T0297074

Solid State Physics

An Introduction to Theory

Solid State Physics

An Introduction to Theory

Joginder Singh Galsin
Department of Mathematics
Statistics and Physics
Punjab Agricultural University
Ludhiana, India

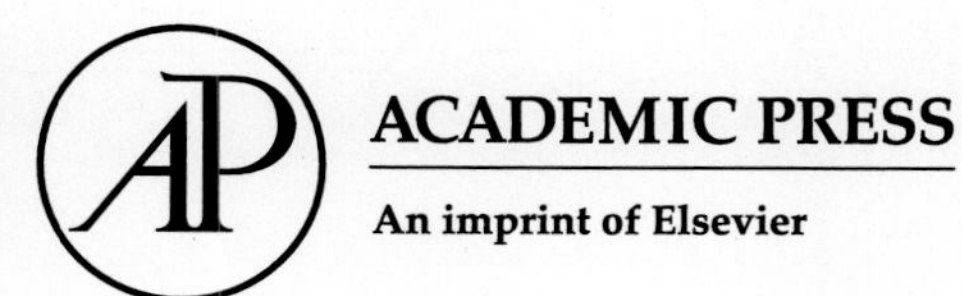

Academic Press is an imprint of Elsevier
125 London Wall, London EC2Y 5AS, United Kingdom
525 B Street, Suite 1650, San Diego, CA 92101, United States
50 Hampshire Street, 5th Floor, Cambridge, MA 02139, United States
The Boulevard, Langford Lane, Kidlington, Oxford OX5 1GB, United Kingdom

Library of Congress Cataloging-in-Publication Data
A catalog record for this book is available from the Library of Congress

British Library Cataloguing-in-Publication Data
A catalogue record for this book is available from the British Library

ISBN 978-0-12-817103-5

For information on all Academic Press publications visit our website at https://www.elsevier.com/books-and-journals

Publisher: Joseph P. Hayton (ELS-CMA)
Acquisition Editor: Katey Birtcher (ELS-CMA)
Editorial Project Manager: Andrea Gallego Ortiz (ELS-CBG)
Production Project Manager: Omer Mukthar
Cover Designer: Mark Rogers

Typeset by SPi Global, India

Dedication

I dedicate this book to my Mom and late Dad with love.

Contents

14. Semiconductors

15. Dielectric Properties of Nonconducting Solids

16. Ferroelectric Solids

17. Optical Properties of Solids

18. Magnetism

19. Ferromagnetism

About the Author

Dr. Joginder Singh Galsin is a physicist who was born in the north Indian city of Ludhiana, Punjab. After graduating in science, he went on to acquire an MSc (Honors School) degree and a PhD in theoretical solid state physics from Punjab University, Chandigarh. He later worked as a postdoctoral fellow for 1 year in the same department.

He started his professional career in 1977 as an assistant professor of physics in Punjab Agricultural University, Ludhiana and, over the next 30 years, became a powerhouse in the Sciences Department there. In 2007, he retired from the University at the age of 60 as professor and head, Department of Mathematics, Statistics and Physics, with a brief stint as reader in physics at Guru Nanak Dev University, Amritsar, from 1984 to 1986. After his retirement, he served for another 7 years in three institutions in various capacities: as head, Department of Physics, Lovely Professional University, Jalandhar; as director, Gulzar Institute of Engineering & Technology, Khanna; and as professor, Ludhiana Institute of Engineering & Technology, Katani Kalan, Ludhiana. He eventually retired from service in 2014 at the age of 67 after 37 long years of committed and dedicated educative service in various educational institutions.

Over the course of his academic years, he was an external expert on various academic/professional committees, including the Board of Studies in Physics, Punjabi University, Patiala; the Faculty of Physical Sciences of Punjabi University, Patiala, and M.D. University, Rohtak; and the Research Degree Committee of Guru Nanak Dev University, Amritsar. He was a member of the Academic Councils of Punjab Technical University, Jalandhar, and Lovely Professional University, Jalandhar.

He was awarded the Best Teacher Award in 1982 by The Punjab Agricultural University Teachers Association. He has more than 80 research papers in journals of national/international repute to his credit (41 in international and 39 in national journals/conferences). The areas of his professional and personal experience and interest include the lattice dynamics of transition metals, band magnetism in metals, and the electronic structure of metallic alloys. He has supervised a number of MSc and MPhil students and jointly supervised PhD students in the above-mentioned fields. He attended a number of national and international conferences in the above-mentioned fields and delivered invited talks on various teaching and research topics, including nanotechnology.

He authored a book called *Impurity Scattering in Metallic Alloys*, which was published by Kluwer Academic/Plenum Publishers, New York, in 2002 (now with Springer). It is a fulfilling moment to mention that the present book entitled *Solid State Physics: An Introduction to Theory*," the outcome of 16 committed years, is sure to be of immense value to the physics community.

Preface

For the past three decades, many scientists have been jumping onto the bandwagon of applied science, thereby hampering the development of basic science. If this emerging trend is permitted to persist over a long period of time, research in applied science will find itself at a crossroad. Recent years have been characterized by debates at the international level over attracting intelligent people to the basic sciences. During my entire professional career, which spans more than 40 years in the field of theoretical solid state physics, I have found that textbooks on solid state physics greatly outnumber books on theoretical solid state physics. This unfortunate trend motivated me to write an elementary textbook on theoretical solid state physics. A major portion of this book has been derived from lectures I delivered on solid state physics at various Indian universities over a period of three decades. I began writing this book in 2000 and it took me almost 17 years of concentrated effort to accomplish a task of such magnitude. Needless to say, the collection of material commenced much earlier.

Solid state physics is such a diverse field that it cannot be covered in a single book. Further, the theory of solids is progressing at a very fast pace and is reaching an increased level of sophistication, greatly complicating the task of providing up-to-date knowledge of the whole subject. Therefore, I have tried to concentrate on the fundamentals of the theoretical aspects of those topics that are required in a first course for undergraduate students of physics, chemistry, materials science, and engineering at various universities across the globe. There are two approaches involved in the development of a book on solid state physics. First is the phenomenological approach, which includes hypotheses and models that are important in the development of the subject. Second is the fundamental approach, based on quantum mechanics and statistical mechanics, which provides greater insight into the actual processes responsible for the various properties of solids. I have tried to present a unified quantum mechanical treatment for the different properties of solids, touching upon phenomenological models wherever necessary. Some of the salient features of the book are discussed later.

For the study of the various properties of solids, a general formalism for the fundamentals has been derived wherever possible. Detailed mathematical steps are presented to make it comprehensible even to students with a minimal mathematical background. The results for simple structures in one-, two-, and three-dimensional solids are derived for particular cases. All of the chapters of the book are coherently interrelated. Elementary courses in quantum mechanics and statistical mechanics may be considered prerequisites for understanding the subject matter.

Dirac's notation has been used, which highlights the physics contained in the mathematics in a befitting and compact manner.

More than 400 diagrams and geometrical constructions of the elementary processes present in solids have been used to enable students to easily comprehend the subject matter.

A considerable number of problems have been inserted at appropriate places in all the chapters with the aim of providing deeper insight into the subject. Throughout the text, bold letters represent vector quantities. Greek letters with arrows also represent vector quantities.

The book contains an elementary account of some recent topics, such as the quantum Hall effect, high-T_c superconductivity, and nanomaterials. The topics of elasticity in solids, dislocations, polymers, point defects, and nanomaterials are of special interest for engineering students. The inclusion of abstract methods of quantum field theory, though important in many-body problems, have been deliberately avoided as they may not be very relevant to the diverse student communities for whom this book was written.

At the end of the book, some elementary textbooks on solid state physics are listed for supplementary reading. Advanced books on the topics covered in the present text are also included in the list, which may be helpful to advanced learners in carrying out further work.

I am indebted to Professor K.N. Pathak, former Vice Chancellor of Panjab University, Chandigarh, for fostering and nurturing my interest in the subject of solid state physics while I was a student. I am thankful to my daughters Amardeep Galsin, Manveen Galsin, my son-in-law Dr. Nirjhar Hore, and my son Damanjit Singh Galsin, who have been a constant source of encouragement and support for me during the completion of this work. I am very grateful to my wife, Professor

Surinder Kaur, for encouraging me to liberally devote time to the writing of this book and also for editing the technical aspects of the English language. I am grateful to Mr. Rakesh Kumar (Somalya Printers, Ludhiana) for undertaking the artwork for this book so diligently and efficiently. I am also thankful to all my loved ones, colleagues, and well-wishers especially Dr. Jagtar Singh Dhiman, Dr. Nathi Singh and Dr. Paramjit Singh, who silently urged me to move on toward the successful completion of this momentous project. Last but not least, my journey with the Elsevier team, from the submission of the manuscript to the finished product, has been very pleasant. The book has not been read by any subject expert, therefore, any omission or error is my sole responsibility. I would welcome and appreciate comments/suggestions/ feedback for the improvement of the book in the near future. **A big thanks to Lord Almighty-our creator.**

Joginder Singh Galsin

Chapter 1

Crystal Structure of Solids

Chapter Outline

Matter exists in three states: solid, liquid, and gas. At very low temperatures, all forms of matter condense to form a solid. Matter consists of very small particles called atoms that can exist independently. The most remarkable property of the solid state is that the atoms of most of the solids, in the pure form, arrange themselves in a periodic fashion. Such materials are called *crystalline solids*. The word crystal comes from the Greek word meaning "clear ice." This term was used for transparent quartz material because for a long period in ancient times only quartz was known to be a crystalline material. The modern theory of solids is founded on the science of crystallography, which is concerned with the enumeration and classification of the actual structures exhibited by various crystalline solids. There are 103 stable elements in the periodic table and the majority of these exist in the solid state. Today metallic solids play an indispensable role in engineering, technology, and industry. Tools and machines ranging from sewing needles to automobiles and aircraft are made of metallic solids with required properties. Thus, the study of various physical properties of solids is very important. In this chapter we shall give an introductory account of the various periodic arrangements of atoms in solids.

1.1 CLOSE PACKING OF ATOMS IN SOLIDS

There exist forces of attraction and repulsion among the atoms in a solid. But the net force between any two atoms must be attractive for a solid to exist. In solids, each atom is attracted approximately equally and indiscriminately to all of its neighboring atoms. As a result, in a crystalline solid the atoms have the tendency to settle in a close-packed structure. In an ideal close-packed structure, atoms touch one another just like peas placed in a vessel. The packing of atoms into a minimum total volume is called *close packing*. If the atoms are assumed to have a spherical shape, then a close-packed layer of atoms of an element with centers at positions A appears as shown in Fig. 1.1A. Above this layer, there are two types of voids, labelled B and C. Therefore, in the second layer, above the first one, the atoms can settle down with their centers at either of the positions B or C. If the atoms in the second layer go over the B positions then there are two nonequivalent choices for the third layer. The atoms in the third layer can have their centers at either the A or C positions and so on. Therefore, the most common close-packed structures that are obtained have a layer stacking given by ABABA… (or BCBCB… or CACAC…) and ABCABCA… The stacking of layers given by ABABA (or BCBCB or CACAC) gives a hexagonal close-packed (hcp) structure while the second type of stacking, ABCABCA, gives a face-centered cubic (fcc) structure. Therefore, the most common close-packed structures exhibit either cubic or hexagonal symmetry: the basic symmetries of crystal structure. The details of the geometry of these close-packed structures will be discussed in the coming sections.

Solid State Physics. https://doi.org/10.1016/B978-0-12-817103-5.00001-3

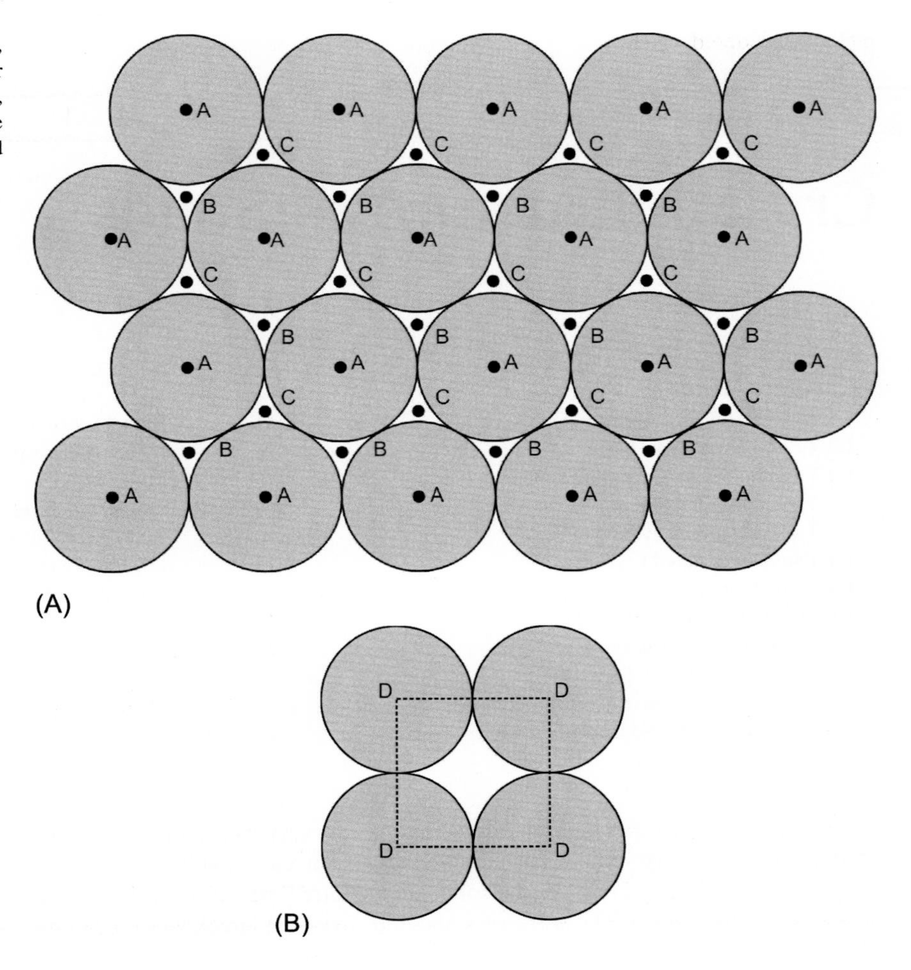

FIG. 1.1 (A) A close-packed layer of atoms, which are assumed to be hard spheres, with their centers at the points marked A. Above this layer, voids exist at points B and C. (B) The close packing of atoms with centers at points marked D in the *bottom* layer of the sc structure.

Another close-packed structure exhibited by some elements is a simple cubic (sc) structure. The bottom layer of an sc structure is shown in Fig. 1.1B, in which the centers of the atoms are shown by the points D. If in all of the layers above the first layer the atoms settle down with their centers at the positions D then an sc structure is formed.

There can also be other sequences of layer stacking that show either a lower order or no order. Such structures are called faulted close-packed structures. For example, some rare-earth elements exhibit structures possessing layer stacking ABACA. This corresponds to a *stacking fault* appearing in every fourth layer and leads to a doubling of the hexagonal structure along the vertical axis (double hexagonal structure). Samarium has a unique structure, which has stacking sequence ABABCBCAC.

A quantitative measurement of the degree of close packing is given by a parameter called the *packing fraction*, f_p. It is defined as the ratio of actual volume occupied by an atom V_a to its average volume V_0 in a crystalline structure,

$$f_p = \frac{V_a}{V_0} \tag{1.1}$$

The value of f_p will be calculated for some simple structures later in this chapter. Here we would like to mention two facts about crystal structures. First, crystals with a higher value of f_p are more likely to exist. Second, in real crystals, the atoms may not necessarily touch each other but may instead settle down at some equilibrium distance that depends on the binding force between them.

From the above discussion, it is evident that a crystalline solid is obtained by piling planes of atoms one above the other at regular intervals with the different planes bound together by interplanar electrostatic forces. Each atomic plane consists of periodic arrangement of atoms in two dimensions that are bound together by intraplanar electrostatic forces. A crystalline solid may exhibit one-dimensional, two-dimensional, or three-dimensional behavior depending on the strength of the interplanar and intraplanar forces. If the interplanar forces are much weaker than the intraplanar forces, then each atomic plane can be considered to be independent of the other atomic planes. Such a situation can arise in a solid in which the distance between the atoms of the same plane is much smaller than that of the atoms belonging to different planes. These crystalline solids exhibit the behavior of a two-dimensional solid. Further, each atomic plane can be considered to be made of parallel lines of atoms (atomic lines) and the atoms in the same and different atomic lines are bound together by electrostatic forces. If the forces between the atoms belonging to different atomic lines are much weaker than those among the atoms belonging to the same atomic line, then each atomic line becomes nearly independent of the other atomic lines and the solid behaves as a one-dimensional solid. Such a situation may arise in a two-dimensional solid when the distance between the atoms in the same atomic line is much smaller than the distance between atoms belonging to different atomic lines. Therefore, a crystalline solid will behave as a one-dimensional solid if the interplanar forces and the forces between different atomic lines in a plane are quite weak.

1.2 CRYSTAL LATTICE AND BASIS

An ideal crystal consists of a periodic arrangement of an infinite number of atoms in a three-dimensional space. In order to express the periodicity of a crystal in mathematical language, it is convenient to define a crystal lattice (space lattice), or more commonly a *Bravais lattice*. A Bravais lattice consists of an infinite array of points distributed periodically in three-dimensional space in which each point has surroundings identical to those of every other point. A crystal lattice is an idealized mathematical concept and does not exist in reality. Fig. 1.2 shows a one-dimensional lattice in which the lattice vector is defined as

$$\mathbf{R}_n = n\mathbf{a}_1 \tag{1.2a}$$

$$\mathbf{a}_1 = a\hat{\mathbf{i}}_1 \tag{1.2b}$$

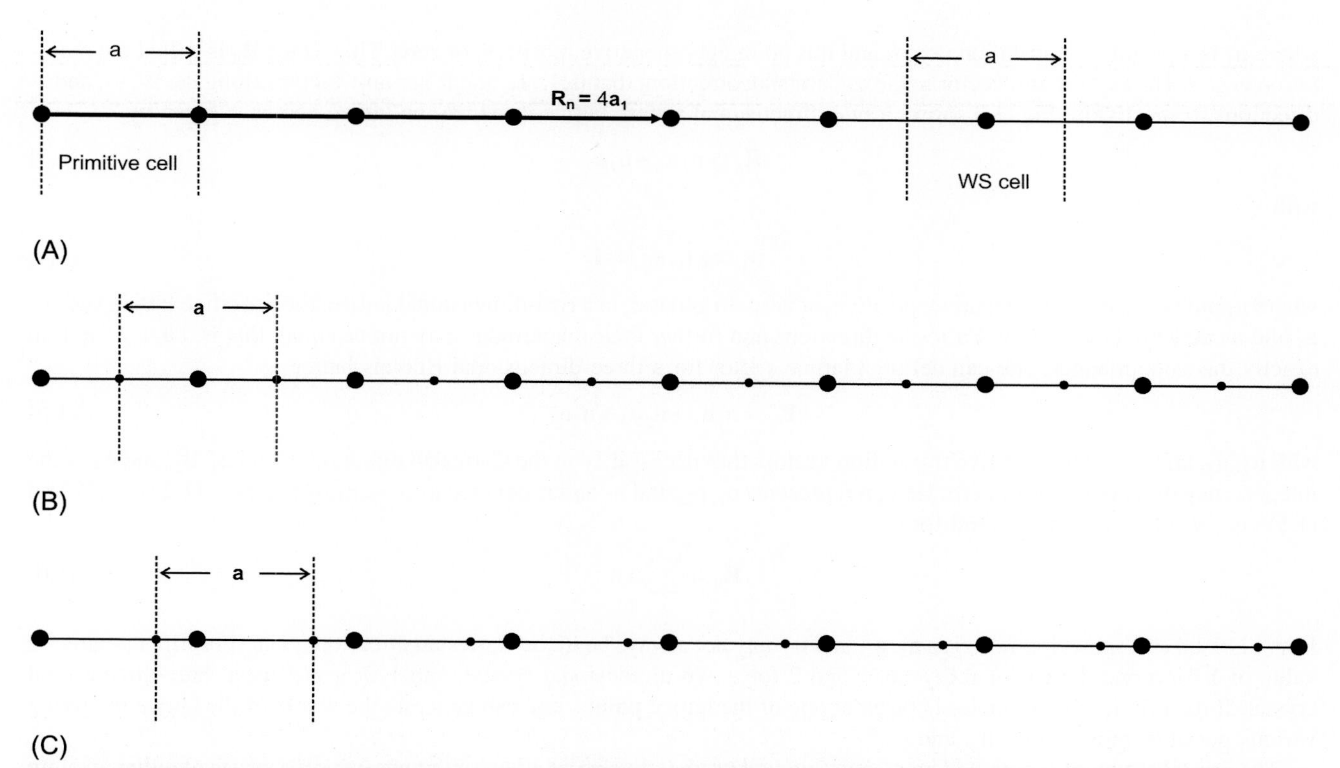

FIG. 1.2 Monatomic linear lattice with periodicity "a." (A) One-dimensional solid with lattice points at the position of the atoms. Here each end of the primitive cell contributes, on average, half a lattice point/atom, thus yielding one lattice point/atom in the primitive cell. (B) One-dimensional solid with lattice point in the middle of the two atoms, that is, at a distance a/2 from the atom. The new primitive cell contains one lattice point/atom. (C) One-dimensional solid with lattice point at a distance a/4 toward the left of the atom. The new primitive cell contains one lattice point/atom.

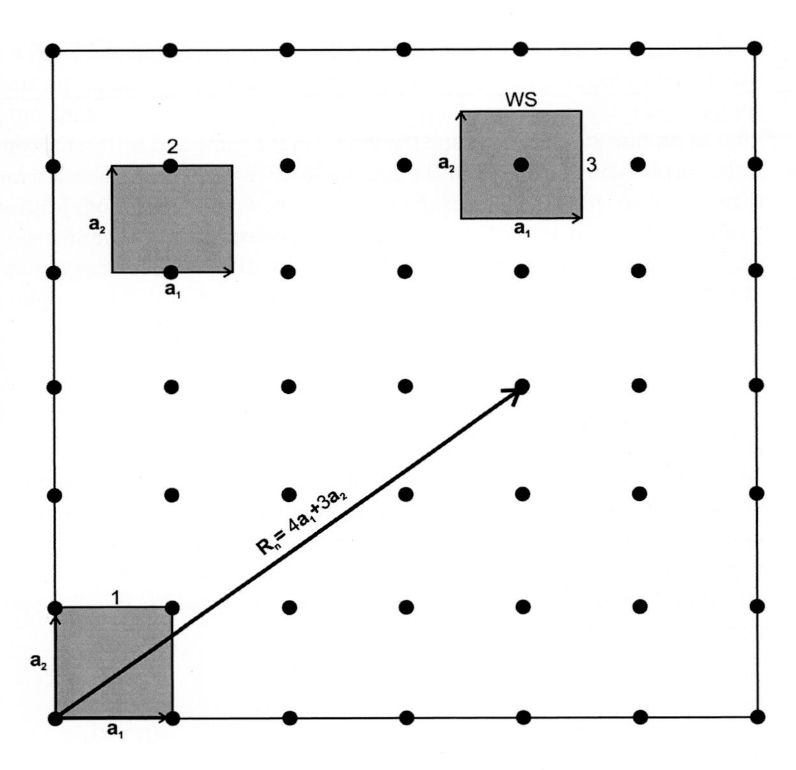

FIG. 1.3 Monatomic square lattice with primitive vectors $\mathbf{a}_1$ and $\mathbf{a}_2$, where $|\mathbf{a}_1| = |\mathbf{a}_2|$. In part 1 of the figure, the atom is assumed to be situated at the position of the lattice point and the primitive cell has atoms at its corners. The lattice points/atoms at the corners contribute, on average one-fourth of the lattice point/atom to the primitive cell, thus yielding one lattice point per primitive cell. In part 2 of the figure, the lattice point is situated in the middle of the two atoms on the horizontal lines of atoms. The primitive cell has one lattice point as each corner contributes 1/4 of a lattice point to the primitive cell. Further, an atom at the middle of the side contributes ½ atom to the cell. In part 3 of the figure, again the atom is assumed to be situated at the position of the lattice point. It shows the Wigner-Seitz (WS) cell of a monatomic square lattice with one lattice point/atom at its center. Further, the area of the WS cell is the same as in cases 1 and 2.

where $\mathbf{a}_1$ is a primitive translation vector and n is an integer: negative, positive, or zero. The vector $\mathbf{R}_n$ is called the *translation vector*. Here $\hat{\mathbf{i}}_\alpha$ is a unit vector in the α-Cartesian direction, that is, $\hat{\mathbf{i}}_1$, $\hat{\mathbf{i}}_2$, and $\hat{\mathbf{i}}_3$ are unit vectors along the x-, y-, and z-directions, respectively. Fig. 1.3 shows a two-dimensional square lattice in which the lattice vector is given by

$$\mathbf{R}_n = n_1\mathbf{a}_1 + n_2\mathbf{a}_2 \tag{1.3}$$

with

$$\mathbf{a}_1 = a\hat{\mathbf{i}}_1, \mathbf{a}_2 = a\hat{\mathbf{i}}_2 \tag{1.4}$$

where $n_1$and n_2 are integers: negative, positive, or zero. In general, in a two-dimensional lattice, the primitive lattice vectors $\mathbf{a}_1$ and $\mathbf{a}_2$ may not be along the Cartesian directions and further their magnitudes may not be equal, that is, $|\mathbf{a}_1| \neq |\mathbf{a}_2|$. In exactly the same manner, one can define a lattice vector for a three-dimensional Bravais lattice as

$$\mathbf{R}_n = n_1\mathbf{a}_1 + n_2\mathbf{a}_2 + n_3\mathbf{a}_3 \tag{1.5}$$

with $\mathbf{a}_1$, $\mathbf{a}_2$, and $\mathbf{a}_3$ as the primitive translation vectors (not necessarily in the Cartesian directions), and n_1, n_2, and n_3 as the integers: negative, positive, or zero. Here, n represents n_1, n_2, and n_3 and is denoted as $n = (n_1,n_2,n_3)$. Eqs. (1.2a), (1.3), and (1.5) can be written in the general form

$$\mathbf{R}_n = \sum_i n_i\mathbf{a}_i \tag{1.6}$$

Here i is used as a subscript (not α) as $\mathbf{a}_1$, $\mathbf{a}_2$, and $\mathbf{a}_3$ may not always be in the Cartesian directions. The subscript i assumes a value of 1 for a one-dimensional crystal, 1 and 2 for a two-dimensional crystal, and 1, 2, and 3 for a three-dimensional crystal. If the origin of coordinates is taken at one of the lattice points, one can generate the whole of the lattice by giving various possible values to n_1, n_2, and n_3.

The crystal structure is obtained by associating with each lattice point a *basis of atoms*, which consists of either an atom or a group of atoms. The basis of atoms associated with every lattice point must be identical, both in composition and orientation. If there is only one atom in the basis, it is usually assumed to be situated at the lattice point itself. However, if there is more than one atom in the basis, one of them can be assumed to be situated at the lattice point and the others can be

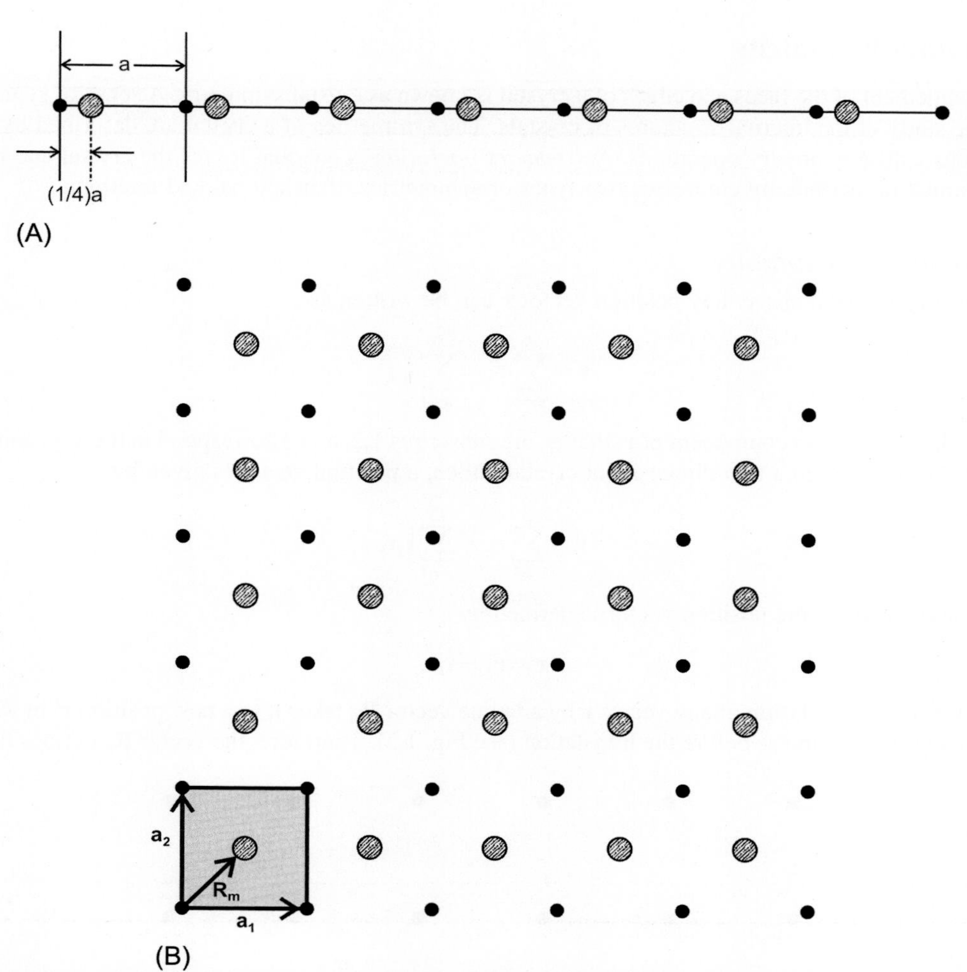

FIG. 1.4 (A) A diatomic linear lattice with lattice constant a. The first atom *(black sphere)* of the crystal is at the origin and the basis atom *(shaded)* with respect to it is at a distance of (1/4)a. (B) A square lattice with a basis of two atoms: the *shaded and black spheres* represent the two types of atoms in the lattice. The lattice points are taken at the positions of the black spheres with the coordinates of the basis atoms given by $\mathbf{R}_m = 0$, $(1/2)\mathbf{a}_1 + (1/2)\mathbf{a}_2$.

specified with respect to it. Fig. 1.4 shows a linear lattice and a square lattice with a basis of two atoms in which one of the atoms is taken at the lattice point. So, in general, the position of the mth basis atom associated with the nth lattice point may be written as

$$\mathbf{R}_{nm} = \mathbf{R}_n + \mathbf{R}_m \tag{1.7}$$

with

$$\mathbf{R}_m = m_1 \mathbf{a}_1 + m_2 \mathbf{a}_2 + m_3 \mathbf{a}_3 \tag{1.8}$$

where m_1, m_2, and m_3 are constants and usually $0 \leq m_1, m_2, m_3 \leq 1$. Here m denotes all three numbers m_1, m_2, m_3 and is usually written as $m = (m_1, m_2, m_3)$. Such a lattice is called a Bravais lattice with a basis. It is worth mentioning here that the choice of the lattice point is not unique, but rather a number of choices are possible. Fig. 1.2A–C shows three possible choices of lattice points in a one-dimensional crystal. Similarly, Fig. 1.3 shows two possible choices, namely, 1, 2, of lattice points in a two-dimensional square lattice. It is evident from the figures that the magnitude and the orientation of the primitive lattice vectors remain the same, although the positions of the basis atoms with respect to the lattice point change. In other words, for all the choices of the lattice points, the crystal lattice exhibits the same periodicity.

1.3 PERIODICITIES IN CRYSTALLINE SOLIDS

In a pure crystalline solid there are basically two types of periodicities: structural and electrostatic. In this chapter, we shall discuss only the structural periodicity, while the electrostatic periodicity will be discussed in Chapter 12.

1.3.1 Structural Periodicity

The ordered arrangement of the faces and edges of a crystal is known as crystal symmetry. A sense of symmetry is a powerful tool for the study of the internal structures of crystals. The symmetries of a crystal are described by certain mathematical operations called symmetry operations. A *symmetry operation* is one that leaves the crystal and its environment invariant. The structural periodicity comprises two types of symmetries: translational and rotational.

1.3.1.1 Translational Symmetry

In a three-dimensional crystal space, any position vector **r** can be written as

$$\mathbf{r} = \sum_{\alpha=1}^{3} \mathbf{r}_\alpha = \sum_{\alpha=1}^{3} \hat{\mathbf{i}}_\alpha \mathrm{r}_\alpha \tag{1.9}$$

where r_α or $\mathbf{r}_\alpha$ is the α-Cartesian component of **r**, that is, the subscripts 1,2, and 3 correspond to the x, y, and z components, respectively, of the vector **r**. In a two-dimensional crystal lattice, a position vector is given by

$$\mathbf{r} = \sum_{\alpha=1}^{2} \mathbf{r}_\alpha = \sum_{\alpha=1}^{2} \hat{\mathbf{i}}_\alpha \mathrm{r}_\alpha \tag{1.10}$$

In a one-dimensional crystal, the position vector is defined by

$$\mathbf{r} = \mathbf{r}_1 = \hat{\mathbf{i}}_1 \mathrm{r}_1 \tag{1.11}$$

In a crystalline solid, the translation of any vector **r** by a lattice vector $\mathbf{R}_n$ takes it to a new position $\mathbf{r}'$ in which the atomic arrangement is exactly the same as before the translation (see Fig. 1.5). Therefore, the vector $\mathbf{R}_n$ defines the *translational*

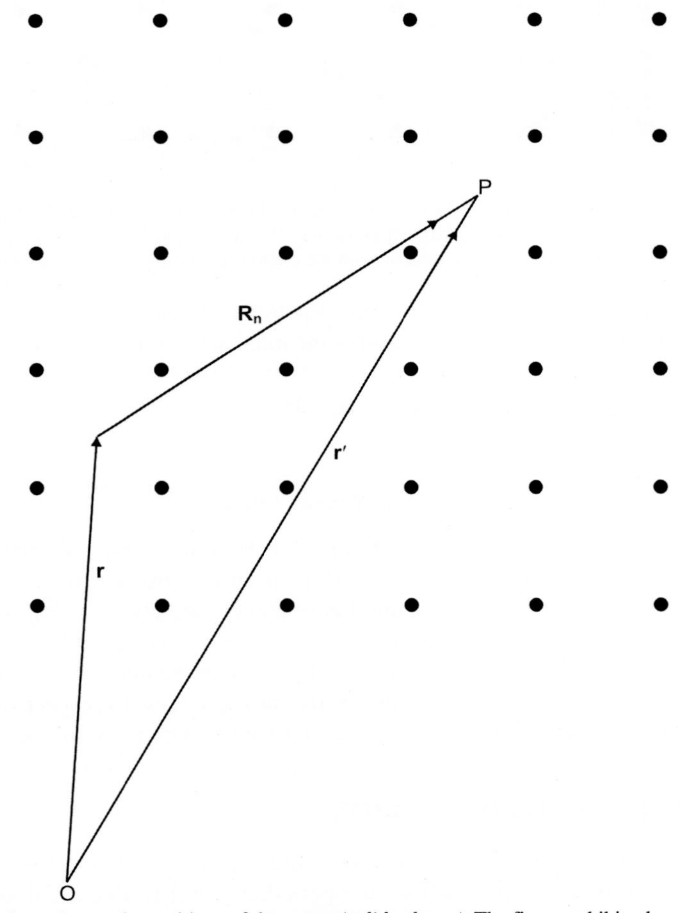

FIG. 1.5 A square lattice with lattice points at the positions of the atoms *(solid spheres)*. The figure exhibits the same distribution of atoms around any two points **r** and $\mathbf{r}' = \mathbf{r} + \mathbf{R}_n$.

symmetry of the crystalline solid. Let the subscript n denote the nth unit cell and m the number of atoms in it. Then, in a crystalline solid with s number of atoms in a unit cell (s atoms associated with a lattice point), the density of atoms, $\rho^a(\mathbf{r})$, is defined as

$$\rho^a(\mathbf{r}) = \frac{1}{V} \sum_n \sum_m \delta(\mathbf{r} - \mathbf{R}_{nm}) \tag{1.12}$$

where the summation nm is over the crystal and V is the volume of the crystal. It can easily be proved that

$$\rho^a(\mathbf{r}) = \rho^a(\mathbf{r} + \mathbf{R}_n) \tag{1.13}$$

Eq. (1.13) shows that the atomic arrangement at $\mathbf{r}$ and $\mathbf{r} + \mathbf{R}_n$ is the same and therefore defines the translational symmetry of the crystal lattice mathematically.

1.3.1.2 Near Neighbors

It has already been noted that the distribution of lattice points and of atoms around any lattice point is the same. To be more specific, the distribution of lattice points can be classified in terms of near neighbors (NNs) of different orders about a given lattice point. In a Bravais lattice, the lattice points closest to a given lattice point are called first nearest neighbors (1NNs) and the number of 1NNs is usually called the coordination number. The next closest lattice points to that particular lattice point are called the second nearest neighbors (2NNs). In this way, one can define third nearest neighbors (3NNs), fourth nearest neighbors (4NNs) and, in general, the nth nearest neighbors (nNNs). As the lattice is periodic, each lattice point in a given crystal structure has the same number of nNNs for all values of n. The number, position, and distance of 1NNs and 2NNs in some simple crystal structures are given in Table 1.1.

1.3.1.3 Primitive Unit Cell

The most important property of structural periodicity is that it allows us to divide the whole of the lattice into the smallest identical cells, called primitive unit cells or simply *primitive cells*. Figs. 1.2 and 1.3 show the primitive cells of one- and two-dimensional lattices, while Fig. 1.6 shows the primitive cell of an sc lattice. In a monatomic linear lattice, the primitive cell is a line segment of length $|\mathbf{a}_1| = a$ with one lattice point in it on average. One can say that each end contributes, on

TABLE 1.1 Positions, Distances, and Numbers of 1NNs and 2NNs in sc, fcc, and bcc Structures

nNN	Position	Number	Distance
sc structure			
1NN	a(±1, 0, 0) a(0, ±1, 0) a(0, 0, ±1)	6	a
2NN	a(±1, ±1, 0) a(0, ±1, ±1) a(±1, 0, ±1)	12	$\sqrt{2}a$
fcc structure			
1NN	a(±1/2, ±1/2, 0) a(±1/2, 0, ±1/2) a(0, ±1/2, ±1/2)	12	$a/\sqrt{2}$
2NN	a(±1, 0, 0) a(0, ±1, 0) a(0, 0, ±1)	6	a
bcc structure			
1NN	a(±1/2, ±1/2, ±1/2)	8	$\sqrt{3}a/2$
2NN	a(±1, 0, 0) a(0, ±1, 0) a(0, 0, ±1)	6	a

Here a is the lattice parameter.

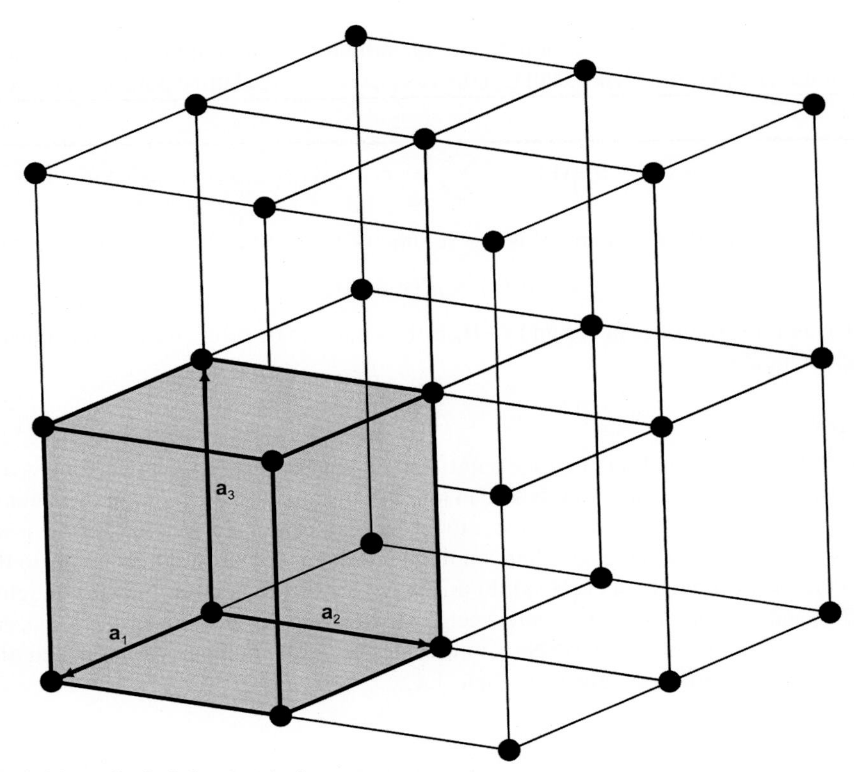

FIG. 1.6 Conventional primitive cell *(shaded region)* in the sc structure.

average, a half lattice point to the primitive cell. Therefore, the number of lattice points per unit length, N_0 (linear density of lattice points), is given by

$$N_0 = \frac{1}{a} \tag{1.14}$$

In a square lattice, the primitive cell is a square bounded by primitive vectors $\mathbf{a}_1$ and $\mathbf{a}_2$ with sides having length $|\mathbf{a}_1| = |\mathbf{a}_2| = a$. There is one lattice point, on average, in a primitive cell: each corner of the square contributes, on average, one-fourth of a lattice point to the primitive cell. In general, in a two-dimensional lattice, the primitive cells are parallelograms bounded by vectors $\mathbf{a}_1$ and $\mathbf{a}_2$ and having area A_0 given by

$$A_0 = |\mathbf{a}_1 \times \mathbf{a}_2| \tag{1.15}$$

Therefore, the number of lattice points per unit area, N_0 (surface density of lattice points), is given by

$$N_0 = \frac{1}{A_0} \tag{1.16}$$

In an sc lattice, the primitive cell is a cube bounded by the primitive vectors $\mathbf{a}_1$, $\mathbf{a}_2$, and $\mathbf{a}_3$ with lattice points (atoms) at the corners and with each corner contributing one-eighth of the lattice point (atom) to the primitive cell (Fig. 1.6). In general, in a three-dimensional lattice, the primitive cell is a parallelepiped bounded by vectors $\mathbf{a}_1$, $\mathbf{a}_2$, $\mathbf{a}_3$ and having volume V_0 given by

$$V_0 = |\mathbf{a}_1 \cdot \mathbf{a}_2 \times \mathbf{a}_3| \tag{1.17}$$

Hence the volume density of lattice points, N_0, in a three-dimensional lattice is given by

$$N_0 = \frac{1}{V_0} \tag{1.18}$$

One should note that in a monatomic crystal the density of lattice points N_0 is equal to the atomic density ρ^a. In many crystals, a primitive cell contains one lattice point with a basis containing more than one atom. If the subscript n is assumed to label the primitive cell, then $\mathbf{R}_{nm}$ gives the position of the mth atom in the nth cell. The translation of a primitive cell by all possible $\mathbf{R}_n$ vectors just fills the crystal space without overlap or voids.

The crystal space can also be filled up without any overlap by the translation of cells larger than the primitive cell, whose volume is usually an integral multiple of the volume of the primitive cell. Such cells are called *unit cells* and their choice is not unique. The shape of a unit cell may be different from that of a primitive cell and it may contain more than one lattice point. For example, in an sc structure, a cube with side 2a (see Fig. 1.6) is one choice for a unit cell. It contains eight primitive cells and hence eight lattice points. Therefore, *a primitive cell can be defined as a unit cell with minimum volume.*

The choice of a primitive cell is also not unique, but its volume is independent of the choice for a particular crystal structure. Wigner and Seitz gave an alternative and elegant method to construct a primitive cell. In a Bravais lattice, a given lattice point is joined by lines to its 1NN, 2NN, 3NN.... lattice points. The smallest polyhedron bounded by perpendicular bisector planes of these lines is called the *Wigner-Seitz* (WS) *cell*. Figs. 1.2A and 1.3 show the WS cells for a monatomic linear lattice and a square lattice, respectively. The WS cell for an sc structure is shown in Fig. 1.7. The WS cell in a monatomic linear lattice is a line segment of length a with the lattice point (atom) at its center. Similarly, the WS cell in a square lattice is a square with area a^2, with a lattice point at the center. In an sc lattice, it is a cube with volume a^3, again having a lattice point at the center. It is evident that in these simple crystal structures both the shape and the volume of the conventional primitive cell and the WS cell are the same. But, in general, the shape of the two types of cells may differ in other crystals. The WS cell exhibits the following characteristic features. First, the WS cell is independent of the choice of primitive lattice vectors. Second, the lattice point lies at the center of the WS cell as a result of which the WS cell is nearly symmetrical about the lattice point, unlike the conventional primitive cell. This symmetry allows us to replace the actual WS cell by a sphere whose volume is equal to that of the WS cell. It is usually called the *WS sphere* and simplifies many of the theoretical calculations.

The translational symmetry of a lattice can be deduced from the concept of the WS cell. Fig. 1.8 shows one of the planes of the WS cell, the equation for which can be written directly as

$$\mathbf{r}\cdot\hat{\mathbf{R}}_n=\frac{1}{2}|\mathbf{R}_n| \tag{1.19}$$

where $\hat{\mathbf{R}}_n=\mathbf{R}_n/|\mathbf{R}_n|$ is a unit vector in the direction of $\mathbf{R}_n$. Eq. (1.19) is equivalent to the relation

$$\mathbf{r}'=\mathbf{r}+\mathbf{R}_n \tag{1.20}$$

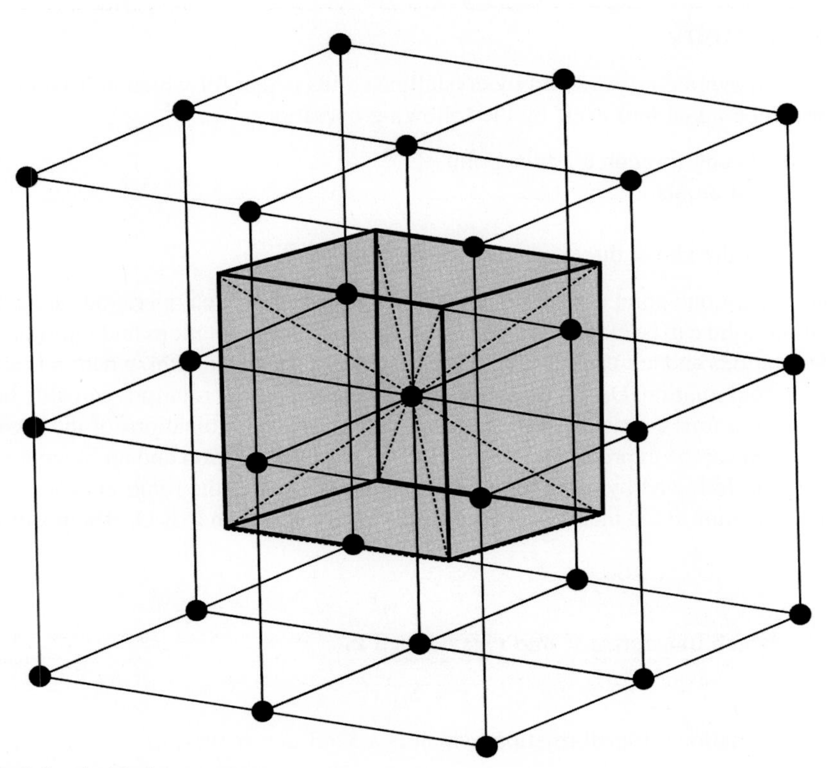

FIG. 1.7 The WS cell *(shaded region)* in the sc structure.

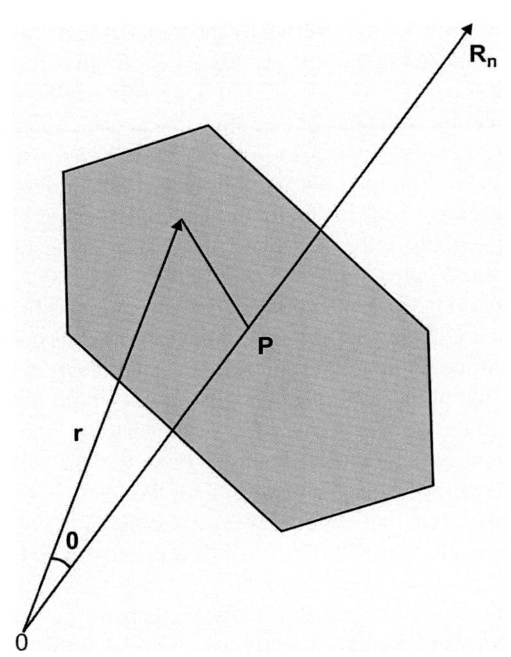

FIG. 1.8 The perpendicular bisector plane of the translation vector $\mathbf{R}_n$, where $\mathbf{r}$ is the position vector of a point in the plane.

where $|\mathbf{r}'| = |\mathbf{r}|$. Fig. 1.5 shows the points $\mathbf{r}$ and $\mathbf{r}'$ defined by Eq. (1.20) and these are found to be equivalent. Therefore, Eq. (1.20) describes the translational symmetry of a Bravais lattice. The translational symmetry allows us to generate the whole lattice by making all possible translations of the WS cell. It can be easily proved that two successive translations are equivalent to a single translation and, moreover, two successive translations commute with each other. Therefore, the collection of lattice translations forms an *Abelian* group.

1.3.2 Rotational Symmetry

The second type of structural symmetry exhibited by crystalline solids is that for which at least one point of the lattice is fixed. A Bravais lattice can be taken into itself by the following operations:

1. Rotation about an axis passing through a lattice point.
2. Reflection about a plane of atoms.
3. Inversion.
4. Different combinations of the above three symmetry operations.

In all of these operations at least one point of the lattice is fixed and therefore such operations are called *point symmetries*. The rotations in a crystalline solid can be classified into two categories: proper rotations and improper rotations. The *proper rotations* are the simple rotations and are usually expressed in terms of the angle $2\pi/n$, where n is an integer. The rotation through $2\pi/n$ is called an n-fold rotation. Detailed analysis shows that the proper rotations can only be through multiples of $\pi/3$ and $\pi/2$. The *improper rotations* consist of inversions, reflections, and combinations of them with rotations. It can be easily proved that a reflection can be expressed as the product of a proper rotation and an inversion. An inversion can be expressed as a 2-fold rotation followed by a reflection in the plane normal to the rotation axis.

Let S_{ni} be a symmetry operator (3×3 matrix) for the n-fold rotation about an axis O_i. The position vector $\mathbf{r}$ after the n-fold rotation becomes

$$\mathbf{r}' = S_{ni}\mathbf{r} \tag{1.21}$$

The inverse operator S_{ni}^{-1}, which transforms $\mathbf{r}'$ into $\mathbf{r}$, is defined as

$$\mathbf{r} = S_{ni}^{-1}\mathbf{r}' \tag{1.22}$$

One can define the identity rotational transformation, which is a 3×3 unit matrix, as

$$\mathbf{r} = I\mathbf{r} \tag{1.23}$$

The collection of all the rotational symmetry operations forms a group, usually known as a *point group*, because two successive rotations are equivalent to a single rotation. The point group is non-Abelian because the two successive rotations do not commute.

1.3.2.1 Space Group

The group of all the translational and rotational symmetry operations that transform a Bravais lattice into itself forms a bigger group known as the *space group* of the Bravais lattice. The general symmetry transformation in a space group is defined as

$$\mathbf{r}' = S_{ni}\,\mathbf{r} + \mathbf{R}_n \tag{1.24}$$

It means that first an n-fold rotation is performed, which is followed by a translation through $\mathbf{R_n}$. For convenience, Eq. (1.24) is written as:

$$\mathbf{r}' = \{S_{ni}\,|\,\mathbf{R}_n\}\,\mathbf{r} \tag{1.25}$$

where $\{S_{ni}\,|\,\mathbf{R}_n\}$ defines the operator corresponding to the transformation (1.24). The inverse transformation corresponding to Eq. (1.25) is defined as

$$\mathbf{r} = \{S_{ni}\,|\,\mathbf{R}_n\}^{-1}\,\mathbf{r}' \tag{1.26}$$

All of the pure lattice translations are given by the collection of symmetry operators $\{I\,|\,\mathbf{R}_n\}$, while all of the pure rotations are given by the collection of symmetry operators $\{S_{ni}\,|\,0\}$, and both of them form the subgroups of the space group.

Problem 1.1

If $\{S_{ni}\,|\,\mathbf{R}_n\}$ and $\{S_{mi}\,|\,\mathbf{R}_{n'}\}$ are two transformations of a space group, prove that

$$\{S_{ni}|\mathbf{R}_n\}\{S_{mi}|\mathbf{R}_{n'}\} = \{S_{ni}S_{mi}|S_{ni}\mathbf{R}_{n'} + \mathbf{R}_n\} \tag{1.27}$$

Problem 1.2

Prove that the inverse transformation of $\{S_{ni}\,|\,\mathbf{R_n}\}$ is given as

$$\{S_{ni}|\mathbf{R}_n\}^{-1} = \{S_{ni}^{-1}|-S_{ni}^{-1}\mathbf{R}_n\} \tag{1.28}$$

The important property of the space group is that the subgroup of pure translations $\{I\,|\,\mathbf{R}_n\}$ is invariant. As a result, in three-dimensional crystals, the only allowed rotations are those that satisfy this invariant property. Let $\{S_{mi}\,|\,\mathbf{R}_{n'}\}$ and $\{I\,|\,\mathbf{R}_n\}$ be the members of the space group of a lattice. The invariance demands that $\{S_{mi}\,|\,\mathbf{R}_{n'}\}\{I\,|\,\mathbf{R}_n\}\{S_{mi}\,|\,\mathbf{R}_{n'}\}^{-1}$ must be a lattice translation. Using Eqs. (1.27), (1.28), it can be readily proved that

$$\{S_{mi}|\mathbf{R}_{n'}\}\{I|\mathbf{R}_n\}\{S_{mi}|\mathbf{R}_{n'}\}^{-1} = \{I|S_{mi}\mathbf{R}_n\} \tag{1.29}$$

According to Eq. (1.29), the lattice translation vector after an m-fold rotation about an axis, that is, $S_{mi}\mathbf{R}_n$, must be a lattice vector that restricts the allowed rotations. With the help of this property the allowed rotations can be found.

1.3.2.2 Allowed Rotations in a Crystal

Consider a row of lattice points in a crystalline solid represented by the line ABCD (see Fig. 1.9). Let $\mathbf{a}$ be the primitive translation vector with magnitude a. The vectors BA and CD are rotated clockwise and counterclockwise, respectively, through an angle $\theta_n = 2\pi/n$ (n-fold rotation) with final positions given by the BE and CF vectors. The rotational symmetry of Eq. (1.29) demands that the points E and F must correspond to lattice points if the crystal lattice is to possess an axis of n-fold rotational symmetry. Clearly, EF must be parallel to AD and the magnitude of EF must be an integral multiple of a, that is, EF = ma where m is an integer. From Fig. 1.9, it is evident that

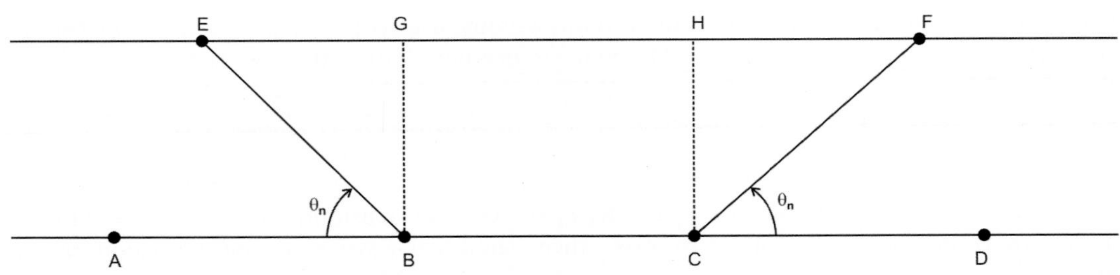

FIG. 1.9 The lattice points A, B, C, and D along a particular line in a crystal. E and F are the positions of the lattice points A and D after n-fold rotation θ_n about the points B and C, respectively.

$$\cos\theta_n = \frac{EG}{BE} = \frac{FH}{CF}$$

and it gives

$$EG = FH = a\cos\theta_n$$

Hence, the rotational symmetry gives

$$EF = a + 2a\cos\theta_n = ma$$

which yields the value of $\cos\theta_n$ given by

$$\cos\theta_n = \frac{m-1}{2} = \frac{N}{2} \tag{1.30}$$

where N is an integer. The allowed values of N are obtained from the fact that $\cos\theta_n$ lies between +1 and −1. Table 1.2 gives the possible values of N, θ_n, and n. It shows that all the allowed rotations are multiples of either $\pi/2$ or $\pi/3$ and that the 5-fold rotation is not allowed, that is, not allowed by the condition of Eq. (1.29). We want to mention here that the 7-fold rotation is also not allowed as it is not a multiple of either $\pi/2$ or $\pi/3$ and therefore is not compatible with the translational symmetry of the three-dimensional lattice. The geometric proof of the fact that 5- and 7-fold symmetries are not allowed is as follows. Fig. 1.10 shows that primitive cells with five-fold rotational symmetry (pentagon) do not fill the space completely but leave voids, which is not allowed. On the other hand, primitive cells with seven-fold symmetry (see Fig. 1.10) overlap when translated to fill the space, which again is not allowed. Therefore, both the 5-fold and 7-fold rotational symmetries are not allowed in a three-dimensional crystal lattice. Before we proceed further, we state a few theorems for the student to prove.

- **Theorem 1**: If a Bravais lattice has a line of symmetry, it has a second line of symmetry at right angles to the first.
- **Theorem 2**: There is a two-fold axis passing through every lattice point of a Bravais lattice and every midpoint between two lattice points.
- **Theorem 3**: If a Bravais lattice has a twofold axis, it also has a plane of symmetry at right angles to that axis, and vice versa.
- **Theorem 4**: If a Bravais lattice has an axis of n-fold symmetry, it also has n-fold symmetry about any lattice point.

TABLE 1.2 Allowed Rotations in a Three-Dimensional Crystal

N	$\cos\theta_n$	θ_n	n
−2	−1	π	2
−1	−1/2	$2\pi/3$	3
0	0	$\pi/2$	4
1	½	$\pi/3$	6
2	1	2π	1

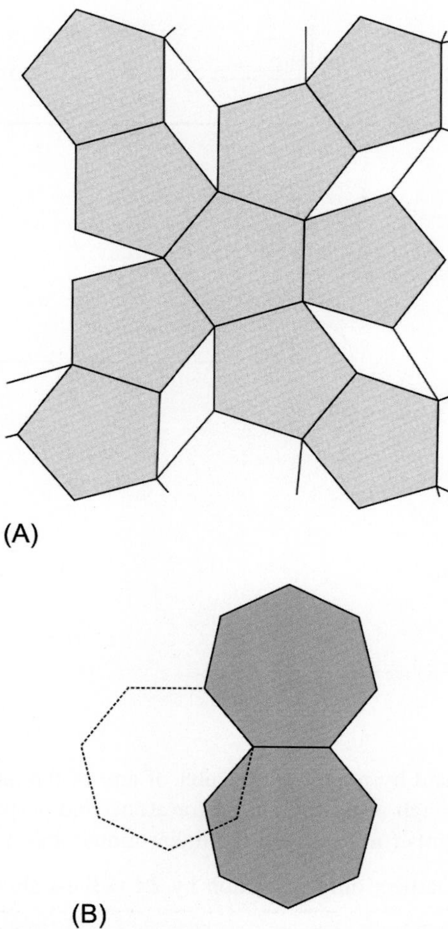

FIG. 1.10 (A) Five-fold rotational symmetry cannot exist in a lattice as it is not possible to fill the whole of the space. Voids are created with a connected array of pentagons. (B) A seven-fold rotational symmetry cannot exist in a lattice as the connected polygons with seven sides overlap with one another (Kepler's demonstration).

- **Theorem 5**: If a lattice has two lines of symmetry making an angle θ, it also has rotational symmetry about their intersection with an angle 2θ.
- **Theorem 6**: If a Bravais lattice has two planes of symmetry making an angle θ, the intersection of the two planes is a rotational axis of period 2θ.

The consideration of a space group for a particular solid yields a number of crystallographic point groups. Once we know the point group corresponding to a particular class of crystals, information can be obtained about the primitive translations $\{I|\mathbf{R}_n\}$, which are invariant under the operations of its point group. It is sufficient to put restrictions on the basic primitive vectors.

To discuss crystals with different dimensions, we first represent the primitive vectors $\mathbf{a}_1$, $\mathbf{a}_2$, and $\mathbf{a}_3$ and the angles between them α, β, and γ (see Fig. 1.11). The values of $\mathbf{a}_1$, $\mathbf{a}_2$, $\mathbf{a}_3$, α, β, and γ are chosen in such a way that the invariance of the lattice under the point symmetry group is satisfied.

1.4 ONE-DIMENSIONAL CRYSTALS

In one-dimensional crystals, there is one primitive vector $\mathbf{a}_1$ and the translation vector is given by Eq. (1.2). In these crystals, there is one translational group and two point symmetry operations or groups. First, point symmetry operation is the identity operation (equivalent to a rotation of 2π about a lattice point) and second is a reflection through a lattice point, which transforms x into $-$x. The total number of space groups, n, is obtained by multiplying the number of translational groups n_T and point symmetry groups n_R, that is, $n = n_T \times n_R$. Therefore, in a one-dimensional monatomic crystal there are two space groups and both satisfy the invariance property under point symmetry groups.

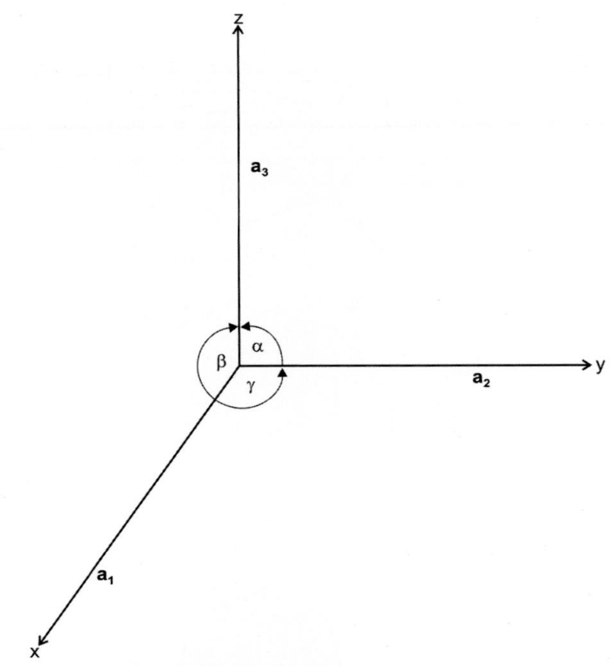

FIG. 1.11 The angles α, β, and γ between the primitive vectors $\mathbf{a}_1$, $\mathbf{a}_2$, and $\mathbf{a}_3$.

1. Rotation by 2π (identity operation) and by π about the center of any of the lattice points (or atoms).
2. Reflection about a plane passing through any lattice point (or atom) and perpendicular to the linear lattice.
3. Inversion about any of the lattice points (or atoms) of the linear monatomic lattice.

On the other hand, in a diatomic linear lattice, only a rotation by 2π radians about any lattice point is allowed and this comprises the point group.

1.5 TWO-DIMENSIONAL CRYSTALS

In two-dimensional crystals there are two primitive vectors $\mathbf{a}_1$ and $\mathbf{a}_2$ with an angle γ between them. In a two-dimensional lattice, it has been found that there are 5 distinct translational groups (Bravais lattices) and 10 crystallographic point groups. Further, it has been established that there are 17 permissible space groups in total. One should note that the total number of permissible space groups is less than the total number of space groups $n = 5 \times 10 = 50$. The five Bravais lattices in two-dimensional crystals are shown in Fig. 1.12 and have the following relation between $\mathbf{a}_1$, $\mathbf{a}_2$, and angle γ.

1. $|\mathbf{a}_1| = |\mathbf{a}_2|$, $\gamma = 90°$ square lattice
2. $|\mathbf{a}_1| = |\mathbf{a}_2|$, $\gamma = 120°$ hexagonal lattice
3. $|\mathbf{a}_1| \neq |\mathbf{a}_2|$, $\gamma = 90°$ rectangular lattice
4. $|\mathbf{a}_1| \neq |\mathbf{a}_2|$, $\gamma = 90°$ centered rectangular lattice
5. $|\mathbf{a}_1| \neq |\mathbf{a}_2|$, $\gamma \neq 90°$ oblique lattice

1.6 THREE-DIMENSIONAL CRYSTALS

In a three-dimensional crystal, the invariance of the lattice under the point symmetry group yields 14 Bravais lattices (translational groups) and 32 crystallographic point groups. Further, it has been established that there are in all 230 permissible space groups, which is less than the total number of space groups $n = n_T \times n_R = 14 \times 32 = 448$. Therefore, in general, each translational group is compatible with a limited number of point groups. It is a common practice to divide the 14 Bravais lattices into seven groups as stated below (see Fig. 1.13). In the seven classes of Bravais lattices, the unit cell may or may not be primitive in nature. Let us discuss some features of these crystals.

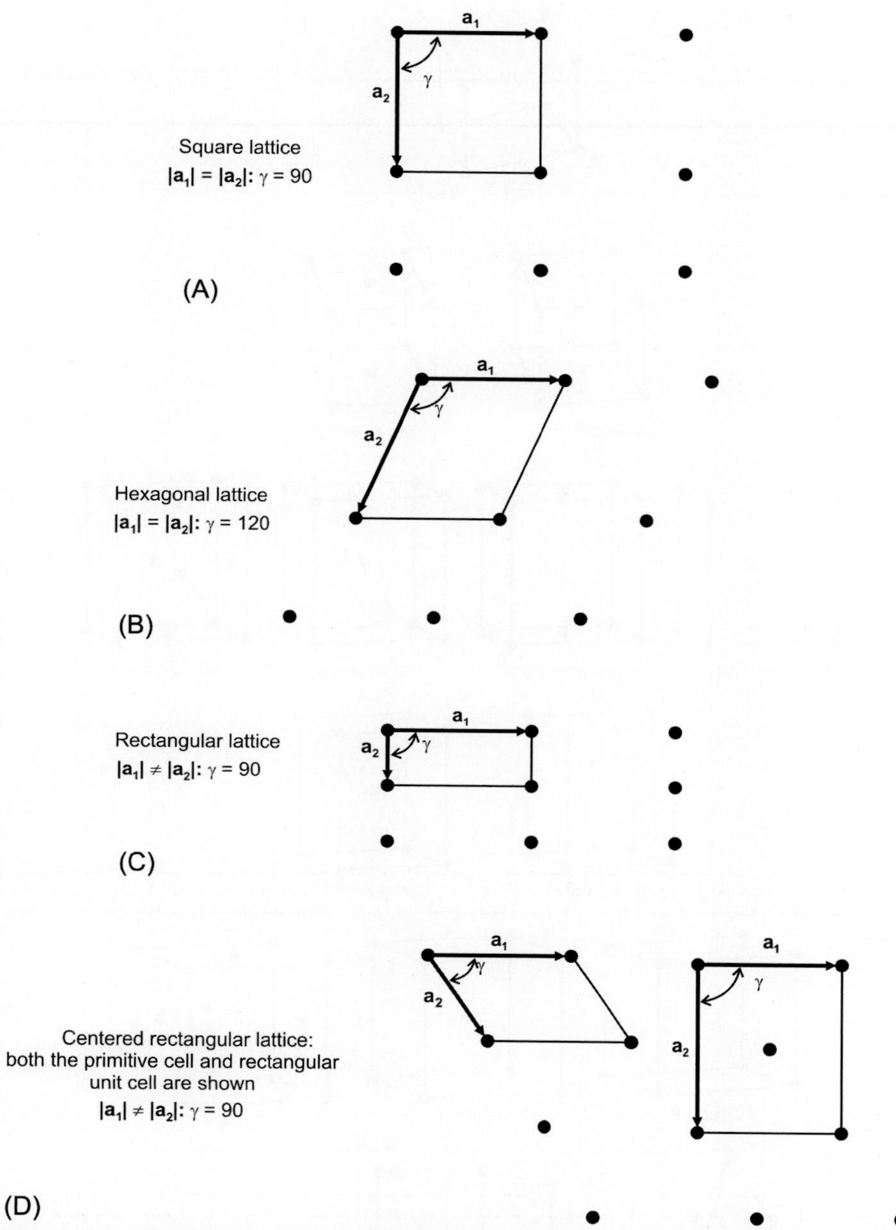

FIG. 1.12 The possible primitive cells of the two-dimensional lattices permitted by the property of invariance under translational and rotational symmetries.

Cubic Crystals In this class of crystals

$$|\mathbf{a}_1| = |\mathbf{a}_2| = |\mathbf{a}_3| \;\&\; \alpha = \beta = \gamma = 90°$$

These are high-symmetry crystals in which the primitive vectors are orthogonal to each other and the repetitive interval is the same along the three axes. The cubic lattices may have sc, bcc, or fcc structures.

Trigonal Crystals In the trigonal symmetry

$$|\mathbf{a}_1| = |\mathbf{a}_2| = |\mathbf{a}_3| \;\&\; \alpha = \beta = \gamma \langle 120°, \neq 90°$$

There is only one type of trigonal crystal in which the unit cell is primitive in nature. Note that the three primitive vectors are equally inclined to each other.

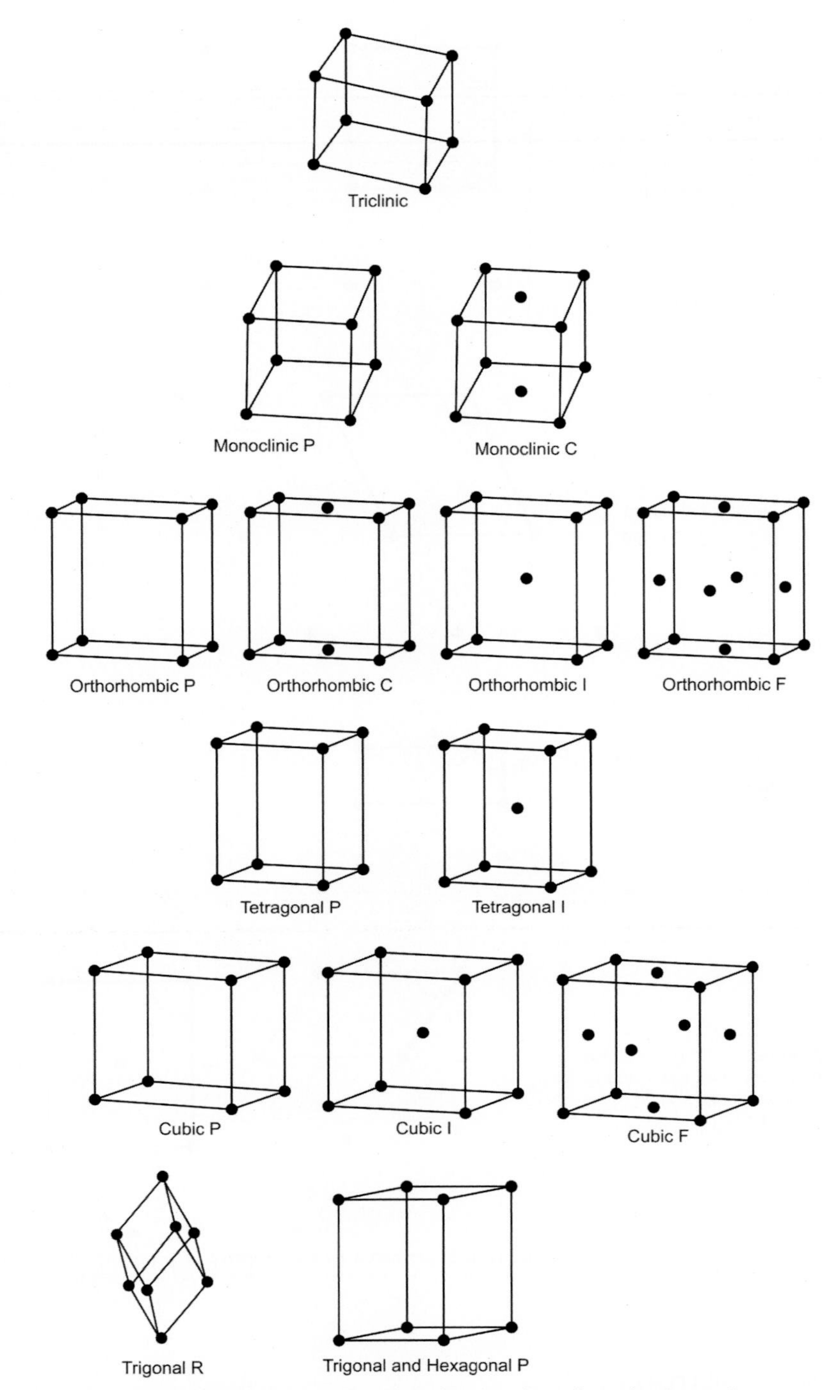

FIG. 1.13 The conventional unit cells of the possible fourteen Bravais lattices in a three-dimensional crystal.

Tetragonal Crystals In tetragonal symmetry

$$|\mathbf{a}_1| = |\mathbf{a}_2| \neq |\mathbf{a}_3| \;\&\; \alpha = \beta = \gamma = 90^\circ$$

There are two crystal lattices. One is simple and the other is a body-centered tetragonal crystal. The simple tetragonal crystal has a primitive unit cell.

Hexagonal Crystals In the hexagonal symmetry

$$|\mathbf{a}_1| = |\mathbf{a}_2| \neq |\mathbf{a}_3| \;\&\; \alpha = \beta = 90^\circ, \gamma = 120^\circ$$

In this class of crystals $\mathbf{a}_1$ and $\mathbf{a}_2$ make $2\pi/3 = 120$ degrees angles, and there is sixfold rotational symmetry: thus the name hexagonal. The third primitive vector $\mathbf{a}_3$ is perpendicular to both $\mathbf{a}_1$ and $\mathbf{a}_2$. The hexagonal lattice is primitive in nature (see Fig. 1.13).

Orthorhombic Crystals In this class of Bravais lattices

$$|\mathbf{a}_1| \neq |\mathbf{a}_2| \neq |\mathbf{a}_3| \; \& \; \alpha = \beta = \gamma = 90°$$

There are four types of orthorhombic crystals. They are simple, base-centered, body-centered, and face-centered orthorhombic crystals. The simple orthorhombic crystal has a primitive cell.

Monoclinic Crystals In this class

$$|\mathbf{a}_1| \neq |\mathbf{a}_2| \neq |\mathbf{a}_3| \; \& \; \alpha = \gamma = 90° \neq \beta$$

There are two lattices. One lattice has a primitive cell with lattice points (atoms) at its corners while the other has a non-primitive cell with base-centered planes formed by $\mathbf{a}_1$ and $\mathbf{a}_2$.

Triclinic Crystals In this class, there is only one lattice with

$$|\mathbf{a}_1| \neq |\mathbf{a}_2| \neq |\mathbf{a}_3| \; \& \; \alpha \neq \beta \neq \gamma$$

The lattice has a primitive cell as there is one lattice point in it.

1.7 SIMPLE CRYSTAL STRUCTURES

Atoms in a crystal have the tendency to settle in close-packed structures. The simplest close-packed structures have either cubic or hexagonal symmetry. Both the translational and the point groups of the full cubic group are of highest symmetry. The operations of the cubic group are as follows:

1. The identity operation $\{I|0\}$.
2. The four-fold rotation $\{S_{4i}|0\}$ about the edge of a cubic unit cell.
3. The two-fold rotation $\{S_{2i}|0\}$ about an edge of a cubic unit cell.
4. The three-fold rotation $\{S_{3i}|0\}$ about the diagonal of a cubic unit cell.
5. The inversion J with respect to the origin.
6. Any of the above-mentioned rotations followed by an inversion about the origin, that is, $J\{S_{4i}|0\}$, $J\{S_{4i}|0\}^2$, $J\{S_{2i}|0\}$, $J\{S_{3i}|0\}$

Note that $\{I|0\}$, $\{S_{4i}|0\}$, $\{S_{4i}|0\}^2$, $\{S_{2i}|0\}$, and $\{S_{3i}|0\}$ form one subgroup while $\{I|0\}$, $\{S_{4i}|0\}^2$, $J\{S_{4i}|0\}$, $J\{S_{2i}|0\}$, and $\{S_{3i}|0\}$ form another subgroup. The compatibility considerations of the translational and point symmetries show that 10 space groups are associated with the full cubic point group. These include sc, bcc, fcc, diamond lattices, and others.

1.7.1 Simple Cubic Structure

The simplest crystal structure is the sc structure in which the primitive translation vectors are given by

$$\mathbf{a}_1 = a\hat{\mathbf{i}}_1, \mathbf{a}_2 = a\hat{\mathbf{i}}_2, \mathbf{a}_3 = a\hat{\mathbf{i}}_3 \tag{1.31}$$

It is a monatomic crystal structure, that is, there is one atom per primitive cell. The volume of the primitive cell is given by

$$V_0 = |\mathbf{a}_1 \cdot \mathbf{a}_2 \times \mathbf{a}_3| = a^3 \tag{1.32}$$

and the density of lattice points (or atoms) in monatomic crystals is given by

$$N_0 = \frac{1}{V_0} = a^{-3} \tag{1.33}$$

In a close-packed sc structure the four atoms in the basal plane of the primitive cell touch each other to form a square. Exactly above these atoms are another four atoms that form the top face of the primitive cell. Hence, in a close-packed sc structure

$$V_a = \frac{4\pi}{3}\left(\frac{a}{2}\right)^3 \tag{1.34}$$

The packing fraction, f_p, in the sc structure is given by

$$f_p = \frac{V_a}{V_0} = \frac{\frac{4\pi}{3}\left(\frac{a}{2}\right)^3}{a^3} = \frac{\pi}{6} = 0.52 \tag{1.35}$$

Therefore, in a crystal with sc structure only 52% of the space is occupied by the atoms.

1.7.2 Body-Centered Cubic Structure

The bcc structure of a pure element is obtained by the penetration of two identical sc structures. Fig. 1.14 shows two identical sc unit cells in which one sc cell is shifted from the other sc cell by a vector.

$$\mathbf{r} = \frac{1}{2}a\left(\hat{\mathbf{i}}_1 + \hat{\mathbf{i}}_2 + \hat{\mathbf{i}}_3\right) \tag{1.36}$$

where a is the magnitude of a side of the cube. Fig. 1.15 shows one of the convenient choices of the primitive translation vectors of the bcc structure. They are given by

$$\begin{aligned} \mathbf{a}_1 &= \frac{1}{2}a\left(\hat{\mathbf{i}}_1 + \hat{\mathbf{i}}_2 - \hat{\mathbf{i}}_3\right) \\ \mathbf{a}_2 &= \frac{1}{2}a\left(-\hat{\mathbf{i}}_1 + \hat{\mathbf{i}}_2 + \hat{\mathbf{i}}_3\right) \\ \mathbf{a}_3 &= \frac{1}{2}a\left(\hat{\mathbf{i}}_1 - \hat{\mathbf{i}}_2 + \hat{\mathbf{i}}_3\right) \end{aligned} \tag{1.37}$$

Problem 1.3

Prove that the average volume per atom in a bcc structure is given by

$$V_0 = a^3/2$$

Problem 1.4

Prove that the angle between the primitive vectors of the bcc structure is 109 degrees, 28′.

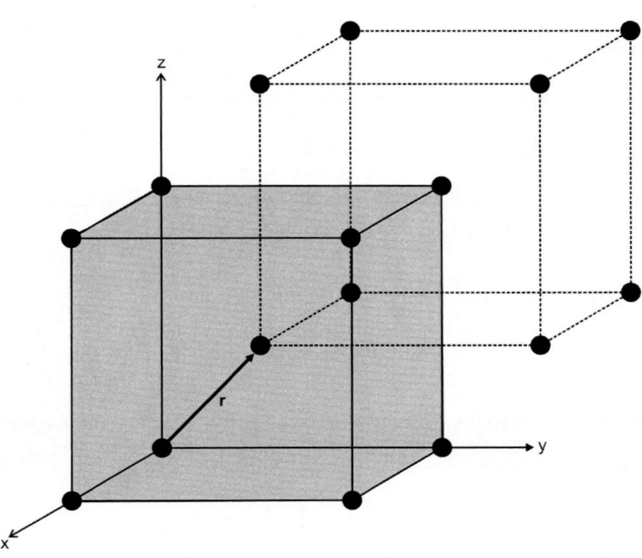

FIG. 1.14 The penetration of two cubic unit cells in the formation of the unit cell of a bcc structure *(shaded region)*.

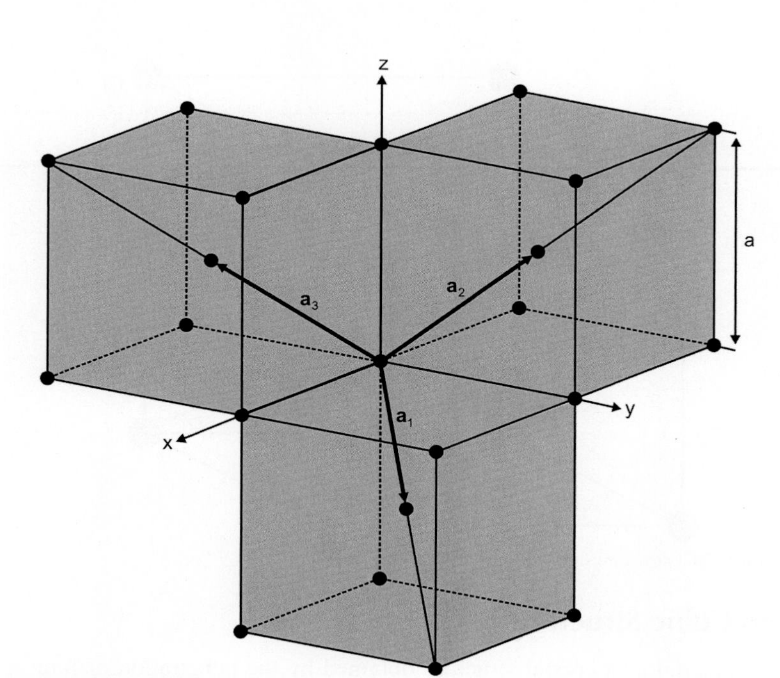

FIG. 1.15 The primitive translation vectors of a bcc structure.

Pure elements, such as V, Cr, Nb, Mo, Ta, and W, possess a bcc structure and are monatomic (one atom per primitive cell). The conventional unit cell and the WS cell for the bcc structure are shown in Fig. 1.16. The WS cell for the bcc structure is a truncated octahedron: the perpendicular bisector planes of the lines joining the 2NNs cut the corners of the octahedron formed by the perpendicular bisector planes of the 1NNs. The WS cell is nearly symmetric about its center. Some compounds, such as CsCl, RbCl, TlBr, and TlI, also exhibit bcc structure. Fig. 1.17 shows the unit cell of CsCl in which two sc lattices of Cs and Cl penetrate into one another. The bcc structure of CsCl has a basis of two atoms and the unit cell contains one molecule of CsCl.

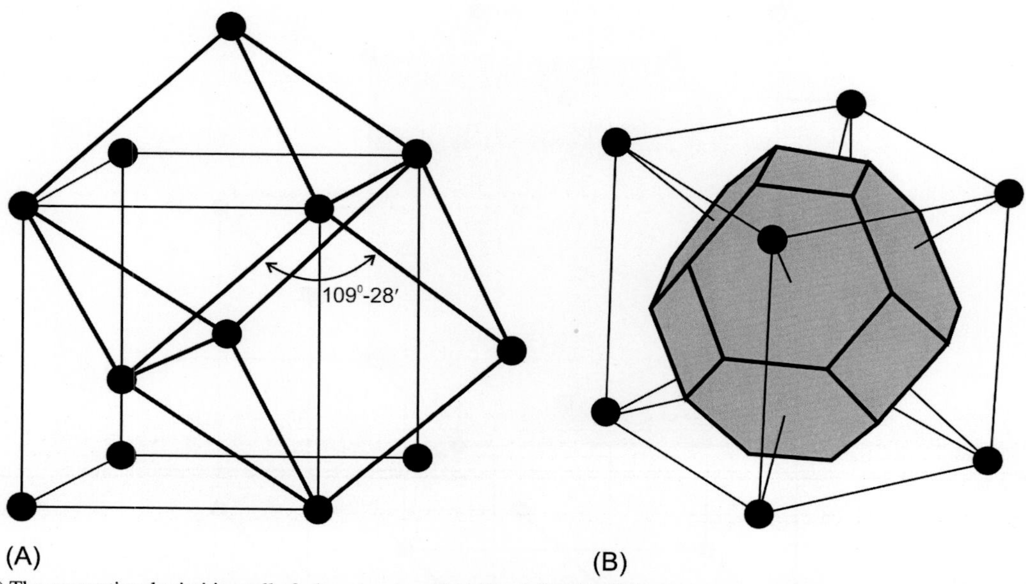

FIG. 1.16 (A) The conventional primitive cell of a bcc structure *(dark lines)*. (B) The WS cell of a bcc structure (*shaded region* enclosed by *dark lines*).

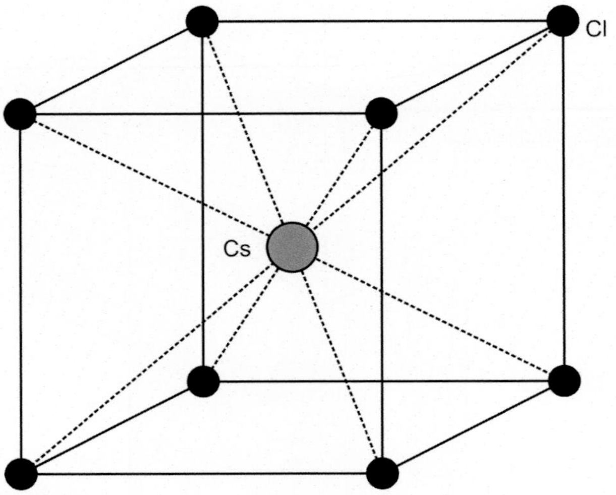

FIG. 1.17 The primitive cell of the CsCl structure.

1.7.3 Face-Centered Cubic Structure

The fcc structure can be considered as a crystal structure obtained by the penetration of four sc structures as shown in Fig. 1.18. Here the black cube shows the primitive cell of the fcc structure with primitive vectors defined by (see Figs. 1.18 and 1.19)

$$\begin{aligned}\mathbf{a}_1 &= \frac{1}{2}a\left(\hat{\mathbf{i}}_1 + \hat{\mathbf{i}}_2\right)\\ \mathbf{a}_2 &= \frac{1}{2}a\left(\hat{\mathbf{i}}_2 + \hat{\mathbf{i}}_3\right)\\ \mathbf{a}_3 &= \frac{1}{2}a\left(\hat{\mathbf{i}}_3 + \hat{\mathbf{i}}_1\right)\end{aligned} \tag{1.38}$$

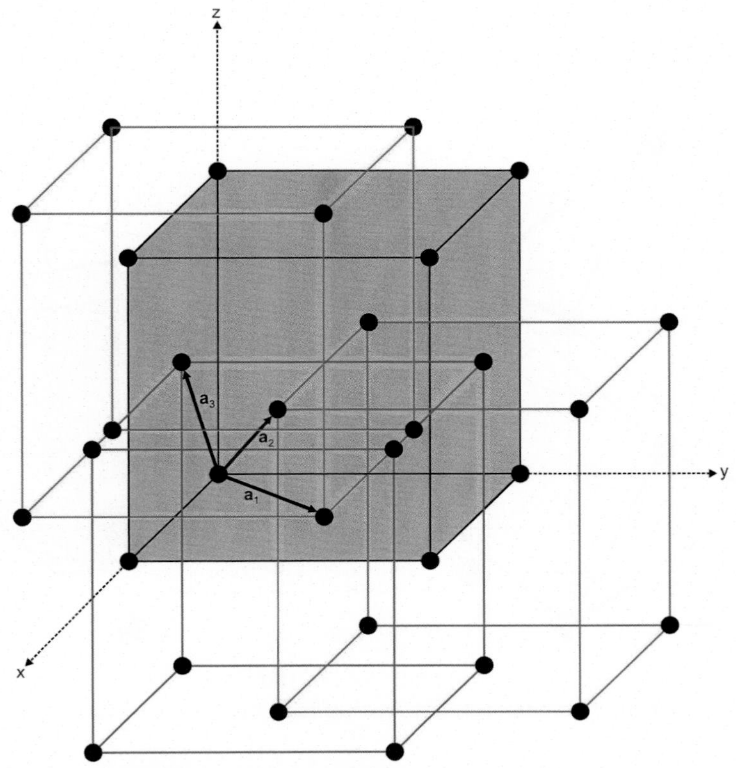

FIG. 1.18 The penetration of four cubic structures in the formation of the unit cell of an fcc structure *(shaded region)*.

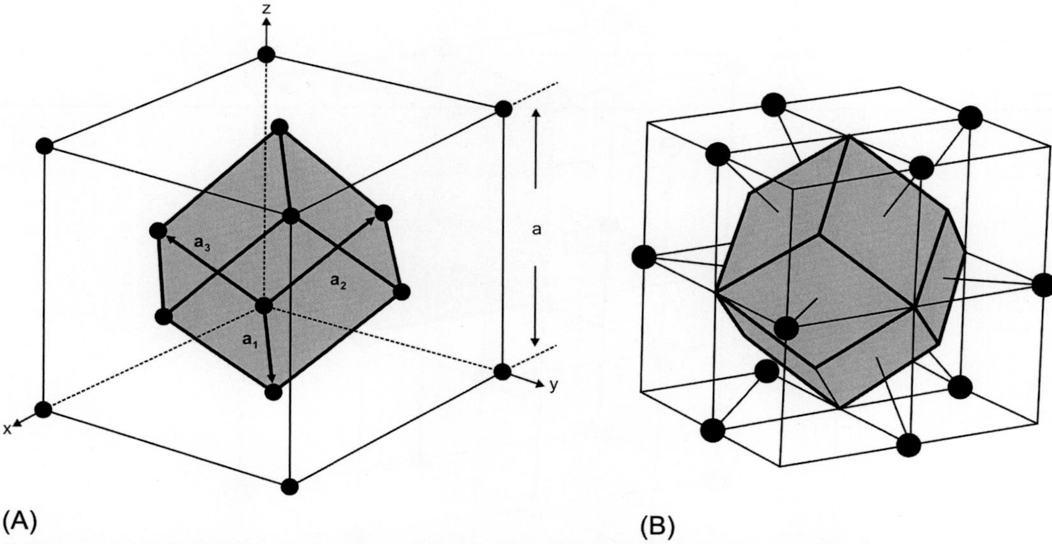

FIG. 1.19 (A) The conventional primitive cell of an fcc structure (*shaded region* enclosed by *dark lines*). (B) The WS cell of an fcc structure (*shaded region* enclosed by *dark lines*).

It is noteworthy that the sc unit cells with green, blue, and red colors are displaced from the black sc unit cell by vectors $-(1/2)[\mathbf{a}_1+\mathbf{a}_2]$, $(1/2)[\mathbf{a}_2-\mathbf{a}_3]$, and $(1/2)[\mathbf{a}_1-\mathbf{a}_3]$, respectively.

Problem 1.5

Prove that the average volume per atom in an fcc structure is given by

$$V_0 = a^3/4 \tag{1.39}$$

Problem 1.6

Prove that the angle between the primitive vectors in an fcc structure is 60 degrees.

Pure elements, such as Cu, Ag, Au, Ni, Pd, and Pt, possess fcc structure and are monatomic in nature. The conventional unit cell and the WS cell for the fcc structure are shown in Fig. 1.19. The WS cell in the fcc structure is a regular 12-faced polyhedron (dodecahedron) and is nearly symmetric about its center.

Many compounds also exhibit fcc structure and NaCl is a good example of this case. Fig. 1.20 shows the fcc structure possessed by a NaCl crystal, which is a Bravais lattice with a basis: the basis consists of one Na atom and one Cl atom and these are separated by one-half of the body diagonal of the unit cube. There are four molecules of NaCl in one unit cell.

The above discussion shows that the bcc crystal structure is obtained by the penetration of two sc structures, while the fcc crystal structure is obtained by the penetration of four sc structures. Therefore, on physical grounds, one expects a maximum packing fraction in the case of fcc crystal structure and a minimum in the sc structure, which is the case in reality.

Problem 1.7

Prove that the packing fraction f_p in

(a) bcc crystal structure is 0.68.

(b) fcc crystal structure is 0.74.

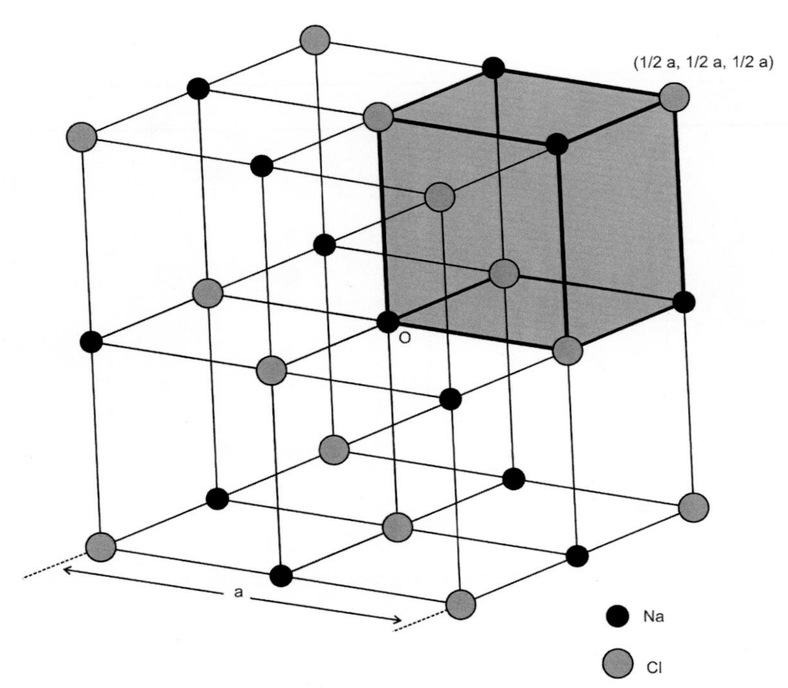

FIG. 1.20 The unit cell of the NaCl structure. The shaded region enclosed by the dark lines shows the primitive cell of the NaCl crystal.

1.7.4 Hexagonal Structure

One of the close-packed structures in crystalline solids is the hcp structure, which is represented by the stacking sequence ABABA … (see Fig. 1.1). We first describe simple hexagonal structure, which is possessed by only a few elements. It exhibits 6-fold symmetry in the basal plane usually called *hexagonal symmetry*. Fig. 1.21 shows the unit and primitive

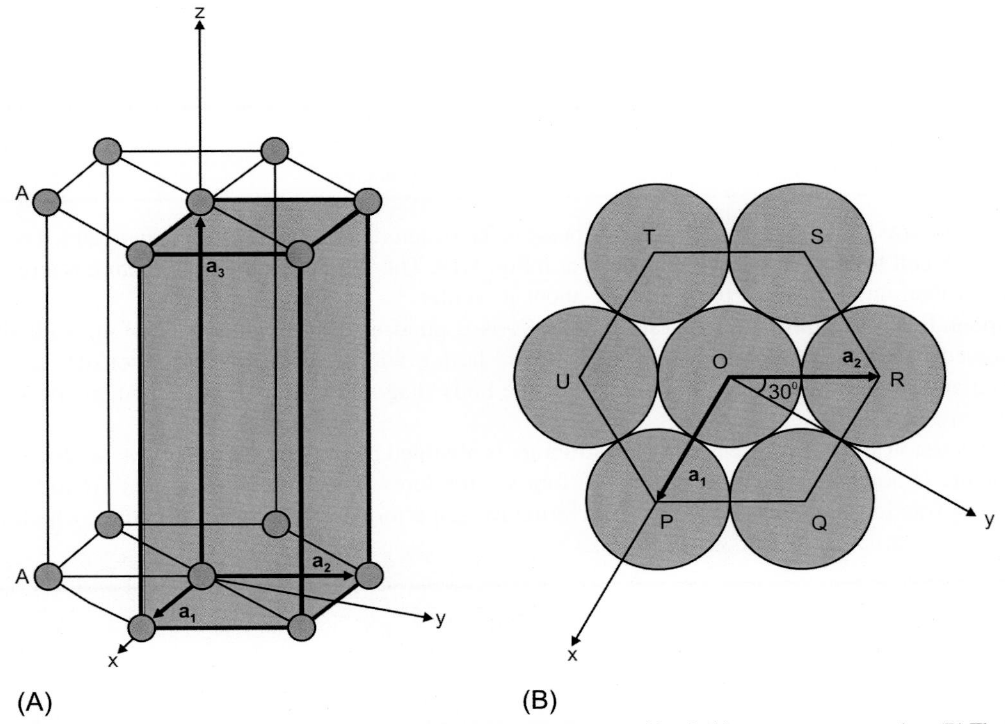

FIG. 1.21 (A) The unit cell and the primitive cell *(shaded region)* of a hexagonal structure with primitive vectors $\mathbf{a}_1$, $\mathbf{a}_2$, and $\mathbf{a}_3$. (B) The basal plane of the hexagonal structure shown in Fig. 1.21A.

cells of the simple hexagonal structure with $\mathbf{a}_1$, $\mathbf{a}_2$, and $\mathbf{a}_3$ as the primitive lattice vectors: $|\mathbf{a}_1| = |\mathbf{a}_2| = a$ and $|\mathbf{a}_3| = c$. It is a monatomic crystal structure. The mathematical forms of the primitive translation vectors vary for different choices of the Cartesian coordinates. Consider the basal plane, shown in Fig. 1.21, with the x-axis along the $\mathbf{a}_1$ primitive vector. In this set of Cartesian coordinates, the primitive vectors are given by

$$\begin{aligned} \mathbf{a}_1 &= \mathbf{a}\hat{\mathbf{i}}_1 \\ \mathbf{a}_2 &= -\frac{1}{2}\mathbf{a}\hat{\mathbf{i}}_1 + \frac{\sqrt{3}}{2}\mathbf{a}\hat{\mathbf{i}}_2 \\ \mathbf{a}_3 &= \mathbf{c}\hat{\mathbf{i}}_3 \end{aligned} \tag{1.40}$$

Another choice for the x- and y-axes in the basal plane of the simple hexagonal structure is shown in Fig. 1.22 for which the primitive translation vectors acquire the form

$$\begin{aligned} \mathbf{a}_1 &= \frac{1}{2}\mathbf{a}\hat{\mathbf{i}}_1 - \frac{\sqrt{3}}{2}\mathbf{a}\hat{\mathbf{i}}_2 \\ \mathbf{a}_2 &= \frac{1}{2}\mathbf{a}\hat{\mathbf{i}}_1 + \frac{\sqrt{3}}{2}\mathbf{a}\hat{\mathbf{i}}_2 \\ \mathbf{a}_3 &= \mathbf{c}\hat{\mathbf{i}}_3 \end{aligned} \tag{1.41}$$

Problem 1.8

Prove that the volume of the primitive cell of the simple hexagonal structure in the crystal space, which is the average volume per atom, obtained either from Eq. (1.40) or (1.41), is given by

$$V_0 = |\mathbf{a}_1 \cdot \mathbf{a}_2 \times \mathbf{a}_3| = \frac{\sqrt{3}}{2}a^2c \tag{1.42}$$

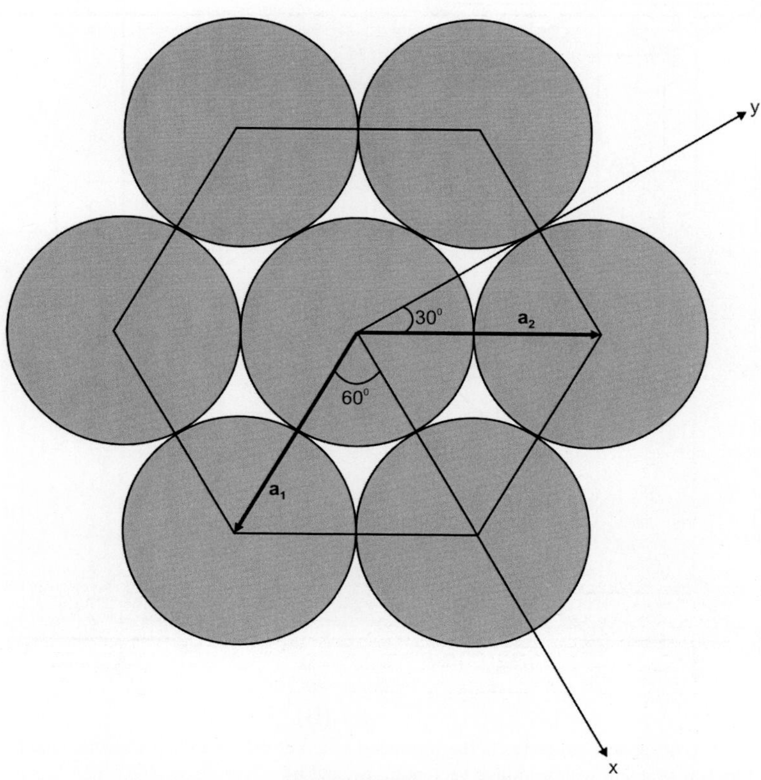

FIG. 1.22 The primitive vectors $\mathbf{a}_1$ and $\mathbf{a}_2$ are depicted with respect to another set of x and y Cartesian coordinates in the basal plane.

1.7.5 Hexagonal Close-packed Structure

Hexagonal close-packed (hcp) structure is obtained by the penetration of one hexagonal structure into another hexagonal structure. The basal plane of one hexagonal unit cell is at half the height of another hexagonal unit cell and is also laterally displaced (see Fig. 1.23A). The unit and primitive cells of hcp structure are shown in Fig. 1.23B. The hcp structure is a diatomic structure: one atom of the basis is taken at the origin while the other at position **r** is defined by

$$\mathbf{r}=\frac{2}{3}\mathbf{a}_1+\frac{1}{3}\mathbf{a}_2+\frac{1}{2}\mathbf{a}_3 \tag{1.43}$$

Hence, the average volume occupied by an atom in an hcp structure is $\left(\sqrt{3}/4\right)a^2c$. In an ideal close packing of the hcp structure, the distance between the two basis atoms in a primitive cell must be equal to a, that is, $|\mathbf{r}|=a$. So

$$\left|\frac{2}{3}\mathbf{a}_1+\frac{1}{3}\mathbf{a}_2+\frac{1}{2}\mathbf{a}_3\right|^2=a^2$$

The above equation can be written as

$$\left[\frac{2}{3}\mathbf{a}_1+\frac{1}{3}\mathbf{a}_2\right]\cdot\left[\frac{2}{3}\mathbf{a}_1+\frac{1}{3}\mathbf{a}_2\right]+\frac{1}{4}c^2=a^2 \tag{1.44}$$

which on simplification gives

$$\frac{c}{a}=\sqrt{8/3}=1.633 \tag{1.45}$$

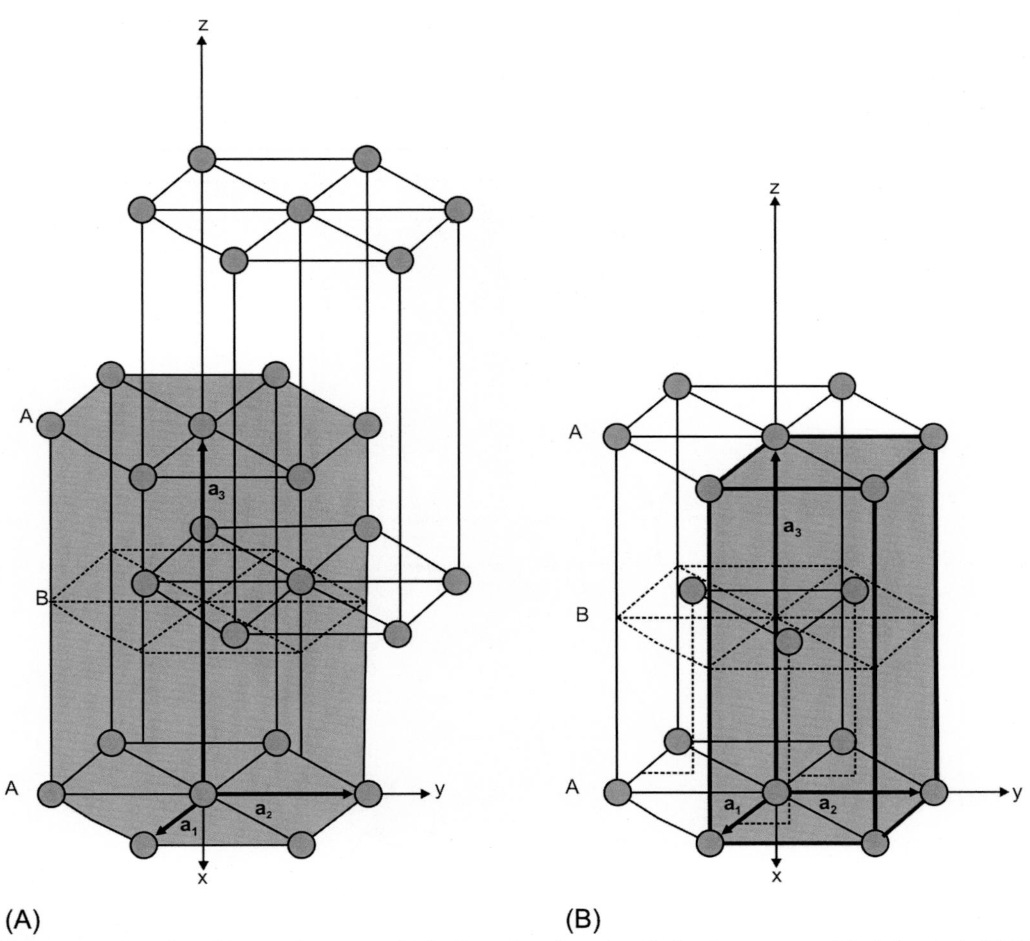

FIG. 1.23 (A) The penetration of two hexagonal structures in the formation of a unit cell of an hcp structure *(shaded region)*. (B) The unit cell and primitive cell *(shaded region)* of an hcp structure with primitive vectors $\mathbf{a}_1$, $\mathbf{a}_2$, and $\mathbf{a}_3$.

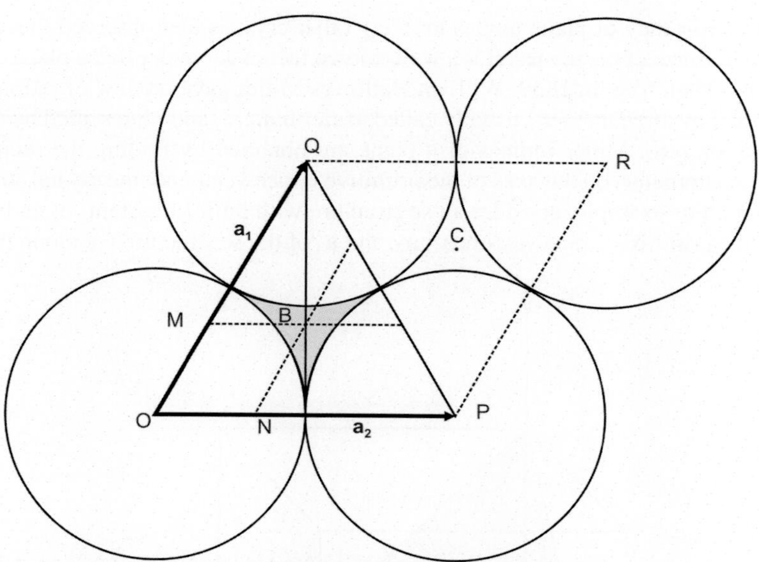

FIG. 1.24 The close packing of four atoms in the basal plane of the primitive cell of an hcp structure.

Eq. (1.45) gives the ideal value of the axial ratio c/a for the hcp structure. It is worth mentioning here that in the real crystals the axial ratio c/a deviates significantly from its ideal value 1.633. It simply means that the atoms do not exhibit ideal close packing in real crystals.

Alternate Method for c/a

Fig. 1.24 shows the close packing of four atoms in the bottom layer of the primitive cell in an hcp structure in which the B and C points represent the centers of the two types of voids between them. The center of an atom in the second layer of the hcp structure is at the point B, which is at a distance of c/2 above the bottom layer. Let $\mathbf{a}_1$ and $\mathbf{a}_2$ be the vectors in the basal plane as shown in the figure, then $|\mathbf{a}_1| = |\mathbf{a}_2| = a$. Let the third vector $\mathbf{a}_3$ be in the vertical direction passing through O such that $|\mathbf{a}_3| = c$. Now to arrive at the center of the atom situated at the point B in the second layer, one has to travel a distance OM ($= |\mathbf{a}_1|/3$) along the vector $\mathbf{a}_1$ and then from M, a distance $|\mathbf{a}_2|/3$ parallel to the vector $\mathbf{a}_2$, and finally a vertical distance of c/2 ($|\mathbf{a}_3|/2$) to reach the point B. Hence the position vector $\vec{\delta}$ of the atom at position B in the second layer with respect to the point O is given by

$$\vec{\delta} = \frac{\mathbf{a}_1 + \mathbf{a}_2}{3} + \frac{\mathbf{a}_3}{2} \tag{1.46}$$

In the case of close packing of the atoms $\left|\vec{\delta}\right| = a$, that is,

$$\left|\frac{\mathbf{a}_1 + \mathbf{a}_2}{3} + \frac{\mathbf{a}_3}{2}\right|^2 = a^2$$

which on simplification gives

$$\frac{c}{a} = \sqrt{8/3} = 1.633.$$

Problem 1.9

Prove that the packing fraction f_p in an hcp structure is 0.74.

From the problems 1.7 and 1.9 it is evident that the packing fraction for both the fcc and hcp structures is the same.

1.8 MILLER INDICES

An actual crystal has a definite shape bounded by a set of planes. In 1669, Niels Stenson found that the angles between the similar faces of a quartz crystal are always the same, no matter how the crystal is prepared. In 1772, Jean Baptiste Rome de l'Isle extended the law of consistency of plane angles to many other crystals and observed that their values depend on a particular crystal and not on the size of the planes. It is a well-known fact in geometry that a plane can be specified uniquely by three noncollinear points. Therefore, in 1839, William Hallowes Miller gave the law of rational numbers according to which a plane can be specified by three integers, usually called *Miller indices*, and represented by (hkl) where h, k, and l are integers, positive, negative, or zero. Miller indices of a plane are obtained by finding the reciprocals of the intercepts (expressed in units of lattice parameter) on the axes of the primitive lattice vectors of the crystal structure and then reducing them to the smallest integers. For example, consider an sc structure with lattice constant 'a' and Cartesian coordinates as shown in Fig. 1.25a. Here the primitive lattice vectors $\mathbf{a}_1$, $\mathbf{a}_2$, and $\mathbf{a}_3$ of the sc structure are along the Cartesian coordinates.

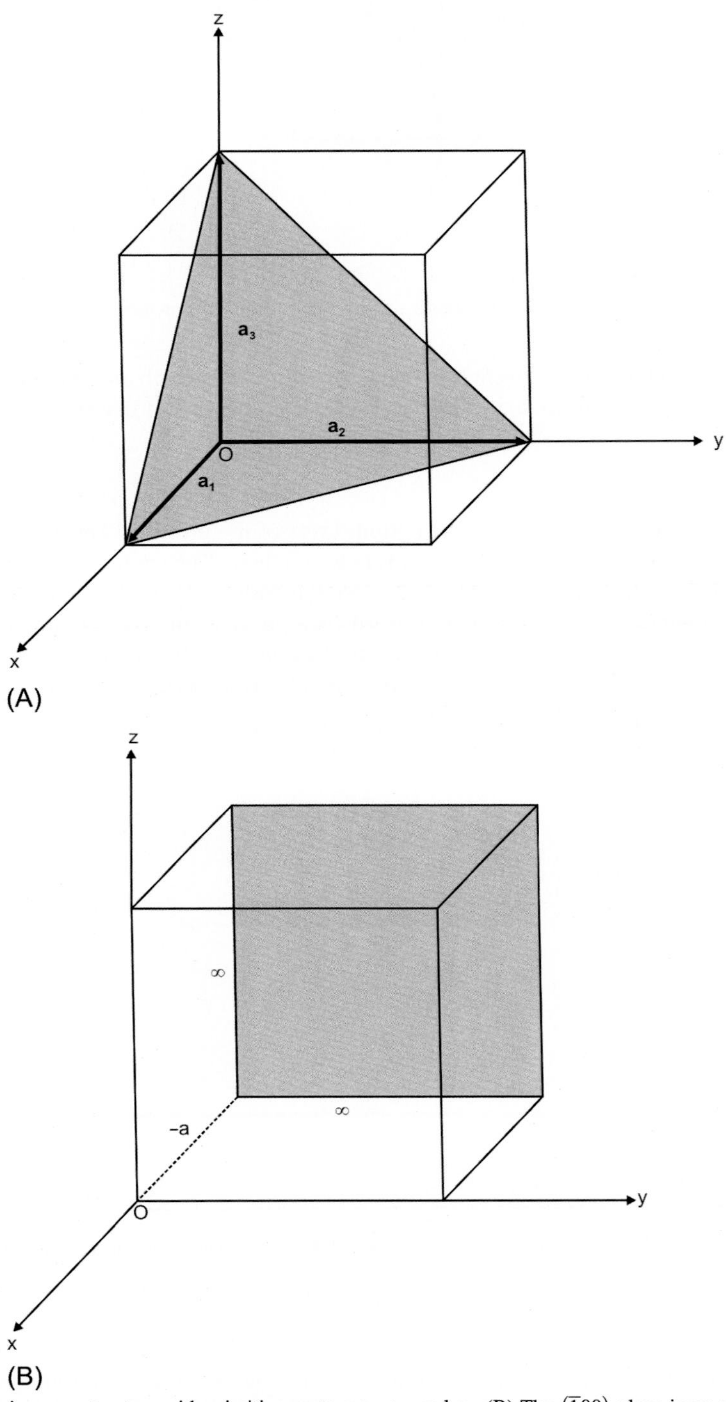

FIG. 1.25 (A) The (111) plane in an sc structure with primitive vectors $\mathbf{a}_1$, $\mathbf{a}_2$, and $\mathbf{a}_3$. (B) The $(\bar{1}00)$ plane in an sc structure.

The magnitudes of the intercepts of the plane shown in the figure by the shaded area along $\mathbf{a}_1$, $\mathbf{a}_2$, and $\mathbf{a}_3$ are equal to a and the reciprocals of the intercepts are 1/a, 1/a, and 1/a. The reduction of these values to the smallest integers yields 1,1,1. Hence the Miller indices of the plane are written as (111). If any of the intercepts of a plane are on the negative side of the axes, then the corresponding smallest integer is also negative and is represented by putting a bar over the number, for example, $-1=\bar{1}$. Thus, the Miller indices of a plane (the shaded area) in Fig. 1.25b are written as ($\bar{1}$00).

Therefore, in general, (hkl) represents a single plane. But in a crystal structure, there can exist a number of equivalent planes corresponding to particular Miller indices (hkl). Such planes are collectively represented by the symbol {hkl}. For example, the Miller indices of the six faces of an sc structure are represented by (100), (010), (001), ($\bar{1}$00), (0$\bar{1}$0), and (00$\bar{1}$). All of the six planes of a cubic crystal are equivalent by symmetry and are therefore denoted by {100}.

Problem 1.10

Draw the planes in an sc structure represented by the Miller indices (100), (200), and (110).

A direction in a crystal structure can also be represented by a set of the smallest three integers written in a square bracket as [hkl]. In order to examine what these three integers represent, consider the sc structure shown in Fig. 1.26. In the direction OA, the first lattice point from the origin is positioned at A. To move from 0 to A one has to move a distance a along the $\mathbf{a}_1$-axis, that is, OC, and then a distance a parallel to the $\mathbf{a}_2$-axis, that is, CA. The coordinates of point A are (a,a,0) and the indices of this direction are obtained by reducing the coordinates to smallest integers. Hence, the direction OA is represented by [110]. Similarly, the direction OB is represented by [111]. Note that in cubic crystals, the direction [hkl] is perpendicular to the plane (hkl) with the same indices. This fact can straightway be proved from Fig. 1.26. Here the x-axis with [100] is perpendicular to the front face of the cubic cell represented by (100). Similarly, the lines [110] and [111] are perpendicular to the planes (110) and (111).

Let us examine the representation of planes in a hexagonal structure having lower symmetry. Fig. 1.27 shows the basal plane of the hexagonal structure with primitive vectors $\mathbf{a}_1$ and $\mathbf{a}_2$ passing through the origin. The third primitive vector $\mathbf{a}_3$ is perpendicular to the basal plane and passes through the origin. The planes passing through AB, BC, CD, DE, EF, and FA, and perpendicular to the basal plane, are called prism faces of the hexagon and comprise a set of similar planes possessing 6-fold rotational symmetry about the origin. With respect to the primitive vectors $\mathbf{a}_1$ and $\mathbf{a}_2$, the Miller indices of the prism

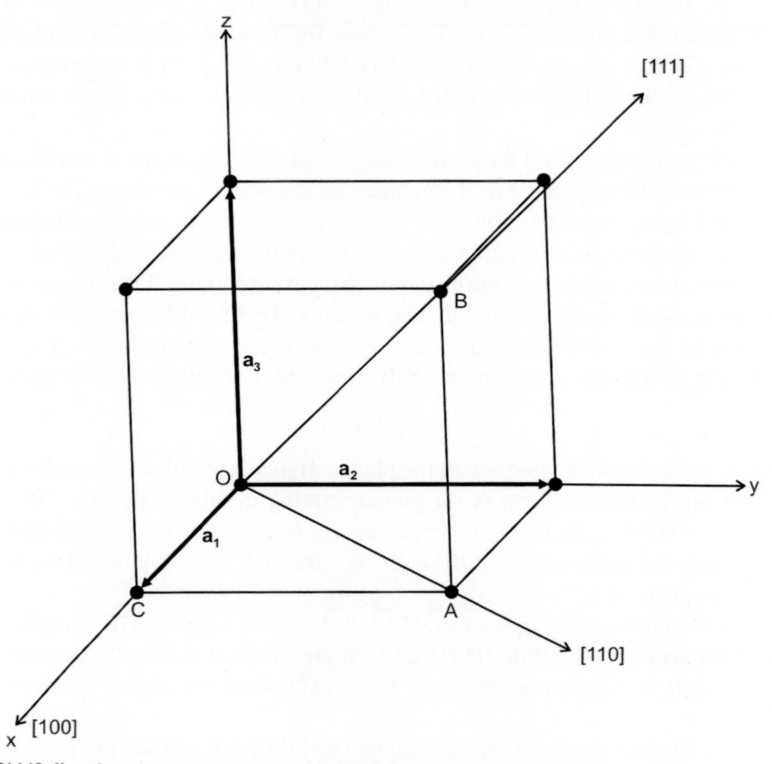

FIG. 1.26 [100], [110], and [111] directions in an sc structure with primitive vectors $\mathbf{a}_1$, $\mathbf{a}_2$, and $\mathbf{a}_3$.

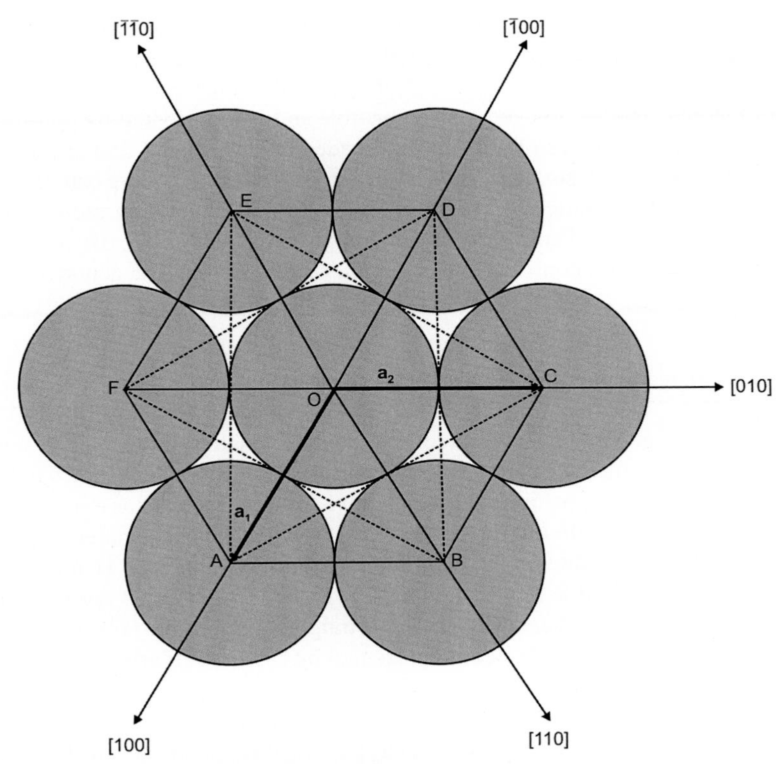

FIG. 1.27 [100], [010], [110] directions in the basal plane of the hexagonal crystal structure.

faces passing through AB, BC, CD, DE, EF, and FA are (100), (010), $(\bar{1}10)$, $(\bar{1}00)$, $(0\bar{1}0)$, and $(1\bar{1}0)$, respectively. These Miller indices form two groups: the first group consists of planes (100), (010), $(\bar{1}00)$, and $(0\bar{1}0)$ and the second group consists of the $(\bar{1}10)$ and $(1\bar{1}0)$ planes. But one expects similar indices for the equivalent planes in a crystal structure. We find the same thing in the situation of planes passing through AC, BD, CE, DF, EA, and FB, and perpendicular to the basal plane, which have Miller indices (110), $(\bar{1}20)$, $(\bar{2}10)$, $(\bar{1}\bar{1}0)$, $(1\bar{2}0)$, $(2\bar{1}0)$, respectively. These again form two groups, that is, (110), $(\bar{1}\bar{1}0)$ and $(\bar{1}20)$, $(\bar{2}10)$, $(1\bar{2}0)$, $(2\bar{1}0)$.

To obtain one group of Miller indices for all the equivalent planes, Bravais adopted a different coordinate system that has four axes, with three being in the basal plane. Fig. 1.28 shows the coordinate axes $\mathbf{a}_1$, $\mathbf{a}_2$, and $\mathbf{a}_4$ in the basal plane, which are inclined at an angle of $2\pi/3$ to one another. Now four indices will represent a plane, which are called *Miller-Bravais indices*: finding the reciprocals of intercepts on the four axes and then reducing them to the smallest integers. Miller-Bravais indices are represented as (hkil), where the index i corresponds to the $\mathbf{a}_4$ axis. For all the planes, the intercept on the $\mathbf{a}_4$ axis has a definite relationship with the intercepts on the $\mathbf{a}_1$ and $\mathbf{a}_2$ axes. To find this relationship, consider a plane passing through AC and perpendicular to the basal plane (Fig. 1.28). For this plane, the intercepts on the $\mathbf{a}_1$ and $\mathbf{a}_2$ axes are equal to a in magnitude, while on the $\mathbf{a}_4$ axis the intercept is $-$ OB/2 = $-$a/2. For this plane it is straightforward to prove that

$$i = -(h + k) \tag{1.47}$$

It can be proved that the above relation holds good for all the planes. Hence, the index i is completely determined by the first two Miller indices. Now the Miller-Bravais indices for planes passing through AB, BC, CD, DE, EF, and FA become $(10\bar{1}0)$, $(01\bar{1}0)$, $(\bar{1}100)$, $(\bar{1}010)$, $(0\bar{1}10)$, and $(1\bar{1}00)$, which evidently form one group in contrast to the Miller indices. Similarly, the Miller-Bravais indices for planes passing through AC, BD, CE, DF, EA, and FB are $(11\bar{2}0)$, $(\bar{1}2\bar{1}0)$, $(\bar{2}110)$, $(\bar{1}\bar{1}20)$, $(1\bar{2}10)$, and $(2\bar{1}\bar{1}0)$, respectively.

A direction can also be represented by four indices as [hkil]. A direction is defined by translations parallel to each of the four axes that give motion in the required direction. These translations are then reduced to the smallest integers. In addition, the first three indices must satisfy the condition given by Eq. (1.47). High-symmetry directions in the basal plane of a hexagon are shown in Fig. 1.29.

Let us examine the direction of primitive vector $\mathbf{a}_1$ represented by [100] in the system of three indices (old system). In the system of four indices (new system) one can obtain the representation in the same way. If one moves directly from the origin 0

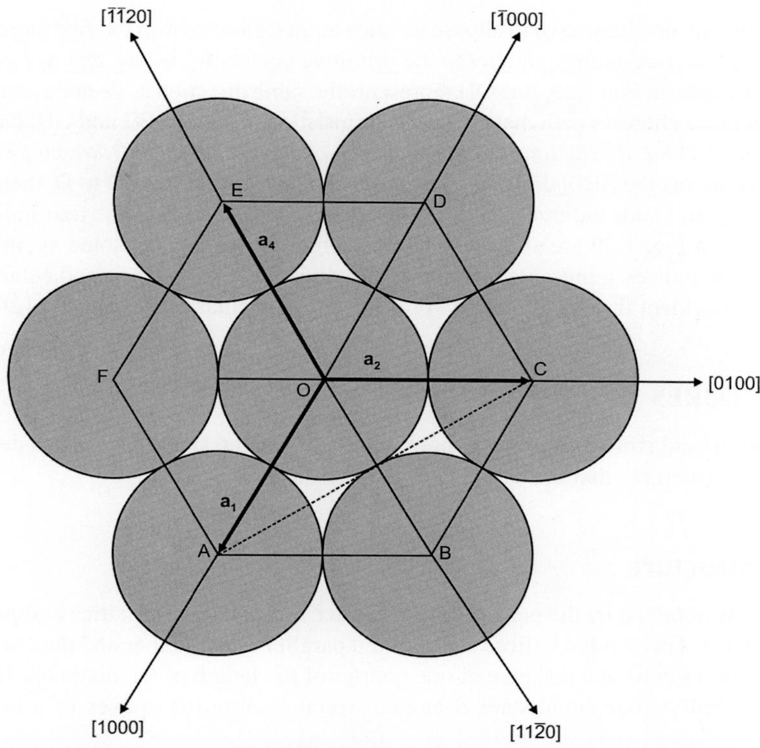

FIG. 1.28 The primitive vectors $\mathbf{a}_1$, $\mathbf{a}_2$, and $\mathbf{a}_4$ in the basal plane of the hexagonal structure. The figure also shows the directions [1000], [0100], $[11\bar{2}0]$, $[\bar{1}000]$, and $[\bar{1}\bar{1}20]$, in terms of Miller-Bravais indices, in the basal plane.

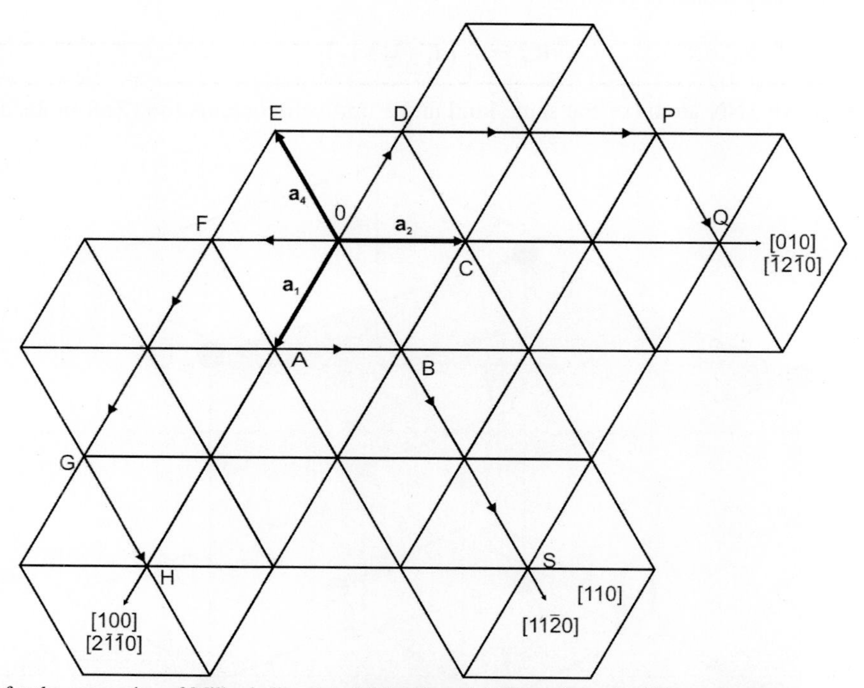

FIG. 1.29 The method for the conversion of Miller indices into Miller-Bravais indices along the different symmetry directions.

to point A, then the translations along the four directions are a, 0, 0, 0 and hence the direction of $\mathbf{a}_1$ is represented by [1000]. Similarly, the direction of $\mathbf{a}_3$ becomes [0001] if the translation is performed directly from the origin to the atom at a distance c perpendicular to the basal plane. But in the [1000] and [0001] representations the property given by Eq. (1.47) is not satisfied.

In order to fulfill this requirement, one may have to choose a lattice atom different from the first atom in a particular direction. Further, to reach this lattice point, translations parallel to the primitive vectors $\mathbf{a}_1$, $\mathbf{a}_2$, $\mathbf{a}_3$, and $\mathbf{a}_4$ have to be chosen such that Eq. (1.47) is satisfied. For example, in Fig. 1.29, line OH represents the same direction as vector $\mathbf{a}_1$, that is, [100]. Starting from the origin, to reach the point H we choose a path that consists of translations OF, then FG and GH; these are parallel to vectors $-\mathbf{a}_2$, $\mathbf{a}_1$, and $-\mathbf{a}_4$, respectively. These translations have magnitudes of 2a, −a, −a, and 0, which give the indices of direction as $[2\bar{1}\bar{1}0]$. Similarly, OQ represents the [010] direction and, to reach Q, we travel from O to D, then from D to P, and finally from P to Q (Fig. 1.29). This path yields indices $[\bar{1}2\bar{1}0]$ for the direction of vector $\mathbf{a}_2$. The four indices representation of the direction satisfies Eq. (1.47). In Fig. 1.29 are shown four indices representations for some symmetry directions. Another important property of the four indices representation is that the direction [hkil] is perpendicular to the plane (hkil). For example, from Fig. 1.28, it is evident that the direction $[11\bar{2}0]$ is perpendicular to the plane $(11\bar{2}0)$.

1.9 OTHER STRUCTURES

There are a number of other crystal structures that are more involved than the simple structures described earlier. We shall briefly describe some of the structures that are relevant to the present text.

1.9.1 Zinc Sulfide Structure

Zinc sulfide (ZnS) structure is obtained by the penetration of two fcc lattices: one fcc lattice composed of Zn atoms and the other of S atoms (see Fig. 1.30). The two fcc lattices are oriented parallel to each other and the corner of one cube is placed on the body diagonal of the other cube at a distance of one quarter of the length of the diagonal. Here the atoms of the two kinds are connected tetrahedrally: four equidistant S atoms, which occupy the apexes of a tetrahedron, surround one Zn atom.

The space lattice for the ZnS crystal structure can be considered as an fcc lattice with a basis of two atoms, one Zn atom and the other an S atom. Let the lattice constant of the ZnS structure be a, that is, the length of the edge of the cube. Then, the vector connecting the two basis atoms is given by

$$\mathbf{R}_m = \frac{a}{4}\left(\hat{\mathbf{i}}_1 + \hat{\mathbf{i}}_2 + \hat{\mathbf{i}}_3\right) \tag{1.48}$$

Note that an atom has twelve 1NN atoms of the same kind and a unit cell contains four ZnS molecules.

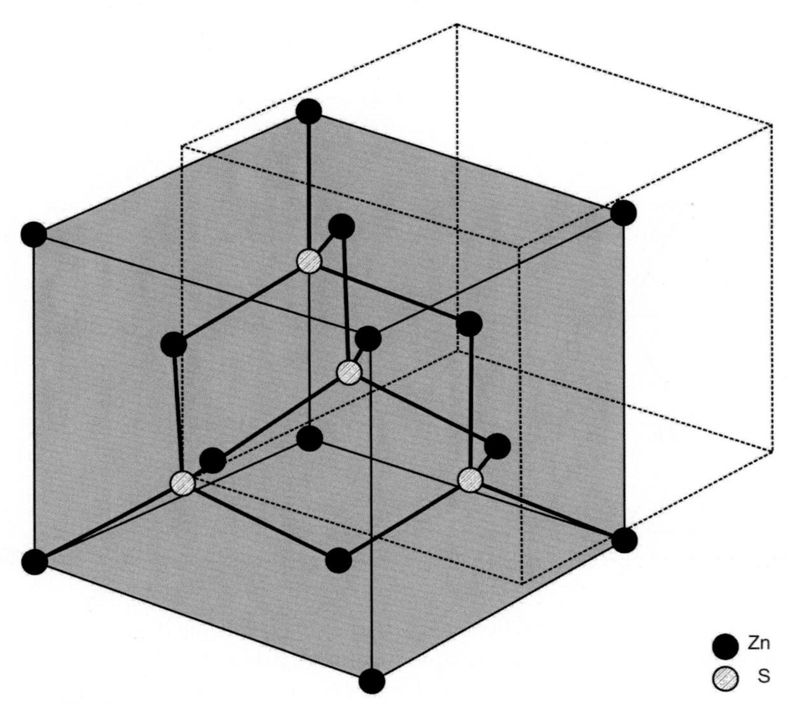

FIG. 1.30 The crystal structure of the ZnS compound.

1.9.2 Diamond Structure

The crystal structure of diamond is exactly similar to that of cubic ZnS, in which both the interpenetrating fcc lattices are made of carbon atoms (see Fig. 1.31). Hence the space lattice of diamond structure is an fcc lattice with a basis of two identical carbon atoms situated at (0, 0, 0) and (a/4, a/4, a/4). Here, each C atom has four 1NNs, which are positioned at the apexes of a tetrahedron, and twelve 2NNs. There are eight carbon atoms in a unit cube.

1.9.3 Wurtzite Structure

Wurtzite structure has basic hexagonal symmetry. It can be considered as being formed by the penetration of two hcp lattices (see Fig. 1.32). The two hcp lattices have the same axis ($\mathbf{a}_3$–axis) but one of them is displaced with respect to the other. The wurtzite structure may be considered as an hcp structure with a basis of two atoms. The primitive vector along $\mathbf{a}_3$ has a length of 3/8 times the $\mathbf{a}_3$ vector, that is, (3/8) $\mathbf{a}_3$. In the wurtzite structure, the atoms are also arranged with tetrahedral symmetry, that is, an atom has four 1NN atoms of another kind and twelve 2NN atoms of the same kind. It is noteworthy that the atomic arrangement along the $\mathbf{a}_3$ axis in the wurtzite structure is similar to the atomic arrangement along the [111] direction in the ZnS structure. In the wurtzite structure, there are four atoms per unit cell. Hence, the average volume per atom in the wurtzite structure is given by $(\sqrt{3}/8)a^2c$. The wurtzite structure has uniaxial symmetry and a number of piezoelectric and pyroelectric crystals possess this structure.

1.9.4 Perovskite Structure

Some naturally occurring minerals have cubic structure with the formula ABO_3, where A and B are cations while O is an oxygen anion, for example, $BaTiO_3$ and $CaTiO_3$. This structure is generally called perovskite structure and is shown in Fig. 1.33. Most of the materials having this structure exhibit ferroelectric behavior and are therefore important from a technological point of view.

In the perovskite structure A^{+m} occupy the corners of the cubic unit cell and the centers of the faces of the cube are occupied by O^{-2} ions. The O^{-2} ions form an octahedron at the center of which is located a small B^{+n} ion. One of the important compounds possessing perovskite structure is $BaTiO_3$. In this compound Ba^{+2} ions occupy the corners of the cubic unit cell while Ti^{+4} is situated at the center of the oxygen octahedron. But the atomic arrangement is not restricted only to divalent and quadrivalent ions as in $BaTiO_3$, $CaTiO_3$, or $SrNbO_3$. In fact, compounds such as $KNbO_3$ and $LaAlO_3$ also possess the same structure: here the ions are K^{+1} and Nb^{+5} in $KNbO_3$ and La^{+3} and Al^{+3} in $LaAlO_3$. It appears that the

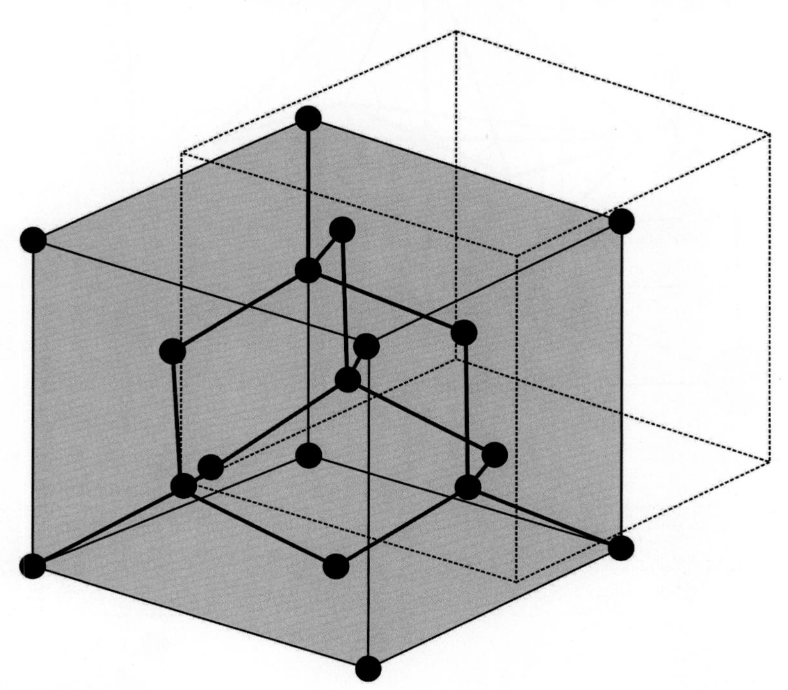

FIG. 1.31 The crystal structure of diamond.

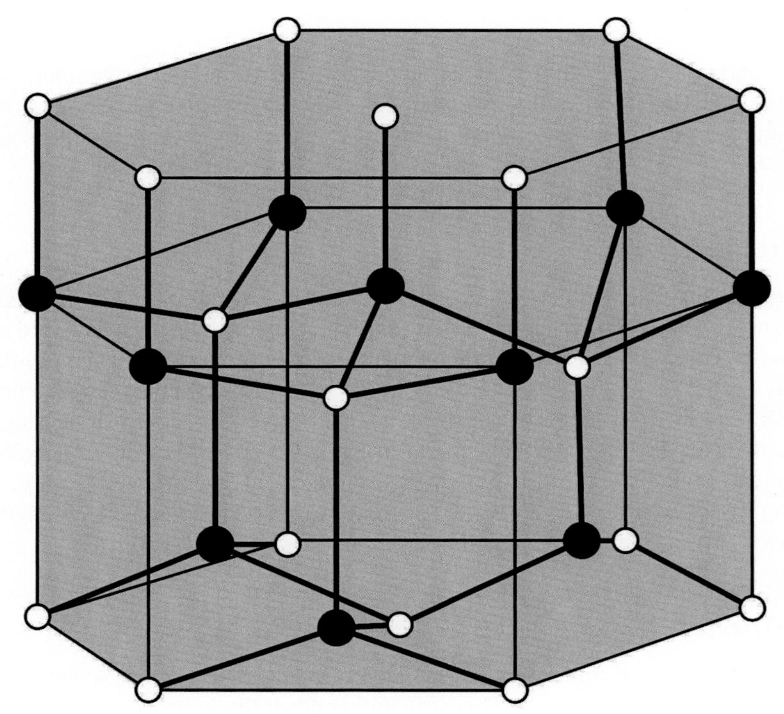

FIG. 1.32 Wurtzite crystal structure.

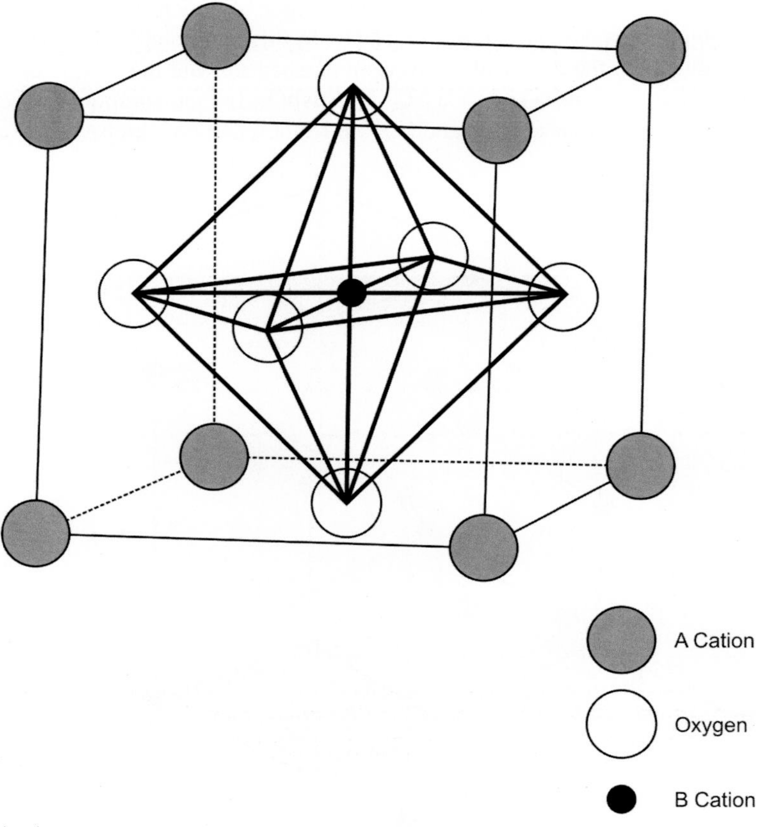

FIG. 1.33 ABO_3 perovskite structure.

size of the atoms is a more important factor than the valency of the atoms in determining the arrangement of the atoms. Hence, in general, perovskite structure can be assigned the formula $A^{+m}B^{+n}O_3$ with (m,n) = (2,4), (3,3), or (1,5). Almost all the ABO_3 compounds are insulators except when A and B are both nontransitional elements. But these metallic compounds generally have low carrier density.

There is an intimate relationship between the atomic arrangement and the ferroelectric properties of a material. For example, in $BaTiO_3$, the Ti^{+4} ion is considerably smaller than the space available inside the oxygen octahedron. As a result, below a certain temperature called the Curie temperature, the structure is slightly deformed, with Ba^{+2} and Ti^{+4} ions displaced relative to the O^{-2} ions, thereby developing a dipole moment. The displacement of Ba^{+2} and Ti^{+4} ions gives rise to spontaneous polarization in $BaTiO_3$.

Another important property of ABO_3 compounds is that A and/or B sites may be partially occupied by cations of a different kind (say C) giving rise to $A_{1-x}C_xBO_3$ or $AB_{1-x}C_xO_3$ compounds. The common examples of such compounds are $K_{1-x}Ba_xBiO_3$, $SrGa_{1-x}Nb_xO_3$, $BaPb_{1-x}Bi_xO_3$ etc. High-temperature superconductivity is observed in the pseudoternary compound $BaPb_{1-x}Bi_xO_3$ with $T_c = 13K$. The compound $K_{1-x}Ba_xBiO_3$ exhibit superconductivity with $T_c = 30K$. Replacing Ba^{+2} by K^{+1} is likely to generate oxygen vacancies. The deficiency of oxygen is responsible for enhanced electron-phonon interactions, which may yield high-T_c behavior. In these compounds, the conductivity depends on the value of parameter x, which is very difficult to control. In this sense, such materials are sometimes called alloys.

1.9.5 High-T_c Superconductors

In 1986, Bednorz and Muller discovered a new class of *superconductors* in which the transition temperature T_c is considerably higher than in normal superconductors; thus, these are commonly known as high-T_c superconductors. They discovered superconductivity in $La_{2-x}Ba_xCuO_4$ compounds with $T_c > 30$ K: $La_{2-x}Ba_xCuO_4$ with $x = 0.15$ exhibits bulk superconductivity with $T_c \sim 35$ K. Later, a large number of superconductors were found with different values of T_c and it is interesting to note that all of the superconductors contain layers of copper oxide in their crystal structures. Therefore, they are also sometimes called oxide superconductors. Historically, one can say that the high-T_c superconductive cuprates have evolved from materials related to the perovskite family ABO_3. All of the high-T_c superconductors can be grouped into three categories.

1. The first group of high-T_c superconductors is represented by the formula $(La, M)_2CuO_4$, which means that some of the positions of La atoms in La_2CuO_4 are occupied by atoms labeled as M. These superconductors have a K_2NiO_4-type structure, as shown in Fig. 1.34A. One should note that Fig. 1.34A shows a schematic ideal structure but the actual structure exhibits some distortions. For a theoretical study, the ideal structure is more suitable as it makes mathematical treatment simple. The highest T_c achieved in these superconductors is 35 K for $La_{1.85}Sr_{0.15}CuO_{4-y}$, where y represents the deficiency of oxygen atoms in the structure. The value of T_c is very sensitive to the value of y, which is difficult to control.
2. The second group can be represented by a general chemical formula $MBa_2Cu_3O_{7-y}$, where y ranges from 0.0 to 0.5. Here, M can be one of the rare-earth elements Sc, Y, La, Nd, Sm, Gd, Tb, Dy, Ho, Er, etc., and binary combinations with Lu, Sc, Y, and La are prototype rare-earth elements. This structure is usually called the 123 structure, as there is 1 atom of M, 2 atoms of Ba, and 3 atoms of Cu. The ideal 123 structure is shown in Fig. 1.34B. In these superconductors, a suitable choice for the rare-earth element M enhances T_c; the highest value of T_c that has been reached is 95 K in the case of $YBa_2Cu_3O_{7-y}$.
3. The third group of high-T_c superconductors can be represented by the chemical formula $Bi_2Sr_2Ca_nCu_{n+1}O_{n+6+y}$. For n = 0, that is, $Bi_2Sr_2CuO_{6+y}$, the highest T_c that has been reached is less than 20 K for some suitable values of y. For n = 1, that is, $Bi_2Sr_2CaCu_2O_{7+y}$, the highest T_c is 85 K. Such superconductors are said to possess 2212 structure, which is shown in Fig. 1.34C in the ideal case. For n = 2, that is, $Bi_2Sr_2Ca_2Cu_3O_{8+y}$, the highest $T_c \approx 110$ K. In the structure of these compounds, perovskite and BiO layers alternate.

$Tl_2Ca_2Ba_nCu_{n+1}O_{n+6+y}$ represents a similar group of superconductors. In these superconductors, for n = 1 (2212 structure), the highest value of $T_c = 105$ K, while for n = 2 the highest value of $T_c = 127$ K.

1.10 QUASICRYSTALS

A solid can be obtained from the molten state of an element (or a mixture of elements) in three ways:

1. If the material in the molten state is cooled slowly, then a crystalline solid or crystal is obtained, which exhibits perfect long-range periodicity. The different structures of crystalline solids have already been discussed.

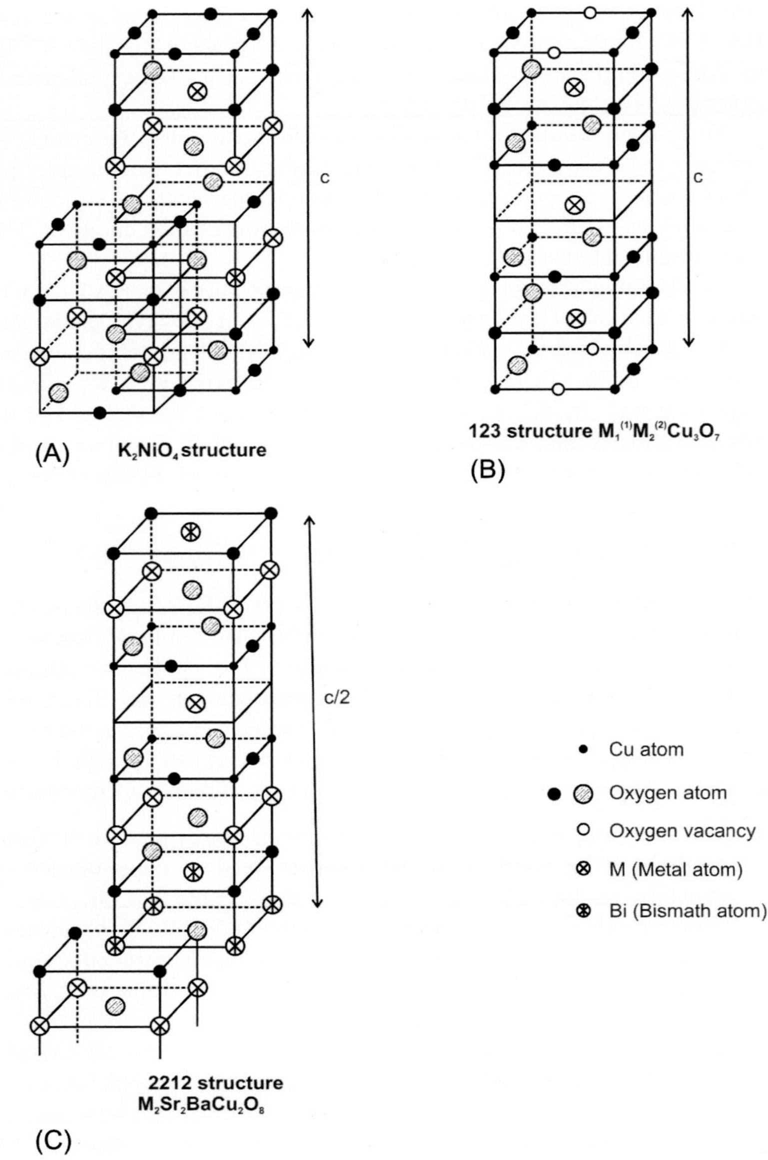

FIG. 1.34 The structure of high-T_c superconductors is divided into three categories: (A) $(La,M)_2CuO_4$ structure: In this class, the high-T_c superconductors have K_2NiO_4-type crystal structure. Here M is a metallic element. (B) $M_1^{(1)}M_1^{(2)}Cu_3O_7$ (123) structure for high-T_c superconductors. Here $M^{(1)}$ and $M^{(2)}$ are mostly rare-earth elements. An example of a 123 structure is $MBa_2Cu_3O_7$. (C) $M_2Sr_2BaCu_2O_8$ (2212) structure for high-T_c superconductors with M as the metallic atom. An example of 2212 structure is $Bi_2Sr_2BaCu_2O_8$.

2. If the material in the molten state is cooled very rapidly, then an amorphous solid is obtained in which periodicity is absent. Glass is an example of an amorphous solid. Actually, there exists a short-range order in an amorphous solid. In the very rapid cooling process the atoms (molecules) do not have time to arrange themselves in a periodic fashion.
3. If the molten state of the material is cooled rapidly (neither slowly nor very rapidly), then a *quasicrystalline solid* (quasicrystal) is produced in some materials. This is because the atoms (molecules) do not have sufficient time to arrange themselves in perfect order. A quasicrystalline solid consisting of small periodic structures, which are bound together in an irregular manner, is obtained. For example, when a mixture of Al and Mn in the molten state is cooled rapidly, a quasicrystalline alloy, Al_6Mn, is obtained. A diffraction study of Al_6Mn shows a sharp diffraction pattern corresponding to icosahedral structure, which exhibits 5-fold (pentagonal) and 10-fold (decagonal) symmetries.

It is easy to study the structure of quasicrystals in two dimensions. In order to fill the two-dimensional space, Penrose constructed two small plane surfaces (usually called tiles) of different shapes from a unit pentagon (Fig. 1.35) and these form rhombi with their areas forming an irrational number $\varphi = 1.618$. He was able to cover the two-dimensional space perfectly in an infinite number of aperiodic ways yielding infinite patterns called Penrose patterns. Each arrangement possesses a long-ranged quasiperiodicity that is responsible for the discrete diffraction pattern. In the same way, one can study the

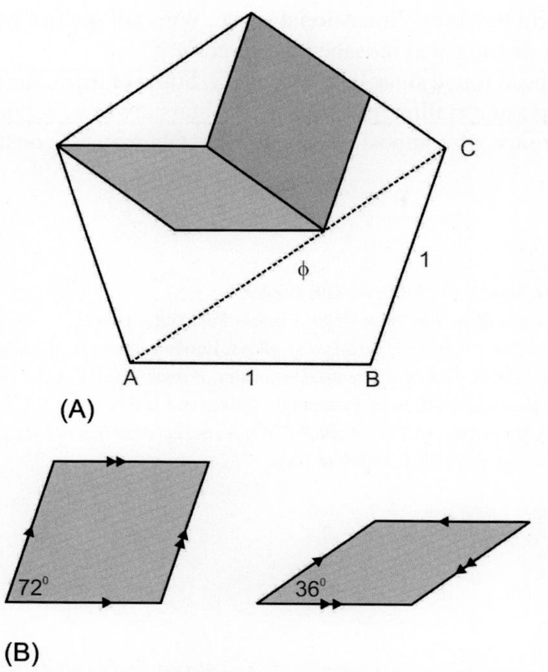

FIG. 1.35 (A) The unit pentagon and the Penrose tiles *(shown shaded)* derived from it. AC is the irrational number φ. (B) The Penrose tiles; they may only be used if matching sides, as indicated by *arrows*, are placed together.

symmetry elements of quasicrystals in three dimensions. In analogy with the two-dimensional tiles, construct two types of rhombohedra. Fill the three-dimensional space with different arrangements of rhombohedra without overlaps and gaps. The rhombohedra produce an irregular but quasiperiodic arrangement of lattice points with icosahedral symmetry. In Fig. 1.36 an icosahedron is shown, which has twenty similar sides forming equilateral triangles. Further, the icosahedron has six 5-fold, ten 3-fold, and fifteen 2-fold axes of rotational symmetry. Just as pentagons cannot cover the two-dimensional space

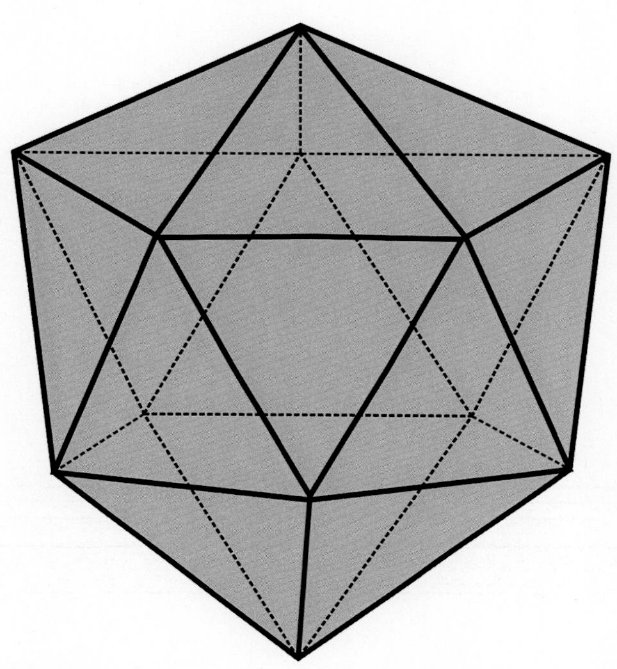

FIG. 1.36 An icosahedron in a quasicrystal.

perfectly, so the icosahedra cannot fill the three-dimensional space. We shall see in Chapter 25 that some clusters of atoms (nanoparticles) possess icosahedral or truncated icosahedral structure.

A number of quasicrystals have been found since their discovery. Some of them, for example, $Al_{65}Cu_{20}Fe_{12}$, do not even require rapid cooling to form the quasicrystalline phase. Quasicrystals always contain two or more components and the quasiperiodicity may arise over a range of compositions. Quasicrystals may be considered a link between the perfectly periodic and amorphous states.

SUGGESTED READING

Azaroff, L. V. (1960). *Introduction to solids*. New York: McGraw-Hill Book Co.

Barrett, C., & Massalski, T. B. (1968). *Structure of metals*. New Delhi: Eurasia Publishing House.

Hume-Rothery, W., & Raynor, G. V. (1956). *The structure of metals and alloys*. London: Institute of Metals.

Moffatt, W. G., Pearsall, G. W., & Wulff, J. (1964). *The structure and properties of materials*. (Vol. 1). New York: J. Wiley & Sons.

Phillips, C. (1971). *An introduction to crystallography* (4th ed.). Edinburgh: Oliver and Boyd.

Verma, A. R. (1971). Crystals and symmetry properties. In F. C. Auluck (Ed.), *A short course in solid state physics* (pp. 3–116). India: Thomson Press.

Wyckoff, R. W. G. (1971). *Crystal structures*. New York: J. Wiley & Sons.

Chapter 2

Crystal Structure in Reciprocal Space

Chapter Outline

In Chapter 1, we saw that atoms in a crystalline solid form a three-dimensional periodic array but nothing has been said about the determination of the structure of a crystalline solid, which involves knowledge of the positions of the atoms. Two types of studies are performed for the determination of crystal structure:

1. Direct imaging method
2. Diffraction method

With the help of high-resolution electron microscopes, such as field-ion microscopes, electron tunneling microscopes, atomic force microscopes, and magnetic force microscopes, one can obtain a direct image of a real crystal structure. This very useful technique is ideal for investigating point defects and dislocations and to study surfaces and interfaces. This method will be discussed in some detail in the last chapter of the book. On the other hand, diffraction studies yield detailed information about the crystal structure. To observe the diffraction pattern from a crystal, one requires a wave having a wavelength comparable to the interatomic spacing, which is on the order of a few angstroms. Visible light cannot be used for the determination of crystal structure because its wavelength is very large (on the order of few thousand angstroms) compared with the interatomic spacing. X-rays, low-energy electrons, and thermal neutrons are best suited for diffraction studies in crystalline solids.

2.1 X-RAY DIFFRACTION

In 1895, Wilhelm Roentgen accidentally discovered X-rays while studying electric discharge through gases at low pressures but was not able to establish their nature. Because of the vast and diversified applications of X-rays, Roentgen was awarded the first Nobel Prize in 1901. Max von Laue thought that the periodic arrangement of atoms in a crystalline solid could satisfy the condition for diffraction of X-rays. On his suggestion, the most important discovery in solid state physics was made by W. Friedrich and P. Knipping in 1912. They showed that interference among X-rays occurs on passing through a crystal. This was the first proof of the wave nature of X-rays and also of the existence of a space lattice in a crystal. Thus, the wave theory of X-rays and the atomic theory of crystals came into the light together. Max von Laue derived the scattering expression, known as Laue scattering, which marked the beginning of scattering studies from crystals. In 1913, William Henry Bragg and his son William Lawrence Bragg studied, for the first time, the structure of rock salt crystal

Solid State Physics. https://doi.org/10.1016/B978-0-12-817103-5.00002-5

with the help of X-ray diffraction experiments. They were able to obtain the value of the wavelength λ of X-rays and the lattice parameter of the crystal and found that these are on the order of a few angstroms. This study marked the beginning of X-ray spectroscopy, which enabled the determination of the structure of a large number of crystalline solids.

2.1.1 Bragg's Law of X-Ray Diffraction

In a crystalline solid, there are a number of sets of parallel planes of atoms and each set acts as a three-dimensional diffraction grating rather than the plane grating used in optics. The beauty of Bragg's study is that he gave a very simple law for the X-ray diffraction from a particular set of parallel planes of a crystal. For simplicity we consider a square lattice as shown in Fig. 2.1, in which a set of parallel planes with Miller indices (120) are shown. A parallel beam of X-rays falls on the crystal and is reflected from the (120) set of parallel planes. At a particular angle of incidence, all the X-rays in a reflected beam are parallel and interfere constructively to yield a bright spot on the photosensitive plate. According to Bragg's law, two incident rays undergoing reflection from the adjoining parallel planes of a particular set interfere constructively if the path difference between them after reflection is an integral multiple of λ, that is,

$$2d \sin\theta = n\lambda \tag{2.1}$$

where n is an integer, d is the spacing between adjoining parallel planes of the crystal, and θ is the angle made by incident X-ray beam with one of the planes (see Fig. 2.1). Note that d may or may not be equal to the lattice parameter of the crystal.

2.2 ELECTRON DIFFRACTION

In 1898, Sir J. J. Thomson discovered the electron as one of the constituent particles of an atom and gave a physical model of an atom. In 1924, Louis de Broglie gave the famous and revolutionary idea of the wave-particle duality in which the wavelength λ associated with a particle having momentum $p = |\mathbf{p}|$ is given by

$$\lambda = \frac{h}{p} \tag{2.2}$$

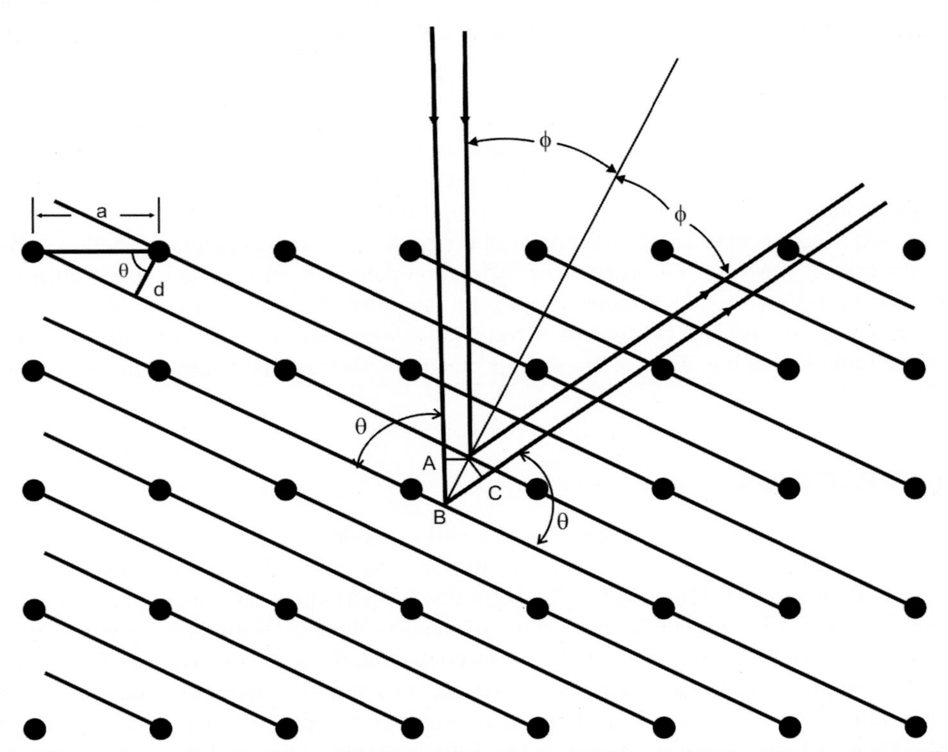

FIG. 2.1 Reflection of X-rays from a set of parallel atomic planes (120) of a crystal. The path difference between two parallel incident rays after reflection from the adjacent planes is AB + BC = 2d sin θ.

where h is Planck's constant. The energy of a free electron E is given by

$$E = \frac{p^2}{2m_e} \tag{2.3}$$

where m_e is the mass of an electron. From Eqs. (2.2), (2.3) one gets

$$\lambda = \frac{h}{\sqrt{2m_e E}} \tag{2.4}$$

Assuming the wave-particle duality to be true, Elsasser in 1925, thought that the waves associated with electrons might interfere on passing through crystals as X-rays do. In 1927, Davison and Germer and G. P. Thomson, independently, obtained the interference pattern produced by electrons passing through a crystal and also measured the wavelength of electron waves. Electrons are light charged particles, so they interact strongly with matter and penetrate relatively short distances into the crystal. Therefore, the low-energy electron diffraction (LEED) technique is very useful for studying the structure of surfaces and thin films and in surface science in general. LEED is also a very important technique for the study of surface structures formed by atoms or molecules adsorbed on metal or semiconductor surfaces.

2.3 NEUTRON DIFFRACTION

In 1932, J. Chadwick found a new particle called a neutron in radioactive processes. According to the wave-particle duality principle, the wavelength of a neutron is also given by Eq. (2.4) by replacing the electron mass by the neutron mass M_n, that is,

$$\lambda = \frac{h}{\sqrt{2M_n E}} \tag{2.5}$$

The energy of a thermal neutron (energy at room temperature T) is given by

$$E = \frac{3}{2} k_B T \tag{2.6}$$

where k_B is the Boltzmann constant. From Eqs. (2.5), (2.6) the wavelength of a thermal neutron is given by

$$\lambda = \frac{h}{\sqrt{3M_n k_B T}} \tag{2.7}$$

At $T = 293\,K$ one gets $\lambda = 1.49 Å$, which is of the order of atomic spacing in crystalline solids, a necessary condition for interference from a crystal lattice. Therefore, a beam of thermal neutrons, having velocity on the order of 3×10^{15} cm/sec, suffers diffraction from the atoms of a crystalline solid and provides information about the crystal structure in the same way as X-ray diffraction does.

An important property of neutrons is that they possess finite magnetic moment even though neutrons have no charge. The magnetic moment of a neutron is $-1.91307 \pm 0.0006\ \mu_{Bp}$ where

$$\mu_{Bp} = \frac{eh}{4\pi\ M_p c} = 5.05 \times 10^{-24}\ \text{erg/gauss} \tag{2.8}$$

is the Bohr magnetron of a free proton having mass M_p and c is the velocity of light. It is noteworthy that the magnetic moment of a neutron is much different in magnitude and sign from that of a proton. In neutron diffraction, the magnetic moments of neutrons interact with the magnetic moments of crystal atoms. Thus, neutron diffraction has an added advantage over X-ray diffraction as it yields information about the magnetic structure (orientation of magnetic moments of atoms of the crystal) in addition to the chemical structure of crystalline solids. Shull and Smart (1949) performed the first neutron diffraction study to investigate the structure of antiferromagnetic MnO crystal. The early studies on neutron diffraction from crystals are nicely reviewed by Becon (1975) and Shull (1995). In the early 1950s, experimental studies on neutron diffraction from crystals received considerable impetus (Koehler & Wollan, 1955; Shull, Strauser, & Wollan, 1951; Shull & Wilkinson, 1955; Shull & Wollan, 1956). Neutron diffraction studies give information about the ferromagnetic and antiferromagnetic ordering of magnetic moments in crystalline solids (Izyumov & Ozerov, 1970). In the case of magnetic binary alloys, neutron diffraction studies yield information about the magnetic moments of the two constituent atoms (Shull & Wollan, 1956). Now it is generally accepted that thermal neutron diffraction is one of the most powerful experimental techniques to study the properties of crystalline solids.

2.4 LAUE SCATTERING THEORY

Consider a crystalline solid on which is incident a parallel beam of radiation. The amplitude of radiation at any point **r** having frequency ω is given by

$$A(\mathbf{r}) = A_0 e^{\imath(\mathbf{K}\cdot\mathbf{r}-\omega t)} \tag{2.9}$$

where A_0 is the amplitude at $\mathbf{r} = 0$ and **K** is the wave vector of the incident radiation. As the frequency ω remains constant in the scattering process so $e^{-\imath\omega t}$ is constant and can be absorbed in amplitude A_0 at a particular instant. Hence, the incident radiation at a particular instant can be written as

$$A(\mathbf{r}) = A_0 e^{\imath\mathbf{K}\cdot\mathbf{r}} \tag{2.10}$$

The radiation is scattered from the atoms in a solid and the geometry of the scattering process is shown in Fig. 2.2. In the case of elastic scattering, $|\mathbf{K}'| = |\mathbf{K}| = K$ and the amplitude of scattered waves depends on the amplitude of the incident radiation at position $\mathbf{R}_n$. Hence, the amplitude of a scattered wave $A_s(\mathbf{r})$ at the observation point P is given by

$$A_s(\mathbf{r}) = C\left(A_0\, e^{\imath\mathbf{K}\cdot\mathbf{R}_n}\right)\frac{e^{\imath Kr}}{r} \tag{2.11}$$

which, in the case of elastic scattering, can be written as

$$A_s(\mathbf{r}) = C\, A_0 e^{\imath\mathbf{K}\cdot\mathbf{R}_n}\frac{e^{\imath\mathbf{K}'\cdot\mathbf{r}}}{r} \tag{2.12}$$

Here C is the constant of proportionality and depends on the details of the scattering center. From the geometry of the scattering process (Fig. 2.2)

$$\mathbf{r} = \mathbf{R}_{n'} - \mathbf{R}_n \tag{2.13}$$

which in the asymptotic limit, with $|\mathbf{R}_n| << |\mathbf{r}|$, $|\mathbf{R}_n| << |\mathbf{R}_{n'}|$ is given by

$$r = R_{n'} - \hat{\mathbf{R}}_{n'}\cdot\mathbf{R}_n = R_{n'}$$

$$K\hat{\mathbf{R}}_{n'} \approx K\hat{\mathbf{r}} = \mathbf{K}' \tag{2.14}$$

where $\hat{\mathbf{R}}_{n'} = \mathbf{R}_{n'}/|\mathbf{R}_{n'}|$ is a unit vector in the direction of $\mathbf{R}_{n'}$ and $\hat{\mathbf{r}}$ is a unit vector in the direction of vector **r**. Here symbols without a vector sign are moduli (absolute values) of their vector forms. Substituting Eqs. (2.13), (2.14) into Eq. (2.12) we get

$$A_s(\mathbf{r}) = C A_0 \frac{e^{-\imath\Delta\mathbf{K}\cdot\mathbf{R}_n}}{R_{n'}} \tag{2.15}$$

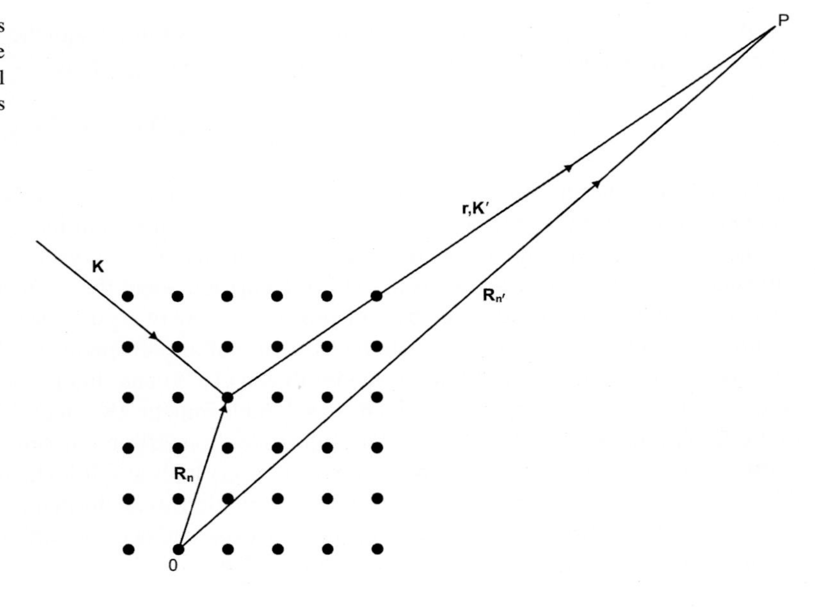

FIG. 2.2 Electromagnetic waves with plane wave fronts having wave vector **K** are incident on an atom at $\mathbf{R}_n$ in the crystal lattice. After scattering, radiation with a spherical wave front having wave vector **K′** is produced and travels in the direction of the position vector **r**.

where

$$\Delta\mathbf{K} = \mathbf{K}' - \mathbf{K} \tag{2.16}$$

Here, the factor $\exp(\imath\mathbf{K}'\cdot\mathbf{R}_{n'}) = \exp(\imath K R_{n'})$ is nearly a constant quantity and is absorbed in A_0. This is the case when $|\mathbf{R}_{n'}|$ is very large compared with the dimensions of the crystal under consideration. The total scattering amplitude from all the atoms of a crystalline material is obtained by summing over all the atoms, that is,

$$A_s(\mathbf{r}) = C A_0 \sum_n \frac{e^{-\imath\Delta\mathbf{K}\cdot\mathbf{R}_n}}{R_{n'}} \tag{2.17}$$

Actually, the incident radiation interacts with the electrons of the atom at a particular site $\mathbf{R}_n$ and is scattered. Therefore, the strength of scattered radiation depends on the electron density $n_e(\mathbf{R}_n)$ at $\mathbf{R}_n$. So, the total scattering amplitude of radiation becomes

$$A_s(\mathbf{r}) = \frac{C A_0}{R_{n'}} \int \sum_n n_e(\mathbf{R}_n) e^{-\imath\Delta\mathbf{K}\cdot\mathbf{R}_n} d^3R_n \tag{2.18}$$

Here, the integral is over the volume of the nth atom. If the material under consideration is monatomic, then the atom can be assumed to be at a lattice point and the scattering strength of all the atoms is the same. But if there is more than one atom in the basis (s number of atoms), then the total scattering amplitude from the whole of the crystal is found to be

$$\begin{aligned} A_s(\mathbf{r}) &= \frac{1}{R_{n'}} \sum_n \sum_{m=1}^{s} f_m(\Delta\mathbf{K}) e^{-\imath\Delta\mathbf{K}\cdot\mathbf{R}_{nm}} \\ &= \frac{1}{R_{n'}} A_L(\Delta\mathbf{K})\, F_S(\Delta\mathbf{K}) \end{aligned} \tag{2.19}$$

where

$$A_L(\Delta\mathbf{K}) = \sum_n e^{-\imath\Delta\mathbf{K}\cdot\mathbf{R}_n} \tag{2.20}$$

$$F_S(\Delta\mathbf{K}) = \sum_{m=1}^{s} f_m(\Delta\mathbf{K}) e^{-\imath\Delta\mathbf{K}\cdot\mathbf{R}_m} \tag{2.21}$$

Here $A_L(\Delta\mathbf{K})$ gives the scattering amplitude from the lattice of a crystalline solid and depends on the periodicity of the lattice. On the other hand, $F_S(\Delta\mathbf{K})$ gives the amplitude of scattering from a unit cell and depends upon the distribution of atoms in the unit cell and hence is called the geometrical structure factor of the basis. The product of $A_L(\Delta\mathbf{K})$ and $F_S(\Delta\mathbf{K})$ determines the total scattering amplitude from the crystalline structure. Here f_m measures the amplitude of scattered waves from the mth atom of the unit cell and its value depends on the electron distribution of the mth atom in the nth unit cell. In terms of the electron density of the atom at the $\mathbf{R}_m$ site, that is, $n_e(\mathbf{R}_m)$, f_m can be written as

$$f_m(\Delta\mathbf{K}) = \int n_e(\mathbf{R}_m) e^{\imath\Delta\mathbf{K}\cdot\mathbf{R}_m} d^3R_m$$

It can also be written as

$$f_m(\Delta\mathbf{K}) = \int n_e(\mathbf{r}) e^{\imath\Delta\mathbf{K}\cdot\mathbf{r}} d^3r \tag{2.22}$$

where the volume integral is over the volume of the mth atom. The value of f_m for the mth atom is the same for each lattice point $\mathbf{R}_n$. Thus, every lattice point $\mathbf{R}_n$ has associated with it a unit cell and every unit cell yields the same amplitude of scattered radiation.

Further, f_m also depends on the nature of the incident radiation. If X-ray diffraction is studied from a crystalline material, then $f_m = f_m^X$ is called the *X-ray scattering amplitude* of the mth atom in the unit cell. In the case of a neutron diffraction study, $f_m = f_m^N$ is called the *neutron scattering amplitude* from the mth nonmagnetic atom of the unit cell. On the other hand, in the case of neutron diffraction from a magnetic atom (magnetic material) $f_m = f_m^M$ is called the *magnetic scattering amplitude*.

Problem 2.1

Substitute for $\mathbf{R}_n$ from Eq. (1.5) in Eq. (2.20) and prove that the intensity of scattered waves is maximum when the following conditions are satisfied:

$$\begin{aligned} \mathbf{a}_1 \cdot \Delta\mathbf{K} &= 2\pi N_1 \\ \mathbf{a}_2 \cdot \Delta\mathbf{K} &= 2\pi N_2 \\ \mathbf{a}_3 \cdot \Delta\mathbf{K} &= 2\pi N_3 \end{aligned} \tag{2.23}$$

where N_1, N_2, and N_3 are integers. The conditions given by Eq. (2.23) are called *Laue's diffraction conditions.* Further, prove that the intensity of the scattered beam is directly proportional to the number of atoms in the crystal.

2.5 RECIPROCAL LATTICE

An X-ray or neutron diffraction pattern is in a space that has dimensions of reciprocal length and such a space is called reciprocal space. Therefore, to interpret the X-ray diffraction pattern and then relate it to the crystal (direct) space, one has to define a reciprocal lattice. Ewald originally proposed the formal concept of the reciprocal lattice in 1921 and Bernal studied its applications in 1927.

Consider a Bravais lattice in the reciprocal space with primitive translation vectors $\mathbf{b}_1$, $\mathbf{b}_2$, and $\mathbf{b}_3$. One of the ways to construct a reciprocal lattice is to make its primitive vectors orthogonal to the primitive vectors of the direct space, which mathematically can be written as

$$\mathbf{a}_\alpha \cdot \mathbf{b}_\beta = 2\pi\delta_{\alpha\beta} \tag{2.24}$$

where α and β take values 1, 2, and 3. The factor 2π in Eq. (2.24) arises due to geometrical reasons. We have seen in Chapter 1 that $\mathbf{a}_1$, $\mathbf{a}_2$, and $\mathbf{a}_3$ are not generally orthogonal to one another and so are the vectors $\mathbf{b}_1$, $\mathbf{b}_2$, and $\mathbf{b}_3$. From Eq. (2.24) it is evident that $\mathbf{b}_1$ is perpendicular to $\mathbf{a}_2$ and $\mathbf{a}_3$, $\mathbf{b}_2$ is perpendicular to $\mathbf{a}_3$ and $\mathbf{a}_1$, and $\mathbf{b}_3$ is perpendicular to $\mathbf{a}_1$ and $\mathbf{a}_2$. Therefore, one can write

$$\begin{gathered} \mathbf{b}_1 \propto \mathbf{a}_2 \times \mathbf{a}_3 \\ \text{or } \mathbf{b}_1 = K_1 \mathbf{a}_2 \times \mathbf{a}_3 \end{gathered} \tag{2.25}$$

where K_1 is a constant of proportionality which can be determined from the condition

$$\mathbf{a}_1 \cdot \mathbf{b}_1 = 2\pi \tag{2.26}$$

Substituting $\mathbf{b}_1$ from Eq. (2.25) into Eq. (2.26) we get

$$K_1 = \frac{2\pi}{\mathbf{a}_1 \cdot \mathbf{a}_2 \times \mathbf{a}_3} \tag{2.27}$$

Hence, the vector $\mathbf{b}_1$ from Eq. (2.25), (2.27) becomes

$$\mathbf{b}_1 = 2\pi \frac{\mathbf{a}_2 \times \mathbf{a}_3}{\mathbf{a}_1 \cdot \mathbf{a}_2 \times \mathbf{a}_3} \tag{2.28a}$$

Similarly, expressions for $\mathbf{b}_2$ and $\mathbf{b}_3$ can be obtained and are given by

$$\mathbf{b}_2 = 2\pi \frac{\mathbf{a}_3 \times \mathbf{a}_1}{\mathbf{a}_1 \cdot \mathbf{a}_2 \times \mathbf{a}_3} \tag{2.28b}$$

$$\mathbf{b}_3 = 2\pi \frac{\mathbf{a}_1 \times \mathbf{a}_2}{\mathbf{a}_1 \cdot \mathbf{a}_2 \times \mathbf{a}_3} \tag{2.28c}$$

Problem 2.2

If the primitive translation vectors $\mathbf{a}_1$, $\mathbf{a}_2$, and $\mathbf{a}_3$ are constructed from $\mathbf{b}_1$, $\mathbf{b}_2$, and $\mathbf{b}_3$ in the same way as $\mathbf{b}_1$, $\mathbf{b}_2$, and $\mathbf{b}_3$ are constructed from $\mathbf{a}_1$, $\mathbf{a}_2$, and $\mathbf{a}_3$ [see Eq. (2.28)], prove that

$$\mathbf{a}_1 = 2\pi \frac{\mathbf{b}_2 \times \mathbf{b}_3}{\mathbf{b}_1 \cdot \mathbf{b}_2 \times \mathbf{b}_3}, \mathbf{a}_2 = 2\pi \frac{\mathbf{b}_3 \times \mathbf{b}_1}{\mathbf{b}_1 \cdot \mathbf{b}_2 \times \mathbf{b}_3}, \mathbf{a}_3 = 2\pi \frac{\mathbf{b}_1 \times \mathbf{b}_2}{\mathbf{b}_1 \cdot \mathbf{b}_2 \times \mathbf{b}_3} \tag{2.29}$$

From Eqs. (2.24) and (2.28), it is evident that vectors $\mathbf{b}_1$, $\mathbf{b}_2$, and $\mathbf{b}_3$ have dimensions of reciprocal of length and are therefore, called primitive reciprocal lattice vectors. The periodic repetition of $\mathbf{b}_1$, $\mathbf{b}_2$, and $\mathbf{b}_3$ generates a new lattice called the reciprocal lattice in a space called the reciprocal space or the Fourier space. A general reciprocal lattice vector $\mathbf{G}_p$ is defined in the conventional way as

$$\mathbf{G}_p = p_1\mathbf{b}_1 + p_2\mathbf{b}_2 + p_3\mathbf{b}_3 \tag{2.30}$$

where p_1, p_2, and p_3 are integers: negative, positive, or zero. The set of integers p_1, p_2, and p_3 is, collectively, represented by an integer p, that is, $p = (p_1, p_2, p_3)$. The important property of the reciprocal lattice is that

$$e^{\imath \mathbf{G}_p \cdot \mathbf{R}_n} = 1 \tag{2.31}$$

The reciprocal lattice can also be constructed by making use of the periodicity of the electron and atomic densities.

2.5.1 Periodicity of Electron Density

The concept of reciprocal lattice can also be derived from the electronic band theory of solids (see Chapter 12). The general wave function for an electron in a periodic crystalline solid is the Bloch wave function defined as

$$|\psi_{\mathbf{k}}(\mathbf{r})\rangle = e^{\imath \mathbf{k}\cdot\mathbf{r}} u_{\mathbf{k}}(\mathbf{r}) \tag{2.32}$$

where $\mathbf{k}$ is the electron wave vector and $u_{\mathbf{k}}(\mathbf{r})$ is a scalar complex function that satisfies the periodicity of the crystal, that is,

$$u_{\mathbf{k}}(\mathbf{r}) = u_{\mathbf{k}}(\mathbf{r} + \mathbf{R}_n) \tag{2.33}$$

If $T(\mathbf{R}_n)$ is the translation operator, then the Bloch wave function satisfies the following condition

$$T(\mathbf{R}_n)|\psi_{\mathbf{k}}(\mathbf{r})\rangle = |\psi_{\mathbf{k}}(\mathbf{r} + \mathbf{R}_n)\rangle = e^{\imath \mathbf{k}\cdot\mathbf{R}_n}|\psi_{\mathbf{k}}(\mathbf{r})\rangle \tag{2.34}$$

Eq. (2.34) is usually called the Bloch condition [see Eq. (12.33)]. Let us consider some vector $\mathbf{k} = \mathbf{G}_p$ such that

$$e^{\imath \mathbf{G}_p \cdot \mathbf{R}_n} = 1 \tag{2.31}$$

then Eq. (2.34) reduces to

$$T(\mathbf{R}_n)\left|\psi_{\mathbf{G}_p}(\mathbf{r})\right\rangle = \left|\psi_{\mathbf{G}_p}(\mathbf{r} + \mathbf{R}_n)\right\rangle = \left|\psi_{\mathbf{G}_p}(\mathbf{r})\right\rangle \tag{2.35}$$

Eq. (2.35) shows that the choice of $\mathbf{k} = \mathbf{G}_p$ makes $|\psi_{\mathbf{k}}(\mathbf{r})\rangle$ invariant under translation (translational periodicity). Eq. (2.31) is satisfied if

$$\mathbf{G}_p \cdot \mathbf{R}_n = 2\pi p \tag{2.36}$$

where p is an integer: negative, positive, or zero. Using Eq. (1.5) for $\mathbf{R}_n$ in Eq. (2.36) we get

$$n_1\mathbf{G}_p \cdot \mathbf{a}_1 + n_2\mathbf{G}_p \cdot \mathbf{a}_2 + n_3\mathbf{G}_p \cdot \mathbf{a}_3 = 2\pi p \tag{2.37}$$

For each of the integers (n_1, n_2, n_3), there exists some integer p for which the above equation holds. In other words, the value of p depends on the set $\{n_1, n_2, n_3\}$. As a simple case, consider $n_2 = n_3 = 0$, then Eq. (2.37) gives

$$\mathbf{G}_p \cdot \mathbf{a}_1 = 2\pi \frac{p}{n_1} = 2\pi p_1 \tag{2.38a}$$

where $p/n_1 = p_1$ is another integer. Similarly, one can get

$$\mathbf{G}_p \cdot \mathbf{a}_2 = 2\pi p_2 \tag{2.38b}$$

$$\mathbf{G}_p \cdot \mathbf{a}_3 = 2\pi p_3 \tag{2.38c}$$

From Eqs. (2.38a), (2.38b), and (2.38c) the general form of vector $\mathbf{G}_p$ can be written as

$$\mathbf{G}_p = p_1\mathbf{b}_1 + p_2\mathbf{b}_2 + p_3\mathbf{b}_3 \tag{2.39}$$

which is the same as Eq. (2.30). It is noteworthy that $\mathbf{G}_p$ is defined by satisfying Eq. (2.31). Using Eq. (1.5) for $\mathbf{R}_n$ and Eq. (2.39) for $\mathbf{G}_p$ in Eq. (2.31) one can prove the condition defined by Eq. (2.24). This shows that the factor of 2π comes from the geometry of the crystal.

2.5.2 Periodicity of Atomic Density

2.5.2.1 Monatomic Linear Lattice

In the direct crystal space, the atomic density is periodic [Eq. (1.13)]. For a monatomic linear lattice, the periodicity of the atomic density $\rho^a(\mathbf{r}_1)$ can be written as

$$\rho^a(\mathbf{r}_1) = \rho^a(\mathbf{r}_1 + \mathbf{R}_n) \tag{2.40}$$

where

$$\mathbf{R}_n = n\mathbf{a}_1 \tag{2.41}$$

Here n is an integer: negative, positive, or zero. A periodic function can always be expanded in the Fourier space. Therefore, $\rho^a(\mathbf{r}_1)$ can be written as

$$\rho^a(\mathbf{r}_1) = \sum_p \rho_p^a e^{\imath\left(\frac{2\pi}{a}p\hat{\mathbf{i}}_1\right)\cdot\mathbf{r}_1} \tag{2.42}$$

where p is an integer. Let us define a vector $\mathbf{G}_p$ in the Fourier space by

$$\mathbf{G}_p = \frac{2\pi}{a}p\hat{\mathbf{i}}_1 = p\mathbf{b}_1 \tag{2.43}$$

then Eq. (2.42) becomes

$$\rho^a(\mathbf{r}_1) = \sum_p \rho_{\mathbf{G}_p}^a e^{\imath\mathbf{G}_p\cdot\mathbf{r}_1} \tag{2.44}$$

Here $\mathbf{G}_p$ is called the reciprocal lattice vector in one dimension. The translation of $\mathbf{G}_p$, by varying all possible values of p, forms a periodic lattice called the reciprocal lattice, which in a one-dimensional crystal is also along the $\hat{\mathbf{i}}_1$ direction (see Fig. 2.3). Here and hereafter the small dots represent the reciprocal lattice points. From Eq. (2.41) and Eq. (2.43) one can write

$$\mathbf{G}_p \cdot \mathbf{R}_n = 2\pi \times \text{integer} \tag{2.45}$$

and therefore,

$$e^{\imath\mathbf{G}_p\cdot\mathbf{R}_n} = 1 \tag{2.46}$$

is an important property of a reciprocal lattice.

2.5.2.2 Two-Dimensional Square Lattice

In a square lattice the periodic atomic density is given by Eq. (1.13), which for completeness is written again as

$$\rho^a(\mathbf{r}) = \rho^a(\mathbf{r} + \mathbf{R}_n) \tag{2.47}$$

Here $\mathbf{r}$ and $\mathbf{R}_n$ for a two-dimensional crystal are given by Eqs. (1.10) and (1.3), respectively. The expansion for $\rho^a(\mathbf{r})$ in the Fourier space becomes

$$\rho^a(\mathbf{r}) = \sum_{p_1,p_2} \rho_{p_1,p_2}^a e^{\imath\left(\frac{2\pi}{a}p_1\hat{\mathbf{i}}_1 + \frac{2\pi}{a}p_2\hat{\mathbf{i}}_2\right)\cdot(\mathbf{r}_1+\mathbf{r}_2)} \tag{2.48}$$

$$= \sum_{\mathbf{G}_p} \rho_{\mathbf{G}_p}^a e^{\imath\mathbf{G}_p\cdot\mathbf{r}} \tag{2.49}$$

FIG. 2.3 Reciprocal lattice of a monatomic one-dimensional crystal having primitive lattice vector $\mathbf{a}_1$ such that $|\mathbf{a}_1| = a$. Here, dots represent the reciprocal lattice points and $\mathbf{b}_1 = (2\pi/a)\hat{\mathbf{i}}_1$ is the primitive reciprocal lattice vector.

where

$$\mathbf{G}_p = \frac{2\pi}{a}\left(p_1\hat{\mathbf{i}}_1 + p_2\hat{\mathbf{i}}_2\right) = p_1\mathbf{b}_1 + p_2\mathbf{b}_2 \tag{2.50}$$

Here we have used Eq. (1.10) for $\mathbf{r}$. The reciprocal lattice of a square lattice is shown in Fig. 2.4.

2.5.2.3 Three-Dimensional Cubic Lattice

In a three-dimensional cubic crystal, the periodic atomic density is given by Eq. (1.13), where $\mathbf{r}$ and $\mathbf{R}_n$ are the position vector and lattice vector given by Eqs. (1.9) and (1.5). Just as in the two-dimensional crystal, one can prove that in a three-dimensional crystal the atomic density can be represented in the form

$$\rho^a(\mathbf{r}) = \sum_{\mathbf{G}_p} \rho^a_{\mathbf{G}_p} e^{\imath \mathbf{G}_p \cdot \mathbf{r}} \tag{2.51}$$

where

$$\mathbf{G}_p = \frac{2\pi}{a}\left(p_1\hat{\mathbf{i}}_1 + p_2\hat{\mathbf{i}}_2 + p_3\hat{\mathbf{i}}_3\right) = p_1\mathbf{b}_1 + p_2\mathbf{b}_2 + p_3\mathbf{b}_3 \tag{2.52}$$

In the reciprocal lattice, one can introduce physical concepts corresponding to those already defined in the direct space. For simplicity consider a square lattice in the reciprocal space as shown in Fig. 2.4. Here $\mathbf{K}$ is any wave vector in the reciprocal space, which is related to any other wave vector $\mathbf{K}'$ through a reciprocal lattice vector $\mathbf{G}_p$, that is,

$$\mathbf{K}' = \mathbf{K} + \mathbf{G}_p \tag{2.53}$$

(see Fig. 2.4). Note that the points corresponding to vectors $\mathbf{K}$ and $\mathbf{K}'$ are equivalent in the reciprocal space. The equivalence between $\mathbf{K}$ and $\mathbf{K}'$ can be proved using the key property of the reciprocal lattice defined by Eq. (2.31). The phase factor of $\mathbf{K}'$ is given by

$$\begin{aligned} e^{\imath \mathbf{K}' \cdot \mathbf{R}_n} &= e^{\imath(\mathbf{K} + \mathbf{G}_p) \cdot \mathbf{R}_n} \\ &= e^{\imath \mathbf{K} \cdot \mathbf{R}_n} \end{aligned} \tag{2.54}$$

which is the phase factor for vector $\mathbf{K}$. Eq. (2.54) shows that vectors $\mathbf{K}$ and $\mathbf{K}'$ in Eq. (2.53) are equivalent. From Eq. (2.53) one can generate an infinite number of points that are equivalent to vector $\mathbf{K}$. Hence one can say that *a reciprocal lattice is a*

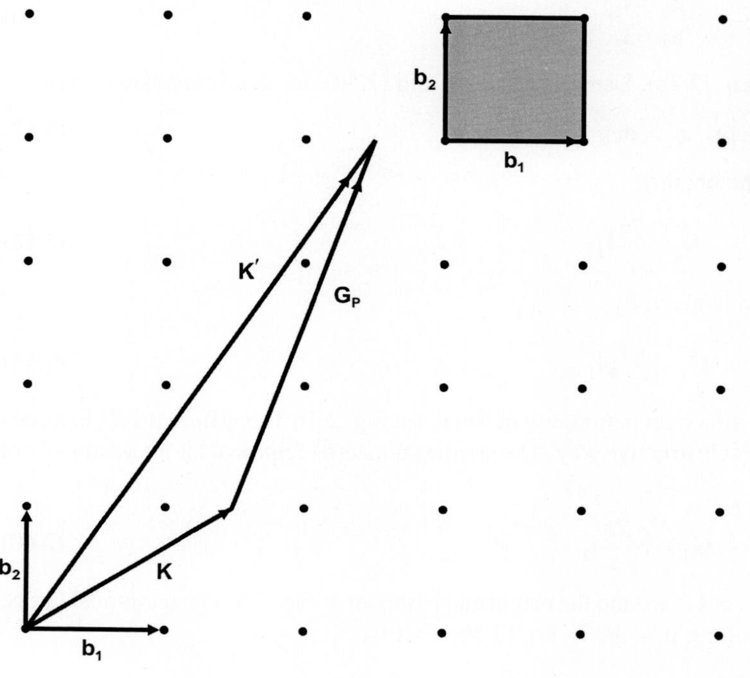

FIG. 2.4 Square lattice in the reciprocal space having primitive reciprocal lattice vectors $\mathbf{b}_1$and $\mathbf{b}_2$ (dots represent the reciprocal lattice points). Let $\mathbf{K}$ be any vector in the reciprocal space. The vectors $\mathbf{K}$ and $\mathbf{K}'$ are related to one another through a reciprocal lattice vector $\mathbf{G}_p = \mathbf{b}_1 + 3\mathbf{b}_2$, where $|\mathbf{b}_1| = |\mathbf{b}_2| = 2\pi/a$ and a is the lattice constant of the square lattice. The shaded square represents the conventional primitive cell of the lattice.

mathematical abstraction and consists of an array of an infinite number of points in the reciprocal space in which each point has identical surroundings. Further, Eq. (2.53) in the reciprocal space is equivalent to Eq. (1.20) in the direct space: Eq. (1.20) defines periodicity of the lattice in the direct space while Eq. (2.53) defines the periodicity of the lattice in the reciprocal space.

2.6 PRIMITIVE CELL IN RECIPROCAL SPACE

The primitive cell in the reciprocal space can be defined in the conventional way as a polyhedron bounded by the primitive vectors $\mathbf{b}_1$, $\mathbf{b}_2$, and $\mathbf{b}_3$ with volume given by

$$\Omega_0 = \left|\mathbf{b}_1 \cdot \mathbf{b}_2 \times \mathbf{b}_3\right| = \frac{(2\pi)^3}{V_0} \tag{2.55}$$

In the conventional primitive cell, the reciprocal lattice points are at the corners of the polyhedron, which yield on average one reciprocal lattice point per cell. However, the common practice is to define WS cells in the reciprocal lattice in exactly the same way as in the crystal space. The WS cells in the reciprocal space are known as *Brillouin zones* (BZs). In constructing a BZ, we draw the perpendicular bisector planes to the lines joining the reciprocal lattice point under consideration (assumed to be at the origin) to the reciprocal lattice points corresponding to 1NNs, 2NNs, 3NNs, and higher order NNs from the origin. We start from the origin and proceed away from it until the first set of bisector planes is encountered. The region inside these planes is called the first BZ (1BZ) with a reciprocal lattice point at its center. The surfaces of the intersecting planes define the surface of the 1BZ. We then start from the surface of the 1BZ and move away from it until we encounter the next new bisector planes. The surface of these planes defines the surface of the second BZ (2BZ) and the volume between the surfaces of the 1BZ and the 2BZ gives the volume of the 2BZ. In general, if we start from the surface of the $(n\text{-}1)^{th}$ BZ and move away from it until the next bisector planes are encountered, then the surface of these planes define the surface of the nth BZ (nBZ). The volume between the surfaces of the $(n\text{-}1)^{th}$ and n^{th} BZs gives the volume of the nBZ. We take some simple examples for obtaining the reciprocal lattices and constructing their BZs.

2.6.1 Linear Monatomic Lattice

Consider a linear monatomic lattice, along the x-axis, with lattice constant a (see Fig. 1.2A). The primitive translation vector $\mathbf{a}_1$ is given by Eq. (1.2b). The primitive vector $\mathbf{b}_1$ can be calculated in terms of $\mathbf{a}_1$. To do so, we introduce temporary unit lattice vectors $\mathbf{a}_2$ and $\mathbf{a}_3$ along y- and z-directions defined as

$$\begin{aligned}\mathbf{a}_2 &= \hat{\mathbf{i}}_2 \\ \mathbf{a}_3 &= \hat{\mathbf{i}}_3\end{aligned} \tag{2.56}$$

The vectors $\mathbf{a}_2$ and $\mathbf{a}_3$ are introduced so as to use Eq. (2.28). Using Eqs. (1.2b) and (2.56) one can immediately write

$$\left|\mathbf{a}_1 \cdot \mathbf{a}_2 \times \mathbf{a}_3\right| = a \tag{2.57}$$

Further substituting $\mathbf{a}_1$, $\mathbf{a}_2$, and $\mathbf{a}_3$ in Eq. (2.28a) one obtains

$$\mathbf{b}_1 = \frac{2\pi}{a}\hat{\mathbf{i}}_1 \tag{2.58}$$

Now, the reciprocal lattice vector in one dimension is given by

$$\mathbf{G}_p = \frac{2\pi}{a} p\hat{\mathbf{i}}_1 \tag{2.59}$$

Hence, the reciprocal lattice is also along the x-direction with periodicity of $2\pi/a$ (see Fig. 2.3). The different BZs in a one-dimensional crystal can be shown in a simple but most instructive way. The shortest nonzero reciprocal lattice vectors from Eq. (2.59) are for $p = \pm 1$, that is,

$$\mathbf{G}_1 = \pm\frac{2\pi}{a}\hat{\mathbf{i}}_1 \tag{2.60}$$

The 1BZ is subtended by the perpendicular bisectors of $\mathbf{G}_1$ around the origin and is shown in Fig. 2.5. The reciprocal lattice vectors next to the shortest ones are obtained by putting $p = \pm 2$ in Eq. (2.59), that is,

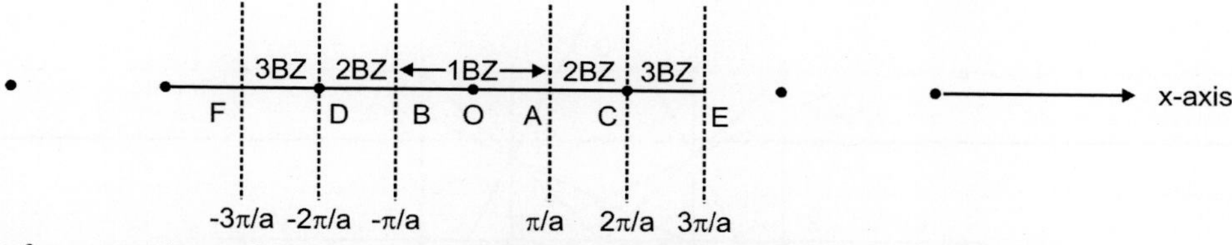

FIG. 2.5 The BZs of a monatomic one-dimensional lattice having lattice constant a. The dots represent the reciprocal lattice points.

$$\mathbf{G}_2 = \pm\frac{4\pi}{a}\hat{\mathbf{i}}_1 \tag{2.61}$$

The 2BZ formed by the perpendicular bisectors of $\mathbf{G}_2$ is shown in Fig. 2.5 and it has two segments: one on the left and the other on the right of the 1BZ, but the total length of the 2BZ is the same as that of the 1BZ. Similarly, one can draw the 3BZ, 4BZ, … in the reciprocal lattice and all will have the same length, that is, $2\pi/a$.

2.6.2 Square Lattice

The square lattice in the crystal space is shown in Fig. 1.3 and the translation vector and the primitive translation vectors are given by Eqs. (1.3) and (1.4), respectively. The reciprocal lattice vector in two dimensions is given by

$$\mathbf{G}_p = p_1\mathbf{b}_1 + p_2\mathbf{b}_2 \tag{2.62}$$

The primitive vectors $\mathbf{b}_1$and $\mathbf{b}_2$ can be evaluated from Eq. (2.28) by introducing a third temporary unit vector $\mathbf{a}_3$ along the z-direction defined as

$$\mathbf{a}_3 = \hat{\mathbf{i}}_3 \tag{2.63}$$

Now, it is straightforward to prove that

$$|\mathbf{a}_1 \cdot \mathbf{a}_2 \times \mathbf{a}_3| = a^2 \tag{2.64}$$

From Eqs (2.28) one can immediately write

$$\mathbf{b}_1 = \frac{2\pi}{a}\hat{\mathbf{i}}_1, \quad \mathbf{b}_2 = \frac{2\pi}{a}\hat{\mathbf{i}}_2 \tag{2.65}$$

From Eqs. (2.62), (2.65) the general reciprocal lattice vector becomes:

$$\mathbf{G}_p = \frac{2\pi}{a}\left(p_1\hat{\mathbf{i}}_1 + p_2\hat{\mathbf{i}}_2\right) \tag{2.66}$$

which is the same as Eq. (2.50). Here $\mathbf{b}_1$ and $\mathbf{b}_2$ are perpendicular to each other with the same magnitude. It is noteworthy that the directions of $\mathbf{b}_1$ and $\mathbf{b}_2$ are the same as those of $\mathbf{a}_1$ and $\mathbf{a}_2$ but with different magnitude. Hence, the direct and reciprocal lattices of a square lattice exhibit the same symmetry. From Eq. (2.66) the shortest nonzero reciprocal lattice vectors are given by

$$\mathbf{G}_1 = \pm\frac{2\pi}{a}\hat{\mathbf{i}}_1, \quad \pm\frac{2\pi}{a}\hat{\mathbf{i}}_2 \tag{2.67}$$

The perpendicular bisectors of four $\mathbf{G}_1$ vectors form the 1BZ, which is a square having side $2\pi/a$ with a reciprocal lattice point at its center (see Fig. 2.6). The construction of the 2BZ and 3BZ is also shown in Fig. 2.6. The 2BZ has four parts, namely, 2a, 2b, 2c, and 2d while the 3BZ has eight parts, namely, 3a, 3b, 3c, 3d, 3e, 3f, 3g, and 3h. When we fold back the different parts of the 2BZ or 3BZ, they form squares with area equal to that of the 1BZ.

2.6.3 sc Lattice

Consider an sc lattice with primitive lattice vectors given by Eq. (1.31). It is straightforward to prove that

$$|\mathbf{a}_1 \cdot \mathbf{a}_2 \times \mathbf{a}_3| = a^3 \tag{2.68}$$

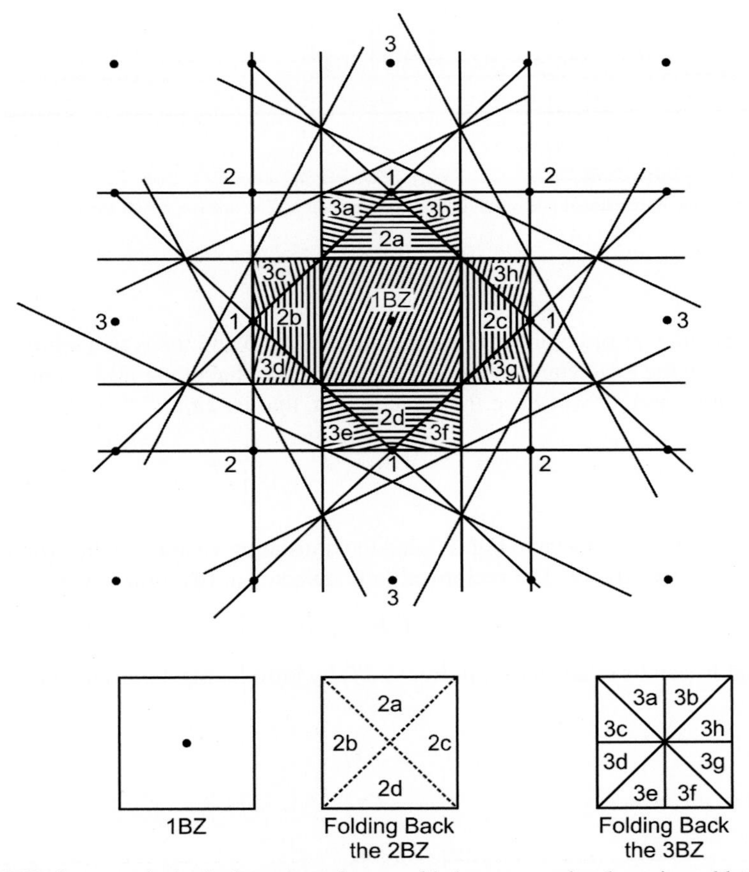

FIG. 2.6 The 1BZ, 2BZ, and 3BZ of a square lattice in the reciprocal space with dots representing the reciprocal lattice points.

and

$$\mathbf{b}_1 = \frac{2\pi}{a}\hat{\mathbf{i}}_1$$
$$\mathbf{b}_2 = \frac{2\pi}{a}\hat{\mathbf{i}}_2 \tag{2.69}$$
$$\mathbf{b}_3 = \frac{2\pi}{a}\hat{\mathbf{i}}_3$$

Hence, the general reciprocal lattice can be written as

$$\mathbf{G}_p = \frac{2\pi}{a}\left(p_1\hat{\mathbf{i}}_1 + p_2\hat{\mathbf{i}}_2 + p_3\hat{\mathbf{i}}_3\right) \tag{2.70}$$

which is the same as Eq. (2.52). From the above equation, the shortest nonzero reciprocal lattice vectors are six in number and are given by

$$\mathbf{G}_1 = \pm\frac{2\pi}{a}\hat{\mathbf{i}}_1, \quad \pm\frac{2\pi}{a}\hat{\mathbf{i}}_2, \quad \pm\frac{2\pi}{a}\hat{\mathbf{i}}_3 \tag{2.71}$$

Hence, the reciprocal lattice of an sc crystal structure is also an sc lattice with primitive vectors given by Eq. (2.69). The volume of the primitive cell is given by $(2\pi/a)^3$, which is consistent with Eq. (2.55). The 1BZ of the sc lattice, formed by the perpendicular bisector planes of the $\mathbf{G}_1$ vectors, is a cube with side $2\pi/a$ in magnitude (see Fig. 2.7). It is noteworthy that the reciprocal lattice point is situated at the center of the 1BZ, which is symmetrical about the lattice point. The 2BZ of the sc reciprocal lattice is a dodecahedron and is shown in Fig. 2.8.

In all of the above examples, the direct and reciprocal primitive vectors lie along the same directions with the same symmetry, but this may not be always true. We consider some examples below in which the direct and reciprocal lattices exhibit different symmetries.

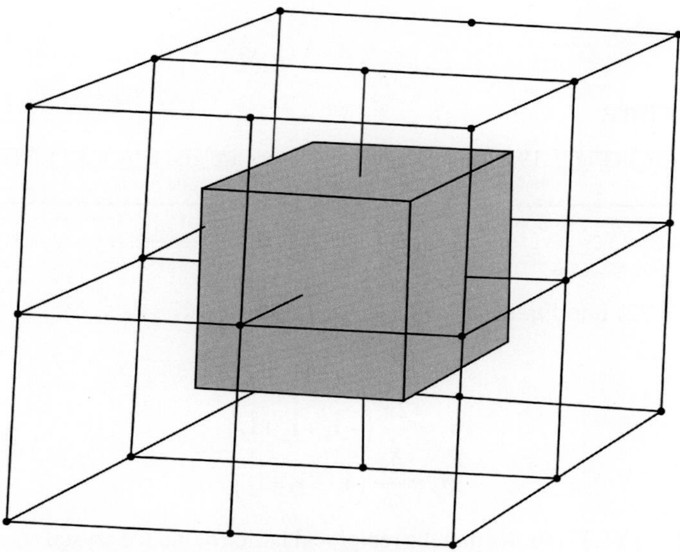

FIG. 2.7 The reciprocal lattice and 1BZ of sc crystal structure having lattice constant a.

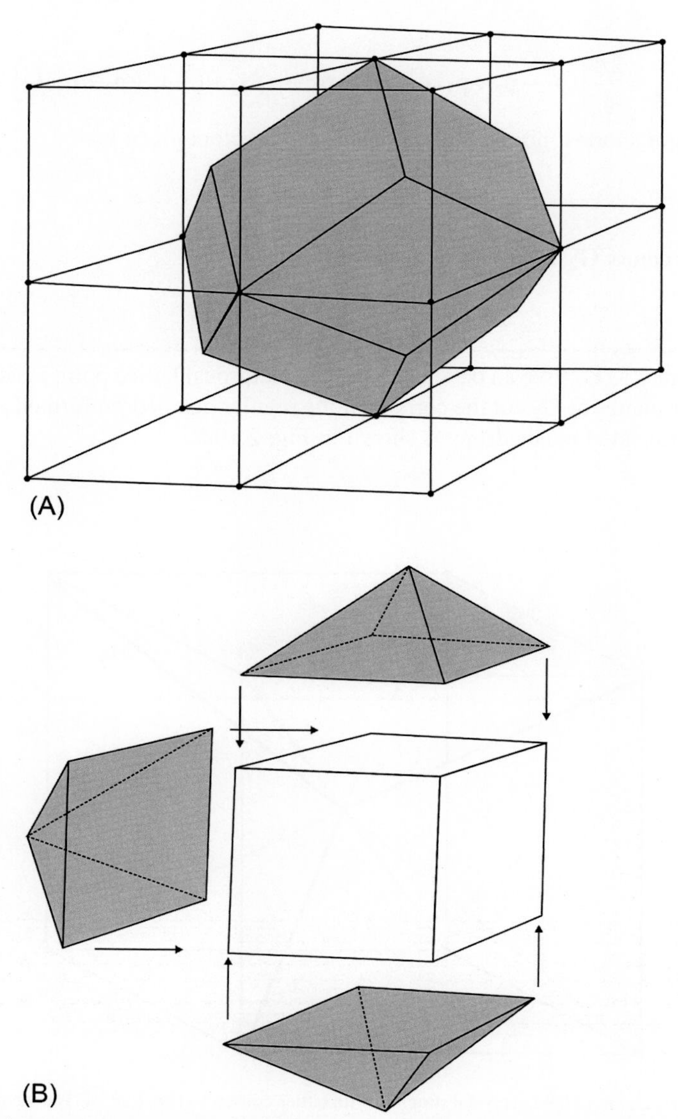

FIG. 2.8 (A) The 2BZ of an sc lattice (omitting the interior cube, which is the 1BZ). (B) Three typical segments of the 2BZ and the interior cube that forms the 1BZ.

2.6.4 fcc Crystal Structure

The fcc crystal structure is shown in Fig. 1.19a and its primitive vectors are given by Eq. (1.38). The volume of the primitive cell from Eq. (1.39) is given by

$$V_0 = |\mathbf{a}_1 \cdot \mathbf{a}_2 \times \mathbf{a}_3| = \frac{a^3}{4} \tag{2.72}$$

Substituting Eqs. (1.38) and (2.72) into Eq. (2.28), we can prove that

$$\begin{aligned} \mathbf{b}_1 &= \frac{2\pi}{a}\left(\hat{\mathbf{i}}_1 + \hat{\mathbf{i}}_2 - \hat{\mathbf{i}}_3\right) \\ \mathbf{b}_2 &= \frac{2\pi}{a}\left(-\hat{\mathbf{i}}_1 + \hat{\mathbf{i}}_2 + \hat{\mathbf{i}}_3\right) \\ \mathbf{b}_3 &= \frac{2\pi}{a}\left(\hat{\mathbf{i}}_1 - \hat{\mathbf{i}}_2 + \hat{\mathbf{i}}_3\right) \end{aligned} \tag{2.73}$$

Comparison of Eq. (2.73) with Eq. (1.37) shows that the reciprocal lattice of the fcc crystal structure exhibits bcc symmetry. The reciprocal lattice vectors given by Eq. (2.73) are shown in Fig. 2.9. Here the origin is chosen to be at the center of the cube while the x-, y-, and z-directions are parallel to the edges of the cube. So, the reciprocal lattice of the fcc crystal lattice is a bcc lattice with a primitive cell of volume $4(2\pi/a)^3$, using Eq. (2.55). Substituting Eqs. (2.73) into Eq. (2.30), the general reciprocal lattice is given by

$$\mathbf{G}_p = \frac{2\pi}{a}\left[(p_1 - p_2 + p_3)\hat{\mathbf{i}}_1 + (p_1 + p_2 - p_3)\hat{\mathbf{i}}_2 + (-p_1 + p_2 + p_3)\hat{\mathbf{i}}_3\right] \tag{2.74}$$

The above equation yields eight shortest nonzero reciprocal lattice vectors given by

$$\mathbf{G}_1 = \frac{2\pi}{a}\left(\pm\hat{\mathbf{i}}_1, \pm\hat{\mathbf{i}}_2, \pm\hat{\mathbf{i}}_3\right) \tag{2.75}$$

The next shortest reciprocal vectors $\mathbf{G}_2$ are six in number and are given by

$$\mathbf{G}_2 = \pm\frac{4\pi}{a}\hat{\mathbf{i}}_1, \quad \pm\frac{4\pi}{a}\hat{\mathbf{i}}_2, \quad \pm\frac{4\pi}{a}\hat{\mathbf{i}}_3 \tag{2.76}$$

The perpendicular bisector planes to $\mathbf{G}_1$ give an octahedron with a reciprocal lattice point at its center. However, it is found that the perpendicular bisector planes of $\mathbf{G}_2$ cut the corners of the regular octahedron formed above. Therefore, the 1BZ of the fcc crystal structure is a truncated octahedron, as shown in Fig. 2.10A.

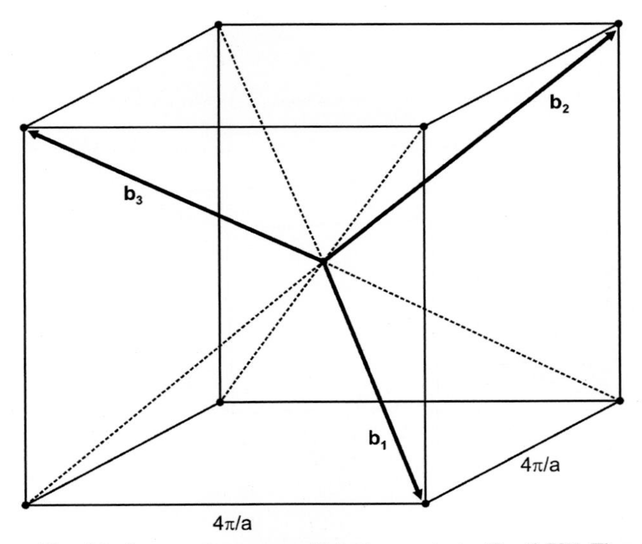

FIG. 2.9 The primitive vectors $\mathbf{b}_1$, $\mathbf{b}_2$, and $\mathbf{b}_3$ of the fcc crystal structure with lattice constant a [Eq. (2.73)]. The reciprocal lattice exhibits bcc symmetry.

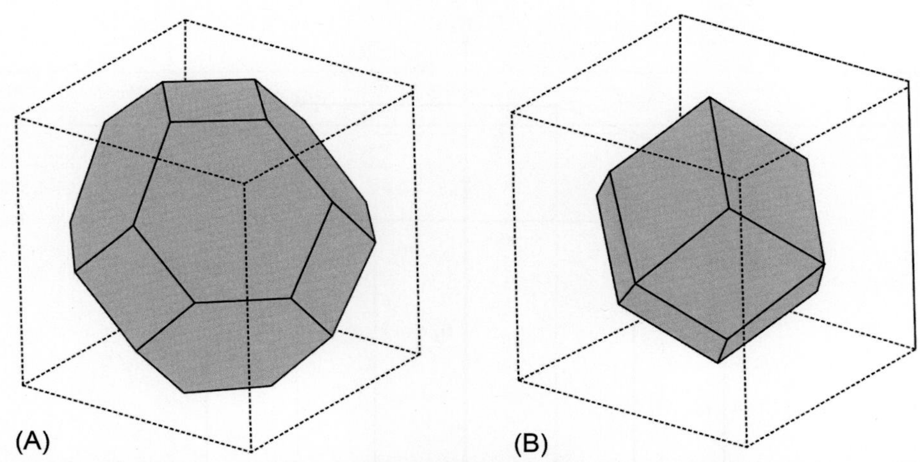

FIG. 2.10 The 1BZs for (A) fcc and (B) bcc crystal structures. Each BZ is inscribed in a cube of edge $4\pi/a$, where a is the edge of the cubic cell in the crystal space.

Problem 2.3

Consider a crystal with bcc structure whose primitive vectors are given by Eq. (1.37). Prove that its reciprocal lattice has fcc symmetry with primitive vectors given by

$$\begin{aligned}\mathbf{b}_1 &= \frac{2\pi}{a}\left(\hat{\mathbf{i}}_1 + \hat{\mathbf{i}}_2\right)\\ \mathbf{b}_2 &= \frac{2\pi}{a}\left(\hat{\mathbf{i}}_2 + \hat{\mathbf{i}}_3\right)\\ \mathbf{b}_3 &= \frac{2\pi}{a}\left(\hat{\mathbf{i}}_3 + \hat{\mathbf{i}}_1\right)\end{aligned} \tag{2.77}$$

Draw the primitive vectors $\mathbf{b}_1$, $\mathbf{b}_2$, and $\mathbf{b}_3$ in the reciprocal space.

The general reciprocal lattice vector obtained from Eqs. (2.77) and (2.30) is given by

$$\mathbf{G}_p = \frac{2\pi}{a}\left[(p_1 + p_3)\hat{\mathbf{i}}_1 + (p_1 + p_2)\hat{\mathbf{i}}_2 + (p_2 + p_3)\hat{\mathbf{i}}_3\right] \tag{2.78}$$

The above equation yields twelve shortest nonzero reciprocal lattice vectors given by

$$\mathbf{G}_1 = \frac{2\pi}{a}\left(\pm\hat{\mathbf{i}}_1, \pm\hat{\mathbf{i}}_2\right), \ \frac{2\pi}{a}\left(\pm\hat{\mathbf{i}}_2, \pm\hat{\mathbf{i}}_3\right), \ \frac{2\pi}{a}\left(\pm\hat{\mathbf{i}}_3, \pm\hat{\mathbf{i}}_1\right) \tag{2.79}$$

The 1BZ is obtained by the perpendicular bisector planes of $\mathbf{G}_1$, defined by Eq. (2.79), and is a regular rhombic dodecahedron, as shown in Fig. 2.10B.

2.6.5 Hexagonal Crystal Structure

The primitive translation vectors of the hexagonal structure in direct space, as shown in the basal plane of Fig. 1.22, are given by Eq. (1.41). Using Eq. (2.28), one can find the primitive lattice vectors in the reciprocal space that are given by

$$\begin{aligned}\mathbf{b}_1 &= \frac{2\pi}{a}\left(\hat{\mathbf{i}}_1 - \frac{1}{\sqrt{3}}\hat{\mathbf{i}}_2\right)\\ \mathbf{b}_2 &= \frac{2\pi}{a}\left(\hat{\mathbf{i}}_1 + \frac{1}{\sqrt{3}}\hat{\mathbf{i}}_2\right)\\ \mathbf{b}_3 &= \frac{2\pi}{c}\hat{\mathbf{i}}_3\end{aligned} \tag{2.80}$$

Here we have used Eq. (1.42) for the volume of the primitive cell of the hexagonal structure. Fig. 2.11 shows the unit cell formed by vectors $\mathbf{b}_1$, $\mathbf{b}_2$, and $\mathbf{b}_3$. Hence, the reciprocal lattice of hexagonal crystal structure also exhibits hexagonal

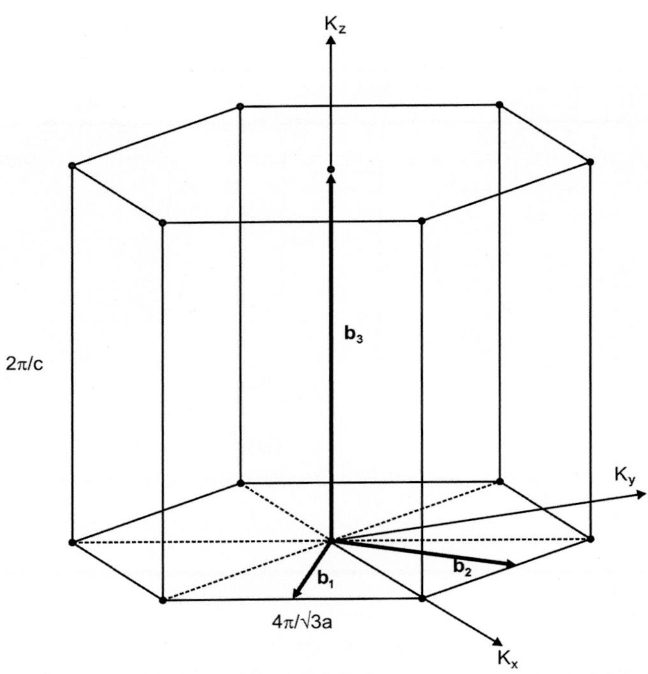

FIG. 2.11 The unit cell in the reciprocal space formed by primitive lattice vectors $\mathbf{b}_1$, $\mathbf{b}_2$, and $\mathbf{b}_3$ given by Eq. (2.80) for the hexagonal crystal structure.

symmetry. The general reciprocal lattice vector for the hexagonal structure can be written from Eqs. (2.30), (2.80) and is given by

$$\mathbf{G}_p = \frac{2\pi}{a}\left[(p_1 + p_2)\hat{\mathbf{i}}_1 - \frac{1}{\sqrt{3}}(p_1 - p_2)\hat{\mathbf{i}}_2 + \frac{a}{c}p_3\hat{\mathbf{i}}_3\right] \tag{2.81}$$

Problem 2.4

If the primitive lattice vectors in the crystal space in a hexagonal structure are given by

$$\begin{aligned} \mathbf{a}_1 &= a\hat{\mathbf{i}}_1 \\ \mathbf{a}_2 &= -\frac{a}{2}\hat{\mathbf{i}}_1 + \frac{\sqrt{3}}{2}a\hat{\mathbf{i}}_2 \\ \mathbf{a}_3 &= c\hat{\mathbf{i}}_3 \end{aligned} \tag{2.82}$$

(see Fig. 1.21), prove that the primitive lattice vectors in the reciprocal space are given by

$$\begin{aligned} \mathbf{b}_1 &= \frac{2\pi}{a}\left(\hat{\mathbf{i}}_1 + \frac{1}{\sqrt{3}}\hat{\mathbf{i}}_2\right) \\ \mathbf{b}_2 &= \frac{2\pi}{a}\frac{2}{\sqrt{3}}\hat{\mathbf{i}}_2 \\ \mathbf{b}_3 &= \frac{2\pi}{c}\hat{\mathbf{i}}_3 \end{aligned} \tag{2.83}$$

The primitive vectors in the reciprocal space are shown in Fig. 2.12. Further, show that the general reciprocal lattice vector $\mathbf{G}_p$ from Eqs. (2.30), (2.83) is given by[1]

$$\mathbf{G}_p = \frac{2\pi}{a}\left[p_1\hat{\mathbf{i}}_1 + \frac{1}{\sqrt{3}}(p_1 + 2p_2)\hat{\mathbf{i}}_2 + \frac{a}{c}p_3\hat{\mathbf{i}}_3\right] \tag{2.84}$$

---

1. If $I_1 = p_1$, $I_2 = p_1 + 2p_2$, and $I_3 = p_3$, then $I_1 + I_2 = 2(p_1 + p_2)$. Because p_1, p_2 and p_3 are integers, therefore $I_1 + I_2$ is always an even integer. While generating reciprocal lattice vectors of hexagonal structure, one should incorporate the above condition ($I_1 + I_2$ is always even) in the variation of p_1, p_2 and p_3.

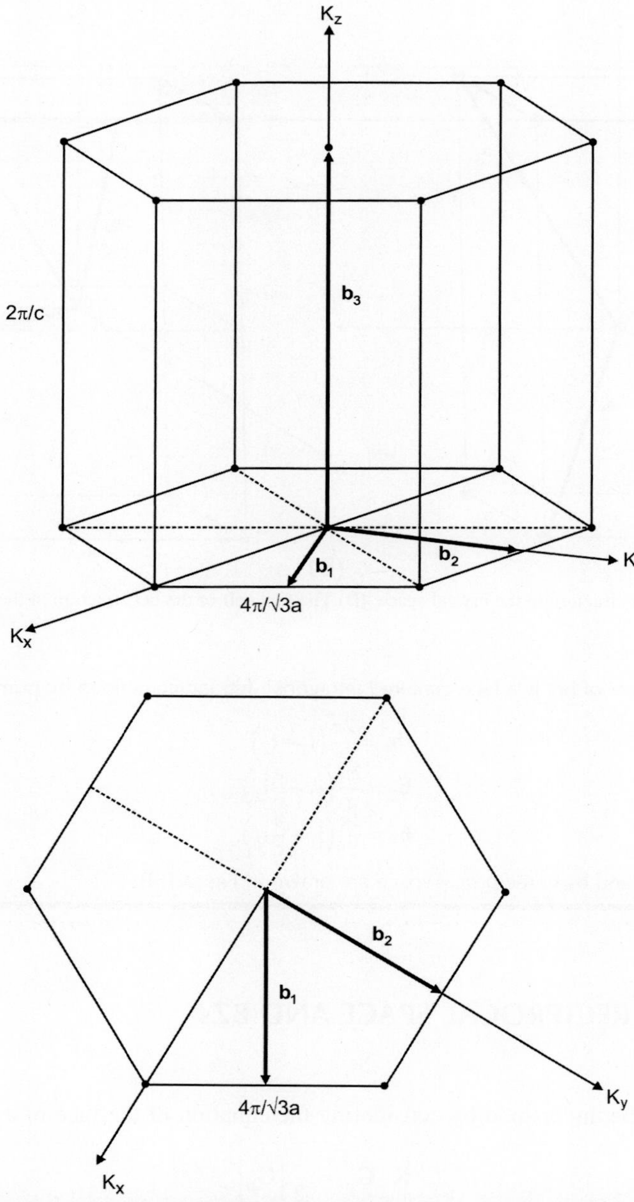

FIG. 2.12 The unit cell in the reciprocal space formed by primitive lattice vectors $\mathbf{b}_1$, $\mathbf{b}_2$, and $\mathbf{b}_3$ given by Eq. (2.83) of the hexagonal structure. The figure also shows separately the primitive vectors $\mathbf{b}_1$ and $\mathbf{b}_2$ in the basal plane of the unit cell.

Problem 2.5

Consider a body-centered tetragonal (bct) structure (see Fig. 2.13A) with primitive vectors given by

$$\begin{aligned}\mathbf{a}_1 &= \frac{a}{2}\hat{\mathbf{i}}_1 + \frac{a}{2}\hat{\mathbf{i}}_2 - \frac{c}{2}\hat{\mathbf{i}}_3 \\ \mathbf{a}_2 &= -\frac{a}{2}\hat{\mathbf{i}}_1 + \frac{a}{2}\hat{\mathbf{i}}_2 + \frac{c}{2}\hat{\mathbf{i}}_3 \\ \mathbf{a}_3 &= \frac{a}{2}\hat{\mathbf{i}}_1 - \frac{a}{2}\hat{\mathbf{i}}_2 + \frac{c}{2}\hat{\mathbf{i}}_3\end{aligned} \tag{2.85}$$

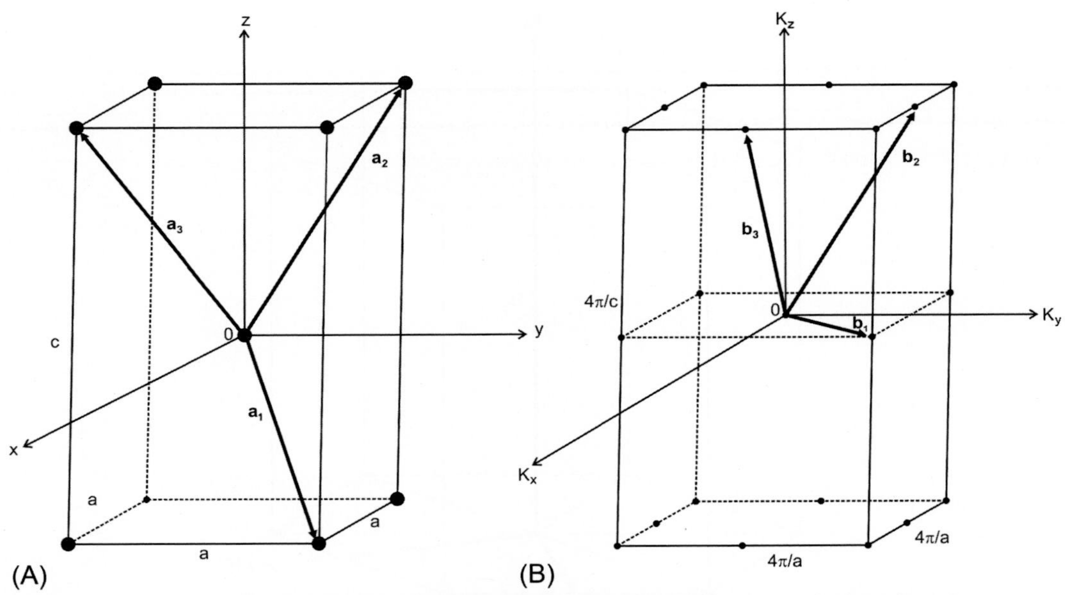

FIG. 2.13 (A) The unit cell of the bct structure in the crystal space. (B) The unit cell of the bct structure in the reciprocal space.

Prove that the reciprocal lattice of bct is a face-centered tetragonal (fct) lattice defined by primitive reciprocal lattice vectors

$$\begin{aligned}\mathbf{b}_1 &= \frac{2\pi}{a}\left(\hat{\mathbf{i}}_1 + \hat{\mathbf{i}}_2\right)\\ \mathbf{b}_2 &= \frac{2\pi}{a}\left(\hat{\mathbf{i}}_2 + \frac{a}{c}\hat{\mathbf{i}}_3\right)\\ \mathbf{b}_3 &= \frac{2\pi}{a}\left(\hat{\mathbf{i}}_1 + \frac{a}{c}\hat{\mathbf{i}}_3\right)\end{aligned} \tag{2.86}$$

The primitive vectors $\mathbf{b}_1$, $\mathbf{b}_2$, and $\mathbf{b}_3$ in reciprocal space are drawn in Fig. 2.13B.

2.7 IMPORTANCE OF RECIPROCAL SPACE AND BZs

2.7.1 Bragg Reflection

The importance of the BZ can be understood by considering the equation of the face of a BZ (see Fig. 2.14) given by

$$\mathbf{K}\cdot\hat{\mathbf{G}}_p = \frac{1}{2}\left|\mathbf{G}_p\right| \tag{2.87}$$

where $\mathbf{K}$ is any wave vector ending on the face of the BZ and $\hat{\mathbf{G}}_p$ is a unit vector along $\mathbf{G}_p$. If $\mathbf{G}_p$ is the shortest reciprocal lattice vector, then Eq. (2.87) gives the face of the 1BZ. Eq. (2.87) is equivalent to the equation

$$\mathbf{K}' = \mathbf{K} + \mathbf{G}_p \tag{2.88}$$

with $|\mathbf{K}'| = |\mathbf{K}|$. Eq. (2.88) may be viewed as an expression for the conservation of momentum in a crystal. Note that Eqs. (2.87), (2.88) in the reciprocal space are equivalent to Eqs. (1.19) and (1.20), respectively, in the crystal space. Eq. (2.88) is just the same as Eq. (2.53) and gives the Bragg reflection condition and can be represented by Ewald's construction (Fig. 2.15). Eq. (2.87) can also be written as

$$2\mathbf{K}\cdot\mathbf{G}_p + G_p^2 = 0 \tag{2.89}$$

It is noteworthy that the Eqs. (2.87), (2.88), and (2.89) are all equivalent and represent the Bragg reflection condition. According to Fig. 2.15, Bragg reflection occurs if the wave vectors before ($\mathbf{K}$) and after ($\mathbf{K}'$) reflection end at the reciprocal lattice points. Hence, the Bragg reflection condition is satisfied at all points on the surface of the BZ, which is an important property of the reciprocal lattice.

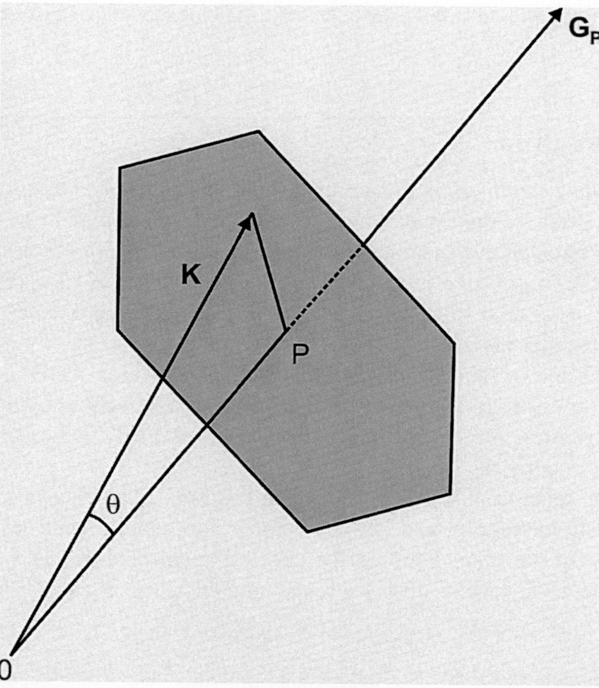

FIG. 2.14 A plane bisects perpendicularly the reciprocal vector $\mathbf{G}_p$ with P as the mid point. The vector **K** represent the position vector of any point on the plane.

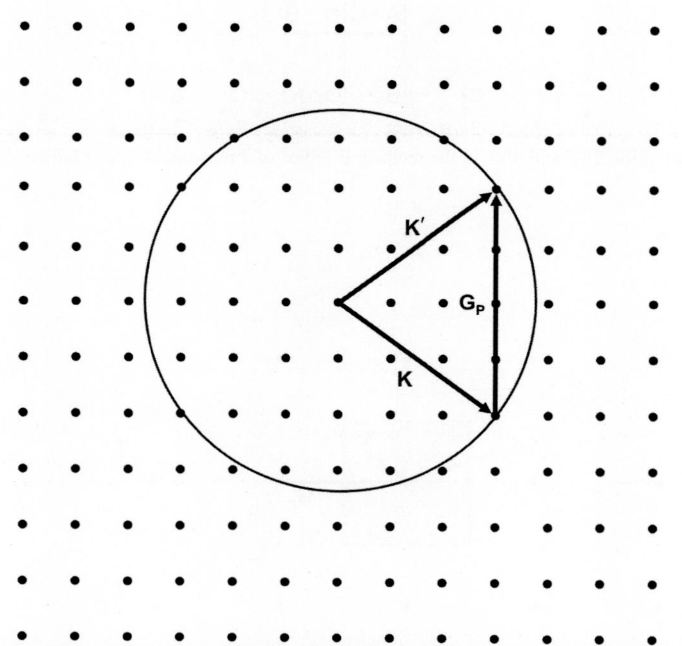

FIG. 2.15 Ewald's construction for a square reciprocal lattice with dots representing the lattice points. The wave vectors **K** and **K′** are such that they join lattice points and $|\mathbf{K}|=|\mathbf{K}'|$.

Problem 2.6

Prove that Eq. (2.88) or Eq. (2.89) can be written as

$$2a\sin\theta = n\lambda \tag{2.90}$$

for cubic crystals. Here a and λ are the lattice parameter and the wavelength, n is an integer, and angle θ is defined in the same way as in Fig. 2.1.

Note that Eq. (2.90) is the Bragg reflection condition, except that the distance between the two planes d [Eq. (2.1)] is replaced by a.

2.7.2 Significant Wave Vectors

Eqs. (2.53), (2.54) show that a wave vector **K** is not unique. An infinite number of equivalent wave vectors **K**′ can be generated using Eq. (2.53). Let us denote the set of wave vectors that are equivalent to **K** by {**K**′}. Any wave vector in {**K**′} is related to **K** by some reciprocal lattice vector $\mathbf{G}_p$. In other words, every **K** possesses an infinite number of equivalent wave vectors in the reciprocal space. In a set of equivalent wave vectors {**K**′}, if **K** forms the smallest vector, it is called the significant wave vector. Let us investigate if there exists any particular region in the reciprocal space that contains all the significant wave vectors for a particular crystal.

Consider a one-dimensional lattice in reciprocal space (Fig. 2.16) in which AOB is the 1BZ. Now add the shortest $\mathbf{G}_p(=2\pi/a)$ to the portion DB of the 2BZ. It will occupy the position in the 1BZ shown in Fig. 2.16. Similarly, subtract the shortest $\mathbf{G}_p$ from the second portion AC of the 2BZ to bring it to the 1BZ (Fig. 2.16). Similarly, the 3BZ can also be brought to the 1BZ and it will occupy the position shown in Fig. 2.16.

Consider a square lattice in the reciprocal space, as shown in Fig. 2.17. The wave vector $\mathbf{K}_1$ lies inside, while $\mathbf{K}_2$ lies outside the 1BZ. When even the shortest reciprocal lattice vector is subtracted from (or added to) $\mathbf{K}_1$, the resultant wave vector $\mathbf{K}_1'$ lies outside the 1BZ. On the other hand, if the shortest reciprocal lattice vector is subtracted from $\mathbf{K}_2$, the resultant vector $\mathbf{K}_2'$ lies inside the 1BZ. In both the cases, the smaller wave vector (significant wave vector) lies inside

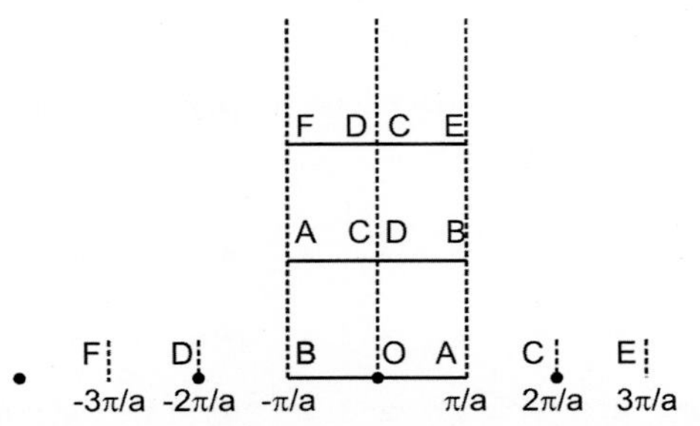

FIG. 2.16 Schematic representation of the 2BZ and 3BZ in the reciprocal lattice of a monatomic linear lattice, when transferred to the 1BZ.

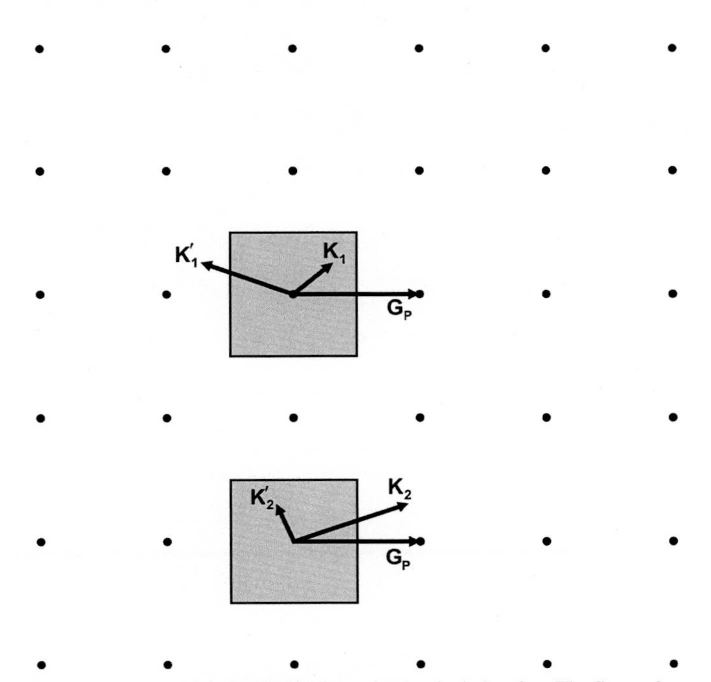

FIG. 2.17 A square lattice in the reciprocal space in which the 1BZ is shown by the shaded region. The figure shows two cases: one with wave vector $\mathbf{K}_1$ lying in the 1BZ and the second with wave vector $\mathbf{K}_2$ outside the 1BZ.

the 1BZ. One can generalize this fact by saying that a significant wave vector **K**, corresponding to a set of equivalent wave vectors $\{\mathbf{K}'\}$, always lies inside the 1BZ. In general, all the significant wave vectors for a particular reciprocal lattice lie inside the 1BZ. Sometimes the significant wave vectors are called reduced wave vectors and are usually represented by vector **q**. It is for this reason that the electronic properties, particularly the energy band structure, are calculated in the 1BZ and these get repeated in the higher BZs.

2.7.3 Construction of Reciprocal Lattice

The 1BZ of a crystal structure is a primitive unit cell in the reciprocal space. By translating the 1BZ by all possible reciprocal lattice vectors one can fill the whole of the reciprocal space and can generate the reciprocal lattice. From Fig. 2.10, it is evident that the 1BZ is symmetrical about the lattice point situated at its center. Therefore, for convenience, in the theoretical investigations of the properties of a crystal, one can replace the actual 1BZ by a sphere of equal volume to that of the 1BZ, generally called the WS sphere. Hence, the reciprocal lattice is a very important concept in the study of crystalline materials because:

1. The diffraction pattern of a crystal (Laue spots) is a picture of the reciprocal lattice of the crystal. In principle, the primitive translation vectors and their orientation in the crystal space can be obtained by making a transformation from the reciprocal to the crystal space with the help of Eq. (2.29).
2. The mathematical solution of many physical problems in solid state physics is very difficult in the crystal space. But if one transforms the problem into reciprocal space, using Eq. (2.28), the mathematical solution becomes easy.

2.8 ATOMIC SCATTERING FACTOR

The atomic scattering factor measures the amplitude of the wave scattered from a particular atom of a unit cell. The scattering of X-rays depends on the electron distribution in an atom. The atomic scattering factor of the mth atom of a unit cell from Eq. (2.22) is given by

$$f_m(\Delta\mathbf{K}) = \int n_e(\mathbf{r})\, e^{\imath\Delta\mathbf{K}\cdot\mathbf{r}} d^3r \tag{2.91}$$

The integral is over the electron density of the atom under consideration. Eq. (2.91) is nothing but the Fourier transform of the atomic electron density. The plane wave can be expanded in terms of Legendre polynomials $P_\ell(\cos\theta)$ as

$$e^{\imath\Delta\mathbf{K}\cdot\mathbf{r}} = \sum_{\ell=0}^{\infty}(2\ell+1)\,\imath^\ell\, j_\ell(\Delta K r)\, P_\ell\left(\cos\theta_{\Delta\mathbf{K},\mathbf{r}}\right) \tag{2.92}$$

$j_\ell(\Delta K\, r)$ is a spherical Bessel function of order ℓ (orbital quantum number) and $\theta_{\Delta\mathbf{K},\,\mathbf{r}}$ is the angle between $\Delta\mathbf{K}$ and **r**. Assuming the electron density of an atom to be spherically symmetric, only the $\ell = 0$ component contributes to Eq. (2.92). Therefore,

$$e^{\imath\Delta\mathbf{K}\cdot\mathbf{r}} = j_0(\Delta K r) = \frac{\sin(\Delta K r)}{\Delta K r} \tag{2.93}$$

Hence, Eqs. (2.93), (2.91) give

$$f_m(\Delta K) = 4\pi\int n_e(\mathbf{r})\,\frac{\sin(\Delta K r)}{\Delta K r}\, r^2 dr \tag{2.94}$$

As the angle θ between **K** and $\mathbf{K}'$ goes to zero (see Fig. 2.18), $\Delta\mathbf{K}$ also reduces to zero, which gives

$$f_m(0) = 4\pi\int n_e(\mathbf{r})\, r^2 dr = Z \tag{2.95}$$

where Z is the valence of atom under consideration. If $\Delta\mathbf{K}$ is nonzero, then according to Bragg's reflection $\Delta\mathbf{K} = \mathbf{G}_p$ [see Fig. (2.15)]. Therefore, Eq. (2.94) reduces to

$$f_m\left(\mathbf{G}_p\right) = 4\pi\int n_e(\mathbf{r})\frac{\sin\left(G_p r\right)}{G_p r} r^2 dr \tag{2.96}$$

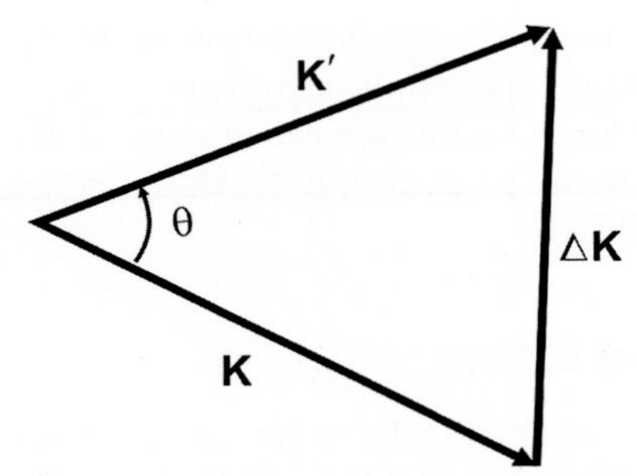

FIG. 2.18 The change in wave vector $\Delta\mathbf{K}$ after the scattering process. Here, $\mathbf{K}$ is the wave vector before scattering and $\mathbf{K}'$ is the wave vector after scattering.

If the whole of the electron distribution of an atom, situated at the origin, is concentrated at the origin, then the atom can be treated as a point charge and Eq. (2.96) reduces to

$$f_m\left(\mathbf{G}_p\right) = 4\pi \int n_e(\mathbf{r})\, r^2\, dr = Z \tag{2.97}$$

Therefore, in both the approximations of $\mathbf{G}_p \rightarrow 0$ and $\mathbf{r} \rightarrow 0$, the total scattering amplitude reduces to the same value.

2.9 GEOMETRICAL STRUCTURE FACTOR

The geometrical structure factor from Eq. (2.21) is given by

$$F_S(\Delta\mathbf{K}) = \sum_{m=1}^{s} f_m(\Delta\mathbf{K})\, e^{-\imath \Delta\mathbf{K}\cdot\mathbf{R}_m} \tag{2.98}$$

$F_S(\mathbf{G}_p)$ gives the amplitude of scattering from a unit cell of the crystal. Bragg reflection occurs at various atoms in the unit cell when $\Delta\mathbf{K} = \mathbf{G}_p$, which reduces $F_S(\Delta\mathbf{K})$ to

$$F_S\left(\mathbf{G}_p\right) = \sum_{m=1}^{s} f_m\left(\mathbf{G}_p\right) e^{-\imath \mathbf{G}_p\cdot\mathbf{R}_m} \tag{2.99}$$

Substituting the values of $\mathbf{G}_p$ and $\mathbf{R}_\mathbf{m}$ from Eqs. (2.30) and (1.8), respectively, one gets

$$F_S\left(\mathbf{G}_p\right) = \sum_{m=1}^{s} f_m\left(\mathbf{G}_p\right) e^{-2\pi\imath(p_1 m_1 + p_2 m_2 + p_3 m_3)} \tag{2.100}$$

It is evident from Eq. (2.100) that $F_S(\mathbf{G}_p)$ may not be real, but the scattered wave intensity, given by $F_S^*(\mathbf{G}_p)F_S(\mathbf{G}_p)$, is real. With the help of Eq. (2.100), one can find directions in which the Bragg's reflection has a maximum and others in which it is a minimum. Such a study will help us in the determination of the structure of a crystalline solid.

2.9.1 sc Crystal Structure

In a crystal with sc structure there is one basis atom, which is assumed to be situated at the lattice point. If this lattice point is at the origin, then, $m = (m_1, m_2, m_3) = (0,0,0) = 0$ and Eq. (2.100) reduces to

$$F_S\left(\mathbf{G}_p\right) = f_0\left(\mathbf{G}_p\right) \tag{2.101}$$

which is real. The intensity of scattered wave I_S becomes

$$I_S = F_S^*\left(\mathbf{G}_p\right) F_S\left(\mathbf{G}_p\right) = f_0^2\left(\mathbf{G}_p\right) \tag{2.102}$$

2.9.2 fcc Crystal Structure

In the unit cell of the fcc structure, there are four atoms at the positions a(0,0,0), a(½,½,0), a(½,0, ½), and a(0, ½, ½), see Fig. 1.19. Substituting these coordinates into Eq. (2.100) one gets

$$F_S\left(\mathbf{G}_p\right) = f_0\left(\mathbf{G}_p\right)\left[1 + e^{-\imath\pi(p_1+p_2)} + e^{-\imath\pi(p_2+p_3)} + e^{-\imath\pi(p_3+p_1)}\right] \quad (2.103)$$

Here, it is assumed that all of the four atoms are identical (pure crystalline solid), each having atomic form factor $f_0(\mathbf{G}_p)$. The maximum value of $F_S(\mathbf{G}_p)$ from Eq. (2.103) becomes

$$F_S\left(\mathbf{G}_p\right) = 4f_0\left(\mathbf{G}_p\right) \quad (2.104)$$

when the integers p_1, p_2, and p_3 are all even or all odd. The maximum intensity of Bragg reflection then becomes

$$I_S = 16f_0^2\left(\mathbf{G}_p\right) \quad (2.105)$$

On the other hand, if one of the integers p_1, p_2, and p_3 is odd and the others are even, or if one integer is even and others odd, then $F_S(\mathbf{G}_p)$ goes to zero; in such directions, there is no Bragg reflection. All of the four atoms in a unit cell may not be identical if the crystal is a compound or an alloy. In these materials, one obtains minima in the Bragg reflection instead of no reflection because the atomic form factors f_m have different values for different atoms.

2.9.3 bcc Crystal Structure

In a pure crystal with bcc structure, the basis consists of two identical atoms at a(0,0,0) and a(½,½,½) (see Fig. 1.14). Substituting the coordinates of the basis atoms into Eq. (2.100), we obtain

$$F_S\left(\mathbf{G}_p\right) = f_0\left(\mathbf{G}_p\right)\left[1 + e^{-\imath\pi(p_1+p_2+p_3)}\right] \quad (2.106)$$

If $p_1 + p_2 + p_3$ is an odd integer, then

$$F_S\left(\mathbf{G}_p\right) = 0 \quad (2.107)$$

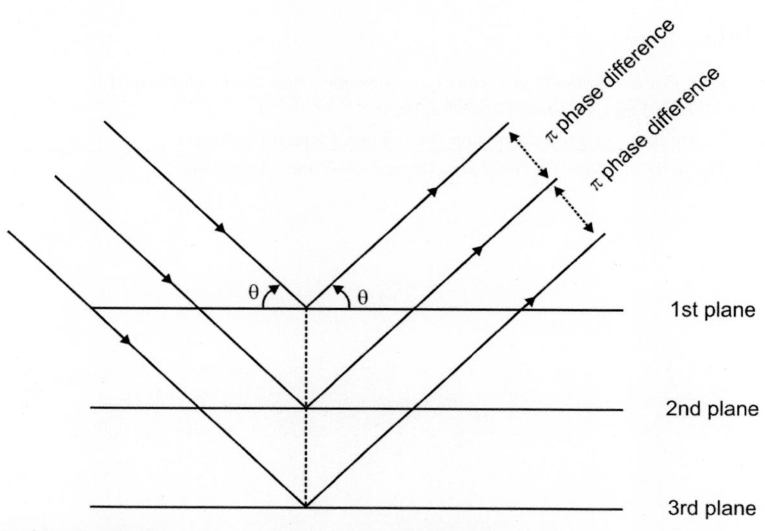

FIG. 2.19 The X-ray reflection from the first three atomic planes in a bcc crystal structure.

and yields zero for the value of the scattered amplitude. Therefore, the intensity of the Bragg reflection in these directions of the unit cell is zero. If $p_1 + p_2 + p_3$ is an even integer, then

$$F_S\left(\mathbf{G}_p\right) = 2f_0\left(\mathbf{G}_p\right) \tag{2.108}$$

Eq. (2.108) yields the maximum value of the intensity of scattered waves in these directions and is given by

$$I_S = 4f_0^2\left(\mathbf{G}_p\right) \tag{2.109}$$

Eqs. (2.107), 2.108) show that the diffraction pattern does not contain lines such as [100], [111], [300], or [221], but does contain lines such as [110], [200], [211], and [222].

One can explain physically why the [100] reflection is missing in the diffraction pattern from a bcc structure of a pure element. The [100] reflection occurs when the reflections from the first and third planes differ in phase by 2π (see Fig. 2.19) and these planes bind the unit cell. The reflection from the first and the intervening second plane are out of phase, thereby canceling exactly the effect of each other due to the identical nature of the two planes. On the other hand, if the bcc structure is of a compound, such as CsCl (see Fig. 1.17), then the atoms in the first and second planes are of a different kind: the first plane consists of Cl atoms, while the second plane consists of Cs atoms and the atomic scattering factors of the two types of atoms are different say f_0 and f_1. From Eq. (2.100) one obtains

$$F_S\left(\mathbf{G}_p\right) = f_0\left(\mathbf{G}_p\right) + f_1\left(\mathbf{G}_p\right)e^{-\imath\pi(p_1+p_2+p_3)} \tag{2.110}$$

In this case $F_S(\mathbf{G}_p)$ will never be zero, but it will oscillate from a minimum $f_0 - f_1$ to a maximum value of $f_0 + f_1$. Hence, the intensities of lines will vary from a maximum value of $|f_0 + f_1|^2$ to a minimum value $|f_0 - f_1|^2$.

REFERENCES

Becon, G. E. (1975). *Neutron diffraction* (3rd ed.). Oxford: Clarendon Press.

Izyumov, Y. A., & Ozerov, R. P. (1970). *Magnetic neutron diffraction.* New York: Plenum Press.

Koehler, W. C., & Wollan, E. O. (1955). Neutron diffraction by metallic erbium. *Physical Review*, *97*, 1177–1178.

Shull, C. G. (1995). Early development of neutron scattering. *Reviews of Modern Physics*, *67*, 753–757.

Shull, C. G., & Smart, J. S. (1949). Detection of antiferromagnetism by neutron diffraction. *Physical Review*, *76*, 1256–1259.

Shull, C. G., Strauser, W. A., & Wollan, E. O. (1951). Neutron diffraction by paramagnetic and antiferromagnetic substances. *Physical Review*, *83*, 333–345.

Shull, C. G., & Wilkinson, M. K. (1955). Neutron diffraction studies of the magnetic structure of alloys of transition elements. *Physical Review*, *97*, 304–310.

Shull, C. G., & Wollan, E. O. (1956). Applications of neutron diffraction to solid state problems. F. Seitz, & D. Turnbull (Eds.), *Solid state physics*. (pp. 137–217). Vol. 2. New York: Academic Press.

SUGGESTED READING

Azaroff, L. V., & Buerger, M. J. (1958). *Powder methods in X-ray crystallography.* New York: McGraw-Hill.

Becon, G. E. (1975). *Neutron diffraction* (3rd ed.). London: Clarendon Press.

Izyumov, Y. A., & Ozerov, R. (1970). *Magnetic neutron diffraction.* New York: Plenum Publishers.

Lipson, H., & Cochran, W. (1954). *The determination of crystal structures.* New York: Macmillan.

Chapter 3

Approximations in the Study of Solids

Chapter Outline

A pure crystalline solid consists of a periodic array of atoms in three dimensions and each atom consists of a nucleus with electrons revolving around it. In a solid, one talks either about the electronic properties, such as electron states, electronic band structure, and electrical conductivity, or about the lattice properties, such as phonons, lattice specific heat, and thermal conductivity. Solids can be classified broadly as insulators, conductors, and semiconductors. In an insulator, there are no free electrons and the atoms as such form the periodic array. But in a conducting solid, each atom provides some free electrons, leaving an ion behind. Therefore, in a conductor there are free electrons with ions fixed at the lattice positions. An intrinsic semiconductor behaves as an insulator at low temperatures, but at reasonably high temperatures some electrons are excited to the conduction band to become free. On the other hand, in an extrinsic semiconductor, there are a few free electrons or holes even at room temperature, which give rise to finite electrical conductivity. Therefore, in general, a crystalline solid consists of a large number of electrons and ions, which give rise to a finite electrostatic field inside it, usually called the *crystal field*. The crystal field plays a central role in the theoretical study of the various physical properties of crystalline solids but its exact determination is very difficult due to the many-body nature of the problem. To simplify the theoretical study, some approximations are made in the estimation of the crystal field of a crystalline solid and these are discussed in this chapter.

3.1 SEPARATION OF ION-CORE AND VALENCE ELECTRONS

In most solids, except the inert gas crystals, the outermost electron orbit of each atom is partially filled and the electrons belonging to it are called valence electrons. Below the outermost electron orbit, there is the ion core in which all the electron orbits are completely filled. The valence electrons play an important role in the study of various physical properties of solids. In a metallic solid the valence electrons are loosely bound to the nucleus of an atom. The valence electrons experience the crystal field as a result of which they get detached from the atom and become free to move anywhere in the solid. So, a metallic solid can be considered to be a sea of free (conduction) electrons in which the ions are embedded at the lattice positions. It is usually assumed that there is a sharp distinction between the valence electrons and the ion core and these can be dealt with separately in a theoretical study. This approximation works well in most solids. If the ion core is small and spherical in shape, which is the case in light elements, it can be treated as a point with charge Ze, where Z is the valency of the atom/ion.

3.2 RIGID ION-CORE APPROXIMATION

The electrons belonging to an ion core, called core electrons, are assumed to move rigidly along with the nucleus and cannot be excited at available energies in a solid. This is called the *rigid ion approximation*. Further, the core states in a solid are assumed to be the same as in an isolated atom, which means that the crystal field does not affect (distort) the core states. This approximation works reasonably well in solids with small ion core size, such as simple metals. But in solids with large ion core size, such as d- and f-band metals, this is not a very good approximation as the core states are affected significantly by the crystal field.

Solid State Physics. https://doi.org/10.1016/B978-0-12-817103-5.00003-7

3.3 SELF-CONSISTENT POTENTIAL APPROXIMATION

The exact evaluation of the crystal potential $V(\mathbf{r})$ is not possible. This is because $V(\mathbf{r})$ depends on the electron states, which in turn depend on $V(\mathbf{r})$. Therefore, $V(\mathbf{r})$ must be calculated self-consistently and such a potential is called a self-consistent potential. This approximation is called the self-consistent approximation.

3.4 THE BORN-OPPENHEIMER APPROXIMATION

In a solid the electrons, being lighter particles, move much more quickly than ions. Further, calculations show that the average speed of an electron in the hydrogen molecule-ion is approximately 1000 times that of a proton. This means that an electron can complete an orbit before the nuclei of the molecule-ion have moved significantly. This fact enables the electrons to adjust their orbitals almost instantaneously in response to any change in the positions of the two nuclei. Therefore, the motion of the nuclei of the hydrogen molecule-ion (representing translation, vibration, and rotation) can be separated from the electron motion. This fact can be generalized to all the crystalline solids in which the electron motion can be considered to be independent from the motion of the ions. This is known as the *Born-Oppenheimer approximation* or the *adiabatic approximation* because the electrons follow the motion of the ions adiabatically in a solid.

The adiabatic approximation can be explained further by considering the Hamiltonian of a crystalline solid given by

$$\widehat{H} = T_e + T_i + V_{ii} + V_{ee} + V_{ei} + V_{xc} + V_c \tag{3.1}$$

where

$$T_e = -\sum_i \frac{\hbar^2}{2m_e}\nabla_i^2 \tag{3.2}$$

$$T_i = -\sum_n \frac{\hbar^2}{2M_n}\nabla_n^2 \tag{3.3}$$

$$V_{ee} = \frac{1}{2}\sum_{i,j\,(i\neq j)} \frac{e^2}{|\mathbf{r}_i - \mathbf{r}_j|} \tag{3.4}$$

$$V_{ii} = \frac{1}{2}\sum_{n,n'\,(n\neq n')} \frac{Z_n Z_{n'} e^2}{|\mathbf{R}_n - \mathbf{R}_{n'}|} \tag{3.5}$$

$$V_{ei} = -\sum_{n,i} \frac{Z_n e^2}{|\mathbf{r}_i - \mathbf{R}_n|} \tag{3.6}$$

Here $\mathbf{r}_i$ and $\mathbf{R}_n$ are the positions of ith electron and n^{th} ion. M_n is the mass of the nth ion, having charge $Z_n e$, while m_e is the mass of an electron having charge $-e$. T_e and T_i are the kinetic energies of all the electrons and ions, respectively. V_{ee}, V_{ii}, and V_{ei} represent potentials due to electron-electron, ion-ion, and electron-ion interactions in a solid. The factor of ½ in V_{ee} (V_{ii}) avoids the occurrence of a pair of electrons (ions) twice in the summation. V_{xc} and V_c represent the potentials arising from the exchange interactions and correlation interactions among the electrons, respectively, and will be discussed in detail later in this chapter.

The Schrodinger wave equation for a solid can be written as

$$\widehat{H}\,|\Psi(\{\mathbf{r}\},\{\mathbf{R}\})\rangle = E\,|\Psi(\{\mathbf{r}\},\{\mathbf{R}\})\rangle \tag{3.7}$$

The coordinates of all the electrons are collectively written as $\{\mathbf{r}\}$ and those of ions as $\{\mathbf{R}\}$. $|\Psi(\{\mathbf{r}\},\{\mathbf{R}\})\rangle$ is the wave function of the solid, which is a function of the coordinates of all the ions and electrons. In the adiabatic approximation, the total wave function of the solid can be written as the product of an electronic wave function $|\Psi(\{\mathbf{r}\})\rangle$, which is a function only of the electron coordinates for the fixed positions of ions, and an ionic wave function $|\Phi(\{\mathbf{R}\})\rangle$, which is a function only of the ionic coordinates, that is,

$$|\Psi(\{\mathbf{r}\},\{\mathbf{R}\})\rangle = |\Psi(\{\mathbf{r}\})\rangle|\Phi(\{\mathbf{R}\})\rangle \tag{3.8}$$

In order to study separately the electronic and lattice properties of a solid, one should split the total Hamiltonian into two parts: the electronic part $\widehat{H}_e$ and the ionic part $\widehat{H}_i$. From Eq. (3.6), we see that V_{ei} involves the coordinates of all the electrons and ions, therefore, it should be included in both the electronic and ionic parts of the Hamiltonian. The electronic part of the Hamiltonian is defined as

$$\widehat{H}_e = T_e + V_e(\{\mathbf{r}\}) \tag{3.9}$$

where

$$V_e(\{\mathbf{r}\}) = V_{ee} + V_{ei} + V_{xc} + V_c \tag{3.10}$$

It is noteworthy that $V_e(\{\mathbf{r}\})$ is a function of the coordinates of all the electrons for fixed ion positions. The ionic part of the Hamiltonian is defined as

$$\widehat{H}_i = T_i + V_i(\{\mathbf{R}\}) \tag{3.11}$$

where

$$V_i(\{\mathbf{R}\}) = V_{ii} + V_{ei} \tag{3.12}$$

$V_i(\{\mathbf{R}\})$ is a function of the positions of all the ions in the solid. One should note that $V_i(\{\mathbf{R}\})$ also includes the direct ion-ion overlap interaction, not written here, in addition to the ion-ion and electron-ion Coulomb interactions. Eqs. (3.9), (3.11) show that the solutions of $\widehat{H}_e$ and $\widehat{H}_i$ are many-body problems.

3.5 ONE-ELECTRON APPROXIMATION

In studying the electronic properties of a solid, one has to solve the electronic part of the Schrodinger wave equation defined as

$$\widehat{H}_e\left(\mathbf{r}_1, \mathbf{r}_2, \ldots, \mathbf{r}_{N_e}\right) \left|\Psi\left(\mathbf{r}_1, \mathbf{r}_2, \ldots, \mathbf{r}_{N_e}\right)\right\rangle = E\left|\Psi\left(\mathbf{r}_1, \mathbf{r}_2, \ldots, \mathbf{r}_{N_e}\right)\right\rangle \tag{3.13}$$

Here $|\Psi(\mathbf{r}_1, \mathbf{r}_2, \ldots, \mathbf{r}_{N_e})\rangle$ is an orthonormal wave function, which is a function of the positions of all the electrons, and E is the energy of the composite electron system. The exact solution of Eq. (3.13) is not possible as it is a many-body problem with a large number of electrons in a solid and, therefore, one has to resort to some simplifying assumption. The one-electron approximation is usually adopted in which an electron is assumed to move in some average potential $V(\mathbf{r})$ due to all the ions and the remaining electrons in a solid. In this approximation, one replaces the real system by a system of N_e independent electrons with the effective Hamiltonian of the ith electron given by

$$\widehat{H}_e(\mathbf{r}_i) = -\frac{\hbar^2}{2m_e}\nabla_i^2 + V(\mathbf{r}_i) \tag{3.14}$$

$\widehat{H}_e(\mathbf{r}_i)$ satisfies the one-electron Schrodinger equation defined by

$$\widehat{H}_e(\mathbf{r}_i)|\psi_i(\mathbf{r}_i)\rangle = E_i|\psi_i(\mathbf{r}_i)\rangle \tag{3.15}$$

where $|\psi_i(\mathbf{r}_i)\rangle$ and E_i are the one-electron orthonormal wave function and energy of the ith electron. The orthonormality condition demands

$$\langle\psi_i(\mathbf{r}_i)|\psi_j\left(\mathbf{r}_j\right)\rangle = \delta_{ij} \tag{3.16}$$

Now, the total energy of the composite system will be the sum of the energies of the individual electrons, that is,

$$E = \sum_{i=1}^{N_e} E_i \tag{3.17}$$

and correspondingly the total Hamiltonian will be the sum of the Hamiltonians of the individual electrons, that is,

$$\widehat{H}_e\left(\mathbf{r}_1, \mathbf{r}_2, \ldots, \mathbf{r}_{N_e}\right) = \sum_{i=1}^{N_e} \widehat{H}_e(\mathbf{r}_i) \tag{3.18}$$

The wave function of the composite system can be proved, from Eqs. (3.14)–(3.18), to be the product of the individual electron wave functions as

$$\left|\Psi\left(\mathbf{r}_1, \mathbf{r}_2, \ldots, \mathbf{r}_{N_e}\right)\right\rangle = |\psi_1(\mathbf{r}_1)\rangle\,|\psi_2(\mathbf{r}_2)\rangle \ldots \left|\psi_{N_e}\left(\mathbf{r}_{N_e}\right)\right\rangle \tag{3.19}$$

Therefore, in the one-electron approximation, the solution of the one-electron Schrodinger equation (3.15) is employed to find the wave function and energy of the composite system. From Eqs. (3.15), (3.17) the total energy of the system is given by

$$E = \sum_{i=1}^{N_e} \langle \psi_i(\mathbf{r}_i)|\widehat{H}_e(\mathbf{r}_i)\,|\psi_i(\mathbf{r}_i)\rangle \tag{3.20}$$

The solution of Eqs. (3.15), (3.20) requires knowledge of the one-electron potential, which must be calculated self-consistently. Hartree (1928) gave a self-consistent method for the determination of the ground state energy of a system. It was extended by Fock (1930) by incorporating symmetry in the wave function and this is usually called the Hartree-Fock self-consistent field theory.

3.6 ELECTRON EXCHANGE AND CORRELATION INTERACTIONS

In addition to the usual Coulomb interactions, which vary as 1/r, there are many-body electron interactions in a solid that can be classified into two categories:

1. Electron exchange interactions
2. Electron correlation interactions

Pines (1963) has described in detail the physics of exchange and correlation effects.

3.6.1 Electron Exchange Interactions

In a system of two or more indistinguishable electrons, the exchange of any two electrons gives rise to an additional contribution to the energy, usually called exchange energy, and the possible interactions are called exchange interactions. Therefore, the exchange interactions are many-body interactions that play an important role in understanding the electronic properties of solids. These interactions have purely quantum mechanical origin and have no classical analogue. The various exchange interactions in a solid are described below.

3.6.1.1 Intra-Atomic Exchange Interactions

Intra-atomic exchange interactions are the interactions between the electrons in the same or different orbits of a particular atom. These exchange interactions tend to align the spins of the electrons parallel to each other so as to give the maximum value of spin permitted by the available states (Hund's rule). In the transition metals (TMs), p-d and d-d exchange interactions are important, while in the rare earth metals (REMs) the important exchange interactions are d-f and f-f interactions.

3.6.1.2 Interatomic Exchange Interactions

This is an exchange interaction between the electrons belonging to different atoms or ion cores in a solid. The most important contributions are the s-d and d-d exchange interactions in TMs and the s-f and f-f exchange interactions in REMs. One can further classify the interatomic exchange interactions into direct and indirect interatomic exchange interactions, as discussed below.

Direct Interatomic Exchange Interaction

The electrons around different nuclei are more separated compared with those around the same nucleus. Therefore, the strength of the direct interatomic exchange interaction is much less than for an intra-atomic exchange interaction. But at the same time, the strength of the attractive electron-ion interaction is quite large. In the interatomic exchange interaction, the antiparallel arrangement of the spins is more probable (preferred). The most important of such exchange interactions are the d-d (f-f) interactions in the TMs (REMs).

Indirect Interatomic Exchange Interaction

In the indirect interatomic exchange interaction, the spins of the two electrons belonging to two different ions interact with one another via the conduction electrons, favoring a parallel alignment of the spins. Zener (1951a, 1951b) proposed the s-d interaction in TMs for the first time.

In a TM, the s-conduction electrons around a d-shell ion get polarized due to the ionic magnetic moment. These spin-polarized s-electrons interact with the spin of any other neighboring magnetic d-shell ion. This is called the s-d interaction, which favors a parallel alignment of the spins of the d-electrons. The s-d interaction, therefore, is responsible for the existence of ferromagnetism in solids. In the TMs, d-electrons are quasilocalized and possess an itinerant character to a significant extent. Therefore, it cannot be said with certainty that only the s-d interaction is responsible for the origin of ferromagnetism in TMs. On the other hand, in the REMs, the f-electrons are highly localized and therefore the s-f exchange interaction is mainly responsible for ferromagnetism in many of these metals. But some of the REMs exhibit spiral magnetic ordering of various complexities, which cannot be understood in terms of the s-f interaction as it favors ferromagnetic ordering. The spiral ordering can be accounted for in terms of the competition between the ferromagnetic and antiferromagnetic interactions between the adjacent spins.

A quantitative analysis shows that the s-d (s-f) interaction between the spins is independent of their separation, which is physically an unreasonable result because an interaction cannot have an infinite extent. Therefore, the higher-order terms must be included in the Zener's s-d interaction. Ruderman and Kittel (1954), Kasuya (1956), and Yosida (1957) proposed another interaction, usually called the Ruderman-Kittel-Kasuya-Yosida (RKKY) interaction, which includes the higher-order terms in the s-d (s-f) interaction. In the RKKY interaction, the s-conduction electrons around a magnetic ion are spin polarized (yielding finite magnetization) and the spin polarization is oscillatory in nature. The spin polarization of the conduction electrons tends to align more and more of the s-conduction electrons in a particular direction. One can therefore talk about the spin density $n_S(\mathbf{r})$, defined as

$$n_S(\mathbf{r}) = n_\uparrow(\mathbf{r}) - n_\downarrow(\mathbf{r}) \tag{3.21}$$

where $n_\uparrow(\mathbf{r})$ and $n_\downarrow(\mathbf{r})$ are the densities of the s-conduction electrons with up and down spins. The quantity $n_S(\mathbf{r})$ is related to the magnetization $M(\mathbf{r})$ as

$$M(\mathbf{r}) = \mu_B n_S(\mathbf{r}) \tag{3.22}$$

where μ_B is the Bohr magnetron. It has been shown that $n_S(\mathbf{r})$ and hence $M(\mathbf{r})$ varies as $1/r^3$ where r is the distance from the magnetic ion. Fig. 3.1 shows $M(\mathbf{r})$, produced by the spin polarization of the s-electrons, as a function of r. Such an oscillatory spin polarization can couple the ionic spins in a pure ferromagnetic arrangement, pure antiferromagnetic arrangement, or a partially ferromagnetic and partially antiferromagnetic arrangement depending on the ionic separation. Therefore, the RKKY interaction is capable of explaining the various types of spiral orderings at suitable distances.

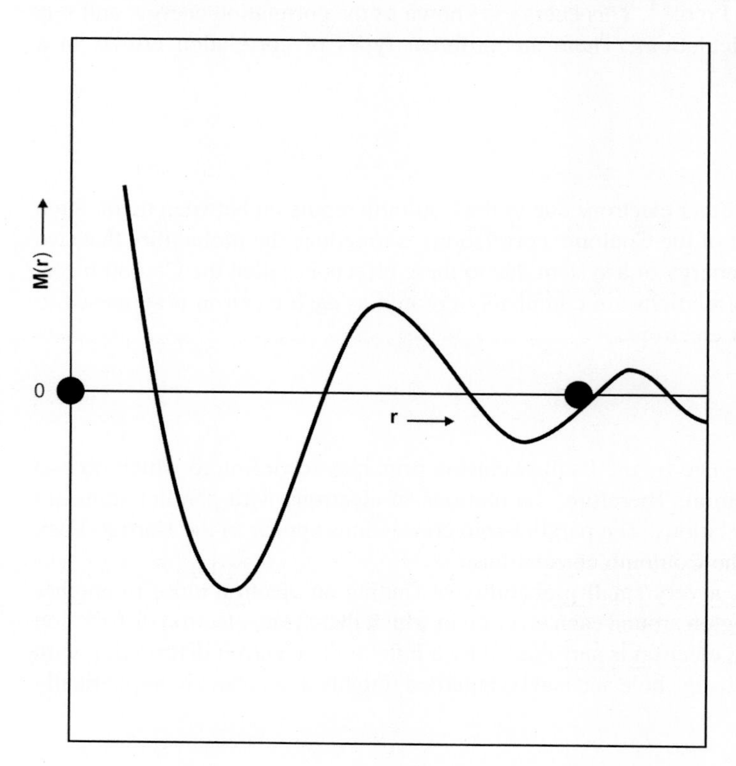

FIG. 3.1 Magnetization $M(\mathbf{r})$ as a function of **r** around an ion at the origin with finite spin. The solid spheres represent ions.

Further, the rapid decrease in the amplitude of the spin polarization or magnetization means that the interionic spin coupling is a localized effect and falls off rapidly with an increase in the separation between the ions.

In some materials, such as MnO, MnSe, and MnTe, the spins of the Mn ions exhibit antiferromagnetic order even though they are well separated by the nonmagnetic ions O, Se, and Te, respectively. Interatomic exchange is a highly localized effect and, therefore, is unable to explain the spin-spin interaction in these materials that have a large separation between the magnetic ions. To explain the antiferromagnetic behavior of such materials, another interaction, usually called the superexchange interaction, has been proposed. It has been suggested that a superexchange involves the transfer of an electron from a nonmagnetic ion to a vacancy in the orbit of a magnetic ion. As a result, the nonmagnetic ion with an unbalanced spin will become paramagnetic and then its spin will be able to interact with the spin of another nearby magnetic ion through direct interatomic coupling. It is this indirect interaction of two magnetic ions via a nonmagnetic ion that produces antiferromagnetic order among the spins of the magnetic ions. It must be emphasized that the superexchange mechanism relates to nonmetallic systems.

3.6.1.3 Conduction Electron-Conduction Electron Exchange Interaction

There exists a spin-spin exchange interaction between the conduction electrons in a metal. In all of the SMs, TMs, and REMs it is named the s-s exchange interaction as the s-electrons in the outermost orbit of the atoms behave as itinerant electrons. Many authors (Hubbard, 1963, 1964a, 1964b; Kohn & Sham, 1965; Lindgren & Schwarz, 1972; Sham, 1961; Singwi, Sjolander, Tosi, & Land, 1970; Toigo & Woodruff, 1970) have put forward the exchange interaction in a free electron gas. Many of these authors gave a parameterized form of the exchange interactions for a paramagnetic electron gas.

3.6.2 Electron Correlation Interactions

In a free electron system, the electrons move independently of each other, but no real system exhibits the ideal free-electron behavior. There always exists a finite repulsion between the electrons, which causes them to avoid each other. In other words, the motions of the electrons are correlated in the presence of repulsive interactions. Such effects are called correlation effects and affect the energy of an electron system. The neglect of electron correlations causes the total energy of a typical atom to be overestimated by an amount ≈ 100 kJ mol^{-1}. This energy is known as the correlation energy, and it is very difficult to make proper allowance for it in calculations. There are different types of correlation effects in a crystalline solid.

3.6.2.1 Coulomb Correlations

The motion of one electron is affected by the motion of other electrons due to the Coulomb repulsion between them. Such correlations are called Coulomb correlations. The effect of the Coulomb correlations is to reduce the probability that two electrons approach closely to one other. The change in energy of a system due to these effects is called the Coulomb correlation energy. In the Hartree theory, the Coulomb correlations are completely ignored as each electron is supposed to move in the average charge distribution of all the other electrons.

3.6.2.2 Parallel-Spin Correlations

The motion of two electrons with parallel spins is governed by the Pauli exclusion principle according to which no two electrons with parallel spins can occupy the same position. Therefore, the motions of electrons with parallel spins are coupled together and these are called parallel-spin correlations. The parallel-spin correlations appear in the Hartree-Fock theory and they affect the energy in the same way as the Coulomb correlations.

According to the Pauli exclusion principle, there is a very small probability of finding an electron close to another electron with parallel spin. As a result, there is a finite region around each electron in which there is no electron distribution with parallel spin. In other words, one can say that each electron is surrounded by a hole in the electron distribution with parallel spin. This is usually called the Fermi hole or exchange hole and can be regarded roughly as a sphere in a spherically symmetric system.

3.6.2.3 Antiparallel-Spin Correlations

There are finite antiparallel-spin correlation effects. Each electron might be expected to be surrounded by a hole, similar to the exchange hole, in the electron distribution with antiparallel-spin also. Such a hole is called a correlation hole in the electron distribution.

REFERENCES

Fock, V. A. (1930). A method for the solution of many-body problems in quantum mechanics. *Zeitschrift fur Physik, 61*, 126–148.

Hartree, D. R. (1928). The wave mechanics of an atom with a non-Coulomb central field. *Proceedings of Cambridge Philosophical Society, 24*, 89–132.

Hubbard, J. (1963). Electron correlations in narrow energy bands. *Proceedings of the Royal Society of London, A266*, 238–257.

Hubbard, J. (1964a). Electron correlations in narrow energy bands II-The degenerate band case. *Proceedings of the Royal Society of London, A277*, 237–259.

Hubbard, J. (1964b). Electron correlations in narrow energy bands III- An improved solution. *Proceedings of the Royal Society of London, A281*, 401–419.

Kasuya, T. (1956). A theory of metallic ferro- and antiferromagnetism on Zener model. *Progress of Theoretical Physics (Kyoto), 16*, 45–57.

Kohn, W., & Sham, L. J. (1965). Self-consistent equations including exchange and correlation effects. *Physical Review A, 140*, 1133–1138.

Lindgren, I., & Schwarz, K. (1972). Analysis of the electron exchange in atoms. *Physical Review A, 5*, 542–550.

Pines, D. (1963). *Elementary excitations in solids.* New York: Benjamin Inc.

Ruderman, M. A., & Kittel, C. (1954). Indirect exchange coupling of nuclear magnetic moments by conduction electrons. *Physical Review, 96*, 99–102.

Sham, L. J. (1961). Electron-phonon interaction in the method of pseudo potentials. *Proceedings of Physical Society (London), 78*, 895–902.

Singwi, K. S., Sjolander, A., Tosi, M. P., & Land, R. H. (1970). Electron correlations at metallic densities. *Physical Review B, 1*, 1044–1053.

Toigo, F., & Woodruff, T. O. (1970). Calculation of dielectric function for a degenerate electron gas with interactions I- Static limit. *Physical Review B, 2*, 3958–3966.

Yosida, K. (1957). Magnetic properties of Cu-Mn alloys. *Physical Review, 106*, 893–898.

Zener, C. (1951a). Interaction between the d-shells in the transition metals. *Physical Review, 81*, 440–444.

Zener, C. (1951b). Interaction between the d-shells in the transition metals III- Calculation of Weiss factors in Fe, Co and Ni. *Physical Review, 83*, 299–301.

SUGGESTED READING

Anderson, P. W. (1963). Theory of magnetic exchange interactions: exchange in insulators and semiconductors. In F. Seitz, & D. Turnbull (Eds.), *Solid state physics, Vol. 14* (pp. 99–214). New York: Academic Press.

Enz, C. P. (1991). *A course on many-body theory applied to solid state physics*. Singapore: World Scientific.

Kittel, C. (1968). Indirect exchange interactions in metals. In F. Seitz, D. Turnbull, & H. Ehrenreich (Eds.), *Solid state physics, Vol. 22* (pp. 1–26). New York: Academic Press.

Pines, D. (1955). Electron interactions in metals. In F. Seitz & D. Turnbull (Eds.), *Solid state physics, Vol. 1* (pp. 367–450). New York: Academic Press.

Pines, D. (1961). *The many-body problem.* Reading: W A Benjamin Inc.

Reitz, J. R. (1955). Methods of one-electron theory of solids. In F. Seitz, & D. Turnbull (Eds.), *Solid state physics, Vol. 1* (pp. 1–95). New York: Academic Press.

Singwi, K. S., & Tosi, M. P. (1981). Correlations in electron liquids. In H. Ehrenreich, F. Seitz, & D. Turnbull (Eds.), *Solid state physics, Vol. 36* (pp. 177–266). New York: Academic Press.

Chapter 4

Bonding in Solids

Chapter Outline

To understand the formation of a crystalline solid in the form of a three-dimensional periodic array, one has to consider interactions between the various atoms. Any two atoms in a solid interact with each other via the repulsive and attractive interactions that oppose each other (Chapter 3) but the net interaction between them is attractive. The attractive interaction should be sufficiently strong to form a stable aggregate of atoms at temperatures of interest. The phenomenon of holding the atoms together is known as *bonding* or, more appropriately, *chemical bonding*. Further, the different elements crystallize in different structures as explained in Chapter 1. A particular arrangement of atoms in a crystalline solid is determined by the character, strength, and directionality of the chemical bonding and cohesive forces. As regards the nature of chemical bonding, the crystalline solids can be classified as follows:

1. Inert gas crystals
2. Ionic crystals
3. Covalent crystals
4. Metallic crystals

In addition to these, there are hydrogen-bonded crystals. In this chapter we shall briefly describe the various types of bonding and the related properties in crystalline solids.

4.1 INTERACTIONS BETWEEN ATOMS

Consider two atoms labeled 1 and 2, separated by an infinitely large distance. There will be no interaction between the atoms and they will behave as free particles. As the atoms are brought closer, they start interacting with each other via electrostatic forces. The net interaction energy U(R) between the atoms may be attractive or repulsive, as is shown in Fig. 4.1. If the interaction is attractive, the two atoms bind together and such a state is called a *bonding state*. On the other hand, if the interaction is repulsive, the atoms do not bind together and the state is called an *antibonding state*.

Suppose two atoms interact via attractive forces. When the atoms come very close to each other, their electron clouds begin to overlap (Fig. 4.2), which gives rise to an additional repulsive interaction due to the Pauli exclusion principle. As a result, the electron states split up causing promotion of electrons to higher unoccupied states of the atom. The repulsive contribution to the interaction potential increases the energy of the system. The repulsive interaction comes into existence only when the distance is on the order of atomic dimensions and increases very quickly as the distance decreases, ultimately

Solid State Physics. https://doi.org/10.1016/B978-0-12-817103-5.00004-9

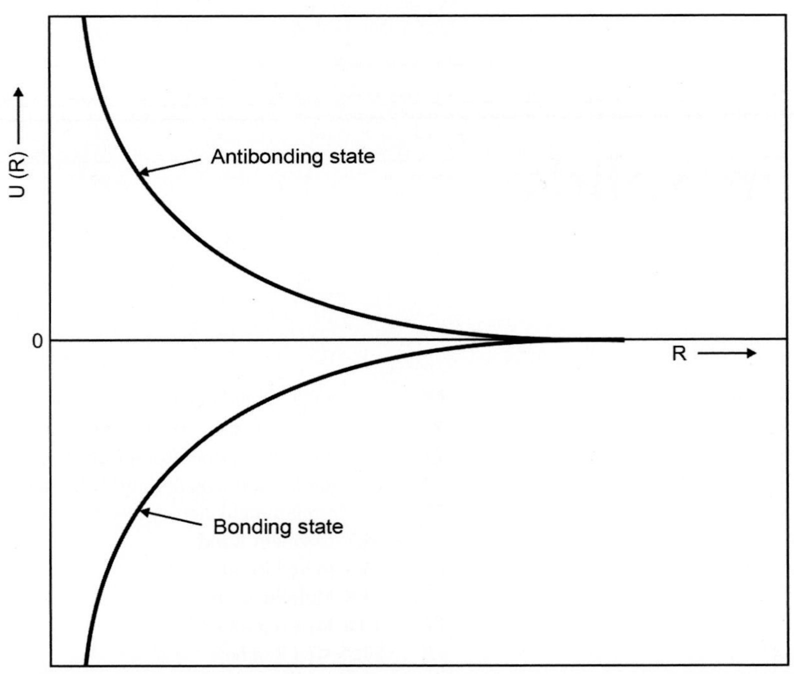

FIG. 4.1 The interaction potential U(R) as a function of the distance R between two atoms for the bonding and antibonding states. Here the repulsive interaction due to the direct overlap of electron distributions of two atoms is not taken into account.

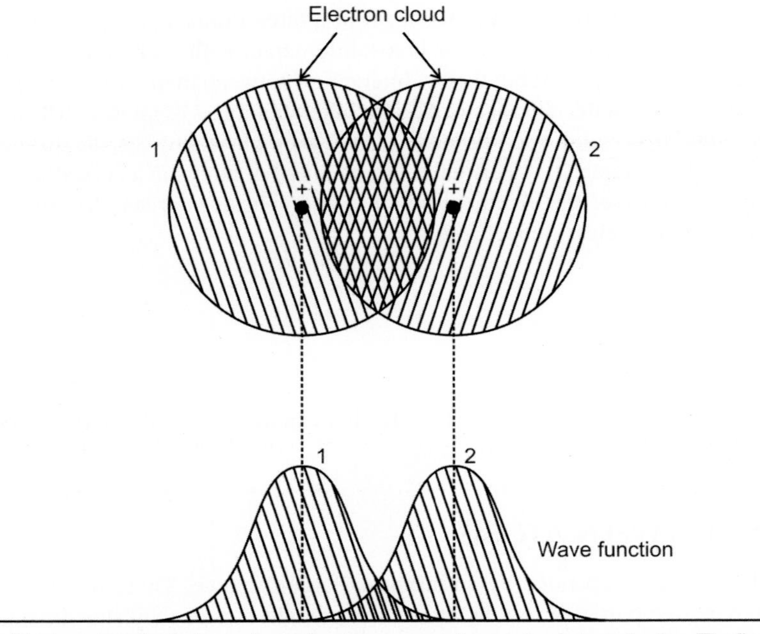

FIG. 4.2 The electron clouds of the two atoms begin to overlap as the atoms approach very closely to each other. The figure also depicts the overlapping of the electronic wave functions of the two atoms.

overpowering the attractive interaction. One should note that the attractive interaction is a long-ranged one, while the repulsive interaction operates over short range. The net interaction energy in a stable state of two atoms, which exhibits a minimum at a particular distance R_0, is shown in Fig. 4.3. Here R_0 represents the equilibrium distance between the two atoms. In a solid, there are a large number (N) of atoms and the above description can be generalized to all of the atoms in it. In a bonding state, all of the atoms of a solid attract each other and the total energy of the solid is the sum of the energies of the individual atoms in addition to their interaction energy.

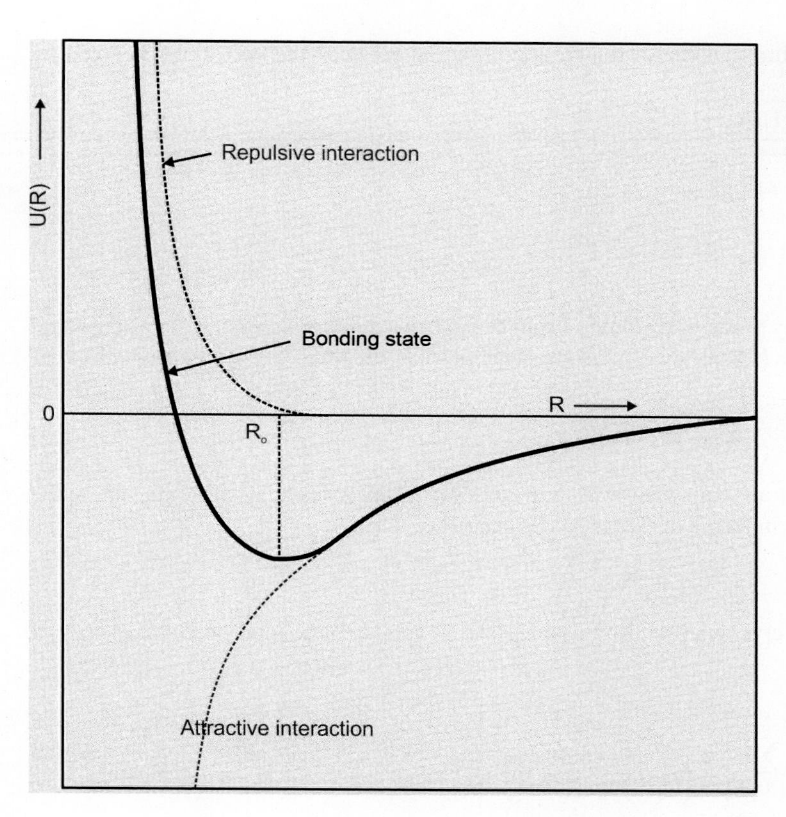

FIG. 4.3 The repulsive and attractive interaction potentials between the two atoms are shown separately as a function of distance R between the two atoms. The net interaction potential U(R) obtained by adding the two contributions is plotted, where R_0 gives the equilibrium distance.

The general form of the interaction potential energy between two atoms can be written as

$$U(R) = -\frac{A}{R^m} + \frac{B}{R^n} \tag{4.1}$$

where R is the distance between the atoms. Here A and B are constants and m and n are integers to be determined. The first term in Eq. (4.1) gives the attractive interaction, while the second term gives the repulsive interaction. The equilibrium distance R_0 between two atoms can be evaluated by minimizing U(R) as

$$\left.\frac{dU}{dR}\right|_{R=R_0} = 0 \tag{4.2}$$

Substitute Eq. (4.1) in Eq. (4.2) to get

$$R_0 = \left(\frac{n}{m}\frac{B}{A}\right)^{\frac{1}{n-m}} \tag{4.3}$$

For the energy to be minimum, the double derivative of U(R) must be positive, that is,

$$\left.\frac{d^2U}{dR^2}\right|_{R=R_0} \rangle\, 0 \tag{4.4}$$

Eqs. (4.1), (4.4) yield

$$\frac{n(n+1)B}{R_0^{n+2}} - \frac{m(m+1)A}{R_0^{m+2}} \rangle\, 0$$

Simplifying the above equation, one gets

$$n \rangle\, m \tag{4.5}$$

Here we have used Eq. (4.3) for R_0. Now the minimum value of interaction potential between the two atoms is given by

$$U(R_0) = -\frac{A}{R_0^m} + \frac{B}{R_0^n} \tag{4.6}$$

Substituting the value of R_0 from Eq. (4.3) in Eq. (4.6), one gets

$$U(R_0) = \frac{A}{R_0^m}\left[1 - \frac{m}{n}\right] \tag{4.7}$$

Eq. (4.7) represents the interaction potential energy when one atom is brought closer to another atom. The total potential energy in a solid is obtained by summing over all the N atoms. Considering the ith atom as the reference atom, the position of the jth atom is given by

$$\mathbf{R}_{ij} = \mathbf{R}_i - \mathbf{R}_j \tag{4.8}$$

where $\mathbf{R}_i$ and $\mathbf{R}_j$ are the positions of the ith and j^{th} atoms. In a solid with one atom per primitive cell, all the atoms are at the lattice positions; therefore, $\mathbf{R}_{ij}$ can be represented in terms of the 1NN distance $\mathbf{R}$ as

$$\mathbf{R}_{ij} = p_{ij}\mathbf{R} \tag{4.9}$$

where p_{ij} is a number whose value depends on the crystal structure. From Eqs. (4.1) and (4.9) the total potential energy of the ith atom is given by

$$U_i(R) = \sum_{j(i\neq j)}\left[-\frac{A}{\left(p_{ij}R\right)^m} + \frac{B}{\left(p_{ij}R\right)^n}\right] \tag{4.10}$$

The potential energy of the ith atom in the equilibrium state is given by substituting $R = R_0$ in Eq. (4.10), that is,

$$U_i(R_0) = \sum_{j(i\neq j)}\left[-\frac{A}{\left(p_{ij}R_0\right)^m} + \frac{B}{\left(p_{ij}R_0\right)^n}\right] \tag{4.11}$$

The total interaction potential of a solid with N atoms, in the equilibrium state, is given by

$$U(R_0) = \sum_i U_i(R_0) = \frac{1}{2}\sum_{i,j(i\neq j)}\left[-\frac{A}{\left(p_{ij}R_0\right)^m} + \frac{B}{\left(p_{ij}R_0\right)^n}\right]$$

$$= \frac{1}{2}N\sum_{j(i\neq j)}\left[-\frac{A}{\left(p_{ij}R_0\right)^m} + \frac{B}{\left(p_{ij}R_0\right)^n}\right] \tag{4.12}$$

The factor of one-half appears in order to avoid double counting a pair of atoms. Here we have used the fact that every atom yields the same amount of potential energy due to the periodicity of the crystal structure. Another form of the interaction potential of a solid that appears in the literature assumes

$$A = 4\kappa\eta^m, B = 4\kappa\eta^n \tag{4.13}$$

where κ and η are new constants. From Eqs. (4.12), (4.13) one can write

$$U(R_0) = 2N\kappa\sum_{j(i\neq j)}\left[-\left(\frac{\eta}{p_{ij}R_0}\right)^m + \left(\frac{\eta}{p_{ij}R_0}\right)^n\right] \tag{4.14}$$

Eqs. (4.12), (4.14) are the two forms of potential energy of a solid that appear in the literature.

4.2 COHESIVE ENERGY

Let E_T be the total energy of N free atoms. When the atoms are brought closer they start interacting with one another and in doing so some of the energy is used in binding the atoms together. Let E_T' be the energy of the N atoms when bound together to form a solid. Naturally, E_T' is the sum of the kinetic and potential energies of the atoms of the solid and $E_T' \langle E_T$ for a stable solid. The *cohesive energy* E_C of a solid is the difference between E_T and E_T' and is usually defined per atom as

$$E_C = \frac{1}{N}\left(E_T - E_T'\right) \tag{4.15}$$

The cohesive energy is usually defined for one kilomole (Kmol) of the solid. Therefore, the cohesive energy or *binding energy* is defined as the energy of formation of one Kmol of a solid from its atoms or ions. It is equal, but opposite in sign, to the energy of dissociation of a solid. The cohesive energy can be obtained from a knowledge of the thermodynamics and spectroscopic data. Inert gas crystals are weakly bound, while alkali metal crystals have intermediate values of cohesive energy. On the other hand, TMs exhibit strong binding, yielding high values of cohesive energy. Ionic solids also exhibit large values of cohesive energy.

4.3 EQUILIBRIUM DISTANCE

The total interaction potential of a solid composed of N atoms is given by Eq. (4.14). Let us define the following parameters

$$\sum_{j(j\neq i)}\left(\frac{1}{p_{ij}}\right)^m = p_{mi} \quad \text{and} \quad \sum_{j(j\neq i)}\left(\frac{1}{p_{ij}}\right)^n = p_{ni} \tag{4.16}$$

Substituting Eq. (4.16) in Eq. (4.14), we can write

$$U(R) = 2N\kappa\left[p_{ni}\left(\frac{\eta}{R}\right)^n - p_{mi}\left(\frac{\eta}{R}\right)^m\right] \tag{4.17}$$

The equilibrium distance, from Eqs. (4.2), (4.17), is given by

$$\left(\frac{R_0}{\eta}\right)^{n-m} = \frac{np_{ni}}{mp_{mi}} \tag{4.18}$$

From Eqs. (4.17), (4.18) the interaction potential in the equilibrium position is given by

$$U(R_0) = 2N\kappa p_{mi}\left(\frac{mp_{mi}}{np_{ni}}\right)^{\frac{m}{n-m}}\left[\frac{m}{n} - 1\right] \tag{4.19}$$

Eq. (4.19) gives the cohesive energy of a solid and is always negative as $n \rangle m$.

Problem 4.1

From Eq. (4.12) prove that the equilibrium distance is given by

$$R_0^{n-m} = \frac{n}{m}\frac{B}{A}\frac{p_{ni}}{p_{mi}} \tag{4.20}$$

Further, the total interaction potential in the equilibrium state is given by

$$U(R_0) = \frac{1}{2}N\frac{p_{mi}A}{R_0^m}\left[\frac{m}{n} - 1\right] \tag{4.21}$$

4.4 BULK MODULUS AND COMPRESSIBILITY

The bulk modulus B_M is defined as the ratio of stress to strain in three dimensions and can be written as

$$B_M = -\frac{dP}{dV/V} = -V\frac{dP}{dV} \tag{4.22}$$

Here dP is the change in pressure and dV is the corresponding change in volume. The negative sign represents the fact that the increase in pressure decreases the volume. According to the first law of thermodynamics

$$dQ = dU + dW \tag{4.23}$$

where dQ is the change in heat energy, dU is the change in internal energy, and dW is the work done. At constant pressure the work done is given by

$$dW = P\,dV \tag{4.24}$$

We know that

$$dQ = TdS \tag{4.25}$$

where T and S are temperature and entropy, respectively. From Eqs. (4.23)–(4.25) one can write

$$dU = TdS - PdV \tag{4.26}$$

At constant entropy, the pressure from Eq. (4.26) becomes

$$P = -\left(\frac{dU}{dV}\right)_S \tag{4.27}$$

Substituting Eq. (4.27) in Eq. (4.22), the bulk modulus is given by

$$B_M = \frac{1}{K_C} = V\left(\frac{d^2U}{dV^2}\right) \tag{4.28}$$

Here K_C denotes the compressibility, which is the reciprocal of the bulk modulus.

Let us evaluate the bulk modulus for a solid with an fcc structure. If a is the lattice constant, the average volume per atom is given by $(1/4)a^3$. The volume of a solid with N atoms becomes

$$V = \frac{1}{4}Na^3 \tag{4.29}$$

The 1NN distance R in an fcc structure is given by

$$R = \frac{a}{\sqrt{2}} \tag{4.30}$$

From Eqs. (4.29), (4.30) one can write

$$V = \frac{1}{\sqrt{2}}NR^3 \tag{4.31}$$

From the above equation one can write

$$\frac{1}{R} = \frac{N^{1/3}}{2^{1/6}V^{1/3}} \tag{4.32}$$

Substituting Eq. (4.32) into Eq. (4.17) allows us to write

$$U(V) = 2N\kappa\left[\frac{P_{ni}}{V^{n/3}} - \frac{P_{mi}}{V^{m/3}}\right] \tag{4.33}$$

where

$$P_{ni} = \frac{N^{n/3}p_{ni}\eta^n}{2^{n/6}} \quad \text{and} \quad P_{mi} = \frac{N^{m/3}p_{mi}\eta^m}{2^{m/6}} \tag{4.34}$$

In the equilibrium state the interaction potential must be minimum, that is,

$$\left.\frac{dU}{dV}\right|_{V_0} = 0 \tag{4.35}$$

From Eqs. (4.33), (4.35) one can find the equilibrium volume V_0, which is given by

$$V_0 = \left(\frac{nP_{ni}}{mP_{mi}}\right)^{\frac{3}{n-m}} \tag{4.36}$$

From Eqs. (4.33), (4.36) the double derivative of the interaction potential U(V) in the equilibrium state is given by

$$\left.\frac{d^2U}{dV^2}\right|_{V=V_0} = 2N\kappa \frac{nP_{ni}}{V_0^{\frac{n}{3}+2}} \frac{n-m}{9} \tag{4.37}$$

Substituting Eq. (4.37) into Eq. (4.28), the bulk modulus in the equilibrium state is given by

$$B_M = \frac{1}{K_C} = 2N\kappa \frac{nP_{ni}}{V_0^{\frac{n}{3}+1}} \frac{n-m}{9} \tag{4.38}$$

B_M is a positive quantity as n is always greater than m.

4.5 INERT GAS CRYSTALS

The inert gas crystals constitute the 8th column of the periodic table and their outermost electron orbit is completely filled with a nearly spherical distribution of electronic charge. Such atoms exhibit maximum stability compared with others having incomplete outermost electron orbits and, therefore, have large values of ionization energy. The inert gas solids crystallize at low temperatures, forming transparent insulators. As the inert gas atoms are nearly spherical in shape, their crystals exhibit fcc structure: a close-packed structure with cubic symmetry. A schematic diagram of the arrangement of atoms in an inert gas crystal is shown in Fig. 4.4. The inert gas crystals are weakly bound with a low melting point. Therefore, the electron distribution in an inert gas atom in a crystal is not much different from that of the free atom except for some distortion. Two types of interactions are present between the atoms in an inert gas crystal.

1. Van der Waals-London interaction
2. Repulsive interaction

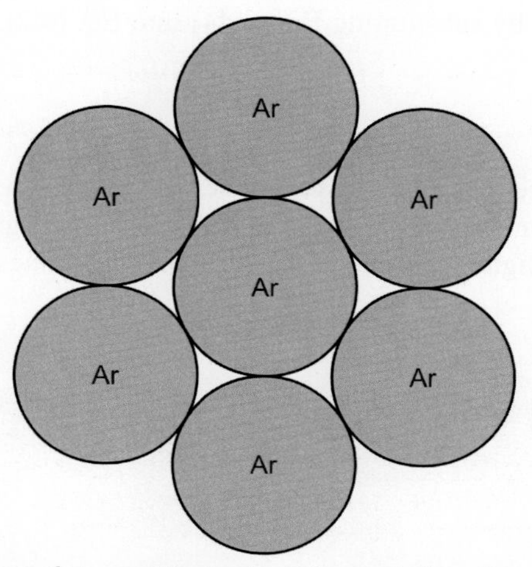

FIG. 4.4 A schematic diagram for the arrangement of atoms of the inert gas element Ar in a plane.

Both the interactions are discussed in Appendix A. The total interaction potential between two inert gas atoms is obtained from Eq. (4.1) by substituting m = 6 and n = 12 and is written as

$$U(R) = -\frac{A}{R^6} + \frac{B}{R^{12}} \tag{4.39}$$

The constants A and B from Eq. (4.13) are given as

$$A = 4\kappa\eta^6, B = 4\kappa\eta^{12} \tag{4.40}$$

Therefore, the interaction potential (4.39) becomes

$$U(R) = 4\kappa\left[-\left(\frac{\eta}{R}\right)^6 + \left(\frac{\eta}{R}\right)^{12}\right] \tag{4.41}$$

This is known as the *Lennard-Jones potential*, which is attractive at large distances but repulsive in nature at small distances.

4.5.1 Equilibrium Lattice Constant

Neglecting the kinetic energy, the cohesive energy of an inert gas crystal is obtained by summing over all the pairs of atoms in the crystal. The total potential of a solid containing N atoms can be obtained from Eq. (4.14) by substituting m = 6 and n = 12 to write

$$U(R) = 2N\kappa \sum_{j(i \neq j)} \left[-\left(\frac{\eta}{p_{ij}R}\right)^6 + \left(\frac{\eta}{p_{ij}R}\right)^{12}\right] \tag{4.42}$$

The summation in Eq. (4.42) is to be evaluated assuming fcc structure for the inert gas crystal. It can be shown that

$$\sum_{j(j \neq i)} \left(\frac{1}{p_{ij}}\right)^6 = 14.454 \text{ and } \sum_{j(j \neq i)} \left(\frac{1}{p_{ij}}\right)^{12} = 12.131 \tag{4.43}$$

One should note that there are twelve 1NNs in the fcc structure and the above summations yield values near twelve. This shows that the 1NNs contribute the most to the interaction potential. So, Eq. (4.42) gives

$$U(R) = 2N\kappa\left[12.131\left(\frac{\eta}{R}\right)^{12} - 14.454\left(\frac{\eta}{R}\right)^6\right] \tag{4.44}$$

The equilibrium distance R_0 is obtained by substituting Eq. (4.44) into Eq. (4.2), which, after simplification, gives

$$\frac{R_0}{\eta} = 1.09 \tag{4.45}$$

The above derivation gives the same value of R_0/η for all of the inert gas crystals with fcc structure. But the actual values of R_0/η vary a little from the above value and these are listed in Table 4.1. The agreement between the calculated and the experimental values is quite good. The slight departure of the experimental values from 1.09 can be attributed to quantum mechanical effects.

TABLE 4.1 Equilibrium Distance for the Inert Gas Crystals

Element	Ne	Ar	Kr	Xe
$\frac{R_0}{\eta}$	1.14	1.11	1.10	1.09

4.5.2 Cohesive Energy of Inert Gas Crystals

The cohesive energy of a solid with N atoms is given by Eq. (4.44) by replacing R by R_0, that is,

$$U(R_0) = 2N\kappa\left[12.131\left(\frac{\eta}{R_0}\right)^{12} - 14.454\left(\frac{\eta}{R_0}\right)^{6}\right] \tag{4.46}$$

Substituting the value of η/R_0 from Eq. (4.45) into Eq. (4.46), one gets

$$U(R_0) = -4.30(2N\kappa) \tag{4.47}$$

The cohesive energy also comes out to be the same for all of the inert gas elements. In the above derivation, the cohesive energy is calculated assuming zero kinetic energy. But the kinetic energy is always finite and its inclusion reduces the cohesive energy. Further, the value of the kinetic energy is different for different elements, which makes the cohesive energy different also. For example, in heavier elements the kinetic energy is small and hence the reduction in cohesive energy is small.

Problem 4.2

Prove that the result given by Eq. (4.47) can be obtained from Eq. (4.19) by substituting $m = 6$ and $n = 12$.

4.5.3 Bulk Modulus

The bulk modulus for inert gas crystals can be calculated from the general expression given by Eq. (4.38) by substituting $m = 6$ and $n = 12$ to write

$$B_M = \frac{1}{K_C} = 2N\kappa\frac{12P_{12i}}{V_0^5}\frac{2}{3} \tag{4.48}$$

$$\text{where } P_{12i} = \frac{1}{4}N^4\eta^{12}p_{12i} \tag{4.49}$$

Here we have used Eq. (4.34) for P_{ni}. Substituting Eq. (4.49) into Eq. (4.48), one gets

$$B_M = \frac{1}{K_C} = 4N\kappa p_{12i}\frac{N^4\eta^{12}}{V_0^5} \tag{4.50}$$

Substituting the value of V_0 from Eq. (4.36) allows us to write

$$B_M = \frac{1}{K_C} = \frac{4\kappa}{\eta^3}\frac{(p_{6i})^{5/2}}{(p_{12i})^{3/2}} \tag{4.51}$$

From the above equation it is evident that B_M is on the order of κ/η^3 in inert gas crystals.

Problem 4.3

From Eq. (4.33) the total interaction potential for $m = 6$ and $n = 12$ can be written in terms of volume as

$$U(V) = \frac{b_{12}}{V^4} - \frac{b_6}{V^2} \tag{4.52}$$

Here b_6 and b_{12} are new constants that can be expressed in terms of κ and η. Prove that for the interaction potential given by Eq. (4.52) the bulk modulus is given by

$$B_M = \sqrt{2}\frac{b_6^{5/2}}{b_{12}^{3/2}} \tag{4.53}$$

Further, show that B_M is on the order of κ/η^3.

4.6 IONIC BONDING

An element that acquires a positive charge by giving an electron is called an electropositive element, while one that acquires negative charge by receiving an electron is called an electronegative element. In other words, an electronegative element has an affinity for negative charge. The electropositive elements lie on the left side of the periodic table, while the electronegative elements lie on the right side. *Electronegativity* is a measure of the tendency of an atom or a radical to attract electrons in the formation of an ionic bond. Pauling gave the most commonly used scale, called the Pauling scale of electronegativity. The higher the associated electronegativity number, the more attracted is an element or compound toward an electron (negative charge). For example, fluorine (F) is the most electronegative element and is assigned the value 4.0, while cesium (Cs) and francium (Fr) are the least electronegative elements with values of 0.7. Ionic bonding is the simplest type of bonding between the electropositive and electronegative elements. For this reason, the ionic bond is also called the heteropolar bond. Crystalline solids in which the atoms exhibit ionic bonds are called ionic crystals. The ionic crystals are made of positive and negative ions and these are arranged in such a way that the Coulomb attraction between the oppositely charged ions is stronger than the Coulomb repulsion between the ions of the same charge. Therefore, the ionic bond is purely electrostatic in nature. NaCl and CsCl are common examples of ionic crystals and their crystal structures are shown in Figs. 1.20 and 1.17.

Let us study the formation of an ionic solid. Consider a NaCl crystal in which the Na and Cl atoms have the following electronic configuration:

$$\begin{aligned} &\mathrm{Na}: 1s^2 2s^2 2p^6 3s^1 \\ &\mathrm{Cl}: 1s^2 2s^2 2p^6 3s^2 3p^5 \end{aligned} \tag{4.54}$$

In an atom of Na, there is one electron in the outermost electron orbit that is loosely bound to the nucleus. So, an atom of Na can easily lose this electron to the surroundings so as to acquire the most stable inert gas configuration with all of the electron orbits completely filled. On the other hand, an atom of Cl has one electron deficit in its outermost orbit and, therefore, can easily accept an electron to acquire the most stable inert gas electron configuration. This process leads to the formation of positively and negatively charged ions with the following configurations:

$$\begin{aligned} &\mathrm{Na}^{+1}: 1s^2 2s^2 2p^6 \\ &\mathrm{Cl}^{-1}: 1s^2 2s^2 2p^6 3s^2 3p^6 \end{aligned} \tag{4.55}$$

Na^{+1} and Cl^{-1} ions attract each other to form a molecule that is electrically neutral. The formation of a NaCl molecule can be described by the following equation

$$\mathrm{Na} + \mathrm{Cl} \rightarrow \mathrm{Na}^{+1} + \mathrm{Cl}^{-1} \rightarrow \mathrm{NaCl} \tag{4.56}$$

Because chlorine is a gas with Cl_2 as its molecule, the above equation is modified as

$$2\mathrm{Na} + \mathrm{Cl}_2 \rightarrow 2\mathrm{Na}^{+1} + 2\mathrm{Cl}^{-1} \rightarrow 2\mathrm{NaCl} \tag{4.57}$$

Some energy is required to carry out the above reaction. But once the reaction starts it proceeds vigorously with the evolution of light and heat from the ionic bond formation causing a sizable decrease in energy.

4.6.1 Ionic-Bond Energy

Let us estimate the ionic-bond energy of a NaCl crystal. The ionization energy of an atom of Na (= 5.1 eV) is required to remove an electron from the Na atom, yielding a Na^{+1} ion and an electron. Mathematically one can write

$$\mathrm{Na} + 5.1\,\mathrm{eV} = \mathrm{Na}^{+1} + e^{-1} \tag{4.58}$$

The electron affinity of Cl is 3.6 eV, so in the formation of a chlorine ion one can write

$$\mathrm{Cl} + e^{-1} = \mathrm{Cl}^{-1} + 3.6\,\mathrm{eV} \tag{4.59}$$

Adding Eqs. (4.58), (4.59) allows us to write

$$\mathrm{Na} + \mathrm{Cl} + 1.5\,\mathrm{eV} = \mathrm{Na}^{+1} + \mathrm{Cl}^{-1} \tag{4.60}$$

Thus, an energy of 1.5 eV is needed to create Na^{+1} and Cl^{-1} ions from the corresponding atoms. The Na^{+1} and Cl^{-1} ions join together by attractive forces to form a NaCl molecule with minimum potential energy. In the formation of a molecule from the Na^{+1} and Cl^{-1} ions, some energy is released, usually called the bond energy. It can be calculated as follows.

The equilibrium distance between the Na^{+1} and Cl^{-1} ions is $R_0 = 2.4 \times 10^{-8}$ cm. So the potential energy of the NaCl molecule becomes

$$V(R_0) = -\frac{e^2}{R_0} = -\frac{(4.8 \times 10^{-10})^2}{2.4 \times 10^{-8} \times 1.6 \times 10^{-12}} = -6.0\,\text{eV}$$

Thus, in the formation of a sodium chloride molecule one can write

$$Na^{+1} + Cl^{-1} = NaCl + 6\,\text{eV} \tag{4.61}$$

From Eqs. (4.60), (4.61) one can write

$$Na + Cl = NaCl + 4.5\,\text{eV} \tag{4.62}$$

Eq. (4.62) shows that in the formation of a NaCl molecule from the Na and Cl atoms, an energy of 4.5 eV is released. In other words, when a molecule of NaCl dissociates into Na and Cl atoms it requires an energy of 4.5 eV.

4.6.2 Lattice Energy

Lattice energy is defined as the energy released when the constituent atoms are placed in their respective positions on the crystal lattice. It can also be defined as the amount of energy that is spent to separate an ionic crystal into its constituent ions. It can be evaluated considering different contributions to the potential energy. In an ionic solid there are present three types of interactions:

1. The electrostatic interactions between the ions, which yield the overall very strong attractive interaction.
2. Van der Waals interactions, which fall off as the 6th power of the distance. These are attractive but very weak and are usually neglected in comparison with the electrostatic interactions.
3. The ion-ion overlap interactions, which are repulsive in nature and fall off as the 12th power of the distance. These interactions are significant at very small distances.

Therefore, the interaction energy between the ith and j^{th} ions $U_{ij}(R_{ij})$ is given by

$$U_{ij}\left(R_{ij}\right) = \frac{B}{R_{ij}^n} \pm \frac{Z^2 e^2}{R_{ij}} \tag{4.63}$$

Here Ze is the charge on the ion and the charge on both types of ions is assumed to be equal and opposite. The first term represents the repulsive overlap interaction between the ions, while the second term represents the electrostatic interaction between them. In the above equation, the plus sign is for ions with like charges, while the negative sign is for ions with unlike charges. With the help of Eq. (4.9), Eq. (4.63) can be written in terms of the 1NN distance R as

$$U_{ij}(R) = \frac{B}{\left(p_{ij} R\right)^n} \pm \frac{Z^2 e^2}{p_{ij} R} \tag{4.64}$$

The net interaction energy of the ith ion in the lattice becomes

$$U_i(R) = \sum_{j(j \neq i)} U_{ij}(R) = \sum_{j(j \neq i)} \frac{B}{\left(p_{ij} R\right)^n} \pm \sum_{j(j \neq i)} \frac{Z^2 e^2}{p_{ij} R} \tag{4.65}$$

The total interaction energy of a crystal with N ions, which gives the lattice energy, becomes

$$U(R) = N U_i(R) = N \sum_{j(j \neq i)} \frac{B}{\left(p_{ij} R\right)^n} \pm N \sum_{j(j \neq i)} \frac{Z^2 e^2}{p_{ij} R} \tag{4.66}$$

In writing the above equation, it is assumed that the interaction energy of all the ions is the same. This is true because the surroundings of all the ions are similar due to the symmetry of the lattice. The calculation of U(R) is difficult as it involves a sum over the whole of the lattice. The problem is simplified if one assumes that the repulsive interaction, being very short-ranged, is appreciable only up to the 1NNs of an ion. In this approximation, Eq. (4.66) is simplified to

$$U(R) = N\left[n_0 \frac{B}{R^n} - \alpha_M \frac{Z^2 e^2}{R}\right] \tag{4.67}$$

where n_0 is the number of 1NNs and α_M is the Madelung constant defined as

$$\alpha_M = \pm \sum_{j\,(j \neq i)} \frac{1}{p_{ij}} \tag{4.68}$$

In Eq. (4.67) we have taken the negative sign before the electrostatic term as it yields a positive value of α_M. In α_M, the positive sign is for charges of opposite sign, while the negative sign is for charges of the same sign. From Eqs. (4.2), (4.67) one gets, in the equilibrium state, the constant B as

$$B = \alpha_M \frac{Z^2 e^2}{n n_0} R_0^{n-1} \tag{4.69}$$

So, the total interaction potential in the equilibrium state is obtained by substituting the value of constant B from Eq. (4.69) in Eq. (4.67), that is,

$$U(R_0) = -\frac{N \alpha_M Z^2 e^2}{R_0}\left[1 - \frac{1}{n}\right] \tag{4.70}$$

The term $-N\alpha_M Z^2 e^2/R_0$ is called the Madelung energy. Being very short ranged, the repulsive interaction is very small at the equilibrium distance R_0. Therefore, the lattice energy $U(R_0)$ is mainly electrostatic in nature.

4.6.3 Difference Between Bond Energy, Cohesive Energy, and Lattice Energy

The bond energy is the energy by which the ions/atoms of a molecule are held together and, therefore, it involves the ions/atoms of a single molecule. The lattice energy of an ionic crystal is the energy released when a crystal is formed by placing the ions on the lattice and, therefore, involves all the ions of the crystal. The cohesive energy of an ionic solid is the energy that would be liberated in the formation of the ionic solid from the individual neutral atoms. The cohesive energy is also called the binding energy of the solid and involves the interaction of all the atoms present in the solid.

4.6.4 Bulk Modulus of Ionic Crystals

From Eq. (4.28) it is evident that the bulk modulus depends on the double derivative of potential with respect to volume, which can be calculated from the interaction potential as follows:

$$\frac{dU}{dV} = \frac{dU}{dR}\frac{dR}{dV} \tag{4.71}$$

Differentiating Eq. (4.71) again allows us to write

$$\frac{d^2U}{dV^2} = \frac{d^2U}{dR^2}\left(\frac{dR}{dV}\right)^2 + \frac{dU}{dR}\frac{d^2R}{dV^2} \tag{4.72}$$

In the equilibrium state, $dU/dR = 0$. Therefore, Eq. (4.72) reduces to

$$\frac{d^2U}{dV^2} = \frac{d^2U}{dR^2}\left(\frac{dR}{dV}\right)^2\Bigg|_{V=V_0} \tag{4.73}$$

Substituting Eq. (4.73) into Eq. (4.28), the bulk modulus is given by

$$B_M = V\frac{d^2U}{dR^2}\left(\frac{dR}{dV}\right)^2\Bigg|_{V=V_0} \tag{4.74}$$

Let us evaluate the bulk modulus for a NaCl crystal having fcc structure. If a is the lattice parameter, then $(1/4)a^3$ is the volume per molecule. The total volume of a NaCl crystal with N number of molecules becomes (see Fig. 1.20)

$$V = \frac{1}{4}Na^3 = \frac{1}{4}N(2R)^3 = 2NR^3 \tag{4.75}$$

From the above equation

$$\frac{dV}{dR} = 6NR^2 \tag{4.76}$$

Substituting Eqs. (4.75), (4.76) into Eq. (4.74), the bulk modulus becomes

$$B_M = \frac{1}{18NR_0} \frac{d^2U}{dR^2}\bigg|_{R=R_0} \tag{4.77}$$

From Eqs. (4.67), (4.69) one can write

$$\frac{d^2U}{dR^2}\bigg|_{R=R_0} = \frac{N\alpha_M Z^2 e^2}{R_0^3}(n-1) \tag{4.78}$$

Substituting Eq. (4.78) into Eq. (4.77), one gets

$$B_M = \frac{\alpha_M Z^2 e^2}{18R_0^4}(n-1) \tag{4.79}$$

Problem 4.4

Assuming the interaction potential to be of the form given by Eq. (4.67), prove that the bulk modulus for the CsCl structure (Fig. 1.17) is given by[1]

$$B_M = \frac{\alpha_M Z^2 e^2}{8\sqrt{3}R_0^4}(n-1) \tag{4.80}$$

One of the requirements of a reliable theory of cohesion is that it should be able to predict the correct structure of the solid that yields the minimum value of the interaction energy given by Eq. (4.70). It has been found that the energy given by Eq. (4.70) is not adequate for this purpose and the reason may be two-fold. First, the repulsive interaction may not be reliable, and second, the small but nevertheless finite Van der Waals interaction must be considered. The improved interaction potential suggested by Born and Mayer is given by

$$U_i(R) = \sum_{j(j\neq i)} \frac{B}{(p_{ij}R)^n} - \alpha_M \frac{Z^2 e^2}{R} - \sum_{j(j\neq i)} \frac{A}{(p_{ij}R)^m} + \sum_{j(j\neq i)} D_{ij} \tag{4.81}$$

The third term on the right side of the above equation is the Van der Waals interaction and the last term contains other corrective interactions.

The expressions derived for the bulk modulus of a solid can be used to find the value of n, the exponent in the repulsive potential. The equilibrium distance R_0 can be determined from X-ray studies, the bulk modulus B_M can be measured experimentally, and the values of Ze, α_M, and n_0 can be calculated from the structure of the solid. Knowing these quantities, one can determine the value of n from Eq. (4.79). Let us take the case of KCl, in which

$$B_M = 1.97 \times 10^{11} \text{ dynes/cm}^2$$

$$R_0 = 3.14 \times 10^{-8} \text{ cm}$$

$$Z = 1$$

$$\alpha_M = 1.75$$

---

1. It should be noted that the difference in the bulk modulus arises due to the different unit cell volume for different structures. One can attempt the general case in which the volume of the crystal can be assumed to be $V = CNR^3$ where C is a constant whose value depends on the structure of the solid.

From Eq. (4.79) one can calculate the value of the exponent n, which is given by

$$n = \frac{18R_0^4B_M}{\alpha_M Z^2 e^2} + 1 = 9.4$$

Hence, in KCl, the power of the repulsive interaction $n \approx 9 - 10$.

4.6.5 Exponential Repulsive Potential

The interaction potential $U_{ij}(R_{ij})$, representing the repulsive potential in exponential form, can be written as

$$U_{ij}\left(R_{ij}\right) = \lambda_R e^{-R_{ij}/\rho} \pm \frac{Z^2 e^2}{R_{ij}} \tag{4.82}$$

where the parameter λ_R gives the strength of the repulsive potential and ρ is the decay factor. Considering only the 1NN interactions in the repulsive potential, the total interaction potential of the crystal is given by

$$U(R) = NU_i(R) = N\left[n_0 \lambda_R e^{-R/\rho} - \alpha_M \frac{Z^2 e^2}{R}\right] \tag{4.83}$$

The improved form of the interaction potential given by Eq. (4.83) must include the Van der Waals interactions and the residual interactions that are included in the potential given by Eq. (4.81).

Problem 4.5

Using Eq. (4.83) prove that the total interaction potential in the equilibrium state is given by

$$U(R_0) = -N\alpha_M \frac{Z^2 e^2}{R_0}\left[1 - \frac{\rho}{R_0}\right] \tag{4.84}$$

where the 1NN equilibrium distance R_0 is given by

$$R_0^2 e^{-R_0/\rho} = \rho \alpha_M \frac{Z^2 e^2}{\lambda_R n_0} \tag{4.85}$$

Problem 4.6

Using the interaction potential given by Eq. (4.83), show that the bulk modulus for the NaCl structure (see Fig. 1.20) is given by

$$B_M = \frac{\alpha_M Z^2 e^2}{18R_0^4}\left[\frac{R_0}{\rho} - 2\right] \tag{4.86}$$

One can calculate the values of ρ and λ_R for the exponential form of the repulsive potential using Eqs. (4.85), (4.86). Substituting the values of B_M, R_0, Z, and α_M for KCl, given above, into Eq. (4.86) yields

$$\frac{R_0}{\rho} = \frac{18R_0^4 B_M}{\alpha_M Z^2 e^2} + 2 = 10.4$$

The above equation yields

$$\rho = \frac{R_0}{10.4} = \frac{3.14 \times 10^{-8}}{10.4} = 0.30 \times 10^{-8}\text{cm}$$

Knowing the value of ρ one can calculate the value of λ_R from Eq. (4.85), that is,

$$\lambda_R = \frac{\alpha_M Z^2 e^2}{n_0 R_0^2} \rho e^{R_0/\rho} = \frac{1.75 \times (4.8 \times 10^{-10})^2}{6 \times (3.14 \times 10^{-8})^2} 0.30 \times 10^{-8} \times e^{10.4}$$
$$= 0.67 \times 10^{-8} \text{cm}$$

With these values of R_0 and ρ the cohesive energy from Eq. (4.84) is given by

$$U(R_0) = -1.16 \times 10^{-11} \text{ erg}$$
$$= -7.26 \text{ eV}$$

The observed value of the cohesive energy for KCl is −7.397 eV. Therefore, the theoretical value agrees reasonably well with the observed value.

4.6.6 Calculation of the Madelung Constant

The Madelung constant was defined in Eq. (4.68), which, when divided by R, gives

$$\frac{\alpha_M}{R} = \pm \sum_{j(j \neq i)} \frac{1}{p_{ij} R} \tag{4.87}$$

Eq. (4.87) shows that the Madelung constant depends on the crystal structure and, therefore, has different values for different ionic solids. Further, the value of α_M for a particular structure depends on the way it is defined: whether it is defined in terms of the 1NN distance R or the lattice parameter a, or in some other way. In the calculation of α_M, if one takes the negative ion as the reference ion, then a positive sign will be used for positive ions and a negative sign for negative ions.

4.6.6.1 First Method

To illustrate the evaluation of α_M, consider the simplest case of a monatomic linear lattice of ions (see Fig. 4.5). Let us consider a negatively charged ion as the reference ion, then Eq. (4.87) yields

$$\frac{\alpha_M}{R} = \frac{2}{R} - \frac{2}{2R} + \frac{2}{3R} - \frac{2}{4R} + \ldots\ldots\ldots \tag{4.88}$$

Here the factor of 2 in the first term is due to the fact that there are two 1NNs, one on the left and the other on the right, and the same is true for all other NNs. Simplifying the above equation, we get

$$\frac{\alpha_M}{R} = \frac{2}{R} \ln 2 \tag{4.89}$$

Therefore, the Madelung constant for a linear chain of ions becomes

$$\alpha_M = 2 \ln 2 \tag{4.90}$$

4.6.6.2 Second Method

The interaction potential of a negatively charged reference ion with its two 1NNs is $-2e^2/R$, with its 2NNs, $2e^2/2R$, and so on. Here each ion is assumed to be monovalent. Hence the total interaction potential seen by the reference ion due to whole of the linear chain of ions is given by

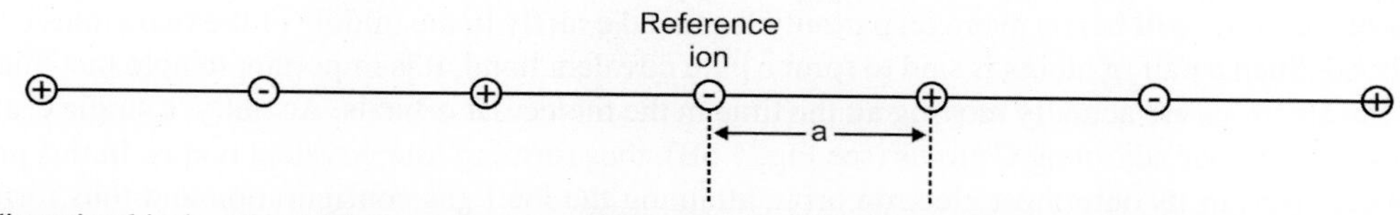

FIG. 4.5 One-dimensional ionic solid with lattice parameter a. The negatively and positively charged ions occur alternately.

$$U(R) = -\frac{2e^2}{R} + \frac{2e^2}{2R} - \frac{2e^2}{3R} + \frac{2e^2}{4R} - \ldots\ldots$$
$$= -\frac{e^2}{R}[2\ \ln 2] \tag{4.91}$$

Eq. (4.91) can be written more conveniently as (Appendix B)

$$U(R) = -\alpha_M \frac{e^2}{R} \tag{4.92}$$

where α_M is given by Eq. (4.90).

4.6.6.3 Madelung Constant for NaCl Structure

It is much more difficult to perform the summation in a three-dimensional crystal. Let us consider the case of NaCl crystal shown in Fig. 1.20. The Na^{+1} ion at position O is taken as the reference ion. The Na^{+1} ion has six Cl^{-1} ions as its 1NNs at a distance R, so the interaction potential seen by the reference ion due to all its 1NNs becomes $-6e^2/R$. There are twelve 2NN Na^{+1} ions each at a distance of $\sqrt{2}R$ which contribute potential energy of $12e^2/\sqrt{2}R$ and there are eight 3NNs Cl^{-1} ions each at a distance of $\sqrt{3}R$ with potential energy contribution of $-8e^2/\sqrt{3}R$. Further, there are six 4NNs, each at a distance of 2R, which gives a potential energy of $6e^2/2R$ and so on. Summing over all the NNs, the total potential energy becomes

$$U(R) = -\frac{6e^2}{R} + \frac{12e^2}{\sqrt{2}R} - \frac{8e^2}{\sqrt{3}R} + \frac{6e^2}{2R} - \ldots.$$
$$= -\alpha_M \frac{e^2}{R} \tag{4.93}$$

where

$$\alpha_M = 6 - \frac{12}{\sqrt{2}} + \frac{8}{\sqrt{3}} - \frac{6}{\sqrt{4}} + \ldots.$$
$$= 1.75 \tag{4.94}$$

Therefore, the Madelung constant for NaCl structure is 1.75.

4.7 COVALENT BOND

A covalent bond is a homopolar bond that is formed by the sharing of an even number of electrons between atoms. A single covalent bond involves the sharing of two electrons, one from each atom. Two bonds involve the sharing of four electrons, and so on. In general, a multiple covalent bond involves the sharing of 2n electrons where n is the number of bonds. Familiar examples of covalently bonded crystals are found among the group IV semiconductors, which comprise the elements C, Si, Ge, and Sn. Covalent bonding also exists in crystals of the form $A^N B^{8-N}$ made of elements A and B with N and $8-N$ valence electrons per atom, respectively. This gives eight s or p valence electrons per pair of atoms.

To understand covalent bonding, let us take the case of the element carbon (C). A C atom possesses the electronic configuration given below:

$$C: 1s^2 2s^2 2p^2 \tag{4.95}$$

Consider the covalent bond between two C atoms. The first C atom shares its electron with the second C atom and the second C atom in turn shares its electron with the first C atom (see Fig. 4.6A). In this process of sharing, both the C atoms possess 5 electrons in their outermost orbit. The electrons forming the bond tend to localize partially in the region between the two atoms joined by the bond. The spins of the two electrons participating in the bond should be antiparallel. If the two atoms are equally electronegative, then both have the tendency to attract the bonding pair of electrons equally. Therefore, the bonding pair of electrons will be, on average, partially localized exactly in the middle of the two atoms or in the middle of the covalent bond. Such a pair of atoms is said to form a pure covalent bond. It is important to note that this is an average picture because the electrons are actually moving all the time in the molecular orbitals. Actually, a single C atom can share its valence electrons with four adjoining C atoms (see Fig. 4.6B), thus forming four covalent bonds. In this process, each C atom acquires 8 electrons in its outermost electron orbit, attaining the inert gas configuration and thus forming the most stable state.

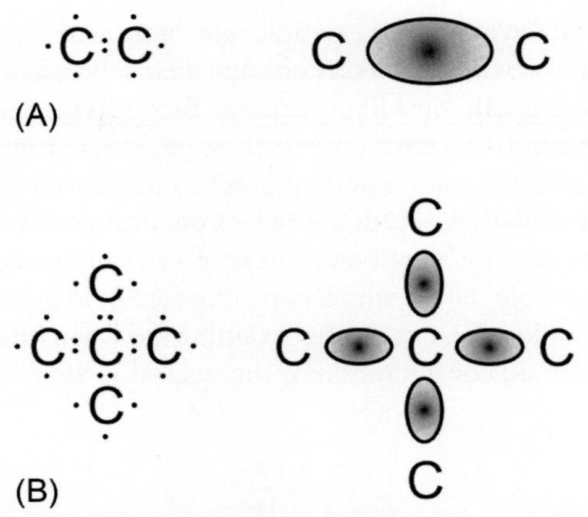

FIG. 4.6 (A) Covalent bond between two carbon (C) atoms. The two electrons in the middle of the C atoms are shared by them. The covalent bond is represented by the electron charge in the shaded region between the two C atoms. (B) Covalent bond of an atom of C, in the center, with four C atoms forming the 1NNs. The description of the figure is the same as in part (A).

The covalent bond is a strong bond, which is evident from the unusual hardness of diamond. The bond strength of a covalent crystal is comparable with that of an ionic crystal. A typical value of the binding energy of a covalent bond is a few electron volts per bond. The most striking feature of a covalent bond is its highly directional properties. One should note that C, Si, and Ge all have the diamond structure in which each atom makes covalent bonds with four atoms at tetrahedral angles to one another (see Fig. 4.7). In the pure elements mentioned above, identical atoms form covalent bonds. The binding energy (E_B) of a covalent bond arises mainly from the following three contributions:

1. The formation of tetragonal bonds gives a negative contribution to E_B as some energy is required to form tetragonal sp^3 hybrid orbits from the s- and p- quantum orbits in a free atom.
2. The Coulomb repulsion between the ions or between the inner orbit electrons, such as $1s^22s^22p^6$ orbits.
3. Exchange interaction potential.

In general, the covalent bonds repel each other for the simple reason that the clouds of negatively charged electrons have a repulsive interaction. Therefore, it has been observed that, in an element, the covalent bonds formed by the same atoms

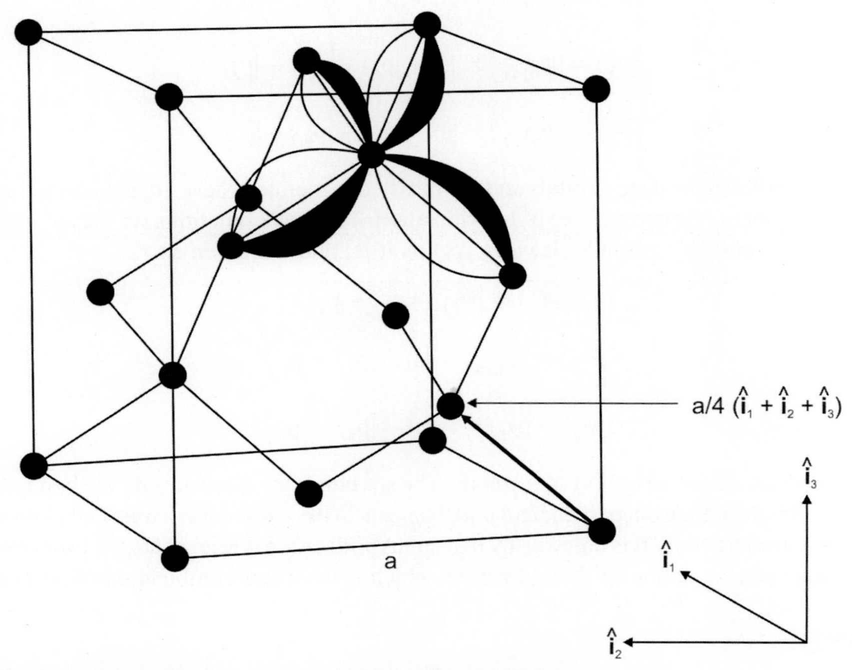

FIG. 4.7 Four covalent bonds around an atom in a pure crystal with the diamond structure.

arrange themselves symmetrically around the atom. For example, one atom will form two covalent bonds in a straight line. If the atom forms three covalent bonds, these will most likely arrange themselves in a plane at an angle of 120° to each other. If the atom forms four covalent bonds, these will most likely arrange themselves in such a way that the four atoms form the corners of a tetrahedron. The tetrahedral bonds in a covalent crystal are shown in Fig. 4.7. The tetrahedral bonds naturally fit into cubic symmetry because the four nonadjacent corners of a cube correspond to the corners of a tetrahedron.

This plausible account still does not explain why a double-electron arrangement produces a bond, that is, an attractive interatomic interaction. The explanation of a covalent bond can be given only through quantum mechanics. The simplest known example is that of a hydrogen molecule, H_2, in which two atoms are held together by covalent bonds by sharing their electrons (Heitler-London theory of H_2 molecule). Let us now examine the wave functions of the tetragonal covalent bonds formed in a crystal of carbon. The electronic configuration of the neutral C atom is given by Eq. (4.95). In this configuration, one has the following orbitals:

$$|s\rangle = \frac{1}{4\pi} \tag{4.96}$$

$$|p_x\rangle = \sqrt{\frac{1}{2}}\left(Y_1^1 + Y_1^{-1}\right) = \left(\frac{3\pi}{4}\right)^{1/2} \sin\theta\cos\varphi \tag{4.97}$$

$$\left|p_y\right\rangle = -\sqrt{\frac{1}{2}}\,\imath\left(Y_1^1 - Y_1^{-1}\right) = \left(\frac{3\pi}{4}\right)^{1/2} \sin\theta\sin\varphi \tag{4.98}$$

$$|p_z\rangle = \left(\frac{3\pi}{4}\right)^{1/2} \cos\theta \tag{4.99}$$

But the electronic configuration of an atom of C in the valence state having the tetragonal bonding angles is known to be

$$C: 1s^2 2s 2p^3 \tag{4.100}$$

This is usually called the sp^3 configuration. The four wave functions that determine the sp^3 configuration are the linear combinations of the orbitals given by Eqs. (4.96)–(4.99) and are given below:

$$|\psi_1\rangle = |\psi_{0111}\rangle = |s\rangle + |p_x\rangle + \left|p_y\right\rangle + |p_z\rangle \tag{4.101}$$

$$|\psi_2\rangle = |\psi_{01\bar{1}\bar{1}}\rangle = |s\rangle + |p_x\rangle - \left|p_y\right\rangle - |p_z\rangle \tag{4.102}$$

$$|\psi_3\rangle = |\psi_{0\bar{1}1\bar{1}}\rangle = |s\rangle - |p_x\rangle + \left|p_y\right\rangle - |p_z\rangle \tag{4.103}$$

$$|\psi_4\rangle = |\psi_{0\bar{1}\bar{1}1}\rangle = |s\rangle - |p_x\rangle - \left|p_y\right\rangle + |p_z\rangle \tag{4.104}$$

The above four orbitals are called hybrid sp^3 orbitals and are nearly degenerate. These are also sometimes called tetrahedral orbits. In exactly the same manner one can define sp^2 hybridization. The wave functions for the sp^2 configuration consist of the superposition of $|s\rangle$, $|p_x\rangle$, and $|p_y\rangle$ orbitals (leaving $|p_z\rangle$ as it is) and are given as

$$|\psi_1\rangle = |\psi_{01}\rangle = |s\rangle + |p_x\rangle \tag{4.105}$$

$$|\psi_2\rangle = |\psi_{0\bar{1}1}\rangle = |s\rangle - |p_x\rangle + \left|p_y\right\rangle \tag{4.106}$$

$$|\psi_2\rangle = |\psi_{0\bar{1}\bar{1}}\rangle = |s\rangle - |p_x\rangle - \left|p_y\right\rangle \tag{4.107}$$

The above three wave functions are called hybrid sp^2 orbitals. The sp^2 bonds are three degenerate bonds 120 degrees apart in a plane, with the leftover p-orbital sticking out perpendicular to the plane in the z-direction (commonly known as the c-direction). The sp^2 bonds give a hexagonal structure. It is noteworthy that sp^2 hybridization is relevant to the formation of graphene planes and carbon nanotubes. One can also define sp^1 hybridization, which is the linear combination of $|s\rangle$ and $|p_x\rangle$ orbitals.

4.8 MIXED BOND

In the above discussion, we have considered only the pure bonds: either purely ionic or purely covalent in nature. There are, however, many crystals that exhibit mixed bonds, that is, the mixture of ionic and covalent bonds. Let us understand the formation of a mixed bond from the concept of sharing of electrons in a chemical bond between two atoms A and B. We have seen that if two atoms forming a bond are equally electronegative, a pure covalent bond is formed. But in general, the two atoms forming the bond may possess different electronegativities, for example, atom B may be more electronegative than atom A. In this case, the bonding pair of electrons, on average, will be partially localized near the atom B as compared with atom A. One can interpret this by saying that some electronic charge is transferred from A (cation) to B (anion). In other words, one can say that the bond exhibits a mixed character: it is partially ionic ($A^+ - B^-$) and partially covalent (A : B). Depending on the difference in electronegativities, the extent of the covalent and ionic nature of the bond may vary: the greater the difference, the greater the ionic nature of the bond. In the extreme case in which the electronegativity of atom B is very large as compared with that of atom A, the bonding pair of electrons is dragged very near atom B. So, for all practical purposes, the atom A has lost control on its electron while the atom B has complete control over both the bonding electrons. In other words, one can say that ions are formed that give rise to pure ionic bonding. GaAs presents a good example of a mixed bond. In GaAs charge transfer does take place but it is not complete: only 0.46 of an electron is transferred on average from the Ga to the As atom. This transfer accounts for a part of the binding force in GaAs but the major part comes from the electron sharing (covalent bond) between Ga and its neighboring As atoms. Other examples that exhibit the mixed bond character are InP, InAs, GaSb and, SiC.

The extent (in percent) of the ionic and covalent characters is described quantum mechanically. The valence electron would, then, probabilistically speaking, spend a part of its time in an ionic state and a part in a covalent state. Such alternation is often called a resonance in analogy with the harmonic oscillator, which alternately stores its energy in kinetic and potential forms.

The simplest example of covalent bonding is between two hydrogen atoms in the H_2 molecule. When two hydrogen atoms are far apart their wave functions do not overlap (Fig. 4.8A). On coming closer, their electronic wave functions overlap and the net wave function is the linear combination of the two wave functions given by

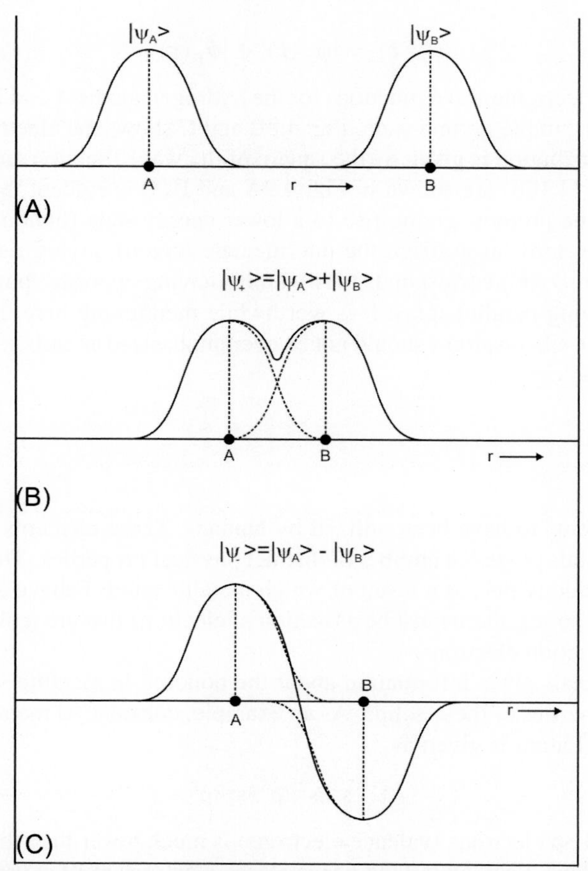

FIG. 4.8 (A) The wave functions of two atoms separated by a large distance. Here $|\psi_A(\mathbf{r})\rangle$ and $|\psi_B(\mathbf{r})\rangle$ are the wave functions of the individual atoms. (B) Overlapping of the wave functions of two atoms having antiparallel spins that form a bonding state defined by $|\psi_+(\mathbf{r})\rangle = |\psi_A(\mathbf{r})\rangle + |\psi_B(\mathbf{r})\rangle$. (C) Overlapping of the wave functions of two atoms having parallel spins that form an antibonding state defined by $|\psi_-(\mathbf{r})\rangle = |\psi_A(\mathbf{r})\rangle - |\psi_B(\mathbf{r})\rangle$.

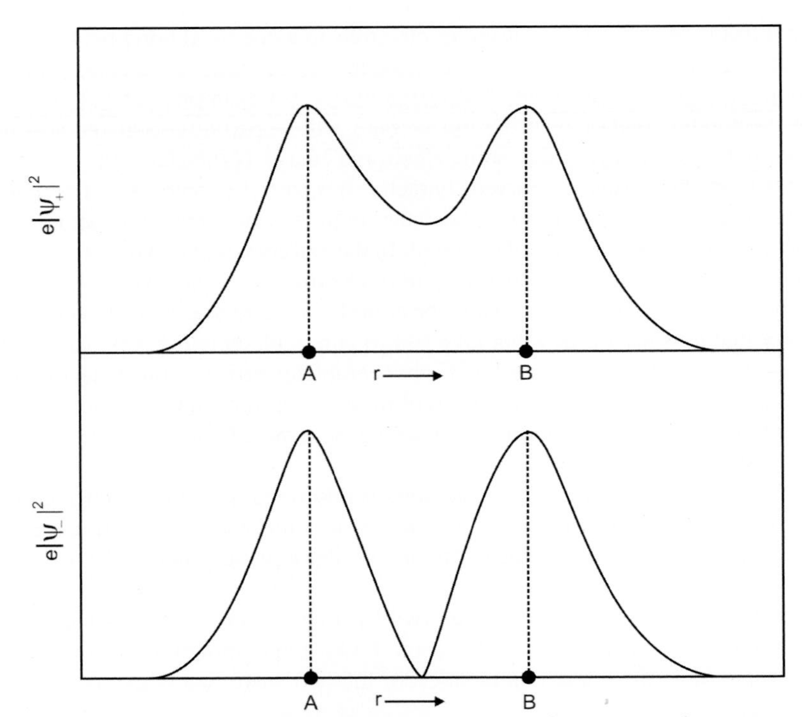

FIG. 4.9 The charge densities of the bonding and antibonding states defined by e $|\psi_+(\mathbf{r})|^2$ and e $|\psi_-(\mathbf{r})|^2$, respectively, as a function of distance.

$$|\psi_\pm(\mathbf{r})\rangle = |\psi_A(\mathbf{r})\rangle \pm |\psi_B(\mathbf{r})\rangle \tag{4.108}$$

Here $|\psi_A(\mathbf{r})\rangle$ and $|\psi_B(\mathbf{r})\rangle$ are the electronic wave functions for the hydrogen atoms A and B. The electrons in both the wave functions are in the 1s state, that is, in the ground state. Fig. 4.8B and C shows the electronic wave functions $|\psi_+(\mathbf{r})\rangle$ and $|\psi\_(\mathbf{r})\rangle$. The electronic charge distribution is given by the square of the wave function and the charge distributions for the two wave functions given by Eq. (4.108) are shown in Fig. 4.9A and B. It is evident that $|\psi_+(\mathbf{r})\rangle$ deposits the electrons primarily in the region between the protons, giving rise to a lower energy state (bonding state), while $|\psi_-(\mathbf{r})\rangle$ deposits electrons around the individual protons (away from the intermediate region), giving rise to a higher energy state (antibonding state). The wave function $|\psi_+(\mathbf{r})\rangle$ corresponds to two atoms having opposite spins, while the wave function $|\psi_-(\mathbf{r})\rangle$ corresponds to two atoms having parallel spins. It is worthwhile mentioning here that the similarity of the bonding between carbon atoms and between silicon atoms should not be overemphasized as carbon gives biology while silicon gives geology and semiconductor technology.

4.9 METALLIC BOND

Metals are perhaps the first elements to have been utilized by humans. These elements constitute more than half of the elements in the periodic table. Metals possess a number of distinct physical properties. The most important are high values for the electrical and thermal conductivities, as a result of which metallic solids behave as good conductors of electricity. This fact indicates that in metallic solids, there must be a fraction of electrons that are mobile and conduct electricity. Such electrons are usually called conduction electrons.

The electronic structure of metals gives information about the bonding in metallic solids, which is central in understanding the various physical properties of these solids. As an example, consider Al metal composed of N atoms in which the electronic structure of each Al atom is given as

$$\mathrm{Al}: 1s^2 2s^2 2p^6 3s^2 3p^1 \tag{4.109}$$

The ionization potential of 3s and 3p electrons (valence electrons) is much lower than that of 2p electrons and, therefore, they are loosely bound to the nucleus. Each atom (and hence all the electrons in it) experiences the electrostatic potential due to all of the other atoms, as a result of which the 3s and 3p electrons are detached from the atom. These electrons roam about more or less freely in the whole of the crystal and are thus called free electrons. As these are the electrons largely

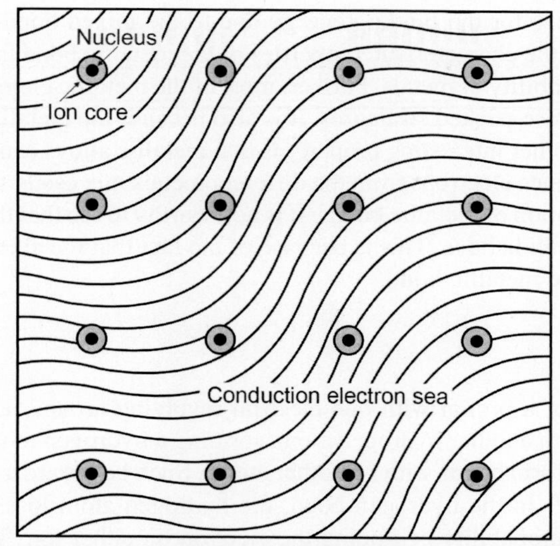

FIG. 4.10 Model representation of a simple metal or free-electron metal. Spheres represent the ion cores, which are assumed to be hard.

responsible for the conduction of charge in a metal, they are also called conduction electrons. The ion cores left behind are somewhat distorted by the crystal field of the metallic solid and have the electronic structure

$$Al^{+3} : 1s^2 2s^2 2p^6 \tag{4.110}$$

Metallic crystals in which the conduction electrons are of type s or p are usually called simple metals. In Al metal with N atoms, there will be 3N conduction electrons and N positively charged Al^{+3} ions at the lattice positions. As the simplest approximation, *a simple metal can be represented as a sea of conduction electrons in which are embedded the ions at the lattice positions*. Such a representation is shown in Fig. 4.10. The electrostatic potential seen by an electron due to all the ions and the rest of the electrons is called the crystal potential $V(\mathbf{r})$. In a metallic solid, the different contributions to $V(\mathbf{r})$, in the one-electron approximation, have already been discussed in Chapter 3 and it should be calculated self-consistently. For a metallic crystal to form, the attractive interaction potential V_{ei} must dominate the sum of the repulsive interaction potentials V_{ii} and V_{ee}. Three main terms contribute to the binding energy E_B of a metallic solid:

1. *Kinetic energy*: It gives a positive contribution to the binding energy. But in metallic solids the interatomic distance is comparatively large, which gives rise to lower values of kinetic energy for the conduction electrons.[2]
2. *Exchange interaction potential*: The exchange interaction potential V_{xc} is negative in a free-electron gas.
3. *Coulomb interaction potential*: The Coulomb repulsion and attraction both increase in metals and contribute toward E_B. The attractive interaction potential V_{ei} is large for large electron-ion distances, but the kinetic energy decreases.[2]

Pure metallic bonds are not strong chemical bonds. The binding energy in metallic solids ranges from 0.7eV for Hg to 8.8 eV for W. The alkali metals Li, Na, K, Rb, and Cs fall in the lower limit of binding energy and hence have relatively low melting and boiling points due to the presence of pure metallic bonds. On the other hand, TMs, such as W, fall in the upper limit and exhibit extremely high melting and boiling points. In the TMs, the conduction electrons possess both s- and d-characters. The s-conduction electrons are nearly free and form metallic bonds. The d-conduction electrons are below the s-conduction electrons and are quasilocalized. The d-electrons form covalent bonds and make an additional contribution to the binding energy.

---

2. The uncertainty principle states that

$$\Delta x \Delta p \geq h/2\pi$$
$$\text{Or } \Delta p \geq h/2\pi\Delta x$$

So, the kinetic energy K.E. is given by

$$\text{K.E.} = \frac{(\Delta p)^2}{2m} \approx \frac{\hbar^2}{2m(\Delta x)^2}$$

So, for larger interatomic distance Δx, the value of K.E. is smaller.

Thus, the TMs exhibit high values for the binding energy due to the mixed bonding character: the bonds are partly metallic and partly covalent. One of the characteristic properties of the metallic bond is that it is isotropic in nature, which gives rise to the ductility and machinability of metals. Further, most of the metallic elements crystallize into fcc, bcc, or hcp structures, which are basically the close-packed structures. It is also because the metallic bonds are isotropic in nature and the ions are hard spherical balls. Another interesting property is that metallic alloys can be easily formed by mixing two or more metals. This is because the valence electrons from the different metals mix easily together to form a sea of conduction electrons and participate in the formation of metallic bonds. It is noteworthy to see that the binding energy of alkali metals is considerably less than that of the alkali halides. This is because of the fact that the alkali halides are ionic crystals and the ionic bond is much stronger than the metallic bond.

4.10 HYDROGEN BOND

A neutral hydrogen atom consists of a proton with one electron revolving around it. Therefore, one hydrogen atom is expected to form a covalent bond with another hydrogen atom, forming a hydrogen molecule (H_2). But, under certain conditions, a hydrogen atom is found to form bonds with two other atoms. Such bonds are usually called hydrogen bonds with a bond energy on the order of 0.1 eV. In the hydrogen bond, the hydrogen atom loses one electron to either of the two adjoining atoms and there is equal probability of finding the electron on either ion. The hydrogen bond is largely ionic in nature and is formed with the most electronegative elements, such as F, O, and N. Fig. 4.11 shows a hydrogen bond with F atoms in which hydrogen difluoride (HF_2^{-1}) is formed. Here a proton forms a hydrogen bond with the two F^- ions. The proton tends to draw the two anions more closely together than their normal spacing so that the shortening of their interatomic spacing serves to indicate the presence of a hydrogen bond. The proton, being very small in size, can accommodate only two F atoms on either side.

Water molecules (H_2O) interact with each other via hydrogen bonds, which is responsible for the electrostatic attraction between the electric dipole moments of water molecules. This fact gives rise to the remarkable physical properties of water and ice. The hydrogen bond is also important in some ferroelectric crystals.

In the above discussion and also in most of books, the bulk modulus is evaluated from Eq. (4.14). But one can derive the expression for the bulk modulus from Eq. (4.12) in exactly the same manner. Here we give this as a problem for the students.

Problem 4.7

With the help of Eq. (4.32), the interaction potential given by Eq. (4.12) can be written as

$$U(V) = \frac{1}{2}N\left[\frac{BP'_{ni}}{V^{n/3}} - \frac{AP'_{mi}}{V^{m/3}}\right] \tag{4.111}$$

$$\text{where } P'_{ni} = \frac{N^{n/3}P_{ni}}{2^{n/6}},\quad P'_{mi} = \frac{N^{m/3}P_{mi}}{2^{m/6}} \tag{4.112}$$

In Eq. (4.12) R_0 has been replaced by R. From Eq. (4.111), prove that the equilibrium volume is given by

$$V_0 = \left(\frac{n}{m}\frac{BP'_{ni}}{AP'_{mi}}\right)^{\frac{3}{n-m}} \tag{4.113}$$

Further, the bulk modulus in terms of volume, from Eq. (4.111), is given by

$$B_M = \frac{1}{2}NB\frac{nP'_{ni}}{V^{\frac{n}{3}+1}}\frac{n-m}{9} \tag{4.114}$$

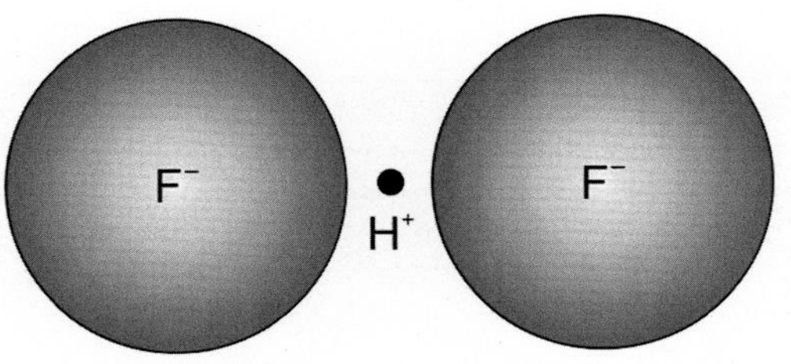

FIG. 4.11 Hydrogen difluoride ion HF_2^{-1} represents a stable hydrogen bond. It is a schematic representation of the ion as proton is shown to be a base particle.

It is easy to prove that the expressions for V_0 and B_M, given by Eqs. (4.113), (4.114) are precisely the same as given in Eqs. (4.36), (4.38).

SUGGESTED READING

Jaswon, M. A. (1954). *Metal physics and physical metallurgy: The theory of cohesion*. London: Pergamon Press.

Pettifor, D. M. (1995). *Bonding and structure of molecules and solids*. London: Oxford University Press.

Tosi, M. P. (1964). Cohesion of ionic solids in the Born model. F. Seitz, & D. Turnbull (Eds.), *Solid state physics* (pp. 1–120). *Vol. 16*. New York: Academic Press.

Chapter 5

Elastic Properties of Solids

Chapter Outline

The theoretical study of the various properties of crystalline solids requires a knowledge of the crystal potential, the reliable determination of which is difficult. One of the oldest and simplest methods is to view the crystal as an isotropic continuous elastic material with uniform density instead of as a discrete periodic array of atoms. The various properties of the solids, such as elastic constants, lattice vibrations, and thermal properties, can be studied using the continuum elasticity theory of solids. The study of elastic constants is of immense importance because they give information about the nature of binding forces in solids: a central physical quantity in studying various properties of solids. The continuum elasticity theory is valid for low-frequency waves, such as elastic waves with wavelength $\lambda > 10^{-6}$ cm or frequency $\nu < 10^{11}$ Hz, as these waves are not able to see the atomic structure of a crystalline solid. The present chapter presents the calculation of elastic constants in the isotropic linear elasticity approximation in which Hooke's law is valid.

5.1 STRAIN TENSOR

Consider a solid with Cartesian coordinate axes having unit vectors $\hat{\mathbf{i}}_\alpha$ with $\alpha = 1, 2,$ and 3. The position vector of a point or a particle in the solid (Fig. 5.1) is defined as

$$\mathbf{r} = \sum_\alpha \hat{\mathbf{i}}_\alpha \mathrm{r}_\alpha \tag{5.1}$$

where r_α are the Cartesian components of the vector $\mathbf{r}$. The application of a weak external force produces a small deformation in the solid that changes the orientation and magnitude of both the position vector and the unit vectors. The strain produced in the body can be studied in two approaches. In the first approach the unit vectors are kept unchanged and the orientation and magnitude of the Cartesian components of the position vector are changed. In the second approach the Cartesian components of the position vector are kept unchanged and the orientation and magnitude of the unit vectors are changed. It is more convenient to adopt the second approach as is done here. In a strained body, a particle at point $\mathbf{r}$ moves to $\mathbf{r}'$ (Fig. 5.1) given by

$$\mathbf{r}' = \mathbf{r} + \mathbf{u}(\mathbf{r}) \tag{5.2}$$

where $\mathbf{u}(\mathbf{r})$ is the displacement produced. If u_α is the Cartesian component of $\mathbf{u}(\mathbf{r})$ along the unit vector $\hat{\mathbf{i}}_\alpha$, then

$$\mathbf{u}(\mathbf{r}) = \sum_\alpha \mathrm{u}_\alpha \hat{\mathbf{i}}_\alpha \tag{5.3}$$

Solid State Physics. https://doi.org/10.1016/B978-0-12-817103-5.00005-0

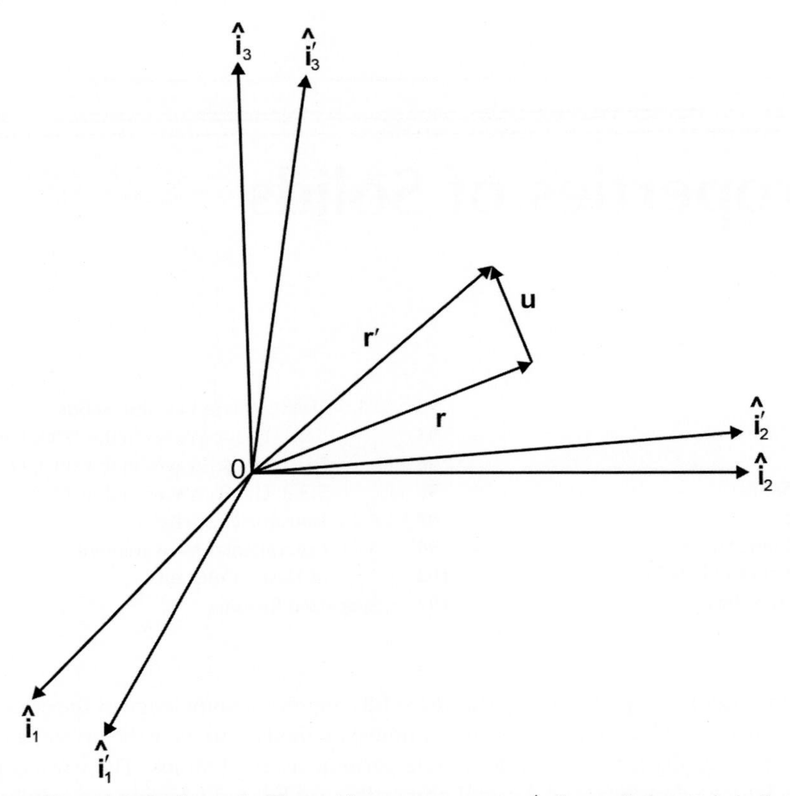

FIG. 5.1 Coordinate axes in perfect and strained crystals. The position vector **r** goes to **r′** after producing strain in the crystal.

Note that **u(r)** is not a constant quantity but is a function of the position vector. Therefore, **u(r)** varies continuously throughout the solid and forms a vector field usually called the *strain field or displacement field.* The components of the unit vector after displacement, denoted as $\hat{\mathbf{i}}'_\alpha$, are given as

$$\begin{aligned}\hat{\mathbf{i}}'_\alpha &= \frac{\partial}{\partial \mathrm{r}_\alpha}(\mathbf{r}+\mathbf{u}) = \frac{\partial}{\partial \mathrm{r}_\alpha}\sum_\beta \left[\mathrm{r}_\beta \hat{\mathbf{i}}_\beta + \mathrm{u}_\beta \hat{\mathbf{i}}_\beta\right] \\ &= \sum_\beta \left(\delta_{\alpha\beta} + e_{\alpha\beta}\right)\hat{\mathbf{i}}_\beta\end{aligned} \tag{5.4}$$

where

$$e_{\alpha\beta} = \partial \mathrm{u}_\beta / \partial \mathrm{r}_\alpha \tag{5.5}$$

The position vector **r′** can be written, in terms of $\hat{\mathbf{i}}'_\alpha$, as

$$\mathbf{r}' = \sum_\alpha \mathrm{r}_\alpha \hat{\mathbf{i}}'_\alpha \tag{5.6}$$

In the above equation the Cartesian components of **r′** remain unchanged after deformation. To make the measurement of deformation free from the orientation of the coordinate axes, it is convenient to define a scalar product between the new unit vectors as

$$g_{\alpha\beta} = \hat{\mathbf{i}}'_\alpha \cdot \hat{\mathbf{i}}'_\beta = \delta_{\alpha\beta} + e_{\alpha\beta} + e_{\beta\alpha} + \sum_\gamma e_{\alpha\gamma} e_{\beta\gamma} \tag{5.7}$$

From the above equation the angle between the two deformed axes is

$$\theta_{\alpha\beta} = \cos^{-1}\left(g_{\alpha\beta} / \left[|\hat{\mathbf{i}}'_\alpha||\hat{\mathbf{i}}'_\beta|\right]\right) \tag{5.8}$$

In terms of $g_{\alpha\beta}$ the strain components $\varepsilon_{\alpha\beta}$ are defined as

$$\begin{aligned}\varepsilon_{\alpha\beta} &= \frac{1}{2}\left(g_{\alpha\beta} - \delta_{\alpha\beta}\right) = \frac{1}{2}\left(e_{\alpha\beta} + e_{\beta\alpha}\right) \\ &= \frac{1}{2}\left[\frac{\partial u_\beta}{\partial r_\alpha} + \frac{\partial u_\alpha}{\partial r_\beta}\right]\end{aligned} \tag{5.9}$$

In defining the above equation, the second order terms in $e_{\alpha\beta}$ are neglected (the linear elasticity approximation). Eq. (5.9) shows that the strain tensor $\overleftrightarrow{\varepsilon}$ is symmetric, that is, $\varepsilon_{\alpha\beta} = \varepsilon_{\beta\alpha}$. The diagonal components of the strain tensor from Eq. (5.9) are given as

$$\varepsilon_{\alpha\alpha} = \frac{\partial u_\alpha}{\partial r_\alpha} = e_{\alpha\alpha} \tag{5.10}$$

When the direction of the axes is reversed, that is, α goes to $-\alpha$, then under the reversal of the coordinate axes one can write

$$r_{-\alpha} = -r_\alpha, \quad u_{-\alpha} = -u_\alpha \tag{5.11}$$

From Eqs. (5.9), (5.10) one can write

$$\varepsilon_{-\alpha-\alpha} = \frac{\partial u_{-\alpha}}{\partial r_{-\alpha}} = \varepsilon_{\alpha\alpha} \tag{5.12}$$

$$\varepsilon_{\alpha-\beta} = \frac{1}{2}\left[\frac{\partial u_{-\beta}}{\partial r_\alpha} + \frac{\partial u_\alpha}{\partial r_{-\beta}}\right] = -\varepsilon_{\alpha\beta} \tag{5.13}$$

Eqs. (5.12), (5.13) show that with the reversal of direction of one of the Cartesian coordinates, the nondiagonal components of the strain tensor change their sign while the diagonal components remain unchanged.

Ideally speaking, two types of strain can exist in a solid. The first is *hydrostatic pressure* in which the volume of the solid changes without any change in its shape and is represented by the diagonal components of the strain tensor $\overleftrightarrow{\varepsilon}$. The second is pure shear in which the volume of the body remains unchanged by the deformation and only the shape changes. Pure shear is represented by the nondiagonal components of $\overleftrightarrow{\varepsilon}$. But in an actual strain, both the volume and shape of the solid may change. Therefore, the general strain may be represented as the linear combination of pure hydrostatic pressure and pure shear. To do so, the strain components are defined as

$$\varepsilon_{\alpha\beta} = \left[\varepsilon_{\alpha\beta} - \frac{1}{3}\delta_{\alpha\beta}\sum_\gamma \varepsilon_{\gamma\gamma}\right] + \frac{1}{3}\delta_{\alpha\beta}\sum_\gamma \varepsilon_{\gamma\gamma} \tag{5.14}$$

The first term on the right side is evidently a pure shear as the sum of its diagonal terms is zero, while the second term is a pure hydrostatic pressure. Hydrostatic pressure acts perpendicular to the surface of the solid, while pure shear acts along the surface. Therefore, these two scalar components are independent of each other and the square of the strain component $\varepsilon_{\alpha\beta}$ can be obtained by the addition of the squares of these two components, that is,

$$\left[\varepsilon_{\alpha\beta}\right]^2 = \left[\varepsilon_{\alpha\beta} - \frac{1}{3}\delta_{\alpha\beta}\sum_\gamma \varepsilon_{\gamma\gamma}\right]^2 + \frac{1}{3}\left[\sum_\gamma \varepsilon_{\gamma\gamma}\right]^2 \tag{5.15}$$

The strain field, from Eqs. (5.1), (5.2), (5.4), and (5.6), can be written as

$$\mathbf{u}(\mathbf{r}) = \mathbf{r}' - \mathbf{r} = \sum_{\alpha,\beta}\left[r_\alpha e_{\alpha\beta}\right]\hat{\mathbf{i}}_\beta \tag{5.16}$$

From Eqs. (5.3), (5.16) the components of the strain field are given by

$$u_\beta = \sum_\alpha r_\alpha e_{\alpha\beta} \tag{5.17}$$

5.2 DILATION

Dilation is defined as the fractional change in volume of a solid due to the deformation produced by the application of an external force. It can be expressed in terms of the strain field components $\varepsilon_{\alpha\beta}$. Consider a solid in the form of a cube with unit edges $\hat{\mathbf{i}}_1$, $\hat{\mathbf{i}}_2$, and $\hat{\mathbf{i}}_3$. The volume of the solid V is given by

$$V = \hat{\mathbf{i}}_1 \cdot \hat{\mathbf{i}}_2 \times \hat{\mathbf{i}}_3 = 1 \tag{5.18}$$

After deformation the cube becomes a parallelepiped with edges $\hat{\mathbf{i}}'_1$, $\hat{\mathbf{i}}'_2$, and $\hat{\mathbf{i}}'_3$, having volume V′ given by

$$\mathrm{V}' = \hat{\mathbf{i}}'_1 \cdot \hat{\mathbf{i}}'_2 \times \hat{\mathbf{i}}'_3 = 1 + \sum_{\alpha} \varepsilon_{\alpha\alpha} \tag{5.19}$$

in the linear elasticity approximation. The dilation produced by the deformation becomes

$$\delta_{\mathrm{D}} = \frac{\mathrm{V}' - \mathrm{V}}{\mathrm{V}} = \frac{\Delta \mathrm{V}}{\mathrm{V}} = \sum_{\alpha} \varepsilon_{\alpha\alpha} \tag{5.20}$$

The symmetric strain tensor has six independent elements. For simplicity of notation, it is very convenient to use the Voigt notation for the strain components, according to which the subscripts are defined as

$$\begin{aligned} 1 &\equiv 11 = \mathrm{xx}; \quad 2 \equiv 22 = \mathrm{yy}; \quad 3 \equiv 33 = \mathrm{zz} \\ 4 &\equiv 23 = \mathrm{yz}; \quad 5 \equiv 31 = \mathrm{zx}; \quad 6 \equiv 12 = \mathrm{xy} \end{aligned} \tag{5.21}$$

In the Voigt notation $\varepsilon_{\alpha\beta}$ can be written as ε_{m} where m takes the values 1, 2, 3, 4, 5, and 6.

5.3 STRESS TENSOR

There are two types of forces acting on a solid body. The first type is called the body force, which acts throughout the body and exerts influence on the whole of the mass distribution. The inertial and gravitational forces are examples of the body force. The body force is expressed per unit mass or per unit volume. If $\mathbf{f}(\mathbf{r})$ is the body force per unit volume, it can be written as

$$\mathbf{f}(\mathbf{r}) = \sum_{\alpha} \mathrm{f}_{\alpha} \hat{\mathbf{i}}_{\alpha} \tag{5.22}$$

Here f_{α} is the α-component of the body force density. The second type is the surface force, which acts on the surface of the body. A surface force is expressed in units of force per unit area and is called stress. The stress is further divided into two categories: stress acting normal to the surface (normal stress) and stress acting along the surface (shear stress). As the normal stress and the shear stress act perpendicularly to each other, therefore, they are orthogonal stresses. Stress is represented by a tensor $\overleftrightarrow{\sigma}$ with components $\sigma_{\alpha\beta}$ where the first subscript indicates the direction of the stress and the second subscript the direction of the normal to the surface on which the stress is acting. Fig. 5.2 shows the different components of stress acting on a cubic solid. It can be easily shown that the stress tensor is symmetric, that is,

$$\sigma_{\alpha\beta} = \sigma_{\beta\alpha} \tag{5.23}$$

which is a consequence of the fact that shear stress does not cause angular rotation. The symmetric stress tensor $\overleftrightarrow{\sigma}$ has six independent elements, which, in the Voigt notation, can be written as σ_{m} where m can take values 1, 2, 3, 4, 5, and 6. If a body under stress is in the equilibrium state, its equation of motion is

$$\sum_{\beta} \frac{\partial \sigma_{\alpha\beta}}{\partial \mathrm{r}_{\beta}} + \mathrm{f}_{\alpha} = 0 \tag{5.24}$$

The first term gives the body force per unit volume applied externally on the body and the second term is the internal body force per unit volume.

5.4 ELASTIC CONSTANTS OF SOLIDS

In the linear elasticity approximation, valid for very small deformations only, the strain is linearly proportional to stress and vice versa. In mathematical language, we can write

$$\sigma_{\alpha\beta} = \sum_{\mu,\nu} \mathrm{C}_{\alpha\beta\mu\nu} \varepsilon_{\mu\nu} \tag{5.25a}$$

$$\varepsilon_{\alpha\beta} = \sum_{\mu,\nu} \mathrm{C}^{\mathrm{s}}_{\alpha\beta\mu\nu} \sigma_{\mu\nu} \tag{5.25b}$$

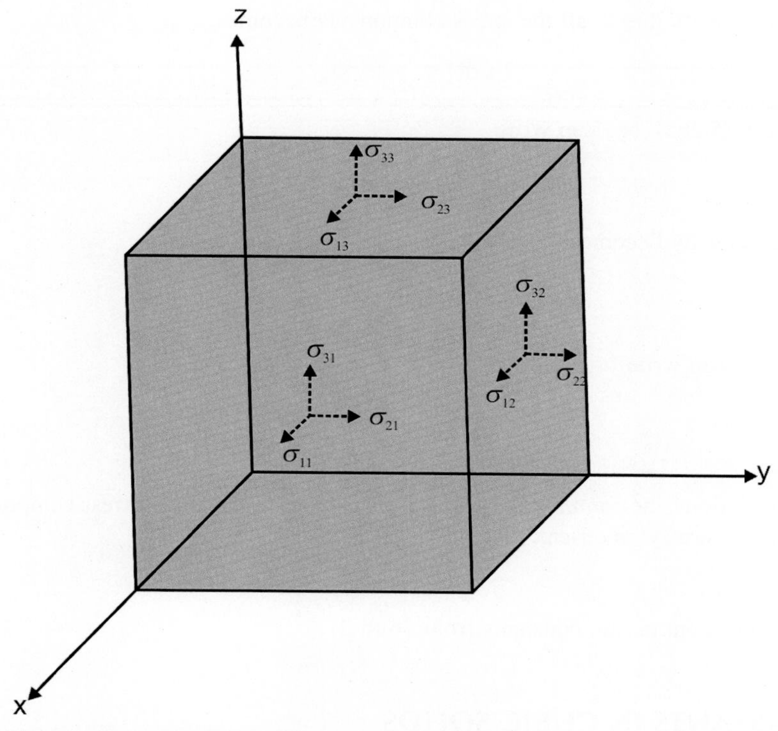

FIG. 5.2 Components $\sigma_{\alpha\beta}$ of the stress tensor in a cubic crystal.

where $C_{\alpha\beta\mu\nu}$ and $C^{s}_{\alpha\beta\mu\nu}$ are called the *elastic stiffness constants* and *elastic compliance constants*. The quantity $C_{\alpha\beta\mu\nu}$ has dimensions of energy/volume, while $C^{s}_{\alpha\beta\mu\nu}$ has dimensions of volume/energy. In the Voigt notation Eqs. (5.25a), (5.25b) become

$$\sigma_{m} = \sum_{n} C_{mn}\,\varepsilon_{n} \tag{5.26a}$$

$$\varepsilon_{m} = \sum_{n} C^{s}_{mn}\,\sigma_{n} \tag{5.26b}$$

where m and n have integral values from 1 to 6. The constants C_{mn} are usually called the *elastic constants* of the crystalline solid.

5.5 ELASTIC ENERGY DENSITY

If U is the elastic energy density per unit volume, the stress components in terms of it are given by

$$\sigma_{\alpha\beta} = -\frac{\partial U}{\partial \varepsilon_{\alpha\beta}} \tag{5.27a}$$

or

$$dU = -\sigma_{\alpha\beta} d\varepsilon_{\alpha\beta} \tag{5.27b}$$

The minus sign indicates that U is the work done on the system. In Voigt notation one can write (magnitude)

$$dU = \sigma_{n} d\varepsilon_{n} \tag{5.28}$$

From the above equation the stress components can be written in terms of energy density as

$$\sigma_{n} = \frac{dU}{d\varepsilon_{n}} \tag{5.29}$$

The total change in energy density due to all the stress components becomes

$$dU = \sum_{n} \sigma_n d\varepsilon_n \tag{5.30}$$

Substituting for σ_n from Eq. (5.26a), one can write

$$dU = \sum_{n,m} C_{nm} \varepsilon_m d\varepsilon_n \tag{5.31}$$

Therefore, the total energy density becomes

$$U = \int dU = \frac{1}{2} \sum_{n,m} C_{nm} \varepsilon_m \varepsilon_n \tag{5.32}$$

From Eqs. (5.29), (5.32) one can write

$$\sigma_1 = \frac{dU}{d\varepsilon_1} = C_{11}\varepsilon_1 + \frac{1}{2} \sum_{n=2}^{6} (C_{1n} + C_{n1})\varepsilon_n \tag{5.33}$$

Hence, in the stress-strain relations, the combination $1/2(C_{mn}+C_{nm})$ appears in all the stress components. It follows that the elastic stiffness constants C_{mn} are symmetrical, that is,

$$C_{mn} = C_{nm} \tag{5.34}$$

Eq. (5.34) reduces the independent elastic constants from 36 to 21.

5.6 ELASTIC CONSTANTS IN CUBIC SOLIDS

In the linear elasticity approximation, the stress-strain relation given by Eq. (5.26a) allows us to write the different components of stress as

$$\begin{aligned}
\sigma_{xx} &= C_{11}\varepsilon_{xx} + C_{12}\varepsilon_{yy} + C_{13}\varepsilon_{zz} + C_{14}\varepsilon_{yz} + C_{15}\varepsilon_{zx} + C_{16}\varepsilon_{xy} \\
\sigma_{yy} &= C_{21}\varepsilon_{xx} + C_{22}\varepsilon_{yy} + C_{23}\varepsilon_{zz} + C_{24}\varepsilon_{yz} + C_{25}\varepsilon_{zx} + C_{26}\varepsilon_{xy} \\
\sigma_{zz} &= C_{31}\varepsilon_{xx} + C_{32}\varepsilon_{yy} + C_{33}\varepsilon_{zz} + C_{34}\varepsilon_{yz} + C_{35}\varepsilon_{zx} + C_{36}\varepsilon_{xy} \\
\sigma_{yz} &= C_{41}\varepsilon_{xx} + C_{42}\varepsilon_{yy} + C_{43}\varepsilon_{zz} + C_{44}\varepsilon_{yz} + C_{45}\varepsilon_{zx} + C_{46}\varepsilon_{xy} \\
\sigma_{zx} &= C_{51}\varepsilon_{xx} + C_{52}\varepsilon_{yy} + C_{53}\varepsilon_{zz} + C_{54}\varepsilon_{yz} + C_{55}\varepsilon_{zx} + C_{56}\varepsilon_{xy} \\
\sigma_{xy} &= C_{61}\varepsilon_{xx} + C_{62}\varepsilon_{yy} + C_{63}\varepsilon_{zz} + C_{64}\varepsilon_{yz} + C_{65}\varepsilon_{zx} + C_{66}\varepsilon_{xy}
\end{aligned} \tag{5.35}$$

The above equation can be written in matrix form as

$$\begin{pmatrix} \sigma_{xx} \\ \sigma_{yy} \\ \sigma_{zz} \\ \sigma_{yz} \\ \sigma_{zx} \\ \sigma_{xy} \end{pmatrix} = \begin{pmatrix} C_{11} & C_{12} & C_{13} & C_{14} & C_{15} & C_{16} \\ C_{21} & C_{22} & C_{23} & C_{24} & C_{25} & C_{26} \\ C_{31} & C_{32} & C_{33} & C_{34} & C_{35} & C_{36} \\ C_{41} & C_{42} & C_{43} & C_{44} & C_{45} & C_{46} \\ C_{51} & C_{52} & C_{53} & C_{54} & C_{55} & C_{56} \\ C_{61} & C_{62} & C_{63} & C_{64} & C_{65} & C_{66} \end{pmatrix} \begin{pmatrix} \varepsilon_{xx} \\ \varepsilon_{yy} \\ \varepsilon_{zz} \\ \varepsilon_{yz} \\ \varepsilon_{zx} \\ \varepsilon_{xy} \end{pmatrix} \tag{5.36}$$

Consider an isotropic solid in which the physical properties do not alter under the symmetry operations of a cubic solid. The number of elastic constants can be reduced by considering the different symmetry relations of the cubic solid. Following are the important symmetry operations in a cubic solid.

Fourfold $(2\pi/4)$ rotation about one of the edges of a cube (Fig. 5.3), which changes the sign of the coordinates as follows (anticlockwise rotation)

$$\begin{aligned}
&\text{About z} - \text{axis}: \quad x \to y, \quad y \to -x \\
&\text{About y} - \text{axis}: \quad z \to x, \quad x \to -z \\
&\text{About x} - \text{axis}: \quad y \to z, \quad z \to -y
\end{aligned} \tag{5.37}$$

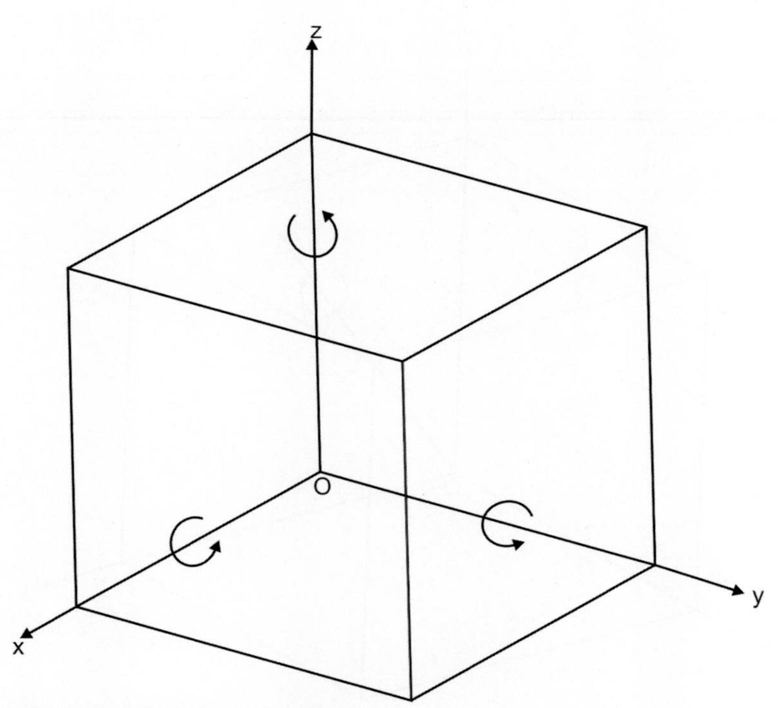

FIG. 5.3 Twofold and fourfold rotations about the edges of a cube.

Twofold ($2\pi/2$) rotation about one of the edges of a cube, which changes the axes as follows (anticlockwise rotation):

$$\begin{aligned}&\text{About z}-\text{axis}: \; x \to -x, \; y \to -y\\&\text{About y}-\text{axis}: \; z \to -z, \; x \to -x\\&\text{About x}-\text{axis}: \; y \to -y, \; z \to -z\end{aligned} \tag{5.38}$$

Threefold ($2\pi/3$) rotation about the diagonals of a cube (Fig. 5.4). There are four diagonals about which the transformation of the axes is as follows:

$$\begin{aligned}&x \to y \to z \to x\\&-x \to z \to -y \to -x\\&x \to z \to -y \to x\\&-x \to y \to z \to -x\end{aligned} \tag{5.39}$$

We see from the above symmetry relations that the rotations either interchange the axes or reverse their sign. Let us apply the operation of a twofold rotation to the cubic solids in which the signs of the axes change. If the sign of the y-axis is reversed, then the left-hand side of the first expression of Eq. (5.35) is given as

$$\begin{aligned}\sigma_{xx} &= C_{11}\varepsilon_{xx} + C_{12}\varepsilon_{-y-y} + C_{13}\varepsilon_{zz} + C_{14}\varepsilon_{-yz} + C_{15}\varepsilon_{zx} + C_{16}\varepsilon_{x-y}\\&= C_{11}\varepsilon_{xx} + C_{12}\varepsilon_{yy} + C_{13}\varepsilon_{zz} - C_{14}\varepsilon_{yz} + C_{15}\varepsilon_{zx} - C_{16}\varepsilon_{xy}\end{aligned} \tag{5.40}$$

In the above equation we have used the properties (5.12), (5.13) of $\varepsilon_{\alpha\beta}$ and we see that all the terms remain unchanged except the 4th and 6th terms of Eq. (5.40). Therefore, σ_{xx} is invariant under the transformation $y \to -y$ only if

$$C_{14} = C_{16} = 0 \tag{5.41}$$

The symmetry property of the elastic constants yields

$$C_{41} = C_{61} = 0 \tag{5.42}$$

Similarly, if we make the transformation $z \to -z$ in the first expression of Eq. (5.35), we get

$$C_{14} = C_{41} = C_{15} = C_{51} = 0 \tag{5.43}$$

Applying the transformation $x \to -x$ in the second expression of Eq. (5.35), we find

$$\sigma_{yy} = C_{21}\varepsilon_{xx} + C_{22}\varepsilon_{yy} + C_{23}\varepsilon_{zz} + C_{24}\varepsilon_{yz} - C_{25}\varepsilon_{zx} - C_{26}\varepsilon_{xy} \tag{5.44}$$

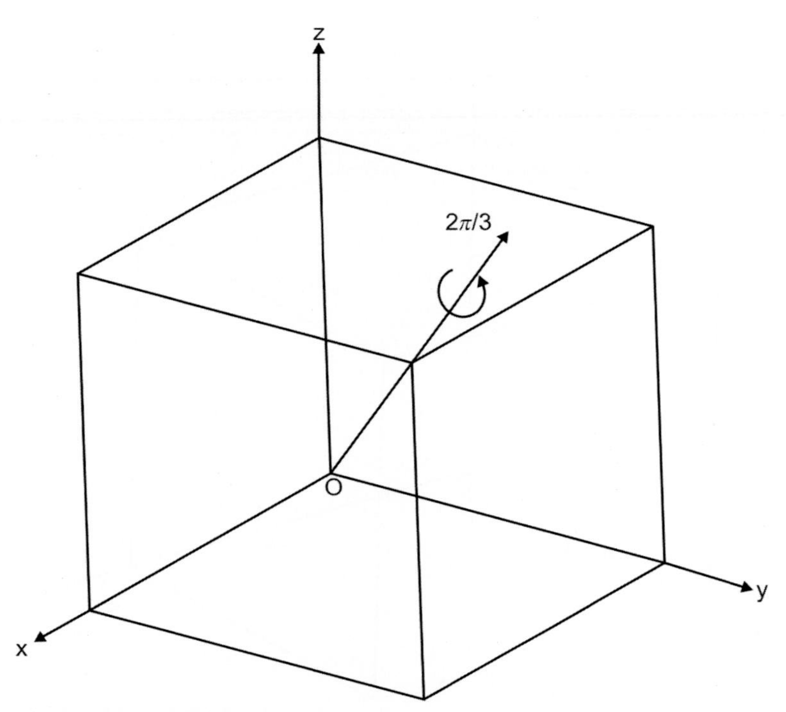

FIG. 5.4 Threefold rotation about the diagonal of a cubic structure.

and the transformation $z \to -z$ in the second expression of Eq. (5.35) yields

$$\sigma_{yy} = C_{21}\varepsilon_{xx} + C_{22}\varepsilon_{yy} + C_{23}\varepsilon_{zz} - C_{24}\varepsilon_{yz} - C_{25}\varepsilon_{zx} + C_{26}\varepsilon_{xy} \tag{5.45}$$

The property of invariance yields, from Eqs. (5.44), (5.45),

$$C_{24} = C_{25} = C_{26} = C_{42} = C_{52} = C_{62} = 0 \tag{5.46}$$

The transformations $x \to -x$ and $y \to -y$, when applied to the third expression of Eq. (5.35), reduce the following elastic constants to zero

$$C_{34} = C_{35} = C_{36} = C_{43} = C_{53} = C_{63} = 0 \tag{5.47}$$

In the above description we have considered the invariance of only the normal stress components σ_{xx}, σ_{yy}, and σ_{zz} in cubic crystals. Using Eqs. (5.41)–(5.43), (5.46), and (5.47) in Eq. (5.35), one can write

$$\begin{aligned}
\sigma_{xx} &= C_{11}\varepsilon_{xx} + C_{12}\varepsilon_{yy} + C_{13}\varepsilon_{zz} \\
\sigma_{yy} &= C_{21}\varepsilon_{xx} + C_{22}\varepsilon_{yy} + C_{23}\varepsilon_{zz} \\
\sigma_{zz} &= C_{31}\varepsilon_{xx} + C_{32}\varepsilon_{yy} + C_{33}\varepsilon_{zz} \\
\sigma_{yz} &= C_{44}\varepsilon_{yz} + C_{45}\varepsilon_{zx} + C_{46}\varepsilon_{xy} \\
\sigma_{zx} &= C_{54}\varepsilon_{yz} + C_{55}\varepsilon_{zx} + C_{56}\varepsilon_{xy} \\
\sigma_{xy} &= C_{64}\varepsilon_{yz} + C_{65}\varepsilon_{zx} + C_{66}\varepsilon_{xy}
\end{aligned} \tag{5.48}$$

Let us now consider the shear stress defined by the last three expressions of Eq. (5.48). If we apply the transformation $x \to -x$ to the fourth expression of Eq. (5.48), the invariance property yields

$$C_{45} = C_{46} = C_{54} = C_{64} = 0 \tag{5.49}$$

We have used the fact that σ_{yz} does not change under the transformation $x \to -x$ as it does not involve the x-coordinate. Similarly, the application of transformation $y \to -y$ to the fifth expression and of $z \to -z$ to the sixth expression of Eq. (5.48) give

$$C_{54} = C_{56} = C_{45} = C_{65} = 0 \tag{5.50}$$

$$C_{64} = C_{65} = C_{46} = C_{56} = 0 \tag{5.51}$$

Substituting Eqs. (5.49)–(5.51) into Eq. (5.48), one gets

$$\begin{aligned}\sigma_{xx} &= C_{11}\varepsilon_{xx} + C_{12}\varepsilon_{yy} + C_{13}\varepsilon_{zz}\\ \sigma_{yy} &= C_{21}\varepsilon_{xx} + C_{22}\varepsilon_{yy} + C_{23}\varepsilon_{zz}\\ \sigma_{zz} &= C_{31}\varepsilon_{xx} + C_{32}\varepsilon_{yy} + C_{33}\varepsilon_{zz}\\ \sigma_{yz} &= C_{44}\varepsilon_{yz}\\ \sigma_{zx} &= C_{55}\varepsilon_{zx}\\ \sigma_{xy} &= C_{66}\varepsilon_{xy}\end{aligned} \tag{5.52}$$

In the above discussion we have applied a twofold rotation about one of the edges of the cubic structure. Let us apply a fourfold rotation about one of the edges of the cube. In a fourfold rotation either the transformation $x \to y$, $y \to z$, $z \to x$ or $x \to -y$, $y \to -z$, $z \to -x$ can take place. In both of these transformations, invariance is required. When fourfold rotation is applied about the x-axis, then the allowed transformations are $y \to z$ and $z \to -y$ and in these transformations the first expression of Eq. (5.52) changes to

$$\sigma_{xx} = C_{11}\varepsilon_{xx} + C_{12}\varepsilon_{zz} + C_{13}\varepsilon_{yy} \tag{5.53}$$

The first expression of Eqs. (5.52), (5.53) must be the same (invariance property) under the transformation, which gives

$$C_{12} = C_{13} \tag{5.54}$$

Similarly, when fourfold rotation is applied to the second expression of Eq. (5.52) about the y-axis and to the third expression of Eq. (5.52) about the z-axis, one gets

$$C_{21} = C_{23} = C_{31} = C_{32} \tag{5.55}$$

Eqs. (5.54), (5.55) collectively can be written as

$$C_{12} = C_{21} = C_{31} = C_{32} = C_{13} = C_{23} \tag{5.56}$$

Use of Eq. (5.56) in Eq. (5.52) yields

$$\begin{aligned}\sigma_{xx} &= C_{11}\varepsilon_{xx} + C_{12}\left(\varepsilon_{yy} + \varepsilon_{zz}\right)\\ \sigma_{yy} &= C_{22}\varepsilon_{yy} + C_{12}\left(\varepsilon_{zz} + \varepsilon_{xx}\right)\\ \sigma_{zz} &= C_{33}\varepsilon_{zz} + C_{12}\left(\varepsilon_{xx} + \varepsilon_{yy}\right)\\ \sigma_{yz} &= C_{44}\varepsilon_{yz}\\ \sigma_{zx} &= C_{55}\varepsilon_{zx}\\ \sigma_{xy} &= C_{66}\varepsilon_{xy}\end{aligned} \tag{5.57}$$

Let us apply a threefold rotation to Eq. (5.57) in which $x \to y \to z \to x$. The application of the transformation $x \to y$ to the first expression of Eq. (5.57) yields

$$\sigma_{yy} = C_{11}\varepsilon_{yy} + C_{12}\left(\varepsilon_{xx} + \varepsilon_{zz}\right) \tag{5.58}$$

Comparing Eq. (5.58) with the second expression of Eq. (5.57), the invariance of σ_{yy} demands

$$C_{11} = C_{22} \tag{5.59}$$

Similarly, when the transformations $y \to z$ and $z \to x$ are applied to the second and third terms, respectively, of Eq. (5.57) these give

$$C_{22} = C_{33}, \quad C_{33} = C_{11} \tag{5.60}$$

The application of the transformation $x \to y \to z \to x$ to the last three expressions of Eq. (5.57) yields

$$C_{44} = C_{55} = C_{66} \tag{5.61}$$

Using Eqs. (5.59)–(5.61) in Eq. (5.57) one gets

$$\begin{aligned}\sigma_{xx} &= C_{11}\varepsilon_{xx} + C_{12}\left(\varepsilon_{yy} + \varepsilon_{zz}\right)\\ \sigma_{yy} &= C_{11}\varepsilon_{yy} + C_{12}\left(\varepsilon_{zz} + \varepsilon_{xx}\right)\\ \sigma_{zz} &= C_{11}\varepsilon_{zz} + C_{12}\left(\varepsilon_{xx} + \varepsilon_{yy}\right)\\ \sigma_{yz} &= C_{44}\varepsilon_{yz}\\ \sigma_{zx} &= C_{44}\varepsilon_{zx}\\ \sigma_{xy} &= C_{44}\varepsilon_{xy}\end{aligned} \tag{5.62}$$

Eq. (5.62) can be written in matrix form as

$$\begin{pmatrix} \sigma_{xx} \\ \sigma_{yy} \\ \sigma_{zz} \\ \sigma_{yz} \\ \sigma_{zx} \\ \sigma_{xy} \end{pmatrix} = \begin{pmatrix} C_{11} & C_{12} & C_{12} & 0 & 0 & 0 \\ C_{12} & C_{11} & C_{12} & 0 & 0 & 0 \\ C_{12} & C_{12} & C_{11} & 0 & 0 & 0 \\ 0 & 0 & 0 & C_{44} & 0 & 0 \\ 0 & 0 & 0 & 0 & C_{44} & 0 \\ 0 & 0 & 0 & 0 & 0 & C_{44} \end{pmatrix} \begin{pmatrix} \varepsilon_{xx} \\ \varepsilon_{yy} \\ \varepsilon_{zz} \\ \varepsilon_{yz} \\ \varepsilon_{zx} \\ \varepsilon_{xy} \end{pmatrix} \tag{5.63}$$

Eq. (5.62) or Eq. (5.63) shows that in a solid with a cubic structure there are three independent elastic constants, C_{11}, C_{12}, and C_{44}, the determination of which explains all of the elastic properties of these solids. From Eq. (5.63) one can derive the relation between the elastic stiffness and the elastic compliance constants for a cubic crystal. Table 5.1 presents the elastic stiffness constants for Al, Cu, Ag, and Au metals.

5.7 ELASTIC ENERGY DENSITY IN CUBIC SOLIDS

The elastic energy density, given by Eq. (5.32), is a quadratic function of the elastic strain. In cubic crystals there are only three nonzero elastic constants, C_{11}, C_{12}, and C_{44}. Therefore, the nonzero terms in Eq. (5.32), with the help of Eq. (5.63), can be written as

$$U = \frac{1}{2} C_{11} \left(\varepsilon_1^2 + \varepsilon_2^2 + \varepsilon_3^2\right) + \frac{1}{2} C_{44} \left(\varepsilon_4^2 + \varepsilon_5^2 + \varepsilon_6^2\right) + C_{12} \left(\varepsilon_1 \varepsilon_2 + \varepsilon_1 \varepsilon_3 + \varepsilon_2 \varepsilon_3\right) \tag{5.64}$$

The above equation can be written in terms of the x, y, and z axes as

$$U = \frac{1}{2} C_{11} \left(\varepsilon_{xx}^2 + \varepsilon_{yy}^2 + \varepsilon_{zz}^2\right) + \frac{1}{2} C_{44} \left(\varepsilon_{yz}^2 + \varepsilon_{zx}^2 + \varepsilon_{xy}^2\right) + C_{12} \left(\varepsilon_{xx} \varepsilon_{yy} + \varepsilon_{yy} \varepsilon_{zz} + \varepsilon_{zz} \varepsilon_{xx}\right) \tag{5.65}$$

5.8 BULK MODULUS IN CUBIC SOLIDS

The bulk modulus B_M is defined, in terms of the energy density U, as

$$U = \frac{1}{2} B_M \delta_D^2 \tag{5.66}$$

where δ_D is the dilation. From Eq. (5.65), U can be calculated in terms of δ_D. For uniform dilation

$$\varepsilon_{xx} = \varepsilon_{yy} = \varepsilon_{zz} = \frac{1}{3} \delta_D \tag{5.67}$$

and

$$\varepsilon_{yz} = \varepsilon_{zx} = \varepsilon_{xy} = 0 \tag{5.68}$$

TABLE 5.1 Adiabatic Elastic Stiffness Constants (10^{12} dyne/cm^2) at Room Temperature (300 K) for Al, Cu, Ag, and Au[a]

Metal	C_{11}	C_{12}	C_{44}	A_{as}
Al	1.068	0.607	0.282	0.611
Cu	1.684	1.214	0.754	1.604
Ag	1.240	0.937	0.461	1.521
Au	1.923	1.631	0.420	1.438

[a]*Kittel, C. (1971).* Introduction to solid state physics *(4th ed.). New York: J. Wiley & Sons.*

Substituting Eqs. (5.67), (5.68) into Eq. (5.65), one can write

$$U = \frac{1}{6}(C_{11} + 2C_{12})\delta_D^2 \tag{5.69}$$

Comparing Eqs.(5.66), (5.69), the bulk modulus becomes

$$B_M = \frac{1}{3}(C_{11} + 2C_{12}) \tag{5.70}$$

and the compressibility K_C can be written as

$$K_C = \frac{1}{B_M} = \frac{3}{C_{11} + 2C_{12}} \tag{5.71}$$

With the knowledge of the elastic constants, the bulk modulus and compressibility of a cubic solid can be estimated.

5.9 ELASTIC WAVES IN CUBIC SOLIDS

The propagation of elastic waves in a solid can be understood by considering the mechanical system shown in Fig. 5.5. Here P, Q, and R are wooden blocks joined together with springs A and B along the x-direction. Elastic waves are produced by stretching the springs along the x-direction. If the springs A and B are stretched equally in opposite directions, the force acting on the block Q is zero. Further, if uniform stress σ_{xx} is applied, the force acting on the blocks is again zero. If spring B is stretched more than spring A, nonuniform stress is produced in the system and the net stress acting on the block Q becomes

$$\Delta\sigma_{xx} = \sigma_{xx}(B) - \sigma_{xx}(A) \tag{5.72}$$

$\Delta\sigma_{xx}$ will make the block Q move in the x-direction, thereby producing an elastic wave. Let us now consider a small cubic solid with sides Δx, Δy, and Δz (Fig. 5.6). The application of a nonuniform stress $\sigma_{xx}(x)$ in the x-direction produces a nonuniform strain in the cube. Let $\sigma_{xx}(x+\Delta x)$ and $\sigma_{xx}(x)$ be the stress applied to the opposite faces of the cube (see Fig. 5.6), then

$$\sigma_{xx}(x + \Delta x) = \sigma_{xx}(x) + \frac{\partial \sigma_{xx}}{\partial x}\Delta x \tag{5.73}$$

The net stress acting on the face ABCD is given by

$$\Delta\sigma_{xx}(\Delta x) = \sigma_{xx}(x + \Delta x) - \sigma_{xx}(x) = \frac{\partial \sigma_{xx}}{\partial x}\Delta x \tag{5.74}$$

Therefore, the force $\Delta F(\Delta x)$ acting on the face ABCD becomes

$$\Delta F(\Delta x) = \Delta\sigma_{xx}\Delta y\Delta z = \frac{\partial \sigma_{xx}}{\partial x}\Delta x\Delta y\Delta z \tag{5.75}$$

The application of stress $\Delta\sigma_{xx}(\Delta x)$ makes the cube move along the x-direction, thereby producing an elastic wave in the same direction. The stress $\Delta\sigma_{xx}$ will also produce strain (deformation) in the y- (in the faces BCGF and ADHE) and z- (in the faces DCGH and ABFE) directions. The forces acting along the y- and z-directions are given by

$$\Delta F(\Delta y) = F(y + \Delta y) - F(y) = \frac{\partial \sigma_{xy}}{\partial y}\Delta x\Delta y\Delta z \tag{5.76}$$

$$\Delta F(\Delta z) = F(z + \Delta z) - F(z) = \frac{\partial \sigma_{xz}}{\partial z}\Delta x\Delta y\Delta z \tag{5.77}$$

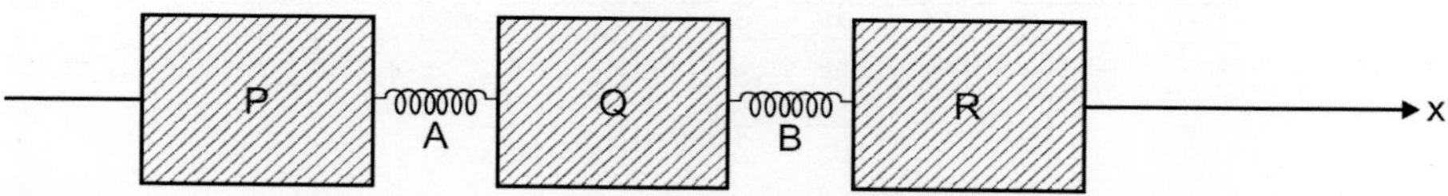

FIG. 5.5 A mechanical system of three wooden blocks P, Q, and R joined together with springs A and B along the x-direction.

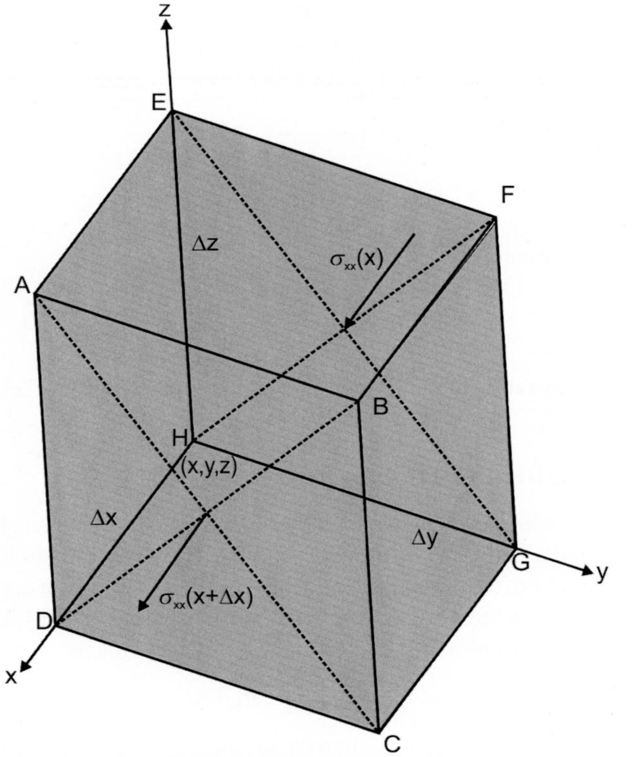

FIG. 5.6 A small cubic solid with sides Δx, Δy, and Δz along the three Cartesian directions. A nonuniform stress $\sigma_{xx}(x)$ is applied along the x-direction.

Hence the total force acting on the cube due to the stress applied in the x-direction becomes

$$\begin{aligned}\Delta F &= \Delta F(\Delta x) + \Delta F(\Delta y) + \Delta F(\Delta z) \\ &= \left(\frac{\partial \sigma_{xx}}{\partial x} + \frac{\partial \sigma_{xy}}{\partial y} + \frac{\partial \sigma_{xz}}{\partial z}\right)\Delta x \Delta y \Delta z\end{aligned} \tag{5.78}$$

Let ρ_m be the mass density of the homogeneous isotropic material, then according to Newton's second law

$$(\rho_m \Delta x \Delta y \Delta z)\frac{d^2 u_1}{dt^2} = \Delta F \tag{5.79}$$

where u_1 is the displacement in the x-direction. From Eqs. (5.78), (5.79) one can write

$$\rho_m \frac{d^2 u_1}{dt^2} = \frac{\partial \sigma_{xx}}{\partial x} + \frac{\partial \sigma_{xy}}{\partial y} + \frac{\partial \sigma_{xz}}{\partial z} \tag{5.80}$$

The solution of Eq. (5.80) requires knowledge of the elements of the stress tensor of a cubic solid. Differentiating the expressions for σ_{xx}, σ_{xy}, and σ_{xz} given by Eq. (5.62), and using Eqs. (5.9), (5.10) for the elements of the strain tensor, one gets

$$\frac{\partial \sigma_{xx}}{\partial x} = C_{11}\frac{\partial \varepsilon_{xx}}{\partial x} + C_{12}\left(\frac{\partial \varepsilon_{yy}}{\partial x} + \frac{\partial \varepsilon_{zz}}{\partial x}\right) = C_{11}\frac{\partial^2 u_1}{\partial x^2} + C_{12}\left[\frac{\partial^2 u_2}{\partial x \partial y} + \frac{\partial^2 u_3}{\partial x \partial z}\right] \tag{5.81}$$

$$\frac{\partial \sigma_{xy}}{\partial y} = C_{44}\frac{\partial \varepsilon_{xy}}{\partial y} = \frac{1}{2}C_{44}\left[\frac{\partial^2 u_1}{\partial y^2} + \frac{\partial^2 u_2}{\partial x \partial y}\right] \tag{5.82}$$

$$\frac{\partial \sigma_{xz}}{\partial z} = C_{44}\frac{\partial \varepsilon_{xz}}{\partial z} = \frac{1}{2}C_{44}\left[\frac{\partial^2 u_1}{\partial z^2} + \frac{\partial^2 u_3}{\partial x \partial z}\right] \tag{5.83}$$

Substituting Eqs. (5.81)–(5.83) into Eq. (5.80), one can write

$$\rho_m \frac{\partial^2 u_1}{\partial t^2} = C_{11}\frac{\partial^2 u_1}{\partial x^2} + \frac{1}{2}C_{44}\left[\frac{\partial^2 u_1}{\partial y^2} + \frac{\partial^2 u_1}{\partial z^2}\right] + \left(C_{12} + \frac{1}{2}C_{44}\right)\left[\frac{\partial^2 u_2}{\partial x \partial y} + \frac{\partial^2 u_3}{\partial z \partial x}\right] \tag{5.84}$$

Eq. (5.84) gives the equation of motion of the elastic wave produced in a cubic solid when stress is applied externally in the x-direction. In exactly the same manner, one can apply the nonhomogeneous stress in the y- and z-directions and can obtain the corresponding equations of motion for u_2 and u_3, which are given by

$$\rho \frac{\partial^2 u_2}{\partial t^2} = C_{11}\frac{\partial^2 u_2}{\partial y^2} + \frac{1}{2}C_{44}\left[\frac{\partial^2 u_2}{\partial x^2} + \frac{\partial^2 u_2}{\partial z^2}\right] + \left(C_{12} + \frac{1}{2}C_{44}\right)\left[\frac{\partial^2 u_1}{\partial x \partial y} + \frac{\partial^2 u_3}{\partial y \partial z}\right] \tag{5.85}$$

$$\rho_m \frac{\partial^2 u_3}{\partial t^2} = C_{11}\frac{\partial^2 u_3}{\partial z^2} + \frac{1}{2}C_{44}\left[\frac{\partial^2 u_3}{\partial x^2} + \frac{\partial^2 u_3}{\partial y^2}\right] + \left(C_{12} + \frac{1}{2}C_{44}\right)\left[\frac{\partial^2 u_1}{\partial x \partial z} + \frac{\partial^2 u_2}{\partial y \partial z}\right] \tag{5.86}$$

Eqs. (5.84)–(5.86) can be written in a general form as

$$\rho_m \frac{\partial^2 u_\alpha}{\partial t^2} = C_{11}\frac{\partial^2 u_\alpha}{\partial r_\alpha^2} + \frac{1}{2}C_{44}\sum_{\beta(\beta\neq\alpha)}\frac{\partial^2 u_\alpha}{\partial r_\beta^2} + \left(C_{12} + \frac{1}{2}C_{44}\right)\sum_{\beta(\beta\neq\alpha)}\frac{\partial^2 u_\beta}{\partial r_\alpha \partial r_\beta} \tag{5.87}$$

To examine the actual elastic waves produced in a cubic solid, let us apply Eq. (5.87) in different directions.

5.9.1 Elastic Waves in the [100] Direction

The elastic waves along the [100] direction may be longitudinal or transverse in nature. The displacement of longitudinal (L) elastic waves in the x-direction u_{1L} is given by

$$u_{1L} = u_{0L}\exp[-\imath(Kr_1 - \omega_L t)] = u_{0L}\exp[-\imath(Kx - \omega_L t)] \tag{5.88}$$

Here u_{0L} and K are the amplitude and propagation wave vector along the x-direction and ω_L is the frequency of the L elastic wave. Substituting Eq. (5.88) into Eq. (5.84), the frequency of the elastic wave is given by

$$\omega_L = \sqrt{\frac{C_{11}}{\rho_m}}K \tag{5.89}$$

Eq. (5.89) is called the *dispersion relation* because it relates the frequency to the wave vector. It immediately gives the linear velocity as

$$v_L = \frac{\omega_L}{K} = \sqrt{\frac{C_{11}}{\rho_m}} \tag{5.90}$$

Here the group velocity and the phase velocity are the same.

There are two transverse (T) elastic waves with displacements along the y- and z-directions but with K along the x-direction and they are defined as

$$u_{2T_1} = u_{02}\exp\left[-\imath\left(Kr_1 - \omega_{T_1}t\right)\right] \tag{5.91}$$

$$u_{3T_2} = u_{03}\exp\left[-\imath\left(Kr_1 - \omega_{T_2}t\right)\right] \tag{5.92}$$

where ω_{T_1} and ω_{T_2} are the frequencies of the T elastic waves. Substituting Eq. (5.91) into Eq. (5.85) and Eq. (5.92) into Eq. (5.86), one can solve for the velocities of the waves v_{T_1} and v_{T_2} to get

$$v_{T_1} = \frac{\omega_{T_1}}{K} = \sqrt{\frac{C_{44}}{2\rho_m}} \tag{5.93a}$$

$$v_{T_2} = \frac{\omega_{T_2}}{K} = \sqrt{\frac{C_{44}}{2\rho_m}} \tag{5.93b}$$

Eqs. (5.93a), (5.93b) show that the velocities of the two T elastic waves are the same, but differ from that of the L elastic wave. The experimental measurement of the velocities of the L and T elastic waves in a cubic crystal allows us to find the

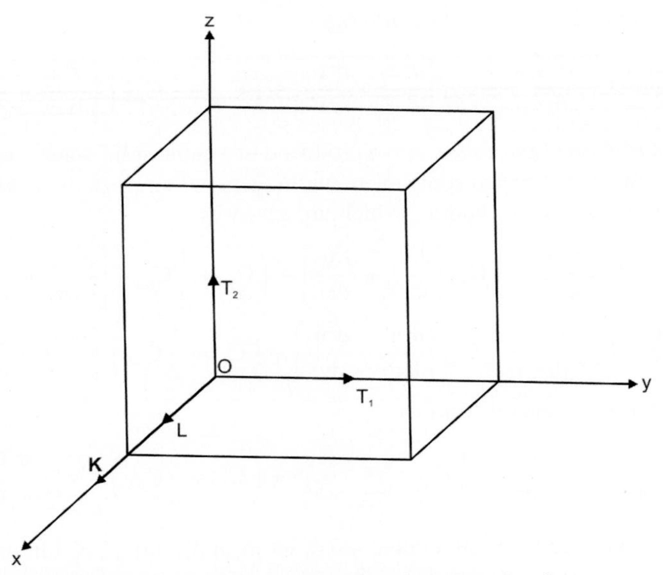

FIG. 5.7 Longitudinal and transverse polarizations of the elastic wave traveling along the x-direction.

elastic constants C_{11} and C_{44} using Eqs. (5.90), (5.93a), (5.93b). Fig. 5.7 shows the L and T polarizations of the elastic wave with propagation wave vector K along the x-direction.

5.9.2 Elastic Waves in the [110] Direction

The propagation of elastic waves in the [100] direction gives information only about the elastic constants C_{11} and C_{44}, but the elastic constant C_{12} remains undetermined. Let us consider the propagation of elastic waves along the [110] symmetry direction as shown in Fig. 5.8. The propagation wavevector **K** in the [110] direction is given by

$$\mathbf{K} = \hat{\mathbf{i}}_1 K_1 + \hat{\mathbf{i}}_2 K_2 = \frac{K}{\sqrt{2}}\left(\hat{\mathbf{i}}_1 + \hat{\mathbf{i}}_2\right) \tag{5.94}$$

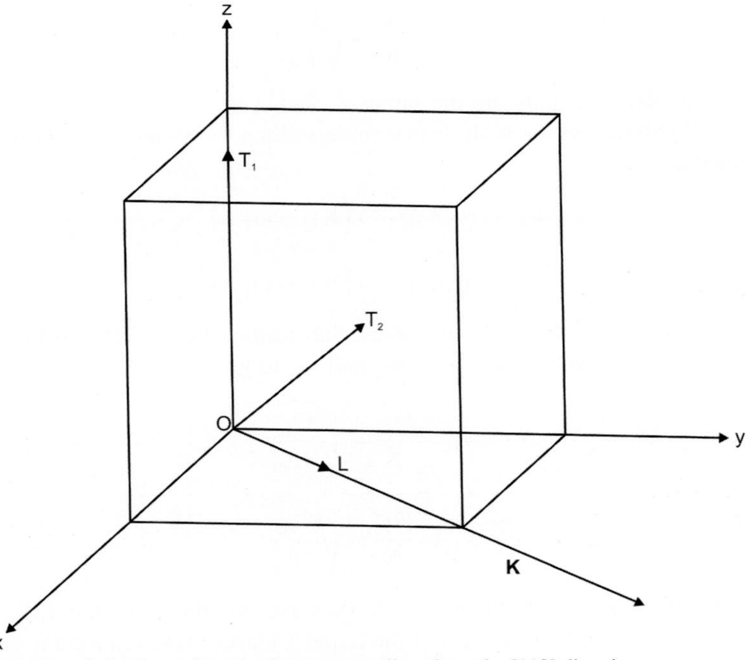

FIG. 5.8 Longitudinal and transverse polarizations of an elastic wave traveling along the [110] direction.

The components $K_1 = K_x$ and $K_2 = K_y$ have the same magnitude. Fig. 5.8 shows the different polarizations (modes) of the elastic wave. It is evident from the figure that one of the transverse elastic waves T_1 propagates in the xy-plane with particle displacement along the z-direction defined as

$$u_{3T_1} = u_{03} \exp\left[-\imath\left\{\frac{K}{\sqrt{2}}(x+y) - \omega_{T_1} t\right\}\right] \tag{5.95}$$

Substituting Eq. (5.95) into Eq. (5.86), one gets

$$\omega_{T_1} = \sqrt{\frac{C_{44}}{2\rho_m}} K \tag{5.96}$$

Therefore, the velocity of the T_1 elastic wave becomes

$$v_{T_1} = \sqrt{\frac{C_{44}}{2\rho_m}} \tag{5.97}$$

The second transverse elastic wave T_2 and the L elastic wave propagate in the xy-plane with particle displacement also in the xy-plane and they are defined as

$$u_1 = u_{01} \exp\left[-\imath\left\{\frac{K}{\sqrt{2}}(x+y) - \omega t\right\}\right] \tag{5.98}$$

$$u_2 = u_{02} \exp\left[-\imath\left\{\frac{K}{\sqrt{2}}(x+y) - \omega t\right\}\right] \tag{5.99}$$

Substituting u_1 and u_2 into Eqs. (5.84), (5.85), one gets

$$\omega^2 \rho_m u_1 = \left(C_{11} + \frac{1}{2}C_{44}\right)\frac{K^2}{2} u_1 + \left(C_{12} + \frac{1}{2}C_{44}\right)\frac{K^2}{2} u_2 \tag{5.100}$$

$$\omega^2 \rho_m u_2 = \left(C_{11} + \frac{1}{2}C_{44}\right)\frac{K^2}{2} u_2 + \left(C_{12} + \frac{1}{2}C_{44}\right)\frac{K^2}{2} u_1 \tag{5.101}$$

Eqs. (5.100), (5.101) have nontrivial solutions only if the determinant of the coefficients of u_1 and u_2 is zero, that is,

$$\begin{vmatrix} -\omega^2\rho_m + \left(C_{11} + \frac{1}{2}C_{44}\right)\frac{K^2}{2} & \left(C_{12} + \frac{1}{2}C_{44}\right)\frac{K^2}{2} \\ \left(C_{12} + \frac{1}{2}C_{44}\right)\frac{K^2}{2} & -\omega^2\rho_m + \left(C_{11} + \frac{1}{2}C_{44}\right)\frac{K^2}{2} \end{vmatrix} = 0 \tag{5.102}$$

The above determinant gives a quadratic equation in ω^2 with the following solutions

$$\omega_1^2 \rho_m = \frac{1}{2}(C_{11} + C_{12} + C_{44})K^2 \tag{5.103}$$

$$\omega_2^2 \rho_m = \frac{1}{2}(C_{11} - C_{12})K^2 \tag{5.104}$$

So, the velocities of these elastic waves are given by

$$v_1 = \sqrt{\frac{C_{11} + C_{12} + C_{44}}{2\rho_m}} \tag{5.105}$$

$$v_2 = \sqrt{\frac{C_{11} - C_{12}}{2\rho_m}} \tag{5.106}$$

Let us examine the nature of the two waves described by Eqs. (5.103), (5.104). Substituting Eq. (5.103) into Eq. (5.100) for the frequency $\omega = \omega_1$, one can prove that

$$u_1 = u_2 \tag{5.107}$$

Thus, corresponding to the frequency ω_1, the displacement of the particle is in the [110] direction, which gives the L elastic wave. Similarly, substituting Eq. (5.104) into Eq. (5.100) for the frequency $\omega = \omega_2$, one can get

$$u_1 = -u_2 \tag{5.108}$$

According to Eq. (5.108) the displacement of the particle is along the $[1\bar{1}0]$ direction, which is perpendicular to the direction of propagation [1 1 0]. Hence Eq. (5.104) gives the frequency of the T_2 elastic wave. Now, the velocities of the L and T_2 elastic waves are given by

$$v_L = \sqrt{\frac{C_{11} + C_{12} + C_{44}}{2\rho_m}} \tag{5.109}$$

$$v_{T_2} = \sqrt{\frac{C_{11} - C_{12}}{2\rho_m}} \tag{5.110}$$

Hence, from the experimental measurements of the velocities of the L and T elastic waves, given by Eqs. (5.97), (5.109), and (5.110), one can determine all of the elastic constants C_{11}, C_{12} and C_{44} of a cubic crystal.

5.9.3 Elastic Waves in the [111] Direction

Another high-symmetry direction of interest in cubic crystals is the [111] direction. Consider an elastic wave propagating in the [111] direction with wave vector **K** (Fig. 5.9) defined as

$$\begin{aligned}\mathbf{K} &= \hat{\mathbf{i}}_1 K_1 + \hat{\mathbf{i}}_2 K_2 + \hat{\mathbf{i}}_3 K_3 = \hat{\mathbf{i}}_1 K_x + \hat{\mathbf{i}}_2 K_y + \hat{\mathbf{i}}_3 K_z \\ &= \frac{K}{\sqrt{3}}\left(\hat{\mathbf{i}}_1 + \hat{\mathbf{i}}_2 + \hat{\mathbf{i}}_3\right)\end{aligned} \tag{5.111}$$

Here we have used the fact that, in the [111] direction, the magnitude of all the Cartesian components of **K** is equal. Therefore, the Cartesian components of displacement are given by

$$u_1(\mathbf{r}) = u_{01} \exp\left[-\imath\left\{\frac{K}{\sqrt{3}}(x + y + z) - \omega t\right\}\right] \tag{5.112}$$

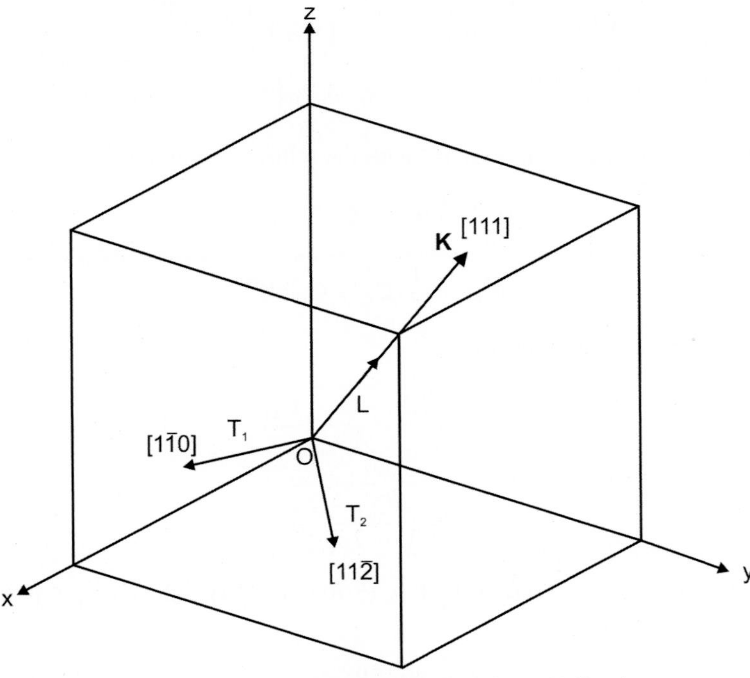

FIG. 5.9 Longitudinal and transverse polarizations of an elastic wave traveling along the [111] direction.

$$u_2(\mathbf{r}) = u_{02} \exp\left[-\imath\left\{\frac{K}{\sqrt{3}}(x+y+z) - \omega t\right\}\right] \tag{5.113}$$

$$u_3(\mathbf{r}) = u_{03} \exp\left[-\imath\left\{\frac{K}{\sqrt{3}}(x+y+z) - \omega t\right\}\right] \tag{5.114}$$

The total displacement $\mathbf{u}(\mathbf{r})$ becomes

$$\mathbf{u}(\mathbf{r}) = \left(u_{01}\hat{\mathbf{i}}_1 + u_{02}\hat{\mathbf{i}}_2 + u_{03}\hat{\mathbf{i}}_3\right) \exp\left[-\imath\left\{\frac{K}{\sqrt{3}}(x+y+z) - \omega t\right\}\right] \tag{5.115}$$

One is interested in finding the velocities of the different modes of the elastic wave propagating in the [111] direction. Substituting Eqs. (5.112)–(5.114) into Eq. (5.84) and simplifying, we obtain

$$(C_{11} + C_{44} - 3\Lambda)u_1 + \left(C_{12} + \frac{1}{2}C_{44}\right)u_2 + \left(C_{12} + \frac{1}{2}C_{44}\right)u_3 = 0 \tag{5.116}$$

where

$$\Lambda = \rho_m \frac{\omega^2}{K^2} \tag{5.117}$$

Similarly, substituting Eqs. (5.112)–(5.114) into Eqs. (5.85), (5.86), one can write

$$\left(C_{12} + \frac{1}{2}C_{44}\right)u_1 + (C_{11} + C_{44} - 3\Lambda)u_2 + \left(C_{12} + \frac{1}{2}C_{44}\right)u_3 = 0 \tag{5.118}$$

$$\left(C_{12} + \frac{1}{2}C_{44}\right)u_1 + \left(C_{12} + \frac{1}{2}C_{44}\right)u_2 + (C_{11} + C_{44} - 3\Lambda)u_3 = 0 \tag{5.119}$$

Eqs. (5.116), (5.118), and (5.119) have nontrivial solutions only if the determinant of the coefficient of u_1, u_2, and u_3 is zero, that is,

$$\begin{vmatrix} A & B & B \\ B & A & B \\ B & B & A \end{vmatrix} = 0 \tag{5.120}$$

where

$$A = C_{11} + C_{44} - 3\Lambda \tag{5.121}$$

and

$$B = C_{12} + \frac{1}{2}C_{44} \tag{5.122}$$

Expanding the determinant of Eq. (5.120), we get

$$(A - B)^2(A + 2B) = 0 \tag{5.123}$$

The above equation gives three solutions (out of which one solution is doubly degenerate) given by

$$A = B \tag{5.124}$$

$$A = -2B \tag{5.125}$$

Eqs. (5.124), (5.125), with the help of Eqs. (5.121), (5.122), yield

$$\omega_1 = \sqrt{\frac{C_{11} - C_{12} + \frac{1}{2}C_{44}}{3\rho_m}}\,K \tag{5.126}$$

$$\omega_2 = \sqrt{\frac{C_{11} + 2C_{12} + 2C_{44}}{3\rho_m}}\,K \tag{5.127}$$

Let us determine the nature of the polarization of the elastic waves with frequencies ω_1 and ω_2. Eq. (5.125) can be written as

$$3\Lambda_2 = C_{11} + 2C_{12} + 2C_{44} \tag{5.128}$$

Substituting Eq. (5.128) into Eq. (5.116), one gets

$$2u_1 = u_2 + u_3 \tag{5.129}$$

Similarly, substituting Eq. (5.128) into Eqs. (5.118), (5.119), we obtain

$$2u_2 = u_3 + u_1 \tag{5.130}$$

$$2u_3 = u_1 + u_2 \tag{5.131}$$

From Eqs. (5.129)–(5.131) one can immediately write

$$u_1 = u_2 = u_3 \tag{5.132}$$

Hence, for the eigenvalue Λ_2, the three components of displacement are equal, which yields the displacement **u** along the [111] direction. Therefore, the elastic wave corresponding to the eigenvalue Λ_2 or ω_2 represents the L elastic wave. If ω_L represents the frequency of the L wave, then from Eq. (5.127) its velocity is given by

$$v_L = \frac{\omega_L}{K} = \sqrt{\frac{C_{11} + 2C_{12} + 2C_{44}}{3\rho_m}} \tag{5.133}$$

Let us now consider the wave with frequency ω_1, which from Eqs. (5.124), (5.121), and (5.122) gives

$$3\Lambda_1 = C_{11} - C_{12} + \frac{1}{2}C_{44} \tag{5.134}$$

Substituting Eq. (5.134) into Eq. (5.116), one immediately gets

$$u_1 + u_2 + u_3 = 0 \tag{5.135}$$

If we substitute Eq. (5.134) into Eqs. (5.118), (5.119) we obtain the same expression as given by Eq. (5.135). Therefore, Eq. (5.134) represents two degenerate T modes of the elastic wave propagating in the [111] direction and these must be orthogonal to each other and also to the L elastic wave. Let us take $[1\bar{1}0]$ as the direction of displacement of one of the transverse waves (say T_1). Then the third displacement vector $\mathbf{X}_3$, representing the second transverse wave (say T_2), must be perpendicular to both $X_1 = [111]$ and $\mathbf{X}_2 = [1\bar{1}0]$, that is,

$$\begin{aligned}\mathbf{X}_3 = \mathbf{X}_1 \times \mathbf{X}_2 &= \left(\hat{\mathbf{i}}_1 + \hat{\mathbf{i}}_2 + \hat{\mathbf{i}}_3\right) \times \left(\hat{\mathbf{i}}_1 - \hat{\mathbf{i}}_2 + 0\hat{\mathbf{i}}_3\right) \\ &= \hat{\mathbf{i}}_1 + \hat{\mathbf{i}}_2 - 2\hat{\mathbf{i}}_3\end{aligned} \tag{5.136}$$

Hence the displacement vector $\mathbf{X}_3$ of the T_2 elastic wave is in the $[11\bar{2}]$ direction. The polarizations of three elastic waves are shown in Fig. 5.9. From Eq. (5.126) the velocities of both the T elastic waves are the same and are given by

$$v_{T_1} = v_{T_2} = v_T = \frac{\omega_T}{K} = \sqrt{\frac{C_{11} - C_{12} + \frac{1}{2}C_{44}}{3\rho_m}} \tag{5.137}$$

An alternate method for determining the eigenvectors of the elastic wave is given in Appendix C.

The above discussion yields only two velocity equations, that is, Eqs. (5.133), (5.137), but there are three elastic constants. Therefore, one cannot determine all the elastic constants of a cubic solid in this case. It is worthwhile to note that, in general, the elastic waves in an isotropic medium are mixtures of both the L and T polarizations, depending on the direction of propagation of the wave: only in high-symmetry directions do these possess pure L or T polarization. Further, the two T elastic waves may not, in general, have the same velocity.

5.10 ISOTROPIC ELASTICITY

The substitution

$$C_{11} - C_{12} = C_{44} \tag{5.138}$$

into Eqs. (5.90), (5.109), and (5.133) gives the same velocity of the L elastic waves in the different symmetry directions [100], [110], and [111]. Further, the velocities of the T elastic waves, given by Eqs. (5.93a), (5.93b), (5.97), (5.110), and (5.137) along the different directions [100], [110], and [111], also become the same, although different from that of the L elastic waves. In other words, regardless of the direction of propagation, the velocity of an elastic wave with a particular polarization becomes the same, subject to the condition (5.138). Therefore, this is known as the *elastic isotropy condition.* The anisotropy factor A_{as} in cubic crystals is defined as the square of the ratio of the velocities of the T elastic waves propagating in the [100] and [110] directions, that is,

$$A_{as} = \frac{C_{44}}{C_{11} - C_{12}} \tag{5.139}$$

A_{as} is also equal to the ratio of the squares of the velocities of the two T elastic waves propagating in the [110] direction. A_{as} is unity for elastic isotropy. The departure of the values of A_{as} from unity is a measure of the anisotropy in cubic crystals (see Table 5.1).

5.11 EXPERIMENTAL MEASUREMENT OF ELASTIC CONSTANTS

Ultrasonic waves are elastic waves whose velocity in solids can be measured experimentally. The elastic constants can be evaluated from the experimentally measured velocities of ultrasonic waves with different polarizations propagating in different symmetry directions in cubic solids. One of the most commonly used methods to measure the velocities is the ultrasonic pulse method, a schematic setup of which is shown in Fig. 5.10. In this method, ultrasonic pulses at regular intervals are produced by a quartz crystal fixed at one end of the specimen crystal and these are allowed to travel through it, as shown. The pulses are reflected back at the opposite end of the specimen, which ultimately reach the quartz crystal again. The time t taken by an ultrasonic pulse to travel the forward and backward journey in the crystal is measured experimentally. If d is the length of the specimen, then the velocity of the ultrasonic waves is given by

$$v = \frac{2d}{t} \tag{5.140}$$

Actually, the ultrasonic waves are allowed to travel along one of the symmetry directions, say [110], and the velocity of the waves with different polarizations is measured. Then the elastic constants of the cubic solid are evaluated using Eqs. (5.97), (5.109), and (5.110).

In the above discussion we have defined the second-order elastic constants assuming Hooke's law to be valid. In this approximation, the elastic energy density is a quadratic function of the strain. But for large stresses and strains, the Hooke's law is not valid and one has to consider higher-order terms in strain and stress-strain relations (Eqs. 5.26a, 5.26b) and the energy density expression (Eq. 5.32). The elastic energy density involving cubic terms of the strain elements should be considered and these are manifestations of nonlinear effects, such as the interaction of phonons and thermal expansion. Therefore, one can define the third-order elastic constants from the energy density involving cubic terms of the strain elements.

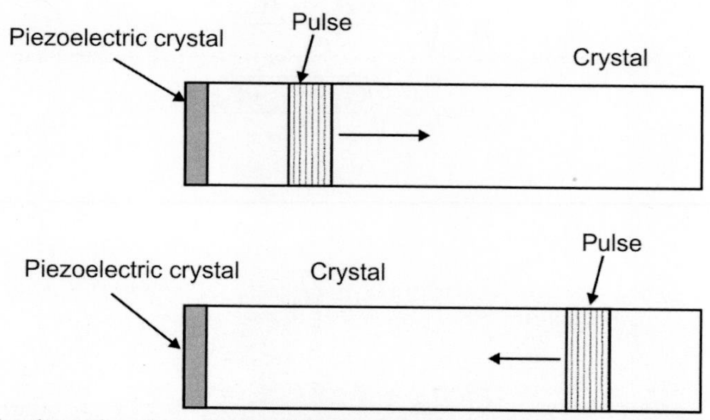

FIG. 5.10 Schematic diagram for the ultrasonic pulse method used for determining velocities of ultrasonic waves in solids.

Problem 5.1

If the factor of one-half is neglected in Eq. (5.9) and the strain components are defined as

$$\varepsilon_{\alpha\beta} = \frac{\partial u_\beta}{\partial r_\alpha} + \frac{\partial u_\alpha}{\partial r_\beta}, \varepsilon_{\alpha\alpha} = \frac{\partial u_\alpha}{\partial r_\alpha} \tag{5.141}$$

prove that the equation of motion for any component u_α of the displacement field is given by

$$\rho_m \frac{\partial^2 u_\alpha}{\partial t^2} = C_{11} \frac{\partial^2 u_\alpha}{\partial r_\alpha^2} + C_{44} \sum_{\beta(\beta \neq \alpha)} \frac{\partial^2 u_\alpha}{\partial r_\beta^2} + (C_{12} + C_{44}) \sum_{\beta(\beta \neq \alpha)} \frac{\partial^2 u_\beta}{\partial r_\alpha \partial r_\beta} \tag{5.142}$$

In some books, the equation of motion given by Eq. (5.142) is used.

Problem 5.2

Assume the running wave-like solution for the displacement $\mathbf{u}(\mathbf{r})$ defined as

$$\mathbf{u}(\mathbf{r}) = \mathbf{u}_0 e^{-\imath(\mathbf{K}\cdot\mathbf{r} - \omega t)} \tag{5.143}$$

where $\mathbf{K}$ is the propagation wave vector for the elastic wave.

(a) If the elastic wave is traveling in the [100] direction, then from Eq. (5.142) prove that the velocities for the longitudinal and transverse waves are given by

$$v_L = \sqrt{\frac{C_{11}}{\rho_m}}, v_{T_1} = v_{T_2} = \sqrt{\frac{C_{44}}{\rho_m}} \tag{5.144}$$

(b) If the elastic wave is traveling in the [110] direction, then from Eq. (5.142) prove that the velocities for the longitudinal and transverse waves are given by

$$v_L = \sqrt{\frac{C_{11} + C_{12} + 2C_{44}}{2\rho_m}} \tag{5.145}$$

$$v_{T_1} = \sqrt{\frac{C_{44}}{\rho_m}}, v_{T_2} = \sqrt{\frac{C_{11} - C_{12}}{2\rho_m}} \tag{5.146}$$

(c) If the elastic wave is traveling in the [111] direction, then from Eq. (5.142) prove that the velocities for the longitudinal and transverse waves are given by

$$v_L = \sqrt{\frac{C_{11} + 2C_{12} + 4C_{44}}{3\rho_m}} \tag{5.147}$$

$$v_{T_1} = v_{T_2} = \sqrt{\frac{C_{11} - C_{12} + C_{44}}{3\rho_m}} \tag{5.148}$$

Problem 5.3

Using the Newton second law of force in equation

$$f_\alpha = \sum_\beta \frac{\partial \sigma_{\alpha\beta}}{\partial r_\beta} \tag{5.149}$$

prove that for a wave-like solution, the equation of motion is given by

$$\sum_{\mu,\nu,\beta} \left[C_{\alpha\beta\mu\nu} K_\beta K_\nu - \omega^2 \rho_m \delta_{\alpha\mu} \right] u_0^\mu = 0 \tag{5.150}$$

Here $\mathbf{u}_0$ is the amplitude of the elastic wave and the other symbols have their usual meanings. Eq. (5.150) is called the *Christoffel equation.*

Problem 5.4

The free energy of a deformed body is defined as

$$F = F_0 + \frac{1}{2}\lambda_L \left[\sum_\alpha \varepsilon_{\alpha\alpha} \right]^2 + \mu_L \varepsilon_{\alpha\beta}^2 \tag{5.151}$$

where F_0 is a constant quantity and λ_L and μ_L are called the Lame's coefficients. Express the free energy as a sum of the pure shear strain and pure hydrostatic compression. Further, find the stress components from the free energy.

SUGGESTED READING

Eshelby, J. D. (1956). The continuum theory of lattice defects. F. Seitz, & D. Turnbull (Eds.), *Solid state physics* (pp. 79–144). Vol. 3. New York: Academic Press.

Huntington, H. B. (1958). The elastic constants of crystals. F. Seitz, & D. Turnbull (Eds.), *Solid state physics* (pp. 213–351). Vol. 7. New York: Academic Press.

Landau, L. D., & Lifshitz, E. M. (1970). *Theory of elasticity*. London: Pergamon Press.

Timoshenko, S., & Doodier, J. N. (1970). *Theory of elasticity* (3rd ed.). New York: McGraw-Hill Book Co.

Chapter 6

Lattice Vibrations-1

Chapter Outline

The study of the thermal properties of solids is an important field in the subject of solid-state physics. There are a number of thermal properties, such as lattice vibrations, specific heat, thermal conductivity, and thermal expansion. The temperature variation of lattice vibrations, magnetism, and superconductivity form another class of thermal properties. At absolute zero, all of the atoms in insulators and dielectrics are at rest at the lattice positions. With an increase in temperature, the atoms acquire thermal energy given by

$$E_{TH} = k_B T \tag{6.1}$$

where k_B is the Boltzmann constant and T is the temperature in degrees Kelvin. By gaining finite thermal energy, each atom starts vibrating about its equilibrium position with finite frequency. The amplitude and frequency of atomic vibrations increase with an increase in temperature. As the atoms in a solid are bound together, the vibrations of one atom are handed over to the next atom and so on. Therefore, all of the atoms vibrate collectively in the form of an elastic wave. Such a collective motion is called the normal mode of vibration of the lattice. The total number of normal modes of vibration is equal to the number of degrees of freedom, which is 3N if there are N atoms in a solid. In a metal, there are ions at the lattice positions and the conduction electrons are free to move in it. Therefore, in a metal at finite temperature, the ions vibrate about the lattice positions as in insulators, while the conduction electrons move with some finite velocity. In this chapter and Chapter 7, we study the lattice vibrations at finite temperatures, which play a central role in the study of a number of lattice properties, such as lattice specific heat, lattice conduction, and thermal expansion. In this chapter, the lattice vibrations in one-dimensional solids will be studied using the classical approach.

6.1 VIBRATIONS IN A HOMOGENEOUS ELASTIC MEDIUM

For simplicity, consider a one-dimensional homogeneous elastic solid along the x-direction, having great length but a small uniform area of cross section (Fig. 6.1). Let ρ_m be the mass per unit length (linear mass density) of the solid, then

$$\rho_m = \frac{dM}{dx} \tag{6.2}$$

where M denotes the mass. When opposing external forces F_1 and F_2 are applied at the points x and $x+\Delta x$ along the x-direction, the net force acting on the small element Δx is

$$\Delta F = F_2 - F_1 \tag{6.3}$$

The force ΔF produces strain in the elemental length Δx. Let $\varepsilon_{xx}(x)$ and $\varepsilon_{xx}(x+\Delta x)$ represent the x-component of strain produced at the points x and $x+\Delta x$. The strain component $\varepsilon_{xx}(x)$ is defined as

$$\varepsilon_{xx}(x) = \frac{\partial u}{\partial x} \tag{6.4}$$

Solid State Physics. https://doi.org/10.1016/B978-0-12-817103-5.00006-2

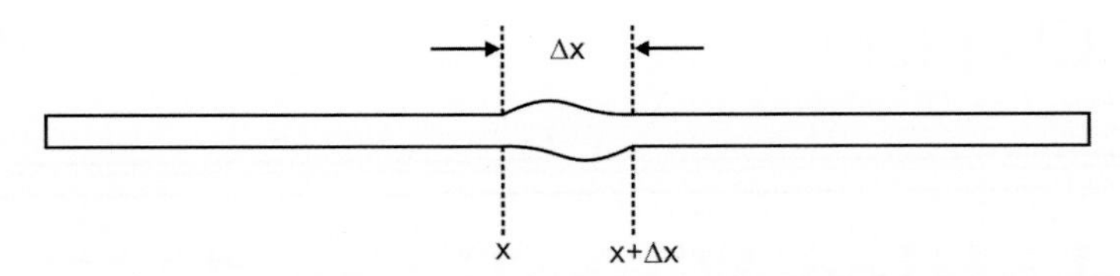

FIG. 6.1 One-dimensional homogeneous line with strain produced between x and x + Δx.

where u(x) is the displacement at point x due to the strain. In a one-dimensional solid, the stress cannot be defined in the same way as in a three-dimensional solid. But one can talk about the force acting at a point, which is assumed to be linearly proportional to the strain at that point. Therefore, one can write

$$F_1 = C_{11}\,\varepsilon_{xx}(x) \tag{6.5}$$

$$F_2 = C_{11}\,\varepsilon_{xx}(x + \Delta x) \tag{6.6}$$

where C_{11} is the proportionality constant and is the modulus of elasticity. From Eqs. (6.5), (6.6) it is easy to write

$$\Delta F = C_{11}\left[\varepsilon_{xx}(x + \Delta x) - \varepsilon_{xx}(x)\right] \tag{6.7}$$

If the strain $\varepsilon_{xx}(x + \Delta x)$ is varying slowly, it can be expanded as

$$\varepsilon_{xx}(x + \Delta x) = \varepsilon_{xx}(x) + \frac{\partial \varepsilon_{xx}}{\partial x}\Delta x \tag{6.8}$$

Substituting Eq. (6.4) into Eq. (6.8), the net strain acting on the element Δx is given by

$$\varepsilon_{xx}(x + \Delta x) - \varepsilon_{xx}(x) = \frac{\partial^2 u}{\partial x^2}\Delta x \tag{6.9}$$

with the help of which Eq. (6.7) becomes

$$\Delta F = C_{11}\frac{\partial^2 u}{\partial x^2}\Delta x \tag{6.10}$$

From Newton's second law of motion, ΔF can immediately be written as

$$\Delta F = \rho_m \Delta x \frac{\partial^2 u}{\partial t^2} \tag{6.11}$$

From Eqs. (6.10), (6.11) one gets

$$\frac{\partial^2 u}{\partial x^2} = \frac{1}{v_x^2}\frac{\partial^2 u}{\partial t^2} \tag{6.12}$$

where

$$v_x = \sqrt{\frac{C_{11}}{\rho_m}} \tag{6.13}$$

Here v_x gives the velocity of the elastic wave. Eq. (6.12) is the well-known Newton's formula for the velocity of sound waves. For Eq. (6.12) we are seeking a wave solution of the form

$$u(x) = u_0 e^{\imath(Kx - \omega t)} \tag{6.14}$$

where u_0 is the amplitude, K is the propagation wave vector, and ω is the frequency of the wave. Substituting Eq. (6.14) into Eq. (6.12), one obtains

$$\omega = v_x K \tag{6.15}$$

This is called the dispersion relation as it relates the frequency to the wave vector K. Eq. (6.15) shows that ω is linearly proportional to K, which implies that velocity v_x is independent of the wavelength. Further, there is no upper limit to the frequency of vibration in a homogeneous medium. For an infinitely long one-dimensional solid, the values of K vary continuously and so does the frequency. Fig. 6.2 shows the dispersion relation for an infinite homogeneous line. If the

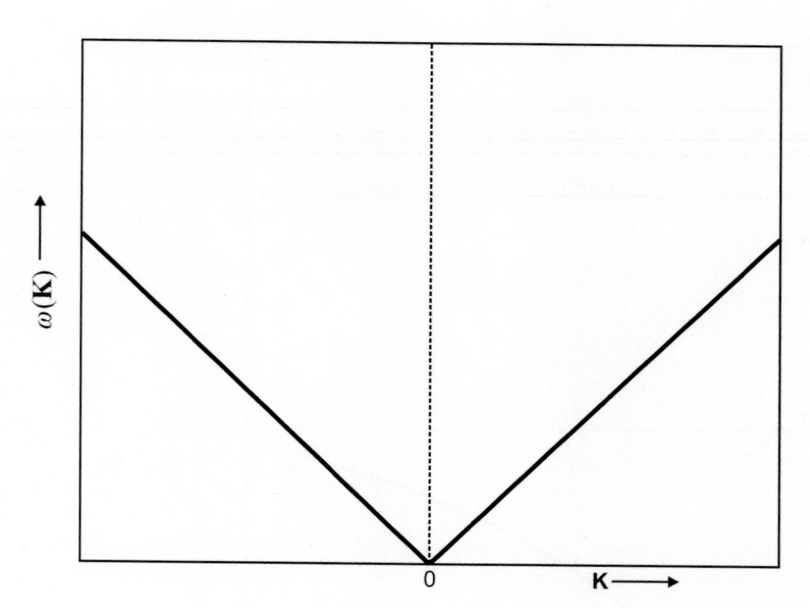

FIG. 6.2 Dispersion curve for a homogeneous line. The slope of the curve gives v_p or v_g.

one-dimensional solid has a finite length L, then the solution must satisfy the boundary conditions: the displacement must be zero at the fixed boundaries, that is,

$$u(0) = u(L) = 0 \tag{6.16}$$

Substituting Eq. (6.14) into Eq. (6.16), one obtains discrete values of K given by

$$K = \frac{2\pi n}{L} \tag{6.17}$$

where n is an integer: negative, positive, or zero. The different values of K give different modes of vibration. The dispersion relation gives the phase velocity v_p as

$$v_p = \frac{\omega}{K} = v_x \tag{6.18}$$

while the group velocity v_g is given as

$$v_g = \frac{d\omega}{dK} = v_x \tag{6.19}$$

Hence the phase and group velocities of the elastic wave in a homogeneous medium are the same.

6.2 INTERATOMIC POTENTIAL IN SOLIDS

The interaction potential V(**r**) between two atoms, denoted by the numerals 1 and 2, of a crystalline solid is shown by the curve PQR in Fig. 6.3. At absolute zero the equilibrium distance between the atoms is R_0, which corresponds to the interaction potential

$$V(R_0) = -V_0 \tag{6.20}$$

In the equilibrium position, atom 2 occupies the position O. It is evident from Fig. 6.3 that the interaction potential energy curve PQR is asymmetrical about the point Q (or O) and can be evaluated in the following manner.

Let atom 2 be displaced by a distance **u**(t) at the time t from its equilibrium position $\mathbf{R}_0$. So, the displaced position of atom 2, keeping atom 1 fixed, becomes

$$\mathbf{r} = \mathbf{R}_0 + \mathbf{u}(t) \tag{6.21}$$

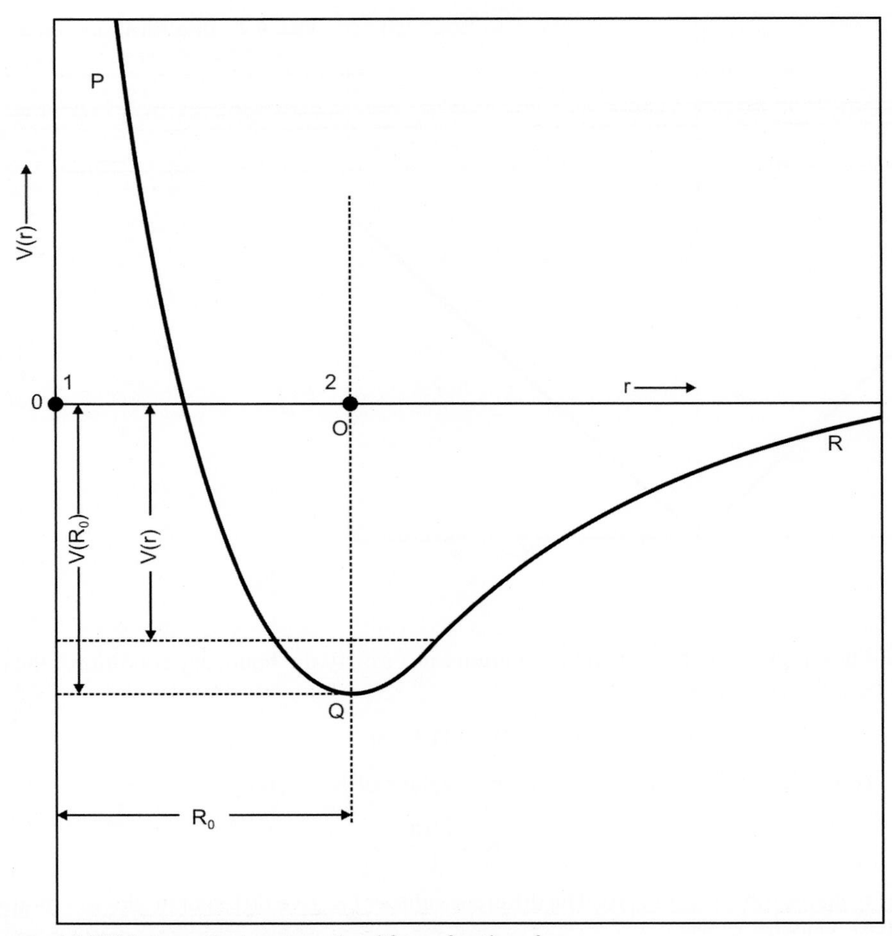

FIG. 6.3 Interatomic potential V(r) between the two atoms 1 and 2 as a function of r.

If the displacement is small as compared with the interatomic distance, then $V(\mathbf{r})$ can be expanded about the mean position $\mathbf{R}_0$ as

$$V(\mathbf{r}) = V(\mathbf{R}_0) + \sum_{\alpha} u_\alpha \left.\frac{\partial V}{\partial r_\alpha}\right|_{\mathbf{r}=\mathbf{R}_0} + \frac{1}{2}\sum_{\alpha,\beta} u_\alpha u_\beta \left.\frac{\partial^2 V}{\partial r_\alpha \partial r_\beta}\right|_{\mathbf{r}=\mathbf{R}_0} + \frac{1}{6}\sum_{\alpha,\beta,\gamma} u_\alpha u_\beta u_\gamma \left.\frac{\partial^3 V}{\partial r_\alpha \partial r_\beta \partial r_\gamma}\right|_{\mathbf{r}=\mathbf{R}_0} + \frac{1}{24}\sum_{\alpha,\beta,\gamma,\delta} u_\alpha u_\beta u_\gamma u_\delta \left.\frac{\partial^4 V}{\partial r_\alpha \partial r_\beta \partial r_\gamma \partial r_\delta}\right|_{\mathbf{r}=\mathbf{R}_0} + \cdots \tag{6.22}$$

If the solid is isotropic, the potential becomes independent of direction, at least in the high-symmetry directions (e.g., solid with cubic structure). In such a solid, $u_\alpha = u_\beta = u_\gamma = u_\delta = u$ and Eq. (6.22) reduces to

$$V(\mathbf{r}) = V(\mathbf{R}_0) + u\left.\frac{\partial V}{\partial r}\right|_{\mathbf{r}=\mathbf{R}_0} + \frac{1}{2}u^2\left.\frac{\partial^2 V}{\partial r^2}\right|_{\mathbf{r}=\mathbf{R}_0} + \frac{1}{6}u^3\left.\frac{\partial^3 V}{\partial r^3}\right|_{\mathbf{r}=\mathbf{R}_0} + \frac{1}{24}u^4\left.\frac{\partial^4 V}{\partial r^4}\right|_{\mathbf{r}=\mathbf{R}_0} + \cdots \tag{6.23}$$

In the equilibrium position the force acting on atom 2 vanishes, thereby reducing the term with the first derivative of $V(\mathbf{r})$ to zero. So, Eq. (6.23) becomes

$$V(\mathbf{r}) = V(\mathbf{R}_0) + \frac{1}{2}\alpha_F u^2 - \frac{1}{3}\gamma_F u^3 - \frac{1}{4}\delta_F u^4 + \cdots \tag{6.24}$$

where

$$\alpha_F = \left.\frac{\partial^2 V}{\partial r^2}\right|_{\mathbf{r}=\mathbf{R}_0} \tag{6.25}$$

$$\gamma_F = -\frac{1}{2}\left.\frac{\partial^3 V}{\partial r^3}\right|_{\mathbf{r}=\mathbf{R}_0}, \quad \delta_F = -\frac{1}{6}\left.\frac{\partial^4 V}{\partial r^4}\right|_{\mathbf{r}=\mathbf{R}_0} \tag{6.26}$$

Here α_F is called the force constant and is a measure of the rigidity of the bond between the two atoms. γ_F and δ_F are the derivatives of the force constants and represent higher-order force constants. For simplicity, α_F, γ_F, and δ_F can be taken as proportionality constants. Corresponding to $V(\mathbf{r})$, and given by Eq. (6.24), the force between atoms 1 and 2 is given by

$$F = -\frac{\partial V(\mathbf{r})}{\partial r} = -\alpha_F u + \gamma_F u^2 + \delta_F u^3 + \cdots \tag{6.27}$$

Eqs. (6.24), (6.27) form an infinite series and are exact expressions for the potential and force. The exact evaluation of $V(\mathbf{r})$ is very difficult but it can be estimated in various approximations.

6.2.1 Square-Well Potential

$V(\mathbf{r})$ is, sometimes, approximated by a square-well potential centered on the point O (see Fig. 6.4) and is defined as

$$\begin{aligned} V(\mathbf{r}) &= -V_0 \quad \text{for } |\mathbf{r}| \le a_0 \\ &= 0 \quad \text{for } |\mathbf{r}| > a_0 \end{aligned} \tag{6.28}$$

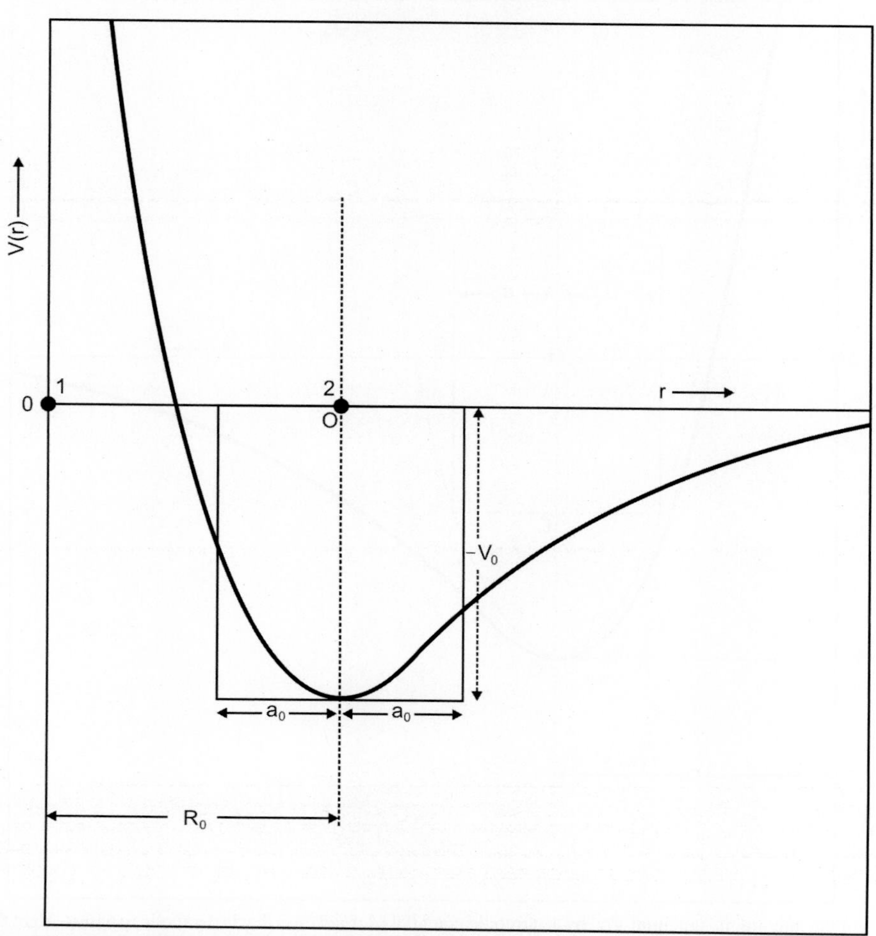

FIG. 6.4 Approximation of the interatomic potential $V(r)$ by a square-well potential with depth $-V_0$ and width $2a_0$.

But the square-well potential is in no way close to the exact interaction potential, except that its value corresponds to the actual value at R_0.

6.2.2 Harmonic Interaction Potential

As the lowest-order approximation, one can retain terms up to the second power of displacement in Eq. (6.24), which gives

$$V(\mathbf{r}) = -V_0 + \frac{1}{2}\alpha_F u^2 \tag{6.29}$$

In defining the above equation, the reference level of potential is assumed to be at $V(\mathbf{R}_0) = -V_0$. Therefore, the force from Eq. (6.27) is given by

$$F = -\alpha_F u \tag{6.30}$$

In this approximation, the force acting on atom 2 is proportional to its displacement and is directed toward its equilibrium position. It is wellknown that an atom acted upon by such a force oscillates harmonically about its equilibrium position and hence the name harmonic force. Therefore, Eq. (6.29) defines the harmonic potential, which is parabolic in nature. It is evident from Fig. 6.5 that atom 2 oscillates with equal amplitude on both sides of its equilibrium position. The third-, fourth-, and higher-order terms in Eq. (6.24) give an anharmonic contribution to the interaction potential, which is assumed to be negligible here.

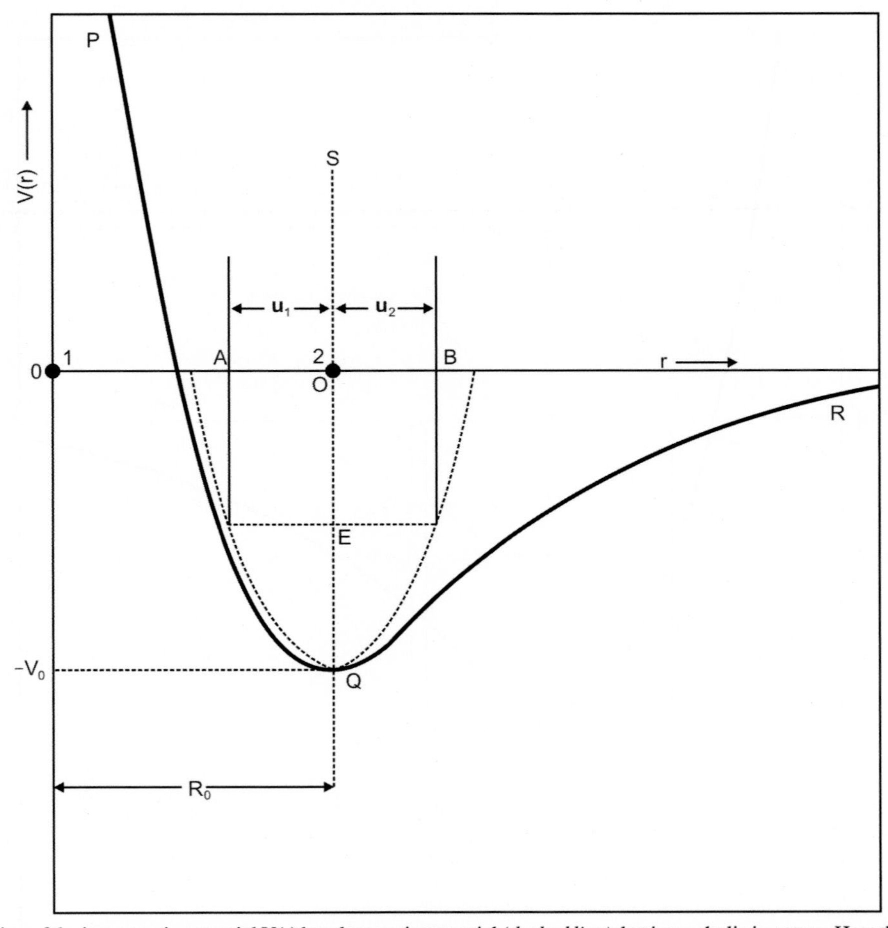

FIG. 6.5 Approximation of the interatomic potential V(r) by a harmonic potential *(dashed line)* that is parabolic in nature. Here the amplitude of atom 2 is the same on both sides, that is, $|\mathbf{u}_1| = |\mathbf{u}_2| = u$.

6.3 LATTICE VIBRATIONS IN A DISCRETE ONE-DIMENSIONAL LATTICE

It has already been discussed in Chapter 1 that a one-dimensional solid consists of a periodic arrangement of atoms (molecules) in a particular direction and that its primitive cell may contain one or more atoms. In this section, we shall study lattice vibrations in monatomic and diatomic linear lattices.

6.3.1 Monatomic Linear Lattice

Fig. 6.6 shows a monatomic linear lattice along the x-direction with distance "a" between the consecutive atoms (lattice points). At finite temperature, the atoms start vibrating about their mean positions along the x-direction. The position of the nth atom at time t is given by

$$R(n,t) = R_n + u(n,t) = na + u(n,t) \tag{6.31}$$

R_n is the lattice vector and $u(n,t)$ is the displacement of the nth atom at time t and both are along the same direction, that is, the x-direction. To find the frequencies of vibration of the atoms we make the approximations described below.

First, the force F acting on a vibrating atom is assumed to obey Hooke's law, according to which F is linearly proportional to the displacement u of the atom (harmonic force), that is,

$$F = -\alpha_F u \tag{6.32}$$

The parameter α_F is the force constant. Because a vibrating spring executes a simple harmonic motion about its mean position, the atoms can be assumed to be connected via massless springs. It is for this reason that α_F is sometimes called the spring constant. Thus, a linear solid can be replaced by a mechanical system of the form shown in Fig. 6.6.

Let us consider the equation of motion of the nth atom of the lattice. The force $F_{n,s}$ acting on the nth atom due to (n+s)th atom depends on the relative displacement of these two atoms and is written as

$$F_{n,s} = -\alpha_{Fs}[u(n) - u(n+s)] \tag{6.33}$$

where u(n) and u(n+s) are the displacements of the nth and (n+s)th atoms from their mean positions and their time dependence is assumed to be understood. The parameter α_{Fs} represents the force constant for the sNN of the nth atom under consideration. In general, an atom interacts with all of the other atoms of the solid, therefore, the total force acting on the nth atom is given by

$$F_n = \sum_s F_{n,s} = -\sum_s \alpha_{Fs}[u(n) - u(n+s)] \tag{6.34}$$

where $s = 0, \pm 1, \pm 2, \ldots$ Secondly, we assume that an atom interacts with its 1NNs only. The force constant α_{Fs} for both the 1NNs is the same (say α_F). In this case, the subscript s takes the values 1 and –1 in Eq. (6.34) to give

$$F_n = \alpha_F[u(n+1) + u(n-1) - 2u(n)] \tag{6.35}$$

According to Newton's second law of motion

$$F_n = M\frac{\partial^2 u(n)}{\partial t^2} \tag{6.36}$$

where M is the mass of an atom. From Eqs. (6.35), (6.36), the equation of motion for the nth atom becomes

$$M\frac{\partial^2 u(n)}{\partial t^2} = \alpha_F[u(n+1) + u(n-1) - 2u(n)] \tag{6.37}$$

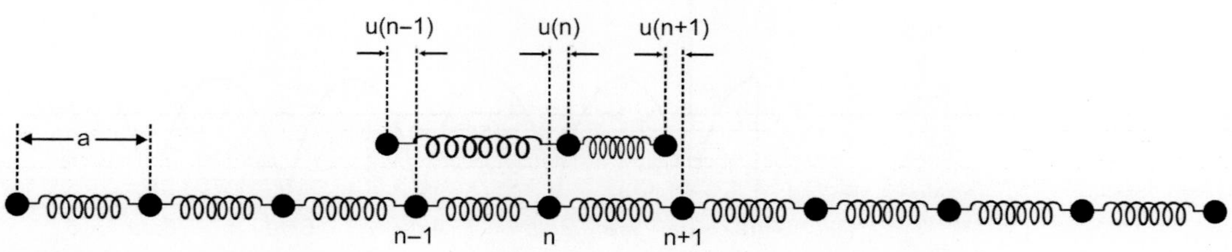

FIG. 6.6 Monatomic linear lattice with distance "a" between the consecutive atoms. At finite temperature, u(n − 1), u(n), and u(n+1) are the instantaneous displacements of the (n − 1)th, nth, and (n+1)th atoms, respectively, from their mean positions.

The wave-like solution of Eq. (6.37) can be written as

$$u(n) = u_0 e^{\imath(KR_n - \omega t)} = u_0 e^{\imath(Kna - \omega t)} \tag{6.38}$$

Similarly, we can write

$$u(n+1) = u_0 e^{\imath[K(n+1)a - \omega t]} \tag{6.39}$$

$$u(n-1) = u_0 e^{\imath[K(n-1)a - \omega t]} \tag{6.40}$$

Substituting Eqs. (6.38)–(6.40) into Eq. (6.37) and simplifying, we get

$$M\omega^2 = 4\alpha_F \sin^2\left(\frac{Ka}{2}\right) \tag{6.41}$$

Therefore, the frequency of vibration of atoms is given by

$$\omega = \pm\sqrt{\frac{4\alpha_F}{M}} \sin\left(\frac{Ka}{2}\right) \tag{6.42}$$

Because the frequency of vibration is always positive for a stable lattice, therefore, in Eq. (6.42), we should take the positive square root and modulus of the sine function to write

$$\omega = \sqrt{\frac{4\alpha_F}{M}} \left|\sin\left(\frac{Ka}{2}\right)\right| \tag{6.43}$$

Eq. (6.43) shows that ω depends nonlinearly on the wave vector K and gives the dispersion relation. The sine function is a periodic function and its modulus varies from 0 to 1, therefore, the value of K varies from 0 to π/a (see Fig. 6.7). Beyond this value the sine function is repeated, therefore, the independent values of K range from 0 to π/a on the positive side. The wave can propagate either to the right or to the left, therefore, K can have both positive and negative values. Hence the range of independent values of K is

$$-\frac{\pi}{a} \le K \le \frac{\pi}{a} \tag{6.44}$$

This range of independent values of K is called the 1BZ (see Chapter 2) of the one-dimensional lattice. Fig. 6.8 shows the frequencies of vibration of atoms (phonon frequencies) as a function of K in the 1BZ. Eq. (6.43) can also be written in terms of the elastic constant C_{11} and the density per unit length ρ_m. One can easily write

$$\alpha_F = \frac{C_{11}}{a} \quad \text{and} \quad M = a\rho_m \tag{6.45}$$

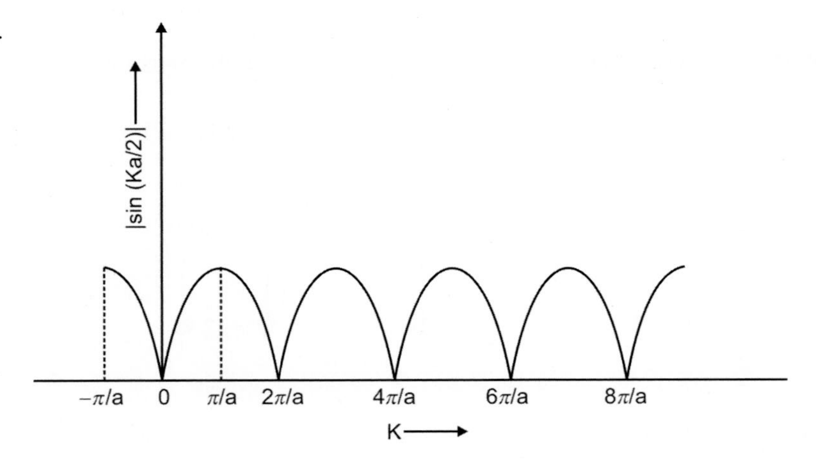

FIG. 6.7 Plot of | sin(Ka/2) | as a function of K.

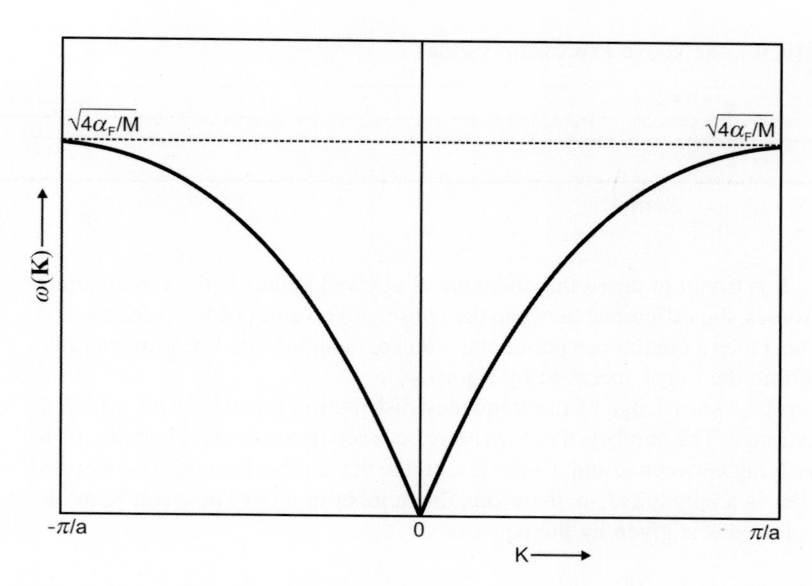

FIG. 6.8 Vibrational frequencies ω as a function of the wave vector K for a linear monatomic lattice in the 1BZ.

Using Eq. (6.45) in Eq. (6.43), one can write

$$\omega = \frac{2}{a} v_x \sin\left(\frac{Ka}{2}\right) \tag{6.46}$$

where

$$v_x = \sqrt{\frac{C_{11}}{\rho_m}} \tag{6.47}$$

Eq. (6.47) is the same as Eq. (6.13). Let us calculate the phase and group velocities for a linear monatomic lattice. The phase velocity, from Eq. (6.46), is given by

$$v_p = \frac{\omega}{K} = \frac{2v_x}{Ka} \sin\left(\frac{Ka}{2}\right) \tag{6.48}$$

and the group velocity is given by

$$v_g = \frac{d\omega}{dK} = v_x \cos\left(\frac{Ka}{2}\right) \tag{6.49}$$

It is evident that v_p and v_g are different for a linear monatomic lattice. At the 1BZ boundary with $K = \pi/a$, the phase and group velocities are given as

$$v_p = \frac{2v_x}{\pi}, \quad v_g = 0 \tag{6.50}$$

The group velocity represents the transfer of signal or energy, therefore, at the 1BZ there is no transfer of energy and the wave is a standing wave. At low frequencies, that is, in the limit of $K \to 0$, Eq. (6.46) reduces to

$$\omega = v_x K \tag{6.51}$$

which is the same dispersion relation as that obtained for a homogeneous line (see Eq. 6.15). Such behavior is expected because long wavelengths would not be sensitive to the discreteness of the lattice. Further, at long wavelengths (low frequencies), the group and phase velocities become the same ($=v_x$).

Let us calculate the values of the wave vector K in a discrete monatomic linear lattice. If the crystal is finite with a number N of atoms in it, the periodic boundary condition demands

$$u(n) = u(n + N) \tag{6.52}$$

Substituting the value of displacement from Eq. (6.38), the above expression yields

$$e^{\imath KNa}=1 \tag{6.53}$$

which is satisfied if

$$K=\frac{2\pi m}{Na} \tag{6.54}$$

Here $m=0,\pm 1,\pm 2,\ldots$ From Eqs. (6.38), (6.54), it is trivial to prove that there are N allowed values K for a monatomic linear lattice with N atoms. As the value of N increases, the difference between the consecutive values of K decreases, and finally, in the limit of very large N, the vector K becomes a continuous parameter. Hence, in an infinite linear monatomic crystal, the wave vector K varies continuously within the range specified by Eq. (6.44).

An estimation of the specific heat of solids involves knowledge of the frequency distribution function $g(\omega)$, which is defined as the number of modes per unit frequency range. The number of modes lying between frequencies ω and $\omega+d\omega$ is given by $g(\omega)\,d\omega$. Therefore, the number of modes lying between ω and $\omega+d\omega$ is equal to the number of modes between K and $K+dK$. From Eq. (6.54) one K state (mode) lies in a length $2\pi/Na$, therefore, the number of modes between K and K $+dK$ is given by $(Na/2\pi)\,dK$. The total number of modes is given by the equation

$$\int g(\omega)\,d\omega=\int 2\frac{Na}{2\pi}\,dK=N \tag{6.55}$$

According to Eq. (6.44) or Eq. (6.54) every positive value of K has a corresponding negative value. The factor of two takes into account both the positive and negative values of K. The above equation allows us to write

$$g(\omega)=\frac{Na}{\pi}\frac{dK}{d\omega} \tag{6.56}$$

From Eq. (6.43) it is straightforward to write

$$\frac{dK}{d\omega}=\frac{2}{a}\frac{1}{\sqrt{\omega_{max}^2-\omega^2}} \tag{6.57}$$

where

$$\omega_{max}=\sqrt{\frac{4\alpha_F}{M}} \tag{6.58}$$

Substituting Eq. (6.57) into Eq. (6.56), one gets

$$g(\omega)=\frac{2N/\pi}{\sqrt{\omega_{max}^2-\omega^2}} \tag{6.59}$$

The function $g(\omega)$ is shown in Fig. 6.9 and has the maximum frequency ω_{max}. The figure shows that most of the modes of vibration lie near ω_{max}. In the above treatment, only the interactions with the 1NNs have been included, but in an exact treatment one should include the interactions with all the NNs.

Problem 6.1

With the help of Eq. (6.59) prove that the total number of modes in a one-dimensional monatomic solid is equal to the total number of atoms.

6.3.2 Diatomic Linear Lattice

Consider a one-dimensional solid with two different types of atoms in the basis (see Fig. 6.10) with masses M_1 and M_2, where $M_1>M_2$. Note that the masses of the two types of atoms may not necessarily be different. If "a" is the distance

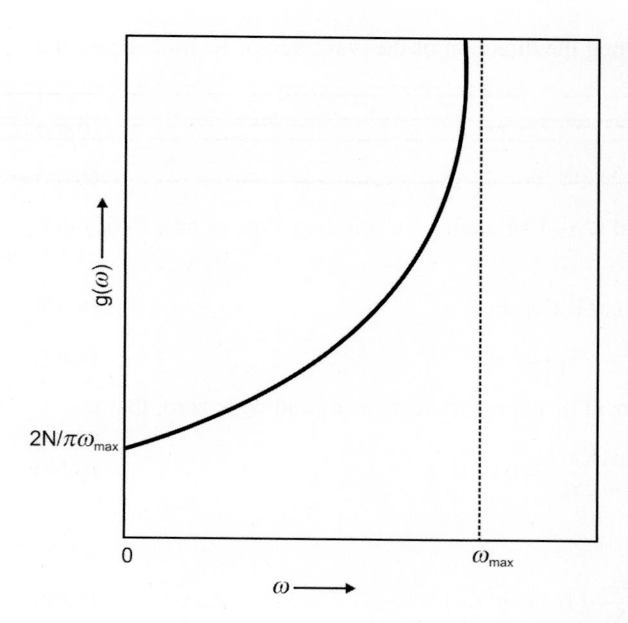

FIG. 6.9 Frequency distribution function $g(\omega)$ as a function of ω for a one-dimensional monatomic lattice.

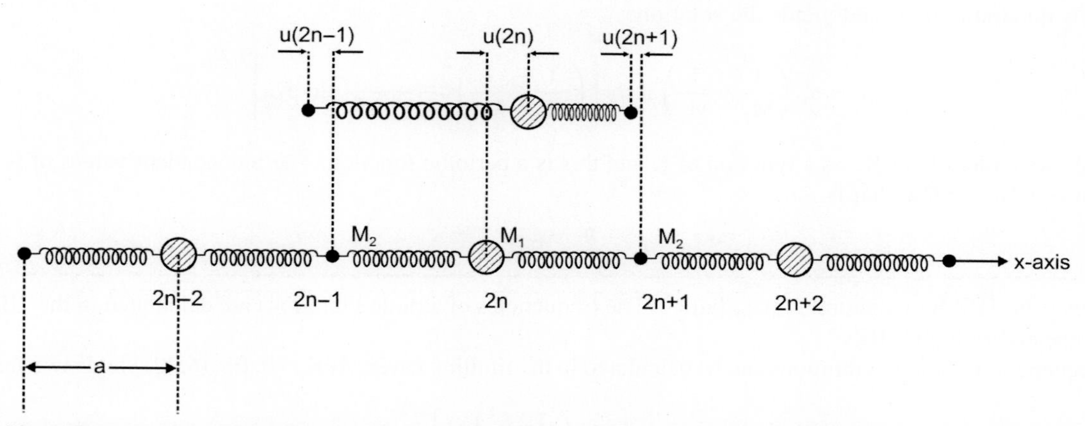

FIG. 6.10 Linear diatomic lattice with even-numbered atoms having mass M_1 and the odd-numbered atoms having mass M_2. Here, a is the distance between two consecutive atoms.

between consecutive atoms, then the repeat distance is 2a. The position of the nth atom at time t is given by Eq. (6.31). Considering the harmonic forces, the equations of motion for the 2nth and (2n+1)th atoms are given by

$$M_1 \ddot{u}(2n) = -\sum_s \alpha_{Fs}[u(2n) - u(2n+s)] \tag{6.60}$$

$$M_2 \ddot{u}(2n+1) = -\sum_s \alpha_{Fs}[u(2n+1) - u(2n+1+s)] \tag{6.61}$$

where $s = 0, \pm 1, \pm 2, \ldots$ For further simplification of the equations of motion of the atoms, only the 1NN interactions are retained. The force constants between an atom and its 1NNs are the same (say α_F) as the forces are identical. Therefore, the equations of motion of the 2nth and (2n+1)th atoms become

$$M_1 \ddot{u}(2n) = \alpha_F[u(2n+1) + u(2n-1) - 2u(2n)] \tag{6.62}$$

$$M_2 \ddot{u}(2n+1) = \alpha_F[u(2n+2) + u(2n) - 2u(2n+1)] \tag{6.63}$$

Let us consider a longitudinal wave in which the atoms vibrate along the direction of the wave vector K. In this case the wave-like solutions of Eqs. (6.62), (6.63) are given by

$$u(2n) = u_1 e^{\imath[2Kna - \omega t]} \tag{6.64}$$

$$u(2n+1) = u_2 e^{\imath[K(2n+1)a - \omega t]} \tag{6.65}$$

where u_1 and u_2 are the amplitudes of the waves for the 2nth and (2n+1)th atoms. Substituting Eqs. (6.64), (6.65) into Eqs. (6.62), (6.63), one gets

$$\left(M_1\omega^2 - 2\alpha_F\right)u_1 + (2\alpha_F \cos Ka)\,u_2 = 0 \tag{6.66}$$

$$(2\alpha_F \cos Ka)\,u_1 + \left(M_2\omega^2 - 2\alpha_F\right)u_2 = 0 \tag{6.67}$$

The above equations have nontrivial solution only if the determinant of the coefficients of u_1 and u_2 is zero, that is,

$$\begin{vmatrix} M_1\omega^2 - 2\alpha_F & 2\alpha_F \cos Ka \\ 2\alpha_F \cos Ka & M_2\omega^2 - 2\alpha_F \end{vmatrix} = 0 \tag{6.68}$$

The expansion of the above determinant yields the expression

$$\omega^4 - 2\alpha_F\left(\frac{1}{M_1} + \frac{1}{M_2}\right)\omega^2 + \frac{4\alpha_F^2}{M_1 M_2}\left(1 - \cos^2 Ka\right) = 0 \tag{6.69}$$

Eq. (6.69) is quadratic in ω^2 and yields the solutions

$$\omega^2 = \alpha_F\left(\frac{1}{M_1} + \frac{1}{M_2}\right) \pm \alpha_F\left[\left(\frac{1}{M_1} + \frac{1}{M_2}\right)^2 - \frac{4}{M_1 M_2}\sin^2 Ka\right]^{1/2} \tag{6.70}$$

Fig. 6.11 shows a plot of $\sin^2 Ka$ as a function of K and this is a periodic function. The independent values of K lie in the range from $-\pi/2a$ to $\pi/2a$, that is,

$$-\frac{\pi}{2a} \le K \le \frac{\pi}{2a} \tag{6.71}$$

which defines the 1BZ for a diatomic linear lattice. The frequencies of atomic vibrations are calculated in the 1BZ only as these are repeated in higher BZs.

The frequencies of atomic vibrations can be calculated in the limiting cases. At K = 0, Eq. (6.70) yields two frequencies

$$\omega_+ = \left[2\alpha_F\left(\frac{1}{M_1} + \frac{1}{M_2}\right)\right]^{1/2} \tag{6.72a}$$

$$\omega_- = 0 \tag{6.72b}$$

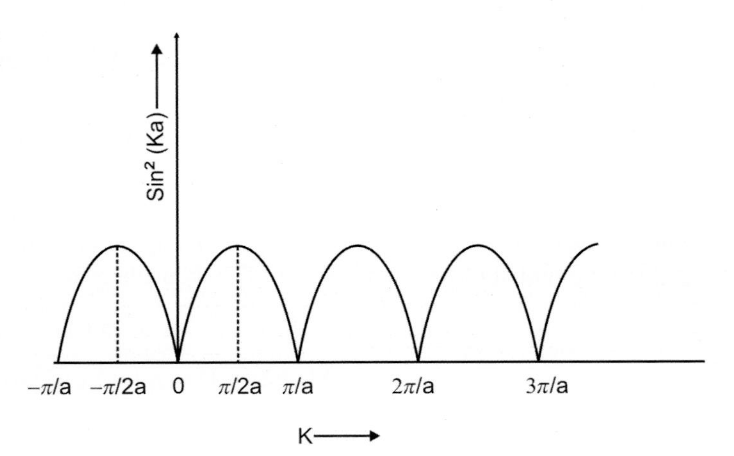

FIG. 6.11 Plot of $\sin^2 Ka$ as a function of K for a diatomic linear lattice with a as the distance between two consecutive atoms.

ω_+ and ω_- are the atomic vibration frequencies corresponding to the plus and minus signs in Eq. (6.70). For very small values of K, sin Ka = Ka, therefore, the atomic vibration frequencies from Eq. (6.70) are given by

$$\omega_+ = \left[2\alpha_F\left(\frac{1}{M_1}+\frac{1}{M_2}\right) - \frac{2\alpha_F(Ka)^2}{(M_1+M_2)}\right]^{1/2} \tag{6.73a}$$

$$\omega_- = \left[\frac{2\alpha_F(Ka)^2}{(M_1+M_2)}\right]^{1/2} \tag{6.73b}$$

At the 1BZ boundary, that is, at $K=\pi/2a$, $\sin Ka = 1$, therefore, Eq. (6.70) gives the frequency of atomic vibrations as

$$\omega_+ = \left(\frac{2\alpha_F}{M_2}\right)^{1/2} \tag{6.74a}$$

and

$$\omega_- = \left(\frac{2\alpha_F}{M_1}\right)^{1/2} \tag{6.74b}$$

With the help of Eqs. (6.70), (6.72a), (6.72b)–(6.74a), (6.74b), ω can be plotted as a function of K, which is shown in Fig. 6.12. The curve for ω_- as a function of K is called the acoustical branch and that for ω_+ is called the optical branch of the dispersion relations for the longitudinal waves. Fig. 6.12 shows that wave-like solutions do not exist for frequencies between $(2\alpha_F/M_2)^{1/2}$ and $(2\alpha_F/M_1)^{1/2}$ at the 1BZ boundary. This frequency gap is a characteristic feature of the elastic waves in a diatomic lattice. If one looks for a solution with real ω in this gap, then the wave vector K will be complex, which means that the wave is damped in space.

To study the nature of acoustical and optical waves, let us find the amplitudes of these waves. Substituting the value of ω_- from Eq. (6.73b) into Eq. (6.66), we find

$$2\alpha_F u_1\left(\frac{M_1(Ka)^2}{M_1+M_2} - 1\right) + 2\alpha_F u_2 \cos Ka = 0 \tag{6.75}$$

For very small values of K ($Ka \ll 1$) the above expression yields

$$u_1 = u_2 \tag{6.76}$$

Hence, for very small values of K or for very large values of wavelength, the amplitudes of the two types of atoms are the same. The corresponding wave is shown in Fig. 6.13A. Such a wave can be stimulated by some kind of force that makes all of the atoms move in the same direction, such as a compressional wave or a sound wave; that is why it is called the

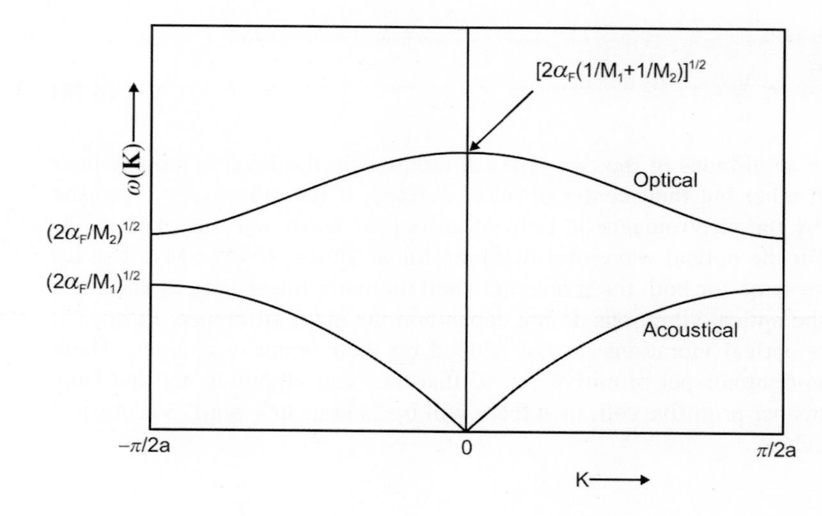

FIG. 6.12 Dispersion relations for longitudinal waves in a linear diatomic lattice with a as the distance between two consecutive atoms.

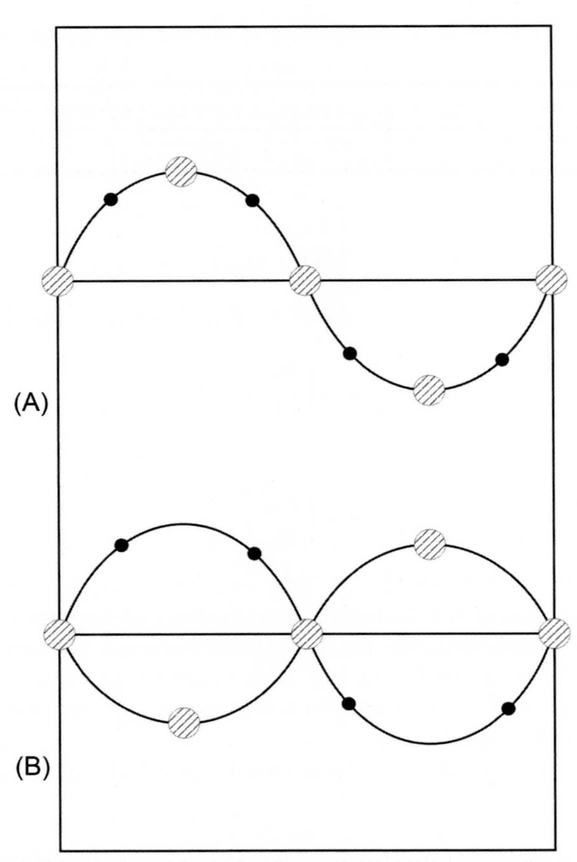

FIG. 6.13 (A) Acoustic wave in a diatomic linear lattice with a as the distance between two consecutive atoms. (B) Optical wave in a diatomic linear lattice with a as the distance between two consecutive atoms.

acoustical branch. A monatomic lattice can respond only to this kind of excitation. Substituting the value of ω_+ from Eq. (6.73a) into Eq. (6.66), we find

$$\frac{u_1}{u_2} = -\frac{1}{M_1} \frac{\cos Ka}{\frac{1}{M_2} - \frac{(Ka)^2}{M_1 + M_2}} \tag{6.77}$$

For very small values of $K\,(Ka \ll 1)$, we obtain

$$\frac{u_1}{u_2} = -\frac{M_2}{M_1} \tag{6.78}$$

Therefore, in the optical branch, the ratio of the amplitudes of the two types of atoms is in the inverse ratio of their masses. Further, the atoms vibrate against each other but their center of mass is fixed. If the atoms carry opposite charges, a motion of this type can be excited by the electromagnetic field of an optical wave and this explains its name. Fig. 6.13B shows a schematic diagram for the optical wave of a diatomic linear lattice. If $M_1 = M_2$, then $u_1/u_2 = -1$. In this case the frequency range is the same for both the monatomic and diatomic linear lattices and there will be no forbidden gap. It is noteworthy that the optical vibrations do not depend on the mass difference. Further, if the diatomic linear lattice consists of ions, the optical vibrations do not depend on their opposite charges. These depend only on the fact that there are two or more atoms per primitive cell so that they can vibrate in and out from the center of mass. In general, if there are s atoms per primitive cell, then there will be 3s branches in all: 3 acoustical branches and the remaining 3s − 3 optical branches.

From the discussion of lattice vibrations of discrete monatomic and diatomic linear lattices, the following features emerge:

1. The frequency ω is no longer linearly proportional to the wave vector K but is a periodic function. This fact imposes an upper limit on ω in contrast with the wave propagation in a homogeneous medium.
2. There exist allowed frequency (energy) bands separated by a forbidden frequency (energy) band. The forbidden band is related to Bragg's reflection in crystalline solids.

Problem 6.2

Consider a diatomic linear lattice as shown in Fig. 6.14 where the unit cell, having length a, contains two atoms named 1 and 2 with masses M_1 and M_2, respectively. Assuming a wave-like solution for the two atoms of the form

$$u(nm) = u(m)\, e^{i(\mathbf{K}\cdot\mathbf{R}_n - \omega t)} \tag{6.79}$$

prove that the phonon frequency of the lattice is given by

$$\omega^2 = \alpha_F\left(\frac{1}{M_1}+\frac{1}{M_2}\right) \pm \alpha_F\left[\left(\frac{1}{M_1}+\frac{1}{M_2}\right)^2 - \frac{4}{M_1M_2}\sin^2\left(\frac{Ka}{2}\right)\right]^{1/2} \tag{6.80}$$

Further prove that

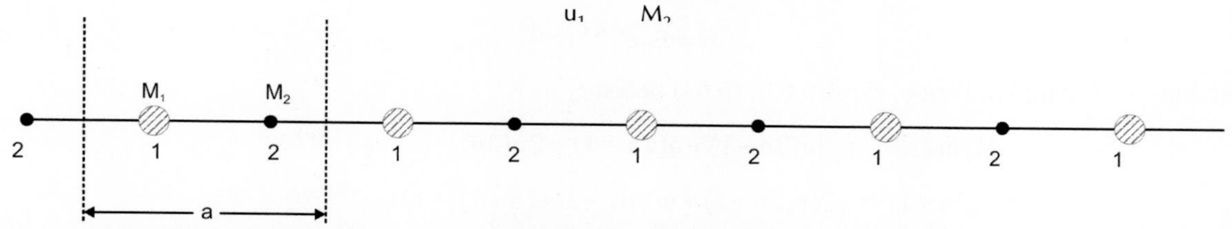

FIG. 6.14 Linear diatomic lattice with repeat distance a. The figure shows that each unit cell contains two atoms of type 1 and 2.

The expressions for the phonon frequencies obtained in Section 6.3.2 and in the above problem are the same except for the different values of the repeat distance (the repeat distance in Section 6.3.2 is 2a, while in Problem 6.1 it is a). A plot of the vibrational frequencies as a function of K obtained from Eq. (6.80) is shown in Fig. 6.15. The only difference in Figs. 6.12 and 6.15 is due to the different boundaries of the 1BZ arising from the different values of the repeat distance.

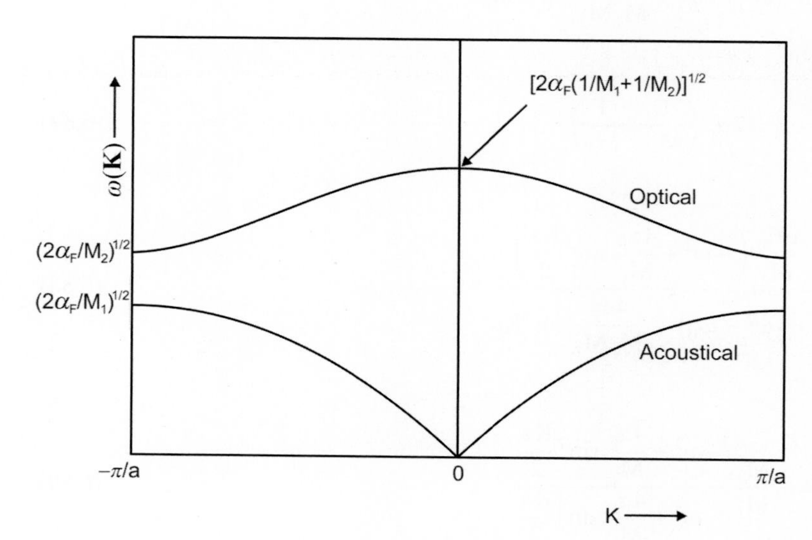

FIG. 6.15 Vibrational frequencies ω as a function of the wave vector K in the 1BZ of the linear diatomic lattice with repeat distance a and for $M_1 > M_2$.

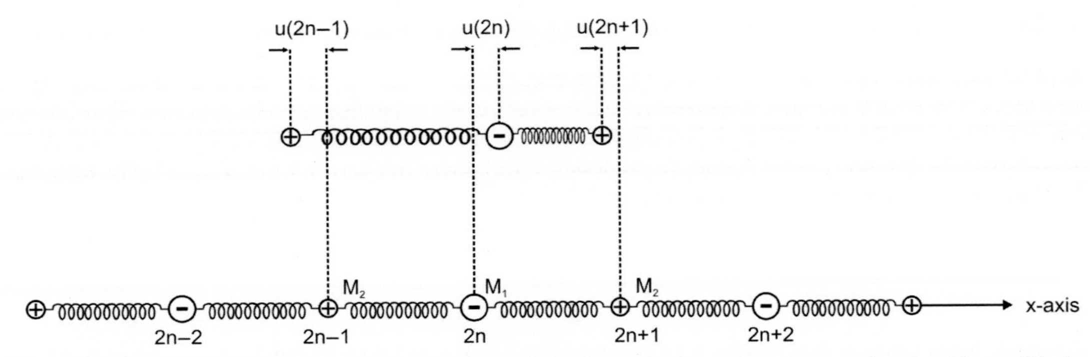

FIG. 6.16 One-dimensional ionic lattice with charges ±e on alternate ions. The negative ions occupy even-numbered positions, while the positive ions occupy odd-numbered positions.

6.4 EXCITATION OF IONIC LATTICE IN INFRARED REGION

Consider a one-dimensional ionic lattice with charges ±e on the ions. Let the negative ions occupy even-numbered positions and the positive ions the odd-numbered positions (see Fig. 6.16). A field of electromagnetic waves can affect the optical branches of a vibrating ionic lattice. To examine this, allow infrared radiation to fall on an ionic crystal for which the electric field can be defined as

$$E = E_0 e^{\imath(Kx - \omega t)} \tag{6.81}$$

In the presence of infrared radiation, Eqs. (6.62), (6.63) become

$$M_1 \ddot{u}(2n) = \alpha_F [u(2n+1) + u(2n-1) - 2u(2n)] - eE_0 e^{\imath(2Kna - \omega t)} \tag{6.82}$$

$$M_2 \ddot{u}(2n+1) = \alpha_F [u(2n+2) + u(2n) - 2u(2n+1)] + eE_0 e^{\imath[K(2n+1)a - \omega t]} \tag{6.83}$$

Substituting Eqs. (6.64), (6.65) into the above expressions, we get

$$\left(-M_1\omega^2 + 2\alpha_F\right) u_1 - (2\alpha_F \cos Ka)\, u_2 + eE_0 = 0 \tag{6.84}$$

$$-(2\alpha_F \cos Ka)\, u_1 + \left(-M_2\omega^2 + 2\alpha_F\right) u_2 - eE_0 = 0 \tag{6.85}$$

Multiplying Eq. (6.84) by $(-M_2\omega^2 + 2\alpha_F)$ and Eq. (6.85) by $2\alpha_F \cos Ka$ and adding, we get

$$u_1 = \frac{-\dfrac{eE_0}{M_1}\left(-\omega^2 + \dfrac{4\alpha_F}{M_2}\sin^2\dfrac{Ka}{2}\right)}{\omega^2\left(\omega^2 - \omega_0^2\right) + \dfrac{4\alpha_F^2}{M_1 M_2}\sin^2 Ka} \tag{6.86}$$

where

$$\omega_0^2 = 2\alpha_F\left(\frac{1}{M_1} + \frac{1}{M_2}\right) \tag{6.87}$$

Similarly, one can get

$$u_2 = \frac{\dfrac{eE_0}{M_2}\left(-\omega^2 + \dfrac{4\alpha_F}{M_1}\sin^2\dfrac{Ka}{2}\right)}{\omega^2\left(\omega^2 - \omega_0^2\right) + \dfrac{4\alpha_F^2}{M_1 M_2}\sin^2 Ka} \tag{6.88}$$

Hence the ratio of two amplitudes is given by

$$\frac{u_1}{u_2} = -\frac{M_2}{M_1}\,\frac{-\omega^2 + \dfrac{4\alpha_F}{M_2}\sin^2\dfrac{Ka}{2}}{-\omega^2 + \dfrac{4\alpha_F}{M_1}\sin^2\dfrac{Ka}{2}} \tag{6.89}$$

In the long wavelength limit ($Ka \ll 1$), the individual amplitudes from Eqs. (6.86), (6.88) are given by

$$u_1 = \frac{\frac{e}{M_1} E_0}{\omega^2 - \omega_0^2} \tag{6.90}$$

$$u_2 = -\frac{\frac{e}{M_2} E_0}{\omega^2 - \omega_0^2} \tag{6.91}$$

Therefore, the amplitudes u_1 and u_2 exhibit resonance behavior at ω_0. From Eqs. (6.90), (6.91) the ratio of amplitudes become

$$\frac{u_1}{u_2} = -\frac{M_2}{M_1} \tag{6.92}$$

which is the same result as that given by Eq. (6.78) for the optical branch of the diatomic lattice. The general Eq. (6.89) also reduces to Eq. (6.92) for $Ka \ll 1$. Comparing Eqs. (6.77), (6.89), it is evident that the infrared radiation affects the amplitudes of the two types of ions forming the optical wave in an ionic lattice. In other words, one can say that the infrared radiation can excite optical vibrations in an ionic lattice.

Problem 6.3

In the harmonic force approximation, the equation of motion for a lattice is given by

$$M\ddot{\mathbf{u}}(n_1, n_2, n_3) = -\sum_s \alpha_{Fs} [\mathbf{u}(n_1, n_2, n_3) - \mathbf{u}(n_1 + s_1, n_2 + s_2, n_3 + s_3)] \tag{6.93}$$

where $n = (n_1, n_2, n_3)$ represents the nth atom in a lattice and $s = (s_1, s_2, s_3)$ represents the sth nearest neighbor of the nth atom. Further, M is the mass of an atom and α_{Fs} is the force constant for the sth nearest neighbor of the nth atom. Assuming 1NN interactions with force constant α_F, show that the dispersion relation for the square lattice is given by

$$\omega^2 M = 2\alpha_F \left(2 - \cos K_x a - \cos K_y a\right) \tag{6.94}$$

Problem 6.4

The general equation of motion of the lattice in the harmonic approximation is given by Eq. (6.93). Assuming 1NN interactions with force constant α_F, derive the expression for the dispersion relation for an sc lattice.

Chapter 7

Lattice Vibrations-2

Chapter Outline

In the last chapter, we found that the frequency of atomic vibrations is determined by the force constant, which is the derivative of the force acting on an atom or the double derivative of the lattice potential. A determination of the exact lattice potential is undoubtedly very difficult (see Eq. 3.12) but can be estimated within some simplifying assumptions described in Chapter 3. In the adiabatic approximation, electrons are considered to move in a field of ions regarded as instantaneously stationary. The electronic energy eigenvalue so determined serves as the potential energy for ionic motion. To study the lattice properties, one needs to consider only the lattice part of the Hamiltonian as the motion of the ions can be separated from that of the electrons.

In this chapter, we present the general theory of lattice vibrations in a three-dimensional metallic solid, usually called the lattice dynamics of metals. Consider a crystal composed of an infinite number of unit cells, each of which is a parallelepiped bounded by three noncoplanar vectors $\mathbf{a}_1$, $\mathbf{a}_2$, and $\mathbf{a}_3$. Here we consider Bravais crystals with a basis in which each unit cell contains "s" ions. Such crystals are also called nonprimitive crystals. If there is only one ion in the unit cell, the crystals are called primitive crystals or Bravais crystals.

7.1 EQUATION OF MOTION OF THE LATTICE

We restrict our attention to stable crystals in which, as a result of thermal energy, each ion is displaced from its equilibrium position by an amount $\mathbf{u}(\mathrm{nm})$ at a specific point of time. So, the instantaneous position of the ion can be written as

$$\mathbf{R}(\mathrm{nm}) = \mathbf{R}_{\mathrm{nm}} + \mathbf{u}(\mathrm{nm}) \tag{7.1}$$

where $\mathbf{R}_{\mathrm{nm}}$ is the equilibrium position vector. Now, the total Hamiltonian of the lattice becomes

$$\widehat{\mathrm{H}}_{\mathrm{i}} = \frac{1}{2}\sum_{\mathrm{nm}\alpha} \mathrm{M}_{\mathrm{m}}\dot{\mathrm{u}}_{\alpha}^{2}(\mathrm{nm}) + \mathrm{V}_{\mathrm{i}}(\mathbf{R}) \tag{7.2}$$

where M_{m} is the mass of the mth type of ion. $\mathrm{V}_{\mathrm{i}}(\mathbf{R})$ is the total lattice potential energy given by Eq. (3.12) and is a function of the instantaneous position-coordinates of all the ions denoted by $\mathbf{R}$. Note that $\mathrm{V}_{\mathrm{i}}(\mathbf{R})$ also includes the direct overlap interaction between the electron distributions of the ions. It is convenient to expand $\mathrm{V}_{\mathrm{i}}(\mathbf{R})$ in powers of displacements $\mathbf{u}(\mathrm{nm})$ of the ions about their equilibrium positions (Taylor expansion) as

Solid State Physics. https://doi.org/10.1016/B978-0-12-817103-5.00007-4

$$
\begin{aligned}
V_i(\mathbf{R}) = V_0 &+ \sum_{nm\alpha} V_\alpha(nm)\, u_\alpha(nm) \\
&+ \frac{1}{2}\sum_{nm\alpha}\sum_{n'm'\beta} V_{\alpha\beta}(nm, n'm')\, u_\alpha(nm)\, u_\beta(n'm') \\
&+ \frac{1}{6}\sum_{nm\alpha}\sum_{n'm'\beta}\sum_{n''m''\gamma} V_{\alpha\beta\gamma}(nm, n'm', n''m'')\, u_\alpha(nm) u_\beta(n'm') u_\gamma(n''m'') + \cdots
\end{aligned}
\tag{7.3}
$$

where

$$
\begin{aligned}
&V_\alpha(nm) = \left.\frac{\partial V_i(\mathbf{R})}{\partial u_\alpha(nm)}\right|_{\mathbf{u}=0}, \quad V_{\alpha\beta}(nm, n'm') = \left.\frac{\partial^2 V_i(\mathbf{R})}{\partial u_\alpha(nm)\,\partial u_\beta(n'm')}\right|_{\mathbf{u}=0}, \\
&V_{\alpha\beta\gamma}(nm, n'm', n''m'') = \left.\frac{\partial^3 V_i(\mathbf{R})}{\partial u_\alpha(nm)\,\partial u_\beta(n'm')\,\partial u_\gamma(n''m'')}\right|_{\mathbf{u}=0}
\end{aligned}
\tag{7.4}
$$

V_0 is the potential energy of a perfect crystal. The subscript $u = 0$ indicates that the derivatives are evaluated in the equilibrium configuration of the crystal. Let us first interpret physically all the quantities occurring in Eq. (7.3). The coefficient $V_\alpha(nm)$ is the negative of the force acting in the α-direction on the atom at $\mathbf{R}_{nm}$. The coefficient $V_{\alpha\beta}(nm, n'm')$, in the first approximation, is the negative of the force exerted in the α-direction on the atom (nm) when the atom $(n'm')$ is displaced by a unit distance in the β-direction, all other atoms being kept fixed at their equilibrium positions: it is just the force constant as defined in Chapter 6. In exactly the same way, one can interpret the coefficients $V_{\alpha\beta\gamma}(nm, n'm', n''m'')$, which gives us the bond-bending forces.

The motion of ions is described by Eqs. (7.2), (7.3), which can be regarded as a set of coupled anharmonic oscillators. The potential energy expansion up to the second order gives harmonic terms, while the higher-order terms are anharmonic in nature. If the ionic displacements are small compared with the interatomic spacing, the series (7.3) is expected to converge rapidly. Therefore, to a good approximation, one can retain terms up to the second order only and neglect the higher-order terms, treating them as small. This is called the *harmonic approximation* in which the forces obey Hooke's law. In the equilibrium position, the net force acting on an ion is zero, that is, $V_\alpha(nm) = 0$. Therefore, the potential in the harmonic approximation becomes

$$
V_i(\mathbf{R}) = V_0 + \frac{1}{2}\sum_{nm\alpha}\sum_{n'm'\beta} V_{\alpha\beta}(nm, n'm')\, u_\alpha(nm)\, u_\beta(n'm') \tag{7.5}
$$

From Eqs. (7.2), (7.5), the lattice Hamiltonian in the equilibrium position is given as

$$
\begin{aligned}
\widehat{H}_i = &\frac{1}{2}\sum_{nm\alpha} M_m \dot{u}_\alpha^2(nm) + V_0 \\
&+ \frac{1}{2}\sum_{nm\alpha}\sum_{n'm'\beta} V_{\alpha\beta}(nm, n'm')\, u_\alpha(nm)\, u_\beta(n'm')
\end{aligned}
\tag{7.6}
$$

The equations of motion of the lattice are then easily found to be

$$
F_\gamma(n''m'') = M_{m''}\ddot{u}_\gamma(n''m'') = -\frac{\partial V_i(\mathbf{R})}{\partial u_\gamma(n''m'')} \tag{7.7}
$$

Substituting $V_i(\mathbf{R})$ from Eq. (7.5) into Eq. (7.7), one gets

$$
F_\gamma(n''m'') = -\frac{1}{2}\sum_{n'm'\beta} V_{\gamma\beta}(n''m'', n'm')\, u_\beta(n'm') - \frac{1}{2}\sum_{nm\alpha} V_{\alpha\gamma}(nm, n''m'')\, u_\alpha(nm) \tag{7.8}
$$

As the force constants $V_{\alpha\beta}(nm, n'm')$ are the second-order partial derivatives of $V_i(\mathbf{R})$, therefore,

$$
V_{\alpha\beta}(nm, n'm') = V_{\beta\alpha}(n'm', nm) \tag{7.9}
$$

Using the above relation in Eq. (7.8), one can write

$$
M_{m''}\ddot{\mathbf{u}}_\gamma(n''m'') = -\frac{1}{2}\sum_{n'm'\beta} V_{\gamma\beta}(n''m'', n'm')\, u_\beta(n'm') - \frac{1}{2}\sum_{nm\alpha} V_{\gamma\alpha}(n''m'', nm)\, u_\alpha(nm)
$$

The two terms on the right side of the above equation are the same, therefore,

$$M_{m''}\ddot{u}_{\gamma}(n''m'') = -\sum_{nm\alpha} V_{\gamma\alpha}(n''m'', nm)u_{\alpha}(nm)$$

which can also be written as

$$M_m\ddot{u}_{\alpha}(nm) = -\sum_{n'm'\beta} V_{\alpha\beta}(nm, n'm')u_{\beta}(n'm') \tag{7.10}$$

7.1.1 Restrictions on Atomic Force Constants

The atomic force constants satisfy several conditions due to the infinitesimal translational and rotational invariance. However, we shall prove only the translational invariance conditions, which are of use here. Let us replace each displacement vector $\mathbf{u}(nm)$ by an arbitrary constant vector $\mathbf{u}^0$, which is independent of n and m. Such a displacement field describes the rigid body translation of the whole of the crystal by an amount $\mathbf{u}^0$ and, therefore, the potential energy of the lattice should remain unchanged. For rigid body displacement, the potential from Eq. (7.3), in the harmonic approximation, becomes

$$V_i(\mathbf{R}) = V_0 + \sum_{nm\,\alpha} V_{\alpha}(nm)\,u_{\alpha}^0 + \frac{1}{2}\sum_{nm\alpha}\sum_{n'm'\beta} V_{\alpha\beta}(nm, n'm')\,u_{\alpha}^0 u_{\beta}^0 \tag{7.11}$$

The apparent change in $V_i(\mathbf{R})$, described by the last two terms on the right-hand side of Eq. (7.11), must vanish. Because $\mathbf{u}^0$ is an arbitrary vector, we must equate the coefficients of each power of u_{α}^0 to zero. In this way, we obtain

$$\sum_{nm} V_{\alpha}(nm) = 0 \tag{7.12a}$$

$$\sum_{nm}\sum_{n'm'} V_{\alpha\beta}(nm, n'm') = 0 \tag{7.12b}$$

According to the first equation the net force on the whole of the lattice vanishes. This condition is automatically satisfied if all of the atoms of the crystal are in their equilibrium positions as the net force on each atom vanishes. If the atoms are not in the equilibrium positions, then the net force on each atom is finite, but still the net force acting on whole of the lattice vanishes.

A more restrictive condition on $V_{\alpha\beta}(nm, n'm')$ follows from the behavior of the force on each atom $F_{\alpha}(nm)$ under rigid body translation of the crystal by $\mathbf{u}^0$. The force in general from Eq. (7.7) is given as

$$F_{\alpha}(nm) = -\frac{\partial V_i(\mathbf{R})}{\partial u_{\alpha}(nm)}$$

$$= -V_{\alpha}(nm) - \sum_{n'm'\beta} V_{\alpha\beta}(nm, n'm')\,u_{\beta}(n'm') + \cdots$$

using Eq. (7.3). When the rigid body translation is performed, the force becomes

$$F_{\alpha}(nm) = -V_{\alpha}(nm) - \sum_{n'm'\beta} V_{\alpha\beta}(nm, n'm')\,u_{\beta}^0 + \cdots$$

Under rigid body translation, there should be no change in the force acting on an atom, which is so only if the coefficient of u_{β}^0 is zero, that is,

$$\sum_{n'm'} V_{\alpha\beta}(nm, n'm') = 0 \tag{7.12c}$$

It is evident that Eq. (7.12b) implies Eq. (7.12c) but not vice versa as Eq. (7.12c) is more restrictive than Eq. (7.12b).

In a pure crystalline solid, there is perfect periodicity as a result of which the environments of each lattice point are the same in the crystal. Therefore, if we change the origin of the coordinate system from one lattice point to another, it does not affect the physical properties, such as $V_{\alpha\beta}(nm, n'm')$. This is equivalent to the fact that $V_{\alpha\beta}(nm, n'm')$ depends on $\mathbf{R}_n$ and $\mathbf{R}_{n'}$ only through their vector difference $\mathbf{R}_n - \mathbf{R}_{n'}$. Mathematically, we can write

$$V_{\alpha\beta}(nm, n'm') = V_{\alpha\beta}(n - n'm, 0m') = V_{\alpha\beta}(0m, n' - nm') \tag{7.12d}$$

$$V_{\alpha}(nm) = V_{\alpha}(0m) = V_{\alpha}(m) \tag{7.12e}$$

Eq. (7.12e) says that the force acting on an atom does not depend on the value of n or the unit cell as explained above.

7.2 NORMAL COORDINATE TRANSFORMATION

To study the lattice vibrations, one has to solve Eq. (7.10), which represents a set of an infinite number of coupled linear differential equations. The problem may be solved by making a transformation to new coordinates called *normal coordinates*, which diagonalizes the Hamiltonian and reduces the problem to that of uncoupled oscillators. Making use of the periodicity of the lattice, the wave solution can be written in the form

$$u_{\alpha}(nm) = \frac{1}{\sqrt{M_m}} u_{\alpha}(m)\, e^{\imath[\mathbf{q}\cdot\mathbf{R}_{nm} - \omega t]} \tag{7.13}$$

where

$$\mathbf{K} = \mathbf{q} - \mathbf{G} \tag{7.14}$$

Here $u_{\alpha}(m)$ is the α-component of amplitude $\mathbf{u}(m)$. The wave vector $\mathbf{q}$ of the vibrational wave is restricted to the 1BZ with $|\mathbf{q}| = 2\pi/\lambda$ where λ is the wavelength. Because the vibrational frequencies are calculated in the 1BZ, only the reduced wave vector $\mathbf{q}$ is introduced. Substituting Eq. (7.13) into Eq. (7.10), we write

$$\omega^2 u_{\alpha}(m) = \sum_{m'\beta} D_{\alpha\beta}(\mathbf{q}, mm')\, u_{\beta}(m') \tag{7.15}$$

where

$$D_{\alpha\beta}(\mathbf{q}, mm') = \frac{1}{\sqrt{M_m M_{m'}}} \sum_{n'} V_{\alpha\beta}(nm, n'm')\, e^{-\imath\mathbf{q}\cdot[\mathbf{R}_{nm} - \mathbf{R}_{n'm'}]} \tag{7.16}$$

$D_{\alpha\beta}(\mathbf{q}, mm')$ are the elements of a dynamical matrix. One can write Eq. (7.15) in the following form

$$\sum_{m'\beta} \left[\omega^2 \delta_{\alpha\beta}\delta_{mm'} - D_{\alpha\beta}(\mathbf{q}, mm')\right] u_{\beta}(m') = 0 \tag{7.17}$$

As $u_{\beta}(m')$ is an arbitrary amplitude, therefore, the above equation has a nontrivial solution only if the determinant of the coefficients of $u_{\beta}(m')$ is zero, that is,

$$\det\left|\omega^2 \delta_{\alpha\beta}\delta_{mm'} - D_{\alpha\beta}(\mathbf{q}, mm')\right| = 0 \tag{7.18}$$

For a given value of $\mathbf{q}$, Eq. (7.15) constitutes a set of 3s linear homogeneous algebraic equations. The frequency ω is a function of both $\mathbf{q}$ and j where j is called the branch index. Therefore, one can write

$$\omega = \omega(\mathbf{q}j), \quad j = 1, 2, 3, \ldots. 3s \tag{7.19}$$

The relation given by Eq. (7.19) is known as the dispersion relation and $\omega(\mathbf{q}j)$ is the frequency of the normal mode of vibrations of the lattice. For a stable crystal it is necessary that $\omega^2(\mathbf{q}j)$ be positive for every normal mode, otherwise $\omega(\mathbf{q}j)$ will be imaginary, which is not allowed. For each of the 3s values of $\omega(\mathbf{q}j)$ for a given $\mathbf{q}$, there are eigenvectors of the dynamical matrix $D_{\alpha\beta}(\mathbf{q}, mm')$ denoted as $\mathbf{e}(\mathbf{q}j, m)$ such that

$$\omega^2(\mathbf{q}j)\, e_{\alpha}(\mathbf{q}j, m) = \sum_{m'\beta} D_{\alpha\beta}(\mathbf{q}, mm')\, e_{\beta}(\mathbf{q}j, m') \tag{7.20}$$

The eigenvectors $e_\alpha(\mathbf{q}j, m)$ are in fact elements of a unitary matrix, which diagonalize $D_{\alpha\beta}(\mathbf{q}, mm')$. As a result, the eigenvectors satisfy the following orthonormality and completeness relations

$$\sum_{m\,\alpha} e^*_\alpha(\mathbf{q}j, m)\, e_\alpha(\mathbf{q}j', m) = \delta_{jj'} \tag{7.21a}$$

$$\sum_{j} e^*_\alpha(\mathbf{q}j, m)\, e_\beta(\mathbf{q}j, m') = \delta_{mm'}\,\delta_{\alpha\beta} \tag{7.21b}$$

7.3 PROPERTIES OF DYNAMICAL MATRIX AND EIGENVECTORS

The dynamical matrix defined by Eq. (7.16) depends on both n and n′. Because of the periodicity of the lattice, the force constants depend only on the vector difference $\mathbf{R}_n - \mathbf{R}_{n'}$ (Eq. 7.12d). Assuming the nth cell to be at the origin of the coordinate axes, the expression for the dynamical matrix (Eq. 7.16) becomes

$$D_{\alpha\beta}(\mathbf{q}, mm') = \frac{1}{\sqrt{M_m M_{m'}}} \sum_{n'} V_{\alpha\beta}(0m, n'm')\, e^{-\imath\mathbf{q}\cdot[\mathbf{R}_m - \mathbf{R}_{m'} - \mathbf{R}_{n'}]} \tag{7.22}$$

The dynamical matrix can be separated into two parts as:

(i) For $m \neq m'$, the dynamical matrix (7.22) is given by

$$D_{\alpha\beta}(\mathbf{q}, mm') = \frac{1}{\sqrt{M_m M_{m'}}} \sum_{n} V_{\alpha\beta}(0m, nm')\, e^{-\imath\mathbf{q}\cdot[\mathbf{R}_m - \mathbf{R}_{m'} - \mathbf{R}_{n}]} \tag{7.23a}$$

(ii) For $m = m'$, $\mathbf{R}_m - \mathbf{R}_{m'} = 0$ and therefore, in the equilibrium position, Eq. (7.22) becomes

$$\begin{aligned} D_{\alpha\beta}(\mathbf{q}, mm) &= \frac{1}{M_m} \sum_{n} V_{\alpha\beta}(0m, nm)\, e^{\imath\mathbf{q}\cdot\mathbf{R}_n} \\ &= \frac{1}{M_m}\left[\sum_{n\neq 0} V_{\alpha\beta}(0m, nm) e^{\imath\mathbf{q}\cdot\mathbf{R}_n} + V_{\alpha\beta}(0m, 0m)\right] \end{aligned} \tag{7.23b}$$

The rigid displacement of the lattice as a whole does not change the force acting on an atom, which gives condition (7.12c). If we take $n = 0$, then Eq. (7.12c) becomes

$$\sum_{nm'} V_{\alpha\beta}(0m, nm') = 0$$

which on expansion can be written as

$$V_{\alpha\beta}(0m, 0m) = -\sum_{n\neq 0} V_{\alpha\beta}(0m, nm) - \sum_{n} \sum_{m'(m'\neq m)} V_{\alpha\beta}(0m, nm') \tag{7.23c}$$

Substituting Eq. (7.23c) into Eq. (7.23b), we get

$$D_{\alpha\beta}(\mathbf{q}, mm) = \frac{1}{M_m}\left[\sum_{n\neq 0} V_{\alpha\beta}(0m, nm)\, e^{\imath\mathbf{q}\cdot\mathbf{R}_n} - \sum_{n\neq 0} V_{\alpha\beta}(0m, nm) - \sum_{n} \sum_{m'(m'\neq m)} V_{\alpha\beta}(0m, nm')\right]$$

Adding and subtracting a term $V_{\alpha\beta}(0m, 0m)$, that is, the term for $n = 0$, and absorbing these in the first and second terms of the above equation, we have

$$D_{\alpha\beta}(\mathbf{q}, mm) = \frac{1}{M_m}\left[\sum_{n} V_{\alpha\beta}(0m, nm) e^{\imath\mathbf{q}\cdot\mathbf{R}_n} - \sum_{n} V_{\alpha\beta}(0m, nm) - \sum_{n} \sum_{m'(m'\neq m)} V_{\alpha\beta}(0m, nm')\right] \tag{7.23d}$$

Now, the general expression for the dynamical matrix can be written as

$$D_{\alpha\beta}(\mathbf{q}, mm') = \overline{D}_{\alpha\beta}(\mathbf{q}, mm') - \text{Lim}_{\mathbf{q}\to 0}\,\delta_{mm'} \sum_{m''} \overline{D}_{\alpha\beta}(\mathbf{q}, mm'') \tag{7.24}$$

where

$$\overline{D}_{\alpha\beta}(\mathbf{q}, mm')\frac{1}{\sqrt{M_m M_{m'}}}\sum_n V_{\alpha\beta}(0m, nm')\, e^{i\mathbf{q}\cdot[\mathbf{R}_n-\mathbf{R}_m+\mathbf{R}_{m'}]} \tag{7.25}$$

The dynamical matrix is fully known from Eqs. (7.24), (7.25) if $V_{\alpha\beta}(0m, nm')$ are known. Therefore, the basic problem is reduced to the determination of the atomic force constants. From Eq. (7.16) it is straightforward to prove the following properties of the dynamical matrix

$$D_{\alpha\beta}(-\mathbf{q}, mm') = D^*_{\alpha\beta}(\mathbf{q}, mm') \tag{7.26}$$

$$D^*_{\alpha\beta}(\mathbf{q}, mm') = D_{\beta\alpha}(\mathbf{q}, m'm) \tag{7.27}$$

Here we have assumed the force constants to be real. Substituting $-\mathbf{q}$ for $\mathbf{q}$ in Eq. (7.20), we get

$$\omega^2(-\mathbf{q}j)\, e_\alpha(-\mathbf{q}j, m) = \sum_{m'\beta} D_{\alpha\beta}(-\mathbf{q}, mm')\, e_\beta(-\mathbf{q}j, m') \tag{7.28}$$

Taking the complex conjugate of the above equation and using Eq. (7.26), we have

$$\omega^2(-\mathbf{q}j)\, e^*_\alpha(-\mathbf{q}j, m) = \sum_{m'\beta} D_{\alpha\beta}(\mathbf{q}, mm')\, e^*_\beta(-\mathbf{q}j, m') \tag{7.29}$$

As $D_{\alpha\beta}(\mathbf{q}, mm')$ is Hermitian, its eigenvalues $\omega^2(-\mathbf{q}j)$ must be real. Further, we note that $\mathbf{e}^*(-\mathbf{q}j, m)$ are also the eigenfunctions of the same dynamical matrix, therefore, if $\mathbf{q}$ is not in the 1BZ at which $D_{\alpha\beta}(\mathbf{q}, mm')$ has degenerate values, we can write

$$\omega^2(-\mathbf{q}j) = \omega^2(\mathbf{q}j) \tag{7.30}$$

We see from Eqs. (7.20), (7.29) that both $e_\alpha(\mathbf{q}j, m)$ and $e^*_\alpha(-\mathbf{q}j, m)$ are the eigenfunctions of the dynamical matrix with the same eigenvalue. Therefore, the two eigenfunctions differ only by a constant factor whose modulus is unity and one can write

$$e^*_\alpha(-\mathbf{q}j, m) = C\, e_\alpha(\mathbf{q}j, m) \tag{7.31}$$

Sometimes, the constant C is written as a phase factor $e^{i\phi}$. There are two common choices of C, 1 or -1. As the choice of the constant does not affect any physical property, it is chosen to be unity, giving

$$e^*_\alpha(-\mathbf{q}j, m) = e_\alpha(\mathbf{q}j, m) \tag{7.32}$$

Now we are in a position to make a normal coordinate transformation. The transformation from the original displacements $u_\alpha(nm)$ to the normal coordinates $Q(\mathbf{q}j, t)$ is given by

$$u_\alpha(nm) = \frac{1}{\sqrt{NM_m}}\sum_{\mathbf{q}j} e_\alpha(\mathbf{q}j, m)\, Q(\mathbf{q}j, t)\, e^{i\mathbf{q}\cdot\mathbf{R}_n} \tag{7.33}$$

In defining the above equation, $\mathbf{q}$ is assumed to possess discrete values. Because the displacement is always real, therefore

$$u^*_\alpha(nm) = u_\alpha(nm) \tag{7.34}$$

Substituting Eq. (7.33) into Eq. (7.34) and using Eq. (7.32), one gets

$$Q^*(\mathbf{q}j, t) = Q(-\mathbf{q}j, t) \tag{7.35}$$

Using Eq. (7.33), the expression for kinetic energy becomes

$$\begin{aligned} T &= \frac{1}{2}\sum_{nm\alpha} M_m \dot{u}^2_\alpha(nm) \\ &= \frac{1}{2}\sum_{m\alpha}\sum_{\mathbf{q}jj'} e_\alpha(\mathbf{q}j, m)\, e^*_\alpha(\mathbf{q}j', m)\, \dot{Q}(\mathbf{q}j, t)\, \dot{Q}^*(\mathbf{q}j', t) \end{aligned} \tag{7.36}$$

Here we have used Dirac's delta function defined as

$$\delta(\mathbf{q}) = \frac{1}{N}\sum_{n} e^{\imath \mathbf{q}\cdot\mathbf{R}_n} \tag{7.37}$$

along with the properties given by Eqs. (7.32), (7.35). Further, using the orthogonality relation defined by Eq. (7.21a) in Eq. (7.36), we get

$$T = \frac{1}{2}\sum_{\mathbf{q}j} \dot{Q}(\mathbf{q}j, t)\,\dot{Q}^*(\mathbf{q}j, t) \tag{7.38}$$

In the crystal potential, defined by Eq. (7.5), V_0 is a constant potential of a perfectly periodic crystal so one can take the reference point such that $V_0 = 0$. Therefore, the lattice potential from Eq. (7.5) can be written as

$$V_i(\mathbf{R}) = \frac{1}{2}\sum_{nm\alpha}\sum_{n'm'\beta} V_{\alpha\beta}(nm, n'm')\,u_\alpha(nm)u_\beta(n'm') \tag{7.39}$$

Substituting the value of $u_\alpha(nm)$ in terms of normal coordinates from Eq. (7.33) and simplifying, we write

$$V_i = \frac{1}{2}\sum_{\mathbf{q}j} \omega^2(\mathbf{q}j)\,Q(\mathbf{q}j, t)\,Q^*(\mathbf{q}j, t) \tag{7.40}$$

Eq. (7.20), along with the properties of $e_\alpha(\mathbf{q}j, m)$, have been used in obtaining the above expression. Further, in obtaining Eq. (7.40) the dynamical matrix is defined as

$$D_{\alpha\beta}(\mathbf{q}, mm') = \frac{1}{\sqrt{M_m M_{m'}}}\sum_{n-n'} V_{\alpha\beta}(n-n'm, 0m')\,e^{-\imath\mathbf{q}\cdot[\mathbf{R}_{n'}-\mathbf{R}_n]} \tag{7.41}$$

Knowing the kinetic and potential energies, the Hamiltonian and Lagrangian of the lattice can be written as

$$\widehat{H}_i = T + V_i = \frac{1}{2}\sum_{\mathbf{q}j}\left[\dot{Q}(\mathbf{q}j, t)\dot{Q}^*(\mathbf{q}j, t) + \omega^2(\mathbf{q}j)Q(\mathbf{q}j, t)Q^*(\mathbf{q}j, t)\right] \tag{7.42}$$

$$L = T - V_i = \frac{1}{2}\sum_{\mathbf{q}j}\left[\dot{Q}(\mathbf{q}j, t)\dot{Q}^*(\mathbf{q}j, t) - \omega^2(\mathbf{q}j)Q(\mathbf{q}j, t)Q^*(\mathbf{q}j, t)\right] \tag{7.43}$$

The momentum conjugate to $Q^*(\mathbf{q}j, t)$ is defined as

$$P(\mathbf{q}j, t) = \frac{\partial L}{\partial \dot{Q}^*(\mathbf{q}j, t)} = \dot{Q}(\mathbf{q}j, t) \tag{7.44}$$

Hamilton's equations give

$$\dot{P}(\mathbf{q}j, t) = -\frac{\partial \widehat{H}_i}{\partial Q^*(\mathbf{q}j, t)} = -\omega^2(\mathbf{q}j)\,Q(\mathbf{q}j, t) \tag{7.45}$$

Substituting the value of $P(\mathbf{q}j, t)$ from Eq. (7.44) into Eq. (7.45), we get

$$\ddot{Q}(\mathbf{q}j, t) + \omega^2(\mathbf{q}j)Q(\mathbf{q}j, t) = 0 \tag{7.46}$$

which is the equation of motion of a simple harmonic oscillator. We see that each of the new coordinates is simply a periodic function that involves only one of the frequencies $\omega(\mathbf{q}j)$. In the theory of dynamical systems, such coordinates are customarily called *normal coordinates*. Each normal coordinate describes an independent mode of vibration of the crystal with only one frequency and such vibration modes are referred to as *normal modes*. Every atom in each normal mode vibrates with the same frequency and with the same phase and we see that there are as many normal modes as there are degrees of freedom (3sN) in the crystal. From Eq. (7.33) we see that the general motion of the crystal as a whole is a superposition of the normal mode motions, each weighted appropriately by the coefficients $e_\alpha(\mathbf{q}j, m)\exp(\imath\mathbf{q}\cdot\mathbf{R}_n)$.

7.4 QUANTIZATION OF LATTICE HAMILTONIAN

In the previous sections our discussion has been classical. A transition to quantum mechanics is made by regarding Q(**q**j, t) and P(**q**j, t) as equal time operators, which will be discussed below. From the displacement $u_\alpha(nm)$, given by Eq. (7.33), one can define the momentum $p_\alpha(nm)$ as

$$\begin{aligned} p_\alpha(nm) &= M_m \dot{u}_\alpha(nm) \\ &= \sqrt{\frac{M_m}{N}} \sum_{\mathbf{q}j} e_\alpha(\mathbf{q}j, m) P(\mathbf{q}j)\, e^{\imath \mathbf{q}\cdot \mathbf{R}_n} \end{aligned} \tag{7.47}$$

The lattice field is quantized if the displacement $u_\alpha(nm)$ and momentum $p_\alpha(nm)$ are replaced by the corresponding operators that satisfy the following commutation relations

$$\begin{aligned} \left[u_\alpha(nm), u_\beta^+(n'm')\right] &= \left[p_\alpha(nm), p_\beta^+(n'm')\right] = 0 \\ \left[u_\alpha(nm), p_\beta^+(n'm')\right] &= \imath\hbar\, \delta_{nn'}\, \delta_{mm'}\, \delta_{\alpha\beta} \end{aligned} \tag{7.48}$$

Here $u_\beta^+(nm)$ denotes the transpose conjugate of $u_\beta(nm)$ and so on. Substituting the values of the displacement $u_\alpha(nm)$ in terms of the normal coordinates from Eq. (7.33) and the momentum $p_\alpha(nm)$ from Eq. (7.47), one can show that

$$[Q^+(\mathbf{q}j), P(\mathbf{q}'j')] = [Q(\mathbf{q}j), P^+(\mathbf{q}'j')] = \imath\hbar\, \delta_{\mathbf{q}\mathbf{q}'}\, \delta_{jj'} \tag{7.49a}$$

$$[Q(\mathbf{q}j), Q^+(\mathbf{q}'j')] = [P(\mathbf{q}j), P^+(\mathbf{q}'j')] = [Q(\mathbf{q}j), P(\mathbf{q}'j')] = 0 \tag{7.49b}$$

Let us define Q(**q**j) and P(**q**j) in terms of the creation and destruction operators denoted by $a_{\mathbf{q}j}^+$ and $a_{\mathbf{q}j}$, respectively, as follows

$$Q(\mathbf{q}j) = \left[\frac{\hbar}{2\omega(\mathbf{q}j)}\right]^{1/2} \left[a_{\mathbf{q}j} + a_{-\mathbf{q}j}^+\right] \tag{7.50a}$$

$$P(\mathbf{q}j) = -\imath \left[\frac{\hbar\omega(\mathbf{q}j)}{2}\right]^{1/2} \left[a_{\mathbf{q}j} - a_{-\mathbf{q}j}^+\right] \tag{7.50b}$$

Substituting Eqs. (7.50a), (7.50b) into Eqs. (7.49a), (7.49b) one can further show that the annihilation and creation operators satisfy the following commutation relations

$$\begin{aligned} \left[a_{\mathbf{q}j}, a_{\mathbf{q}'j'}^+\right] &= \delta_{\mathbf{q}\mathbf{q}'}\, \delta_{jj'} \\ \left[a_{\mathbf{q}j}, a_{\mathbf{q}'j'}\right] &= \left[a_{\mathbf{q}j}^+, a_{\mathbf{q}'j'}^+\right] = 0 \end{aligned} \tag{7.51}$$

Substituting the values of Q(**q**j) and P(**q**j) from Eqs. (7.50a), (7.50b) in the Hamiltonian given by Eq. (7.42), we can write

$$\widehat{H}_i = \frac{1}{2} \sum_{\mathbf{q}j} \hbar\omega(\mathbf{q}j) \left[a_{\mathbf{q}j}^+ a_{\mathbf{q}j} + a_{-\mathbf{q}j} a_{-\mathbf{q}j}^+\right] \tag{7.52}$$

Using the first commutation relation of Eq. (7.51) in Eq. (7.52), we can get

$$\widehat{H}_i = \sum_{\mathbf{q}j} \hbar\omega(\mathbf{q}j) \left[a_{\mathbf{q}j}^+ a_{\mathbf{q}j} + \frac{1}{2}\right] \tag{7.53}$$

Here we have used the fact that $\omega^2(\mathbf{q}j) = \omega^2(-\mathbf{q}j)$. Eq. (7.53) is the standard form of the Hamiltonian for a collection of independent harmonic oscillators. To obtain the eigenvalues of $\widehat{H}_i$, let us first describe the properties of the creation and destruction operators. Let $|n\rangle$ denote the state having n particles in it. When the operator a^+ acts on the state $|n\rangle$ it increases the number of particles by one (or creates a particle) and its eigenvalue equation is given by

$$a^+|n\rangle = \sqrt{n+1}\,|n+1\rangle \tag{7.54}$$

Similarly, the operator a annihilates one particle and satisfies the equation

$$a\,|n\rangle = \sqrt{n}\,|n-1\rangle \tag{7.55}$$

Using the above two equations, it is easy to prove that

$$a^{+}a\,|n\rangle = n\,|n\rangle \tag{7.56}$$

The above equation shows that when the operator $a^{+}a$ acts on a state, it gives the total number of particles in that state and therefore $a^{+}a$ is called the *number operator*. Now when the Hamiltonian defined by Eq. (7.53) acts on the state $|n\rangle$, it gives the energy, that is

$$\widehat{H}_i\,|n\rangle = E\,|n\rangle \tag{7.57}$$

where[1]

$$E = \sum_{\mathbf{q}j}\left[n_{\mathbf{q}j} + \frac{1}{2}\right]\hbar\omega(\mathbf{q}j) \tag{7.58}$$

Here $n_{\mathbf{q}j}$ takes any integral value from zero to infinity. Eq. (7.58) says that the energy eigenvalues are quantized and the quantum of excitation energy of the lattice vibrations is called a *phonon* having energy $\hbar\omega(\mathbf{q}j)$. The frequency $\omega(\mathbf{q}j)$ is called the *phonon frequency* and the plot of $\omega(\mathbf{q}j)$ as a function of $\mathbf{q}$ is called the *phonon dispersion relation.*

7.5 SIMPLE APPLICATIONS

7.5.1 Linear Monatomic Lattice

Let us consider a one-dimensional array of atoms, each having the same mass M as shown in Fig. 6.6. Further, assume that the atoms interact with one another through the 1NN harmonic forces. The general expression for the equation of motion of a lattice is

$$M_m\ddot{u}_\alpha(nm) = -\sum_{n'm'\beta} V_{\alpha\beta}(nm, n'm')\,u_\beta(n'm') \tag{7.59}$$

which is the same as Eq. (7.10) and is written here for completeness. Here the labels m and m′ have no meaning and become redundant as the lattice is monatomic. Further, in a one-dimensional lattice (say along the x-axis), $\alpha = \beta = x$. Under these approximations Eq. (7.59) reduces to

$$M\ddot{u}_x(n) = -\sum_{n'} V_{xx}(n, n')\,u_x(n') \tag{7.60}$$

As the interactions are assumed to be finite with the 1NNs only, therefore, the force constants $V_{xx}(n,n)$, $V_{xx}(n,n+1)$ and $V_{xx}(n,n-1)$ are finite, with the others being zero. Further, as the atoms are identical, the interactions of the nth atom with the (n − 1)th and (n + 1)th atoms are the same and hence so are the force constants, that is,

$$V_{xx}(n, n+1) = V_{xx}(n, n-1) = V_i'' = -\alpha_F \tag{7.61}$$

V_i'' is the second derivative of the lattice potential $V_i(\mathbf{R})$. As the nth atom is attracted by both the (n − 1)th and (n + 1)th atoms, so the potential $V_i(\mathbf{R})$, and hence the force constant α_F, are attractive. From Eq. (7.12c), the force constants of a one-dimensional monatomic crystal satisfy the relation

$$\sum_{n'} V_{xx}(n, n') = 0 \tag{7.62}$$

Considering only the 1NN interactions, the above expression gives $V_{xx}(n,n)$ as

$$V_{xx}(n, n) = -[V_{xx}(n, n+1) + V_{xx}(n, n-1)] = 2\alpha_F \tag{7.63}$$

Substituting Eqs. (7.61), (7.63) into Eq. (7.60), we get

$$M\ddot{u}_x(n) = \alpha_F\,[u_x(n+1) + u_x(n-1) - 2u_x(n)] \tag{7.64}$$

---

1. In obtaining Eq. (7.57) we have taken the state $|n\rangle$ as
$|n\rangle = |n_{\mathbf{q}_1 j_1}, n_{\mathbf{q}_1 j_2} \cdots, n_{\mathbf{q}_1 j_{3s}}, n_{\mathbf{q}_2 j_1}, n_{\mathbf{q}_2 j_2} \cdots, n_{\mathbf{q}_2 j_{3s}}, \ldots, n_{\mathbf{q}_N j_1}, n_{\mathbf{q}_N j_2} \cdots, n_{\mathbf{q}_N j_{3s}}\rangle$

The displacement $u_x(n)$ satisfies the periodicity of the lattice, so it can be written as

$$u_x(n) = u_{0x} e^{\imath(\mathbf{q}\cdot\mathbf{R}_n - \omega t)} \tag{7.65}$$

u_{0x} is the amplitude of vibration in the x-direction. In one dimension, $\mathbf{R}_n = n\mathbf{a}$, so the above equation becomes

$$u_x(n) = u_{0x} e^{\imath(qna - \omega t)} \tag{7.66}$$

Substituting Eq. (7.66) into Eq. (7.64), we get

$$\omega = \sqrt{\frac{4\alpha_F}{M}} \sin\left(\frac{qa}{2}\right) \tag{7.67}$$

From the periodic boundary condition the allowed values of the wave vector are given as $q = 2\pi n/Na$, where n is an integer and N is the number of cells in the crystal. In the limit $N \to \infty$, the wave vector q, and hence the frequencies of the normal modes, form a quasicontinuum. Eq. (7.67) is the same as Eq. (6.43) with q replaced by K and hence the plot of phonon frequencies is the same as that shown in Fig. 6.8. At small values of q, Eq. (7.67) yields

$$\omega = \sqrt{\frac{4\alpha_F}{M}} \frac{qa}{2} \tag{7.68}$$

Therefore, at small wave vectors there is a linear relationship between ω and q with a slope $\sqrt{\alpha_F/M}\,a$.

7.5.2 Linear Diatomic Lattice

Let us now consider a linear lattice with two different kinds of atoms, named 1 and 2, per primitive cell, having masses M_1 and M_2 $(M_1 \rangle M_2)$ (see Fig. 7.1). The figure shows three consecutive cells of the diatomic lattice. Here again we assume that the atoms interact with one another through the 1NN harmonic forces. So atom 1 in the nth cell interacts with atom 2 in the nth and (n – 1)th cells with the same interaction. Similarly, atom 2 in the nth cell interacts with atom 1 in the nth and (n+1)th cells with the same interaction potential as in the case of the interaction of atom 1 with atom 2. Therefore, one can write

$$\begin{aligned} V_{xx}(n1, n2) = V_{xx}(n1, n-12) = V_{xx}(n2, n1) = V_{xx}(n2, n+11) \\ = V_i'' = -\alpha_F \end{aligned} \tag{7.69}$$

Considering only the 1NN interactions, Eq. (7.12c) for 1 and 2 types of atoms of the nth cell can be written as

$$V_{xx}(n1, n1) + V_{xx}(n1, n2) + V_{xx}(n1, n-12) = 0 \tag{7.70a}$$

$$V_{xx}(n2, n2) + V_{xx}(n2, n+11) + V_{xx}(n2, n1) = 0 \tag{7.70b}$$

Using Eq. (7.69) in Eqs. (7.70a), (7.70b), one can write

$$V_{xx}(n1, n1) = V_{xx}(n2, n2) = 2\alpha_F \tag{7.71}$$

The equations of motion for the atoms 1 and 2 in the nth cell can straightway be written from Eq. (7.59) as

$$M_1 \ddot{u}_x(n1) = \alpha_F [u_x(n2) + u_x(n-12) - 2u_x(n1)] \tag{7.72a}$$

$$M_2 \ddot{u}_x(n2) = \alpha_F [u_x(n1) + u_x(n+11) - 2u_x(n2)] \tag{7.72b}$$

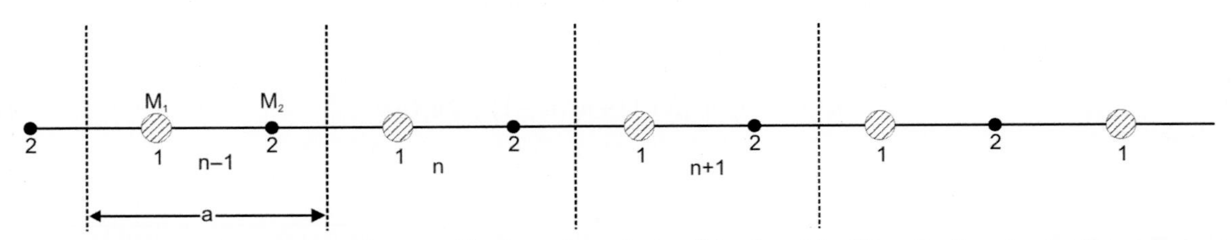

FIG. 7.1 Linear diatomic lattice with repeat distance a. The figure shows the (n – 1)th, nth, and (n+1)th unit cells and each cell contains two atoms of types 1 and 2.

In writing the above equations, Eqs. (7.69), (7.71) have been used. Assume the solution to be a running plane wave defined by

$$\mathbf{u}(nm) = \mathbf{u}(m)\, e^{\imath(\mathbf{q}\cdot\mathbf{R}_n - \omega t)} \tag{7.73}$$

Substituting Eq. (7.73) into Eqs. (7.72a), (7.72b) and using the fact that $\mathbf{R}_n = n\mathbf{a}$, we get

$$\left(\omega^2 M_1 - 2\alpha_F\right) u_x(1) + \alpha_F\left(1 + e^{-\imath qa}\right) u_x(2) = 0 \tag{7.74a}$$

$$\alpha_F\left(1 + e^{\imath qa}\right) u_x(1) + \left(\omega^2 M_2 - 2\alpha_F\right) u_x(2) = 0 \tag{7.74b}$$

Eqs. (7.74a), (7.74b) possess nontrivial solutions only if the determinant of the coefficients of $u_x(1)$ and $u_x(2)$ is zero, that is,

$$\begin{vmatrix} \omega^2 M_1 - 2\alpha_F & \alpha_F\left(1 + e^{-\imath qa}\right) \\ \alpha_F\left(1 + e^{\imath qa}\right) & \omega^2 M_2 - 2\alpha_F \end{vmatrix} = 0$$

Expanding the determinant one can write

$$M_1 M_2 \omega^4 - 2\alpha_F (M_1 + M_2)\omega^2 + 2\alpha_F^2 (1 - \cos qa) = 0 \tag{7.75}$$

The above equation can be solved for ω^2 to get

$$\omega^2 = \alpha_F\left(\frac{1}{M_1} + \frac{1}{M_2}\right) \pm \alpha_F\left[\left(\frac{1}{M_1} + \frac{1}{M_2}\right)^2 - \frac{4}{M_1 M_2}\sin^2\left(\frac{qa}{2}\right)\right]^{1/2} \tag{7.76}$$

which is the same equation as given by Eq. (6.80). From the periodic boundary conditions the wave vector is given by $q = 2\pi n/Na$, where the integer n lies in the range $-N/2 \le n \le N/2$ and N is assumed to be an even integer. As N approaches infinity, q becomes quasicontinuous. At $q = 0$, Eq. (7.76) gives two values of ω as

$$\omega_- = 0$$
$$\omega_+ = \left[2\alpha_F\left(\frac{1}{M_1} + \frac{1}{M_2}\right)\right]^{1/2} \tag{7.77}$$

Here ω_+ and ω_- are the values of ω for the plus and minus signs in Eq. (7.76). The values of ω^2 at the 1BZ boundary are obtained by putting $q = \pm\pi/a$ in Eq. (7.76), giving

$$\omega_+ = \left(\frac{2\alpha_F}{M_1}\right)^{1/2}, \quad \omega_- = \left(\frac{2\alpha_F}{M_2}\right)^{1/2} \tag{7.78}$$

The plot of phonon frequencies ω as a function of q in the 1BZ is the same as that shown in Fig. 6.15 (replacing K by q). Fig. 7.2 shows the phonon frequency curves in both the 1BZ and 2BZ, that is, in the extended zone scheme. The same diagram can be extended to the higher BZs. The phonon frequencies exhibit periodic behavior due to the periodicity of the lattice.

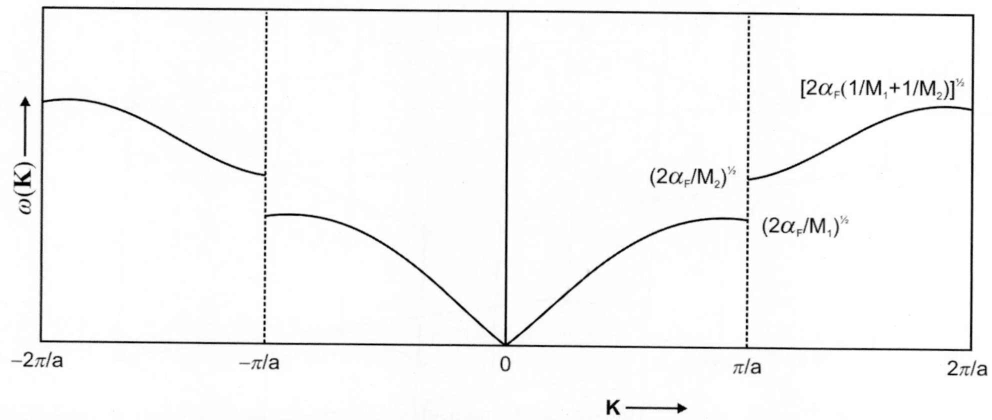

FIG. 7.2 Phonon frequencies as a function of wave vector **K** for the normal modes of a diatomic linear lattice in the extended zone scheme. The wave vector **K** can have any value ranging from zero to infinity. The phonon dispersion relations of the 2BZ can be brought to the 1BZ by adding or subtracting the smallest lattice vector $\mathbf{G}_1$ to **K**, that is, $\mathbf{q} = \mathbf{K} \pm \mathbf{G}_1$, thus obtaining the dispersion relations shown in Fig. 6.15. It is for this reason that **q** is called the reduced wave vector.

7.5.3 Simple Cubic Lattice

The general formalism developed in this chapter can be applied to study the phonon frequencies in three-dimensional crystals having different structures. Here we consider an sc lattice, which is the simplest among the three-dimensional crystals. To further simplify the equations of motion, we assume central forces between the 1NN and 2NN atoms only (see Fig. 7.3). The equation of motion along the x-direction, for an sc lattice is given by

$$\begin{aligned}
M\ddot{u}_x(\mathbf{R}_n) = {} & \alpha_{F1} \sum_{n_1'=\pm 1} \left[u_x(n_1+n_1', n_2, n_3) - u_x(n_1, n_2, n_3)\right] \\
& + \alpha_{F2} \sum_{n_1'=\pm 1} \sum_{n_2'=\pm 1} \Big[u_x(n_1+n_1', n_2+n_2', n_3) - u_x(n_1, n_2, n_3) \\
& + n_1' n_2' \left\{u_y(n_1+n_1', n_2+n_2', n_3) - u_y(n_1, n_2, n_3)\right\}\Big] \\
& + \alpha_{F2} \sum_{n_1'=\pm 1} \sum_{n_3'=\pm 1} \Big[u_x(n_1+n_1', n_2, n_3+n_3') - u_x(n_1, n_2, n_3) \\
& + n_1' n_3' \left\{u_z(n_1+n_1', n_2, n_3+n_3') - u_z(n_1, n_2, n_3)\right\}\Big]
\end{aligned} \tag{7.79}$$

where α_{F1} and α_{F2} are the 1NN and 2NN force constants. To make the representation explicit, the lattice sites, which are also the atomic sites in this case, are specified by three integers n_1, n_2, n_3, and these are collectively written as n in $\mathbf{R}_n$ (see Chapter 1). The equations of motion along the y- and z-directions can also be written by the cyclic permutations of x, y, z and the increments of n_1, n_2, n_3. We are interested in the plane wave solutions of the form

$$u_\alpha(\mathbf{R}_n) = u_\alpha^0 e^{\imath(\mathbf{q}\cdot\mathbf{R}_n - \omega t)} \tag{7.80}$$

Substituting Eq. (7.80) into Eq. (7.79), one can immediately write the equation of motion along the x-direction as

$$\omega^2 M u_x^0 = \left[2\alpha_{F1}(1-C_x) + 4\alpha_{F2}\left(2 - C_x C_y - C_x C_z\right)\right] u_x^0 + 4\alpha_{F2} S_x S_y u_y^0 + 4\alpha_{F2} S_x S_z u_z^0 \tag{7.81}$$

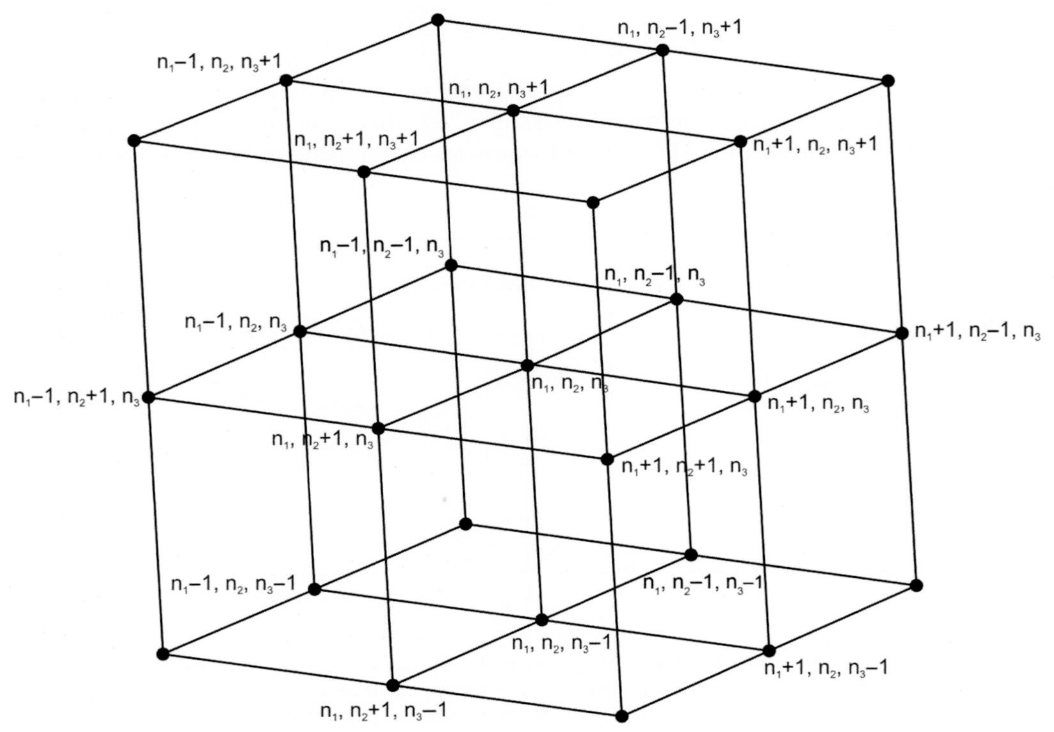

FIG. 7.3 The 1NNs and 2NNs of an atom (n_1, n_2, n_3) in an sc structure.

where

$$C_\alpha = \cos(q_\alpha a), \quad S_\alpha = \sin(q_\alpha a) \tag{7.82}$$

Similarly, the equations of motion along the y- and z-directions can be written as

$$\omega^2 M u_y^0 = 4\alpha_{F2} S_x S_y u_x^0 + \left[2\alpha_{F1}\left(1 - C_y\right) + 4\alpha_{F2}\left(2 - C_x C_y - C_y C_z\right)\right] u_y^0 + 4\alpha_{F2} S_y S_z u_z^0 \tag{7.83}$$

$$\omega^2 M u_z^0 = 4\alpha_{F2} S_x S_z u_x^0 + 4\alpha_{F2} S_y S_z u_y^0 + \left[2\alpha_{F1}(1 - C_z) + 4\alpha_{F2}\left(2 - C_x C_z - C_y C_z\right)\right] u_z^0 \tag{7.84}$$

The nontrivial solution of Eqs. (7.81), (7.83), (7.84) is obtained by equating to zero the determinant of the coefficients of u_x^0, u_y^0, u_z^0, that is,

$$\begin{vmatrix} 2\alpha_{F1}(1 - C_x) + 4\alpha_{F2}\left(2 - C_x C_y - C_x C_z\right) - \omega^2 M & 4\alpha_{F2} S_x S_y & 4\alpha_{F2} S_x S_z \\ 4\alpha_{F2} S_x S_y & 2\alpha_{F1}\left(1 - C_y\right) + 4\alpha_{F2}\left(2 - C_x C_y - C_y C_z\right) - \omega^2 M & 4\alpha_{F2} S_y S_z \\ 4\alpha_{F2} S_x S_z & 4\alpha_{F2} S_y S_z & 2\alpha_{F1}(1 - C_z) + 4\alpha_{F2}\left(2 - C_x C_z - C_y C_z\right) - \omega^2 M \end{vmatrix} = 0 \tag{7.85}$$

Eq. (7.85) is a cubic equation in ω^2 and in a general direction its solution is difficult. But in high-symmetry directions, such as [100], [110], and [111], its solution becomes quite simple. Consider the case of a wave propagating in the [100] direction in which the wave vector $\mathbf{q}$ = [q00], for which

$$S_y = S_z = 0, \quad C_y = C_z = 1$$

So Eq. (7.85) reduces to

$$\begin{vmatrix} 2\alpha_{F1}(1 - C_x) + 8\alpha_{F2}(1 - C_x) - \omega^2 M & 0 & 0 \\ 0 & 4\alpha_{F2}(1 - C_x) - \omega^2 M & 0 \\ 0 & 0 & 4\alpha_{F2}(1 - C_x) - \omega^2 M \end{vmatrix} = 0 \tag{7.86}$$

The solutions of the above equation are given as

$$\omega_1^2 = \frac{4\alpha_{F1}}{M}\left(1 + 4\frac{\alpha_{F2}}{\alpha_{F1}}\right)\sin^2\frac{qa}{2} \tag{7.87}$$

$$\omega_2^2 = \frac{8\alpha_{F2}}{M}\sin^2\frac{qa}{2} \tag{7.88}$$

Eq. (7.87) gives a single solution, while Eq. (7.88) gives two equal solutions. Using the standard technique, it is straightforward to find the eigenvectors corresponding to the frequencies given by Eqs. (7.87), (7.88). Eq. (7.87) corresponds to longitudinal waves in which the displacements are parallel to the x-axis, that is, in the direction of propagation. The two solutions given by Eq. (7.88) correspond to degenerate transverse waves in which the displacement is perpendicular to the direction of propagation, that is, in the y- and z-directions. Therefore, Eqs. (7.87), (7.88) yield

$$\omega_L = \sqrt{\frac{4\alpha_{F1}}{M}\left(1 + 4\frac{\alpha_{F2}}{\alpha_{F1}}\right)}\sin\frac{qa}{2} \tag{7.89}$$

$$\omega_T = \sqrt{\frac{8\alpha_{F2}}{M}}\sin\frac{qa}{2} \tag{7.90}$$

The dispersion curves for longitudinal and transverse waves with $\mathbf{q}$ along the [100] direction are shown in Fig. 7.4. In exactly the same manner the phonon frequencies can be evaluated for wave vector $\mathbf{q}$ along the [110], [111] directions.

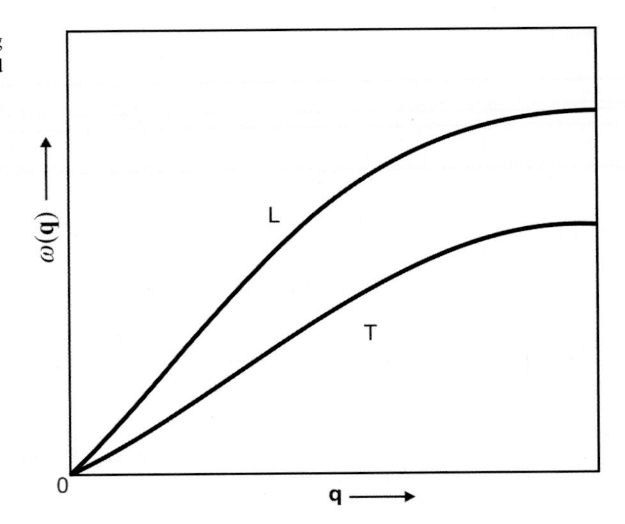

FIG. 7.4 Schematic representation of phonon frequencies as a function of **q** along the [100] direction for the normal modes of an sc solid. Here **q** is the reduced wave vector.

It is noteworthy that purely longitudinal and purely transverse waves can be obtained only in the high-symmetry directions. In a general direction of propagation, the normal modes of vibration are mixtures of the two types of wave and it is quite cumbersome to obtain the solution.

7.6 EXPERIMENTAL DETERMINATION OF PHONON FREQUENCIES

Scattering phenomena play a vital role in the determination of different properties of crystalline solids and these can be divided into two broad categories.

1. *Elastic scattering* is scattering in which the energy of the incident particles (radiation) remains unchanged during the scattering process, with a familiar example being Bragg reflection.

2. *Inelastic scattering* is scattering in which the incident particles (radiation) either gain or lose energy during the scattering process. For example, when incident radiation interacts with a lattice, the energy of the radiation changes either by the emission or absorption of a phonon.

The elastic and inelastic scattering phenomena are further divided into two categories: coherent and incoherent scattering processes. In coherent scattering from a lattice, the scattering of radiation takes place collectively from the ions, which thus interfere with one another. Elastic coherent scattering provides information about the structure of crystalline solids and has been discussed in Chapter 2. Inelastic coherent scattering provides information about the phonon frequencies of a crystalline solid. In incoherent scattering, the ions on the lattice scatter the radiation independently and, therefore, do not interfere with one another. Elastic incoherent scattering does not give any useful information, but inelastic incoherent scattering determines the frequency distribution function directly.

The phonon frequencies are determined either by using the X-ray diffraction or neutron diffraction techniques. The neutron diffraction technique is more powerful as it also provides information about the magnetic state of a crystalline solid. In the present text we shall briefly describe the neutron diffraction technique.

7.6.1 Neutron Diffraction Technique

Two methods are used to determine the phonon frequencies in crystalline solids.

7.6.1.1 Time-of-Flight Method

Neutrons are produced in a nuclear reactor with an average energy of 2 MeV. They are slowed down by passing through a material (moderator) to acquire thermal energy (≈ 0.025 eV for $T = 300$ K). The thermal neutrons are then Bragg reflected from a single large crystal, for example of Al or Pb, to produce a monochromatic beam of neutrons. In the time-of-flight method a monochromatic beam of neutrons is allowed to fall on the specimen crystal. The beam after reflection through an angle ϕ is collected by a counter and the time of flight is measured. In this method neutrons scattered at two or more than two angles can be analyzed. The main disadvantage of this method is that the phonon wave vector **q** is chosen randomly.

7.6.1.2 Constant Momentum Method

Brockhouse (1995) provided a constant **q** method for neutron diffraction, which is now widely used for measuring the phonon frequencies in crystalline solids. Let **K** be the wave vector of incident neutrons and **K**′ the wave vector after scattering. The conservation of momentum demands

$$\mathbf{K}' - \mathbf{K} = \Delta\mathbf{K} = \pm\mathbf{q} + \mathbf{G} \tag{7.91}$$

where **q** is the reduced wave vector. The plus (minus) sign on the right side of the above equation gives the absorption (creation) of a phonon during the neutron scattering process. The conservation of energy of the neutrons gives

$$\frac{\hbar^2}{2M_n}\left(K'^2 - K^2\right) = \pm\hbar\omega(\mathbf{q}) \tag{7.92}$$

M_n is the mass of the neutron and $\omega(\mathbf{q})$ is the frequency of vibration of the lattice with wave vector **q**. Again, the plus (minus) sign on the right side of Eq. (7.92) gives the absorption (creation) of a phonon. Thus, accurate measurements of **K** and **K**′ and the energy loss of the neutron beam provide the relation between $\omega(\mathbf{q})$ and **q**, that is, the phonon dispersion relation.

The experimental set up used to measure the phonon frequencies, which is usually called *triple axis spectrometer*, is shown in Fig. 7.5. A parallel beam of thermal neutrons from a nuclear reactor is allowed to fall on a single crystal X_M, usually called the monochromating crystal. The Bragg reflected neutrons from X_M constitute a monochromatic beam of thermal neutrons, which is allowed to pass through a collimator. By changing the angle θ_M it is possible to obtain the desired value of **K** or the momentum of incident neutrons. The monochromatic narrow beam of neutrons with the desired value of the wave vector **K** is allowed to fall on the specimen crystal S whose phonon frequencies are to be studied. The direction of incidence of the neutrons on the specimen crystal plane corresponding to a particular reciprocal lattice vector can be varied by changing the value of the angle θ_R. The neutron beam scattered through an angle ϕ is allowed to fall on the crystal X_A, called the analyzing crystal. By rotating the second arm, the angle ϕ between **K** and **K**′ can be varied. The scattered neutron beam is Bragg reflected from the crystal X_A in the direction θ_A and analyzed by the detector D. Knowing the values of θ_M and θ_A, and using Bragg's law, one can determine the wave vector of the incident neutron, **K**, and that of the scattered neutron, **K**′. Hence the main advantage of the triple axis spectrometer is that one can choose the desired values of wave vector **q** in a particular symmetry direction.

The triple axis spectrometer allows us to measure the energy of scattered neutrons by varying θ_A but keeping ϕ constant. In this way one obtains a peak in the scattered neutron beam for a particular wave vector **q**. The same experiment is repeated for different values of the scattering angle ϕ. The frequency $\omega(\mathbf{q})$ can be calculated from the energy conservation condition (Eq. 7.92). The measured values of the phonon frequencies of Al metal along the high-symmetry directions are shown in

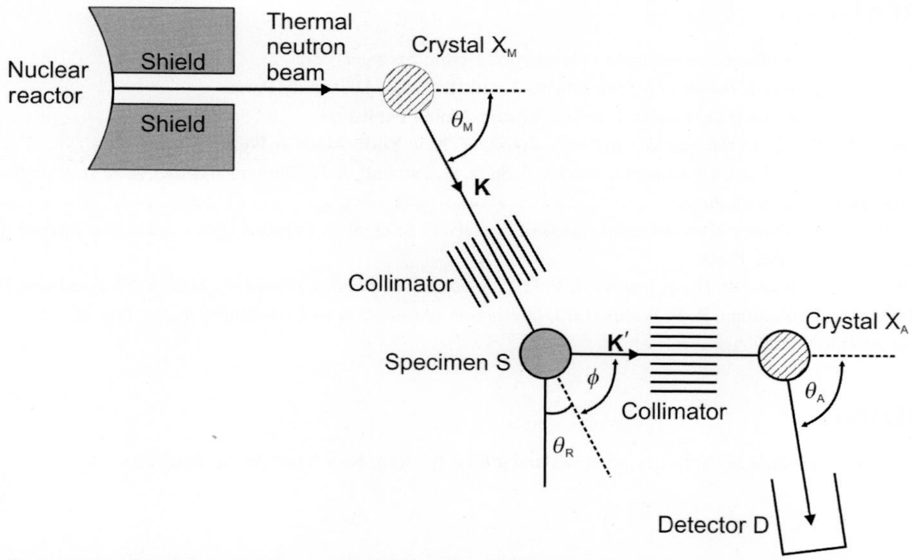

FIG. 7.5 Schematic diagram of a triple axis spectrometer. The angle θ_R is made by the incident direction with the vertical direction. *(Modified from Ghatak, A. K., & Kothari, L. S. (1972). An introduction to lattice dynamics. New York: Addison-Wesley.)*

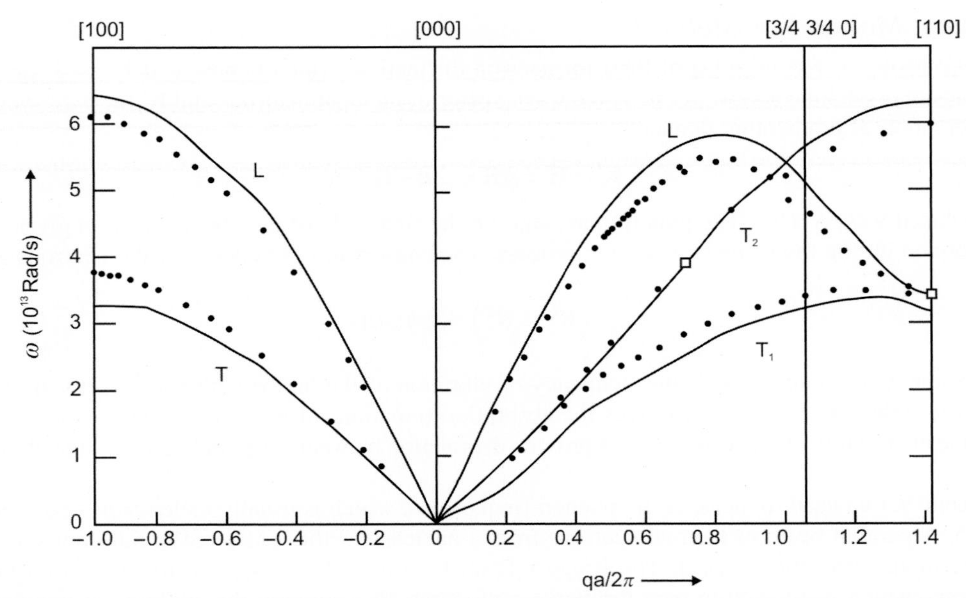

FIG. 7.6 The phonon frequencies of Al metal. *Dots* represent the experimental values of the phonon frequencies due to Yarnell et al. (1965), while the *curves* represent the theoretical results obtained by using the Harrison model potential. The points at which the pseudopotential was fit to the experiment are indicated by *squares. (Modified from Harrison, W. A. (1966). Pseudopotentials in the theory of metals and alloys 300. New York: W. A. Benjamin)*

Fig. 7.6 (Yarnell, Warren, & Koenig, 1965). In the [100] symmetry direction one obtains phonon frequencies of a longitudinal (L) wave, usually called the L branch, and one transverse (T) wave called the T branch. But actually, there are two T branches that overlap each other as these travel with the same velocity. In the [110] direction there is an L branch and two distinct T branches, namely T_1 and T_2.

REFERENCES

Brockhouse, B. N. (1995). Slow neutron spectroscopy and the grand atlas of the physical world. *Reviews of Modern Physics, 67*, 735–751.

Yarnell, J. L., Warren, J. L., & Koenig, S. H. (1965). Experimental dispersion curves for phonons in aluminium. In R. F. Wallis (Ed.), *Lattice dynamics* (pp. 57–61). Oxford: Pergamon.

SUGGESTED READING

Bilz, H., & Kress, W. (1979). *Phonon dispersion relations in insulators*. Berlin: Springer-Verlag.

Born, M., & Huang, K. (1954). *Dynamical theory of crystal lattices*. London: Oxford University Press.

Cochran, W. (1973). *The dynamics of atoms in crystals*. London: Edward Arnold Publishers.

Ghatak, A. K., & Kothari, L. S. (1972). *An introduction to lattice dynamics*. New York: Addison-Wesley.

Joshi, S. K., & Rajagopal, A. K. (1968). Lattice dynamics of metals. F. Seitz, D. Turnbull, & H. Ehrenreich (Eds.), *Solid state physics* (pp. 159–312). Vol. 22(pp. 159–312). New York: Academic Press.

Liebfried, G., & Ludwig, W. (1961). Theory of anharmonic effects in crystals. F. Seitz, & D. Turnbull (Eds.), *Solid state physics* (pp. 275–444). Vol. 12 (pp. 275–444). New York: Academic Press.

Maradudin, A. A., Montroll, E. W., Weiss, F. H., & Ipatova, I. P. (1971). *Solid state physics, (Suppl. 3)*. New York: Academic Press.

Wallis, R. F. (1977). Phonons and polaritons. R. F. Wallis (Ed.), *Interaction of radiation with condensed matter* (pp. 163–215). Vol. 1(pp. 163–215). Vienna: International Atomic Energy Agency.

FURTHER READING

Harrison, W. A. (1966). *Pseudopotentials in the theory of metals and alloys*: (p. 300). New York: W. A. Benjamin.

Chapter 8

Specific Heat of Solids

Chapter Outline

Specific heat is the most extensively studied thermal property of solids and is a measure of the capacity to store heat energy in a solid. Because of this it is sometimes called the heat capacity and is usually expressed per unit mass of the solid. Therefore, the specific heat of a solid is the heat energy required to raise the temperature of one gram (gram mole) of the solid by one degree centigrade. It is usually measured in cal/gram K (cal/mole K) of solid unless otherwise stated. The specific heat comprises two contributions:

1. Lattice specific heat

It is the contribution to the specific heat from the lattice of a solid.

2. Electronic specific heat

It is the contribution to the specific heat arising from the conduction electron distribution and is finite in conductors and semiconductors.

In this chapter we shall describe the lattice specific heat, which can be defined in two ways. First is the lattice specific heat at constant pressure C_P, which can be measured experimentally. Second is the lattice specific heat at constant volume C_V, which is easy to deal with theoretically. C_V quantitatively measures the ability of a system to absorb energy into its internal degrees of freedom, which, in turn, are related to the atomic and molecular characteristics of the system. Thus, the specific heat can provide an important link between the observed macroscopic behavior of a solid and its detailed atomic and molecular structure. The two types of specific heat are related to one another through the relation

$$C_P - C_V = 9\Gamma_{TH}^2 B_M V T \tag{8.1}$$

where Γ_{TH} is the coefficient of linear expansion. B_M, V, and T represent the bulk modulus, volume, and temperature (in degrees Kelvin), respectively. One can also write

$$C_P - C_V = R \tag{8.2}$$

where R is the gas constant.

Solid State Physics. https://doi.org/10.1016/B978-0-12-817103-5.00008-6

8.1 EXPERIMENTAL FACTS

For the first time in 1819, Dulong and Petit measured the specific heat of a solid at room temperature experimentally and found it to be 3R (≈ 6 calories per mole). This value is below the melting point of a solid. At and above the melting point, the specific heat starts increasing due to the change in phase. Actually, the specific heat is a function of temperature: at very low temperatures, it varies as T^3 and approaches zero at absolute zero (see Fig. 8.1). At intermediate temperatures, the specific heat varies linearly with T and approaches the Dulong and Petit's value at reasonably large temperatures (≈ room temperature). In a superconducting solid, the specific heat decreases exponentially below the superconducting transition temperature and goes to zero as the temperature approaches absolute zero. Further, in magnetic substances, the specific heat becomes large over the temperature range in which magnetic moments are ordered. This may be because the ordering of the magnetic moments decreases the entropy of the solid faster, thereby increasing the specific heat. Below 0.1K the ordering of magnetic (nuclear) moments may give very large values of specific heat.

8.2 THERMODYNAMICAL DEFINITION

The specific heat C in mathematical language can be written as

$$C = \frac{dQ}{dT} \tag{8.3}$$

where Q is the heat energy. When a solid is heated by imparting a small amount of energy dQ, then, a part of it is used in increasing the internal energy of the solid dE and the rest is used in doing work. So, one can write

$$dQ = dE + PdV \tag{8.4}$$

where P is pressure and dV is the volume change in a solid. Dividing Eq. (8.4) by dT we get

$$C = \frac{dQ}{dT} = \frac{dE}{dT} + P\frac{dV}{dT} \tag{8.5}$$

The energy E depends on both T and V. The specific heat at constant volume V, from Eq. (8.5), becomes

$$C_V = \left(\frac{dQ}{dT}\right)_V = \left(\frac{dE}{dT}\right)_V \tag{8.6}$$

From the second law of thermodynamics, for reversible processes,

$$dQ = TdS \tag{8.7}$$

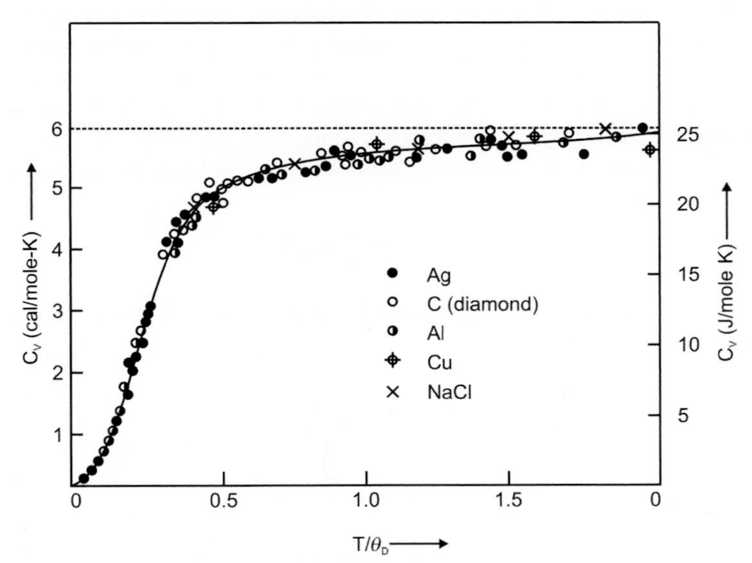

FIG. 8.1 Temperature dependence of experimentally measured values of specific heat for some elements. The *solid line* represents the theoretical values obtained for C_V from the Debye theory. *(Modified from Epifanov, G. I. (1979).* Solid state physics *(p. 119). Moscow: Mir Publishers.)*

where S is entropy of the system. So

$$C_V = T\left(\frac{dS}{dT}\right)_V \tag{8.8}$$

In theoretical investigations, we usually consider C_V as it can easily be obtained from E and S of a thermodynamic system. Therefore, in this chapter, we discuss the evaluation of C_V from which C_P can be obtained by using Eq. (8.2).

8.3 PHASE SPACE

A space that gives both the position and velocity of a particle, as a function of time, is known as phase space (or configuration space). Consider a physical system with a large number of particles in the phase space. The state of the system at a given instant of time may be defined by specifying the position and velocity of each particle at that time. The simplest example of phase space is one with a one-dimensional velocity distribution, say along the x-direction. Let x and v_x represent the position and velocity of a particle in the x-direction. Such a phase space is two-dimensional and the elemental volume reduces to an elemental area (see Fig. 8.2) given by

$$d\tau_2 = dx\, dv_x \tag{8.9}$$

In a realistic system the velocity distribution is three dimensional and the position and velocity of a particle at a particular time t is specified by the corresponding coordinates as

$$\mathbf{r} = \sum_\alpha \hat{\mathbf{i}}_\alpha r_\alpha \tag{8.10}$$

$$\mathbf{v} = \sum_\alpha \hat{\mathbf{i}}_\alpha v_\alpha \tag{8.11}$$

The phase space for a single particle is six dimensional: three components for position and three components for velocity. The volume element in this space is given by

$$d\tau_6 = d^3r d^3v \tag{8.12}$$

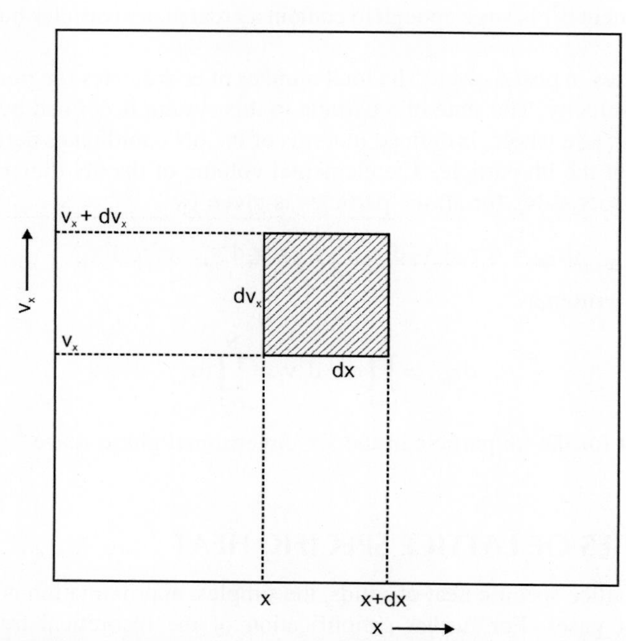

FIG. 8.2 Two-dimensional phase space with one-dimensional velocity distribution function $f(x, v_x, t)$. The *shaded region* shows the elemental area for particles having positions between x and x+dx and velocity between v_x and v_x+dv_x.

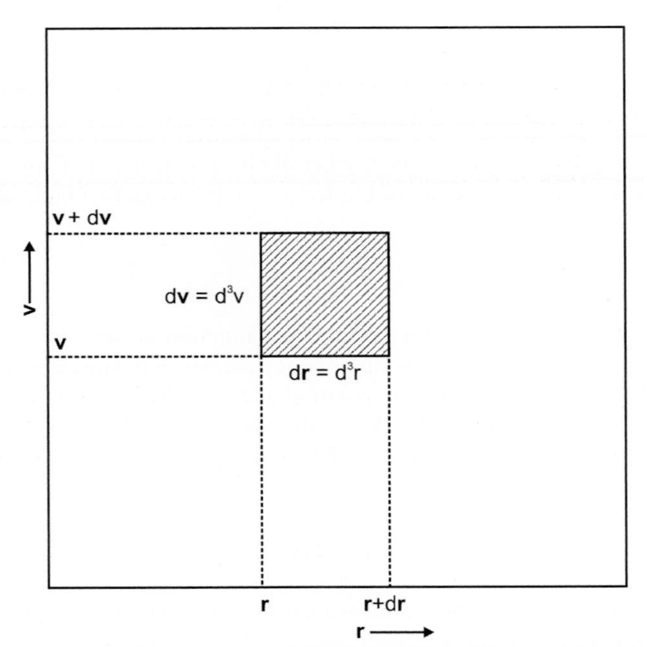

FIG. 8.3 Six-dimensional phase space with three-dimensional velocity distribution function f(**r**, **v**, t). The *shaded region* shows the elemental volume for particles having positions between **r** and **r**+**dr** and velocity between **v** and **v**+**dv** .

where

$$d^3r = dr_1 dr_2 dr_3 = dx\,dy\,dz \tag{8.13}$$

$$d^3v = dv_1 dv_2 dv_3 = dv_x dv_y dv_z \tag{8.14}$$

Here d^3r is the volume element around the terminal point **r** in the position space and d^3v is the volume element around the terminal point **v** in the velocity space. Fig. 8.3 shows the volume element for a particle in six-dimensional phase space. Here it is assumed that the volume element d^3r is large enough to contain a great many particles but small enough compared with the dimensions of the system.

Consider a system of N particles in phase space. The total number of coordinates (degrees of freedom) of the system is 6N: 3N for position and 3N for velocity. The state of a particle in this system is defined by six coordinates as mentioned above. But the state of the system, as a whole, is defined in terms of the 6N coordinates defining a 6N-dimensional space. Let $\mathbf{r}_i$ and $\mathbf{v}_i$ be the coordinates of the ith particle. The elemental volume of the 6N-dimensional space with coordinates ranging from $\mathbf{r}_i$ to $\mathbf{r}_i + \mathbf{dr}_i$ and $\mathbf{v}_i$ to $\mathbf{v}_i + \mathbf{dv}_i$, for all the particles is given by

$$d\tau_{6N} = d^3r_1 d^3v_1 d^3r_2 d^3v_2 \ldots d^3r_i d^3v_i \ldots d^3r_N d^3v_N \tag{8.15}$$

The above equation can also be written as

$$d\tau_{6N} = \prod_{i=1}^{N} d^3r_i d^3v_i = \prod_{i=1}^{N} d\tau_6^i \tag{8.16}$$

where $d\tau_6^i$ is the volume element for the ith particle in the six-dimensional phase space.

8.4 CLASSICAL THEORIES OF LATTICE SPECIFIC HEAT

In the classical treatment of the lattice specific heat of solids, the simplest approximation is to treat atoms as free particles and apply the kinetic theory of gases. For further simplification of the theoretical treatment we consider an ideal monatomic gas.

8.4.1 Free Atom Model

8.4.1.1 One-Dimensional Solid

Consider a one-dimensional ideal gas of atoms in which the atoms move freely along the x-direction. The energy of a free atom is given by

$$E_x = \frac{p_x^2}{2M} \tag{8.17}$$

where M and p_x are the mass and momentum of an atom along the x-direction. If $f(E,T)$ is the distribution function for the atoms, the average energy of an atom can be written as

$$\overline{E} = \frac{\int E\, f(E,T)\, d\tau}{\int f(E,T)\, d\tau} \tag{8.18}$$

Here $d\tau$ is the volume element in the phase space, given in one dimension by $d\tau = dx\, dp_x$. The Maxwell-Boltzmann distribution, usually called the classical distribution of atoms, is given by

$$f(E,T) = e^{-\frac{E}{k_B T}} \tag{8.19}$$

where k_B is the Boltzmann constant. Using Eqs. (8.17), (8.19) in Eq. (8.18), the average energy of an atom in a one-dimensional solid is given by

$$\overline{E} = \frac{\int \frac{p_x^2}{2M} e^{\frac{-p_x^2/2M}{k_B T}} dx\, dp_x}{\int e^{\frac{-p_x^2/2M}{k_B T}} dx\, dp_x} \tag{8.20}$$

The integral over x gives L_x, the length of the one-dimensional solid, while the integral over p_x will be from $-\infty$ to ∞. Hence Eq. (8.20) reduces to

$$\overline{E} = \frac{\int_{-\infty}^{\infty} \frac{p_x^2}{2M} e^{-\beta_0 p_x^2/2M} dp_x}{\int_{-\infty}^{\infty} e^{-\beta_0 p_x^2/2M} dp_x} \tag{8.21}$$

where

$$\beta_0 = \frac{1}{k_B T} \tag{8.22}$$

The integrands of the integrals in Eq. (8.21) are an even function of p_x so the integration can be performed from 0 to ∞ and multiplied by a factor of two. Therefore

$$\overline{E} = \frac{\int_{0}^{\infty} \frac{p_x^2}{2M} e^{-\beta_0 p_x^2/2M} dp_x}{\int_{0}^{\infty} e^{-\beta_0 p_x^2/2M} dp_x}$$

With the help of Eq. (8.17) the above integral can be written as

$$\overline{E} = \frac{\int_0^\infty E_x^{1/2} e^{-\beta_0 E_x} dE_x}{\int_0^\infty E_x^{-1/2} e^{-\beta_0 E_x} dE_x} \tag{8.23}$$

Integrating by parts, one finally gets

$$\overline{E} = \frac{1}{2} k_B T \tag{8.24}$$

Eq. (8.24) gives the energy associated with one degree of freedom. The average energy of a one-dimensional solid with N atoms becomes

$$E = N\overline{E} = \frac{1}{2} N k_B T \tag{8.25}$$

From Eqs. (8.6), (8.25) C_V is given by

$$C_V = \frac{1}{2} N k_B = \frac{1}{2} R \tag{8.26}$$

where the gas constant R is given by

$$R = N k_B \tag{8.27}$$

8.4.1.2 Two-Dimensional Solid

Consider a two-dimensional solid with atoms moving freely in the xy-plane (two-dimensional ideal gas of atoms). The energy of a free atom can be written as

$$E = \frac{p^2}{2M} \tag{8.28}$$

with

$$p^2 = p_x^2 + p_y^2 \tag{8.29}$$

Substituting Eqs. (8.19), (8.28) into Eq. (8.18), the average energy of an atom becomes

$$\overline{E} = \frac{\int \frac{p^2}{2M} e^{-\beta_0 p^2/2M} d\tau}{\int e^{-\beta_0 p^2/2M} d\tau} \tag{8.30}$$

In two dimensions

$$d\tau = d^2 r\, d^2 p = (2\pi r dr)(2\pi p dp) \tag{8.31}$$

Substituting Eq. (8.31) into Eq. (8.30), one can write

$$\overline{E} = \frac{\int_0^\infty \frac{p^2}{2M} e^{-\beta_0 p^2/2M} p dp}{\int_0^\infty e^{-\beta_0 p^2/2M} p dp} \tag{8.32}$$

The integral over r gives the area A of the two-dimensional solid. Converting momentum into energy with the help of Eq. (8.28), we get

$$\overline{E} = \frac{\int_0^\infty E e^{-\beta_0 E} dE}{\int_0^\infty e^{-\beta_0 E} dE} \tag{8.33}$$

Solving the integral by parts, one gets

$$\overline{E} = \frac{1}{\beta_0} = k_B T \tag{8.34}$$

$\overline{E}$ is the energy associated with two degrees of freedom and therefore is double the energy of an atom in a one-dimensional solid. Hence the average energy of a two-dimensional solid with N atoms becomes

$$E = N\overline{E} = Nk_B T \tag{8.35}$$

The specific heat at constant volume C_V is given by

$$C_V = Nk_B = R \tag{8.36}$$

which is twice of the value obtained in a one-dimensional solid.

8.4.1.3 Three-Dimensional Solid

Consider a three-dimensional solid in the form of a cube of edge L. The atoms move freely in all possible directions and the energy of a free atom is given by

$$E = \frac{p^2}{2M} \tag{8.37}$$

with

$$p^2 = p_x^2 + p_y^2 + p_z^2 \tag{8.38}$$

The average energy of an atom in the phase space is given by

$$\overline{E} = \frac{\int \frac{p^2}{2M} e^{-\beta_0 p^2/2M} d^3r\, d^3p}{\int e^{-\beta_0 p^2/2M} d^3r\, d^3p} \tag{8.39}$$

The integration over the direct space gives the volume of the solid $V = L^3$. Using spherical coordinates, the integral over momentum becomes

$$\overline{E} = \frac{\int_0^\infty \frac{p^2}{2M} e^{-\beta_0 p^2/2M} p^2 dp}{\int_0^\infty e^{-\beta_0 p^2/2M} p^2 dp} \tag{8.40}$$

Substituting Eq. (8.37) for the energy E into Eq. (8.40), one can write

$$\overline{E} = \frac{\int_0^\infty E^{3/2} e^{-\beta_0 E} dE}{\int_0^\infty E^{1/2} e^{-\beta_0 E} dE} \tag{8.41}$$

Solving the above integral by parts, we get

$$\overline{E}=\frac{3}{2}\frac{1}{\beta_0}=\frac{3}{2}k_BT \tag{8.42}$$

which is the sum of energies of the three degrees of freedom of an atom. If there are N atoms in the solid, the total energy becomes

$$E=N\overline{E}=\frac{3}{2}Nk_BT \tag{8.43}$$

The specific heat C_V, therefore, is given by

$$C_V=\frac{3}{2}Nk_B=\frac{3}{2}R \tag{8.44}$$

Eq. (8.44) gives only half the value observed by Dulong and Petit. From the above discussion it is evident that the average energy of an atom is equal to the energy associated with one degree of freedom multiplied by the number of degrees of freedom n of the atom. So, in general, one can write the average energy of an atom as

$$\overline{E}=\frac{1}{2}nk_BT \tag{8.45}$$

Note that n is also equal to the dimensionality of the solid. The specific heat C_V, then, becomes

$$C_V=\frac{1}{2}nNk_B=\frac{1}{2}nR \tag{8.46}$$

8.4.2 Fixed Classical Harmonic Oscillator Model

In a solid, the atoms cannot move from one place to another but can oscillate about their mean positions. So, the first improvement over the free-atom model is to assume each atom as a harmonic oscillator with a fixed equilibrium position.

8.4.2.1 One-Dimensional Solid

In a one-dimensional solid, an atom will execute simple harmonic motion along the x-direction. So, the energy of an atom is given by

$$E=\frac{p_x^2}{2M}+\frac{1}{2}\alpha_F x^2=\frac{p_x^2}{2M}+\frac{1}{2}M\omega^2x^2 \tag{8.47}$$

where

$$\omega=\sqrt{\frac{\alpha_F}{M}} \tag{8.48}$$

Here α_F is the force constant, M is the mass, and ω is the natural frequency of vibration of the atom. The average energy of an atom, from Eq. (8.18), becomes

$$\overline{E}=\frac{\int_{-\infty}^{\infty}\left(\frac{p_x^2}{2M}+\frac{1}{2}M\omega^2x^2\right)e^{-\beta_0\left(\frac{p_x^2}{2M}+\frac{1}{2}M\omega^2x^2\right)}dx\,dp_x}{\int_{-\infty}^{\infty}e^{-\beta_0\left(\frac{p_x^2}{2M}+\frac{1}{2}M\omega^2x^2\right)}dx\,dp_x} \tag{8.49}$$

The above integral can be simplified to write

$$\overline{E}=\frac{\int_{-\infty}^{\infty}\frac{p_x^2}{2M}e^{-\beta_0\frac{p_x^2}{2M}}dp_x}{\int_{-\infty}^{\infty}e^{-\beta_0\frac{p_x^2}{2M}}dp_x}+\frac{1}{2}M\omega^2\frac{\int_{-\infty}^{\infty}x^2e^{-\frac{1}{2}\beta_0M\omega^2x^2}dx}{\int_{-\infty}^{\infty}e^{-\frac{1}{2}\beta_0M\omega^2x^2}dx} \tag{8.50}$$

Putting $E_x = p_x^2/2M$ and $q = \frac{1}{2}\beta_0 M\omega^2 x^2$ into Eq. (8.50), we get

$$\overline{E} = \frac{\int_0^\infty E_x^{1/2} e^{-\beta_0 E_x} dE_x}{\int_0^\infty E_x^{-1/2} e^{-\beta_0 E_x} dE_x} + \frac{1}{\beta_0} \frac{\int_0^\infty q^{1/2} e^{-q} dq}{\int_0^\infty q^{-1/2} e^{-q} dq}$$

The first integral is exactly the same as in the case of the one-dimensional free atom model and the second integral can be solved by integrating it by parts to give

$$\overline{E} = \frac{1}{2} k_B T + \frac{1}{2} k_B T = k_B T \tag{8.51}$$

If there are N atoms in the one-dimensional solid, then the total energy E becomes

$$E = N\overline{E} = Nk_B T \tag{8.52}$$

Therefore, the specific heat C_V is given by

$$C_V = Nk_B = R \tag{8.53}$$

Problem 8.1 Two-Dimensional Solid

Consider a two-dimensional solid with N atoms in which each atom is considered to be a fixed harmonic oscillator. The energy of an atom is given by

$$E = \frac{p^2}{2M} + \frac{1}{2}\alpha_F r^2 \tag{8.54}$$

where

$$p^2 = p_x^2 + p_y^2$$
$$r^2 = x^2 + y^2$$

and

$$\omega = \sqrt{\frac{\alpha_F}{M}} \tag{8.55}$$

Here M is the mass of an atom. Prove that the total energy of the solid and the specific heat are given by

$$E = 2Nk_B T \tag{8.56}$$
$$C_V = 2Nk_B = 2R \tag{8.57}$$

8.4.2.2 Three-Dimensional Solid

Consider a three-dimensional solid with N atoms in which each atom is considered to be a fixed harmonic oscillator with energy

$$E = \frac{p^2}{2M} + \frac{1}{2}\alpha_F r^2 \tag{8.58}$$

where

$$r^2 = x^2 + y^2 + z^2 \tag{8.59}$$

The average energy of an atom is found by substituting Eq. (8.58) into Eq. (8.18), that is,

$$\overline{E}=\frac{\int_{-\infty}^{\infty}\left(\frac{p^2}{2M}+\frac{1}{2}M\omega^2r^2\right)e^{-\beta_0\left(\frac{p^2}{2M}+\frac{1}{2}M\omega^2r^2\right)}d^3r\,d^3p}{\int_{-\infty}^{\infty}e^{-\beta_0\left(\frac{p^2}{2M}+\frac{1}{2}M\omega^2r^2\right)}d^3r\,d^3p} \tag{8.60}$$

Simplifying the above integral, we find

$$\overline{E}=\frac{\int_{-\infty}^{\infty}\frac{p^2}{2M}e^{-\frac{\beta_0p^2}{2M}}d^3p}{\int_{-\infty}^{\infty}e^{-\frac{\beta_0p^2}{2M}}d^3p}+\frac{1}{2}M\omega^2\frac{\int_{-\infty}^{\infty}r^2e^{-\frac{1}{2}\beta_0M\omega^2r^2}d^3r}{\int_{-\infty}^{\infty}e^{-\frac{1}{2}\beta_0M\omega^2r^2}d^3r} \tag{8.61}$$

Using spherical coordinates, the above equation reduces to

$$\overline{E}=\frac{\int_{0}^{\infty}\frac{p^2}{2M}e^{-\frac{\beta_0p^2}{2M}}p^2dp}{\int_{0}^{\infty}e^{-\frac{\beta_0p^2}{2M}}p^2dp}+\frac{1}{2}M\omega^2\frac{\int_{0}^{\infty}r^4e^{-\frac{1}{2}\beta_0M\omega^2r^2}dr}{\int_{0}^{\infty}r^2e^{-\frac{1}{2}\beta_0M\omega^2r^2}dr} \tag{8.62}$$

The first integral is the same as in Eq. (8.40). The second integral can be solved by parts, finally yielding

$$\overline{E}=\frac{3}{2}k_BT+\frac{3}{2}k_BT=3k_BT \tag{8.63}$$

Hence the total energy of the three-dimensional solid becomes

$$E=N\overline{E}=3Nk_BT \tag{8.64}$$

The specific heat is given by

$$C_V=3Nk_B=3R\quad\approx 6\,\text{calories/degree/mole} \tag{8.65}$$

This model reproduces the Dulong and Petit's value but does not explain the temperature variation of C_V at low temperatures.

8.5 QUANTUM MECHANICAL THEORIES

8.5.1 Einstein Theory of Specific Heat

The classical theory yields a constant value of specific heat, but the experimental results exhibit temperature variation at low temperatures. To explain the temperature variation of specific heat, Einstein assumed that the atoms in a solid are identical independent quantum harmonic oscillators vibrating with the same natural frequency ω_E, usually called the Einstein frequency. The energy of a quantum harmonic oscillator is given by

$$E_n=\left(n+\frac{1}{2}\right)\hbar\omega_E \tag{8.66}$$

where n = 0, 1, 2, ... As the temperature decreases, the amplitude and energy of a harmonic oscillator decrease and, at absolute zero, the energy becomes $(1/2)\hbar\omega_E$, the zero-point energy. Einstein assumed that the oscillators obey the Maxwell-Boltzmann distribution and, therefore, the average energy of an oscillator is given by

$$\bar{E}(\omega_E) = \frac{\sum_{n=0}^{\infty}\left(n+\frac{1}{2}\right)\hbar\omega_E e^{-\beta_0(n+1/2)\hbar\omega_E}}{\sum_{n=0}^{\infty} e^{-\beta_0(n+1/2)\hbar\omega_E}} = \frac{\sum_{n=0}^{\infty} n\hbar\omega_E e^{-\beta_0 n\hbar\omega_E}}{\sum_{n=0}^{\infty} e^{-\beta_0 n\hbar\omega_E}} + \frac{1}{2}\hbar\omega_E \tag{8.67}$$

Solving Eq. (8.67) the average energy of a mode of vibration is given by

$$\bar{E}(\omega_E) = \frac{\hbar\omega_E}{e^{\frac{\hbar\omega_E}{k_B T}} - 1} + \frac{1}{2}\hbar\omega_E \tag{8.68}$$

Consider a solid comprising N atoms. Let D_n denotes the total number of degrees of freedom (total number of modes of vibration) of all the atoms in the solid: it has a value of 1 N, 2N, and 3 N for one-, two-, and three-dimensional solids, respectively. Therefore, the total energy of the solid is given by

$$E_T(\omega_E) = D_n \bar{E}(\omega_E) = D_n \frac{\hbar\omega_E}{e^{\frac{\hbar\omega_E}{k_B T}} - 1} + \frac{1}{2} D_n \hbar\omega_E \tag{8.69}$$

The specific heat at constant volume is given by

$$C_V = \left(\frac{\partial E_T}{\partial T}\right)_V = D_n \left(\frac{\partial \bar{E}}{\partial T}\right)_V \tag{8.70}$$

Substituting Eq. (8.68) into Eq. (8.70), C_V is given by

$$C_V = D_n \frac{\partial}{\partial T}\left(\frac{\hbar\omega_E}{e^{\frac{\hbar\omega_E}{k_B T}} - 1}\right) \tag{8.71}$$

The above expression shows that the zero-point energy of the solid does not contribute toward C_V and therefore may be disregarded in the forthcoming discussion on the specific heat. Let us define the Einstein temperature θ_E by

$$\theta_E = \frac{\hbar\omega_E}{k_B} \tag{8.72}$$

In terms of θ_E, Eq. (8.71) can be written as

$$C_V = D_n k_B \left(\frac{\theta_E}{T}\right)^2 \frac{e^{\theta_E/T}}{\left(e^{\theta_E/T} - 1\right)^2} \tag{8.73}$$

This is the general expression for C_V and one can study its limiting cases. At high temperatures $T >> \theta_E$ or $\theta_E/T \ll 1$, Eq. (8.73) reduces to

$$C_V = D_n k_B e^{\theta_E/T} \cong D_n k_B \tag{8.74}$$

Eq. (8.74) gives values $1 Nk_B$, $2 Nk_B$, and $3 Nk_B$ for one-, two-, and three-dimensional solids, respectively, and this agrees with the Dulong-Petit's value at high temperatures. At very low temperatures $T << \theta_E$ or $\theta_E/T \gg 1$, Eq. (8.73) reduces to

$$C_V = D_n k_B \left(\frac{\theta_E}{T}\right)^2 e^{-\theta_E/T} \tag{8.75}$$

Eq. (8.75) shows that the temperature variation of C_V in one-, two-, and three-dimensional solids is the same and approaches zero exponentially as the temperature goes to absolute zero. Fig. 8.4 shows the temperature variation of C_V for a three-dimensional solid obtained in the Einstein model. Experimentally it has been observed that C_V varies as T^3 at very low temperatures. Therefore, although the Einstein theory exhibits the temperature variation of specific heat, the decrease is much faster than that observed experimentally.

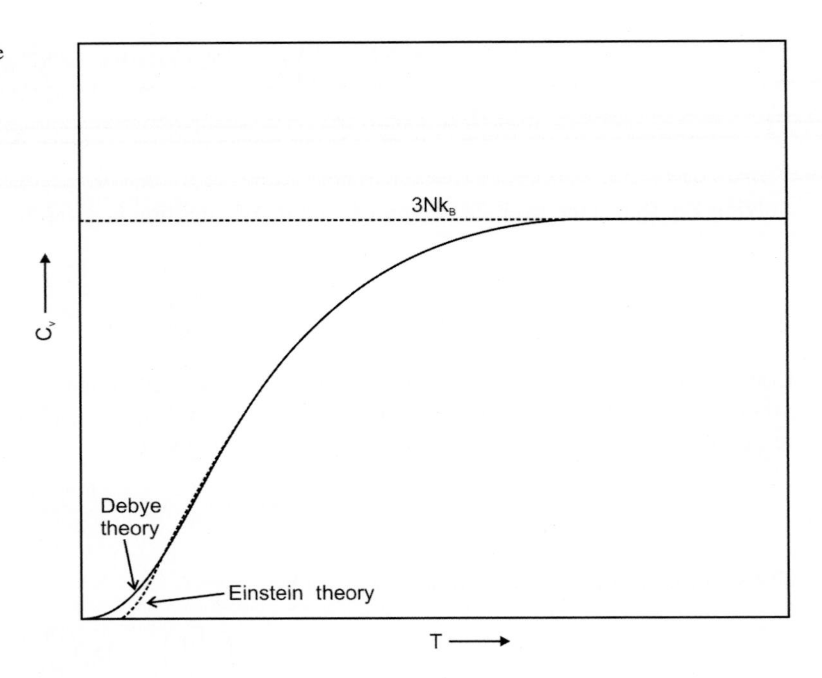

FIG. 8.4 The specific heat C_V as a function of temperature T due to the Einstein and Debye theories.

8.5.2 Debye Theory of Specific Heat

Debye pointed out a possible reason for the discrepancy in the Einstein theory of specific heat. He said that all of the atoms in a solid do not vibrate with the same frequency. Rather, there must be some spread in the allowed frequencies of vibration of the atoms. He assumed that the frequencies of atomic vibrations vary from zero to some maximum frequency ω_D, usually called the Debye frequency.

From Eq. (8.68) it is evident that, at low temperatures, the average energy per mode $\overline{E}(\omega)$ becomes small for large values of frequency ω. Therefore, at low temperatures, a significant contribution to the internal energy comes from the long wavelength (small ω) modes of vibration. For long wavelengths, it is reasonable to assume that the solid is a homogeneous continuous medium in which the dispersion relation is given by

$$\omega(\mathbf{K}) = vK \tag{8.76}$$

where v is the phase velocity, which is also equal to the group velocity.

8.5.2.1 Linear Monatomic Lattice

In a one-dimensional solid the number of modes of vibration in a wave vector K is given by

$$n = \frac{L}{2\pi}K \tag{8.77}$$

Therefore, the number of vibration modes in the range from K to K+dK is given by

$$dn = \frac{L}{\pi}dK \tag{8.78}$$

It is well known that every positive value of the wave vector K has the corresponding negative value −K. Therefore, in Eq. (8.78), we have multiplied by a factor of 2 to get the total number of vibration modes. From Eq. (8.76) one can write

$$d\omega = vdK \tag{8.79}$$

Substituting the value of dK from Eq. (8.79) into Eq. (8.78), one gets the number of vibration modes in the frequency range ω and $\omega + d\omega$ as

$$g(\omega)\,d\omega = \frac{L}{\pi v}\,d\omega = A\,d\omega \tag{8.80}$$

where $A = g(\omega) = L/\pi v$ is a constant. The function $g(\omega)$ gives the phonon density of states and should be defined in such a way that the total number of modes of vibration given by it must be equal to the total number of degrees of freedom N, that is,

$$\int_0^{\omega_D} g(\omega)\,d\omega = N \tag{8.81}$$

Substituting Eq. (8.80) into Eq. (8.81), one can immediately write

$$g(\omega) = A = \frac{N}{\omega_D} \tag{8.82}$$

Assuming $g(\omega)$ to be a continuous function of ω, the total energy of the solid becomes

$$E_T = \int \overline{E}\,g(\omega)\,d\omega = \int_0^{\infty} \frac{\hbar\omega}{e^{\frac{\hbar\omega}{k_B T}} - 1}\,g(\omega)\,d\omega \tag{8.83}$$

In writing the above expression, the zero-point energy has been neglected, as it does not contribute toward the lattice specific heat. Substituting

$$x = \frac{\hbar\omega}{k_B T} \quad \text{and} \quad \theta_D = \frac{\hbar\omega_D}{k_B} \tag{8.84}$$

along with Eq. (8.82) into Eq. (8.83), we obtain

$$E_T = N k_B \frac{T^2}{\theta_D} \int_0^{\theta_D/T} \frac{x\,dx}{e^x - 1} \tag{8.85}$$

Here θ_D is the Debye temperature. Eq. (8.85) gives the general expression for the total energy of a one-dimensional solid, which gives analytical expressions only in the limiting cases. At high temperatures, $x \ll 1$, Eq. (8.85) reduces to

$$E_T = N k_B T \tag{8.86}$$

Therefore, at high temperatures, one obtains the classical value of C_V, that is,

$$C_V = N k_B \tag{8.87}$$

At low temperatures, $x \gg 1$ and $\theta_D/T \to \infty$, Eq. (8.85) yields

$$E_T = N k_B \frac{T^2}{\theta_D} \int_0^{\infty} \frac{x\,dx}{e^x - 1} \tag{8.88}$$

The integral in the above expression is a standard one and has a value of $\pi^2/6$, therefore, the total energy is given by

$$E_T = \frac{\pi^2}{6} N k_B \frac{T^2}{\theta_D} \tag{8.89}$$

From the above expression, C_V is immediately written as

$$C_V = \frac{\pi^2}{3} N k_B \frac{T}{\theta_D} \tag{8.90}$$

Therefore, in a one-dimensional solid, C_V varies linearly with temperature.

The lattice specific heat can also be estimated using the frequency distribution function of the lattice vibrations for a one-dimensional solid given by Eq. (6.59). Substituting the value of $g(\omega)$ from Eq. (6.59) into Eq. (8.83), one gets

$$E_T = \frac{2N\hbar}{\pi} \int_0^{\omega_{max}} \frac{\omega d\omega}{\left(e^{\hbar\omega/k_BT} - 1\right)\left(\omega_{max}^2 - \omega^2\right)^{1/2}} \tag{8.91}$$

At low temperatures, the states with small energy (low frequency) are occupied or one can say that the high-frequency modes are effectively frozen. Therefore, at low temperatures, $\omega/\omega_{max} \langle\langle 1$, in this approximation, Eq. (8.91) becomes

$$E_T = \frac{2N\hbar}{\pi} \cdot \frac{1}{\omega_{max}} \int_0^{\omega_{max}} \frac{\omega d\omega}{e^{\hbar\omega/k_BT} - 1} \tag{8.92}$$

Making the substitution

$$x = \frac{\hbar\omega}{k_BT} \text{ and } \theta_D = \frac{\hbar\omega_{max}}{k_B} \tag{8.93}$$

Eq. (8.92) becomes

$$E_T = \frac{2Nk_B}{\pi} \frac{T^2}{\theta_D} \int_0^{\theta_D/T} \frac{x dx}{e^x - 1} \tag{8.94}$$

The integral in the above expression is the same as in Eq. (8.85), so it gives

$$E_T = \frac{\pi}{3} Nk_B \frac{T^2}{\theta_D} \tag{8.95}$$

The heat capacity at constant volume is given by

$$C_V = \frac{2\pi}{3} Nk_B \left(\frac{T}{\theta_D}\right) \tag{8.96}$$

Eq. (8.96) is similar to Eq. (8.90) except for the constant factor. At high temperature, $\hbar\omega \ll k_BT$, Eq. (8.91) reduces to

$$E_T = k_BT \int_0^{\omega_{max}} g(\omega) d\omega = Nk_BT \tag{8.97}$$

which is the same result as Eq. (8.86). The value of C_V from the above equation is given by

$$C_V = Nk_B \tag{8.98}$$

which is the Dulong and Petit's value for a one-dimensional solid.

8.5.2.2 Two-Dimensional Lattice

The phonon density of states in a two-dimensional solid can be calculated from the periodicity of the lattice. Consider a two-dimensional solid in the form of a square of side L. The number of phonon states n in a circle of radius K is given by

$$n = \pi K^2 \left(\frac{L}{2\pi}\right)^2 \tag{8.99}$$

So, the number of phonon states in a ring of radius K and thickness dK is obtained by differentiating Eq. (8.99) to obtain

$$dn = \frac{A_0}{2\pi} K dK \tag{8.100}$$

Here A_0 is the area of the two-dimensional solid. If we change the wave vector K into the frequency ω using the Debye approximation given by Eq. (8.76), we get

$$dn = \frac{A_0}{2\pi v^2} \omega \, d\omega \tag{8.101}$$

The above expression gives the number of phonon states between two circles of constant energy ω and $\omega + d\omega$, that is, $g(\omega)$ $d\omega$. Therefore, one can write

$$g(\omega) = \frac{A_0}{2\pi v^2}\omega = C\omega \tag{8.102}$$

where C is a constant given by

$$C = \frac{A_0}{2\pi v^2} \tag{8.103}$$

Eq. (8.102) can also be obtained from the general expression for the density of phonon states in two-dimensional solids (Appendix E). The total number of modes of vibration must be equal to the total number of degrees of freedom, that is,

$$\int_0^{\omega_D} g(\omega)\, d\omega = 2N = C\int_0^{\omega_D} \omega\, d\omega \tag{8.104}$$

The above equation gives C as

$$C = \frac{4N}{\omega_D^2} \tag{8.105}$$

Eqs. (8.102), (8.105) give the phonon density of states

$$g(\omega) = \frac{4N}{\omega_D^2}\omega \tag{8.106}$$

The total energy of the solid is given by

$$E_T = \int_0^{\omega_D} \overline{E}(\omega)\, g(\omega)\, d\omega \tag{8.107}$$

Substituting the values of $\overline{E}(\omega)$ from Eq. (8.68) and neglecting the zero-point energy and substituting $g(\omega)$ from Eq. (8.106) into Eq. (8.107), we write

$$E_T = \frac{4N}{\omega_D^2}\hbar \int_0^{\omega_D} \frac{\omega^2\, d\omega}{e^{\frac{\hbar\omega}{k_B T}} - 1} \tag{8.108}$$

Making the substitution given by Eq. (8.84) into Eq. (8.108), one gets

$$E_T = 4Nk_B \frac{T^3}{\theta_D^2} \int_0^{\theta_D/T} \frac{x^2\, dx}{e^x - 1} \tag{8.109}$$

It is the general expression for the total energy of a two-dimensional solid with N atoms. At high temperatures $x \ll 1$, Eq. (8.109) reduces to

$$E_T = 4Nk_B \frac{T^3}{\theta_D^2} \int_0^{\theta_D/T} x\, dx = 2Nk_B T \tag{8.110}$$

Eq. (8.110) yields the classical value of C_V as

$$C_V = 2Nk_B \tag{8.111}$$

It gives the Dulong and Petit's value for a two-dimensional solid. At low temperatures, $x \gg 1$, Eq. (8.109) reduces to

$$E_T = 4Nk_B \frac{T^3}{\theta_D^2} \int_0^{\infty} e^{-x} x^2\, dx = 8Nk_B \frac{T^3}{\theta_D^2} \tag{8.112}$$

Hence C_V from Eq. (8.112) becomes

$$C_V = 24Nk_B\left(\frac{T}{\theta_D}\right)^2 \tag{8.113}$$

Eq. (8.113) shows that in a two-dimensional solid C_V varies as T^2 at very low temperatures.

8.5.2.3 Three-Dimensional Lattice

The phonon density of states can be calculated from the periodicity of the lattice as in the two-dimensional solid. If the solid is in the form of a cube with edges of length L, then the number of states per unit volume in the K-space is $(L/2\pi)^3$. The number of phonon states n in a sphere of radius K is given by

$$n = \frac{4\pi}{3}K^3\left(\frac{L}{2\pi}\right)^3 \tag{8.114}$$

Therefore, the number of phonon states in a spherical shell of radius K and thickness dK is obtained by differentiating the above equation, that is,

$$dn = 4\pi K^2 dK \frac{V}{(2\pi)^3} \tag{8.115}$$

where V is the volume of the solid. The number of modes of vibration between K and K+dK, given by Eq. (8.115), corresponds to the number of modes between frequency ω and $\omega + d\omega$, that is,

$$g(\omega)d\omega = \frac{V}{2\pi^2 v^3}\omega^2 d\omega \tag{8.116}$$

Every atom possesses three modes of vibration, therefore, $g(\omega)$ from the above equation can be written as

$$g(\omega) = \frac{3V}{2\pi^2 v^3}\omega^2 = A\omega^2 \tag{8.117}$$

where

$$A = \frac{3V}{2\pi^2 v^3} \tag{8.118}$$

Eq. (8.117) can be obtained from the general expression for the density of phonon states for the three-dimensional solid as given in Appendix E. Fig. 8.5 shows the frequency distribution function for a three-dimensional solid in the Debye approximation. The figure also shows the Einstein frequency ω_E, which is greater than the Debye frequency ω_D. The function $g(\omega)$ must satisfy the relation

$$\int_0^{\omega_D} g(\omega)d\omega = 3N \tag{8.119}$$

Substituting Eq. (8.117) into Eq. (8.119), the constant A is given by

$$A = \frac{9N}{\omega_D^3} \tag{8.120}$$

From the above equation the Debye frequency is given by

$$\omega_D = \left(\frac{9N}{A}\right)^{1/3} = v\left(\frac{6\pi^2 N}{V}\right)^{1/3} = vK_D \tag{8.121}$$

Eq. (8.121) allows us to write the Debye wave vector K_D as

$$K_D = \left(6\pi^2\frac{N}{V}\right)^{1/3} \tag{8.122}$$

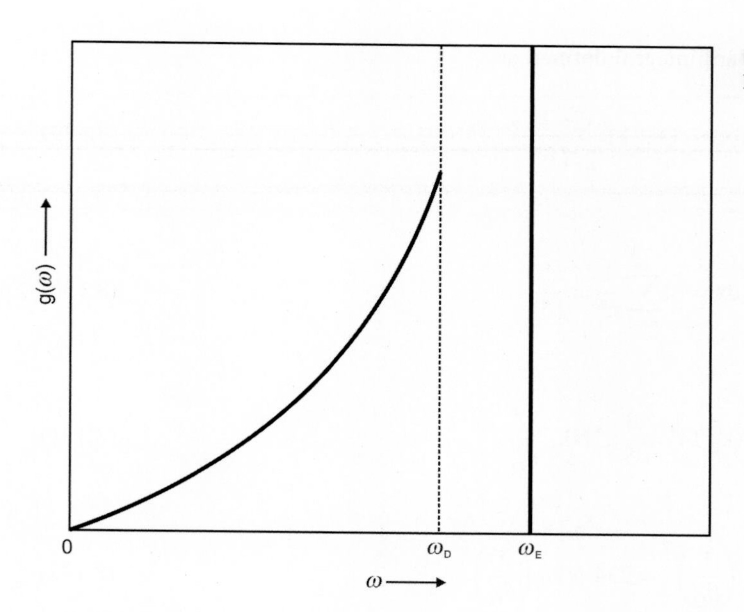

FIG. 8.5 The density of phonon states $g(\omega)$ as a function of ω. The Einstein frequency ω_E is greater than the Debye frequency ω_D.

The total energy of the lattice is given by

$$E_T = \int_0^{\omega_D} \overline{E}(\omega)\, g(\omega)\, d\omega \tag{8.123}$$

Substituting Eq. (8.117) for $g(\omega)$ and Eq. (8.68) for the average energy (neglecting the zero-point energy) into the above equation, we get

$$E_T = A\hbar \int_0^{\omega_D} \frac{\omega^3\, d\omega}{e^{\frac{\hbar\omega}{k_B T}} - 1} \tag{8.124}$$

Making the substitution given by Eq. (8.84), the above equation can be written as

$$E_T = \frac{A}{\hbar^3}(k_B T)^4 \int_0^{\theta_D/T} \frac{x^3}{e^x - 1} dx \tag{8.125}$$

The integral in Eq. (8.125) cannot be solved analytically but one can study the special cases for E_T. At very high temperatures, that is, $(T \gg \theta_D)$ $x \ll 1$, Eq. (8.125) reduces to

$$\begin{aligned} E_T &= \frac{A}{\hbar^3}(k_B T)^4 \int_0^{\theta_D/T} x^2 dx \\ &= 3Nk_B T \end{aligned} \tag{8.126}$$

Hence the specific heat at constant volume becomes

$$C_V = 3Nk_B \tag{8.127}$$

which is the Dulong and Petit's value for C_V (classical result). At low temperatures, that is, $(T \ll \theta_D)$ $x \gg 1$, Eq. (8.125) can be written as

$$E_T = \frac{A}{\hbar^3}(k_B T)^4 \int_0^{\infty} \frac{x^3}{e^x - 1} dx \tag{8.128}$$

The above integral can be evaluated by using the standard integral defined as

$$\int_0^\infty \frac{x^{s-1}}{e^x - 1} dx = (s-1)! \sum_{n=1}^{\infty} \frac{1}{n^s} \tag{8.129}$$

For s = 4 the above equation gives

$$\int_0^\infty \frac{x^3}{e^x - 1} dx = 3! \sum_{n=1}^{\infty} \frac{1}{n^4} = \frac{\pi^4}{15} \tag{8.130}$$

Substituting Eq. (8.130) into Eq. (8.128), one gets

$$E_T = \frac{A\pi^4}{15\hbar^3} (k_B T)^4 = \frac{3}{5} \pi^4 N k_B \frac{T^4}{\theta_D^3} \tag{8.131}$$

Hence the specific heat at constant volume becomes

$$C_V = \frac{12}{5} \pi^4 N k_B \left(\frac{T}{\theta_D}\right)^3 = 234\ N k_B \left(\frac{T}{\theta_D}\right)^3 \tag{8.132}$$

The above expression shows that at very low temperatures, the lattice specific heat varies as T^3, which agrees with the experimental observations (see Fig. 8.1). It is usually known as the Debye T^3 law. Fig. 8.6 shows C_V for one-, two-, and three-dimensional solids. It is evident from the figure that the temperature variation of C_V in the three types of solids is different at very low temperatures in contrast with that found in the Einstein model. The theoretical results obtained in the Einstein and Debye models are compared in Fig. 8.4. It is noteworthy that the Debye approximation works well at sufficiently low temperatures because at these temperatures only long-wavelength acoustic modes are excited. The energy of short-wavelength modes (e.g., the optical modes) is too high to allow these to be populated at low temperatures. Further, according to the dispersion relation (8.76), $\omega = 0$ at $K = 0$, but for the optical modes the frequency of vibration is finite at

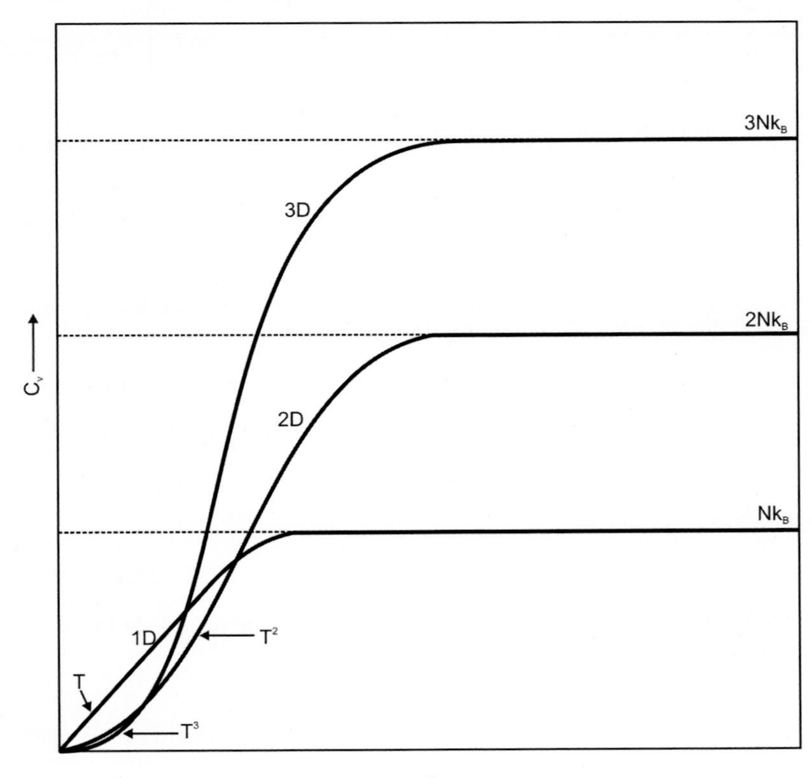

FIG. 8.6 Schematic representation of specific heat C_V as a function of temperature T for one-, two-, and three-dimensional solids in the Debye theory.

K = 0 (see Chapter 6). Therefore, the Debye theory does not account for the optical modes. We want to mention here that the Debye approximation includes only the linear term in the dispersion relation (Eq. 8.76), which is the lowest order approximation. In fact, one should consider the dispersion relations calculated from the experimentally measured phonon frequencies and then calculate the lattice specific heat.

8.6 EFFECT OF ELECTRONS ON SPECIFIC HEAT

One distinctive property of metals (also of semiconductors) is their electrical conductivity. It might be expected that the conduction electrons make a significant contribution to the specific heat of a metal. It will be shown in Chapter 9 that the electronic contribution to the specific heat varies linearly with temperature. Therefore, the total specific heat in a metal becomes

$$C_{metal} = \beta_\ell T^3 + \gamma_e T \tag{8.133}$$

Here β_ℓ and γ_e are constants. At room temperature the electronic contribution is very small compared with the lattice contribution to the specific heat (see Fig. 8.7A). But at low temperatures the electrons make an appreciable contribution to the specific heat, as is evident from Fig. 8.7B.

8.7 IDEAL PHONON GAS

Consider a crystal containing N atoms that are vibrating about their equilibrium positions at finite temperature. It has been shown in Chapter 7 that the energy of a crystal, in the harmonic approximation, is equivalent to the energy of 3N independent harmonic oscillators. The energy of each harmonic oscillator is quantized and is given from Eq. (7.58) by

$$E_{n_q} = \left(n_q + \frac{1}{2}\right)\hbar\omega(q) \tag{8.134}$$

with $n_q = 0, 1, 2 \ldots$ and $\omega(q)$ is the angular frequency of the normal mode. In the above equation we have omitted the index of polarization. Hence the difference in the energies of two consecutive states n_q and $n_q + 1$ is given by

$$E_{ph} = E_{n_q+1} - E_{n_q} = \hbar\omega(q) = h\nu(q) \tag{8.135}$$

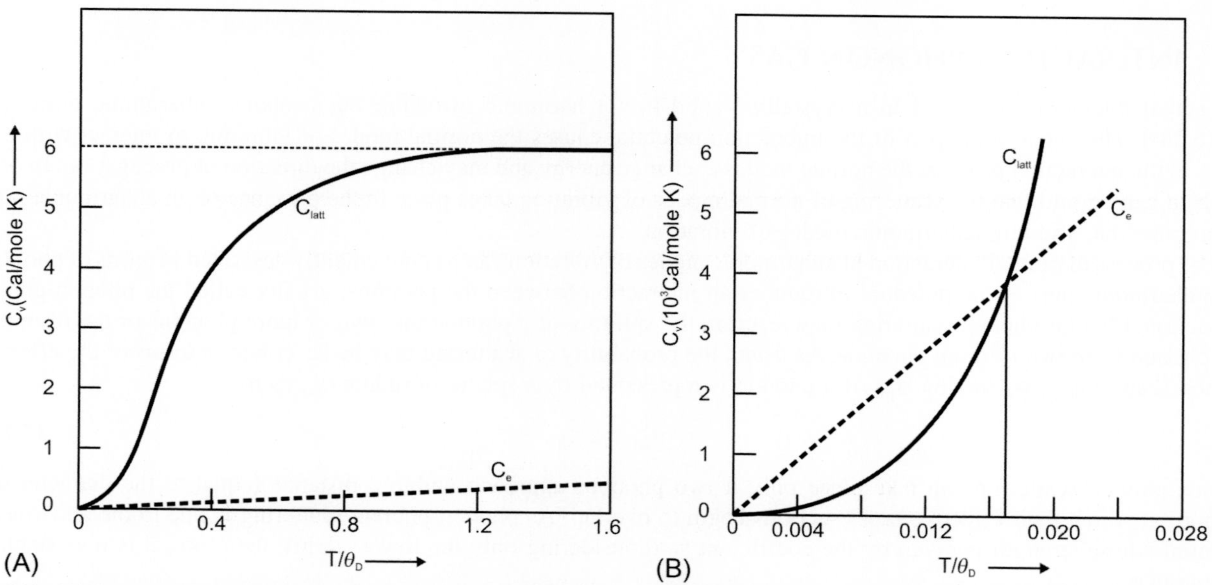

FIG. 8.7 (A) Temperature dependence of the lattice contribution C_{latt} and the electronic contribution C_e to the specific heat for $Cr_{0.8}V_{0.2}$ alloy. (B) C_{latt} and C_e for $Cr_{0.8}V_{0.2}$ alloy at very low temperatures. $C_e > C_{latt}$ up to $T \approx 8.5$ K. Here $\theta_D = 500$ K. *(Modified from Epifanov, G. I. (1979).* Solid state physics. *(p. 122). Moscow: Mir Publishers.)*

Here $\nu(q)$ is the frequency of thermal vibrations propagating through the crystal in the form of elastic waves. Eq. (8.135) gives the quantum of energy of elastic waves, called a phonon. The field of elastic waves may be treated as a gas of phonons (or a gas of quanta of normal modes of the lattice), having energy given by Eq. (8.135) and momentum p_{ph} given by

$$p_{ph} = \frac{\hbar\omega(q)}{v} = \frac{h}{\lambda} = \hbar q \tag{8.136}$$

where $q = 2\pi/\lambda$ is the phonon wave vector restricted to the 1 BZ. Here v is the velocity and λ is the wavelength of the elastic waves. The phonon density n_{ph} is defined as the number of phonons excited per unit volume of the crystal and is equal to the lattice energy per unit volume $E_{lattice}$ divided by the energy of a phonon, that is,

$$n_{ph} = \frac{E_{lattice}}{\hbar\omega(q)} \tag{8.137}$$

In the harmonic approximation, the atomic vibrations are strictly noninteracting harmonic elastic waves, and therefore, travel through the lattice without scattering and without meeting any resistance. So, in the harmonic approximation, phonons make up an ideal phonon gas. Because the waves do not meet any resistance, considerable heat flux flows, even for an infinitesimally small temperature difference, yielding an infinitely large thermal conductivity for a solid.

The behavior of n_{ph} can be studied in the limiting cases. In the Debye approximation at low temperatures, the lattice energy is given by (see Eq. 8.131)

$$E_{lattice} = \frac{E_T}{V} \propto T^4 \tag{8.138}$$

and the phonon energy $\hbar\omega(q) \approx k_B T \approx T$. Hence from Eq. (8.137)

$$n_{ph} \propto T^3 \tag{8.139}$$

On the other hand, in the high-temperature limit, the Debye approximation gives (see Eq. 8.126)

$$E_{lattice} = \frac{E_T}{V} \propto T \tag{8.140}$$

But at high temperatures the phonon energy attains the maximum value of $\hbar\omega_D \approx k_B\theta_D$ and is independent of temperature. Therefore, in the high-temperature limit, from Eq. (8.137), one gets

$$n_{ph} \propto T \tag{8.141}$$

8.8 INTERACTING PHONON GAS

The actual interaction potential in a crystalline solid is not harmonic in nature but contains anharmonic terms [see Eq. (6.24)]. The anharmonic part of the interaction potential causes the normal modes of vibration to interact with each other. In the interaction process, the normal modes exchange energy and may change the direction of propagation. In other words, it can be said that the scattering of normal modes of vibration takes place in the presence of an anharmonic interaction potential, yielding anharmonic modes of vibration.

The process of mutual interaction of anharmonic modes of vibration can be conveniently described in terms of phonons. The anharmonic part of the potential introduces an interaction between the phonons, usually called the phonon-phonon interaction. Phonon-phonon scattering may result in the splitting of a phonon into two or more phonons or the formation of a phonon from two or more phonons. As usual, the probability of scattering may be described in terms of the effective phonon scattering cross section σ_{ph}. If a phonon is represented by a sphere of radius r_{ph}, then

$$\sigma_{ph} = \pi r_{ph}^2 \tag{8.142}$$

Phonon-phonon scattering can take place only if two phonons approach within a distance (equal to the diameter of a phonon) at which their effective cross sections begin to overlap. As phonon-phonon scattering is due to the anharmonic potential whose strength is given by the coefficient γ_F (considering only the lowest term), therefore, it is reasonable to assume that

$$\sigma_{ph} \cdot \propto \gamma_F^2 \tag{8.143}$$

Knowing σ_{ph}, one can calculate the phonon's mean free path l_{ph}, which is the average distance traveled by a phonon between two consecutive scattering events. Calculations show that

$$l_{ph} = \frac{1}{n_{ph}\sigma_{ph}} \propto \frac{1}{n_{ph}\gamma_F^2} \tag{8.144}$$

8.9 THERMAL EXPANSION OF SOLIDS

In the harmonic approximation the interaction potential is parabolic in nature, which is symmetric about its axis, that is, about the line QS at a distance R_0 from atom 1 (see Fig. 6.5). Therefore, the displacements $\mathbf{u}_1$ and $\mathbf{u}_2$ are equal, yielding the mean position of atom 2 as R_0. Assuming atom 1 to be fixed at its position, the increase in temperature excites atom 2 to the higher energy state, thereby increasing its amplitude of vibration. But the vibration remains symmetric, yielding the same mean position O for atom 2. Hence the harmonic interaction potential is unable to explain thermal expansion in solids. The anharmonic potential from Eq. (6.24) is written as

$$V(u) = V(R_0) + \frac{1}{2}\alpha_F u^2 - \frac{1}{3}\gamma_F u^3 - \frac{1}{4}\delta_F u^4 + \cdots \tag{8.145}$$

The first term in the above equation is constant. If the reference level of the potential is taken as $V(R_0)$ and only the first anharmonic term is retained in Eq. (8.145), then one can write

$$V(u) = \frac{1}{2}\alpha_F u^2 - \frac{1}{3}\gamma_F u^3 \tag{8.146}$$

When atom 2 is displaced to the positive u values (toward the right of O) or to increasing values of r, the second term in Eq. (8.146) is subtracted from the first term, thereby decreasing the slope of part QR of the interaction potential curve (see Fig. 8.8). On the other hand, for negative u values (toward the left of O) or for decreasing values of r, the second term in Eq. (8.146) is positive and is added to the first term, thereby increasing the slope of part PQ of the interaction potential curve. Thus, anharmonicity in the potential makes the interaction potential asymmetric about the line QS, as shown in Fig. 8.8. Because of the asymmetric nature of the potential, the amplitudes of vibration of atom 2 toward left and right are different; the amplitude in the former direction being less than that in the latter direction. As a result, the mean position of atom 2 no longer coincides with the previous equilibrium position O but shifts by a distance $\bar{u}$ toward the right to the position O_1. Let E be the energy of atom 2 at temperature T. Then, thermal vibrations of atom 2 increase the distance between the two atoms 1 and 2 by $\bar{u}$ on average and this causes thermal expansion of the solid. The above explanation shows that the cause of thermal expansion is the anharmonic nature of atomic vibrations in a solid. Heating to higher temperature T' increases the energy of atom 2 further to E'. With the increase in energy, the mean position of atom 2 shifts more to the right to O_1', thereby further increasing the mean distance between the atoms (Fig 8.8). The above explanation can be extended to a solid in which the equilibrium distance between all the adjoining atoms increases with an increase in the temperature. Hence thermal expansion increases with an increase in the temperature of a solid.

Let us estimate the coefficient of linear thermal expansion of a solid. The force acting on an atom from Eq. (8.146) becomes

$$F = -\alpha_F u + \gamma_F u^2 \tag{8.147}$$

The average value of the force $\bar{F}$ caused by the displacement of atom 2 from its equilibrium position is

$$\bar{F} = -\alpha_F \bar{u} + \gamma_F u^2 \tag{8.148}$$

In the equilibrium position of the particle, the average force vanishes to give

$$\bar{u} = \frac{\gamma_F}{\alpha_F} u^2 \tag{8.149}$$

The average value of the potential from Eq. (8.146) is

$$\bar{V}(u) = \frac{1}{2}\alpha_F u^2 \tag{8.150}$$

In addition to the potential energy, an atom also possesses finite kinetic energy. It is well known that the average kinetic energy of a vibrating atom is equal to its average potential energy. Therefore, the total average energy $\bar{E}$ of atom 2 becomes

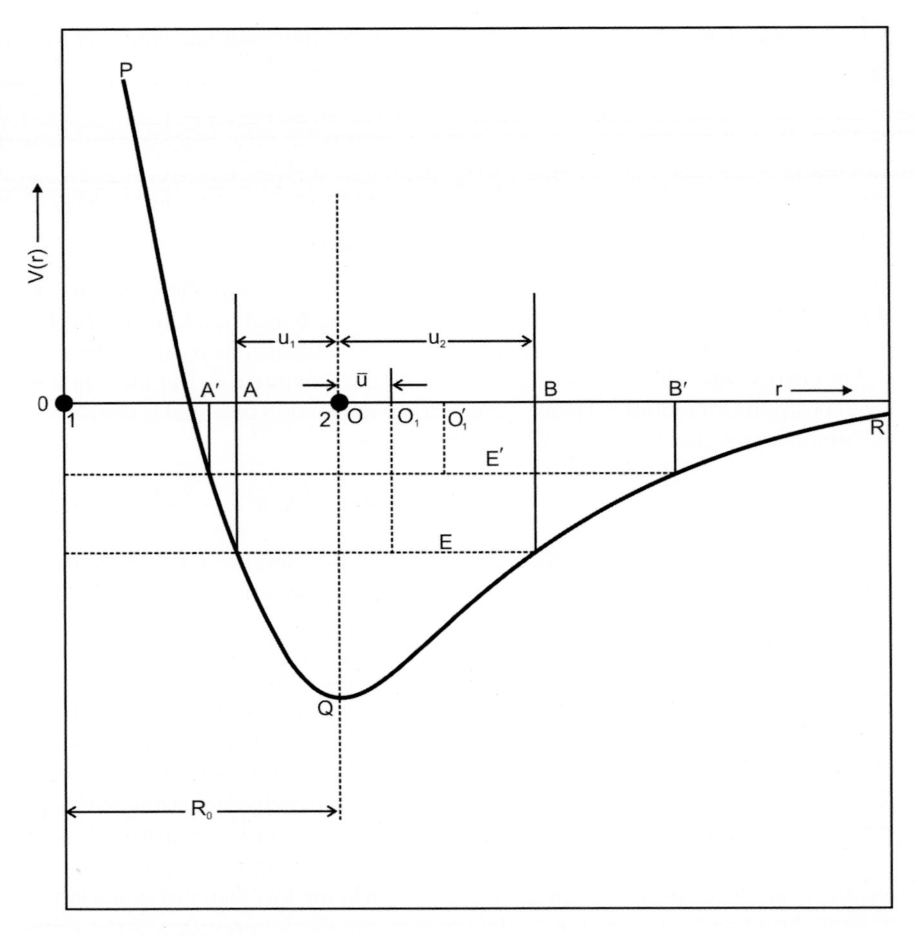

FIG. 8.8 Interatomic potential V(r) as a function of r between two atoms 1 and 2. Here u_1 and u_2 represent the displacements of atom 2, at finite temperature T, to the left and right of the mean position R_0 and is asymmetrical about R_0. The new mean position of atom 2 becomes $R_0+\bar{u}$ where $\bar{u}$ is the average shift in the mean position.

$$\overline{E}=2\overline{V}(u)=\alpha_F u^2 \tag{8.151}$$

From Eqs. (8.149), (8.151), the average displacement $\bar{u}$ in terms of average total energy becomes

$$\bar{u}=\frac{\gamma_F}{\alpha_F^2}\overline{E} \tag{8.152}$$

An alternate method for the calculation of average displacement is described in Appendix G. The coefficient of linear thermal expansion Γ_{TH} is defined as the increase in length per unit original length per degree rise in temperature. In the present case of two atoms it can be written as

$$\Gamma_{TH}=\frac{1}{R_0}\frac{d\bar{u}}{dT} \tag{8.153}$$

Substituting $\bar{u}$ from Eq. (8.152) into Eq. (8.153), we get

$$\Gamma_{TH}=K_P C_V \tag{8.154}$$

where

$$K_P=\frac{\gamma_F}{\alpha_F^2 R_0} \tag{8.155}$$

$$C_V=\frac{d\overline{E}}{dT} \tag{8.156}$$

Fig. 8.9 shows the temperature dependence of both Γ_{TH} and C_V. It is evident that both are interrelated (proportional to each other). In the high-temperature limit, the average energy of an atom $\overline{E}=k_BT$ and therefore $C_V=k_B$, which is temperature independent. Hence in the high-temperature limit,

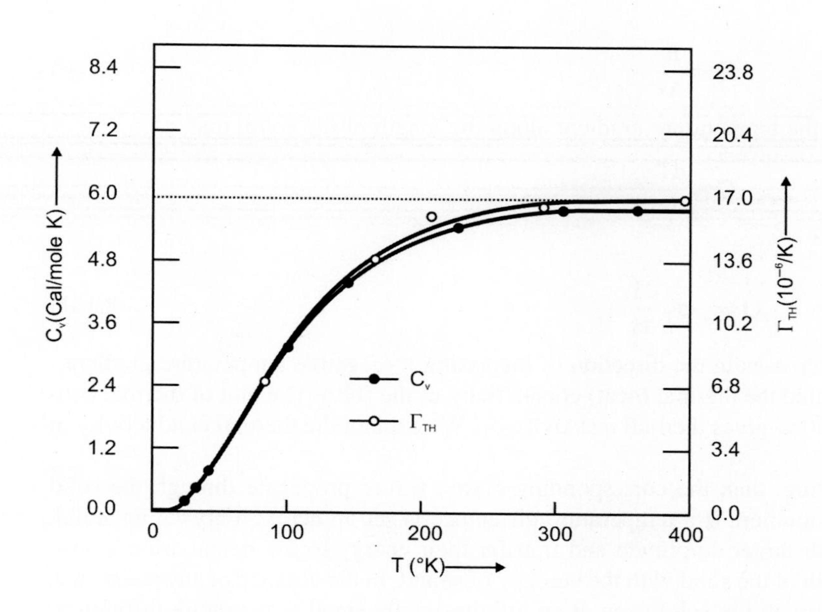

FIG. 8.9 Temperature dependence of the specific heat C_V and the coefficient of linear expansion Γ_{TH}. *(Modified from Epifanov, G. I. (1979).* Solid state physics *(p. 126). Moscow: Mir Publishers.)*

$$\Gamma_{TH} = \frac{\gamma_F k_B}{\alpha_F^2 R_0} \tag{8.157}$$

Substituting the values of γ_F, k_B, α_F, and R_0 for various solids, Γ_{TH} is found to be on the order of $10^{-4} - 10^{-5}$, which is in agreement with experiment. In the low-temperature limit, Γ_{TH} behaves in a way similar to that of C_V, that is, $\Gamma_{TH} \propto T^3$.

In the case of metals, Gruneisen proposed a formula for Γ_{TH}, which is given by

$$\Gamma_{TH} = \frac{\gamma_G K_M}{3V_0} C_V \tag{8.158}$$

where K_M is the metal compressibility, V_0 is the atomic volume, and γ_G is the Gruneisen constant whose value is equal to 1.5–2.5 depending on the metal. It is noteworthy that both Eqs. (8.154), (8.158) are similar as for as the temperature dependence is concerned.

8.10 THERMAL CONDUCTIVITY OF SOLIDS

Consider a solid with length L_x along the x-direction and rectangular opposite faces having area A perpendicular to the length (see Fig. 8.10). Let T_1 and T_2 ($T_1 > T_2$) be the temperatures on the opposite faces. The gradient of temperature along the length of the solid dT/dx is

$$\frac{dT}{dx} = \frac{T_1 - T_2}{L_x} \tag{8.159}$$

Let E be the heat energy, which flows from the face PQRS to the face UVWY in time t. The flux of heat energy Q (heat energy transmitted per unit area per unit time) is given by

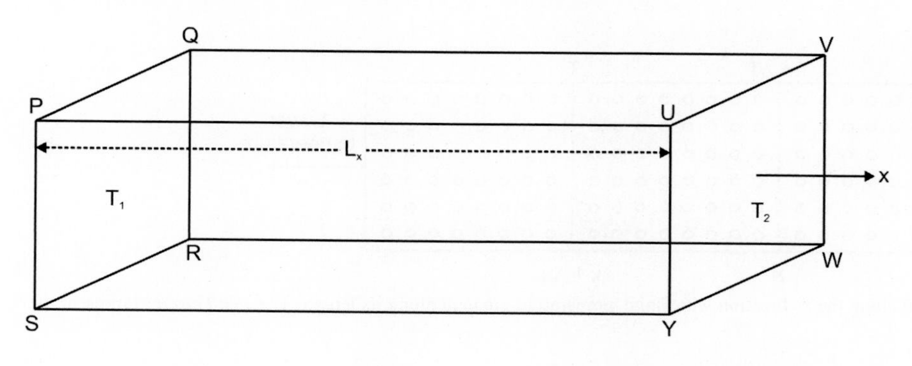

FIG. 8.10 A solid with length L_x along the x-direction. The opposite faces PQRS and UVWY are *rectangular* in shape, each having area of cross section A and at temperatures T_1 and T_2 with $T_1 > T_2$.

$$Q = \frac{E}{At} \tag{8.160}$$

The flow of heat flux is directly proportional to the temperature gradient along the length of the solid, that is,

$$Q \propto \frac{dT}{dx}$$

This gives

$$Q = -\sigma_T \frac{dT}{dx} \tag{8.161}$$

The negative sign shows that the temperature decreases in the direction of increasing x (negative temperature gradient). Here σ_T is a constant of proportionality and is called the thermal (heat) conductivity of the solid. The unit of thermal conductivity is J/s cm K or W/cm K. The reciprocal of σ_T gives thermal resistivity ρ_T. We explain the thermal conductivity in what follows.

If the atomic vibrations are harmonic in nature, then the corresponding elastic waves propagate through the solid without experiencing any resistance. In such a situation, if a temperature difference is set up across a crystalline solid, the atoms at the hot end will start vibrating with larger amplitude and transfer their energy to the neighboring atoms. As a result, a heat wave will travel along the length of the solid with the velocity of sound. In the absence of any resistance, a considerable amount of heat flux will flow through the solid even at an infinitesimally small temperature difference. Hence, in the harmonic approximation, the thermal conductivity of the solid will be infinitely large. But the real potential in a solid is anharmonic in nature, which produces anharmonic modes of vibration. These modes cause phonon-phonon interactions and offer finite resistance to the flow of the heat wave along the solid. Hence anharmonic waves are responsible for the finite thermal conductivity in a solid.

8.10.1 Thermal Conductivity for an Ideal Gas of Atoms

The expression for thermal conductivity can be derived assuming the solid to consist of free atoms forming an ideal gas of atoms. Let n_a be the density of free atoms per unit volume in a rectangular-shaped rod of a solid along the x-direction (see Fig. 8.11). Further, assume that the temperature gradient is produced along the length of the rod. The flow of heat from the end at higher temperature to the other at lower temperature is due to the motion of atoms along the x-direction. The atoms move with equal probability in both the positive and negative x-directions, so on average half of the atoms, that is, $(1/2)n_a$ atoms, move in the positive x-direction with average velocity $\langle v_x \rangle$, while the other half move along the negative x-direction with the same average velocity. Hence the flux of atoms in both the positive and negative x-directions is $(1/2)n_a\langle v_x \rangle$ in magnitude. Let c_V be the specific heat of an atom at constant volume. Then in moving from region x with temperature T $+\Delta T$ to region $x+\Delta x$ with temperature T (see Fig. 8.11), the energy given up by an atom is $c_V \Delta T$. The temperature difference between the two ends of the mean free path $l_a = \langle v_x \rangle \tau_a$ of an atom is given by

$$\Delta T = -\frac{dT}{dx}\langle v_x \rangle \tau_a \tag{8.162}$$

where τ_a is the time between two consecutive scatterings of atoms, usually called the relaxation time. The net flux of energy from atoms moving in both the positive and negative directions is given by

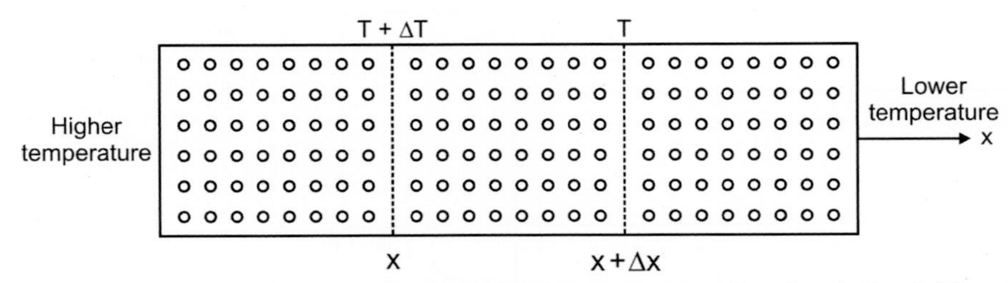

FIG. 8.11 A *rectangular-shaped rod* of a solid along the x-direction with finite temperature gradient along its length. The *small circles* represent the atoms in the solid.

$$Q = n_a \langle v_x \rangle c_V \Delta T = -n_a \langle v_x \rangle^2 c_V \tau_a \frac{dT}{dx} \tag{8.163}$$

If the velocity is constant and equal in all three Cartesian directions, then

$$v^2 = \langle v^2 \rangle = 3 \langle v_x \rangle^2 \tag{8.164}$$

Therefore,

$$Q = -\frac{1}{3} n_a \langle v^2 \rangle c_V \tau_a \frac{dT}{dx} = -\frac{1}{3} n_a v^2 c_V \tau_a \frac{dT}{dx} \tag{8.165}$$

If C_V is the specific heat per unit volume, then $C_V = n_a c_V$. Further, the mean free path of an atom is $l_a = v\, \tau_a$, therefore, one can straightway write

$$Q = -\frac{1}{3} C_V v l_a \frac{dT}{dx} \tag{8.166}$$

Comparing Eqs. (8.161), (8.166) we get

$$\sigma_T = \frac{1}{3} C_V v l_a \tag{8.167}$$

8.10.2 Thermal Conductivity in Insulators and Dielectrics

In insulators atoms vibrate about their equilibrium positions at finite temperature. Therefore, the atoms can be considered to be anharmonic oscillators, which give rise to an interacting phonon gas. Eq. (8.167) for an ideal gas of atoms can also be applied to a phonon gas by replacing l_a by the mean free path of phonons l_{ph}, v by the velocity of sound, and C_V by the specific heat of the solid. Therefore, the lattice thermal conductivity is given by

$$\sigma_{latt} = \frac{1}{3} C_V v l_{ph} \tag{8.168}$$

Substituting the value of l_{ph} from Eq. (8.144) into the above equation, we find

$$\sigma_{latt} \propto \frac{C_V v}{n_{ph} \gamma_F^2} \tag{8.169}$$

In the above expression, γ_F and v determine the magnitude of the thermal conductivity, while C_V and n_{ph} determine its temperature dependence. The parameters v and γ_F depend strongly on the rigidity of the bonds between the atoms in a solid: bonds of higher rigidity yield higher values of v and lower values of γ_F because the strengthening of bonds reduces both anharmonicity and the thermal vibration amplitude of atoms. Therefore, σ_{latt} increases with an increase in the rigidity of the bonds. This conclusion is supported by the experimental results. Further, a detailed analysis shows that σ_{latt} also depends strongly on the mass M of the atoms: σ_{latt} is larger for smaller values of M. Experimentally it is found that for light elements, such as B, C, and Si, σ_{latt} is on the order of a few tens or hundreds of W/cm K. For elements in the middle of the periodic table, σ_{latt} is on the order of several W/cm K, but for heavy elements, σ_{latt} is on the order of several tenths of W/cm K. This trend may be due to the fact that with an increase in M, the value of the rigidity of the bonds decreases as the bond length increases.

The temperature dependence of σ_{latt} depends on C_V and n_{ph} (or l_{ph}). In the high-temperature limit, C_V is independent of temperature (Dulong and Petit's law) but n_{ph} is proportional to T (see Eq. 8.141). Therefore, Eq. (8.169) gives

$$\sigma_{latt} \cdot \propto \frac{1}{T} \tag{8.170}$$

which is consistent with the experimental findings. For T values below the Debye temperature θ_D, n_{ph} decreases strongly with a decrease in T, leading to a sharp increase in the mean free path l_{ph}. At sufficiently low temperatures, l_{ph} becomes comparable with the dimensions of the solid, therefore, any further decrease in temperature does not lead to any increase in l_{ph}. Hence at very low T values, the temperature variation of σ_{latt} is determined by the behavior of C_V only. It has already been seen that, at very low temperatures, $C_V \propto T^3$ (Debye law), therefore,

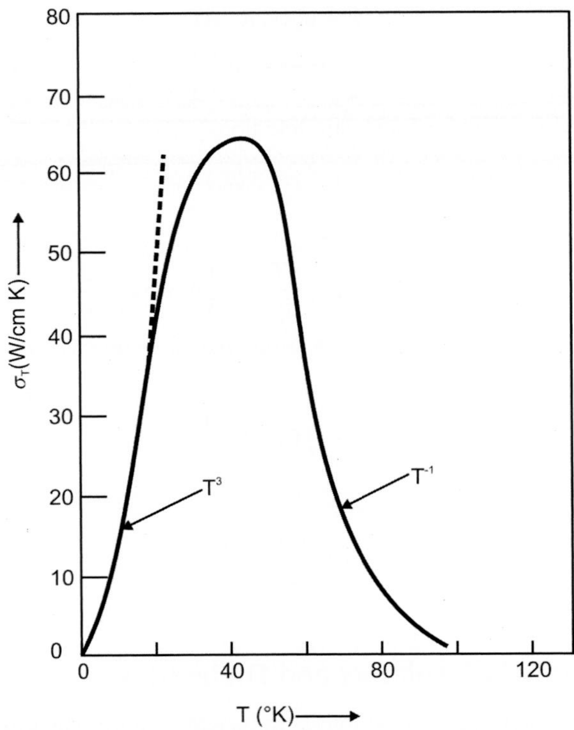

FIG. 8.12 Temperature dependence of thermal conductivity σ_T of synthetic sapphire (Al_2O_3). *(Modified from Berman, R. (1958).* Z. f. Phys. Chem. *(Neue Folge)* ***16****, 10.)*

$$\sigma_{latt} \cdot \propto T^3 \tag{8.171}$$

The behavior of σ_{latt} predicted by Eq. (8.171) is also in qualitative agreement with the experimental results. From the discussion above it is evident that, with an increase in temperature, σ_{latt} increases at very low temperatures (Eq. 8.171), while at high temperatures, σ_{latt} decreases (Eq. 8.170). Hence at some temperature in between, σ_{latt} should exhibit a peak (maximum). Fig. 8.12 shows the experimental results for the temperature dependence of the thermal conductivity σ_T for synthetic sapphire.

8.10.3 Thermal Conductivity of Metals

In metals there are ions at the lattice positions and electrons move freely in the crystal. Therefore, heat is conducted by both the vibrating ions (phonons) and the free electrons and the total thermal conductivity σ_T is the sum of the two contributions, that is,

$$\sigma_T = \sigma_{latt} + \sigma_{el} \tag{8.172}$$

where σ_{el} is the electronic contribution to the thermal conductivity. The main contribution to σ_{el} comes from the electrons at the Fermi surface having velocity v_F. σ_{el} can be estimated from Eq. (8.167) where C_V is replaced by the electronic specific heat C_e, v is replaced by v_F, and l_a is replaced by the mean free path of electrons l_e. Therefore, σ_{el} can be written as

$$\sigma_{el} = \frac{1}{3} C_e v_F l_e \tag{8.173}$$

The mean free path of electrons depends on the following scattering processes:

1. l_e depends on the scattering of electrons from the ions (e-p scattering) and is inversely proportional to the phonon density n_{ph}, that is,

$$l_e \cdot \propto \cdot \frac{1}{n_{ph}} \tag{8.174}$$

2. l_e depends on the scattering of electrons from the impurities present in the crystal and is inversely proportional to the impurity concentration n_i, that is,

$$l_e \cdot \propto \frac{1}{n_i} \tag{8.175}$$

The electronic specific heat C_e is given by (see Eq. 9.107)

$$C_e = \frac{1}{2}\pi^2 N_e k_B \frac{k_B T}{E_F} \tag{8.176}$$

N_e is the total number of free electrons in the metal. Substituting Eq. (8.176) into Eq. (8.173), we obtain

$$\sigma_{el} = \frac{1}{3}\pi^2 \frac{N_e k_B^2}{m_e v_F} l_e T \tag{8.177}$$

Here m_e is the mass of a free electron. In Eq. (8.177), only l_e depends on the temperature. It is interesting to study the behavior of σ_{el} in the limiting cases.

At very low temperatures, n_{ph} in a metal becomes very small, therefore, the electrons are scattered mainly from the impurities present in the metal. As the electron scattering from the impurities is temperature independent, so is l_e. Hence, at very low temperatures, Eq. (8.177) yields

$$\sigma_{el} \cdot \propto \cdot T \tag{8.178}$$

But at very low temperatures, σ_{latt} is proportional to T^3 (see Eq. 8.171). Hence, at very low temperatures, the main contribution to σ_T comes from σ_{el} and is proportional to T, an experimental fact. At low temperatures, n_{ph} and hence l_e is finite and therefore contributes significantly toward σ_{el}. Substituting the value of l_e from Eq. (8.174) into Eq. (8.177), we get

$$\sigma_{el} \cdot \propto \cdot \frac{\pi^2}{3} \frac{N_e k_B^2}{m_e v_F} \frac{T}{n_{ph}} \tag{8.179}$$

At low temperatures, n_{ph} is given by Eq. (8.139), which when substituted in Eq. (8.179) gives

$$\sigma_{el} \cdot \propto T^{-2} \tag{8.180}$$

Hence in the low temperature range, σ_{el} in metals is inversely proportional to the square of the absolute temperature. In the high-temperature limit, n_{ph} is proportional to T (Eq. 8.141), which when substituted in Eq. (8.179) yields

$$\sigma_{el} = \text{Constant} \tag{8.181}$$

Hence σ_{el} becomes constant in pure metals at very high temperatures.

It is interesting to compare the lattice and electronic contributions to the thermal conductivity of metals. From Eqs. (8.168), (8.173) one can write

$$\frac{\sigma_{el}}{\sigma_{latt}} = \frac{C_e v_F l_e}{C_V v\, l_{ph}} \tag{8.182}$$

In pure metals, $C_e/C_V \approx 0.01$, $v_F = 10^8$ cm/s, $v = 5 \times 10^5$ cm/s, $l_e \approx 10^{-6}$ cm, and $l_{ph} \approx 10^{-7}$ cm. Substituting these values in Eq. (8.182) we get

$$\frac{\sigma_{el}}{\sigma_{latt}} \approx 0.2 \times 10^2 \tag{8.183}$$

This shows that σ_T in metals is determined mainly by σ_{el}, with σ_{latt} being negligible. The magnitude of σ_{el} can also be found by substituting the values of various quantities in Eq. (8.173). Fig. 8.13 shows the experimental results for the temperature variation of the thermal conductivity of Cu metal. It is found that the behavior of the experimental results is the same as that of σ_{el} in the whole of the range. Table 8.1 gives the values of σ_T for some particular metals at room temperature.

In the case of alloys, the situation may be entirely different as the impurities form strong scattering centers for the conduction electrons. Therefore, the mean free path l_e is mainly determined by the impurity scattering processes. In metallic alloys, l_e and l_{ph} may come out to be of the same order of magnitude, making both the lattice and electronic contributions to the thermal conductivity equally important.

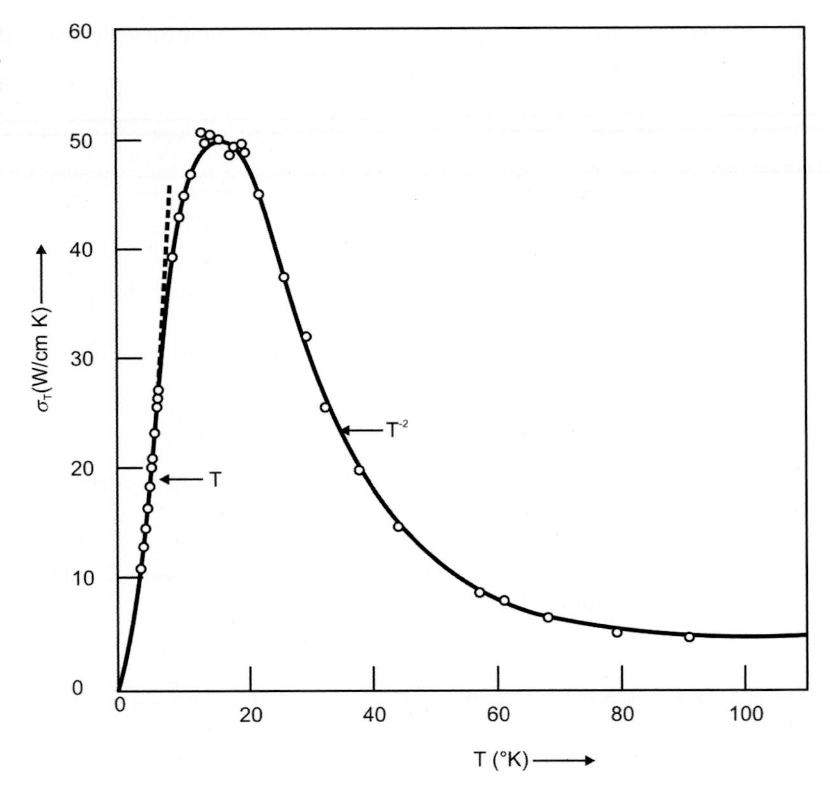

FIG. 8.13 Experimental values of thermal conductivity σ_T as a function of temperature T for Cu metal. σ_T becomes constant at high temperatures. *(Modified from Epifanov, G. I. (1979).* Solid state physics *(p. 131). Moscow: Mir Publishers.)*

TABLE 8.1 Experimental Values of Thermal Conductivity σ_T of Some Selected Metals

Metal	σ_T (W/cm K)
Cu	3.84
Ag	4.03
Au	2.96
Al	2.10
Ni	0.60

FURTHER READING

Adkins, C. J. (1986). *An introduction to thermal physics*. London: Cambridge University Press.

Launay, J. D. (1956). The theory of specific heats and lattice vibrations. In F. Seitz, & D. Turnbull (Eds.), *Solid state physics*, Vol. 2. (pp. 219–303) New York: Academic Press.

Pamplin, B. R. (1980). *Thermal expansion of solids*. New York: Pergamon Press.

Parrot, J. E., & Stuckes, A. D. (1975). *Thermal conductivity of solids*. London: Pion.

Swalin, R. A. (1962). *Thermodynamics of solids*. New York: J. Wiley & Sons.

Chapter 9

Free-Electron Theory of Metals

Chapter Outline

A metallic solid consists of a periodic array of atoms and each atom consists of a nucleus with electrons revolving around it. The electrons in the outermost orbit of an atom are called valence electrons and are loosely bound to the nucleus. Each atom (and hence the electrons in it) experiences the crystal potential $V(\mathbf{r})$, as a result of which the valence electrons get detached from the atom and are able to move more or less freely anywhere in the crystal. These electrons determine the conduction properties of metallic solids and thus the name conduction electrons. Therefore, a metal can be represented by a sea of conduction electrons in which the ions are embedded at the lattice positions (Fig. 4.10). In simple metals the conduction electrons are s- or p-electrons, which are nearly free, yielding a nearly uniform electron charge density. In a more simplified representation of a simple metal, the discrete positive ionic charge is assumed to be smeared out to form a uniform positive background. In this approximation a simple metal consists of a nearly uniform conduction electron gas moving in a uniform positive background. It is usually called the *jellium model* of metal.

9.1 FREE-ELECTRON APPROXIMATION

In Chapter 3, it has been discussed that the electronic properties of crystalline solids are studied in the one-electron approximation in which an electron experiences a self-consistent potential. The self-consistent potential contains both the repulsive part due to the electron-electron and the ion-ion interactions and the attractive part due to the electron-ion interactions. In simple metals it is usually assumed that the average repulsive part of the potential cancels exactly the attractive part, giving rise to a net zero potential. Therefore, the conduction electrons experience zero potential and are free to move anywhere in the metal. This is called the *free-electron approximation*. In this approximation the conduction electrons in a metal form a free-electron gas. We offer no detailed justification for this approximation except to say that it yields reasonably good results for some simple metals, such as Na, K, and Al. We want to mention here that in the jellium model the negative charge density due to the electrons is equal and opposite to the positive charge density of the background, thus yielding net zero charge density. Therefore, the net potential is zero due the vanishing charge density in the jellium model.

9.2 THREE-DIMENSIONAL FREE-ELECTRON GAS

Consider a three-dimensional free-electron gas, having N_e electrons, confined to a cube with sides of length L. The Schrodinger wave equation for a free electron is given by

$$\widehat{H}_e |\psi_{\mathbf{k}}(\mathbf{r})\rangle = E_{\mathbf{k}} |\psi_{\mathbf{k}}(\mathbf{r})\rangle \tag{9.1}$$

Solid State Physics. https://doi.org/10.1016/B978-0-12-817103-5.00009-8

where $\widehat{H}_e$ is the Hamiltonian of an electron. $|\psi_\mathbf{k}(\mathbf{r})\rangle$ and $E_\mathbf{k}$ are the wave function and energy of a state with wave vector $\mathbf{k}$. In the free-electron approximation, the potential energy is zero so the Hamiltonian $\widehat{H}_e$ contains only the kinetic energy $p^2/2m_e$ where m_e is the mass of an electron. Therefore, the Schrodinger wave equation for a free electron becomes

$$-\frac{\hbar^2}{2m_e}\nabla^2|\psi_\mathbf{k}(\mathbf{r})\rangle = E_\mathbf{k}|\psi_\mathbf{k}(\mathbf{r})\rangle \tag{9.2}$$

Here ∇^2 is the three-dimensional Laplacian operator and $\mathbf{k}$ and $\mathbf{r}$ are given by

$$\mathbf{k} = \hat{\mathbf{i}}_1 k_x + \hat{\mathbf{i}}_2 k_y + \hat{\mathbf{i}}_3 k_z \tag{9.3}$$

$$\mathbf{r} = \hat{\mathbf{i}}_1 x + \hat{\mathbf{i}}_2 y + \hat{\mathbf{i}}_3 z \tag{9.4}$$

The probability of finding a free electron at any point in the system is the same, therefore, the wave function can be written in the exponential form as

$$|\psi_\mathbf{k}(\mathbf{r})\rangle = C e^{\imath\mathbf{k}\cdot\mathbf{r}} \tag{9.5}$$

where C is a constant and its value can be obtained by normalizing the wave function to unity. The normalization condition for the wave function is defined as

$$\langle\psi_\mathbf{k}(\mathbf{r})|\psi_\mathbf{k}(\mathbf{r})\rangle = 1 \tag{9.6}$$

Substituting Eq. (9.5) into Eq. (9.6), one gets

$$C = \left(\frac{1}{V}\right)^{1/2} \tag{9.7}$$

where $V = L^3$ is the volume of the free-electron gas. Hence the normalized wave function for the three-dimensional free-electron gas is given by

$$|\psi_\mathbf{k}(\mathbf{r})\rangle = \left(\frac{1}{V}\right)^{1/2} e^{\imath\mathbf{k}\cdot\mathbf{r}} = |\mathbf{k}\rangle \tag{9.8}$$

Here $|\mathbf{k}\rangle$ or $|\psi_\mathbf{k}(\mathbf{r})\rangle$ represents a normalized plane wave and these form a complete orthonormal set of wave functions. For a finite system of the free-electron gas the wave function satisfies the cyclic boundary condition according to which

$$|\psi_\mathbf{k}(\mathbf{r})\rangle = |\psi_\mathbf{k}(\mathbf{r}+\mathbf{L})\rangle \tag{9.9}$$

Substituting Eq. (9.8) into Eq. (9.9), one can write

$$e^{\imath\mathbf{k}\cdot\mathbf{L}} = 1 \tag{9.10}$$

which is equivalent to

$$e^{\imath k_x L} = e^{\imath k_y L} = e^{\imath k_z L} = 1 \tag{9.11}$$

The above equation is satisfied for the following values of k_x, k_y, and k_z

$$k_x = \frac{2\pi n_x}{L},\quad k_y = \frac{2\pi n_y}{L},\quad k_z = \frac{2\pi n_z}{L} \tag{9.12}$$

where $n_x = n_y = n_z = 0, \pm1, \pm2, \ldots$. Substituting Eq. (9.8) into Eq. (9.2) and operating on the equation by $\langle\mathbf{k}|$ from the left side, the energy of the $\mathbf{k}$-state is given by

$$E_\mathbf{k} = \langle\mathbf{k}| -\frac{\hbar^2}{2m_e}\nabla^2|\mathbf{k}\rangle = \frac{\hbar^2 k^2}{2m_e} \tag{9.13}$$

The above equation gives a parabolic energy band, as shown in Fig. 9.1. The linear momentum of an electron can be obtained from the wave function as

$$\mathbf{p}|\psi_\mathbf{k}(\mathbf{r})\rangle = -\imath\hbar\nabla|\psi_\mathbf{k}(\mathbf{r})\rangle = \hbar\mathbf{k}|\psi_\mathbf{k}(\mathbf{r})\rangle \tag{9.14}$$

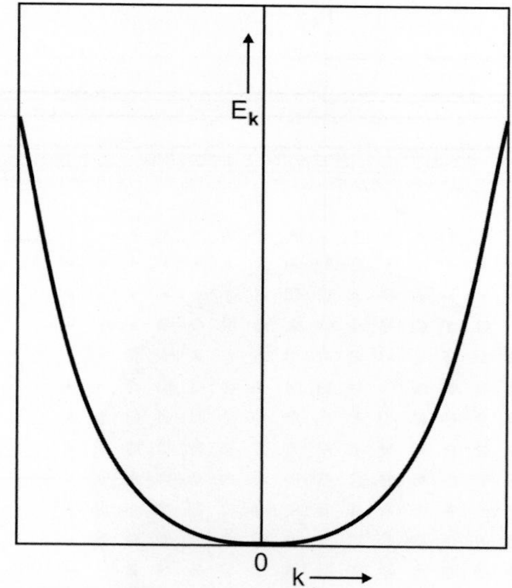

FIG. 9.1 E_k as a function of $k = |\mathbf{k}|$ for a three-dimensional free-electron gas.

Therefore, the velocity **v** of an electron with momentum **p** is given by

$$\mathbf{v} = \frac{\hbar \mathbf{k}}{m_e} \tag{9.15}$$

Points in the **k**-space represented by each set of k_x, k_y, k_z values from Eq. (9.12) represent the allowed states. N_e electrons can be accommodated in different **k**-states. As the electrons obey the Pauli exclusion principle, two electrons possessing opposite spins can be accommodated in each **k**-state (spin degeneracy) until all the electrons are exhausted (see Fig. 9.2). The highest filled state is the Fermi state with wave vector k_F, called the Fermi wave vector. The energy of the highest filled state E_F is called the Fermi energy and is given by

$$E_F = \frac{\hbar^2 k_F^2}{2m_e} \tag{9.16}$$

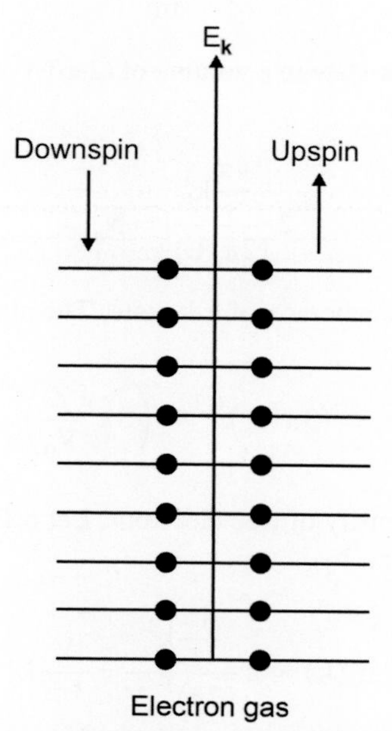

FIG. 9.2 Distribution of electrons, represented by *dots*, in up- and downspin electron states in accordance with the Pauli exclusion principle.

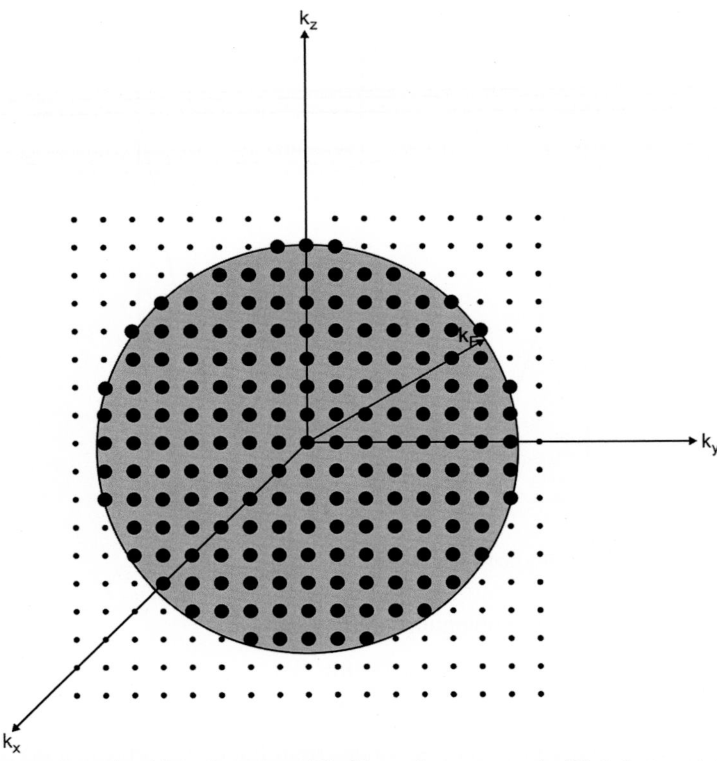

FIG. 9.3 The Fermi sphere in a three-dimensional free-electron gas. The *bigger dots* represent the filled electron states and lie within the Fermi sphere. Outside the Fermi sphere all the electron states (*represented by smaller dots*) are empty.

If the three-dimensional free-electron gas has large dimensions, then the **k**-points lie very close to each other and become quasicontinuous. The occupied states lie approximately in a sphere called the Fermi sphere (see Fig. 9.3). The **k**-states outside the Fermi sphere are all empty. The radius k_F of this sphere is called the Fermi radius or Fermi wave vector. The velocity of electrons on the surface of the Fermi sphere is called the Fermi velocity and is given by

$$v_F = \frac{\hbar k_F}{m_e} \tag{9.17}$$

From Eq. (9.12) it is evident that there is one **k**-state in a volume of $(2\pi/L)^3$ of the **k**-space. Therefore, the number of states in the Fermi sphere is given by

$$2\frac{\frac{4\pi}{3}k_F^3}{(2\pi/L)^3} = N_e \tag{9.18}$$

The factor of 2 takes into account the spin degeneracy of a **k**-state. The above equation yields the Fermi wave vector as

$$k_F = \left(3\pi^2 n_e\right)^{1/3} = \left(3\pi^2 \frac{Z}{V_0}\right)^{1/3} \tag{9.19}$$

where $n_e = N_e/V = Z/V_0$ gives the volume density of free electrons. Let $n_e(\mathbf{k})$ be the number of electron states with wave vector less than or equal to $k = |\mathbf{k}|$. It is given by

$$n_e(\mathbf{k}) = 2\frac{\frac{4\pi}{3}k^3}{(2\pi)^3/V} = \frac{V}{3\pi^2}k^3 \tag{9.20}$$

The density of electron states per unit energy $N_e(E_\mathbf{k})$ in three dimensions is given by

$$N_e(E_\mathbf{k}) = \frac{dn_e(\mathbf{k})}{dE_\mathbf{k}} \tag{9.21}$$

Substituting Eqs. (9.13), (9.20) into Eq. (9.21), we get

$$N_e(E_\mathbf{k}) = \frac{V}{2\pi^2}\left(\frac{2m_e}{\hbar^2}\right)^{3/2} E_\mathbf{k}^{1/2} \tag{9.22}$$

The density of electron states per unit energy per unit volume becomes

$$g_e(E_\mathbf{k}) = \frac{1}{2\pi^2}\left(\frac{2m_e}{\hbar^2}\right)^{3/2} E_\mathbf{k}^{1/2} \tag{9.23}$$

The general expression for the density of electron states $g_e(E_\mathbf{k})$ is given in Appendix F. A plot of $g_e(E_\mathbf{k})$ as a function of $E_\mathbf{k}$ is shown in Fig. 9.4 for the ground state of the free-electron gas. At absolute zero all the states below E_F are filled, while those above are completely empty. The total energy of the system is given by

$$\begin{aligned} E_T &= \int_0^{E_F} E_\mathbf{k} N_e(E_\mathbf{k})\, dE_\mathbf{k} \\ &= \frac{V}{5\pi^2}\left(\frac{2m_e}{\hbar^2}\right)^{3/2} E_F^{5/2} \end{aligned} \tag{9.24}$$

The above equation can also be written as

$$E_T = \frac{3}{5} N_e E_F \tag{9.25}$$

Hence the average energy of an electron in a three-dimensional free-electron gas is $(3/5)E_F$.

The physical quantities of the free-electron gas can also be expressed in terms of the interelectronic distance r_e. In the free-electron gas under consideration, the average volume per electron is

$$\frac{1}{n_e} = \frac{V}{N_e} = \frac{4\pi}{3} r_e^3 \tag{9.26}$$

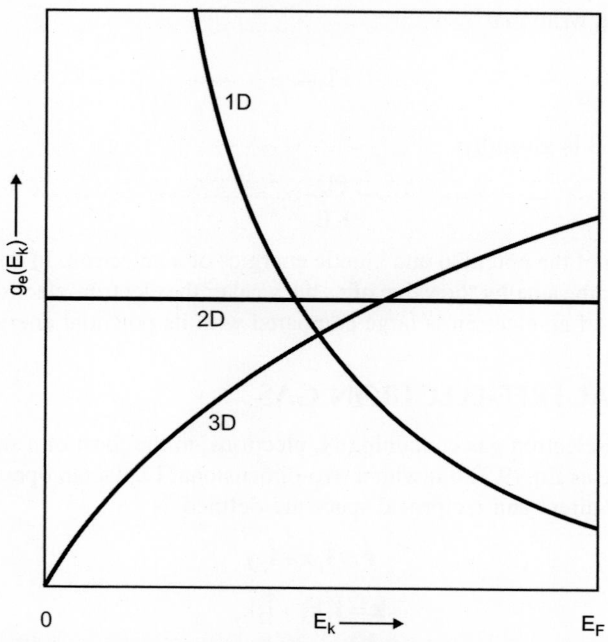

FIG. 9.4 The density of electron states per unit volume per unit energy $g_e(E_\mathbf{k})$ as a function of energy $E_\mathbf{k}$ for a one- (1D), two- (2D), and three-dimensional (3D) free-electron gas at absolute zero.

From the above equation the interelectronic distance r_e is given by

$$r_e = \left(\frac{3}{4\pi n_e}\right)^{1/3} \tag{9.27}$$

The Fermi wave vector k_F is obtained by substituting the value of n_e from Eq. (9.26) into Eq. (9.19) to write

$$k_F = \left(\frac{9\pi}{4}\right)^{1/3} \frac{1}{r_s a_0} \tag{9.28}$$

where

$$r_e = r_s a_0 \tag{9.29}$$

Here $a_0 = \hbar^2/m_e e^2 = 0.529$ Å is the Bohr radius. The Fermi energy, in terms of r_s, is obtained from Eqs. (9.16), (9.28) as

$$\begin{aligned} E_F &= \frac{m_e e^4}{2\hbar^2}\left(\frac{9\pi}{4}\right)^{2/3} \frac{1}{r_s^2} \\ &= \left(\frac{9\pi}{4}\right)^{2/3} \frac{1}{r_s^2} = \frac{3.68}{r_s^2} \text{Ryd} \end{aligned} \tag{9.30}$$

which is the Fermi energy for the free-electron gas at absolute zero in the ground state. From Eqs. (9.25), (9.30) the average energy per electron becomes

$$\frac{E_T}{N_e} = \frac{2.21}{r_s^2} \text{ Ryd} \tag{9.31}$$

The value of r_s is different for different metals (in the range of 1.8–5.8 Å approximately) and is maximum in the case of monovalent metals. The Fermi energy is basically the kinetic energy (KE) of the electrons, so from Eq. (9.30) we can write

$$\text{KE} \propto \frac{1}{r_s^2} \tag{9.32}$$

The potential energy (PE) of an electron is given by

$$V(r) = \frac{e^2}{r_e}$$

From the above equation one can write that

$$\text{PE} \propto \frac{1}{r_s} \tag{9.33}$$

Therefore, the ratio of PE and KE is given by

$$\frac{\text{PE}}{\text{KE}} = r_s \tag{9.34}$$

Hence r_s is a measure of the ratio of the potential and kinetic energies of an electron. In other words, r_s is a measure of the interaction between the electrons: the smaller the value of r_s, the weaker the electron-electron interaction. This is equivalent to saying that the kinetic energy of an electron is large compared with its potential energy for $r_s < 1$.

9.3 TWO-DIMENSIONAL FREE-ELECTRON GAS

Consider a two-dimensional free-electron gas containing N_e electrons, in the form of a square of side L. Here the Schrodinger wave equation is the same as Eq. (9.2) but with a two-dimensional Laplacian operator ∇^2. In the two-dimensional space the position vectors in the direct and reciprocal space are defined as

$$\mathbf{r} = \hat{\mathbf{i}}_1 x + \hat{\mathbf{i}}_2 y \tag{9.35}$$

$$\mathbf{k} = \hat{\mathbf{i}}_1 k_x + \hat{\mathbf{i}}_2 k_y \tag{9.36}$$

The normalized wave function for the two-dimensional free-electron gas is given as

$$|\psi_{\mathbf{k}}(\mathbf{r})\rangle = \frac{1}{\sqrt{A_0}} e^{i\mathbf{k}\cdot\mathbf{r}} = |\mathbf{k}\rangle \tag{9.37}$$

where $A_0 = L^2$ is the area of the free-electron gas. The cyclic boundary condition defined by Eq. (9.9) gives discrete values of k_x and k_y as

$$k_x = \frac{2\pi n_x}{L}, \quad k_y = \frac{2\pi n_y}{L} \tag{9.38}$$

where $n_x = n_y = 0, \pm 1, \pm 2, \ldots$ The points in the **k**-space represented by each set of k_x and k_y give the allowed states. In a two-dimensional free-electron gas the Fermi surface is circular, having radius k_F and area πk_F^2. All the **k**-states within the Fermi circle are occupied but are empty outside. From Eq. (9.38) it is evident that there is one electron state in an area of $(2\pi/L)^2$ in the **k**-space. The total number of occupied states in the Fermi circle is given by

$$2\frac{\pi k_F^2}{(2\pi/L)^2} = N_e \tag{9.39}$$

which gives the value of k_F as

$$k_F = (2\pi n_e)^{1/2} \tag{9.40}$$

where $n_e = N_e/A_0$ gives the number of electrons per unit area, that is, the surface density of the electrons. The Fermi energy is obtained by substituting Eq. (9.40) into Eq. (9.16) to give

$$E_F = \frac{\pi\hbar^2}{m_e} n_e \tag{9.41}$$

The Fermi velocity v_F can be obtained by substituting k_F from Eq. (9.40) into Eq. (9.17) giving

$$v_F = \frac{\hbar}{m_e}(2\pi n_e)^{1/2} \tag{9.42}$$

Let $n_e(\mathbf{k})$ be the number of electron states with a wave vector less than or equal to **k**, then

$$n_e(\mathbf{k}) = \pi k^2 \left(\frac{L}{2\pi}\right)^2 \tag{9.43}$$

One can calculate $N_e(E_{\mathbf{k}})$ and $g_e(E_{\mathbf{k}})$ in exactly the same manner as for the three-dimensional electron gas. They are given by

$$N_e(E_{\mathbf{k}}) = \frac{m_e A_0}{\pi\hbar^2} \tag{9.44}$$

$$g_e(E_{\mathbf{k}}) = \frac{m_e}{\pi\hbar^2} \tag{9.45}$$

It is evident that $N_e(E_{\mathbf{k}})$ and $g_e(E_{\mathbf{k}})$ for the two-dimensional gas are independent of energy. Fig. 9.4 shows $g_e(E_{\mathbf{k}})$ as a function of energy for a two-dimensional free-electron gas. The total energy of the system can be written as

$$\begin{aligned} E_T &= \int_0^{E_F} N_e(E_{\mathbf{k}}) E_{\mathbf{k}} \, dE_{\mathbf{k}} \\ &= \frac{1}{2} N_e E_F \end{aligned} \tag{9.46}$$

Hence the average energy of an electron in a two-dimensional free-electron gas is half of the Fermi energy of the system.

Problem 9.1

In a one-dimensional free-electron gas with N_e electrons, and having length L, the Schrodinger wave equation is defined as

$$-\frac{\hbar^2}{2m_e}\frac{d^2}{dx^2}|\psi_k(x)\rangle = E_k|\psi_k(x)\rangle \tag{9.47}$$

where the normalized wave function $|\psi_k(x)\rangle$ is given by

$$|\psi_k(x)\rangle = \frac{1}{\sqrt{L}}e^{\iota kx} = |k\rangle \tag{9.48}$$

Prove that the Fermi wavevector, Fermi energy, and the density of electron states are given by

$$k_F = \pi\frac{N_e}{L} = \pi n_e \tag{9.49}$$

$$E_F = \frac{\hbar^2 k_F^2}{2m_e} = \frac{\pi^2\hbar^2 n_e^2}{2m_e} \tag{9.50}$$

$$N_e(E_k) = \frac{L}{2\pi}\left(\frac{2m_e}{\hbar^2}\right)^{1/2} E_k^{-1/2} \tag{9.51}$$

$$g_e(E_k) = \frac{1}{2\pi}\left(\frac{2m_e}{\hbar^2}\right)^{1/2} E_k^{-1/2} \tag{9.52}$$

Further prove that the total energy of the free-electron gas is given by

$$E_T = \frac{1}{3}N_e E_F \tag{9.53}$$

The function $g_e(E_\mathbf{k})$ as a function of energy for a one-dimensional free-electron gas is shown in Fig. 9.4.

9.4 COHESIVE ENERGY AND INTERATOMIC SPACING OF IDEAL METAL

In a simple metal there are two contributions to the Coulomb energy: the repulsive part of the energy is given by the kinetic energy of the free electron gas and the attractive part is given by the interaction of electrons with the positive ions situated at the lattice positions. Consider a metal with valence Z and interatomic distance R. The electronic charge $-Ze$ on an atom can be considered to be distributed uniformly in a sphere of radius R. Therefore, the electronic charge density $-en_e$ is given by

$$-en_e = -\frac{Ze}{\frac{4\pi}{3}R^3} \tag{9.54}$$

where n_e is the electron density. Let the nucleus be assumed to be a point with positive charge Ze and situated at a distance r from the center of the sphere of electronic charge (see Fig. 9.5). If we draw a concentric sphere with radius r, then the electronic charge inside this sphere q is given by

$$q = -Ze\left(\frac{r}{R}\right)^3 \tag{9.55}$$

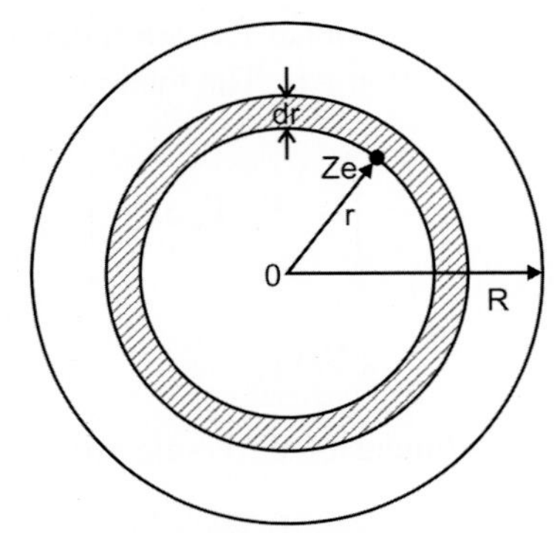

FIG. 9.5 Schematic representation of an atom: electronic charge is uniformly distributed in a sphere of radius R. The nucleus with positive charge Ze, represented by a *dot*, is situated at a distance r from the center of the electronic charge distribution.

So, the net charge inside the sphere of radius r, which also contains positive charge of magnitude Ze, becomes

$$q(r) = Ze - Ze\left(\frac{r}{R}\right)^3 \tag{9.56}$$

The Coulomb potential at a distance r from the center of the sphere becomes

$$\phi(r) = \frac{q(r)}{r} = \frac{Ze}{r}\left[1 - \left(\frac{r}{R}\right)^3\right] \tag{9.57}$$

The electronic charge dq(r) in a spherical shell of radius r and thickness dr is given by

$$dq(r) = -\frac{Ze}{\frac{4\pi}{3}R^3}4\pi r^2 dr \tag{9.58}$$

The contribution to potential energy by this spherical shell is given by

$$dE_C = dq(r)\,\phi(r) = -\frac{3Z^2e^2}{R^3}r\left[1 - \left(\frac{r}{R}\right)^3\right]dr \tag{9.59}$$

Hence the Coulomb potential energy per atom becomes

$$E_C = \int_0^R dE_C = -\frac{9}{10}\frac{Z^2e^2}{R} \tag{9.60}$$

Now the average kinetic energy per electron E_K in a free electron gas is given by

$$E_K = \frac{3}{5}E_F = \frac{3}{5}\frac{\hbar^2 k_F^2}{2m_e} \cdot = \frac{3}{5}\frac{\hbar^2}{2m_e}\left(3\pi^2 n_e\right)^{2/3} \tag{9.61}$$

where we have used Eq. (9.19). Substituting the value of n_e from Eq. (9.54), one gets

$$E_K = \frac{3}{5}\frac{\hbar^2}{2m_e}\left(\frac{9\pi}{4}Z\right)^{2/3}\frac{1}{R^2} \tag{9.62}$$

Therefore, the kinetic energy per atom, which contains Z electrons, becomes

$$E_K = \frac{3Z}{5}\frac{\hbar^2}{2m_e}\left(\frac{9\pi}{4}Z\right)^{2/3}\frac{1}{R^2} \tag{9.63}$$

The total energy per atom is the sum of the potential and kinetic energies given by Eqs. (9.60), (9.63), that is,

$$\begin{aligned} E(R) &= E_C + E_K \\ &= -\frac{9}{10}\frac{Z^2e^2}{R} + \frac{3Z}{5}\frac{\hbar^2}{2m_e}\left(\frac{9\pi}{4}Z\right)^{2/3}\frac{1}{R^2} \end{aligned} \tag{9.64}$$

The equilibrium value of interatomic distance R_0 is obtained by minimizing the total energy per atom, that is,

$$\left.\frac{dE(R)}{dR}\right|_{R=R_0} = 0 \tag{9.65}$$

Substituting Eq. (9.64) into Eq. (9.65), the value of R_0 is given by

$$\begin{aligned} R_0 &= \frac{2}{3}\left(\frac{9\pi}{4}\right)^{2/3}\frac{\hbar^2}{m_e e^2}Z^{-1/3} \\ &= 1.30Z^{-1/3}\text{Å} \end{aligned} \tag{9.66}$$

The value of energy per atom in the equilibrium state is obtained by substituting R_0, from Eq. (9.66) for R, into Eq. (9.64) to obtain

$$\begin{aligned} E(R_0) &= -\frac{9}{20}\frac{Z^2e^2}{R_0} \\ &= -0.3\,Z^2\,\text{Ryd} \approx -5.0\,Z^2\text{eV} \end{aligned} \tag{9.67}$$

We know that the binding energy of the hydrogen atom ($Z = 1$) is -13.6 eV. Therefore, the above model gives only the order of energy per atom.

9.5 THE FERMI-DIRAC DISTRIBUTION FUNCTION

The description of the free-electron theory presented so far has been temperature independent, but a number of electronic properties depend on temperature. The effect of temperature on various electronic properties of solids can be studied with the help of the Fermi-Dirac distribution function. It has been seen that at absolute zero all of the electron states are filled up to the Fermi energy E_F, which represents the ground state of the system. When the temperature is increased, the kinetic energy of the electrons increases and the electrons in the vicinity of E_F cross the Fermi energy. As a result, some of the electron states close to but above E_F are occupied. In other words, some energy levels above E_F are occupied that were vacant at absolute zero, and some levels below E_F become partially vacant that were fully occupied at absolute zero.

The Fermi-Dirac distribution function $f(E_{\mathbf{k}} - \mu)$ correctly describes the distribution of electrons at finite temperatures and is defined as

$$f(E_k - \mu) = \frac{1}{e^{\frac{E_k - \mu}{k_B T}} + 1} \tag{9.68}$$

Here $\mu(T)$ is the chemical potential and is chosen in such a way that the total number of electrons N_e is conserved. It is noteworthy that $\mu(T)$ is equal to the Fermi energy E_F at absolute zero. The Fermi-Dirac distribution function $f(E_{\mathbf{k}} - \mu)$ gives the probability of occupation of an electron state with energy $E_{\mathbf{k}}$ in thermal equilibrium. At absolute zero we know that $E_{\mathbf{k}} \leq \mu(T)$, so from Eq. (9.68), $f(E_{\mathbf{k}} - \mu)$ is unity at absolute zero. But for $E_{\mathbf{k}} > \mu(T)$, $E_{\mathbf{k}} - \mu(T)$ is positive and therefore $f(E_{\mathbf{k}} - \mu)$ is zero at absolute zero. On the whole, one can say that at absolute zero the probability of occupation of all the states below E_F is one (occupied) and for states above E_F is zero (unoccupied). The dashed line in Fig. 9.6A shows $f(E_{\mathbf{k}} - \mu)$ as a function of $E_{\mathbf{k}}$ at absolute zero, which changes discontinuously from the value 1 to 0 at E_F.

To have an idea about the variation of the Fermi-Dirac distribution function at finite temperature, let us calculate it at some points of interest. Here we take $\mu = E_F$ (the value at absolute zero). If $E_{\mathbf{k}} = E_F - k_B T$, then from Eq. (9.68) we get

$$f(E_k - \mu) = \frac{1}{e^{-1} + 1} = 0.73$$

At $E_{\mathbf{k}} = E_F$, Eq. (9.68) gives

$f(E_{\mathbf{k}} - \mu) = 0.50$

At $E_{\mathbf{k}} = E_F + k_B T$, one gets

$f(E_{\mathbf{k}} - \mu) = 0.27$

The dark line in Fig. 9.6A shows the variation of $f(E_{\mathbf{k}} - \mu)$ as a function of $E_{\mathbf{k}}$ at a finite T value. It shows that as the temperature is raised from absolute zero, only a few electrons having energy close to E_F (within a range $\approx k_B T$) are raised to higher states to give free electrons and all other electrons are tightly bound. At $E_{\mathbf{k}} = E_F$, $f(E_{\mathbf{k}} - \mu)$ has the value one-half at all temperatures. Hence the Fermi energy can also be defined as the electron state whose probability of occupation is always one-half. From Fig. 9.6A it is evident that the slope of the Fermi-Distribution function, at finite temperature, is finite only around E_F (approximately within an energy range of $k_B T$) and is zero at all other values of energy. Fig. 9.6B shows the derivative of $f(E_{\mathbf{k}} - \mu)$ as a function of $E_{\mathbf{k}}$. At finite temperature, it is a peaked function around $E_{\mathbf{k}} = E_F$ whose width increases with an increase in temperature. But at absolute zero it becomes the Dirac delta function.

If $E_{\mathbf{k}} - \mu >> k_B T$, then the exponential term in Eq. (9.68) dominates and, hence, the distribution function can be written as

$$f(E_k - \mu) = e^{\frac{(\mu - E_k)}{k_B T}} \tag{9.69}$$

This is essentially the Maxwell-Boltzmann distribution.

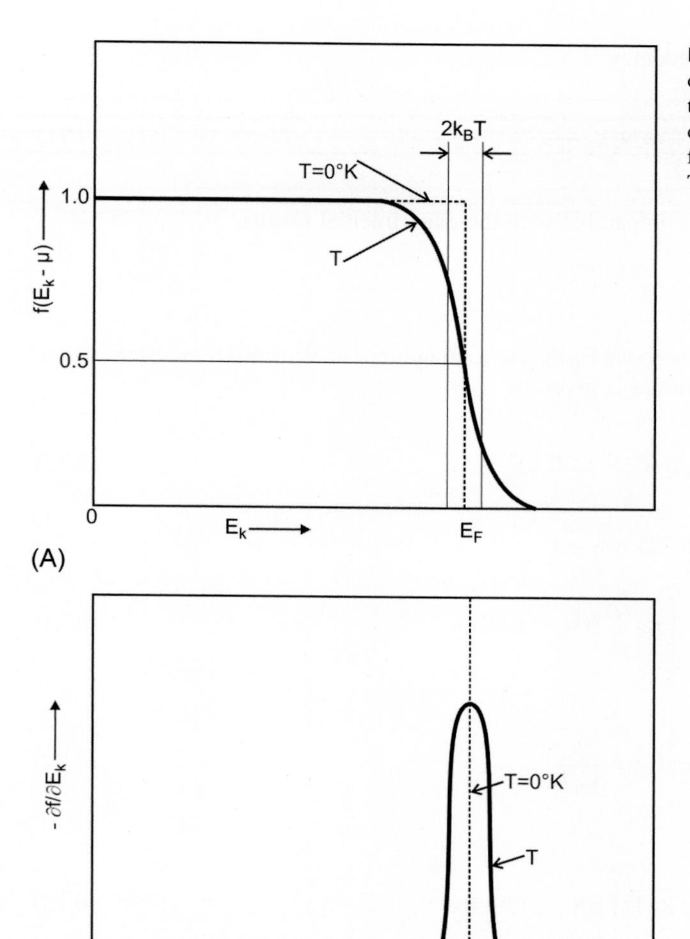

FIG. 9.6 (A) The Fermi distribution function $f(E_k - \mu)$ as a function of energy E_k. The *dashed line* represents $f(E_k - \mu)$ at absolute zero, while the *dark line* represents $f(E_k - \mu)$ at a finite temperature T. (B) The energy derivative of the Fermi-Dirac distribution function, that is, $-\partial f/\partial E_k$ as a function of energy E_k at absolute zero (*dashed line*) and at finite temperature T (*dark line*).

9.6 SPECIFIC HEAT OF ELECTRON GAS

The early development of the theory of specific heat of electrons, which was based on classical mechanics, caused great difficulty. For example, consider a solid with N atoms, each contributing one conduction electron. If the conduction electrons are considered to be free electrons, then the solid has a free electron gas with $N_e (=N)$ electrons. It has already been proved in Chapter 8 that each free electron possesses thermal energy of $(3/2)k_BT$, giving rise to a contribution of $(3/2)k_B$ to the specific heat. Hence the total electronic contribution to the specific heat at constant volume C_e becomes $(3/2)N_ek_B$. But at room temperature, the observed value of C_e is not more than 0.01 times the value predicted above. This happened because all the free electrons contributed only to the electrical conduction and not to the specific heat.

The fact that electrons obey the Fermi-Dirac distribution solved this problem. In the previous section, it was found that, as the temperature is increased from absolute zero to T, not every electron gains thermal energy ($\approx k_BT$) as expected classically. Rather, only those electrons that lie in states around E_F, within an energy range of k_BT, gain energy, and, hence, participate in the thermal properties. Therefore, only a fraction on the order of T/T_F of the total number of electrons N_e can be excited thermally at temperature T because only these electrons lie within an energy range on the order of k_BT. The $N_e(T/T_F)$ electrons, each having thermal energy k_BT, give a total electronic energy E as

$$E = N_e\left(\frac{T}{T_F}\right)k_BT \tag{9.70}$$

Therefore, the electronic contribution to the specific heat becomes

$$C_e = \left(\frac{\partial E}{\partial T}\right)_V \approx N_e k_B \frac{T}{T_F} \tag{9.71}$$

which is very small compared with the classical value, but comparable with the experimental results.

9.6.1 One-Dimensional Free-Electron Gas

At finite temperature some of the electrons are excited to states above E_F but the total number of electrons remains the same.

Therefore, the total number of electrons at finite temperature is given by

$$N_e = \int_0^\infty N_e(E_\mathbf{k}) f(E_\mathbf{k} - \mu) dE_\mathbf{k} \tag{9.72}$$

Substituting the value of $N_e(E_\mathbf{k})$ from Eq. (9.51) into Eq. (9.72), we get

$$N_e = \frac{L}{2\pi}\left(\frac{2m_e}{\hbar^2}\right)^{1/2} \int_0^\infty E_k^{-1/2} f(E_k - \mu) dE_k \tag{9.73}$$

Making the substitution

$$x = \frac{E_k}{k_B T} \text{ and } y = \frac{\mu}{k_B T} \tag{9.74}$$

in the above integral, one can write

$$N_e = \frac{L}{2\pi}\left(\frac{2m_e}{\hbar^2}\right)^{1/2} \sqrt{k_B T} \int_0^\infty f(x-y) \frac{1}{\sqrt{x}} dx \tag{9.75}$$

The integral in Eq. (9.75) is the Fermi distribution function integral (see Appendix I) with

$$h(x) = \frac{1}{\sqrt{x}} \tag{9.76}$$

Therefore, H(x), from Eqs. (9.76), (I.5) of Appendix I, is given by

$$H(x) = \int_0^x \frac{1}{\sqrt{x}} dx = 2x^{1/2} \tag{9.77}$$

Using Eq. (I.17) of Appendix I, the integral in Eq. (9.75) can be solved, giving

$$N_e = \frac{L}{2\pi}\left(\frac{2m_e}{\hbar^2}\right)^{1/2} \sqrt{k_B T}\left[1 + \frac{\pi^2}{6}\frac{\partial^2}{\partial y^2} + \cdots\right] H(y) \tag{9.78}$$

Substituting H(y) from Eq. (9.77) into Eq. (9.78) and simplifying, we obtain

$$N_e = \frac{L}{\pi}\left(\frac{2m_e}{\hbar^2}\right)^{1/2} \mu^{1/2}\left[1 - \frac{\pi^2}{24}\left(\frac{k_B T}{\mu}\right)^2 + \cdots\right] \tag{9.79}$$

At absolute zero, N_e can be obtained from Eq. (9.49) by substituting the value of k_F from Eq. (9.50) and is given by

$$N_e = \frac{L}{\pi}\left(\frac{2m_e}{\hbar^2}\right)^{1/2} E_F^{1/2} \tag{9.80}$$

Equating Eqs. (9.79), (9.80) one obtains the chemical potential $\mu(T)$ as

$$E_F(T) = \mu(T) = E_F\left[1 + \frac{\pi^2}{12}\left(\frac{k_B T}{E_F}\right)^2 + \cdots\right] \tag{9.81}$$

In writing the above expression, we have substituted E_F (at absolute zero) in place of $\mu(T)$ in the denominator of the second term in the square brackets on the right side of Eq. (9.81): an approximate expression. It shows that the Fermi energy increases with an increase in temperature. The total energy of the free-electron gas at temperature T is given by

$$E_T = \int_0^\infty E_{\mathbf{k}} N_e(E_{\mathbf{k}}) f(E_{\mathbf{k}} - \mu)\, dE_{\mathbf{k}} \tag{9.82}$$

Substituting the value of $N_e(E_{\mathbf{k}})$ from Eq. (9.51) and making the substitution given by Eq. (9.74), one can write

$$E_T = \frac{L}{2\pi}\left(\frac{2m_e}{\hbar^2}\right)^{1/2} (k_B T)^{3/2} \int_0^\infty f(x-y)\, x^{1/2}\, dx \tag{9.83}$$

Here

$$H(x) = \int_0^x x^{1/2}\, dx = \frac{2}{3} x^{3/2} \tag{9.84}$$

The total energy can be calculated by solving the Fermi distribution function integral in Eq. (9.83) and is given by

$$E_T = \frac{L}{2\pi}\left(\frac{2m_e}{\hbar^2}\right)^{1/2} (k_B T)^{3/2}\left[1 + \frac{\pi^2}{6}\frac{\partial^2}{\partial y^2} + \cdots\right]\left(\frac{2}{3} y^{3/2}\right) \tag{9.85}$$

Solving the above equation, one gets, at low temperatures,

$$E_T = \frac{N_e}{2E_F^{1/2}}\left[\frac{2}{3}\mu^{3/2} + \frac{\pi^2}{12}\frac{(k_B T)^2}{\mu^{1/2}}\right] \tag{9.86}$$

In writing the above expression, the terms with the lowest powers of T are retained. Substituting the value of $\mu(T) = E_F(T)$ from Eq.(9.81) into Eq. (9.86), we finally obtain

$$E_T = \frac{1}{3} N_e E_F\left[1 + \frac{\pi^2}{4}\left(\frac{k_B T}{E_F}\right)^2\right] \tag{9.87}$$

retaining the terms up to T^2. The electronic specific heat at constant volume C_e can be calculated from Eq. (9.87) and is given by

$$C_e = \left(\frac{\partial E_T}{\partial T}\right)_V = \frac{\pi^2}{6} N_e k_B \left(\frac{T}{T_F}\right) \tag{9.88}$$

This shows that C_e depends linearly on the temperature.

9.6.2 Two-Dimensional Free-Electron Gas

Substituting $N_e(E_{\mathbf{k}})$ for a two-dimensional electron gas from Eq. (9.44) into Eq. (9.72) and making the substitution given by Eq. (9.74), we obtain

$$N_e = \frac{m_e A_0}{\pi\hbar^2} k_B T \int_0^\infty f(x-y)\, dx \tag{9.89}$$

Here $h(x) = 1$ in the above integral, so

$$H(x) = \int_0^x dx = x \tag{9.90}$$

N_e can be evaluated by solving the Fermi distribution function integral in Eq. (9.89) and is given as

$$\begin{aligned} N_e &= \frac{m_e A_0}{\pi \hbar^2} k_B T \left[1 + \frac{\pi^2}{6} \frac{\partial^2}{\partial y^2} + \cdots \right] y \\ &= \frac{m_e A_0}{\pi \hbar^2} \mu(T) = \frac{m_e A_0}{\pi \hbar^2} E_F(T) \end{aligned} \tag{9.91}$$

At $T = 0$ K, N_e can also be obtained by integrating $N_e(E_{\mathbf{k}})$, given by Eq. (9.44), from zero to the Fermi energy, yielding

$$N_e = \frac{m_e A_0}{\pi \hbar^2} E_F \tag{9.92}$$

The total number of electrons in the free-electron gas is constant at all temperatures, therefore, equating Eqs. (9.91), (9.92), we obtain

$$E_F(T) = E_F \tag{9.93}$$

This shows that the Fermi energy in a two-dimensional free-electron gas is independent of temperature. The total energy of the two-dimensional free-electron gas is obtained by substituting $N_e(E_{\mathbf{k}})$ from Eq. (9.44) into Eq. (9.82) to get

$$E_T = \frac{m_e A_0}{\pi \hbar^2} (k_B T)^2 \cdot \int_0^\infty x f(x - y) dx \tag{9.94}$$

Here

$$H(x) = \frac{x^2}{2} \tag{9.95}$$

The Fermi distribution integral in Eq. (9.94) can be solved easily to get

$$E_T = \frac{1}{2} N_e E_F \left[1 + \frac{\pi^2}{3} \left(\frac{T}{T_F} \right)^2 \right] \tag{9.96}$$

The electronic specific heat at constant volume becomes

$$C_e = \left(\frac{\partial E_T}{\partial T} \right)_V = \frac{\pi^2}{3} N_e k_B \left(\frac{T}{T_F} \right) \tag{9.97}$$

From Eqs. (9.88), (9.97) it is evident that C_e in a two-dimensional free-electron gas is double the value in a one-dimensional free-electron gas.

9.6.3 Three-Dimensional Free-Electron Gas

The total number of electrons in a three-dimensional free-electron gas at finite temperature is obtained by substituting the value of $N_e(E_{\mathbf{k}})$ from Eq. (9.22) into Eq. (9.72), which gives

$$N_e = \frac{V}{2\pi^2} \left(\frac{2m_e}{\hbar^2} \right)^{3/2} \cdot \int_0^\infty E_{\mathbf{k}}^{1/2} f(E_{\mathbf{k}} - \mu) dE_{\mathbf{k}} \tag{9.98}$$

With the help of Eq. (9.74), the above equation can be written as

$$N_e = \frac{V}{2\pi^2}\left(\frac{2m_e}{\hbar^2}\right)^{3/2}(k_BT)^{3/2}\cdot\int_0^\infty x^{1/2}f(x-y)\,dx \tag{9.99}$$

The Fermi distribution function integral above can be solved easily to get

$$N_e = \frac{V}{3\pi^2}\left(\frac{2m_e}{\hbar^2}\right)^{3/2}\cdot\mu^{3/2}\left[1+\frac{\pi^2}{8}\left(\frac{k_BT}{\mu}\right)^2+\cdots\right] \tag{9.100}$$

At absolute zero the total number of electrons in the three-dimensional gas, from Eqs. (9.16), (9.19) is given by

$$N_e = \frac{V}{3\pi^2}\left(\frac{2m_e}{\hbar^2}\right)^{3/2}E_F^{3/2} \tag{9.101}$$

Equating Eqs. (9.100), (9.101), the temperature dependence of the Fermi energy becomes

$$E_F(T) = \mu(T) = E_F\left[1-\frac{\pi^2}{12}\left(\frac{k_BT}{E_F}\right)^2+\cdots\right] \tag{9.102}$$

This shows that E_F decreases with an increase in temperature. Fig. 9.7 shows the behavior of E_F as a function of temperature for one-, two-, and three-dimensional solids. Now the total energy of the three-dimensional gas is obtained by substituting the value of $N_e(E_k)$ from Eq. (9.22) into Eq. (9.82) and further using the substitution (Eq. 9.74) to get

$$E_T = \frac{V}{2\pi^2}\left(\frac{2m_e}{\hbar^2}\right)^{3/2}(K_BT)^{5/2}\int_0^\infty x^{3/2}f(x-y)\,dx \tag{9.103}$$

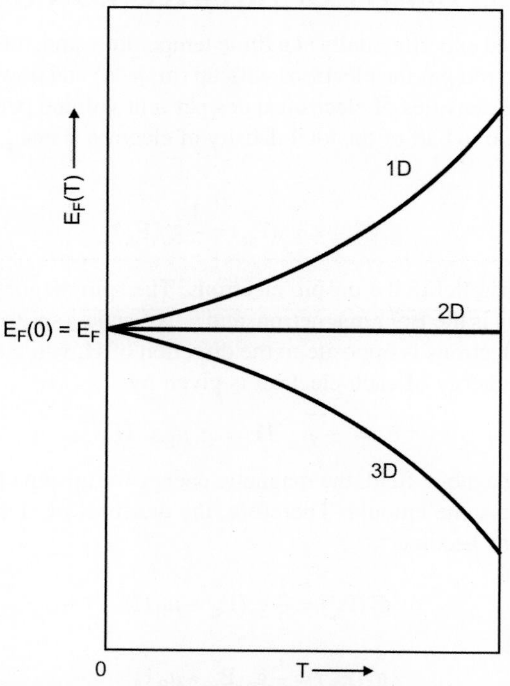

FIG. 9.7 Schematic representation of the temperature variation of the Fermi energy $E_F(T)$ for one-, two-, and three-dimensional free-electron gases.

From the above integral

$$H(x) = \int_0^x x^{3/2} dx = \frac{2}{5} x^{5/2} \tag{9.104}$$

The Fermi distribution function integral in Eq. (9.103) can be easily solved to write

$$E_T = \frac{V}{5\pi^2} \left(\frac{2m_e}{\hbar^2} \right)^{3/2} \mu^{5/2} \left[1 + \frac{5\pi^2}{8} \left(\frac{k_B T}{\mu} \right)^2 + \cdots \right] \tag{9.105}$$

Substituting $\mu(T)$ from Eq. (9.102) into the above expression and retaining terms only up to T^2 we get

$$E_T = \frac{3}{5} N_e E_F \left[1 + \frac{5\pi^2}{12} \left(\frac{k_B T}{E_F} \right)^2 + \cdots \right] \tag{9.106}$$

In writing the above expression we have used Eqs. (9.24), (9.25). With the help of the above expression, the electronic specific heat at constant volume becomes

$$\begin{aligned} C_e &= \left(\frac{\partial E_T}{\partial T} \right)_V = \frac{1}{2} \pi^2 N_e k_B \left(\frac{T}{T_F} \right) \\ &= \frac{1}{3} \pi^2 N_e(E_F) k_B^2 T \end{aligned} \tag{9.107}$$

It is noteworthy, from Eqs. (9.88), (9.97), (9.107) that the magnitude of the electronic specific heat is in the ratio of 1:2:3 for one-, two-, and three-dimensional solids. The temperature variation of C_e is the same regardless of the dimensionality of the solid, which is due to the same temperature variation of the total energy in these solids. In the above derivation the electron-electron and electron-ion interactions, which are finite in a solid, have been neglected. If these interactions are included in the derivation of C_e, then the form of the expression will remain the same except that the density of electron states $N_e(E_\mathbf{k})$ is modified.

9.7 PARAMAGNETIC SUSCEPTIBILITY OF FREE-ELECTRON GAS

Paramagnetic susceptibility is measured experimentally at a finite temperature and, therefore, temperature effects should be included in the theory. In the free-electron gas the electrons with up ($m_s = ½$) and down ($m_s = -½$) spins are distributed in parabolic bands. At absolute zero, the densities of electron states per unit volume per unit energy for up $g_\uparrow(E_\mathbf{k})$ and down $g_\downarrow(E_\mathbf{k})$ spins are equal and each is equal to half of the total density of electron states $g_e(E_\mathbf{k})$ (see Fig. 9.8A). Therefore, one can write

$$g_\uparrow(E_\mathbf{k}) = g_\downarrow(E_\mathbf{k}) = \frac{1}{2} g_e(E_\mathbf{k}) \tag{9.108}$$

Let the magnetic field **H** be applied parallel to the upspin electrons. The spin magnetic moment is given as $\vec{\mu}_s = -g_s \mu_B \mathbf{s}$ (see Chapter 18) where **s** is the spin, μ_B is the Bohr magnetron, and g_s is Lande's splitting factor for spin ($g_s = 2$). Therefore, the magnetic moment of the upspin electrons is opposite to the direction of **H**, while that of the downspin electrons is parallel to **H**. The magnetic interaction energy of each electron is given by

$$E = -\vec{\mu}_s \cdot \mathbf{H} = g_s \mu_B \mathbf{s} \cdot \mathbf{H} \tag{9.109}$$

Therefore, after the application of a magnetic field, the magnetic energy for upspin electrons is raised by $\mu_B H$ and that for downspin electrons is lowered by the same amount. Therefore, the densities of electron states per unit volume per unit energy for up- and downspin electrons become

$$g_\uparrow(E_\mathbf{k}) = \frac{1}{2} g_e(E_\mathbf{k} - \mu_B H) \tag{9.110}$$

$$g_\downarrow(E_\mathbf{k}) = \frac{1}{2} g_e(E_\mathbf{k} + \mu_B H) \tag{9.111}$$

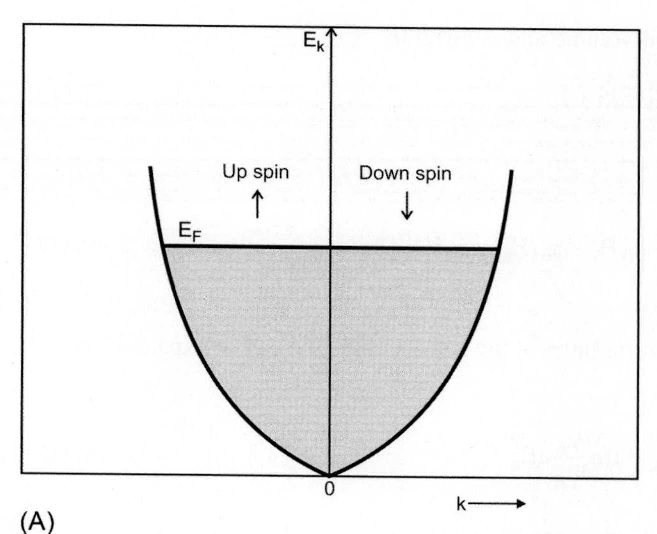

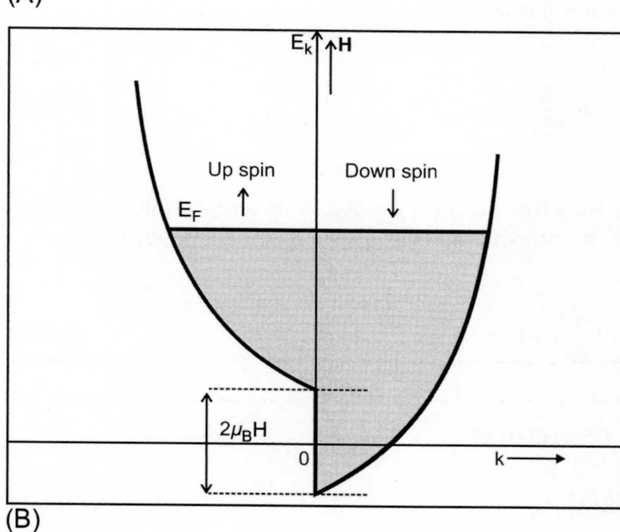

FIG. 9.8 (A) The parabolic energy bands for up- and downspin electrons in the absence of a magnetic field. The Fermi energy E_F is the same for both the spin states. (B) The parabolic energy bands for up- and downspin electrons in the presence of magnetic field **H** applied in the vertical direction.

To obtain the equilibrium state, the upspin electrons from the higher energy states will shift to the lower energy states with spin down until the Fermi level of the two distributions become the same (Fig. 9.8B). Hence the densities of electrons per unit volume with up $n_\uparrow$ and down spins $n_\downarrow$ become

$$n_\uparrow = \frac{1}{2}\int_{\mu_B H}^{\infty} f(E_\mathbf{k} - \mu)\, g_e(E_\mathbf{k} - \mu_B H)\, dE_\mathbf{k} \tag{9.112}$$

$$n_\downarrow = \frac{1}{2}\int_{-\mu_B H}^{\infty} f(E_\mathbf{k} - \mu)\, g_e(E_\mathbf{k} + \mu_B H)\, dE_\mathbf{k} \tag{9.113}$$

For a weak magnetic field, $\mu_B H$ is small and, therefore, the lower limit in the above integrals can be taken to be zero, allowing us to write

$$n_\uparrow = \frac{1}{2}\int_{0}^{\infty} f(E_\mathbf{k} - \mu)\, g_e(E_\mathbf{k} - \mu_B H)\, dE_\mathbf{k} \tag{9.114}$$

$$n_\downarrow = \frac{1}{2}\int_{0}^{\infty} f(E_\mathbf{k} - \mu)\, g_e(E_\mathbf{k} + \mu_B H)\, dE_\mathbf{k} \tag{9.115}$$

The magnetization is defined as the magnetic moment per unit volume and is given by

$$M = \mu_B \left(n_\downarrow - n_\uparrow \right) \tag{9.116}$$

Substituting Eqs. (9.114), (9.115) into Eq. (9.116), we get

$$M = \frac{1}{2} \mu_B \int_0^\infty f(E_\mathbf{k} - \mu) \left[g_e(E_\mathbf{k} + \mu_B H) - g_e(E_\mathbf{k} - \mu_B H) \right] dE_\mathbf{k} \tag{9.117}$$

Further, for a weak H field, the functions for the density of electron states in the above expression can be expanded around $E_\mathbf{k}$ to write (retaining terms linear in H)

$$M = \mu_B^2 H \int_0^\infty f(E_\mathbf{k} - \mu) \frac{\partial g_e}{\partial E_\mathbf{k}} dE_\mathbf{k} \tag{9.118}$$

Now the magnetic susceptibility per unit volume of the free-electron gas becomes

$$\chi_M = \frac{M}{H} = \mu_B^2 \int_0^\infty f(E_\mathbf{k} - \mu) \frac{\partial g_e}{\partial E_\mathbf{k}} dE_\mathbf{k} \tag{9.119}$$

This is the general expression for the paramagnetic susceptibility for a free-electron gas at a finite temperature and is also called the Pauli spin susceptibility. Eq. (9.119), with the help of the substitution from Eq. (9.74), can be written as

$$\chi_M = \mu_B^2 \int_0^\infty dx\, f(x - y) \frac{\partial}{\partial x} g_e(E_\mathbf{k}) \tag{9.120}$$

The density of electron states $g_e(E_\mathbf{k})$ for a free-electron gas can be written as

$$g_e(E_\mathbf{k}) = C\, g_e(x) \tag{9.121}$$

where the constant C depends on the dimensionality of the free-electron gas under consideration. Eq. (9.120) becomes

$$\chi_M = \mu_B^2 C \int_0^\infty dx\, f(x - y) \frac{\partial}{\partial x} g_e(x) \tag{9.122}$$

Eq. (9.122) is the Fermi distribution function integral which allows us to write

$$h(x) = \frac{\partial}{\partial x} g_e(x) \tag{9.123}$$

and hence

$$H_{FD}(x) = \int_0^x h(x)\, dx = g_e(x) \tag{9.124}$$

After solving the integral in Eq. (9.122), χ_M is given by

$$\begin{aligned} \chi_M &= \mu_B^2 C \left[1 + \frac{\pi^2}{6} \frac{\partial^2}{\partial y^2} + \cdots \right] H_{FD}(y) \\ &= \mu_B^2 C\, g_e(y) \left[1 + \frac{\pi^2}{6} \frac{g_e''(y)}{g_e(y)} + \cdots \right] \end{aligned} \tag{9.125}$$

where

$$g_e''(y) = \frac{\partial^2}{\partial y^2} g_e(y) \tag{9.126}$$

Eq. (9.125) is the general expression and shows that the temperature dependence of χ_M depends on the density of electron states. Let us calculate χ_M at absolute zero. Integrating Eq. (9.119) by parts, one gets

$$\chi_M = \chi_P = \mu_B^2 \int_0^{\infty} g_e(E_{\mathbf{k}}) \left(-\frac{\partial f}{\partial E_{\mathbf{k}}} \right) dE_{\mathbf{k}} \tag{9.127}$$

At absolute zero, the slope of $f(E_{\mathbf{k}} - \mu)$ is finite only at E_F and is the Dirac delta function (see Fig. 9.6B), that is,

$$-\left(\frac{\partial f}{\partial E_{\mathbf{k}}} \right) = \delta(E_{\mathbf{k}} - E_F) \tag{9.128}$$

Substituting Eq. (9.128) into Eq. (9.127), one gets the familiar expression for χ_M as

$$\chi_M = \chi_P = \mu_B^2 g_e(E_F) \tag{9.129}$$

The same expression can be obtained by putting $f(E_{\mathbf{k}} - \mu) = 1$ in Eq. (9.119).

9.7.1 One-Dimensional Free-Electron Gas

In a one-dimensional free-electron gas, $g_e(E_{\mathbf{k}})$ is given by Eq. (9.52), which can be written as

$$g_e(E_{\mathbf{k}}) = C\, g_e(x) \tag{9.130}$$

where

$$C = (k_B T)^{-1/2} \tag{9.131}$$

and

$$g_e(x) = \frac{1}{2\pi} \left(\frac{2m_e}{\hbar^2} \right)^{1/2} x^{-1/2} \tag{9.132}$$

Substituting the values of C and $g_e(y)$ from Eqs. (9.131), (9.132) into Eq. (9.125), one gets

$$\chi_P = \mu_B^2 \frac{1}{2\pi} \left(\frac{2m_e}{\hbar^2} \right)^{1/2} \mu^{-1/2} \left[1 + \frac{\pi^2}{8} \left(\frac{k_B T}{\mu} \right)^2 + \cdots \right] \tag{9.133}$$

Substituting $\mu(T)$ from Eq. (9.81) into the above expression and retaining terms up to the second power of T, one gets

$$\chi_P = \mu_B^2 g_e(E_F) \left[1 + \frac{\pi^2}{12} \left(\frac{k_B T}{E_F} \right)^2 + \cdots \right] \tag{9.134}$$

According to Eq. (9.134), at low temperatures, Pauli's spin susceptibility increases with an increase in temperature. At absolute zero it gives the same expression as Eq. (9.129).

9.7.2 Two-Dimensional Free-Electron Gas

In a two-dimensional free-electron gas, $g_e(E_{\mathbf{k}})$ is given by Eq. (9.45) and is a constant quantity. Therefore,

$$g_e(E_{\mathbf{k}}) = \frac{m_e}{\pi \hbar^2} = g_e(y) \tag{9.135}$$

and

$$g_e''(y) = 0, \quad C = 1 \tag{9.136}$$

Substituting Eqs. (9.135), (9.136) into Eq. (9.125), one immediately writes

$$\chi_P = \mu_B^2 g_e(E_F) \tag{9.137}$$

Therefore, in a two-dimensional free-electron gas, Pauli's paramagnetic spin susceptibility is independent of temperature.

9.7.3 Three-Dimensional Free-Electron Gas

In a three-dimensional free-electron gas, $g_e(E_\mathbf{k})$ is given by Eq. (9.23), which can also be written as

$$g_e(E_\mathbf{k}) = (k_B T)^{1/2} g_e(x) \tag{9.138}$$

where

$$g_e(x) = \frac{1}{2\pi^2}\left(\frac{2m_e}{\hbar^2}\right)^{3/2} x^{1/2} \tag{9.139}$$

From Eq. (9.138) we can write

$$C = (k_B T)^{1/2} \tag{9.140}$$

Now it is straight forward to prove (from Eq. 9.139) that

$$\frac{g_e''(y)}{g_e(y)} = -\frac{1}{4} y^{-2} \tag{9.141}$$

Substituting Eqs. (9.139), (9.141) into Eq. (9.125) and simplifying, we get

$$\chi_P = \mu_B^2 \frac{1}{2\pi^2}\left(\frac{2m_e}{\hbar^2}\right)^{3/2} \mu^{1/2}\left[1 - \frac{\pi^2}{24}\left(\frac{k_B T}{\mu}\right)^2 + \cdots\right] \tag{9.142}$$

Substituting $\mu(T)$ from Eq. (9.102) into the above equation and simplifying, we get

$$\chi_P = \mu_B^2 g_e(E_F)\left[1 - \frac{\pi^2}{12}\left(\frac{k_B T}{E_F}\right)^2 + \cdots\right] \tag{9.143}$$

The above expression shows that Pauli's spin susceptibility decreases with an increase in temperature. The temperature dependences of the paramagnetic susceptibilities for one-, two-, and three-dimensional free-electron gases are shown in Fig. 9.9 and they exhibit different trends for different dimensionalities. Further, comparing Eqs. (9.81), (9.93), (9.102) for $E_F(T)$ with Eqs. (9.134), (9.137), (9.143) for χ_P, it is evident that the temperature variation of χ_P is the same as that of the corresponding $E_F(T)$. It is also evident from Figs. 9.7 and 9.9.

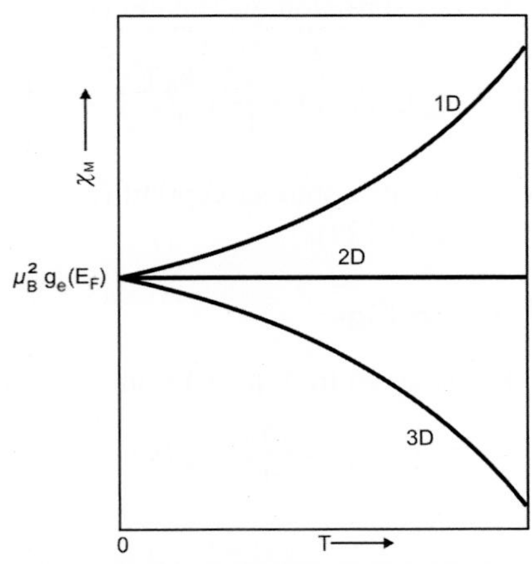

FIG. 9.9 Schematic representation of the temperature variation of paramagnetic susceptibility (the Pauli spin susceptibility) of one-, two-, and three-dimensional free-electron gases.

9.8 CLASSICAL SPIN SUSCEPTIBILITY

The classical expression for spin susceptibility can be derived from the general expression in the limit of very high temperatures. In the very high temperature limit, the Fermi-Dirac distribution function, defined by Eq. (9.68), becomes

$$f(E_{\mathbf{k}} - \mu) = e^{\frac{\mu - E_{\mathbf{k}}}{k_B T}} \tag{9.144}$$

So, in the very high temperature limit, the magnetic susceptibility of the free-electron gas, given by Eq. (9.119), becomes

$$\chi_M = \mu_B^2 e^{\frac{\mu}{k_B T}} \int_0^{\infty} e^{\frac{-E_{\mathbf{k}}}{k_B T}} \frac{\partial g_e}{\partial E_{\mathbf{k}}} dE_k \tag{9.145}$$

Integrating the above equation by parts, one gets

$$\chi_M = \frac{\mu_B^2}{k_B T} e^{\frac{\mu}{k_B T}} \int_0^{\infty} e^{\frac{-E_{\mathbf{k}}}{k_B T}} g_e(E_{\mathbf{k}}) dE_{\mathbf{k}} \tag{9.146}$$

In writing the above expression we have used the fact that $g_e(E_{\mathbf{k}}) \to 0$ as $E_{\mathbf{k}} \to 0$. From Eqs. (9.114), (9.115) the total number of electrons per unit volume can be written as

$$n_e = n_\uparrow + n_\downarrow = \int_0^{\infty} f(E_{\mathbf{k}} - \mu)\, g_e(E_{\mathbf{k}})\, dE_{\mathbf{k}} \tag{9.147}$$

which, in the very high temperature limit, becomes

$$n_e = e^{\frac{\mu}{k_B T}} \int_0^{\infty} e^{\frac{-E_{\mathbf{k}}}{k_B T}} g_e(E_k)\, dE_{\mathbf{k}} \tag{9.148}$$

Substituting Eq. (9.148) into Eq. (9.146), one can immediately obtain

$$\chi_M = \frac{\mu_B^2 n_e}{k_B T} \tag{9.149}$$

which is the classical expression for the spin magnetic susceptibility per unit volume.

In this chapter we have concentrated on simple metals that are characterized by free conduction electrons contained in broad s- (p-) bands. But in the periodic table of the elements, there exist a large number of d- and f-band metals in which the d-band and f-band are localized. In a d-band metal each atom possesses s- or p-electrons in the outermost shell, which are loosely bound to the nucleus and can be regarded as free electrons. Just below the outermost shell is the d-shell containing electrons that are neither tightly bound to the nucleus nor free like the s-electrons. Therefore, a d-band metal can be regarded as a sea of s-conduction electrons, with nearly uniform density, in which the ions with a quasilocalized (or deformable) shell are embedded at the lattice positions. The study of the electronic structure and electronic properties of these metals is more involved and is not within the scope of this book, but interested readers may consult Galsin (2002) for further study.

REFERENCE

Galsin, J. S. (2002). *Impurity scattering in metallic alloys*. New York: Kluwer Academic/Plenum Publishers.

SUGGESTED READING

Donovan, B. (1967). *Elementary theory of metals.* New York: Pergamon Press.
Hume-Rothery, W. (1931). *The metallic state: Electrical properties and theories.* London: Clarendon Press.
Hurd, C. M. (1975). *Electrons in metals.* New York: J. Wiley & Sons.
Mott, N. F., & Jones, H. (1936). *The theory of the properties of metals and alloys.* New York: Dover Publishers.
Seitz, F. (1943). *The physics of metals.* New York: McGraw-Hill Book Co.
Wilson, A. H. (1954). *The theory of metals* (2nd ed.). London: Cambridge University Press.
Wilson, A. W. (1939). *Semiconductors and metals: An introduction to the electron theory of metals.* London: Cambridge University Press.

Chapter 10

Electrons in Electric and Magnetic Fields

Chapter Outline

A number of properties of solids, such as the electrical properties, dielectric properties, and magnetic properties, depend on the electric and magnetic fields. Therefore, it is of great interest to study, in general, the effects of electric and magnetic fields on crystalline solids. In this Chapter we consider simple metals in which the ions are situated at the lattice positions, while the conduction electrons interact with each other.

10.1 EQUATION OF MOTION

Consider a system of interacting electrons confined to a cubical box of finite size. An electron, during its motion, interacts with other electrons and gets scattered. Therefore, one can define a relaxation time τ_e, which is some sort of average time between two consecutive scattering processes (collisions) of an electron. The equation of motion of an interacting electron in the presence of an applied force **F** is given by

$$\frac{d\mathbf{p}}{dt} + \frac{\mathbf{p}}{\tau_e} = \mathbf{F} \tag{10.1}$$

where $\mathbf{p} = m_e\mathbf{v}$ is the momentum of an electron moving with velocity **v**. The first term is due to Newton's second law of motion and the second due to the collision processes. The above equation can be written in terms of **v** as

$$m_e\left[\frac{d}{dt} + \frac{1}{\tau_e}\right]\mathbf{v} = \mathbf{F} \tag{10.2}$$

From wave mechanics we know that

$$\mathbf{p} = m_e\mathbf{v} = \hbar\mathbf{k} \tag{10.3}$$

where **k** is the electron wave vector in the reciprocal space. Therefore, Eq. (10.1) can also be written as

$$\hbar\left[\frac{d}{dt} + \frac{1}{\tau_e}\right]\mathbf{k} = \mathbf{F} \tag{10.4}$$

Solid State Physics. https://doi.org/10.1016/B978-0-12-817103-5.00010-4

If **F** is the Lorentz force experienced by an electron due to the application of an electromagnetic field, it is given by

$$\mathrm{F} = -e\left[\mathbf{E} + \frac{1}{c}\mathbf{v}\times\mathbf{H}\right] \tag{10.5}$$

Here **E** and **H** are electric and magnetic fields and $-e$ is the electronic charge. Substituting Eq. (10.5) into Eqs. (10.2), (10.4), the general equation of motion becomes

$$m_e\left[\frac{d}{dt} + \frac{1}{\tau_e}\right]\mathbf{v} = -e\left[\mathbf{E} + \frac{1}{c}\mathbf{v}\times\mathbf{H}\right] \tag{10.6}$$

$$\hbar\left[\frac{d}{dt} + \frac{1}{\tau_e}\right]\mathbf{k} = -e\left[\mathbf{E} + \frac{1}{c}\mathbf{v}\times\mathbf{H}\right] \tag{10.7}$$

For free electrons the relaxation time goes to infinity, that is, $\tau_e \rightarrow \infty$. Therefore, the equation of motion for a free electron becomes

$$m_e\frac{d\mathbf{v}}{dt} = -e\left[\mathbf{E} + \frac{1}{c}\mathbf{v}\times\mathbf{H}\right] \tag{10.8}$$

$$\hbar\frac{d\mathbf{k}}{dt} = -e\left[\mathbf{E} + \frac{1}{c}\mathbf{v}\times\mathbf{H}\right] \tag{10.9}$$

10.2 FREE ELECTRONS IN A STATIC ELECTRIC FIELD

Let a static electric field **E** be applied to the free electron gas with electron density n_e. The equation of motion of an electron from Eq. (10.8) reduces to

$$m_e\frac{d\mathbf{v}}{dt} = -e\mathbf{E} \tag{10.10}$$

In the presence of an electric field, the electrons move in the direction opposite to that of the electric field. From Eq. (10.10), a small change in velocity $\delta\mathbf{v}(\delta t)$ in time δt is given by

$$\delta\mathbf{v}(\delta t) = -\frac{e\mathbf{E}}{m_e}\delta t \tag{10.11}$$

The free electrons move continuously without any hindrance in the presence of the applied field **E**. But in the presence of collisions, the electrons acquire new equilibrium positions after the relaxation time τ_e. The change in velocity in time τ_e is given by

$$\delta\mathbf{v}(\tau_e) = -\frac{e\mathbf{E}}{m_e}\tau_e \tag{10.12}$$

The electric current density **J** becomes

$$\mathbf{J} = -n_e e\,\delta\mathbf{v}(\tau_e) = \sigma_0\mathbf{E} \tag{10.13}$$

where

$$\sigma_0 = \frac{n_e e^2 \tau_e}{m_e} \tag{10.14}$$

Here we have substituted the value of $\delta\mathbf{v}(\tau_e)$ from Eq. (10.12). Eq. (10.13) is nothing but the Ohm's law and σ_0 gives the Drude's conductivity for an electron gas. The resistivity of the electron gas ρ_0 is given by

$$\rho_0 = \frac{1}{\sigma_0} = \frac{m_e}{n_e e^2 \tau_e} \tag{10.15}$$

One can also derive the expression for **J** in terms of the wave vector **k**. For a free electron, Eq. (10.9) in the presence of **E** reduces to

$$\hbar \frac{d\mathbf{k}}{dt} = -e\mathbf{E} \tag{10.16}$$

In the presence of electron collisions, the change in wave vector $\delta\mathbf{k}$ in time τ_e is given by

$$\delta\mathbf{k}(\tau_e) = -\frac{e\mathbf{E}}{\hbar}\tau_e \tag{10.17}$$

After time τ_e the electrons acquire their new equilibrium positions, as a result of which the Fermi sphere is displaced by $\delta\mathbf{k}(\tau_e)$ (see Fig. 10.1). From Eqs. (10.3), (10.13) the current density in terms of $\delta\mathbf{k}(\tau_e)$ can be written as

$$\mathbf{J} = -\frac{n_e e\hbar}{m_e}\delta\mathbf{k} \tag{10.18}$$

Substituting the value of $\delta\mathbf{k}$ from Eq. (10.17) into Eq. (10.18), one can immediately get the expression for σ_0 given by Eq. (10.14).

10.3 FREE ELECTRONS IN A STATIC MAGNETIC FIELD

If a free electron moves in the presence of an applied static magnetic field $\mathbf{H}$, then Eq. (10.8) reduces to

$$m_e \frac{d\mathbf{v}}{dt} = -\frac{e}{c}\mathbf{v} \times \mathbf{H} \tag{10.19}$$

If $\mathbf{H}$ is in the z-direction, that is, $\mathbf{H} = \hat{\mathbf{i}}_3 H$, then the Cartesian components of Eq. (10.19) become

$$m_e \frac{dv_x}{dt} = -\frac{eH}{c}v_y \tag{10.20}$$

$$m_e \frac{dv_y}{dt} = \frac{eH}{c}v_x \tag{10.21}$$

$$\frac{dv_z}{dt} = 0 \tag{10.22}$$

Eq. (10.22) shows that the velocity v_z in the direction of the applied magnetic field is constant in time. On the other hand, Eqs. (10.20), (10.21) show that v_x and v_y are functions of time and their solution can be obtained as described below. Differentiating Eq. (10.20) with respect to time and using Eq. (10.21), we get

$$\frac{d^2v_x}{dt^2} + \omega_c^2 v_x = 0 \tag{10.23}$$

where

$$\omega_c = \frac{eH}{m_e c} \tag{10.24}$$

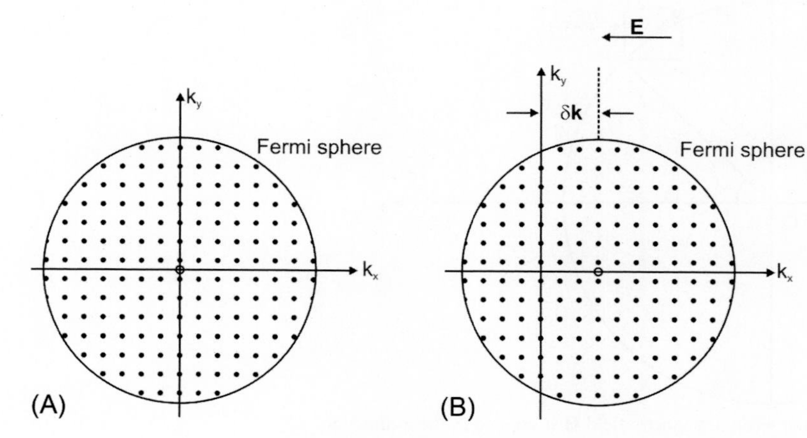

FIG. 10.1 (A) The Fermi sphere of a free-electron gas in the ground state (at absolute zero) in the absence of an electric field at zero time. The *dots* inside the Fermi sphere represent the filled electronic states. (B) The Fermi sphere of the electron gas in the presence of an electric field $\mathbf{E}$. It is displaced by a wave vector $\delta\mathbf{k}$ in time δt after the application of the electric field. Here the shape of the Fermi surface is assumed to be unaffected in the presence of the electric field.

Here ω_c is called the cyclotron frequency. In exactly the same manner, the differential equation for v_y can also be obtained, given by

$$\frac{d^2v_y}{dt^2}+\omega_c^2 v_y=0 \tag{10.25}$$

The applied magnetic field does not change the energy of an electron. Therefore, the solution of Eqs. (10.22), (10.23), (10.25) should be such that the energy of the electron remains unchanged. But v_z is already constant, therefore, the velocity of the electron in the xy-plane should also be constant. If v_0 is the magnitude of constant velocity in the xy-plane, then

$$v_x^2+v_y^2=v_0^2 \tag{10.26}$$

Eq. (10.26) shows that the electron moves in a circular path with frequency ω_c in the xy-plane, that is, perpendicular to the direction of magnetic field (see Fig. 10.2). Therefore, v_x and v_y from Eqs. (10.23), (10.25) are given by

$$v_x=-v_0\sin\omega_c t \tag{10.27}$$

$$v_y=v_0\cos\omega_c t \tag{10.28}$$

10.4 ELECTRONS IN STATIC ELECTRIC AND MAGNETIC FIELDS

From Eqs. (10.6) one can write

$$\dot{\mathbf{v}}=-\frac{e}{m_e}\left[\mathbf{E}+\frac{1}{c}\mathbf{v}\times\mathbf{H}\right]-\frac{\mathbf{v}}{\tau_e} \tag{10.29}$$

If the magnetic field **H** is in the z-direction, then the components of Eq. (10.29) are given by

$$\dot{v}_x=-\frac{e}{m_e}E_x-\frac{e}{m_e c}v_y H-\frac{v_x}{\tau_e} \tag{10.30}$$

$$\dot{v}_y=-\frac{e}{m_e}E_y-\frac{e}{m_e c}(-v_x H)-\frac{v_y}{\tau_e} \tag{10.31}$$

$$\dot{v}_z=-\frac{e}{m_e}E_z-\frac{v_z}{\tau_e} \tag{10.32}$$

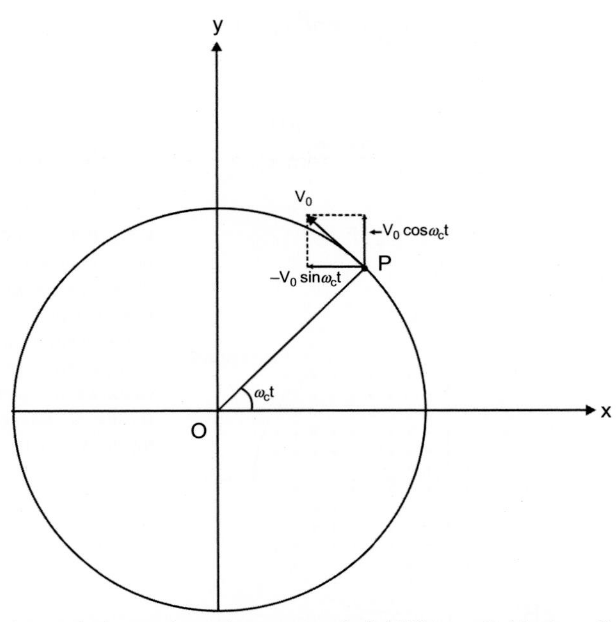

FIG. 10.2 The circular motion of an electron in the xy-plane when a magnetic field **H** is applied in the z-direction.

Multiplying Eqs. (10.30), (10.31), (10.32) by $n_e e\ \tau_e$, we obtain

$$-\tau_e \dot{J}_x = -\sigma_0 E_x + \omega_c \tau_e J_y + J_x \quad (10.33)$$

$$-\tau_e \dot{J}_y = -\sigma_0 E_y - \omega_c \tau_e J_x + J_y \quad (10.34)$$

$$-\tau_e \dot{J}_z = -\sigma_0 E_z + J_z \quad (10.35)$$

Here Eqs. (10.13), (10.14) for **J** and σ_0, respectively, have been used. In the equilibrium condition

$$\dot{J}_x = \dot{J}_y = \dot{J}_z = 0 \quad (10.36)$$

Therefore, for a system in the equilibrium state, Eqs. (10.33), (10.34), (10.35) reduce to

$$J_x + \omega_c \tau_e J_y = \sigma_0 E_x \quad (10.37)$$

$$-(\omega_c \tau_e) J_x + J_y = \sigma_0 E_y \quad (10.38)$$

$$J_z = \sigma_0 E_z \quad (10.39)$$

Eqs. (10.37), (10.38), (10.39) can be written in matrix form as

$$\begin{pmatrix} E_x \\ E_y \\ E_z \end{pmatrix} = \begin{pmatrix} \frac{1}{\sigma_0} & \frac{\omega_c \tau_e}{\sigma_0} & 0 \\ -\frac{\omega_c \tau_e}{\sigma_0} & \frac{1}{\sigma_0} & 0 \\ 0 & 0 & \frac{1}{\sigma_0} \end{pmatrix} \begin{pmatrix} J_x \\ J_y \\ J_z \end{pmatrix} \quad (10.40)$$

which can also be written as

$$E_\alpha = \sum_\beta \rho_{\alpha\beta} J_\beta \quad (10.41)$$

Here $\rho_{\alpha\beta}$ are the Cartesian components of the magnetoresistivity tensor $\overleftrightarrow{\rho}$. From Eqs. (10.40), (10.41) it is evident that the diagonal and nondiagonal components of $\overleftrightarrow{\rho}$ are given by

$$\rho_{xx} = \rho_{yy} = \rho_{zz} = \frac{1}{\sigma_0} \quad (10.42)$$

$$\rho_{xy} = -\rho_{yx} = \frac{\omega_c \tau_e}{\sigma_0}, \quad \rho_{xz} = \rho_{zx} = \rho_{yz} = \rho_{zy} = 0 \quad (10.43)$$

The diagonal components $\rho_{\alpha\alpha}$ are called the magnetoresistivity and are scalar quantities independent of the magnetic field. ρ_{zz} is called the longitudinal magnetoresistivity as the applied magnetic field is parallel to the current. The nondiagonal component ρ_{xy} (or ρ_{yx}) depends on the magnetic field and gives the Hall resistivity. These are also called the transverse magnetoresistivity as the magnetic field is perpendicular to the current or electric field. In the presence of electric and magnetic fields perpendicular to each other, the electron executes a helical path, as shown in Fig. 10.3. One should note that if the magnetic field or the relaxation time is zero, then ρ_{xy} and ρ_{yx} go to zero. The diagonal and nondiagonal components of resistivity in a cubical solid with side L are related to the resistance R as follows:

$$\rho_{xx} = \frac{E_x}{J_x} = \frac{V_x/L}{I_x/L^2} = R_{xx} L \quad (10.44)$$

$$\rho_{yx} = \frac{E_y}{J_x} = \frac{V_y/L}{I_x/L^2} = R_{yx} L \quad (10.45)$$

V_x and V_y are the components of voltage along the x- and y-directions and I_x and I_y are the corresponding current components. Eqs. (10.44), (10.45) show that the resistance in three dimensions depends on the size of the solid. Eqs. (10.37), (10.38), (10.39) can be solved for the components of the current density to write

$$J_x = \frac{\sigma_0}{1 + (\omega_c \tau_e)^2} E_x + \frac{(-\omega_c \tau_e)\sigma_0}{1 + (\omega_c \tau_e)^2} E_y \quad (10.46)$$

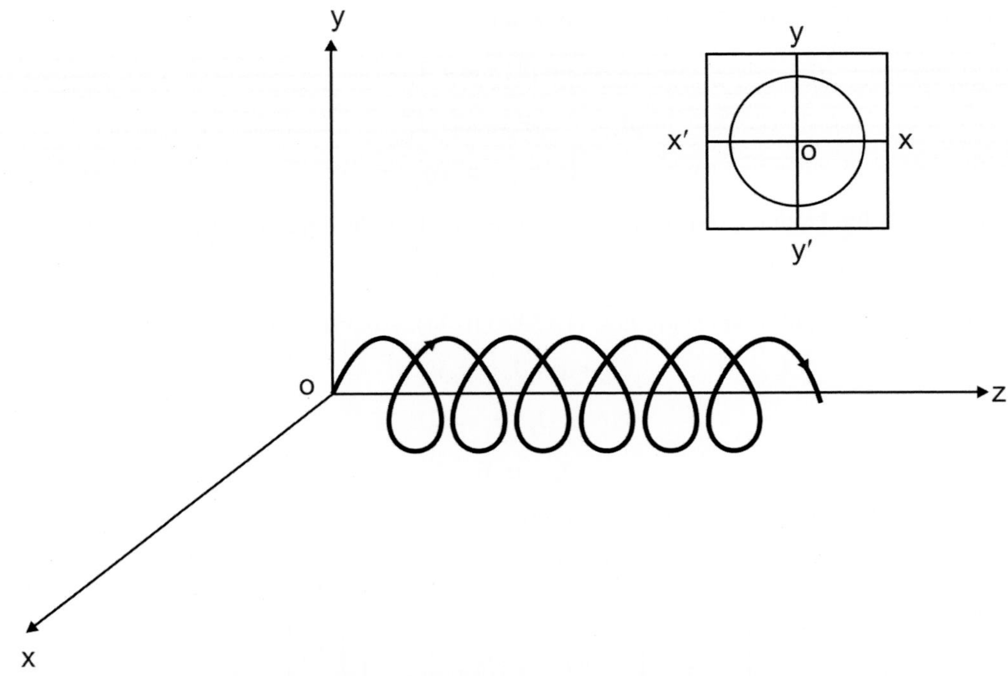

FIG. 10.3 Spiral motion of an electron about the direction of a magnetic field, that is, about the z-axis. The inset diagram shows the circular motion of the electron perpendicular to the magnetic field, that is, in the xy-plane.

$$J_y = \frac{\omega_c \tau_e \sigma_0}{1+(\omega_c \tau_e)^2} E_x + \frac{\sigma_0}{1+(\omega_c \tau_e)^2} E_y \tag{10.47}$$

$$J_z = \sigma_0 E_z \tag{10.48}$$

The above equations can be written in matrix form as

$$\begin{pmatrix} J_x \\ J_y \\ J_z \end{pmatrix} = \begin{pmatrix} \frac{\sigma_0}{1+(\omega_c \tau_e)^2} & \frac{-(\omega_c \tau_e)\sigma_0}{1+(\omega_c \tau_e)^2} & 0 \\ \frac{(\omega_c \tau_e)\sigma_0}{1+(\omega_c \tau_e)^2} & \frac{\sigma_0}{1+(\omega_c \tau_e)^2} & 0 \\ 0 & 0 & \sigma_0 \end{pmatrix} \cdot \begin{pmatrix} E_x \\ E_y \\ E_z \end{pmatrix} \tag{10.49}$$

which can also be written as

$$J_\alpha = \sum_\beta \sigma_{\alpha\beta} E_\beta \tag{10.50}$$

$\sigma_{\alpha\beta}$ are the components of magnetoconductivity tensor $\overleftrightarrow{\sigma}$. It is evident from Eq. (10.49) that

$$\sigma_{xx} = \sigma_{yy} (\neq \sigma_{zz}), \ \sigma_{xy} = -\sigma_{yx}$$
$$\sigma_{yz} = \sigma_{zy} = \sigma_{zx} = \sigma_{xz} = 0 \tag{10.51}$$

The diagonal components σ_{xx} and σ_{yy} give the magnetoconductivity, while the nondiagonal components σ_{xy} and σ_{yx} give the Hall conductivity. Further, if the relaxation time or the magnetic field is zero, then $\sigma_{xy} = \sigma_{yx} = 0$ and the diagonal components give the Drude conductivity.

10.5 THE HALL EFFECT IN METALS

Consider a slab of metallic material subjected to a uniform magnetic field **H** along the z-direction, that is, $\mathbf{H} = \hat{\mathbf{z}}H$, and with a current density J_x passed through the material along the x-direction (see Fig. 10.4). Due to the magnetic force acting on the charge carriers (Fleming's left-hand rule), they will be deflected and a current in the y-direction is set up. Soon after, an

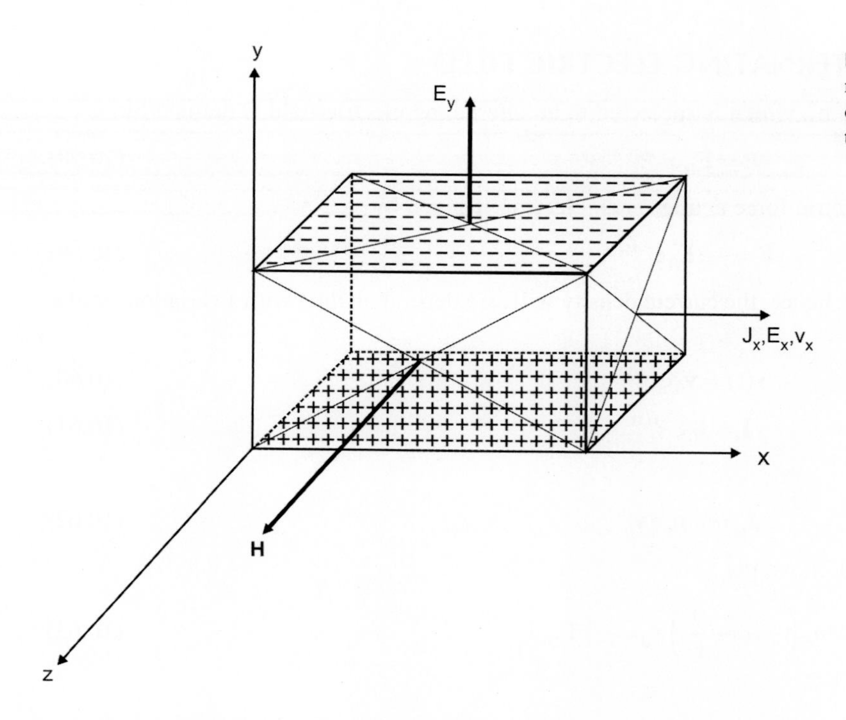

FIG. 10.4 The Hall voltage E_y in a rectangular slab of a metallic solid when a magnetic field **H** is applied in the z-direction. The applied electric field E_x and the velocity of the charge carriers v_x are in the x-direction.

equilibrium state is achieved, which makes the current in the y-direction zero, but develops an electric field E_y called the Hall field. In a metallic solid the current is due to the flow of electrons; therefore, the bottom surface will become positively charged as the electrons collect on the top surface. But if the current in a solid is due to the positive charges, then the bottom surface will become negative relative to the top surface. The Hall coefficient R_{HC} is defined as the electric field E_y developed per unit magnetic field H per unit current density J_x, that is,

$$R_{HC} = \frac{E_y}{J_x H} \tag{10.52}$$

The coefficient R_{HC} can be calculated for a general charge carrier q. The Lorenz force acting on a charge q is given by

$$\mathbf{F} = q\left(\mathbf{E} + \frac{1}{c}\mathbf{v} \times \mathbf{H}\right) \tag{10.53}$$

where **v** is the velocity of the charge carrier. For the magnetic field in the z-direction, the force acting on the charge carrier along the y-direction is given by

$$F_y = q\left[E_y + \frac{1}{c}(-v_x H)\right] \tag{10.54}$$

In the equilibrium position, the force F_y must be zero, which gives E_y as

$$E_y = \frac{1}{c} v_x H \tag{10.55}$$

But the current density J_x is given by

$$J_x = nqv_x \tag{10.56}$$

Substituting Eqs. (10.55), (10.56) into Eq. (10.52), R_{HC} is given by

$$R_{HC} = \frac{1}{nqc} \tag{10.57}$$

One noteworthy feature of the Hall effect is that R_{HC} is negative if the charge carriers are negatively charged (say electrons with charge −e) and positive if the charge carriers are positively charged (say holes with charge e). Therefore, the nature of charge carriers in a solid can be inferred from the sign of R_{HC} determined from the experimental study of the Hall effect.

10.6 FREE ELECTRONS IN AN ALTERNATING ELECTRIC FIELD

Consider an electron gas with electron density n_e, which is subjected to an alternating electric field $\mathbf{E}$ defined by

$$\mathbf{E} = \mathbf{E}_0 e^{-\imath\omega t} \tag{10.58}$$

where ω is the frequency of the field. The electric force acting on an electron is given by

$$\mathbf{F} = -e\mathbf{E}_0 e^{-\imath\omega t} \tag{10.59}$$

In this situation the velocity of the electron and, hence, the current density will also depend on time with a variation similar to that of $\mathbf{E}$, that is,

$$\mathbf{v}(t) = \mathbf{v}_0 e^{-\imath\omega t} \tag{10.60}$$

$$\mathbf{J} = \mathbf{J}_0 e^{-\imath\omega t} \tag{10.61}$$

where

$$\mathbf{J}_0 = -n_e e\mathbf{v}_0 \tag{10.62}$$

Substituting Eqs. (10.59), (10.60) into Eq. (10.2), we get

$$m_e\left(-\imath\omega + \frac{1}{\tau_e}\right)\mathbf{v}_0 = -e\mathbf{E}_0 \tag{10.63}$$

From the above equation $\mathbf{v}_0$ is given by

$$\mathbf{v}_0 = -\frac{e\tau_e/m_e}{1-\imath\omega\tau_e}\mathbf{E}_0 \tag{10.64}$$

Using Eq. (10.64), the amplitude of the current density $\mathbf{J}_0$, defined by Eq. (10.62), becomes

$$\mathbf{J}_0 = \frac{\sigma_0}{1-\imath\omega\tau_e}\mathbf{E}_0 \tag{10.65}$$

The frequency-dependent electrical conductivity $\sigma(\omega)$ is defined as

$$\mathbf{J}_0 = \sigma(\omega)\mathbf{E}_0 \tag{10.66}$$

From Eqs. (10.65), (10.66) one can write

$$\sigma(\omega) = \frac{\sigma_0}{1-\imath\omega\tau_e} \tag{10.67}$$

It is evident that $\sigma(\omega)$ is a complex quantity having both real and imaginary parts, denoted by $\sigma_1(\omega)$ and $\sigma_2(\omega)$, respectively. Therefore, one can write

$$\sigma(\omega) = \sigma_1(\omega) + \imath\,\sigma_2(\omega) \tag{10.68}$$

$$\sigma_1(\omega) = \frac{\sigma_0}{1+(\omega\tau_e)^2} \tag{10.69}$$

$$\sigma_2(\omega) = \frac{(\omega\tau_e)\sigma_0}{1+(\omega\tau_e)^2} \tag{10.70}$$

From Eq. (10.69), it is clear that $\sigma_1(\omega)$ has a maximum value σ_0 at $\omega\tau_e = 0$ and thereafter decreases continuously with an increase in $\omega\tau_e$. On the other hand, $\sigma_2(\omega)$ shows resonant behavior with a resonance at $\omega\tau_e = 1$ having maximum value of $(1/2)\sigma_0$. It is interesting to study the behavior of $\sigma(\omega)$ at high frequencies. For $\omega\tau_e >> 1$, Eqs. (10.69), (10.70) reduce to

$$\sigma_1(\omega) = \frac{\sigma_0}{(\omega\tau_e)^2} = \frac{n_e e^2}{m_e\omega^2\tau_e} \tag{10.71}$$

$$\sigma_2(\omega) = \frac{\sigma_0}{\omega\tau_e} = \frac{n_e e^2}{m_e\omega} \tag{10.72}$$

It is evident from Eqs. (10.71), (10.72) that $\sigma_2(\omega)$ is independent of the relaxation time and dominates over $\sigma_1(\omega)$.

It is always convenient to express the results in terms of a complex dielectric function $\varepsilon(\omega)$ that is defined as

$$\varepsilon(\omega) = 1 + 4\pi\chi(\omega) \tag{10.73}$$

where

$$\chi(\omega) = \frac{\mathbf{P}_0}{\mathbf{E}_0} \tag{10.74}$$

where $\mathbf{P}_0$ is the complex amplitude of the polarization $\mathbf{P}$ defined as

$$\mathbf{P} = \mathbf{P}_0 e^{-\imath\omega t} \tag{10.75}$$

To calculate the polarization $\mathbf{P}_0$, Eq. (10.2) can be written as

$$m_e\left(\frac{d^2}{dt^2} + \frac{1}{\tau_e}\frac{d}{dt}\right)\mathbf{r} = \mathbf{F} \tag{10.76}$$

Here $\mathbf{r}$ is the position vector and $\mathbf{v} = d\mathbf{r}/dt$. The time dependence of $\mathbf{r}$ is the same as that of $\mathbf{v}$, that is,

$$\mathbf{r} = \mathbf{r}_0 e^{-\imath\omega t} \tag{10.77}$$

Substituting Eq. (10.77) into Eq. (10.76), we get

$$\mathbf{r}_0 = \frac{(e/m_e)\mathbf{E}_0}{\omega^2 + \imath\omega/\tau_e} \tag{10.78}$$

The dipole moment of an electron is given by $-e\mathbf{r}_0$. Therefore, $\mathbf{P}_0$ for the electron gas is given by

$$\mathbf{P}_0 = -n_e e\mathbf{r}_0 = -\frac{(n_e e^2/m_e)\mathbf{E}_0}{\omega^2 + \imath\omega/\tau_e} \tag{10.79}$$

$\mathbf{P}_0$ is a complex quantity and its real and imaginary parts can be separated to write

$$\mathbf{P}_0 = \mathbf{P}_0' + \imath\mathbf{P}_0'' \tag{10.80}$$

where

$$\mathbf{P}_0' = -\frac{(n_0 e^2/m_e)\mathbf{E}_0}{\omega^2\left[1 + 1/(\omega\tau_e)^2\right]} \tag{10.81}$$

$$\mathbf{P}_0'' = \frac{(n_0 e^2/m_e)\mathbf{E}_0}{\omega^2\left[\omega\tau_e + 1/\omega\tau_e\right]} \tag{10.82}$$

Substituting $\mathbf{P}_0$ from Eqs. (10.80)–(10.82) in Eq. (10.74) and then in Eq. (10.73), the complex dielectric matrix $\varepsilon(\omega)$ becomes

$$\varepsilon(\omega) = \varepsilon_1(\omega) + \imath\,\varepsilon_2(\omega) \tag{10.83}$$

where

$$\varepsilon_1(\omega) = 1 - \frac{\omega_P^2}{\omega^2}\left[\frac{1}{1 + (1/\omega\tau_e)^2}\right] \tag{10.84}$$

$$\varepsilon_2(\omega) = \frac{\omega_P^2}{\omega^2}\left[\frac{\omega/\tau_e}{\omega^2 + 1/\tau_e^2}\right] \tag{10.85}$$

and

$$\omega_P^2 = \frac{4\pi n_e e^2}{m_e} \tag{10.86}$$

ω_P is a constant frequency and is usually called the plasma frequency for reasons to be described below. The wavelength λ_P associated with ω_P is given by $\lambda_P = 2\pi c/\omega_P$. Eqs. (10.83)–(10.85) give the general expression for the dielectric function.

The limiting cases of $\varepsilon(\omega)$ are of more interest. For example, when $\tau_e \to \infty$, $\varepsilon_2(\omega)$ reduces to zero, while $\varepsilon_1(\omega)$ is finite. Therefore, in the free-electron approximation, $\varepsilon(\omega)$ reduces to $\varepsilon_1(\omega)$ given by

$$\varepsilon(\omega) = \varepsilon_1(\omega) = 1 - \frac{\omega_P^2}{\omega^2} \tag{10.87}$$

The dielectric function $\varepsilon(\omega)$ is negative for $\omega^2 < \omega_P^2$. The waves satisfying this condition (with wavelength λ greater than λ_P) possess an imaginary wave vector and decay exponentially. Therefore, waves with $\omega^2 < \omega_P^2$ are totally reflected from the free-electron gas. On the other side, $\varepsilon(\omega)$ is positive for $\omega^2 > \omega_P^2$ and such high-frequency waves (with λ less than λ_P) can pass through the free-electron gas. For $\omega = \omega_P$, $\varepsilon(\omega)$ is zero and, therefore, ω_P acts as a cutoff frequency. The above discussion shows that the free-electron gas acts as a high-pass filter with a cutoff frequency ω_P (or wavelength λ_P) that depends on the electron density. It has been found that the alkali metals are transparent to ultraviolet light.

10.7 QUANTUM MECHANICAL THEORY OF ELECTRONS IN STATIC ELECTRIC AND MAGNETIC FIELDS

Consider an electron gas in which an electron experiences an electric field E_0 in the x-direction. The force experienced by an electron is $F(x) = -e E_0$. As a result, an electron is subject to the potential $V(x)$ given by

$$V(x) = -\int F(x)\,dx = e E_0 x \tag{10.88}$$

The Schrodinger equation for an electron is given by

$$\left(\frac{p^2}{2m_e} + V(x)\right)|\psi(\mathbf{r})\rangle = E\,|\psi(\mathbf{r})\rangle \tag{10.89}$$

The momentum operator $\mathbf{p} = -\imath\hbar\nabla$. To introduce the spin, we consider

$$p^2 = \mathbf{p}\cdot\mathbf{p} = (\mathbf{S}^P\cdot\mathbf{p})(\mathbf{S}^P\cdot\mathbf{p}) = (\mathbf{S}^P\cdot\mathbf{p})^2 \tag{10.90}$$

Here $\mathbf{S}^P = (S_x^P, S_y^P, S_z^P)$ denotes the Pauli spin matrices and $\mathbf{S}^P = 2\mathbf{s}$ where $\mathbf{s}$ denotes the spin matrices. In writing the above equation, the following identity has been used

$$(\mathbf{S}^P\cdot\mathbf{a})(\mathbf{S}^P\cdot\mathbf{b}) = \mathbf{a}\cdot\mathbf{b} + \imath\mathbf{S}^P\cdot(\mathbf{a}\times\mathbf{b}) \tag{10.91}$$

Here $\mathbf{a}$ and $\mathbf{b}$ are any two vectors. Using Eq. (10.90) in Eq. (10.89), we write

$$\left[\frac{1}{2m_e}(\mathbf{S}^P\cdot\mathbf{p})^2 + V(x)\right]|\psi(\mathbf{r})\rangle = E|\psi(\mathbf{r})\rangle \tag{10.92}$$

If a magnetic field $\mathbf{H}$ is applied on the electron gas, then the momentum changes as follows:

$$\mathbf{p} \to \mathbf{p} - \frac{e}{c}\mathbf{A} \tag{10.93}$$

Here $\mathbf{A}$ is the vector potential defined as follows

$$\mathbf{H} = \nabla\times\mathbf{A} \tag{10.94}$$

Hence the Schrodinger equation in the presence of electric and magnetic fields is given by

$$\left[\frac{1}{2m_e}\left\{\mathbf{S}^P\cdot\left(\mathbf{p} - \frac{e}{c}\mathbf{A}\right)\right\}^2 + V(x)\right]|\psi(\mathbf{r})\rangle = E|\psi(\mathbf{r})\rangle \tag{10.95}$$

Using the identity (10.91)

$$\begin{aligned}\left[\mathbf{S}^{\mathrm{P}}\cdot\left(\mathbf{p}-\frac{\mathrm{e}}{\mathrm{c}}\mathbf{A}\right)\right]^2&=\left[\mathbf{S}^{\mathrm{P}}\cdot\left(\mathbf{p}-\frac{\mathrm{e}}{\mathrm{c}}\mathbf{A}\right)\right]\left[\mathbf{S}^{\mathrm{P}}\cdot\left(\mathbf{p}-\frac{\mathrm{e}}{\mathrm{c}}\mathbf{A}\right)\right]\\&=\left(\mathbf{p}-\frac{\mathrm{e}}{\mathrm{c}}\mathbf{A}\right)^2+\imath\mathbf{S}^{\mathrm{P}}\cdot\left(\mathbf{p}-\frac{\mathrm{e}}{\mathrm{c}}\mathbf{A}\right)\times\left(\mathbf{p}-\frac{\mathrm{e}}{\mathrm{c}}\mathbf{A}\right)\\&=\left(\mathbf{p}-\frac{\mathrm{e}}{\mathrm{c}}\mathbf{A}\right)^2-\imath\frac{\mathrm{e}}{\mathrm{c}}\mathbf{S}^{\mathrm{P}}\cdot(\mathbf{p}\times\mathbf{A}+\mathbf{A}\times\mathbf{p})\\&=\left(\mathbf{p}-\frac{\mathrm{e}}{\mathrm{c}}\mathbf{A}\right)^2-\imath\frac{\mathrm{e}}{\mathrm{c}}\mathbf{S}^{\mathrm{P}}\cdot(-\imath\hbar\mathbf{H})\\&=\left(\mathbf{p}-\frac{\mathrm{e}}{\mathrm{c}}\mathbf{A}\right)^2-\frac{2\mathrm{e}\hbar}{\mathrm{c}}\mathbf{s}\cdot\mathbf{H}\end{aligned}\tag{10.96}$$

Substituting Eq. (10.96) into Eq. (10.95), one gets

$$\left[\frac{1}{2\mathrm{m_e}}\left(\mathbf{p}-\frac{\mathrm{e}}{\mathrm{c}}\mathbf{A}\right)^2+\vec{\mu}_{\mathrm{s}}\cdot\mathbf{H}+\mathrm{V}(\mathrm{x})\right]|\psi(\mathbf{r})\rangle=\mathrm{E}|\psi(\mathbf{r})\rangle\tag{10.97}$$

where $\vec{\mu}_s = -(e\hbar/m_e c)\,\mathbf{s}$ is the operator for the spin magnetic moment. To evaluate the stationary states for the electron system, the magnetic field is assumed to be in the z-direction, that is, $\mathbf{H}=\hat{\mathbf{z}}H$. The vector potential $\mathbf{A}$ is chosen by using the gauge transformation

$$\mathbf{A}=(0,\mathrm{Hx},0)\tag{10.98}$$

Eq. (10.97) can be expanded to write

$$\frac{1}{2\mathrm{m_e}}\left(\mathrm{p}^2-\frac{2\mathrm{e}}{\mathrm{c}}\mathbf{A}\cdot\mathbf{p}+\frac{\mathrm{e}^2}{\mathrm{c}^2}\mathrm{A}^2\right)|\psi(\mathbf{r})\rangle+\mathrm{eE_0x}|\psi(\mathbf{r})\rangle=[\mathrm{E}-(\pm\mu_{\mathrm{B}}\mathrm{H})]|\psi(\mathbf{r})\rangle\tag{10.99}$$

Here we have substituted $s_z = \pm 1/2$. Using Eq. (10.98) for $\mathbf{A}$ and the operator form of $\mathbf{p}$, the above equation becomes

$$\begin{aligned}\left[\frac{1}{2\mathrm{m_e}}\left\{-\hbar^2\left(\frac{\partial^2}{\partial x^2}+\frac{\partial^2}{\partial y^2}+\frac{\partial^2}{\partial z^2}\right)+\frac{2\imath\hbar\mathrm{eH}}{\mathrm{c}}\mathrm{x}\frac{\partial}{\partial y}+\frac{\mathrm{e}^2\mathrm{H}^2}{\mathrm{c}^2}\mathrm{x}^2\right\}+\mathrm{eE_0x}\right]|\psi(\mathbf{r})\rangle\\=[\mathrm{E}-(\pm\mu_{\mathrm{B}}\mathrm{H})]|\psi(\mathbf{r})\end{aligned}\tag{10.100}$$

$$\begin{aligned}-\frac{\hbar^2}{2\mathrm{m_e}}\left(\frac{\partial^2}{\partial x^2}+\frac{\partial^2}{\partial z^2}\right)|\psi(\mathbf{r})\rangle-\frac{\hbar^2}{2\mathrm{m_e}}\left(\frac{\partial^2}{\partial y^2}-\frac{2\imath\mathrm{eHx}}{\hbar\mathrm{c}}\frac{\partial}{\partial y}+\frac{\imath^2\mathrm{e}^2\mathrm{H}^2}{\hbar^2\mathrm{c}^2}\mathrm{x}^2\right)|\psi(\mathbf{r})\rangle\\+\mathrm{eE_0}x|\psi(\mathbf{r})\rangle=[\mathrm{E}-(\pm\mu_{\mathrm{B}}\mathrm{H})]|\psi(\mathbf{r})\rangle\end{aligned}\tag{10.101}$$

Eq. (10.101) obviously has a solution of the form

$$|\psi(\mathbf{r})\rangle=\mathrm{e}^{\imath\mathrm{k_y y}}\mathrm{e}^{\imath\mathrm{k_z z}}\mathrm{u(x)}\tag{10.102}$$

Substituting Eq. (10.102) into Eq. (10.101) and rearranging the terms, we get

$$\begin{aligned}\left[-\frac{\hbar^2}{2\mathrm{m_e}}\frac{\mathrm{d}^2}{\mathrm{dx}^2}+\frac{1}{2}\mathrm{m_e}\omega_{\mathrm{c}}^2(\mathrm{x}-\mathrm{X})^2+\mathrm{eE_0X}+\frac{1}{2}\mathrm{m_e}\left(\frac{\mathrm{cE_0}}{\mathrm{H}}\right)^2\right]\mathrm{u(x)}\\=\left[\mathrm{E}-\frac{\hbar^2\mathrm{k_z^2}}{2\mathrm{m_e}}-(\pm\mu_{\mathrm{B}}\mathrm{H})\right]\mathrm{u(x)}\end{aligned}\tag{10.103}$$

where

$$\mathrm{X}=\frac{\mathrm{c}\hbar\mathrm{k_y}}{\mathrm{eH}}-\frac{\mathrm{eE_0}}{\mathrm{m_e}\omega_{\mathrm{c}}^2}\tag{10.104}$$

The electron has linear velocity v_z in the z-direction due to the applied electric field, but it executes a circular motion in the xy-plane due to the application of the magnetic field along the z-direction. Therefore, according to Eq. (10.103), an electron, in the presence of both electric and magnetic fields, executes a spiral motion about the center X. If only the magnetic field H is acting on the system ($E_0 = 0$), then Eq. (10.103) yields

$$\left[-\frac{\hbar^2}{2m_e}\frac{d^2}{dx^2}+\frac{1}{2}m_e\omega_c^2(x-x_0)^2\right]u(x)=\left[E-\frac{\hbar^2k_z^2}{2m_e}-(\pm\mu_B H)\right]u(x) \tag{10.105}$$

Eq. (10.105) is the equation of motion of a harmonic oscillator having natural frequency ω_c and centered at

$$x_0=\frac{c\hbar k_y}{eH}=\frac{1}{\omega_c}\frac{\hbar k_y}{m_e}=\frac{v_y}{\omega_c} \tag{10.106}$$

The electron moves in a spiral motion about the center x_0. Thus, the energy eigenvalues of Eq. (10.105) are

$$E_s=\left(n_s+\frac{1}{2}\right)\hbar\omega_c \tag{10.107}$$

where n_s is an integer. Therefore,

$$E-\frac{\hbar^2k_z^2}{2m_e}-(\pm\mu_B H)=E_s \tag{10.108}$$

which gives

$$E=\frac{\hbar^2k_z^2}{2m_e}+E_s\pm\mu_B H \tag{10.109}$$

In Eq. (10.109) the first term corresponds to the free motion of electrons along the direction of the magnetic field **H** and yields parabolic bands due to the continuous values of k_z in a solid. The second term gives a discrete set of eigenvalues due to the harmonic oscillations of electrons in a plane perpendicular to **H** and the last term gives the spin splitting of the energy states. Fig. 10.5A shows the energy bands in the presence of a magnetic field, neglecting the effect of spin splitting. It shows that the motion of the electrons gets quantized in the crystal plane perpendicular to **H**. If the electric field is also finite, then from Eq. (10.103) the energy eigenvalues are given by

$$E=\frac{\hbar^2k_z^2}{2m_e}+E_s\pm\mu_B H+eE_0X+\frac{1}{2}m_e\left(\frac{cE_0}{H}\right)^2 \tag{10.110}$$

The last two terms give the effect of the electric field E_0.

Let us investigate the degeneracy of the energy given by Eq. (10.109). From Eq. (10.102) it is evident that the allowed values of k_y are given as

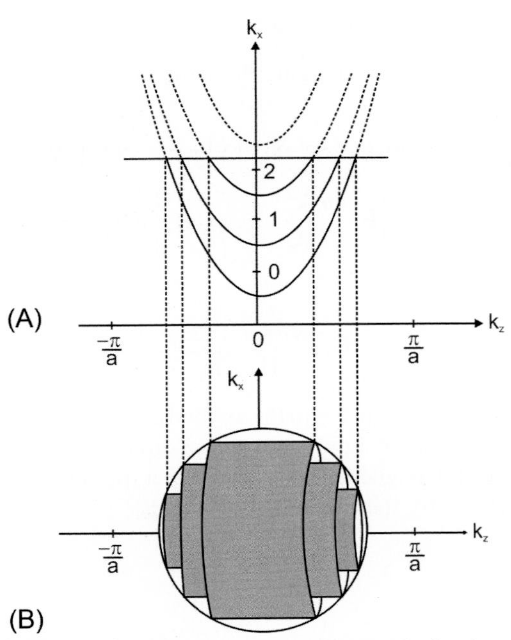

FIG. 10.5 (A) The quantized parabolic energy bands in the presence of a magnetic field neglecting the spin splitting of the energy bands. (B) The quantized Landau cylindrical levels within the Fermi sphere in the presence of the magnetic field.

$$k_y = \frac{2\pi n_y}{L_y} \tag{10.111}$$

where $n_y = 0, \pm 1, \pm 2, \ldots$ This means k_y is quantized in units of $2\pi/L_y$. But the energy is independent of k_y [Eq. (10.103)]. Therefore, for a particular value of n_s, one may think that k_y has any value out of the infinite series of its allowed values. But actually, it is not so as will be clear from the following arguments. As Eq. (10.105) represents a linear oscillator centered about x_0 (dependent on k_y), therefore, the function u(x) also depends on k_y. This means that if an electron starts off in the y-direction with velocity v_y it will move in a circular path in the magnetic field, with center at x_0 (see Fig. 10.6). This path must not be too big and lies inside the xy-plane with

$$0 < x_0 < L_x \tag{10.112}$$

In terms of x_0 the values of k_y are given by (see Eq. 10.106)

$$k_y = \frac{m_e \omega_c}{\hbar} x_0 \tag{10.113}$$

The restriction on the allowed values of x_0 also restricts the values of k_y, which are given by

$$0 < k_y < \frac{m_e \omega_c}{\hbar} L_x \left(= \frac{eH}{\hbar c} L_x \right) \tag{10.114}$$

There is one value of k_y in distance $2\pi/L_y$, so the maximum number of its values is

$$n_{k_y} = \frac{L_y}{2\pi} \frac{m_e \omega_c}{\hbar} L_x \tag{10.115}$$

Hence, for a particular value of n_s, there are n_{k_y} states corresponding to each value of k_y. In other words, an energy state for a particular value of n_s is n_{k_y}-fold degenerate. As k_z has continuous values for a bulk solid, the **k**-states lie on Landau cylinders (see Fig. 10.5B) in which the limit of occupancy is set by the original Fermi surface. Fig. 10.5B shows that the electron energy varies with k_z up to E_F on each cylinder, which is a one-dimensional magnetic subband, called a Landau level, rather than a constant energy surface. Each circle around any cylinder, with both k_z and n_s fixed, is a line of constant energy, generally called a Landau circle or Landau level. Fig. 10.7 shows the Landau circles for different values of n_s in the presence of a magnetic field.

One should note that in the absence of a magnetic field, k_x and k_y are also quantized, having values

$$k_x = \frac{2\pi n_x}{L_x} \tag{10.116}$$

$$k_y = \frac{2\pi n_y}{L_y} \tag{10.117}$$

where $n_x = n_y = 0, \pm 1, \pm 2, \ldots$ With the application of a magnetic field, the quantization described by Eqs. (10.116), (10.117) is broken. The wave function (Eq. 10.102) gives states with energy defined by Eq. (10.109) and these are n_{k_y}-fold

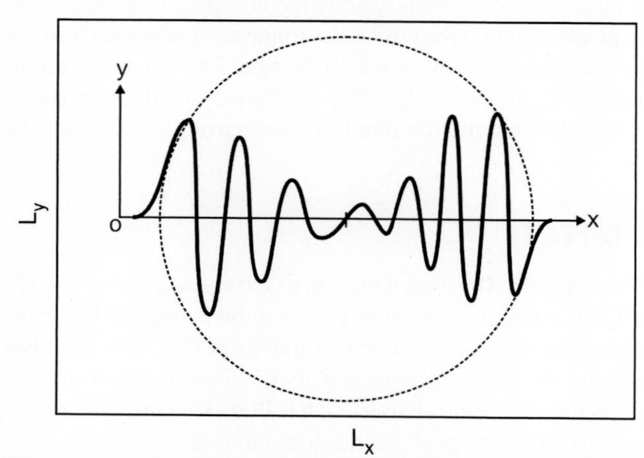

FIG. 10.6 The solution of the Schrodinger wave equation in the xy-plane in the presence of a magnetic field **H**. The electron moves in a circular path (*dashed line*) perpendicular to the z-direction.

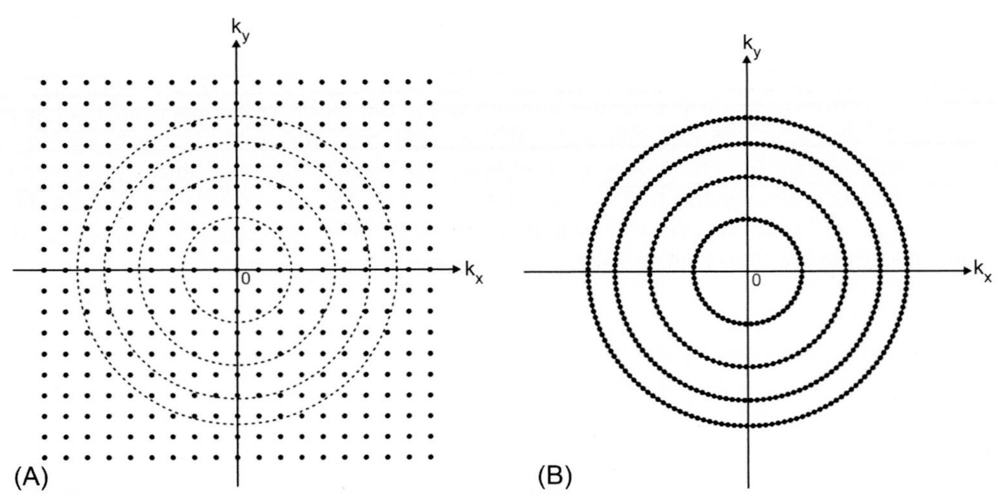

FIG. 10.7 (A) Quantization of the free electron states, shown by *dots*, in the absence of a magnetic field. The *circles (dashed lines)* show the Landau levels, which appear only in the presence of a magnetic field, as shown in part (B) of the figure. (B) Quantization of the electron states, represented by *dots*, in the Landau orbits in the presence of a magnetic field.

degenerate. Consider two energy surfaces with energies E and E + δE. The area between the two energy surfaces separated by energy δE can be written as [see Appendix J, Eq. (J.10)]

$$\delta A = \frac{2\pi m_c}{\hbar^2}\delta E \tag{10.118}$$

Here the electronic mass m_e is replaced by the cyclotron mass m_c. Let us assume that δE is the quantum of energy due to the cyclotron frequency, that is, $\delta E = \hbar\omega_c$. We know that in the free electron case the density of allowed states per unit area is given as $L_xL_y/(2\pi)^2$ in the (k_x, k_y) space. Therefore, the number of electron states in area δA is

$$\frac{L_xL_y}{(2\pi)^2}\delta A = \frac{2\pi m_c}{\hbar^2}(\hbar\omega_c)\frac{L_xL_y}{(2\pi)^2} \tag{10.119}$$

which, after simplification, gives

$$\frac{L_xL_y}{(2\pi)^2}\delta A = \frac{L_xL_y}{2\pi}\frac{m_c\omega_c}{\hbar} = n_{k_y} \tag{10.120}$$

Eq. (10.120) gives the number of allowed states between two quantized orbits. Therefore, the effect of the magnetic field is to create these quantized states (orbits) in **k**-space (see Fig. 10.7) and to cause the free electron states to "condense" onto the nearest such orbit. The number of states in each orbit is exactly the number of allowed states in the annulus in which it lies. The new states are not really fixed at any point on the circle, but rotate around it with frequency ω_c. In the magnetic field we can classify the various levels by naming the circles on which they lie. The quantized circular orbits, with energy given by Eq. (10.107), are called Landau orbits (Landau levels). The degeneracy of the Landau orbits depends upon the applied magnetic field (see Eq. 10.115). In other words, the number of electrons that occupy Landau orbits is proportional to the magnetic field.

10.8 QUANTUM HALL EFFECT

The Quantum Hall Effect (QHE) was observed by Von Klitzing, a German scientist (Von Klitzing, Dorda, & Pepper, 1980). The most interesting aspect of the QHE is that it is observed in two-dimensional (2D) electron systems only, for example, in an inversion layer of a metal-oxide-semiconductor field-effect transistor (MOSFET). It has been observed that for certain combinations of the magnetic field and the surface (areal) density of electrons, the Hall conductance has plateaus at values that are integral multiples of e^2/h: e is the electronic charge and h is Planck's constant. These plateaus extend over a range of the electron density. The QHE is observed under special conditions quite different from those of the ordinary Hall effect: the sample is kept at liquid He temperature ($\approx$ 4 K) and subjected to very high magnetic fields ($\approx$ 10 Tesla). Actually, Ando, Matsumoto, and Uemura (1975) were the first to observe this effect, but the results lacked sufficient precision to make sound conclusions. But now the results are of high precision (at least a few parts in 10^8).

10.8.1 Two-Dimensional Electron System

A 2D system consists of electrons confined to a thin layer, of about 100 Å thickness, near an interface between two dissimilar materials. An important question is how a layer of electrons with a finite thickness behaves as a 2D system. One can argue that if the thickness of the layer is smaller than the thermal wavelength of electrons at low temperatures, the motion of the electrons perpendicular to the layer becomes quantized. The excitation energies of electrons in the perpendicular direction are then much larger than the excitation energies in the plane of the layer and also much larger than the thermal energy. Under such conditions, the motion of electrons in the perpendicular direction is frozen, but they can move easily in the plane of the layer and hence it behaves like a 2D system. This fact has been confirmed experimentally, which shows that the density of states is indeed two dimensional. According to Ohm's law

$$\mathbf{E} = \rho \mathbf{J} \tag{10.121}$$

In a two-dimensional crystal, **J** is the current per unit width and **E** is the voltage per unit length. Hence, in a two-dimensional crystal, from Eq. (10.121),

$$\rho = \frac{\text{voltage}}{\text{current}} = \text{R} \tag{10.122}$$

where R is the resistance. Hence, in a two-dimensional crystal, the resistivity is equal to the resistance of the system, in contrast to a 3D system. In Ohm's law, **J** and **E** are in the same direction, yielding scalar resistivity. But when magnetic a field **H** is applied perpendicular to the surface of a two-dimensional system (and hence perpendicular to the current), an electric field perpendicular to both **H** and **J** is generated, which is called the Hall field. For a 2D system, the resistivity matrix can be written as

$$\begin{pmatrix} E_x \\ E_y \end{pmatrix} = \begin{pmatrix} \rho_{xx} & \rho_{xy} \\ \rho_{yx} & \rho_{yy} \end{pmatrix} \begin{pmatrix} J_x \\ J_y \end{pmatrix} \tag{10.123}$$

The off-diagonal components $\rho_{\alpha\beta}$ ($\alpha \neq \beta$) yield the Hall resistivity. In order to make measurements of the Hall effect, one has to specify the direction of the current when a magnetic field **H** is applied perpendicular to the plane of the 2D electron system having length L_x and breadth L_y (Fig. 10.8). Let the current flow along the x-direction ($J_y = 0$), then from Eq. (10.123) one can write

$$E_x = \rho_{xx} J_x \tag{10.124}$$

$$E_y = \rho_{yx} J_x \tag{10.125}$$

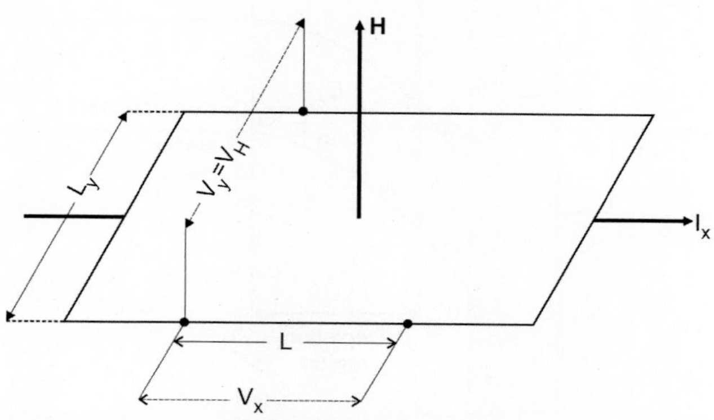

FIG. 10.8 Schematic diagram of a 2D system for measuring the quantum Hall effect. The length of the crystal is L_x and the breadth is L_y. The magnetic field **H** is applied in the z-direction.

We can relate the resistivity components with the components of resistance. From Eqs. (10.124), (10.125)

$$\rho_{xx}=\frac{E_x}{J_x}=\frac{V_x/L}{I_x/L_y}=\frac{V_x}{I_x}\frac{L_y}{L}=R_{xx}\frac{L_y}{L} \tag{10.126}$$

and

$$\rho_{yx}=\frac{E_y}{J_x}=\frac{V_y/L_y}{I_x/L_y}=\frac{V_y}{I_x}=R_{yx} \tag{10.127}$$

Here L is the distance between two points, along the direction of I_x, between which the potential difference V_x is measured. L_y is the distance, perpendicular to both I_x and **H**, across which the potential difference V_y is measured (see Fig. 10.9). It is noteworthy that ρ_{yx} and R_{yx} are equal, without any geometric factor, in contrast with the 3D crystal. Hence, a measurement of the Hall resistivity ρ_{yx} of a 2D system is independent of the dimensions of the crystal. It is this fact that makes possible the measurement of the Quantum Hall resistivity with a high degree of accuracy. In general, the conductivity matrix $\overleftrightarrow{\sigma}$ in two dimensions can be written as

$$\begin{pmatrix} J_x \\ J_y \end{pmatrix} = \begin{pmatrix} \sigma_{xx} & \sigma_{xy} \\ \sigma_{yx} & \sigma_{yy} \end{pmatrix} \begin{pmatrix} E_x \\ E_y \end{pmatrix} \tag{10.128}$$

10.8.2 Classical Theory of Conductivity in a Magnetic Field

The motion of an electron in the presence of electric and magnetic fields was described in Section 10.4. From Eq. (10.40) one can straightway write the expression for resistivity in two dimensions as

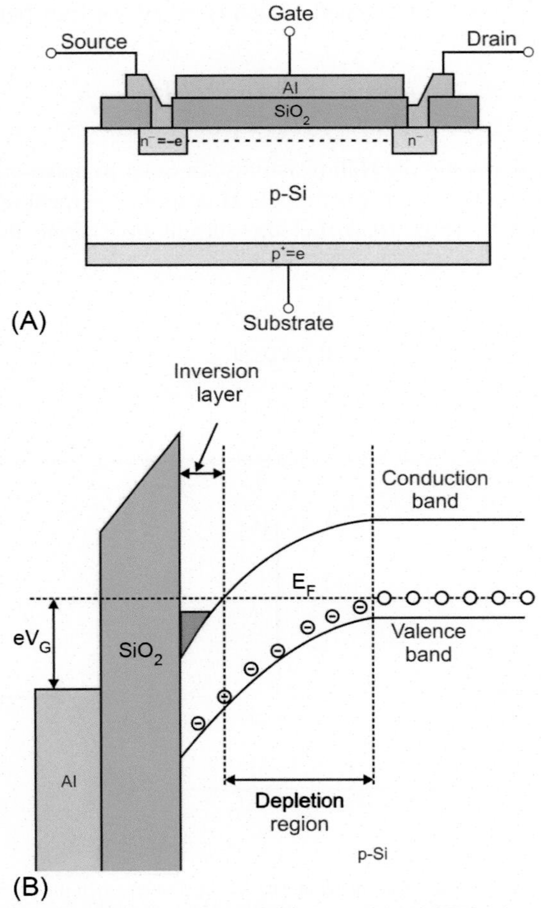

FIG. 10.9 (A) Schematic diagram of Si-MOSFET. (B) The energy band diagram of Si-MOSFET when positive gate voltage V_G is applied.

$$\begin{pmatrix} E_x \\ E_y \end{pmatrix} = \begin{pmatrix} \frac{1}{\sigma_0} & \frac{\omega_c \tau_e}{\sigma_0} \\ -\frac{\omega_c \tau_e}{\sigma_0} & \frac{1}{\sigma_0} \end{pmatrix} \begin{pmatrix} J_x \\ J_y \end{pmatrix} \tag{10.129}$$

Comparing Eq. (10.129) with Eq. (10.123), one can write

$$\rho_{xx} = \rho_{yy} = \frac{1}{\sigma_0}; \rho_{xy} = -\rho_{yx} = \frac{\omega_c \tau_e}{\sigma_0} \tag{10.130}$$

From Eq. (10.49), the conductivity of a two-dimensional solid can be written as

$$\begin{pmatrix} J_x \\ J_y \end{pmatrix} = \begin{pmatrix} \frac{\sigma_0}{1+(\omega_c\tau_e)^2} & \frac{-(\omega_c\tau_e)\sigma_0}{1+(\omega_c\tau_e)^2} \\ \frac{(\omega_c\tau_e)\sigma_0}{1+(\omega_c\tau_e)^2} & \frac{\sigma_0}{1+(\omega_c\tau_e)^2} \end{pmatrix} \begin{pmatrix} E_x \\ E_y \end{pmatrix} \tag{10.131}$$

Comparing Eqs. (10.131), (10.128), one immediately gets

$$\sigma_{xx} = \sigma_{yy} = \frac{\sigma_0}{1+(\omega_c\tau_e)^2}, \quad \sigma_{xy} = -\sigma_{yx} = -\frac{(\omega_c\tau_e)\sigma_0}{1+(\omega_c\tau_e)^2} \tag{10.132}$$

The relation between the components of the conductivity and resistivity matrices in a two-dimensional crystal can be obtained by dividing both the numerator and denominator on the right side of Eq. (10.132) by $(1/\sigma_0)^2$, which gives

$$\sigma_{xx} = \sigma_{yy} = \frac{\rho_{xx}}{\rho_{xx}^2 + \rho_{xy}^2}; \quad \sigma_{xy} = -\sigma_{yx} = -\frac{\rho_{xy}}{\rho_{xx}^2 + \rho_{xy}^2} \tag{10.133}$$

In studying the Hall effect, we are mainly interested in the calculation of σ_{xy}, which from Eq. (10.132) can be written as

$$-\sigma_{xy} = \frac{n_e ec}{H} \frac{(\omega_c\tau_e)^2}{1+(\omega_c\tau_e)^2} \tag{10.134}$$

In writing the above expression we have used Eqs. (10.14), (10.24) for σ_0 and ω_c, respectively. Further manipulation of the above equation will yield

$$-\sigma_{xy} = \frac{n_e ec}{H} - \frac{\sigma_{xx}}{\omega_c\tau_e} \tag{10.135}$$

If σ_{xx} vanishes, the Hall resistivity σ_{xy} is given by

$$\sigma_{xy} = -\frac{n_e ec}{H} \tag{10.136}$$

Eq. (10.136) is purely a classical result.

10.8.3 Quantum Theory of a 2D Free-Electron Gas in a Magnetic Field

Consider a two-dimensional free-electron gas in the xy-plane in which a magnetic field **H** is applied along the z-direction. The free electrons in a magnetic field will satisfy the Schrodinger wave equation

$$\frac{1}{2m_e}\left(\mathbf{p} - \frac{e}{c}\mathbf{A}\right)^2 |\psi\rangle = E|\psi\rangle \tag{10.137}$$

where **A** is the vector potential and $|\psi(x,y)\rangle$ is the wave function. We want to evaluate stationary states of the two-dimensional free-electron gas. Let us choose the vector potential using the gauze transformation as

$$\mathbf{A} = (0, Hx, 0) \tag{10.138}$$

which gives **H** along the z-direction. Using the two-dimensional Laplacian operator and the vector potential **A** from Eq. (10.138) and simplifying in exactly the same manner as we did in Section 10.7, we find

$$\frac{\partial^2}{\partial x^2}|\psi\rangle + \left(\frac{\partial}{\partial y} - \frac{\imath eH}{\hbar c}x\right)^2|\psi\rangle + \frac{2m_e E}{\hbar^2}|\psi\rangle = 0 \tag{10.139}$$

Eq. (10.139) obviously has a solution of the form

$$|\psi(x, y)\rangle = e^{\imath k_y y}u(x) \tag{10.140}$$

Substituting Eq. (10.140) into Eq. (10.139) and simplifying, we write

$$-\frac{\hbar^2}{2m_e}\frac{d^2u}{dx^2} + \frac{1}{2}m_e\omega_c^2(x - x_0)^2u(x) = Eu(x) \tag{10.141}$$

where x_0 is given by Eq. (10.106). Eq. (10.105) reduces to Eq. (10.141) if one substitutes

$$\frac{\hbar^2k_z^2}{2m_e} = \mu_B H = 0 \tag{10.142}$$

The energy in the z-direction is zero because the system is a two-dimensional free-electron gas and the magnetic interaction between the electron magnetic moment and the magnetic field is neglected. Eq. (10.141) is the equation of motion of a one-dimensional harmonic oscillator, having natural frequency ω_c, and centered at x_0. The electron makes a spiral motion about the center x_0. Thus, the energy eigenvalues of Eq. (10.141) are given by

$$E = \left(n_s + \frac{1}{2}\right)\hbar\omega_c \tag{10.143}$$

where n_s is an integer. This shows that the motion of electrons gets quantized in the plane of a two-dimensional free-electron gas.

The degeneracy in energy E, given by Eq. (10.143), can be calculated in exactly the same way as was done in Section 10.7. For a particular value of s, there are n_{k_y} states corresponding to each value of k_y given by

$$n_{k_y} = \frac{L_x L_y}{2\pi}\frac{m_e\omega_c}{\hbar} = \frac{A_0}{2\pi}\frac{m_e\omega_c}{\hbar} \tag{10.144}$$

where $A_0 = L_x L_y$ is the area of the 2D system. n_{k_y} gives the number of allowed states between two quantized orbits. The quantized circular orbits with energy given by Eq. (10.143) are called Landau orbits. The degeneracy of the Landau orbits depends upon the applied magnetic field (see Eq. 10.144). In other words, the number of electrons that occupy Landau levels is proportional to the magnetic field.

For the QHE to be observed, the temperature has to be low and the magnetic field has to be high enough so that the separation between the Landau levels (equal to magnetic energy $\hbar\omega_c$) is much larger than the thermal energy $k_B T$. With these conditions, the lower lying Landau levels will be completely filled with electrons and the higher levels completely empty. Under these circumstances, it is found that the Hall resistance R_{xy} is quantized and is given as

$$R_{xy} = \frac{h}{s_0 e^2} \tag{10.145}$$

where s_0 is the number of completely filled Landau levels, generally called the filling factor of the Landau levels. The filling factor can be changed either by changing the charge carrier density or by adjusting the magnetic field **H**. In either case, the position of the Fermi level is shifted relative to the position of the Landau levels. The plateaus in the Hall resistance are observed for integral values of the filling factor s_0, as mentioned above, and, therefore, it is usually called the integral QHE. The quantized Hall resistance can also be written in terms of the fine structure constant a_{fs} as

$$R_{xy} = \frac{\mu_0 c}{2 s_0 a_{fs}} \tag{10.146}$$

where μ_0 is the permeability of vacuum and has a value of 4×10^{-7} H/m. From Eqs. (10.145), (10.146) we have

$$a_{fs} = \frac{\mu_0 e^2 c}{2h} \tag{10.147}$$

Problem 10.1

The vector potential **A**, yielding a magnetic field in the z-direction, can also be given by the gauze

$$\mathbf{A}=(-yH,0,0) \tag{10.148}$$

Prove that the Schrodinger wave equation for a two-dimensional system with **A** given by Eq. (10.148) becomes

$$-\frac{\hbar^2}{2m_e}\frac{\partial^2\psi}{\partial y^2}-\frac{\hbar^2}{2m_e}\left(\frac{\partial}{\partial x}+\imath\frac{eH}{c\hbar}y\right)^2|\psi\rangle=E|\psi\rangle \tag{10.149}$$

Solve the Schrodinger wave equation to obtain the energy eigenvalues.

Problem 10.2

The general gauze for the vector potential, which yields a magnetic field along the z-direction, is written as

$$A=\left(-\frac{1}{2}yH,\frac{1}{2}xH,0\right) \tag{10.150}$$

Show that the Schrodinger wave equation, for the two-dimensional system with **A** given by Eq. (10.150) becomes

$$-\frac{\hbar^2}{2m_e}\left(\frac{\partial}{\partial x}+\imath\frac{eH}{2c\hbar}y\right)^2|\psi\rangle-\frac{\hbar^2}{2m_e}\left(\frac{\partial}{\partial y}-\imath\frac{eH}{2c\hbar}x\right)^2|\psi\rangle=E|\psi\rangle \tag{10.151}$$

10.8.4 Experimental Setup for QHE

There are two types of 2D systems used for observing the QHE:

1. Silicon MOSFETs (Si-MOSFETs)
2. Semiconductor heterojunctions

Both of these devices create 2D electron systems with very small thickness.

10.8.4.1 Silicon MOSFETs

A systematic diagram of a Si-MOSFET is shown in Fig. 10.9A. It consists of a Si base doped with a p-type material. An SiO_2 layer of about 1000 $\mathring{A}$ thickness is grown over the substrate. Above the SiO_2 layer is an Al metal layer for making good electrical contacts. A positive gate voltage is applied to the Al layer, which generates an electric field on the order of 10^6 volts/cm across the oxide layer. This field separates the electron-hole pairs in p-Si. The electrons move toward the SiO_2 interface, while the holes move away into the bulk p-Si substrate. The band picture of Si-MOSFET is shown in Fig. 10.9B. The gate voltage bends the energy bands of p-Si near the oxide interface. At sufficiently high gate voltage, the conduction band in Si bends so much near the interface that it crosses the Fermi energy. The electrons of the acceptor level and conduction band are, therefore, held by the energy barrier between the conduction band of Si and the oxide layer. Thus, the surface layer near the interface contains more electrons than holes even though the material is doped with acceptors and, therefore, is called an inversion layer. The electrons are confined to a narrow region of about 50 Å near the interface. The energies of excitation in a direction perpendicular to the interface are on the order of 20 meV, which are much larger than the excitation energy in the plane of the interface.

A very useful feature of the Si-MOSFET is that the carrier density is directly proportional to the gate voltage and can be adjusted continuously. The source and the drain provide contacts to the 2D system of electrons and a current source can be connected across these contacts.

10.8.4.2 Semiconductor Heterojunctions

A 2D electron system can be created with the help of a differentially doped semiconductor heterojunction interface. The most commonly used devices use a GaAs – (Al_xGa_{1-x})As interface, as shown in Fig. 10.10. It consists of a substrate of GaAs doped with Cr^+. Above this is a GaAs layer of 1 μm thickness and then layers of (AlGa)As and (AlGaAs) : Si(doped with Si). At the top again, we have a layer of GaAs. These heterojunctions are synthesized with the help of sophisticated techniques, such as Molecular Beam Epitaxy (MBE), which ensure abrupt steps in the conduction band at the interface. The experimental setup for the QHE with heterojunctions is also shown in Fig. 10.10.

The conduction band of GaAs is lower than the conduction band of (AlGa)As by about 300 meV. To maintain a constant Fermi level throughout the junction, the Si donors of the (AlGa)As side of the interface get ionized and transfer electrons to the GaAs side of the interface. The charge transfer produces strong electric fields and band bending near the interface, as a result of which a 2D electron system is formed, which is analogous to that formed in Si-MOSFET. Ohmic contacts can be made in the 2D system by alloying indium (IN) with the epilayer.

The mobility of the charge carriers in these devices is extremely high ($\approx 2 \times 10^6$ cm^2/V s.). This corresponds to a mean free path as large as 10 μm for an elastic event. In general, the low-temperature mobility is limited by the scattering of charge carriers from the ionized impurities. But in heterojunctions, such a scattering is small as the spatial separation between the charge carriers and the parent donors is sufficiently large. The introduction of a thin layer of undoped (AlGa)As ($\approx$ 100 Å) between doped (AlGaAs) : Si and pure GaAs layers further separates the charge carriers and the donors and hence enhances the mobility.

It is interesting to compare the merits and demerits of the two devices. In heterojunctions the charge carrier density is fixed and depends upon the Si donor concentration and the thickness of the undoped (AlGa)As layer. Hence, to change the position of the Fermi level relative to the position of the Landau levels, as is required in the QHE, the magnetic field has to be adjusted. This is a disadvantage because it is relatively more difficult to change the magnetic field than to change the charge carrier density by adjusting the gate voltage in MOSFETS. An important advantage of heterojunctions is that in them the effective mass of electrons is small ($\approx$0.068 m_e) as compared with that in MOSFETs in which it is on the order of 0.2 m_e for Si. The smaller effective mass leads to a larger characteristic magnetic energy ($=\hbar\omega_c$), which allows one to work at smaller magnetic fields ($\approx$ 8 Tesla versus 13 Tesla for Si) and at comparatively higher temperatures.

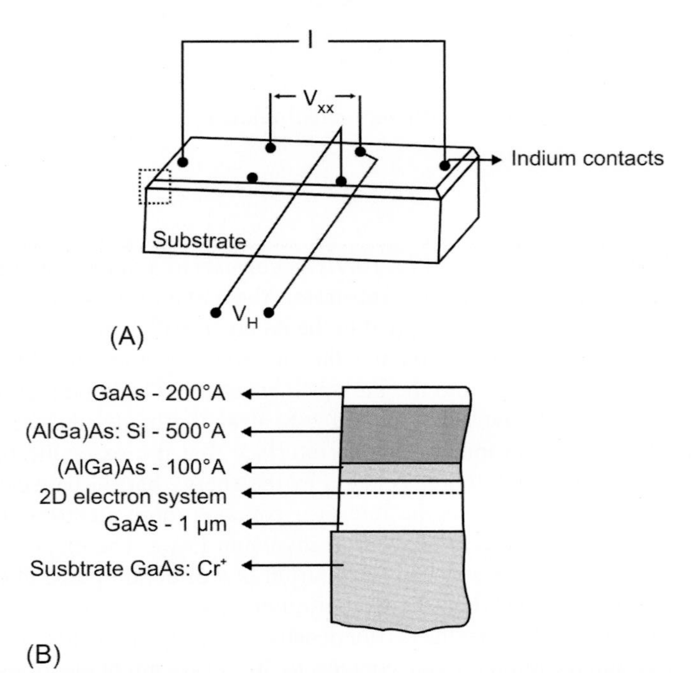

FIG. 10.10 (A) Experimental setup for measurements of the integral QHE with heterojunctions. (B) Magnified view of the edge of the heterojunction shown by a *small square (dashed line)* in part (A) of the figure.

10.8.5 Integral Quantum Hall Effect

When an integral numbers of Landau levels are completely filled, it is usually called the integral QHE. The experimental setup for measuring the Hall effect is shown in Fig. 10.9. A current of 1–10 μA is passed through the sample at the He temperature in the presence of a magnetic field **H**. The linear voltage V_{xx} and the Hall voltage V_H are measured. The linear resistance R_{xx} and the Hall resistance R_{xy} are given by the ratio of the respective voltages to the current I_x.

Typical measurements of Von Klitzing et al. (1980) on Si-MOSFET are shown in Fig. 10.11 in which the Hall voltage and the linear voltage are plotted as a function of the gate voltage V_G for a given source-drain current of 1 μA. A series of plateaus referred to as "Hall steps" are observed in which R_{xy} appears to be essentially constant and independent of V_G. The value of $R_{xy}(R_H)$ at these steps is given by Eq. (10.145) to a high degree of accuracy. The resistance R_{xx} appears to vanish in the region of these plateaus: ρ_{xx} as low as $5 \times 10^{-7}\,\Omega$ cm has been measured. This value is almost an order of magnitude lower than the resistivity of any nonsuperconducting material at any temperature: ρ_{xx} further decreases with temperature.

There are two main features of the QHE that require a physical understanding: one is the existence of Hall steps and the dissipationless current flow in the regions of Hall steps and the second is the high precision to which the Hall steps are quantized. A simple explanation is as follows: In a high magnetic field, the electrons occupy Landau levels having discrete energy. The Landau levels are highly degenerate with degeneracy given by Eq. (10.144). Therefore, the number of electrons per unit area n_e in a Landau level from Eq. (10.144) is given by

$$n_e = \frac{1}{2\pi}\frac{m_e\omega_c}{\hbar} = \frac{eH}{hc} \tag{10.152}$$

If s_0 is the number of completely filled Landau levels, the total carrier density n_e due to all these levels is

$$n_e = s_0\frac{eH}{hc} \tag{10.153}$$

In the Hall effect, the Hall constant is given as

$$R_{HC} = \frac{1}{n_e ec} \tag{10.154}$$

and the Hall resistivity is given as

$$\rho_{xy} = R_{HC}H = \frac{H}{n_e ec} \tag{10.155}$$

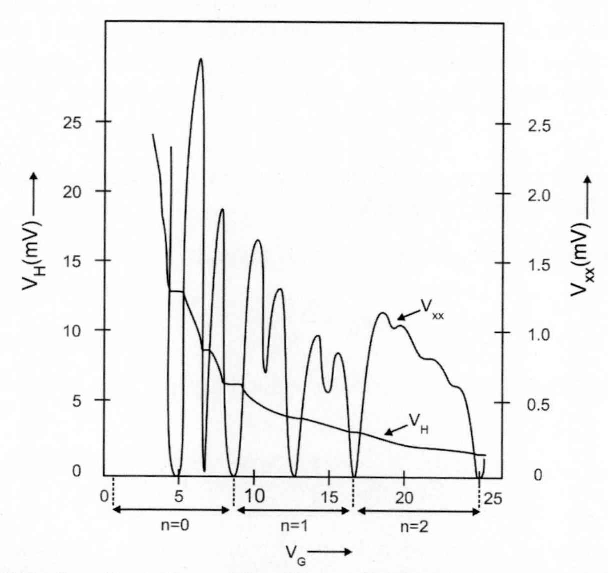

FIG. 10.11 The measurements of the Hall voltage V_H and linear voltage V_{xx} as a function of gate voltage V_G in a Si-MOSFET at T = 1.5 K and H = 18 Tesla. Here the source-drain current is kept fixed at 1 μA. *(Modified from Von Klitzing, K., Dorda, G., & Pepper, M. (1980). New method for high-accuracy determination of the fine-structure constant based on quantized Hall resistance.* Physical Review Letters, 45, *494.)*

Substituting the value of n_e from Eq. (10.153) into Eq. (10.155), we get

$$R_{xy} = \rho_{xy} = \frac{h}{s_0 e^2} \tag{10.156}$$

which gives Eq. (10.145) for the QHE. The QHE is observed at very low temperatures and high magnetic fields so that the thermal excitation of electrons from the last full Landau level s_0 to the next higher Landau level can be neglected.

In reality the above explanation does not hold. The probability that s_0 Landau levels will be completely filled in a real physical system is zero because the charge carrier density is specified. As soon as an extra electron is added to a 2D system with s_0 Landau levels completely full, the extra electron goes to the (s_0+1)th Landau level and the Fermi level jumps discontinuously. What is required, then, is a reservoir of electrons, which would keep s_0 Landau levels completely full by transferring electrons to and from the 2D system when the electron density is varied over a limited range. Note that the above explanation ignores any impurities present in the system.

The integral QHE has been measured (especially the Hall resistance R_{xy}) with an accuracy on the order of 0.1 ppm. The integral QHE can be used in defining a primary resistance standard similar to using the Josephson effect in defining a primary voltage standard. This is because of the fact that R_{xy} is expressed in terms of universal constants (Eq. 10.156). This would make the international comparison of resistance easier and more reliable. The integral QHE can also be used to find the universal constant h or e if one measures R_{xy}.

10.8.6 Fractional Quantum Hall Effect

Soon after the discovery of the integral QHE, Tsui, Stormer, and Gossard (1982) found that the Hall conductance has plateaus at fractional filling of the Landau levels also. They found that plateaus exist at filling factors with 1/3 and 2/3 values. Later on, plateaus were found at values of filling factors equal to simple fractions of the type p/q where q is an odd integer (Stormer et al., 1983).

The fractional QHE is observed in the extreme quantum limit (at very high magnetic fields) and only in samples with a very high mobility of charge carriers. This is in contrast to the case of the integral QHE in which the width of the plateaus increases with an increase in impurity concentration, which decreases the charge carrier mobility. This observation shows that the origins of the integral and fractional QHE are different. The two effects actually compete with each other. Many explanations have been put forward for the fractional QHE based on the ideas of Wigner crystallization of electrons into a solid, condensation of electrons into a quantum liquid, and others. The most appealing explanation is due to Laughlin (1983). According to this explanation, the Coulomb interactions between the electrons cause the electron system to condense into a highly correlated incompressible quantum liquid. Laughlin has given a variational wave function for such a quantum state. An energy gap separates the ground state from the excited state. The quasiparticles of this quantum liquid, which carry the current, behave as particles having a fractional charge of 1/q where q is an odd integer. These fractional charges lead to the plateaus at fractional quantum numbers. On changing the filling factor from the exact fractional quantum number, excess quasiparticles are created. These excess quasiparticles are trapped by the impurities, which lead to the plateaus.

10.9 WIEDEMANN-FRANZ-LORENTZ LAW

It has already been proved that the electrical conductivity σ_0 is given by

$$\sigma_0 = \frac{n_e e^2 \tau_e}{m_e} \tag{10.157}$$

The corresponding electronic contribution to the thermal conductivity per unit volume σ_{el} in metals is given by (see Eq. 8.177)

$$\sigma_T = \sigma_{el} = \frac{1}{3}\pi^2 \frac{n_e k_B^2}{m_e v_F} l_e T \tag{10.158}$$

It may be noted that in the above equation σ_{el} has been defined for unit volume. The mean free path at the Fermi energy is given as

$$l_e = v_F \tau_e \tag{10.159}$$

From Eqs. (10.158), (10.159) one gets

$$\sigma_T = \sigma_{el} = \frac{1}{3}\pi^2 \frac{n_e k_B^2 \tau_e}{m_e} T \tag{10.160}$$

Dividing Eq. (10.160) by Eq. (10.157), we find

$$\frac{\sigma_T}{\sigma_0} = L_N T \tag{10.161}$$

where

$$L_N = \frac{\pi^2}{3}\left(\frac{k_B}{e}\right)^2 \tag{10.162}$$

L_N is called the Lorentz number and is independent of the particular metal under consideration. Eq. (10.161) is the Wiedemann-Franz-Lorentz law, which states that the ratio of the thermal conductivity to the electrical conductivity is proportional to the temperature and the constant of proportionality (Lorentz number) is independent of the metal. In the derivation of Eq. (10.161) it is assumed that the relaxation times for the electrical and thermal processes are the same. If these are different, then the Lorentz number gets modified.

Problem 10.3

Magnetoresistance is defined as the fractional change in resistivity in the presence of a magnetic field. Prove that the magnetoresistance is proportional to the square of the magnetic field intensity.

REFERENCES

Ando, T., Matsumoto, Y., & Uemura, Y. (1975). High temperature relaxation of the uniaxial anisotropy of $Eu_{2.2}Gd_{0.8}Al_{0.6}Fe_{4.4}O_{12}$ garnet. *Journal of the Physical Society of Japan, 39*, 279–288.

Laughlin, R. B. (1983). Anomalous quantum Hall effect: an incompressible quantum fluid with fractionally charged excitations. *Physical Review Letters, 50*, 1395–1398.

Stormer, H. L., Chang, A. M., Tsui, D. S., Huang, J. C. M., Gossard, A. C., & Weigmann, W. (1983). Fractional quantization in the Hall effect. *Physical Review Letters, 50*, 1953–1956.

Tsui, D. C., Stormer, H. L., & Gossard, A. C. (1982). Two dimensional magnetotransport in extreme quantum limit. *Physical Review Letters, 48*, 1562.

Von Klitzing, K., Dorda, G., & Pepper, M. (1980). New method for high-accuracy determination of the fine-structure constant based on quantized Hall resistance. *Physical Review Letters, 45*, 494.

SUGGESTED READING

Blatt, F. J. (1968). *Physics of electronic conduction in solids*. New York: McGraw-Hill.

Pippard, A. B. (1988). *Magnetoresistance in metals*. London: Cambridge University Press.

Prange, R. E., & Girvin, S. M. (1987). *The quantum Hall effect*. New York: Springer-Verlag.

Rossiter, P. L. (1987). *Electrical resistivity of metals and alloys*. London: Cambridge University Press.

Stanley, J. K. (1963). *Electrical and magnetic properties of metals*. Ohio: American Society of Metals.

Chapter 11

Transport Phenomena

Chapter Outline

The most important phenomena are the motion of particles and the flow of energy in solids, liquids, and gases. For example, in metallic solids electric current is produced by the motion of conduction electrons in the presence of an externally applied electric field. On the other hand, heat current arises due to the flow of thermal energy, which is handed over from one atom (or electron) to another in a solid. In liquids electric current is produced by the motion of both the electrons and ions and the heat current arises from the convection currents set up by the actual motion of the atoms or ions. Further, plasma consists of a gas of moving electrons, ions, or neutral particles and it flows in a particular direction. In all of these examples there is transport of either charge or energy. In Chapter 10, we discussed the motion of electrons in the presence of electric and magnetic fields using the Newton laws of motion and some related properties were explained. In the present chapter, we shall describe the motion of electrons and, in general, of charged particles in the presence of electric and magnetic fields using the Boltzmann transport equation, which is an entirely different approach from the previous one. Some basic properties, such as the conduction of electricity and heat in solids, will be presented.

11.1 VELOCITY DISTRIBUTION FUNCTION

In a crystalline solid there are large numbers of particles (electrons or ions) that are distributed in space and possess velocity ranging from zero to some maximum value. The state of a moving particle at a particular time is precisely defined if both the position and velocity are known at that time. Hence it is useful to define a distribution function $f(\mathbf{r},\mathbf{v},t)$ that gives the number of particles per unit volume at position $\mathbf{r}$ that have velocity $\mathbf{v}$ at time t in phase space. Hence $f(\mathbf{r},\mathbf{v},t)\,d^3r\,d^3v$ gives the number of particles in a volume element $d^3r\,d^3v$ of phase space at time t.

11.2 ELECTRIC CURRENT AND ELECTRICAL CONDUCTIVITY

In a metallic solid, electric current is produced by the motion of conduction electrons and is given by

$$\mathbf{J} = -e\int \mathbf{v}\,f(\mathbf{r},\mathbf{v},t)\,d^3v \tag{11.1}$$

The motion of conduction electrons in a crystalline solid originates from two types of interactions.

1. Electrostatic interactions
2. Collision interactions

Solid State Physics. https://doi.org/10.1016/B978-0-12-817103-5.00011-6

11.2.1 Electrostatic Interactions

Two types of electrostatic fields are responsible for the motion of conduction electrons in a crystalline solid:

1. Crystal field
2. External electric field

The crystal potential is the sum of the atomic potentials of the lattice atoms and exhibits the periodicity of the lattice. The interaction of electrons with the crystal potential produces a uniform motion of electrons throughout the crystal. Hence, in the equilibrium state, the motion of an electron in any particular direction is equally probable. In other words, for each electron moving with velocity $\mathbf{v}$ there is a corresponding electron with velocity $-\mathbf{v}$. Mathematically, one can write

$$f_0(\mathbf{r},\mathbf{v},t)=f_0(\mathbf{r},-\mathbf{v},t) \tag{11.2}$$

Here $f_0(\mathbf{r},\mathbf{v},t)$ is the distribution function in the equilibrium state and is an even function of velocity. Therefore, the electric current in the equilibrium state $\mathbf{J}_0$ is given by

$$\mathbf{J}_0=-e\int \mathbf{v}\,f_0(\mathbf{r},\mathbf{v},t)\,d^3v=0 \tag{11.3}$$

$\mathbf{J}_0$ is zero because $\mathbf{v}\,f_0(\mathbf{r},\mathbf{v},t)$ is an odd function of velocity. According to the Bloch theorem, in a crystal with periodic crystal potential, no spontaneous current is possible, which is equivalent to the uniform motion of conduction electrons in all directions in a solid. When an external electric field is applied, the conduction electrons tend to move along the direction of the field. Hence, the application of the field produces changes in both the position and velocity of the electrons, thus causing a change in the distribution function $f(\mathbf{r},\mathbf{v},t)$. Hence, in the presence of an external electric field, the current density $\mathbf{J}$ becomes finite and is given by Eq. (11.1). Bloch has proved that in a periodic potential there is no hindrance to the flow of conduction electrons in a crystalline solid in the presence of an electric field, yielding zero resistance.

11.2.2 Collision Interactions

In a metallic crystal the ions start vibrating about their mean positions by gaining thermal energy k_BT, thereby breaking the periodicity of the lattice and hence of the crystal potential. Because of the breaking of the periodicity of the crystal potential, the conduction electrons are scattered from the ions by exchanging energy in the form of phonons [electron-phonon (e-p) interactions]. The e-p interactions offer hindrance to the flow of conduction electrons in the presence of an external field, giving rise to finite electrical resistance. From the above argument, it is evident that the external electric field and the collision interactions oppose each other. Hence, to calculate the electric current density or conductivity, we must study the combined effect of both the collision interactions and the external electric field. Note that the lattice periodicity is also destroyed by the presence of impurities (lattice defects), thus making an additional contribution to the resistivity or conductivity, especially in alloys.

For weak electric fields, the current density in a metallic solid is linearly proportional to the applied electric field, that is,

$$\mathbf{J}_\alpha=\sum_{\beta=1}^{3}\sigma_{\alpha\beta}\mathbf{E}_\beta \tag{11.4}$$

In a nonhomogeneous and anisotropic solid, an electric field $\mathbf{E}_\beta$ applied in the β-Cartesian direction produces a current density $\mathbf{J}_\alpha$ in the α-Cartesian direction. In this situation the conductivity becomes a tensor and its elements are represented by $\sigma_{\alpha\beta}$. In a homogeneous and isotropic solid, the current density and the applied field are in the same direction and, therefore, the conductivity becomes a scalar, that is,

$$\sigma_{\alpha\beta}=\sigma\,\delta_{\alpha\beta} \tag{11.5}$$

Substituting Eq. (11.5) into Eq. (11.4), we get

$$\mathbf{J}=\sigma\mathbf{E} \tag{11.6}$$

In cubic crystals all three Cartesian directions are equivalent and thus the solid behaves as an isotropic medium. It is noteworthy that Eq. (11.4) does not hold for strong electric fields because then one has to include higher-order terms in $\mathbf{E}$, which incorporate nonlinearity in the conduction phenomenon. In metallic crystals the electrons are responsible for the electrical conduction, while in semiconductors both the electrons and holes are responsible for it.

11.3 HEAT CURRENT AND THERMAL CONDUCTIVITY

Consider a solid in the form of a rod with one end at a higher temperature than the other (see Fig. 8.10). It is well known that heat flows from the higher temperature end to the lower temperature end of the rod. It is reasonable to assume that the heat current density **Q** is linearly proportional to the temperature gradient. Mathematically, one can write

$$\mathbf{Q} = -\sigma_T \nabla_{\mathbf{r}} T \tag{11.7}$$

where σ_T is a constant of proportionality and is called the thermal conductivity. The negative sign indicates that heat current flows in the direction of negative temperature gradient. The heat current originates from the transfer of kinetic energy of electrons in a finite temperature gradient. Let $E_{\mathbf{k}}$ be the band energy and $\mu(T)$ denote the chemical potential of a solid with finite temperature gradient, then the difference $E_{\mathbf{k}} - \mu(T)$ appears as kinetic energy and is responsible for the flow of heat current density.

11.4 THE BOLTZMANN TRANSPORT EQUATION

11.4.1 Classical Formulation

Consider a system of electrons in a solid having a one-dimensional velocity distribution function $f(x, v_x, t)$ with position and velocity in the x-direction. The electrons are assumed to be point charges and, upon moving closer, interact with one another over a finite time.[1] The average time of interaction "dt" between two electrons, or between an electron and an ion, is assumed to be very small compared with the average time τ_e between two consecutive interactions. The number of electrons at any time t in a small elemental area $dx\,dv_x$ of the phase space (see Fig. 8.2) is given by

$$dN_e = f(x, v_x, t)\,dx\,dv_x \tag{11.8}$$

The application of an electric field to the solid changes the distribution for two reasons:

1. The function $f(x, v_x, t)$ changes due to the drift velocity produced by the application of the electric field.
2. The function $f(x, v_x, t)$ changes due to the collision interactions.

Hence the total rate of change in the distribution function $\partial f/\partial t$ is given by

$$\frac{\partial f(x, v_x, t)}{\partial t} = \left.\frac{\partial f(x, v_x, t)}{\partial t}\right|_{drift} + \left.\frac{\partial f(x, v_x, t)}{\partial t}\right|_{coll} \tag{11.9}$$

In the equilibrium state of the system the distribution function becomes constant with respect to time. Therefore, Eq. (11.9) reduces to

$$\left.\frac{\partial f(x, v_x, t)}{\partial t}\right|_{drift} + \left.\frac{\partial f(x, v_x, t)}{\partial t}\right|_{coll} = 0 \tag{11.10}$$

This is the basic form of the Boltzmann equation. Eq. (11.10) can be written as

$$\left.\frac{\partial f(x, v_x, t)}{\partial t}\right|_{drift} = -\left.\frac{\partial f(x, v_x, t)}{\partial t}\right|_{coll} \tag{11.11}$$

Eq. (11.11) shows that the rates of change of the distribution function due to the electric field and the collision processes are equal and opposite. In other words, the collision processes oppose the effect of an external electric field.

To start with, let us assume that the collision processes are absent. The application of an electric field produces drift in the position and velocity of the electrons. The number of electrons dN_e' at time $t+dt$, in the cell $dx\,dv_x$, at position and velocity coordinates $x+dx$ and v_x+dv_x, respectively, is given by

$$dN_e' = f(x+dx, v_x+dv_x, t+dt)\,dx\,dv_x \tag{11.12}$$

1. If the particles are not charged, then they are assumed to be elastic and impenetrable spheres and the interaction between them takes place instantaneously.

In the equilibrium state, the number of electrons in the elemental volume $dx\,dv_x$ must be constant with time, that is,

$$dN_e = dN'_e \tag{11.13}$$

Hence from Eqs. (11.8), (11.12), and (11.13) one can write

$$f(x+dx, v_x+dv_x, t+dt) = f(x, v_x, t) \tag{11.14}$$

Expanding the left side of Eq. (11.14) around x, v_x, and t, we find.

$$\frac{\partial f(x, v_x, t)}{\partial x}dx + \frac{\partial f(x, v_x, t)}{\partial v_x}dv_x + \left.\frac{\partial f(x, v_x, t)}{\partial t}\right|_{drift} dt = 0$$

which can also be written as

$$\left.\frac{\partial f(x, v_x, t)}{\partial t}\right|_{drift} = -v_x\frac{\partial f(x, v_x, t)}{\partial x} - a_x\frac{\partial f(x, v_x, t)}{\partial v_x} \tag{11.15}$$

where $a_x = dv_x/dt$ gives the acceleration in the x-direction produced by the applied electric field. If F_x is the force produced by the electric field in the x-direction, then the acceleration can also be written as $a_x = F_x/m_e$. Substituting Eq. (11.15) into Eq. (11.9), the total change in the distribution function becomes

$$\frac{\partial f(x, v_x, t)}{\partial t} = -v_x\frac{\partial f(x, v_x, t)}{\partial x} - a_x\frac{\partial f(x, v_x, t)}{\partial v_x} + \left.\frac{\partial f(x, v_x, t)}{\partial t}\right|_{coll} \tag{11.16}$$

In the equilibrium state of the system we get

$$\left.\frac{\partial f(x, v_x, t)}{\partial t}\right|_{coll} = -\left.\frac{\partial f(x, v_x, t)}{\partial t}\right|_{drift} = v_x\frac{\partial f(x, v_x, t)}{\partial x} + a_x\frac{\partial f(x, v_x, t)}{\partial v_x} \tag{11.17}$$

Eq. (11.17) gives the Boltzmann transport equation in one dimension. One can derive similar equations for the velocity distributions along the y- and z-directions separately. Therefore, the Boltzmann transport equation in three dimensions is given by

$$\left.\frac{\partial f(\mathbf{r}, \mathbf{v}, t)}{\partial t}\right|_{coll} = -\left.\frac{\partial f(\mathbf{v}, \mathbf{r}, t)}{\partial t}\right|_{drift} = \mathbf{v}\cdot\nabla_{\mathbf{r}} f + \frac{\mathbf{F}}{m_e}\cdot\nabla_{\mathbf{v}} f \tag{11.18}$$

where $\nabla_{\mathbf{r}}$ and $\nabla_{\mathbf{v}}$ are gradient operators with respect to **r** and **v**, respectively.

Appendix K describes an alternate method for the derivation of the Boltzmann transport equation.

Let us now suppose that collision processes are also present in the system. The rate of change of the distribution function due to collision processes is given by

$$\left.\frac{\partial f}{\partial t}\right|_{coll} = -\frac{f - f_0}{\tau_e} \tag{11.19}$$

where τ_e is the electron relaxation time. The negative sign ensures the decay of the distribution function to its equilibrium form $f_0(\mathbf{r}, \mathbf{v}, t)$. Because $f_0(\mathbf{r}, \mathbf{v}, t)$ is constant in time, Eq. (11.19) can be written as

$$\left.\frac{\partial (f - f_0)}{\partial t}\right|_{coll} = -\frac{f - f_0}{\tau_e} \tag{11.20}$$

The solution of the above equation gives

$$(f - f_0)_t = (f - f_0)_{t=0}\, e^{-t/\tau_e} \tag{11.21}$$

which gives an exponential decay of the change in the distribution function from its equilibrium value. From Eq. (11.21) the relaxation time is defined as the time in which the perturbed distribution function $f - f_0$ decreases by a factor of e. Substituting Eq. (11.19) into Eq. (11.18), the distribution function $f(\mathbf{r}, \mathbf{v}, t)$ becomes

$$f(\mathbf{r}, \mathbf{v}, t) = f_0(\mathbf{r}, \mathbf{v}, t) - \tau_e\left[\mathbf{v}\cdot\nabla_{\mathbf{r}} f + \frac{\mathbf{F}}{m_e}\cdot\nabla_{\mathbf{v}} f\right] \tag{11.22}$$

The concept of relaxation time in the collision process has been introduced in an arbitrary fashion. The different relaxation times are defined for different physical phenomena and they reduce to the ordinary differential equation given by

Eq. (11.22). For example, any type of current decays to its equilibrium value in elastic scattering processes. In an inhomogeneous system, the density of electrons is different at different points and, therefore, one should define different distribution functions at different points. In such systems, one should define the equilibrium function as a function of the local density $n(\mathbf{r})$, that is, $f[n(\mathbf{r})]$.

11.4.2 Quantum Formulation

Electrons in a crystalline solid form a quantum mechanical system in which the wave vector of an electron is a good quantum number. Therefore, the state of an electron is characterized by its wave vector $\mathbf{k}$ and energy $E_{\mathbf{k}}$. The quantum mechanical expressions for the velocity and force on an electron are given by

$$\mathbf{v} = \frac{1}{\hbar}\nabla_{\mathbf{k}}E_{\mathbf{k}} \tag{11.23}$$

$$\mathbf{F} = \hbar\dot{\mathbf{k}} \tag{11.24}$$

Treating the electrons as free, the energy can be approximated by a parabolic band, that is,

$$E_{\mathbf{k}} = \frac{\hbar^2 k^2}{2m_e} \tag{11.25}$$

Substituting Eq. (11.25) into Eq. (11.23), we get

$$\mathbf{v} = \frac{\hbar\mathbf{k}}{m_e} \tag{11.26}$$

Further, substituting Eq. (11.24), (11.26) into Eq. (11.18), we get.

$$\left.\frac{\partial f(\mathbf{v},\mathbf{r},t)}{\partial t}\right|_{coll} = -\left.\frac{\partial f(\mathbf{v},\mathbf{r},t)}{\partial t}\right|_{drift} = \mathbf{v}\cdot\nabla_{\mathbf{r}}f + \frac{\mathbf{F}}{\hbar}\cdot\nabla_{\mathbf{k}}f \tag{11.27}$$

Eq. (11.27) gives the quantum mechanical expression for the Boltzmann transport equation.

The solution of the Boltzmann equation is not simple because it is an integro-differential equation on account of the collision processes. To solve for the distribution function arising from the collision processes, it is convenient to make the relaxation time approximation. Consider a system of electrons in a crystalline solid with equilibrium distribution function $f_0(\mathbf{r},\mathbf{k},t)$. The application of an external electric field shifts the distribution of electrons by $\delta\mathbf{k}$ (say) because the electrons collide with the lattice ions, thereby producing finite drift. After some time τ_e, usually called the electron relaxation time, the system comes a steady state with distribution function $f(\mathbf{r},\mathbf{k},t)$ in the presence of the external field. If the external field is switched off, the system will again come to the original state, in time τ_e, with distribution function $f_0(\mathbf{r},\mathbf{k},t)$.

11.5 LINEARIZATION OF BOLTZMANN EQUATION

The Boltzmann transport equation is a nonlinear equation and its solution is thus a difficult problem. Therefore, one has to resort to some simplification and one of the obvious simplifications is to linearize the Boltzmann transport equation. Consider a system of free electrons in the equilibrium state with distribution function $f_0(\mathbf{r},\mathbf{v},t)$. A weak electric field $\mathbf{E}$ is applied to the system, which changes the distribution function to $f(\mathbf{r},\mathbf{v},t)$ defined as

$$f(\mathbf{r},\mathbf{v},t) = f_0(\mathbf{r},\mathbf{v},t) + \Delta f(\mathbf{r},\mathbf{v},t) \tag{11.28}$$

$\Delta f(\mathbf{r},\mathbf{v},t)$ is the small change in the distribution function caused by the application of the $\mathbf{E}$ field. Hence $\Delta f(\mathbf{r},\mathbf{v},t)$ can be treated as a perturbation in the solution of the Boltzmann equation for any physical property. The linear Boltzmann equation is obtained by substituting Eq. (11.28) into Eq. (11.27) and retaining terms only up to the first order. It gives

$$\left.\frac{\partial f}{\partial t}\right|_{coll} = -\left.\frac{\partial f}{\partial t}\right|_{drift} = \mathbf{v}\cdot\nabla_{\mathbf{r}}f_0(\mathbf{r},\mathbf{v},t) - \frac{e\mathbf{E}}{\hbar}\cdot\nabla_{\mathbf{k}}f_0(\mathbf{r},\mathbf{v},t) \tag{11.29}$$

where it is assumed that

$$|\mathbf{v}\cdot\nabla_{\mathbf{r}}\Delta f(\mathbf{r},\mathbf{v},t)| \ll |\mathbf{v}\cdot\nabla_{\mathbf{r}}f_0(\mathbf{r},\mathbf{v},t)| \tag{11.30}$$

$$|\mathbf{E}\cdot\nabla_{\mathbf{k}}\Delta f(\mathbf{r},\mathbf{v},t)| \ll |\mathbf{E}\cdot\nabla_{\mathbf{k}}f_0(\mathbf{r},\mathbf{v},t)| \tag{11.31}$$

The linear Boltzmann equation can be solved for a system of free electrons in which the equilibrium distribution function can be taken to be the Fermi-Dirac distribution function given by

$$f_0(E_{\mathbf{k}}, \mu, T) = \frac{1}{e^{[E_{\mathbf{k}} - \mu(T)]/k_B T} + 1} \tag{11.32}$$

Now it is straightforward to prove that.

$$\nabla_{\mathbf{k}} f_0(E_{\mathbf{k}}) = \frac{\partial f_0}{\partial E_{\mathbf{k}}} \nabla_{\mathbf{k}} E_{\mathbf{k}} = \frac{\partial f_0}{\partial E_{\mathbf{k}}} \hbar \mathbf{v} \tag{11.33}$$

$$\begin{aligned} \nabla_{\mathbf{r}} f_0(E_{\mathbf{k}}) &= \frac{\partial f_0}{\partial T} \nabla_{\mathbf{r}} T \\ &= -\frac{\partial f_0}{\partial E_{\mathbf{k}}} \left(\frac{E_{\mathbf{k}} - \mu(T)}{T} + \frac{\partial \mu}{\partial T} \right) \nabla_{\mathbf{r}} T \end{aligned} \tag{11.34}$$

In writing Eq. (11.33) we have used Eq. (11.23). Substituting Eqs. (11.33), (11.34) into Eq. (11.29), the linear Boltzmann equation for a system of electrons becomes

$$\left.\frac{\partial f}{\partial t}\right|_{coll} = -\left.\frac{\partial f}{\partial t}\right|_{drift} = -\mathbf{v} \cdot \left(\frac{\partial f_0}{\partial E_{\mathbf{k}}} \mathbf{A} \right) \tag{11.35}$$

where

$$\begin{aligned} \mathbf{A} &= e\mathbf{E} + \left(\frac{E_{\mathbf{k}} - \mu(T)}{T} + \frac{\partial \mu}{\partial T} \right) \nabla_{\mathbf{r}} T \\ &= e\mathbf{E} + \frac{E_{\mathbf{k}} - \mu(T)}{T} \nabla_{\mathbf{r}} T + \nabla_{\mathbf{r}} \mu \end{aligned} \tag{11.36}$$

Here $\mathbf{A}$ is the vector field, which combines the actions of electric and thermal fields.

11.6 ELECTRICAL CONDUCTIVITY

Lorentz investigated the problem of electrical conductivity with the Boltzmann transport equation, treating electrons as classical particles and using a simplified model for the collision process between the electrons and ions in the lattice. His treatment led to some serious difficulties. Later, Sommerfeld calculated the electrical conductivity of metals using the Boltzmann transport equation and treating the electrons as quantum particles. He, further, assumed that the relaxation time is a function of energy without investigating the actual mechanism of interaction between the electrons and ions.

Consider a crystalline solid to which an electric field $\mathbf{E}$ is applied in the x-direction, that is, $\mathbf{E} = \hat{\mathbf{i}}_1 E_x$. The force acting on an electron in the direction of the field is given by

$$\mathbf{F} = -eE_x \hat{\mathbf{i}}_1 \tag{11.37}$$

and the acceleration produced by this force is

$$\mathbf{a} = \frac{\mathbf{F}}{m_e} = -\frac{eE_x}{m_e} \hat{\mathbf{i}}_1 \tag{11.38}$$

Substituting Eq. (11.38) into Eq. (11.18), one gets the one-dimensional Boltzmann equation as

$$\left.\frac{\partial f}{\partial t}\right|_{coll} = -\left.\frac{\partial f}{\partial t}\right|_{drift} = v_x \frac{\partial f}{\partial x} - \frac{eE_x}{m_e} \frac{\partial f}{\partial v_x} \tag{11.39}$$

Substituting Eq. (11.19) for the rate of change of the distribution function arising from the collision interactions into Eq. (11.39), we obtain

$$f(x, v_x, t) = f_0(x, v_x, t) - \tau_e \left[v_x \frac{\partial f}{\partial x} - \frac{eE_x}{m_e} \frac{\partial f}{\partial v_x} \right] \tag{11.40}$$

According to Eq. (11.40), $f(x, v_x, t)$ depends on itself through its derivatives and, therefore, its solution can be obtained through the successive approximation method (iteration method). If the applied electric field is weak, the deviation of

$f(x, v_x, t)$ from its equilibrium value $f_0(x, v_x, t)$ is small. Hence, in the lowest order approximation, one can substitute $f(x, v_x, t) \approx f_0(x, v_x, t)$ on the right side of Eq. (11.40) to write

$$f(x, v_x, t) = f_0(x, v_x, t) - \tau_e \left[v_x \frac{\partial f_0}{\partial x} - \frac{eE_x}{m_e} \frac{\partial f_0}{\partial v_x} \right] \tag{11.41}$$

In the equilibrium state, $f_0(x, v_x, t)$ for an electron is nothing but the Fermi-Dirac distribution function $f_0(E, \mu, T)$ defined by Eq. (11.32). In the equilibrium state, the energy E is constant and, therefore, one can write

$$\frac{\partial f_0(E, \mu, T)}{\partial x} = \frac{\partial f_0}{\partial \mu} \frac{\partial \mu}{\partial T} \frac{\partial T}{\partial x} + \frac{\partial f_0}{\partial T} \frac{\partial T}{\partial x} \tag{11.42}$$

We are interested in the electrical conductivity of a crystalline solid at a particular temperature yielding $\partial T/\partial x = 0$, which gives $\partial f_0/\partial x = 0$. Therefore, Eq. (11.41) yields

$$f(x, v_x, t) = f_0(x, v_x, t) + \frac{eE_x \tau_e}{m_e} \frac{\partial f_0}{\partial v_x} \tag{11.43}$$

If the conduction electrons in a solid are assumed to be free, they possess only kinetic energy, that is,

$$E = \frac{1}{2} m_e v^2 \tag{11.44}$$

The velocity derivative of the distribution function can be written as

$$\frac{\partial f_0}{\partial v_x} = \frac{\partial f_0}{\partial E} \frac{\partial E}{\partial v_x} = m_e v_x \frac{\partial f_0}{\partial E} \tag{11.45}$$

Therefore, the distribution function in the free-electron approximation is given by substituting Eq. (11.45) into Eq. (11.43) to obtain

$$f(x, v_x, t) = f_0(x, v_x, t) + e\tau_e v_x E_x \frac{\partial f_0}{\partial E} \tag{11.46}$$

With the knowledge of the distribution function, the current density in the x-direction (see Eq. 11.1) is given by

$$J_x = -e \int v_x f(x, v_x, t)\, d^3v \tag{11.47}$$

Substituting $f(x, v_x, t)$ from Eq. (11.43) into Eq. (11.47), we get

$$J_x = -e \int v_x f_0(x, v_x, t)\, d^3v - \frac{e^2 E_x}{m_e} \int v_x \frac{\partial f_0}{\partial v_x} \tau_e(\mathbf{v})\, d^3v \tag{11.48}$$

Here the relaxation time is assumed to be a function of the velocity or energy of the electron. The first integral in Eq. (11.48) is zero in the equilibrium state (see Eq. 11.3), therefore, one can write

$$J_x = -\frac{e^2 E_x}{m_e} \int v_x \frac{\partial f_0}{\partial v_x} \tau_e(\mathbf{v})\, d^3v \tag{11.49}$$

Comparing Eqs. (11.6), (11.49), the expression for electrical conductivity can be written as

$$\sigma = -\frac{e^2}{m_e} \int v_x \frac{\partial f_0}{\partial v_x} \tau_e(\mathbf{v})\, d^3v \tag{11.50}$$

This is the general expression for conductivity when the current density is in the x-direction. If the conduction electrons in a crystalline solid are assumed to be free, then from Eqs. (11.45), (11.50), σ becomes

$$\sigma = -e^2 \int v_x^2 \frac{\partial f_0}{\partial E} \tau_e(\mathbf{v})\, d^3v \tag{11.51}$$

In the derivation of Eq. (11.51), it is assumed that the applied electric field is in the x-direction, as a result of which the electrons move with velocity v_x. Similar expressions for σ can also be obtained for electron motions along the y- and z-directions. The general expression for σ is obtained by substituting the average of the square of velocities in the Cartesian directions, which are related as

$$\langle v_x^2 \rangle = \langle v_y^2 \rangle = \langle v_z^2 \rangle = v^2/3 \tag{11.52}$$

where v is the velocity of an electron in a general direction. Substitution of Eq. (11.52) into Eq. (11.51) yields

$$\sigma = -\frac{e^2}{3}\int v^2 \frac{\partial f_0}{\partial E}\tau_e(\mathbf{v})\, d^3v \tag{11.53}$$

11.6.1 Classical Theory

Consider a finite system containing N_e conduction electrons and having volume V. In the equilibrium state one can write

$$\int f_0(\mathbf{r}, \mathbf{v}, t)\, d^3r d^3v = N_e \tag{11.54}$$

If, in the equilibrium state, the electrons are distributed uniformly, say with density n_e, then $f_0(\mathbf{r}, \mathbf{v}, t)$ becomes independent of position $\mathbf{r}$. Therefore, Eq. (11.54) reduces to

$$\int f_0(\mathbf{v}, t)\, d^3v = \frac{N_e}{V} = n_e \tag{11.55}$$

Lorentz treated the electrons as classical particles obeying the Maxwell-Boltzmann distribution defined by

$$f_0(E, T) = n_e \left(\frac{m_e}{2\pi k_B T}\right)^{3/2} e^{-E/k_B T} \tag{11.56}$$

where E is the energy of the electron state. From Eq. (11.56) one can write

$$\frac{\partial f_0}{\partial E} = -\frac{1}{k_B T} f_0 \tag{11.57}$$

Lorentz, further, assumed the relaxation time of the conduction electrons to be a constant. Substituting Eqs. (11.56), (11.57) into Eq. (11.53), the electrical conductivity in polar coordinates becomes

$$\sigma = \frac{4\pi n_e e^2}{3k_B T}\tau_e \left(\frac{m_e}{2\pi k_B T}\right)^{3/2} \int_0^\infty e^{-E/k_B T} v^4\, dv \tag{11.58}$$

If the conduction electrons are treated as free, then their energies are given by Eq. (11.44). Therefore, substituting Eq. (11.44) into Eq. (11.58), one gets

$$\sigma = \frac{4\pi n_e e^2}{3k_B T}\tau_e \left(\frac{m_e}{2\pi k_B T}\right)^{3/2} \int_0^\infty e^{-\frac{m_e v^2}{2k_B T}} v^4\, dv \tag{11.59}$$

Making the substitution

$$x = \frac{m_e v^2}{2 k_B T} \tag{11.60}$$

in Eq. (11.59) and simplifying, one gets

$$\sigma = \frac{4 n_e e^2}{3\sqrt{\pi}} \frac{\tau_e}{m_e} \int_0^\infty x^{3/2} e^{-x}\, dx \tag{11.61}$$

The integral in the above equation is a standard gamma integral, so

$$\sigma = \frac{4 n_e e^2}{3\sqrt{\pi}} \frac{\tau_e}{m_e} \Gamma(5/2) = \frac{n_e e^2 \tau_e}{m_e} = \sigma_0 \tag{11.62}$$

which is nothing but the Drude conductivity. Here $\Gamma(5/2)$ is a gamma function given by

$$\Gamma\left(\frac{5}{2}\right)=\frac{3}{2}\cdot\frac{1}{2}\cdot\sqrt{\pi} \tag{11.63}$$

11.6.2 Quantum Theory

Sommerfeld treated electrons as quantum particles (fermions) with the Fermi-Dirac function $f_0(E,\mu,T)$, defined by Eq. (11.32), as the distribution function. We know that the number of electron states per unit volume in **k**-space is given by $V/(2\pi)^3$. Hence the number of states in an elemental volume d^3k becomes $V/(2\pi)^3\,d^3k$. Taking account of the spin degeneracy, the total number of electrons is given by

$$N_e=\frac{2V}{(2\pi)^3}\int f_0(E,\mu,T)\,d^3k=\frac{2V}{(2\pi)^3}\int\frac{1}{e^{(E-\mu)/k_BT}+1}\,d^3k \tag{11.64}$$

We know that $d^3p=\hbar^3d^3k=m_e^3d^3v$. Changing the variable of integration from **k** to **v**, Eq. (11.64) reduces to

$$n_e=\frac{N_e}{V}=2\left(\frac{m_e}{2\pi\hbar}\right)^3\int\frac{1}{e^{(E-\mu)/k_BT}+1}\,d^3v \tag{11.65}$$

Comparing Eq. (11.65) with Eq. (11.55) the distribution function in the equilibrium state is given by

$$f_0(\mathbf{v},t)=f_0(\mathbf{v},T)=2\left(\frac{m_e}{2\pi\hbar}\right)^3\frac{1}{e^{(E-\mu)/k_BT}+1} \tag{11.66}$$

Substituting $f_0(\mathbf{v},T)$ from Eq. (11.66) into Eq. (11.53), we get

$$\sigma=-\frac{2e^2}{3}\left(\frac{m_e}{2\pi\hbar}\right)^3\int v^2\frac{\partial}{\partial E}\left(\frac{1}{e^{(E-\mu)/k_BT}+1}\right)\tau_e(\mathbf{v})\,d^3v \tag{11.67}$$

Changing the variable from **v** to E using Eq. (11.44) and simplifying, we obtain

$$\sigma=-\frac{8\pi e^2}{3m_e}\left(\frac{2}{m_e}\right)^{3/2}\left(\frac{m_e}{2\pi\hbar}\right)^3\int E^{3/2}\frac{\partial}{\partial E}\left(\frac{1}{e^{(E-\mu)/k_BT}+1}\right)\tau_e(E)\,dE \tag{11.68}$$

The conductivity is usually measured at quite low temperatures. At such temperatures the Fermi-Dirac distribution function can be approximated by its value at absolute zero, that is, $f_0(E,\mu,T)$ is one for $E\le E_F$ and zero for $E>E_F$. At absolute zero the energy derivative of $f_0(E,\mu,T)$ is zero everywhere except at $E=E_F$ where it is minus infinity, that is,

$$-\frac{\partial}{\partial E}\left(\frac{1}{e^{(E-\mu)/k_BT}+1}\right)=\delta(E-E_F) \tag{11.69}$$

Substituting Eq. (11.69) into Eq. (11.68), we get

$$\sigma=\frac{e^2}{3\pi^2m_e}\left(\frac{2m_e}{\hbar^2}\right)^{3/2}E_F^{3/2}\tau_e(E_F)=\frac{2e^2}{3m_e}g_e(E_F)E_F\tau_e(E_F) \tag{11.70}$$

where

$$g_e(E_F)=\frac{1}{2\pi^2}\left(\frac{2m_e}{\hbar^2}\right)^{3/2}E_F^{1/2}=\frac{3n_e}{2E_F} \tag{11.71}$$

Hence the conductivity can finally be written as

$$\sigma=\sigma_0=\frac{n_ee^2\tau_e(E_F)}{m_e} \tag{11.72}$$

It is noteworthy that Eqs. (11.62), (11.72) are the same, except that in Sommerfeld theory the relaxation time is evaluated at E_F, while in the Lorentz theory it is constant and independent of energy. In the derivation of electrical conductivity, $\partial f_0/\partial E$ has been taken to be a Dirac delta function, which is true only at absolute zero. Actually, the electrical conductivity is measured at a finite temperature at which $\partial f_0/\partial E$ is a sharply peaked function with some spread in energy, which is on the order of k_BT around E_F (see Fig. 9.6B). Hence only those electrons that occupy states in the range of k_BT around E_F contribute to conduction and are the actual conduction electrons. Hence, the Sommerfeld theory gives a clear-cut definition of the conduction electrons. The expression for σ can be improved by including the actual peaked function $\partial f_0/\partial E$ with finite spread (it will yield an additional contribution to the value given by Eq. 11.72).

11.7 THERMAL CONDUCTIVITY

It has already been discussed that the heat current arises due to the flow of energy (kinetic energy) from one end of a solid to the other in the presence of a finite temperature gradient. From Eq. (11.7) it is evident that the thermal conductivity σ_T is a measure of the flow of heat and is a property of the solid under consideration. In a metallic solid most of the heat is carried by the conduction electrons and a very little by the lattice vibrations. In a semiconductor, both the electrons and lattice vibrations transport heat. But in dielectrics, it is only the lattice vibrations that carry heat. The flow of heat can be studied by solving the Boltzmann transport equation in the presence of a finite temperature gradient.

Let an electric field $\mathbf{E}$ be applied to a crystalline solid in the x-direction, that is, $\mathbf{E} = \hat{\mathbf{i}}_1 E_1 = \hat{\mathbf{i}}_1 E_x$, in the presence of a finite temperature gradient dT/dx, again along the x-direction. Then the electric current and heat current densities flow in the x-direction and are given by

$$J_x = -e\int v_x f(x, v_x, t)\, d^3v \tag{11.73}$$

$$Q_x = \int v_x E f(x, v_x, t)\, d^3v \tag{11.74}$$

Substituting the value of $f(x, v_x, t)$ from Eq. (11.41) into Eqs. (11.73), (11.74), we write

$$J_x = e\int \tau_e(\mathbf{v})\left[v_x^2\frac{\partial f_0}{\partial x} - v_x\frac{eE_x}{m_e}\frac{\partial f_0}{\partial v_x}\right] d^3v \tag{11.75}$$

$$Q_x = -\int \tau_e(\mathbf{v})\, E\left[v_x^2\frac{\partial f_0}{\partial x} - v_x\frac{eE_x}{m_e}\frac{\partial f_0}{\partial v_x}\right] d^3v \tag{11.76}$$

In writing the above expressions for J_x and Q_x, the equilibrium condition given by Eq. (11.3) is used.

11.7.1 Classical Theory

If the electrons are treated as classical particles, they obey the Maxwell-Boltzmann distribution $f_0(E, T)$, given by Eq. (11.56). The derivatives of $f_0(E, T)$ are given by

$$\frac{\partial f_0}{\partial x} = \frac{\partial f_0}{\partial T}\frac{dT}{dx} \tag{11.77}$$

$$\frac{\partial f_0}{\partial v_x} = \frac{\partial f_0}{\partial E}\frac{dE}{dv_x} = m_e v_x \frac{\partial f_0}{\partial E} \tag{11.78}$$

In writing Eq. (11.78) the electrons are assumed to be free particles with energy given by Eq. (11.44). Substituting Eq. (11.56) for $f_0(E, T)$ in the above equations, we get

$$\frac{\partial f_0}{\partial x} = \left(\beta_0 E - \frac{3}{2}\right) f_0(E, T)\frac{1}{T}\frac{dT}{dx} \tag{11.79}$$

$$\frac{\partial f_0}{\partial v_x} = -\beta_0 m_e v_x f_0(E, T) \tag{11.80}$$

where

$$\beta_0 = \frac{1}{k_B T} \tag{11.81}$$

Substituting Eqs. (11.79), (11.80) into Eqs. (11.75), (11.76), we get

$$J_x = e\int \tau_e(\mathbf{v})\, v_x^2 \left[\left(\beta_0 E - \frac{3}{2}\right)\frac{1}{T}\frac{dT}{dx} + eE_x\beta_0\right] f_0(E,T)\, d^3v \tag{11.82}$$

$$Q_x = -\int \tau_e(\mathbf{v})\, E\, v_x^2 \left[\left(\beta_0 E - \frac{3}{2}\right)\frac{1}{T}\frac{dT}{dx} + eE_x\beta_0\right] f_0(E,T)\, d^3v \tag{11.83}$$

The expressions for J_x and Q_x are simplified to a great extent if one assumes the relaxation time to be independent of velocity, that is, a constant parameter. Then

$$J_x = e\tau_e \frac{1}{T}\frac{dT}{dx}\left(\frac{1}{2}m_e\beta_0 I_2 - \frac{3}{2}I_1\right) + e^2\tau_e E_x \beta_0 I_1 \tag{11.84}$$

$$Q_x = -\frac{m_e\tau_e}{2}\left[\frac{1}{T}\frac{dT}{dx}\left(\frac{m_e\beta_0}{2}I_3 - \frac{3}{2}I_2\right) + eE_x\beta_0 I_2\right] \tag{11.85}$$

where

$$I_1 = \int v_x^2 f_0(E,T)\, d^3v \tag{11.86}$$

$$I_2 = \int v_x^2 v^2 f_0(E,T)\, d^3v \tag{11.87}$$

$$I_3 = \int v_x^2 v^4 f_0(E,T)\, d^3v \tag{11.88}$$

The integrals I_1, I_2, and I_3 can be solved analytically for the Maxwell-Boltzmann distribution in polar coordinates assuming $v_x^2 = v_y^2 = v_z^2 = v^2/3$ and they are given by

$$I_1 = \frac{n_e}{m_e\beta_0} \tag{11.89}$$

$$I_2 = \frac{5n_e}{m_e^2\beta_0^2} \tag{11.90}$$

$$I_3 = \frac{35n_e}{m_e^3\beta_0^3} \tag{11.91}$$

Substituting the values of the integrals from Eqs. (11.89)–(11.91) into Eqs. (11.84), (11.85), we find

$$J_x = \frac{n_e e^2\tau_e}{m_e}E_x + \frac{n_e e\tau_e}{m_e\beta_0}\frac{1}{T}\frac{dT}{dx} \tag{11.92}$$

$$Q_x = -\frac{5n_e\tau_e}{2m_e\beta_0^2}\left[eE_x\beta_0 + 2\frac{1}{T}\frac{dT}{dx}\right] \tag{11.93}$$

The thermal conductivity is measured when no electric current passes through the solid, that is, $J_x = 0$. So, for $J_x = 0$, Eq. (11.92) gives

$$eE_x\beta_0 = -\frac{1}{T}\frac{dT}{dx} \tag{11.94}$$

Substituting Eq. (11.94) into Eq. (11.93), the heat current density becomes

$$Q_x = -\sigma_T \frac{dT}{dx} \tag{11.95}$$

$$\sigma_T = \frac{5n_e\tau_e}{2m_e\beta_0^2 T} \tag{11.96}$$

Here σ_T gives the thermal conductivity of a solid. In metallic solids Lorentz defined a parameter, usually called the Lorentz number, as

$$L_N = \frac{\sigma_T}{\sigma_0 T} \tag{11.97}$$

Substituting the values of σ_0 and σ_T from Eqs. (11.62), (11.96), the classical theory yields the value as

$$L_N = \frac{5}{2}\left(\frac{k_B}{e}\right)^2 \tag{11.98}$$

Eq. (11.98) shows that the electrical and thermal conductivities, determined by using the Boltzmann transport equation, satisfy the Wiedemann-Franz-Lorentz law. But the Lorentz number L_N given by Eq. (11.98) is not in agreement with the experimental results, which shows that the electrons cannot be assumed to be classical particles, but rather that they should be treated as quantum particles.

11.7.2 Quantum Theory

Sommerfeld treated electrons as quantum particles obeying the Fermi-Dirac distribution function defined by Eq. (11.66). Therefore, the general expression for J_x and Q_x given by Eqs. (11.75), (11.76) involve the Fermi-Dirac distribution function $f_0(E, \mu, T)$ and its derivatives. In the equilibrium state of the system, the energy is constant in space, that is, $dE/dx = 0$. Hence the derivatives of $f_0(E, \mu, T)$ can be written as

$$\frac{\partial f_0}{\partial x} = \frac{\partial f_0}{\partial \mu}\frac{d\mu}{dx} + \frac{\partial f_0}{\partial T}\frac{dT}{dx} \tag{11.99}$$

$$\frac{\partial f_0}{\partial v_x} = \frac{\partial f_0}{\partial E}\frac{\partial E}{\partial v_x} = m_e v_x \frac{\partial f_0}{\partial E} \tag{11.100}$$

Eq. (11.100) is the same as Eq. (11.78). From Eq. (11.66) the derivatives of $f_0(E, \mu, T)$ are given by

$$\frac{\partial f_0}{\partial E} = -2\beta_0\left(\frac{m_e}{2\pi\hbar}\right)^3 \frac{e^{\beta_0(E-\mu)}}{\left[e^{\beta_0(E-\mu)} + 1\right]^2} \tag{11.101}$$

$$\frac{\partial f_0}{\partial \mu} = -\frac{\partial f_0}{\partial E} \tag{11.102}$$

$$\frac{\partial f_0}{\partial T} = -\frac{E-\mu}{T}\frac{\partial f_0}{\partial E} \tag{11.103}$$

Substituting Eqs. (11.102), (11.103) into Eq. (11.99), we get

$$\frac{\partial f_0}{\partial x} = -\frac{\partial f_0}{\partial E}\left[T\frac{d}{dx}\left(\frac{\mu}{T}\right) + \frac{E}{T}\frac{dT}{dx}\right] \tag{11.104}$$

Substituting the values of $\partial f_0/\partial x$ and $\partial f_0/\partial v_x$ from Eqs. (11.104), (11.100) into Eqs. (11.75), (11.76), we can write

$$J_x = -e\int d^3v\, \tau_e(\mathbf{v})\, v_x^2 \frac{\partial f_0}{\partial E}\left[eE_x + T\frac{d}{dx}\left(\frac{\mu}{T}\right) + \frac{E}{T}\frac{dT}{dx}\right] \tag{11.105}$$

$$Q_x = \int d^3v\, \tau_e(\mathbf{v})\, v_x^2 E \frac{\partial f_0}{\partial E}\left[eE_x + T\frac{d}{dx}\left(\frac{\mu}{T}\right) + \frac{E}{T}\frac{dT}{dx}\right] \tag{11.106}$$

Changing the variable of integration from v to E using Eq. (11.44), we can write Eqs. (11.105), (11.106) in polar coordinates as

$$J_x = -\frac{8\pi}{3m_e^2}\left(\frac{2}{m_e}\right)^{1/2}\int_0^\infty dE E^{3/2}\tau_e(E)\frac{\partial f_0}{\partial E}\left[e^2E_x + eT\frac{d}{dx}\left(\frac{\mu}{T}\right) + \frac{eE}{T}\frac{dT}{dx}\right] \tag{11.107}$$

$$Q_x = \frac{8\pi}{3m_e^2}\left(\frac{2}{m_e}\right)^{1/2}\int_0^\infty dE E^{5/2}\tau_e(E)\frac{\partial f_0}{\partial E}\left[eE_x + T\frac{d}{dx}\left(\frac{\mu}{T}\right) + \frac{E}{T}\frac{dT}{dx}\right] \tag{11.108}$$

To solve for J_x and Q_x we define an integral

$$I_n = -\frac{8\pi}{3m_e^2}\left(\frac{2}{m_e}\right)^{1/2}\int_0^\infty dE E^{n+\frac{1}{2}}\tau_e(E)\frac{\partial f_0}{\partial E} \tag{11.109}$$

where n is an integer. Therefore, in terms of I_n, we can write expressions for J_x and Q_x as follows:

$$J_x = \left[e^2 E_x + e T\frac{d}{dx}\left(\frac{\mu}{T}\right)\right] I_1 + \frac{e}{T}\frac{dT}{dx} I_2 \tag{11.110}$$

$$Q_x = -\left[e E_x + T\frac{d}{dx}\left(\frac{\mu}{T}\right)\right] I_2 - \frac{1}{T}\frac{dT}{dx} I_3 \tag{11.111}$$

To calculate the values of J_x and Q_x we have to evaluate the integral I_n. Substituting $f_0(\mathbf{v}, T)$ from Eq. (11.66) into Eq. (11.109) and integrating by parts, we obtain

$$I_n = C\int_0^\infty dE\, f_0(E, \mu, T)\frac{d}{dE}\left[E^{n+\frac{1}{2}}\tau_e(E)\right] \tag{11.112}$$

where

$$C = \frac{16\pi}{3m_e^2}\left(\frac{2}{m_e}\right)^{1/2}\left(\frac{m_e}{2\pi\hbar}\right)^3 \tag{11.113}$$

Here $f_0(E, \mu, T)$ is given by Eq. (11.32). If we make the substitution

$$h(x) = \frac{d}{dx}\left[x^{n+\frac{1}{2}}\tau_e(x)\right] \tag{11.114}$$

then

$$H(E) = \int_0^E h(x)\,dx = E^{n+\frac{1}{2}}\tau_e(E) \tag{11.115}$$

Now the integral I_n can be written as

$$I_n = C\int_0^E dE f_0(E, \mu, T)\, h(E) \tag{11.116}$$

I_n is the Fermi distribution function integral (see Appendix I) and its solution is given by

$$I_n = C\left[1 + \frac{\pi^2}{6}\frac{\partial^2}{\partial y^2} + \cdots\right] H(\mu) \tag{11.117}$$

where

$$y = \frac{\mu}{k_B T} = \beta_0 \mu \tag{11.118}$$

So Eqs. (11.115), (11.117), and (11.118) give

$$I_n = C\left[\mu^{n+\frac{1}{2}}\tau_e(\mu) + \frac{(\pi k_B T)^2}{6}\frac{d^2}{d\mu^2}\left\{\mu^{n+\frac{1}{2}}\tau_e(\mu)\right\} + \cdots\right] \tag{11.119}$$

From Eq. (11.119) the integrals I_1, I_2, and I_3 are given by

$$I_1 = \frac{16\pi}{3m_e^2}\left(\frac{2}{m_e}\right)^{1/2}\left(\frac{m_e}{2\pi\hbar}\right)^3\left[\mu^{3/2}\tau_e(\mu) + \frac{(\pi k_B T)^2}{6}\frac{d^2}{d\mu^2}\left\{\mu^{3/2}\tau_e(\mu)\right\} + \cdots\right] \tag{11.120}$$

$$I_2 = \frac{16\pi}{3m_e^2}\left(\frac{2}{m_e}\right)^{1/2}\left(\frac{m_e}{2\pi\hbar}\right)^3\left[\mu^{5/2}\tau_e(\mu) + \frac{(\pi k_B T)^2}{6}\frac{d^2}{d\mu^2}\left\{\mu^{5/2}\tau_e(\mu)\right\} + \cdots\right] \tag{11.121}$$

$$I_3 = \frac{16\pi}{3m_e^2}\left(\frac{2}{m_e}\right)^{1/2}\left(\frac{m_e}{2\pi\hbar}\right)^3\left[\mu^{7/2}\tau_e(\mu) + \frac{(\pi k_B T)^2}{6}\frac{d^2}{d\mu^2}\left\{\mu^{7/2}\tau_e(\mu)\right\} + \cdots\right] \quad (11.122)$$

The electrical conductivity is measured at a particular temperature; therefore, $dT/dx = d\mu/dx = 0$. Hence from Eq. (11.110) J_x is given by

$$J_x = e^2 I_1 E_x \quad (11.123)$$

which gives the electrical conductivity as

$$\sigma = e^2 I_1 = \frac{16\pi e^2}{3m_e^2}\left(\frac{2}{m_e}\right)^{1/2}\left(\frac{m_e}{2\pi\hbar}\right)^3\left[\mu^{3/2}\tau_e(\mu) + \frac{(\pi k_B T)^2}{6}\frac{d^2}{d\mu^2}\left\{\mu^{3/2}\tau_e(\mu)\right\} + \cdots\right] \quad (11.124)$$

Substituting $\mu = E_F$ and considering $\tau_e(E_F)$ to be a constant, one can solve the second term of Eq. (11.124), which, after simplification gives

$$\sigma = \frac{n_e e^2 \tau_e(E_F)}{m_e}\left[1 + \frac{1}{8}\pi^2\left(\frac{k_B T}{E_F}\right)^2 + \cdots\right] \quad (11.125)$$

It should be noted that the first term of Eq. (11.125) gives the same value as in Eq. (11.72) and the second term is a correction to it. But at room temperature, $k_B T \ll E_F$ and hence the correction term is very small and can be neglected in comparison with the first term to give

$$\sigma = \sigma_0 = \frac{n_e e^2 \tau_e(E_F)}{m_e} \quad (11.126)$$

The thermal conductivity is measured in the absence of electric current. Hence, by putting $J_x = 0$, Eq. (11.110) yields the condition

$$eE_x + T\frac{d}{dx}\left(\frac{\mu}{T}\right) = -\frac{1}{T}\frac{dT}{dx}\frac{I_2}{I_1} \quad (11.127)$$

Substituting Eq. (11.127) into Eq. (11.111), the thermal current becomes

$$Q_x = -\sigma_T \frac{dT}{dx} \quad (11.128)$$

with

$$\sigma_T = \frac{1}{T}\frac{I_1 I_3 - I_2^2}{I_1} \quad (11.129)$$

The value of σ_T can be obtained by substituting the values of I_1, I_2, and I_3 from Eqs. (11.120), (11.121), and (11.122). If one retains only the first terms in I_1, I_2, and I_3, then the term $(I_1 I_3 - I_2^2)/I_1$ vanishes. The finite value of $(I_1 I_3 - I_2^2)/I_1$ is obtained by retaining at least the first two terms in each of the integrals I_1, I_2, and I_3, which gives

$$\frac{I_1 I_3 - I_2^2}{I_1} = \frac{n_e \pi^2}{3m_e}\tau_e(\mu)(k_B T)^2 \quad (11.130)$$

Substituting Eq. (11.130) into Eq. (11.129), the thermal conductivity becomes

$$\sigma_T = \frac{n_e \pi^2}{3m_e}\tau_e(\mu) k_B^2 T \quad (11.131)$$

Hence, the thermal conductivity at $\mu = E_F$ is given by

$$\sigma_T = \frac{n_e \pi^2}{3m_e}\tau_e(E_F) k_B^2 T \quad (11.132)$$

The value of the Lorentz number, in the quantum theory of Sommerfeld, is obtained by substituting the values of σ_0 and σ_T from Eqs. (11.126), (11.132) into Eq. (11.97), giving

$$L_N = \frac{\pi^2}{3}\left(\frac{k_B}{e}\right)^2 \tag{11.133}$$

Eq. (11.133) gives the value of L_N that agrees with the experimental results of the Wiedemann-Franz-Lorentz law.

11.8 HALL EFFECT

The conductivity measurements do not give any information about the sign of the charges responsible for conduction in solids. To study the nature of the charge carriers responsible for the electric current and hence the conductivity, one has to study the Hall effect. Consider a metallic solid in the form of a slab (see Fig. 10.4) with electric and magnetic fields applied in the x- and z-directions, that is,

$$\mathbf{E} = \hat{\mathbf{i}}_1 E_x \tag{11.134}$$

$$\mathbf{H} = \hat{\mathbf{i}}_3 H_z \tag{11.135}$$

If q and m_q are the charge and mass of the charge carriers responsible for conduction, then the Lorentz force acting on them is given by Eq. (10.53). The charge carriers, in a metallic material, are electrons, but in a semiconductor the charge carriers consist of both electrons and holes.

In the equilibrium state, the distribution function is independent of position, that is, $\nabla_\mathbf{r} f = 0$, therefore, the Boltzmann transport equation, given by Eq. (11.18), becomes

$$\left.\frac{\partial f}{\partial t}\right|_{coll} = \frac{\mathbf{F}}{m_q}\cdot\nabla_\mathbf{v} f \tag{11.136}$$

The Lorentz force produces a perturbation f_1 in the distribution function and hence the perturbed distribution function f becomes

$$f = f_0 + f_1 \tag{11.137}$$

Substituting Eq. (11.137) into Eq. (11.136), we find

$$\left.\frac{\partial f}{\partial t}\right|_{coll} = \frac{\mathbf{F}}{m_q}(\nabla_\mathbf{v} f_0 + \nabla_\mathbf{v} f_1) \tag{11.138}$$

In the relaxation time approximation, the rate of change of the distribution function due to collisions is given by Eq. (11.19), which, when substituted in Eq. (11.138), can be written as

$$f = f_0 - \tau_q \frac{\mathbf{F}}{m_q}\cdot\nabla_\mathbf{v} f_0 - \frac{\tau_q}{m_q}\mathbf{F}\cdot\nabla_\mathbf{v} f_1 \tag{11.139}$$

where τ_q is the relaxation time for the charge carriers. But

$$\nabla_\mathbf{v} f_0 = \hat{\mathbf{i}}_\mathbf{v}\frac{\partial f_0}{\partial v} = \hat{\mathbf{i}}_\mathbf{v}\frac{\partial f_0}{\partial E}\frac{\partial E}{\partial v} = m_q\mathbf{v}\frac{\partial f_0}{\partial E} \tag{11.140}$$

Here $\hat{\imath}_v$ is the unit vector in the direction of velocity **v**. Therefore, the distribution function in the presence of the Lorentz force is given by

$$f = f_0 - \tau_q(\mathbf{v}\cdot\mathbf{F})\frac{\partial f_0}{\partial E} - \frac{\tau_q}{m_q}\mathbf{F}\cdot\nabla_\mathbf{v} f_1 \tag{11.141}$$

Substituting Eq. (10.53) into Eq. (11.141), we obtain the distribution function up to the first order,

$$f = f_0 - q\tau_q(\mathbf{v}\cdot\mathbf{E})\frac{\partial f_0}{\partial E} - \frac{q\tau_q}{m_q c}(\mathbf{v}\times\mathbf{H})\cdot\nabla_\mathbf{v} f_1 \tag{11.142}$$

Here the term containing $\mathbf{E}\cdot\nabla_{\mathbf{v}}f_1$ is neglected, as it is a second-order term. Comparing Eqs. (11.137), (11.142), the change in distribution function due to the Lorentz force is given by

$$f_1 = -q\tau_q(\mathbf{v}\cdot\mathbf{E})\frac{\partial f_0}{\partial E} - \frac{q\tau_q}{m_q c}(\mathbf{v}\times\mathbf{H})\cdot\nabla_{\mathbf{v}}f_1 \tag{11.143}$$

We want to find the current density due to the distribution function given by Eq. (11.142). To use our previous knowledge for calculating the current density due to an electric field (see Section 11.6), we reduce the distribution function to a form similar to that given by Eq. (11.46). To do so we define an effective electric field $\mathbf{E}_{eff}$ such that f_1, given by Eq. (11.143), reduces to

$$f_1 = -q\tau_q(\mathbf{v}\cdot\mathbf{E}_{eff})\frac{\partial f_0}{\partial E} \tag{11.144}$$

and hence the distribution function f can be written as

$$f = f_0 + f_1 = f_0 - q\tau_q(\mathbf{v}\cdot\mathbf{E}_{eff})\frac{\partial f_0}{\partial E} \tag{11.145}$$

To find the relationship between $\mathbf{E}$ and $\mathbf{E}_{eff}$ we calculate the velocity derivative of f_1 from Eq. (11.144) given by

$$\nabla_{\mathbf{v}}f_1 = -q\tau_q\left[m_q\mathbf{v}(\mathbf{v}\cdot\mathbf{E}_{eff})\frac{\partial^2 f_0}{\partial E^2} + \mathbf{E}_{eff}\frac{\partial f_0}{\partial E}\right] \tag{11.146}$$

Substituting Eq. (11.146) into Eq. (11.142), we get

$$f = f_0 - q\tau_q(\mathbf{v}\cdot\mathbf{E})\frac{\partial f_0}{\partial E} + \frac{(q\tau_q)^2}{m_q c}(\mathbf{v}\times\mathbf{H})\cdot\mathbf{E}_{eff}\frac{\partial f_0}{\partial E} \tag{11.147}$$

The last two terms of Eq. (11.147) give the perturbation f_1 in the distribution function, which, when equated to f_1 given by Eq. (11.144), yields

$$\mathbf{E} = \mathbf{E}_{eff} + \frac{q\tau_q}{m_q c}(\mathbf{H}\times\mathbf{E}_{eff}) \tag{11.148}$$

This equation gives the relationship between $\mathbf{E}$ and $\mathbf{E}_{eff}$.

The current density due to the charge carriers with charge q and distribution function $f(\mathbf{v},E)$ is given by

$$\mathbf{J} = q\int \mathbf{v}f(\mathbf{v},E)\,d^3v \tag{11.149}$$

Substituting Eq. (11.145) into Eq. (11.149) and using the equilibrium state condition given by Eq. (11.3), we get

$$\mathbf{J} = -q^2\int \mathbf{v}(\mathbf{v}\cdot\mathbf{E}_{eff})\,\tau_q(\mathbf{v})\frac{\partial f_0}{\partial E}\,d^3v \tag{11.150}$$

If the electric field is applied in the x-direction, then the charged particles move along the x-direction with velocity $\mathbf{v} = \hat{\mathbf{i}}_1 v_x$. Therefore, Eq. (11.150) can be written as

$$\begin{aligned} J &= -q^2\int v_x^2 E_{eff}\,\tau_q(\mathbf{v})\frac{\partial f_0}{\partial E}\,d^3v \\ &= -\frac{q^2}{3}\int v^2 E_{eff}\,\tau_q(\mathbf{v})\frac{\partial f_0}{\partial E}\,d^3v \end{aligned} \tag{11.151}$$

Here E_{eff} is the component of $\mathbf{E}_{eff}$ along the x-direction. In writing Eq. (11.151) we have substituted the average of the square of velocities in the Cartesian directions (see Eq. 11.52). Now one can solve Eq. (11.151) in the same way as in Section 11.6 and can straightway obtain the relation

$$\mathbf{J} = \sigma_0\mathbf{E}_{eff} \tag{11.152}$$

with

$$\sigma_0 = \frac{n_q q^2 \tau_q(E_F)}{m_q} \tag{11.153}$$

Substituting the value of $\mathbf{E}_{eff}$ from Eq. (11.152) into Eq. (11.148), we get

$$\begin{aligned}\mathbf{E} &= \frac{\mathbf{J}}{\sigma_0} + \frac{q\tau_q}{m_q c \sigma_0}\mathbf{H} \times \mathbf{J} \\ &= \frac{\mathbf{J}}{\sigma_0} + \frac{1}{n_q q c}\mathbf{H} \times \mathbf{J}\end{aligned} \tag{11.154}$$

If the magnetic field is zero the second term in Eq. (11.154) goes to zero yielding the familiar expression for conductivity. The second term gives the component of the electric field perpendicular to both **J** and **H**, that is, the transverse electric field usually called the Hall field, which is produced by the combined effect of the electric and magnetic fields. The coefficient of the second term is called the Hall constant and is given by

$$R_{HC} = \frac{1}{n_q q c} \tag{11.155}$$

(see Eq. 10.57). In metals the conductivity is due to the flow of electrons ($q = -e$), which yields a negative value of the Hall coefficient, that is,

$$R_{HC} = -\frac{1}{n_e e c} \tag{11.156}$$

In n-type semiconductors the Hall coefficient is also negative as the majority carriers are electrons. But in p-type semiconductors the Hall coefficient is positive as the majority carriers are holes with positive charge, that is, $q = e$, and

$$R_{HC} = \frac{1}{n_h e c} \tag{11.157}$$

Here n_h is the density of holes.

11.9 MOBILITY OF CHARGE CARRIERS IN SOLIDS

In a crystalline solid the conduction is due to the motion of free charges. Suppose that the magnitude of the charges is q. If an electric field **E** is applied to the solid, the charges experience a force, which produces an acceleration in their motion. During the accelerated motion charges suffer collisions with the lattice ions and impurities that are present in the solid and continuously lose kinetic energy. Ultimately, the charges acquire some constant average velocity $\mathbf{v}_d$, usually called the drift velocity. The expression for $\mathbf{v}_d$ can be obtained from simple arguments. When a particle with charge q and mass m_q moves in an electric field **E**, the acceleration produced is given by

$$\mathbf{a} = \frac{\mathbf{F}}{m_q} = \frac{q\mathbf{E}}{m_q} \tag{11.158}$$

If a constant average velocity $\mathbf{v}_d$ is achieved in the relaxation time τ_q for the charge q, then one can write

$$\mathbf{v}_d = \mathbf{a}\tau_q = \frac{q\tau_q}{m_q}\mathbf{E} \tag{11.159}$$

The above expression can be written as

$$\mathbf{v}_d = \mu_q \mathbf{E} \tag{11.160}$$

where

$$\mu_q = \frac{q\tau_q}{m_q} \tag{11.161}$$

Eq. (11.160) shows that the drift velocity is linearly proportional to the electric field and the constant of proportionality μ_q is called the mobility of the charge carriers. According to Eq. (11.160) the mobility of a charge carrier is defined as the average drift velocity per unit field.

The expression for μ_q can also be obtained from the Boltzmann transport equation. The current density **J** produced by the charges is given by

$$\mathbf{J} = n_q q \mathbf{v}_d \tag{11.162}$$

where n_q is the density of charges. From Eqs. (11.160), (11.162) we get

$$\mu_q = \frac{\mathbf{J}}{n_q q \mathbf{E}} \tag{11.163}$$

With the help of Eq. (11.53), the current density is given by

$$\mathbf{J} = \sigma \mathbf{E} = -\frac{q^2 \mathbf{E}}{3} \int v^2 \frac{\partial f_0}{\partial E} \tau_q(\mathbf{v}) \, d^3v \tag{11.164}$$

Therefore, from Eqs. (11.163), (11.164), the general expression for the mobility of charge q becomes

$$\mu_q = -\frac{1}{3} q \frac{\int v^2 \frac{\partial f_0}{\partial E} \tau_q(\mathbf{v}) \, d^3v}{\int f_0(\mathbf{v}, t) \, d^3v} \tag{11.165}$$

In the above expression the density of charges n_q has been taken from Eq. (11.55). The integrals in Eq. (11.165) can be evaluated using the Maxwell-Boltzmann or the Fermi-Dirac distributions in the same way as was done earlier. Using the Maxwell-Boltzmann distribution given by Eq. (11.56), μ_q reduces to

$$\mu_q = \frac{q \tau_q}{m_q} \tag{11.166}$$

In the above derivation, the relaxation time τ_q is assumed to be a constant, but actually it is a function of the velocity or energy. Assuming the charges to be quantum particles (e.g., electrons or positrons), the Fermi-Dirac distribution given by Eq. (11.66) should be used. Solving the integrals in Eq. (11.165) using the Fermi-Dirac distribution (see Section 11.6), one obtains

$$\mu_q = \frac{q \tau_q(E_F)}{m_q} \tag{11.167}$$

which is the familiar result. Note that Eqs. (11.166), (11.167) involve the magnitude of the charge, that is, $q = |q|$.

Problem 11.1

Let the relaxation time be a function of the energy E:

$$\tau_q(E) = \tau_0 E^p \tag{11.168}$$

where p is a rational number. Assuming the Maxwell-Boltzmann distribution for $f_0(\mathbf{v}, t)$ in Eq. (11.165), prove that the mobility is given by

$$\mu_q = \frac{4}{3\sqrt{\pi}} (k_B T)^p \frac{q \tau_0}{m_q} \Gamma\left(p + \frac{5}{2}\right) \tag{11.169}$$

where $\Gamma(p + 5/2)$ is the gamma function.

To estimate the average relaxation time and mobility, consider a physical example with $p = -1$. Then, from Eq. (11.169), we get

$$\mu_q = \frac{4}{3\sqrt{\pi}} \frac{1}{k_B T} \frac{q \tau_0}{m_q} \Gamma\left(\frac{3}{2}\right) = \frac{2}{3} \frac{q \tau_q(E)}{m_q} \tag{11.170}$$

with

$$\tau_q(E) = \frac{\tau_0}{k_B T} = \frac{\tau_0}{E} \tag{11.171}$$

If we write the mobility as

$$\mu_q = \frac{q\left\langle \tau_q(E) \right\rangle}{m_q} \tag{11.172}$$

then the average relaxation time $\langle \tau_q(E) \rangle$ can be written as

$$\left\langle \tau_q(E) \right\rangle = \frac{2}{3}\tau_q(E) \tag{11.173}$$

Thus, the Boltzmann equation gives the same result as given by Eqs. (11.166) or (11.167), but here the average relaxation time is used. Further, the numerical factor 2/3 is not much different from unity and the deviation from unity is comparable with the experimental error due to the inhomogeneity of the sample material used.

In metallic solids, conduction occurs because of the motion of free electrons. Hence the mobility of the electrons is given by

$$\mu_e = \frac{e\,\tau_e(E_F)}{m_e} \tag{11.174}$$

But in semiconductors, there are both electrons and holes and they both contribute to the mobility. One can easily obtain the mobility contribution of the holes along similar lines to what was done above to obtain

$$\mu_h = \frac{e\,\tau_h(E_F)}{m_h} \tag{11.175}$$

Assuming the relaxation times for both electrons and holes to be the same, that is, $\tau_e(E_F) = \tau_h(E_F)$, the net mobility μ in a semiconductor is given by

$$\mu = \mu_e + \mu_h = e\,\tau_e(E_F)\left[\frac{1}{m_e} + \frac{1}{m_h}\right] \tag{11.176}$$

In n-type semiconductors the majority carriers are electrons, while in p-type semiconductors they are holes. Substituting the value of $\tau_e(E_F)$ from Eq. (11.174) into Eq. (11.72), one gets the expression for conductivity due to electrons as

$$\sigma_e = n_e e \mu_e \tag{11.177}$$

Similarly, one can obtain the expression for the conductivity due to the holes as

$$\sigma_h = n_h e \mu_h \tag{11.178}$$

where n_h is the density of holes. Therefore, the total electrical conductivity σ in a semiconductor is the sum of the electron and hole contributions, that is,

$$\sigma = \sigma_e + \sigma_h = n_e e \mu_e + n_h e \mu_h \tag{11.179}$$

In an intrinsic semiconductor the density of electrons and holes is the same, that is, $n_e = n_h$, but in an extrinsic semiconductor the two are unequal.

Problem 11.2

The resistivity, electron density, and effective mass of the electrons in Cu metal are given as:

$$m_e^* = 1.01\ m_e$$

$$n_e = 8.5 \times 10^{28}/m^3$$

$$\rho = 1.7 \times 10^{-8}\,\Omega\ m$$

Find the relaxation time at the Fermi surface $\tau_e(E_F)$. If the Fermi velocity of the electrons is $v_F = 1.55 \times 10^6$ m/s, find the mean free path of the electrons in Cu metal at the Fermi surface.

SUGGESTED READING

Haug, A. (1972). *Theoretical solid state physics*. (Vols. 1 & 2). New York: Pergamon Press.

Nag, B. R. (1971). Electron transport properties. F. C. Auluck (Ed.), *A short course in solid state physics* (pp. 395–424). Vol. 1(pp. 395–424). New Delhi: Thomson Press.

Sreedhar, A. K. (1971). Electron transport properties (magnetoresistance). F. C. Auluck (Ed.), *A short course in solid state physics* (pp. 425–455). Vol. 1 (pp. 425–455). New Delhi: Thomson Press.

Wilson, A. H. (1954). *The theory of metals* (2nd ed.). London: Cambridge University Press.

Chapter 12

Energy Bands in Crystalline Solids

Chapter Outline

The free-electron theory of metals was developed in Chapter 9. It explained, in a satisfactory manner, a number of electronic properties, such as electrical and thermal conductivities, specific heat, thermionic emission, and paramagnetic susceptibility. The most important achievement of the free-electron approximation is the precise definition of conduction electrons. Unlike the classical theory, it shows that only those free electrons in the vicinity of the Fermi energy participate in conduction. But the free-electron theory could not explain why some solids behave as metals, others as insulators, and still others as semiconductors. It also could not explain other properties, such as the transport properties. The free-electron theory yields only negative values of the Hall coefficient as the charge carriers are simply the electrons. Further, experimental studies show that the Fermi surface is nonspherical in most of the metals, which contradicts the free-electron approximation.

In solids there are electrons (core or valence electrons) and nuclei, which interact among themselves. As a result, a finite crystal potential arises that obeys the periodicity of the lattice (Chapter 1). Therefore, in explaining the various properties of crystalline solids, the crystal potential must be incorporated in the theory. As mentioned earlier in Chapter 3, an exact calculation of the crystal potential is not possible due to the many-body nature of the problem. Therefore, to study the electronic properties, particularly the electronic energy bands in crystalline solids, some simplifying approximations are made: the one-electron approximation is made for the wave function and energy and the crystal potential is estimated in the self-consistent approximation. Let us first examine the nature of the electron wave function in a crystalline solid.

12.1 BLOCH THEOREM

12.1.1 One-Dimensional Solid

Consider a one-dimensional monatomic crystalline solid with "a" as the periodicity and length $L = Na$ (see Fig. 12.1). Let $V(x)$ be the self-consistent crystal potential in the one-electron approximation that satisfies the periodicity of the crystal, that is,

$$V(x+a) = V(x) \tag{12.1}$$

Solid State Physics. https://doi.org/10.1016/B978-0-12-817103-5.00012-8

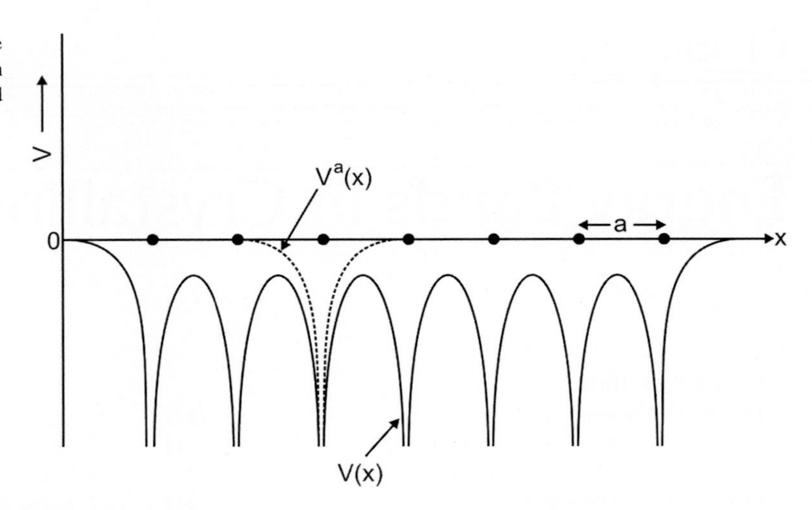

FIG. 12.1 The attractive atomic potential $V^a(x)$ and the crystal potential V(x) for a monatomic linear lattice with "a" as the periodicity. Here V(x), for simplicity, is obtained by the linear combination of atomic potentials.

The Schrodinger wave equation for a one-dimensional solid is written as

$$\widehat{H}_e|\psi_k(x)\rangle = E_k|\psi_k(x)\rangle \tag{12.2}$$

where $|\psi_k(x)\rangle$ and E_k represent the one-electron wave function and energy, respectively, for the electron state with wave vector k. $\widehat{H}_e$ represents the one-electron Hamiltonian defined as

$$\widehat{H}_e = -\frac{\hbar^2}{2m_e}\frac{d^2}{dx^2} + V(x) \tag{12.3}$$

The Bloch theorem states that the wave functions for a wave equation with periodic potential V(x) are of the form

$$|\psi_k(x)\rangle = e^{\iota kx}u_k(x) \tag{12.4}$$

Here $u_k(x)$ is a scalar function that satisfies the periodicity of the lattice, that is,

$$u_k(x+a) = u_k(x) \tag{12.5}$$

The wave function given by Eq. (12.4) is generally called the Bloch function. It is evident from Eq. (12.4) that the Bloch function is a plane wave modified by the periodic potential of the lattice. It is noteworthy that the Bloch functions are the general one-electron wave functions for an ideal crystalline solid.

Consider a translation operator $T(a) = \{I|a\}$ (see Chapter 1), which, when it acts on the wave function, translates it by the distance a so that

$$T(a)|\psi_k(x)\rangle = |\psi_k(x+a)\rangle \tag{12.6}$$

Applying the translation operator on the left side of the Schrodinger wave equation given by Eq. (12.2), we write

$$T(a)\widehat{H}_e|\psi_k(x)\rangle = \left[-\frac{\hbar^2}{2m_e}\frac{d^2}{d(x+a)^2} + V(x+a)\right]|\psi_k(x+a)\rangle = \widehat{H}_e T(a)|\psi_k(x)\rangle \tag{12.7}$$

Eq. (12.7) shows that T(a) commutes with $\widehat{H}_e$, that is,

$$\left[\widehat{H}_e, T(a)\right] = 0 \tag{12.8}$$

where [ ] represent the commutation brackets. From elementary quantum mechanics we know that the commutating operators T(a) and $\widehat{H}_e$ possess simultaneous eigenfunctions. Because $\widehat{H}_e$ is a constant of motion, T(a) is also a constant of motion. So, one can write

$$T(a)|\psi_k(x)\rangle = C_0|\psi_k(x)\rangle \tag{12.9}$$

where C_0 is a constant. From Eqs. (12.6), (12.9) one can write

$$|\psi_k(x+a)\rangle = C_0|\psi_k(x)\rangle \tag{12.10}$$

Similarly, one can show that

$$|\psi_k(x+2a)\rangle = C_0^2|\psi_k(x)\rangle \tag{12.11}$$

and so on. In general, one can show that

$$|\psi_k(x+L)\rangle = |\psi_k(x+Na)\rangle = C_0^N|\psi_k(x)\rangle \tag{12.12}$$

The cyclic boundary condition of the finite crystal demands

$$|\psi_k(x+L)\rangle = |\psi_k(x)\rangle \tag{12.13}$$

From Eqs. (12.12), (12.13) one gets

$$C_0^N = 1 = e^{\imath 2\pi n} \tag{12.14}$$

where $n = 0, \pm 1, \pm 2, \ldots$ Eq. (12.14) gives the value of C_0 as

$$C_0 = e^{2\pi\imath\frac{n}{N}} \tag{12.15}$$

Substituting the value of C_0 in Eq. (12.10), we write

$$|\psi_k(x+a)\rangle = e^{2\pi\imath\frac{n}{N}}|\psi_k(x)\rangle \tag{12.16}$$

The above equation immediately yields

$$|\psi_k(x+a)|^2 = |\psi_k(x)|^2 \tag{12.17}$$

Eq. (12.17) shows that the probability density also satisfies the periodicity of the lattice. But we know that k has got discrete values given by (see Eq. 9.12)

$$k = \frac{2\pi n}{Na} \tag{12.18}$$

From Eqs. (12.16), (12.18) one gets

$$|\psi_k(x+a)\rangle = e^{\imath ka}|\psi_k(x)\rangle \tag{12.19}$$

Eq. (12.19) is usually called the Bloch condition. It can be easily shown that Eqs. (12.4), (12.19) are equivalent if the function $u_k(x)$ satisfies the periodicity of the lattice. From Eq. (12.4) one can write

$$|\psi_k(x+a)\rangle = e^{\imath k(x+a)}u_k(x+a)$$

or

$$|\psi_k(x+a)\rangle = e^{\imath ka}|\psi_k(x)\rangle \tag{12.20}$$

which is nothing but the Bloch condition. Hence the general wave functions for a one-dimensional crystal with periodic potential are the Bloch wave functions defined by Eq. (12.4).

12.1.2 Three-dimensional Solid

Consider a three-dimensional lattice with $\mathbf{a}_1$, $\mathbf{a}_2$, and $\mathbf{a}_3$ as the primitive translation vectors. Let the dimensions of the crystal along the three Cartesian directions be given by $L_1 = N_1\mathbf{a}_1$, $L_2 = N_2\mathbf{a}_2$, and $L_3 = N_3\mathbf{a}_3$ where N_1, N_2, and N_3 are integers. The general lattice vector $\mathbf{R}_n$ in the direct space is given by Eq. (1.5). In a perfect crystalline solid the crystal potential exhibits the periodicity of the lattice, that is,

$$V(\mathbf{r}) = V(\mathbf{r}+\mathbf{R}_n) \tag{12.21}$$

In the presence of the periodic potential, all the physical properties of a solid remain the same when a translation in made through a direct lattice vector. The Schrodinger wave equation in three dimensions is given by

$$\widehat{H}_e|\psi_\mathbf{k}(\mathbf{r})\rangle = E_\mathbf{k}|\psi_\mathbf{k}(\mathbf{r})\rangle \tag{12.22}$$

where the Hamiltonian is given by

$$\widehat{H}_e = -\frac{\hbar^2}{2m_e}\nabla^2 + V(\mathbf{r}) = -\frac{\hbar^2}{2m_e}\frac{d^2}{d\mathbf{r}^2} + V(\mathbf{r}) \tag{12.23}$$

Let us define the translation operator as $T(\mathbf{R}_n) = \{I \mid \mathbf{R}_n\}$, then one can show that

$$\left[T(\mathbf{R}_n), \widehat{H}_e\right] = 0 \tag{12.24}$$

Therefore, $T(\mathbf{R}_n)$ is a constant of motion and both $T(\mathbf{R}_n)$ and $\widehat{H}_e$ possess simultaneous eigenfunctions. Operating $T(\mathbf{R}_n)$ on Eq. (12.22) from left-hand side and using Eq. (12.24), we get

$$\widehat{H}_e T(\mathbf{R}_n)|\psi_\mathbf{k}(\mathbf{r})\rangle = E_\mathbf{k} T(\mathbf{R}_n)|\psi_\mathbf{k}(\mathbf{r})\rangle \tag{12.25}$$

From Eqs. (12.22), (12.25) it is evident that both $|\psi_\mathbf{k}(\mathbf{r})\rangle$ and $T(\mathbf{R}_n)|\psi_\mathbf{k}(\mathbf{r})\rangle$ are eigenfunctions of $\widehat{H}_e$ with the same energy eigenvalue. Therefore, these eigenfunctions differ only by a constant and can be written as

$$T(\mathbf{R}_n)|\psi_\mathbf{k}(\mathbf{r})\rangle = C_0|\psi_\mathbf{k}(\mathbf{r})\rangle \tag{12.26}$$

or

$$|\psi_\mathbf{k}(\mathbf{r}+\mathbf{R}_n)\rangle = C_0|\psi_\mathbf{k}(\mathbf{r})\rangle \tag{12.27}$$

According to the translational symmetry of the crystal, the translation of a position vector by a direct translation vector changes only the origin, but the environment of the new position vector remains the same. Therefore, the probability density must be the same at both the position vectors, that is,

$$|\psi_\mathbf{k}(\mathbf{r}+\mathbf{R}_n)|^2 = |\psi_\mathbf{k}(\mathbf{r})|^2 \tag{12.28}$$

By using Eq. (12.27) the above equation yields

$$|C_0|^2 = 1 \tag{12.29}$$

To find the parameter C_0 let us use the following properties of the translation operator:

$$T(\mathbf{R}_n)T(\mathbf{R}_{n'})|\psi_\mathbf{k}(\mathbf{r})\rangle = |\psi_\mathbf{k}(\mathbf{r}+\mathbf{R}_n+\mathbf{R}_{n'})\rangle = T(\mathbf{R}_n+\mathbf{R}_{n'})|\psi_\mathbf{k}(\mathbf{r})\rangle \tag{12.30}$$

$$T(\mathbf{R}_n)T(\mathbf{R}_{n'})|\psi_\mathbf{k}(\mathbf{r})\rangle = C_0C_0'|\psi_\mathbf{k}(\mathbf{r})\rangle \tag{12.31}$$

The properties given by Eqs. (12.30), (12.31) are satisfied if one assumes

$$C_0 = e^{\imath\mathbf{k}\cdot\mathbf{R}_n} \tag{12.32}$$

Substituting the value of C_0 into Eq. (12.27), one gets

$$|\psi_\mathbf{k}(\mathbf{r}+\mathbf{R}_n)\rangle = e^{\imath\mathbf{k}\cdot\mathbf{R}_n}|\psi_\mathbf{k}(\mathbf{r})\rangle \tag{12.33}$$

which is nothing but the Bloch condition. Eq. (12.33) is satisfied by the Bloch functions, defined as

$$|\psi_\mathbf{k}(\mathbf{r})\rangle = e^{\imath\mathbf{k}\cdot\mathbf{r}}u_\mathbf{k}(\mathbf{r}) \tag{12.34}$$

Here $u_\mathbf{k}(\mathbf{r})$ is a scalar function that satisfies the periodicity of the lattice, that is,

$$u_\mathbf{k}(\mathbf{r}+\mathbf{R}_n) = u_\mathbf{k}(\mathbf{r}) \tag{12.35}$$

The Bloch functions define the general wave functions for the electrons in a crystalline solid. The possible values of the electron wave vector **k** in the Bloch functions are obtained from the cyclic boundary condition on the wave function, which is given as

$$|\psi_\mathbf{k}(\mathbf{r})\rangle = |\psi_\mathbf{k}(\mathbf{r}+\mathbf{L})\rangle = |\psi_\mathbf{k}(\mathbf{r}+N_1\mathbf{a}_1+N_2\mathbf{a}_2+N_3\mathbf{a}_3)\rangle \tag{12.36}$$

Using Eq. (12.34) for the wave function in Eq. (12.36), we write

$$e^{\imath\mathbf{k}\cdot(N_1\mathbf{a}_1+N_2\mathbf{a}_2+N_3\mathbf{a}_3)} = 1 \tag{12.37}$$

Here we have used the fact that $u_{\mathbf{k}}(\mathbf{r})$ satisfies the periodicity of the crystal. As the primitive vectors $\mathbf{a}_1$, $\mathbf{a}_2$, and $\mathbf{a}_3$ are independent, so the exponential terms involving these vectors will separately be unity, that is,

$$e^{\imath \mathbf{k}\cdot N_1 \mathbf{a}_1} = 1 = e^{\imath 2\pi n_1}$$

$$e^{\imath \mathbf{k}\cdot N_2 \mathbf{a}_2} = 1 = e^{\imath 2\pi n_2}$$

$$e^{\imath \mathbf{k}\cdot N_3 \mathbf{a}_3} = 1 = e^{\imath 2\pi n_3}$$

From the above equations one gets

$$\mathbf{k}\cdot\mathbf{a}_1 = \frac{2\pi n_1}{N_1} \quad (12.38)$$

$$\mathbf{k}\cdot\mathbf{a}_2 = \frac{2\pi n_2}{N_2} \quad (12.39)$$

$$\mathbf{k}\cdot\mathbf{a}_3 = \frac{2\pi n_3}{N_3} \quad (12.40)$$

For an sc structure $\mathbf{a}_1 = a\hat{\mathbf{i}}_1$, $\mathbf{a}_2 = a\hat{\mathbf{i}}_2$, and $\mathbf{a}_3 = a\hat{\mathbf{i}}_3$, therefore, one can write the wave vector $\mathbf{k}$ as

$$\mathbf{k} = \frac{2\pi}{a}\left(\frac{n_1}{N_1}\hat{\mathbf{i}}_1 + \frac{n_2}{N_2}\hat{\mathbf{i}}_2 + \frac{n_3}{N_3}\hat{\mathbf{i}}_3\right) \quad (12.41)$$

In the same way, one can find the expression for $\mathbf{k}$ in different structures.

12.2 THE KRONIG-PENNEY MODEL

With knowledge of the electron wave function, one can study the behavior of conduction electrons in a crystalline solid in the presence of a periodic crystal potential. The one-dimensional Kronig-Penney model is a very simple mathematical exercise, which explains beautifully the nature of the energy bands in a crystalline solid. In this model, the periodic potential is assumed to be an array of square-well potentials (see Fig. 12.2) defined as follows:

$$\begin{aligned} V(x) &= V_0 \ \text{for} \ -b < x < 0 \\ &= 0 \ \text{for} \ 0 < x < a \end{aligned} \quad (12.42)$$

Here, as a rough approximation, the potential in the vicinity of an atom is approximated by the potential energy well. The Schrodinger wave equation for the electron state with wave vector $\mathbf{k}$ in the two regions is given by

$$\frac{d^2}{dx^2}|\psi_k^a(x)\rangle + \frac{2m_e E_k}{\hbar^2}|\psi_k^a(x)\rangle = 0 \ \text{for} \ 0 < x < a \quad (12.43)$$

$$\frac{d^2}{dx^2}|\psi_k^b(x)\rangle + \frac{2m_e}{\hbar^2}(E_k - V_0)|\psi_k^b(x)\rangle = 0 \ \text{for} \ -b < x < 0 \quad (12.44)$$

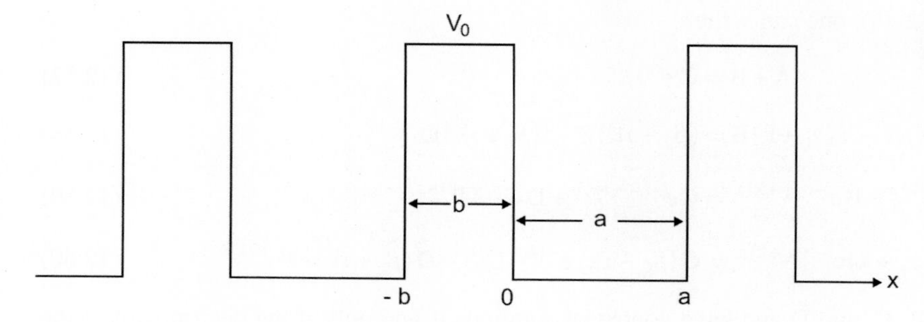

FIG. 12.2 Schematic representation of the periodic square-well potential with periodicity a+b for a monatomic lattice.

If the energy E_k of the electrons is smaller than the potential V_0, then one can define real quantities as

$$\alpha_k^2 = \frac{2m_e E_k}{\hbar^2}, \quad \beta_k^2 = \frac{2m_e}{\hbar^2}(V_0 - E_k) \tag{12.45}$$

Substituting Eq. (12.45) into Eqs. (12.43), (12.44), one can write

$$\frac{d^2}{dx^2}|\psi_k^a\rangle + \alpha_k^2|\psi_k^a\rangle = 0 \text{ for } 0 < x < a \tag{12.46}$$

$$\frac{d^2}{dx^2}|\psi_k^b\rangle - \beta_k^2|\psi_k^b\rangle = 0 \text{ for } -b < x < 0 \tag{12.47}$$

Let us suppose that the solutions of the above equations are the Bloch functions defined as

$$|\psi_k^a(x)\rangle = e^{\imath kx} u_k^a(x) \tag{12.48}$$

$$|\psi_k^b(x)\rangle = e^{\imath kx} u_k^b(x) \tag{12.49}$$

Substituting Eqs. (12.48), (12.49) into Eqs. (12.46), (12.47), one can write

$$\left[D^2 + 2\imath kD + \left(\alpha_k^2 - k^2\right)\right] u_k^a(x) = 0 \text{ for } 0 < x < a \tag{12.50}$$

$$\left[D^2 + 2\imath kD - \left(\beta_k^2 + k^2\right)\right] u_k^b(x) = 0 \text{ for } -b < x < 0 \tag{12.51}$$

where

$$D = \frac{d}{dx} \tag{12.52}$$

Eq. (12.50) has a nontrivial solution if the coefficient of $u_k^a(x)$ is zero, which gives

$$D = \imath(\alpha_k - k) \text{ and } -\imath(\alpha_k + k)$$

Hence the solution of Eq. (12.50) for $u_k^a(x)$ becomes

$$u_k^a(x) = A\,e^{\imath(\alpha_k - k)x} + B\,e^{-\imath(\alpha_k + k)x} \text{ for } 0 < x < a \tag{12.53}$$

Similarly, the solution of Eq. (12.51) can be obtained and is given by

$$u_k^b(x) = C\,e^{(\beta_k - \imath k)x} + De^{-(\beta_k + \imath k)x} \text{ for } -b < x < 0 \tag{12.54}$$

To find the solutions for $u_k^a(x)$ and $u_k^b(x)$, one needs to know the constants A, B, C, and D appearing in Eqs. (12.53), (12.54). These constants can be determined by using the continuity conditions given below:

$$u_k^a(0) = u_k^b(0), \quad u_k^a(a) = u_k^b(-b) \tag{12.55}$$

$$\left(\frac{du_k^a}{dx}\right)_{x=0} = \left(\frac{du_k^b}{dx}\right)_{x=0}, \quad \left(\frac{du_k^a}{dx}\right)_{x=a} = \left(\frac{du_k^b}{dx}\right)_{x=-b} \tag{12.56}$$

The conditions on the left side of Eqs. (12.55), (12.56) represent the requirement of continuity of the wave functions at the origin, while those on the right side represent the periodicity of the wave functions and their derivatives. Substituting Eqs. (12.53), (12.54) into Eqs. (12.55), (12.56), one can write

$$A + B = C + D \tag{12.57}$$

$$\imath(\alpha_k - k)A - \imath(\alpha_k + k)B = (\beta_k - \imath k)C - (\beta_k + \imath k)D \tag{12.58}$$

$$Ae^{\imath(\alpha_k - k)a} + Be^{-\imath(\alpha_k + k)a} = Ce^{-(\beta_k - \imath k)b} + De^{(\beta_k + \imath k)b} \tag{12.59}$$

$$A\imath(\alpha_k - k)e^{\imath(\alpha_k - k)a} - B\imath(\alpha_k + k)e^{-\imath(\alpha_k + k)a} = C(\beta_k - \imath k)\,e^{-(\beta_k - \imath k)b} - D(\beta_k + \imath k)\,e^{(\beta_k + \imath k)b} \tag{12.60}$$

These are homogeneous equations in A, B, C, and D and have nontrivial solutions if and only if the determinant of the coefficients of A, B, C, and D is zero, that is,

$$\begin{vmatrix} 1 & 1 & -1 & -1 \\ \imath(\alpha_k - k) & -\imath(\alpha_k + k) & -(\beta_k - \imath k) & (\beta_k + \imath k) \\ e^{\imath(\alpha_k - k)a} & e^{-\imath(\alpha_k + k)a} & -e^{-(\beta_k - \imath k)b} & -e^{(\beta_k + \imath k)b} \\ \imath(\alpha_k - k)e^{\imath(\alpha_k - k)a} & -\imath(\alpha_k + k)e^{-\imath(\alpha_k + k)a} & -(\beta_k - \imath k)e^{-(\beta_k - \imath k)b} & (\beta_k + \imath k)e^{(\beta_k + \imath k)b} \end{vmatrix} = 0 \quad (12.61)$$

Solving the above determinant, one gets

$$\frac{\beta_k^2 - \alpha_k^2}{2\alpha_k \beta_k} \sinh \beta_k b \sin \alpha_k a + \cosh \beta_k b \cos \alpha_k a = \cos k(a+b) \quad (12.62)$$

To obtain the energy band structure one has to solve Eq. (12.62), which is complex in nature. Kronig and Penney simplified the problem by assuming the potential barrier to be a delta function potential, that is, V_0 approaches infinity as b approaches zero such that $V_0 b$ is finite and much less than unity. In this approximation

$$\beta_k = \left(\frac{2m_e}{\hbar^2} V_0\right)^{1/2}, \quad \beta_k b \ll 1, \quad b \ll a, \quad \text{and} \quad \alpha_k \ll \beta_k \quad (12.63)$$

Using Eq. (12.63) to simplify Eq. (12.62), one can write

$$P \frac{\sin \alpha_k a}{\alpha_k a} + \cos \alpha_k a = \cos ka \quad (12.64)$$

where

$$P = \frac{m_e a V_0 b}{\hbar^2} \quad (12.65)$$

Therefore, P is proportional to the area under the potential barrier, that is, $V_0 b$, and is the measure of the strength of the barrier potential. In other words, P gives the binding of an electron to the potential well: the greater the value of P, the greater the binding of the electron to the potential well.

Fig. 12.3 shows the plot of the left side of Eq. (12.64) as a function of $\alpha_k a$ for $P = 3\pi/2$. As α_k is proportional to E_k, the abscissa is the measure of energy. Now the right side of Eq. (12.64) has values ranging from -1 to $+1$. Therefore, only those values of energy that satisfy Eq. (12.64) are allowed. The values of energy corresponding to the shaded regions in Fig. 12.3 are allowed. From Fig. 12.3 the following interesting conclusions can be drawn:

1. The energy spectrum of the electrons consists of a number of allowed energy bands separated by forbidden energy gaps.
2. According to Eq. (12.64) the energies are allowed if cos ka ranges from -1 to $+1$. Therefore, the range of k values is as follows:

$$k = -\frac{n\pi}{a} \rightarrow \frac{n\pi}{a}$$

for the nth band. Hence for the 1BZ, the k values range from $-\pi/a$ to π/a and for the 2BZ these range from $-\pi/a$ to $-2\pi/a$ and from π/a to $2\pi/a$ and so on.

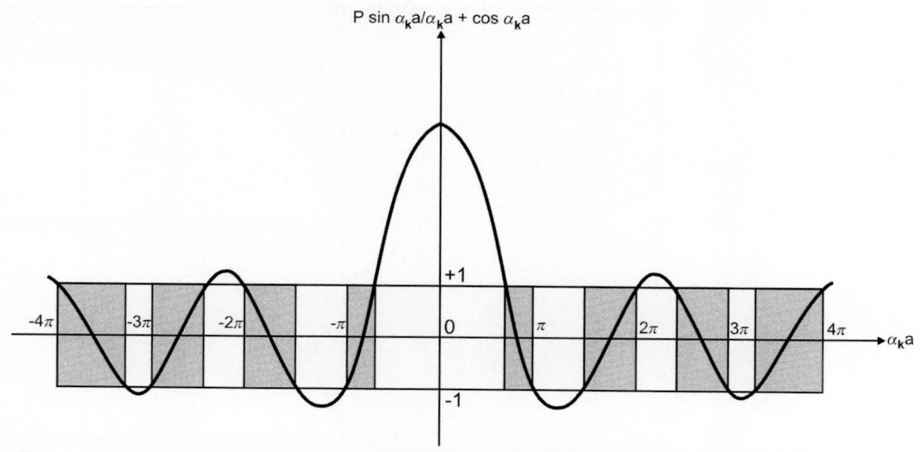

FIG. 12.3 A graph of P $\sin \alpha_k a/\alpha_k a + \cos \alpha_k a$ as a function of $\alpha_k a$ with $P = 3\pi/2$ for a linear monatomic lattice. The *shaded regions* represent the allowed energy bands, while the white spaces between the shaded regions represent the forbidden energy gaps.

3. The width of the allowed energy band increases while that of the forbidden energy gap decreases with an increase in the value of $\alpha_k a$, that is, with an increase in the value of energy. This is a consequence of the fact that the first term of Eq. (12.64) decreases on average with an increase in the value of $\alpha_k a$.
4. The width of a particular allowed energy band decreases with increasing P value, that is, with increasing binding energy of the electrons. One can study Eq. (12.64) in the limiting cases. If the electrons are not bound (free electron case), then $P = 0$ and Eq. (12.64) reduces to

$$\cos \alpha_k a = \cos ka$$

or

$$\alpha_k a = ka \tag{12.66}$$

Substituting the value of α_k from Eq. (12.45) into Eq. (12.66), one gets

$$E_{\mathbf{k}} = \frac{\hbar^2 k^2}{2m_e} \tag{12.67}$$

which is nothing but the energy of a free electron. On the other hand, if the electrons are tightly bound, then $P = \infty$. In this case the first term of Eq. (12.64) becomes finite only if

$$\sin \alpha_k a = 0$$

which gives

$$\alpha_k a = n\pi \tag{12.68}$$

Substituting the value of α_k in the above equation, one gets

$$E_{\mathbf{k}} = E_n = \frac{n^2 \pi^2 \hbar^2}{2m_e a^2} \tag{12.69}$$

These are the energy levels of a particle in a box of atomic dimensions and with finite potential. When the potential becomes infinite ($P = \infty$), the tunneling of an electron through the barrier becomes impossible and the allowed energy spectrum becomes a line spectrum. Fig. 12.4 shows a plot of the energy as a function of P (binding) and we see that for $P = 0$ there are no bands, that is, the energy is quasicontinuous. Also, for $P = \infty$, there are no bands but the line spectrum.

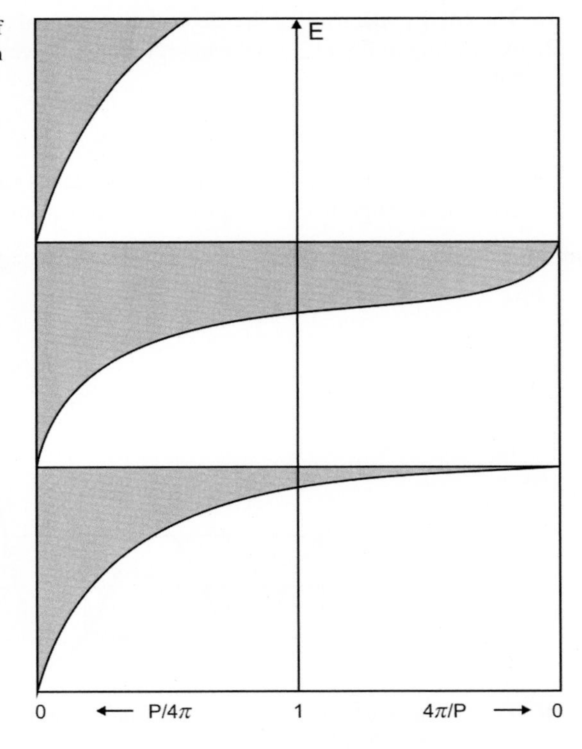

FIG. 12.4 A plot of energy E as a function of parameter P, which gives the strength of the barrier potential. On the left side of 1, the value of P decreases, while it increases on the right side.

12.3 NEARLY FREE-ELECTRON THEORY

The Kronig-Penney model does not represent a physical system, except that the electron wave functions are assumed to be the Bloch functions in the presence of a periodic square-well potential. In this section we study the nature of energy bands in a crystalline solid with self-consistent periodic crystal potential $V(\mathbf{r})$. The crystal potential $V(\mathbf{r})$ can be taken to be the linear combination of self-consistent atomic potentials $V^a(\mathbf{r})$ as

$$V(\mathbf{r}) = \sum_n V^a(\mathbf{r} - \mathbf{R}_n) \tag{12.70}$$

The Schrodinger wave equation for a three-dimensional solid is given by

$$\left[-\frac{\hbar^2}{2m_e}\nabla^2 + V(\mathbf{r})\right] |\psi_{\mathbf{k}}(\mathbf{r})\rangle = E_{\mathbf{k}} |\psi_{\mathbf{k}}(\mathbf{r})\rangle \tag{12.71}$$

As the simplest approximation, one can assume $V(\mathbf{r})$ to be a weak potential. In this approximation, the conduction electrons in the crystalline solid behave as nearly free electrons and, therefore, this approximation is called the nearly free-electron approximation. One can use perturbation theory to evaluate $E_{\mathbf{k}}$ and $|\psi_{\mathbf{k}}(\mathbf{r})\rangle$ in the presence of a weak periodic potential. In the absence of a potential, Eq. (12.71) reduces to

$$\left[-\frac{\hbar^2}{2m_e}\nabla^2\right] |\psi^0_{\mathbf{k}}(\mathbf{r})\rangle = E^0_{\mathbf{k}} |\psi^0_{\mathbf{k}}(\mathbf{r})\rangle \tag{12.72}$$

which is the wave equation for the free electrons. The unperturbed wave function is a plane wave $|\mathbf{k}\rangle$ defined as

$$|\psi^0_{\mathbf{k}}(\mathbf{r})\rangle = \frac{1}{\sqrt{V}} e^{\imath\mathbf{k}\cdot\mathbf{r}} = |\mathbf{k}\rangle \tag{12.73}$$

where

$$V = NV_0 \tag{12.74}$$

Here V_0 and V are the atomic volume and crystal volume and N is the total number of atoms. Operating on Eq. (12.72) with $\langle\psi_{\mathbf{k}'}{}^0(\mathbf{r})|$ from the left side and simplifying, one can write

$$E^0_{\mathbf{k}} = \langle\mathbf{k}| -\frac{\hbar^2}{2m_e}\nabla^2 |\mathbf{k}\rangle = \frac{\hbar^2 k^2}{2m_e} \tag{12.75}$$

In the presence of a weak potential the plane waves are no longer independent of each other. Therefore, the perturbed wave function up to the first order $|\psi_{\mathbf{k}}(\mathbf{r})\rangle$ is obtained by taking a linear combination of plane waves as

$$|\psi_{\mathbf{k}}(\mathbf{r})\rangle = |\psi^0_{\mathbf{k}}(\mathbf{r})\rangle + \sum_{\mathbf{k}'(\neq\mathbf{k})} A_{\mathbf{k}'\mathbf{k}} |\psi^0_{\mathbf{k}'}(\mathbf{r})\rangle \tag{12.76}$$

where the constants $A_{\mathbf{k}'\mathbf{k}}$ are given by

$$A_{\mathbf{k}'\mathbf{k}} = \frac{V_{\mathbf{k}'\mathbf{k}}}{E^0_{\mathbf{k}} - E^0_{\mathbf{k}'}} \tag{12.77}$$

$V_{\mathbf{k}'\mathbf{k}}$ are the matrix elements of $V(\mathbf{r})$ between the electron states with wave vectors $\mathbf{k}'$ and $\mathbf{k}$ and they are given by

$$V_{\mathbf{k}'\mathbf{k}} = \langle\mathbf{k}'|V(\mathbf{r})|\mathbf{k}\rangle = \frac{1}{V}\int e^{-\imath\mathbf{k}'\cdot\mathbf{r}} V(\mathbf{r}) e^{\imath\mathbf{k}\cdot\mathbf{r}} d^3r \tag{12.78}$$

The perturbed energy, correct up to the second order, is given by

$$E_{\mathbf{k}} = E^0_{\mathbf{k}} + V_{\mathbf{k}\mathbf{k}} + \sum_{\mathbf{k}'(\neq\mathbf{k})} \frac{|V_{\mathbf{k}'\mathbf{k}}|^2}{E^0_{\mathbf{k}} - E^0_{\mathbf{k}'}} \tag{12.79}$$

To calculate $|\psi_{\mathbf{k}}(\mathbf{r})\rangle$ and $E_{\mathbf{k}}$ for a solid, one has to evaluate the matrix elements $V_{\mathbf{k}'\mathbf{k}}$. The diagonal matrix element $V_{\mathbf{k}\mathbf{k}}$ gives the average crystal potential in the electron state $|\mathbf{k}\rangle$ and, as a suitable reference of energy, it can be taken as zero, that is,

$$V_{\mathbf{kk}} = \langle \mathbf{k}|V(\mathbf{r})|\mathbf{k}\rangle = \frac{1}{V}\int V(\mathbf{r})\,d^3r = 0 \tag{12.80}$$

Therefore, the perturbed energy from Eq. (12.79) becomes

$$E_{\mathbf{k}} = E_{\mathbf{k}}^0 + \sum_{\mathbf{k}'(\neq \mathbf{k})} \frac{|V_{\mathbf{k}'\mathbf{k}}|^2}{E_{\mathbf{k}}^0 - E_{\mathbf{k}'}^0} \tag{12.81}$$

Substituting the Eq. (12.70) for the crystal potential in Eq. (12.78) and rearranging the terms, one can write

$$V_{\mathbf{k}'\mathbf{k}} = \frac{1}{V}\sum_n e^{-\imath(\mathbf{k}'-\mathbf{k})\cdot\mathbf{R}_n}\int e^{-\imath\mathbf{k}'\cdot(\mathbf{r}-\mathbf{R}_n)}V^a(\mathbf{r}-\mathbf{R}_n)\,e^{\imath\mathbf{k}\cdot(\mathbf{r}-\mathbf{R}_n)}\,d^3(r-R_n) \tag{12.82}$$

Here we have used the fact that $d^3r = d^3(r - R_n)$. As $V^a(\mathbf{r})$ is the atomic potential, so its normalization constant will be the atomic volume. Therefore, Eq. (12.82) can be written as

$$V_{\mathbf{k}'\mathbf{k}} = S_{\mathbf{k}'\mathbf{k}}V_{\mathbf{k}'\mathbf{k}}^a \tag{12.83}$$

where

$$S_{\mathbf{k}'\mathbf{k}} = \frac{1}{N}\sum_n e^{-\imath(\mathbf{k}'-\mathbf{k})\cdot\mathbf{R}_n} \tag{12.84}$$

$$V_{\mathbf{k}'\mathbf{k}}^a = \frac{1}{V_0}\int e^{-\imath(\mathbf{k}'-\mathbf{k})\cdot\mathbf{r}}V^a(\mathbf{r})\,d^3r \tag{12.85}$$

$V_{\mathbf{k}'\mathbf{k}}{}^a$ is nothing but the Fourier transform of the atomic potential, generally called the atomic form factor. $S_{\mathbf{k}'\mathbf{k}}$ describes the structure of the solid in the reciprocal space and it can be shown that

$$\begin{aligned} S_{\mathbf{k}'\mathbf{k}} &= 1 \quad \text{if } \mathbf{k}' = \mathbf{k} - \mathbf{G}_p \\ &= 0 \quad \text{otherwise} \end{aligned} \tag{12.86}$$

Here $\mathbf{G}_p$ is the reciprocal lattice vector (see Eq. 2.30). Therefore, $S_{\mathbf{k}'\mathbf{k}}$ is called the structure factor of the crystal. From Eqs. (12.83), (12.86) one can write

$$V_{\mathbf{k}'\mathbf{k}} = V_{\mathbf{k}-\mathbf{G}_p\,\mathbf{k}} = V_{\mathbf{G}_p}^a \tag{12.87}$$

where

$$V_{\mathbf{G}_p}^a = \frac{1}{V_0}\int e^{\imath\mathbf{G}_p\cdot\mathbf{r}}V^a(\mathbf{r})\,d^3r \tag{12.88}$$

Substituting Eq. (12.87) into Eqs. (12.76), (12.81), we can write

$$|\psi_{\mathbf{k}}(\mathbf{r})\rangle = |\psi_{\mathbf{k}}^0(\mathbf{r})\rangle + \sum_{\mathbf{G}_p(\neq 0)} \frac{V_{\mathbf{G}_p}^a}{E_{\mathbf{k}}^0 - E_{\mathbf{k}-\mathbf{G}_p}^0}\left|\psi_{\mathbf{k}-\mathbf{G}_p}^0(\mathbf{r})\right\rangle \tag{12.89}$$

$$E_{\mathbf{k}} = E_{\mathbf{k}}^0 + \sum_{\mathbf{G}_p(\neq 0)} \frac{\left|V_{\mathbf{G}_p}^a\right|^2}{E_{\mathbf{k}}^0 - E_{\mathbf{k}-\mathbf{G}_p}^0} \tag{12.90}$$

Eqs. (12.89), (12.90) are valid under the following conditions:

1. $V_{G_p}{}^a$ rapidly approaches zero as $\mathbf{G}_p$ increases.
2. The states $\mathbf{k}$ and $\mathbf{k} - \mathbf{G}_p$ are nondegenerate because, for degenerate states, the wave function and energy both blow up and hence perturbation theory is not valid. For degenerate states

$$E_{\mathbf{k}}^0 = E_{\mathbf{k}-\mathbf{G}_p}^0$$

which gives

$$|\mathbf{k}|^2 = \left|\mathbf{k} - \mathbf{G}_p\right|^2 \tag{12.91}$$

The above equation is equivalent to

$$\left|\mathbf{G}_p\right|^2 - 2\mathbf{k}\cdot\mathbf{G}_p = 0 \tag{12.92}$$

Eq. (12.92) is the Bragg reflection condition (Ewald's construction). Therefore, Eqs. (12.89), (12.90) are not valid at or near the BZ boundary.

At the BZ boundary, one has to use degenerate perturbation theory in which the wave function is represented as a linear combination of wave functions $|\psi^0_{\mathbf{k}}(\mathbf{r})\rangle$ and $|\psi^0_{\mathbf{k}-\mathbf{G}}(\mathbf{r})\rangle$, that is,

$$|\psi_{\mathbf{k}}(\mathbf{r})\rangle = A_0\,|\psi^0_{\mathbf{k}}(\mathbf{r})\rangle + A_{\mathbf{G}_p}\left|\psi^0_{\mathbf{k}-\mathbf{G}_p}\right\rangle \tag{12.93}$$

Substituting Eq. (12.93) into Eq. (12.71), one gets

$$A_0 E^0_{\mathbf{k}}|\psi^0_{\mathbf{k}}(\mathbf{r})\rangle + A_{\mathbf{G}_p}E^0_{\mathbf{k}-\mathbf{G}_p}\left|\psi^0_{\mathbf{k}-\mathbf{G}_p}(\mathbf{r})\right\rangle + V(\mathbf{r})\left[A_0|\psi^0_{\mathbf{k}}(\mathbf{r})\rangle + A_{\mathbf{G}_p}\left|\psi^0_{\mathbf{k}-\mathbf{G}_p}\right\rangle\right] = E_{\mathbf{k}}\left[A_0|\psi^0_{\mathbf{k}}(\mathbf{r})\rangle + A_{\mathbf{G}_p}\left|\psi^0_{\mathbf{k}-\mathbf{G}_p}(\mathbf{r})\right\rangle\right] \tag{12.94}$$

Operating $\langle\psi^0_{\mathbf{k}}(\mathbf{r})|$ on Eq. (12.94) from the left side and simplifying, we get

$$A_0\left[E^0_{\mathbf{k}} - E_{\mathbf{k}}\right] + A_{\mathbf{G}_p}\left[V^a_{\mathbf{G}_p}\right]^* = 0 \tag{12.95}$$

Similarly, operating $\langle\psi_{\mathbf{k}-\mathbf{G}_p}{}^0(\mathbf{r})|$ on Eq. (12.94) from the left side and simplifying, we get

$$A_0 V^a_{\mathbf{G}_p} + \left[E^0_{\mathbf{k}-\mathbf{G}_p} - E_{\mathbf{k}}\right]A_{\mathbf{G}_p} = 0 \tag{12.96}$$

Eqs. (12.95), (12.96) have a nontrivial solution if and only if the determinant of the coefficients of A_0 and $A_{\mathbf{G}_p}$ is zero, that is,

$$\begin{vmatrix} E^0_{\mathbf{k}} - E_{\mathbf{k}} & \left[V^a_{\mathbf{G}_p}\right]^* \\ V^a_{\mathbf{G}_p} & E^0_{\mathbf{k}-\mathbf{G}_p} - E_{\mathbf{k}} \end{vmatrix} = 0 \tag{12.97}$$

Solving the above determinant, the energy is given by

$$E_{\mathbf{k}} = \frac{1}{2}\left[\left(E^0_{\mathbf{k}} + E^0_{\mathbf{k}-\mathbf{G}_p}\right) \pm \left\{\left(E^0_{\mathbf{k}} - E^0_{\mathbf{k}-\mathbf{G}_p}\right)^2 + 4\left|V^a_{\mathbf{G}_p}\right|^2\right\}^{1/2}\right] \tag{12.98}$$

If $[E_{\mathbf{k}} - E_{\mathbf{k}-\mathbf{G}_p}]^2 \gg 4\,|V_{\mathbf{G}_p}{}^a|^2$, then one is quite far away from the BZ boundary and Eq. (12.98) yields

$$E_{\mathbf{k}} = E^o_{\mathbf{k}} \tag{12.99}$$

$$E_{\mathbf{k}} = E^o_{\mathbf{k}-\mathbf{G}} \tag{12.100}$$

which are nothing but the free-electron energy bands. But if $[E_{\mathbf{k}} - E_{\mathbf{k}-\mathbf{G}_p}]^2 \ll 4\,|V_{\mathbf{G}_p}{}^a|^2$, one is very near the BZ boundary and Eq. (12.98) reduces to

$$E_{\mathbf{k}} = \frac{1}{2}\left(E^0_{\mathbf{k}} + E^0_{\mathbf{k}-\mathbf{G}_p}\right) \pm \left|V^a_{\mathbf{G}_p}\right| \tag{12.101}$$

At the BZ boundary with $k = (1/2)\,|\mathbf{G}_p|$, the energies from Eq. (12.101) are given by

$$E^+_{\frac{1}{2}\left|\mathbf{G}_p\right|} = E^0_{\frac{1}{2}\left|\mathbf{G}_p\right|} + \left|V^a_{\mathbf{G}_p}\right| \tag{12.102}$$

$$E^-_{\frac{1}{2}\left|\mathbf{G}_p\right|} = E^0_{\frac{1}{2}\left|\mathbf{G}_p\right|} - \left|V^a_{\mathbf{G}_p}\right| \tag{12.103}$$

Eqs. (12.102), (12.103) show that the two bands are separated by an energy gap of magnitude $E_g = 2\,|V_{\mathbf{G}_p}{}^a|$ at the BZ boundary, corresponding to the vector $\mathbf{G}_p$. It is noteworthy that the magnitude of the band gap at the boundary of a particular BZ depends on the atomic potential. Further, its magnitude is different at the boundaries of different BZs, even for the same atomic potential.

To study the nature of the waveform at the BZ boundary, one has to calculate the coefficients A_0 and A_{G_p}. From Eq. (12.95) one can write

$$\frac{A_0}{A_{Gp}} = \frac{\left[V^a_{G_p}\right]^*}{E_k - E^0_k} \tag{12.104}$$

At the BZ boundary the ratio of coefficients for E^+ and E^- becomes

$$\left(\frac{A_0}{A_{G_p}}\right)_+ = \frac{\left[V^a_{G_p}\right]^*}{E^+_{\frac{1}{2}\left|G_p\right|} - E^0_{\frac{1}{2}\left|G_p\right|}} = \sqrt{\frac{\left[V^a_{G_p}\right]^*}{V^a_{G_p}}} \tag{12.105}$$

$$\left(\frac{A_0}{A_{G_p}}\right)_- = \frac{\left[V^a_{G_p}\right]^*}{E^-_{\frac{1}{2}\left|G_p\right|} - E^0_{\frac{1}{2}\left|G_p\right|}} = -\sqrt{\frac{\left[V^a_{G_p}\right]^*}{V^a_{G_p}}} \tag{12.106}$$

If the potential is real, then

$$\left[V^a_{G_p}\right]^* = V^a_{G_p} = V^a_{-G_p} \tag{12.107}$$

From Eqs. (12.105), (12.106), and (12.107) one gets

$$\left(\frac{A_0}{A_{G_p}}\right)_+ = -\left(\frac{A_0}{A_{G_p}}\right)_- = 1 \tag{12.108}$$

From Eqs. (12.93), (12.108) the wave functions at the BZ boundary are given by

$$\left|\psi^{\pm}_{\frac{1}{2}G_p}(\mathbf{r})\right\rangle = A_0\left[\left|\psi^0_{\frac{1}{2}G_p}(\mathbf{r})\right\rangle \pm \left|\psi^0_{-\frac{1}{2}G_p}(\mathbf{r})\right\rangle\right] \tag{12.109}$$

Substituting $|\psi^0_{\mathbf{k}}(\mathbf{r})\rangle$ for the required wave vector, from Eq. (12.73), into the above equation, we obtain

$$\left|\psi^{\pm}_{\frac{1}{2}G_p}(\mathbf{r})\right\rangle = \frac{A_0}{\sqrt{V}}\left[e^{\imath\frac{1}{2}G_p\cdot\mathbf{r}} \pm e^{-\imath\frac{1}{2}G_p\cdot\mathbf{r}}\right] \tag{12.110}$$

Therefore, the two wave functions are given by

$$\left|\psi^{+}_{\frac{1}{2}G_p}(\mathbf{r})\right\rangle = \frac{2A_0}{\sqrt{V}}\cos\left(\frac{1}{2}\mathbf{G}_p\cdot\mathbf{r}\right) \tag{12.111}$$

$$\left|\psi^{-}_{\frac{1}{2}G_p}(\mathbf{r})\right\rangle = \frac{2\imath A_0}{\sqrt{V}}\sin\left(\frac{1}{2}\mathbf{G}_p\cdot\mathbf{r}\right) \tag{12.112}$$

Eqs. (12.111), (12.112) represent standing waves at the BZ boundary, as two plane waves that are exactly the same are moving in opposite directions. The corresponding probability densities at the BZ boundary are given by

$$\rho^{+}_{\frac{1}{2}G_p}(\mathbf{r}) = \left|\psi^{+}_{\frac{1}{2}G_p}(\mathbf{r})\right|^2 = \frac{4A_0^2}{V}\cos^2\left(\frac{1}{2}\mathbf{G}_p\cdot\mathbf{r}\right) \tag{12.113}$$

$$\rho^{-}_{\frac{1}{2}G_p}(\mathbf{r}) = \left|\psi^{-}_{\frac{1}{2}G_p}(\mathbf{r})\right|^2 = \frac{4A_0^2}{V}\sin^2\left(\frac{1}{2}\mathbf{G}_p\cdot\mathbf{r}\right) \tag{12.114}$$

We know that $\mathbf{G}_p\cdot\mathbf{R}_n = 2\pi n'$, where n′ is an integer, therefore,

$$\frac{1}{2}\mathbf{G}_p\cdot\mathbf{R}_n = n'\pi \tag{12.115}$$

From Eqs. (12.113), (12.114) it is evident that $\rho^{+}_{\frac{1}{2}\mathbf{G}_p}(\mathbf{r})$ is maximum while $\rho^{-}_{\frac{1}{2}\mathbf{G}_p}(\mathbf{r})$ is minimum at the lattice positions defined by $\mathbf{R}_n$. In order to study the behavior of $\rho^{+}_{\frac{1}{2}\mathbf{G}_p}(\mathbf{r})$ and $\rho^{-}_{\frac{1}{2}\mathbf{G}_p}(\mathbf{r})$ in detail, one can apply the general expressions to one-, two-, and three-dimensional crystalline solids.

12.3.1 Application to One-Dimensional Solid

Consider a monatomic linear lattice along the x-direction with periodicity "a" and length L = Na (see Fig. 12.1). The position vector can be written as

$$\mathbf{r} = x\hat{\mathbf{i}}_1 \tag{12.116}$$

The reciprocal lattice vector of the linear lattice is given by

$$\mathbf{G}_p = \frac{2\pi p}{a}\hat{\mathbf{i}}_1 \tag{12.117}$$

Therefore,

$$\mathbf{G}_p \cdot \mathbf{r} = \frac{2\pi p}{a} x \tag{12.118}$$

So, the unperturbed wave function is a one-dimensional plane wave defined as

$$|\psi^0_k(x)\rangle = \frac{1}{\sqrt{L}} e^{\imath kx} \tag{12.119}$$

Substituting Eq. (12.117) into Eqs. (12.102), (12.103), we get

$$E^{+}_{p\pi/a} = E^{0}_{p\pi/a} + V^{a}_{\mathbf{G}_p} \tag{12.120}$$

$$E^{-}_{p\pi/a} = E^{0}_{p\pi/a} - V^{a}_{\mathbf{G}_p} \tag{12.121}$$

where $V_{\mathbf{G}_p}{}^{a}$ is the Fourier transform of the real atomic potential $V^a(x)$ in one dimension and is given by

$$V^{a}_{\mathbf{G}_p} = \frac{1}{a}\int_0^a V^a(x)\, e^{2\pi\imath px/a} dx \tag{12.122}$$

Hence, at the boundary of the pBZ, there are two energy eigenvalues, namely $E^{+}_{p\pi/a}$ and $E^{-}_{p\pi/a}$, which are separated by the forbidden energy having value $E_g = 2\,|\,V_{\mathbf{G}_p}{}^{a}|$.

In the free-electron approximation, the energy bands, shown by dashed lines in Fig. 12.5, are parabolic and are degenerate at the BZ boundary. With the introduction of a weak periodic potential, the parabolic distribution of electrons is modified to remove the degeneracy and the energy bands at the BZ boundary are separated by an energy gap of the order $E_g = 2\,|\,V_{\mathbf{G}_p}{}^{a}\,|$. Fig. 12.5 shows the energy bands in the first three BZs in the nearly free-electron approximation. Substituting Eq. (12.119) into Eq. (12.109), one gets

$$\left|\psi^{+}_{p\pi/a}(x)\right\rangle = \frac{2A_0}{L}\cos\left(\frac{p\pi}{a}x\right) \tag{12.123}$$

$$\left|\psi^{-}_{p\pi/a}(x)\right\rangle = \frac{2\imath A_0}{L}\sin\left(\frac{p\pi}{a}x\right) \tag{12.124}$$

The wave functions $|\,\psi^{+}_{p\pi/a}\rangle$ and $|\,\psi^{-}_{p\pi/a}\rangle$ represent standing waves. If the atomic potential $V^a(x)$ is attractive (Fig. 12.1), that is, the matrix elements $V_{\mathbf{G}_p}{}^{a}$ are negative, then, from Eqs. (12.120), (12.121), it is evident that $E^{+}_{p\pi/a}$ has a lower value than $E^{-}_{p\pi/a}$. Hence, the wave function $|\,\psi^{+}_{p\pi/a}\rangle$ corresponds to the lower energy state. The above fact can also be explained with the help of the electron probability densities, which, for the states $|\,\psi^{+}_{p\pi/a}\rangle$ and $|\,\psi^{-}_{p\pi/a}\rangle$, are given by

$$\rho^{+}_{p\pi/a}(x) = \frac{4A_0^2}{L}\cos^2\left(\frac{p\pi}{a}x\right) \tag{12.125}$$

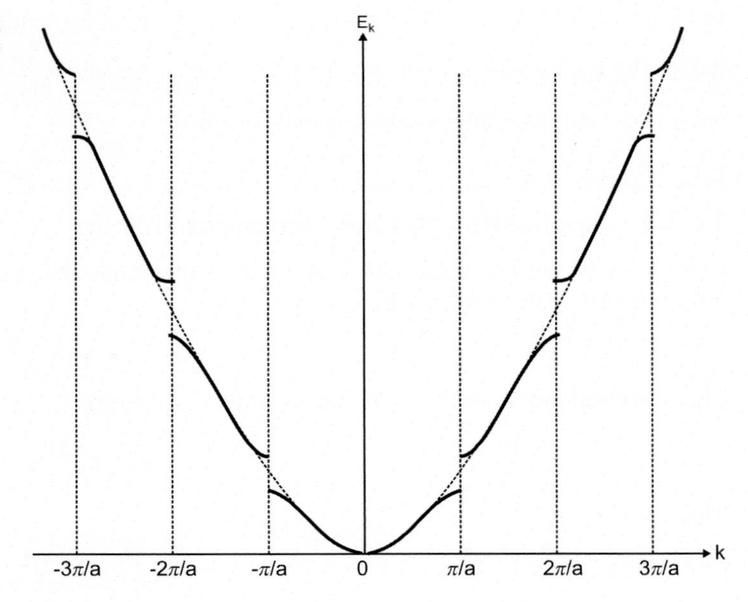

FIG. 12.5 The band energy E_k as a function of wave vector k in various BZs of a linear lattice (extended zone scheme) in the nearly free-electron approximation.

$$\rho^{-}_{p\pi/a}(x) = \frac{4A_0^2}{L}\sin^2\left(\frac{p\pi}{a}x\right) \tag{12.126}$$

Fig. 12.6 shows that $\rho^{+}_{p\pi/a}(x)$ is maximum at the atomic positions and minimum midway between the atoms. Therefore, the electronic screening of the atoms is maximum in the case of $|\psi^{+}_{p\pi/a}\rangle$, thereby lowering the potential energy in comparison with the average potential energy. On the other hand, $\rho^{-}_{p\pi/a}(x)$ has its minimum value at the atomic sites but is maximum midway between the atoms. In this case the potential energy is maximum due to the minimum electronic screening. The wave function $|\psi^{-}_{p\pi/a}(x)\rangle$ corresponds to the higher energy eigenvalue.

The standing waves are formed only if the travelling wave is reflected back in the opposite direction. It can be shown that the reflection at the BZ boundary is nothing but the Bragg's reflection. The wave will be Bragg reflected only if the Bragg reflection condition is satisfied, that is,

$$2d\sin\theta = p\lambda \tag{12.127}$$

Here p is the order of reflection. For normal incidence the above equation reduces to

$$2d = p\,\lambda \tag{12.128}$$

In the case of a linear monatomic lattice, d = a and, therefore, the above equation gives

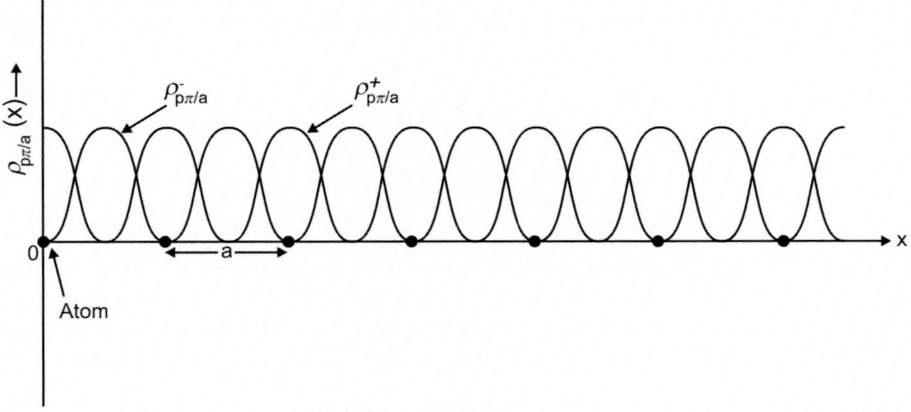

FIG. 12.6 The electron probability densities $\rho^{+}_{p\pi/a}$ and $\rho^{-}_{p\ \pi/a}$ for the two types of standing waves formed in a linear lattice in the nearly free-electron approximation.

$$\lambda = \frac{2a}{p} \tag{12.129}$$

From the definition of wave vector $k(=2\pi/\lambda)$ one gets

$$k = \frac{p\pi}{a} \tag{12.130}$$

which defines the boundary of the pBZ. Therefore, a travelling plane wave with wave vector $k = \pm p\pi/a$ is reflected back at the BZ boundary forming the standing wave. In other words, one can say that in a linear lattice the waves, corresponding to $k = \pm p\pi/a$, reflected from one atom interfere constructively with those reflected from the adjacent atoms. The energies of the two standing waves are different, causing a finite energy gap at the BZ boundary (see Fig. 12.5). The first energy gap occurs at $k = \pm\pi/a$, corresponding to the 1BZ boundary, while the others occur at the boundaries of the higher-order BZs.

The number of electron states in a band can be found using the plane wave function. From the periodicity of the wave function defined by Eq. (12.119), the allowed values of k in a one-dimensional solid are given by

$$k = 0, \pm\frac{2\pi}{L}, \pm\frac{4\pi}{L}, \ldots, \pm\frac{N\pi}{L} \tag{12.131}$$

The series has been cut at $N\pi/L$ as it gives the boundary of the 1BZ. According to Eq. (12.131), in a wave vector of length of $2\pi/L$ in k-space, there is one energy state. Hence the total number of electron states in one band is given by

$$\int_{-\pi/a}^{\pi/a} \frac{L}{2\pi} dk = \frac{L}{2\pi}\frac{2\pi}{a} = N \tag{12.132}$$

Therefore, the total number of states in one band is equal to the number of primitive cells or the number of lattice points in the solid, which is also equal to the number of k-points in one BZ. One should note that each primitive cell contributes one independent value of k to one energy band. If account is also taken of the spin orientation, then there are 2N independent states in a band. We can remark here that if in a linear solid the atoms are monovalent, then the valence band will be half filled with a total of N electrons at absolute zero. But if the atoms are divalent, then the valence band will be completely filled with 2N electrons.

Problem 12.1

Let the atomic potential $V^a(r)$ seen by an electron be represented by a Coulomb potential of the form

$$V^a(r) = -\frac{Ze^2}{r}$$

Find the Fourier transform $V_{G_p}{}^a$ of the atomic potential.

12.4 DIFFERENT ENERGY ZONE SCHEMES

The electronic energy bands in a crystalline solid calculated either in the Kronig-Penney model or in the nearly free-electron approximation possess the same main features as shown in Figs. 12.3 and 12.5. The energy bands are represented in three zone schemes as described below.

12.4.1 Extended Zone Scheme

The representation of the nondegenerate energy bands in different BZs, as shown in Fig. 12.5, is usually called the extended zone scheme. It is an actual representation of the energy bands in which the first band lies in the 1BZ, the second band lies in the 2BZ, and so on. Further, the value of the energy increases with an increase in the order of the band or the order of the BZ.

12.4.2 Periodic Zone Scheme

The various physical properties of crystalline solids, especially the electronic energy bands, show periodic behavior due to the periodic nature of the Bloch wave functions. The Bloch wave function is defined as

$$|\psi_{\mathbf{k}'}(\mathbf{r})\rangle = e^{\imath \mathbf{k}'\cdot\mathbf{r}} u_{\mathbf{k}'}(\mathbf{r})$$

If we substitute $\mathbf{k}' = \mathbf{k} + \mathbf{G}_p$ into the equation above, then

$$\begin{aligned} |\psi_{\mathbf{k}'}(\mathbf{r})\rangle &= e^{\imath \mathbf{k}\cdot\mathbf{r}} \left[e^{\imath \mathbf{G}_p\cdot\mathbf{r}} u_{\mathbf{k}+\mathbf{G}_p}(\mathbf{r}) \right] \\ &= e^{\imath \mathbf{k}\cdot\mathbf{r}} u_{\mathbf{k}}(\mathbf{r}) = |\psi_{\mathbf{k}}(\mathbf{r})\rangle \end{aligned} \tag{12.133}$$

where

$$u_{\mathbf{k}}(\mathbf{r}) = e^{\imath \mathbf{G}_p\cdot\mathbf{r}} u_{\mathbf{k}+\mathbf{G}_p}(\mathbf{r}) \tag{12.134}$$

is a periodic function. Eq. (12.134) shows that the values of $\mathbf{k}$ are not uniquely defined. According to Eq. (12.133) the Bloch wave function gets repeated after every reciprocal lattice vector, as is $E_{\mathbf{k}}$: an energy band is periodic with $\mathbf{G}_p$ as the periodicity, that is,

$$E_{\mathbf{k}} = E_{\mathbf{k}+\mathbf{G}_p} \tag{12.135}$$

The above physical property has also been proved in the nearly free-electron theory. Therefore, every energy band in the extended zone scheme can be repeated in all the BZs. Such a representation of the energy bands in a one-dimensional solid is shown in Fig. 12.7 and is usually called the periodic zone scheme.

12.4.3 Reduced Zone Scheme

In the periodic zone scheme, the representation of each energy band is translated to different BZs by adding suitable reciprocal lattice vectors. The reverse can also be done, that is, one can bring all of the energy bands to the 1BZ by adding or subtracting suitable reciprocal lattice vectors. Such a representation of the bands is called the reduced zone scheme. Fig. 12.8 shows the energy bands of Fig. 12.5 in the reduced zone scheme. This scheme is widely used in the literature as one can represent a number of bands in a compact way.

From Eqs. (12.89), (12.90) it is evident that the electron energy bands depend on two physical quantities, the crystal potential $V(\mathbf{r})$ and the electron wave function $|\psi_{\mathbf{k}}(\mathbf{r})\rangle$. Therefore, the use of different wave functions and crystal potentials yield different methods for determining the energy bands, such as the orthogonalized plane wave method and the augmented plane wave method. Another extreme state occurs when the electrons are considered tightly bound to the nucleus. The determination of energy bands in the tight-binding approximation is of interest here as the Schrodinger wave equation in this approximation can be solved analytically.

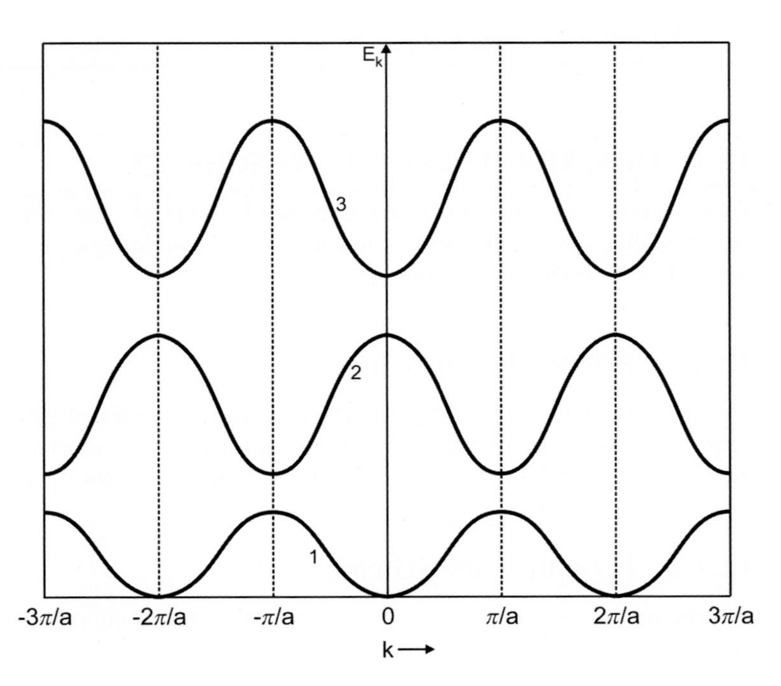

FIG. 12.7 The band energy E_k as a function of wave vector k for the first three bands, labeled as 1, 2, and 3, in all of the BZs of the linear lattice (periodic zone scheme) in the nearly free-electron theory.

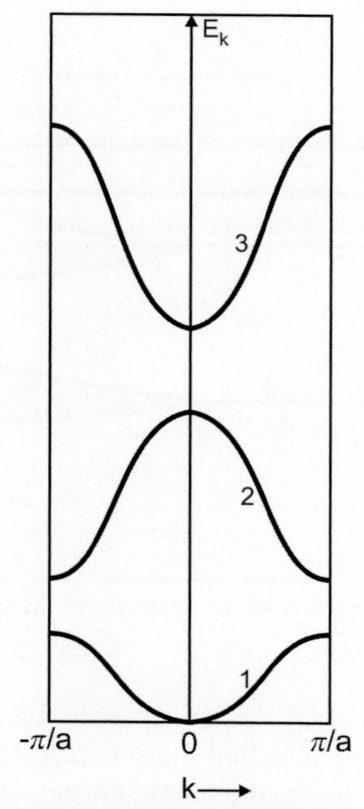

FIG. 12.8 The band energy $E_\mathbf{k}$ as a function of wave vector k for the first three bands, labeled as 1, 2, and 3, in the 1BZ of a linear lattice (reduced zone scheme) in the nearly free-electron theory.

Problem 12.2

Let the atomic potential $V^a(\mathbf{r})$ be represented by a screened Coulomb potential of the type

$$V^a(\mathbf{r}) = -\frac{Ze^2}{\varepsilon r}$$

where ε is the dielectric screening constant. Prove that the Fourier transform $V_{\mathbf{G}_p}{}^a$ of this potential is given by

$$V^a_{\mathbf{G}_p} = -\frac{4\pi Ze^2}{\varepsilon V_0 G_p^2}$$

Problem 12.3

Draw the free-electron energy bands in the reduced band scheme.

12.5 TIGHT-BINDING THEORY

An atom is associated with a localized wave function, which has maximum amplitude and, hence, maximum probability density, at the atomic position. The wave functions of two neutral hydrogen (H) atoms separated by a large distance do not overlap (Fig. 4.8A). As the H atoms are brought closer, their wave functions start to overlap and so do their charge distributions. The overlap can be described by a linear combination of the wave functions as follows:

$$|\psi_+\rangle = |\psi_A\rangle + |\psi_B\rangle \tag{12.136}$$

$$|\psi_-\rangle = |\psi_A\rangle - |\psi_B\rangle \tag{12.137}$$

The wave functions for the above combinations are shown in Fig. 4.8B and C. Each combination shares electrons equally between the two protons. An electron in the state $|\psi_A\rangle + |\psi_B\rangle$ will possess somewhat lower energy than in the state $|\psi_A\rangle - |\psi_B\rangle$ for the following reason: In the state $|\psi_A\rangle + |\psi_B\rangle$, an electron spends part of its time in the region midway between the protons and is under the influence of the finite attractive potential of both protons, thereby increasing the binding energy of the state. But in the state $|\psi_A\rangle - |\psi_B\rangle$, the potential is zero midway between the protons and, hence, the extra contribution to the binding energy does not appear. We know that for greater binding, the energy eigenvalue is

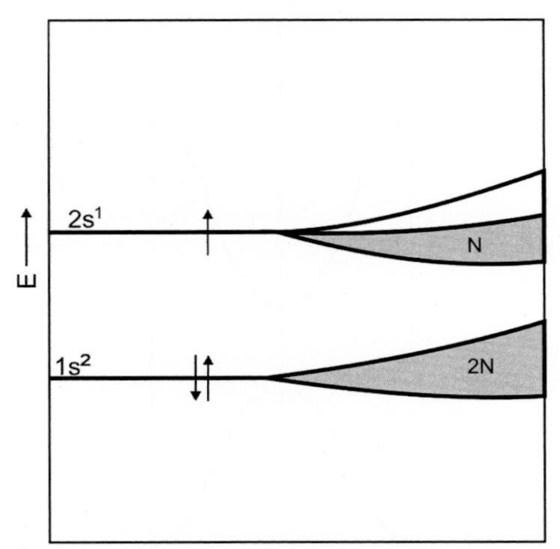

FIG. 12.9 The electronic band structure of a $_3Li^7$ atom. The left-hand side of the figure shows the sharp energy states of a $_3Li^7$ atom, while the right side shows the energy bands in a solid made of $_3Li^7$ atoms.

lower and, hence, the state is more stable. So $|\psi_A\rangle + |\psi_B\rangle$ represents a more stable state as compared with the state $|\psi_A\rangle - |\psi_B\rangle$. It is for this reason that $|\psi_A\rangle + |\psi_B\rangle$ is called the *bonding state*, while $|\psi_A\rangle - |\psi_B\rangle$ is called the *antibonding* state. Thus, as the two atoms are brought closer, the single energy state is split up into two states $|\psi_A\rangle + |\psi_B\rangle$ and $|\psi_A\rangle - |\psi_B\rangle$ with different energies. In general, when N atoms are brought closer to form a solid, each energy level of an isolated atom is split into N closely spaced levels, which collectively form an energy band. Thus, all of the energy levels of an atom become energy bands in a solid. In other words, a state with quantum number n of a free atom is spread out into a band of energies in a solid. The s-, p-, d-, and higher states of an atom form energy bands and the width of these bands is proportional to the strength of the overlap interaction between the neighboring atoms.

Consider the case of a lithium ($_3Li^7$) atom in the ground state. The sharp energy levels of a free $_3Li^7$ atom are shown in the left part of Fig. 12.9. In the $1s^2$ state of a $_3Li^7$ atom, there are two electrons with opposite spins. Therefore, the $1s^2$ state forms a degenerate state and the degeneracy can be broken if a magnetic field is applied. The $2s^1$ state contains only a single electron. Let there be a number N of $_3Li^7$ atoms with no interaction between them. In this case, the $1s^2$ state each independent $_3Li^7$ atom has the same energy and, hence, is N-fold degenerate. Similarly, the $2s^1$ state of the independent atoms is also N-fold degenerate. As the atoms are brought closer, their charge distributions start to overlap and the interaction energy comes into play. The N-fold degeneracy is broken in each of the $1s^2$ and $2s^1$ states due to the Pauli exclusion principle and they get split into N states. The $1s^2$ band will have 2N electrons (each atom contributes two electrons) and will be completely filled, while the $2s^1$ band has only got N electrons and is half filled (see Fig. 12.9).

The nature of the energy bands of electrons tightly bound to the nuclei of atoms in a solid form another limiting case. The wave function of an electron in a free atom is usually called an atomic orbital and the atomic orbitals with different energies and belonging to different atoms are orthonormal. The Bloch wave function, which fully describes an electron in the periodic field of the crystal, can be constructed by taking the linear combination of atomic orbitals (LCAO) belonging to different atoms. This is called the LCAO method and is more suitable for electrons in the inner shells of an atom.

Let $V^a(\mathbf{r})$ be the self-consistent atomic potential experienced by an atomic electron at a distance $\mathbf{r}$ from the nucleus of the atom to which it belongs (*dashed line* in Fig. 12.1). Let the wave function of the electron in a free atom be represented by the atomic orbital $|\psi^a(\mathbf{r})\rangle$ with energy E_0. The Schrodinger wave equation for the atomic electron is given by

$$\widehat{H}_a |\psi^a(\mathbf{r})\rangle = E_0 |\psi^a(\mathbf{r})\rangle \tag{12.138}$$

where $\widehat{H}_a$ is the Hamiltonian of an atomic electron and is given by

$$\widehat{H}_a = -\frac{\hbar^2}{2m_e}\nabla^2 + V^a(\mathbf{r}) \tag{12.139}$$

Suppose a number N of similar atoms are brought together to form a crystal. The crystal potential is obtained by the superposition of the atomic potentials and is shown by the continuous curve in Fig. 12.1. Assume that the origin is at the position of a particular atom, then $\mathbf{R}_n$ gives the position of the nth atom. In the tight-binding approximation, it is assumed that an electron belonging to the nth atom is only slightly influenced by the presence of other atoms. In this approximation, the wave function of the electron with position vector $\mathbf{r}$ and belonging to the nth atom (Fig. 12.10) is given by the atomic orbital

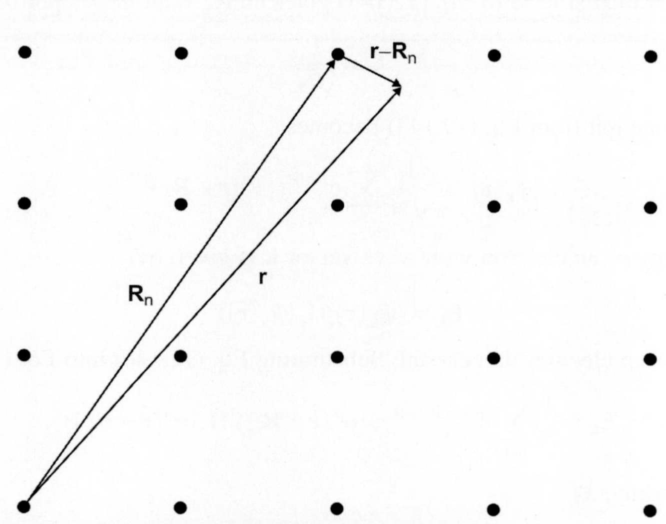

FIG. 12.10 Showing the position vector **r** of an electron belonging to an atom at position $\mathbf{R}_n$.

$|\psi^a(\mathbf{r}-\mathbf{R}_n)\rangle$. The energy of the electron is very close to E_0 and, ideally speaking, it is taken to be E_0. Now the electronic wave function with wave vector **k** in the crystal is the linear combination of the form

$$|\psi_{\mathbf{k}}(\mathbf{r})\rangle = \sum_n C_n(\mathbf{k})\,|\psi^a(\mathbf{r}-\mathbf{R}_n)\rangle \tag{12.140}$$

Here **r** lies very close to $\mathbf{R}_n$, therefore, all the contributions to the sum will be small except for that from $|\psi^a(\mathbf{r}-\mathbf{R}_n)\rangle$ with the smallest value of $\mathbf{r}-\mathbf{R}_n$. In a crystalline solid an electron experiences a periodic potential, so the wave function must be of the form of the Bloch function, which restricts the value of $C_n(\mathbf{k})$ to $e^{\imath\mathbf{k}\cdot\mathbf{R}_n}$. Therefore, the wave function given by Eq. (12.140) becomes

$$|\psi_{\mathbf{k}}(\mathbf{r})\rangle = \sum_n e^{\imath\mathbf{k}\cdot\mathbf{R}_n}\,|\psi^a(\mathbf{r}-\mathbf{R}_n)\rangle \tag{12.141}$$

Eq. (12.141) satisfies the properties of the Bloch functions. This can be realized by translating the above wave function by a lattice vector $\mathbf{R}_m$, that is,

$$|\psi_{\mathbf{k}}(\mathbf{r}+\mathbf{R}_m)\rangle = \sum_n e^{\imath\mathbf{k}\cdot\mathbf{R}_n}|\psi^a(\mathbf{r}+\mathbf{R}_m-\mathbf{R}_n)\rangle$$

which, after simplification, can be written as

$$|\psi_{\mathbf{k}}(\mathbf{r}+\mathbf{R}_m)\rangle = e^{\imath\mathbf{k}\cdot\mathbf{R}_m}\,|\psi_{\mathbf{k}}(\mathbf{r})\rangle \tag{12.142}$$

Eq. (12.142) is nothing but the Bloch condition. The wave function given by Eq. (12.141) can be normalized to unity. Let C be the normalizing factor, then

$$|\psi_{\mathbf{k}}(\mathbf{r})\rangle = C\sum_n e^{\imath\mathbf{k}\cdot\mathbf{R}_n}\,|\psi^a(\mathbf{r}-\mathbf{R}_n)\rangle \tag{12.143}$$

The normalization condition demands that

$$\begin{aligned}\langle\psi_{\mathbf{k}}(\mathbf{r})|\psi_{\mathbf{k}}(\mathbf{r})\rangle = 1 &= C^2\left[\sum_{n,m} e^{\imath\mathbf{k}\cdot(\mathbf{R}_n-\mathbf{R}_m)}\langle\psi^a(\mathbf{r}-\mathbf{R}_m)|\psi^a(\mathbf{r}-\mathbf{R}_n)\rangle\right]\\ &= C^2\left[\sum_n \langle\psi^a(\mathbf{r}-\mathbf{R}_n)|\psi^a(\mathbf{r}-\mathbf{R}_n)\rangle\right]\\ &\quad + C^2\left[\sum_{n,m\,(n\neq m)} e^{\imath\mathbf{k}\cdot(\mathbf{R}_n-\mathbf{R}_m)}\langle\psi^a(\mathbf{r}-\mathbf{R}_m)|\psi^a(\mathbf{r}-\mathbf{R}_n)\rangle\right]\end{aligned} \tag{12.144}$$

As the electrons are assumed to be tightly bound to the nucleus, their wave functions are highly localized about the nucleus. Therefore, the overlap of wave functions of electrons belonging to different nuclei is negligible. To the lowest order approximation, neglecting the overlap of electronic wave functions, the second term on the right-hand side in Eq. (12.144) becomes zero and each integral of first term of Eq. (12.144) gives unity. With these approximations, Eq. (12.144) gives

$$C = \frac{1}{\sqrt{N}} \tag{12.145}$$

Hence the normalized wave function from Eq. (12.143) becomes

$$|\psi_{\mathbf{k}}(\mathbf{r})\rangle = \frac{1}{\sqrt{N}} \sum_{n} e^{\imath \mathbf{k} \cdot \mathbf{R}_n} |\psi^a(\mathbf{r} - \mathbf{R}_n)\rangle \tag{12.146}$$

The expectation value of energy of an electron with wave vector $\mathbf{k}$ is given by

$$E_{\mathbf{k}} = \langle \psi_{\mathbf{k}}(\mathbf{r}) | \widehat{H}_e | \psi_{\mathbf{k}}(\mathbf{r}) \rangle \tag{12.147}$$

Here, $\widehat{H}_e$ is the Hamiltonian of an electron in a crystal. Substituting Eq. (12.146) into Eq. (12.147), one can write

$$E_{\mathbf{k}} = \frac{1}{N} \sum_{n,m} e^{\imath \mathbf{k} \cdot (\mathbf{R}_n - \mathbf{R}_m)} \langle \psi^a(\mathbf{r} - \mathbf{R}_m) | \widehat{H}_e | \psi^a(\mathbf{r} - \mathbf{R}_n) \rangle \tag{12.148}$$

The Hamiltonian $\widehat{H}_e$ can be written as

$$\begin{aligned} \widehat{H}_e &= -\frac{\hbar^2}{2m_e} \nabla^2 + V(\mathbf{r}) \\ &= -\frac{\hbar^2}{2m_e} \nabla^2 + V^a(\mathbf{r} - \mathbf{R}_n) + V'(\mathbf{r} - \mathbf{R}_n) \end{aligned} \tag{12.149}$$

where

$$V'(\mathbf{r} - \mathbf{R}_n) = V(\mathbf{r}) - V^a(\mathbf{r} - \mathbf{R}_n) \tag{12.150}$$

$V^a(\mathbf{r} - \mathbf{R}_n)$ is the potential seen by an electron at position $\mathbf{r}$ due to the atom at position $\mathbf{R}_n$. $V'(\mathbf{r} - \mathbf{R}_n)$ is the difference of self-consistent potential seen by an electron when all atoms are present and the atomic potential due to the single atom at position $\mathbf{R}_n$. In other words, $V'(\mathbf{r} - \mathbf{R}_n)$ represents the potential seen by the electron at $\mathbf{r}$ resulting from all of the atoms except the one located at position $\mathbf{R}_n$. Therefore, $V'(\mathbf{r} - \mathbf{R}_n)$ is a weak potential as the overlap is negligible in the LCAO approximation and can be treated as a perturbation. Further, $V'(\mathbf{r} - \mathbf{R}_n)$ is a negative quantity as the crystal potential is smaller than the atomic potential (see Fig. 12.1). Substituting Eq. (12.149) into Eq. (12.148) and using Eq. (12.138), we get

$$E_{\mathbf{k}} = E_0 - \frac{1}{N} \sum_{n,m} e^{\imath \mathbf{k} \cdot (\mathbf{R}_n - \mathbf{R}_m)} \gamma_{mn} \tag{12.151}$$

where

$$\gamma_{mn} = -\langle \psi^a(\mathbf{r} - \mathbf{R}_m) | V'(\mathbf{r} - \mathbf{R}_n) | \psi^a(\mathbf{r} - \mathbf{R}_n) \rangle \tag{12.152}$$

The summation in Eq. (12.151) can be split into two parts as

$$E_{\mathbf{k}} = E_0 - \frac{1}{N} \sum_{n} \gamma_{nn} - \frac{1}{N} \sum_{n,m\,(n \neq m)} e^{\imath \mathbf{k} \cdot (\mathbf{R}_n - \mathbf{R}_m)} \gamma_{mn} \tag{12.153}$$

In order to solve the above equation, we make some simplifying approximations. First, we assume that only the 1NN interactions are significant and so we sum over the 1NNs in the last term of Eq. (12.153). Second, it is assumed that the atomic orbitals are spherically symmetric, that is, that they depend on the magnitude of $\mathbf{r} - \mathbf{R}_n$. In this approximation, all the integrals in the last term of Eq. (12.153) become equal and so this term is equal to the magnitude of the single integral multiplied by the number of the 1NNs. To simplify the notation, we write

$$\gamma_0 = \gamma_{nn} \text{ and } \gamma_{mn} = \gamma_1 \tag{12.154}$$

which allows us to write Eq. (12.153) as

$$E_{\mathbf{k}} = E_0 - \gamma_0 - \gamma_1 \sum_m e^{\imath \mathbf{k}\cdot(\mathbf{R}_n - \mathbf{R}_m)} \tag{12.155}$$

The quantities γ_0 and γ_1 are positive because $V'(\mathbf{r} - \mathbf{R}_n)$ is always negative. From Eq. (12.155) it is evident that the energy of an electron in a crystal differs from its energy in a free atom by a constant factor γ_0 plus a term that depends on the wave vector and the crystal structure. It is the last contribution in Eq. (12.155) that transforms the discrete energy levels into energy bands in a solid. Eq. (12.155) gives the general expression for energy in the tight-binding approximation and one can apply it to different structures.

12.5.1 Linear Monatomic Lattice

Consider a linear monatomic lattice along the x-direction with "a" as its periodicity. Therefore, an atom at $\mathbf{R}_n$ (assumed to be the origin) has two 1NNs with coordinates given by

$$\mathbf{R}_n - \mathbf{R}_m = (\pm a, 0, 0) \tag{12.156}$$

Substituting the coordinates of the 1NNs in Eq. (12.155), we get

$$E_{k_x} = E_0 - \gamma_0 - \gamma_1 \left(e^{\imath k_x a} + e^{-\imath k_x a}\right)$$

which can be written as

$$E_{k_x} = E_0 - \gamma_0 - 2\gamma_1 \cos k_x a \tag{12.157}$$

The parameters γ_0 and γ_1 are given by

$$\gamma_0 = -\langle \psi^a(x) | V'(x) | \psi^a(x) \rangle \tag{12.158}$$

$$\gamma_1 = -\langle \psi^a(x \pm a) | V'(x) | \psi^a(x) \rangle \tag{12.159}$$

The slope of the energy band is given by

$$\frac{dE_{k_x}}{dk_x} = 2\gamma_1 a \sin k_x a \tag{12.160}$$

The value of γ_1 depends on both the magnitude and sign of $|\psi^a(x)\rangle$ and $|\psi^a(x \pm a)\rangle$. First, we consider the case when both $|\psi^a(x)\rangle$ and $|\psi^a(x \pm a)\rangle$ have the same sign, which yields a positive value of γ_1. The energy band given by Eq. (12.157) is plotted in Fig. 12.11A in the 1BZ. In this band, the energy is minimum at $k_x = 0$ and maximum at $k_x = \pi/a$. As the value of

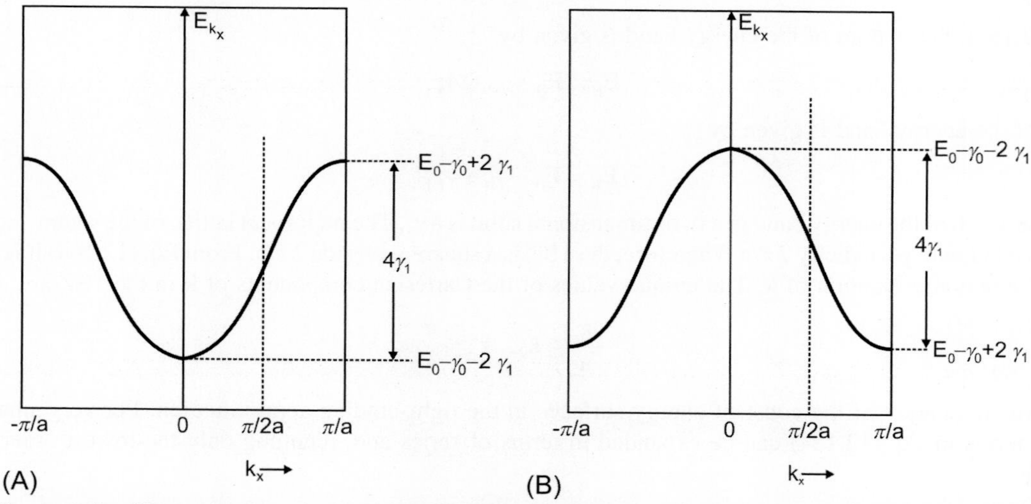

FIG. 12.11 The energy band E_{k_x} as a function of k_x for a monatomic linear lattice with periodicity "a" in the tight-binding approximation. In (A) the parameter γ_1 is positive, but in (B) γ_1 is negative.

cos $k_x a$ ranges from -1 to 1, therefore, the width of the energy band is $4\gamma_1$. In this case the energy is lower so it corresponds to the state given by Eq. (12.136). The details of the energy band can be investigated from Eqs. (12.157), (12.160). At small values of k_x, the cosine term of Eq. (12.157) can be expanded in a series to write

$$\begin{aligned} E_{k_x} &= E_0 - \gamma_0 - 2\gamma_1\left(1 - \frac{(k_x a)^2}{2} + \cdots\right) \\ &= E_0 - \gamma_0 - 2\gamma_1 + \gamma_1 a^2 k_x^2 \end{aligned} \tag{12.161}$$

At small values of k_x the band is a parabola with a positive slope, which increases with an increase in k_x, but becomes maximum at $k_x = \pi/2a$. After this, the slope becomes negative and decreases, going to zero at the BZ boundary with $k_x = \pi/a$ (see Fig. 12.11A). This shows that there is a point of inflection in the energy curve at the midpoint of the 1BZ with $k_x = \pi/2a$. Further, the energy E_{k_x} is periodic with a periodicity of $2\pi/a$. Hence the unique values of k_x are defined only in the 1BZ. It is noteworthy that apart from the position of zero of energy, Eq. (12.161) is similar to the free electron energy. To point out the similarity we sometimes write

$$\gamma_1 a^2 = \frac{\hbar^2}{2m_e^*} \tag{12.162}$$

where m_e^* is the effective mass of the electron near $k_x = 0$.

In the second case, $|\psi^a(x)\rangle$ and $|\psi^a(x \pm a)\rangle$ have opposite signs, so γ_1 becomes negative. In this case the parabola is inverted, that is, the energy is maximum at $k_x = 0$ and minimum at the BZ boundary with $k_x = \pi/a$ (see Fig. 12.11B). Here the energy band corresponds to the state given by Eq. (12.137), which has an energy a little more than that of the state described by Eq. (12.136). There is an important difference between the bands obtained in the nearly free-electron and tight-binding approximations. In the nearly free-electron approximation, the band is almost a parabola except near the BZ boundary where it is an inverted parabola. But in the tight-binding approximation, the band has the same symmetry around the midpoint, at which it exhibits a point of inflection.

12.5.2 Two-Dimensional Square Lattice

In a square lattice the atoms are arranged on a square matrix with "a" as the periodicity. The 1NNs along the x- and y-direction are $(\pm a, 0, 0)$ and $(0, \pm a, 0)$, respectively. So, one can write

$$\mathbf{R}_n - \mathbf{R}_m = (\pm a, 0, 0), (0, \pm a, 0) \tag{12.163}$$

Substituting Eq. (12.163) into Eq. (12.155) and simplifying, one gets

$$E_{\mathbf{k}} = E_0 - \gamma_0 - 2\gamma_1\left[\cos k_x a + \cos k_y a\right] \tag{12.164}$$

From Eq. (12.164), the bottom of the energy band is given by

$$E_{\mathbf{k}} = E_0 - \gamma_0 - 4\gamma_1 \tag{12.165}$$

and the top of the energy band is given by

$$E_{\mathbf{k}} = E_0 - \gamma_0 + 4\gamma_1 \tag{12.166}$$

Therefore, the width of the energy band in a two-dimensional solid is $8\gamma_1$. The reciprocal lattice of the square lattice is again a square lattice but with periodicity $2\pi/a$. Therefore, the 1BZ is a square with side $2\pi/a$. From Eq. (12.164) it is evident that the energy is a periodic function of **k**. The unique values of the Cartesian components of **k** in the 1BZ are given as

$$-\frac{\pi}{a} \le k_x,\ k_y \le \frac{\pi}{a} \tag{12.167}$$

Let us examine the nature of the constant energy surfaces in the tight-binding approximation. For very small values of **k** the cosine terms in Eq. (12.164) can be expanded in terms of series and, retaining only the lowest order terms in k, one can write

$$E_{\mathbf{k}} = E_0 - \gamma_0 - 4\gamma_1 + \gamma_1 a^2 k^2 \tag{12.168}$$

Again, near the bottom of the band, the energy bands are parabolic, as in the case of free electrons. Therefore, for small **k** values the constant energy surfaces are circular in nature. In this region the electrons can be assumed to be free with an effective mass m_e^* From Eq. (12.168) one can write

$$m_e^* = \frac{\hbar^2}{2\gamma_1 a^2} \tag{12.169}$$

which is the same result as in the case of a one-dimensional solid. In the tight-binding approximation, the value of m_e^* is large because of the small value of γ_1, which is a measure of the overlap of the wave functions of the 1NNs. In other words, the electrons cannot move freely in the bands with small bandwidths. At the top of the first band (1BZ boundary), $k_x = k_y = \pm\pi/a$ and this gives

$$\cos k_x a = \cos k_y a = -1 \tag{12.170}$$

In the reduced zone scheme, the corners correspond to the states at the top of the band. In the vicinity of these corners one can expand the cosine term in a series. The expansion of the $\cos k_x a$ term around a corner can be written as

$$\cos k_x a = \cos\left[\pi - \left(\frac{\pi}{a} - k_x\right)a\right] = \cos\left(\pi - k_x' a\right) = -\cos k_x' a \tag{12.171}$$

where

$$k_x' = \frac{\pi}{a} - k_x \tag{12.172}$$

Here k_x' is measured relative to the corner. Therefore, for small values of k_x', one can expand the cosine term as

$$\cos k_x a = -\cos k_x' a = -1 + \frac{(k_x' a)^2}{2} + \cdots \tag{12.173}$$

Similarly, one can expand $\cos k_y a$. Substituting these expansions into Eq. (12.164), one can write

$$E_{\mathbf{k}'} = E_0 - \gamma_0 + 4\gamma_1 - \gamma_1 a^2 k'^2 \tag{12.174}$$

where

$$k'^2 = k_x'^2 + k_y'^2 \tag{12.175}$$

Eq. (12.174) shows that the energy bands near the top are also parabolic in shape, giving rise to circular constant energy surfaces. The nature of the bands away from the bottom and top of the band can also be studied. Suppose we are interested in the constant energy bands having energy

$$E_{\mathbf{k}} = E_0 - \gamma_0 \tag{12.176}$$

Then, from Eq. (12.164), we have

$$\cos k_x a + \cos k_y a = 0$$

From the above equation k_x can be written in terms of k_y and vice versa as follows:

$$\cos k_x a = -\cos k_y a = \cos\left\{\pm\left(\pi - k_y a\right)\right\} \tag{12.177}$$

$$\cos k_y a = \cos\{\pm(\pi - k_x a)\} \tag{12.178}$$

Eqs. (12.177), (12.178) yield

$$k_x = \pm\frac{\pi}{a} \mp k_y \tag{12.179}$$

$$k_y = \pm\frac{\pi}{a} \mp k_x \tag{12.180}$$

Note that Eq. (12.179) represents a straight line passing through the points $(\pi/a, 0)$ and $(0, \pi/a)$ and the second set of points are $(-\pi/a, 0)$ and $(0, \pi/a)$. Similarly, Eq. (12.180) represents straight lines passing through the two sets of points $(0, \pi/a)$, $(\pi/a, 0)$ and $(0, -\pi/a)$, $(\pi/a, 0)$. With the knowledge of these points, one can draw constant energy straight lines having value $E_0 - \gamma_0$.

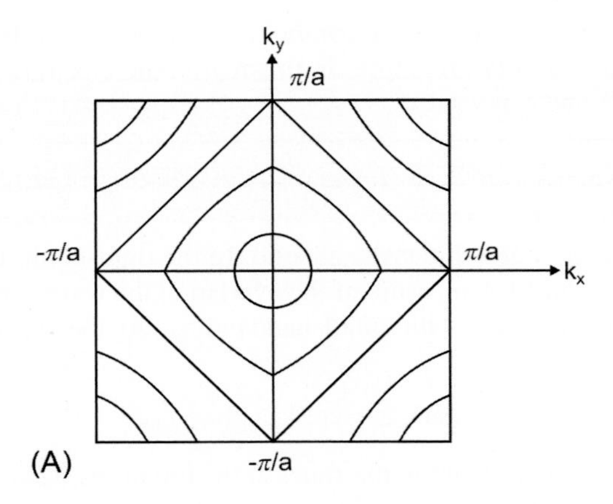

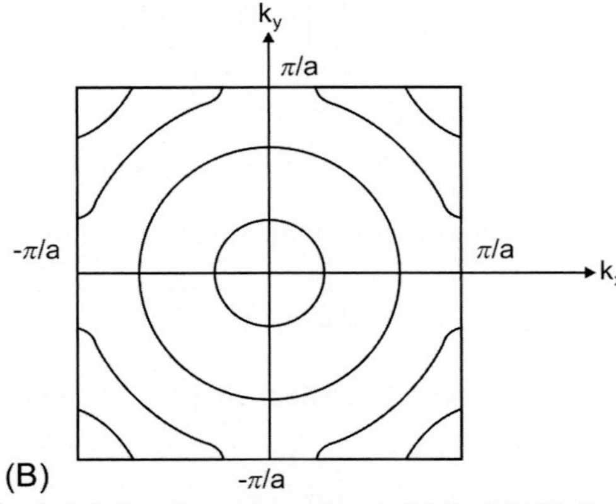

FIG. 12.12 The constant energy surfaces in the (k_x, k_y) plane for a square lattice in (A) the tight-binding approximation; and (B) the nearly free-electron approximation. *(Modified from Dekker, A. J. (1971). Solid state physics (p. 262). London: Macmillan Press.)*

The constant energy curves in the 1BZ of the square lattice are shown in Fig. 12.12A in the tight-binding approximation. This clearly shows that the energy bands are circular near the bottom and top of the band, but become flat as we move toward the center of the band. Fig. 12.12B shows the constant energy curves obtained in the nearly free-electron theory. It is noteworthy that in the nearly free-electron approximation, the k^2 dependence of the bands (and hence the circular constant energy lines) extend to much larger values of the wave vector than in the tight-binding approximation. The circular constant energy lines near the top of the band (near the corners of the BZ) extend to only small values of the wave vectors.

12.5.3 Three-Dimensional sc Lattice

In a solid with sc structure, there are six 1NNs with coordinates given by

$$\mathbf{R}_n - \mathbf{R}_m = (\pm a, 0, 0), (0, \pm a, 0), (0, 0, \pm a) \tag{12.181}$$

Substituting the coordinates of 1NNs from Eq. (12.181) into Eq. (12.155), one can write

$$E_{\mathbf{k}} = E_0 - \gamma_0 - 2\gamma_1 \left[\cos k_x a + \cos k_y a + \cos k_z a\right] \tag{12.182}$$

The width of the energy band in a three-dimensional solid is $12\gamma_1$ and all the energy levels are contained in it. In the tight-binding approximation, the value of γ_1 is small due to the small overlap of the wave functions, which yields narrow bands (bands with small bandwidths). The inner electron energy levels give rise to very narrow bands in a solid because of further decreases in the overlap. The reciprocal lattice of an sc structure with periodicity "a" is again an sc lattice but with periodicity $2\pi/a$. Therefore, the 1BZ is a cube of edge $2\pi/a$. The energy is a periodic function of **k** (Eq. 12.182) and its values are uniquely defined in the 1BZ as

$$-\frac{\pi}{a} \le k_x, k_y, k_z \le \frac{\pi}{a} \tag{12.183}$$

The nature of the constant energy surfaces in a three-dimensional solid can be examined in the same way as in a two-dimensional solid. For very small values of **k**, one can expand the cosine terms in series and can retain only the lowest order terms in k to write

$$E_{\mathbf{k}} = E_0 - \gamma_0 - 6\gamma_1 + \gamma_1 a^2 k^2 \tag{12.184}$$

So, near the bottom of the band, the energy bands are parabolic, as in the case of free electrons, and the constant energy surfaces are spherical. In this region electrons can be assumed to be free with an effective mass m_e^* given by (Eq. 12.184)

$$m_e^* = \frac{\hbar^2}{2\gamma_1 a^2} \tag{12.185}$$

The value of m_e^* is large because of the small value of γ_1. At the top of the first band, $k_x = k_y = k_z = \pm\pi/a$, giving

$$\cos k_x a = \cos k_y a = \cos k_z a = -1 \tag{12.186}$$

By defining the Cartesian components of the wave vector relative to the corners of the 1BZ, it can easily be proved that

$$E_{\mathbf{k}'} = E_0 - \gamma_0 + 6\gamma_1 - \gamma_1 a^2 k'^2 \tag{12.187}$$

where

$$k'^2 = k_x'^2 + k_y'^2 + k_z'^2 \tag{12.188}$$

and

$$k_x' = \frac{\pi}{a} - k_x \tag{12.189}$$

$$k_y' = \frac{\pi}{a} - k_y \tag{12.190}$$

$$k_z' = \frac{\pi}{a} - k_z \tag{12.191}$$

Eq. (12.187) shows that near the top the energy bands are also parabolic in shape, giving rise to spherical constant energy surfaces. The nature of all of the bands and the constant energy surfaces in an sc solid can be calculated in the xy-, yz-, and zx-planes of the 1BZ in exactly the same manner as in the two-dimensional solid. It is found that the energy bands are parabolic near the bottom and top of the band, but become flat as we move toward the midpoint of the band. In other words, one can say that the constant energy surfaces are spherical near the center and the corners of the 1BZ and become flat as one moves away from them.

Problem 12.4

Show that in the tight-binding approximation, the energy $E_{\mathbf{k}}$ for

(a) a bcc lattice is given by

$$E_{\mathbf{k}} = E_0 - \gamma_0 - 8\gamma_1 \cos(k_x a)\cos(k_y a)\cos(k_z a) \tag{12.192}$$

(b) an fcc lattice is given by

$$E_{\mathbf{k}} = E_0 - \gamma_0 - 4\gamma_1\left[\cos(k_x a)\cos(k_y a) + \cos(k_y a)\cos(k_z a) + \cos(k_z a)\cos(k_x a)\right] \tag{12.193}$$

Here 2a is the cube edge. Also show that for small values of **k** the energy is proportional to k^2. Further, discuss the shape of the constant energy surfaces in the **k**-space.

The real wave function in a crystalline solid is shown in Fig. 12.13. It is evident from the figure that the real wave function is oscillatory in nature near an atom (more precisely within the ion core) due to the localized core states, but behaves like a plane wave midway between the atoms. This shows that an electron is neither free nor tightly bound. Further, Fig. 12.13 suggests that the conduction electron wave function can be obtained by taking a linear combination of the plane waves and core states. There are different ways to have a linear combination of the plane waves and the core states, thus yielding different methods for the determination of the energy bands in solids.

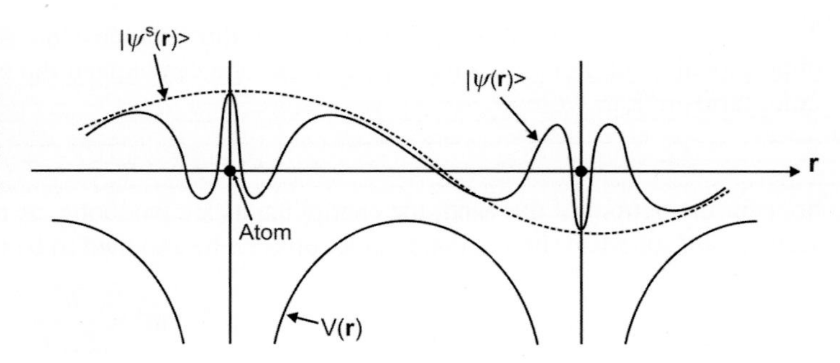

FIG. 12.13 The Bloch wave function $|\psi_{\mathbf{k}}(\mathbf{r})\rangle$ oscillates near the atomic sites due to the presence of localized core states, while, between the atomic sites, the wave function behaves like a plane wave. Near the atomic sites the crystal potential V(**r**) is strong and attractive, but between the atomic sites the potential is weak and flat in nature. The dashed line shows the pseudowave function $|\psi^s(\mathbf{r})\rangle$.

12.6 ORTHOGONALIZED PLANE WAVE (OPW) METHOD

In a metallic solid there are conduction states, represented by plane waves $|\mathbf{k}\rangle$ and core states $|\phi_c(\mathbf{r})\rangle$. The subscript c denotes the quantum numbers along with the position of the ion. The core states are orthogonal to each other, which is represented mathematically as

$$\langle\phi_c(\mathbf{r})|\phi_{c'}(\mathbf{r})\rangle = \delta_{cc'} \tag{12.194}$$

Both the conduction and core states are eigenfunctions of the same one-electron Hamiltonian $\widehat{H}_e$, so

$$\widehat{H}_e|\mathbf{k}\rangle = E_{\mathbf{k}}|\mathbf{k}\rangle \tag{12.195}$$

$$\widehat{H}_e|\phi_c(\mathbf{r})\rangle = E_c|\phi_c(\mathbf{r})\rangle \tag{12.196}$$

where E_c is the energy of a core state in the crystalline solid, which is different from the core energy in an isolated ion. The real conduction electron wave function is constructed by taking a linear combination of the plane waves and core states in such a way that it is orthogonal to the core states. Such a conduction state is called an orthogonalized plane wave (OPW) and is defined as

$$|\mathbf{k}_{OPW}\rangle = |\mathbf{k}\rangle - \sum_c |\phi_c(\mathbf{r})\rangle\langle\phi_c(\mathbf{r})|\mathbf{k}\rangle \tag{12.197}$$

Eq. (12.197) can be written as

$$|\mathbf{k}_{OPW}\rangle = (1-P)|\mathbf{k}\rangle \tag{12.198}$$

where

$$P = \sum_c |\phi_c(\mathbf{r})\rangle\langle\phi_c(\mathbf{r})| \tag{12.199}$$

P is the projection operator onto the core states. Now it is straightforward to prove the orthogonality of the OPWs and the core states, that is,

$$\langle\mathbf{k}_{OPW}|\phi_c(\mathbf{r})\rangle = 0 \tag{12.200}$$

But the OPWs are not orthogonal to each other. One can easily prove that

$$\langle\mathbf{k}'_{OPW}|\mathbf{k}_{OPW}\rangle = \delta_{\mathbf{k}'\mathbf{k}} - \sum_c \langle\mathbf{k}'|\phi_c(\mathbf{r})\rangle\langle\phi_c(\mathbf{r})|\mathbf{k}\rangle \tag{12.201}$$

In a crystalline solid the general conduction electron wave function $|\psi(\mathbf{r})\rangle$ is taken as a linear combination of OPWs, that is,

$$|\psi(\mathbf{r})\rangle = \sum_{\mathbf{k}} a_{\mathbf{k}}|\mathbf{k}_{OPW}\rangle = (1-P)|\psi^s(\mathbf{r})\rangle \tag{12.202}$$

where

$$|\psi^s(\mathbf{r})\rangle = \sum_{\mathbf{k}} a_{\mathbf{k}}|\mathbf{k}\rangle \tag{12.203}$$

$|\psi^s(\mathbf{r})\rangle$ is a linear combination of plane waves and is called a pseudowave function. Due to the presence of core states, an OPW decreases much faster with an increase in the value of **k** and, therefore, gives a significant contribution only at small values of $\mathbf{k}$. $|\psi^s(\mathbf{r})\rangle$ is a smooth wave function and is shown in Fig. 12.13 by a dashed line. Note that $|\psi^s(\mathbf{r})\rangle$ is equal in magnitude to the true wave function (except possibly for normalization) outside the core because the operator P is zero there. So, the real wave function may be obtained simply by orthogonalizing the pseudowave function to the core states with the operator $1-P$ and then renormalizing it. The one-electron Schrodinger wave equation satisfied by $|\psi(\mathbf{r})\rangle$ is written as

$$\left[-\frac{\hbar^2}{2m_e}\nabla^2+V(\mathbf{r})\right]|\psi(\mathbf{r})\rangle=E|\psi(\mathbf{r})\rangle \tag{12.204}$$

Substituting Eq. (12.202) into Eq. (12.204) and rearranging the terms, we get

$$\left[-\frac{\hbar^2}{2m_e}\nabla^2+W(\mathbf{r})\right]|\psi^s(\mathbf{r})\rangle=E|\psi^s(\mathbf{r})\rangle \tag{12.205}$$

where

$$W(\mathbf{r})=V(\mathbf{r})+\sum_c(E-E_c)|\phi_c(\mathbf{r})\rangle\langle\phi_c(\mathbf{r})|=V(\mathbf{r})+\left(E-\widehat{H}_e\right)P \tag{12.206}$$

$W(\mathbf{r})$ is called the pseudopotential, which is nonlocal in nature, and Eq. (12.205) is called the pseudowave equation, with $|\psi^s(\mathbf{r})\rangle$ as the pseudowave function. In writing Eq. (12.205) the effect of orthogonality has been transferred from the wave function to the crystal potential. The energy of the core states E_c is always negative; therefore, the second term of $W(\mathbf{r})$ is always positive and represents the repulsive contribution to the potential. The repulsive term originates from the orthogonality condition, which is frequently ascribed to the Pauli exclusion principle. Therefore, the pseudopotential $W(\mathbf{r})$, which is the sum of attractive and repulsive contributions, is a weak potential. The noteworthy feature here is that Eq. (12.205) is the Schrodinger wave equation with a weak potential as in the nearly free-electron theory. Hence, perturbation theory can be applied to study the electronic properties of solids.

Problem 12.5: Nonuniqueness in Pseudowave Function

The pseudowave function $|\psi^s(\mathbf{r})\rangle$, given by Eq. (12.203), is the eigenfunction of the pseudowave equation given by Eq. (12.205). Add a linear combination of core states $|\phi_{c'}\rangle$ to $|\psi^s(\mathbf{r})\rangle$ to obtain a new pseudowave function $|\psi^{s'}(\mathbf{r})\rangle$ written as

$$\left|\psi^{s'}(\mathbf{r})\right\rangle=|\psi^s(\mathbf{r})\rangle+\sum_{c'}a_{c'}|\phi_{c'}\rangle \tag{12.207}$$

where $a_{c'}$ is a constant. Prove that $|\psi^{s'}(\mathbf{r})\rangle$ is also the eigenfunction of Eq. (12.205) with the same energy.

Problem 12.6

Operate $\langle\mathbf{k}'_{OPW}|$ from the left side on Eq. (12.205) and show that

$$\sum_{\mathbf{k}}a_{\mathbf{k}}\left[\left(\frac{\hbar^2k^2}{2m_e}-E\right)\left\{\delta_{\mathbf{k}'\mathbf{k}}-\sum_{c'}\langle\mathbf{k}'|\phi_{c'}(\mathbf{r})\rangle\langle\phi_{c'}(\mathbf{r})|\mathbf{k}\rangle\right\}+\langle\mathbf{k}'|W(\mathbf{r})|\mathbf{k}\rangle\right]=0 \tag{12.208}$$

From Eq. (12.206) it is evident that the pseudopotential $W(\mathbf{r})$ depends on energy E and vice versa. Therefore, $W(\mathbf{r})$ must be obtained self-consistently. Solving Eq. (12.205) or Eq. (12.208) one can obtain the energy bands in the OPW method. It has already been pointed out that the OPWs decay much faster with an increase in $\mathbf{k}$ value. Therefore, in this method, one uses either one OPW or two OPWs for the evaluation of the electronic energy bands.

Because of nonuniqueness in the pseudowave function there can be many forms of the pseudopotential. If in Eq. (12.206) the positive quantity $E-E_c$ is replaced by any function $f(E,c)$, then the pseudopotential $W(\mathbf{r})$ can be written as

$$W(\mathbf{r})=V(\mathbf{r})+\sum_c f(E,c)|\phi_c(\mathbf{r})\rangle\langle\phi_c(\mathbf{r})| \tag{12.209}$$

With this new pseudopotential, the pseudowave equation (12.205) becomes

$$\left[-\frac{\hbar^2}{2m_e}\nabla^2+V(\mathbf{r})\right]|\psi^s(\mathbf{r})\rangle+\sum_c f(E,c)\,|\phi_c(\mathbf{r})\rangle\langle\phi_c(\mathbf{r})|\psi^s(\mathbf{r})\rangle=E'\,|\psi^s(\mathbf{r})\rangle \tag{12.210}$$

where E' is the energy eigenvalue of the new pseudopotential equation. Operating the true wave function $\langle\psi(\mathbf{r})|$ from the left side on Eq. (12.210), one can easily prove that

$$E'\langle\psi(\mathbf{r})|\psi^s(\mathbf{r})\rangle=E\langle\psi(\mathbf{r})|\psi^s(\mathbf{r})\rangle \tag{12.211}$$

In order for Eq. (12.211) to hold, either the true wave function should be orthogonal to the pseudowave function, which is not true, or their energies should be equal, which is the case here. It is important to note that all forms of the repulsive part f(E, c) give the correct energy eigenvalues and eigenfunctions if the pseudowave equation is solved exactly. In other words, it means that there is no correct pseudopotential, but there are many valid forms of it. This leads to the formulation of a number of self-consistent pseudopotentials, which are used to study the electronic properties of the crystalline solids. Another class of pseudopotentials comprises model potentials in which the fact that the repulsive part f(E, c) can have any form is exploited. There exist a number of model potentials for metals with different forms of the repulsive contribution and the reader is referred to Harrison (1966) and Galsin (2002) for further study.

12.7 AUGMENTED PLANE WAVE (APW) METHOD

The actual crystal potential $V(\mathbf{r})$ shown in Fig. 12.13 can be accurately described by a screened Coulomb potential near an ion, which can be assumed to be spherically symmetric in nature. But, between the ions, $V(\mathbf{r})$ is flat and weak. It should be noted that $V(\mathbf{r})$ is consistent with the nature of the real wave function, which is oscillatory within the ion core and a plane wave outside. It has already been pointed out in Chapter 3 that the exact evaluation of $V(\mathbf{r})$ in a crystalline solid is difficult; therefore, it is usually estimated in some justifiable approximation.

It was discussed in Chapter 1 that a crystalline solid can be divided into identical WS cells with an atom (ion) at the center. One can draw a sphere of radius R_{MT}, with its center at the center of the WS cell, so that it lies well within the WS cell (see Fig. 12.14A). Such a sphere is called a muffin-tin (MT) sphere. The radius R_{MT} is such that, within the MT sphere, the potential can be approximated by a screened Coulomb potential $V_{ion}(\mathbf{r})$ of an ion. But outside the MT sphere, $V(\mathbf{r})$ is assumed to be constant and preferably zero (by shifting the zero of the energy scale). Such a potential is called an MT potential. Mathematically, one can write

$$\begin{aligned} V(\mathbf{r}) &= V_{ion}(\mathbf{r}) \quad \text{for } r\le R_{MT} \\ &= 0 \qquad\quad\ \ \text{for } r\rangle R_{MT} \end{aligned} \tag{12.212}$$

The MT potential is shown in Fig. 12.14B. In the literature there exist more refined and sophisticated forms of the MT potential. The wave function $|\psi_\mathbf{k}(\mathbf{r})\rangle$ is assumed to be spherically symmetric inside the MT sphere, but outside it is in the form of a plane wave. Both of these wave functions must be continuous on the surface of the MT sphere, that is, the wave functions and their derivatives must be the same at $r=R_{MT}$. In the muffin-tin approximation for the potential, the wave function is given by

$$\begin{aligned} |\psi_\mathbf{k}(\mathbf{r})\rangle &= \sum_{\ell,m} A_{\ell m} R_\ell(kr)\, Y_\ell^m(\theta,\phi) \quad \text{for } r\le R_{MT} \\ &= e^{\imath\mathbf{k}\cdot\mathbf{r}} \qquad\qquad\qquad\quad\ \text{for } r\rangle R_{MT} \end{aligned} \tag{12.213}$$

Here $R_\ell(kr)$ is the radial wave function and $Y_\ell^m(\theta,\phi)$ are spherical harmonics. The wave function defined by Eq. (12.213) is usually called the augmented plane wave (APW) function and the method of determining the energy eigenvalues using this wave function is called the APW method. The constants $A_{\ell m}$ can be obtained from the expansion of a plane wave in terms of spherical harmonics $Y_\ell^m(\theta,\phi)$ given as

$$e^{\imath\mathbf{k}\cdot\mathbf{r}}=4\pi\sum_{\ell,m}\imath^\ell j_\ell(kr)\,Y_\ell^m(\theta,\phi)\,Y_\ell^{m*}(\theta',\phi') \tag{12.214}$$

where $j_\ell(kr)$ is the spherical Bessel function for orbital quantum number ℓ. The continuity condition of the wave functions, given by Eqs. (12.213), (2.214), at $r=R_{MT}$ gives

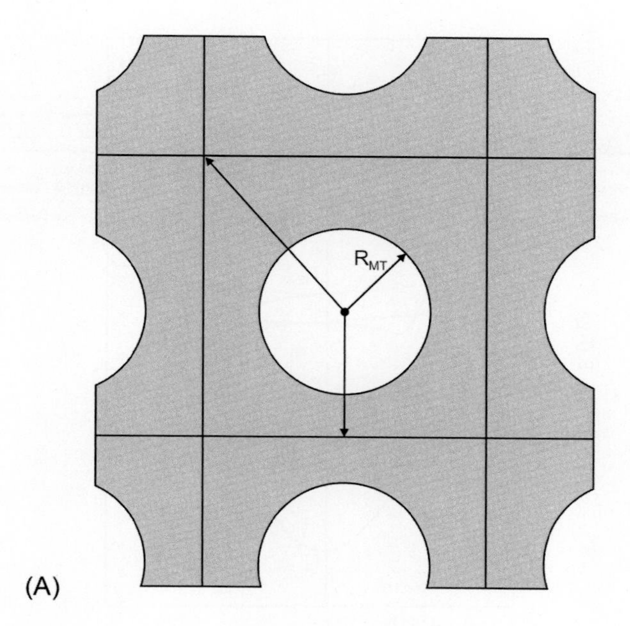

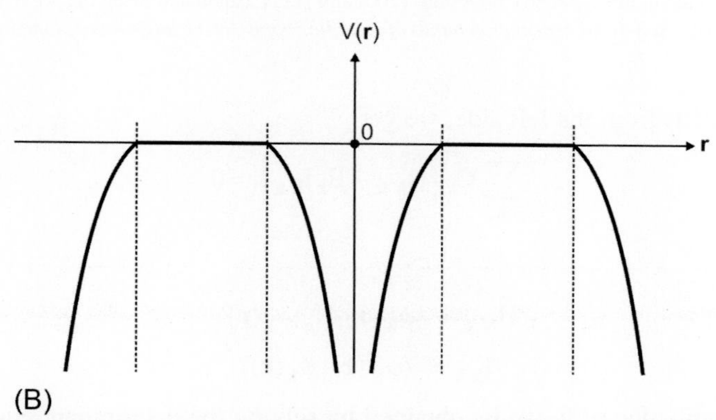

FIG. 12.14 (A) The square lattice is divided into WS cells having an atom at their centers. The MT sphere is shown in each WS cell with the center of the MT sphere coinciding with the center of the WS cell. (B) The MT potential for a crystalline solid.

$$A_{\ell m} = 4\pi\imath^{\ell} \frac{j_{\ell}(kr)}{R_{\ell}(kr)} Y_{\ell}^{m*}(\theta', \phi') \tag{12.215}$$

The radial wave function $R_{\ell}(kr)$ is obtained by solving the radial part of the Schrodinger wave equation given by

$$-\frac{1}{r^2}\frac{d}{dr}\left(r^2\frac{dR_{\ell}}{dr}\right) + \left[\frac{\ell(\ell+1)}{r^2} + \frac{2m_e}{\hbar^2}V(\mathbf{r})\right]R_{\ell}(kr) = \frac{2m_e E}{\hbar^2}R_{\ell}(kr) \tag{12.216}$$

for a given energy E. The electron wave function is obtained by taking a linear combination of $|\psi_{\mathbf{k}}(\mathbf{r})\rangle$, that is,

$$|\psi(\mathbf{r})\rangle = \sum_{\mathbf{k}} C_{\mathbf{k}} |\psi_{\mathbf{k}}(\mathbf{r})\rangle \tag{12.217}$$

The coefficients $C_{\mathbf{k}}$ are chosen to minimize the energy of $|\psi(\mathbf{r})\rangle$. The energy eigenvalues can be obtained by solving the Schrodinger wave equation

$$\widehat{H}_e |\psi(\mathbf{r})\rangle = E |\psi(\mathbf{r})\rangle \tag{12.218}$$

Substituting Eq. (12.217) into Eq. (12.218), we obtain

$$\sum_{\mathbf{k}} C_{\mathbf{k}} \left[\widehat{H}_e |\psi_{\mathbf{k}}(\mathbf{r})\rangle - E_{\mathbf{k}} |\psi_{\mathbf{k}}(\mathbf{r})\rangle\right] = 0 \tag{12.219}$$

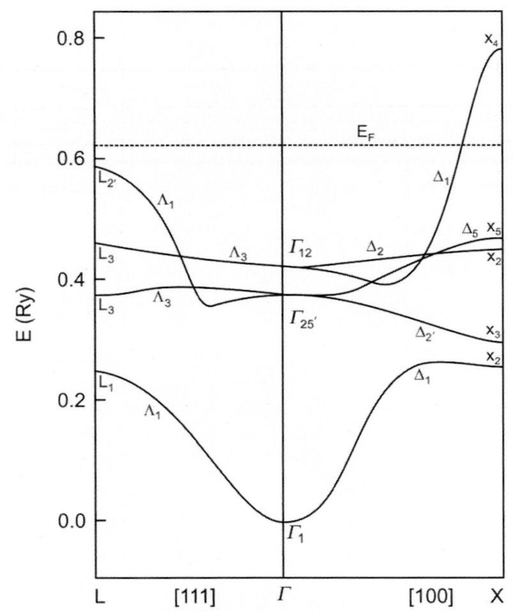

FIG. 12.15 Energy bands in Cu metal along the symmetry directions [100] and [111], calculated using the APW method. The *dashed line* denotes the Fermi energy E_F. *(After Snow, E. C. (1968). Self-consistent energy bands of metallic copper by the augmented plane wave method II.* Physical Review, 171, *785–789.)*

Operating $\langle\psi_{\mathbf{k}'}(\mathbf{r})|$ on Eq. (12.219) from the left side, we get

$$\sum_{\mathbf{k}} C_{\mathbf{k}}\left[\widehat{H}_{\mathbf{k}'\mathbf{k}} - E_{\mathbf{k}} I_{\mathbf{k}'\mathbf{k}}\right] = 0 \tag{12.220}$$

where

$$\widehat{H}_{\mathbf{k}'\mathbf{k}} = \langle\psi_{\mathbf{k}'}(\mathbf{r})|\widehat{H}_e|\psi_{\mathbf{k}}(\mathbf{r})\rangle \tag{12.221}$$

$$I_{\mathbf{k}'\mathbf{k}} = \langle\psi_{\mathbf{k}'}(\mathbf{r})\,|\,\psi_{\mathbf{k}}(\mathbf{r})\rangle \tag{12.222}$$

The energy eigenvalues for each value of **k** can be obtained by solving the determinant equation

$$\det\left|\widehat{H}_{\mathbf{k}'\mathbf{k}} - E_{\mathbf{k}} I_{\mathbf{k}'\mathbf{k}}\right| = 0 \tag{12.223}$$

Knowing the values of the energy eigenvalues $E_{\mathbf{k}}$, the coefficients $C_{\mathbf{k}}$ can be determined from Eq. (12.220). Note that the matrix elements $\widehat{H}_{\mathbf{k}'\mathbf{k}}$ and $I_{\mathbf{k}'\mathbf{k}}$ depend on $E_{\mathbf{k}}$ and vice versa. Therefore, one has to start with some trial energy eigenvalue $E_{\mathbf{k}}^0$ and then determine the energy self-consistently. Fig. 12.15 shows energy bands in Cu metal obtained using the APW method (Snow 1968). It is evident from the figure that the shape of the energy bands is very much different from those obtained in the nearly free-electron and tight-binding theories. Further, the energy bands along the two symmetry directions have different values and different shapes.

There are a number of other methods for determining the energy eigenvalues in crystalline solids. One method worth mentioning here is the Korringa-Kohn-Rostoker (KKR) method. The KKR method makes use of the Green's function technique for determining the energy bands in a solid, but it is not in the scope of this book. The interested reader may consult Galsin (2002) and Callaway (1974) for a brief account of this method.

12.8 DYNAMICS OF ELECTRONS IN ENERGY BANDS

An electron in a band may not be free to move, but rather its motion depends on the energy state occupied by it in a particular energy band. In other words, the velocity and effective mass of the electron may not remain constant throughout the energy band. In the previous sections, the band energy $E_{\mathbf{k}}$ has been calculated using different methods, but here we calculate the effective mass of an electron using a simple form of energy band, as shown in Fig. 12.16. For simplicity it is assumed that the BZ under consideration contains only one electron. From the wave mechanical theory, the velocity of an electron is equal to the group velocity of the wave packet that is given by the standard relation as

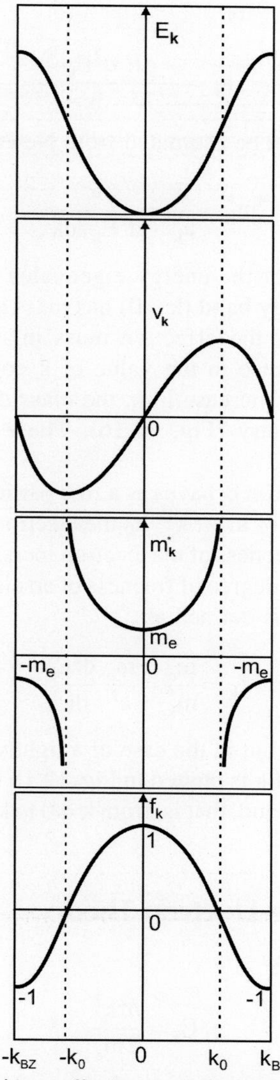

FIG. 12.16 Schematic representation of energy E_k, velocity v_k, effective mass m^*_k, and the parameter f_k as a function of wave vector k. Here k_0 is the wave vector at the point of inflection in the E_k curves and k_{BZ} is the wave vector at the 1BZ boundary.

$$v_{\mathbf{k}} = \frac{d\omega(\mathbf{k})}{dk} = \frac{1}{\hbar}\frac{dE_{\mathbf{k}}}{dk} \tag{12.224}$$

where $\omega(\mathbf{k})$ is the angular frequency for the wave vector $\mathbf{k}$. From Eq. (12.224) it is evident that $v_{\mathbf{k}}$ depends on the first derivative of the energy curve and acceleration, therefore, on the second derivative. Fig. 12.16 shows $v_{\mathbf{k}}$ as a function of k for a one-dimensional solid. The velocity v_k of an electron increases with an increase in the value of E_k and becomes maximum at the point of inflection occurring at $k = k_0$. With a further increase in the k value, the velocity decreases and goes to zero at the BZ boundary. It is evident that v_k is zero either at the center of the band or at the BZ boundaries.

The effective mass of the electron m^*_k can be calculated as a function of k from an energy band. If an electric field $\mathbf{E}$ is applied to the material, then the energy gained by an electron in an energy band is given by

$$dE_k = \mathbf{F} \cdot d\mathbf{S} = -eE(v_k dt) \tag{12.225}$$

where $\mathbf{F} = -e\mathbf{E}$. Using Eq. (12.224) for $v_{\mathbf{k}}$ in Eq. (12.225), one obtains

$$\frac{dk}{dt} = -\frac{eE}{\hbar} \tag{12.226}$$

Now the acceleration $a^*_{\mathbf{k}}$ can be written as

$$a^*_{\mathbf{k}} = \frac{dv_{\mathbf{k}}}{dt} = \frac{1}{\hbar}\frac{d^2E_{\mathbf{k}}}{dk^2}\frac{dk}{dt} \tag{12.227}$$

Substituting Eq. (12.226) into Eq. (12.227), one gets

$$a_{\mathbf{k}}^{*}=-\frac{eE}{\hbar^{2}}\frac{d^{2}E_{\mathbf{k}}}{dk^{2}} \tag{12.228}$$

With the knowledge of $a_{\mathbf{k}}^{*}$, $m_{\mathbf{k}}^{*}$ of an electron can be estimated from Newton's law as

$$m_{\mathbf{k}}^{*}=\frac{F}{a_{\mathbf{k}}^{*}}=\frac{\hbar^{2}}{d^{2}E_{\mathbf{k}}/dk^{2}} \tag{12.229}$$

Eqs. (12.228), (12.229) show the importance of the energy eigenvalue curves derived from band theory. Fig. 12.16 shows that the electron at the center of an energy band ($k=0$) has mass equal to the mass of a free electron m_e, that is, $m_{\mathbf{k}}^{*}=m_e$. With an increase in the value of k, the effective mass $m_{\mathbf{k}}^{*}$ increases and becomes infinity at the point of inflection at wave vector k_0. With an increase in the value of k compared with k_0, the effective mass becomes very large but negative in sign. With a further increase in k, the mass decreases and reaches the mass of a free electron but with negative sign at the BZ boundary (Fig. 12.16). These results can be interpreted according to the discussion below.

From the foregoing, it is evident that an electron behaves as a free particle at the center of a band and that the degree of freeness decreases with an increase in the value of k. At $k=k_0$ the electron becomes tightly bound to the nucleus. With a further increase in the value of k, the degree of freeness of the electron increases again and it becomes completely free at the BZ boundary. A factor of $f_{\mathbf{k}}$, which measures the degree of freeness of an electron in an energy state with wave vector **k** in a three-dimensional solid, can be introduced and is defined as

$$f_{\mathbf{k}}=\frac{m_e}{m_{\mathbf{k}}^{*}}=\frac{m_e}{\hbar^{2}}\frac{d^{2}E_{\mathbf{k}}}{dk^{2}} \tag{12.230}$$

For a free electron $m_{\mathbf{k}}^{*}=m_e$, which gives $f_{\mathbf{k}}=1$, but in the case of a tightly bound electron $m_{\mathbf{k}}^{*}=\infty$ and $f_{\mathbf{k}}=0$. For a one-dimensional solid, the value of $f_{\mathbf{k}}$ as a function of k is plotted in Fig. 12.16 for the given energy band. It is evident from the figure that $f_{\mathbf{k}}$ is positive for the lower part of the band, that is, from $k=0$ to $k=k_0$, and becomes negative in the upper part of the band.

12.8.1 Behavior of Electrons in Free-Electron Theory

In the free-electron theory the energy is given by

$$E_{\mathbf{k}}=\frac{\hbar^{2}k^{2}}{2m_e} \tag{12.231}$$

Using Eq. (12.231) in Eqs. (12.224), (12.229), one gets

$$v_{\mathbf{k}}=\frac{\hbar k}{m_e}=\frac{p}{m_e} \tag{12.232}$$

$$m_{\mathbf{k}}^{*}=m_e \tag{12.233}$$

Therefore, in the free-electron approximation, an electron behaves as a free particle for all values of k, but the velocity increases linearly with wave vector k. In this approximation

$$f_{\mathbf{k}}=1 \tag{12.234}$$

12.8.2 Behavior of Electrons in Tight-Binding Approximation

In the tight-binding approximation (one-dimensional case), the energy is given by

$$E_k=E_0-\gamma_0-2\gamma_1\cos ka \tag{12.235}$$

Now the velocity in this approximation becomes

$$v_k=\frac{2\gamma_1 a}{\hbar}\sin ka \tag{12.236}$$

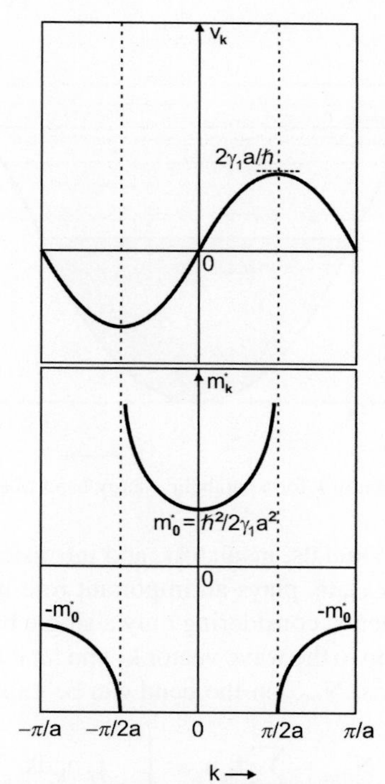

FIG. 12.17 Variation of the velocity v_k and effective mass m^*_k in the tight-binding approximation (one-dimensional case).

So, the velocity varies sinusoidally as shown in Fig. 12.17. The effective mass is given by

$$m^*_k = \frac{\hbar^2}{2\gamma_1 a^2 \cos ka} \tag{12.237}$$

Eq. (12.237) gives the variation of m^*_k as a function of k, which is similar to that shown in Fig. 12.16 with the value of m^*_k at $k=0$ as

$$m^*_0 = \frac{\hbar^2}{2\gamma_1 a^2} \tag{12.238}$$

The variation of v_k and m^*_k in the tight-binding approximation is shown in Fig. 12.17. The only difference in the Figs. 12.16 and 12.17 is that the mass at $k = 0$ is different in the two cases. It is noteworthy that in a realistic energy band structure the point of inflection may not be exactly in the middle of the BZ as it is in the tight-binding approximation. The factor f_k, in the tight-binding approximation, is given as

$$f_k = \frac{m_e}{m^*_0} \cos ka \tag{12.239}$$

Here we have used Eqs. (12.230), (12.237), and (12.238). If $m^*_0 = m_e$, then

$$f_k = \cos ka \tag{12.240}$$

Eq. (12.240) gives the variation of f_k with k, which is similar to that shown in Fig. 12.16. The above treatment of effective mass can be extended to the three-dimensional case in which m^*_k is given by (see Eq. 12.229)

$$\frac{1}{m^*_k} = \frac{1}{\hbar^2} \nabla_k \cdot \nabla_k E_k \tag{12.241}$$

Eq. (12.241) gives a tensor with nine components of the general form $\partial^2 E_k / \partial k_\alpha \partial k_\beta$.

12.9 DISTINCTION BETWEEN METALS, INSULATORS, AND SEMICONDUCTORS

The classification of elements as metals, insulators, and semiconductors is done on the basis of electrical conductivity, which in turn depends on the number of free electrons present. The energy band theory employing the periodic potential

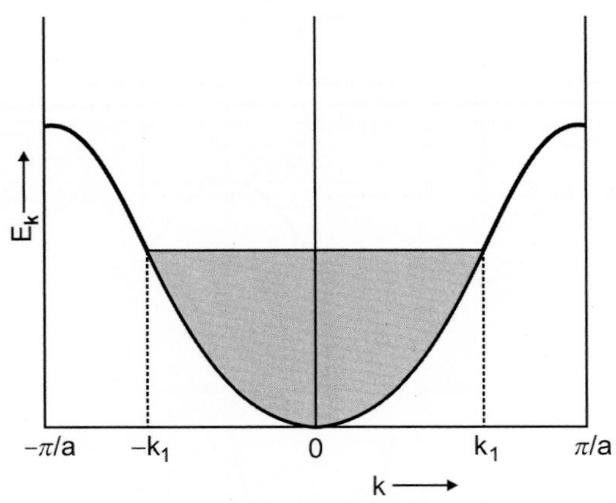

FIG. 12.18 Schematic representation of E_k as a function of k for a parabolic energy band filled up to the wave vector k_1 in the 1BZ.

of the lattice allows us to distinguish between metals, insulators, and intrinsic semiconductors. The factor $f_{\mathbf{k}}$, which gives the degree of freeness of an electron in the **k** state, plays an important role in making this distinction.

We demonstrate the classification of elements, considering only a single band in one dimension, in Fig. 12.18. Let this energy band be partially filled with electrons up to the wave vector k_1 and let it contain a number N_e of electrons, which may be partially free. The number of free electrons, N_{free}, in the band can be expressed in terms of f_k as follows:

$$N_{free} = \sum_k f_k n_k = \int_{-k_1}^{k_1} f_k n_k \, dk \tag{12.242}$$

where the summation is over all the occupied states in the band and n_k is the density of electron states with wave vector k. The number of k-states per unit length in the reciprocal space is $n_k = L/2\pi$. So, the number of k-states lying between k and k +dk is given by

$$n_k \, dk = 2 \frac{L}{2\pi} dk \tag{12.243}$$

Here the factor of 2 takes care of the spin degeneracy of the electron states. Hence, the number of free electrons is obtained by substituting Eqs. (12.230), (12.243) in Eq. (12.242), that is,

$$N_{free} = \frac{L}{\pi} \frac{m_e}{\hbar^2} \int_{-k_1}^{k_1} \frac{d^2 E_k}{dk^2} dk \tag{12.244}$$

The above integral is even; therefore, one can write

$$N_{free} = \frac{2 m_e L}{\pi \hbar^2} \int_0^{k_1} \frac{d^2 E_k}{dk^2} dk = \frac{2 m_e L}{\pi \hbar^2} \frac{dE_k}{dk}\bigg|_{k_1} \tag{12.245}$$

From Eq. (12.245) it is evident that N_{free} depends on the first derivative of E_k at the topmost filled state in the band. This result allows us to draw the following conclusions:

1. There are no free electrons in a completely filled (completely empty) band because $dE_{\mathbf{k}}/dk$ vanishes at the top (bottom) of the band. Consider a solid in which some energy bands are completely filled while others are completely empty. The topmost filled band is called the valence band as it contains the valence electrons. The next higher band is empty and represents the first excited state, usually called the conduction band. In such a solid the conduction and valence bands are separated by an energy gap E_g, as shown in Fig. 12.19. There are two equivalent representations of the energy bands, the choice of which is based on convenience. In Fig. 12.19A the energy E_k of the parabolic bands is plotted as a function of k. Such a representation is useful when the maximum of the valence band and the minimum of the conduction band occur at different values of k in the BZ. But Fig. 12.19B shows only the energy E_k of the bands and does not give information about the positions of the maximum and minimum of the bands in the BZ. One should note that the situation shown in Fig. 12.19 is true only at 0 K. At finite temperature, some electrons from the valence band may get excited to the conduction band, resulting in finite conductivity. If the forbidden energy gap E_g is of the order of several electron volts (e.g., in diamond $E_g \cong 7.0$ eV), then the electrons cannot jump from the valence to the conduction band, yielding

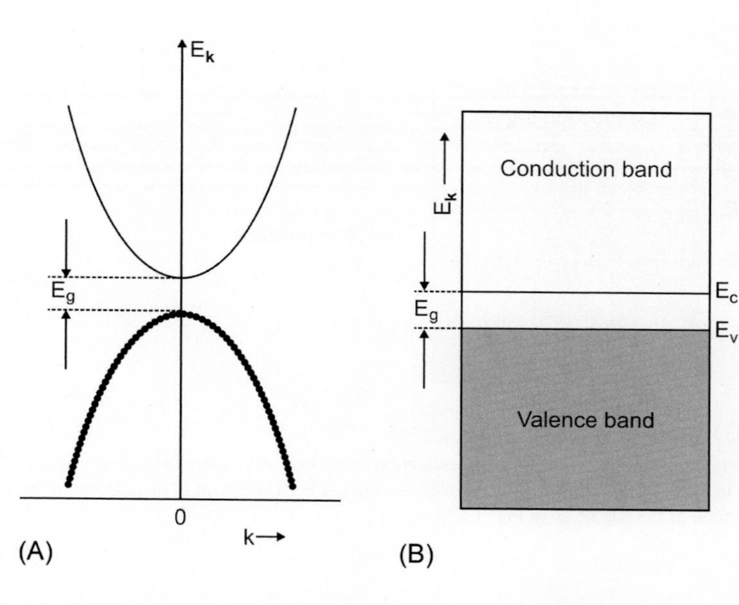

FIG. 12.19 (A) The energy E_k is plotted as a function of k for parabolic valence and conduction bands in a solid separated by a band gap E_g. The valence band is completely full, while the conduction band is completely empty. (B) The valence band with energy E_v at its top and the conduction band with energy E_c at its bottom. The valence band is completely full (shown by the shaded region), while the conduction band is empty. E_g is the energy band gap between them.

zero conductivity. Such a solid is an insulator for all practical purposes. But for small values of the energy gap, the number of thermally excited electrons may become appreciable, which is the case in intrinsic semiconductors, such as Ge ($E_g=0.7$ eV) and Si ($E_g=1.1$ eV). Therefore, the distinction between insulators and semiconductors is only a quantitative one. In fact, all of the semiconductors are insulators at $T=0\,K$, whereas all the insulators may behave as semiconductors at finite temperatures.

2. The value of N_{free} increases with an increase in k-value and acquires a maximum value for a band filled up to the point of inflection, as dE_k/dk has its maximum value at this point. Therefore, a solid with a partially filled band exhibits metallic character. In such solids the electrons may get excited to the higher energy states in the same partially filled band at finite temperatures (see Fig. 12.20). Alkali metals, such as Na and K, in which the valence band is half filled, are examples of such solids. It is well known that Ca, Ba, and Sr are also good conductors in which the energy bands are completely filled. It is because the valence and conduction bands overlap in such solids (see Fig. 12.21) and that the electrons can move without any hindrance from the valence to the conduction band by receiving even a very small thermal energy. The overlap of the valence and conduction bands may occur in one or more directions, which will become clear only by studying from ab initio the energy band structure of a solid in three dimensions. Hence, there is a possibility that a solid that is an insulator in a one-dimensional study may turn out to be a metal in a three-dimensional energy band study. It is interesting to note that the conductivity of semiconductors increases with an increase in temperature, whereas that of metals decreases with an increase in temperature.

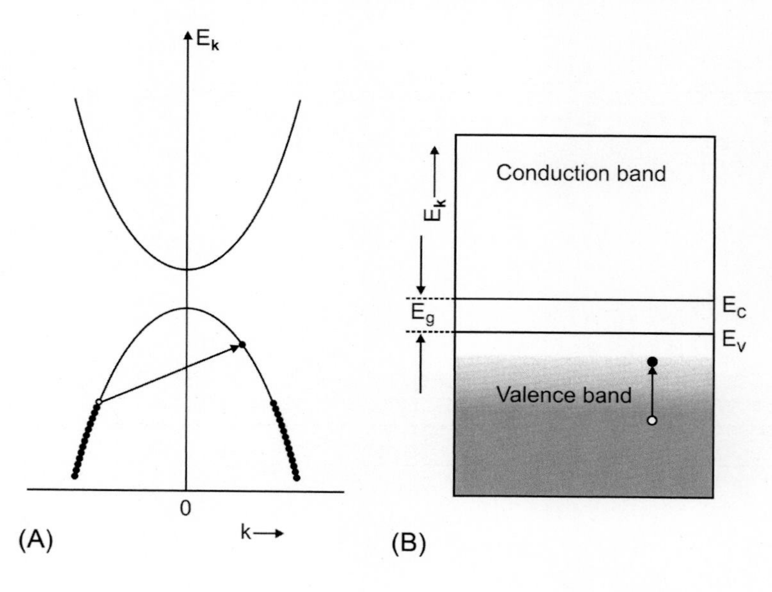

FIG. 12.20 (A) The energy E_k is plotted as a function of k for parabolic valence and conduction bands in a solid separated by a band gap E_g. Here the valence band is partially filled, while the conduction band is empty. (B) E_v and E_c are the energies at the top and bottom of the valence and conduction bands, respectively, with band gap E_g between them. The electrons in the partially filled valence band make transitions from the filled states to the empty states. The occupancy of the states in the valence band is given by the tone of the shade (occupancy increases with an increase in the tone of the shade). The conduction band is completely empty.

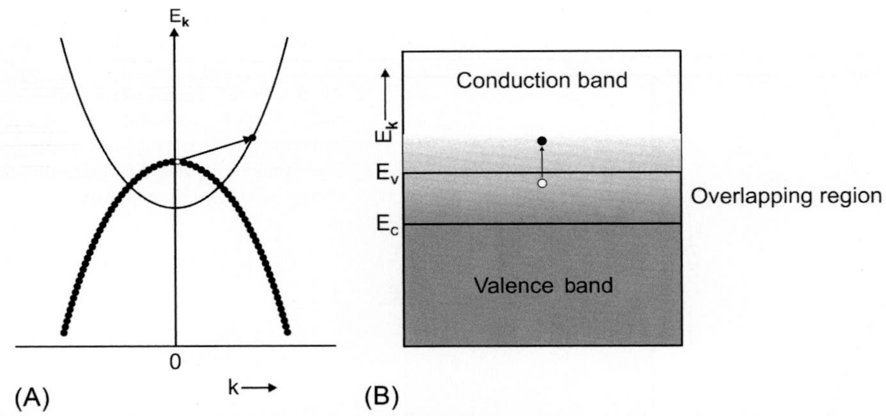

FIG. 12.21 (A) The energy E_k is plotted as a function of k for overlapping parabolic valence and conduction bands in a solid. Here the valence band is completely full. (B) The overlapping valence and conduction bands in a solid. Here the energy E_v crosses the energy E_c. The extent of occupation of the electron states in the valence and conduction bands is given by the tone of the shade, as in Fig. 12.20.

REFERENCES

Callaway, J. (1974). *Quantum theory of the solid state (parts A & B)*. New York: Academic Press.

Galsin, J. S. (2002). *Impurity scattering in metallic alloys*. New York: Kluwer Academic/Plenum Publishers.

Harrison, W. A. (1966). *Pseudopotentials in the theory of metals and alloys*. New York: W. A. Benjamin.

Snow, E. C. (1968). Self-consistent energy bands of metallic copper by the augmented plane wave method II. *Physical Review*, *171*, 785–789.

SUGGESTED READING

Altmann, S. L. (1970). *Band theory of metals*. New York: Pergamon Press.

Blount, E. I. (1962). Formalism of band theory. F. Seitz, & D. Turnbull (Eds.), *Solid state physics* (pp. 306–373). Vol. 13(pp. 306–373). New York: Academic Press.

Callaway, J. (1958). Electron energy bands in solids. F. Seitz, & D. Turnbull (Eds.), *Solid state physics* (pp. 100–212). Vol. 7(pp. 100–212). New York: Academic Press.

Cornwell, J. F. (1969). *Selected topics in solid state physics: Group theory and electronic energy bands in solids*. Amsterdam: North-Holland Publ. Co.

Lehmann, G., & Ziesche, P. (1991). *Electronic properties of metals*. Amsterdam: Elsevier Publishing Co.

Loucks, T. L. (1967). *Augmented plane wave method*. New York: W. A. Benjamin.

Marcus, P. M., Janak, J. F., & Williams, A. R. (1971). *Computational methods in band theory*. New York: Plenum Publishers.

Moruzzi, V. L., Janak, J. F., & Williams, A. R. (1978). *Calculated electronic properties of metals*. New York: Pergamon Press.

Seitz, F. (1940). *Modern theory of solids*. New York: McGraw-Hill Book Co.

Chapter 13

The Fermi Surfaces

Chapter Outline

13.1 CONSTANT ENERGY SURFACES

The locus of all the points at which the energy $E_{\mathbf{k}}$ has a constant value is called a constant energy surface. The shape of the constant energy surface depends on the nature of the energy bands. One can define any number of constant energy surfaces for different values of $E_{\mathbf{k}}$. In a one-dimensional free-electron gas the energy bands are parabolic in nature and are given by

$$E_{\mathbf{k}} = \frac{\hbar^2 k^2}{2m_e} \tag{13.1}$$

In this case one can define points with constant energy and Fig. 13.1A shows two equidistant points with constant energy $E_{\mathbf{k}}$. In a two-dimensional solid, one can define constant energy contours. Fig. 13.1B shows *circles* as the constant energy contours in a two-dimensional free-electron gas. But in a three-dimensional solid one can define constant energy surfaces. In a three-dimensional free-electron gas, the energy bands are parabolic, just as defined by Eq. (13.1), and they yield spherical constant energy surfaces (see Fig. 13.1C).

13.2 THE FERMI SURFACES

The Fermi surface (FS) is a special constant energy surface and is defined as the locus of all of the points at which the value of energy is E_F. In the preceding chapters, we have discussed the fact that it is the electron states very near the FS that determine most of the electronic properties of crystalline solids. This is because the electrons in these states can be easily excited to the vacant states above the FS by the application of a small external field. Hence the nature of energy bands in the neighborhood of the FS is of primary importance and requires our special attention.

13.3 THE FERMI SURFACE IN THE FREE-ELECTRON APPROXIMATION

In a one-dimensional free-electron gas, the Fermi energy is given by (Eq. 9.50)

$$E_F = \frac{\hbar^2 k_F^2}{2m_e} = \frac{\pi^2 \hbar^2 n_e^2}{2m_e} \tag{13.2}$$

where n_e is the linear density of electrons. The points at which energy has value E_F are equidistant from the center of the 1BZ and are symmetrically spaced on both sides of it (see Fig. 13.1A). In a two-dimensional free-electron gas the Fermi energy is given by (see Eq. 9.41)

Solid State Physics. https://doi.org/10.1016/B978-0-12-817103-5.00013-X

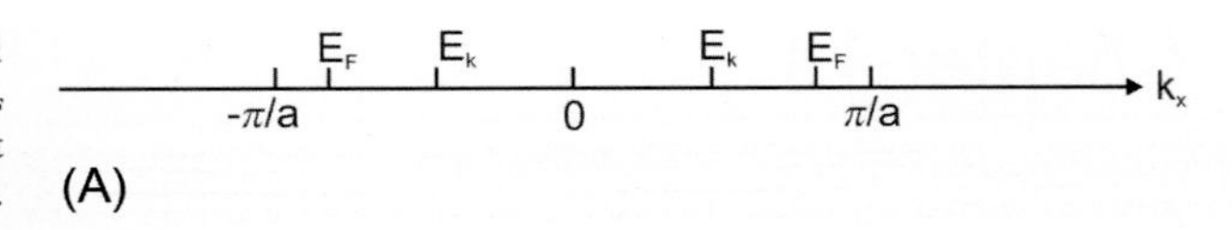

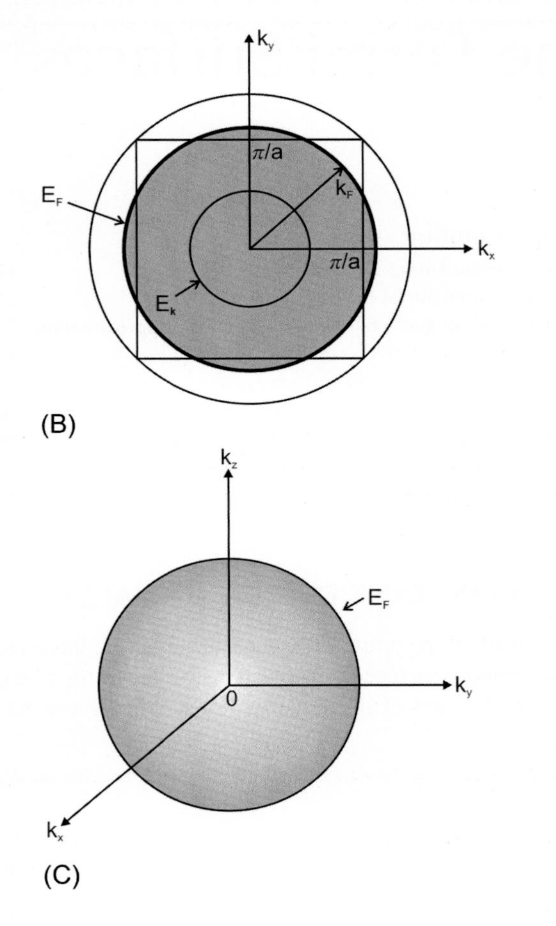

FIG. 13.1 (A) Two points having constant energy E_k in a one-dimensional free-electron gas. The figure also shows two points having Fermi energy E_F lying within the 1BZ. (B) The constant energy circular contours with different values of energy $E_{\mathbf{k}}$ in the square lattice of a two-dimensional free-electron gas. The figure also shows the circular contour having Fermi energy E_F. (C) The spherical Fermi surface with constant energy E_F in a three-dimensional free-electron gas.

$$E_F = \frac{\hbar^2 k_F^2}{2m_e} = \frac{\pi \hbar^2 n_e}{m_e} \tag{13.3}$$

where n_e is the surface electron density. The Fermi circle with its center as the center of the 1BZ and radius k_F is shown in Fig. 13.1B. In the three-dimensional free-electron gas the Fermi energy, from Eqs. (9.16), (9.19), is given by

$$E_F = \frac{\hbar^2}{2m_e}\left(3\pi^2 \frac{Z}{V_0}\right)^{2/3} \tag{13.4}$$

The components of the wave vector $\mathbf{k}$ at the 1BZ boundary are given by

$$k_X = k_y = k_Z = \pm\frac{\pi}{a} \tag{13.5}$$

The FS is a sphere with radius k_F having center at the center of the 1BZ of the lattice and is shown in Fig. 13.1C. In the free-electron approximation, three distinct classes of the Fermi surface are observed.

13.3.1 Type I Fermi Surface

It is evident from Eq. (13.4) that in the free-electron approximation, the value of k_F depends upon the electron concentration. In solids with a low electron concentration, the entire FS may lie in the 1BZ. In this case, only the first band (which

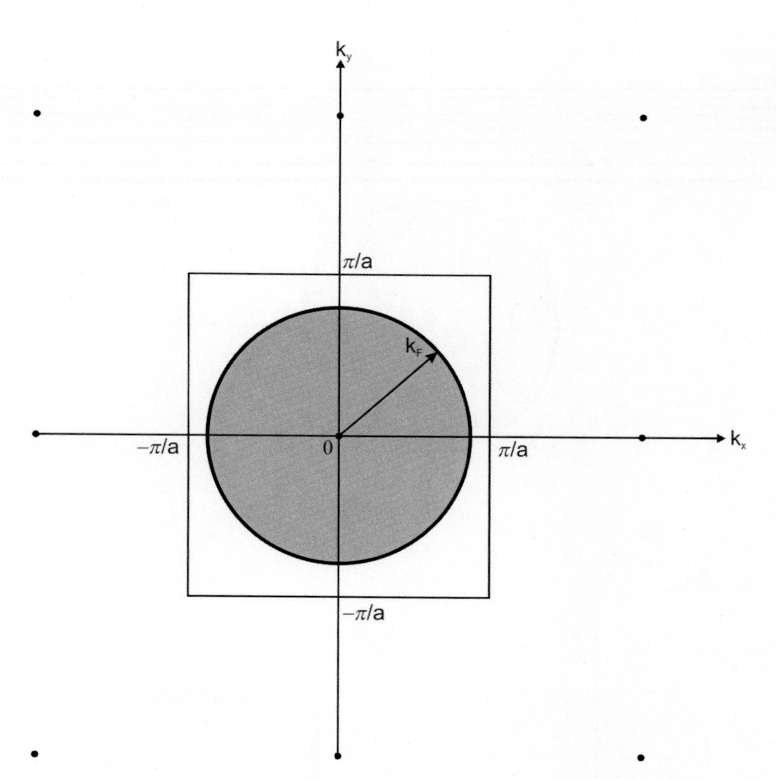

FIG. 13.2 The Fermi circle of a gas of low-electron-density free electrons in a square lattice. The *dots* represent the lattice points in the reciprocal lattice of a square lattice with periodicity a. The 1BZ is a square with side $2\pi/a$.

lies in the 1BZ) is partially filled, the others are empty. Fig. 13.2 shows, for simplicity, a circular FS for a two-dimensional square lattice with low electron density.

13.3.2 Type II Fermi Surface

In crystalline solids with a reasonably large concentration of electrons, the FS may extend to the higher order BZs. Fig. 13.3A shows the FS of a square lattice, which extends to the 2BZ. Here the FS exhibits two partially filled bands: the first band lies in the 1BZ and the second in the 2BZ. It has already been discussed that all of the significant values of **k** lie in the 1BZ. Therefore, it is the usual practice to reduce the FS to the 1BZ (reduced zone scheme).

Fig. 13.3B and C shows the first and second bands in the reduced zone scheme, that is, in the 1BZ. Fig. 13.4A and B shows the first and second bands of the square lattice in the periodic zone scheme, which are repeated periodically. From Fig. 13.4A it is evident that the FS for the first band consists of pockets of holes at the corners of the 1BZ, while the FS of the second band (Fig. 13.4B) exhibits electron pockets at the middle of the sides of the 1BZ. It should be noted that both the bands in Fig. 13.4 are plotted in the 1BZ and the pieces of the FS that stick out of the zone must be interpreted in the periodic zone scheme.

13.3.2.1 Electron Orbits

For a changing electron state, the **k** point with a particular energy changes its position with time. The closed path of the **k** point in the reciprocal space is called an *orbit* and the periodic zone scheme is more suitable (not essential) for its representation. It is of particular interest to consider orbits along which the energy is constant and has a value equal to E_F. These orbits provide very useful information about the shape of the FS as they lie on the FS. Moreover, such orbits are experimentally accessible. In two-dimensional crystals such orbits just coincide with the FS itself. Fig. 13.5A shows the part of the FS for the first energy band in the 1BZ of a two-dimensional electron gas. The orbit corresponding to the hole pockets of the first band in the reduced zone scheme is a discontinuous one. The motion of an electron in this orbit can be described as follows:

1. Let an electron start from the state represented by point A and go from state A to state B. The electron is then Bragg reflected at B and goes to state C. The state C is identical to the state B as it differs from state B by a reciprocal lattice vector, $\mathbf{k}=\mathbf{G}$.

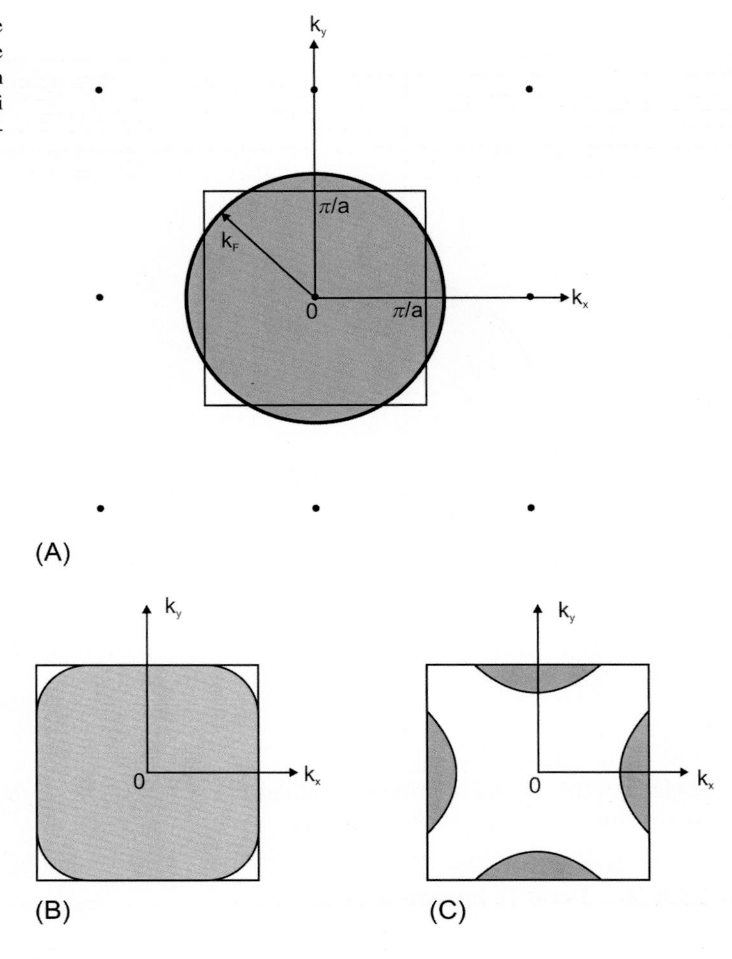

FIG. 13.3 (A) The Fermi surface extending to the 2BZ in the square lattice of a two-dimensional free-electron gas with reasonably large electron density. (B) The Fermi surface of the first energy band in the 1BZ of a two-dimensional free-electron gas. (C) The Fermi surface of the second energy band reduced to the 1BZ in a two-dimensional free-electron gas.

2. The electron goes from state C to state D and is then Bragg reflected at D and goes to state E.
3. The electron from state E goes to state F and is then Bragg reflected at F and goes to state G.
4. Finally, the electron moves from state G to state H and is then Bragg reflected from here to state A.

The same process can be shown in the periodic zone scheme in which the hole orbit is a continuous one (see Fig. 13.5B). The reader is also referred to Fig. 13.4A. It is evident that the continuous orbits in the periodic zone scheme are more appealing and easier to understand than the discontinuous orbits in the reduced zone scheme.

Fig. 13.6A and B shows the electron orbits for the second band in the reduced zone and periodic zone schemes. Again, the discontinuous orbit in the reduced zone scheme can be visualized as a continuous electron orbit in the periodic zone scheme. Further, the representation of the electron pockets (situated at the midpoint of each side of the 1BZ) is more appealing in the periodic zone scheme (also see Fig. 13.4B).

13.3.3 Type III Fermi Surface

In crystalline solids with higher electron concentration, the FS may extend to the higher order BZs. Fig. 13.7A shows a square lattice in **k**-space with its FS extending to the 4BZ. The portions of the FS in the 1BZ, 2BZ, 3BZ, and 4BZ, which show, respectively, the FS sections due to the first, second, third, and fourth bands, can be reduced to the 1BZ (Fig. 13.7B). It is evident that the FS due to the second band exhibits the orbit of a hole pocket around the center of the 1BZ. The FS corresponding to the third and fourth bands is disconnected in the reduced zone scheme. The discontinuous electron orbits for the third and fourth energy bands, in the reduced zone scheme, are shown in Fig. 13.8.

Any band can be plotted in the periodic zone scheme by translating its plot in the reduced zone scheme by all possible reciprocal lattice vectors. The plot of the second band and its hole orbit in the periodic zone scheme remain the same, as shown in Fig. 13.7B. Plots of the third and fourth bands are shown in Fig. 13.9, in which the discontinuous electron orbits

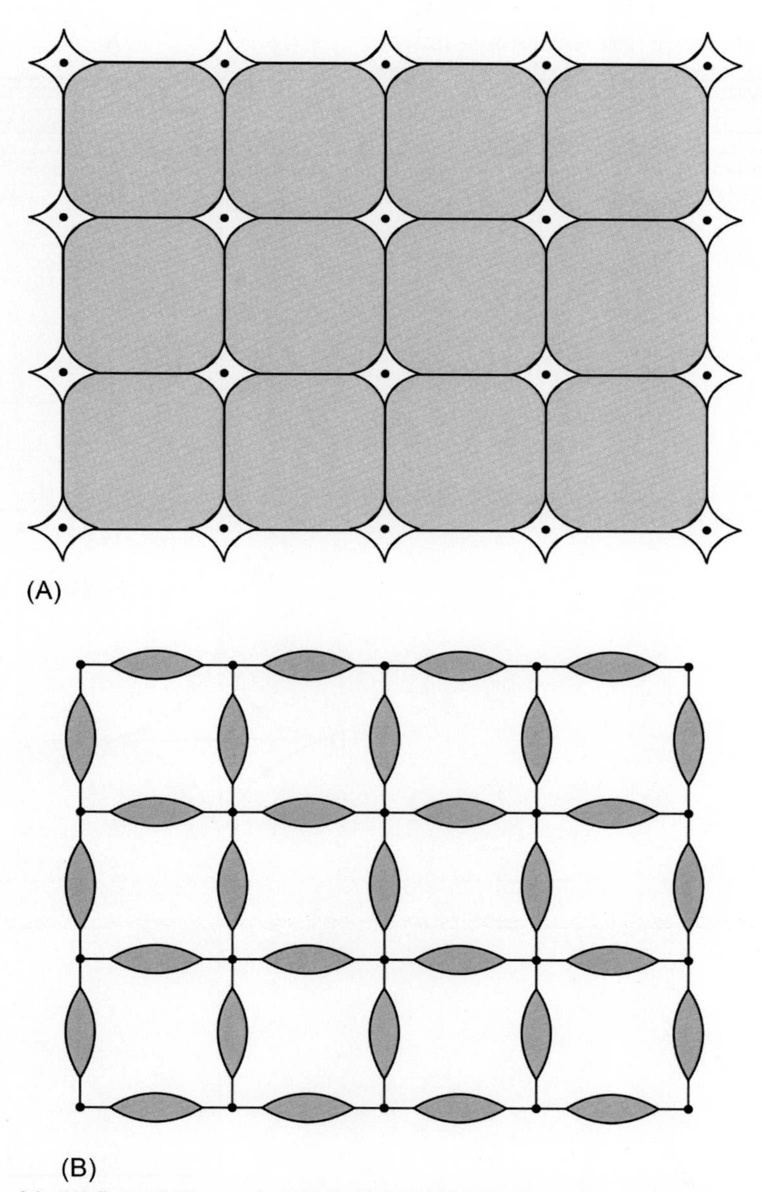

FIG. 13.4 The Fermi surface of the (A) first and (B) second energy bands in the periodic zone scheme in a two-dimensional free-electron gas having a square lattice.

are represented by continuous electron orbits. The third and fourth band Fermi surfaces exhibit electron pockets centered at the corners of the 1BZ. It is noteworthy that the hole orbit corresponding to the second band FS is continuous, even in the reduced band scheme, in contrast to the third and fourth band FS sections.

13.4 HARRISON'S CONSTRUCTION OF THE FERMI SURFACE

Harrison (1960) gave an elegant method for the construction of an FS corresponding to different bands of a crystalline solid in the periodic zone scheme. In this method, the reciprocal lattice corresponding to the crystal structure is determined and the free electron Fermi sphere is drawn around each lattice point. The problem is how to assign the various segments to the bands. Within each Fermi sphere all the states are occupied, while the states are empty outside. Now if some point, in the reduced zone scheme, lies within n spheres, then there are n occupied energy bands at that point in the reduced zone scheme and these bands are ordered in increasing energy. Therefore, the spherical segments, which separate regions of the reduced zone within n spheres from regions within (n+1) spheres are segments of the FS arising from the nth band, that is, these

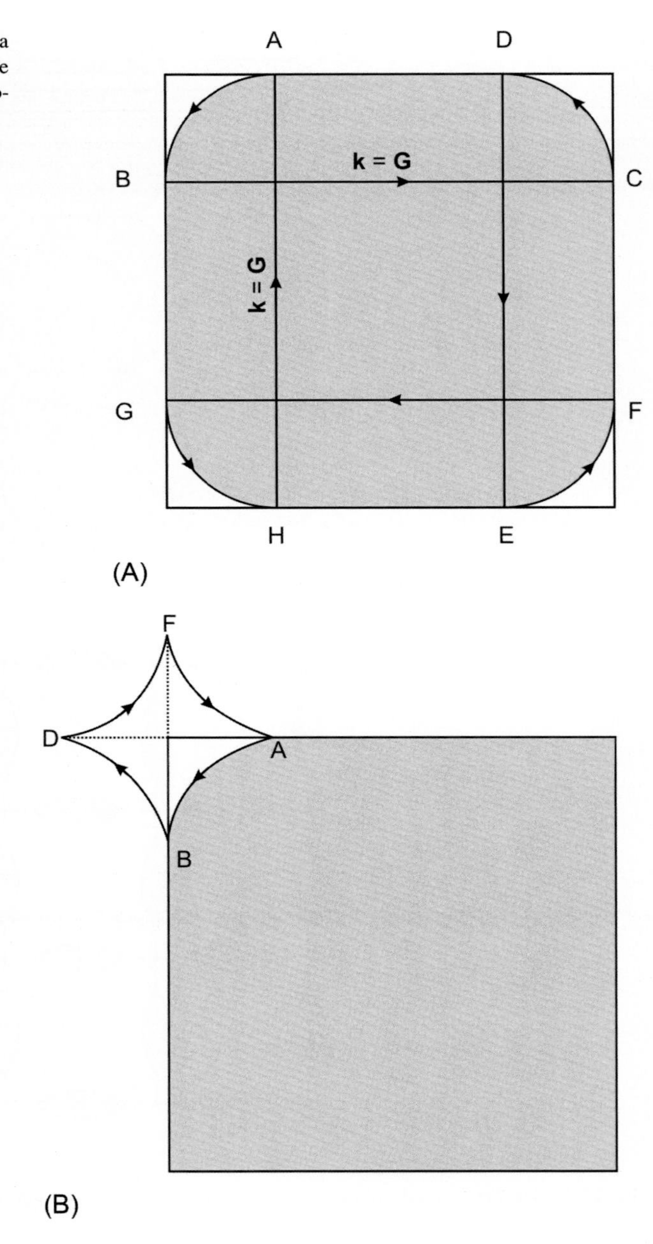

FIG. 13.5 (A) The discontinuous hole orbit of the first band in the 1BZ of a square lattice in a two-dimensional free-electron gas. (B) The continuous hole orbit of the first band in the periodic zone scheme of a square lattice in a two-dimensional free-electron gas.

segments separate occupied regions from unoccupied ones in the nth band. For example, points in the reciprocal space that lie within at least one sphere correspond to occupied states in the first zone (1st band). Points within at least two spheres correspond to occupied states in the second zone (2nd band), with similar results for points in three or more spheres. Hence, the construction of the FS arising from the various bands is reduced to the construction of Fermi spheres and counting them at a particular point or region.

Fig. 13.10 shows Harrison's construction of the FS for a square lattice in the free-electron approximation when the Fermi sphere extends to the 4BZ. It shows the second, third, and fourth bands in the periodic zone scheme. Here the 1BZ has been constructed in two ways. The first is the usual 1BZ, which contains one lattice point at its center and is called the 1BZ of type (a) for convenience. The second construction of the 1BZ has lattice points at the corners and is called the 1BZ of type (b). The first band fills the whole of the 1BZ. The FS of the second band in the 1BZ of type (a) represents a hole pocket with a lattice point at its center, while in the 1BZ of type (b) it consists of hole pockets at the corners of the 1BZ (see Fig. 13.10). The FS of the third band consists of four electron pockets, each at the corner of the 1BZ of type (a), but forming a rosette at the center of the 1BZ of type (b). The FS of the fourth band consists of four electron pockets at the corners of the

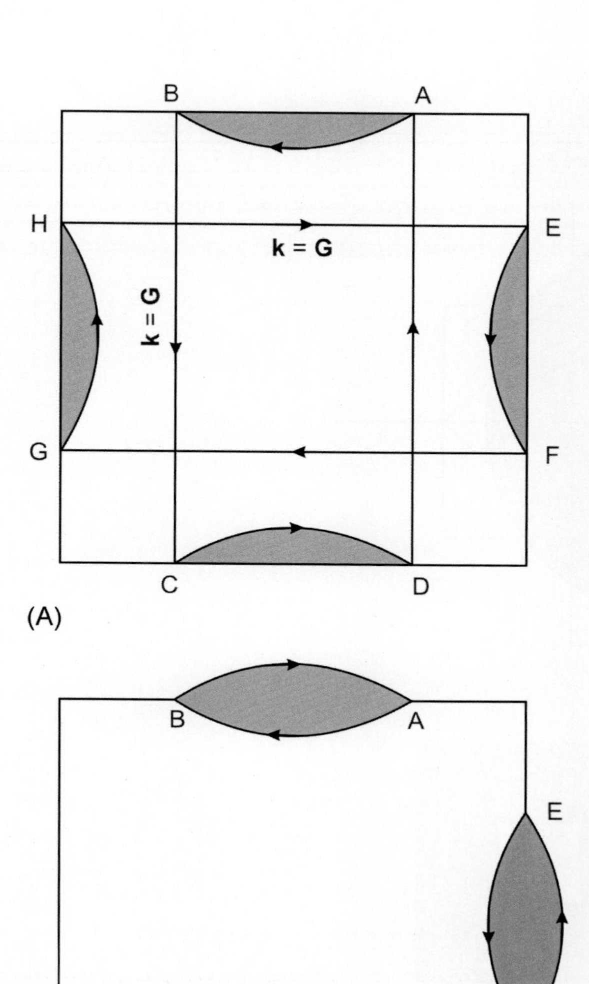

FIG. 13.6 (A) The discontinuous electron orbit of the second band, in the reduced zone scheme, of a square lattice in a two-dimensional free-electron gas. (B) The continuous electron orbit of the second band, in the periodic zone scheme, of a square lattice in a two-dimensional free-electron gas.

1BZ of type (a) but forms a single electron pocket at the center of the 1BZ of type (b). Fig. 13.11 shows the FS of the first four bands drawn in the 1BZ of type (a), while Fig. 13.12 shows the FS of the third and fourth bands drawn in the 1BZ of type (b). It is evident from Fig. 13.11 that one gets the same FS as shown in Fig. 13.7.

13.5 NEARLY FREE-ELECTRON APPROXIMATION

In the previous section the effect of the periodic lattice potential on the motion of an electron and on the FS has been neglected. In reality, an electron experiences the periodic potential of the lattice, which modifies its motion significantly, and hence modifies the electron energy bands and the Fermi surface in a crystalline solid. In Chapter 12, the nearly free-electron approximation, which yielded the lowest order improvement in the energy bands, was discussed. Therefore, it is worthwhile to construct the FS in this approximation and compare the results with those obtained in the free-electron approximation. In the nearly free-electron approximation the energy bands are modified as the zone boundary is reached in the following fashion:

1. The energy bands cut the BZ boundary in a perpendicular direction.
2. The bands exhibit an energy band gap at the zone boundary due to the rounding off of the energy bands.

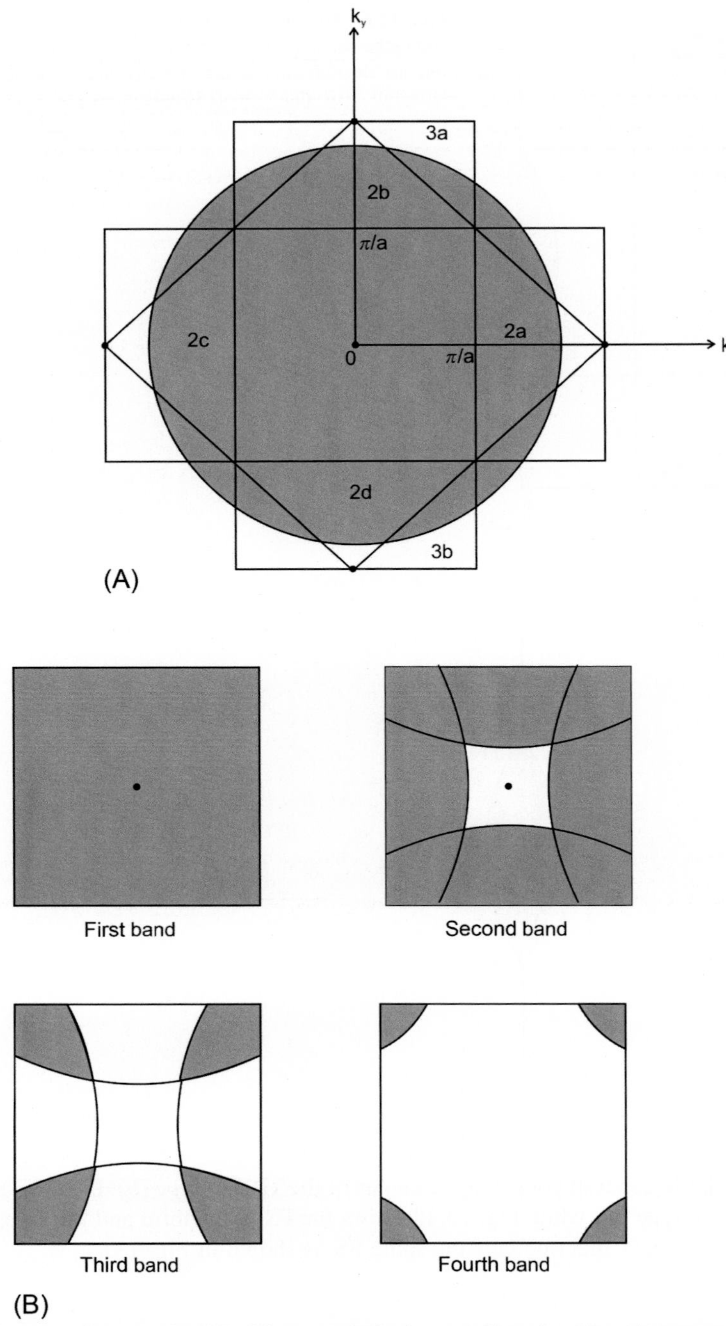

FIG. 13.7 (A) The Fermi surface extending up to the 4BZ of the square lattice in a two-dimensional free-electron gas. (B) The Fermi surfaces of the first four bands, in the reduced zone scheme, of a square lattice in a two-dimensional free-electron gas.

The FS in the nearly free-electron approximation, which extends to the 2BZ, is shown in Fig. 13.13. The FS in the reduced zone scheme is shown in Fig. 13.14 and in the periodic zone scheme in Fig. 13.15. A comparison of Figs. 13.4 and 13.15 shows that the Fermi surfaces for both the first and second bands, in the nearly free-electron approximation, are rounded off at the corners. Further, the FS intersects the BZ boundary in the perpendicular direction. It is noteworthy that the total volume enclosed by the FS depends only on the electron concentration and is independent of the details of the periodic potential.

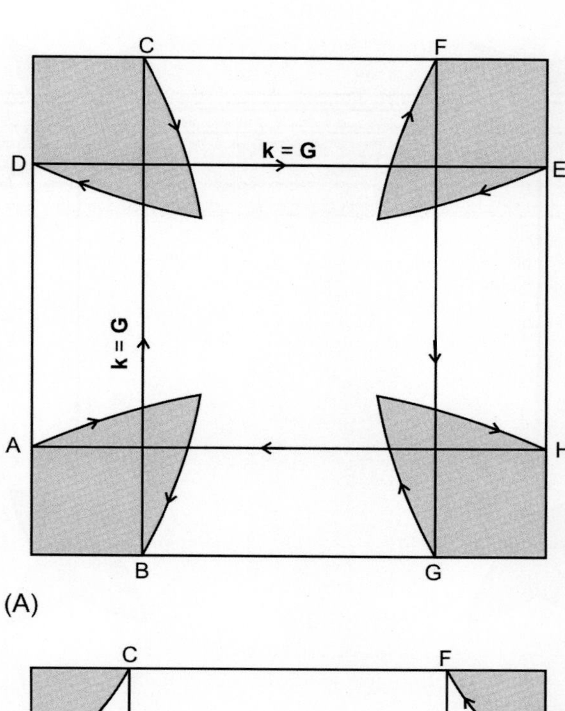

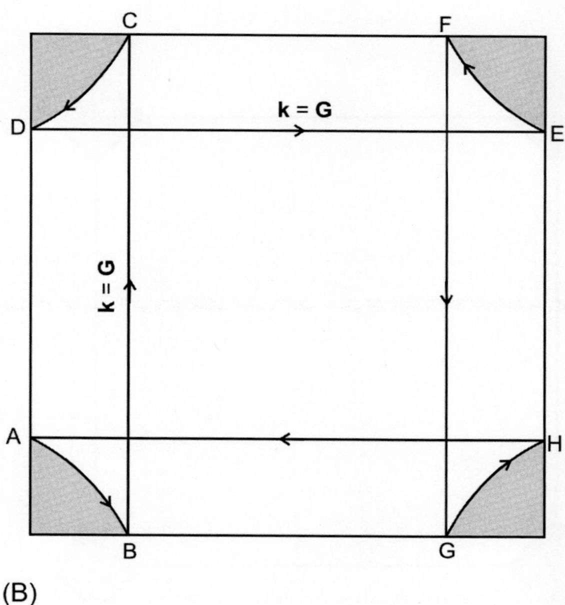

FIG. 13.8 The discontinuous electron orbits for the (A) third and (B) fourth bands, in the reduced zone scheme, of a square lattice in a free-electron gas.

13.6 THE ACTUAL FERMI SURFACES

Metal is a three-dimensional crystalline solid in which there can be more than one electron energy band lying on or crossing the FS. Each of these bands will yield a part of the FS, which separates the occupied states from the unoccupied ones. Here we shall consider only simple metals, which exhibit simple Fermi surfaces.

13.6.1 Monovalent Metals

Monovalent free-electron metals with conduction electrons possessing the s-character constitute the simplest of the metals: the alkali metals, such as Na and K, constitute such a category. Na metal has a bcc crystal structure with the 1NN distance as $\sqrt{3}a/2$ and the volume per atom $V_0 = a^3/2$. The reciprocal lattice of Na metal has fcc symmetry with the shortest reciprocal lattice vectors given by $(2\pi/a)(\pm 1, \pm 1, 0)$, $(2\pi/a)(\pm 1, 0, \pm 1)$, $(2\pi/a)(0, \pm 1, \pm 1)$. The boundary of the 1BZ is at a distance

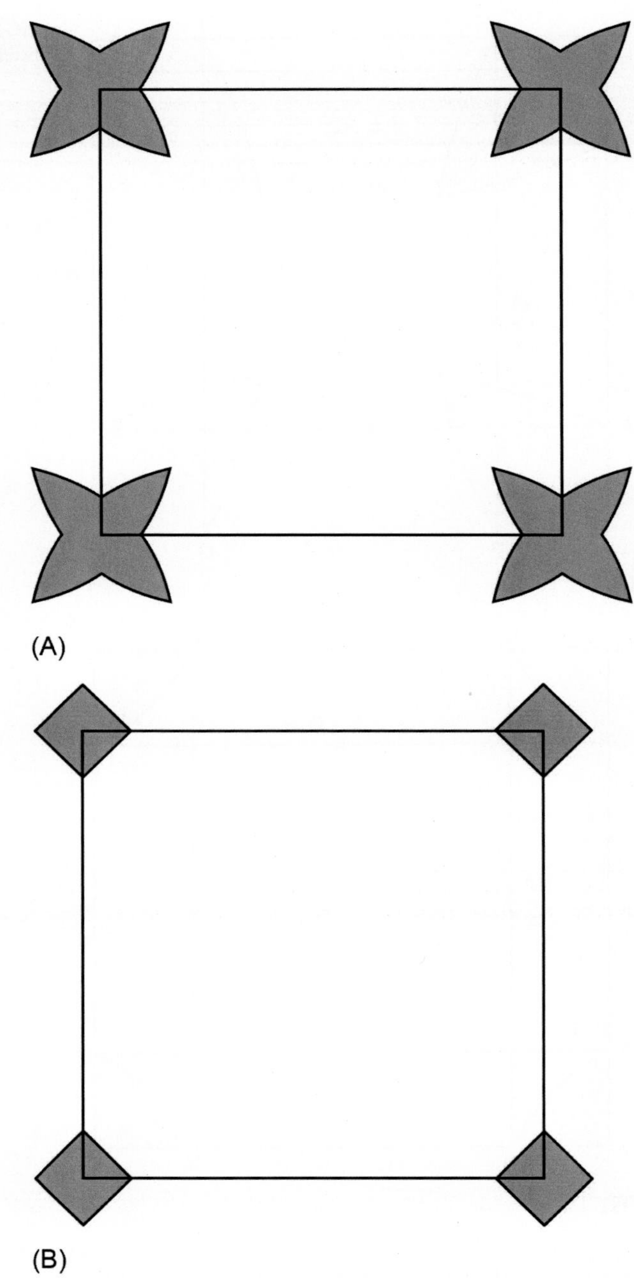

FIG. 13.9 The continuous electron orbits for the (A) third and (B) fourth bands, in the periodic zone scheme, of a square lattice in the free-electron approximation.

$d = \sqrt{2}\pi/a$, which is half of the distance of the 1NN in the reciprocal lattice. For a monovalent metal one can get from Eq. (9.19)

$$k_F = \left(\frac{3\pi^2}{V_0}\right)^{1/3} \tag{13.6}$$

For Na metal with a bcc structure one can write, from Eq. (13.6),

$$k_F = \frac{(6\pi^2)^{1/3}}{a} \tag{13.7}$$

Hence,

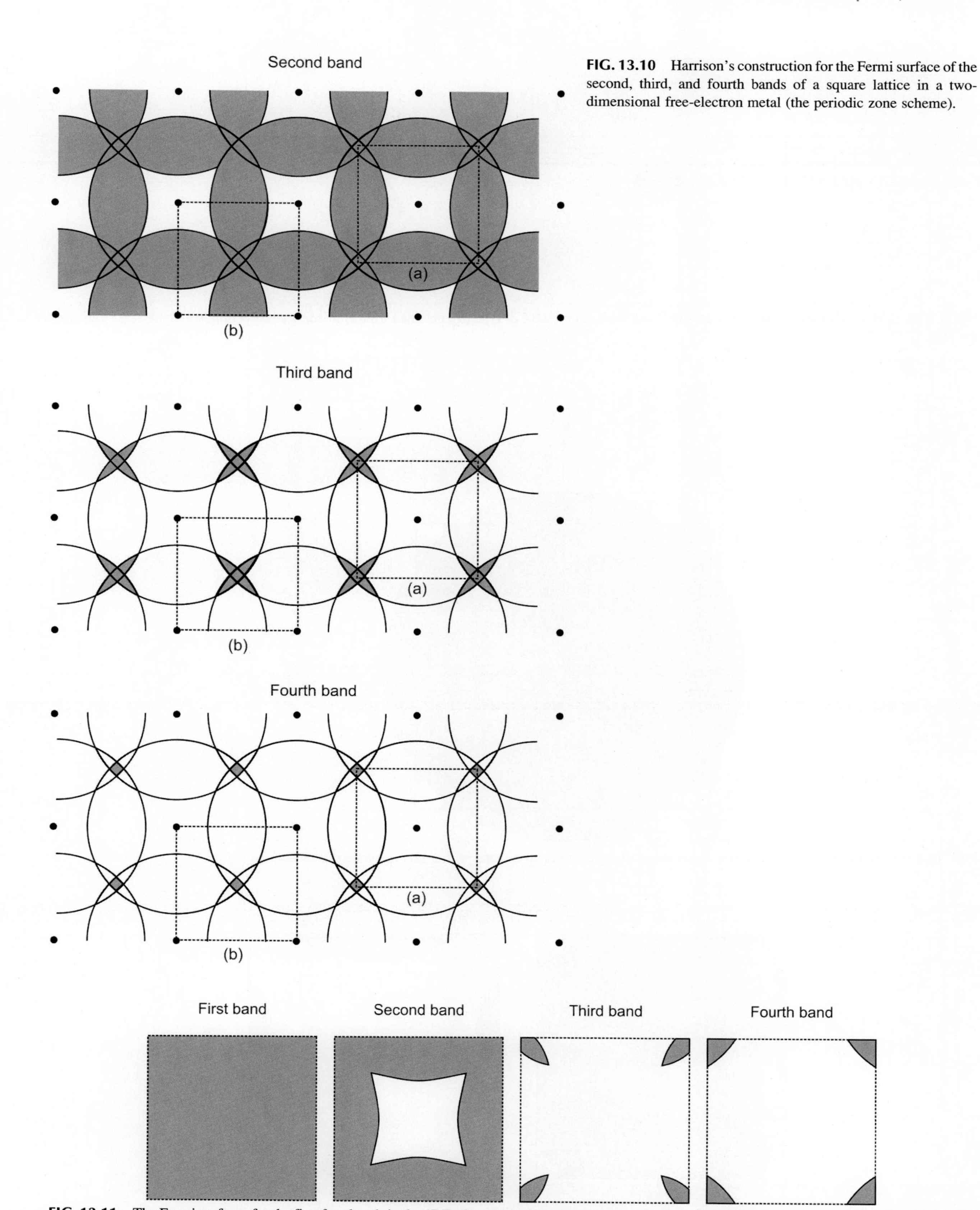

FIG. 13.10 Harrison's construction for the Fermi surface of the second, third, and fourth bands of a square lattice in a two-dimensional free-electron metal (the periodic zone scheme).

FIG. 13.11 The Fermi surfaces for the first four bands in the 1BZ of type (a) of the square lattice obtained from Harrison's construction in a free-electron metal (the reduced zone scheme).

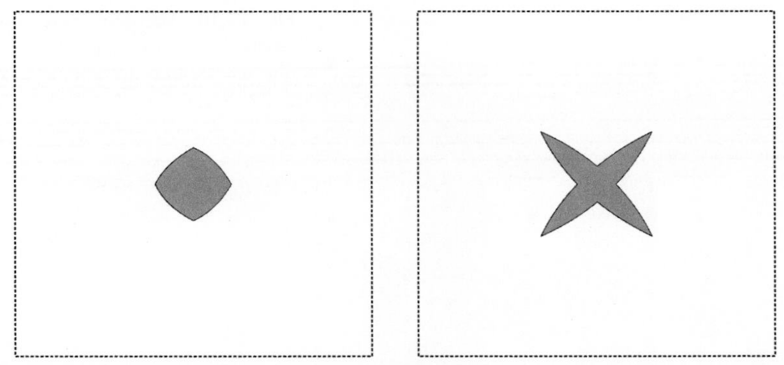

FIG. 13.12 The Fermi surfaces for the third and fourth bands in the 1BZ of type (b), obtained from Harrison's construction, which lie at the center of the 1BZ.

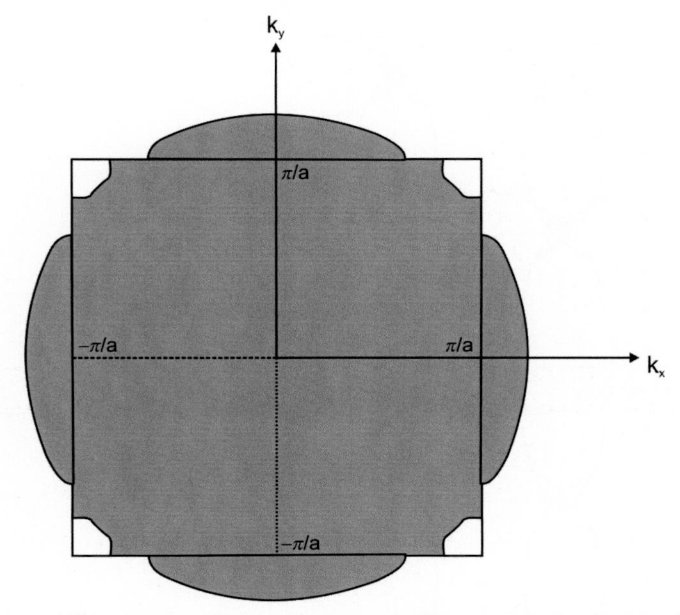

FIG. 13.13 The Fermi surface for a two-dimensional nearly free-electron metal with a square lattice that extends to the 2BZ.

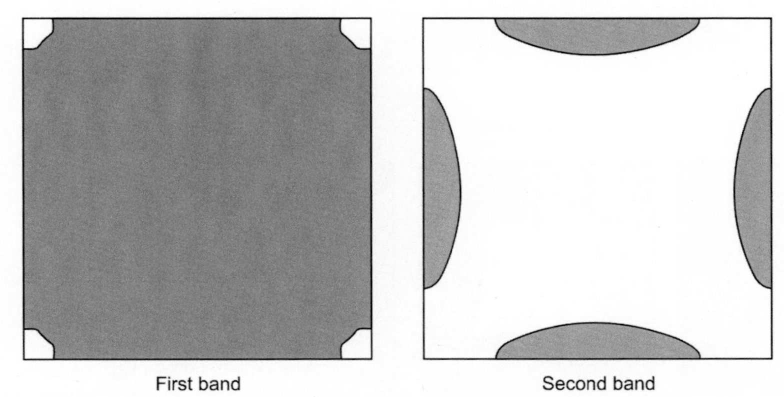

FIG. 13.14 The Fermi surface of the first and second energy bands in a square lattice for a two-dimensional nearly free-electron metal (the reduced zone scheme).

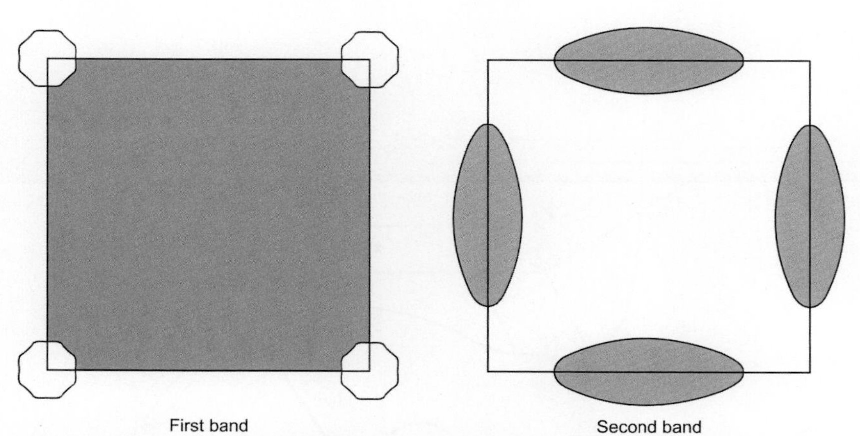

FIG. 13.15 The Fermi surface of the first and second energy bands in a square lattice for a two-dimensional nearly free-electron metal (in the periodic zone scheme).

$$\frac{k_F}{d} = \left(\frac{3}{\sqrt{2}\pi}\right)^{1/3} = 0.876 \tag{13.8}$$

From Eq. (13.8), it is evident that $k_F < d$. Hence, if the Na metal is treated as a free-electron metal, then the whole of the Fermi sphere will be well within the 1BZ as in Fig. 13.2. But, if Na is assumed to be a nearly free-electron metal, then there are zone boundary effects, that is, the bands bulge in the outward direction near the zone boundary and intersect it perpendicularly. Because of zone boundary effects, the FS of Na exhibits necks at the centers of the BZ faces (the points at which the FS reaches out to the zone boundary).

Consider nearly free-electron monovalent metals with fcc crystal structure, such as Cu, Ag, and others. Here the 1NN distance is $\sqrt{2}a$ and the atomic volume is $V_0 = a^3/4$. The reciprocal lattice of fcc is bcc and, therefore, the shortest reciprocal lattice vectors of Cu metal are $(2\pi/a)(\pm1, \pm1, \pm1)$ and the faces of the 1BZ lie midway to the points $(2\pi/a)(\pm1, \pm1, \pm1)$, etc. The distance of the zone face from the center of the 1BZ becomes $d = \sqrt{3}\pi/a$. Hence for monovalent Cu metal

$$\frac{k_F}{d} = \left(\frac{4}{\sqrt{3}\pi}\right)^{1/3} = 0.91 \tag{13.9}$$

Again in Cu, $k_F < d$ and the FS exhibits necks at the center of the BZ hexagonal faces due to the zone boundary effects. The FS for Cu in the reduced zone scheme is shown in Fig. 13.16. The figure shows two constant energy orbits: the first one is a belly orbit labeled as H_{100}, which should appear in the de Haas-van Alphen effect when a magnetic field **H** is applied along the [100] direction, and the second is a neck orbit labeled as N, which should appear with **H** along the [111] direction. Note that in Cu, d-electrons are well below E_F except for one d-subband. Therefore, the effect of the d-electrons on the FS is negligible. But, in general, the d-electrons give rise to strong and anisotropic localized potentials, which make the periodic lattice potential strong and anisotropic. As a result, the FS suffers large deviations from the nearly free-electron shape.

Problem 13.1

Show that in a monovalent metal with sc structure

$$\frac{k_F}{d} = \left(\frac{3}{\pi}\right)^{1/3} \tag{13.10}$$

Discuss the FS for such a metal.

13.6.2 Polyvalent Metals

Consider a divalent metal in the free-electron approximation. Using Eq. (9.19), the ratio k_F/d is given by

$$\frac{k_F}{d} = \frac{1}{d}\left(\frac{6\pi^2}{V_0}\right)^{1/3} \tag{13.11}$$

Therefore, for divalent metals with different structures, one can write

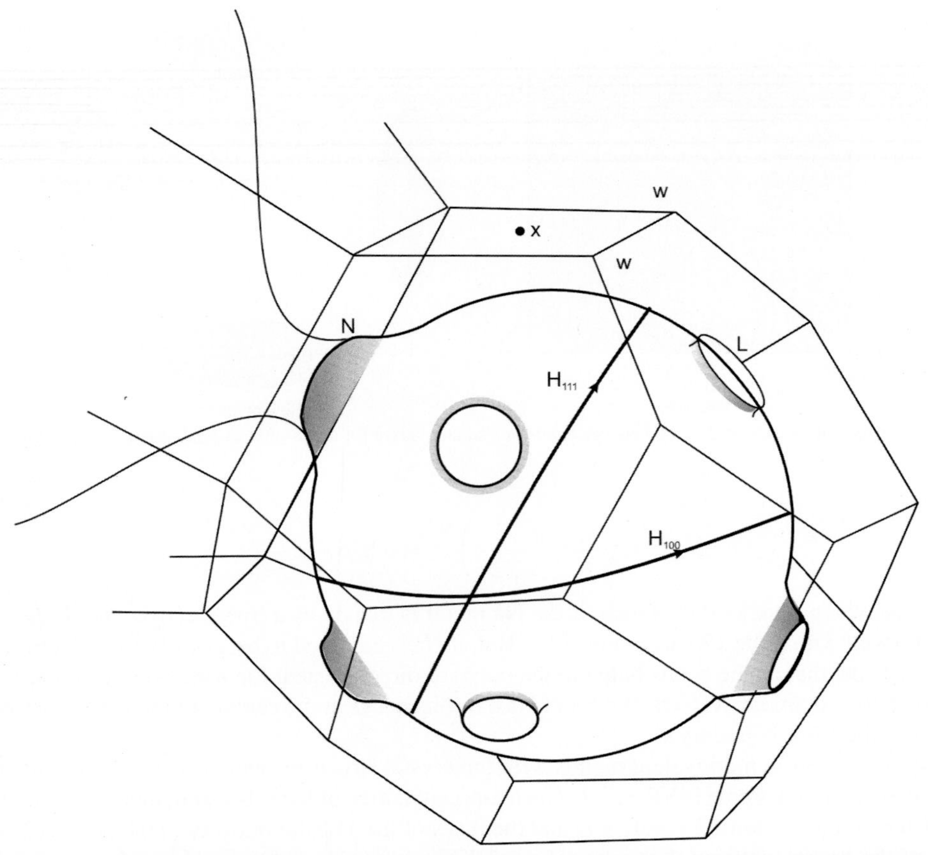

FIG. 13.16 The Fermi surface of the nearly free-electron metal Cu showing necks at the center of the zone faces. X and L are the symmetry points at the centers of the square and hexagonal faces, respectively, and W is the symmetry point at the corner of the face. *(After Burdick, G. A. (1963). Energy band structure of copper.* Physical Review, 129*, 138–150.)*

$$\begin{aligned}\frac{k_F}{d} &= \left(\frac{8}{\sqrt{3}\pi}\right)^{1/3} \text{ for fcc structure}\\ &= \left(\frac{3\sqrt{2}}{\pi}\right)^{1/3} \text{ for bcc structure}\\ &= \left(\frac{6}{\pi}\right)^{1/3} \text{ for sc stucture}\end{aligned} \tag{13.12}$$

Similarly, one can find the ratio k_F/d for trivalent metals in the free-electron approximation:

$$\begin{aligned}\frac{k_F}{d} &= \left(\frac{4\sqrt{3}}{\pi}\right)^{1/3} \text{ for fcc structure}\\ \frac{k_F}{d} &= \left(\frac{9}{\sqrt{2}\pi}\right)^{1/3} \text{ for bcc structure}\\ \frac{k_F}{d} &= \left(\frac{9}{\pi}\right)^{1/3} \text{ for sc structure}\end{aligned} \tag{13.13}$$

From Eqs. (13.12), (13.13) it is evident that $k_F > d$; therefore, in the polyvalent metals, the occupied bands extend to the higher-order BZs. As a result, the FS of polyvalent metals extends to the higher-order BZs. For example, Al, which is trivalent, is a nearly free-electron metal as is evident from its electron energy bands. The lower band exhibits 3s character, while the upper valence bands exhibit 3p character. Further, in Al, the lower 3s band is completely filled with 2N electrons, while the rest of the N valence electrons are distributed in the 3p bands.

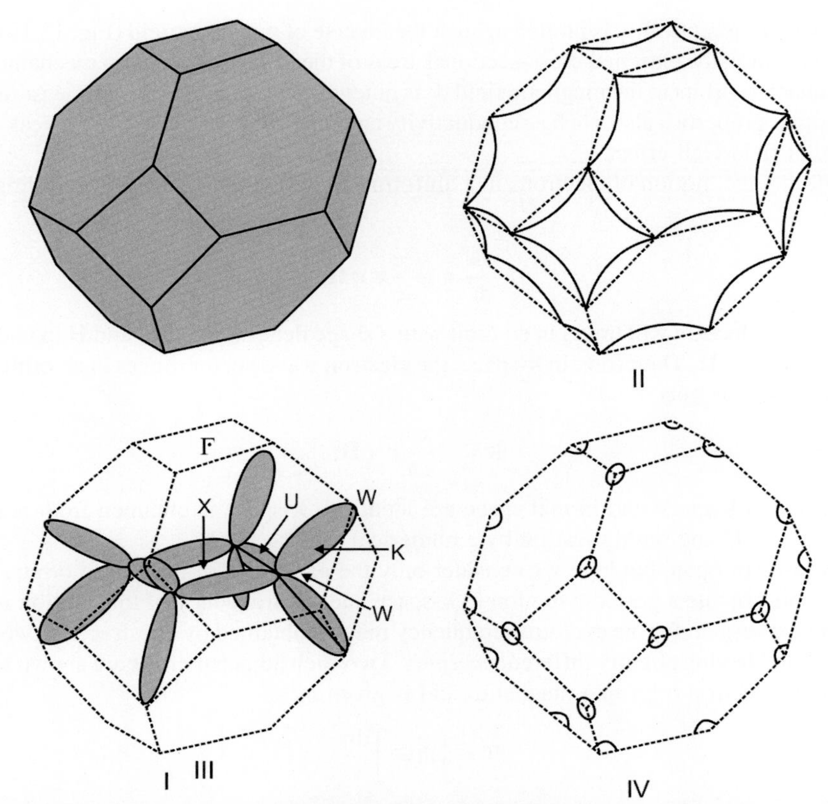

FIG. 13.17 The Fermi surface of Al metal in the free-electron model (Harrison, 1960). The first (I) band is completely full. In the second band (II) the region around the faces is full. In the third band (III) a complex region around the edges with many narrow areas is full. These narrow regions may be translated and put together to form a monster. In the fourth band there are small pockets around the W points that are full.

Harrison (1960) studied the FS of Al in the free-electron approximation, which extends to the 4BZ in the extended zone scheme (Fig. 13.17). It is evident that the first band is completely full with 2N electrons in it. The rest of the N valence electrons are distributed in the second, third, and fourth bands: in the second band the region around the zone faces is full, while in the third and fourth bands small electron pockets are formed around the edges and corners, respectively. The occupied region of the third band, when reduced to the 1BZ, forms a monster-like shape (see Fig. 13.17). The three-dimensional view of the FS of metals is complex and is still more complex for metals with geometry of low symmetry.

13.7 EXPERIMENTAL METHODS IN FERMI SURFACE STUDIES

A number of powerful experimental techniques have been developed for the study of the FS of metallic solids. Some of these methods are:

1. de Haas-van Alphen effect
2. Cyclotron resonance
3. Anomalous skin effect
4. Magnetoresistance
5. Ultrasonic propagation in magnetic fields

The de Haas-van Alphen effect and cyclotron resonance exhibit quantization of electronic states in the presence of a magnetic field. In these methods the effect of a uniform magnetic field on the electronic motion in **k**-space can be visualized. It is from this insight that the shape of the FS in **k**-space can be determined.

13.7.1 de Haas-van Alphen Effect

De Haas and Van Alphen, in 1931, discovered that, at low temperatures, the diamagnetic susceptibility χ_M of pure Bi shows periodic oscillations when plotted against the high values of an applied magnetic field **H**. These oscillations display a remarkable periodicity when susceptibility is plotted against the inverse of magnetic field (Fig. 13.18). This effect has been used successfully in determining the extremal cross-sectional areas of the FS. It is a quantum mechanical effect arising from the quantization of the electron orbits in the magnetic field. It is noteworthy that more precise measurements exhibit similar oscillatory behavior in other properties also, such as conductivity and magnetic resistance. Very weak oscillations have also been observed in the high-field Hall effect.

We begin by considering the motion of electrons in a uniform magnetic field **H,** which according to Newton's second law of motion gives

$$\hbar \frac{d\mathbf{k}}{dt} = -\frac{e}{c} \mathbf{v} \times \mathbf{H} \tag{13.14}$$

According to Eq. (13.14) the electron will travel in an orbit with a shape determined by **v** and **H** in real space and the rate of change of **k** is a vector normal to **H**. Therefore, in **k**-space, the electron wave vector moves in an orbit with its plane normal to **H**. Integrating Eq. (13.14) one gets

$$\mathbf{k} = -\frac{e}{c\hbar} \mathbf{r} \times \mathbf{H} \tag{13.15}$$

This implies that the orbits in **k**-space and in real space are identical: **k**-space is obtained from real space by a rotation through $\pi/2$ about the axis of **H** and multiplication by a numerical factor $eH/c\hbar$.

The orbits may be closed or open, but here we consider only the properties of the closed orbits. In a closed orbit, the electron wave vector **k** will execute a periodic motion in **k**-space and the frequency of this motion is called the cyclotron frequency. A convenient expression for the cyclotron frequency may be obtained by constructing two orbits in **k**-space in a plane perpendicular to **H** and having slightly different energies. Two such adjacent orbits are shown in Fig. J2 of Appendix J. The time period T of an electron orbit in a magnetic field is given by

$$T = \oint dt = \oint \frac{d\mathbf{r}}{\mathbf{v}} \tag{13.16}$$

where **v** is the velocity of the electron in a band with energy E in **k**-space and is given by

$$v = \frac{dr}{dt} = \frac{1}{\hbar}\frac{dE}{dk} = \frac{1}{\hbar}\frac{dE}{dk_\perp} \tag{13.17}$$

where $d\mathbf{k}_\perp$ is the normal distance in **k**-space between constant energy surfaces of energy E and E + dE. Substituting **v** from Eq. (13.17) into Eq. (13.16), we find

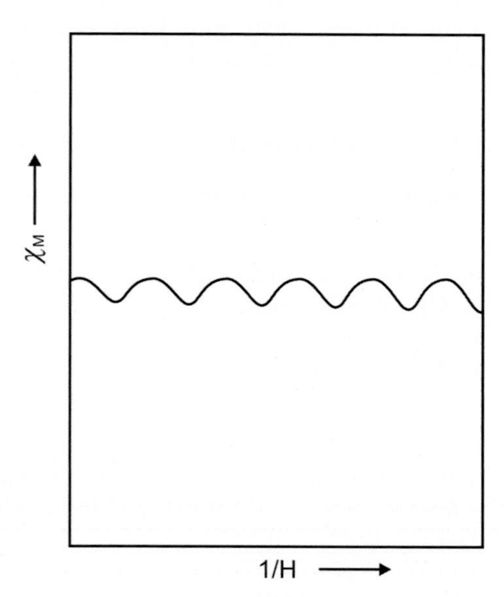

FIG. 13.18 The diamagnetic susceptibility of Bi as a function of 1/H at high magnetic field H.

$$T = \hbar\oint\frac{\mathbf{dr}\times\mathbf{dk}_\perp}{dE} = \frac{c\hbar^2}{eH}\oint\frac{|\mathbf{dk}\times\mathbf{dk}_\perp|}{dE} = \frac{c\hbar^2}{eH}\frac{dA_e}{dE} \tag{13.18}$$

Here **dk** is an infinitesimal change in the wave vector along the orbit in an infinitesimal time dt. Therefore, the term $\oint|d\mathbf{k}\times\mathbf{dk}_\perp|$ is simply the area dA_e between the two orbits in wave vector space. The cyclotron frequency is given by

$$\omega_c = \frac{2\pi}{T} = \frac{2\pi eH}{c\hbar^2}\frac{dE}{dA_e} \tag{13.19}$$

The derivative of the orbital area with respect to energy is taken at a constant component of **k** parallel to the magnetic field. The cyclotron frequency from Eq. (10.24) can be written as

$$\omega_c = \frac{eH}{m_c^* c} \tag{13.20}$$

where m_c^* is the effective cyclotron mass. Comparing Eqs. (13.19), (13.20), m_c^* can be written as

$$m_c^* = \frac{\hbar^2}{2\pi}\frac{dA_e}{dE} \tag{13.21}$$

So far, the discussion of electron motion has been classical. Even under such conditions, the Bohr correspondence principle provides the quantization condition as

$$\oint\mathbf{p}\cdot\mathbf{dr} = 2\pi\hbar\left(n+\frac{1}{2}\right) \tag{13.22}$$

The momentum of a free electron in the presence of a magnetic field changes as

$$\mathbf{p}\rightarrow\mathbf{p}-\frac{e}{c}\mathbf{A} \tag{13.23}$$

where **A** is the vector potential defined as

$$\mathbf{H} = \nabla\times\mathbf{A} \tag{13.24}$$

Substituting for **k** from Eq. (13.15) in Eq. (13.23) and then **p** in Eq. (13.22), we get

$$\oint\mathbf{H}\cdot(\mathbf{r}\times\mathbf{dr}) - \oint\mathbf{A}\cdot\mathbf{dr} = \frac{2\pi\hbar c}{e}\left(n+\frac{1}{2}\right)$$

Here the integral $\oint\mathbf{r}\times d\mathbf{r}$ gives twice the area of the orbit, so that $\oint\mathbf{H}\cdot(\mathbf{r}\times\mathbf{dr})$ is twice the magnetic flux Φ. The term $\oint\mathbf{A}\cdot d\mathbf{r}$ also gives magnetic flux Φ. Therefore, the above equation, in terms of the magnetic flux Φ, can be written as

$$\Phi = \frac{2\pi\hbar c}{e}\left(n+\frac{1}{2}\right) \tag{13.25}$$

Eq. (13.25) implies that the magnetic flux through an electron orbit in real space is quantized in units of $2\pi\hbar c/e$. The magnetic flux is given by $\Phi = HA_r$ where A_r is the area of the orbit in real space. We have seen before, in this section itself, that the radius of the orbit in **k**-space is $eH/c\hbar$ larger than that of the orbit in **r**-space. Therefore

$$A_e = \left(\frac{eH}{c\hbar}\right)^2 A_r \tag{13.26}$$

One should note that A_e represents the extremal cross-sectional area of the FS in a plane normal to the magnetic field **H**. The extremal area is either a maximum or minimum area of the cross section of the FS; therefore, the derivative of A_e with respect to **k** must be zero at that point. Substituting the value of A_r, one gets the magnetic flux as

$$\Phi = \frac{c^2\hbar^2}{e^2H}A_e = \frac{2\pi\hbar c}{e}\left(n+\frac{1}{2}\right)$$

From the above expression one gets

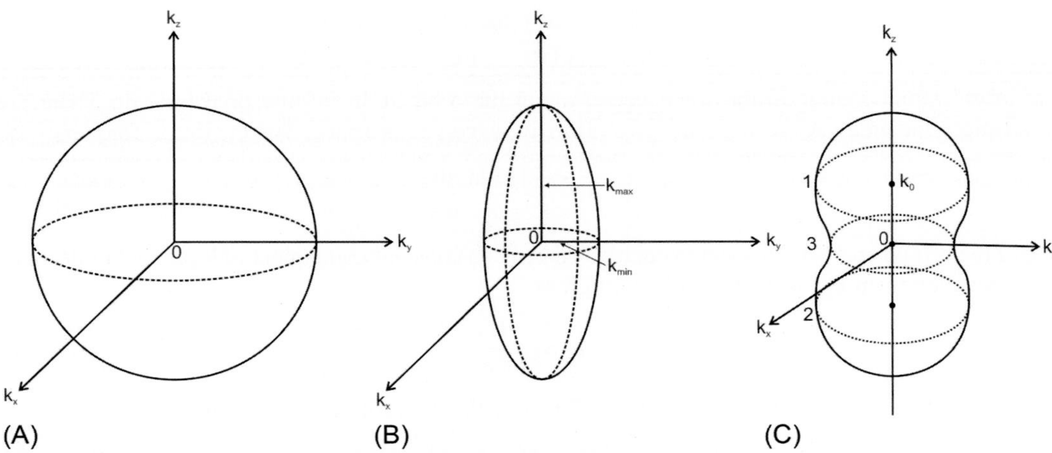

FIG. 13.19 The extremal areas of cross section in spherical, ellipsoidal, and dumbbell shaped Fermi surfaces.

$$A_e = \frac{2\pi e H}{c\hbar}\left(n + \frac{1}{2}\right) \tag{13.27}$$

This is known as the Onsager-Lifshitz quantization condition and this is the basis of the de Haas-van Alphen effect. Thus, the quantum condition allows only a certain discrete set of orbital areas in **k**-space. The size of these orbits is directly proportional to the magnetic field. The effect of a magnetic field has already been described in Chapter 10. In the absence of a magnetic field, the allowed states in two dimensions are shown in Fig. 10.7A, while they are shown in Fig. 10.7B in the presence of a magnetic field. Evidently the effect of the magnetic field is to create quantized circles in the **k**-space and cause the free electron states to condense into the nearest circle. These are the familiar Landau levels, as discussed in Chapter 10. The successive Landau levels correspond to successive values for the quantum number n. The reciprocal of the magnetic field, from Eq. (13.27), is given by

$$\frac{1}{H} = \frac{2\pi e}{A_e \hbar c}\left(n + \frac{1}{2}\right) \tag{13.28}$$

The reciprocal of the magnetic field induces fluctuations in the magnetic susceptibility. The period of oscillation is inversely proportional to the cross-sectional area of the FS. In three dimensions the allowed states lie in tubes in **k**-space each of which has constant cross section in the planes perpendicular to the magnetic field. Such sets of tubes or cylinders are shown in Fig. 10.5B. Each tube has been cut off at a constant energy surface corresponding to the FS.

In the determination of the FS, a magnetic field is applied at different angles to the axis of the single crystal and time period T is measured as a function of **H**. It then becomes possible to measure the extremal area of the FS normal to the direction of **H** using Eq. (13.28). The extremal areas for simple shapes of the Fermi surface are shown in Fig. 13.19. In a spherical Fermi surface there is only one extremal (maximum) area for all the directions of the **H** field and that is a circle having area πk_F^2 (Fig. 13.19A). In the case of an ellipsoidal Fermi surface with k_{max} and k_{min} as the major and minor axes, the magnitude and shape of the extremal area depends on the direction of the **H** field. If the **H** field is applied along the z-direction, then the extremal area is a circle with area πk_{min}^2 (see Fig. 13.19B). But if the **H** field is along the y-direction, the extremal area is an ellipse with area $\pi k_{max} k_{min}$. For a dumbbell-shaped Fermi surface, the extremal areas are shown in Fig. 13.19C. If the **H** field is applied in the z-direction, then there are three extremal areas: two circular orbits with maximum area (labeled as 1 and 2) and one circular orbit (labeled as 3) with minimum area. But if the **H** field is in the y-direction, then there is one extremal area having a dumbbell shape.

The oscillatory behavior of the magnetic susceptibility and other related properties can be explained as follows. The quantum condition produces a sharp oscillatory structure in the electron density of states, with its peak occurring at the energy corresponding to the extremal orbit satisfying the quantum condition. At the extremal orbit, the area of the portion of the tube is enormously enhanced as a result of the slow variation of the energy along the tube near the given orbit.

13.7.2 Cyclotron Resonance

The cyclotron resonance method makes use of the fact that, if an rf electric field is applied to a metallic solid, it penetrates at the surface by a small distance (skin depth). The Azbel-Kaner geometry is often employed for the study of cyclotron

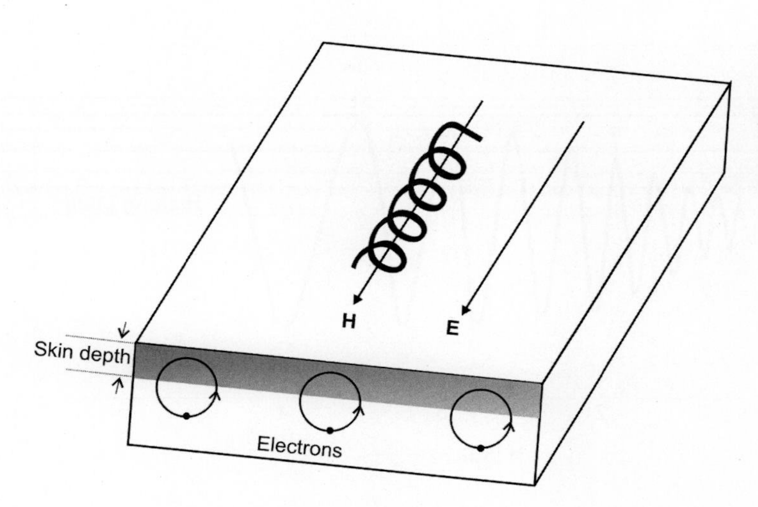

FIG. 13.20 The Azbel-Kaner geometry for cyclotron resonance in a metallic slab with **H** and **E** fields parallel to each other and in the same direction (longitudinal geometry). The circular motion of the electrons is shown in the front face of the metallic slab, which is perpendicular to **H** field. The shaded region shows the extent of penetration of the field inside the slab.

resonances in metals: here an rf electric field **E** and static magnetic field **H** are applied parallel to the surface of a metallic slab. In this geometry two types of studies are performed: first when **E** and **H** are perpendicular to each other (transverse geometry) and second when **E** and **H** are parallel to each other (longitudinal geometry). Fig. 13.20 shows the longitudinal geometry in which **E** and **H** are in the same direction. Note that if **E** and **H** are parallel but in opposite directions, then the direction of motion of the electrons is reversed. The shading in the figure near the surface of the solid indicates the penetration depth of the rf field. Under the combined effect of both the **E** and **H** fields, the electron moves in a helical path. Fig. 13.20 also shows the circular path of the electrons perpendicular to the direction of **H** in the front face of the metallic slab. The frequency of circular motion, called the cyclotron frequency, is given by Eq. (13.20). The time period T of the electron orbit is defined by

$$T = \frac{2\pi}{\omega_c} = \frac{2\pi m_c^* c}{eH} \tag{13.29}$$

The radius of the orbit of an electron in a magnetic field of 10 kilogauss is on the order of 10^{-3} cm, which is much larger than the skin depth at radio frequencies in a pure metal at low temperatures. Electrons in orbits, such as those shown in Fig. 13.20, will see the rf field only for a small part (near the top) of each cycle of their motion. The electrons are accelerated in each cycle if the phase of the electrons when they arrive in the skin depth part of each cycle is the same as that of the rf field. This will happen only when the frequency of the rf field ω is equal to an integral multiple of the cyclotron frequency, that is,

$$\omega = n\omega_c \tag{13.30}$$

Here n is an integer 1, 2, 3, … and it defines the harmonics of the cyclotron frequency. This is called the Azbel-Kaner resonance or cyclotron resonance and in this condition the electrons absorb the maximum energy. Substituting the value of ω_c from Eq. (13.20) into Eq. (13.30), we find

$$\omega = \frac{neH}{m_c^* c} \tag{13.31}$$

which gives

$$H = \frac{\omega m_c^* c}{ne} \tag{13.32}$$

One can express this resonance condition in terms of the extremal areas of cross section by substituting m_c^* from Eq. (13.21) into Eq. (13.32) to get

$$\frac{1}{H} = \frac{2\pi ne}{c\hbar^2\omega}\frac{dE}{dA_e} \tag{13.33}$$

Fig. 13.21 represents the Azbel-Kaner cyclotron resonance spectrum for Cu metal at 4.2 K. The ordinate of the curve represents the derivative of the surface resistivity with respect to the field. In general, the electrons in different regions of the surface have different cyclotron frequencies. But the frequency that is most pronounced in absorption is the frequency

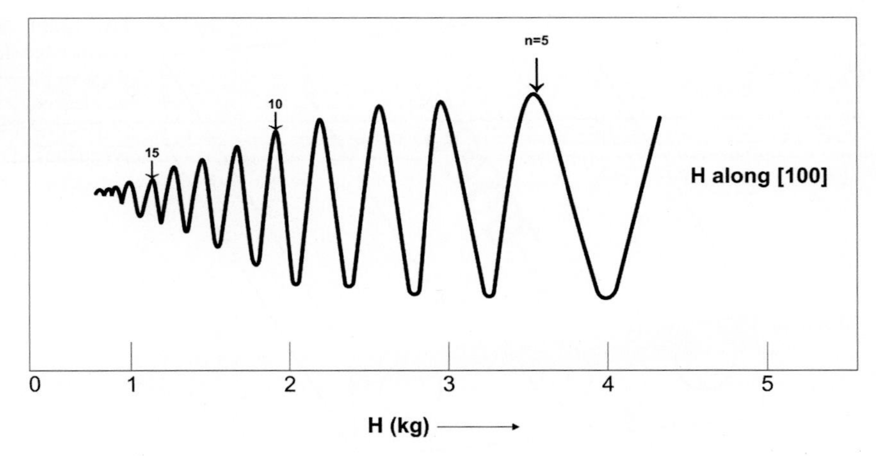

FIG. 13.21 Azbel-Kaner cyclotron resonance spectrum for Cu metal at 4 K. The upper crystal surface is cut along the (100) plane. *(After Haussler, P. & Welles, S. J. (1966). Determination of relaxation times in cyclotron resonance in copper.* Physical Review, 152, *675.)*

appropriate to the extremal orbit in which the FS cross section perpendicular to **H** is maximum or minimum. Therefore, by varying the orientation of **H**, one can measure the extremal sections in various directions and reconstruct the FS. One can determine the various electronic properties and their oscillatory behavior in the same way as described in the de Haas-van Alphen effect.

REFERENCES

Harrison, W. A. (1960). Electronic structure of polyvalent metals. *Physical Review*, *118*, 1190–1208.

SUGGESTED READING

Altmann, S. L. (1970). *Band theory of metals*. New York: Pergamon Press.

Callaway, J. (1958). Electron energy bands in solids. F. Seitz, & D. Turnbull (Eds.), *Solid state physics* (pp. 100–212). Vol. 7(pp. 100–212). New York: Academic Press.

Cornwell, J. F. (1969). *Selected topics in solid state physics: Group theory and electronic energy bands in solids*. Amsterdam: North-Holland Publ. Co.

Cracknell, A. P., & Wong, K. C. (1973). *The Fermi surface: Its concept, determination and use in the physics of metals*. London: Clarendon Press.

FURTHER READING

Burdick, G. A. (1963). Energy band structure of copper. *Physical Review*, *129*, 138–150.

Chapter 14

Semiconductors

Chapter Outline

Semiconductors are of immense value from both the technological and industrial point of view and the band gap in them is usually less than 2 eV. They can be classified into two categories. The first category comprises pure semiconductors called *intrinsic semiconductors*. In these semiconductors most of the properties are structure dependent. At absolute zero the intrinsic semiconductors behave as insulators and their conductivity increases with an increase in temperature. But these semiconductors remain poor conductors of electricity at temperatures of interest. The most common examples of intrinsic semiconductors are Si and Ge, which possess energy band gaps of the order of 1.1 and 0.72 eV, respectively. The properties of semiconductors change drastically with the presence of even very small amounts of impurities and other imperfections ($\approx$100–1000 ppm). For example, the electrical conductivity of Si and Ge increases many-fold with the addition of a very small As impurity. The second category constitutes semiconductors with impurities, which are called *extrinsic semiconductors*, and these have brought a revolution in electronics and condensed matter physics. Common examples of extrinsic semiconductors are SiAs, SiIn, GeAs, and GeIn, in which the first element represents the host semiconductor and the second element the impurity. In this chapter, we will present the basics of semiconductors and their properties in terms of the energy band theory.

14.1 INTRINSIC SEMICONDUCTORS

The intrinsic semiconductor Si possesses the following electronic structure:

$$\mathrm{Si}: 1s^2 2s^2 2p^6 3s^2 3p^2$$

A Si atom exhibits valency 4 and forms pure covalent bonds by sharing its four valence electrons with its four neighboring Si atoms. Fig. 14.1A presents a schematic representation for the formation of four covalent bonds. One should note that the four neighboring Si atoms are not coplanar with the given Si atom, as shown in Fig. 14.1A. In reality the Si atom forms four hybrid sp^3 orbitals with its four neighboring Si atoms situated at the corners of a tetrahedron, as shown in Fig. 14.1B. With the formation of the covalent bonds, the valence band of each Si atom becomes full with 8 electrons in it. The binding energy of each electron in Si is 1.1 eV. The energy band diagram of Si is shown in Fig. 14.2. At absolute zero the valence band is completely full, while the conduction band is completely empty, and both are separated by an energy band gap E_g of 1.1 eV. Further, at absolute zero, the Fermi energy E_F lies exactly in the middle of E_g. With an increase in temperature, a small fraction of the electrons in the valence band gets excited to the conduction band, giving rise to finite conductivity. At room temperature (300 K), the thermal energy E_T has a value of 0.026 eV, which is very small compared with the band gap.

Solid State Physics. https://doi.org/10.1016/B978-0-12-817103-5.00014-1

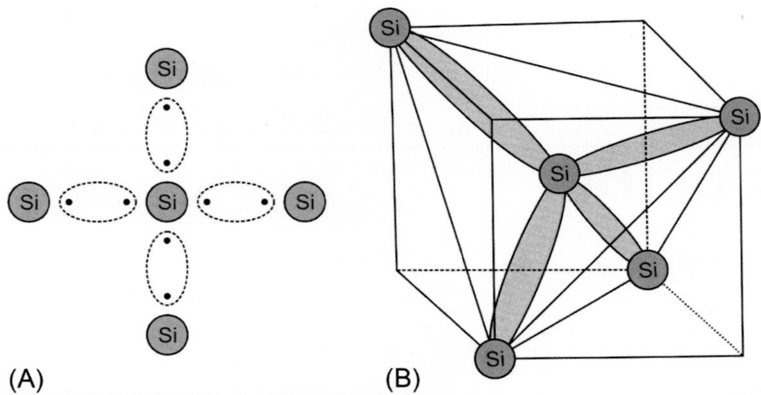

FIG. 14.1 (A) A schematic representation of the covalent bond formation of a Si atom with its four 1NN Si atoms in an intrinsic semiconductor. The two dots between the two Si atoms represent the shared electrons forming a covalent bond. (B) The four sp^3-hybrid bonding orbitals in a crystalline solid of Si.

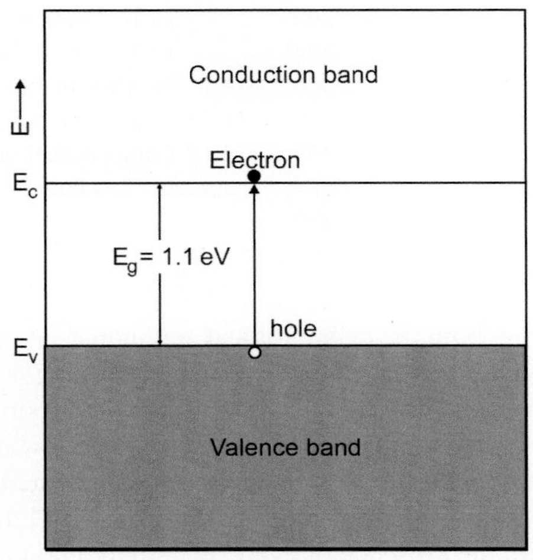

FIG. 14.2 Schematic energy band model of a crystalline solid of Si. The valence band is completely filled, with E_v as the energy of its topmost filled state. The conduction band is completely empty, with E_c as its lowest state.

So, at room temperature an intrinsic Si semiconductor behaves nearly as an insulator. With a further increase in temperature, more and more electrons go to the conduction band, thereby increasing the conductivity of the intrinsic semiconductor. The excitation of an electron from the valence to conduction band leaves behind a hole (absence of an electron) in the valence band (Fig. 14.2). For all practical purposes the hole acts as a particle with negative mass and having a charge equal and opposite to that of an electron. Therefore, in an intrinsic semiconductor, the number density of free electrons n_e in the conduction band is equal to the number density of holes n_h in the valence, that is,

$$n_e = n_h \tag{14.1}$$

In the presence of an external electric field, the electrons move in the conduction band, while the holes move in the valence band but in the opposite direction, and both contribute toward the conductivity.

There are two categories of intrinsic semiconductors: direct band gap and indirect band gap semiconductors. A semiconductor with a valence band maximum and conduction band minimum at the center of the BZ (see Fig. 14.3A) is called a *direct band gap semiconductor*. Fig. 14.3B shows explicitly that the motion of the hole in the valence band is equivalent to the motion of the valence electron in the reverse direction. But there is no physical reason why both the valence band maximum and the conduction band minimum should lie at the center of the BZ. The minimum of the conduction band could lie at any point inside the BZ (even at the boundary of the BZ), as shown in Fig. 14.4. A semiconductor with valence band maximum and conduction band minimum situated at different values of the wave vector in the 1BZ is called an *indirect band gap semiconductor*. In this class of semiconductors an electron can jump from the valence band to the conduction band in two ways. First, an electron can follow the path PQR (path 1) in which both the momentum and energy of the electron change: the momentum of the electron in the valence band increases by $\hbar \mathbf{k}'$ in going from P to Q and then it

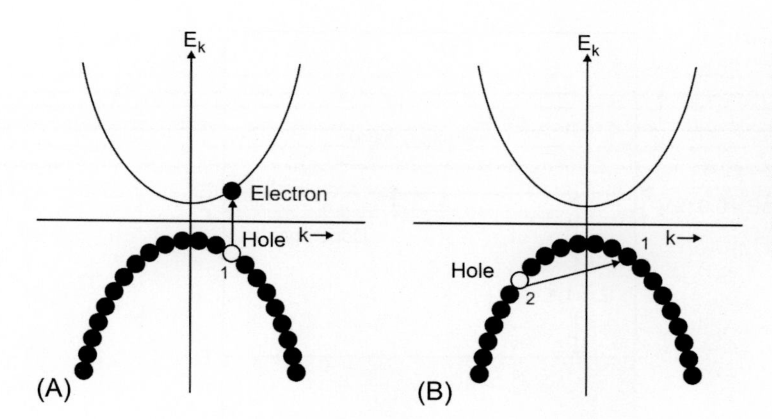

FIG. 14.3 Schematic representation of the band structure of a semiconductor near the zone center. (A) The electron at position 1 makes a jump from the completely full valence band to the conduction band leaving behind a hole (empty space) (B) In the valence band the motion of a hole in the **k**-space in one direction is equivalent to the motion of an electron in the opposite direction. Here an electron from position 2 moves to position 1, thereby filling the hole in position 1. It is equivalent to the motion of the hole from position 1 to position 2.

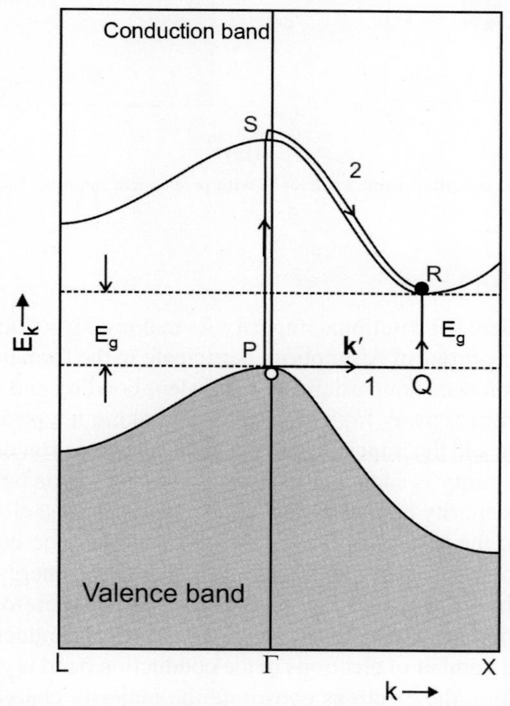

FIG. 14.4 Schematic representation of the band structure of an indirect gap semiconductor. Two paths are shown, labeled 1 and 2, for the excitation of an electron from the valence band to the conduction band.

gains energy (=E_g) in going from Q to R. Second, an electron in the valence band can be promoted directly upward to a higher energy state at the same momentum along the path PS, then the excited electron falls down into the conduction band minimum along the path SR by gaining momentum but losing energy (path 2), as seen in Fig. 14.4.

14.2 EXTRINSIC SEMICONDUCTORS

One of the methods to increase conductivity is to introduce a substitutional impurity in small amounts in the lattice of an intrinsic semiconductor. The introduction of a substitutional impurity in an otherwise pure semiconductor is called *doping*. It has been observed that even a very small level of impurity (100–1000 ppm) drastically changes the conductivity of a semiconductor. Either a pentavalent or a trivalent impurity is introduced in a semiconductor, which creates an imbalance between electrons in the conduction band and holes in the valence band. The extrinsic semiconductors are of two types:

1. Semiconductors with pentavalent impurities are called *n-type semiconductors*. Common examples are SiAs and GeAs.
2. Semiconductors with trivalent impurities are called *p-type semiconductors*. Common examples are SiIn and GeIn.

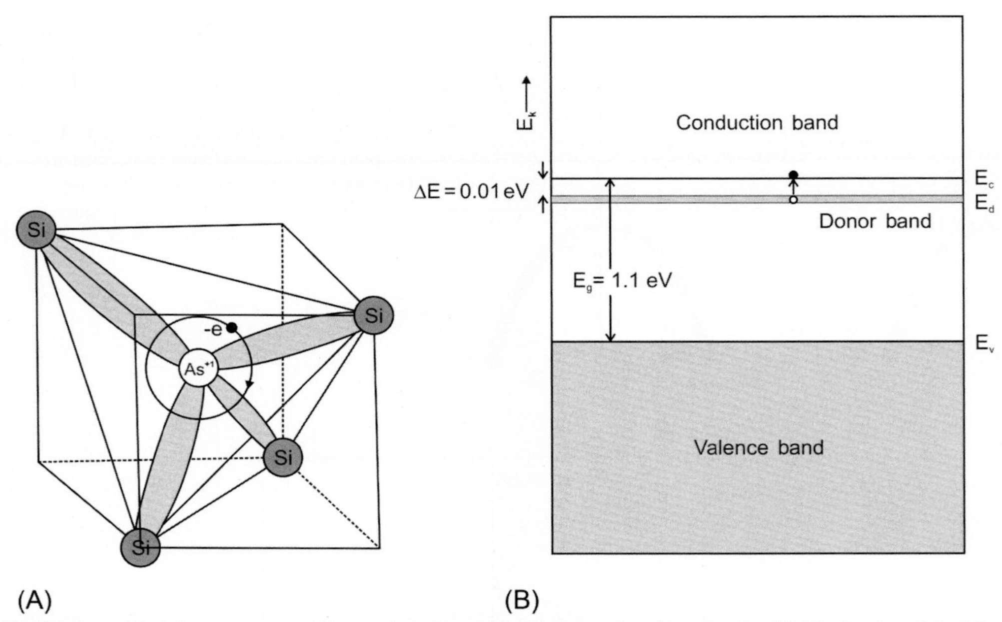

FIG. 14.5 (A) The bond model of the n-type crystalline semiconductor Si with pentavalent impurity As. (B) The band model of the n-type crystalline semiconductor Si with pentavalent impurity As.

14.2.1 n-Type Semiconductors

Consider a Si crystal with the pentavalent substitutional impurity As making SiAs semiconductor. The valence bond model of SiAs is shown in Fig. 14.5A. Four electrons of As impurity participate in the formation of covalent bonds with the neighboring Si atoms, while the fifth electron does not participate in covalent bonding and is loosely bound to the As atom. As a result of weak binding, this electron moves away from the impurity, making it a positively charged ion, that is, As^{+1}. The force of attraction between the electron and the impurity ion As^{+1} makes the electron orbit around the As^{+1} ion. The distance of the fifth electron from the As^{+1} impurity is such that its ground state energy is below but very close to the conduction band of Si (see Fig. 14.5B). Such an impurity band is called a *donor band* and the electrons in it are called the *donor electrons*. The energy difference between the bottom of the conduction band and the donor band is $\Delta E \approx 0.01$ eV, which is smaller than the thermal energy at room temperature. On the other hand, the energy difference between the donor band and the top of the valence band is on the order of 1.09 eV, which is quite large. Therefore, at room temperature, the electrons in the donor band jump to the conduction band, yielding appreciable electrical conductivity. Such a semiconductor is called an n-type semiconductor, in which the number of electrons in the conduction band is greater than the number of holes in the valence band, that is, $n_e > n_h$. Therefore, the electrons constitute the majority charge carriers and the holes the minority charge carriers in n-type semiconductors.

14.2.2 p-Type Semiconductors

Trivalent impurities, such as B, Al, Ga, and In, can also be substituted in a semiconductor. Fig. 14.6A shows the bond model of an In impurity in a Si semiconductor. The three valence electrons of an In atom form three covalent bonds with three Si atoms, while the fourth bond contains one electron and one hole forming an unsaturated bond. The hole moves away from the In impurity when the neighboring electron occupies its position. In this process the fourth bond also becomes saturated and the In impurity acquires negative charge, that is, In^{-1}, thereby attracting the hole toward it. The force of attraction between the hole and the In^{-1} impurity ion makes the hole move around the In^{-1} in an orbit. The distance of the hole from the In^{-1} ion is such that its energy band lies above, but very near, the valence band. The band model of SiIn semiconductor is shown in Fig. 14.6B. At room temperature the electrons from the top of the valence band jump to fill the empty states of the impurity band, thereby creating holes in the valence band. The impurity band, therefore, is called the *accepter band*. Such a semiconductor is called a p-type semiconductor as the number of holes in the valence band is greater than the number of electrons in the conduction band, that is, $n_h > n_e$. Therefore, the holes constitute the majority carriers and the electrons the minority carriers in a p-type semiconductor. Both the electrons in the conduction band and the holes in the valence band

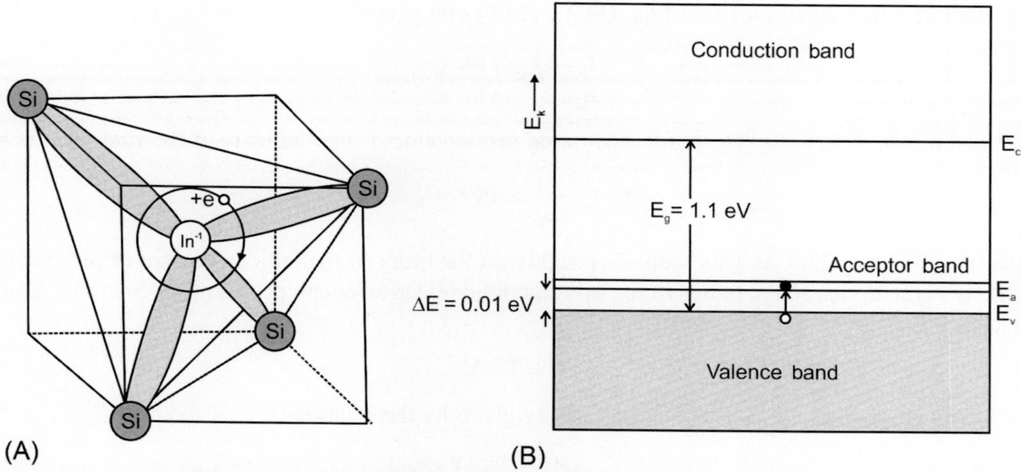

FIG. 14.6 (A) The bond model of the p-type crystalline semiconductor Si with trivalent impurity In. (B) The band model of the p-type crystalline semiconductor Si with trivalent impurity In.

contribute toward the conduction properties, such as the electrical conductivity, of the extrinsic semiconductors. From the above discussion we conclude that the extrinsic semiconductors are materials with mixed bonding: partly ionic and partly covalent.

It is important to note that donor impurities, such as P, Sb, and As, are used in doping because the donor bands formed by these impurities are close to the conduction band and, therefore, yield a significant carrier concentration at room temperature. Similarly, B, Al, Ga, and In are used as acceptor impurities as the acceptor bands formed by these impurities lie close to the valence band. It is noteworthy that transition metals, such as Fe, Ni, Co, and Cu, if added to Si or Ge, form impurity bands far removed from the edges of the valence and conduction bands and, therefore, are not suitable doping materials.

14.3 IONIZATION ENERGY OF IMPURITY

In Section 14.2.1 it was assumed that the donor electron orbits in a circle around the donor ion embedded in a semiconductor. The energy of the donor electron depends on the radius of the orbit. The energy required to raise the donor electron to the conduction band is usually called the ionization energy of the donor atom or impurity. As the simplest approximation, one can make use of the Bohr model to estimate the ionization energy of the donor atom. One should keep in mind that the use of the Bohr model is reasonably justified if the radius of the orbit of the donor electron is on the order of the Bohr radius a_0.

If the donor electron is assumed to move in a circular orbit of radius r_n around the donor ion, then its electrostatic attractive force must be balanced by the centripetal force, that is,

$$\frac{e^2}{\varepsilon r_n^2} = \frac{m_e^* v^2}{r_n} \tag{14.2}$$

where m_e^* and v are the effective mass and velocity of the donor electron. As the donor atom is embedded in a semiconductor with dielectric constant ε, the electrostatic force is decreased by a factor of ε. In the Bohr model the angular momentum of the donor electron is quantized and is given as

$$m_e^* v r_n = n\hbar \tag{14.3}$$

where n is a positive integer. Substituting the value of v from Eq. (14.3) into Eq. (14.2) and simplifying, one gets

$$r_n = \varepsilon \frac{n^2 \hbar^2}{m_e^* e^2} \tag{14.4}$$

The Bohr radius for a free electron in a hydrogen atom is given by

$$a_0 = \frac{\hbar^2}{m_e e^2} \tag{14.5}$$

Here m_e is the mass of a free electron. From Eqs. (14.4), (14.5) one gets

$$\frac{r_n}{a_0} = \varepsilon n^2 \left(\frac{m_e}{m_e^*}\right) \tag{14.6}$$

For Si, $\varepsilon = 11.7$ and $m_e^*/m_e = 0.26$, therefore, for a Si-based semiconductor the radius r_1 of the first orbit is given by

$$\frac{r_1}{a_0} = 45 \tag{14.7}$$

The above equation gives $r_1 = 23.9$ Å. This radius is roughly on the order of the lattice constant of Si. One unit cell of Si contains effectively eight atoms, so the radius of the orbiting donor electron encompasses many Si atoms. The total energy of the donor electron E is given by

$$E = T + V \tag{14.8}$$

where T is the kinetic energy and V is the potential energy given by the relations

$$T = \frac{1}{2} m_e^* v^2 \tag{14.9}$$

$$V = -\frac{e^2}{\varepsilon r_n} \tag{14.10}$$

Substituting the value of v from Eq. (14.3) into Eq. (14.9), one can write

$$T = \frac{m_e^* e^4}{2\varepsilon^2 n^2 \hbar^2} \tag{14.11}$$

Similarly, substituting the value of r_n from Eq. (14.4) into Eq. (14.10), we get

$$V = -\frac{m_e^* e^4}{\varepsilon^2 n^2 \hbar^2} \tag{14.12}$$

From Eqs. (14.8), (14.11), and (14.12), the total energy of the donor electron is given by

$$E = -\frac{m_e^* e^4}{2\varepsilon^2 n^2 \hbar^2} \tag{14.13}$$

For a hydrogen atom, one can write $m_e^* = m_e$ and $\varepsilon = 1$ for vacuum. The energy of the lowest orbit of the hydrogen atom is given by

$$E_0 = -\frac{m_e e^4}{2\hbar^2} = -13.6 \text{ eV} \tag{14.14}$$

The energy of the lowest state of Si is obtained by substituting its values for ε and m_e^* to give

$$E_0(\text{Si}) = -0.0258 \text{ eV} \tag{14.15}$$

which is much less than the energy gap E_g in Si. So, a small energy of 0.0258 eV is required by the electron in the donor band to make the transition to the conduction band. This simple calculation based on the Bohr model shows two facts:

1. The donor energy band is very close to the conduction band.
2. The donor electron can be excited to the conduction band at room temperature (300 K)

The Bohr model can also be applied to evaluate the ionization energy for the acceptor impurities in Si and Ge. The experimental values of the ionization energies of some of the donor and acceptor impurities in Si and Ge are given in Table 14.1.

14.4 CARRIER MOBILITY

In an intrinsic or extrinsic semiconductor, there are two types of charge carriers: electrons in the conduction band and holes in the valence band. When an electric field **E** is applied, both the charge carriers move in opposite directions. Both the electrons and holes lose energy through collisions among themselves and acquire some constant velocity, usually called

TABLE 14.1 Impurity Ionization Energies in Semiconductors Si and Ge

Impurity	Ionization Energy (eV)	
	Si	Ge
Donors		
P	0.045	0.012
As	0.050	0.0127
Acceptors		
B	0.045	0.0104
Al	0.060	0.0102

the drift velocity. In Chapter 11, expressions were derived for the drift velocity and mobility of both the electrons and holes. The total conductivity in a semiconductor is the sum of electronic and hole contributions and is given by (see Eq. 11.179)

$$\sigma = \sigma_e + \sigma_h = n_e e \mu_e + n_h e \mu_h \tag{14.16}$$

where μ_e and μ_h are the mobilities of the electrons and holes, respectively. The reciprocal of conductivity is resistivity ρ, that is,

$$\rho = \frac{1}{\sigma} = \frac{1}{(n_e \mu_e + n_h \mu_h) e} \tag{14.17}$$

and is measured in Ω cm. It is evident from Eqs. (14.16), (14.17) that knowledge of n_e, μ_e, n_h, and μ_h is required to evaluate the conductivity or resistivity of a semiconductor. It has been observed experimentally that resistivity ρ is not a linear function of the impurity concentration in extrinsic semiconductors.

In the laboratory we measure current I and voltage V; therefore, one should express Eqs. (14.16), (14.17) in terms of I and V. Consider a bar of semiconducting material (Fig. 14.7) with A as the area of cross section and L as the length, then

$$J = \frac{I}{A}, \quad E = \frac{V}{L} \tag{14.18}$$

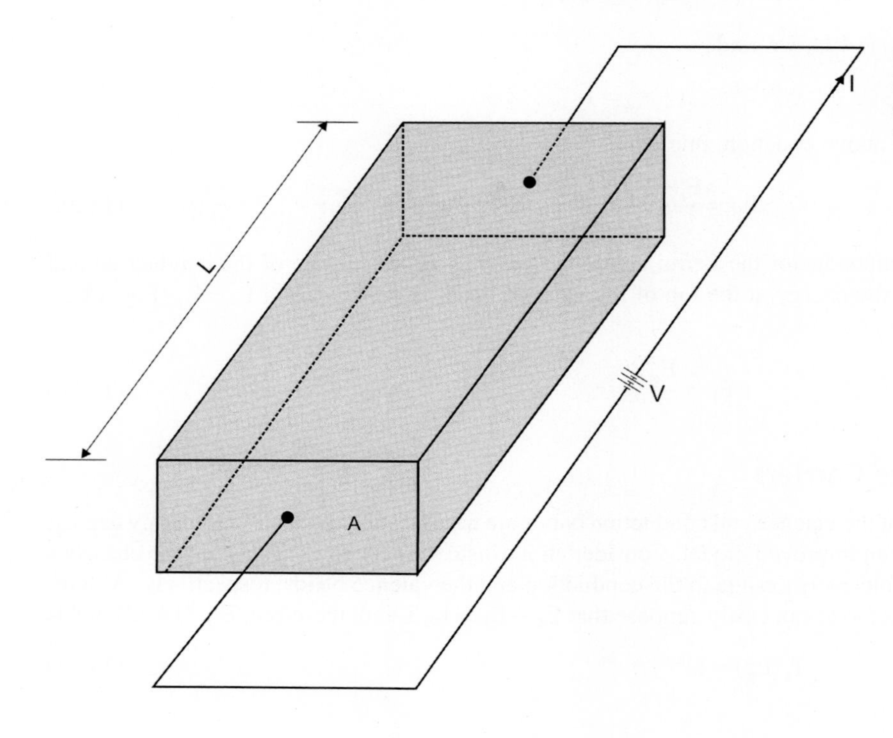

FIG. 14.7 A bar of a semiconducting material, having length L and area of cross section A, acts as a resistor. The current I flows through the material when a voltage V is applied across it.

The expression for the current density is given by

$$\mathbf{J}=\sigma\mathbf{E} \tag{14.19}$$

Substituting Eq. (14.18) into Eq. (14.19), one can write

$$V=IR \tag{14.20}$$

where

$$R=\rho\frac{L}{A}=\frac{L}{A\sigma} \tag{14.21}$$

Eq. (14.20) is the famous Ohm's law. Eq. (14.21) shows that the resistance is a function of resistivity (conductivity) as well as the geometry of the sample of the semiconductor.

14.5 THEORY OF INTRINSIC SEMICONDUCTORS

Consider a semiconductor with a number N electron states per unit volume. At absolute zero, in an intrinsic semiconductor, the conduction band is completely empty and the valence band is completely filled (see Fig. 14.2). At finite temperature, some electrons from the valence band get excited to the conduction band and the number of electrons in the conduction band is equal to the number of holes in the valence band in an intrinsic semiconductor. Further, the holes are created near the top of the valence band in a small energy range, while the electrons accumulate near the bottom of the conduction band and their energy range is considered to be much smaller than the energy gap E_g. Let E_c be the energy at the bottom of the conduction band and E_v the energy at the top of the valence band. The Fermi energy E_F lies somewhere in the middle of E_c and E_v.The probability of occupation of an electron state with energy $E_{\mathbf{k}}$ is given by the Fermi distribution function defined as

$$f_e(E_{\mathbf{k}})=\frac{1}{e^{(E_{\mathbf{k}}-E_F)/k_B T}+1} \tag{14.22}$$

The Fermi-Dirac distribution function for the holes becomes $f_h(E_{\mathbf{k}})=1-f_e(E_{\mathbf{k}})$. The density of electrons n_e at E_c and of holes n_h at E_v are given by

$$n_e=Nf_e(E_c)=\frac{N}{e^{(E_c-E_F)/k_B T}+1} \tag{14.23}$$

$$n_h=Nf_h(E_v)=\frac{N}{e^{-(E_v-E_F)/k_B T}+1} \tag{14.24}$$

Substituting Eqs. (14.23), (14.24) into Eq. (14.1), we find

$$e^{(E_c+E_v-2E_F)/k_B T}=1$$

Taking the logarithm of both sides of the above equation, one gets

$$E_F=\frac{E_c+E_v}{2} \tag{14.25}$$

Eq. (14.25) shows that in an intrinsic semiconductor the Fermi energy lies exactly in the middle of the conduction and valence bands. If zero energy is taken as the energy at the top of the valence band, then $E_v=0$ and $E_c=E_g$ (Fig. 14.2); therefore, one can write

$$E_F=\frac{E_g}{2} \tag{14.26}$$

14.5.1 Concentration of Charge Carriers

In the preceding section it was assumed that the valence and conduction bands are narrow compared with the energy gap E_g. If they are not narrow, then one has to use an improved model. Consider an intrinsic semiconductor in which the electrons and holes are distributed over an appreciable energy range in the conduction and the valence bands, respectively. At temperatures of interest (say room temperature), one can easily suppose that $E_{\mathbf{k}}-E_F \gg k_B T$ and, therefore, Eq. (14.22) yields

$$f_e(E_{\mathbf{k}})=e^{(E_F-E_{\mathbf{k}})/k_B T} \tag{14.27}$$

Near the bottom of the conduction band the energy of the electrons is a quadratic function of the wave vector **k** and is therefore given by

$$E_{\mathbf{k}} = E_c + \frac{\hbar k^2}{2m_e^*} \tag{14.28}$$

where m_e^* is the effective mass of an electron. The density of electron states per unit energy per unit volume at absolute zero is given by (Eq. 9.23)

$$g_e(E_{\mathbf{k}}) = \frac{1}{2\pi^2}\left(\frac{2m_e^*}{\hbar^2}\right)^{3/2}(E_{\mathbf{k}} - E_c)^{1/2} \tag{14.29}$$

Therefore, the density of electrons per unit volume in the conduction band at finite temperature T is given by

$$n_e = \int_{E_c}^{top} dE_{\mathbf{k}} f_e(E_{\mathbf{k}}) g_e(E_{\mathbf{k}}) \tag{14.30}$$

Substituting Eq. (14.27), (14.29) into Eq. (14.30), one gets

$$n_e = \frac{1}{2\pi^2}\left(\frac{2m_e^*}{\hbar^2}\right)^{3/2} e^{E_F/k_B T}\int_{E_c}^{\infty} dE_{\mathbf{k}} (E_{\mathbf{k}} - E_c)^{1/2} e^{-E_{\mathbf{k}}/k_B T} \tag{14.31}$$

Here we have used the fact that $f_e(E_{\mathbf{k}})$ decreases very quickly with an increase in $E_{\mathbf{k}}$; therefore, in Eq. (14.30) the upper limit can safely be taken as infinity. Now substituting

$$E_{\mathbf{k}} - E_c = x \tag{14.32}$$

one can write Eq. (14.31) as

$$n_e = \frac{1}{2\pi^2}\left(\frac{2m_e^*}{\hbar^2}\right)^{3/2} e^{(E_F - E_c)/k_B T}\int_0^{\infty} dx\, x^{1/2} e^{-x/k_B T} \tag{14.33}$$

Again substituting

$$y = \frac{x}{k_B T} \tag{14.34}$$

in Eq. (14.33), one can write

$$n_e = \frac{1}{2\pi^2}\left(\frac{2m_e^*}{\hbar^2}\right)^{3/2} e^{(E_F - E_c)/k_B T}(k_B T)^{3/2}\int_0^{\infty} dy\, y^{1/2} e^{-y} \tag{14.35}$$

The integral in Eq. (14.35) is a standard integral whose value is $\sqrt{\pi}/2$. Therefore, Eq. (14.35) finally gives the density of electrons in the conduction band as

$$n_e = N_e e^{(E_F - E_c)/k_B T} \tag{14.36}$$

where

$$N_e = 2\left(\frac{m_e^* k_B T}{2\pi\hbar^2}\right)^{3/2} \tag{14.37}$$

Here N_e gives the density of the electron states in the conduction band.

One can also calculate the density of holes in a similar way. The probability of occupation of a hole state with energy $E_{\mathbf{k}}$, at the temperatures of interest, is given by

$$f_h(E_{\mathbf{k}}) = 1 - f_e(E_{\mathbf{k}}) = e^{(E_{\mathbf{k}} - E_F)/k_B T} \tag{14.38}$$

The holes near the top of the valence band behave as particles with effective mass m_h^*. The density of hole states per unit energy per unit volume near the top of the valence band is given by

$$g_h(E_{\mathbf{k}}) = \frac{1}{2\pi^2}\left(\frac{2m_h^*}{\hbar^2}\right)^{3/2}(E_v - E_{\mathbf{k}})^{1/2} \tag{14.39}$$

The density of holes per unit volume in the valence band, therefore, is given by

$$n_h = \int_{\text{bottom}}^{E_v} f_h(E_{\mathbf{k}})\, g_h(E_{\mathbf{k}})\, dE_{\mathbf{k}} \tag{14.40}$$

Further, $f_h(E_{\mathbf{k}})$ decreases rapidly as one goes down below the top of the valence band and, therefore, the lower limit in Eq. (14.40) can safely be taken as $-\infty$, that is,

$$\begin{aligned} n_h &= \int_{-\infty}^{E_v} f_h(E_{\mathbf{k}})\, g_h(E_{\mathbf{k}})\, dE_{\mathbf{k}} \\ &= \frac{1}{2\pi^2}\left(\frac{2m_h^*}{\hbar^2}\right)^{3/2} e^{-E_F/k_B T} \int_{-\infty}^{E_v} (E_v - E_{\mathbf{k}})^{1/2} e^{E_{\mathbf{k}}/k_B T} dE_{\mathbf{k}} \end{aligned} \tag{14.41}$$

Putting $E_v - E_{\mathbf{k}} = x$, the integral becomes

$$n_h = \frac{1}{2\pi^2}\left(\frac{2m_h^*}{\hbar^2}\right)^{3/2} e^{(E_v - E_F)/k_B T} \int_0^{\infty} x^{1/2} e^{-x/k_B T} dx$$

which can be solved to write

$$n_h = N_h\, e^{(E_v - E_F)/k_B T} \tag{14.42}$$

where

$$N_h = 2\left(\frac{m_h^* k_B T}{2\pi\hbar^2}\right)^{3/2} \tag{14.43}$$

Here N_h gives the density of hole states in the valence band. Now multiplying Eqs. (14.36), (14.42), one can write

$$n_e n_h = 4\left(m_e^* m_h^*\right)^{3/2}\left(\frac{k_B T}{2\pi\hbar^2}\right)^3 e^{-(E_c - E_v)/k_B T} \tag{14.44}$$

The above equation can also be written as

$$n_e n_h = N_e N_h\, e^{-(E_c - E_v)/k_B T} \tag{14.45}$$

Note that Eq. (14.45) does not involve the Fermi energy, which means that it does not depend on the particular substance. This expression is sometimes called the *law of mass action*. The only assumption that we have made here is that the Fermi energy is well away from the conduction and valence bands, which is true in an intrinsic semiconductor. In an intrinsic semiconductor $n_e = n_h$, therefore, the density of conduction electrons and holes is given from Eq. (14.44) as

$$n_e = n_h = 2\left(\frac{k_B T}{2\pi\hbar^2}\right)^{3/2}\left(m_e^* m_h^*\right)^{3/4} e^{-\frac{E_c - E_v}{2k_B T}} \tag{14.46}$$

The position of E_F can be evaluated from the values of n_e and n_h. Dividing Eq. (14.36) by Eq. (14.42) and taking the logarithm of both sides, one gets

$$E_F = \frac{E_c + E_v}{2} + \frac{1}{2} k_B T\left[\frac{3}{2}\ \ln\frac{m_h^*}{m_e^*} + \ \ln\frac{n_e}{n_h}\right] \tag{14.47}$$

In an intrinsic semiconductor $n_e = n_h$, so the last term in Eq. (14.47) goes to zero reducing it to

$$E_F = \frac{E_c + E_v}{2} + \frac{3}{4} k_B T\ \ln\left(\frac{m_h^*}{m_e^*}\right) \tag{14.48}$$

If the masses of an electron in the conduction band and a hole in the valence band are also equal, that is, $m_e^* = m_h^*$, then Eq. (14.48) reduces to

$$E_F = \frac{E_c + E_v}{2} \tag{14.49}$$

In this case the Fermi energy lies exactly in the middle of the conduction and valence bands. As a special case, if the zero of energy is taken at the top of the valence band, that is, $E_v = 0$, then Eqs. (14.44), (14.46), and (14.48) become

$$n_e n_h = 4\left(m_e^* m_h^*\right)^{3/2}\left(\frac{k_B T}{2\pi\hbar^2}\right)^3 e^{-E_g/k_b T} \tag{14.50}$$

$$n_e = n_h = 2\left(\frac{k_B T}{2\pi\hbar^2}\right)^{3/2}\left(m_e^* m_h^*\right)^{3/4} e^{-\frac{E_g}{2k_B T}} \tag{14.51}$$

$$E_F = \frac{E_g}{2} + \frac{3}{4} k_B T \ln\left(\frac{m_h^*}{m_e^*}\right) \tag{14.52}$$

14.6 MODEL FOR EXTRINSIC SEMICONDUCTORS

14.6.1 n-Type Semiconductors

The band model for an n-type semiconductor is shown in Fig. 14.5B, in which the donor band having energy E_d lies close to the conduction band. Let N_d be the density of donor states per unit volume in the donor band. At absolute zero the donor band is assumed to be full while the conduction band is completely empty. But at low temperatures a fraction of the electrons from the donor band jumps to the conduction band. At such temperatures the Fermi energy E_F lies about halfway between the donor band and the bottom of the conduction band. We further assume that E_F of the n-type semiconductor lies below the bottom of the conduction band and above the donor band by about an energy of a few $k_B T$. In this approximation, the density of electrons in the conduction band of an n-type semiconductor is given by Eq. (14.36), that is,

$$n_e = N_e\, e^{(E_F - E_c)/k_B T} \tag{14.53}$$

Eq. (14.53) also gives the number of holes in the donor band n_h, which is equal to the number of ionized donors. Now the number of holes in the donor band is given by

$$n_h = N_d f_h(E_d) = N_d\left[1 - f_e(E_d)\right] = N_d e^{(E_d - E_F)/k_B T} \tag{14.54}$$

In writing the above equation we have used Eq. (14.38). If no electron is excited from the valence to the conduction band, which is possible only at low temperatures, then the number of electrons in the conduction band must be equal to the number of holes in the donor band, that is, $n_e = n_h$. So

$$N_d e^{(E_d - E_F)/k_B T} = N_e\, e^{(E_F - E_c)/k_B T} \tag{14.55}$$

Rearranging the terms, we get

$$e^{(2E_F - E_c - E_d)/k_B T} = \frac{N_d}{N_e} \tag{14.56}$$

Taking the logarithm of the above equation and solving for E_F, we get

$$E_F = \frac{E_c + E_d}{2} + \frac{k_B T}{2} \ln\left[\frac{N_d}{N_e}\right] \tag{14.57}$$

The above equation describes the variation of E_F with temperature. At $T = 0$, Eq. (14.57) gives

$$E_F = \frac{E_c + E_d}{2} \tag{14.58}$$

which is exactly in the middle of the bottom of the conduction band and the donor band. As the temperature increases, E_F falls and approaches the donor band. With a further increase in temperature, the Fermi energy continuously goes down and at very high temperatures it approaches the middle of the valence and conduction bands (Fig. 4.8).

From Eq. (14.53) it is evident that the density of electrons in the conduction band depends on E_F, which in turn depends on the temperature and density of the donor states (Eq. 14.57). To evaluate n_e in an extrinsic semiconductor, we substitute E_F from Eq. (14.57) into Eq. (14.53) and simplify to get

$$n_e = \sqrt{N_d N_e} e^{-\Delta E_{cd}/k_B T} \tag{14.59}$$

where

$$\Delta E_{cd} = E_c - E_d \tag{14.60}$$

Here ΔE_{cd} represents the ionization energy (excitation energy) of the donor electron from the donor band to the conduction band. We see that the number of electrons in the conduction band depends on the square root of the density of donor states in the impurity band. Therefore, n_h is also proportional to the square root of the donor concentration.

14.6.2 p-Type Semiconductors

In a p-type semiconductor the acceptor band is empty and lies close to the completely filled valence band (see Fig. 14.6B). At low temperatures some of the electrons from the valence band jump to the acceptor band, thereby creating holes in the valence band. At low temperatures the Fermi energy lies between the valence band and the acceptor band. Further, we assume that E_F lies below the acceptor band and above the valence band by an energy of a few k_BT. In this approximation the number of holes in the valence band is given by Eq. (14.42) and is written as

$$n_h = N_h e^{(E_v - E_F)/k_B T} \tag{14.61}$$

This number must be equal to the number of electrons in the acceptor band. If N_a is the density of acceptor states per unit volume, then the number of electrons in the acceptor band is given by

$$n_e = N_a f_e(E_a) \cong N_a e^{(E_F - E_a)/k_B T} \tag{14.62}$$

where E_a is the energy of the acceptor band. If no electron is excited from the valence band to the conduction band, which is possible only at low temperatures, then the number of electrons in the accepter band must be equal to the number of holes in the valence band. Therefore, from Eqs. (14.61), (14.62), one can write

$$N_a e^{(E_F - E_a)/k_B T} = N_h e^{(E_v - E_F)/k_B T} \tag{14.63}$$

Solving the above equation for E_F, we get

$$E_F = \frac{E_a + E_v}{2} - \frac{k_B T}{2} \ln \left[\frac{N_a}{N_h}\right] \tag{14.64}$$

At absolute zero, Eq. (14.64) gives

$$E_F = \frac{E_a + E_v}{2} \tag{14.65}$$

Therefore, at absolute zero, E_F lies exactly in the middle between the top of the valence band and the acceptor band. With an increase in temperature, E_F increases and ultimately, at very high temperatures, it approaches the middle of the valence and conduction bands. The variation of E_F with temperature for n- and p-type semiconductors is shown in Fig. 14.8.

At a particular temperature, the number of holes in the valence band is obtained by substituting Eq. (14.64) into Eq. (14.61) and simplifying to get

$$n_h = \sqrt{N_a N_h} e^{-\Delta E_{av}/k_B T} \tag{14.66}$$

where

$$\Delta E_{av} = E_a - E_v \tag{14.67}$$

ΔE_{av} represents the excitation energy of an electron from the valence band to the acceptor band. It is evident that the number of holes in the valence band depends on the square root of the density of acceptor states in the acceptor band. In other words, n_h is proportional to the square root of the acceptor concentration.

It is noteworthy that, within the model considered above, the expressions for electrons in n-type semiconductors and holes in p-type semiconductors are similar and depend on the concentration of the doping.

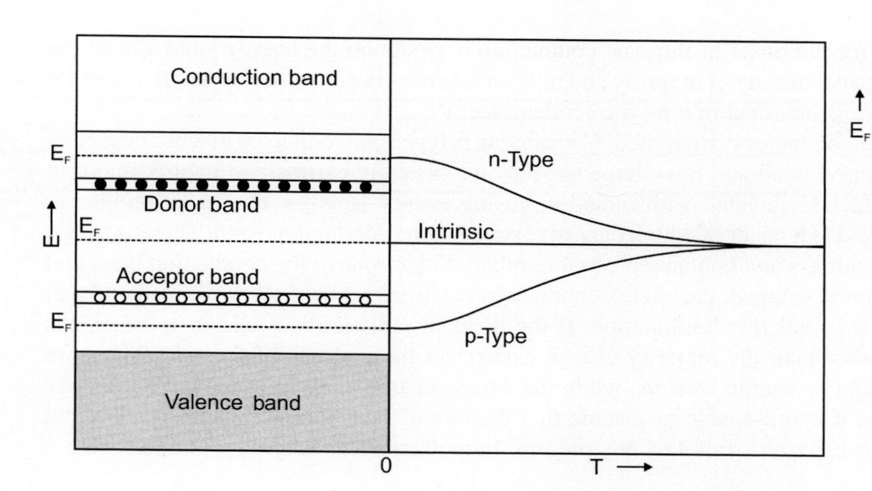

FIG. 14.8 Schematic representation of the variation of E_F with temperature T in intrinsic, n-type and p-type semiconductors. The dashed lines in the left part of the figure show the Fermi energy in the intrinsic, n-type and p-type semiconductors at 0 K. The right part of the figure shows the variation of the Fermi energy with temperature in the three types of semiconductors.

14.7 EFFECT OF TEMPERATURE ON CARRIER DENSITY

At absolute zero the extrinsic semiconductors behave as insulators as there are no free charge carriers in the conduction band. In an n-type semiconductor, with an increase in temperature some electrons get excited from the donor to the conduction band, yielding a finite electron density n_e. At room temperature the thermal energy is sufficient to excite most of the electrons from the donor band to the conduction band. In this temperature range the charge carriers are extrinsic in nature and this is usually called the extrinsic range or impurity range (Fig. 14.9). For example, in Ge the upper limit of the extrinsic range is 100°C, while in Si it is about 200°C. Ultimately, the donor band gets exhausted and remains exhausted for a certain range of temperature (exhaustion range). In the exhaustion range the direct excitation of an electron from the valence band to the conduction band is almost zero and the value of n_e becomes independent of temperature. It is noteworthy that in the extrinsic and exhaustion ranges the concentration of majority carriers is far greater than that of minority carriers. With a further increase in temperature, the electrons start getting excited from the valence to the conduction band in large numbers. At these temperatures an equal number of electrons and holes are liberated, which ultimately exceed the number of extrinsic carriers. The carrier density n_e increases much faster and, therefore, the conduction becomes intrinsic in nature. This is

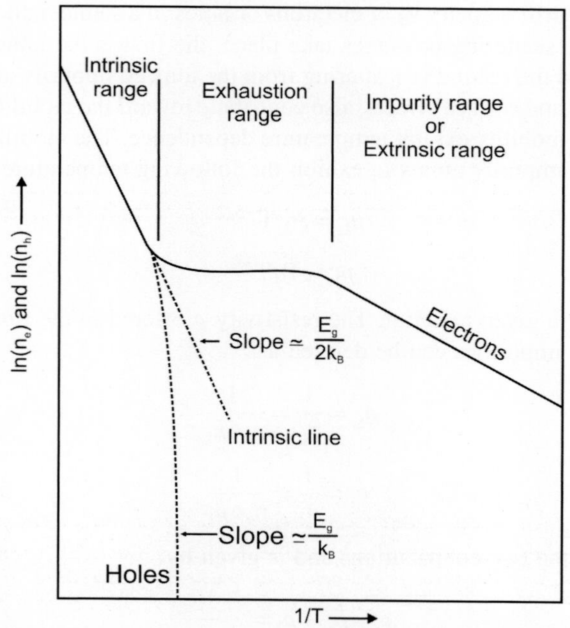

FIG. 14.9 Variation of the density of electrons (holes) as a function of temperature showing the extrinsic, exhaustion, and intrinsic ranges in an n- (p-) type semiconductor.

called the intrinsic range. The temperature for the onset of intrinsic conduction depends on the energy band gap of the semiconductor. Note that in the intrinsic range the density of majority and minority carriers becomes nearly equal. A similar variation of hole density n_h with temperature is obtained in a p-type semiconductor.

Let us examine what happens when a semiconductor is irradiated. Consider an n-type semiconductor in which the electrons from the donor band go to the conduction band and the charge carriers are basically extrinsic in nature: majority charge carriers. Let an n-type semiconductor be irradiated with radiation having energy $E_g = h\nu$. The radiation energy is absorbed by the electrons in the valence band as a result of which they get excited to the conduction band. These intrinsic charge carriers are called the excess charge carriers and contain an equal number of electrons in the conduction band and holes in the valence band. When the radiation is stopped, the excess conduction electrons return to the valence band and combine with the excess holes. This process is called recombination. If the incident radiation is weak, then the excess charge carriers will be much fewer in number than the majority charge carriers in thermal equilibrium. In this sense the extrinsic charge carriers become the majority charge carriers, while the excess charge carriers become the minority charge carriers. For weak incident radiation, it is reasonable to assume that the rate of recombination is proportional to the number of excess charge carriers (excess electrons or holes) N_{ex} present. In mathematical language, one can write

$$\frac{dN_{ex}}{dt} = -\frac{N_{ex}}{\tau_m} \tag{14.68}$$

The factor $1/\tau_m$ is the proportionality constant and τ_m is called the minority carrier lifetime. If the radiation stops at $t=0$, the above equation gives

$$\int \frac{dN_{ex}}{N_{ex}} = -\frac{1}{\tau_m}\int dt$$

Solving the above integral, one gets

$$N_{ex}(t) = N_{ex}(0)\, e^{-t/\tau_m} \tag{14.69}$$

Eq. (14.69) describes the exponential decay of the excess minority carrier density. In time τ_m the density is reduced by a factor e^{-1}.

14.8 TEMPERATURE DEPENDENCE OF MOBILITY

The mobility μ is a measure of the drift velocity v_d of electrons or holes in a semiconductor and depends on the scattering processes. In a semiconductor two scattering processes take place: the first is the lattice scattering (or the scattering of charge carriers due to phonons) and the second is scattering from the ionized impurity atoms (donors or acceptors). Other imperfections, such as dislocations and surface effects, also contribute toward the mobility but to a much lesser extent. The scattering processes and hence the mobility exhibit temperature dependence. The mobility due to lattice scattering μ_L and due to the scattering from ionized impurity atoms μ_I exhibit the following temperature dependence:

$$\mu_L = A_L T^{-3/2} \tag{14.70}$$

$$\mu_I = B_I T^{3/2} \tag{14.71}$$

where A_L and B_I are constants for a given material. The resistivity ρ offered to the flow of electrons in an n-type semiconductor due to the phonons and impurities can be defined as

$$\rho_L = \frac{1}{\sigma_L} = \frac{1}{n_e e \mu_L} \tag{14.72}$$

$$\rho_I = \frac{1}{\sigma_I} = \frac{1}{n_e e \mu_I} \tag{14.73}$$

The total resistivity is the sum of the two contributions and is given by

$$\rho = \rho_L + \rho_I = \frac{1}{n_e e \mu} \tag{14.74}$$

where μ is the total mobility in a semiconductor. Substituting Eqs. (14.72), (14.73) into Eq. (14.74), one gets

$$\frac{1}{\mu}=\frac{1}{\mu_L}+\frac{1}{\mu_I} \tag{14.75}$$

From Eqs. (14.70), (14.71), and (14.75) one can immediately write

$$\mu=\frac{1}{\frac{1}{B_I}T^{-3/2}+\frac{1}{A_L}T^{3/2}} \tag{14.76}$$

At low values of T the term $T^{-3/2}$ dominates in the denominator of Eq. (14.76) and, therefore, reduces the mobility to

$$\mu=B_I\,T^{3/2} \tag{14.77}$$

Therefore, at low temperatures, the impurity scattering dominates due to the fact that the number of phonons is very small at low temperatures. But at high values of T, the term $T^{3/2}$ dominates in the denominator of Eq. (14.76), which gives

$$\mu=A_L\,T^{-3/2} \tag{14.78}$$

Hence, phonon scattering is mainly responsible for the mobility at high temperatures due to the presence of a large number of phonons. In an intrinsic semiconductor, impurity scattering is absent; therefore, Eq. (14.70) applies directly. Fig. 14.10 shows the variation of mobility μ as a function of T.

14.9 THE HALL EFFECT

The Hall effect in semiconductors describes the effect of a magnetic field on moving charge carriers. As an example, let us consider a p-type semiconducting slab in which the current density J_x is passed along the x-direction (Fig. 14.11A). The current density J_x is constituted by the motion of holes in the positive x-direction. Let a magnetic field **H** be applied along the z-direction, then the magnetic force acting on the hole is given by

$$\mathbf{F}^m=\frac{e}{c}\mathbf{v}\times\mathbf{H} \tag{14.79}$$

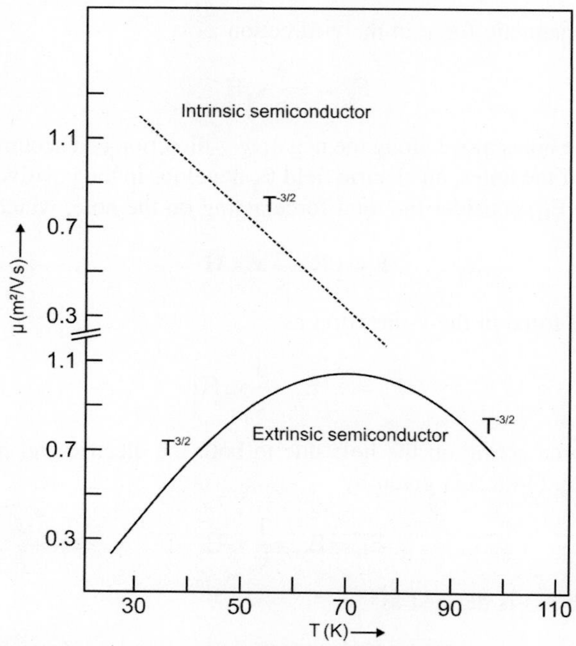

FIG. 14.10 The variation of mobility μ as a function of temperature T in an extrinsic semiconductor is shown by the continuous line. The dashed line shows the variation of mobility with T in an intrinsic semiconductor.

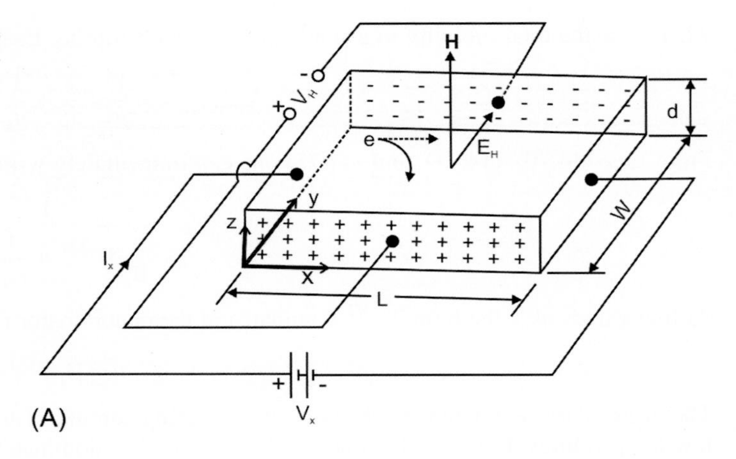

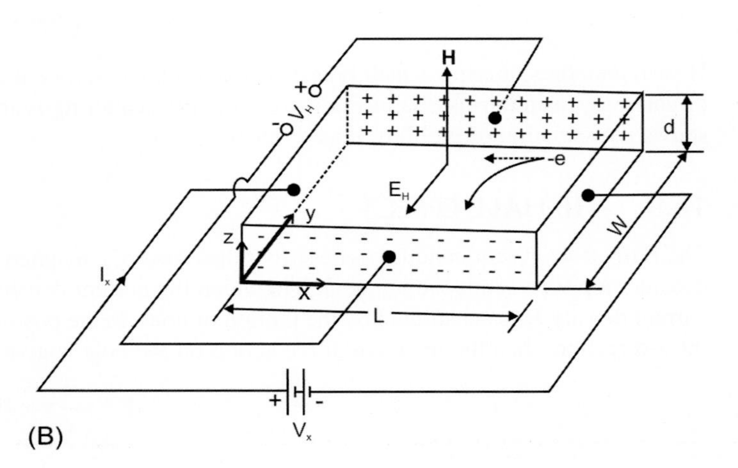

FIG. 14.11 (A) Geometry for measuring the Hall effect in a p-type semiconductor. (B) Geometry for measuring the Hall effect in an n-type semiconductor.

The above equation gives the magnetic force in the y-direction as

$$F_y^m = -\frac{e}{c} v_x H \tag{14.80}$$

As a result of the magnetic force the holes move along the negative y-direction and accumulate on the front face of the slab (Fig. 14.11A). Due to the shifting of the holes, an electric field E_y develops in the positive y-direction, generally called the Hall field E_H. In order to evaluate E_H, consider the total force acting on the hole, which is given by

$$\mathbf{F} = e\mathbf{E} + \frac{e}{c}\,\mathbf{v} \times \mathbf{H} \tag{14.81}$$

The above equation gives the total force in the y-direction as

$$F_y = e\left[E_y - \frac{1}{c} v_x H\right] \tag{14.82}$$

In the equilibrium state the net force acting on the hole due to both the electric and magnetic fields must vanish and, therefore, the Hall voltage from Eq. (14.82) is given by

$$E_H = E_y = \frac{1}{c} v_x H \tag{14.83}$$

The current density J_x due to the holes is defined as

$$J_x = n_h e v_d \tag{14.84}$$

The drift velocity v_d of the holes is equal to v_x in the present situation. So,

$$v_x = v_d = \frac{J_x}{n_h e} \tag{14.85}$$

From Eqs. (14.83), (14.85) the Hall field becomes

$$E_H = \frac{J_x H}{n_h ec} \tag{14.86}$$

Hence, the density of holes from the above equation becomes

$$n_h = \frac{J_x H}{ecE_H} \tag{14.87}$$

The Hall coefficient is defined as the Hall field per unit current density and per unit magnetic field and is given by

$$R_H = \frac{E_H}{J_x H} = \frac{1}{n_h ec} \tag{14.88}$$

The mobility of charge carriers can be estimated from the geometry of the Hall effect setup for a p-type semiconductor. The drift velocity is given by

$$v_d = \mu_h E_x \tag{14.89}$$

Substituting Eq. (14.89) into Eq. (14.84), the mobility of the holes is given by

$$\mu_h = \frac{J_x}{n_h e E_x} \tag{14.90}$$

The mobility of charge carriers can also be calculated if one knows the conductivity and the Hall coefficient of a semiconductor. Consider a semiconductor with one type of charge carrier, say a p-type semiconductor with holes as the majority charge carriers. The conductivity of the semiconductor is given as

$$\sigma_h = n_h e \mu_h \tag{14.91}$$

The measurement of σ_h of the semiconductor gives information about the product $n_h \mu_h$, but it does not allow the separate determination of n_h and μ_h. From Eqs. (14.88), (14.91) one can write

$$\mu_h = c\sigma_h R_H \tag{14.92}$$

If σ_h and R_H of a semiconductor are measured experimentally, one can find μ_h and, knowing μ_h, one can calculate n_h from Eq. (14.91). The above discussion shows that for the individual determination of n_h and μ_h in a semiconductor, the experimental measurement of both σ_h and R_H is required.

From Eqs. (14.86), (14.88) we see that the Hall field and the Hall coefficient are positive for a p-type semiconductor. It is important to note that in an actual experiment the Hall voltage V_H and current I_x are measured instead of the Hall field and current density. Therefore, one should express the results in terms of these measurable physical quantities. Let A be the area of the face at which the current enters the semiconducting slab (see Fig. 14.11A), then

$$E_H = \frac{V_H}{W}, \quad J_x = \frac{I_x}{A} = \frac{I_x}{Wd}, \quad E_x = \frac{V_x}{L} \tag{14.93}$$

Substituting the values of J_x and E_H, in terms of I_x and V_H, in Eqs. (14.86), (14.87) one can write

$$V_H = \frac{1}{n_h ec} \frac{I_x H}{d} \tag{14.94}$$

$$n_h = \frac{1}{ec} \frac{I_x H}{V_H d} \tag{14.95}$$

In order to calculate the mobility in terms of the measurable quantities, we substitute E_x and J_x from Eq. (14.93) in Eq. (14.90) to get

$$\mu_h = \frac{I_x L}{n_h e V_x W d} \tag{14.96}$$

One can also study the Hall effect in an n-type semiconductor. Proceeding in exactly the same manner, one can write the Lorentz force as

$$\mathbf{F} = -e\left[\mathbf{E} + \frac{1}{c}\mathbf{v}\times\mathbf{H}\right] \tag{14.97}$$

In the equilibrium state one gets

$$\mathbf{E} = -\frac{1}{c}\mathbf{v}\times\mathbf{H} \tag{14.98}$$

The geometry of the setup for studying the Hall effect is shown in Fig. 14.11B. The electrons accumulate on the front face of the slab and the Hall field is set up in the negative y-direction. From Eq. (14.98) one can write

$$E_H = E_y = \frac{1}{c}v_x H \tag{14.99}$$

Now the current density J_x due to the flow of electrons is given as

$$J_x = -n_e e v_d \tag{14.100}$$

The drift velocity of the electrons is equal to v_x and is given by

$$v_x = v_d = -\frac{J_x}{n_e e} \tag{14.101}$$

Substituting Eq. (14.101) into Eq. (14.99), the Hall field is given by

$$E_H = -\frac{J_x H}{n_e e c} \tag{14.102}$$

From this equation the electron density can be written as

$$n_e = -\frac{J_x H}{e c E_H} \tag{14.103}$$

The Hall coefficient for an n-type semiconductor is given by

$$R_H = \frac{E_H}{J_x H} = -\frac{1}{n_e e c} \tag{14.104}$$

From Eqs. (14.102), (14.104) it is evident that the Hall field and Hall coefficient are negative for an n-type semiconductor in which the majority carriers are the electrons. Use Eq. (14.93) in Eqs. (14.102), (14.103), one gets

$$V_H = -\frac{1}{n_e e c}\frac{I_x H}{d} \tag{14.105}$$

$$n_e = -\frac{1}{ec}\frac{I_x H}{V_H d} \tag{14.106}$$

One can calculate the mobility μ_e for an n-type semiconductor in exactly the same way as for a p-type semiconductor. It is given by

$$\mu_e = \frac{I_x L}{n_e e V_x W d} \tag{14.107}$$

Here we have neglected the negative sign. The mobility of charge carriers can also be calculated if one knows the conductivity and the Hall coefficient of a semiconductor. In an n-type semiconductor, the conductivity is given as

$$\sigma_e = n_e e \mu_e \tag{14.108}$$

If one measures σ_e, it gives information about the product $n_e \mu_e$. Eqs. (14.104), (14.108) allow us to write (neglecting the negative sign)

$$\mu_e = c\sigma_e R_H \tag{14.109}$$

If σ_e and R_H of a semiconductor are measured experimentally, one can find μ_e and n_e from Eqs. (14.108), (14.109).

14.10 ELECTRICAL CONDUCTIVITY IN SEMICONDUCTORS

After describing the temperature dependence of the mobility of charge carriers, one can obtain explicit expressions for the electrical conductivity of semiconductors.

14.10.1 Intrinsic Semiconductors

The general expression for the conductivity of a semiconductor is given by Eq. (14.16). In an intrinsic semiconductor $n_e = n_h$, therefore, Eq. (14.16) reduces to

$$\sigma = n_e e (\mu_e + \mu_h) \tag{14.110}$$

Substituting for n_e from Eq. (14.51), we find

$$\sigma = \sigma_0 e^{-\frac{E_g}{2k_B T}} \tag{14.111}$$

where

$$\sigma_0 = 2e(\mu_e + \mu_h)\left(\frac{k_B T}{2\pi\hbar^2}\right)^{3/2} \left(m_e^* m_h^*\right)^{3/4} \tag{14.112}$$

From Eq. (14.76) it is evident that in an intrinsic semiconductor the mobility μ decreases with an increase in temperature as

$$\mu (= \mu_e + \mu_h) = A_L T^{-3/2} \tag{14.113}$$

Substituting Eq. (14.113) into Eq. (14.112), one gets

$$\sigma_0 = 2eA_L \left(\frac{k_B}{2\pi\hbar^2}\right)^{3/2} \left(m_e^* m_h^*\right)^{3/4} \tag{14.114}$$

Therefore, σ_0 is independent of temperature. Taking the logarithm of Eq. (14.111), one gets

$$\ln \sigma = \ln \sigma_0 - \frac{E_g}{2k_B T} \tag{14.115}$$

Taking the reciprocal of Eq. (14.111) one can straightway write the expression for resistivity ρ of an intrinsic semiconductor as

$$\rho = \rho_0 e^{\frac{E_g}{2k_B T}} \tag{14.116}$$

where $\rho = 1/\sigma$ and $\rho_0 = 1/\sigma_0$. The logarithm of Eq. (14.116) yields

$$\ln \rho = \ln \rho_0 + \frac{E_g}{2k_B T} \tag{14.117}$$

The plot of $\ln \sigma$ or $\ln \rho$ as a function of T is a straight line whose slope gives the energy band gap E_g. Therefore, if one measures the conductivity σ or resistivity ρ as a function of temperature T, the band gap E_g of the intrinsic semiconductor can be determined.

14.10.2 Extrinsic Semiconductors

The derivation of the general expression for conductivity σ of an extrinsic semiconductor is difficult, but one can calculate σ in limiting cases. At low temperatures the electrons from the valence band cannot be excited to the conduction band. On the other hand, the electrons from the donor band get excited to the conduction band and behave as the charge carriers. Hence, at low temperatures, the conductivity is given as

$$\sigma = n_e e \mu_e \tag{14.118}$$

where n_e is given by Eq. (14.59). From Eqs. (14.59), (14.118) one can write

$$\sigma = e\mu_e \sqrt{N_d N_e}\, e^{-\Delta E_{cd}/k_B T} \tag{14.119}$$

Substituting the value of N_e from Eq. (14.37) into the above equation, we get

$$\sigma = \sigma_{0e} e^{-\Delta E_{cd}/k_B T} \tag{14.120}$$

where

$$\sigma_{0e} = e\mu_e \sqrt{2N_d}\left(\frac{m_e^* k_B T}{2\pi\hbar^2}\right)^{3/4} \tag{14.121}$$

The mobility μ_e is here due to the impurity ions, which exhibit $T^{3/2}$ dependence (Eq. 14.77); therefore, σ_{0e} is temperature dependent.

At high temperatures, most of the electrons in the conduction band are excited from the valence band as the impurity concentration in a semiconductor is low. Therefore, at such temperatures, an n-type semiconductor behaves like an intrinsic semiconductor with $n_e = n_h$ given by Eq. (14.51) and conductivity given by Eq. (14.111).

Problem 14.1

Show that at low temperatures, the electrical conductivity of a p-type semiconductor is given by

$$\sigma = \sigma_{0h} e^{-\Delta E_{av}/k_B T} \tag{14.122}$$

where

$$\sigma_{0h} = e\mu_h \sqrt{2N_a}\left(\frac{m_h^* k_B T}{2\pi\hbar^2}\right)^{3/4} \tag{14.123}$$

Problem 14.2

A Ge crystal has the dimensions of $L = 1.0\,cm$, $W = 0.12\,cm$, and $d = 0.2\,cm$ (see Fig. 14.11A). In the Hall effect measurements, a current of $I_x = 2.4\,mA$ flows when voltage $V_x = 1.0\,V$ and magnetic field $H_z = 5000\,gauss$ is applied. The Hall voltage developed in the crystal is $V = 10\,mV$. Calculate the conductivity of the crystal and the mobility of the charge carriers.

14.11 NONDEGENERATE SEMICONDUCTORS

In an extrinsic semiconductor the concentration of impurity (trivalent or pentavalent) is generally very small. Therefore, in a conventional extrinsic semiconductor, the separation between the impurity atoms is large, as a result of which the impurity-impurity interaction is negligible. Therefore, in an ordinary n- or p-type semiconductor, the impurity (donor or acceptor) states are discrete and one can talk about impurity levels instead of bands. Such types of extrinsic semiconductors are called *nondegenerate semiconductors*.

14.12 DEGENERATE SEMICONDUCTORS

With an increase in the impurity concentration in an extrinsic semiconductor, the impurity-impurity interaction becomes finite. This interaction splits the discrete impurity states, forming an impurity band. With a further increase in the impurity concentration, the width of the impurity band increases and, ultimately, it may overlap either with the conduction band in an n-type semiconductor or with the valence band in a p-type semiconductor (see Fig. 14.12). The overlap occurs when the impurity concentration becomes comparable with the effective density of states, or rather increases compared with the density of states. Such types of semiconductors are called *degenerate semiconductors*. In an n-type degenerate semiconductor the Fermi energy E_F lies in the conduction band, while in a p-type degenerate semiconductor, E_F lies inside the valence band (see Fig. 14.12). In an n-type degenerate semiconductor, the energy states below E_F are mostly filled and above E_F they are mostly empty. Therefore, in an n-type degenerate semiconductor, the energy states between E_F and E_c are mostly filled with electrons; thus, the electron concentration in the conduction band is very large. On the other hand, in a p-type degenerate semiconductor, the energy states between E_v and E_F are mostly empty (have holes in them) and the hole concentration in the valence band is very large.

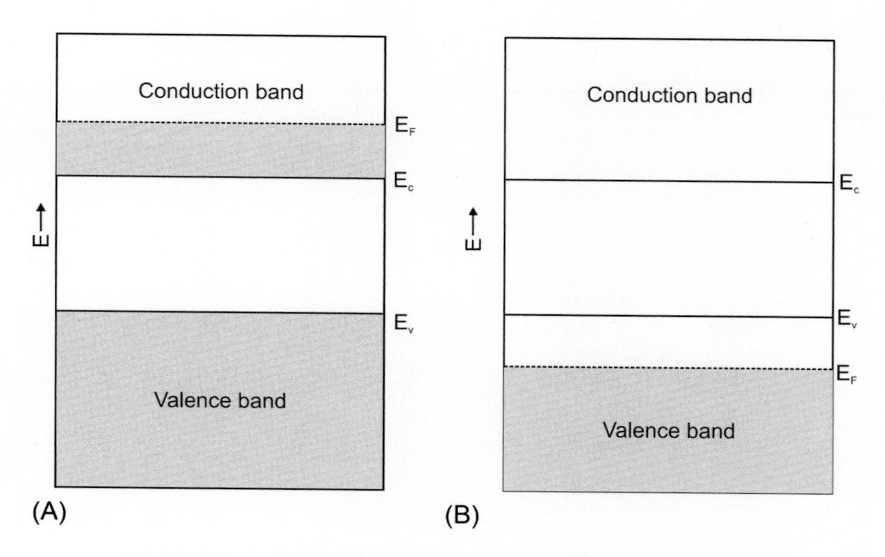

FIG. 14.12 Simplified energy band diagram for degenerately doped (A) n-type and (B) p-type semiconductors. The *shaded region* shows the filled electron states.

14.13 COMPENSATED SEMICONDUCTORS

A *compensated semiconductor* is one that contains both donor and acceptor impurity atoms in the same region. A compensated semiconductor can be formed either by diffusing acceptor impurities into an n-type semiconductor or by diffusing donor impurities into a p-type semiconductor. Let N_d and N_a be the density of states in the donor and acceptor levels, respectively. In an n-type compensated semiconductor $N_a < N_d$ and in a p-type compensated semiconductor $N_d < N_a$. If $N_a = N_d$ one gets a completely compensated semiconductor, which exhibits the characteristics of an intrinsic semiconductor. Compensated semiconductors are created quite naturally during device fabrication.

Problem 14.3

In an intrinsic semiconductor the energy bands for both the conduction and valence bands are ellipsoidal in shape and are defined as

$$E = E_c + \frac{\hbar^2}{2}\left(\frac{k_x^2}{m_x^*} + \frac{k_y^2}{m_y^*} + \frac{k_z^2}{m_z^*}\right) \tag{14.124}$$

$$E = E_v - \frac{\hbar^2}{2}\left(\frac{k_x^2}{m_x^*} + \frac{k_y^2}{m_y^*} + \frac{k_z^2}{m_z^*}\right) \tag{14.125}$$

where E_c is the energy value at the bottom of the conduction band and E_v is the energy at the top of the valence band. Here m_x^*, m_y^*, and m_z^* are the effective component masses along the x-, y-, and z-directions. Find expressions for the density of states in the valence and conduction bands.

Problem 14.4

Suppose that the effective mass of holes in an intrinsic semiconductor with $E_g = 1\,\text{eV}$ is four times that of the electrons. Find the temperature at which the Fermi energy will be shifted by 5% from the middle of the energy gap.

SUGGESTED READING

McKelvey, J. P. (1966). *Solid state and semiconductor physics*. New York: Harper and Row.

Neamen, D. A. (2009). *Semiconductor physics and devices*. New Delhi: Tata-McGraw Hill Publ. Co, Ltd.

Rose, R. M., Shepard, L. A., & Wulff, J. (1966). *The structure and properties of materials: Electronic properties*. (Vol. IV). New York: J. Wiley & Sons.

Shockley, W. (1950). *Electrons and holes in semiconductors*. New York: Van Nostrand Co. Inc.

Snoke, D. W. (2009). *Solid state physics: Essential concepts*. New Delhi: Pearson Education & Dorling Kindersley.

Chapter 15

Dielectric Properties of Nonconducting Solids

Chapter Outline

The properties of conducting and semiconducting solids have already been studied in the preceding chapters. In this chapter, both the macroscopic and microscopic descriptions of the properties of nonconducting solids, particularly the dielectric solids, will be presented. A microscopic study provides more insight into the dielectric properties of solids. The nonconducting solids can be classified into two categories:

1. Nonpolar solids
2. Polar solids

15.1 NONPOLAR SOLIDS

In a nonpolar solid, an atom/molecule does not possess an intrinsic electric dipole moment. In such a solid the centers of negative and positive charges in each atom/molecule coincide (Fig. 15.1A). If an external electric field $\mathbf{E}_0$ is applied to a nonpolar solid, then two situations may arise:

1. In the presence of an external electric field $\mathbf{E}_0$, each atom/molecule may become polarized due to the shifting of the centers of positive and negative charges (Fig. 15.1B). In other words, the electric field induces an electric dipole moment in each atom/molecule. Such solids are called *dielectric solids* in analogy with diamagnetic solids.
2. The atoms/molecules do not suffer any polarization at all and such solids are called *ideal insulators*.

15.2 POLAR SOLIDS

A solid in which each atom/molecule possesses intrinsic electric dipole moment is called a *polar solid*. In such solids the centers of positive and negative charge of an atom/molecule do not coincide, yielding a finite electric dipole moment (Fig. 15.2). These solids can further be classified into two categories.

1. Solids that exhibit net zero electric dipole moment in the absence of an external electric field but a finite dipole moment in the presence of an applied electric field. Such solids are called *paraelectric solids*.
2. Solids that exhibit net finite electric dipole moment even in the absence of an external electric field are called *ferro-electric solids*.

Solid State Physics. https://doi.org/10.1016/B978-0-12-817103-5.00015-3

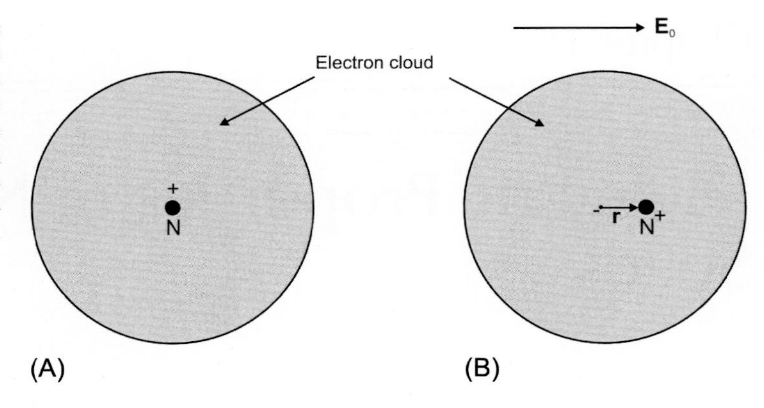

FIG. 15.1 (A) Atom with spherical shape: The electron cloud is spherical in shape with the nucleus N represented by a dot at its center. Here the centers of negative and positive charges coincide. (B) Spherical atom in the presence of an applied electric field $\mathbf{E}_0$. It is assumed that there is no distortion in the shape of the electron cloud in the presence of $\mathbf{E}_0$. The electron cloud as such moves in a direction opposite to that of the field $\mathbf{E}_0$, while the nucleus N moves in the direction of the field, causing displacement $\mathbf{r}$ in the centers of the negative and positive charges.

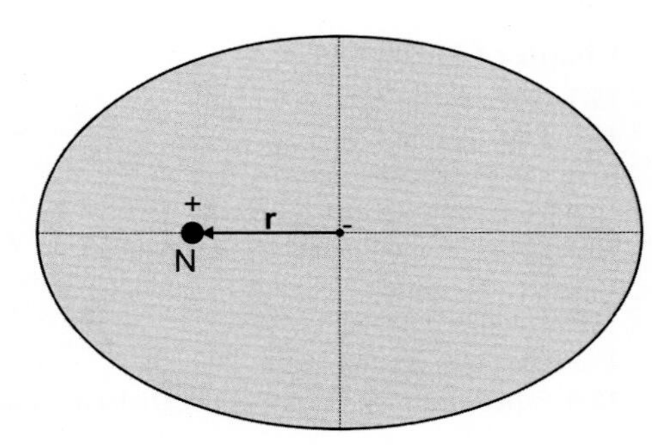

FIG. 15.2 An atom with an elliptical electron cloud has the nucleus at one of its foci. The centers of negative and positive charges are separated by distance $\mathbf{r}$.

15.3 ELECTRIC DIPOLE MOMENT

An electric dipole consists of two equal and opposite charges separated by a finite distance. The moment of an electric dipole $\mathbf{p}$ is defined as the product of the magnitude of the charge and the distance between the two charges (Fig. 15.3). The direction of $\mathbf{p}$ is from the negative to the positive charge. Therefore

$$\mathbf{p} = q\mathbf{r} \tag{15.1}$$

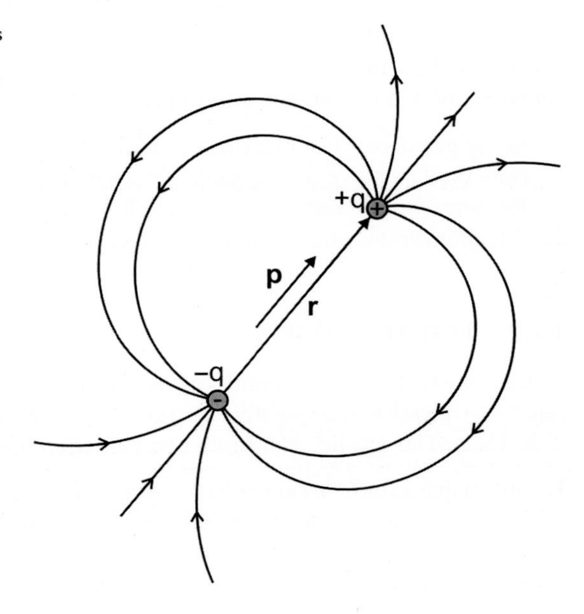

FIG. 15.3 Electric field due to an electric dipole with dipole moment $\mathbf{p}$, which is directed from negative to positive charge.

Here **r** is a vector directed from the negative to the positive charge. The electric field of the electric dipole is directed from the positive to the negative charge, as shown in Fig. 15.3. If there are a number of discrete charges present in the system, then the total electric dipole moment is given by

$$\mathbf{p} = \sum_i q_i \mathbf{r}_i \tag{15.2}$$

In the presence of a continuous electron charge distribution, with charge density $\rho_e(\mathbf{r})$, the electric dipole moment of the system is defined as

$$\mathbf{p} = \int \mathbf{r}\, \rho_e(\mathbf{r})\, d^3r \tag{15.3}$$

Polarization **P** in a solid is defined as the electric dipole moment per unit volume and is written as

$$\mathbf{P} = \frac{\mathbf{p}}{V} \tag{15.4}$$

where V is the volume of the solid. The polarization is a macroscopic property so the average is taken over the macroscopic volume of the solid. The electric field due to a dipole moment **p** varies inversely as the cube of the distance and is given by

$$\mathbf{E}(\mathbf{r}) = \frac{3(\mathbf{p}\cdot\hat{\mathbf{r}})\hat{\mathbf{r}} - \mathbf{p}}{r^3} \tag{15.5}$$

Here $\hat{\mathbf{r}}$ is a unit vector in the direction of **r**.

15.4 MACROSCOPIC ELECTRIC FIELD

Experimental measurements of different physical properties, such as electric field, magnetic field, and electrical resistivity, of a bulk material yield some sort of average value of the property. On the other hand, at the position of an atom/molecule inside the solid, the physical property may be significantly different, both in magnitude and direction, from the measured value for the bulk material. Here we consider a macroscopic electric field inside a dielectric material. Fig. 15.4 shows the bulk material in the presence of an externally applied electric field $\mathbf{E}_0$. The dielectric material gets polarized with **P** as the polarization and $\mathbf{E}_P$ as the electric field resulting from it. The macroscopic (average) field inside the material becomes

$$\mathbf{E} = \mathbf{E}_0 + \mathbf{E}_P \tag{15.6}$$

Let $\mathbf{E}_{loc}(\mathbf{r})$ be the local (microscopic) electric field at an atom/molecule situated at **r**, which may have different values at different atoms/molecules. One can express the macroscopic electric field per unit volume at the position $\mathbf{r}_0$, i.e., $\mathbf{E}(\mathbf{r}_0)$, in terms of the local field as

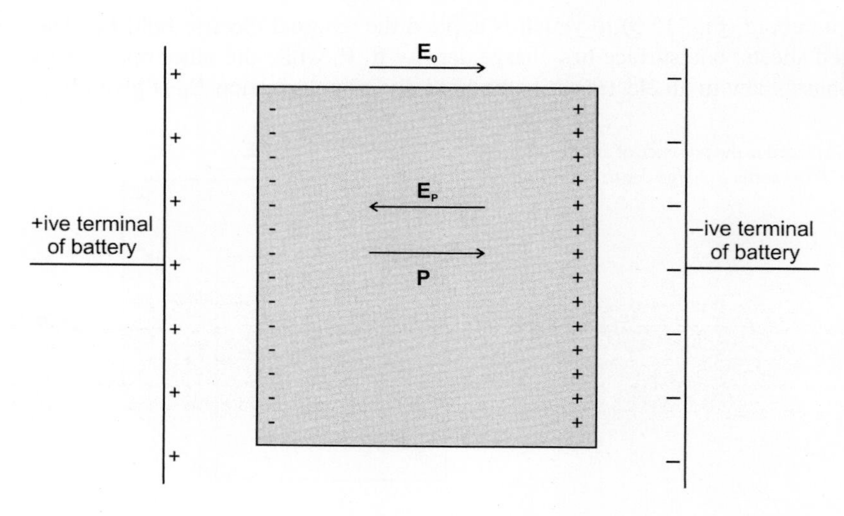

FIG. 15.4 Bulk dielectric material placed in the external electric field $\mathbf{E}_0$. $\mathbf{E}_P$ is the electric field due to the polarization **P** produced inside the material.

$$\mathbf{E}(\mathbf{r}_0) = \frac{1}{V}\int \mathbf{E}_{\text{loc}}(\mathbf{r}-\mathbf{r}_0)\, d^3r \tag{15.7}$$

The field $\mathbf{E}(\mathbf{r})$ is quite smooth compared with $\mathbf{E}_{\text{loc}}(\mathbf{r})$.

15.5 POTENTIAL DUE TO AN ELECTRIC DIPOLE

From elementary electrostatics it is well known that the potential due to an electric dipole moment distribution with polarization $\mathbf{P}$ is given by

$$V(r) = \int d^3r\mathbf{P}\cdot\nabla\frac{1}{r} \tag{15.8}$$

Applying the vector identity

$$\nabla\cdot\left(\frac{1}{r}\mathbf{P}\right) = \nabla\left(\frac{1}{r}\right)\cdot\mathbf{P} + \frac{1}{r}\nabla\cdot\mathbf{P} \tag{15.9}$$

one can write Eq. (15.8) as

$$V(r) = \int \nabla\cdot\left(\frac{1}{r}\mathbf{P}\right) d^3r \tag{15.10}$$

Here we have assumed constant polarization, so that $\nabla\cdot\mathbf{P} = \mathbf{0}$. Applying the divergence theorem to Eq. (15.10) one gets

$$V(r) = \int dS\frac{\hat{\mathbf{n}}\cdot\mathbf{P}}{r} = \int dS\frac{P_n}{r} \tag{15.11}$$

Here dS is an infinitesimal surface element and $\hat{\mathbf{n}}$ is a unit vector normal to the surface and is directed away from the dielectric material. Comparing the above equation with.

$$V(r) = \int dS\frac{\sigma_n}{r} \tag{15.12}$$

the surface charge density normal to the surface σ_n becomes

$$\sigma_n = \hat{\mathbf{n}}\cdot\mathbf{P} = P_n \tag{15.13}$$

Fig. 15.5 shows $\mathbf{E}_0$, $\mathbf{E}_P$, $\mathbf{P}$, and $\hat{\mathbf{n}}\cdot\mathbf{P}$ in a dielectric material in the form of a cuboid. With the help of Eq. (15.13), the electric field due to the polarization of the dielectric material can be evaluated.

15.6 DEPOLARIZATION FIELD DUE TO CUBOID

Consider a dielectric material in the form of a cuboid (Fig. 15.5) to which is applied the external electric field $\mathbf{E}_0$. The dielectric material is equivalent to two charged sheets: one surface has charge density $\hat{\mathbf{n}}\cdot\mathbf{P}$, while the other opposite to it has $-\hat{\mathbf{n}}\cdot\mathbf{P}$, as shown in the figure. From Gauss's law in an electrostatic, the field due to polarization $\mathbf{E}_P$ is given by

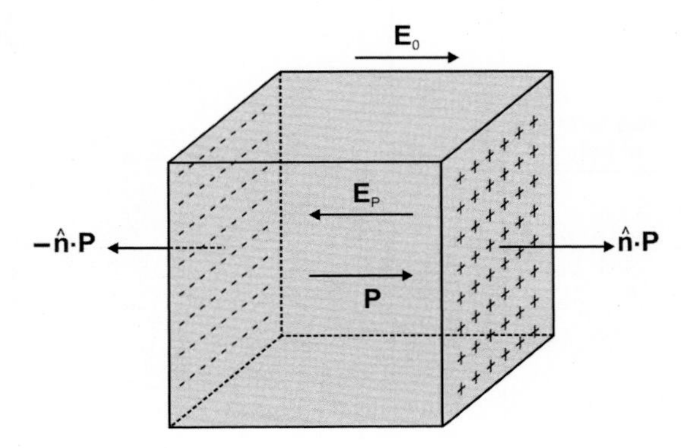

FIG. 15.5 Bulk dielectric material in the form of a cuboid placed in the presence of an external electric field $\mathbf{E}_0$. $\mathbf{E}_P$ gives the electric field and $\hat{\mathbf{n}}\cdot\mathbf{P}$ the surface charge density due to the polarization produced inside the material.

$$E_P = -2\pi\hat{\mathbf{n}}\cdot\mathbf{P} - 2\pi\hat{\mathbf{n}}\cdot\mathbf{P} = -4\pi\hat{\mathbf{n}}\cdot\mathbf{P}$$
$$= -4\pi P_n = -4\pi\sigma_n$$

In the vector form one can write.

$$\mathbf{E}_P = -4\pi\mathbf{P} \tag{15.14}$$

$\mathbf{E}_P$ is called the *depolarization field* because it opposes the applied electric field $\mathbf{E}_0$.

Problem 15.1

Prove that the depolarization field $\mathbf{E}_P$ in a bulk dielectric material in the form of a sphere is given by

$$\mathbf{E}_P = -\frac{4\pi}{3}\mathbf{P}. \tag{15.15}$$

Hence, the macroscopic field in a dielectric material in the form of a cuboid, from Eqs. (15.6) and (15.14), is given by

$$\mathbf{E} = \mathbf{E}_0 - 4\pi\mathbf{P} \tag{15.16}$$

In the above derivation the discrete lattice of the dipoles has been replaced by a smoothly varying polarization $\mathbf{P}$, therefore, $\mathbf{E}_P$ and $\mathbf{E}$ are smoothly varying fields. From Eq. (15.14) it is evident that the constant factor in the depolarization field depends on the shape of the bulk material, therefore, in general the Cartesian components of the depolarization field can be written as

$$E_{P_x} = -N_x P_x, \quad E_{P_y} = -N_y P_y, \quad E_{P_z} = -N_z P_z \tag{15.17}$$

Here N_x, N_y, and N_z are the depolarization factors and their values depend on the shape of the bulk material. For a solid in the form of an ellipsoid, the values of N_x, N_y, and N_z depend on the ratio of the major to minor axes and satisfy the following condition

$$N_x + N_y + N_z = 4\pi \tag{15.18}$$

$\mathbf{E}_0$ and $\mathbf{P}$ are parallel to each other and in the same direction. Further, $\mathbf{E}_P$ is parallel to $\mathbf{P}$ but in the opposite direction, therefore, one can write

$$\mathbf{E}_P = -N\mathbf{P} \tag{15.19}$$

Hence, the macroscopic field in this case becomes

$$\mathbf{E} = \mathbf{E}_0 - N\mathbf{P} \tag{15.20}$$

15.7 POLARIZATION

In general, a crystalline solid is anisotropic in nature. In such solids, when a macroscopic electric field $\mathbf{E}(\mathbf{r},t)$ is applied at position $\mathbf{r}$ at time t, the polarization $\mathbf{P}(\mathbf{r}',t')$ may be produced at all possible values of position $\mathbf{r}'$ and time t'. The reverse is also true, that is, the polarization $\mathbf{P}(\mathbf{r},t)$ at position $\mathbf{r}$ and time t is produced when macroscopic field $\mathbf{E}(\mathbf{r}',t')$ is applied at all possible positions $\mathbf{r}'$ at time t'. In the linear response approximation, this fact can be represented mathematically as follows

$$\mathbf{P}(\mathbf{r},t) = \sum_{\mathbf{r}',t'} \chi_E(\mathbf{r},t;\mathbf{r}',t')\mathbf{E}(\mathbf{r}',t') \tag{15.21}$$

$\chi_E(\mathbf{r},t;\mathbf{r}',t')$ is the linear response function and is usually called the electric susceptibility matrix. In reciprocal space Eq. (15.21) is given by

$$\mathbf{P}(\mathbf{K},\omega) = \sum_{\mathbf{K}',\omega'} \chi_E(\mathbf{K},\omega;\mathbf{K}',\omega')\mathbf{E}(\mathbf{K}',\omega') \tag{15.22}$$

One is usually interested in the equal-time response function, i.e., for $t=t'$ or $\omega=\omega'$. Therefore, for equal time, Eqs. (15.21) and (15.22) can be written as

$$\mathbf{P}(\mathbf{r},t) = \sum_{\mathbf{r}'} \chi_E(\mathbf{r},\mathbf{r}',t)\mathbf{E}(\mathbf{r}',t) \tag{15.23}$$

$$\mathbf{P}(\mathbf{K},\omega) = \sum_{\mathbf{K}'} \chi_E(\mathbf{K},\mathbf{K}',\omega)\mathbf{E}(\mathbf{K}',\omega) \tag{15.24}$$

Here we have written $\chi_E(\mathbf{r}, t; \mathbf{r}', t) = \chi_E(\mathbf{r}, \mathbf{r}', t)$ and $\chi_E(\mathbf{K}, \omega; \mathbf{K}', \omega) = \chi_E(\mathbf{K}, \mathbf{K}', \omega)$. The frequency-dependent susceptibility $\chi_E(\mathbf{K}, \mathbf{K}', \omega)$ is generally called the dynamical susceptibility matrix. In a homogeneous and isotropic solid, the polarization is produced in the direction of the applied field and in this case Eqs. (15.23) and (15.24) reduce to

$$\mathbf{P}(\mathbf{r}, t) = \chi_E(\mathbf{r}, t)\, \mathbf{E}(\mathbf{r}, t) \tag{15.25}$$

$$\mathbf{P}(\mathbf{K}, \omega) = \chi_E(\mathbf{K}, \omega)\, \mathbf{E}(\mathbf{K}, \omega) \tag{15.26}$$

Here the dynamical electric susceptibility $\chi_E(\mathbf{K}, \omega)$ is a scalar quantity. The time-independent polarization (static polarization) for a homogeneous material is obtained by substituting $t = 0$ and $\omega = 0$ in Eqs. (15.25) and (15.26) and is given by

$$\mathbf{P}(\mathbf{r}) = \chi_E(\mathbf{r})\, \mathbf{E}(\mathbf{r}) \tag{15.27}$$

$$\mathbf{P}(\mathbf{K}) = \chi_E(\mathbf{K})\, \mathbf{E}(\mathbf{K}) \tag{15.28}$$

$\chi_E(\mathbf{K})$ is the static electric susceptibility function. If a constant electric field is applied to the solid, the above equation can be written as

$$\mathbf{P} = \chi_E \mathbf{E} \tag{15.29}$$

The electric susceptibility χ_E becomes a constant. With the help of Eq. (15.20) the polarization $\mathbf{P}$ can also be expressed in terms of the applied field $\mathbf{E}_0$. Substituting Eq. (15.20) into Eq. (15.29), one can write

$$\mathbf{P} = \frac{\chi_E}{1 + N\chi_E}\mathbf{E}_0 \tag{15.30}$$

In the limiting case of very large χ_E the above equation reduces to

$$\mathbf{P} \approx \frac{\mathbf{E}_0}{N} \tag{15.31}$$

Hence, the polarization is determined purely by the shape of the material in the limit of very large χ_E. Such a situation must be avoided in the experimental determination of the electrical susceptibility of a solid.

15.8 DIELECTRIC MATRIX

From elementary electricity, the displacement field $\mathbf{D}(\mathbf{r}, t)$ is defined as

$$\mathbf{D}(\mathbf{r}, t) = \mathbf{E}(\mathbf{r}, t) + 4\pi \mathbf{P}(\mathbf{r}, t) \tag{15.32}$$

Substituting the value of $\mathbf{P}(\mathbf{r}, t)$ from Eq. (15.23) into Eq. (15.32), one writes.

$$\mathbf{D}(\mathbf{r}, t) = \mathbf{E}(\mathbf{r}, t) + 4\pi \sum_{\mathbf{r}'} \chi_E(\mathbf{r}, \mathbf{r}', t)\, \mathbf{E}(\mathbf{r}', t) \tag{15.33}$$

Here we have used the equal-time expression for $\mathbf{P}(\mathbf{r}, t)$. The above equation can be written as

$$\mathbf{D}(\mathbf{r}, t) = \sum_{\mathbf{r}'} \varepsilon(\mathbf{r}, \mathbf{r}', t)\, \mathbf{E}(\mathbf{r}', t) \tag{15.34}$$

where

$$\varepsilon(\mathbf{r}, \mathbf{r}', t) = \delta_{\mathbf{r},\mathbf{r}'} + 4\pi \chi_E(\mathbf{r}, \mathbf{r}', t) \tag{15.35}$$

$\varepsilon(\mathbf{r}, \mathbf{r}', t)$ is the response function and is called the dielectric matrix. In reciprocal space Eqs. (15.34) and (15.35) can be written as

$$\mathbf{D}(\mathbf{K}, \omega) = \sum_{\mathbf{K}'} \varepsilon(\mathbf{K}, \mathbf{K}', \omega)\, \mathbf{E}(\mathbf{K}', \omega) \tag{15.36}$$

$$\varepsilon(\mathbf{K}, \mathbf{K}', \omega) = \delta_{\mathbf{K},\mathbf{K}'} + 4\pi \chi_E(\mathbf{K}, \mathbf{K}', \omega) \tag{15.37}$$

For a homogeneous and isotropic solid, Eqs. (15.35) and (15.37) reduce to scalar equations given by

$$\varepsilon(\mathbf{r}, t) = 1 + 4\pi \chi_E(\mathbf{r}, t) \tag{15.38}$$

$$\varepsilon(\mathbf{K},\omega) = 1 + 4\pi\chi_E(\mathbf{K},\omega) \tag{15.39}$$

$\varepsilon(\mathbf{r},t)$ and $\varepsilon(\mathbf{K},\omega)$ represent the scalar dynamical dielectric function in the direct and reciprocal spaces. The time-independent (static) dielectric function for a homogeneous material is obtained by substituting $t=0$ and $\omega=0$ in Eqs. (15.38) and (15.39), respectively, and is given by

$$\varepsilon(\mathbf{r}) = 1 + 4\pi\chi_E(\mathbf{r}) \tag{15.40}$$

$$\varepsilon(\mathbf{K}) = 1 + 4\pi\chi_E(\mathbf{K}) \tag{15.41}$$

The frequency-dependent dielectric function at $\mathbf{K}=0$, i.e., $\varepsilon(\omega)$, from Eq. (15.39) becomes

$$\varepsilon(\omega) = 1 + 4\pi\chi_E(\omega) \tag{15.42}$$

$\varepsilon(\omega)$ gives the dielectric function at very long wavelengths. For a uniform electric field, Eqs. (15.40) and (15.41) reduce to

$$\varepsilon = 1 + 4\pi\chi_E \tag{15.43}$$

ε defines the dielectric constant of a material.

15.9 EXPERIMENTAL MEASUREMENT OF DIELECTRIC CONSTANT

The dielectric constant can be defined as the ratio of the capacity of a capacitor with dielectric material C_{dielc} to the capacity of the same capacitor when empty, C_{vac}, i.e.,

$$\varepsilon = \frac{C_{dielc}}{C_{vac}} \tag{15.44}$$

Experimental measurements of the dielectric constant can be performed using Eq. (15.44). The capacity of a capacitor can be calculated with the help of an LC resonant circuit, as shown in Fig. 15.6. Here L is an inductor, C is an experimental capacitor, and C_G is a variable (gang) capacitor. It should be noted that Fig. 15.6 illustrates just the principle, but there are a number of actual circuits employed to determine the dielectric constant. The resonance frequency of an LC circuit is given by

$$\nu_0 = \frac{1}{2\pi}\sqrt{\frac{1}{L(C+C_G)}} \tag{15.45}$$

By varying the capacity C_G, the resonance frequency of the LC circuit is determined when the capacitor C is empty. From the total capacity, $C_{vac}+C_G$, one can find C_{vac}, i.e., the capacity when there is vacuum inside the experimental capacitor. Now the dielectric material is placed inside the experimental capacitor and C_G is again varied to obtain the same resonance frequency, yielding the total capacity of the circuit as $C_{dielc}+C_G$. From the total capacity one can calculate C_{dielc}, i.e., the capacity of the experimental capacitor with the dielectric inside it. With knowledge of C_{dielc} and C_{vac}, one can find the dielectric constant of the material.

In this chapter we shall present an atomic view (microscopic description) of the properties of a dielectric material. To study the dielectric properties on the atomic scale, one should know that the electric field experienced by an atom may be much different from the macroscopic field.

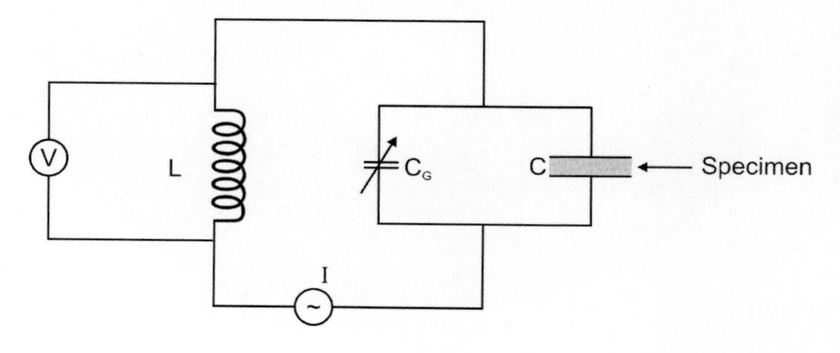

FIG. 15.6 The LC circuit used to measure the dielectric constant of a solid. Here C is a parallel plate capacitor in which the specimen of the given solid is used as a dielectric and C_G is a variable (gang) capacitor.

15.10 LOCAL ELECTRIC FIELD AT AN ATOM

The actual electric field at the site of an atom is called the local electric field $\mathbf{E}_{loc}$, which may be significantly different from the macroscopic electric field **E**. Let us find the general expression for $\mathbf{E}_{loc}$ at the atomic site as it forms the backbone of a microscopic study of the dielectric properties of solids. Consider a solid in ellipsoidal form (Fig. 15.7) in which the external electric field $\mathbf{E}_0$ is applied along the major axis. It is assumed that all the electric dipoles are oriented parallel to the external field. Let us evaluate $\mathbf{E}_{loc}$ at an atom situated at the center of the ellipsoidal solid. The exact value of $\mathbf{E}_{loc}$ at the position of the atom is the sum of the external electric field and the field arising from all the atomic dipoles of the solid, i.e.,

$$\mathbf{E}_{loc} = \mathbf{E}_0 + \sum_i \frac{3(\mathbf{p}_i \cdot \hat{\mathbf{r}}_i)\hat{\mathbf{r}}_i - \mathbf{p}_i}{r_i^3} \tag{15.46}$$

Here the summation is over all the atomic dipoles of the solid, an operation that may be difficult to perform. Therefore, the following simplified procedure is adopted to estimate $\mathbf{E}_{loc}$. The maximum contribution to $\mathbf{E}_{loc}$ at the atom is expected to come from its neighboring atomic dipoles and their contribution should be accounted for exactly. The contribution from the rest of the atomic dipoles of the solid can be averaged out. To achieve this, a small hypothetical spherical solid with the atom under consideration at its center is cut from the solid (see Figs. 15.7 and 15.8). The spherical solid is large enough to contain very many atomic dipoles in it. After removing this spherical portion, a cavity, usually called the Lorentz cavity, is created at the center of the ellipsoidal solid. Leaving aside the spherical portion, one is left with an ellipsoidal dielectric solid having finite charge density at its outer surface and also finite charge density on the inner surface of the Lorentz cavity (see Fig. 15.8). Now $\mathbf{E}_{loc}$ at the center of the cavity can be split into four contributions as follows:

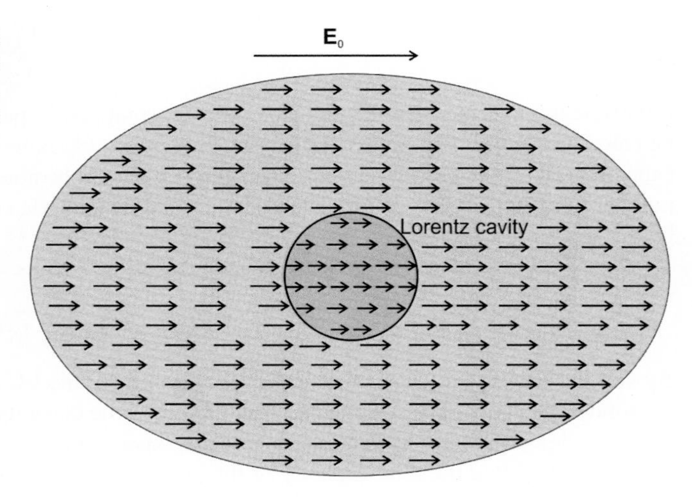

FIG. 15.7 Polarization produced in a solid in the presence of an applied electric field $\mathbf{E}_0$. The sphere in the center of the solid shows a small portion of the solid, which contains a sufficiently large number of atomic dipoles. By removing this small portion of the solid a cavity is produced, usually called the Lorentz cavity.

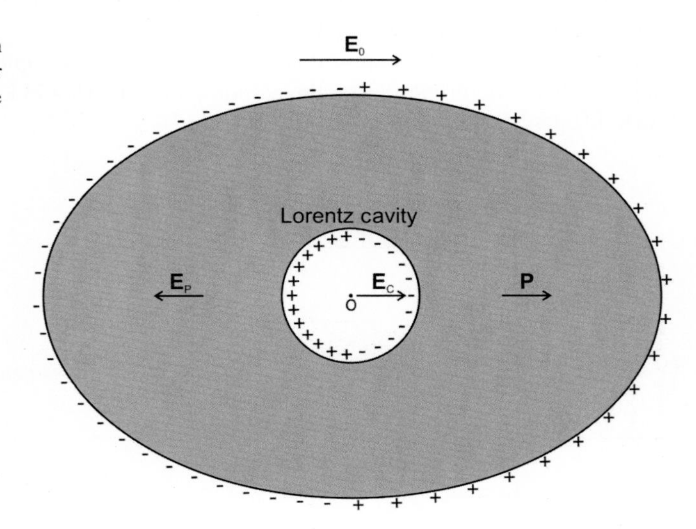

FIG. 15.8 The dipole moments inside the solid cancel the effects of each other; therefore, the surface of the solid is charged as shown. The inner surface of the Lorentz cavity is also charged but opposite to that of the outer surface of the solid.

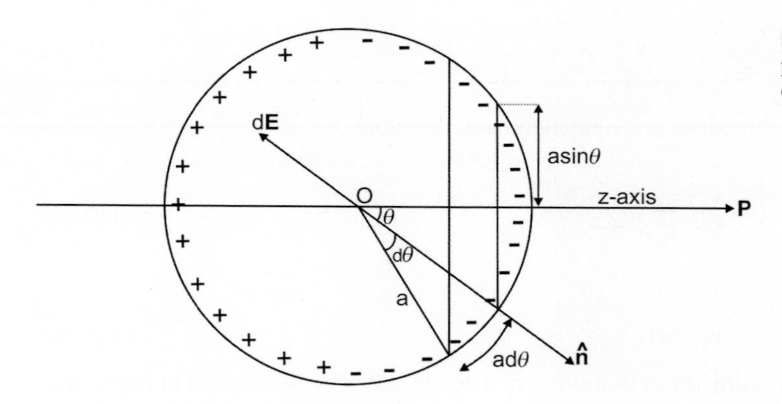

FIG. 15.9 The Lorentz cavity with origin at its center O. The polarization and electric field produced by the charge density on the inner surface of the cavity are shown.

$$\mathbf{E}_{loc} = \mathbf{E}_0 + \mathbf{E}_P + \mathbf{E}_C + \mathbf{E}_D \tag{15.47}$$

$\mathbf{E}_D$ is the net electric field at the center of the Lorentz cavity due to the fields of all of the electric dipoles inside the spherical portion (Lorentz cavity) and is given by

$$\mathbf{E}_D = \sum_{j\,(cavity)} \frac{3\left(\mathbf{p}_j \cdot \hat{\mathbf{r}}_j\right)\hat{\mathbf{r}}_j - \mathbf{p}_j}{r_j^3} \tag{15.48}$$

Here $\mathbf{E}_P$ is the depolarization field due to the charge density on the outer surface of the solid and its value depends on the shape of the solid (Section 15.6). $\mathbf{E}_C$ is the field arising from the charge density appearing on the inner surface of the Lorentz cavity and is calculated according to what follows.

The Lorentz cavity with radius "a" has a finite charge density $-\hat{\mathbf{n}} \cdot \mathbf{P} = -P\cos\theta$ on its inner surface (Fig. 15.9). Let the charge on the circular ring be dq, which, from the figure, is given by

$$dq = 2\pi a \sin\theta\,(a d\theta)\,(-P\cos\theta) \tag{15.49}$$

The electric field due to the circular ring at the center O of the cavity $\mathbf{dE}$ is given by

$$\mathbf{dE} = \frac{dq}{a^2} = -2\pi P \cos\theta \sin\theta\, d\theta \tag{15.50}$$

The component of the electric field in the direction of polarization becomes

$$\mathbf{dE}_C = \mathbf{dE}\cos(\pi - \theta) = -2\pi P\cos\theta \sin\theta d\theta \cos(180 - \theta)$$

Hence the total field at the center of the cavity becomes

$$E_C = 2\pi P \int_0^{\pi} \cos^2\theta \sin\theta\, d\theta = \frac{4\pi}{3} P \tag{15.51}$$

Using Eqs. (15.48) and (15.51) in Eq. (15.47), one gets

$$\mathbf{E}_{loc} = \mathbf{E}_0 + \mathbf{E}_P + \frac{4\pi}{3}\mathbf{P} + \sum_{j\,(cavity)} \frac{3\left(\mathbf{p}_j \cdot \hat{\mathbf{r}}_j\right)\hat{\mathbf{r}}_j - \mathbf{p}_j}{r_j^3} \tag{15.52}$$

The equation above gives the local electric field for any structure. The evaluation of the last term of Eq. (15.52) is difficult for a complex structure, so here we evaluate $\mathbf{E}_{loc}$ for an sc structure. From Eq. (15.48), $\mathbf{E}_D$ can be written as

$$\mathbf{E}_D = \sum_{j\,(cavity)} \frac{3\left(\mathbf{p}_j \cdot \mathbf{r}_j\right)\mathbf{r}_j - r_j^2 \mathbf{p}_j}{r_j^5} \tag{15.53}$$

If all of the electric dipoles are oriented in the z-direction, i.e., $\mathbf{p}_j = p_j\hat{z}$, then $\mathbf{E}_D (= E_D^z)$ in the z-direction is given by

$$E_D^z = \sum_{j\,(\text{cavity})} p_j \frac{3z_j^2 - r_j^2}{r_j^5} = \sum_{j\,(\text{cavity})} p_j \frac{2z_j^2 - x_j^2 - y_j^2}{r_j^5} \tag{15.54}$$

For a cubic crystal with a spherical cavity all three directions are equivalent and, therefore,

$$\sum_j \frac{x_j^2}{r_j^5} = \sum_j \frac{y_j^2}{r_j^5} = \sum_j \frac{z_j^2}{r_j^5} \tag{15.55}$$

Substituting Eq. (15.55) into Eq. (15.54), one gets.

$$E_D^z = 0 \tag{15.56}$$

The result above can also be proved from Eq. (15.48) using the definition of $\hat{\mathbf{r}}_j$. Therefore, for a cubic crystal of any shape, the local field at the site of an atom is given by

$$\mathbf{E}_{loc} = \mathbf{E}_0 + \mathbf{E}_P + \frac{4\pi}{3}\mathbf{P} \tag{15.57}$$

The equation above can be written in terms of the macroscopic field **E** as

$$\mathbf{E}_{loc} = \mathbf{E} + \frac{4\pi}{3}\mathbf{P} \tag{15.58}$$

If the cubic crystal has a spherical shape instead of an ellipsoidal one, then the depolarization field $\mathbf{E}_P$ [given by Eq. (15.15)] cancels the Lorentz cavity field $\mathbf{E}_C$ and, therefore, the local field reduces to the applied field, i.e.,

$$\mathbf{E}_{loc} = \mathbf{E}_0 \tag{15.59}$$

15.11 POLARIZABILITY

The electric dipole moment of an atom/molecule **p** depends on the local electric field experienced by it. Therefore, in the linear response approximation, **p** is given by

$$\mathbf{p} = \alpha^a \mathbf{E}_{loc} \tag{15.60}$$

where α^a is a constant of proportionally called the *atomic polarizability*. The dimensions of α^a are L^3.

15.12 POLARIZATION

For a microscopic description of dielectric properties, the polarization should be expressed in terms of the atomic dipole moments. In general, a solid may contain more than one type of atom. Let ρ_i^a be the number of the ith type of atom per unit volume in the solid, with $\mathbf{p}_i$ as the atomic dipole moment. The polarization of such a solid is given by

$$\mathbf{P} = \sum_i \rho_i^a \mathbf{p}_i \tag{15.61}$$

From Eqs. (15.60) and (15.61) one can write.

$$\mathbf{P} = \sum_i \rho_i^a \alpha_i^a \mathbf{E}_{loc}^i \tag{15.62}$$

Here α_i^a and $\mathbf{E}_{loc}^i$ are the polarizability and local electric field experienced by the ith type of atom. Substituting the value of $\mathbf{E}_{loc}^i$ from Eq. (15.58) into Eq. (15.62), we get

$$\mathbf{P} = \sum_i \rho_i^a \alpha_i^a \left(\mathbf{E} + \frac{4\pi}{3}\mathbf{P}\right) \tag{15.63}$$

From the above equation the electric susceptibility becomes

$$\chi_E = \frac{\mathbf{P}}{\mathbf{E}} = \frac{\sum_i \rho_i^a \alpha_i^a}{1 - \frac{4\pi}{3}\sum_i \rho_i^a \alpha_i^a} \tag{15.64}$$

Using Eq. (15.43) in Eq. (15.64), one obtains a relation between ε and α_i^a as

$$\frac{\varepsilon - 1}{\varepsilon + 2} = \frac{4\pi}{3}\sum_i \rho_i^a \alpha_i^a \tag{15.65}$$

Eq. (15.65) is called the Clausius-Mossotti relation and represents the macroscopic physical quantities in terms of microscopic quantities. This relation can be written in terms of conventional quantities. Let n_i^a be the number of the ith type of atom in a unit cell of volume V_0. Then the number of the ith type of atom per unit volume becomes $\rho_i^a = n_i^a/V_0$. Substituting the value of ρ_i^a in Eq. (15.65), we find

$$\sum_i n_i^a \alpha_i^a = \frac{3V_0}{4\pi}\frac{\varepsilon - 1}{\varepsilon + 2} \tag{15.66}$$

Eq. (15.66) can be used to find the polarizability of the solid. For example, consider a solid with n^a identical atoms in a unit cell. The polarizability α^a from Eq. (15.66) becomes

$$\alpha^a = \frac{3V_0}{4\pi n^a}\frac{\varepsilon - 1}{\varepsilon + 2} \tag{15.67}$$

If, in a solid, there are two types of atoms with polarizabilities α_1^a and α_2^a, but with the same number of atoms per unit cell, i.e., $n_1^a = n_2^a = n^a$, then

$$\sum_i n_i^a \alpha_i^a = n^a\left(\alpha_1^a + \alpha_2^a\right) \tag{15.68}$$

Eqs. (15.66) and (15.68) give the sum of the polarizabilities of two types of atoms by

$$\alpha_1^a + \alpha_2^a = \frac{3V_0}{4\pi n^a}\frac{\varepsilon - 1}{\varepsilon + 2} \tag{15.69}$$

15.13 TYPES OF POLARIZABILITIES

The polarizability of an atom comprises three possible contributions, namely

1. Electronic polarizability α_E^a
2. Ionic polarizability α_I^a
3. Dipolar or orientational polarizability α_D^a

The total atomic polarizability is the sum of the three contributions and is given by

$$\alpha^a = \alpha_E^a + \alpha_I^a + \alpha_D^a \tag{15.70}$$

Fig. 15.10A shows a free spherical atom. When an electric field $\mathbf{E}_0$ is applied, the electron cloud of the atom not only shifts toward the positive side but also suffers distortion (Fig. 15.10B). This causes the centers of negative and positive charges to shift away from each other, thereby producing a finite electric dipole moment. This physical effect contributes to the polarizability, generally called the *electronic* polarizability.

Fig. 15.11A shows a linear ionic solid in which the distance between the nearest neighbors is the same. In the ionic solid the adjacent electric dipole moments are equal and opposite, yielding net zero polarization. When an electric field is applied, the positive and negative ions move in opposite directions (Fig. 15.11B), thereby producing oppositely directed unequal electric dipole moments causing finite ionic polarization and hence *ionic polarizability*.

Consider a solid in which each atom possesses a finite intrinsic electric dipole moment. In the absence of an electric field, all of the dipoles are randomly oriented, yielding zero polarization. When an electric field is applied, the dipole

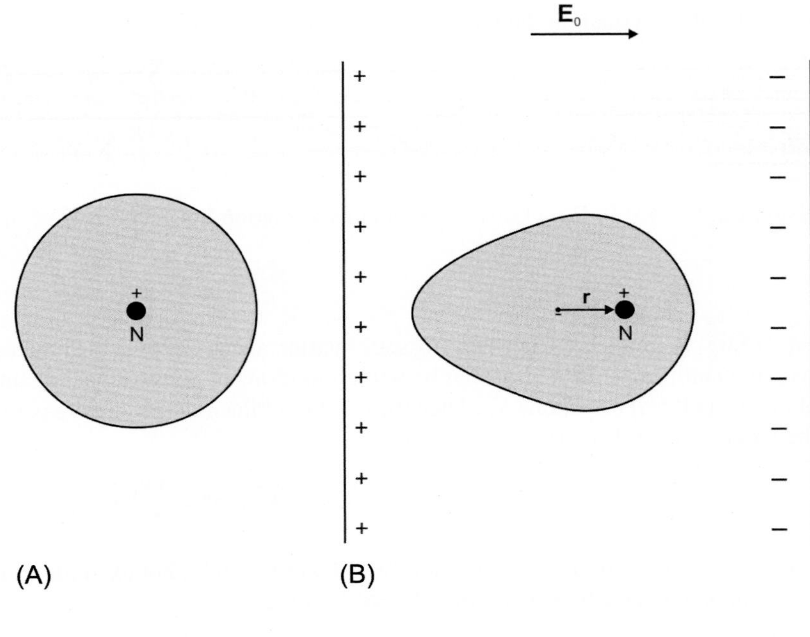

FIG. 15.10 (A) A free spherical atom. (B) An atom in the presence of an external electric field $\mathbf{E}_0$. In the presence of $\mathbf{E}_0$ the shape of the electron cloud gets distorted in addition to the shifting of the centers of negative and positive charges.

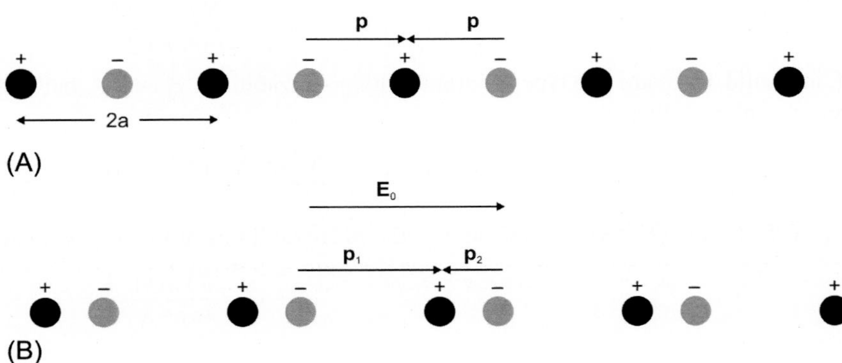

FIG. 15.11 (A) A linear ionic solid with periodicity 2a yields no polarization. (B) A linear ionic solid in the presence of an applied electric field $\mathbf{E}_0$ produces finite ionic polarization.

moments tend to align in the direction of the field, producing finite polarization. The polarizability thus produced is called *dipolar* or *orientational polarizability*. The electronic and ionic polarizabilities are produced in both polar and nonpolar solids, but the dipolar polarizability is produced only in polar solids.

15.14 VARIATION OF POLARIZABILITY WITH FREQUENCY

At very low frequencies the ions, electrons, and intrinsic dipole moments all respond to the applied electric field; therefore, the total polarizability is the sum of the three contributions (Eq. 15.70). With an increase in frequency the reorientation of the intrinsic dipole moments becomes difficult as atoms are heavy particles that interact with one another. Ultimately, the atomic dipole moments stop responding at some particular frequency. With a further increase in frequency, the response of the ions also decreases due to the large inertia and goes to zero at some higher frequency. At very high frequencies only electrons, being very light particles, respond to the applied electric field, yielding finite electronic polarizability. At optical frequencies, only the electronic polarizability is finite. Fig. 15.12 shows the variation of α^a as a function of frequency. The peaks appearing in the figure are due to resonance absorption at certain frequencies and are not of interest to us. At optical frequencies $\varepsilon = n^2$ where n is the refractive index of the material. Therefore, in the optical range, Eq. (15.65) reduces to.

$$\frac{n^2 - 1}{n^2 + 2} = \frac{4\pi}{3} \sum_i \rho_i^a \alpha_i^a \tag{15.71}$$

Here α_i^a corresponds to the electronic polarizability of the i-th type of atom. This equation can be used to evaluate the electronic polarizability as the refractive index of most materials is known to a fair degree of accuracy.

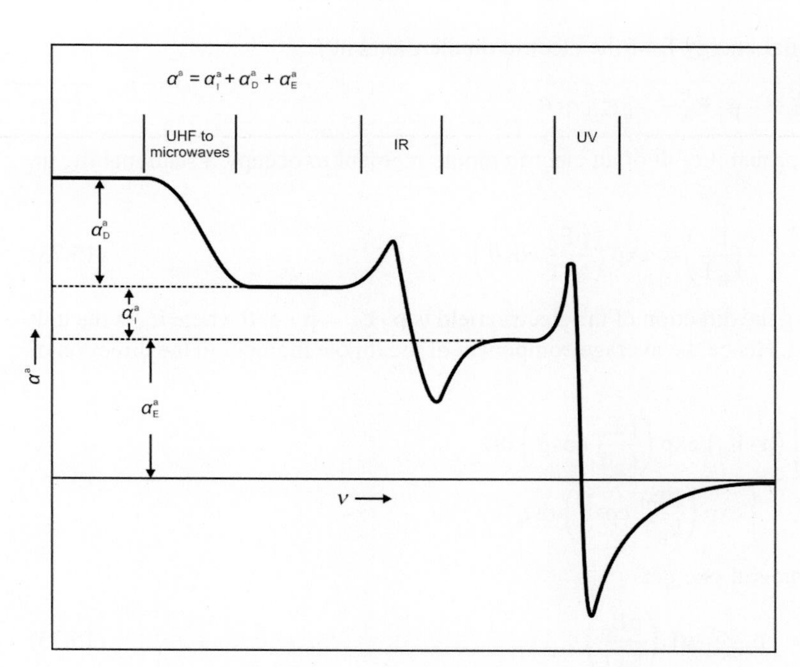

FIG. 15.12 Variation of dipolar, ionic, and electronic polarizabilities of an atom/molecule as a function of frequency.

15.15 ORIENTATIONAL POLARIZABILITY

The orientational or dipolar polarizability is shown by materials in which each atom/molecule possesses an intrinsic electric dipole moment. At finite temperature, in the presence of an external electric field $\mathbf{E}_0$, there are two forces acting on each electric dipole:

1. First is the electric force, which tends to align the dipole moments along the direction of the electric field.
2. Second is the thermal force, which tries to destroy the alignment.

Under the action of these two competing forces, some electric dipole moments align in the direction of the applied electric field and others make some angle θ, which may vary from 0 to π radians for different dipole moments (see Fig. 15.13). Therefore, the specimen shows finite polarization in the direction of the electric field. The polarization increases either with an increase in the strength of the electric field or with a decrease in temperature. The saturation polarization is reached when all the dipole moments align along the direction of the electric field.

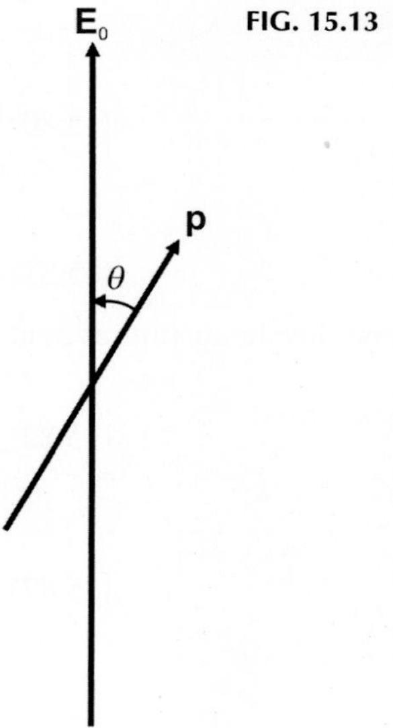

FIG. 15.13 The orientation of the electric dipole moment **p** with respect to an applied electric field $\mathbf{E}_0$.

In the presence of an electric field, the potential energy E of the electric dipole moment is

$$E = -\mathbf{p} \cdot \mathbf{E}_0 = -pE_0 \cos\theta \tag{15.72}$$

According to classical statistical mechanics, the probability Φ of an electric dipole moment to occupy a state making an angle θ with the electric field is given by

$$\Phi \propto \exp\left(-\frac{E}{k_B T}\right) \propto \exp\left(\frac{pE_0}{k_B T}\cos\theta\right) \tag{15.73}$$

The component of an electric dipole moment along the direction of the electric field is $\mathbf{p} \cdot \hat{\mathbf{E}}_0 = p \cos\theta$ where $\hat{\mathbf{E}}_0$ is the unit vector in the direction of the external electric field. Hence the average component of the dipole moment in the direction of the electric field is given by

$$p_{avg} = \frac{\int (\mathbf{p} \cdot \hat{\mathbf{E}}_0) \exp\left(\frac{pE_0}{k_B T}\cos\theta\right) d\Omega_s}{\int \exp\left(\frac{pE_0}{k_B T}\cos\theta\right) d\Omega_s} \tag{15.74}$$

Here $d\Omega_s$ is the solid angle. Solving the above integral one gets

$$p_{avg} = pL\left(\frac{pE_0}{k_B T}\right) \tag{15.75}$$

where

$$L(y) = \coth y - \frac{1}{y} \tag{15.76}$$

Here L(y) is called the Langevin function. If ρ^a is the number of atoms per unit volume, then the electric polarization is given by

$$P(T) = \rho^a p L\left(\frac{pE_0}{k_B T}\right) \tag{15.77}$$

The electric susceptibility, in the linear response approximation, becomes

$$\chi_E(T) = \frac{P(T)}{E_0} = \frac{\rho^a p}{E_0} L\left(\frac{pE_0}{k_B T}\right) \tag{15.78}$$

Simple expressions can be obtained for polarization and susceptibility in the limiting cases. If

$$pE_0 \gg k_B T \tag{15.79}$$

then

$$L\left(\frac{pE_0}{k_B T}\right) = 1 \tag{15.80}$$

Therefore,

$$P = \rho^a p \tag{15.81}$$

which gives the saturation polarization. Hence, the saturation polarization is obtained either at very low temperatures or at very high electric field. If

$$pE_0 \ll k_B T \tag{15.82}$$

then, for small values of y, coth y can be expanded as

$$\coth y = \frac{1}{y} + \frac{y}{3} - \frac{y^3}{45} + \cdots. \tag{15.83}$$

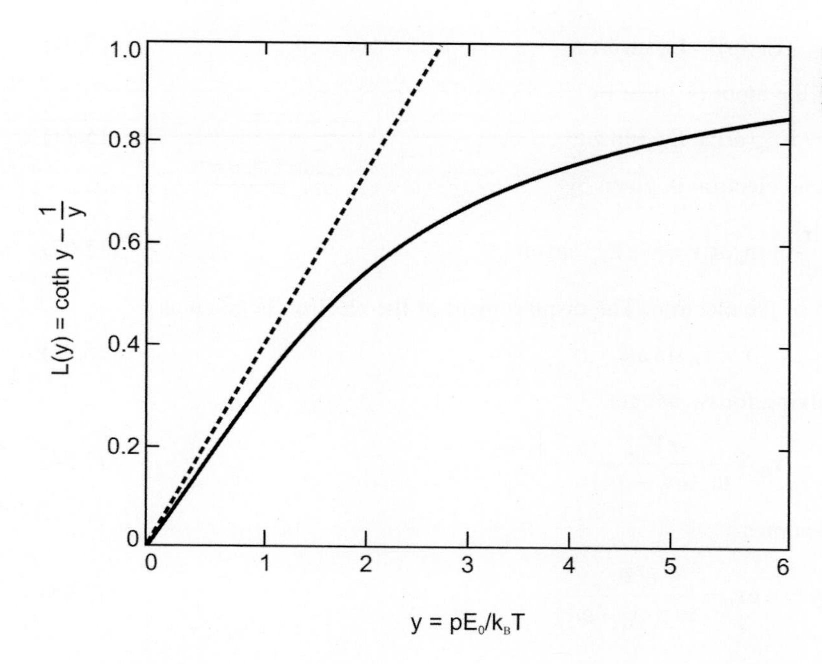

FIG. 15.14 Plot of the Langevin function L(y) as a function of the parameter $y = pE_0/k_BT$. The *dashed line* is tangent to the curve at the origin and its slope is 1/3.

Therefore, for small y, L(y) can be written as

$$L(y) = \left(\frac{1}{y} + \frac{y}{3} + \cdots\cdots\right) - \frac{1}{y} \approx \frac{y}{3} \tag{15.84}$$

The behavior of the Langevin function L(y) is shown in Fig. 15.14. L(y) has a slope of 1/3 at $y = 0$ and attains the saturation value of one at very large y. Hence, at small values of y, the polarization and electric susceptibility are given by

$$P(T) = \frac{\rho^a p^2}{3k_B T} E_0 \tag{15.85}$$

$$\chi_E(T) = \frac{C_E}{T} \tag{15.86}$$

where C_E is called the Curie constant and is given by

$$C_E = \frac{\rho^a p^2}{3k_B} \tag{15.87}$$

Eq. (15.86) is usually called the Curie law. From Eqs. (15.60) and (15.75) the dipolar contribution to the polarizability is given by

$$\alpha_D^a(T) = \frac{\mathbf{p}_{avg}}{\mathbf{E}_{loc}} = \frac{p}{E_0} L\left(\frac{pE_0}{k_B T}\right) \tag{15.88}$$

At weak field E_0 and high temperature T (paramagnetic region), Eq. (15.88) reduces to

$$\alpha_D^a(T) = \frac{p^2}{3k_B T} \tag{15.89}$$

15.16 CLASSICAL THEORY OF ELECTRONIC POLARIZABILITY

Suppose an electron is bound harmonically to the nucleus of an atom in a dielectric material. A frequency-dependent electric field $\mathbf{E}_0(\omega)$ is applied to the material, given by

$$\mathbf{E}_0(\omega)=\mathbf{E}_0 \sin \omega t \tag{15.90}$$

The local field $\mathbf{E}_{loc}(\omega)$ acting on the electron of the atom is given by

$$\mathbf{E}_{loc}(\omega)=\mathbf{E}_{loc} \sin \omega t \tag{15.91}$$

The equation of motion of the harmonically bound electron is given by

$$m_e \frac{d\mathbf{r}^2}{dt}+m_e \omega_0^2 \mathbf{r}=-e\mathbf{E}_{loc} \sin \omega t \tag{15.92}$$

Here ω_0 is the resonance frequency of vibration of the electron. The displacement of the electron is given as

$$\mathbf{r}=\mathbf{r}_0 \sin \omega t \tag{15.93}$$

Substituting Eq. (15.93) into Eq. (15.92) and solving for $\mathbf{r}_0$, we get.

$$\mathbf{r}_0=\frac{-e\mathbf{E}_{loc}}{m_e\left(\omega_0^2-\omega^2\right)} \tag{15.94}$$

Therefore, the dipole moment of the electron becomes

$$\mathbf{p}=-e\mathbf{r}_0=\frac{e^2\mathbf{E}_{loc}}{m_e\left(\omega_0^2-\omega^2\right)} \tag{15.95}$$

The electronic polarizability $\alpha_E^a(\omega)$ becomes

$$\alpha_E^a(\omega)=\frac{\mathbf{p}}{\mathbf{E}_{loc}}=\frac{e^2}{m_e\left(\omega_0^2-\omega^2\right)} \tag{15.96}$$

The static polarizability is obtained by applying a static electric field to the material. Therefore, from Eq. (15.96), the static polarizability becomes

$$\alpha_E^a=\alpha_E^a(0)=\frac{e^2}{m_e\omega_0^2} \tag{15.97}$$

The electronic polarizability can also be treated quantum mechanically and the expression is given by

$$\alpha_E^a(\omega)=\frac{e^2}{m_e}\sum_j \frac{f_{ij}}{\omega_{ij}^2-\omega^2} \tag{15.98}$$

where f_{ij} is called the oscillator strength of the electric dipole transition between the atomic states i and j. Eq. (15.98) is derived for atoms and must be modified for dielectric solids. Note that Eq. (15.98) is quite similar to the classical result given by Eq. (15.96).

Problem 15.2

The Hartree dielectric screening function $\varepsilon(\mathbf{K})$ for wave vector $\mathbf{K}$ is defined as

$$\varepsilon(\mathbf{K})=1-v(\mathbf{K})\,\chi_E(\mathbf{K}) \tag{15.99}$$

where

$$v(\mathbf{K})=\frac{4\pi e^2}{VK^2} \tag{15.100}$$

and

$$\chi_E(\mathbf{K})=\sum_{\mathbf{k}} \frac{f(E_{\mathbf{k}},T)-f(E_{\mathbf{k}+\mathbf{K}},T)}{E_{\mathbf{k}}-E_{\mathbf{k}+\mathbf{K}}} \tag{15.101}$$

Here V is the volume of the solid and $v(\mathbf{K})$ is the electron–electron interaction potential in the reciprocal space. Derive the expressions for $\chi_E(\mathbf{K})$ and $\varepsilon(\mathbf{K})$ at absolute zero in the free-electron approximation.

Problem 15.3

The bare-ion Coulomb potential in the reciprocal space as seen by an electron is given by

$$V^b(K) = -\frac{4\pi Ze^2}{V_0 K^2} \quad (15.102)$$

If the screened ion potential is defined as

$$V(K) = \frac{V^b(K)}{\varepsilon(K)} \quad (15.103)$$

where $\varepsilon(\mathbf{K})$ is the Hartree dielectric function (Problem 5.2). Prove that in the limit $K \to 0$, the screened potential reduces to.

$$\lim_{K\to 0} V(K) = -\frac{2}{3} E_F \quad (15.104)$$

Problem 15.4

The dielectric function in the Thomas-Fermi approximation is given by

$$\varepsilon(\mathbf{K}) = 1 + K_{TF}^2/K^2 \quad (15.105)$$

where K_{TF} is the Thomas-Fermi wave vector defined as

$$K_{TF} = \left(\frac{6\pi n_e e^2}{E_F}\right)^{1/2} \quad (15.106)$$

The bare-ion potential as seen by an electron is given as

$$V^b(K) = -\frac{4\pi Ze^2}{V_0 K^2} \quad (15.107)$$

Find the expression for the screened potential in the crystal space.

Problem 15.5

Find the expression for the electronic polarizability of an atom.

Problem 15.6

Consider the one-dimensional ionic lattice shown in Fig. 15.11. If the external electric field $\mathbf{E}_0$ is applied in the positive x-direction, derive an expression for the ionic polarizability assuming harmonic forces between the ions.

SUGGESTED READING

Coelho, R. (1979). *Physics of dielectrics, Parts I, II, III*. Amsterdam: North-Holland Publishing Co.

Debye, P. (1945). *Polar molecules*. New York: Dover Publishers.

Tareev, B. (1979). *Physics of dielectric materials*. Moscow: Mir Publishers.

Van Vleck, J. H. (1932). *The theory of electric and magnetic susceptibilities*. London: Oxford University Press.

Chapter 16

Ferroelectric Solids

Chapter Outline

A number of dielectric materials exist in which the electric polarization depends nonlinearly on the applied electric field and that exhibit a hysteresis effect. Such materials are called ferroelectric materials and the phenomenon is called ferroelectricity. Ferroelectric crystal can be divided into a large number of regions, called ferroelectric domains. In each domain all the electric dipoles are directed in one direction, yielding a finite net electric dipole within it. The spontaneous polarization $\mathbf{P}_s$ is the polarization in the absence of an electric field. The magnitude of $\mathbf{P}_s$ is determined by the vector sum of the electric dipole moments in an individual domain and, therefore, is a cooperative phenomenon. The different domains have polarizations in different directions. In the absence of an applied electric field all the domains are randomly oriented, yielding net zero polarization. When an electric field $\mathbf{E}_0$ is applied, the domains with polarization in the direction of the field grow, while those with polarization opposite to the direction of the field diminish in size and, ultimately, vanish. When all the dipoles are oriented in the direction of the applied field, then a single domain exists in the solid and saturation polarization is obtained. A plot of polarization $\mathbf{P}$ as a function of the applied field $\mathbf{E}_0$ is shown in Fig. 16.1. From the figure it is evident that the polarization $\mathbf{P}$ always lags behind the applied electric field $\mathbf{E}_0$ and this phenomenon is called hysteresis. The figure shows that the polarization does not retrace its path, resulting in a hysteresis loop whose area gives the loss of electrostatic energy. When the applied electric field is switched off, there exists a finite polarization in the solid, which is called the remnant polarization $\mathbf{P}_r$. To reduce the polarization to zero, one has to apply the electric field to the solid in the reverse direction; this is called the coercive field $\mathbf{E}_c$. The value of $\mathbf{P}_s$ is obtained by extrapolating the linear part BC of the hysteresis loop to zero electric field. The spontaneous polarization $\mathbf{P}_s$, remnant polarization $\mathbf{P}_r$, and coercive field $\mathbf{E}_c$ are shown in Fig. 16.1. The material shows ferroelectric properties only at low temperatures and makes a transition from the ferroelectric state to the paraelectric state at a particular temperature T_C called the transition temperature. We should remark here that the spontaneous polarization in a ferroelectric material arises due to electrostrictive strains in the crystal; therefore, the ferroelectric state has lower symmetry than the paraelectric state.

The dielectric constant ε is given by the slope of the curve OA (see Fig. 16.1) at very small fields so that no motion of the domain boundaries occurs. The dielectric function ε is a function of temperature and above T_C, it satisfies the relation

$$\varepsilon = \frac{C'}{T-\theta} + \varepsilon_0 \tag{16.1}$$

C' is a constant and θ is the characteristic temperature, which is usually less than T_C. Near T_C, the electric susceptibility χ_E is given by

$$\chi_E = \frac{C}{T-\theta} \tag{16.2}$$

Solid State Physics. https://doi.org/10.1016/B978-0-12-817103-5.00016-5

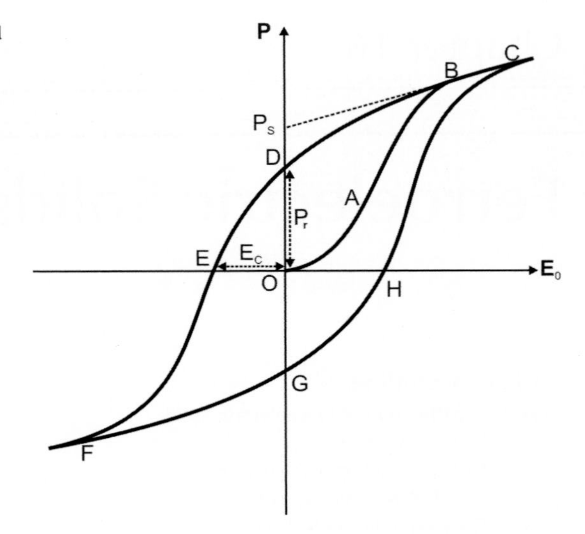

FIG. 16.1 Schematic representation of electric polarization **P** as a function of applied electric field **E** in ferroelectric solids (hysteresis loop).

where

$$C = \frac{C'}{4\pi} \tag{16.3}$$

Eq. (16.2) is usually called the Curie-Weiss law.

16.1 CLASSIFICATION OF FERROELECTRIC SOLIDS

The ferroelectric solids can be classified broadly into three categories.

16.1.1 Tartrate Group

Rochelle salt is a sodium potassium salt of tartaric acid with the formula $NaKC_4H_4O_6 \cdot 4H_2O$ and is one of the first materials in which ferroelectricity was observed. The other ferroelectric materials of this group can be prepared by partially replacing K by NH_4, Rb, or Tl. Lithium ammonium tartrate and lithium tantalum tartrate are also members of this group. Rochelle salt exhibits ferroelectric behavior in the temperature range from −18°C (255 K) to 23°C (296 K), which means it has two transition temperatures. Rochelle salt exhibits orthorhombic structure ($a_1 \neq a_2 \neq a_3$, $\alpha = \beta = \gamma = 90°$), but in the ferroelectric state it has monoclinic structure ($a_1 \neq a_2 \neq a_3$, $\alpha = \gamma = 90° \neq \beta$). The spontaneous polarization occurs along the original a-direction, i.e., positive and negative directions along the a-axis. There are three components of dielectric constant ε_1, ε_2, and ε_3 along the three lattice vectors $\mathbf{a}_1$, $\mathbf{a}_2$, and $\mathbf{a}_3$. The dielectric constant ε_1 exhibits two peaks at the transition temperatures, while ε_2 and ε_3 are found to vary smoothly with temperature (Fig. 16.2). The susceptibility function in Rochelle salt obeys the Curie law in the nonferroelectric state as

$$\chi_E^1 = \frac{C_1}{T - T_{C_1}} \quad \text{for}\, T > T_{C_1} = 296\,\text{K} \tag{16.4}$$

and

$$\chi_E^2 = \frac{C_2}{T_{C_2} - T} \quad \text{for}\, T < T_{C_2} = 255\,\text{K} \tag{16.5}$$

where $C_1 = 178$ K and $C_2 = 93.8$ K. The spontaneous polarization $\mathbf{P}_s(T)$ as a function of temperature T is shown in Fig. 16.3 for Rochelle salt and deuterated Rochelle salt. It is noteworthy that the replacement of an H atom by D (deuterium) has a marked influence on both $\mathbf{P}_s(T)$ and T_C, which indicates the important role of H-bonds. But X-ray and neutron studies show that H and D are not at all involved in the mechanism of ferroelectricity.

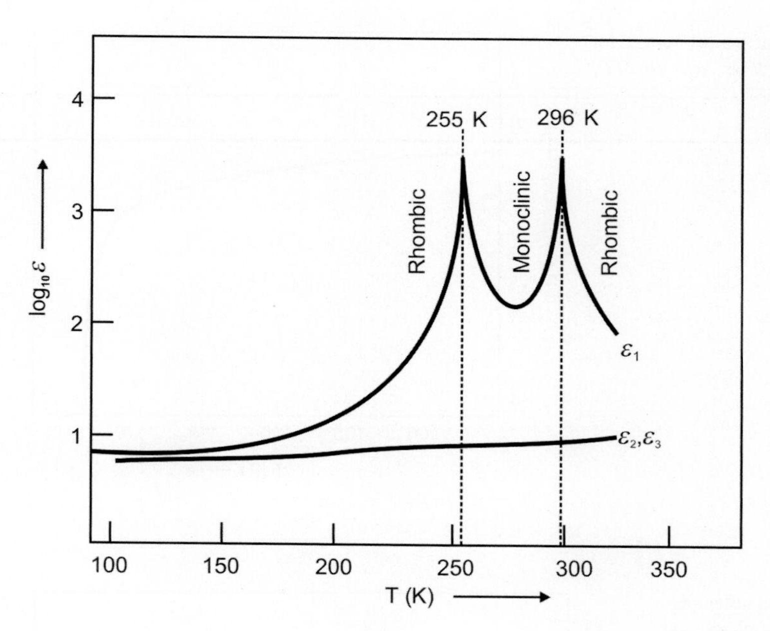

FIG. 16.2 The logarithm of the dielectric constant ε_α of Rochelle salt along the three lattice vectors $\mathbf{a}_\alpha$ ($\alpha = 1$, 2, and 3) as a function of temperature. *(Modified from Halblutzel, J. (1939). Helv. Phys. Acta. 12, 489.)*

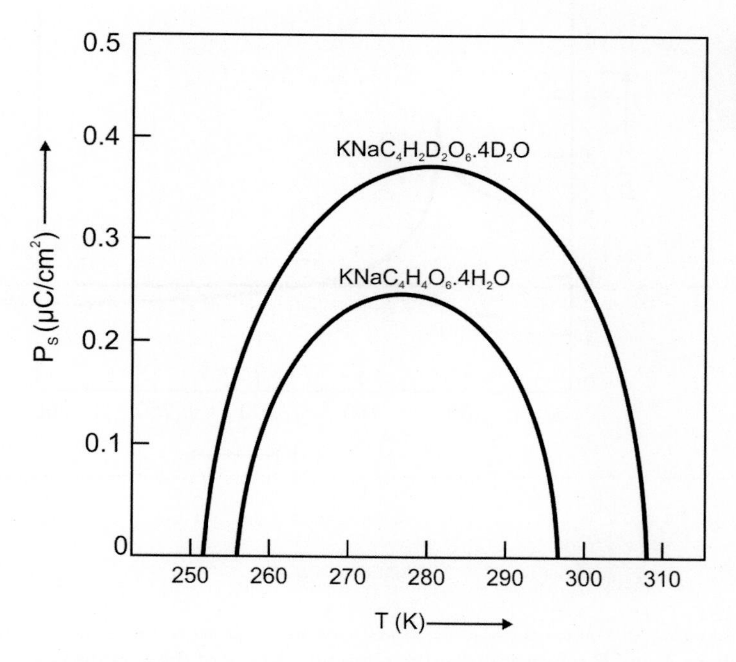

FIG. 16.3 The spontaneous polarization $\mathbf{P}_s$(T) as a function of temperature T. The lower curve corresponds to Rochelle salt, while the upper curve corresponds to deuterated Rochelle salt. *(Modified from Halblutzel, J. (1939). Helv. Phys. Acta. 12, 489.)*

16.1.2 Dihydrophosphates and Arsenates

In the second type of ferroelectrics, the most important is potassium dihydrophosphate with the formula KH_2PO_4, which exhibits ferroelectric behavior below $T_C = 123\,K$. It possesses orthorhombic structure ($a_1 \neq a_2 \neq a_3$, $\alpha = \beta = \gamma = 90^o$) below T_C but tetragonal structure ($a_1 = a_2 \neq a_3$, $\alpha = \beta = \gamma = 90^o$) above T_C. P_s(T) in KH_2PO_4 occurs in the $\mathbf{a_3}$-direction, the only polar direction, and its temperature variation is shown in Fig. 16.4. The temperature variation of ε(T) is shown in Fig. 16.5 and exhibits the same behavior as given by Eq. (16.1) (see Fig. 16.2). In KH_2PO_4, the phosphate group contains four oxygen atoms at the corners of a tetrahedron with a phosphorus atom at its center. The phosphate groups are bound together by the H-bond. It has been observed that if the H-atom is replaced by a D-atom, then T_C changes from 123 K to 213 K, i.e., by 90 K. This again indicates that H-bonds might be playing an important role in ferroelectricity in this class of ferroelectric solids.

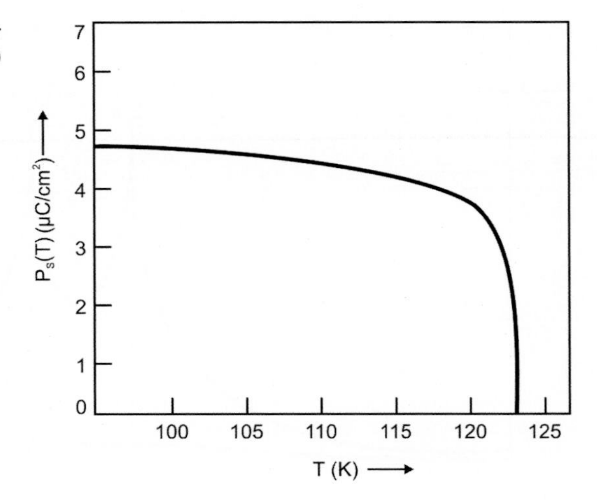

FIG. 16.4 The spontaneous polarization $\mathbf{P}_s(T)$ as a function of temperature T for KH_2PO_4. *(Modified from Von Arx, A. and Bantle, W. (1943). Helv. Phys. Acta. 16, 211.)*

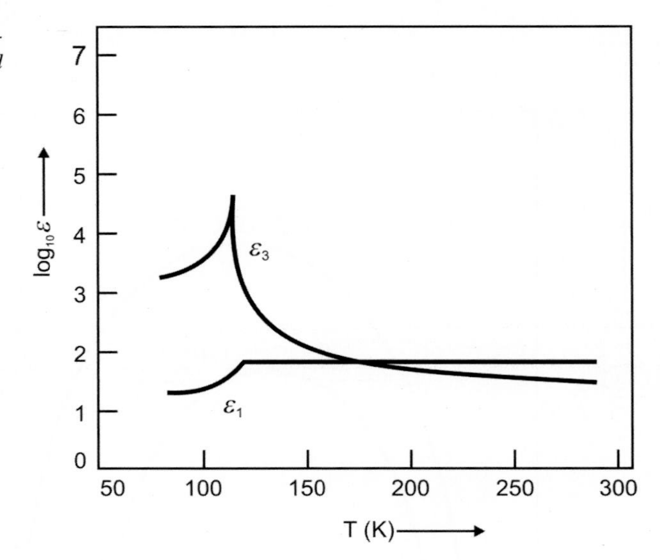

FIG. 16.5 The logarithm of the dielectric constant as a function of temperature T for KH_2PO_4 along the $\mathbf{a}_1$ and $\mathbf{a}_3$ axes. *(Modified from Busch, G. and Scherrer, P. (1935). Naturwiss., 23, 737.)*

16.1.3 Perovskite Structure

The general formula for perovskite structure is ABO_3, where A and B represent metallic elements. Many of these materials are ferroelectric in nature. The salient features of perovskite structure were presented in Chapter 1. The most important material belonging to this class is barium titanate with the formula $BaTiO_3$. The structure of $BaTiO_3$, in which the oxygen atoms form an octahedron, is shown in Fig. 1.33. In the unpolarized state it has cubic symmetry: Ba^{+2} ions occupy the corners of the cubic unit cell, O^{-2} ions are at the centers of the faces, and a Ti^{+4} ion lies at the center of the unit cell. In this structure O^{-2} ions are highly polarizable, producing finite spontaneous polarization. When an electric field is applied, the unit cell of $BaTiO_3$ expands along the $\mathbf{a}_3$-direction and contracts perpendicular to it, thereby producing finite polarization. The spontaneous polarization is measured along the [001] direction. $BaTiO_3$ has a T_C of 120°C (393 K) below which it is ferroelectric with tetragonal structure, but above T_C it is nonferroelectric in nature. As the temperature is decreased, two structural changes take place: at 278 K it acquires orthorhombic structure, while at 193 K its structure changes to a rhombohedral one. Figs. 16.6 and 16.7 show the variation of ε(T) and $P_s(T)$ with T as measured along the [001] direction and the structural transitions are evident from the graphs.

There are other groups of ferroelectric materials but they are not of much interest here.

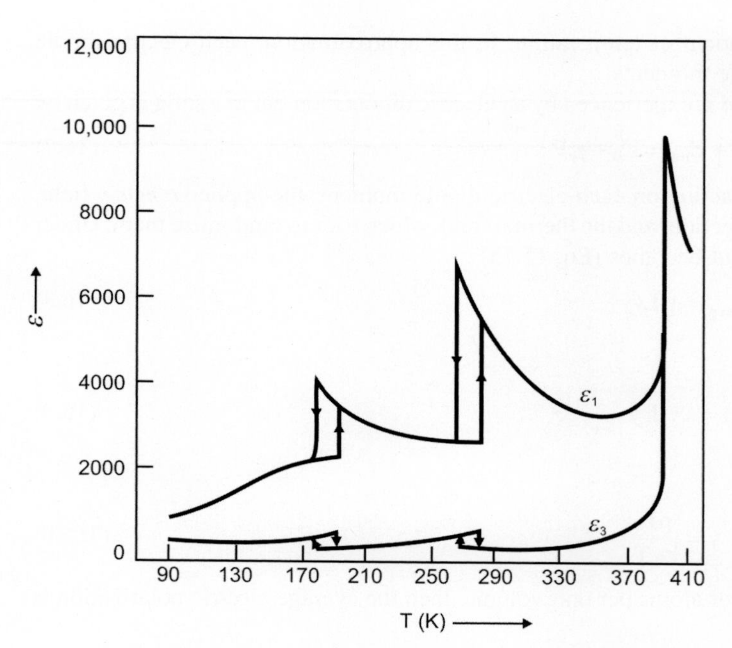

FIG. 16.6 The dielectric constant ε_α of $BaTiO_3$ along the $\mathbf{a}_1$ and $\mathbf{a}_3$ axes as a function of temperature T. *(Modified from Merz, W. J. (1949) Physical Review, 76, 1221.)*

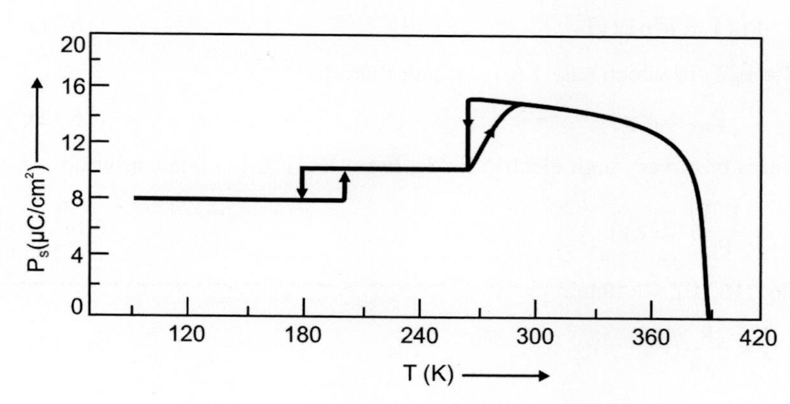

FIG. 16.7 The spontaneous polarization $\mathbf{P}_s(T)$ measured along the [001] direction as a function of temperature T for $BaTiO_3$. *(Modified from Merz, W. J. (1949) Physical Review, 76, 1221.)*

16.2 THEORIES OF FERROELECTRICITY

Two types of theories for ferroelectricity have been put forward:

1. Atomic models, which depend on the crystal structure of the solid.
2. A thermodynamic description, which is independent of the structure of the solid.

16.2.1 Atomic Models

16.2.1.1 Electric Dipole Theory

In the electric dipole theory, the following two assumptions are made:

1. In a ferroelectric substance there exists finite spontaneous polarization $\mathbf{P}(T)$, which is a function of temperature T. Note here that we have used $\mathbf{P}(T)$ as the symbol for spontaneous polarization instead of $\mathbf{P}_s(T)$.
2. Within a ferroelectric solid there exists some internal interaction, which tends to align all the electric dipole moments parallel to each other. This interaction gives rise to a finite temperature-dependent internal electric field $\mathbf{E}_{int}(T)$ and it is assumed to be linearly proportional to the spontaneous polarization, i.e.,

$$\mathbf{E}_{int}(T) = \gamma_i \mathbf{P}(T) \tag{16.6}$$

Here γ_i is the internal field constant, which is independent of temperature. In this approximation, each electric dipole moment sees the average polarization of all other dipole moments.

If $\mathbf{E_0}$ is the applied electric field, the total electric field $\mathbf{E}$ experienced by an electric dipole moment in a solid is given by

$$\mathbf{E} = \mathbf{E_0} + \mathbf{E_{int}} = \mathbf{E_0} + \gamma_i \mathbf{P} \tag{16.7}$$

In a ferroelectric solid there are two competing forces acting on each electric dipole moment: the applied electric field, which tries to align the dipole moments along its own direction, and the thermal field, which tries to randomize them. Under the action of these two fields the average dipole moment becomes (Eq. 15.75)

$$p_{avg} = p\,L(y) \tag{16.8}$$

where

$$L(y) = \coth y - \frac{1}{y} \tag{16.9}$$

and

$$y = \frac{pE}{k_B T} \tag{16.10}$$

Here L(y) is the Langevin function. If ρ^a is the number of atoms per unit volume, then the average electric polarization is given by

$$P(T) = \rho^a p\,L(y) \tag{16.11}$$

The saturation polarization P_{sat} is obtained for $pE \gg k_B T$, in which case $L(y) = 1$ and, therefore,

$$P_{sat} = \rho^a p \tag{16.12}$$

Hence P_{sat} is obtained either at very low temperatures or at very high electric fields. From Eq. (16.11) one can write

$$\frac{P(T)}{P_{sat}} = L(y) \tag{16.13}$$

Substituting the value of E from Eq. (16.7) into Eq. (16.10), we obtain

$$y = \frac{p}{k_B T}(E_0 + \gamma_i P) \tag{16.14}$$

From Eqs. (16.13) and (16.14) one can write

$$\frac{P(T)}{P_{sat}} = L\left(\frac{p}{k_B T}(E_0 + \gamma_i P)\right) \tag{16.15}$$

The spontaneous polarization is obtained by substituting $E_0 = 0$ in the above expression to get

$$\frac{P(T)}{P_{sat}} = L(x) \tag{16.16}$$

where

$$x = \frac{\gamma_i p P}{k_B T} \tag{16.17}$$

A graph of the spontaneous polarization P/P_{sat} as a function of x is shown in Fig. 16.8. The expression for P(T) can also be obtained from Eq. (16.14) and is given by

$$P(T) = \frac{k_B T}{\gamma_i p} y - \frac{E_0}{\gamma_i} \tag{16.18}$$

For spontaneous polarization, we put $E_0 = 0$ in Eq. (16.18) (y goes to x in this case) to get

$$P(T) = \frac{k_B T}{\gamma_i p} x \tag{16.19}$$

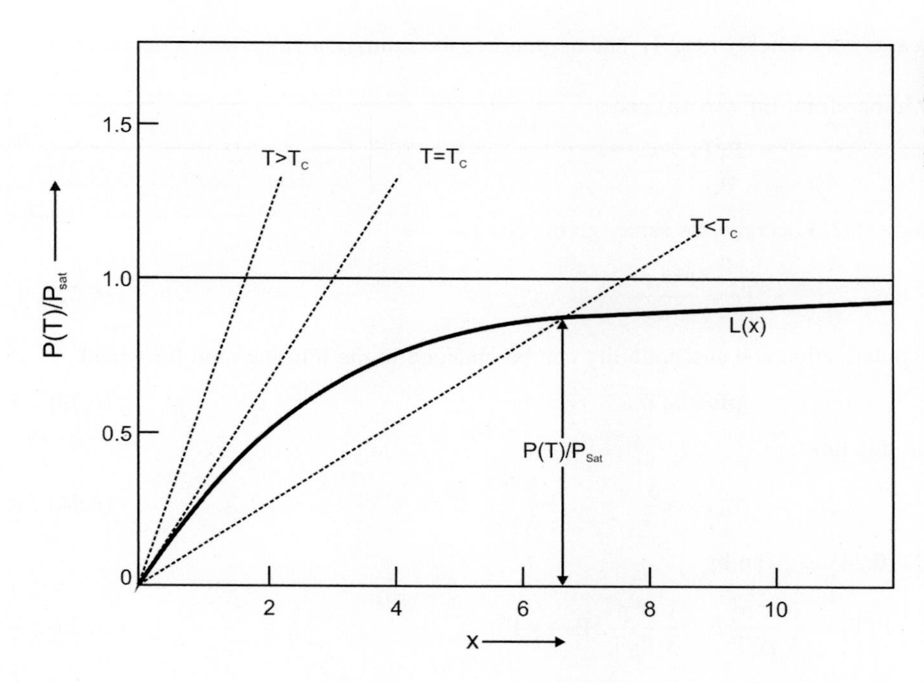

FIG. 16.8 The spontaneous polarization P/P_{sat} is plotted as a function of x where $x = \gamma_i p\, P/k_B T$. The *solid line* represents the Langevin function L (x) as a function of x. The *dashed lines* show plots of Eq. (16.20) for different temperatures.

From the above expression

$$\frac{P(T)}{P_{sat}} = \frac{k_B T}{\rho^a \gamma_i p^2} x \tag{16.20}$$

which represents a straight line passing through the origin. The spontaneous polarization is given by the point of intersection of the curves represented by Eqs. (16.16) and (16.20), as shown in Fig. 16.8. It is evident that the slope of the straight line, represented by Eq. (16.20), increases with an increase in temperature and, at a particular temperature T_C, the straight line becomes a tangent at the origin to the curve represented by Eq. (16.16). From the figure it is evident that for $T < T_C$ the point of intersection yields a finite spontaneous polarization P(T), while for $T > T_C$, P(T) is zero. Therefore, the solid exhibits ferroelectric behavior below T_C, but is paraelectric in nature above T_C. Hence T_C represents the transition temperature between the two states of the solid. For $T \langle T_C$, the spontaneous polarization can be evaluated at different temperatures by the method of intersection as described above. Fig. 16.9 shows the temperature dependence of the spontaneous

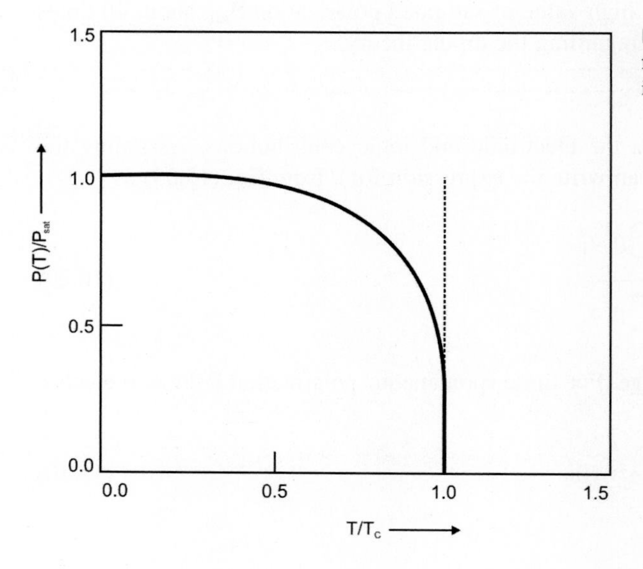

FIG. 16.9 Schematic representation of spontaneous polarization P/P_{sat} as a function of temperature T/T_C. The value of P/P_{sat} corresponds to the point of intersection as explained in Fig. 16.8.

polarization. It is evident that P(T) increases very quickly near T_C and approaches the saturation value with a decrease in temperature.

At very small values of x, $L(x)=x/3$, therefore, Eq. (16.16) gives

$$\frac{\mathbf{P}(T)}{\mathbf{P}_{sat}}\approx\frac{x}{3} \tag{16.21}$$

At $T=T_C$, the slopes of Eqs. (16.20) and (16.21) become the same, giving rise to

$$T_C=\frac{\rho^a\gamma_i p^2}{3k_B} \tag{16.22}$$

A simplified expression for the electric polarization and susceptibility can be obtained in the limiting case for which

$$pE \ll k_B T \tag{16.23}$$

Eq. (16.23) is equivalent to $y \ll 1$ and in this limit

$$L(y)\approx\frac{y}{3} \tag{16.24}$$

The polarization from Eqs. (16.11) and (16.24) is given by

$$P(T)=\frac{1}{3}\frac{\rho^a p^2}{k_B T}E=\frac{1}{3}\frac{\rho^a p^2}{k_B T}(E_0+\gamma_i P) \tag{16.25}$$

The above equation yields the electric susceptibility as

$$\chi_E=\frac{P}{E_0}=\frac{T_C/\gamma_i}{T-T_C}=\frac{C_E}{T-T_C} \tag{16.26}$$

where C_E is the Curie constant and is written as

$$C_E=\frac{T_C}{\gamma_i} \tag{16.27}$$

Using the value of T_C from Eq. (16.22) in the above equation, one gets

$$C_E=\frac{\rho^a p^2}{3k_B} \tag{16.28}$$

Therefore, the electric susceptibility in ferroelectric solids obeys the Curie-Weiss law, as is observed experimentally. There are some objections to the dipole theory. First, the dipole theory is not able to explain the existence of two transition temperatures. Second, if the dipole theory is correct, then most of the polar materials should exhibit ferroelectric behavior, which is contrary to fact. Lastly, the dipole theory yields a very high value of saturated polarization P_{sat}, about 40 times the observed value in Rochelle salt, which creates some doubts regarding the dipole theory.

16.2.1.2 Polarization Catastrophe

In nonpolar solids the dielectric constant ε arises mainly from the electronic and ionic contributions. Assuming the Clausius-Mossotti relation to be valid in a nonpolar solid, one can write the expression for ε from Eq. (15.65) as

$$\varepsilon=\frac{1+\frac{8\pi}{3}\sum_i\rho_i^a\alpha_i^a}{1-\frac{4\pi}{3}\sum_i\rho_i^a\alpha_i^a} \tag{16.29}$$

In ferroelectric solids the value of the dielectric constant ε is large. For finite spontaneous polarization P, in zero electric field, ε goes to infinity if

$$1-\frac{4\pi}{3}\sum_i\rho_i^a\alpha_i^a=0 \tag{16.30}$$

The above equation can be written as

$$\sum_{i} \rho_i^a \alpha_i^a = \frac{3}{4\pi} \tag{16.31}$$

We know that

$$\chi_E = \frac{P}{E} = \frac{\varepsilon - 1}{4\pi} \tag{16.32}$$

For $\varepsilon \gg 1$, the above equation gives

$$P = \frac{\varepsilon}{4\pi} E \tag{16.33}$$

The above equation says that if ε is very large, the polarization P is also very large. The polarization will become infinite when condition (16.31) is satisfied. So, Eq. (16.31) yields the condition for a *polarization catastrophe*. Fig. 16.10 shows the variation of ε with polarizability. For large values of ε the left hand side of Eq. (15.65) approaches unity, therefore, a small change in polarizability $(4\pi/3) \sum_i \rho_i^a \alpha_i^a$ leads to a large change in ε (Eq. 16.29). Consider a very small departure of magnitude S from unity in the value of $(4\pi/3) \sum_i \rho_i^a \alpha_i^a$, which can be written as

$$\frac{4\pi}{3} \sum_{i} \rho_i^a \alpha_i^a = 1 - S \tag{16.34}$$

where $S \ll 1$. Substituting Eq. (16.34) into Eq. (16.29), one gets

$$\varepsilon = \frac{3}{S} - 2$$

which can be written approximately as

$$\varepsilon \approx \frac{3}{S} \tag{16.35}$$

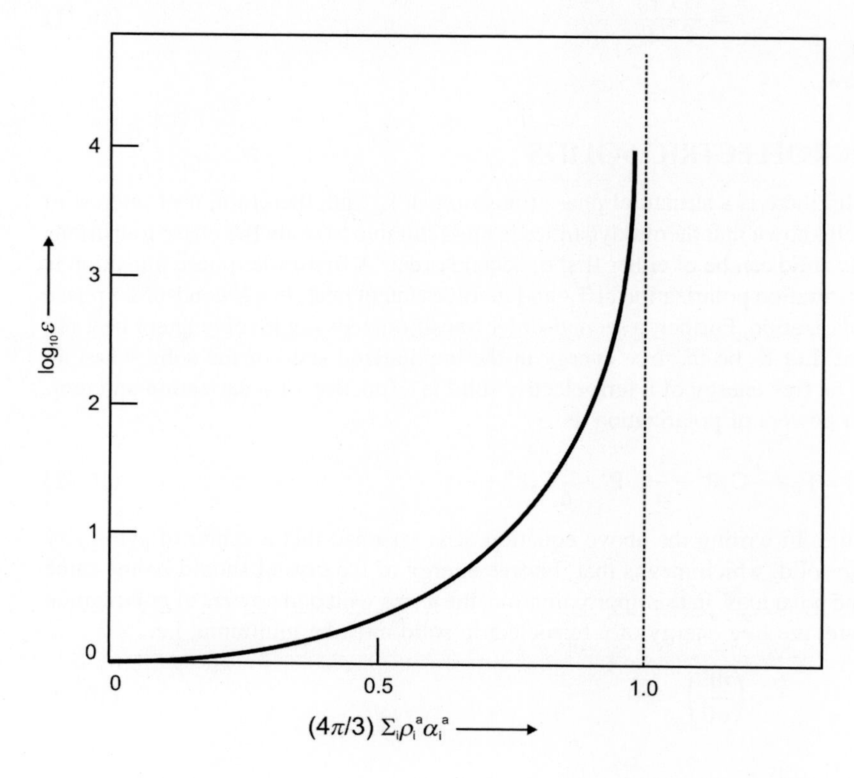

FIG. 16.10 The logarithm of the dielectric function ε as a function of polarizability of the solid.

As S is very small, ε is very large. Note that if $\varepsilon \approx 10$ or smaller, then any change in ρ_i^a resulting from the thermal expansion will not greatly affect the value of ε.

It can be shown that in solids with a large value of ε, one can obtain the Curie-Weiss behavior. Differentiating Eq. (15.65) with respect to T, we get

$$\frac{3}{(\varepsilon+2)^2}\frac{d\varepsilon}{dT}=\frac{4\pi}{3}\sum_i \alpha_i^a \frac{d\rho_i^a}{dT} \tag{16.36}$$

Dividing Eq. (16.36) by Eq. (15.65) and simplifying, we obtain

$$\frac{3}{(\varepsilon-1)(\varepsilon+2)}\frac{d\varepsilon}{dT}=\frac{d}{dT}\left[\ln\left(\sum_i \rho_i^a \alpha_i^a\right)\right] \tag{16.37}$$

It is not easy to solve the above expression. The solution to Eq. (16.37) becomes simple if all of the atoms in the unit cell are assumed to be identical, with atomic density ρ^a and atomic polarizability α^a. Under this approximation Eq. (16.37) reduces to

$$\frac{3}{(\varepsilon-1)(\varepsilon+2)}\frac{d\varepsilon}{dT}=\frac{1}{\rho^a}\frac{d\rho^a}{dT}=-3\Gamma_{TH} \tag{16.38}$$

Γ_{TH} is the coefficient of linear expansion of the solid. For $\varepsilon \gg 1$, the above equation reduces to

$$\frac{1}{\varepsilon^2}\frac{d\varepsilon}{dT}=-\Gamma_{TH} \tag{16.39}$$

Integrating the above equation, we find

$$\frac{1}{\Gamma_{TH}}\int \frac{d\varepsilon}{\varepsilon^2}=-\int dT+\theta \tag{16.40}$$

Here θ is the constant of integration. The solution of the above equation gives

$$\varepsilon=\frac{1/\Gamma_{TH}}{T-\theta} \tag{16.41}$$

which is the usual form of the Curie-Weiss law.

16.3 THERMODYNAMICS OF FERROELECTRIC SOLIDS

It has been observed that in a ferroelectric solid there is a structural phase transition at T_C and, therefore, it of interest to study the behavior of the solid near T_C. It is well known that thermodynamics is most suitable to study the phase transitions in solids. The phase transition in a ferroelectric solid can be of either first or second order. A first-order phase transition is characterized by a discontinuous jump in the saturation polarization at T_C and involves latent heat. In a second-order phase transition there is a continuous variation in polarization. Further, a second-order transition does not involve latent heat but exhibits a discontinuous jump in specific heat. Let F_0 be the free energy in the unpolarized state of the solid when no external pressure or electric field is applied. The free energy of a ferroelectric solid is a function of polarization and temperature, i.e., F(P, T), and can be expanded in powers of polarization as

$$F(P,T)=F_0+\frac{1}{2}C_1P^2+\frac{1}{4}C_2P^4+\frac{1}{6}C_3P^6+\cdots \tag{16.42}$$

Here the coefficients C_n depend on temperature. In writing the above equation, it is assumed that a center of symmetry exists in the crystal structure of a ferroelectric solid, which means that the free energy of the crystal should be the same for positive and negative polarizations along the polar axis. In this approximation, the terms with odd powers of polarization reduce to zero. In the thermal equilibrium state, the free energy of a ferroelectric solid must be minimum, i.e.,

$$\left(\frac{dF}{dP}\right)_T=0 \tag{16.43}$$

Substituting Eq. (16.42) into Eq. (16.43), one gets

$$P_s\left(C_1+C_2P_s^2+C_3P_s^4\right)=0 \tag{16.44}$$

Here P_s is the spontaneous polarization. For $P_s \neq 0$ the above equation gives

$$C_1+C_2P_s^2+C_3P_s^4=0 \tag{16.45}$$

One can calculate the susceptibility function from thermodynamic considerations. Let a weak electric field E_0 be applied to the system. Then the electric susceptibility χ_E is given by

$$\frac{1}{\chi_E}=\frac{\partial E_0}{\partial P} \tag{16.46}$$

From thermodynamics one can write a small change in free energy as

$$dF=-\,SdT-\,pdV+E_0dP \tag{16.47}$$

Here S, p, and V are the entropy, pressure, and volume of the system at temperature T. Under zero pressure the above equation becomes

$$dF=-\,SdT\,+E_0dP \tag{16.48}$$

At constant temperature the applied electric field becomes

$$E_0=\left(\frac{\partial F}{\partial P}\right)_T \tag{16.49}$$

From Eqs. (16.46) and (16.49) one can write

$$\frac{1}{\chi_E}=\left(\frac{\partial^2 F}{\partial P^2}\right)_T \tag{16.50}$$

16.3.1 Second-Order Transition in Ferroelectric Solids

From Eq. (16.42), $F-F_0$ can be plotted as a function of P for different values (negative, positive, or zero) of the constant C_1, the other constants being positive, and the plot is shown in Fig. 16.11A. The spontaneous polarization P_s as a function of C_1 is shown in Fig. 16.11B. It shows that P_s is finite for $C_1<0$ and is zero for $C_1>0$ and, therefore, $C_1=0$ is the value at the transition temperature T_C. The polarization P varies continuously with temperature, but the slope of P and hence P^2 has a

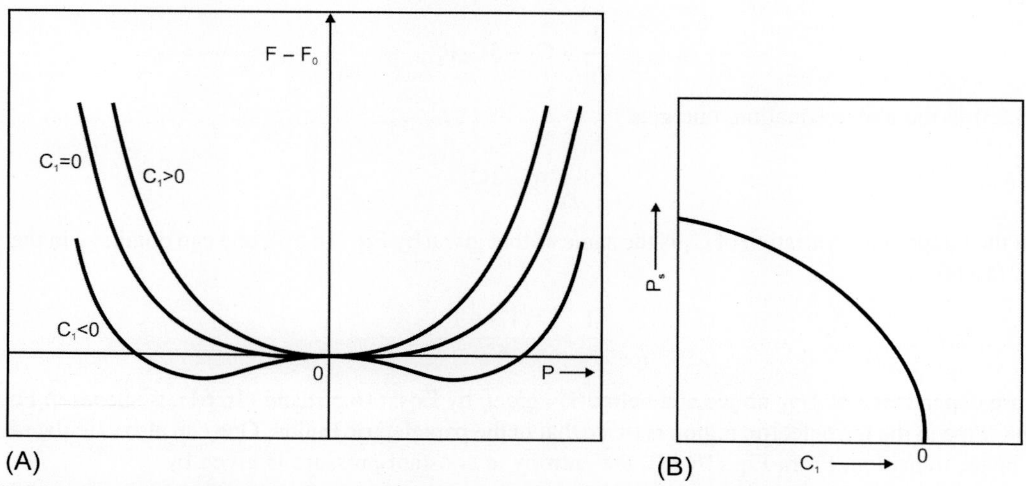

FIG. 16.11 (A) Schematic representation of the free energy $F-F_0$ as a function of the polarization P for various values of C_1, negative, positive, and zero. (B) The spontaneous polarization P_s as a function of the parameter C_1 for a second-order transition.

discontinuity at T_C. Therefore, there should be a discontinuity in the specific heat at T_C resulting in a second-order transition. If, for simplicity, the term with C_3 is assumed to be negligible, then from Eq. (16.42) one can write

$$F = F_0 + \frac{1}{2}C_1P^2 + \frac{1}{4}C_2P^4 \tag{16.51}$$

Therefore, spontaneous polarization satisfies the equation

$$C_1 + C_2P_s^2 = 0 \tag{16.52}$$

which gives

$$P_s^2 = -\frac{C_1}{C_2} \tag{16.53}$$

According to the above equation P_s^2 is positive only if C_1 is negative. Note that P_s is a continuous function of T; therefore, it corresponds to a second-order transition and exhibits a discontinuous jump in specific heat. One can calculate χ_E above and below T_C. Above T_C the polarization is small, so retaining the first two terms in Eq. (16.51) of free energy allows us to write

$$F - F_0 = \frac{1}{2}C_1P^2 \tag{16.54}$$

From Eqs. (16.50) and (16.54) one obtains

$$\frac{1}{\chi_E} = C_1 = \frac{T - T_C}{C_E} \tag{16.55}$$

Here we have used the fact that χ_E above T_C exhibits the Curie behavior given by Eq. (16.26). From the above equation the constant C_1 is given in terms of the Curie constant C_E as

$$C_1 = \frac{T - T_C}{C_E} \tag{16.56}$$

In the ferroelectric phase below T_C the polarization is appreciable, so retaining the first two terms of Eq. (16.42) allows us to write the free energy as

$$F = F_0 + \frac{1}{2}C_1P^2 + \frac{1}{4}C_2P^4 \tag{16.57}$$

From Eqs. (16.50) and (16.57) the electric susceptibility becomes

$$\frac{1}{\chi_E} = C_1 + 3C_2P^2 \tag{16.58}$$

For a weak applied electric field, $P = P_s$, therefore, from the above equation

$$\frac{1}{\chi_E} = C_1 + 3C_2P_s^2 \tag{16.59}$$

Using Eq. (16.53) in the above equation, one gets

$$\frac{1}{\chi_E} = -2C_1 \tag{16.60}$$

Assuming that the temperature variation of C_1 is the same as that given by Eq. (16.56), one can obtain χ_E in the ferroelectric state from Eq. (16.60) as

$$\frac{1}{\chi_E} = \frac{2(T_C - T)}{C_E} \tag{16.61}$$

The temperature dependence of $1/\chi_E$ above and below T_C, given by Eqs. (16.55) and (16.61), is shown in Fig. 16.12. It is evident that the slope in the ferroelectric region is twice that in the paraelectric region. One can also calculate entropy S in a second-order phase transition. From Eq. (16.48), the entropy at constant pressure is given by

$$S = -\left(\frac{\partial F}{\partial T}\right)_P \tag{16.62}$$

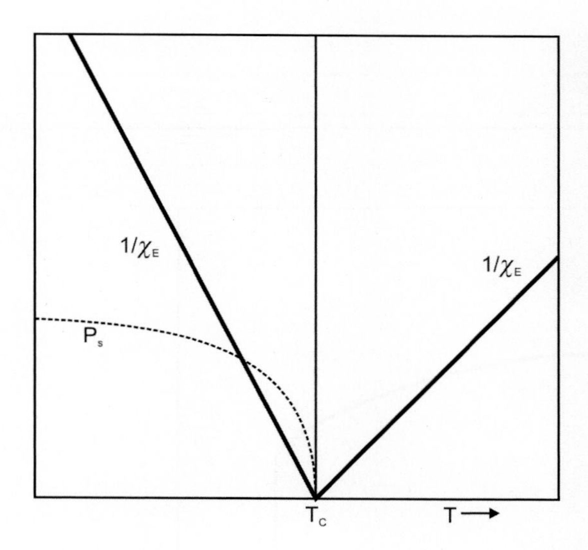

FIG. 16.12 Plot of the reciprocal of susceptibility $1/\chi_E$ as a function of temperature near the critical temperature T_C for a second-order transition. The spontaneous polarization P_s is shown by the *dashed line*.

Substituting Eq. (16.42) into Eq. (16.62), one obtains, to a first-order approximation, the change in entropy as

$$S - S_0 = -\frac{1}{2}\frac{\partial C_1}{\partial T}P^2 \tag{16.63}$$

Here S_0 is the entropy in the unpolarized state. The polarization P varies continuously with temperature, but the slope of P and hence P^2 has a discontinuity at T_C. Therefore, there should be a discontinuity in the specific heat at T_C and the transition should be of second order.

16.3.2 First-Order Transition in Ferroelectric Solids

Consider the case when C_2 is negative but C_3 is positive. In this approximation, the free energy is plotted as a function of P for different values of C_1 ranging from negative to positive values in Fig. 16.13A. In this case, the free energy becomes minimum for finite values of the spontaneous polarization P_s. This corresponds to a transition from the unpolarized to the polarized state. Fig. 16.13B shows a plot of P_s as a function of temperature T and there is a discontinuity (jump) in the value of P_s at T_C. So, according to Eq. (16.63), there is a discontinuity in the entropy S and, therefore, latent heat is involved in the transition; thus, this is a first-order transition. In a first-order transition at T_C,

$$F(T_C) = F_0(T_C) \tag{16.64}$$

Substituting Eq. (16.42) into Eq. (16.64), one can write

$$C_1 + \frac{1}{2}C_2P_s^2(T_C) + \frac{1}{3}C_3P_s^4(T_C) = 0 \tag{16.65}$$

At $T = T_C$, Eq. (16.45) gives

$$C_1 + C_2P_s^2(T_C) + C_3P_s^4(T_C) = 0 \tag{16.66}$$

Subtracting Eq. (16.65) from Eq. (16.66), we get

$$P_s^2(T_C)\left[\frac{1}{2}C_2 + \frac{2}{3}C_3P_s^2(T_C)\right] = 0 \tag{16.67}$$

For nonzero spontaneous polarization, Eq. (16.67) gives

$$\frac{1}{2}C_2 + \frac{2}{3}C_3P_S^2(T_C) = 0 \tag{16.68}$$

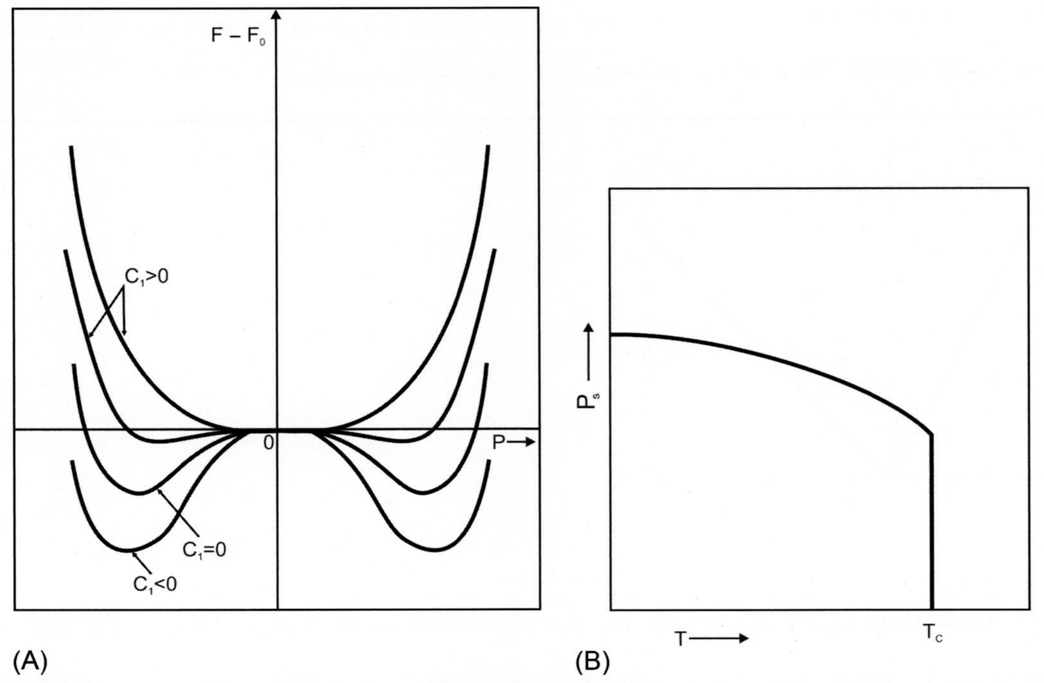

FIG. 16.13 (A) Schematic representation of the free energy $F-F_0$ as a function of polarization P for positive and negative values of C_1 for a first-order transition. (B) The spontaneous polarization P_s as a function of temperature T for a first-order transition.

From this equation $P_s^2(T_C)$ is given by

$$P_s^2(T_C) = -\frac{3}{4}\frac{C_2}{C_3} \tag{16.69}$$

Multiplying Eq. (16.65) by 2 and subtracting from Eq. (16.66), we get

$$P_s^4(T_C) = \frac{3C_1}{C_3} \tag{16.70}$$

Substituting Eqs. (16.69) and (16.70) into Eq. (16.66), one can write

$$C_1 = \frac{3}{16}\frac{C_2^2}{C_3} \tag{16.71}$$

For $T > T_C$, the polarization is very small and the free energy is given by Eq. (16.54). From Eqs. (16.50) and (16.54) one obtains

$$\frac{1}{\chi_E} = C_1 = \frac{T - T_C}{C_E} \tag{16.72}$$

which is the same equation as Eq. (16.55) in the second-order phase transition. For $T < T_C$, the polarization is appreciable and so all the terms should be retained. From Eqs. (16.42) and (16.50) one can write

$$\frac{1}{\chi_E} = C_1 + 3C_2P^2 + 5C_3P^4 \tag{16.73}$$

For a weak electric field $P = P_s$, therefore, the susceptibility becomes

$$\frac{1}{\chi_E} = C_1 + 3C_2P_s^2 + 5C_3P_s^4 \tag{16.74}$$

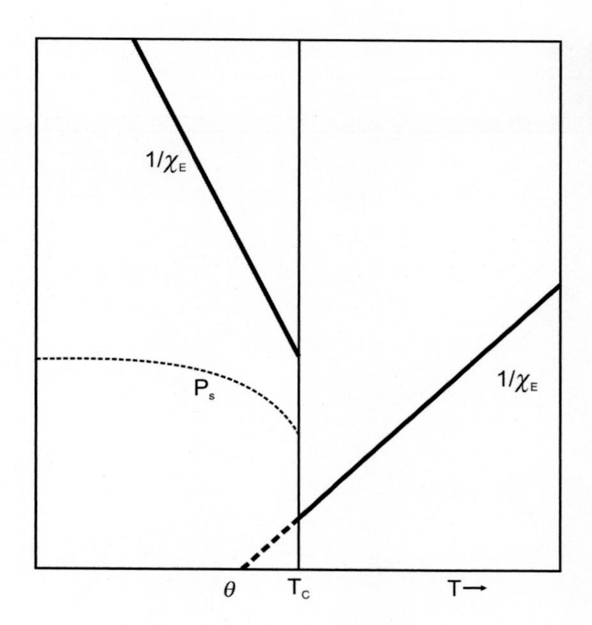

FIG. 16.14 Plot of the reciprocal of susceptibility $1/\chi_E$ as a function of temperature T around the critical temperature T_C for a first-order transition. The spontaneous polarization P_s is shown by the *dashed line*.

Substituting Eqs. (16.69), (16.70), and (16.71) into Eq. (16.74), one obtains

$$\frac{1}{\chi_E} = 4C_1 = 4\frac{T_C - T}{C_E} \tag{16.75}$$

The variation of $1/\chi_E$ as a function of T is shown in Fig. 16.14. From Eqs. (16.72) and (16.75) it is evident that the slope of $1/\chi_E$ in a ferroelectric state is four times the slope in a paraelectric state in a first-order phase transition. Further, the slope of $1/\chi_E$ in a ferroelectric crystal with a first-order phase transition is twice the slope in a ferroelectric crystal with a second-order transition.

A thermodynamic study has some advantages over other approaches used in ferroelectric materials. First, it is independent of the structure of the ferroelectric solid. Second, it is independent of any atomic model. Therefore, a thermodynamic study gives general results in ferroelectric solids. The deficiency of the thermodynamic theory is that it does not provide any physical insight into the mechanism responsible for the existence of a ferroelectric state in a solid.

16.4 FERROELECTRIC DOMAINS

As the ferroelectric material is cooled to below T_C, some regions appear that are polarized along a particular direction, while others are polarized exactly in the opposite direction. In other words, the ferroelectric crystal gets divided into regions that are polarized in only two directions, which are along the same line but opposite in direction. Such regions are called domains and the boundaries between these regions are called domain walls. Further, the direction of polarization of domains is different in different materials. In contrast to ferroelectric materials, the domains in a ferromagnetic material have magnetization in all possible directions (see Chapter 19) and, with a decrease in temperature, there is growth in domains that possess magnetization along the direction of the magnetic field.

Rochelle salt has an orthorhombic structure and some domains are polarized along the positive $\mathbf{a}_1$ direction, while others are polarized along the negative $\mathbf{a}_1$ direction with net polarization along the $\mathbf{a}_1$ axis. If an electric field $\mathbf{E}_0$ is applied along the positive $\mathbf{a}_1$ direction, then the number and size of the domains with polarization along the positive $\mathbf{a}_1$ direction increase until saturation polarization is achieved. If the electric field $\mathbf{E}_0$ is switched off and then applied in the reverse direction, then hysteresis is produced, which is evident from the variation of P with E_0 in Fig. 16.1. The hysteresis gives rise to a dielectric loss, which is proportional to the area of the hysteresis loop. The ferroelectric material KH_2PO_4 possesses tetragonal structure and the domains are polarized either along the positive or negative $\mathbf{a}_3$ direction, yielding a net polarization along the $\mathbf{a}_3$ direction. The material $BaTiO_3$ possesses cubic structure and spontaneous polarization may occur along any of the three Cartesian directions (three edges of the unit cell), leading to six possible directions for spontaneous polarization.

SUGGESTED READING

Cady, W. P. (1946). *Piezoelectricity*. New York: McGraw-Hill Book Co.
Kanzig, W. (1957). Ferroelectrics and antiferroelectrics. F. Seitz, & D. Turnbull (Eds.), *Solid state physics* Vol. 4, pp. 1–196. New York: Academic Press.

Chapter 17

Optical Properties of Solids

Chapter Outline

Luster and color, which have been known to mankind from ancient times, are the most important properties of solids, particularly metals. Because of these properties, metals have been used in jewelry and mirrors for ages. The entire electromagnetic spectrum ranges from radio waves to microwaves, infrared, visible, ultraviolet, and X-rays up to γ-rays (see Fig. 17.1) with wavelengths ranging from zero to infinity. The optical region forms a very small part of the electromagnetic spectrum, with wavelengths ranging from about 4000 to 7000 $\mathring{A}$. The measurable optical properties are the refractive index, reflectance, and transmittance, among others, and they will be dealt with in this chapter both classically and quantum mechanically. Today, optical methods are among the most important tools for studying the electronic structure of solids through their interaction with light. Recently, a number of optical devices, such as lasers, photodetectors, optical fibers, waveguides, light-emitting diodes, and flat-panel displays, have gained considerable technological importance. They are used in communication, medical diagnostics, night viewing, solar applications, optical computing, and for other optoelectronic purposes. Some other traditional uses of optical materials include manufacturing windows, antireflection coatings, lenses, and mirrors.

17.1 PLANE WAVES IN A NONCONDUCTING MEDIUM

The macroscopic Maxwell equations for a nonconducting medium, in the absence of free charges, with magnetic permeability μ and electric permittivity ε are given by

$$\nabla \cdot \mathbf{D} = 0 \tag{17.1}$$

$$\nabla \times \mathbf{H} = \frac{1}{c}\frac{\partial \mathbf{D}}{\partial t} \tag{17.2}$$

$$\nabla \cdot \mathbf{B} = 0 \tag{17.3}$$

$$\nabla \times \mathbf{E} + \frac{1}{c}\frac{\partial \mathbf{B}}{\partial t} = 0 \tag{17.4}$$

where

$$\mathbf{D} = \varepsilon \mathbf{E} \tag{17.5}$$

and

$$\mathbf{B} = \mu \mathbf{H} \tag{17.6}$$

Solid State Physics. https://doi.org/10.1016/B978-0-12-817103-5.00017-7

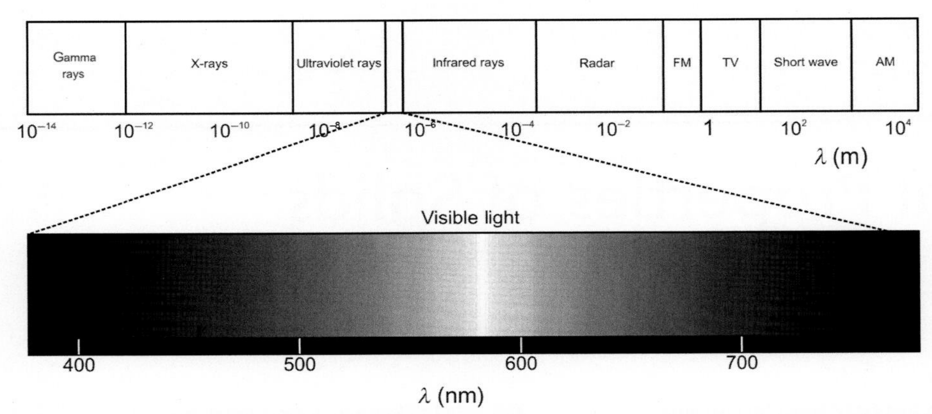

FIG. 17.1 Electromagnetic wave spectrum from gamma rays to radio waves: λ represents the wavelength of the waves. Magnified view of the visible spectrum is shown separately.

For constant values of ε and μ, Eqs. (17.1)–(17.4) reduce to

$$\nabla \cdot \mathbf{E} = 0 \tag{17.7}$$

$$\nabla \times \mathbf{E} + \frac{1}{c}\frac{\partial \mathbf{B}}{\partial t} = 0 \tag{17.8}$$

$$\nabla \cdot \mathbf{B} = 0 \tag{17.9}$$

$$\nabla \times \mathbf{B} = \frac{\mu\varepsilon}{c}\frac{\partial \mathbf{E}}{\partial t} \tag{17.10}$$

Taking the curl of Eq. (17.8) and using Eq. (17.10), one gets

$$\nabla^2 \mathbf{E} - \frac{1}{v^2}\frac{\partial^2 \mathbf{E}}{\partial t^2} = 0 \tag{17.11}$$

where

$$v = \frac{c}{\sqrt{\mu\varepsilon}} = \frac{c}{n} \tag{17.12}$$

with

$$n = \sqrt{\mu\varepsilon} \tag{17.13}$$

Here v is the velocity of electromagnetic waves in the medium and n is its refractive index. Similarly, one can obtain

$$\nabla^2 \mathbf{B} - \frac{1}{v^2}\frac{\partial^2 \mathbf{B}}{\partial t^2} = 0 \tag{17.14}$$

The solutions of Eqs. (17.11) and (17.14) are plane waves defined by

$$\mathbf{E}(\mathbf{r}, t) = \hat{\mathbf{E}} E_0 e^{\imath(\mathbf{K}\cdot\mathbf{r} - \omega t)} \tag{17.15}$$

$$\mathbf{B}(\mathbf{r}, t) = \hat{\mathbf{B}} B_0 e^{\imath(\mathbf{K}\cdot\mathbf{r} - \omega t)} \tag{17.16}$$

where $\hat{\mathbf{E}}$ and $\hat{\mathbf{B}}$ are unit vectors and define the polarization of the electric and magnetic fields. The physical electric and magnetic fields are obtained by taking the real parts of the complex fields. The wave vector **K** gives the direction of propagation of the electromagnetic wave. The orientation of **E**, **B**, and **K** can be obtained from Eqs. (17.7)–(17.10), (17.15), and (17.16). Substituting Eqs. (17.15) and (17.16) into Eqs. (17.7) and (17.9), one gets

$$\mathbf{K} \cdot \hat{\mathbf{E}} = \mathbf{K} \cdot \hat{\mathbf{B}} = 0 \tag{17.17}$$

It is evident from Eq. (17.17) that **K** is perpendicular to both $\hat{\mathbf{E}}$ and $\hat{\mathbf{B}}$ but it does not give any information about the orientation of $\hat{\mathbf{E}}$ and $\hat{\mathbf{B}}$. Substituting Eqs. (17.15) and (17.16) into Eq. (17.8), one gets

$$\mathbf{K} \times \hat{\mathbf{E}}\, \mathrm{E}_0 = \hat{\mathbf{B}} \frac{\omega}{\mathrm{v}\sqrt{\mu\varepsilon}} \mathrm{B}_0 \tag{17.18}$$

We know that

$$|\mathbf{K}| = \frac{\omega}{\mathrm{v}} \tag{17.19}$$

Using Eq. (17.19) one can write Eq. (17.18) as

$$\hat{\mathbf{K}} \times \hat{\mathbf{E}}\, \mathrm{E}_0 = \hat{\mathbf{B}} \frac{\mathrm{B}_0}{\sqrt{\mu\varepsilon}} \tag{17.20}$$

Here $\hat{\mathbf{K}}$ is a unit vector in the direction of the propagation wave vector **K**. The above equation is valid if both the direction and magnitude are the same on both sides, i.e.,

$$\hat{\mathbf{K}} \times \hat{\mathbf{E}} = \hat{\mathbf{B}} \tag{17.21}$$

and

$$\mathrm{B}_0 = \sqrt{\mu\varepsilon}\, \mathrm{E}_0 \tag{17.22}$$

Eq. (17.21) shows that $\hat{\mathbf{B}}$ is perpendicular to both $\hat{\mathbf{K}}$ and $\hat{\mathbf{E}}$. Hence $\hat{\mathbf{E}}, \hat{\mathbf{B}}$, and $\hat{\mathbf{K}}$ form an orthogonal triad and the electromagnetic plane wave is a transverse wave. Eq. (17.22) gives the relation between the electric and magnetic field amplitudes.

17.2 REFLECTION AND REFRACTION AT A PLANE INTERFACE

Reflection and refraction at a plane surface are familiar phenomena and each has two aspects:

1. *Kinematic properties*
   These constitute the laws of reflection and refraction.
2. *Dynamic properties*
   Dynamic properties provide information about the intensities of the reflected and refracted waves. They also give the phase changes and polarizations after reflection and refraction.

17.2.1 Kinematic Properties

Let the media below and above the $z=0$ plane have permeability and permittivity μ, ε and μ', ε' (Fig. 17.2), respectively. A plane wave with wave vector **K** and frequency ω is incident from the medium μ, ε. The reflected and refracted rays have wave vectors $\mathbf{K}''$ and $\mathbf{K}'$, respectively. Let $\hat{\mathbf{n}}$ be a unit vector normal to the $z=0$ plane and directed from the medium μ, ε to the medium μ', ε'. The incident, reflected, and refracted waves are now represented as follows:

Incident wave:

$$\mathbf{E} = \mathbf{E}_0\, \mathrm{e}^{\imath(\mathbf{K}\cdot\mathbf{r}-\omega t)} \tag{17.23}$$

$$\mathbf{B} = \sqrt{\mu\varepsilon}\, \frac{\mathbf{K} \times \mathbf{E}}{\mathrm{K}} \tag{17.24}$$

Refracted wave:

$$\mathbf{E}' = \mathbf{E}'_0\, \mathrm{e}^{\imath(\mathbf{K}'\cdot\mathbf{r}-\omega t)} \tag{17.25}$$

$$\mathbf{B}' = \sqrt{\mu'\varepsilon'}\, \frac{\mathbf{K}' \times \mathbf{E}'}{\mathrm{K}'} \tag{17.26}$$

Reflected wave:

$$\mathbf{E}'' = \mathbf{E}''_0\, \mathrm{e}^{\imath(\mathbf{K}''\cdot\mathbf{r}-\omega t)} \tag{17.27}$$

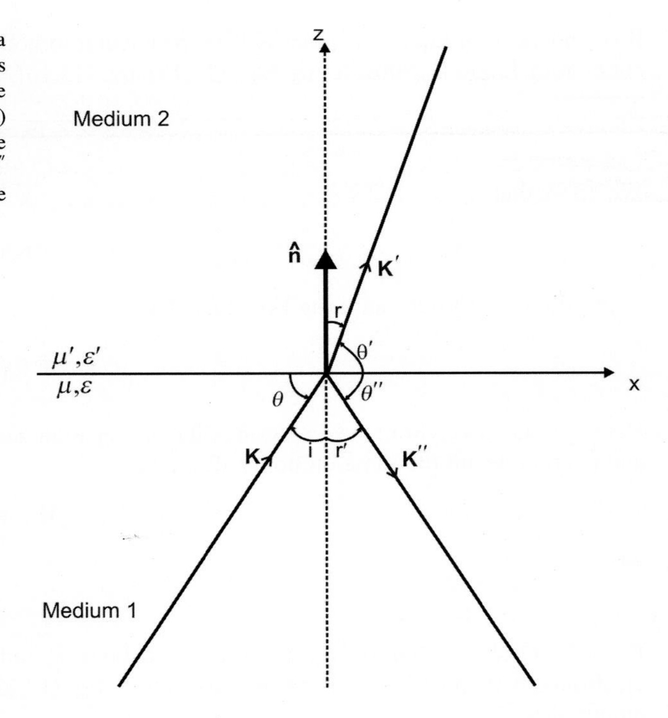

FIG. 17.2 Incident plane electromagnetic wave with wave vector **K** strikes a plane interface between two different media. The incident plane wave gives rise to a reflected wave with wave vector $\mathbf{K}''$ and a refracted wave with wave vector $\mathbf{K}'$. Here $\hat{\mathbf{n}}$ is a unit vector perpendicular to the interface (z=0 plane) separating the two media and is directed from medium 1 to medium 2. The angles of incidence and refraction are represented by i and r. θ, θ', and θ'' are the angles made by the incident, refracted, and reflected rays with the interface.

$$\mathbf{B}'' = \sqrt{\mu\varepsilon}\,\frac{\mathbf{K}'' \times \mathbf{E}''}{\mathrm{K}} \tag{17.28}$$

The wave vector of the reflected wave has the same magnitude as that of the incident wave and can be defined as

$$\mathrm{K} = \frac{\omega}{\mathrm{v}} = \frac{\omega}{\mathrm{c}}\sqrt{\mu\varepsilon} = \frac{\omega}{\mathrm{c}}\,\mathrm{n} \tag{17.29}$$

But the wave vector $\mathbf{K}'$ has the magnitude

$$\mathrm{K}' = |\mathbf{K}'| = \frac{\omega}{\mathrm{c}}\sqrt{\mu'\varepsilon'} = \frac{\omega}{\mathrm{c}}\,\mathrm{n}' \tag{17.30}$$

Here n and n′ are the refractive indices of the two media defined by

$$\mathrm{n} = \sqrt{\mu\varepsilon}, \mathrm{n}' = \sqrt{\mu'\varepsilon'} \tag{17.31}$$

To prove the kinematic properties the boundary conditions at the interface must be satisfied at all points and for all times, which implies that the spatial and time variation of all the fields must be the same at the z=0 plane. It also implies that the phase factors at z=0 must be the same for all the waves, i.e.,

$$\imath(\mathbf{K}\cdot\mathbf{r} - \omega t)_{z=0} = \imath(\mathbf{K}'\cdot\mathbf{r} - \omega t)_{z=0} = \imath(\mathbf{K}''\cdot\mathbf{r} - \omega t)_{z=0} \tag{17.32}$$

or

$$(\mathbf{K}\cdot\mathbf{r})_{z=0} = (\mathbf{K}'\cdot\mathbf{r})_{z=0} = (\mathbf{K}''\cdot\mathbf{r})_{z=0} \tag{17.33}$$

Let us first consider refraction, for which

$$(\mathbf{K}\cdot\mathbf{r})_{z=0} = (\mathbf{K}'\cdot\mathbf{r})_{z=0}$$

This can be written as

$$\mathrm{K\,r}\cos\theta = \mathrm{K'\,r}\cos\theta'$$

From Fig. 17.2, the above equation can be written as

$$\mathrm{K}\cos(90 - \mathrm{i}) = \mathrm{K}'\cos(90 - \mathrm{r})$$

which gives

$$\frac{\sin i}{\sin r}=\frac{K'}{K}=\frac{\lambda}{\lambda'}=\frac{\sqrt{\mu'\varepsilon'}}{\sqrt{\mu\varepsilon}}=\frac{n'}{n} \tag{17.34}$$

Eq. (17.34) is nothing but the Snell's law of refraction. In the case of reflection, the boundary condition from Eq. (17.33) can be written as

$$(\mathbf{K}\cdot\mathbf{r})_{z=0}=(\mathbf{K}''\cdot\mathbf{r})_{z=0} \tag{17.35}$$

From Fig. 17.2 the above equation can be written as

$$K\cos\theta=K\cos\theta''$$

or

$$i=r' \tag{17.36}$$

i.e., the angle of incidence is equal to the angle of reflection (law of reflection). Further, $\mathbf{K}$, $\mathbf{K}'$, and $\mathbf{K}''$ all lie in the same plane and, therefore, the incident, reflected, and refracted rays all lie in the same plane.

17.2.2 Dynamic Properties

The dynamic properties can be studied by considering the boundary conditions satisfied by the fields. The first two boundary conditions are that the normal components of $\mathbf{D}$ and $\mathbf{B}$ should be continuous at the interface, i.e.,

$$(\mathbf{D}\cdot\hat{\mathbf{n}})_1=(\mathbf{D}\cdot\hat{\mathbf{n}})_2 \tag{17.37}$$

$$(\mathbf{B}\cdot\hat{\mathbf{n}})_1=(\mathbf{B}\cdot\hat{\mathbf{n}})_2 \tag{17.38}$$

The subscripts 1 and 2 denote the two media represented by μ, ε and μ', ε'. The other two boundary conditions are that the tangential components of $\mathbf{E}$ and $\mathbf{H}$ should be continuous at the interface, i.e.,

$$(\mathbf{E}\times\hat{\mathbf{n}})_1=(\mathbf{E}\times\hat{\mathbf{n}})_2 \tag{17.39}$$

$$(\mathbf{H}\times\hat{\mathbf{n}})_1=(\mathbf{H}\times\hat{\mathbf{n}})_2 \tag{17.40}$$

In medium 1 the electric field is $\mathbf{E}+\mathbf{E}''$ (of the incident and reflected waves) and in the medium 2 it is $\mathbf{E}'$. The case with magnetic fields is similar, so Eqs. (17.37) and (17.38), in terms of $\mathbf{E}$ and $\mathbf{B}$, can be written as

$$\varepsilon(\mathbf{E}+\mathbf{E}'')\cdot\hat{\mathbf{n}}=\varepsilon'\mathbf{E}'\cdot\hat{\mathbf{n}} \tag{17.41}$$

$$(\mathbf{B}+\mathbf{B}'')\cdot\hat{\mathbf{n}}=\mathbf{B}'\cdot\hat{\mathbf{n}} \tag{17.42}$$

Substituting the values of the $\mathbf{E}$ and $\mathbf{B}$ fields from Eqs. (17.23) to (17.28) and simplifying, one gets

$$\left[\varepsilon(\mathbf{E}_0+\mathbf{E}_0'')-\varepsilon'\mathbf{E}_0'\right]\cdot\hat{\mathbf{n}}=0 \tag{17.43}$$

$$\left[\mathbf{K}\times\mathbf{E}_0+\mathbf{K}''\times\mathbf{E}_0''-\mathbf{K}'\times\mathbf{E}_0'\right]\cdot\hat{\mathbf{n}}=0 \tag{17.44}$$

Now the boundary condition represented by Eqs. (17.39) and (17.40) can be simplified in the same way as Eqs. (17.37) and (17.38) and one can write

$$\left(\mathbf{E}_0+\mathbf{E}_0''-\mathbf{E}_0'\right)\times\hat{\mathbf{n}}=0 \tag{17.45}$$

$$\left[\frac{1}{\mu}\left(\mathbf{K}\times\mathbf{E}_0+\mathbf{K}''\times\mathbf{E}_0''\right)-\frac{1}{\mu'}\mathbf{K}'\times\mathbf{E}_0'\right]\times\hat{\mathbf{n}}=0 \tag{17.46}$$

The boundary conditions represented by Eqs. (17.43)–(17.46) can be applied in two different situations for plane polarized electromagnetic waves:

1. When the electric field vector $\mathbf{E}$ is perpendicular to the plane of incidence.
2. When the electric field vector $\mathbf{E}$ is parallel to the plane of incidence.

17.2.2.1 Electric Field Perpendicular to Plane of Incidence

Fig. 17.3 shows an electric field perpendicular to the plane of incidence and directed away from the viewer. The directions of the **B** fields are chosen to give a positive flow of energy in the direction of propagation, i.e., in the direction of the wave vectors. As the electric field vectors are parallel to the interface, the boundary condition (17.43) becomes redundant (normal components of **E** fields are zero). One can easily show that the boundary conditions (17.44) and (17.45) are equivalent and yield

$$E_0 + E_0'' - E_0' = 0 \tag{17.47}$$

The boundary condition (17.46) can be simplified to finally obtain

$$\sqrt{\frac{\varepsilon}{\mu}}\left(E_0 - E_0''\right)\cos i - \sqrt{\frac{\varepsilon'}{\mu'}}E_0' \cos r = 0 \tag{17.48}$$

In obtaining Eq. (17.48), the angles in the clockwise direction are taken to be positive, while those in the counterclockwise direction are negative. From Eqs. (17.47) and (17.48) one can solve for E_0' in terms of E_0 to get

$$\frac{E_0'}{E_0} = \frac{2\cos i}{\cos i + \dfrac{\mu}{\mu'}\sqrt{\dfrac{\mu'\varepsilon'}{\mu\varepsilon}}\cos r} = \frac{2\cos i}{\cos i + \dfrac{\mu}{\mu'}\dfrac{n'}{n}\cos r} \tag{17.49}$$

For normal incidence ($i = r = 0$) the above equation becomes

$$\frac{E_0'}{E_0} = \frac{2}{1 + \dfrac{\mu}{\mu'}\dfrac{n'}{n}} \tag{17.50}$$

For two media with the same permeability ($\mu = \mu'$) one can write

$$\frac{n'}{n} = \sqrt{\frac{\varepsilon'}{\varepsilon}} \tag{17.51}$$

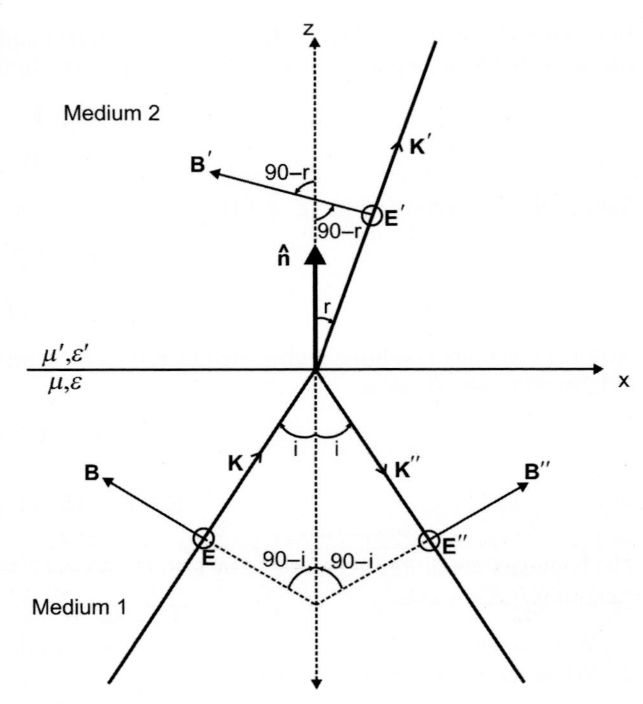

FIG. 17.3 Reflection and refraction of a plane electromagnetic wave with polarization perpendicular the plane of incidence. The rest of the description of the figure is the same as that of Fig. 17.2.

From Eqs. (17.50) and (17.51) one gets

$$\frac{E_0'}{E_0}=\frac{2n}{n+n'}=\frac{2\sqrt{\varepsilon}}{\sqrt{\varepsilon}+\sqrt{\varepsilon'}} \tag{17.52}$$

If $\mu=\mu'=1$, then

$$n=\sqrt{\varepsilon} \text{ and } n'=\sqrt{\varepsilon'} \tag{17.53}$$

Now the amplitude of the electric field vector of the incident wave must be equal to the sum of the amplitudes of the field vectors of the reflected and refracted waves, i.e.,

$$E_0=E_0'+E_0'' \tag{17.54}$$

From Eqs. (17.52) and (17.54) one can write

$$\frac{E_0''}{E_0}=\frac{n'-n}{n'+n} \tag{17.55}$$

Reflectivity R is defined as the ratio of the intensity of the reflected light I_R to the intensity of the incident light I_0 and from Eq. (17.55) is given by

$$R=\frac{I_R}{I_0}=\frac{|E_0''|^2}{|E_0|^2}=\left|\frac{n'-n}{n'+n}\right|^2 \tag{17.56}$$

If medium 1, in which the incident and reflected waves are traveling, is vacuum, then $n=1$ and so Eq. (17.56) reduces to

$$R=\left|\frac{n-1}{n+1}\right|^2 \tag{17.57}$$

Similarly, the ratio of the intensity of transmitted light I_T to the intensity of incident light I_0 is defined as *transmissivity* or *transmittance* T and is given by

$$T=\frac{I_T}{I_0}=\frac{|E_0'|^2}{|E_0|^2}=\frac{4n^2}{(n'+n)^2} \tag{17.58}$$

For $n=1$ (vacuum) one can write

$$T=\frac{4}{(n+1)^2} \tag{17.59}$$

Both R and T are dimensionless quantities and are measured in percent.

Problem 17.1

Consider a plane wave incident at the interface of two media having the same permeability and represented by μ, ε and μ, ε'. If the electric field is parallel to the plane of incidence (see Fig. 17.4) prove that

$$\frac{E_0'}{E_0}=\frac{2\cos i\ \sin r}{\sin(i+r)\cos(i-r)} \tag{17.60}$$

$$\frac{E_0''}{E_0}=\frac{\tan(i-r)}{\tan(i+r)} \tag{17.61}$$

Further, prove that the values of R and T for normal incidence are given by Eqs. (17.57) and (17.59).

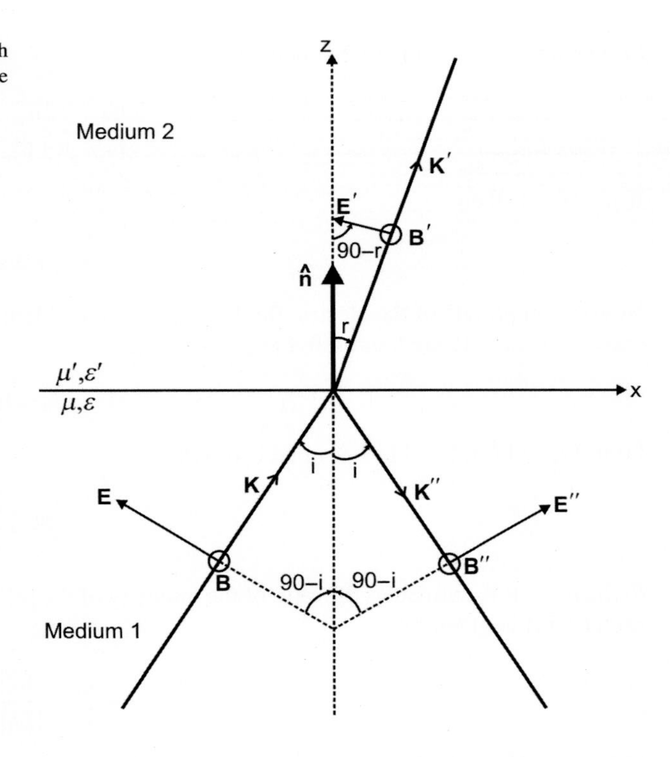

FIG. 17.4 Reflection and refraction of a plane electromagnetic wave with electric polarization parallel to the plane of incidence. The rest of the description of the figure is the same as that of Fig. 17.2.

17.3 ELECTROMAGNETIC WAVES IN A CONDUCTING MEDIUM

Consider a transverse plane electromagnetic wave traveling in a medium with finite conductivity σ. The current density $\mathbf{J}$ is defined by

$$\mathbf{J}=\sigma\mathbf{E} \tag{17.62}$$

The Maxwell equations for a conducting medium in the absence of free charges are given by

$$\nabla\cdot\mathbf{D}=0 \tag{17.63}$$

$$\nabla\times\mathbf{E}+\frac{1}{c}\frac{\partial\mathbf{B}}{\partial t}=0 \tag{17.64}$$

$$\nabla\cdot\mathbf{B}=0 \tag{17.65}$$

$$\nabla\times\mathbf{H}=\frac{1}{c}\frac{\partial\mathbf{D}}{\partial t}+\frac{4\pi}{c}\mathbf{J} \tag{17.66}$$

Substituting the values of $\mathbf{D}$ and $\mathbf{B}$ from Eqs. (17.5) and (17.6), we write the Maxwell equations as

$$\nabla\cdot\mathbf{E}=0 \tag{17.67}$$

$$\nabla\times\mathbf{E}+\frac{\mu}{c}\frac{\partial\mathbf{H}}{\partial t}=0 \tag{17.68}$$

$$\nabla\cdot\mathbf{H}=0 \tag{17.69}$$

$$\nabla\times\mathbf{H}-\frac{\varepsilon}{c}\frac{\partial\mathbf{E}}{\partial t}-\frac{4\pi\sigma}{c}\mathbf{E}=0 \tag{17.70}$$

Taking the curl of Eq. (17.68) and using Eq. (17.70), we get

$$\nabla^2\mathbf{E}-\frac{\mu\varepsilon}{c^2}\frac{\partial^2\mathbf{E}}{\partial t^2}-\frac{4\pi\sigma\mu}{c^2}\frac{\partial\mathbf{E}}{\partial t}=0 \tag{17.71}$$

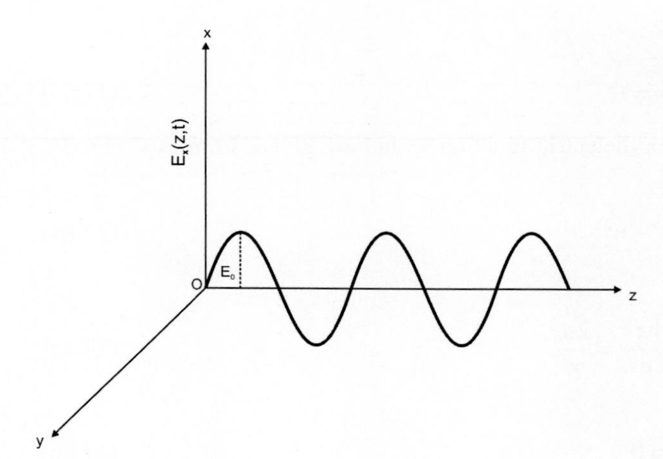

FIG. 17.5 A snapshot of a plane electromagnetic (em) wave $E_x(z,t)$ traveling in the z-direction. The electric field of the wave is in the x-direction with amplitude E_0.

Similarly, we can obtain

$$\nabla^2\mathbf{H} - \frac{\mu\varepsilon}{c^2}\frac{\partial^2\mathbf{H}}{\partial t^2} - \frac{4\pi\sigma\mu}{c^2}\frac{\partial\mathbf{H}}{\partial t} = 0 \tag{17.72}$$

The plane wave solutions of the **E** and **H** fields for Eqs. (17.71) and (17.72) are defined as

$$\mathbf{E}(\mathbf{r},t) = \mathbf{E}_0 \exp[\imath(\mathbf{K}\cdot\mathbf{r} - \omega t)] \tag{17.73}$$

$$\mathbf{H}(\mathbf{r},t) = \mathbf{H}_0 \exp[\imath(\mathbf{K}\cdot\mathbf{r} - \omega t)] \tag{17.74}$$

Consider an electromagnetic wave traveling in the z-direction with its electric field vector in the x-direction (see Fig. 17.5). Such a wave is represented as

$$E_x(z,t) = E_0 \exp[\imath(Kz - \omega t)] = E_0 \exp\left[\imath\omega\left(\frac{n_c}{c}z - t\right)\right] \tag{17.75}$$

Here we have used the symbol n_c for the complex refractive index. Eq. (17.29) has been used in writing the above equation. Then, Eq. (17.71) becomes

$$\frac{\partial^2 E_x}{\partial z^2} - \frac{\mu\varepsilon}{c^2}\frac{\partial^2 E_x}{\partial t^2} - \frac{4\pi\sigma\mu}{c^2}\frac{\partial E_x}{\partial t} = 0 \tag{17.76}$$

Substituting Eq. (17.75) into Eq. (17.76), we obtain

$$n_c^2 = \mu\varepsilon + \imath\frac{4\pi\sigma\mu}{\omega} \tag{17.77}$$

The complex refractive index can, in general, be written as

$$n_c = n_1 + \imath n_2 \tag{17.78}$$

n_1 and n_2 are the real and imaginary components of n_c. The imaginary part of the refractive index n_2 is usually called the damping factor and gives the absorption of electromagnetic waves. The square of n_c, from Eq. (17.78), becomes

$$n_c^2 = n_1^2 - n_2^2 + 2\imath n_1 n_2 \tag{17.79}$$

Comparing Eqs. (17.77) and (17.79), we find

$$n_1^2 - n_2^2 = \mu\varepsilon \tag{17.80}$$

and

$$2n_1 n_2 = \frac{4\pi\sigma\mu}{\omega} \tag{17.81}$$

A system of particular interest is one with $\mu = 1$, for which Eq. (17.77) becomes

$$n_c^2 = \varepsilon + \imath\frac{4\pi\sigma}{\omega} = \varepsilon + \imath\frac{2\sigma}{\nu} \tag{17.82}$$

Here ν is the frequency. In this case n_c^2 can be written as

$$n_c^2 = \varepsilon_c = \varepsilon_1 + \imath\varepsilon_2 \tag{17.83}$$

Here ε_1 and ε_2 are the real and imaginary parts of the complex dielectric function ε_c and are given, from Eqs. (17.79), (17.82), and (17.83), as

$$\varepsilon_1 = \varepsilon = n_1^2 - n_2^2 \tag{17.84}$$

and

$$\varepsilon_2 = 2n_1 n_2 = \frac{4\pi\sigma}{\omega} = \frac{2\sigma}{\nu} \tag{17.85}$$

For an insulator $\sigma = 0$, so from Eq. (17.85) one can write

$$\varepsilon_2(n_2) = 0 \tag{17.86}$$

assuming finite refractive index n_1. Now from Eq. (17.84)

$$n^2 = \varepsilon_1 = \varepsilon \tag{17.87}$$

which is the expected result from the Maxwell equations. In defining Eqs. (17.84) and (17.87) we have written $n_1 = n$, the real part of the refractive index.

Let us investigate the effect of a complex refractive index on an electromagnetic wave. Substituting n_c from Eq. (17.78) into Eq. (17.75), one writes

$$E_x(z, t) = E_0^d \exp\left[\imath\omega\left(\frac{n}{c}z - t\right)\right] \tag{17.88}$$

where

$$E_0^d = E_0 e^{-\frac{\omega n_2}{c}z} \tag{17.89}$$

Eq. (17.88) represents a damped wave in which the amplitude E_0^d decreases exponentially with increasing z (see Fig. 17.6). The decrease in amplitude is determined by both n_2 and ω. For fixed ω, the damping is greater for greater values of n_2 and that is why n_2 is usually called the damping constant. Therefore, high-frequency electromagnetic waves conduct only near the outer surface of the wire (normal skin effect).

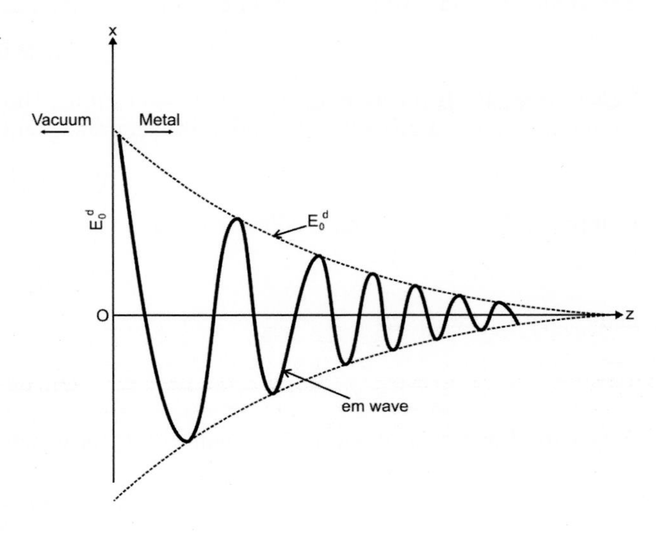

FIG. 17.6 The exponential decrease in amplitude E_0^d of a plane electromagnetic (em) wave traveling in the z-direction in a metal.

It is difficult to measure the amplitude of the electric field, but easier to measure its intensity, which is the square of the modulus of the field amplitude. The intensity I_d corresponding to the damped electric field, from Eqs. (17.88) and (17.89), is given by

$$I_d = \left|E_0^d\right|^2 = I_0 e^{-\frac{2\omega n_2}{c}z} \tag{17.90}$$

where

$$I_0 = |E_0|^2 \tag{17.91}$$

It is customary to define the *characteristic penetration depth* Z_p as the distance at which the intensity of the electric field reduces to e^{-1} times the original value, i.e.,

$$\frac{I_d}{I_0} = e^{-1} \tag{17.92}$$

From Eqs. (17.90) and (17.92) one can write

$$Z_p = \frac{c}{2\omega n_2} = \frac{c}{4\pi\nu n_2} = \frac{\lambda}{4\pi n_2} \tag{17.93}$$

Further, the *absorbance* or *attenuation* α_p is defined as the reciprocal of Z_p, i.e.,

$$\alpha_p = \frac{1}{Z_p} = \frac{4\pi n_2}{\lambda} \tag{17.94}$$

Substituting the value of n_2 from Eq. (17.85), one can write (for $\mu = 1$)

$$\alpha_p = \frac{4\pi\sigma}{nc} \tag{17.95}$$

Substituting the value of n_2 in terms of ε_2, from Eq. (17.85) one gets

$$\alpha_p = \frac{2\pi\varepsilon_2}{\lambda n} \tag{17.96}$$

Hence one can write

$$\alpha_p = \frac{4\pi n_2}{\lambda} = \frac{4\pi\sigma}{nc} = \frac{2\pi\varepsilon_2}{\lambda n} = \frac{2\omega n_2}{c} \tag{17.97}$$

17.4 REFLECTIVITY FROM METALLIC SOLIDS

In a metallic solid the refractive index n_c is a complex quantity. Thus, by replacing n by n_c in Eq. (17.57) and substituting Eq. (17.78) into it, one can write

$$R = \left|\frac{n + \imath n_2 - 1}{n + \imath n_2 + 1}\right|^2 \tag{17.98}$$

The above equation can be written as

$$R = \frac{(n-1)^2 + n_2^2}{(n+1)^2 + n_2^2} \tag{17.99}$$

Eq. (17.99) is called the *Beer equation*. If n_c is imaginary, then its real part goes to zero, i.e., $n = 0$. In this case $R = 1$, which means the reflectivity is 100%. But if n_c is real and positive ($n_2 = 0$), then Eq. (17.98) reduces to

$$R = \left(\frac{n-1}{n+1}\right)^2 \tag{17.100}$$

which is same as Eq. (17.57). In this case the material is essentially transparent to these wavelengths (for perpendicular incidence) and behaves optically like an insulator.

Problem 17.2

For a medium with unit permeability ($\mu = 1$), prove the following relations

$$n^2 + n_2^2 = \sqrt{\varepsilon_1^2 + \varepsilon_2^2} \tag{17.101}$$

$$n^2 = \frac{1}{2}\left[\sqrt{\varepsilon_1^2 + \varepsilon_2^2} + \varepsilon_1\right] \tag{17.102}$$

$$n_2^2 = \frac{1}{2}\left[\sqrt{\varepsilon_1^2 + \varepsilon_2^2} - \varepsilon_1\right] \tag{17.103}$$

$$2n = \sqrt{2\left(\sqrt{\varepsilon_1^2 + \varepsilon_2^2} + \varepsilon_1\right)} \tag{17.104}$$

Problem 17.3

Using Eq. (17.99), prove that for a medium with unit permeability ($\mu = 1$),

$$R = \frac{1 + \sqrt{\varepsilon_1^2 + \varepsilon_2^2} - \left(2\left(\sqrt{\varepsilon_1^2 + \varepsilon_2^2} + \varepsilon_1\right)\right)^{1/2}}{1 + \sqrt{\varepsilon_1^2 + \varepsilon_2^2} + \left(2\left(\sqrt{\varepsilon_1^2 + \varepsilon_2^2} + \varepsilon_1\right)\right)^{1/2}} \tag{17.105}$$

17.5 REFLECTIVITY AND CONDUCTIVITY

In a metallic solid it is interesting to examine the relationship between reflectivity and conductivity. Substituting the value of ε_2 from Eq. (17.85) into Eqs. (17.102) and (17.103), one can write

$$n^2 = \frac{1}{2}\left[\sqrt{\varepsilon_1^2 + \left(\frac{2\sigma}{\nu}\right)^2} + \varepsilon_1\right] \tag{17.106}$$

$$n_2^2 = \frac{1}{2}\left[\sqrt{\varepsilon_1^2 + \left(\frac{2\sigma}{\nu}\right)^2} - \varepsilon_1\right] \tag{17.107}$$

For low frequencies, i.e., $\nu < 10^{13}\,s^{-1}$ (in the infrared region) for a metal with $\sigma \approx 10^{17}\,s^{-1}$ and $\varepsilon_1 \approx 10$, one can write

$$\varepsilon_1^2 + \left(\frac{2\sigma}{\nu}\right)^2 \approx \left(\frac{2\sigma}{\nu}\right)^2$$

$$\approx \left(\frac{10^{17}}{10^{13}}\right)^2 \approx 10^8$$

Hence, from Eqs. (17.106) and (17.107), it is easy to write

$$n^2 = n_2^2 = \frac{\sigma}{\nu} \tag{17.108}$$

The reflectivity from Eq. (17.99) is given by

$$R = 1 - \frac{4n}{2n^2 + 2n + 1} \tag{17.109}$$

Here we have used Eq. (17.108). For metallic solids with $n > 1$, one can neglect the factor $2n + 1$ in comparison with $2n^2$ to yield

$$R = 1 - \frac{2}{n} \tag{17.110}$$

Substituting the value of n from Eq. (17.108), one gets

$$R = 1 - 2\sqrt{\frac{v}{\sigma}} \tag{17.111}$$

For small frequencies (say in the infrared region), σ can be replaced by the dc conductivity σ_0 to write

$$R = 1 - 2\sqrt{\frac{v}{\sigma_0}} \tag{17.112}$$

Eq. (17.112) is usually called the *Hagen-Rubens* equation as this relation was found empirically by Hagen and Rubens from experimental measurements of reflectivity in the infrared region ($\lambda > 30\,\text{nm}$). However. Drude derived Eq. (17.112) theoretically. Further, Eq. (17.112) shows that good metals (with large σ_0) act as very good reflectors in the infrared region as $R \approx 1$.

17.6 KRAMERS-KRONIG RELATIONS

In this and the previous chapters it has been found that, in the linear approximation, the response functions, such as reflectance, absorption, and electrical susceptibility, are functions of variable ω. A general linear response function $N(\omega)$ can be written as

$$N(\omega) = N_1(\omega) + \imath N_2(\omega) \tag{17.113}$$

where $N_1(\omega)$ and $N_2(\omega)$ are the real and imaginary parts of $N(\omega)$. To obtain $N_1(\omega)$ and $N_2(\omega)$ one can apply complex variable calculus. The response function may have one or more resonances (poles) and its total value is the sum of the contributions arising from the various resonances. One of the most important methods to sum the contributions arising from all the resonances is Cauchy's integral theorem in which we calculate the integral in the upper half of the complex plane. The contour in the upper half of the complex plane consists of four parts as shown in Fig. 17.7. Cauchy's integral can be written as

$$N(\omega) = \frac{1}{\pi \imath} P \int_{-\infty}^{\infty} \frac{N(\omega')}{\omega' - \omega} d\omega \tag{17.114}$$

Here P stands for the principal part of the integral. In order to apply Cauchy's integral theorem, the response function must satisfy the following conditions:

1. The poles of $N(\omega)$ lie in the upper half of the complex plane.
2. The integral of $N(\omega)/\omega$ vanishes when calculated around the infinite semicircular contour in the upper half on the complex plane. In this approximation the integral over part 4 of the contour will go to zero.
3. Function $N_1(\omega)$ must be even, but $N_2(\omega)$ must be odd with respect to the variable ω.

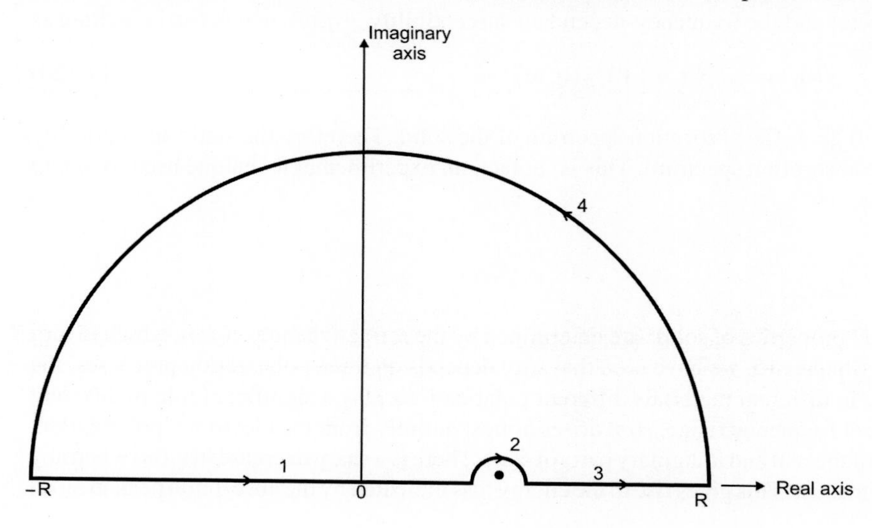

FIG. 17.7 The semicircular contour in the upper half of the complex plane.

Separating the real and imaginary parts in Eq. (17.114), one can write

$$N_1(\omega) = \frac{1}{\pi} P \int_{-\infty}^{\infty} \frac{N_2(\omega')}{\omega' - \omega} d\omega' \tag{17.115}$$

$$N_2(\omega) = -\frac{1}{\pi} P \int_{-\infty}^{\infty} \frac{N_1(\omega')}{\omega' - \omega} d\omega' \tag{17.116}$$

The integral over parts 1, 2, and 3 of the contour, by definition, forms the principal part of the integral. Eq. (17.115) can further be split into two integrals as follows:

$$N_1(\omega) = \frac{1}{\pi} P \int_{0}^{\infty} \frac{N_2(\omega')}{\omega' - \omega} d\omega' + \frac{1}{\pi} P \int_{-\infty}^{0} \frac{N_2(\omega')}{\omega' - \omega} d\omega' = \frac{2}{\pi} P \int_{0}^{\infty} \frac{\omega' N_2(\omega')}{\omega'^2 - \omega^2} d\omega' \tag{17.117}$$

Similarly, from Eq. (17.116) one can obtain

$$N_2(\omega) = -\frac{2\omega}{\pi} P \int_{0}^{\infty} \frac{N_1(\omega')}{\omega'^2 - \omega^2} d\omega' \tag{17.118}$$

Eqs. (17.117) and (17.118) are known as the *Kramers-Kronig* relations. It is clear from Eqs. (17.117) and (17.118) that these relations allow us to find the real part of the response function if its imaginary part is known over all the frequencies and vice versa. As an example, take the case of a complex dielectric function of a material written as

$$\varepsilon(\omega) = \varepsilon_1(\omega) + \imath \varepsilon_2(\omega) \tag{17.119}$$

The imaginary part of the dielectric function $\varepsilon_2(\omega)$ gives the absorption spectrum, while the real part $\varepsilon_1(\omega)$ gives the refractive index $n(\omega)$ of the material. Therefore, if $n(\omega)$ is known over the whole of the frequency range, one can calculate the absorption spectrum without any additional information and vice versa. This shows that refraction and absorption are not independent properties of a material but originate from the same physical effect. But in practice, the refractive index and the absorption spectrum are measured experimentally over a certain finite frequency range. Therefore, the value of the refractive index or the absorption spectrum evaluated using the Kramers-Kronig relations may involve some error. We want to remark here that reflectivity also depends on the refractive index. Therefore, the reflectivity and absorption spectra are also not independent, but rather can be calculated from each other. Note that if the absorption is strong, the transmission through a thick slab may be too weak to measure, but the reflectivity can still be measured.

Other important quantity is the wave vector and the frequency-dependent susceptibility $\chi(\mathbf{q}, \omega)$, which can be written as

$$\chi(\mathbf{q}, \omega) = \chi_1(\mathbf{q}, \omega) + \imath \chi_2(\mathbf{q}, \omega) \tag{17.120}$$

The imaginary part of susceptibility $\chi_2(\mathbf{q}, \omega)$ gives the absorption spectrum of the solid. Therefore,the static susceptibility $\chi_1(\mathbf{q}, 0)$ may be obtained by integrating the absorption spectrum. This is, in fact, an experimental technique used to obtain the static susceptibility of certain materials.

17.7 OPTICAL MODELS

It has become evident by now that the optical properties of solids are determined by the refractive index $n_c(\omega)$, which in turn depends on the dielectric function $\varepsilon(\omega)$. In Chapter 15, we have seen that $\varepsilon(\omega)$ depends on three polarization processes: the electronic, ionic, and dipolar polarizations. In different materials different polarizations play a significant role in different frequency ranges. For example, in the optical frequency range, $\varepsilon(\omega)$ arises almost entirely from the electronic polarization. Fig. 17.8 shows a schematic representation of the real and imaginary parts of $\varepsilon(\omega)$. There is a sharp decrease in $\varepsilon_1(\omega)$ when any one of the polarization contributions ceases to exist. This gives rise to the energy loss indicated by the absorption peak in $\varepsilon_2(\omega)$

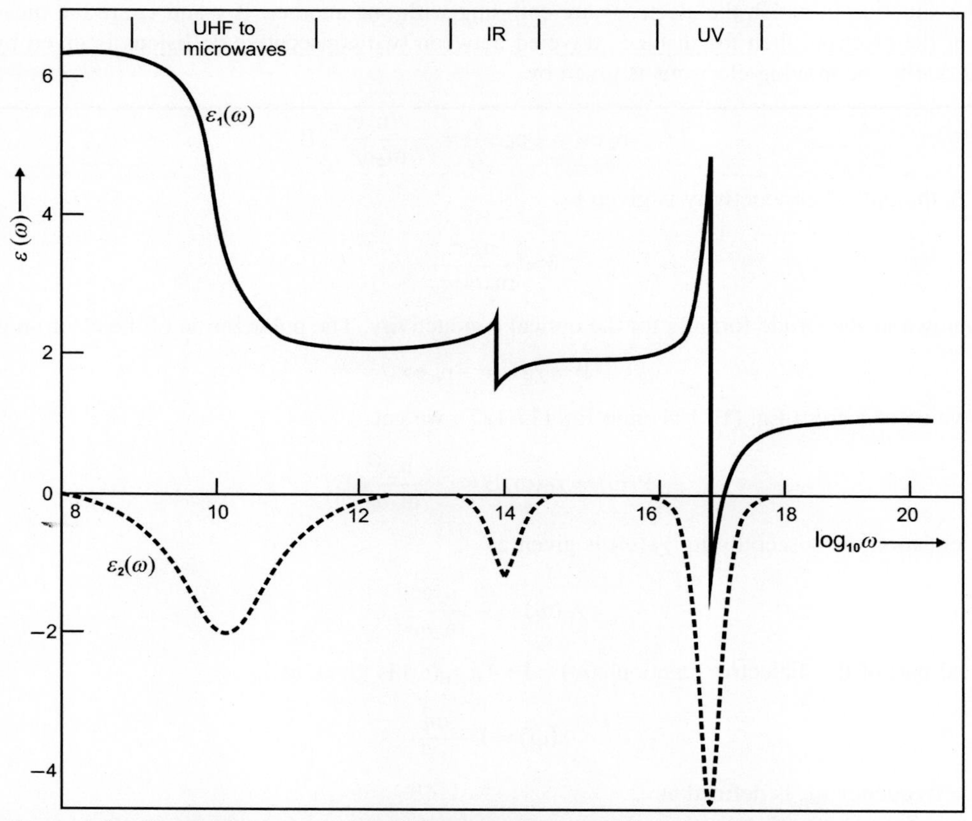

FIG. 17.8 The real part $\varepsilon_1(\omega)$ and imaginary part $\varepsilon_2(\omega)$ of the dielectric function in a solid as a function of frequency ω.

in that frequency range. The first absorption peak occurs in the ultrahigh-frequency to microwave-frequency range, the second peak occurs near the infrared region, and the third sharp peak occurs in the ultraviolet region near the visible frequency range. From a comparison of Figs. 15.12 and 17.8 it is evident that the behavior of $\varepsilon_1(\omega)$ is similar to that of the polarizability $\alpha^a(\omega)$. The optical properties, such as reflectivity, transmission, and absorption, can be evaluated using different models depending on the nature of the solid. In this text, we describe simple models for metals, insulators, and ionic solids.

17.7.1 Drude Model

Drude gave a simple model for the dielectric function and optical conductivity of an electron gas. Let n_e be the electron density of the electron gas. In the absence of any collisions, the electrons are free to move (free-electron gas). The equation of motion of a free electron moving in the x-direction is given by

$$m_e \frac{d^2x}{dt^2} = -eE \tag{17.121}$$

Let both the position x and electric field E have the same time dependence defined as

$$x = x_0 e^{-\imath\omega t} \tag{17.122}$$

$$E = E_0 e^{-\imath\omega t} \tag{17.123}$$

Substituting Eqs. (17.122) and (17.123) into Eq. (17.121), one gets

$$x = \frac{eE}{m_e \omega^2} \tag{17.124}$$

Now consider the situation in which the electrons are colliding with one another. If v and τ_e are the mean velocity and relaxation time of the electron, then the distance traveled between two consecutive collisions is given by $x = v\tau_e$. The current density J due to the moving electrons is given by

$$J = -n_e e v = -n_e e \frac{x}{\tau_e} = -\frac{n_e e^2}{m_e \omega^2 \tau_e} E \tag{17.125}$$

Because $J = \sigma E$, the optical conductivity is given by

$$\sigma = \frac{n_e e^2}{m_e \omega^2 \tau_e} \tag{17.126}$$

Eq. (17.126) is known as the Drude formula for the optical conductivity. The polarization of the electron gas is given by

$$P = \chi_E E = -n_e e x \tag{17.127}$$

Substituting the value of x from Eq. (17.124) into Eq. (17.127), we get

$$P(\omega) = \chi_E(\omega) E = -\frac{n_e e^2}{m_e \omega^2} E \tag{17.128}$$

From the above equation the susceptibility $\chi_E(\omega)$ is given as

$$\chi_E(\omega) = -\frac{n_e e^2}{m_e \omega^2} \tag{17.129}$$

Therefore, the real part of the dielectric function $\varepsilon(\omega) = 1 + 4\pi\chi_E(\omega)$ is given as

$$\varepsilon(\omega) = 1 - \frac{\omega_P^2}{\omega^2} \tag{17.130}$$

where the plasma frequency ω_p is defined as

$$\omega_P = \left(\frac{4\pi n_e e^2}{m_e}\right)^{1/2} \tag{17.131}$$

The dielectric function due to Drude, given by Eq. (17.130), is shown in Fig. 17.9. The refractive index $n(\omega)$ for a material with $\mu = 1$ is given by

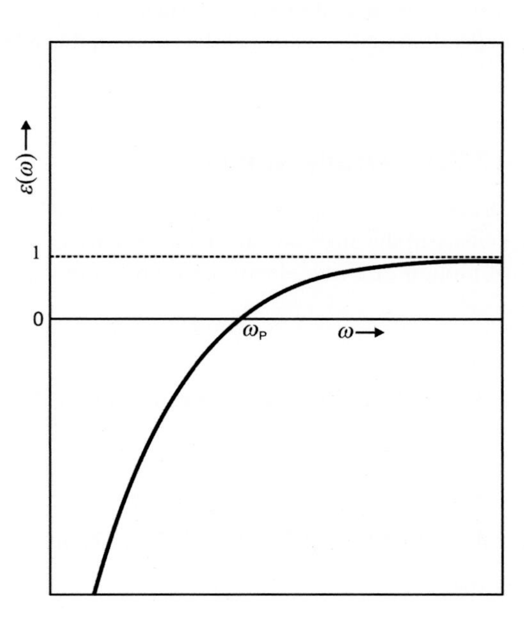

FIG. 17.9 The dielectric function $\varepsilon(\omega)$ as a function of frequency ω for a free-electron gas in the Drude model. The crossover from negative to positive values occurs at the plasma frequency ω_P.

$$n(\omega)=\sqrt{\varepsilon}=\left(1-\frac{\omega_P^2}{\omega^2}\right)^{1/2} \tag{17.132}$$

The behavior of the propagation vector can be studied in a free-electron gas. The equation of motion for the wave is given by Eq. (17.11), which can be written for a nonmagnetic system ($\mu = 1$) as

$$\nabla^2\mathbf{E}-\frac{\varepsilon}{c^2}\frac{\partial^2\mathbf{E}}{\partial t^2}=0 \tag{17.133}$$

We are looking for plane wave solutions of the type

$$\mathbf{E}=\mathbf{E}_0\,e^{\imath(\mathbf{K}\cdot\mathbf{r}-\omega t)} \tag{17.134}$$

Substituting Eq. (17.134) into Eq. (17.133), one gets the dispersion relation

$$\omega^2\varepsilon(\omega) = K^2c^2 \tag{17.135}$$

Substituting the value of $\varepsilon(\omega)$ from Eq. (17.130) into Eq. (17.135), the dispersion for $\varepsilon(\omega)$ becomes

$$\omega^2-\omega_P^2=K^2c^2 \tag{17.136}$$

The dispersion relation defined by Eq. (17.136) is shown in Fig. 17.10 and it describes the transverse electromagnetic waves in a plasma.

The physical meaning of $n(\omega)$ given by Eq. (17.132) is as follows. From Eq. (17.132) it is evident that the behavior of a metal depends on the frequency of the applied field. If $\omega^2\rangle\ \omega_P^2$, the propagation vector is real and the refractive index is real and positive but less than unity. In this case the reflectivity is given by Eq. (17.100) and the metal, therefore, is transparent to normally incident light, but there exists a critical angle of incidence above which total reflection takes place at the surface of the metal. This happen at large frequencies or short wavelengths. On the other hand, if $\omega^2<\omega_P^2$ (large wavelengths), both the propagation vector and refractive index become imaginary so the value of reflectivity R becomes unity and total reflection takes place at all angles of incidence. The reflectivity R as a function of wavelength λ/λ_P is shown in Fig. 17.11A. The plasma wavelength λ_P acts as a dividing line between propagation though the medium and total reflection at the surface of the medium (see Fig. 17.11B). At ω_P the medium becomes transparent to light of wavelength $\lambda_P=2\pi c/\omega_P$. For example, in alkali metals, the electron density is 10^{23} per c.c., which yields an ω_P on the order of 1016 rad/s and a λ_P on the order of 0.1 μm. For this reason, alkali metal reflects visible light but is transparent to ultraviolet light.

Let us examine how Eq. (17.130) is modified when the electromagnetic wave propagates in a metal. In the Jellium model of metal, there is a sea of free electrons in a uniform positive background. The uniform positive background gives a constant contribution to the dielectric constant, denoted as $\varepsilon(\infty)$, up to frequencies well above ω_P. Now the dielectric constant from Eq. (17.130) can be written as

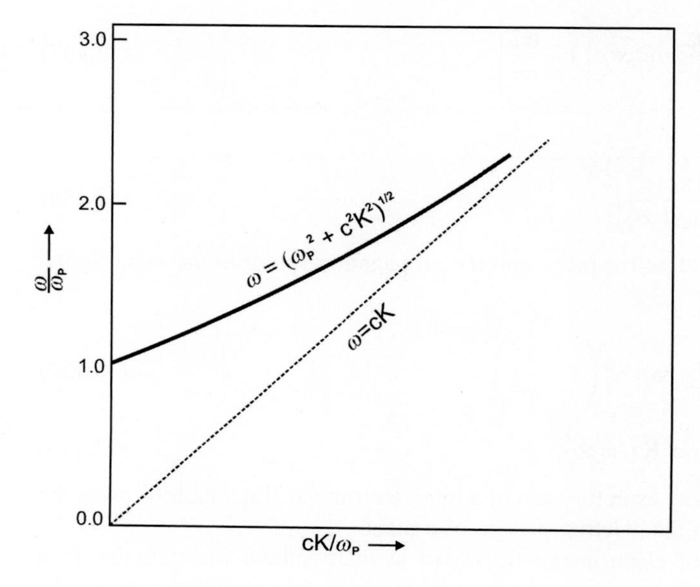

FIG. 17.10 A plot of ω/ω_P as a function of cK/ω_P (dispersion relation) corresponding to the equation $\omega^2=\omega_P^2+c^2K^2$.

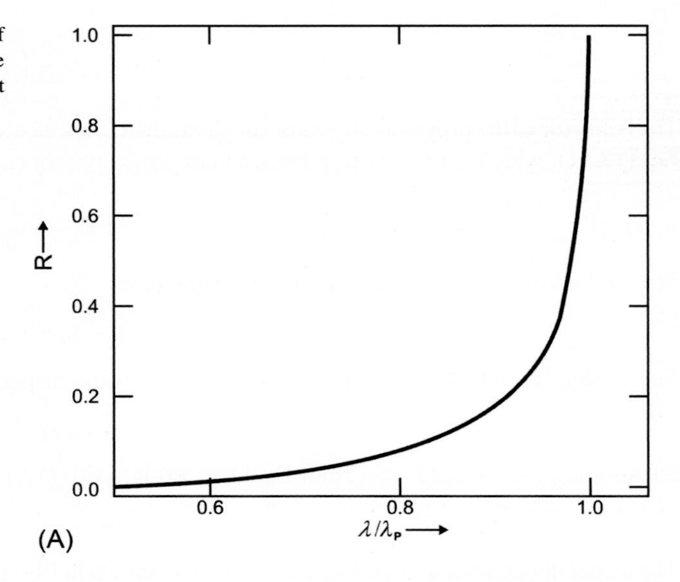

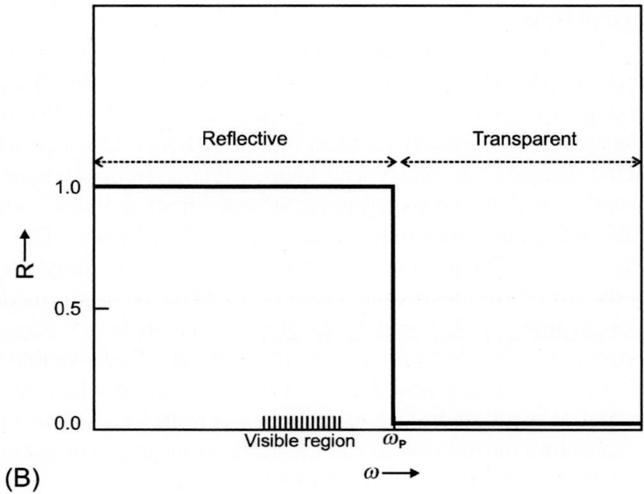

FIG. 17.11 (A) The reflectance R of a free-electron gas as a function of wavelength λ. (B) Schematic representation of the frequency dependence of reflectivity R for an alkali metal in the free-electron theory without damping. *(A: From Zener, C. (1933).* Nature, *23, 968.)*

$$\widetilde{\varepsilon}(\omega)=\varepsilon(\infty)-\frac{\omega_P^2}{\omega^2}=\varepsilon(\infty)\left[1-\frac{\widetilde{\omega}_P^2}{\omega^2}\right] \tag{17.137}$$

where

$$\widetilde{\omega}_P=\left[\frac{4\pi n_e e^2}{m_e\,\varepsilon(\infty)}\right]^{1/2} \tag{17.138}$$

$\widetilde{\omega}_P$ is the normalized plasma frequency. One can find the refractive index and the propagation vector in the metal in the same way as was done above, yielding

$$\widetilde{n}(\omega)=\sqrt{\widetilde{\varepsilon}(\omega)}=[\varepsilon(\infty)]^{1/2}\left(1-\frac{\widetilde{\omega}_P^2}{\omega^2}\right)^{1/2} \tag{17.139}$$

$$\omega^2-\widetilde{\omega}_P^2=c^2K^2/\varepsilon(\infty) \tag{17.140}$$

Eqs. (17.139) and (17.140) can be interpreted in the same way as in the case of a free-electron gas. Eq. (17.140) gives the dispersion relation of a transverse electromagnetic wave in the electron plasma of a metal.

Let us examine the behavior of a metal in the case of electromagnetic waves at intermediate wavelengths. In a metal, the electrons interact with each other and with the ions, giving rise to a finite mean free path and causing the absorption of energy by the metal. The equation of motion of an electron in the presence of a finite mean free path is given by

$$m_e \frac{d^2x}{dt^2} + \frac{m_e}{\tau_e}\frac{dx}{dt} = -eE \tag{17.141}$$

The second term in the above equation is a damping term with $1/\tau_e$ as the damping coefficient. Substituting Eqs. (17.122) and (17.123) into Eq. (17.141), one gets

$$x = \frac{eE}{m_e\omega[\omega + \imath/\tau_e]} \tag{17.142}$$

Further, substituting the value of x from Eq. (17.142) into Eq. (17.127), one can easily obtain

$$\chi_E(\omega) = -\frac{n_e e^2}{m_e\omega[\omega + \imath/\tau_e]} \tag{17.143}$$

Hence the dielectric function becomes

$$\varepsilon(\omega) = 1 - \frac{\omega_P^2}{\omega(\omega + \imath/\tau_e)} \tag{17.144}$$

It is evident that in the limit $\tau_e \rightarrow \infty$, Eq. (17.144) reduces to Eq. (17.130) given by Drude. Separating the real and imaginary parts of Eq. (17.144), we write

$$\varepsilon_1(\omega) = 1 - \frac{\omega_P^2}{\omega^2 + 1/\tau_e^2} \tag{17.145}$$

$$\varepsilon_2(\omega) = \frac{\omega_P^2/\omega\tau_e}{\omega^2 + 1/\tau_e^2} \tag{17.146}$$

From Eq. (17.145) the refractive index of the metal is given by

$$n(\omega) = \sqrt{\varepsilon_1(\omega)} = \left(1 - \frac{\omega_P^2}{\omega^2 + 1/\tau_e^2}\right)^{1/2} \tag{17.147}$$

In the limit $\omega\tau_e \gg 1$, the real part $\varepsilon_1(\omega)$ gives the same expression as Eq. (17.130), while the imaginary part $\varepsilon_2(\omega)$ reduces, from Eq. (17.146), to

$$\varepsilon_2(\omega) = \frac{\omega_P^2}{\omega^3\tau_e} = \frac{4\pi\sigma}{\omega} \tag{17.148}$$

Here we have used Eq. (17.85). From the above equation the magnitude of the photoconductivity is given by

$$\sigma(\omega) = \frac{n_e e^2}{m_e\omega^2\tau_e} = \frac{\omega\varepsilon_2(\omega)}{4\pi}$$

which is the same equation as given by Eq. (17.126). The function $\sigma(\omega)$ decreases with an increase in ω, as shown in Fig. 17.12. Another quantity of interest is $-\text{Im}\,\varepsilon^{-1}$, which is proportional to the absorption spectrum, and is given from Eq. (17.119) as

$$-\text{Im}\,\varepsilon^{-1} = \frac{\varepsilon_2}{\varepsilon_1^2 + \varepsilon_2^2} \tag{17.149}$$

The function $-\text{Im}\,\varepsilon^{-1}$, plotted in Fig. 17.12, is peaked about the plasma frequency. The effect of introducing a finite mean free path or damping coefficient $1/\tau_e$ is that the electrons vibrate freely for a finite time only and then give up their energy to the lattice. For $\omega_P^2/\omega^2 > 1$ the reflecting power becomes slightly less than unity because some energy is absorbed in the surface. On the other hand, for $\omega_P^2/\omega^2 < 1$, the wave in the metal is damped so that the metal is opaque except for relatively thin films. The behavior of the metal in the two cases is shown in Fig. 17.13.

To express Eqs. (17.145) and (17.146) in terms of wavelength, we define the following quantities

$$\omega = 2\pi\frac{c}{\lambda}, \frac{1}{\tau_e} = \frac{2\pi c}{\lambda_T}, \lambda_0 = \left(\frac{\pi m_e c^2}{n_e e^2}\right)^{1/2} \tag{17.150}$$

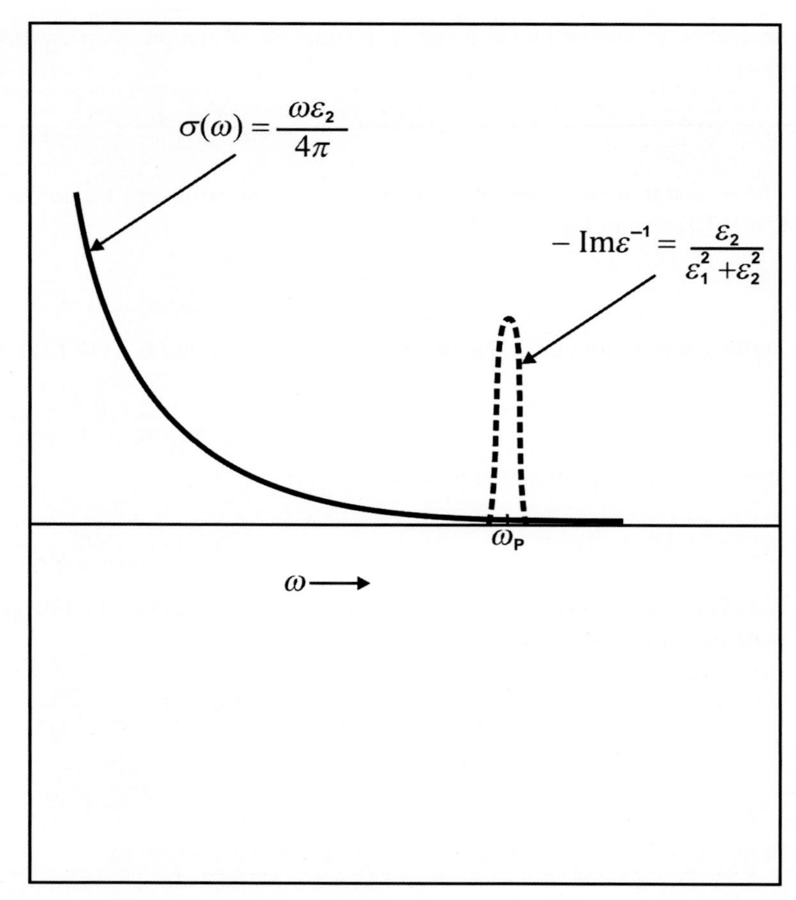

FIG. 17.12 The optical conductivity $\sigma(\omega)$ and the negative of the imaginary part of the inverse of the dielectric function $-\mathrm{Im}\ \varepsilon^{-1}(\omega)$, are plotted as a function of frequency ω.

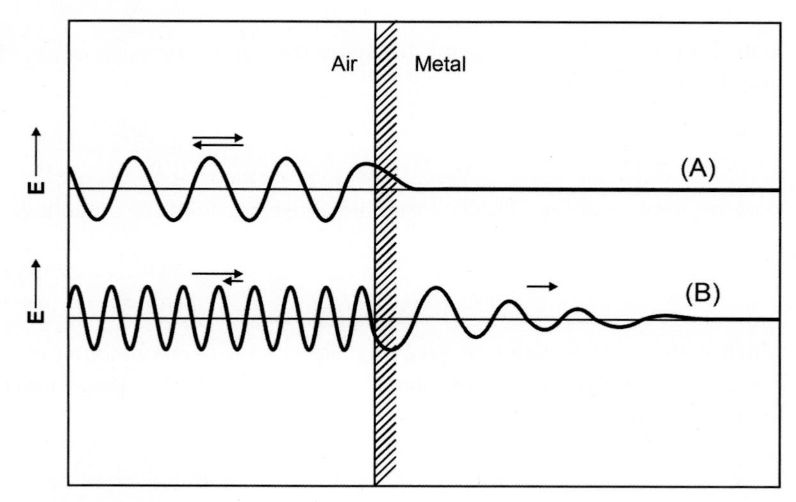

FIG. 17.13 The electric vector **E** of an electromagnetic wave incident on a metal for (A) $\omega_P^2/\omega^2 > 1$ and (B) $\omega_P^2/\omega^2 < 1$.

Using Eq. (17.150) in Eqs. (17.145) and (17.146) one can immediately write

$$\varepsilon_1(\lambda) = 1 - \left(\frac{\lambda}{\lambda_0}\right)^2 \frac{1}{1 + (\lambda/\lambda_T)^2} \tag{17.151}$$

$$\varepsilon_2(\lambda) = \left(\frac{\lambda}{\lambda_0}\right)^2 \frac{\lambda}{\lambda_T} \frac{1}{1 + (\lambda/\lambda_T)^2} \tag{17.152}$$

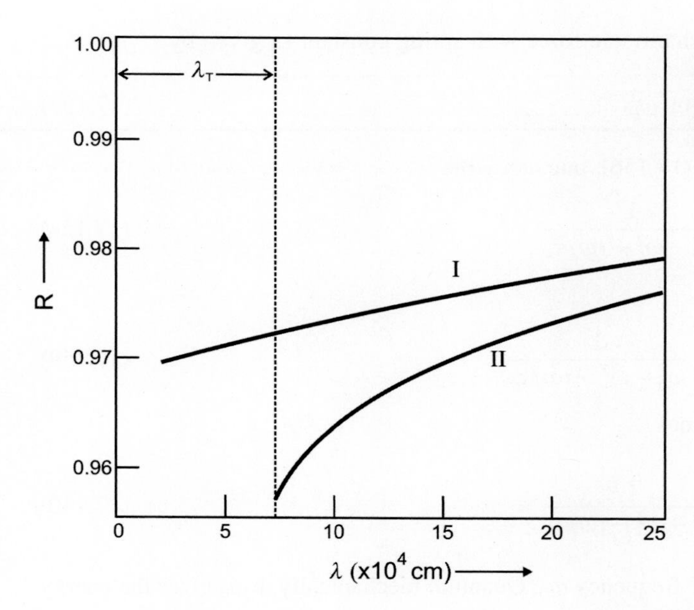

FIG. 17.14 Reflectivity R for Pt metal as a function of wavelength λ in the infrared region. Curve I gives the results obtained from Eqs. (17.151) and (17.152), while Curve II gives the results obtained from the Hagen-Rubens relation given by Eq. (17.112). Here σ_0 has been taken as the observed conductivity at room temperature.

Fig. 17.14 shows the value of reflectivity R, calculated from Eqs. (17.151) and (17.152) in the transition region where $1/\omega \approx \tau_e$, and also from the Hagen-Rubens relation defined by Eq. (17.112). In the limiting case of $1/\omega \ll \tau_e$, i.e., $\lambda/\lambda_T \ll 1$, Eqs. (17.151) reduces to

$$\varepsilon_1(\lambda) = 1 - \left(\frac{\lambda}{\lambda_0}\right)^2 \tag{17.153}$$

which is actually the same as Eq. (17.130) due to Drude.

Problem 17.4

Consider a free-electron metal in which there is a damping of the electron motion due to collisions with the atoms. The damping term is assumed to be $\gamma\,dx/dt$, which is proportional to the electron velocity. The vibration equation is given by

$$m_e \frac{d^2x}{dt^2} + \gamma \frac{dx}{dt} = e\mathbf{E} = eE_0 e^{i\omega t} \tag{17.154}$$

Prove that the damping factor is inversely proportional to the conductivity of the metal σ_0 and is given as

$$\gamma = \frac{n_e e^2}{\sigma_0} \tag{17.155}$$

Here n_e is the electron density.

17.7.2 Lorentz Model for Insulators

Lorentz provided a model to explain the optical properties of nonferromagnetic insulators. He assumed that the major contribution to polarization in an insulator is due to the bound electrons in an atom oscillating with natural frequency ω_0. According to Lorentz, the equation of motion of a bound atomic electron in an insulator is given by

$$m_e \frac{d^2x}{dt^2} + \frac{m_e}{\tau_e}\frac{dx}{dt} + \kappa_s x = -eE \tag{17.156}$$

The third term on the left side of Eq. (17.156) represents a harmonic force with spring constant κ_s given by

$$\kappa_s = m_e \omega_0^2 \tag{17.157}$$

Substituting Eqs. (17.122), (17.123), and (17.157) into Eq. (17.156), one can write

$$x = -\frac{eE}{m_e} \frac{1}{\omega_0^2 - \omega^2 - \imath\omega/\tau_e} \tag{17.158}$$

The dipole moment induced on the atom is given by

$$p = -ex = \frac{e^2 E}{m_e} \frac{1}{\omega_0^2 - \omega^2 - \imath\omega/\tau_e} \tag{17.159}$$

From the above equation the electronic polarizability becomes

$$\alpha_E^a = \frac{e^2}{m_e} \frac{1}{\omega_0^2 - \omega^2 - \imath\omega/\tau_e} \tag{17.160}$$

Eqs. (17.159) and (17.160) exhibit a resonance at the natural frequency ω_0. Quantum mechanically, $\hbar\omega_0$ gives the energy difference before and after the transition.

Consider the material for which the permeability is unity, that is, $\mu = 1$. The dielectric constant in an insulator is given by the expression (see Eq. 16.29)

$$\varepsilon_e(\omega) = \frac{1 + \frac{8\pi}{3}\rho^a \alpha_E^a}{1 - \frac{4\pi}{3}\rho^a \alpha_E^a} \tag{17.161}$$

where ρ^a is the atomic density. Substituting the value of α_E^a from Eq. (17.160) into Eq. (17.161) and separating the real and imaginary parts of the dielectric function, one gets

$$\varepsilon_1(\omega) = \frac{\left(\omega_0^2 - \omega^2 + \frac{2}{3}\omega_P^2\right)\left(\omega_0^2 - \omega^2 - \frac{1}{3}\omega_P^2\right) + \left(\frac{\omega}{\tau_e}\right)^2}{\left(\omega_0^2 - \omega^2 - \frac{1}{3}\omega_P^2\right)^2 + \left(\frac{\omega}{\tau_e}\right)^2} \tag{17.162}$$

$$\varepsilon_2(\omega) = \frac{\omega_P^2\left(\frac{\omega}{\tau_e}\right)}{\left(\omega_0^2 - \omega^2 - \frac{1}{3}\omega_P^2\right)^2 + \left(\frac{\omega}{\tau_e}\right)^2} \tag{17.163}$$

The refractive index can be evaluated from Eq. (17.162) with the help of Eq. (17.87). In covalent Si crystals, the dielectric function $\varepsilon(\omega)$ simply equals the static dielectric function $\varepsilon_0 = \varepsilon_1(0)$ below the frequency of visible light. So, at $\omega = 0$, Eq. (17.162) reduces to

$$\varepsilon_0 = \frac{\omega_0^2 + \frac{2}{3}\omega_P^2}{\omega_0^2 - \frac{1}{3}\omega_P^2} \tag{17.164}$$

In Si, $\omega_0 = 7.2$ and $\omega_P = 11$, which gives $\omega_0/\omega_P = 0.654$. With these values one obtains $\varepsilon_0 = 11$. Fig. 17.15 shows the calculated values of $\varepsilon_1(\omega)$ and $\varepsilon_2(\omega)$ versus the scaled frequency ω/ω_0 for Si.

The Lorentz model can also be applied to explain the polarization in ionic solids or covalent solids with ionic bonding. In the presence of an electric field, the equation of motion of an ionic solid can be written as

$$M\frac{d^2x}{dt^2} + \frac{M}{\Gamma}\frac{dx}{dt} + \kappa_S x = qE \tag{17.165}$$

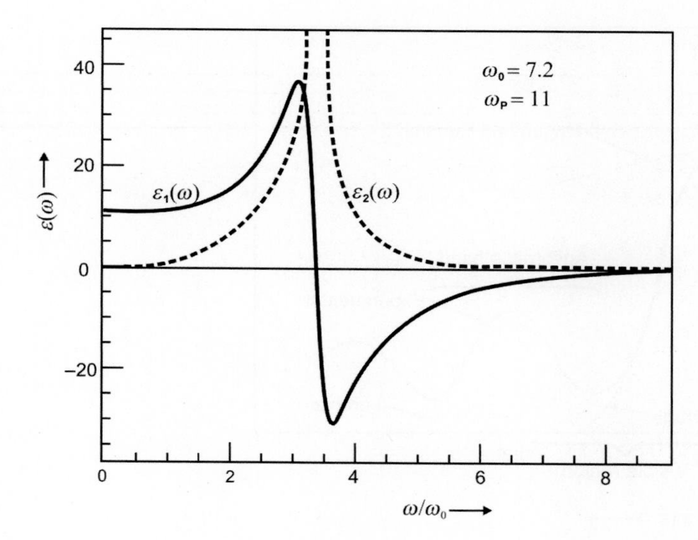

FIG. 17.15 ε_1 and ε_2 are plotted as a function of frequency ω for Si.

where M is the mass of an ion having charge q. Here $1/\Gamma$ is the damping factor and κ_S is the spring constant defined as

$$\kappa_S = M\omega_{0i}^2 \tag{17.166}$$

Here ω_{0i} is the resonance frequency of the ionic polarization. Substituting Eqs. (17.122) and (17.123) into Eq. (17.165), one obtains the displacement as

$$x(\omega) = \frac{q}{M} \frac{1}{\omega_{0i}^2 - \omega^2 - \imath\omega/\Gamma} E \tag{17.167}$$

Hence the ionic polarizability is given by

$$\alpha_I^a(\omega) = \frac{q\,x(\omega)}{E} = \frac{q^2}{M} \frac{1}{\omega_{0i}^2 - \omega^2 - \imath\omega/\Gamma} \tag{17.168}$$

It is known that ω_{0i} is the frequency of the optical branches $\omega_j(0)$ (see Chapter 6), i.e., $\omega_{0i} = \omega_j(0)$. In NaCl, the real part of the dielectric constant $\varepsilon_1(\omega)$ ranges from about 2.8 to 3.0 in the microwave frequency range of 0.4–4 GHz but the static dielectric constant is 5.895. The reduction in $\varepsilon_1(\omega)$ at radio frequency (rf) must be due to the dipolar polarization of defects. But in the covalent crystals, such as diamond or silicon, there is no dipolar contribution as there are very few defects in these crystals. As a result, the refractive index of a covalent crystal is almost constant below infrared frequency.

It is interesting to compare the validity of different models as regards the frequency. Fig. 17.16 shows the frequency dependence of the reflectivity R obtained from different models for metals compared with the experimental results. The theoretical results obtained from the Hagen-Rubens relation (Eq. 17.112) exhibit good agreement with the experimental results in the infrared (IR) region (up to frequency of $10^{13}\ s^{-1}$ and below). It is noteworthy that electromagnetic waves of such low frequencies (long wavelengths) are not able to see the atomic structure of the solid. Therefore, the Hagen-Rubens relation is for a continuous material and is not able to explain the values of R at higher frequencies. With an increase in frequency, the electromagnetic waves start interacting with the atoms of the solid and, therefore, one needs an atomistic model to explain the experimental results for R. The first atomistic model was given by Drude for a free-electron gas, which was then improved by including collisions between the electrons, which cause damping of the electron motion. Drude's model explains the reflectivity in the near-infrared and visible regions (see Fig. 17.16). At still higher frequencies the experimental value of R increases and then decreases, as shown in the figure, giving rise to an absorption band. Lorentz assumed that the electron is bound to an atom and that it oscillates harmonically about its position. The Lorentz model explains the peak in reflectivity R in metals in the ultraviolet (UV) region. Fig. 17.17 shows the results of reflectivity R as a function of frequency in dielectrics in which there are no free electrons. The experimental results exhibit two peaks in the ultraviolet region, which are well explained by the Lorentz model.

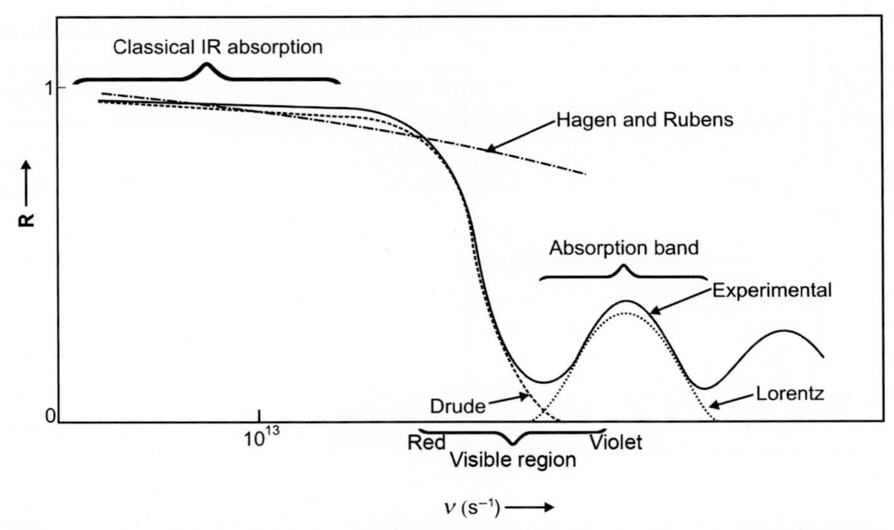

FIG. 17.16 The frequency dependence of reflectivity R for a metal. The *solid line* shows the experimental results, while the *dashed* and *dotted lines* show the theoretical results obtained from the Drude and Lorentz models. The figure also depicts the results obtained from the Hagen-Rubens equation *(dash-dot line)*. *(Modified from Hummel, R. E. (2001) Electronic properties of materials, Springer-Verlag, Berlin.)*

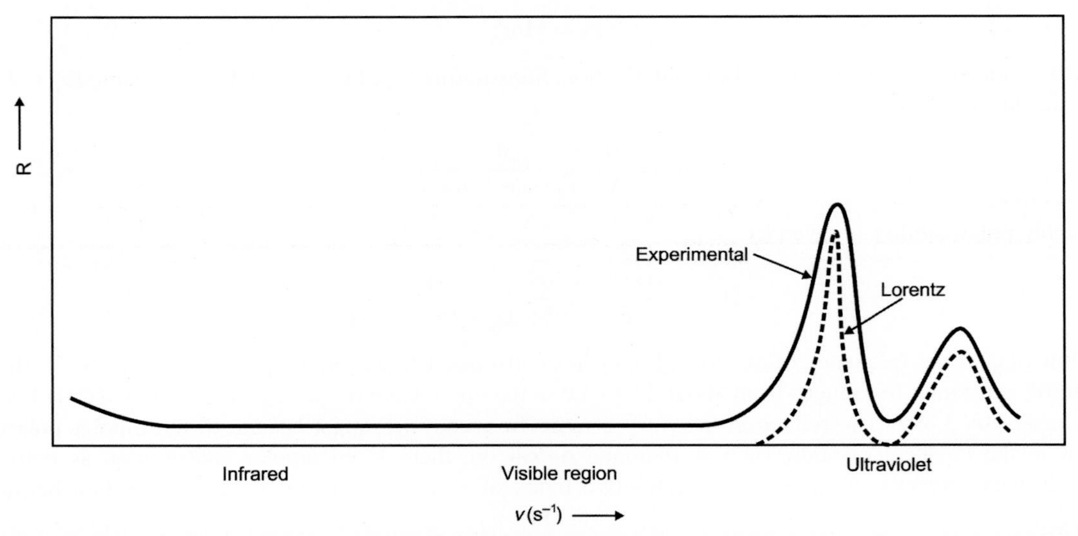

FIG. 17.17 The frequency dependence of reflectivity R for a dielectric solid. The *solid line* shows the experimental results, while the *dashed line* shows the theoretical results obtained from the Lorentz model. *(Modified from Hummel, R. E. (2001) Electronic properties of materials, Springer-Verlag, Berlin.)*

17.8 LYDDANE-SACHS-TELLER RELATION

In Chapter 6 it was explained that the amplitudes of vibration of the two types of ions in an ionic solid are affected by the presence of electromagnetic radiation, particularly IR radiation, that produces a finite dipole moment on the ionic molecule. The polarization and dielectric function of the ionic solid depend on the intensity of the IR radiation. Therefore, it is of interest to study the optical properties of ionic solids in the IR region.

The dipole moment on a molecule of an ionic solid is given by $e(u_2 - u_1)$ where u_1 and u_2 are the amplitudes of vibration of the two types of ions given by Eqs. (6.86) and (6.88). If there are n_i molecules per unit volume, i.e., n_i positive and n_i negative ions, then the ionic polarization is given by

$$P_i = n_i e\,(u_2 - u_1) \tag{17.169}$$

Substituting the values of u_1 and u_2 from Eqs. (6.86) and (6.88) into the above equation, one can write

$$P_i = n_i e^2 E_0 \left[\frac{-\dfrac{\omega^2}{\mu_M} + \dfrac{8\alpha_F}{M_1 M_2}\sin^2(Ka/2)}{\omega^2(\omega^2-\omega_0^2) + \dfrac{4\alpha_F^2}{M_1 M_2}\sin^2 Ka} \right] \tag{17.170}$$

where μ_M is the reduced mass of the ionic molecule and is defined as

$$\frac{1}{\mu_M} = \frac{1}{M_1} + \frac{1}{M_2} \tag{17.171}$$

In the IR region the wave vector K is small, so in the limit of $Ka \ll 1$, Eq. (17.170) reduces to

$$P_i = \frac{n_i e^2}{\mu_M(\omega_0^2 - \omega^2)} E_0 \tag{17.172}$$

We know that the dielectric function $\varepsilon(\omega)$ is given by

$$\varepsilon(\omega) - 1 = 4\pi \frac{P}{E_0} \tag{17.173}$$

In ionic solids the polarization comprises ionic and electronic contributions because the dipolar contribution is negligible. Therefore, for ionic solids the above equation becomes

$$\varepsilon(\omega) - 1 = 4\pi \frac{P_i}{E_0} + 4\pi \frac{P_e}{E_0} \tag{17.174}$$

where P_i and P_e are the ionic and electronic contributions to the polarization. At low frequencies both P_i and P_e are finite, but at high frequencies P_i is zero due to the large ion mass. As $\omega \to \infty$ the above equation can be written as

$$\varepsilon(\infty) - 1 = 4\pi \frac{P_e}{E_0} \tag{17.175}$$

Substituting Eq. (17.175) into Eq. (17.174), one gets

$$\varepsilon(\omega) = \varepsilon(\infty) + 4\pi \frac{P_i}{E_0} \tag{17.176}$$

From Eqs. (17.172) and (17.176) one can write

$$\varepsilon(\omega) = \varepsilon(\infty) - \frac{\omega_{Pi}^2}{\omega^2 - \omega_0^2} \tag{17.177}$$

where

$$\omega_{Pi} = \left(\frac{4\pi n_i e^2}{\mu_M} \right)^{1/2} \tag{17.178}$$

From Eq. (17.177) the dielectric function is positive if

$$\frac{\omega_{Pi}^2}{\omega^2 - \omega_0^2} < \varepsilon(\infty) \tag{17.179}$$

If the above condition is satisfied, then $\varepsilon(\omega)$ and hence the refractive index $n(\omega)$ is positive and the electromagnetic wave passes through the ionic solid. On the other hand, if

$$\frac{\omega_{Pi}^2}{\omega^2 - \omega_0^2} > \varepsilon(\infty) \tag{17.180}$$

then $\varepsilon(\omega)$ is negative and $n(\omega)$ is imaginary and, therefore, the electromagnetic wave is reflected from the solid. The above equation can be written as

$$\omega^2 \left\langle \frac{\omega_{Pi}^2}{\varepsilon(\infty)} + \omega_0^2 \right. \tag{17.181}$$

In writing the above condition we have assumed the second term of Eq. (17.177) to be positive. Therefore, n(ω) is imaginary if the frequency satisfies the following condition

$$\omega_0^2 \langle \omega^2 \langle \frac{\omega_{Pi}^2}{\varepsilon(\infty)} + \omega_0^2 \tag{17.182}$$

Hence the electromagnetic waves cannot propagate through the ionic solid in the range of frequencies defined by Eq. (17.182). In an ionic solid the frequencies of transverse and longitudinal waves are given by

$$\omega_T^2 = \omega_0^2, \omega_L^2 = \omega_0^2 + \frac{\omega_{Pi}^2}{\varepsilon(\infty)} \tag{17.183}$$

Using Eq. (17.183), Eq. (17.182) becomes

$$\omega_T^2 < \omega^2 < \omega_L^2 \tag{17.184}$$

From Eq. (17.177) the static dielectric function $\varepsilon(0)$ is given by

$$\varepsilon(0) = \varepsilon(\infty) + \frac{\omega_{Pi}^2}{\omega_0^2} \tag{17.185}$$

Dividing Eq. (17.185) by $\varepsilon(\infty)$ and rearranging the terms, one gets

$$\frac{\varepsilon(0)}{\varepsilon(\infty)} = \frac{\omega_L^2}{\omega_T^2} \tag{17.186}$$

Eq. (17.186) is known as the *Lyddane-Sachs-Teller relation*. The upper limit for the forbidden band of frequencies (reflected waves) falls at ω_L. Therefore, the dielectric function at ω_L should go to zero (the upper limit for the negative values of ε), i.e.,

$$\varepsilon(\omega_L) = 0. \tag{17.187}$$

Eq. (17.187) can be taken as the definition for the longitudinal optical phonon frequency for small K values.

Problem 17.5

Using Eqs. (17.177) and (17.185) prove that

$$\varepsilon(\omega) - \varepsilon(\infty) = [\varepsilon(0) - \varepsilon(\infty)] \frac{\omega_0^2}{\omega_0^2 - \omega^2} \tag{17.188}$$

Problem 17.6

From Eqs. (17.177), (17.185), and (17.186) prove that

$$\frac{\varepsilon(\omega)}{\varepsilon(\infty)} = \frac{\omega_L^2 - \omega^2}{\omega_T^2 - \omega^2} \tag{17.189}$$

Problem 17.7

The equation of motion of an electron in the presence of finite mean free path is given by (see Eq. 17.141)

$$m_e \frac{d^2x}{dt^2} + \frac{m_e}{\tau_e} \frac{dx}{dt} = -eE$$

Prove that the above equation in terms of current density becomes

$$\frac{dJ}{dt} = \frac{n_e e^2}{m_e} \left(E - \frac{J}{\sigma_0} \right) \tag{17.190}$$

SUGGESTED READING

Abeles, F. (Ed.). (1972). *Optical properties of solids*. Amsterdam: North-Holland Publ. Co..

Givens, M. P. (1958). Optical properties of metals. F. Seitz, & D. Turnbull (Eds.), *Solid state physics* (pp. 313–352). Vol. 6(pp. 313–352). New York: Academic Press.

Hummel, R. E. (2001). *Electronic properties of materials*. Berlin: Springer-Verlag.

Jackson, J. D. (1999). *Classical electrodynamics* (3rd ed.). New York: J. Wiley & Sons.

Nilsson, P. O. (1974). Optical properties of metals and alloys. H. Ehrenreich, F. Seitz, & D. Turnbull (Eds.), *Solid state physics* (pp. 139–234). Vol. 29 (pp. 139–234). New York: Academic Press.

Phillips, J. C. (1966). The fundamentals of optical spectra of solids. F. Seitz, & D. Turnbull (Eds.), *Solid state physics* (pp. 55–164). Vol. 18(pp. 55–164). New York: Academic Press.

Stern, F. (1963). Elementary theory of optical properties in solids. F. Seitz, & D. Turnbull (Eds.), *Solid state physics* (pp. 299–408). Vol. 15(pp. 299–408). New York: Academic Press.

SUGGESTED READINGS

[illegible]

[illegible]

[illegible]

[illegible]

[illegible]

[illegible]

Chapter 18

Magnetism

Chapter Outline

Magnetism was discovered in Magnesia, a place in Greece, around 800 BCE and that is the origin of its name. The writing of Thales, a Greek writer, shows that magnetite or loadstone was known to attract iron pieces. The Chinese made a magnetic compass sometime around 200 BCE. Today, we can observe that most of the elements in the periodic table exhibit magnetism of varying strength. The type of magnetization that occurs when an external magnetic field is applied to an element varies:

1. In some elements of the periodic table, magnetization is induced in a direction opposite to the applied magnetic field. The induced magnetization lasts only for the time the applied magnetic field exists. Such elements are called diamagnetic elements and are repelled by the magnetic field.
2. In many elements, weak magnetization is produced in the direction of the applied magnetic field. Moreover, the magnetization lasts so long as the applied field is finite. Such elements are called paramagnetic elements and are weakly attracted by the magnetic field.
3. In some elements, remarkably strong magnetization is produced in the direction of the applied magnetic field. Further, the magnetization exists even in the absence of the applied field. Such elements are called ferromagnetic elements and are strongly attracted by the magnetic field.

In addition, there exist antiferromagnetic and ferrimagnetic elements, which will be discussed in reasonable detail in the coming chapters. The atomic magnetic dipole moment, induced or intrinsic, is basically responsible for the existence of magnetism in the various elements.

18.1 ATOMIC MAGNETIC DIPOLE MOMENT

In an atom, electrons revolve around the nucleus and the nucleus contains protons and neutrons. An atom as a whole is electrically neutral, but it consists of moving charged particles that may behave as magnetic dipoles. An electron in an atom has two motions: orbital and spin. Similarly, protons and neutrons also possess orbital and spin motions inside the nucleus. Therefore, the magnetic moment of an electron has two principal contributions, which are the orbital and spin magnetic moments. There is also a third contribution to the magnetic moment arising from the spin-orbit interaction. If the spin and orbital motions are assumed to be independent of each other, then the spin-orbit contribution vanishes and the total magnetic moment of the ith electron $\vec{\mu}_{ei}$ is the vector sum of its orbital and spin contributions, i.e.,

$$\vec{\mu}_{ei} = \vec{\mu}_{eil} + \vec{\mu}_{eis} \tag{18.1}$$

Solid State Physics. https://doi.org/10.1016/B978-0-12-817103-5.00018-9

where $\vec{\mu}_{eil}$ and $\vec{\mu}_{eis}$ are the orbital and spin contributions to the magnetic moment of the ith electron. The total electronic contribution to the magnetic moment of an atom $\vec{\mu}_e$, therefore, is the vector sum of the magnetic moments of all the electrons, i.e.,

$$\vec{\mu}_e = \sum_i \vec{\mu}_{ei} \tag{18.2}$$

The protons in a nucleus, being charged particles, possess both orbital and spin magnetic moments, just like electrons. The neutrons, being neutral particles, do not possess an orbital magnetic moment in spite of their orbital motion, but they do possess an intrinsic spin magnetic moment. The total magnetic moment of a nucleus $\vec{\mu}_N$ is the vector sum of the magnetic moments of the neutrons and protons and is given by

$$\mu_N = \sum_j \mu_{pj} + \sum_k \mu_{nk} \tag{18.3}$$

where μ_{pj} and $\vec{\mu}_{nk}$ are the total magnetic moment of the jth proton and kth neutron. From Eqs. (18.2) and (18.3) the magnetic moment of an atom is given by

$$\vec{\mu} = \vec{\mu}_e + \vec{\mu}_N \tag{18.4}$$

We shall see later that the magnetic moment of a nucleus is negligible compared with the electronic contribution (about 2000 times smaller); therefore, the magnetic moment of an atom is determined mainly by the electrons. In the coming discussion the magnetic moment of an atom $\vec{\mu}$ is assumed to include only the electronic contribution.

18.1.1 Orbital Magnetic Moment

Consider an atom in which an electron is moving in an elliptical orbit with a nucleus at one of its foci, say O (Fig. 18.1). Let T be the time period of revolution of the electron around the nucleus. The revolving electron constitutes an electric current I_L given by

$$I_L = -\frac{e}{T} \tag{18.5}$$

The total area of the elliptical orbit swept by the electron in time T is given by

$$A = \frac{1}{2}\int_0^{2\pi} r^2 d\varphi \tag{18.6}$$

where φ is the angle formed by the major axis of the ellipse with the radius vector r (from the focus) of the electron at any time t. From elementary electricity, the orbital magnetic moment arising from the current I_L is given by

$$\mu_L = \frac{I_L A}{c} \tag{18.7}$$

where c is the velocity of light. The angular momentum of the electron is given by

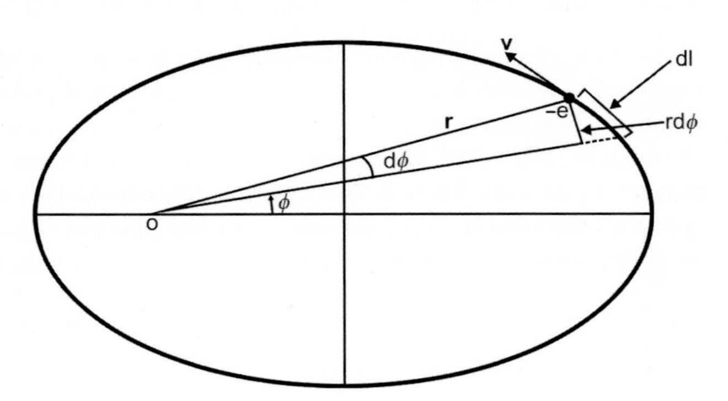

FIG. 18.1 Motion of an atomic electron in an elliptical orbit with a nucleus at one of its foci O. The electron with position vector **r** is moving with velocity **v** in the orbit.

$$p_\varphi = m_e r^2 \omega_\varphi = m_e r^2 \frac{d\varphi}{dt} \tag{18.8}$$

where m_e is the mass and ω_φ is the angular velocity of the electron. Substituting the value of r^2 from Eq. (18.8) into Eq. (18.6), we write

$$A = \frac{1}{2}\int_0^{2\pi} \frac{p_\varphi}{m_e} \frac{1}{{d\varphi}/{dt}} d\varphi = \frac{1}{2}\frac{p_\varphi}{m_e}\int_0^T dt = \frac{1}{2}\frac{p_\varphi T}{m_e} \tag{18.9}$$

Substituting Eqs. (18.5) and (18.9) into Eq. (18.7), we get

$$\mu_L = -\frac{e}{2m_e c} p_\varphi \tag{18.10}$$

From Bohr's quantization rule for orbits, the angular momentum p_φ can be written as

$$p_\varphi = \hbar L \tag{18.11}$$

Here L is called the orbital quantum number and has integral values 1, 2, 3, ... Sometimes L is also called the orbital angular momentum in units of $\hbar = h/2\pi$ where h is the Planck constant. From Eqs. (18.10) and (18.11) one can write

$$\mu_L = -\mu_B L \tag{18.12}$$

where μ_B is called the Bohr magnetron defined as

$$\mu_B = \frac{e\hbar}{2m_e c} \tag{18.13}$$

In vector notation Eq. (18.12) can be written as

$$\vec{\mu}_L = -\mu_B \mathbf{L} \tag{18.14}$$

The negative sign indicates that the orbital magnetic moment is in a direction opposite to the orbital angular momentum and is basically due to the negative charge of the electron. The above expression is valid only for orbital motion. An alternate method for calculating μ_L for an electron moving in a circular orbit is given in Appendix L.

18.1.2 Spin Magnetic Moment

The orbital theory does not explain the multiplicity of atomic spectra, e.g., the doublet of d-states. In addition, it also does not explain the Zeeman levels in some of the elements. These difficulties were resolved by assuming that an electron possesses intrinsic spin angular momentum **S**, which has eigenvalues $\pm(1/2)$ in units of $\hbar$. Note that spin is purely a relativistic property of an electron and arises from quantum effects. The magnetic moment arising from the spin angular momentum is given by

$$\vec{\mu}_S = -2\mu_B \mathbf{S} \tag{18.15}$$

From Eq. (18.15) the value of the spin magnetic moment is numerically equal to the Bohr magnetron. Hence the total magnetic moment of an electron becomes

$$\begin{aligned} \vec{\mu}_J &= \vec{\mu}_L + \vec{\mu}_S \\ &= -\mu_B(\mathbf{J}+\mathbf{S}) \end{aligned} \tag{18.16}$$

where the total angular momentum **J** of an electron is given by

$$\mathbf{J} = \mathbf{L} + \mathbf{S} \tag{18.17}$$

The vector **S** is spinning around the direction of **J** (see Fig. 18.2). So, the average value of the magnetic moment $\vec{\mu}_J$ is obtained by substituting the average value of **S** along the direction of **J**, that is, $\langle \mathbf{S} \rangle$ in Eq. (18.16), allowing us to write

$$\vec{\mu}_J = -\mu_B[\mathbf{J} + \langle \mathbf{S} \rangle] \tag{18.18}$$

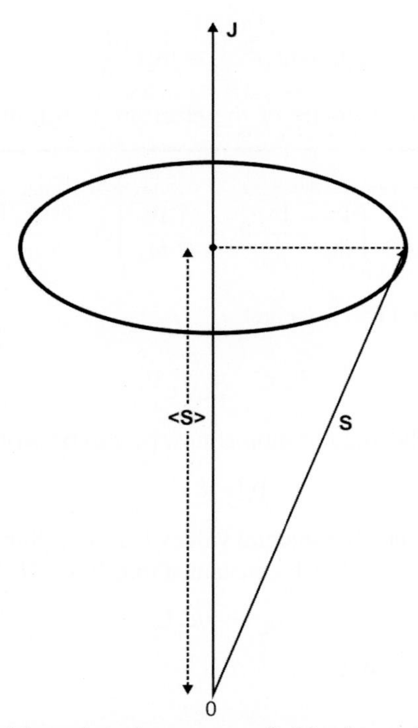

FIG. 18.2 The spinning of an electron spin **S** around the total angular momentum **J** of the electron. The vector $\langle \mathbf{S} \rangle$ gives the average value of spin **S** along the **J** vector.

Here $\mathbf{J}+\langle \mathbf{S} \rangle$ gives the diagonal element of $\mathbf{J}+\mathbf{S}$. The average value of $\langle \mathbf{S} \rangle$ is given by

$$\langle \mathbf{S} \rangle = \frac{\mathbf{J}\cdot\mathbf{S}}{|\mathbf{J}|}\hat{\mathbf{J}} = \frac{\mathbf{J}\cdot\mathbf{S}}{|\mathbf{J}|^2}\mathbf{J} \tag{18.19}$$

where $\hat{\mathbf{J}}$ is a unit vector in the direction of **J**. From Eq. (18.17) we write

$$\mathbf{J}-\mathbf{S}=\mathbf{L} \tag{18.20}$$

Squaring both sides, we find

$$\mathbf{J}\cdot\mathbf{S}=\frac{1}{2}\left(J^2+S^2-L^2\right) \tag{18.21}$$

From Eqs. (18.19) and (18.21) the average value of the spin becomes

$$\langle \mathbf{S} \rangle = \frac{J^2+S^2-L^2}{2J^2}\mathbf{J} \tag{18.22}$$

The eigenvalues of L^2, S^2, and J^2 are $L(L+1)$, $S(S+1)$ and $J(J+1)$. Therefore, the average value of spin along the direction of **J** is given by

$$\langle \mathbf{S} \rangle = \frac{J(J+1)+S(S+1)-L(L+1)}{2J(J+1)}\mathbf{J} \tag{18.23}$$

Substituting the value of $\langle \mathbf{S} \rangle$ from Eq. (18.23) into Eq. (18.18), we get the average value of the magnetic moment as

$$\vec{\mu}_J = -g_J \mu_B \mathbf{J} \tag{18.24}$$

where

$$g_J = 1+\frac{J(J+1)+S(S+1)-L(L+1)}{2J(J+1)} \tag{18.25}$$

The factor g_J is usually called Lande's splitting factor. The above expression gives the magnetic moment of an electron due to its total angular momentum **J**. It can be easily proved from Eq. (18.25) that $g_J = 2$ if there is only the spin motion and that it is equal to 1 if there is only the orbital motion. From experiments, the actual value of g_J for electron spin is found to be 2.0023.

18.1.3 Nuclear Magnetic Moment

One can also calculate the magnetic moment of a proton $\vec{\mu}_p$ and neutron $\vec{\mu}_n$ in exactly the same way as for an electron. The expressions for the magnetic moments are

$$\vec{\mu}_p = \mu_{Bp}\mathbf{I}_p \tag{18.26}$$

and

$$\vec{\mu}_n = \mu_{Bn}\mathbf{I}_n \tag{18.27}$$

where

$$\mu_{Bp} = \frac{e\hbar}{2M_p c}, \quad \mu_{Bn} = \frac{e\hbar}{2M_n c} \tag{18.28}$$

$\mathbf{I}_p$ and $\mathbf{I}_n$ are the total angular momenta for the proton and neutron, respectively. $\mathbf{I}_p$ arises from both the orbital and spin motions, while $\mathbf{I}_n$ arises from the spin motion only. From Eq. (18.26) it is evident that the angular momentum and magnetic moment of a proton are in the same direction, in contrast with an electron, and this is because of the positive charge on the proton. In the case of a neutron, the angular momentum and magnetic moment are also in the same direction, although the neutron is a neutral particle. Further, due to the large mass of the proton, the Bohr magnetron of a proton μ_{Bp} is about 2000 times smaller than the Bohr magnetron of an electron μ_B. The same applies to the neutron Bohr magnetron μ_{Bn}. Therefore, the nuclear magnetic moment is very small compared with the electronic magnetic moment in an atom. In other words, the atomic magnetic moment arises mainly from the electron contribution.

18.2 MAGNETIZATION

When a solid is placed in a magnetic field, it gets magnetized. Therefore, one can talk about the strength of magnetism produced inside the solid, which is determined by a physical quantity called magnetization. *Magnetization is defined as the atomic/molecular magnetic moment per unit volume.* For weak magnetic fields, magnetization $\mathbf{M}(\mathbf{r})$ is linearly proportional to the applied magnetic field $\mathbf{H}(\mathbf{r})$. For inhomogeneous and anisotropic solids.

$$\mathbf{M}(\mathbf{r}) = \sum_{\mathbf{r}'} \chi_M(\mathbf{r}, \mathbf{r}')\mathbf{H}(\mathbf{r}') \tag{18.29}$$

$\mathbf{M}(\mathbf{r})$ is the magnetization produced in the $\mathbf{r}$ direction, while the magnetic field $\mathbf{H}(\mathbf{r}')$ is applied in the $\mathbf{r}'$ direction. Here $\chi_M(\mathbf{r}, \mathbf{r}')$ is the proportionality constant and is, in general, a tensor for an inhomogeneous and anisotropic solid. $\chi_M(\mathbf{r}, \mathbf{r}')$ is usually called the magnetic susceptibility tensor. According to the above expression, the magnetic field applied in all possible directions of $\mathbf{r}'$ contributes to magnetization along the $\mathbf{r}$ direction. If the solid is homogeneous and isotropic, then both the magnetic field and magnetization are in the same direction and one can write

$$\mathbf{M}(\mathbf{r}) = \chi_M(\mathbf{r})\mathbf{H}(\mathbf{r}) \tag{18.30}$$

For such solids the magnetic susceptibility $\chi_M(\mathbf{r})$ becomes a scalar quantity. A uniform magnetic field produces a constant magnetization and, therefore, the magnetic susceptibility χ_M becomes a constant. It can easily be shown from the above expression that the magnetic susceptibility is dimensionless.

18.3 MAGNETIC INDUCTION

In the presence of an externally applied magnetic field, a solid is magnetized. Therefore, the magnetic field inside the solid $\mathbf{B}(\mathbf{r})$, usually called the magnetic induction, is different than the applied field and is given by

$$\mathbf{B}(\mathbf{r}) = \mathbf{H}(\mathbf{r}) + 4\pi\mathbf{M}(\mathbf{r}) \tag{18.31}$$

Substituting the value of **M**(**r**) from Eq. (18.29) into Eq. (18.31), one gets

$$\mathbf{B}(\mathbf{r}) = \mathbf{H}(\mathbf{r}) + 4\pi \sum_{\mathbf{r}'} \chi_M(\mathbf{r}, \mathbf{r}')\, \mathbf{H}(\mathbf{r}') \tag{18.32}$$

The above expression can be written as

$$\mathbf{B}(\mathbf{r}) = \sum_{\mathbf{r}'} \mu(\mathbf{r}, \mathbf{r}')\, \mathbf{H}(\mathbf{r}') \tag{18.33}$$

where

$$\mu(\mathbf{r}, \mathbf{r}') = \delta_{\mathbf{r},\mathbf{r}'} + 4\pi \chi_M(\mathbf{r}, \mathbf{r}') \tag{18.34}$$

Here $\mu(\mathbf{r}, \mathbf{r}')$ is called the magnetic permeability tensor of the material. As already discussed, for a homogeneous and isotropic material the magnetic susceptibility is a scalar, therefore, from Eq. (18.34) the magnetic permeability also becomes a scalar and is given as

$$\mu(\mathbf{r}) = 1 + 4\pi \chi_M(\mathbf{r}) \tag{18.35}$$

As the magnetic susceptibility is dimensionless, the magnetic permeability is also dimensionless.

18.4 POTENTIAL ENERGY OF MAGNETIC DIPOLE MOMENT

Consider an electron moving in an elliptical orbit with its magnetic dipole moment always perpendicular to it. Let a uniform magnetic field **H** be applied in the z-direction, as shown in Fig. 18.3. In the presence of **H**, torque will act on the current loop or the magnetic dipole moment, which is given by

$$\vec{\tau} = \vec{\mu} \times \mathbf{H} \tag{18.36}$$

The magnitude of the torque is given by

$$\tau = \mu \mathrm{H} \sin\theta \tag{18.37}$$

Work will be done by the torque on the magnetic moment, which will change the orientation of the dipole moment. The work done will be stored as the potential energy of the magnetic dipole moment. The zero of the potential energy (reference level) may be taken in any direction of the dipole moment. To be consistent with Eq. (18.37) we usually assume potential energy to be zero when $\vec{\mu}$ and **H** are perpendicular to each other. The potential energy of the magnetic dipole moment in the

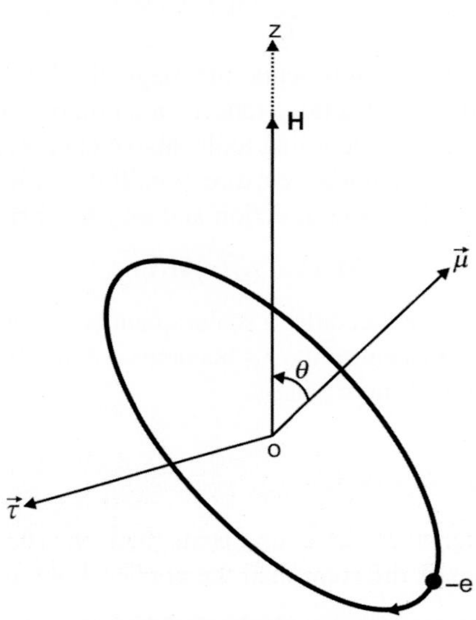

FIG. 18.3 The torque $\vec{\tau}$ acting on the magnetic moment $\vec{\mu}$, arising from a current loop, in the presence of applied magnetic field **H**.

presence of a magnetic field is the work required to rotate the magnetic dipole from the zero energy position ($\theta = 90°$) to an angle θ, i.e.,

$$E = \int_{90}^{\theta} \tau d\theta = \mu H \int_{90}^{\theta} \sin\theta d\theta \tag{18.38}$$

The above integral can easily be solved to get

$$E = -\vec{\mu} \cdot \mathbf{H} \tag{18.39}$$

It should be noted that the choice of the zero energy configuration for E is arbitrary as one is usually interested in the changes in potential energy that occur when a dipole moment is rotated.

18.5 LARMOR PRECESSION

Consider an orbital magnetic moment $\vec{\mu}_L$, associated with an electron, in a uniform magnetic field **H,** as shown in Fig. 18.4. The torque acting on the magnetic moment, from Eqs. (18.14) and (18.36), is given by

$$\vec{\tau}_L = \vec{\mu}_L \times \mathbf{H} = -\mu_B \mathbf{L} \times \mathbf{H} \tag{18.40}$$

So, the magnitude of the torque is given by

$$\tau_L = \mu_B H L \sin\theta \tag{18.41}$$

Depending on the direction of motion, the torque will either accelerate or retard the electron in motion, thereby inducing additional current in the current loop. According to Newton's second law of motion the torque produces a change in the orbital angular momentum **L**, which is at a right angle to itself. Torque can also be defined as the rate of change of angular momentum and is given by

$$\vec{\tau}_L = \frac{d\mathbf{p}_\varphi}{dt} = \hbar \frac{d\mathbf{L}}{dt} \tag{18.42}$$

So, the torque causes **L** to process about the direction of **H** with an angular frequency ω_L. The precession of the orbital angular momentum about the direction of a magnetic field is called the Larmor precession and ω_L is called the Larmor frequency. An alternate simple method for calculating ω_L is presented in Appendix M. From Fig. 18.4, the change in orbital angular momentum **L** in time dt is given by

$$dL = L \sin\theta\, (\omega_L dt)$$

The above equation gives the torque τ_L as

$$\tau_L = \hbar \frac{dL}{dt} = \hbar \omega_L L \sin\theta \tag{18.43}$$

From Eqs. (18.41) and (18.43) one can immediately write

$$\hbar \omega_L = \mu_B H \tag{18.44}$$

From this equation the Larmor precession frequency becomes

$$\omega_L = \frac{eH}{2m_e c} \tag{18.45}$$

Diamagnetism is related to the Larmor precession of the electrons. Diamagnetism is the tendency of electrical charges to partially shield the interior of the solid from the applied magnetic field. The basic principle of diamagnetic behavior can be illustrated with the Lenz law of electricity. Consider an atom with Z electrons revolving around its nucleus in different orbits. When an external magnetic field **H** is applied, the magnetic force acts on every electron. The magnetic force accelerates some of the electrons, while others are retarded depending on the direction of their motion. The change in velocity of

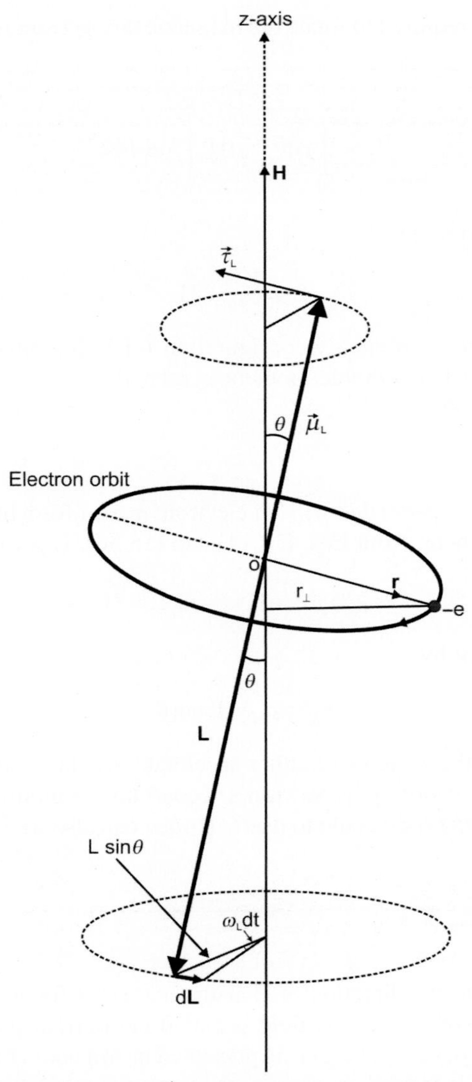

FIG. 18.4 The torque $\vec{\tau}_L$ acting on an orbital magnetic moment $\vec{\mu}_L$ in the presence of an applied magnetic field **H** in the z-direction. The figure also depicts the change in orbital angular momentum **dL** due to the torque.

the electrons gives rise to an induced current that opposes the applied magnetic field (Lenz law). The induced current is responsible for inducing an orbital magnetic moment $\vec{\mu}_L$ on the atom. If T_L is the time period for Larmor precession of the electrons around the magnetic field, the induced current I_L is given by

$$I_L = -\frac{Ze}{T_L} \tag{18.46}$$

But the Larmor frequency is given by

$$\omega_L = \frac{2\pi}{T_L} = \frac{eH}{2m_e c} \tag{18.47}$$

Substituting the value of T_L from Eq. (18.47) into Eq. (18.46), we find

$$I_L = -\frac{Ze^2 H}{4\pi m_e c} \tag{18.48}$$

Let $\langle r_{\perp}^2 \rangle$ be the average of the square of the radius of the electron from the nucleus perpendicular to the direction of the magnetic field. Then the average area of the electron orbit perpendicular to the magnetic field becomes

$$A = \pi \langle r_{\perp}^2 \rangle \tag{18.49}$$

As the magnetic field is in the z-direction, $\langle r_{\perp}^2 \rangle$ is in the xy-plane. One can write

$$\langle r_{\perp}^2 \rangle = \langle x^2 \rangle + \langle y^2 \rangle \tag{18.50}$$

In general, the mean square distance $\langle r^2 \rangle$ of the electrons from the nucleus in three dimensions is given by

$$\langle r^2 \rangle = \langle x^2 \rangle + \langle y^2 \rangle + \langle z^2 \rangle \tag{18.51}$$

In order to estimate the induced magnetic moment, we consider a simple case in which the charge distribution is spherically symmetric, that is,

$$\langle x^2 \rangle = \langle y^2 \rangle = \langle z^2 \rangle \tag{18.52}$$

From Eqs. (18.50), (18.51), and (18.52) one can easily write

$$\langle r_{\perp}^2 \rangle == \frac{2}{3} \langle r^2 \rangle \tag{18.53}$$

Substituting Eqs. (18.48), (18.49), and (18.53) into Eq. (18.7), the induced magnetic moment due to the Larmor precession is given by

$$\mu_L = -\frac{Ze^2 H}{6m_e c^2} \langle r^2 \rangle \tag{18.54}$$

If there are ρ^a atoms per unit volume, the diamagnetic susceptibility is given by

$$\chi_d = \frac{M}{H} = \frac{\rho^a \mu_L}{H} = -\frac{Ze^2 \rho^a}{6m_e c^2} \langle r^2 \rangle \tag{18.55}$$

This is called the Langevin result. From Eq. (18.55) it is evident that the problem of calculating the diamagnetic susceptibility is reduced to the calculation of $\langle r^2 \rangle$ for the atomic electron distribution in an atom, which can be estimated using a quantum mechanical approach.

The units of χ_d can be calculated from Eq. (18.55). Z is a number but ρ^a, as the density of atoms, has dimensions of $1/L^3$ and so, from Eq. (18.55), one can write

$$\chi_d = \frac{1}{L^3} \frac{e^2}{M(LT^{-1})^2} L^2 = \frac{e^2}{L} \frac{1}{ML^2T^{-2}} \tag{18.56}$$

Now e^2/L have the units of energy (work) with dimensions

$$\frac{e^2}{L} = maS = M(LT^{-2})L = ML^2T^{-2} \tag{18.57}$$

From Eqs. (18.56) and (18.57), χ_d is found to be dimensionless. The value of χ_d is specified in the same way as the density ρ^a is defined. If the density ρ^a is defined per unit volume, then the values of χ_d are listed per unit volume, but if ρ^a is taken per gram mole, then χ_d is specified per gram mole.

Problem 18.1

Calculate the diamagnetic susceptibility for a He atom in the ground state, i.e., the 1 s state, taking its radius as the Bohr radius a_0. The density of He atoms is given by $\rho^a = 2.7 \times 10^{24}\,cm^{-3}$.

18.6 QUANTUM THEORY OF DIAMAGNETISM

The Hamiltonian of an electron in an atom (say the Bohr atom) is given by

$$\widehat{H}_0 = \frac{p^2}{2m_e} + V \tag{18.58}$$

where p and m_e are the momentum and mass of the electron, respectively. If the atom is placed in electric and magnetic fields represented by **E** and **H**, respectively, then the Lorentz force acting on the electron is given by

$$\mathbf{F} = -e\mathbf{E} - \frac{e}{c}\mathbf{v} \times \mathbf{H} \tag{18.59}$$

The magnetic field in terms of the vector potential **A** is given by

$$\mathbf{H} = \nabla \times \mathbf{A} \tag{18.60}$$

The momentum of an electron in the presence of an electromagnetic field changes as follows:

$$\mathbf{p} \to \mathbf{p} - \frac{e}{c}\mathbf{A} \tag{18.61}$$

Therefore, the Hamiltonian of an electron in the presence of a magnetic field becomes

$$\widehat{H} = \frac{1}{2m_e}\left(\mathbf{p} - \frac{e}{c}\mathbf{A}\right)^2 + V \tag{18.62}$$

$\widehat{H}$ can be split up into two parts as

$$\widehat{H} = \widehat{H}_0 + \widehat{H}_1 \tag{18.63}$$

where

$$\widehat{H}_0 = \frac{p^2}{2m_e} + V \tag{18.64}$$

$$\widehat{H}_1 = -\frac{e}{2m_e c}(\mathbf{p} \cdot \mathbf{A} + \mathbf{A} \cdot \mathbf{p}) + \frac{e^2}{2m_e c^2}A^2 \tag{18.65}$$

Here $\widehat{H}_0$ is the unperturbed Hamiltonian and $\widehat{H}_1$ is the perturbation. Suppose **H** is uniform and is applied in the z-direction, then the components of the magnetic field from Eq. (18.60) are given as

$$H_x = \frac{\partial A_z}{\partial y} - \frac{\partial A_y}{\partial z} = 0 \tag{18.66}$$

$$H_y = \frac{\partial A_x}{\partial z} - \frac{\partial A_z}{\partial x} = 0 \tag{18.67}$$

$$H_z = \frac{\partial A_y}{\partial x} - \frac{\partial A_x}{\partial y} = H \tag{18.68}$$

The above equations are satisfied if the components of the vector potential are given by

$$A_x = -\frac{1}{2}yH, A_y = \frac{1}{2}xH, A_z = 0 \tag{18.69}$$

This can be written in vector form as

$$\mathbf{A} = \frac{1}{2}\mathbf{H} \times \mathbf{r} \tag{18.70}$$

Substituting $\mathbf{p} = -\imath\hbar\nabla$ into Eq. (18.65), $\widehat{H}_1$ can be written as

$$\widehat{H}_1 = \frac{\imath\hbar e}{2m_e c}(\nabla \cdot \mathbf{A} + \mathbf{A} \cdot \nabla) + \frac{e^2}{2m_e c^2}A^2 \tag{18.71}$$

In terms of Cartesian components $\widehat{H}_1$, from Eq. (18.65), can be written as

$$\widehat{H}_1 = -\frac{e}{2m_e c}\left(p_x A_x + p_y A_y + A_x p_x + A_y p_y\right) + \frac{e^2}{2m_e c^2}\left(A_x^2 + A_y^2\right) \tag{18.72}$$

From Eqs. (18.69) and (18.72) it is straightforward to write

$$\widehat{H}_1 = -\frac{eH}{2m_e c}\left(xp_y - yp_x\right) + \frac{e^2H^2}{8m_e c^2}(x^2 + y^2) \tag{18.73}$$

The orbital angular momentum, defined as $\mathbf{L} = \mathbf{r} \times \mathbf{p}$, can be used to write

$$\widehat{H}_1 = -\frac{eH}{2m_e c}L_z + \frac{e^2H^2}{8m_e c^2}(x^2 + y^2)$$

or

$$\widehat{H}_1 = -\mu_z H + \frac{e^2H^2}{8m_e c^2}(x^2 + y^2)$$

or

$$\widehat{H}_1 = -\vec{\mu} \cdot \mathbf{H} + \frac{e^2H^2}{8m_e c^2}(x^2 + y^2) \tag{18.74}$$

The expectation value of $\widehat{H}_1$ gives us the change in energy due to the application of the magnetic field. The lowest order change in energy is given by the first-order correction in perturbation theory. Let $|\psi_0\rangle = |0\rangle$ represent the ground state of the system. For diamagnetic substances the atomic or molecular magnetic moment is zero in the ground state, therefore,

$$\langle 0|\mu_z|0\rangle = 0 \tag{18.75}$$

Hence the first-order correction to energy in a diamagnetic substance comes from the expectation value of the second term in Eq. (18.74), i.e.,

$$E_1 = \frac{e^2H^2}{8m_e c^2}\langle 0|x^2 + y^2|0\rangle \tag{18.76}$$

$\langle 0|\, x^2 + y^2\, |0\rangle$ is the average value of the area of the electron loop perpendicular to the direction of the magnetic field and is given by

$$\langle 0|x^2 + y^2|0\rangle = \langle r_\perp^2\rangle = \frac{2}{3}\langle r^2\rangle \tag{18.77}$$

Substituting Eq. (18.77) into Eq. (18.76), we obtain

$$E_1 = \frac{e^2H^2}{12m_e c^2}\langle r^2\rangle \tag{18.78}$$

We know that the magnetic energy is given by

$$E = -\vec{\mu} \cdot \mathbf{H} = -\mu_z H \tag{18.79}$$

Therefore, the magnetic moment is given by

$$\mu_z = -\frac{\partial E}{\partial H} \tag{18.80}$$

Substituting Eq. (18.78) into Eq. (18.80), one gets

$$\mu_z = -\frac{e^2H}{6m_e c^2}\langle r^2\rangle \tag{18.81}$$

This is the same result for the magnetic moment as that obtained classically.

Let us find the expectation value of the Hamiltonian of the perturbed ground state. Suppose $|n\rangle$ represents the nth state of the unperturbed system with energy E_n. The matrix element of the magnetic moment between the ground state $|0\rangle$ and the nth state $|n\rangle$ is $\langle n|\mu_z|0\rangle$. When a magnetic field **H** is applied, the perturbed ground state of the system is written as

$$|0'\rangle = |0\rangle + \sum_{n\neq 0} \frac{\langle n|\mu_z H|0\rangle}{E_n - E_0}|n\rangle \tag{18.82}$$

The first-order correction to the magnetic moment with respect to the perturbed ground state of the system, neglecting terms of second and higher order in H, is given by

$$\begin{aligned}\Delta\mu = \langle 0'|\mu_z|0'\rangle = \langle 0|\mu_z|0\rangle + H\sum_{n\neq 0}\frac{\langle 0|\mu_z|n\rangle\langle n|\mu_z|0\rangle}{E_n - E_0}\\ +H\sum_{n'\neq 0}\frac{\langle 0|\mu_z|n'\rangle\langle n'|\mu_z|0\rangle}{E_{n'} - E_0}\end{aligned} \tag{18.83}$$

The first term on the right side of Eq. (18.83) is zero. Further, the second and third terms are equal, yielding

$$\Delta\mu = 2H\sum_{n\neq 0}\frac{|\langle n|\mu_z|0\rangle|^2}{E_n - E_0} \tag{18.84}$$

If there are ρ^a atoms or molecules per unit volume of the solid, then the magnetization produced is given by

$$\Delta M = \rho^a\Delta\mu = 2\rho^a H\sum_{n\neq 0}\frac{|n|\mu_z|0\rangle|^2}{E_n - E_0} \tag{18.85}$$

Therefore, the magnetic susceptibility contribution is given by

$$\Delta\chi_M = \frac{\Delta M}{H} = 2\rho^a\sum_{n\neq 0}\frac{|\langle n|\mu_z|0\rangle|^2}{E_n - E_0} \tag{18.86}$$

Here $E_n > E_0$, therefore, $\Delta\mu$ and hence $\Delta\chi_M$ is positive. With respect to $E_n - E_0$ two cases arise:

1. If $E_n - E_0 \gg k_B T$, i.e., the excited state has energy much greater than the thermal energy, then most of the electrons will be in the ground state. In this case, $\Delta\chi_M$ is positive and independent of temperature. This type of contribution to the magnetic susceptibility of a diamagnetic substance is known as *Van Vleck paramagnetism*.
2. If $E_n - E_0 \ll k_B T$, the excited state has an energy much less than the thermal energy. In this situation, both the ground and excited states are occupied with electrons, but the ground state has a higher population compared with the excited state. The excess population in the ground state is $\rho^a(E_n - E_0)/2k_B T$. Hence the resultant magnetization in the ground state of the system is given by

$$\Delta M = \rho^a\Delta\mu(E_n - E_0)/2k_B T$$

Substituting the value of $\Delta\mu$ from Eq. (18.84) into the above equation, we find

$$\Delta M = \frac{\rho^a H}{k_B T}\sum_{n\neq 0}|\langle n|\mu_z|0\rangle|^2 \tag{18.87}$$

Hence the magnetic susceptibility becomes

$$\Delta\chi_M = \frac{\rho^a}{k_B T}\sum_{n\neq 0}|\langle n|\mu_z|0\rangle|^2 \tag{18.88}$$

$\Delta\chi_M$ has a behavior similar to that of the Curie susceptibility, but the origin of this contribution is entirely different: $\Delta\chi_M$ arises due to the polarization of the states of the system. It should be noted that the energy separation $E_n - E_0$ does not enter in Eq. (18.88). We should also note that if $E_n \rightarrow E_0$, then the electrons become free and the solid becomes a metal; in this case, Eq. (18.88) is not valid.

The above treatment can be generalized for the nth perturbed excited state given by

$$|n'\rangle = |n\rangle - \sum_{n \neq 0} \frac{\langle 0|\mu_z H|n\rangle}{E_n - E_0} |0\rangle$$

The expectation value of the magnetic moment in the perturbed state is given by

$$\Delta\mu' = \langle n'|\mu_z|n'\rangle = -2H \sum_{n \neq 0} \frac{|\langle n|\mu_z|0\rangle|^2}{E_n - E_0}$$

18.7 PARAMAGNETISM

In a paramagnetic substance each atom or molecule possesses an intrinsic magnetic dipole moment $\vec{\mu}$. At finite temperature, all of the magnetic dipole moments are oriented randomly in the form of closed chains yielding zero magnetization. In the presence of an applied magnetic field, two opposing forces act on each atomic dipole moment in a paramagnetic substance:

1. The magnetic field tries to align the dipole moments in the direction of the field, thereby producing finite magnetization along the magnetic field.
2. At finite temperature, the thermal energy tries to randomize the magnetic moments to form closed chains and hence tends to decrease magnetization.

18.7.1 Classical Theory of Paramagnetism

In the classical description, the magnetic dipole moment $\vec{\mu}$ is taken to be a constant physical quantity independent of the quantum numbers. Under the action of the competing forces mentioned above, some dipole moments align in the direction of the applied magnetic field, while others make some angle θ, which is different for different dipole moments. Therefore, a solid shows finite magnetic dipole moment and hence finite magnetization in the direction of the magnetic field. The maximum magnetization is produced when all of the dipole moments align along the direction of the applied field. In the presence of a magnetic field, the potential energy of the magnetic dipole moment is given by

$$E = -\vec{\mu} \cdot \mathbf{H} = -\mu H \cos\theta \tag{18.89}$$

According to classical statistics, the probability P of a dipole moment making an angle θ with the magnetic field is given by

$$P \propto \exp\left(-\frac{E}{k_B T}\right) \propto \exp\left(\frac{\mu H}{k_B T} \cos\theta\right) \tag{18.90}$$

The component of the magnetic moment along the direction of the magnetic field is $\vec{\mu} \cdot \hat{\mathbf{H}} = \mu \cos\theta$ where $\hat{\mathbf{H}}$ is a unit vector in the direction of the field. Hence the average component of magnetic moment in the direction of the magnetic field is given by

$$\mu_{avg} = \frac{\int \left(\vec{\mu} \cdot \hat{\mathbf{H}}\right) \exp\left(\frac{\mu H}{k_B T} \cos\theta\right) d\Omega_s}{\int \exp\left(\frac{\mu H}{k_B T} \cos\theta\right) d\Omega_s} \tag{18.91}$$

Here $d\Omega_s$ is the elemental solid angle. Solving the above integral, one gets

$$\mu_{avg} = \mu L\left(\frac{\mu H}{k_B T}\right) \tag{18.92}$$

where

$$L(y) = \coth y - \frac{1}{y} \tag{18.93}$$

L(y) is the Langevin function (see Section 15.15). If ρ^a is the number of atoms per unit volume, then the magnetization is given by

$$M = \rho^a \mu L\left(\frac{\mu H}{k_B T}\right) \tag{18.94}$$

The magnetic susceptibility χ_M becomes

$$\chi_M = \frac{\rho^a \mu}{H} L\left(\frac{\mu H}{k_B T}\right) \tag{18.95}$$

It is interesting to study M and χ_M in limiting cases. If the magnetic field is very high and the temperature is very low then,

$$\mu H \gg k_B T \tag{18.96}$$

In this limiting case the Langevin function goes to unity, i.e., $L(\mu H/k_B T) = 1$ and therefore

$$M = \rho^a \mu \tag{18.97}$$

which is the saturation magnetization when all the magnetic dipole moments are aligned in the direction of the magnetic field. Hence saturation magnetization is obtained either at very low temperatures or at very high magnetic field values. The other limiting case occurs when the magnetic field is low, but the temperature is high and, according to this.

$$\mu H \ll k_B T \tag{18.98}$$

If y is small, $L(y) \approx y/3$ [see Eq. (15.84)] and hence the magnetization from Eq. (18.94) becomes

$$M(T) = \rho^a \mu \frac{\mu H}{3 k_B T} = \frac{\mu^2 \rho^a}{3 k_B T} H \tag{18.99}$$

The behavior of the magnetization M(y) as a function of y is shown in Fig. 18.5. M(y) acquires the saturated value $\rho^a \mu$ at very large values of y, but the slope of the M(y) curve at $y = 0$ is $\rho^a \mu/3$. From Eq. (18.99) the paramagnetic susceptibility is given by

$$\chi_M(T) = \frac{C_M}{T} \tag{18.100}$$

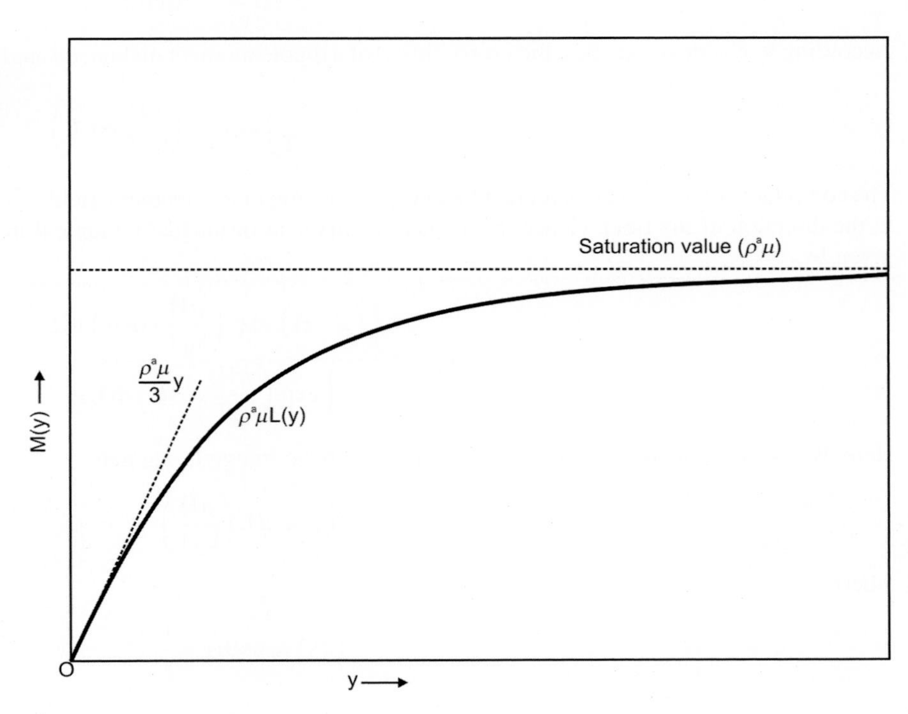

FIG. 18.5 The magnetization M(y) in a paramagnetic solid as a function of parameter $y = \mu H/k_B T$ in the classical theory. The slope of the magnetization curve at the origin is shown by the *dashed line*.

where C_M is the Curie constant and is given by

$$C_M = \frac{\mu^2 \rho^a}{3k_B} \tag{18.101}$$

Eq. (18.100) is the usual Curie law. The limitation of the classical theory is that the distribution of magnetic dipole moments is assumed to be continuous, i.e., all values of θ are allowed. But according to quantum mechanics, the distribution of magnetic dipoles must be discrete.

Problem 18.2

Let the paramagnetic susceptibility be given by

$$\chi_M = \frac{\rho^a \mu_B^2}{3k_B T}$$

where μ_B is the Bohr magnetron. If the density of atoms is $\rho^a = 2 \times 10^{22}$ atoms/cm^3, find the paramagnetic susceptibility at room temperature taken as $T = 300\,K$.

Problem 18.3

If one retains the first two terms in the series expansion of the Langevin theory of paramagnetism, prove that the susceptibility is given by

$$\chi_M = \frac{M}{H} = \frac{\rho^a \mu^2}{3k_B T}\left[1 - \frac{1}{15}\left(\frac{\mu H}{k_B T}\right)^2\right]$$

18.7.2 Quantum Theory of Paramagnetism

Eq. (18.14) yields discrete values for the orbital magnetic moment $\vec{\mu}_L$, which means that it is quantized. Similarly, the spin magnetic moment $\vec{\mu}_S$ is also discrete, having two values [Eq. (18.15)]: μ_B and $-\mu_B$. Therefore, the total magnetic moment $\vec{\mu}_J$ has discrete values. The general expression for the magnetic moment of an atom or an ion in free space is given by

$$\vec{\mu}_J = \gamma_J \hbar \mathbf{J} \tag{18.102}$$

where $\mathbf{J}$ is the total angular momentum. The constant γ_J is the ratio of the magnetic moment to the angular momentum and is called the magneto-mechanical or gyromagnetic ratio. Comparing Eq. (18.102) with Eq. (18.24), one can write

$$g_J \mu_B = -\gamma_J \hbar \tag{18.103}$$

Lande's spectroscopic splitting factor g_J represents the ratio of the number of Bohr magnetrons to the angular momentum in units of ħ.

Suppose a magnetic field $\mathbf{H}$ is applied to a paramagnetic substance along the z-direction. The Hamiltonian of the system is given by

$$\widehat{H} = -\vec{\mu}_J \cdot \mathbf{H} = g_J \mu_B J_z H \tag{18.104}$$

J_z is the z-component of the angular momentum $\mathbf{J}$. If M_J is the eigenvalue of J_z, the interaction energy is given by

$$E = g_J \mu_B H M_J \tag{18.105}$$

M_J is the azimuthal quantum number having the values $-J, -(J-1), \ldots\ldots\ldots -1, 0, 1, \ldots\ldots (J-1), J$, which are $2J+1$ in number. In a paramagnetic substance the occupation probability is given by the Boltzmann distribution as

$$P \propto \exp\left(-\frac{E}{k_B T}\right) \propto \exp\left(-\beta_0 g_J \mu_B H M_J\right) \tag{18.106}$$

The constant β_0 is given by Eq. (8.22). The component of the magnetic moment in the direction of the magnetic field is

$$\mu_z = \vec{\mu}_J \cdot \hat{z} = -g_J \mu_B M_J \tag{18.107}$$

Hence the average magnetic moment in the direction of the magnetic field is given by

$$\mu_{avg} = \frac{\sum_J (-g_J \mu_B M_J) \exp(-\beta_0 g_J \mu_B H M_J)}{\sum_J \exp(-\beta_0 g_J \mu_B H M_J)} \tag{18.108}$$

Substituting

$$y = \beta_0 g_J \mu_B H \tag{18.109}$$

Eq. (18.108) can be written as

$$\begin{aligned} \mu_{avg} &= \frac{-g_J \mu_B \sum_J M_J \exp(-y M_J)}{\sum_J \exp(-y M_J)} \\ &= g_J \mu_B \frac{d}{dy} \ln\left(\sum_J \exp(-y M_J)\right) \end{aligned} \tag{18.110}$$

It can easily be shown that

$$\sum_J \exp(-y M_J) = \frac{\exp\left(\frac{2J+1}{2} y\right) - \exp\left(-\frac{2J+1}{2} y\right)}{\exp\left(\frac{y}{2}\right) - \exp\left(-\frac{y}{2}\right)} \tag{18.111}$$

Substituting Eq. (18.111) into Eq. (18.110) and simplifying, we obtain

$$\mu_{avg} = g_J \mu_B J B_J(x) \tag{18.112}$$

where

$$B_J(x) = \frac{2J+1}{2J} \coth\left(\frac{2J+1}{2J} x\right) - \frac{1}{2J} \coth\left(\frac{1}{2J} x\right) \tag{18.113}$$

and

$$x = yJ = \beta_0 g_J \mu_B H J \tag{18.114}$$

The function $B_J(x)$ is called the Brillouin function. If ρ^a is the number of dipole moments per unit volume, the magnetization is given by

$$M_J(x) = \rho^a g_J \mu_B J B_J(x) \tag{18.115}$$

One can study the particular case in which there is only spin ($L=0$). In the case of spin $J=S=1/2$ and $g_J = g_S = 2$, we find

$$x = \frac{\mu_B H}{k_B T} \tag{18.116}$$

Substituting the above mentioned values, the Brillouin function for spin becomes

$$B_{1/2}\left(\frac{\mu_B H}{k_B T}\right) = 2 \coth\left(2 \frac{\mu_B H}{k_B T}\right) - \coth\left(\frac{\mu_B H}{k_B T}\right) \tag{18.117}$$

which can be simplified to get

$$B_{1/2}\left(\frac{\mu_B H}{k_B T}\right) = \tanh\left(\frac{\mu_B H}{k_B T}\right) \tag{18.118}$$

Substituting Eq. (18.118) into Eq. (18.115) one can write

$$M_{1/2} = \rho^a \mu_B \tanh\left(\frac{\mu_B H}{k_B T}\right) \tag{18.119}$$

The value of the magnetization and magnetic susceptibility can be obtained in a simpler form in the limiting cases. From Eq. (18.114) one can write

$$x = \frac{g_J \mu_B HJ}{k_B T} \tag{18.120}$$

In the limit $x \to 0$, i.e., when the magnetic field is very small or the temperature is very large, one can expand coth x as in Eq. (15.83) and use this in Eq. (18.113) to get

$$B_J(x) = \frac{J+1}{J}\frac{x}{3} \tag{18.121}$$

Therefore, in the limit $x \to 0$, the magnetization from Eq. (18.115) is given by

$$M_J(x) = \frac{\rho^a g_J^2 \mu_B^2 J(J+1)}{3k_B T} H = \rho^a g_J \mu_B \frac{J+1}{3} x \tag{18.122}$$

Hence the magnetic susceptibility becomes

$$\chi_M = \frac{C_J}{T} \tag{18.123}$$

where

$$C_J = \frac{\rho^a g_J^2 \mu_B^2 J(J+1)}{3k_B} \tag{18.124}$$

Eq. (18.123) is just the Curie law with Curie constant C_J, which depends on the total quantum number J. If we compare Eq. (18.124) with Eq. (18.101), we can say that the magnetic moment μ_J associated with an atom having quantum number J is

$$\mu_J = \mu_B p_J \tag{18.125}$$

where

$$p_J = g_J \sqrt{J(J+1)} \tag{18.126}$$

Here p_J gives the effective number of Bohr magnetrons in an atom. The Curie constant in terms of μ_J is given by

$$C_J = \frac{\rho^a \mu_J^2}{3k_B} \tag{18.127}$$

In the limiting case of $x \to \infty$, either the magnetic field is very high or the temperature is very low. In this limit the Brillouin function (Eq. 18.113) goes to unity and, therefore, the magnetization from Eq. (18.115) is given by

$$M_J = \rho^a g_J \mu_B J \tag{18.128}$$

which gives the saturation magnetization of the substance. The variation of $M_J(x)$, given by Eq. (18.115) as a function of x, is shown in Fig. 18.6, which is similar to the magnetization curve obtained in the classical case. $M_J(x)$ increases with an increase in x and approaches the saturation value for large values of magnetic fields.

One can obtain the classical result of paramagnetism from the quantum theory in the limiting case. Let us suppose that the angular momentum **J** makes an angle θ with the direction of **H** (Fig. 18.7). The eigenvalue of **J** is $[J(J+1)]^{1/2}$ and, therefore, the value of the z-component of **J**, i.e., J_z, is given by

$$J_z = \sqrt{J(J+1)}\cos\theta \tag{18.129}$$

FIG. 18.6 The magnetization $M_J(x)$ for a paramagnetic solid as a function of parameter $x = \beta_0 g \mu_B J H$. The magnetization curve is similar to that shown in Fig. 18.5 except that the saturation magnetization and slope of the curve at the origin are different.

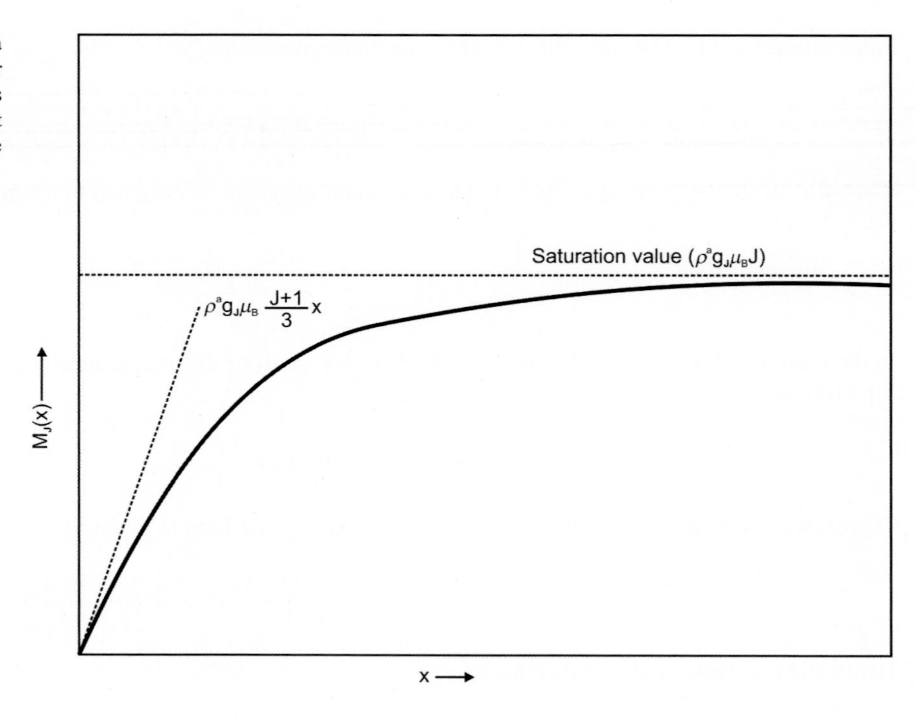

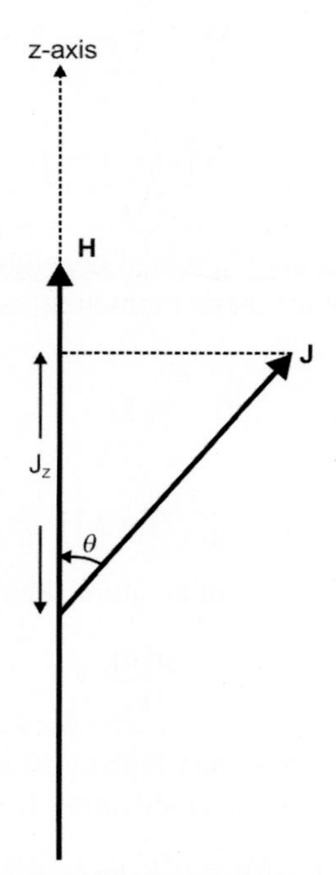

FIG. 18.7 Orientation of total angular momentum **J** with respect to the applied magnetic field **H** in the z-direction.

From quantum mechanics J_z has $2J+1$ eigenvalues ranging from $-J$ to J through zero. Therefore, the values of $\cos\theta$ are given by

$$\cos\theta = \pm\frac{J}{\sqrt{J(J+1)}} = \pm\frac{1}{\left(1+\frac{1}{J}\right)^{1/2}} \tag{18.130}$$

As the values of J are discrete, so are the values of θ. If J has an infinite number of values, then J becomes very large. Hence from Eq. (18.130) $\cos\theta$ has an infinite number of values lying between -1 and $+1$. In other words, the value of θ becomes continuous, that is, the distribution becomes continuous (classical case). In the limit of $J \rightarrow \infty$, it is easy to prove from Eq. (18.113) that

$$\text{Lim}_{J\rightarrow\infty} B_J(x) = \coth x - \frac{1}{x} = L(x) \tag{18.131}$$

Hence the magnetization from Eq. (18.115) in the limit $J \rightarrow \infty$ is given by

$$M_J = \rho^a \mu_J L(x) \tag{18.132}$$

where

$$\mu_J = g_J \mu_B J \tag{18.133}$$

and

$$x = \frac{\mu_J H}{k_B T} \tag{18.134}$$

Eq. (18.132) gives the familiar Langevin paramagnetism.

Problem 18.4

Consider an ion with a partially filled shell of angular momentum J and Z additional electrons in filled shells. Show that the ratio of paramagnetic susceptibility at high temperatures (Curie law) to the diamagnetic susceptibility is given by

$$\frac{\chi_M}{\chi_d} = \frac{g_J^2}{4} \frac{2J(J+1)}{Zk_B T} \frac{\hbar^2}{m_e \langle r^2 \rangle}$$

18.8 HUND'S RULE

The magnetic moment of an atom can be predicted using the knowledge of quantum mechanics in combination with the Pauli exclusion principle and Hund's rule. The Pauli principle says that, in a paramagnetic substance, an electron state can be occupied by two electrons with the same principal (n), orbital (ℓ), and magnetic (m_ℓ) quantum numbers, but with opposite spins (s). In an atom the filled electron states do not contribute to the magnetic moment, but rather its finite value results from the partially filled states.

Hund's rule states that in the ground state of an atom

1. The electron spins add to give the maximum possible total spin S consistent with the Pauli exclusion principle. This rule has its origin in the Coulomb repulsive interaction energy between two electrons.
2. The orbital angular momenta of electrons combine to give the maximum possible total angular momentum L that is consistent with point 1. This rule is based on model calculations of spectral terms.
3. For a partially filled shell, the total angular momentum is given as follows:

$$\begin{aligned} J &= |L - S| \text{ for a shell less than half filled} \\ &= L + S \text{ for a shell more than half filled.} \end{aligned} \tag{18.135}$$

This rule is a consequence of the spin-orbit interaction.

18.8.1 Applications of Hund's Rule

In the paramagnetic elements each atom or molecule has a finite intrinsic magnetic moment. In the periodic table most of the paramagnetic elements are either d-shell or f-shell elements, which possess partially filled electron shells. For example, elements of the iron group, with atomic number Z ranging from 21 to 28, possess incomplete 3d-shells. The elements of the palladium group, with Z ranging from 39 to 46, possess incomplete 4d- shells, while the platinum group elements, with Z ranging from 71 to 78, possess incomplete 5d-shells. The rare-earth elements, with Z ranging from 57 to 72, possess incomplete 4f-shells. The uranium group elements, with Z ranging from 89 to 103, possess incomplete 5f and 6d-shells. To illustrate the method of calculating the atomic magnetic moment, we consider a few different elements.

18.8.1.1 Rare-Earth Group

The rare-earth element Ce^{58} is paramagnetic in nature and has the following electronic configuration.

$$Ce^{58}: 4f^2 5s^2 5p^6 6s^2$$

In the above representation the electronic configuration starting from the first partially filled shells is written. Here the 4f-shell is partially filled and is responsible for the magnetic moment. The ion of Ce^{58} is trivalent and has the configuration.

$$Ce^{+3}: 4f^1 5s^2 5p^6$$

The valence is contributed by one electron in the 4f-shell and two electrons in 6s-shell. The 4f-shell has 7 subshells with orbital magnetic quantum number m_ℓ from −3 to 3, while the spin quantum number m_s has two values 1/2 and −1/2. The distribution of 4f-electrons in the subshells is given below:

$$\begin{array}{ll} m_\ell & : 3\ 2\ 1\ 0\ -1\ -2\ -3 \\ m_{s\uparrow} & : \frac{1}{2} \end{array}$$

The above distribution gives as a maximum value of the orbital quantum number L = 3 and a maximum value of spin S = 1/2 consistent with Hund's rule. As the 4f-shell is less than half filled, the total angular momentum J is given as

$$J = |L - S| = 3 - ½ = 5/2$$

With these values of J, L, and S, the value of g_J can be calculated using Eq. (18.25) yielding

$$g_J = 1 + \frac{\frac{5}{2}\cdot\frac{7}{2} + \frac{1}{2}\cdot\frac{3}{2} - 3\times 4}{2\cdot\frac{5}{2}\cdot\frac{7}{2}} = \frac{6}{7}$$

Hence the effective number of Bohr magnetrons from Eq. (18.126) becomes

$$p_J = \frac{3}{7}\sqrt{35} \cong 2.5$$

The experimental value of the effective number of Bohr magnetrons is $p_{exp} = 2.4$, which is in good agreement with the calculated value.

Another interesting example of the rare-earth elements is Pr^{59} with the following electronic configuration.

$$Pr^{59}: 4f^2 5s^2 5p^6 6s^2 6p^1$$

Here the 4f-shell is partially filled and is responsible for the magnetic moment in paramagnetic Pr. The ion of Pr^{59} is trivalent and has the configuration.

$$Pr^{+3}: 4f^2 5s^2 5p^6$$

The distribution of electrons of Pr^{+3} in the 4f-subshells is given below:

$$\begin{array}{ll} m_\ell & : 3\ 2\ 1\ 0\ -1\ -2\ -3 \\ m_{s\uparrow} & : \frac{1}{2}\ \frac{1}{2} \end{array}$$

According to Hund's rule, L = 3+2 = 5 and S = ½+½ = 1. As the f-shell is less than half filled, therefore,

$$J = |L - S| = 4$$

The value of Lande's splitting factor becomes

$$g_J = 1 - \frac{1}{5} = \frac{4}{5}$$

The effective number of Bohr magnetrons p_J can be found immediately and has the value

$$p_J = \frac{4}{5}\cdot\sqrt{20} = 3.58$$

The experimental value is $p_{exp} = 3.50$, which is in reasonable agreement with theory. The agreement between theory and experiment in the ionic magnetic moment of Ce and Pr is good, but there is a large discrepancy in the case of Eu^{+3} and Sm^{+3} ions.

18.8.1.2 Iron Group

Mn is an important element of the iron group with its atom having the electronic configuration:

$$Mn : 3d^5 4s^2$$

Here the 3d-shell is incomplete and is expected to contribute to the magnetic moment. The electronic configuration of a divalent Mn ion becomes

$$Mn^{+2} : 3d^5$$

The distribution of electrons among the d-subshells is given below:

$$m_\ell \;:\; 2 \quad 1 \quad 0 \quad -1 \quad -2$$

$$m_{s\uparrow} \;:\; \frac{1}{2} \quad \frac{1}{2} \quad \frac{1}{2} \quad \frac{1}{2} \quad \frac{1}{2}$$

Hund's rule yields the following values for the quantum numbers L and S:

$$L=0, \; S=5/2$$

One should note that the 3d-shell is half filled and the value of J is the same using both formulas: one for a shell less than half filled and the other for a shell more than half filled, that is,

$$J=|L-S|=|L+S| = 5/2.$$

The g_J factor has the value 2 because this is a case with spin only. With the above values one can easily find the value of p_J as

$$p_J = g_J\sqrt{J(J+1)} = g_S\sqrt{S(S+1)} = \sqrt{35} = 5.9$$

The experimental value is also the same, that is, $p_{exp} = 5.9$. Hence both the calculated and experimental values agree with each other.

Another peculiar element of the iron group is Cr^{24} with the following electronic configuration:

$$Cr^{24} : 3d^5 4s^1$$

If the valence of Cr is taken to be three, then the electronic configuration of Cr^{+3} becomes

$$Cr^{+3} : 3d^3$$

Here two d-electrons and one s-electron contribute to the valence. The three electrons in the d-subshells of Cr^{+3} contribute to the magnetic moment and their arrangement is given below:

$$m_\ell \;:\; 2 \quad 1 \quad 0 \quad -1 \quad -2$$

$$m_{s\uparrow} \;:\; \frac{1}{2} \quad \frac{1}{2} \quad \frac{1}{2}$$

The values of L, S, and J become 3, 3/2, and 3/2, which yield a value of $g_J = 2/5$. The value of p_J becomes

$$p_J = \frac{1}{5}\sqrt{15} = 0.77$$

But $p_{exp} = 3.8$, which clearly shows a disagreement between theory and experiment. The disagreement may possibly be due to the valence as the Cr atom exhibits variable valence. Let us take the Cr atom as divalent with electronic configuration

$$Cr^{+2} : 3d^4$$

In this case the distribution of electrons is given below:

$$\begin{array}{l}m_\ell \;:\; 2\;\;1\;\;0\;\;-1\;\;-2\\ m_{s\uparrow} \;:\; \frac{1}{2}\;\frac{1}{2}\;\frac{1}{2}\;\;\frac{1}{2}\end{array}$$

The above distribution yields values of $L=2$, $S=2$ and $J=0$ (d-shell is less than half filled). These values yield zero magnetic moment ($p_J=0$) for the Cr^{+2} ion, which is again in disagreement with the experimental value. This shows that there is some other factor that may yield the correct value of the magnetic moment in Cr. Let us examine the case of Cr^{+3} assuming that only the spin contributes to the magnetic moment of the ion. Then

$$p_S = g_S\sqrt{S(S+1)} = 2\sqrt{\frac{3}{2}\cdot\frac{5}{2}} = \sqrt{15} = 3.87$$

The value of p_S is in reasonable agreement with $p_{exp} = 3.8$. Therefore, a Cr^{+3} ion behaves as if it had zero orbital angular momentum. Similarly, it can be shown that in the Fe^{+3} ion, the magnetic moment turns out to be 5.9 if the orbital angular momentum is assumed to be zero, which agrees with the experimental value. One should note that, in general, the ions from the iron group behave as if there were no orbital angular momentum associated with them. In other words, one can say that the orbital angular momentum is quenched in iron group elements.

18.9 CRYSTAL FIELD SPLITTING

Inside a crystal, every atom or molecule experiences a crystal field, which has a significant effect on the atomic/molecular magnetic moment. The 4f-shell in the rare-earth elements is responsible for paramagnetism and lies deep inside the ion. Therefore, the 4f-shell is well shielded from the crystal field by the 5 s- and 5p-shells. On the other hand, in the iron group elements, the 3d-shell is responsible for paramagnetism. The 3d-shell is the outermost shell in an ion and experiences an intense local crystal field produced by the neighboring ions, which is generally inhomogeneous in nature. The interaction of ions with the inhomogeneous crystal field has two major effects.

1. The coupling of the **L** and **S** vectors (**L** − **S** coupling) is largely broken, so the states can no longer be specified by the total angular momentum **J**.
2. The $2L+1$ sublevels (given by m_ℓ) belonging to a given L value are degenerate in a free ion, but they get split up by the inhomogeneous crystal field. The splitting diminishes the contribution of the orbital magnetic moment.

18.9.1 Quenching of Orbital Angular Momentum

In a central field directed toward the nucleus, the plane of the electron orbit is fixed in space, yielding constant components of orbital angular momentum L_x, L_y, and L_z. According to quantum mechanics, in the central field approximation, $\widehat{H}, L_z$, and L^2 are constants of motion, which means that they commute with one another. On the other hand, in the presence of a noncentral crystal field, the plane of the electron orbit is not fixed, but rather it is moving about its center in all possible directions. As a result, the components of orbital angular momentum are continuously changing and they may average out to zero. In such a situation $\widehat{H}$ and L_z are no longer constants of motion, although L^2 may continue to be a constant of motion. In other words, $\widehat{H}$ and L_z do not commute with each other, i.e.,

$$\left[\widehat{H}, L_z\right] \neq 0 \tag{18.136}$$

In this case L_z may average out to zero, leading to quenching of the orbital angular momentum.

The magnetic moment of an atom or molecule depends on the magnetic moment operator $\mu_B(\mathbf{L}+2\mathbf{S})$. If a magnetic field is applied in the z-direction, the orbital magnetic moment is proportional to the expectation value of L_z. If L_z is quenched, the orbital magnetic moment is also quenched. In such elements, the magnetic moment arises from the spin angular momentum only.

Problem 18.5

Derive the expression for the paramagnetic susceptibility in a metal with free electrons contributing to magnetization.

SUGGESTED READING

Bates, L. F. (1961). *Modern magnetism.* London: Cambridge University Press.
Cracknell, A. P. (1975). *Magnetism in crystalline materials.* New York: Pergamon Press.
Mattis, D. C. (1988). *The theory of magnetism: Static and dynamics.* (Vol. 1). New York: Springer-Verlag.
Stoner, E. C. (1934). *Magnetism and matter*. London: Methuen & Co., Ltd.
Van Vleck, J. H. (1932). *The theory of electric and magnetic susceptibilities.* London: Oxford University Press.

Chapter 19

Ferromagnetism

Chapter Outline

A large number of elements can be magnetized with the application of an external magnetic field. But there exist some elements that show magnetism even when the external magnetic field is switched off. Finite magnetization in the absence of a magnetic field is called *spontaneous magnetization*. The elements that show finite spontaneous magnetization are called ferromagnetic elements and the phenomenon is called *ferromagnetism*. The spontaneous magnetization suggests that the electron spins and hence the magnetic moments are aligned in a regular manner. Therefore, a ferromagnetic state is an ordered state and ferromagnetism is a cooperative phenomenon. Some common ferromagnetic elements are Fe^{26}, Ni^{27}, Co^{28}, Gd^{64}, and Dy^{66}. In addition to these there are a large number of ferromagnetic materials that are either oxides or alloys of the elements listed above.

At very low temperatures, a ferromagnetic material shows spontaneous magnetization $\mathbf{M}(T)$, which is a function of temperature T. With an increase in temperature, $\mathbf{M}(T)$ decreases and at a particular temperature T_c, spontaneous magnetization vanishes. T_c is called the ferromagnetic transition temperature above which the material behaves as a paramagnetic material. Therefore, T_c separates the ordered ferromagnetic state from the disordered paramagnetic state.

19.1 WEISS MOLECULAR FIELD THEORY

Weiss was the first to explain ferromagnetic behavior in solids. He made two assumptions

1. A ferromagnetic substance of macroscopic dimensions contains, in general, a large number of small regions, called ferromagnetic domains, which show spontaneous magnetization $\mathbf{M}(T)$ as a function of temperature T. In one domain all the magnetic moments are aligned in one direction, but the direction of alignment may be different in different domains. The spontaneous magnetization in a domain is defined as the vector sum of the magnetic dipole moments of all the atoms in that domain divided by the volume. From the above facts, it is evident that spontaneous magnetization is a cooperative phenomenon of all the atomic dipoles within a single domain.
2. Within each domain there is some internal interaction tending to align all the magnetic moments parallel to each other. This internal interaction gives rise to a field, which was called the molecular field by Weiss, though perhaps more appropriately it should be called the exchange field $\mathbf{H}_{ex}$ (see Section 19.5). Weiss assumed that the molecular field is proportional to the magnetization, that is,

$$\mathbf{H}_{ex}(T) = \lambda_M \mathbf{M}(T) \tag{19.1}$$

Here λ_M is called the molecular (or Weiss) constant and is independent of temperature. According to Eq. (19.1) the molecular field is a function of temperature. In this approximation, each magnetic moment sees the average magnetization of all other magnetic moments.

Solid State Physics. https://doi.org/10.1016/B978-0-12-817103-5.00019-0

In the absence of an applied magnetic field, all of the domains in a ferromagnetic substance are randomly oriented, thereby yielding zero magnetization. When an external magnetic field **H** is applied the domains try to orient themselves in the direction of the field, yielding finite magnetization. Therefore, in a ferromagnetic substance, the total magnetic field $\mathbf{H}_t$ is given by

$$\mathbf{H}_t = \mathbf{H} + \mathbf{H}_{ex} = \mathbf{H} + \lambda_M \mathbf{M}(T) \tag{19.2}$$

If the applied magnetic field is weak, **M**(T) is linearly proportional to the field $\mathbf{H}_t$ (linear approximation), that is,

$$\mathbf{M}(T) = \chi_M (\mathbf{H} + \mathbf{H}_{ex}) \tag{19.3}$$

where χ_M is called the magnetic susceptibility. One can calculate χ_M of a ferromagnetic substance above T_c (in the paramagnetic region) by assuming the Curie form of the susceptibility for a paramagnetic substance, that is,

$$\chi_M = \frac{C_M}{T} \tag{19.4}$$

Substituting Eq. (19.1), (19.4) into Eq. (19.3), one gets

$$\chi_M = \frac{M(T)}{H} = \frac{C_M}{T - T_c} \tag{19.5}$$

where

$$T_c = \lambda_M C_M \tag{19.6}$$

Eq. (19.5) is called the Weiss-Curie formula. χ_M as a function of T is shown in Fig. 19.1.

19.2 CLASSICAL THEORY OF FERROMAGNETISM

The magnetization produced in a paramagnetic substance, according to the Langevin theory, is given by

$$M(T) = \rho^a \mu L(x) \tag{19.7}$$

where

$$x = \beta_0 \mu H \tag{19.8}$$

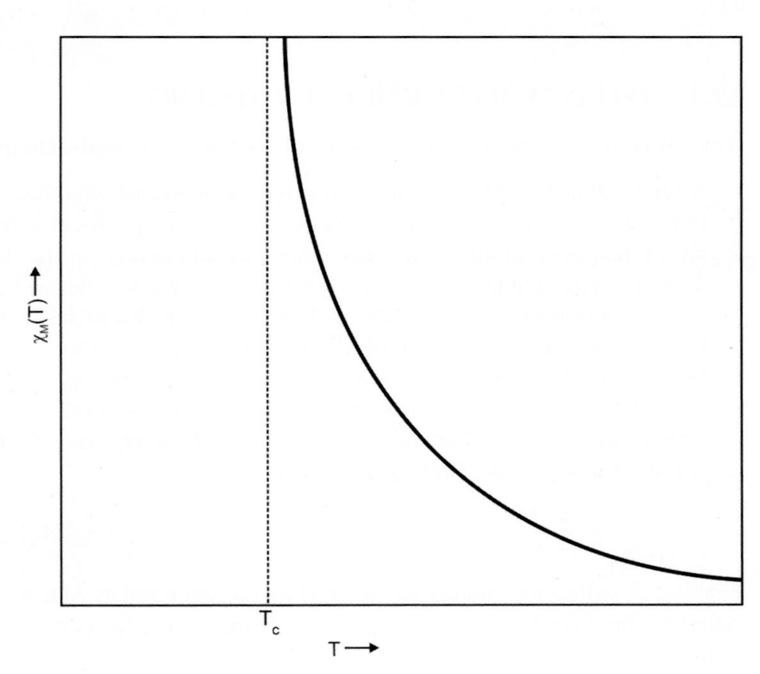

FIG. 19.1 The temperature variation of the magnetic susceptibility χ_M of a ferromagnetic solid for $T \rangle T_c$.

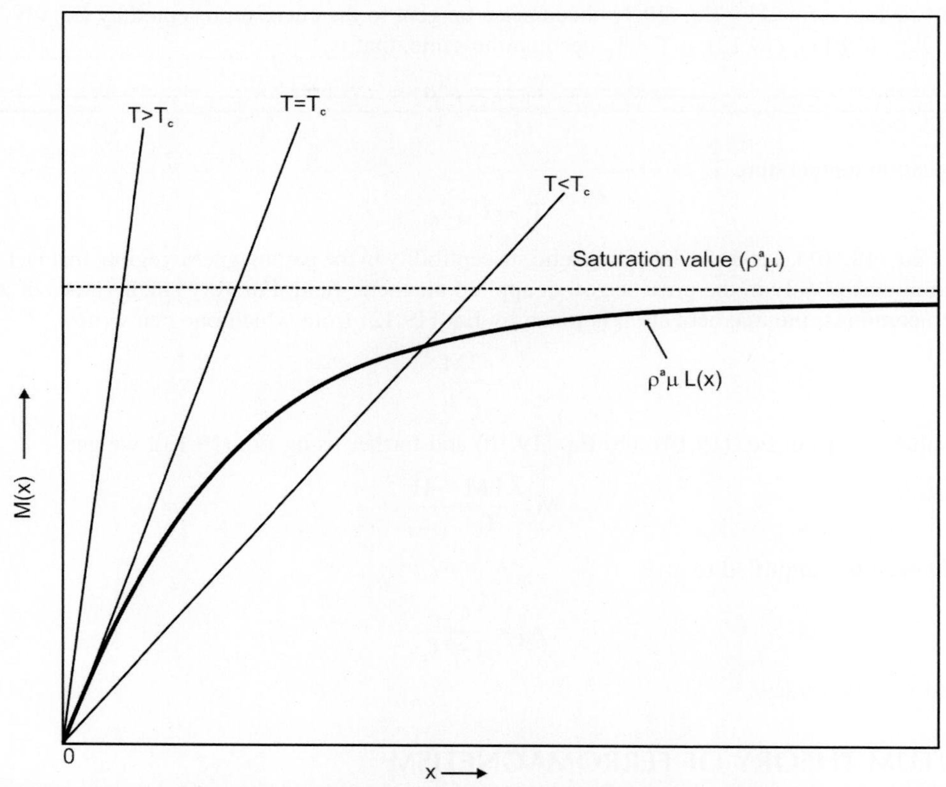

FIG. 19.2 Magnetization M(x) as a function of the parameter $x = \mu H/k_B T$ in the classical theory of a paramagnetic solid. The finite spontaneous magnetization is given by the point of intersection of the curve corresponding to Eq. (19.7) and the straight line corresponding to Eq. (19.11) for $T < T_c$.

A plot of Eq. (19.7) is shown in Fig. 19.2. In a ferromagnetic substance the total field is the sum of the applied magnetic field and the molecular field. One can derive an expression for the magnetization of a ferromagnetic substance by replacing **H** by $\mathbf{H}_t$ in Eq. (19.8). In doing so one gets the same relation as was obtained in Eq. (19.7) in which

$$x = \beta_0 \mu (H + \lambda_M M) \tag{19.9}$$

$$M = \frac{k_B T}{\mu \lambda_M} x - \frac{H}{\lambda_M} \tag{19.10}$$

The plot of Eq. (19.10) gives a straight line with $k_B T/\mu \lambda_M$ as the slope and $(-H/\lambda_M)$ as the intercept. The spontaneous magnetization from Eq. (19.10) is given by

$$M = \frac{k_B T}{\mu \lambda_M} x \tag{19.11}$$

which is a straight line with slope $k_B T/\mu \lambda_M$ and that passes through the origin. In a ferromagnetic substance M must satisfy both Eqs. (19.7), (19.11); therefore, the value of M is given by the point of intersection of these equations (see Fig. 19.2).

The slope of Eq. (19.11) increases with an increase in temperature and, at a particular temperature represented by T_c, the straight line becomes a tangent to the curve represented by Eq. (19.7) at the origin. At $T > T_c$, the straight line intersects the curve only at the origin, yielding a zero value for the spontaneous magnetization M (paramagnetic state). But for $T < T_c$ the straight line intersects the curve at two points: one at the origin and the other at a finite value of M, which indicates that the material shows ferromagnetism in this temperature range. There must be some relation between λ_M and T_c. At very small values of x, Eq. (19.7) gives

$$M = \rho^a \mu \frac{x}{3} \tag{19.12}$$

When the straight line represented by Eq. (19.11) becomes a tangent to the curve represented by Eq. (19.7) at the origin, then the slopes of Eqs. (19.11), (19.12) at $T=T_c$ become the same, that is,

$$\frac{k_B T_c}{\mu \lambda_M} = \frac{\rho^a \mu}{3}$$

This gives the transition temperature T_c as

$$T_c = C_M \lambda_M \tag{19.13}$$

Here we have used Eq. (18.101). Let us find the magnetic susceptibility in the paramagnetic region, that is, for $T \rangle T_c$. In this region magnetization occurs only in the presence of an applied magnetic field. For very small values of x, that is, at low fields and high temperatures, the magnetization is given by Eq. (19.12) from which one can write

$$x = \frac{3M}{\rho^a \mu} \tag{19.14}$$

Substituting the value of x from Eq. (19.14) into Eq. (19.10) and further using Eq. (19.13), we get

$$M = \frac{TM}{T_c} - \frac{H}{\lambda_M} \tag{19.15}$$

The above equation can be simplified to give

$$\chi_M = \frac{C_M}{T - T_c} \tag{19.16}$$

Eq. (19.16) is the Curie-Weiss law.

19.3 QUANTUM THEORY OF FERROMAGNETISM

Consider a ferromagnetic solid with ρ^a number of atoms per unit volume. According to the quantum theory of paramagnetism, the magnetization is given by

$$M_J(x) = \rho^a g_J \mu_B J B_J(x) \tag{19.17}$$

with

$$x = \beta_0 g_J \mu_B J H \tag{19.18}$$

[see Eq. (18.115)]. A plot of $M_J(x)$ as a function of x from Eq. (19.17), which approaches its saturation value $\rho^a g_J \mu_B J$ at large values of x, is shown by the curve in Fig. 19.3. One obtains different curves at different temperatures but the form of the curve remains the same. For a ferromagnetic substance the expression for $M_J(x)$ remains the same except that H is replaced by H_t in the expression for x, that is,

$$x = \beta_0 g_J \mu_B J H_t \tag{19.19}$$

Substituting the value of H_t from Eq. (19.2) into Eq. (19.19) and solving for $M_J(x)$, we get

$$M_J(x) = \frac{k_B T}{\lambda_M g_J \mu_B J} x - \frac{H}{\lambda_M} \tag{19.20}$$

Therefore, the magnetization in a ferromagnetic substance, given by Eq. (19.20), is a straight line. The spontaneous magnetization $M_J(x)$ from Eq. (19.20) is given by

$$M_J(x) = \frac{k_B T}{\lambda_M g_J \mu_B J} x \tag{19.21}$$

Eq. (19.21) represents a straight line passing through the origin whose slope increases with an increase in T. Because $M_J(x)$ must satisfy both Eqs. (19.17), (19.21), the value of $M_J(x)$ is given by the points of intersection of the curve and the straight line (see Fig. 19.3). With a decrease in T the slope of the straight line decreases and at a particular value, say T_c, it becomes a tangent to the curve at the origin. For $T \rangle T_c$, the straight line does not intersect the curve except at the origin, yielding a zero value for $M_J(x)$. But for $T \langle T_c$, the straight line intersects the curve at two points, one at the origin and the other at finite magnetization. Therefore, a finite value of magnetization in the absence of a magnetic field is obtained only for $T \langle T_c$.

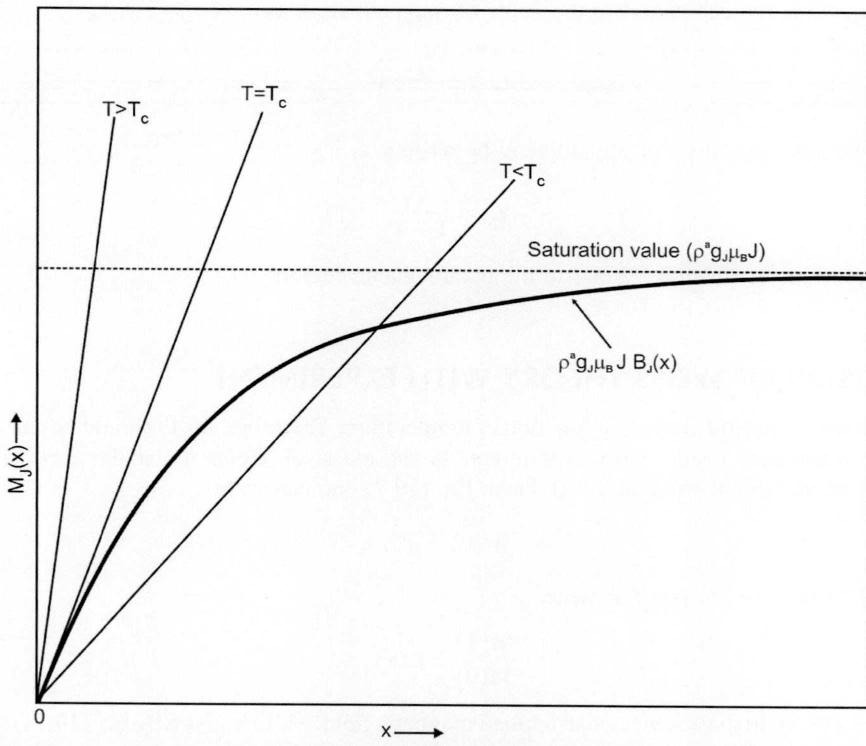

FIG. 19.3 Magnetization M(x) as a function of the parameter $x = g_J\mu_B J H/k_B T$ in the quantum mechanical theory of a paramagnetic solid. The finite spontaneous magnetization is given by the point of intersection of the curve, corresponding to Eq. (19.17), and the straight line, corresponding to Eq. (19.21), for T ⟨ T_c.

Let us investigate the relationship between T_c and λ_M. For very small values of x, the magnetization from Eq. (19.17) is given by

$$M_J(x) = \rho^a g_J \mu_B \frac{J+1}{3} x \tag{19.22}$$

At $T = T_c$, the slopes of both Eqs. (19.21), (19.22), representing straight lines, become the same, therefore

$$\frac{k_B T_c}{\lambda_M g_J \mu_B J} = \rho^a g_J \mu_B \frac{J+1}{3}$$

which gives

$$T_c = C_J \lambda_M \tag{19.23}$$

Here C_J is the Curie constant already defined by Eq. (18.124). Eq. (19.23) says that T_c is large if λ_M is large, that is, the molecular field is large. If there is spin only, then $J = 1/2$ and $g_J = 2$ and in that case $C_J = C_{1/2}$. Therefore, for the case of spin only, Eq. (19.23) reduces to

$$T_c = \frac{\rho^a \mu_B^2}{k_B} \lambda_M \tag{19.24}$$

As a limiting case, the magnetic susceptibility can be calculated in the paramagnetic region, which occurs either at high temperatures or at low magnetic fields, that is, at very small values of x. In this limit, substituting the value of x from Eq. (19.22) into Eq. (19.20), we find

$$M_J = \frac{T M_J}{T_c} - \frac{H}{\lambda_M} \tag{19.25}$$

From the above equation one can calculate the value of magnetization as

$$M_J = \frac{C_J}{T - T_c} H \tag{19.26}$$

Therefore, the magnetic susceptibility can immediately be written as

$$\chi_M = \frac{C_J}{T - T_c} \tag{19.27}$$

which is the same expression as given by Weiss in the paramagnetic region.

19.4 COMPARISON OF WEISS THEORY WITH EXPERIMENT

Magnetization is measured experimentally as a function of temperature. Therefore, M(T) should be calculated as a function of T to compare the theoretical results with experiment. In the classical Weiss molecular theory, M(T) is given by Eq. (19.7), which has its maximum value at T = 0. From Eq. (19.7) one can write

$$M(0) = \rho^a \mu \tag{19.28}$$

Hence, from Eqs. (19.7) and (19.28) one can write

$$\frac{M(T)}{M(0)} = L(x) \tag{19.29}$$

where x is given by Eq. (19.8). In the absence of an applied magnetic field, M(T) is given by Eq. (19.11) and, therefore, from Eqs. (19.28), (19.11) one can write

$$\frac{M(T)}{M(0)} = \frac{k_B T}{\rho^a \mu^2 \lambda_M} x = \frac{T}{3T_c} x \tag{19.30}$$

The quantity M(T)/M(0) for a particular value of T can be calculated from the point of intersection of the curves represented by Eqs. (19.29), (19.30). A plot of magnetization as a function of temperature is shown in Fig. 19.4 for the classical case. It is evident from the figure that the classical theory is not able to explain the experimental results of spontaneous magnetization.

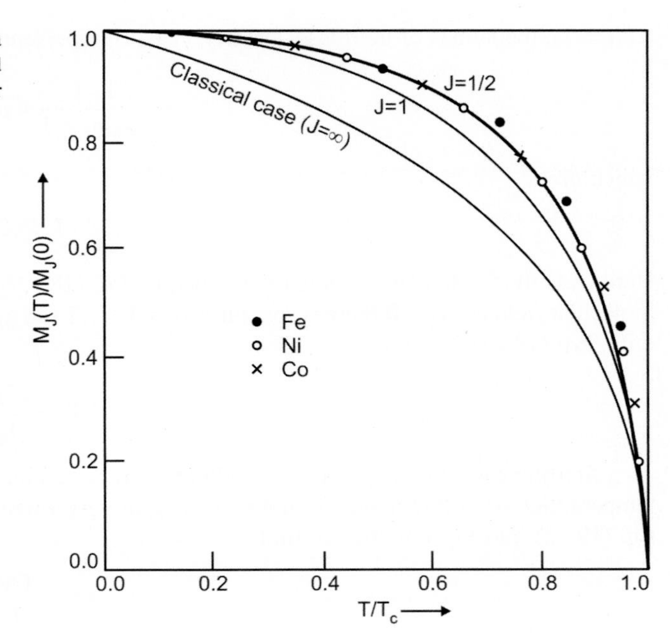

FIG. 19.4 The theoretical values of magnetization M(T)/M(0) as a function of temperature T/T_c for total angular momentum J = ½, 1, and ∞ along with the experimental values. The values of magnetization for J = ∞ correspond to the classical Langevin theory.

Let us apply the quantum theory of ferromagnetism to examine the behavior of $M_J(T)$. From Eq. (19.17) the saturation magnetization is given as

$$M_J(0) = \rho^a g_J \mu_B J \tag{19.31}$$

From Eqs. (19.17), (19.31) one can write

$$\frac{M_J(T)}{M_J(0)} = B_J(x) \tag{19.32}$$

The magnetization in a ferromagnetic substance in the absence of an applied magnetic field is given by Eq. (19.21). So, from Eqs. (19.21), (19.31), one can write

$$\frac{M_J(T)}{M_J(0)} = \frac{T(J+1)}{3T_c J} x = \left(1 + \frac{1}{J}\right) \frac{T}{3T_c} x \tag{19.33}$$

Eq. (19.33) shows that $M_J(T)$ depends on the total angular momentum quantum number J, which is contrary to the classical Weiss theory. As a result, one obtains different curves of $M_J(T)$ as a function of T for different values of J. The value of the quantity $M_J(T)/M_J(0)$ is obtained from the point of intersection of Eqs. (19.32), (19.33). Fig. 19.4 shows graphs for $M_J(T)/M_J(0)$ as a function of T/T_c for $J = 1/2$, 1, and ∞ along with the experimental values. The graph for $J = \infty$ assumes a continuous distribution of magnetic dipoles (classical case). Fig. 19.4 shows that $M_J(T)/M_J(0)$ decreases smoothly with an increase in T and becomes zero at $T = T_c$. This behavior shows that a ferromagnetic-to-paramagnetic transition or vice versa is of second order. From the figure it is evident that the curve for $J = 1/2$ fits the experimental data best, which shows that the magnetization in Fe, Ni, and Co basically arises from the electron spins.

The above result is also confirmed by gyromagnetic experiments. In these experiments one either reverses the magnetization of a freely suspended magnetic material and observes the resulting rotation or one rotates the specimen and observes the resulting magnetization: the former is called the Einstein-de Haas method and the latter the Barnett method. From such experiments the value of the gyromagnetic ratio is found to be two, that is, $g_J = 2$, which confirms that the magnetization is largely due to electron spin.

As a limiting case, one can study the variation of magnetization at very low temperatures. Consider the variation for $J = 1/2$ for which

$$B_{1/2}(x) = \tanh x \quad \text{and} \quad x = \frac{\mu_B H}{k_B T} \tag{19.34}$$

Therefore, from Eq. (19.32) one gets, for ½ spin,

$$\frac{M_{1/2}(T)}{M_{1/2}(0)} = \tanh x \tag{19.35}$$

From Eq. (19.33) for $J = 1/2$, one can write

$$x = \frac{T_c}{T} \frac{M_{1/2}(T)}{M_{1/2}(0)} \tag{19.36}$$

Substituting the value of x from Eq. (19.36) into Eq. (19.35), we find

$$\frac{M_{1/2}(T)}{M_{1/2}(0)} \approx 1 - 2 \exp\left(-\frac{2T_c}{T} \frac{M_{1/2}(T)}{M_{1/2}(0)}\right) \tag{19.37}$$

At very low temperatures the second term on the right side of Eq. (19.37) is very small compared with unity, therefore, $M_{1/2}(T)$ is independent of temperature and is nearly equal to the saturation value, that is,

$$M_{1/2}(T) \approx M_{1/2}(0) \tag{19.38}$$

On the other hand, the experimental results exhibit $T^{3/2}$ dependence as follows:

$$\frac{M(T)}{M(0)} = 1 - A T^{3/2} \tag{19.39}$$

where A is a constant. Eq. (19.39) is known as the Bloch $T^{3/2}$ law. The change in magnetization with temperature is defined as

$$\Delta M(T) = M(0) - M(T) \tag{19.40}$$

From Eq. (19.37) one can write

$$\frac{\Delta M_{1/2}(T)}{M_{1/2}(0)} \approx 2\exp\left(-\frac{2T_c}{T}\frac{M_{1/2}(T)}{M_{1/2}(0)}\right) \tag{19.41}$$

In the case of spin, one can write from Eq. (19.35)

$$\frac{M_{1/2}(T)}{M_{1/2}(0)} = \tanh\left(\frac{\mu_B H}{k_B T}\right) = 1 + \frac{\mu_B H}{k_B T} \tag{19.42}$$

Substituting Eq. (19.42) into the right side of Eq. (19.41), we obtain

$$\frac{\Delta M_{1/2}(T)}{M_{1/2}(0)} \approx 2\exp\left(-\frac{2T_c}{T}\right) \tag{19.43}$$

Order of Magnitude of Molecular Field

From Eqs. (19.23) and (18.124) we write

$$T_c = \frac{\lambda_M \rho^a g_J^2 \mu_B^2 J(J+1)}{3k_B} \tag{19.44}$$

For $J=1$, $g_J=2$, and $T_C = 1000$ K the factor λ_M has the value

$$\lambda_M = 10^4$$

For M = 1700, the molecular field becomes

$$H_{ex} = \lambda_M M \approx 10^7 \text{gauss}$$

We know that the local field due to magnetization is $(4\pi/3)M$, which is very small compared with the molecular field. This shows that the origin of the molecular field is very much different from that of the local magnetic field.

The order of magnitude of the molecular field can also be obtained from a different argument. The energy due to the Bohr magnetron in the molecular field is $\mu_B H_{ex}$ and its maximum value is equal to the thermal energy at T_c, that is, $k_B T_c$. Therefore,

$$\mu_B H_{ex} = k_B T_c \tag{19.45}$$

For a Curie temperature of $T_c \approx 1000$ K, we get $H_{ex} = 10^7$ gauss. The magnetic field arising from the dipole-dipole interaction between neighboring dipoles is

$$H_{dp} = \frac{\mu_B}{a^3} \approx 10^3 \text{gauss} \tag{19.46}$$

where a is the interdipole distance. Therefore, in a ferromagnetic substance, the molecular field constant λ_M is given by

$$\lambda_M = \frac{H_{ex}}{H_{dp}} = 10^4 \tag{19.47}$$

which is the same value as obtained above. Further, it is very large compared with the Lorentz field constant, which is on the order of $4\pi/3$.

19.5 HEISENBERG THEORY OF FERROMAGNETISM

The Weiss molecular field theory explains some aspects of ferromagnetism satisfactorily but does not provide any explanation for the origin of the molecular field. In 1928 Heisenberg showed that the molecular field can be explained in terms of

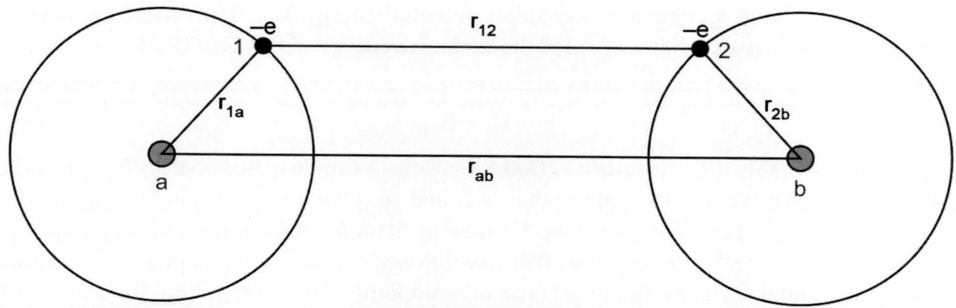

FIG. 19.5 Hydrogen molecule; a and b are nuclei of two hydrogen atoms. Electron 1 belongs to nucleus a, while electron 2 belongs to nucleus b. The distances between the nuclei and electrons are shown in the figure.

the so-called exchange interactions between the electrons. The concept of exchange interactions can be explained by considering a hydrogen molecule, as shown in Fig. 19.5. Let the nuclei in a hydrogen molecule be denoted as a and b and the wave functions of the electrons associated with these nuclei be $|\psi_a\rangle$ and $|\psi_b\rangle$. Here it is assumed that the wave functions of these electrons do not contain orbital angular momentum. In other words, all of the magnetic moments arise from spin angular momentum. The net electrostatic interaction potential between the two atoms is given by

$$V_{ab} = e^2\left(\frac{1}{r_{ab}} + \frac{1}{r_{12}} - \frac{1}{r_{1a}} - \frac{1}{r_{2b}}\right) \tag{19.48}$$

According to the Heitler-London theory the energy of the system can be written as

$$E = E_c \pm E_{ex} \tag{19.49}$$

where

$$E_c = 2E_0 + \langle\psi_a(1)\psi_b(2)|V_{ab}|\psi_a(1)\psi_b(2)\rangle \tag{19.50}$$

$$E_{ex} = \langle\psi_a(1)\psi_b(2)|V_{ab}|\psi_a(2)\psi_b(1)\rangle \tag{19.51}$$

Here E_0 is the energy of a free atom (kinetic energy), E_c is the Coulomb interaction energy between the two atoms, and E_{ex} is the exchange energy. In Eq. (19.49) the plus sign refers to the nonmagnetic state in which the spins of the electrons are antiparallel, while the negative sign corresponds to the magnetic state of the molecule with parallel spins.

To illustrate the formation of magnetic and nonmagnetic states, let us consider the case of C^{12} with an electronic configuration given by

$$C^{12} : 1s^2 2s^2 2p^2$$

In C^{12} there are two possible distributions of electrons in the p-state. The first distribution is given below:

$$\begin{array}{llll} m_\ell : & 1 & 0 & -1 \\ m_{s\uparrow} : & \frac{1}{2} & & \\ m_{s\downarrow} : & -\frac{1}{2} & & \end{array}$$

In the distribution above, the same p-sublevel contains two electrons with opposite spins and it yields net zero spin and, therefore, zero magnetic moment on the carbon atom (nonmagnetic state). Here the spin wave function is antisymmetric, but the orbital wave function is symmetric because the total wave function has to be antisymmetric. In the nonmagnetic state two electrons can come very close to each other giving rise to large potential energy V_{ab}. The second distribution of electrons is as follows:

$$\begin{array}{llll} m_\ell : & 1 & 0 & -1 \\ m_{s\uparrow} : & \frac{1}{2} & \frac{1}{2} & \end{array}$$

Here the two electrons have parallel spins and are in two different sublevels of the p-state, yielding total spin one and hence a finite magnetic moment (magnetic state). In the magnetic state the two electrons cannot come very close to each other due to Pauli's exclusion principle, giving a comparatively small potential energy V_{ab}. The difference between the potential energies of the magnetic and nonmagnetic states is called the exchange energy. From Eq. (19.49) the energy of the magnetic state is given by

$$E = E_c - E_{ex} \tag{19.52}$$

Therefore, the magnetic state is stable if E_{ex} is positive. Bethe has done a simple qualitative analysis of the condition under which E_{ex} is most likely to be positive. Let us suppose that $|\psi_a\rangle$ and $|\psi_b\rangle$ have no nodes in the region where they overlap appreciably, then the product $|\psi_a(1)\,\psi_b(1)\rangle$ or $|\psi_a(1)\,\psi_b(2)\rangle$ may be assumed to be positive everywhere. This condition is always satisfied if $|\psi_a\rangle$ and $|\psi_b\rangle$ are s-wave functions that have no nodes close to the nuclei. This condition may also be satisfied in other cases if the nodal surfaces do not lie near the midpoint of the line joining the centers of the two atoms where there is maximum overlap. Under these conditions the positive terms

$$\frac{e^2}{r_{ab}} + \frac{e^2}{r_{12}}$$

favor magnetism, whereas the negative terms

$$-\frac{e^2}{r_{1a}} - \frac{e^2}{r_{2b}}$$

do not favor magnetism. The exchange energy E_{ex} is most likely positive if

1. The distance r_{ab} is fairly large compared with the orbital radii.
2. The wave functions are comparatively small near the nuclei.

The second condition is most fully satisfied when the orbital quantum number ℓ is high because the wave function varies as r^{ℓ}. Therefore, E_{ex} is expected to be positive for interactions between electrons in the partially filled d- and f-shells when the interatomic distance is large compared with the atomic radius. These conditions are actually satisfied by the pairs of atoms of the iron group metals and rare-earth metals in which the internuclear distances are primarily determined by the s-p valence electrons. The calculated values of E_{ex} are shown in Fig. 19.6. It is found that E_{ex} is negative for small values of r_{ab}/r_0 (r_0 is the atomic radius) but that it becomes positive for reasonably large values. According to Slater, the ratio r_{ab}/r_0 should be larger than 3, but not much larger. E_{ex} is negative for Mn and $\gamma-$Fe (fcc), which are antiferromagnets with $r_{ab}/r_0 \langle 3$, and positive for $\alpha-$Fe(bcc), Co, and Ni, which are ferromagnets with $3.0 \langle r_{ab}/r_0 \langle 5.0$. E_{ex} decreases rapidly

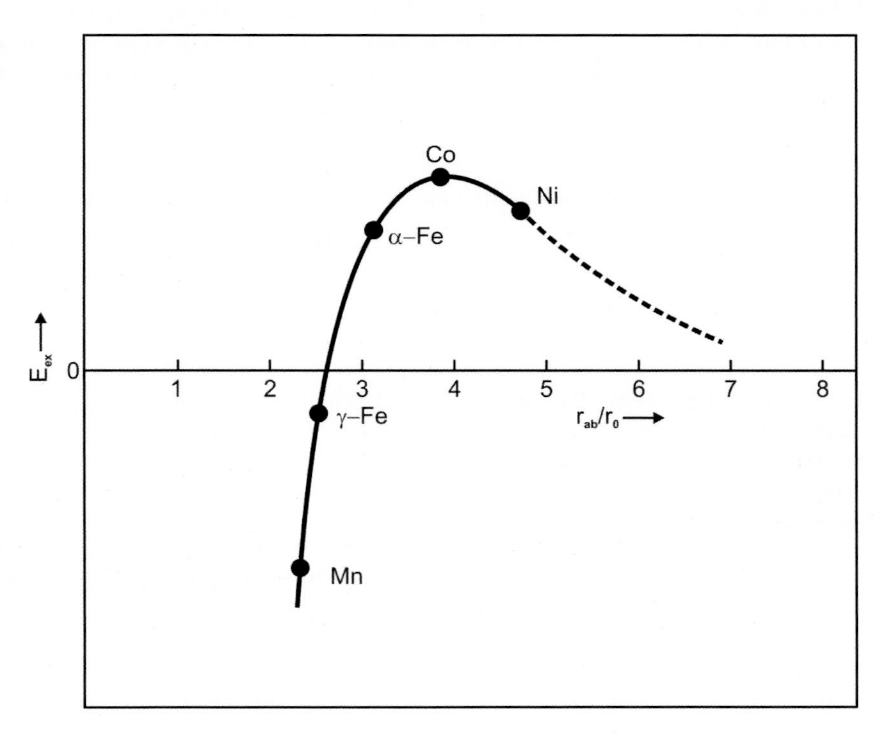

FIG. 19.6 The exchange energy E_{ex}(or exchange integral J), given by Eq. (19.51), is plotted as a function of the internuclear distance r_{ab}/r_0.

for $r_{ab}/r_0 \rangle 5.0$, which indicates that only the interaction between the 1NNs is important for strong ferromagnetism. The values of r_{ab}/r_0 for some of the elements are given below:

Element	Fe	Co	Ni	Cr	Mn	Gd
r_{ab}/r_0	3.26	3.64	3.96	2.60	2.94	3.10

Note that Cr and Mn are not ferromagnets but exhibit antiferromagnetism. Here a question arises as to whether an element with uncompensated spins, which itself is not a ferromagnet because the ratio r_{ab}/r_0 is not favorable, can be combined with another nonferromagnetic element to form a compound for which r_{ab}/r_0 is suitable for ferromagnetism. The answer is yes, it is indeed possible, because MnAs and MnSb are ferromagnetic materials, whereas their components are individually nonferromagnetic in nature.

Let us derive the expression for the exchange interaction theoretically. For simplicity consider two electrons with spins $\mathbf{s}_1$ and $\mathbf{s}_2$. The total spin of the two electrons is given by

$$\mathbf{s} = \mathbf{s}_1 + \mathbf{s}_2 \tag{19.53}$$

Squaring both sides, we write

$$s^2 = s_1^2 + s_2^2 + 2\mathbf{s}_1 \cdot \mathbf{s}_2 \tag{19.54}$$

The eigenvalues of the operator $\mathbf{s}^2$ are s(s+1) in units of $\hbar^2$, with similar relations for $\mathbf{s}_1^2$ and $\mathbf{s}_2^2$. Therefore, from the above equation, the eigenvalues of $\mathbf{s}_1 \cdot \mathbf{s}_2$ are given by

$$2\mathbf{s}_1 \cdot \mathbf{s}_2 = s(s+1) - s_1(s_1+1) - s_2(s_2+1) \tag{19.55}$$

But s_1 and s_2 have the same value (1/2); therefore, one can write

$$2\mathbf{s}_1 \cdot \mathbf{s}_2 = s(s+1) - 2s_1(s_1+1) \tag{19.56}$$

The operators s^2, s_1^2, and s_2^2 are constants of motion and, therefore, $\mathbf{s}_1 \cdot \mathbf{s}_2$ is also a constant of motion. Now s has values 0 and 1, which gives values of 0 and 2 for s(s+1) for antiparallel-spin (nonmagnetic) and parallel-spin (magnetic) states, respectively. The spin s_1 has value ½, which gives $s_1(s_1+1) = ¾$. Substituting these values into Eq. (19.56), one gets

$$2\mathbf{s}_1 \cdot \mathbf{s}_2 = -3/2 \quad \text{for } s = 0 \tag{19.57}$$

and

$$2\mathbf{s}_1 \cdot \mathbf{s}_2 = 1/2 \quad \text{for } s = 2 \tag{19.58}$$

Hence the Hamiltonian corresponding to Eq. (19.49) can is given by

$$\widehat{H} = E_c - \frac{E_{ex}}{2} - 2E_{ex}\mathbf{s}_1 \cdot \mathbf{s}_2 \tag{19.59}$$

The Hamiltonian given by Eq. (19.59) is called the Heisenberg Hamiltonian. The last term in Eq. (19.59) is called the exchange energy and represents the direct coupling between the two spins. It must be emphasized that the exchange interaction is fundamentally electrostatic in nature and the spins enter into the energy expression as a consequence of the Pauli exclusion principle.

The above arguments can be generalized to a system with a large number of electrons. Let there be two atoms each with a number of electrons. The exchange Hamiltonian for the ith electron $\widehat{H}_{ex}^{i}$ is obtained by summing its interaction with other neighboring (jth) spins, so one can write the last term of Eq. (19.59) as

$$\widehat{H}_{ex}^{i} = -2\sum_{j} E_{ex}^{ij}\mathbf{s}_i \cdot \mathbf{s}_j \tag{19.60}$$

Here $\mathbf{s}_i$ and $\mathbf{s}_j$ are the spins of ith and jth electrons. The total exchange Hamiltonian $\widehat{H}_{ex}$ is obtained by summing over all the ith spins, that is,

$$\widehat{H}_{ex} = \frac{1}{2}\sum_{i}\widehat{H}_{ex}^{i} = -\sum_{i,j} E_{ex}^{ij}\mathbf{s}_i \cdot \mathbf{s}_j \tag{19.61}$$

The factor of ½ takes care of the fact that in the summation each pair is counted once only. The exchange energy E_{ex}^{ij} can be replaced by J_{ij} for convenience of notation and it is called the exchange integral. Therefore, Eqs. (19.60), (19.61) can be written as

$$\widehat{H}_{ex}^{i} = -2\sum_{j} J_{ij}\,\mathbf{s}_i \cdot \mathbf{s}_j \tag{19.62}$$

$$\widehat{H}_{ex} = \frac{1}{2}\sum_{i} \widehat{H}_{ex}^{i} = -\sum_{i,j} J_{ij}\,\mathbf{s}_i \cdot \mathbf{s}_j \tag{19.63}$$

Eq. (19.63) also includes the interaction of a spin with itself, which does not give the exchange interaction and, therefore, should be excluded. The total Hamiltonian of the system can be written as

$$\widehat{H}_{ex} = E_0 - \sum_{i,j(i \neq j)} J_{ij}\,\mathbf{s}_i \cdot \mathbf{s}_j \tag{19.64}$$

Here E_0 represent the first two terms of Eq. (19.59). The exact solution of the Heisenberg Hamiltonian is very difficult and most attempts at doing so are directed toward simplified model calculations. One such simplified form is the *Ising model* in which all the spins are directed along the z-direction. In this approximation the exchange Hamiltonian becomes

$$\widehat{H}_{ex} = -\sum_{i,j(i \neq j)} J_{ij}\,s_i^z s_j^z \tag{19.65}$$

where s^z is the z-component of the spin, which is a scalar.

The exchange integral is of central importance in ferromagnetism. Therefore, one would like to relate it to the molecular field constant λ_M and the ferromagnetic transition temperature T_c. For simplicity it is assumed that only the 1NN interactions are dominant in the exchange integral and that it has the same value J_0 for all of the 1NNs. With these assumptions the exchange Hamiltonian $\widehat{H}_{ex}^{i}$ from Eq. (19.62) can be written as

$$\widehat{H}_{ex}^{i} = -2J_0\sum_{j} \mathbf{s}_i \cdot \mathbf{s}_j \tag{19.66}$$

In Eq. (19.66) the summation is over the 1NNs. Assuming the magnetization to be along the z-direction and, hence, assuming the spins to be in the z-direction with magnitude s, Eq. (19.66) becomes

$$\widehat{H}_{ex}^{i} = -2J_0 n\, s^2 \tag{19.67}$$

for the spins aligned parallel to each other. Here n is the number of 1NNs. If the spins are aligned opposite to each other, then

$$\widehat{H}_{ex}^{i} = 2J_0 n\, s^2 \tag{19.68}$$

Eq. (19.67) gives the ground state for ferromagnetism as it yields minimum energy. The energy given by Eq. (19.67) is equivalent to the energy of the magnetic moment $\vec{\mu}$ in the exchange field $\mathbf{H}_{ex}$, that is,

$$-2J_0 n s^2 = -\vec{\mu} \cdot \mathbf{H}_{ex} \tag{19.69}$$

Substituting the value of $\vec{\mu}$ and considering spin only, one writes

$$-2J_0 n s^2 = -g_s \mu_B \mathbf{s} \cdot \mathbf{H}_{ex} \tag{19.70}$$

Substituting $\mathbf{H}_{ex}$ from Eq. (19.1) into the above equation and assuming that $\mathbf{M}$ and $\mathbf{s}$ are in the same direction, one obtains

$$-2J_0 n s^2 = -g_s \mu_B s \lambda_M M \tag{19.71}$$

Therefore, the exchange integral is given by

$$J_0 = \frac{g_s \mu_B \lambda_M M}{2ns} \tag{19.72}$$

The saturation value of magnetization M, the Curie constant C_s, and the transition temperature T_c are given by

$$M = \rho^a g_s \mu_B s \tag{19.73}$$

$$C_s = \frac{\rho^a g_s^2 \mu_B^2 s(s+1)}{3k_B} \tag{19.74}$$

$$T_c = \lambda_M C_s \tag{19.75}$$

Substituting Eqs. (19.73), (19.74), and (19.75) into Eq. (19.72), one obtains

$$\frac{J_0}{k_B T_c} = \frac{3}{2ns(s+1)} \tag{19.76}$$

For an sc lattice, n = 6 and s = ½; therefore, for an sc lattice Eq. (19.76) reduces to

$$\frac{J_0}{k_B T_c} = \frac{1}{3} \approx 0.33 \tag{19.77}$$

But more detailed calculations for an sc lattice give 0.518 and 0.540 values for the above ratio.

19.6 SPIN WAVES

We have discussed in detail that ferromagnetism basically arises from the atomic or molecular spin magnetic moments. Further, the various spins interact with each other (spin-spin interaction), giving rise to an exchange field. Let us examine the effect of the spin-spin interaction on the spin system. Suppose there is a system in which N spins are aligned periodically on a line with a as the distance between the adjacent spins. In the ground state of the system, all of the N spins are parallel to each other, as shown in Fig. 19.7A. The field arising from the spins is given by the Heisenberg Hamiltonian defined as

$$\widehat{H}_{ex} = -2J \sum_i \mathbf{s}_i \cdot \mathbf{s}_{i+1} \tag{19.78}$$

The exchange integral J is assumed to be the same for all the NNs. For simplicity assume that the spin-spin interaction is appreciable only between adjacent spins. At absolute zero all the spins are aligned along one direction, say the z-direction. In this situation $s_x = s_y = 0$ and $s_z = s$. Therefore,

$$\mathbf{s}_i \cdot \mathbf{s}_{i+1} = s^2 \tag{19.79}$$

From Eqs. (19.78), (19.79) the exchange energy becomes

$$E = -2NJs^2 \tag{19.80}$$

If J is positive, then $\widehat{H}_{ex}$ gives the ground state of the system with the lowest energy eigenvalue (negative). But if J is negative, then $\widehat{H}_{ex}$ does not correspond to the ground state as it yields positive energy (higher energy).

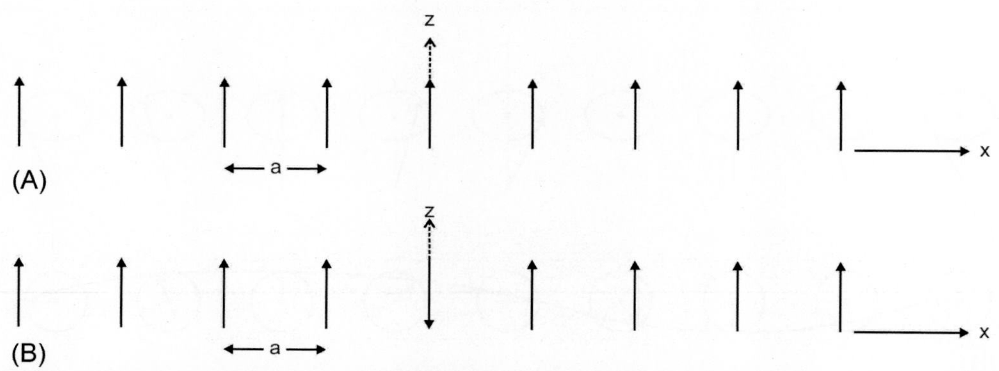

FIG. 19.7 (A) Monatomic linear magnetic solid along the x-direction with periodicity a. The spins of all of the atoms are parallel and are directed along the z-direction. (B) The direction of one of the spins, say the ith spin, is reversed and it represents the excited state of the magnetic linear lattice.

Consider an excited state in which a particular spin, say the ith spin, is reversed, assuming all other spins to remain unaffected, as shown in Fig. 19.7B. In this state the spins adjacent to the ith spin are oppositely directed and, therefore,

$$\mathbf{s}_{i-1} \cdot \mathbf{s}_i = -s^2 = \mathbf{s}_i \cdot \mathbf{s}_{i+1} \tag{19.81}$$

The total energy in this excited state is given by

$$\begin{aligned} E_1 &= -2J\left[(N-2)s^2 - s^2 - s^2\right] \\ &= -2NJs^2 + 8Js^2 \end{aligned} \tag{19.82}$$

From Eqs. (19.80), (19.82) it is evident that to reverse one spin, one requires energy of magnitude $8\,J\,s^2$, which gives the first excited state. Actually, all of the spins interact with each other via the Heisenberg spin-spin interaction. Therefore, when one spin is reversed, the effect will be seen by all other spins. One can show that much lower excitation energy is required if all the spins share in the reversal of a particular spin. In this situation all the spins start precessing about the z-axis, setting up a wave between the spins, as shown in Fig. 19.8. The elementary excitations of a spin system, therefore, have a wave-like form and are called spin waves. The quantization of spin waves yields magnons as the quanta of energy. There are magnon dispersion relations in magnetic materials analogous to the lattice dispersion relations. The spin waves are oscillations in the relative orientations of the spins on the lattice, while the lattice vibrations are oscillations in the relative positions of atoms on a lattice.

19.6.1 Bloch Theory of Spin Waves

The exchange Hamiltonian of a system can be written as

$$\widehat{H}_{ex} = \sum_{\mathbf{R}} \widehat{H}_{\mathbf{R}} \tag{19.83}$$

where

$$\widehat{H}_{\mathbf{R}} = -2J\sum_{\vec{\delta}} \mathbf{s}_{\mathbf{R}} \cdot \mathbf{s}_{\mathbf{R}+\vec{\delta}} \tag{19.84}$$

Here $\mathbf{R}$ denotes the central atom, which is connected to a number n of 1NNs because only the 1NNs are assumed to be significant. The expression for the Hamiltonian of a magnetic dipole $\vec{\mu}_{\mathbf{R}}$ at the position $\mathbf{R}$ that experiences magnetic field $\mathbf{H}_{\mathbf{R}}$ is given by

$$\widehat{H}_{\mathbf{R}} = -\vec{\mu}_{\mathbf{R}} \cdot \mathbf{H}_{\mathbf{R}} \tag{19.85}$$

$$= g_J \mu_B \mathbf{s}_{\mathbf{R}} \cdot \mathbf{H}_{\mathbf{R}} \tag{19.86}$$

where

$$\vec{\mu}_{\mathbf{R}} = -g_J \mu_B \mathbf{s}_{\mathbf{R}} \tag{19.87}$$

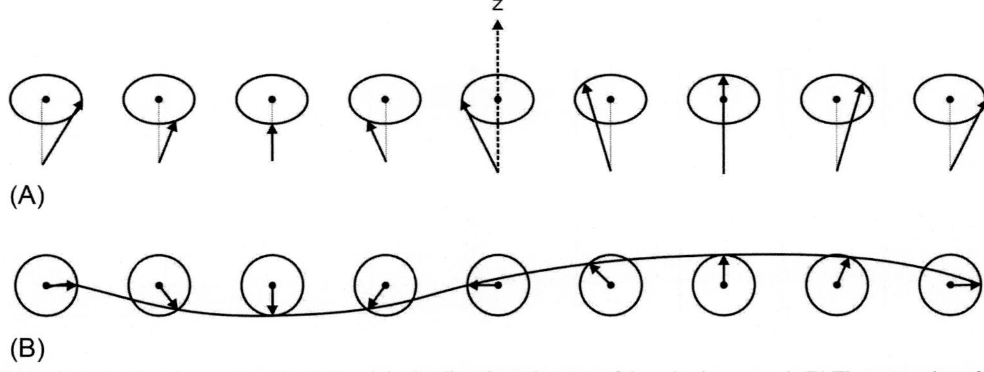

FIG. 19.8 (A) The various atomic spins precess about the original z-direction when one of the spins is reversed. (B) The precession of the spins as viewed from above. The tips of the various precessing spins form a wave called a spin wave.

Eq. (19.84) can be written as

$$\widehat{H}_{\mathbf{R}} = g_J \mu_B \mathbf{s}_{\mathbf{R}} \cdot \left(-\frac{2J}{g_J \mu_B} \sum_{\vec{\delta}} \mathbf{s}_{\mathbf{R}+\vec{\delta}} \right) \tag{19.88}$$

Comparing Eqs. (19.86), (19.88), the effective magnetic field acting on the $\mathbf{R}^{\text{th}}$ atom can be written as

$$\mathbf{H}_{\mathbf{R}} = -\frac{2J}{g_J \mu_B} \sum_{\vec{\delta}} \mathbf{s}_{\mathbf{R}+\vec{\delta}} \tag{19.89}$$

The angular momentum associated with the atom at the $\mathbf{R}^{\text{th}}$ site is $\hbar \mathbf{s}_{\mathbf{R}}$. The rate of change angular momentum gives the torque, that is,

$$\hbar \frac{d\mathbf{s}_{\mathbf{R}}}{dt} = \vec{\mu}_{\mathbf{R}} \times \mathbf{H}_{\mathbf{R}} \tag{19.90}$$

Substituting Eqs. (19.87), (19.89) into Eq. (19.90), one can write

$$\frac{d\mathbf{s}_{\mathbf{R}}}{dt} = \frac{2J}{\hbar} \sum_{\vec{\delta}} \mathbf{s}_{\mathbf{R}} \times \mathbf{s}_{\mathbf{R}+\vec{\delta}} \tag{19.91}$$

From quantum mechanics the equation of motion for spin $\mathbf{s}_{\mathbf{R}}$ is given as

$$\imath\hbar \frac{d\mathbf{s}_{\mathbf{R}}}{dt} = \left[\mathbf{s}_{\mathbf{R}}, \widehat{H}_{\mathbf{R}} \right] \tag{19.92}$$

From Eqs. (19.91), (19.92) one can write

$$\frac{d\mathbf{s}_{\mathbf{R}}}{dt} = \frac{1}{\imath\hbar} \left[\mathbf{s}_{\mathbf{R}}, \widehat{H}_{\mathbf{R}} \right] = \frac{2J}{\hbar} \sum_{\vec{\delta}} \mathbf{s}_{\mathbf{R}} \times \mathbf{s}_{\mathbf{R}+\vec{\delta}} \tag{19.93}$$

From Eq. (19.93), the equations for the components of spin are given as

$$\frac{ds^x_{\mathbf{R}}}{dt} = \frac{2J}{\hbar} \left[s^y_{\mathbf{R}} \sum_{\vec{\delta}} s^z_{\mathbf{R}+\vec{\delta}} - s^z_{\mathbf{R}} \sum_{\vec{\delta}} s^y_{\mathbf{R}+\vec{\delta}} \right] \tag{19.94}$$

$$\frac{ds^y_{\mathbf{R}}}{dt} = \frac{2J}{\hbar} \left[s^z_{\mathbf{R}} \sum_{\vec{\delta}} s^x_{\mathbf{R}+\vec{\delta}} - s^x_{\mathbf{R}} \sum_{\vec{\delta}} s^z_{\mathbf{R}+\vec{\delta}} \right] \tag{19.95}$$

$$\frac{ds^z_{\mathbf{R}}}{dt} = \frac{2J}{\hbar} \left[s^x_{\mathbf{R}} \sum_{\vec{\delta}} s^y_{\mathbf{R}+\vec{\delta}} - s^y_{\mathbf{R}} \sum_{\vec{\delta}} s^x_{\mathbf{R}+\vec{\delta}} \right] \tag{19.96}$$

Eqs. (19.94)–(19.96) represent nonlinear differential equations, which should be linearized in order to obtain solutions. Let us suppose that at absolute zero all the spins are aligned along the z-direction, so

$$s^z_{\mathbf{R}} = s_{\mathrm{R}}, \quad s^x_{\mathbf{R}} = s^y_{\mathbf{R}} = 0 \tag{19.97}$$

But as the temperature increases, there is a deviation in the spin direction and, as a result, the components $s^x_{\mathbf{R}}$ and $s^y_{\mathbf{R}}$ acquire finite values. If the variation in the spin direction is small, then $s^x_{\mathbf{R}}$ and $s^y_{\mathbf{R}}$ are small and their products can be neglected. With these approximations, Eqs. (19.94)–(19.96) give

$$\frac{ds^x_{\mathbf{R}}}{dt} = \frac{2J}{\hbar} \left[n s_{\mathbf{R}} s^y_{\mathbf{R}} - s_{\mathbf{R}} \sum_{\vec{\delta}} s^y_{\mathbf{R}+\vec{\delta}} \right] \tag{19.98}$$

$$\frac{ds_{\mathbf{R}}^{y}}{dt}=\frac{2J}{\hbar}\left[s_{\mathbf{R}}\sum_{\vec{\delta}}s_{\mathbf{R}+\vec{\delta}}^{x}-ns_{\mathbf{R}}s_{\mathbf{R}}^{x}\right] \tag{19.99}$$

$$\frac{ds_{\mathbf{R}}^{z}}{dt}=0 \tag{19.100}$$

The above equations can be written as

$$\frac{ds_{\mathbf{R}}^{x}}{dt}=\frac{2Js_{\mathbf{R}}}{\hbar}\left[ns_{\mathbf{R}}^{y}-\sum_{\vec{\delta}}s_{\mathbf{R}+\vec{\delta}}^{y}\right] \tag{19.101}$$

$$\frac{ds_{\mathbf{R}}^{y}}{dt}=\frac{2Js_{\mathbf{R}}}{\hbar}\left[\sum_{\vec{\delta}}s_{\mathbf{R}+\vec{\delta}}^{x}-ns_{\mathbf{R}}^{x}\right] \tag{19.102}$$

$$\frac{ds_{\mathbf{R}}^{z}}{dt}=0 \tag{19.103}$$

In analogy with the phonon problem, we look for a travelling wave solution of the above equations written as

$$s_{\mathbf{R}+\vec{\delta}}^{x}=u_{\mathbf{K}}\exp\left[\imath\left\{\mathbf{K}\cdot\left(\mathbf{R}+\vec{\delta}\right)-\omega t\right\}\right] \tag{19.104}$$

$$s_{\mathbf{R}+\vec{\delta}}^{y}=v_{\mathbf{K}}\exp\left[\imath\left\{\mathbf{K}\cdot\left(\mathbf{R}+\vec{\delta}\right)-\omega t\right\}\right] \tag{19.105}$$

Here $u_{\mathbf{K}}$ and $v_{\mathbf{K}}$ are the amplitudes, which can be calculated from the boundary conditions of the solutions. Substituting Eqs. (19.104), (19.105) into Eqs. (19.101)–(19.103), we get

$$\imath\omega u_{\mathbf{K}}+\frac{2Js_{\mathbf{R}}}{\hbar}\left[n-\sum_{\vec{\delta}}\exp\left(\imath\mathbf{K}\cdot\vec{\delta}\right)\right]v_{\mathbf{K}}=0 \tag{19.106}$$

$$-\frac{2Js_{\mathbf{R}}}{\hbar}\left[n-\sum_{\vec{\delta}}\exp\left(\imath\mathbf{K}\cdot\vec{\delta}\right)\right]u_{\mathbf{K}}+\imath\omega v_{\mathbf{K}}=0 \tag{19.107}$$

Eqs. (19.106), (19.107) have nontrivial solutions only if the determinant of the coefficients of $u_{\mathbf{K}}$ and $v_{\mathbf{K}}$ is zero, that is,

$$\begin{vmatrix} \imath\omega & \frac{2Js_{\mathbf{R}}}{\hbar}\left[n-\sum_{\vec{\delta}}\exp\left(\imath\mathbf{K}\cdot\vec{\delta}\right)\right] \\ -\frac{2Js_{\mathbf{R}}}{\hbar}\left[n-\sum_{\vec{\delta}}\exp\left(\imath\mathbf{K}\cdot\vec{\delta}\right)\right] & \imath\omega \end{vmatrix}=0 \tag{19.108}$$

The above determinant can be expanded to write

$$\hbar\omega=\hbar\omega_{\mathbf{K}}=2Js_{\mathbf{R}}\left[n-\sum_{\vec{\delta}}\exp\left(\imath\mathbf{K}\cdot\vec{\delta}\right)\right] \tag{19.109}$$

Here we have put ω as $\omega_{\mathbf{K}}$ because the right-hand side of Eq. (19.109) is a function of **K**. Eq. (19.109) is called the magnon dispersion relation and gives the relation between ω and **K**. Substituting Eq. (19.109) into Eq. (19.107), one immediately gets

$$v_{\mathbf{K}}=-\imath u_{\mathbf{K}} \tag{19.110}$$

which corresponds to circular motion of each spin about the z-axis. Let us calculate the magnon dispersion relations in different crystal structures.

19.6.2 Magnons in Monatomic Linear Lattice

Consider a one-dimensional magnetic lattice along the X-axis with periodicity a. Let each atom have spin in the z-direction (see Fig. 19.7A). Here each atom has two 1NNs at $\vec{\delta} = \pm a\hat{\mathbf{i}}_1$. Substituting the values of $\vec{\delta}$ into Eq. (19.109), we find

$$\hbar\omega_{\mathbf{K}} = 2Js\left[2 - e^{\imath K_x a} - e^{-\imath K_x a}\right] \tag{19.111}$$

Here we have assumed $s_{\mathbf{R}} = s$ as all the atoms are identical with the same spin. The above equation can be simplified to get

$$\frac{\hbar\omega_{\mathbf{k}}}{4Js} = 1 - \cos K_x a \tag{19.112}$$

Eq. (19.112) gives the magnon dispersion relation in a one-dimensional lattice. For long wavelengths $K_x a \langle\langle 1$, the above equation reduces to

$$\frac{\hbar\omega_{\mathbf{k}}}{2Js} = K_x^2 a^2 \tag{19.113}$$

Fig. 19.9 shows the magnon frequency as a function of wave vector. It is noteworthy that at large wavelengths the magnon frequency is proportional to K_x^2, whereas the phonon frequency is proportional to K_x (Chapter 6).

19.6.3 Magnons in Square Lattice

A magnetic square lattice with periodicity a in which each atom has four 1NNs with coordinates $\vec{\delta} = \left(\pm a\hat{\mathbf{i}}_1, 0\right)$ and $\left(0, \pm a\hat{\mathbf{i}}_2\right)$ is shown in Fig. 19.10. Substituting the coordinates of the 1NNs in Eq. (19.109) and simplifying, we get

$$\frac{\hbar\omega_{\mathbf{k}}}{4Js} = \left[2 - \cos K_x a - \cos K_y a\right] \tag{19.114}$$

which gives the magnon dispersion relation in a square lattice. For long wavelengths the above relation reduces to

$$\frac{\hbar\omega_{\mathbf{k}}}{2Js} = K^2 a^2 \tag{19.115}$$

where

$$K^2 = K_x^2 + K_y^2 \tag{19.116}$$

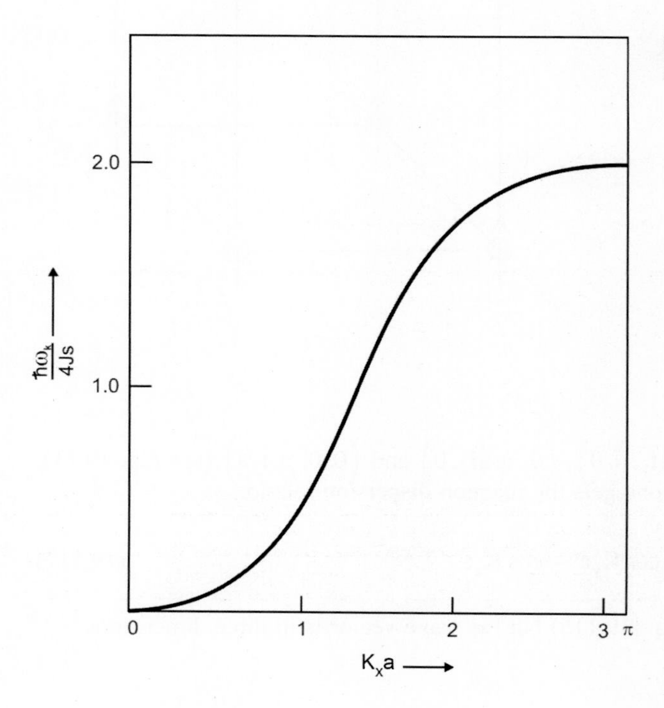

FIG. 19.9 Magnon frequency $\omega_{\mathbf{k}}$ as a function of $K_x a$ in the one-dimensional lattice shown in Fig. 19.7.

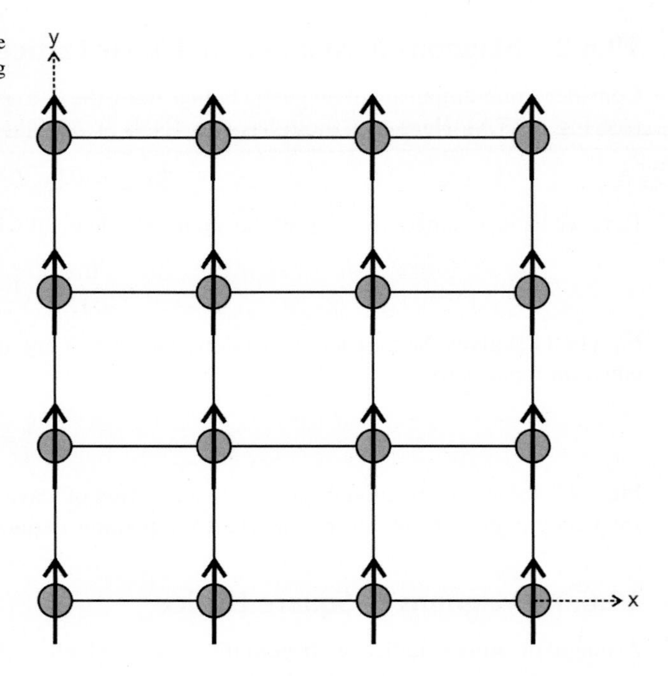

FIG. 19.10 A magnetic square lattice in the xy-plane with periodicity a. The magnetic moments of all of the atoms are parallel to each other and are along the z-direction.

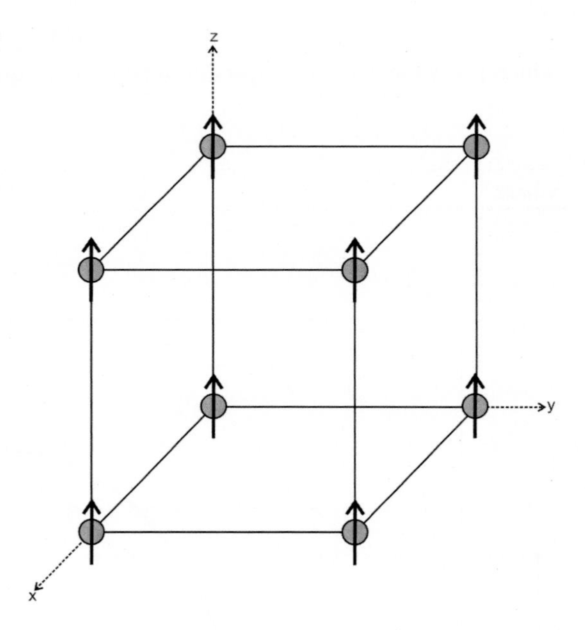

FIG. 19.11 A magnetic sc lattice with periodicity a. The magnetic moments of all of the atoms are parallel to each other and are along the z-direction.

19.6.4 Magnons in sc Lattice

In the sc lattice there are six 1NNs at the positions $\vec{\delta} = \left(\pm a\hat{\mathbf{i}}_1, 0, 0\right)$, $\left(0, \pm a\hat{\mathbf{i}}_2, 0\right)$ and $\left(0, 0, \pm\hat{\mathbf{i}}_3 a\right)$ (see Fig. 19.11). Substituting the values of $\vec{\delta}$ in Eq. (19.109) and simplifying, one gets the magnon dispersion relation as

$$\frac{\hbar\omega_{\mathbf{k}}}{4Js} = 3 - \cos K_x a - \cos K_y a - \cos K_z a \tag{19.117}$$

In the long wavelength limit, one gets the same relation as Eq. (19.115) but the wave vector is in three dimensions.

Problem 19.1

Find the magnon dispersion relations for a crystal with (a) fcc structure and (b) bcc structure. Discuss the results in the long wavelength limit.

19.7 QUANTIZATION OF SPIN WAVES

If there are N spins in a lattice then according to the quantum mechanics of angular momentum the total spin quantum number has the values Ns, Ns − 1,..., 0, ..., −Ns+1, −Ns. In the ferromagnetic ground state, the total spin quantum number is Ns as all the spins are parallel in the ground state. The excitation of a spin wave lowers the total spin because the spins no longer remain parallel. Therefore, there is a relationship between the amplitude of the spin wave and the reduction in the z-component of the total spin quantum number. Substituting Eq. (19.110) into Eqs. (19.104), (19.105), one gets

$$s^x_{\mathbf{R}+\vec{\delta}} = \sum_{\mathbf{K}} u_{\mathbf{K}} \exp\left[\imath\left\{\mathbf{K}\cdot\left(\mathbf{R}+\vec{\delta}\right)-\omega t\right\}\right] \quad (19.118)$$

$$s^y_{\mathbf{R}+\vec{\delta}} = \sum_{\mathbf{K}} (-\imath u_{\mathbf{K}}) \exp\left[\imath\left\{\mathbf{K}\cdot\left(\mathbf{R}+\vec{\delta}\right)-\omega t\right\}\right] \quad (19.119)$$

As the spins are real, we take only the real part of Eqs. (19.118), (19.119) to write

$$s^x_{\mathbf{R}+\vec{\delta}} = \sum_{\mathbf{K}} u_{\mathbf{K}} \cos\left[\mathbf{K}\cdot\left(\mathbf{R}+\vec{\delta}\right)-\omega t\right] \quad (19.120)$$

$$s^y_{\mathbf{R}+\vec{\delta}} = \sum_{\mathbf{K}} u_{\mathbf{K}} \sin\left[\mathbf{K}\cdot\left(\mathbf{R}+\vec{\delta}\right)-\omega t\right] \quad (19.121)$$

Therefore, for the Rth atom, one can write

$$(s^x_{\mathbf{R}})^2 + (s^y_{\mathbf{R}})^2 = \sum_{\mathbf{K}} u^2_{\mathbf{K}} = u^2 \quad (19.122)$$

But

$$s^2_{\mathbf{R}} = (s^x_{\mathbf{R}})^2 + (s^y_{\mathbf{R}})^2 + (s^z_{\mathbf{R}})^2 = \sum_{\mathbf{K}} u^2_{\mathbf{K}} + (s^z_{\mathbf{R}})^2 = u^2 + (s^z_{\mathbf{R}})^2 \quad (19.123)$$

From the above equation one can write

$$s^z_{\mathbf{R}} = \left(s^2_{\mathbf{R}} - \sum_{\mathbf{K}} u^2_{\mathbf{K}}\right)^{1/2} \quad (19.124)$$

For small amplitudes

$$\frac{u_{\mathbf{K}}}{s_{\mathbf{R}}} \langle\langle 1$$

therefore Eq. (19.124) reduces to

$$s^z_{\mathbf{R}} = s_{\mathbf{R}} - \frac{1}{2s_{\mathbf{R}}} \sum_{\mathbf{K}} u^2_{\mathbf{K}} = s_{\mathbf{R}} - \frac{u^2}{2s_{\mathbf{R}}} \quad (19.125)$$

The above equation can be written as

$$s_{\mathbf{R}} - s^z_{\mathbf{R}} = \frac{1}{2s_{\mathbf{R}}} \sum_{\mathbf{K}} u^2_{\mathbf{K}} = \frac{u^2}{2s_{\mathbf{R}}} \quad (19.126)$$

The quantum theory allows only integral values for $s_{\mathbf{R}} - s_{\mathbf{R}}^{z}$. For N spins the above equation becomes

$$Ns_{\mathbf{R}} - Ns_{\mathbf{R}}^{z} = \sum_{\mathbf{K}} n_{\mathbf{K}} \tag{19.127}$$

where

$$n_{\mathbf{K}} = \frac{Nu_{\mathbf{K}}^{2}}{2s_{\mathbf{R}}} \tag{19.128}$$

Eq. (19.127) can also be written as

$$Ns_{\mathbf{R}} - \sum_{\mathbf{K}} n_{\mathbf{K}} = s_{\mathbf{R}}^{z_t} \tag{19.129}$$

where

$$s_{\mathbf{R}}^{z_t} = Ns_{\mathbf{R}}^{z} \tag{19.130}$$

Eq. (19.129) is the quantum condition for N spins in which a spin wave with wave vector **K** is excited; one may write

$$Ns_{\mathbf{R}} - s_{\mathbf{R}}^{z_t} = \sum_{\mathbf{K}} n_{\mathbf{K}} \tag{19.131}$$

Here $n_{\mathbf{K}}$ is an integer and is equal to the number of magnons of wave vector **K** that are excited. Each magnon lowers the z-component of the total spin by unity. Let us find the quantum of energy in the case of a magnetic material. The exchange energy is given by

$$E_{ex} = -J \sum_{\mathbf{R}} \sum_{\vec{\delta}} \mathbf{s}_{\mathbf{R}} \cdot \mathbf{s}_{\mathbf{R}+\vec{\delta}} \tag{19.132}$$

Eq. (19.132) shows that the exchange energy depends on the cosine of the angle between the spins at the adjacent sites represented by **R** and $\mathbf{R} + \vec{\delta}$. Therefore, the difference in phase at time t between the two successive spins is $\mathbf{K} \cdot \vec{\delta}$, as is evident from the equations

$$s_{\mathbf{R}}^{x} = \sum_{\mathbf{K}} u_{\mathbf{K}} e^{\imath(\mathbf{K} \cdot \mathbf{R} - \omega t)} \tag{19.133}$$

$$s_{\mathbf{R}+\vec{\delta}}^{x} = \sum_{\mathbf{K}} u_{\mathbf{K}} e^{\imath\left\{\mathbf{K} \cdot \left(\mathbf{R}+\vec{\delta}\right) - \omega t\right\}} \tag{19.134}$$

The tips of the two spins are separated by a distance $2u_{\mathbf{K}} \sin\left(\mathbf{K} \cdot \vec{\delta}/2\right)$ for a particular wave vector (see Fig. 19.12). Therefore, the angle between the two spin vectors is given by

$$\sin\left(\frac{1}{2}\phi_{\vec{\delta}}\right) = \frac{u_{\mathbf{K}}}{s_{\mathbf{R}}} \sin\left(\frac{1}{2}\mathbf{K} \cdot \vec{\delta}\right) \tag{19.135}$$

Therefore,

$$\begin{aligned} \cos\phi_{\vec{\delta}} &= 1 - 2\sin^2\frac{1}{2}\phi_{\vec{\delta}} \\ &= 1 - 2\frac{u_{\mathbf{K}}^2}{s_{\mathbf{R}}^2}\sin^2\left(\frac{1}{2}\mathbf{K} \cdot \vec{\delta}\right) \end{aligned} \tag{19.136}$$

Now the exchange energy from Eq. (19.132) is given by

$$\begin{aligned} E_{ex} &= -J \sum_{\mathbf{R}} \sum_{\vec{\delta}} |\mathbf{s}_{\mathbf{R}}| \left|\mathbf{s}_{\mathbf{R}+\vec{\delta}}\right| \cos\phi_{\vec{\delta}} \\ &= -J\left(\sum_{\mathbf{R}} s_{\mathbf{R}}^2\right) \sum_{\vec{\delta}} \left[1 - 2\frac{u_{\mathbf{K}}^2}{s_{\mathbf{R}}^2}\sin^2\left(\frac{1}{2}\mathbf{K} \cdot \vec{\delta}\right)\right] \end{aligned} \tag{19.137}$$

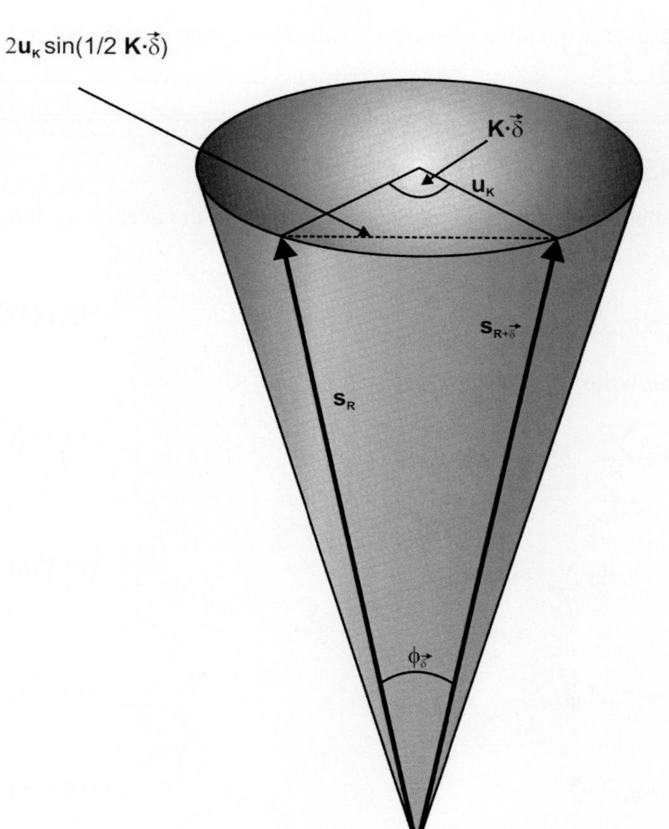

FIG. 19.12 Two consecutive spin vectors $\mathbf{s_R}$ and $\mathbf{s}_{\mathbf{R}+\vec{\delta}}$ with $\varphi_{\vec{\delta}}$ as the angle between them. The figure depicts the relation between $\varphi_{\vec{\delta}}$, spin wave amplitude $u_{\mathbf{K}}$, and the phase angle $\mathbf{K}\cdot\vec{\delta}$.

The summation over all spins gives N times the square of the spin and the summation over the 1NNs gives n in the first term in the square brackets. So, the above equation can be written as

$$E_{ex} = -JNs_{\mathbf{R}}^2\left[n - \frac{2u_{\mathbf{K}}^2}{s_{\mathbf{R}}^2}\sum_{\vec{\delta}}\sin^2\left(\frac{1}{2}\mathbf{K}\cdot\vec{\delta}\right)\right] \tag{19.138}$$

The above equation can be written as

$$E_{ex} = -nNJs^2 + E_{\mathbf{K}} \tag{19.139}$$

where

$$E_{\mathbf{K}} = 2JNu_{\mathbf{K}}^2\sum_{\vec{\delta}}\sin^2\left(\frac{1}{2}\mathbf{K}\cdot\vec{\delta}\right) \tag{19.140}$$

Here $E_{\mathbf{K}}$ is the excitation energy of a spin wave of amplitude $u_{\mathbf{K}}$ and can be written as

$$E_{\mathbf{K}} = 2Js_{\mathbf{R}}n_{\mathbf{K}}\sum_{\vec{\delta}}\left[1-\cos\left(\mathbf{K}\cdot\vec{\delta}\right)\right] = 2Js_{\mathbf{R}}n_{\mathbf{K}}\left[n - \sum_{\vec{\delta}}\cos\left(\mathbf{K}\cdot\vec{\delta}\right)\right] \tag{19.141}$$

Here we have used Eq. (19.128). Using Eq. (19.109) in the above equation one can write

$$E_{\mathbf{K}} = n_{\mathbf{K}}\varepsilon_{\mathbf{K}} \tag{19.142}$$

where

$$\varepsilon_{\mathbf{K}} = \hbar\omega_{\mathbf{K}} = 2Js_{\mathbf{R}}\left[n - \sum_{\vec{\delta}}\cos\left(\mathbf{K}\cdot\vec{\delta}\right)\right] \tag{19.143}$$

The above equation shows that the energy $E_{\mathbf{K}}$ is quantized and the quantum of energy is called a magnon with the energy given by Eq. (19.143).

19.8 THERMAL EXCITATION OF MAGNONS

Magnons are bosons and they obey the Bose-Einstein distribution. Therefore, in thermal equilibrium, the average of the number $n_{\mathbf{K}}$ is given by

$$\langle n_{\mathbf{K}} \rangle = \frac{1}{e^{\beta_0 \hbar \omega_{\mathbf{K}}} - 1} \tag{19.144}$$

Here β_0 is defined by Eq. (8.22). From Eq. (19.129) one can write

$$s_{\mathbf{R}}^{Z_t} = N s_{\mathbf{R}} - \sum_{\mathbf{K}} \langle n_{\mathbf{K}} \rangle \tag{19.145}$$

Multiplying the above equation by $g_J \mu_B$, one can write

$$M(T) = M(0) - g_J \mu_B \sum_{\mathbf{K}} \langle n_{\mathbf{K}} \rangle \tag{19.146}$$

where

$$M(0) = g_J \mu_B N s_{\mathbf{R}} \tag{19.147}$$

and

$$M(T) = g_J \mu_B s_{\mathbf{R}}^{Z_t} \tag{19.148}$$

Eq. (19.147) gives a summation over all parallel spins and, therefore, corresponds to saturation magnetization, that is, magnetization at absolute zero. Eq. (19.148) gives magnetization at T degrees absolute. From Eq. (19.146) it is evident that M (T) can be calculated if $\sum_{\mathbf{K}} \langle n_{\mathbf{K}} \rangle$ is known. From Eq. (19.144) one can write

$$\sum_{\mathbf{K}} \langle n_{\mathbf{K}} \rangle = \int_0^{K_{max}} \frac{d^3K}{(2\pi)^3} \langle n_{\mathbf{K}} \rangle = \frac{1}{2\pi^2} \int_0^{K_{max}} \frac{K^2 dK}{e^{\beta_0 \hbar \omega_{\mathbf{K}}} - 1} \tag{19.149}$$

If T is very small, then the exponential term on the right side of Eq. (19.149) is large, approaching infinity. Hence Eq. (19.149) gives a very small contribution. At very small temperatures and at very small K values, the integral in Eq. (19.149) gives some finite contribution because if $K=0$ then $\omega_{\mathbf{K}}=0$. Hence we see that the integral gives a finite contribution at $T \approx 0$ for very small values of K. At small values of K, Eq. (19.113) or (19.115) gives

$$\hbar \omega_{\mathbf{k}} = 2 J s_{\mathbf{R}} K^2 a^2 \tag{19.150}$$

Substituting Eq. (19.150) into Eq. (19.149), one can write

$$\sum_{\mathbf{K}} \langle n_{\mathbf{K}} \rangle = \frac{1}{2\pi^2} \int_0^{\infty} \frac{K^2 dK}{e^{2\beta_0 J s_{\mathbf{R}} K^2 a^2} - 1} \tag{19.151}$$

In the above integral the limits are taken from 0 to ∞ as we do not know the cutoff value of K. Substituting

$$x = 2\beta_0 J s_{\mathbf{R}} K^2 a^2 \tag{19.152}$$

one can immediately write

$$\sum_{\mathbf{K}} \langle n_{\mathbf{K}} \rangle = \frac{1}{4\pi^2} \left(\frac{1}{2\beta_0 J s_{\mathbf{R}} a^2} \right)^{3/2} \int_0^{\infty} \frac{x^{1/2} dx}{e^x - 1} \tag{19.153}$$

From the standard tables the value of the above integral is $0.0587(4\pi^2)$. Therefore, one gets

$$\sum_{\mathbf{K}} \langle n_{\mathbf{K}} \rangle = 0.0587 \left(\frac{k_B T}{2 J s_{\mathbf{R}} a^2} \right)^{3/2} \tag{19.154}$$

Substituting Eq. (19.154) into Eq. (19.146), we find

$$\Delta M(T) = M(0) - M(T) = 0.0587\, g_J \mu_B \left(\frac{k_B T}{2 J s_{\mathbf{R}} a^2} \right)^{3/2} \tag{19.155}$$

Dividing Eq. (19.155) by M(0) and using Eq. (19.147), one gets

$$\frac{\Delta M(T)}{M(0)} = \frac{0.0587}{N a^3 s_{\mathbf{R}}} \left(\frac{k_B T}{2 J s_{\mathbf{R}}} \right)^{3/2} \tag{19.156}$$

Eq. (19.156) gives the $T^{3/2}$ law for the magnetization as is observed experimentally [see Eq. (19.39)].

19.9 HYSTERESIS CURVE

The application of a magnetic field **H** magnetizes a ferromagnetic material and the magnetization **M** increases in a nonlinear fashion with an increase in **H**. Therefore, the magnetic induction **B**, which is a measure of **M**, increases along the line OA and then reaches a constant value corresponding to the point B (the state of saturation magnetization) in a ferromagnetic material. The variation of magnetization **M** along the line OAB can be explained in terms of the growth of magnetic domains with magnetization along the direction of the applied field **H** in the same way as we did in the case of a ferroelectric material. The **B**-**H** curve does not retrace its path, but instead follows the path BCDE when the field **H** is decreased and the path EFGB when **H** is increased, as in the case of a ferroelectric material (see Fig. 16.1). Therefore, a plot of **B** as a function of **H** forms a loop, as shown in Fig. 19.13. A finite magnetic induction $\mathbf{B}_r$ (residual magnetic induction) corresponding to the point C is obtained when the external field **H** is switched off. This means that a residual magnetization $\mathbf{M}_r$ is obtained in the absence of **H**. $\mathbf{M}_r$ is usually called the remnant magnetization. To reduce **B** or **M** to zero, one has to apply a magnetic field $\mathbf{H}_c$ corresponding to the point D, called the coercive magnetic field, in the reverse direction.

From Fig. 19.13 it is evident that the magnetic induction **B** or the magnetization **M** lags behind the applied field **H** and this phenomenon is called hysteresis. When a ferromagnetic material is taken through the cycle of the **B**-**H** curve, there is a

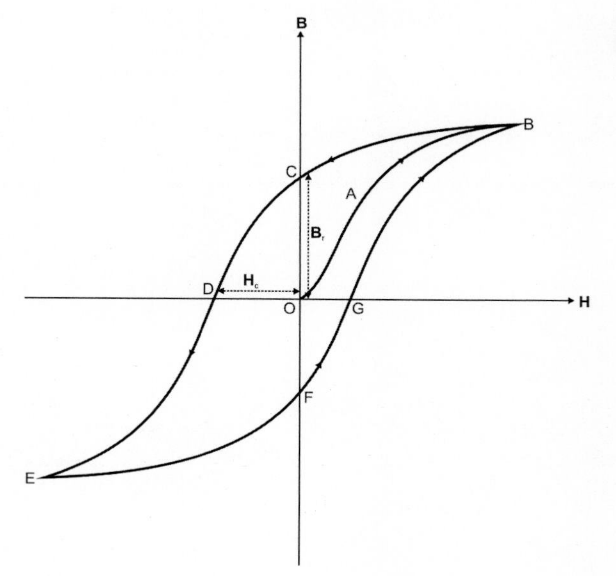

FIG. 19.13 Schematic representation of the magnetic induction **B** in a ferromagnetic material as a function of the applied magnetic field **H**.

loss of energy, generally called the hysteresis loss. The hysteresis loss is proportional to the area of one cycle of the **B**-**H** curve. On the basis of hysteresis loss, ferromagnetic materials are classified into two categories:

1. Ferromagnetic materials in which the area enclosed by the hysteresis loop is small and hence the energy loss is also small. Such materials are called soft ferromagnetic materials as they can easily be magnetized or demagnetized (low coercivity). These materials cannot be permanently magnetized and are used in making transformer cores.
2. The second type of ferromagnetic material is characterized by a hysteresis loop of large area, thus having a large energy loss. Such materials are called hard ferromagnetic materials as they cannot easily be magnetized or demagnetized (high coercivity). These materials retain a considerable amount of magnetization after the field **H** is switched off and are used in making permanent magnets.

SUGGESTED READING

Aharoni, A. (1996). *Introduction to the theory of ferromagnetism*. London: Clarendon Press.
Bozorth, R. M. (1951). *Ferromagnetism*. New York: van Nostrand.
Cracknell, A. P. (1975). *Magnetism in crystalline materials*. New York: Pergamon Press.
Martin, D. H. (1967). *Magnetism in solids*. London: Iliffe Books Ltd.
Moriya, T. (1985). *Spin fluctuations in itinerant electron magnetism*. Berlin: Springer-Verlag.
Rado, G. T., & Suhl, H. (1963). *Magnetism*. (Vols. 1–4). New York: Academic Press.
Wagner, D. (1972). *Introduction to the theory of magnetism*. New York: Pergamon Press.
White, R. M. (1970). *Quantum theory of magnetism*. New York: McGraw-Hill Book Co.
Williams, D. E. G. (1966). *The magnetic properties of matter*. London: Longmans.

Chapter 20

Antiferromagnetism and Ferrimagnetism

Chapter Outline

X-ray diffraction studies of MnO crystal at 80 and 293 K show an fcc structure with lattice constant $a = 4.43$ Å at both temperatures. The unit cell in MnO crystal, obtained from X-ray studies and generally called the chemical unit cell, has a NaCl structure with $a = 4.43$ Å. Neutron diffraction studies on MnO crystal at 293 K yield the same results as those obtained from X-ray diffraction. But one obtains extra neutron reflections at 80 K, which correspond to a unit cell, usually called a magnetic unit cell, with $a = 8.85$ Å (exactly double the value at 293 K). Further, it is observed that the intensity of the extra lines decreases with an increase in temperature and, at some critical temperature T_N (called the Neel temperature), the extra lines disappear. If the ordering were ferromagnetic, then the chemical and magnetic cells would give the same neutron reflection, yielding the same value of a. Therefore, it is argued that the magnetic moments of Mn^{+2} ions are ordered in some nonferromagnetic arrangement. Fig. 20.1 shows the ordering of the spins of the Mn^{+2} ions in MnO crystal as determined from neutron diffraction experiments. It is evident from the figure that all of the spins in a single (111) plane are parallel, but they are antiparallel to the spins in the adjacent (111) plane. Thus, MnO has a spin order that is the opposite of ferromagnetic order. This is, therefore, called antiferromagnetic order and the phenomenon is called antiferromagnetism. A solid that exhibits antiferromagnetic order is called an antiferromagnetic solid. A schematic diagram of antiferromagnetic order is shown in Fig. 20.2. According to the Heisenberg theory, the exchange integral $J > 0$ favors the parallel alignment of spins (ferromagnetism), but $J < 0$ favors the antiparallel alignment of spins (antiferromagnetism). The most interesting property of an antiferromagnetic solid is that its magnetic susceptibility χ_M shows a maximum at the Neel temperature T_N (Fig. 20.3). Such behavior can be explained by using a two-sublattice model.

There is another class of solids called ferrimagnetic solids that contain a mixture of antiferromagnetic and ferromagnetic orders. But on the whole these solids behave as ferromagnetic solids. Ferrimagnetic behavior can also be explained on the basis of the two-sublattice model.

20.1 ANTIFERROMAGNETISM

20.1.1 Two-Sublattice Model

Consider a crystal comprising two interpenetrating lattices called sublattices A and B (see Fig. 20.2): atoms in sublattice A are named A atoms, while those in sublattice B are named B atoms. All of the atoms in sublattice A have up spins, while those in sublattice B have down spins. It is evident from the figure that the two sublattices have the same amount of saturation magnetization in the antiferromagnetic order. Further, let the interaction between the atoms be such that the A spins tend to align antiparallel to the B spins (usually called antiferromagnetic alignment). At very low temperatures the antiferromagnetic interaction is large compared with the thermal energy, which results in nearly perfect antiferromagnetic alignment and yields very small magnetization. With an increase in temperature, the thermal energy increases, which increasingly disturbs the antiferromagnetic alignment, resulting in an increase in magnetization and hence an increase in the magnetic susceptibility χ_M. Finally, at T_N a transition takes place from the antiferromagnetic to the paramagnetic phase and the spins become free. Therefore, above T_N, χ_M decreases with increasing T.

Solid State Physics. https://doi.org/10.1016/B978-0-12-817103-5.00020-7

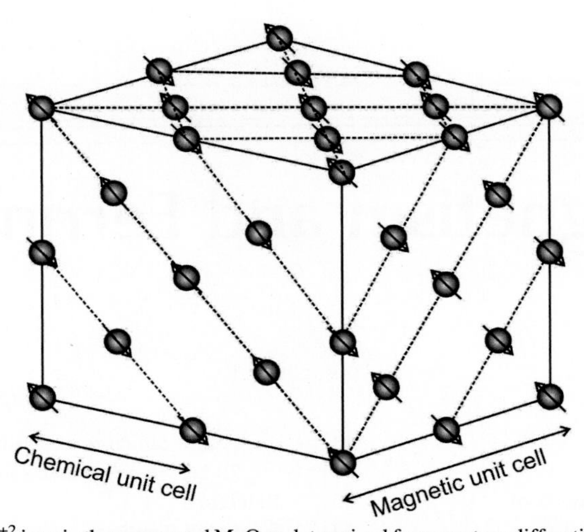

FIG. 20.1 Ordering of the spins of the Mn^{+2} ions in the compound MnO as determined from neutron diffraction experiments. The O^{-2} ions are not shown in the figure.

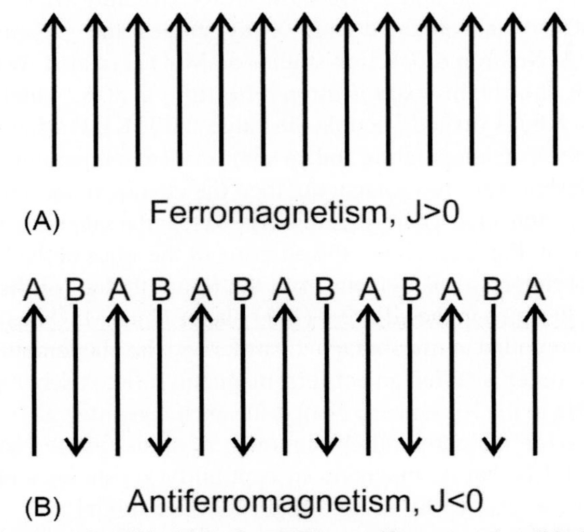

FIG. 20.2 The spin ordering in (A) a ferromagnetic solid with $J>0$ and (B) an antiferromagnetic solid with $J<0$.

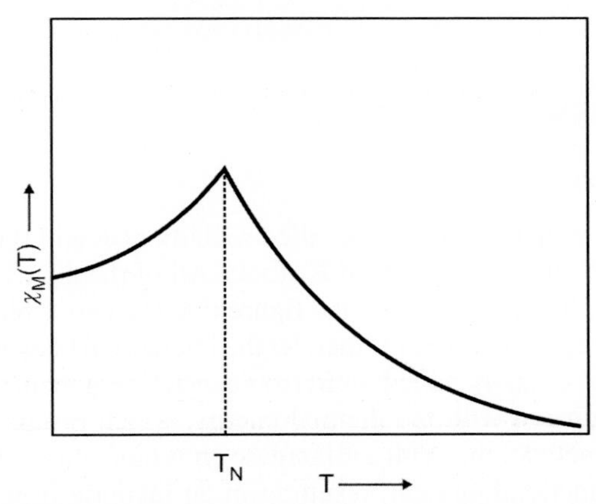

FIG. 20.3 Schematic representation of the experimental magnetic susceptibility χ_M as a function of temperature T in an antiferromagnetic solid. χ_M exhibit a maximum at the Neel temperature T_N.

From Fig. 20.2 it is evident that the 1NNs of an A atom are the B atoms and vice versa. In the lattice, in addition to AB antiferromagnetic interactions, there exist AA and BB ferromagnetic interactions. Further, we assume that

$$J_{AA} = J_{BB} > 0 \text{ and } J_{AB} = J_{BA} < 0 \tag{20.1}$$

where J is the exchange integral. Let $\mathbf{M}_A$ and $\mathbf{M}_B$ be the magnetizations of the sublattices A and B. The net magnetic field acting on sublattice A is denoted by $\mathbf{H}_A$ and is given by

$$\mathbf{H}_A = \mathbf{H} - \lambda_{AA}\mathbf{M}_A - \lambda_{AB}\mathbf{M}_B \tag{20.2}$$

Here $\mathbf{H}$ is the applied magnetic field. An atom in sublattice A experiences the molecular fields due to both sublattices A and B: the second term in Eq. (20.2) gives the molecular field of sublattice A, while the third term gives the molecular field due to sublattice B. Here we have taken the negative sign before the molecular field because J is taken to be positive (actually the value of J is negative in antiferromagnetic solids). Similarly, the net magnetic field acting on sublattice B, denoted by $\mathbf{H}_B$, is given by

$$\mathbf{H}_B = \mathbf{H} - \lambda_{BB}\mathbf{M}_B - \lambda_{BA}\mathbf{M}_A \tag{20.3}$$

Because the two sublattices are identical, it is reasonable to assume that

$$\lambda_{AA} = \lambda_{BB} \text{ and } \lambda_{AB} = \lambda_{BA} \tag{20.4}$$

With these assumptions, Eqs. (20.2), (20.3) reduce to

$$\mathbf{H}_A = \mathbf{H} - \lambda_{AA}\mathbf{M}_A - \lambda_{AB}\mathbf{M}_B \tag{20.5}$$

$$\mathbf{H}_B = \mathbf{H} - \lambda_{AA}\mathbf{M}_B - \lambda_{AB}\mathbf{M}_A \tag{20.6}$$

The magnetization and the magnetic susceptibility are studied in two limiting cases.

20.1.1.1 Susceptibility for $T > T_N$

For T greater than T_N the magnetization shows Curie behavior and, therefore,

$$\mathbf{M}_A = \frac{\rho_A^a g_J^2 \mu_B^2 J(J+1)}{3k_B T}\mathbf{H}_A = \frac{\rho_A^a \mu_J^2}{3k_B T}\mathbf{H}_A \tag{20.7}$$

$$\mathbf{M}_B = \frac{\rho_B^a g_J^2 \mu_B^2 J(J+1)}{3k_B T}\mathbf{H}_B = \frac{\rho_B^a \mu_J^2}{3k_B T}\mathbf{H}_B \tag{20.8}$$

Here ρ_A^a and ρ_B^a are the number of spins (atoms) per unit volume of the A and B sublattices, respectively. In an antiferromagnetic substance $\rho_A^a = \rho_B^a$. Substituting the values of $\mathbf{H}_A$ and $\mathbf{H}_B$ from Eqs. (20.5), (20.6) into Eqs. (20.7), (20.8), one gets

$$\mathbf{M}_A = \frac{\rho_A^a \mu_J^2}{3k_B T}(\mathbf{H} - \lambda_{AA}\mathbf{M}_A - \lambda_{AB}\mathbf{M}_B) \tag{20.9}$$

$$\mathbf{M}_B = \frac{\rho_A^a \mu_J^2}{3k_B T}(\mathbf{H} - \lambda_{AA}\mathbf{M}_B - \lambda_{AB}\mathbf{M}_A) \tag{20.10}$$

From Eqs. (20.9), (20.10) the total magnetization is given by

$$\begin{aligned}\mathbf{M} &= \mathbf{M}_A + \mathbf{M}_B \\ &= \frac{\rho_A^a \mu_J^2}{3k_B T}[2\mathbf{H} - (\lambda_{AA} + \lambda_{AB})\mathbf{M}]\end{aligned} \tag{20.11}$$

From the above equation one can calculate the magnetic susceptibility given by

$$\chi_M = \frac{\mathbf{M}}{\mathbf{H}} = \frac{2C_A}{T + \theta} \tag{20.12}$$

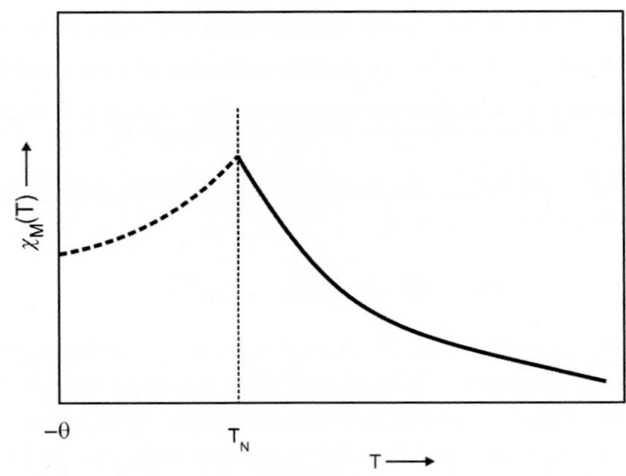

FIG. 20.4 Schematic representation of the theoretical magnetic susceptibility χ_M as a function of temperature T in an antiferromagnetic solid based on the two-sublattice model (*solid line*). The *dashed line* shows the experimental trend below the Neel temperature T_N.

where

$$C_A = \frac{\rho_A^a \mu_J^2}{3k_B} \text{ and } \theta = \frac{\rho_A^a \mu_J^2}{3k_B}(\lambda_{AA} + \lambda_{AB}) = C_A(\lambda_{AA} + \lambda_{AB}) \tag{20.13}$$

Eq. (20.12) shows Curie-type behavior for χ_M and a schematic of this behavior is shown in Fig. 20.4. One can compare Eq. (20.12) with the equation obtained for ferromagnetic behavior (see Eq. 19.5). Let us find the magnetization in the absence of an applied magnetic field. Putting $\mathbf{H} = 0$ in Eqs. (20.9), (20.10), one obtains

$$\mathbf{M}_A = \frac{C_A}{T}(-\lambda_{AA}\mathbf{M}_A - \lambda_{AB}\mathbf{M}_B) \tag{20.14}$$

$$\mathbf{M}_B = \frac{C_A}{T}(-\lambda_{AA}\mathbf{M}_B - \lambda_{AB}\mathbf{M}_A) \tag{20.15}$$

At the Neel temperature T_N the above equations can be arranged to write

$$\left(1 + \frac{C_A}{T_N}\lambda_{AA}\right)\mathbf{M}_A + \frac{C_A}{T_N}\lambda_{AB}\mathbf{M}_B = 0 \tag{20.16}$$

$$\frac{C_A}{T_N}\lambda_{AB}\mathbf{M}_A + \left(1 + \frac{C_A}{T_N}\lambda_{AA}\right)\mathbf{M}_B = 0 \tag{20.17}$$

A nontrivial solution of Eqs. (20.16), (20.17) is obtained by equating the determinant of the coefficients of $\mathbf{M}_A$ and $\mathbf{M}_B$ to zero, that is,

$$\begin{vmatrix} 1 + \frac{C_A}{T_N}\lambda_{AA} & \frac{C_A}{T_N}\lambda_{AB} \\ \frac{C_A}{T_N}\lambda_{AB} & 1 + \frac{C_A}{T_N}\lambda_{AA} \end{vmatrix} = 0 \tag{20.18}$$

Solving the above determinant, one gets

$$T_N = C_A(\lambda_{AB} - \lambda_{AA}) \tag{20.19}$$

which defines the Neel temperature. From Eqs. (20.13), (20.19) one can write

$$\frac{T_N}{\theta} = \frac{\lambda_{AB} - \lambda_{AA}}{\lambda_{AA} + \lambda_{AB}} \tag{20.20}$$

Eqs. (20.13), (20.19), (20.20) show that T_N and θ are different in the high-temperature limit. From experiment it has been found that $T_N \langle \theta$, indicating that λ_{AA} is positive. The above derivation includes ferromagnetic interactions among the spins of sublattice A (also among the spins of sublattice B). If one includes only antiferromagnetic interactions among the spins, that is, $\lambda_{AA} = 0$, then Eqs. (20.19), (20.20) reduce to

$$T_N = \theta = \lambda_{AB} C_A \tag{20.21}$$

20.1.1.2 Susceptibility for $T < T_N$

There are two situations at low temperatures: first when the applied magnetic field is parallel to the axis of the spins (Fig. 20.5A) and second when it is perpendicular to the axis of the spins (Fig. 20.5B). In the first case, when the applied magnetic field is parallel to the spins, the susceptibility obtained is denoted as $\chi_{||}(T)$. At very low temperatures there is perfect antiferromagnetic alignment as a magnetic field parallel to the spins does not disturb the alignment. In other words, the spins do not respond to the magnetic field as they are already in the direction of the field. Therefore, $\chi_{||}(T)$ approaches zero at very low temperatures and will be exactly zero at absolute zero. With an increase in temperature, the thermal energy increases, which disturbs the antiferromagnetic alignment. As a result, the value of $\chi_{||}(T)$ increases. But the calculation of $\chi_{||}(T)$ at a finite temperature is much more complicated and is a function of the total angular momentum **J**. The results of the calculation of Van Vleck (1941) are shown schematically in Fig. 20.6 for different values of spin. They show that $\chi_{||}(T)$ increases with an increase in T.

In the second case the magnetic field is applied in a direction perpendicular to the spins (see Fig. 20.5B). The magnetic force acts on each spin, as a result of which the spins acquire the equilibrium positions determined by the balance of the external magnetic force and the internal restoring force. The angle made by the equilibrium positions of the two types of magnetizations $\mathbf{M}_A$ and $\mathbf{M}_B$ with the original directions (dashed lines) is the same, say ϕ, because $\mathbf{M}_A$ and $\mathbf{M}_B$ have the same magnitude and, therefore, experience the same external force. The net interaction energy per unit volume due to the applied magnetic field is given by

$$E = -\mathbf{M}_A \cdot \mathbf{H}_A - \mathbf{M}_B \cdot \mathbf{H}_B \tag{20.22}$$

Substituting the values of $\mathbf{H}_A$ and $\mathbf{H}_B$ from Eqs. (20.5), (20.6) into Eq. (20.22), we find

$$E = -(\mathbf{M}_A + \mathbf{M}_B) \cdot \mathbf{H} + \lambda_{AA} M_A^2 + \lambda_{AA} M_B^2 + \lambda_{AB} \mathbf{M}_A \cdot \mathbf{M}_B + \lambda_{AB} \mathbf{M}_B \cdot \mathbf{M}_A \tag{20.23}$$

The last two terms in Eq. (20.23) are equal and each represents the interaction energy between the two sublattices. This means that the interaction energy is included twice; therefore, the correct magnetic energy is given by

$$E = -(\mathbf{M}_A + \mathbf{M}_B) \cdot \mathbf{H} + \lambda_{AA} \left(M_A^2 + M_B^2\right) + \lambda_{AB} \mathbf{M}_A \cdot \mathbf{M}_B \tag{20.24}$$

$\lambda_{AA} M_A^2$ and $\lambda_{AA} M_B^2$ are constant terms as they are independent of ϕ. Therefore, Eq. (20.24) can be written as

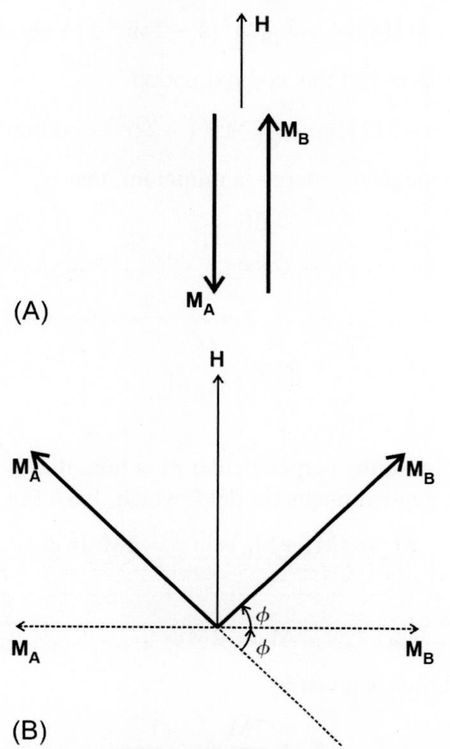

FIG. 20.5 (A) Magnetizations $\mathbf{M}_A$ and $\mathbf{M}_B$ of sublattices A and B are parallel to each other but opposite in direction. Both of the magnetizations are parallel to the applied magnetic field **H**. (B) Magnetizations $\mathbf{M}_A$ and $\mathbf{M}_B$ of sublattices A and B perpendicular to the applied magnetic field **H**.

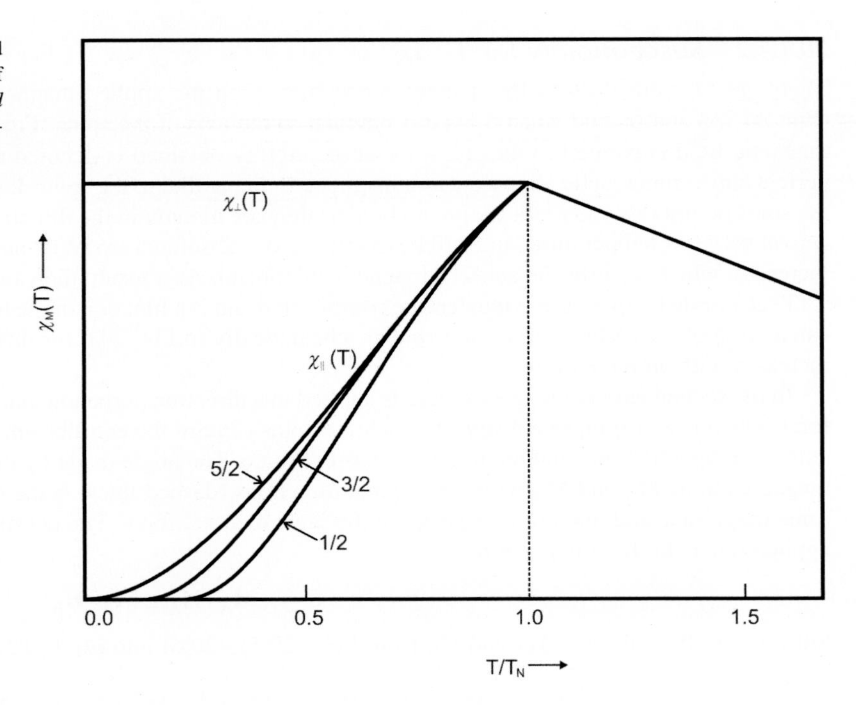

FIG. 20.6 The magnetic susceptibility [both $\chi_{\|}(T)$ and $\chi_{\perp}(T)$] of an antiferromagnetic material as a function of temperature for spin values of ½, 3/2, and 5/2. *Modified from Van Vleck, J. H. (1941). On the theory of antiferromagnetism. The Journal of Chemical Physics, 9, 85.*

$$E = -(\mathbf{M}_A + \mathbf{M}_B)\cdot\mathbf{H} + \lambda_{AB}\mathbf{M}_A\cdot\mathbf{M}_B + \text{constant} \tag{20.25}$$

In an antiferromagnetic solid

$$|\mathbf{M}_A| = |\mathbf{M}_B| = M \tag{20.26}$$

With this fact the magnetic energy from Eq. (20.25) is given by

$$\begin{aligned} E &= -2MH\sin\phi + \lambda_{AB}M^2\cos(180-2\phi) + \text{constant} \\ &= -2MH\sin\phi - \lambda_{AB}M^2\left(1-2\sin^2\phi\right) + \text{constant} \end{aligned} \tag{20.27}$$

If the angle ϕ is very small, then $\sin\phi \approx \phi$ and in this approximation

$$E = -2MH\phi - \lambda_{AB}M^2\left(1-2\phi^2\right) + \text{constant} \tag{20.28}$$

In the equilibrium state of the spins, the magnetic energy is minimum, that is,

$$\frac{dE}{d\phi} = 0 \tag{20.29}$$

From Eqs. (20.28), (20.29) one gets

$$\phi = \frac{H}{2\lambda_{AB}M} \tag{20.30}$$

In order to find the magnetic susceptibility in the perpendicular direction, that is, $\chi_{\perp}(T)$, one requires the component of magnetization along the direction of the applied magnetic field, which, from Fig. 20.5B, is given by

$$M_{\perp} = (M_A + M_B)\sin\phi = 2M\sin\phi \tag{20.31}$$

For small angles

$$M_{\perp} = 2M\phi \tag{20.32}$$

From Eq. (20.32) the magnetic susceptibility is given by

$$\chi_{\perp} = \frac{2M}{H}\phi = \frac{1}{\lambda_{AB}} \tag{20.33}$$

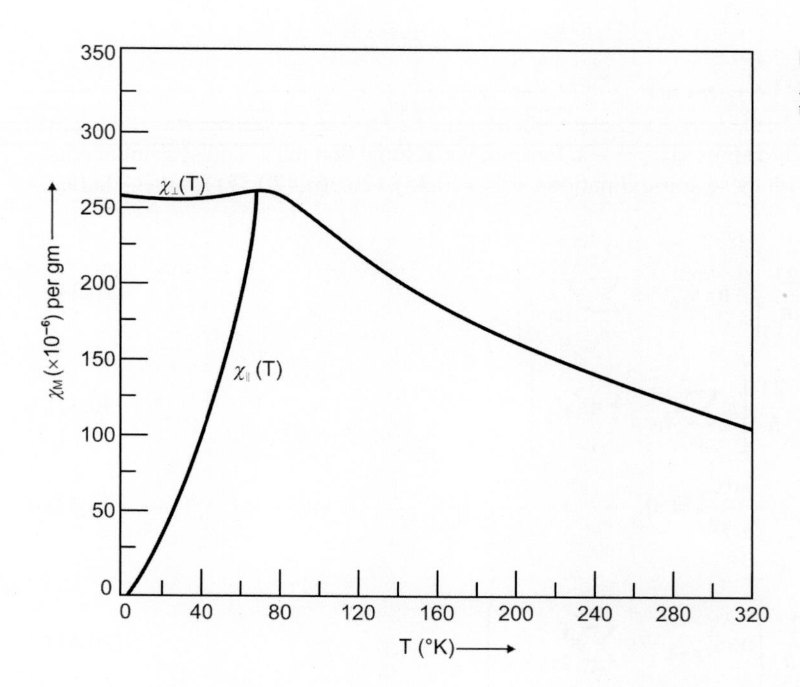

FIG. 20.7 Experimental values of the magnetic susceptibility χ_M of antiferromagnetic MnF_2 parallel and perpendicular to the tetragonal axis.

Eq. (20.33) shows that $\chi_\perp(T)$ is constant below the Neel temperature (see Fig. 20.6). Fig. 20.7 shows experimental measurements for both $\chi_{||}(T)$ and $\chi_\perp(T)$ for MnF_2 and the behavior is found to be similar to that shown in Fig. 20.6. The susceptibility below T_N is given by some average value lying between $\chi_{||}(T)$ and $\chi_\perp(T)$. The simple average, which usually fits the experimental results, is defined as

$$\chi_M(T) = \frac{1}{3}\chi_{||}(T) + \frac{2}{3}\chi_\perp(T) \tag{20.34}$$

20.1.2 Spin Waves in Antiferromagnetism

In an antiferromagnetic substance, suppose that sublattice A consists of up spins (s) with even numbers and sublattice B consists of down spins (−s) with odd numbers. For sublattice A, Eqs. (19.94), (19.95), (19.96) can be written as

$$\frac{ds^x_{2p}}{dt} = \frac{2J}{\hbar}\left[s^y_{2p}\sum_{\vec{\delta}} s^z_{2p,\,\vec{\delta}} - s^z_{2p}\sum_{\vec{\delta}} s^y_{2p,\,\vec{\delta}}\right] \tag{20.35}$$

$$\frac{ds^y_{2p}}{dt} = \frac{2J}{\hbar}\left[s^z_{2p}\sum_{\vec{\delta}} s^x_{2p,\,\vec{\delta}} - s^x_{2p}\sum_{\vec{\delta}} s^z_{2p,\,\vec{\delta}}\right] \tag{20.36}$$

$$\frac{ds^z_{2p}}{dt} = \frac{2J}{\hbar}\left[s^x_{2p}\sum_{\vec{\delta}} s^y_{2p,\,\vec{\delta}} - s^y_{2p}\sum_{\vec{\delta}} s^x_{2p,\,\vec{\delta}}\right] \tag{20.37}$$

For sublattice B one can straightway write

$$\frac{ds^x_{2p+1}}{dt} = \frac{2J}{\hbar}\left[s^y_{2p+1}\sum_{\vec{\delta}} s^z_{2p+1,\,\vec{\delta}} - s^z_{2p+1}\sum_{\vec{\delta}} s^y_{2p+1,\,\vec{\delta}}\right] \tag{20.38}$$

$$\frac{ds^y_{2p+1}}{dt} = \frac{2J}{\hbar}\left[s^z_{2p+1}\sum_{\vec{\delta}} s^x_{2p+1,\,\vec{\delta}} - s^x_{2p+1}\sum_{\vec{\delta}} s^z_{2p+1,\,\vec{\delta}}\right] \tag{20.39}$$

$$\frac{ds^z_{2p+1}}{dt}=\frac{2J}{\hbar}\left[s^x_{2p+1}\sum_{\vec{\delta}}s^y_{2p+1,\vec{\delta}}-s^y_{2p+1}\sum_{\vec{\delta}}s^x_{2p+1,\vec{\delta}}\right] \tag{20.40}$$

For even-numbered spins $s^z_{2p}=s$ and for odd-numbered spins $s^z_{2p+1}=-s$. Further, we assume that the s^x and s^y components are very small, so their product can be neglected. With these approximations, Eqs. (20.35), (20.36), (20.37) for the A lattice become

$$\frac{ds^x_{2p}}{dt}=\frac{2J}{\hbar}\left[-ns\,s^y_{2p}-s\sum_{\vec{\delta}}s^y_{2p,\vec{\delta}}\right] \tag{20.41}$$

$$\frac{ds^y_{2p}}{dt}=\frac{2J}{\hbar}\left[s\sum_{\vec{\delta}}s^x_{2p,\vec{\delta}}+ns\,s^x_{2p}\right] \tag{20.42}$$

$$\frac{ds^z_{2p}}{dt}=0 \tag{20.43}$$

and for lattice B one can write

$$\frac{ds^x_{2p+1}}{dt}=\frac{2J}{\hbar}\left[ns\,s^y_{2p+1}+s\sum_{\vec{\delta}}s^y_{2p+1,\vec{\delta}}\right] \tag{20.44}$$

$$\frac{ds^y_{2p+1}}{dt}=\frac{2J}{\hbar}\left[-s\sum_{\vec{\delta}}s^x_{2p+1,\vec{\delta}}-ns\,s^x_{2p+1}\right] \tag{20.45}$$

$$\frac{ds^z_{2p+1}}{dt}=0 \tag{20.46}$$

Let us define an operator

$$s^+=s^x+\imath s^y \tag{20.47}$$

From Eq. (20.47) the equations of motion of the even and odd spins can be written as

$$\frac{ds^+_{2p}}{dt}=\frac{ds^x_{2p}}{dt}+\imath\frac{ds^y_{2p}}{dt} \tag{20.48}$$

$$\frac{ds^+_{2p+1}}{dt}=\frac{ds^x_{2p+1}}{dt}+\imath\frac{ds^y_{2p+1}}{dt} \tag{20.49}$$

Using Eqs. (20.41)–(20.46) in Eqs. (20.48), (20.49), one can straightway write

$$\frac{ds^+_{2p}}{dt}=\frac{2\imath Js}{\hbar}\left(ns^+_{2p}+\sum_{\vec{\delta}}s^+_{2p,\vec{\delta}}\right) \tag{20.50}$$

$$\frac{ds^+_{2p+1}}{dt}=-\frac{2\imath Js}{\hbar}\left(ns^+_{2p+1}+\sum_{\vec{\delta}}s^+_{2p+1,\vec{\delta}}\right) \tag{20.51}$$

For spin waves one is looking for running-wave solutions of the form

$$s^+_{2p,\vec{\delta}}=u_{\mathbf{K}}e^{\imath\left[\mathbf{K}\cdot\left(\mathbf{R}_{2p}+\vec{\delta}\right)-\omega t\right]} \tag{20.52}$$

$$s^+_{2p+1,\vec{\delta}}=v_{\mathbf{K}}e^{\imath\left[\mathbf{K}\cdot\left(\mathbf{R}_{2p+1}+\vec{\delta}\right)-\omega t\right]} \tag{20.53}$$

Substituting Eqs. (20.52), (20.53) into Eqs. (20.50), (20.51), we obtain

$$\left(\frac{1}{2}n\omega_{ex} - \omega\right) u_{\mathbf{K}} + \left(\frac{1}{2}\omega_{ex} \sum_{\vec{\delta}} e^{\imath \mathbf{K} \cdot \vec{\delta}}\right) v_{\mathbf{K}} = 0 \tag{20.54}$$

$$\left(\frac{1}{2}\omega_{ex} \sum_{\vec{\delta}} e^{\imath \mathbf{K} \cdot \vec{\delta}}\right) u_{\mathbf{K}} + \left(\frac{1}{2}n\omega_{ex} + \omega\right) v_{\mathbf{K}} = 0 \tag{20.55}$$

$$\text{where } \omega_{ex} = -\frac{4Js}{\hbar} \tag{20.56}$$

Eqs. (20.54), (20.55) have a nontrivial solution only if

$$\begin{vmatrix} \frac{1}{2}n\omega_{ex} - \omega & \frac{1}{2}\omega_{ex} \sum_{\vec{\delta}} e^{\imath \mathbf{K}.\vec{\delta}} \\ \frac{1}{2}\omega_{ex} \sum_{\vec{\delta}} e^{\imath \mathbf{K}.\vec{\delta}} & \frac{1}{2}n\omega_{ex} + \omega \end{vmatrix} = 0 \tag{20.57}$$

Expanding the above determinant, we obtain the value of ω^2 as

$$\omega^2 = \frac{1}{4}\omega_{ex}^2 \left[n^2 - \left(\sum_{\vec{\delta}} e^{\imath \mathbf{K} \cdot \vec{\delta}}\right)^2\right] \tag{20.58}$$

This is a general solution of the spin wave for any crystal structure. The spin wave frequency ω can be calculated for simple crystal structures.

20.1.2.1 Linear Monatomic Lattice

In a monatomic linear lattice $n = 2$ and $\vec{\delta} = \pm a\hat{\mathbf{i}}_1$. Substituting these values into Eq. (20.58), one gets

$$\begin{aligned} \omega^2 &= \omega_{ex}^2 - \frac{1}{4}\omega_{ex}^2 \left(e^{\imath K_x a} + e^{-\imath K_x a}\right)^2 \\ &= \omega_{ex}^2 \left(1 - \cos^2 K_x a\right) \end{aligned} \tag{20.59}$$

Therefore, the frequency of the spin waves is given by

$$\omega = \omega_{ex} |\sin K_x a| \tag{20.60}$$

For small values of K_x the above equation yields

$$\omega = \omega_{ex} |K_x a| \tag{20.61}$$

Eq. (20.61) shows that the magnon frequency is linearly proportional to K_x for small values of the wave vector. The magnon dispersion relation for an antiferromagnetic linear lattice is shown in Fig. 20.8 and it is different from that for a ferromagnetic substance.

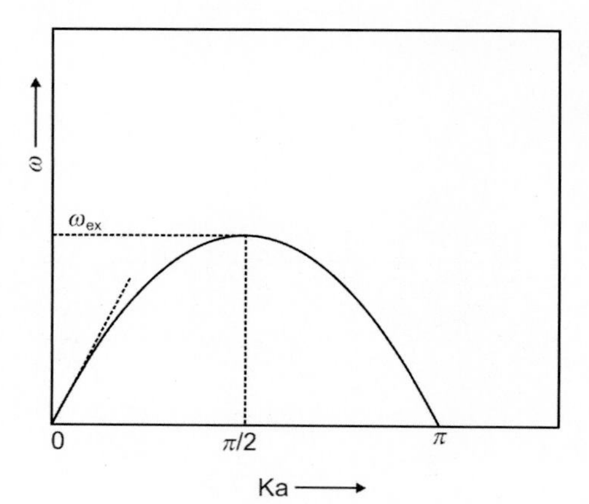

FIG. 20.8 Magnon dispersion relation for a monatomic linear lattice described by Eq. (20.60). In the figure, we have replaced K_x by K for convenience. The *dashed line* is a tangent to the magnon dispersion curve at the origin and shows that the magnon frequency ω, at small values of the wave vector K, is linearly proportional to K (see Eq. 20.61).

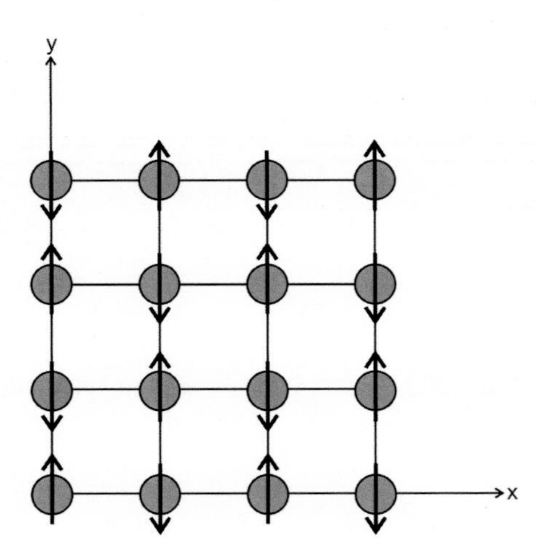

FIG. 20.9 A square lattice with an antiferromagnetic order of atomic spins and having periodicity a.

20.1.2.2 Square Lattice

Consider a square lattice with periodicity a in which $\vec{\delta} = \left(\pm a\hat{\mathbf{i}}_1, 0\right), \left(0, \ \pm a\hat{\mathbf{i}}_2\right)$ (see Fig. 20.9). Substituting the values of $\vec{\delta}$ into Eq. (20.58), one can write

$$\omega^2 = \omega_{ex}^2 \left[4 - \cos^2 K_x a - \cos^2 K_y a \ - 2\cos K_x a \cos K_y a\right] \quad (20.62)$$

Simplifying the above equation, one can write

$$\omega = \omega_{ex} \left[2 + \sin^2 K_x a + \sin^2 K_y a - 2\cos K_x a \cos K_y a\right]^{1/2} \quad (20.63)$$

For very small values of K_x and K_y the trigonometric functions can be expanded to yield

$$\omega = \sqrt{2}\omega_{ex} K a \quad (20.64)$$

Problem 20.1

Find the magnon dispersion relation in an sc structure with antiferromagnetic order (see Fig. 20.10) and show that, at very small values of the wave vector K, it is given by

$$\omega = \sqrt{3}\omega_{ex} K a \quad (20.65)$$

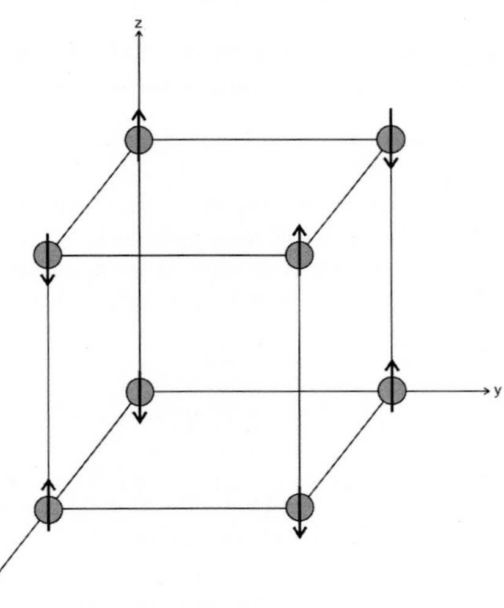

FIG. 20.10 An sc lattice with an antiferromagnetic order of atomic spins and having periodicity a.

Problem 20.2

Find the magnon dispersion relation for an antiferromagnetic material with bcc structure.

From Eqs. (20.61), (20.64), (20.65) it is evident that the magnon dispersion relations in an antiferromagnetic material with cubic structure are linearly proportional to K, while in a ferromagnetic substance they are proportional to K^2. This shows that the magnon dispersion relations in these two types of solids are basically different. Further, from Eqs. (20.61), (20.64), (20.65) we can generalize the expression for the effective magnon frequency as

$$\omega = \sqrt{r}\,\omega_{ex}\,Ka \tag{20.66}$$

where r is the dimensionality of the crystal under consideration.

20.2 FERRIMAGNETISM

Magnetite, generally called load stone, is perhaps the oldest ferromagnetic material known to humans. The chemical formula for magnetite is Fe_3O_4 and, more specifically, $FeO \cdot Fe_2O_3$. The oxide ferromagnetic solids form a special class, generally called ferrimagnetic solids and, more frequently, ferrites. One can produce a number of ferrites by replacing an Fe^{+2} ion by a divalent metal ion M^{+2}, thus yielding the chemical formula $MOFe_2O_3$ or MFe_2O_4, in which M can be any of the metals, such as Mn, Co, Ni, Cu, Mg, Zn , Cd, or others. Ferrites with the formula $MOFe_2O_3$ are called mixed ferrites. One of the important and useful mixed ferrites is $ZnOFe_2O_3$. Zn^{+2} is diamagnetic in nature but $ZnOFe_2O_3$ shows a magnetization that is larger than Fe_3O_4 for small concentrations of Zn.

20.2.1 Structure of Ferrites

The physical properties of ferrites are very much structure dependent. The ferrites are ionic solids and the magnetic moment of one molecule depends on the unpaired spins in the individual ions. The ions Fe^{+2} and Fe^{+3} contribute to the magnetic moment of Fe_3O_4. The electronic configuration of an Fe atom is given by

$$Fe: 3d^6 4s^2$$

Therefore, the electronic configurations of the Fe^{+2} and Fe^{+3} ions are as follows:

$$Fe^{+2}: 3d^6$$

$$Fe^{+3}: 3d^5$$

In an Fe^{+2} ion the two d-electrons are paired, but the rest of the four d-electrons are unpaired, yielding a magnetic moment of $4\,\mu_B$, while in Fe^{+3} all five d-electrons are unpaired, giving rise to a magnetic moment of $5\,\mu_B$. The unit cell of magnetite contains eight FeO. Fe_2O_3 molecules forming a close-packed cubic structure. In other words, the unit cell contains eight Fe^{+2} ions, sixteen Fe^{+3} ions, and twenty-four O^{-2} ions. The twenty four O^{-2} ions form a close-packed cubic structure that has sixteen octahedral interstitial sites and eight tetrahedral interstitial sites. Two types of distributions can occur for the Fe^{+2} and Fe^{+3} ions on the interstitial sites.

Spinel structure: In the first distribution all the Fe^{+3} ions occupy the sixteen octahedral interstitial positions, while the Fe^{+2} ions occupy the eight tetrahedral positions. Such a structure is generally called spinel structure. If the magnetic moments of both the Fe^{+2} and Fe^{+3} ions are aligned in the same direction, then the total magnetic moment on the Fe_3O_4 molecule becomes $5 \times 2 + 4 = 14\,\mu_B$. But if the magnetic moments of the Fe^{+2} and Fe^{+3} ions are aligned opposite to each other, the net magnetic moment on the Fe_3O_4 molecule becomes $10 - 4 = 6\,\mu_B$. Neither of these values agrees with the experimental value of $4 \cdot 08\,\mu_B$.

Inverse spinel structure: In the second distribution the Fe^{+2} ions occupy the eight octahedral interstitial positions, while the Fe^{+3} ions are distributed equally among the octahedral and tetrahedral positions. It is worthwhile pointing out here that FeO. Fe_2O_3 exhibits inverse spinel structure. The schematic representation of the Fe^{+2} and Fe^{+3} ions in a unit cell is shown In Fig. 20.11. In this distribution the magnetic moments of every pair of Fe^{+3} ions (one at the octahedral site and the other at the tetrahedral site) are directed opposite to each other and cancel each other's effect. Therefore, only the Fe^{+2} ions contribute to the magnetic moment of the molecule, which is found to be $4 \cdot 0\,\mu_B$, close to the experimental value.

FIG. 20.11 Schematic representation of the spins of the sixteen Fe^{+3} ions and eight Fe^{+2} ions contained in the unit cell of magnetite Fe_3O_4.

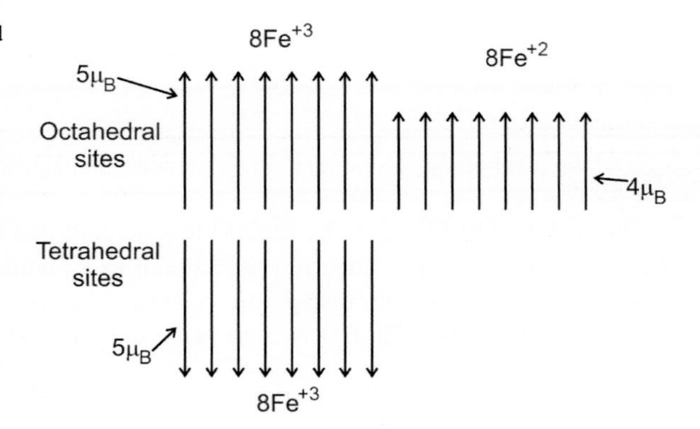

The ferrites exhibit spontaneous magnetization at low temperatures. But at a particular temperature T_c (transition temperature), the solid makes a transition from the ferromagnetic to paramagnetic phase.

20.2.2 Two-Sublattice Model

From Fig. 20.11 it is evident that in a ferrimagnetic solid there are two sublattices with different magnetizations: sublattice A consists of up spins with magnetization $\mathbf{M_A}$ and sublattice B has down spins with magnetization $\mathbf{M_B}$. Therefore, one can use the two-sublattice model for studying the susceptibility and transition temperature in a ferrimagnetic solid. The magnetic field experienced by an atom in the sublattices A and B, from Eqs. (20.2), (20.3), is given by

$$\mathbf{H_A} = \mathbf{H} - \lambda_{AA}\mathbf{M_A} - \lambda_{AB}\mathbf{M_B} \tag{20.67}$$

$$\mathbf{H_B} = \mathbf{H} - \lambda_{BB}\mathbf{M_B} - \lambda_{AB}\mathbf{M_A} \tag{20.68}$$

Here we have assumed that $\lambda_{AB} = \lambda_{BA}$. Above T_c the solid is paramagnetic in nature and the susceptibility obeys the Curie law. Therefore, the magnetizations of sublattices A and B above T_c are given by

$$\mathbf{M_A} = \chi_A \mathbf{H_A} = \frac{C_A}{T}\mathbf{H_A} \tag{20.69}$$

$$\mathbf{M_B} = \chi_B \mathbf{H_B} = \frac{C_B}{T}\mathbf{H_B} \tag{20.70}$$

Substituting Eqs. (20.67), (20.68) into Eqs. (20.69), (20.70), we find

$$(T + C_A\lambda_{AA})\mathbf{M_A} + C_A(\lambda_{AB}\mathbf{M_B} - \mathbf{H}) = 0 \tag{20.71}$$

$$C_B(\lambda_{AB}\mathbf{M_A} - \mathbf{H}) + (T + C_B\lambda_{BB})\mathbf{M_B} = 0 \tag{20.72}$$

At the transition temperature T_c, the spontaneous magnetization is obtained by putting $\mathbf{H} = 0$ in Eqs. (20.71), (20.72) to yield

$$(T_c + C_A\lambda_{AA})\mathbf{M_A} + C_A\lambda_{AB}\mathbf{M_B} = 0 \tag{20.73}$$

$$C_B\lambda_{AB}\mathbf{M_A} + (T_c + C_B\lambda_{BB})\mathbf{M_B} = 0 \tag{20.74}$$

A nontrivial solution of Eqs. (20.73), (20.74) is obtained by setting the determinant of the coefficients of $\mathbf{M_A}$ and $\mathbf{M_B}$ to zero, that is,

$$\begin{vmatrix} T_c + C_A\lambda_{AA} & C_A\lambda_{AB} \\ C_B\lambda_{AB} & T_c + C_B\lambda_{BB} \end{vmatrix} = 0 \tag{20.75}$$

The determinant gives a second order equation in T_c whose value is given by

$$T_c = \frac{1}{2}\left[-(C_A\lambda_{AA} + C_B\lambda_{BB}) \pm \left\{(C_A\lambda_{AA} - C_B\lambda_{BB})^2 + 4C_AC_B\lambda_{AB}^2\right\}^{1/2}\right] \tag{20.76}$$

From Fig. 20.11 it is evident that there are ferromagnetic interactions among the spins of sublattice A. Similarly, there are ferromagnetic interactions among the spins of sublattice B. If only the antiferromagnetic interactions between the spins of sublattices A and B are retained, then $\lambda_{AA} = \lambda_{BB} = 0$. In this approximation Eq. (20.76) reduces to

$$T_c = (C_A C_B)^{1/2} \lambda_{AB} \tag{20.77}$$

The magnetic susceptibility of a ferrimagnetic solid can be calculated by retaining only the antiferromagnetic interactions. Substituting Eqs. (20.67), (20.68) into Eqs. (20.69), (20.70) and retaining only the antiferromagnetic interactions, one can write

$$\mathbf{M}_A T = C_A [\mathbf{H} - \lambda_{AB} \mathbf{M}_B] \tag{20.78}$$

$$\mathbf{M}_B T = C_B [\mathbf{H} - \lambda_{AB} \mathbf{M}_A] \tag{20.79}$$

Multiplying Eq. (20.78) by T and using Eq. (20.79), we get

$$\mathbf{M}_A = \frac{C_A T - C_A C_B \lambda_{AB}}{T^2 - T_c^2} \mathbf{H} \tag{20.80}$$

Similarly, one can obtain from Eq. (20.79) the expression for $\mathbf{M}_B$ as

$$\mathbf{M}_B = \frac{C_B T - C_A C_B \lambda_{AB}}{T^2 - T_c^2} \mathbf{H} \tag{20.81}$$

Adding Eqs. (20.80), (20.81), the magnetic susceptibility is given by

$$\chi_M = \frac{\mathbf{M}_A + \mathbf{M}_B}{\mathbf{H}} = \frac{(C_A + C_B)T - 2C_A C_B \lambda_{AB}}{T^2 - T_c^2} \tag{20.82}$$

One can study antiferromagnetism as a particular case of ferrimagnetism. The Neel temperature in an antiferromagnetic substance can be obtained from Eq. (20.76). In an antiferromagnetic substance the two sublattices are identical and thus have the same magnetizations and Curie constants. Substituting $C_A = C_B$ and $\lambda_{AA} = \lambda_{BB}$ into Eq. (20.76), one gets

$$T_c = T_N = (\lambda_{AB} - \lambda_{AA}) C_A \tag{20.83}$$

which is nothing but the Neel temperature in an antiferromagnetic material (see Eq. 20.19). If only the antiferromagnetic interactions are retained, then $\lambda_{AA} = 0$ and Eq. (20.83) reduces to

$$T_N = \lambda_{AB} C_A \tag{20.84}$$

Substituting $C_A = C_B$ into Eq.(20.82) and using Eq. (20.84), one gets

$$\chi_M = \frac{2C_A}{T + T_N} \tag{20.85}$$

which is same as Eq. (20.12). It is worth mentioning here that in the mixed ferrite $ZnOFe_2O_3$ all the Zn^{+2} ions occupy the tetrahedral interstitial positions, but the Fe^{+3} ions occupy the octahedral interstitial positions. In other words, all the Zn^{+2} ions occupy the positions on one sublattice, say A. But this may not be the case with other divalent metallic ions. In general M^{+2} ions may occupy both the tetrahedral and octahedral positions and, therefore, the mixed ferrite $MOFe_2O_3$ can be written as $Fe_x^{+3} M_{1-x}^{+2} [Fe_{2-x}^{+3} M_x^{+2}] O_4$: here a portion $1 - x$ of the M^{+2} ions occupy positions on sublattice A, while a portion x of the M^{+2} ions occupy positions on sublattice B.

REFERENCE

Van Vleck, J. H. (1941). On the theory of antiferromagnetism. *The Journal of Chemical Physics, 9*, 85.

SUGGESTED READING

Cohen, M. H. (1967). Topics in the theory of magnetic metals. In W. Marshall (Ed.), *Proceedings of the international school of physics "Enrico Fermi" (Course XXXVII)*. New York: Academic Press.

Nagamiya, T., Yosida, K., & Kubo, R. (1955). Antiferromagnetism. *Advances in Physics, 4*, 1.

Spaldin, N. A. (2010). *Magnetic materials: Fundamentals and applications* (2nd ed.). London: Cambridge University Press.

Chapter 21

Magnetic Resonance

Chapter Outline

In solids, both the electrons and nuclei possess intrinsic spin angular momenta, which give rise to spin magnetic dipole moments. We have seen in the previous chapters that a static magnetic field produces magnetization, which yields important information about solids. A great deal of information about magnetic solids can be obtained from the nuclear magnetic resonance (NMR) technique in which an alternating magnetic field is also applied. The basic principle of an NMR method is to apply both dc and rf magnetic fields perpendicular to one another. As a result, the nuclear energy states split up into substates. When the frequency of the applied rf field is equal to the natural frequency in the presence of external magnetic field (which corresponds to the frequency of gyroscopic precession of the magnetic moments), then resonance takes place. There are different resonance techniques with which one can study a number of properties, such as the fine structure of nuclear levels, the electronic structure of an impurity, the interaction between magnetic dipoles, the interaction between magnetic dipoles and the lattice, and the line widths of nuclear transitions. The basic ideas behind nuclear resonance techniques will be described in this chapter.

21.1 NUCLEAR MAGNETIC MOMENT

In analogy with the magnetic moment of an electron, the magnetic moment of a nucleus with nuclear spin **I** is defined as

$$\vec{\mu}_{\mathrm{I}} = \gamma_{\mathrm{I}} \hbar \mathbf{I} \tag{21.1}$$

Here γ_{I} is the ratio of the magnetic moment to the angular momentum and is the gyromagnetic ratio. In the case of a nucleus, γ_{I} satisfies the relation

$$\gamma_{\mathrm{I}} \hbar = g_{\mathrm{I}} \mu_{\mathrm{Bp}} \tag{21.2}$$

where μ_{Bp} is the Bohr magnetron of a proton defined by Eq. (2.8) and g_{I} is the *nuclear spectroscopic splitting factor*. By substituting Eq. (21.2) into Eq. (21.1), one gets

$$\vec{\mu}_{\mathrm{I}} = g_{\mathrm{I}} \mu_{\mathrm{Bp}} \mathbf{I} \tag{21.3}$$

Eq. (21.3) shows that, for a nucleus, $\vec{\mu}_{\mathrm{I}}$ and **I** are in the same direction, in contrast to an electron. Although the gyromagnetic ratio is a more useful quantity for discussing magnetic resonance in many respects, nuclear magnetic moments are usually tabulated in terms of μ_{Bp}, a natural unit with a value of 5.05×10^{-24} erg/gauss (see Chapter 2). One can also define the

Solid State Physics. https://doi.org/10.1016/B978-0-12-817103-5.00021-9

magnetic moment of a nucleus $\vec{\mu}_I$ with nuclear spin **I** as the vector sum of the magnetic moments of the nucleons $\vec{\mu}_n$ that are present in it, that is,

$$\vec{\mu}_I = \sum_n \vec{\mu}_n \tag{21.4}$$

21.2 ZEEMAN EFFECT

The Hamiltonian of a nucleus with magnetic moment $\vec{\mu}_I$ in an external dc magnetic field **H** is

$$\widehat{H}_M = -\vec{\mu}_I \cdot \mathbf{H} = -\gamma_I \hbar \mathbf{I} \cdot \mathbf{H} \tag{21.5}$$

If the magnetic field is applied in the z-direction, that is, $\mathbf{H} = \hat{\mathbf{i}}_3 H$, the Hamiltonian becomes

$$\widehat{H}_M = -\gamma_I \hbar H I_z = -g_I \mu_{Bp} H I_z \tag{21.6}$$

The energy of a nucleus in the state $|m_I\rangle$ is given by

$$E_{m_I} = \langle m_I | \widehat{H}_M | m_I \rangle = -\gamma_I \hbar H m_I = -g_I \mu_{Bp} H m_I \tag{21.7}$$

where m_I is the eigenvalue of I_z defined as

$$\langle m_I | I_z | m_I \rangle = m_I \tag{21.8}$$

Here m_I has $2I+1$ values given as $I, I-1, \ldots \; 1, 0, -1, \ldots \; -(I-1), -I$. Hence, each nuclear state splits up into $2I+1$ substates in the presence of a magnetic field and they are equally spaced. Transitions between the substates are allowed only if the magnetic quantum number m_I changes by unity, that is,

$$\Delta m_I = \pm 1 \tag{21.9}$$

Using Eq. (21.7), we can write

$$\hbar\omega = E_{m_I'} - E_{m_I} = \gamma_I \hbar H \Delta m_I = g_I \mu_{Bp} H \Delta m_I \tag{21.10}$$

where

$$m_I - m_I' = \Delta m_I \tag{21.11}$$

Using the selection rule (21.9) in Eq. (21.10), the transition frequency is given by

$$\omega_L = \gamma_I H = g_I \mu_{Bp} H/\hbar \tag{21.12}$$

Eq. (21.12) yields a single resonance frequency, usually called the Larmor frequency ω_L, which gives the fundamental condition for magnetic resonance absorption. The radiation produced is circularly polarized in a plane perpendicular to the dc magnetic field **H**. Therefore, in order to excite such transitions, it is necessary to supply radiation with a magnetic vector circularly polarized in a plane perpendicular to the dc magnetic field. One can interpret the above result in the following manner: The magnetic dipole moment $\vec{\mu}_I$, when placed in a dc magnetic field **H**, starts precessing about the applied field with the Larmor frequency (see Fig. 21.1).

As an example, consider a solid with $I = 1/2$. In the presence of a magnetic field, $m_I = \pm 1/2$, which gives the energy of the two states as $E_{\pm} = \mp(1/2)\gamma_I \hbar H$ (see Fig. 21.2). If $\hbar\omega_L$ denotes the energy difference between the two energy states, then

$$\omega_L = \gamma_I H = g_I \mu_{Bp} H/\hbar \tag{21.13}$$

which is the same as Eq. (21.12). The Zeeman splitting for a solid with $I = 3/2$ is also shown in Fig. 21.2.

Problem 21.1

Let the population of a Zeeman state N_{m_I} for a particular value of m_I be given as

$$N_{m_I} \propto \exp\left(-E_{m_I}/k_B T\right) \tag{21.14}$$

Here k_B and T are the Boltzmann constant and temperature, respectively. Prove that the average magnetic moment of a nucleus $\langle \mu_I \rangle$ in the direction of the magnetic field is given by

$$\langle \mu_I \rangle = \gamma_I \hbar \left[\frac{2I+1}{2} \coth\left(\frac{2I+1}{2} y\right) - \frac{1}{2} \coth\left(\frac{y}{2}\right)\right] \tag{21.15}$$

where

$$y = \frac{\gamma_I \hbar H}{k_B T} \tag{21.16}$$

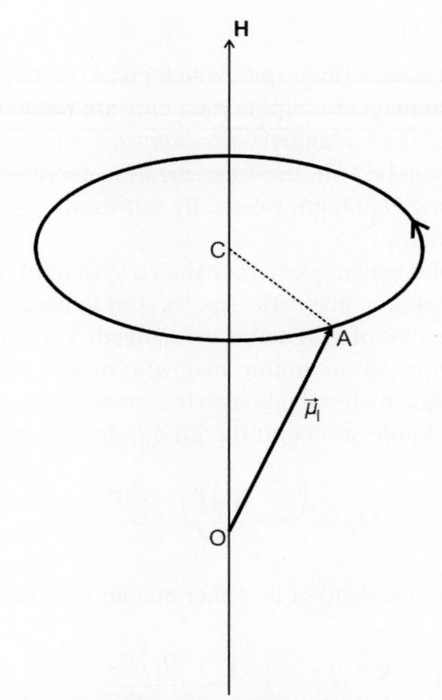

FIG. 21.1 Precession of the magnetic moment $\vec{\mu}_I$ about the applied magnetic field **H**.

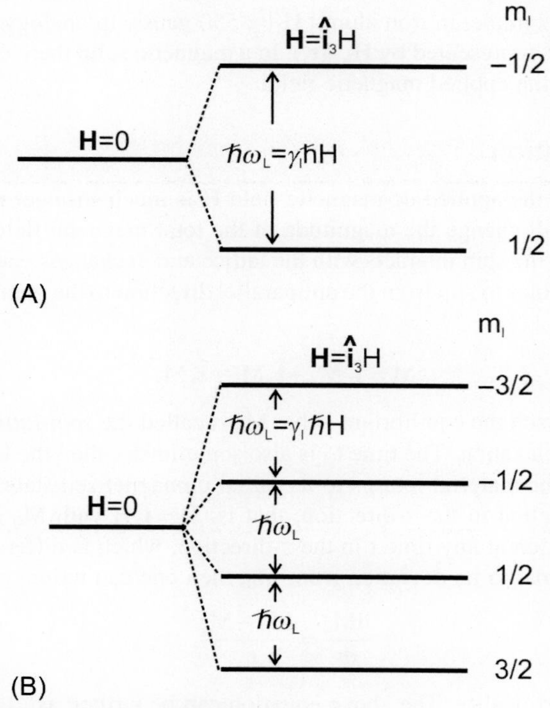

FIG. 21.2 Zeeman splitting in a solid with nuclear spin (A) I=1/2 and (B) I=3/2.

Problem 21.2

Using $\langle \mu_I \rangle$ given by Eq. (21.15), prove that the nuclear magnetic susceptibility in the limit of very high values of T is given by

$$\chi_M = n_{nuc} \gamma_I^2 \hbar^2 \frac{I(I+1)}{3k_B T} \tag{21.17}$$

where n_{nuc} is the number of nuclei per unit volume.

21.3 RELAXATION PHENOMENA

In a paramagnetic substance each atom possesses a finite spin, which gives rise to a finite atomic spin magnetic moment. In the absence of an applied magnetic field the magnetic dipole moments are randomly oriented, giving rise to zero magnetization. With the application of a magnetic field, magnetization appears, which takes some finite time to reach its equilibrium value. When the magnetic field is switched off, the magnetization decreases and goes to zero again in a finite time. Both the increase and decrease in the magnetization are caused by relaxation phenomena and the time taken to reach the equilibrium value is called the relaxation time.

In order to have insight into relaxation phenomena, let us consider a system of free magnetic dipoles oriented randomly. If an external magnetic field is applied, the atomic magnetic dipoles start to precess about it, as shown in Fig. 21.1, but the component of magnetization along the direction of **H** remains unchanged. Therefore, no magnetization will appear. For a finite magnetization to appear, the interaction of an atomic magnetic dipole with other atomic dipoles or with its surroundings is required. Consider a paramagnetic material in which the atoms (magnetic dipoles) form a lattice. The magnetic field at the position $\mathbf{r}_i$ of the ith magnetic dipole $\vec{\mu}_i$ due to the jth dipole $\vec{\mu}_j$ at $\mathbf{r}_j$ is given by

$$\mathbf{H}_{ij} = \frac{\left(\vec{\mu}_j \cdot \mathbf{r}_{ij}\right)\mathbf{r}_{ij} - \vec{\mu}_j r_{ij}^2}{r_{ij}^5} \tag{21.18}$$

The total magnetic field at the position of $\vec{\mu}_i$ due to all of the other magnetic dipoles is obtained by summing over j, that is,

$$\mathbf{H}_i = \sum_j \mathbf{H}_{ij} = \sum_j \frac{\left(\vec{\mu}_j \cdot \mathbf{r}_{ij}\right)\mathbf{r}_{ij} - \vec{\mu}_j r_{ij}^2}{r_{ij}^5} \tag{21.19}$$

$\mathbf{H}_i$ is called the internal magnetic field experienced by an atomic magnetic dipole and is on the order of μ_I/a^3 where a is on the order of a few angstroms. For example, in iron alum $|\mathbf{H}_i| \approx 550$ gauss. In analogy with the local electric field, $\mathbf{H}_i$ is called the local magnetic field and is represented by $\mathbf{H}_{loc}(\mathbf{r}_i)$. In a magnetic solid there can be two types of relaxation phenomena depending on the value of the applied magnetic field.

21.3.1 Spin-Lattice Relaxation

Spin-lattice relaxation occurs when the applied dc magnetic field **H** is much stronger than the local magnetic field $\mathbf{H}_{loc}$. Therefore, a small increase in **H** will change the magnitude of the total magnetic field $\mathbf{H}+\mathbf{H}_{loc}$, but will not change its direction by very much. In this case the spin interacts with the lattice and exchanges energy with it. The lattice interaction may cause some of the magnetic dipoles to slip from the antiparallel direction to the parallel one, thereby producing a finite magnetization **M** defined as

$$\mathbf{M} = \hat{\mathbf{i}}_1 M_1 + \hat{\mathbf{i}}_2 M_2 + \hat{\mathbf{i}}_3 M_3 \tag{21.20}$$

The time taken by the material to reach the equilibrium value M_0 is called the *spin-lattice relaxation time* t_{sl} and the phenomenon is called the spin-lattice relaxation. The time t_{sl} is also sometimes called the longitudinal relaxation time. When the magnetic field is switched off the material returns to its initial unmagnetized state.

Let the magnetic field **H** be applied in the z-direction, that is, $\mathbf{H} = \hat{\mathbf{i}}_3 H$ with M_0 as the equilibrium magnetization. Further, let $M_z(t)$ be the magnetization at any time t in the z-direction, which is different from M_0. If the rate of change of $M_z(t)$ is assumed to be proportional to its deviation from M_0, then one can write

$$\frac{dM_z}{dt} = \frac{M_0 - M_z}{t_{sl}} \tag{21.21}$$

where $1/t_{sl}$ is the constant of proportionality. The above equation can be written as

$$\int_0^{M_z} \frac{dM_z}{M_0 - M_z} = \frac{1}{t_{sl}} \int_0^t dt \tag{21.22}$$

Integrating the above equation, one gets

$$M_z(t) = M_0\left[1 - e^{-t/t_{sl}}\right] \tag{21.23}$$

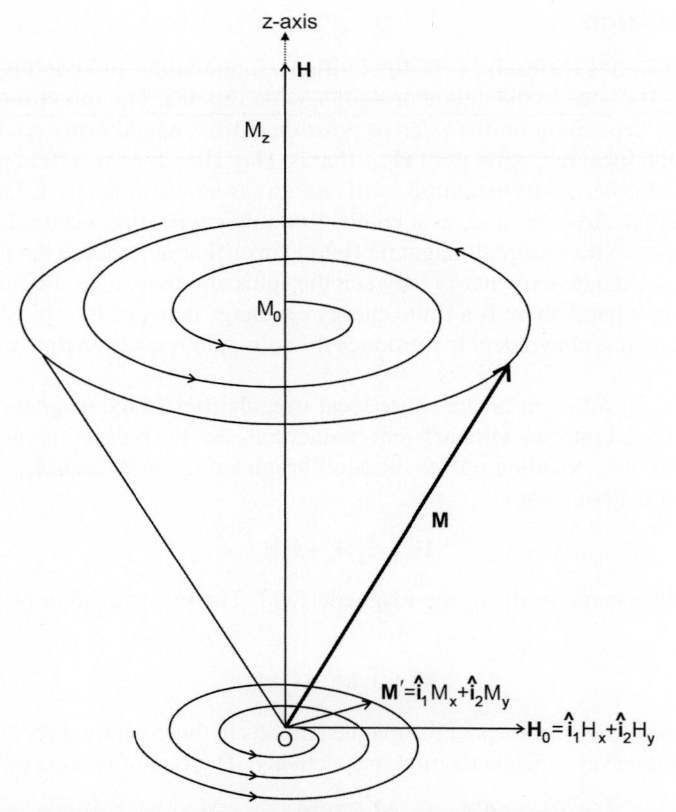

FIG. 21.3 Spin-lattice and spin-spin relaxation processes. In the spin-lattice relaxation process, the magnetization **M** approaches the equilibrium value $\mathbf{M_0}$ along the z-direction. In the spin-spin relaxation process, the magnetization **M′** (in the xy plane) approaches zero, the equilibrium value, in the presence of magnetic field $\mathbf{H_0}$.

Fig. 21.3 shows how a precessing magnetization **M** (about the z-direction) acquires the equilibrium value of magnetization $\mathbf{M_0}$ in the spin-lattice relaxation phenomenon. The variation of $M_z(t)$ with time is shown in Fig. 21.4. The spin-lattice relaxation is measured in strong fields but at low frequencies. It is important to point out that t_{sl} is strongly temperature dependent and decreases with increasing temperature. This variation is due to an increase in the mobility of the magnetic dipoles with an increase in temperature. The time t_{sl}, due to the nuclear moments in solids, may have a value from a few seconds to hours. In solids in which the paramagnetism is due to the electrons, t_{sl} varies between 10^{-11} and 10^{-6} s at room temperature.

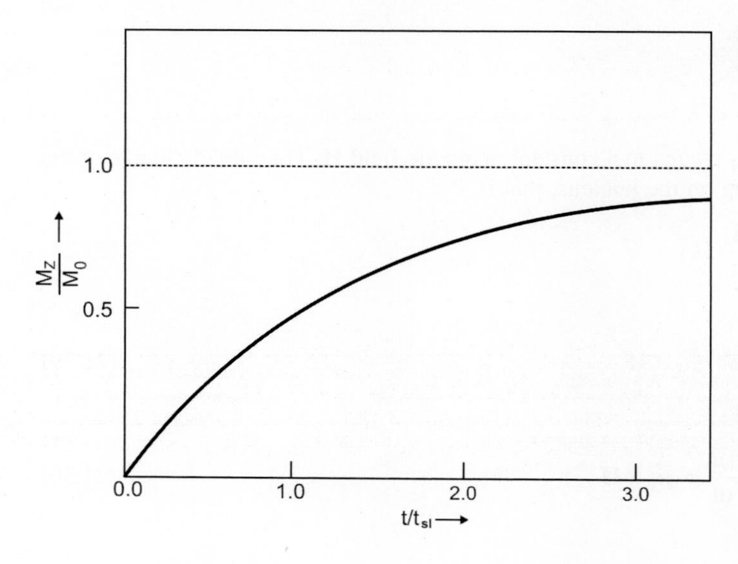

FIG. 21.4 The variation of the z-component of magnetization $\mathbf{M}_z(t)$ with time t in the spin-lattice relaxation phenomenon.

21.3.2 Spin-Spin Relaxation

The nuclear spins also interact among themselves via dipole-dipole interactions. In a paramagnetic solid, each spin experiences a local magnetic field $\mathbf{H}_{loc}$ due to the neighboring magnetic dipoles. The direction and magnitude of $\mathbf{H}_{loc}$ may differ from nucleus to nucleus depending on the relative positions of the neighboring nuclei. If the external magnetic field $\mathbf{H}$ is much smaller than the local magnetic field $\mathbf{H}_{loc}$, that is, $\mathbf{H} \ll \mathbf{H}_{loc}$, then its effect will be to slightly change the direction of the field seen by a dipole, but its magnitude will remain essentially unaltered. The magnetic dipoles will thus precess about a slightly different direction and, as a result, finite magnetization occurs in the direction of $\mathbf{H}$. In this situation the magnetic force due to the external magnetic field is insufficient to cause the magnetic dipoles to flip over and, therefore, there will be no exchange of energy between the spins and the lattice. In other words, spin-lattice relaxation will be absent. On the other hand, there is a finite energy exchange between the spins, giving rise to what is called the spin-spin relaxation process. It is convenient to introduce the spin-spin relaxation time t_{ss} to describe the lifetime of a nuclear spin state.

The different values of $\mathbf{H}_{loc}$ at different nuclei cause local irregularities in the magnetic field. Further, the different values of $\mathbf{H}_{loc}$ make different nuclei precess with different frequencies. So, the precessing nuclei will go out of phase with each other in a time on the order of t_{ss}, yielding zero as the equilibrium values of M_x and M_y. Let an rf magnetic field $\mathbf{H}_0$ be applied in the xy-plane, which is defined as

$$\mathbf{H}_0 = \hat{\mathbf{i}}_1 H_x + \hat{\mathbf{i}}_2 H_y \tag{21.24}$$

Here H_x and H_y are the x- and y-components of the magnetic field. The magnetization produced in the xy-plane $\mathbf{M}'$ is written as

$$\mathbf{M}' = \hat{\mathbf{i}}_1 M_x + \hat{\mathbf{i}}_2 M_y \tag{21.25}$$

where M_x and M_y are the nonequilibrium values of the magnetizations in the x- and y-directions. $\mathbf{H}_0$ and $\mathbf{M}'$ are called the transverse magnetic field and transverse magnetization, respectively. The rate of change of M_x and M_y are given as

$$\frac{dM_x}{dt} = -\frac{M_x}{t_{ss}}, \quad \frac{dM_y}{dt} = -\frac{M_y}{t_{ss}} \tag{21.26}$$

The solution of these differential equations is simple. Solving, we find

$$M_x = M_{0x} e^{-t/t_{ss}}, \quad M_y = M_{0y} e^{-t/t_{ss}} \tag{21.27}$$

Therefore, M_x and M_y approach zero exponentially with time. Fig. 21.3 shows the variation of $\mathbf{M}'$ with time, which ultimately goes to zero (equilibrium value) in the presence of the magnetic field $\mathbf{H}_0$. The time t_{ss} is measured with small magnetic fields that have very high frequencies, on the order of many Mc/s. At such high frequencies the spin-lattice relaxation process is absent. The relaxation time t_{ss} is on the order of 10^{-10} s, which is smaller than the spin-lattice relaxation time t_{sl}. Further, the spin-spin relaxation process is independent of temperature, which makes t_{ss} independent of temperature.

21.4 EQUATION OF MOTION

Consider a nucleus with spin $\mathbf{I}$ and magnetic moment $\vec{\mu}_I$ placed in a constant magnetic field $\mathbf{H}$. The rate of change of the nuclear angular momentum $\mathbf{I}\hbar$ gives the torque $\vec{\tau}$ acting on the nucleus, that is,

$$\frac{d}{dt}(\mathbf{I}\hbar) = \vec{\tau} \tag{21.28}$$

with

$$\vec{\tau} = \vec{\mu}_I \times \mathbf{H} \tag{21.29}$$

Using Eqs. (21.1), (21.29) in Eq. (21.28), one can write

$$\frac{1}{\gamma_I}\frac{d\vec{\mu}_I}{dt} = \vec{\mu}_I \times \mathbf{H} \tag{21.30}$$

We have already seen that $\vec{\mu}_I$ will precess about the magnetic field **H** with the Larmor frequency ω_L. Consider now a magnetic material with weakly interacting nuclear spins. The magnetization **M** is a vector sum of the nuclear magnetic moments per unit volume. If ρ_i is the number of nuclei per unit volume in the ith state, then from Eq. (21.30) one can write

$$\frac{d\mathbf{M}}{dt} = \gamma_I \mathbf{M} \times \mathbf{H} \tag{21.31}$$

where

$$\mathbf{M} = \sum_i \rho_i \vec{\mu}_{Ii} \tag{21.32}$$

Here $\vec{\mu}_{Ii}$ is the nuclear magnetic moment in the ith state of the nucleus having spin **I**. The solution of Eq. (21.31) for a constant magnetic field is the same as that for Eq. (21.30), that is, **M** will precess about **H** with the Larmor frequency ω_L. If **M** is along the z-direction, then in the equilibrium state, $M_z = M_0$ and $M_x = M_y = 0$ and the magnetic susceptibility χ_0 satisfies the Curie law, that is,

$$M_0 = \chi_0 H \tag{21.33}$$

where

$$\chi_0 = \frac{C_M}{T} \tag{21.34}$$

Here C_M is the Curie constant. Hence, in a constant magnetic field H, one observes a resonant Larmor frequency $\nu_L = \omega_L/2\pi$ that is circularly polarized. For $H \approx 10^4$ gauss one obtains ν_L ranging from 1–50 Mc/s.

To observe the resonance absorption of rf energy, one has to apply two magnetic fields:

1. A dc magnetic field **H**, which can be varied along the z-direction.
2. An rf magnetic field of amplitude $\mathbf{H}_0$ ($\ll$**H**) perpendicular to the dc magnetic field.

The reason for the application of the rf field can be understood from the following. Consider a magnetic system with nuclear spin I = 1/2, which gives two energy levels (see Fig. 21.2A). The orientation of the magnetic moment $\vec{\mu}_I$ in the lower energy state is shown in Fig. 21.5 and it precesses about **H** with the Larmor frequency ω_L. For resonance absorption to take place, a

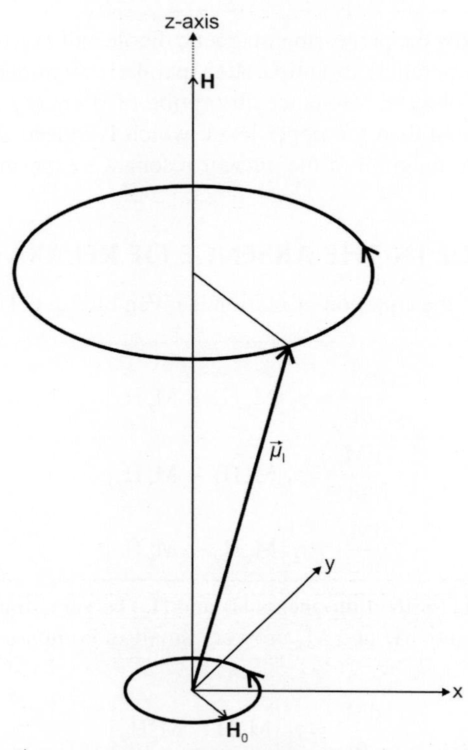

FIG. 21.5 Precession of the magnetic moment $\vec{\mu}_I$ about the dc magnetic field **H** in the presence of rf magnetic field $\mathbf{H}_0$.

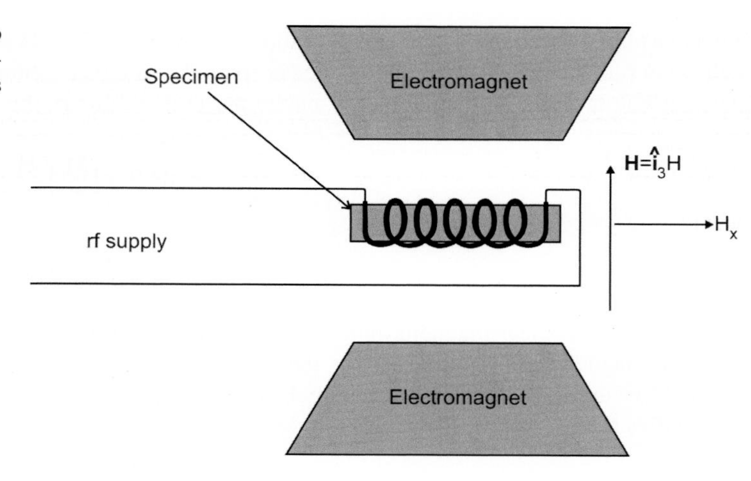

FIG. 21.6 Schematic representation of the experimental setup used for studying the magnetic resonance in a solid. The dc magnetic field is applied in the z-direction, but the rf magnetic field is applied in the x-direction.

transition must occur from the lower state to the higher state (see Fig. 21.2A), which involves a change in spin from ½ to −½. The transition will take place if $\mathbf{H}_0$, applied perpendicular to $\mathbf{H}$, causes the absorption of rf energy at $\omega_L=\gamma_I H$. Further, $\mathbf{H}_0$ reverses the tip of the dipole from parallel ($m_I=1/2$) to antiparallel ($m_I=-1/2$) alignment and vice versa. Let the rf magnetic field be applied in the x-direction. Then, one can write

$$H_x = 2H_0 \cos\omega t$$

$$H_y = H_z = 0 \tag{21.35}$$

But we know that in the Zeeman effect, the resonance frequency emitted or absorbed is circularly polarized. Therefore, it is convenient to represent the rf field as a combination of two circularly polarized magnetic fields in the xy-plane: one rf field rotates clockwise and the other counterclockwise. These two rf fields can be written as

$$\text{Clockwise: } H_x = H_0 \cos\omega t,\ \ H_y = H_0 \sin\omega t \tag{21.36}$$

$$\text{Counter-clockwise: } H_x = H_0 \cos\omega t,\ \ H_y = -H_0 \sin\omega t \tag{21.37}$$

At ω_L, one of the rotating fields will follow the precessing magnetic dipole and exert a constant torque on it. As a result, the magnetic dipole tips from parallel (antiparallel) to antiparallel (parallel) alignment. Therefore, the other rotating field becomes redundant. Hence, in order to observe resonance absorption of rf energy by the spin system, it is essential that the lower level be more heavily populated than the upper level, which is indeed the case at thermal equilibrium due to the Boltzmann distribution. A schematic diagram of the nuclear resonance experimental setup is shown in Fig. 21.6.

21.5 MAGNETIC RESONANCE IN THE ABSENCE OF RELAXATION PHENOMENA

In the absence of relaxation phenomena, the equation of motion is given by Eq. (21.31), in which the three components of magnetization become

$$\frac{dM_x}{dt} = \gamma_I\left[M_y H_z - M_z H_y\right] \tag{21.38}$$

$$\frac{dM_y}{dt} = \gamma_I\left[M_z H_x - M_x H_z\right] \tag{21.39}$$

$$\frac{dM_z}{dt} = \gamma_I\left[M_x H_y - M_y H_x\right] \tag{21.40}$$

Let the magnetic field in the xy-plane $\mathbf{H}_0$ (with components H_x and H_y) be very small in comparison with the dc magnetic field $\mathbf{H}=\hat{\mathbf{i}}_3 H$. As a result, the components M_x and M_y are very small in comparison with M_z. To a first-order approximation, Eqs. (21.38)–(21.40) reduce to

$$\frac{dM_x}{dt} = \gamma_I\left[M_y H - M_z H_y\right] \tag{21.41}$$

$$\frac{dM_y}{dt} = \gamma_I [M_z H_x - M_x H] \tag{21.42}$$

$$\frac{dM_z}{dt} = 0 \tag{21.43}$$

The oscillating components of the magnetic field H_x and H_y give rise to oscillating components of magnetization M_x and M_y, which can be represented as

$$H_x = H_{0x} e^{\imath\omega t}, M_x = M_{0x} e^{\imath\omega t} \tag{21.44}$$

$$H_y = H_{0y} e^{\imath\omega t}, M_y = M_{0y} e^{\imath\omega t} \tag{21.45}$$

Substituting Eqs. (21.44), (21.45) into Eqs. (21.41), (21.42), one gets

$$\imath\omega M_x - \gamma_I H M_y + \gamma_I M_z H_y = 0 \tag{21.46}$$

$$\gamma_I H M_x + \imath\omega M_y - \gamma_I M_z H_x = 0 \tag{21.47}$$

Multiplying Eq. (21.46) by $\gamma_I H$ and Eq. (21.47) by $\imath\omega$ and subtracting, one gets

$$M_y = \frac{\gamma_I M_z}{\omega_L^2 - \omega^2} \left(\omega_L H_y + \imath\omega H_x \right) \tag{21.48}$$

Similarly, one can obtain M_x from Eqs. (21.46), (21.47). We find

$$M_x = \frac{\gamma_I M_z}{\omega_L^2 - \omega^2} \left(\omega_L H_x - \imath\omega H_y \right) \tag{21.49}$$

Eqs. (21.48), (21.49) can be written in matrix form as

$$\begin{pmatrix} M_x \\ \\ M_y \end{pmatrix} = \begin{pmatrix} \dfrac{\omega_L \gamma_I M_z}{\omega_L^2 - \omega^2} & \dfrac{-\imath\omega\gamma_I M_z}{\omega_L^2 - \omega^2} \\ \dfrac{\imath\omega\gamma_I M_z}{\omega_L^2 - \omega^2} & \dfrac{\omega_L \gamma_I M_z}{\omega_L^2 - \omega^2} \end{pmatrix} \begin{pmatrix} H_x \\ \\ H_y \end{pmatrix} \tag{21.50}$$

In matrix form the magnetic susceptibility can be written as

$$\begin{pmatrix} M_x \\ M_y \end{pmatrix} = \begin{pmatrix} \chi_M^{xx} & \chi_M^{xy} \\ \chi_M^{yx} & \chi_M^{yy} \end{pmatrix} \begin{pmatrix} H_x \\ H_y \end{pmatrix} \tag{21.51}$$

Comparing Eqs. (21.50), (21.51), one can write

$$\chi_M^{xx} = \chi_M^{yy} = \frac{\omega_L \gamma_I M_z}{\omega_L^2 - \omega^2} \tag{21.52}$$

$$\chi_M^{xy} = -\chi_M^{yx} = \frac{-\imath\omega\gamma_I M_z}{\omega_L^2 - \omega^2} \tag{21.53}$$

The transverse magnetization $\mathbf{M}'$ and the transverse magnetic field $\mathbf{H}_0$ given by Eqs. (21.25), (21.24) exhibit some interesting features. $\mathbf{M}'$ and $\mathbf{H}_0$ are not in the same direction, rather $\mathbf{M}'$ lags behind $\mathbf{H}_0$. This can be seen by putting $H_x = 0$ in Eqs. (21.48), (21.49) and noticing the phase relation and amplitudes of M_x and M_y. The second interesting feature is that the components of the susceptibility tensor (Eqs. 21.52, 21.53) and magnetization (Eqs. 21.48, 21.49) become infinite at $\omega = \omega_L$, the natural frequency of vibration. But actually, the susceptibility and magnetization become very large at ω_L because, in this state, the applied rf field is synchronous with the precessional motion. This is the condition for *electron paramagnetic resonance*. The infinite values of susceptibility and magnetization suggest that some kind of relaxation phenomenon exists, which makes them finite but large in value.

21.6 BLOCH EQUATIONS

Bloch derived equations of motion for the magnetic response taking account of the spin-lattice and spin-spin relaxation phenomena. From Eqs. (21.21), (21.40) the rate of change of M_z with time is given by

$$\frac{dM_z}{dt}=\gamma_I\left[M_xH_y-M_yH_x\right]+\frac{M_0-M_z}{t_{sl}} \tag{21.54}$$

Similarly, from Eqs. (21.38), (21.39), (21.26) one can write

$$\frac{dM_x}{dt}=\gamma_I\left[M_yH_z-M_zH_y\right]-\frac{M_x}{t_{ss}} \tag{21.55}$$

$$\frac{dM_y}{dt}=\gamma_I[M_zH_x-M_xH_z]-\frac{M_y}{t_{ss}} \tag{21.56}$$

Eqs. (21.54)–(21.56) are called Bloch equations. Substituting the values of H_x and H_y from Eq. (21.37) and putting $H_z=H$ in Eqs. (21.54)–(21.56), one gets

$$\frac{dM_x}{dt}=\gamma_I\left[M_yH+M_zH_0\sin\omega t\right]-\frac{M_x}{t_{ss}} \tag{21.57}$$

$$\frac{dM_y}{dt}=\gamma_I[M_zH_0\cos\omega t-M_xH]-\frac{M_y}{t_{ss}} \tag{21.58}$$

$$\frac{dM_z}{dt}=-\gamma_I\left[M_xH_0\sin\omega t+M_yH_0\cos\omega t\right]+\frac{M_0-M_z}{t_{sl}} \tag{21.59}$$

Eqs. (21.57)–(21.59) can be solved for M_x, M_y, and M_z and the final result is given as

$$M_x=\frac{1}{2}\chi_0\omega_Lt_{ss}\left[\frac{2H_0t_{ss}(\omega_L-\omega)\cos\omega t+2H_0\sin\omega t}{1+(\omega_L-\omega)^2t_{ss}^2+\gamma_I^2H_0^2t_{sl}t_{ss}}\right] \tag{21.60}$$

$$M_y=\frac{1}{2}\chi_0\omega_Lt_{ss}\left[\frac{2H_0\cos\omega t-2H_0t_{ss}(\omega_L-\omega)\sin\omega t}{1+(\omega_L-\omega)^2t_{ss}^2+\gamma_I^2H_0^2t_{sl}t_{ss}}\right] \tag{21.61}$$

$$M_z=\chi_0H\left[\frac{1+(\omega_L-\omega)^2t_{ss}^2}{1+(\omega_L-\omega)^2t_{ss}^2+\gamma_I^2H_0^2t_{sl}t_{ss}}\right] \tag{21.62}$$

In actual experiments the linearly polarized rf magnetic field is applied only in one direction, say in the x-direction (see Fig. 21.6). In this case, $H_x=2H_0\cos\omega t$ and $H_y=0$. For a linearly polarized rf magnetic field, Eqs. (21.60)–(21.62) reduce to

$$M_x=\frac{1}{2}\chi_0\omega_Lt_{ss}\left[\frac{2H_0t_{ss}(\omega_L-\omega)\cos\omega t}{1+(\omega_L-\omega)^2t_{ss}^2+\gamma_I^2H_0^2t_{sl}t_{ss}}\right] \tag{21.63}$$

$$M_y=\frac{1}{2}\chi_0\omega_Lt_{ss}\left[\frac{2H_0\cos\omega t}{1+(\omega_L-\omega)^2t_{ss}^2+\gamma_I^2H_0^2t_{sl}t_{ss}}\right] \tag{21.64}$$

$$M_z=\chi_0H\left[\frac{1+(\omega_L-\omega)^2t_{ss}^2}{1+(\omega_L-\omega)^2t_{ss}^2+\gamma_I^2H_0^2t_{sl}t_{ss}}\right] \tag{21.65}$$

The dynamical magnetic susceptibility can be written as

$$\chi_M(\omega)=\chi_M'(\omega)+\iota\,\chi_M''(\omega) \tag{21.66}$$

$\chi'_M(\omega)$ and $\chi''_M(\omega)$ are the real and imaginary parts, which from Eqs. (21.63), (21.64) are given by

$$\chi'_M(\omega) = \frac{M_x}{H_x} = \frac{M_x}{2H_0 \cos \omega t} = \frac{1}{2}\chi_0 \omega_L t_{ss} \left[\frac{(\omega_L - \omega)\, t_{ss}}{1 + (\omega_L - \omega)^2 t_{ss}^2 + \gamma_I^2 H_0^2 t_{sl} t_{ss}}\right] \tag{21.67}$$

$$\chi''_M(\omega) = \frac{M_y}{H_x} = \frac{M_y}{2H_0 \cos \omega t} = \frac{1}{2}\chi_0 \omega_L t_{ss} \left[\frac{1}{1 + (\omega_L - \omega)^2 t_{ss}^2 + \gamma_I^2 H_0^2 t_{sl} t_{ss}}\right] \tag{21.68}$$

In the denominators of Eqs. (21.63)–(21.65) and Eqs. (21.67)–(21.68), the term $\gamma_I^2 H_0^2 t_{sl} t_{ss}$ is a dimensionless constant that determines the degree of saturation. If H_0 is very small, then saturation is not reached and

$$\gamma_I^2 H_0^2 t_{sl} t_{ss} \ll 1$$

This term can be neglected in the denominators of Eqs. (21.63)–(21.65) and Eqs. (21.67)–(21.68). In this approximation the static magnetic susceptibility from Eq. (21.65) becomes

$$\chi_0 = \frac{M_z}{H} \tag{21.69}$$

The real and imaginary parts of the dynamical magnetic susceptibility, from Eqs. (21.67), (21.68), become

$$\chi'_M(\omega) = \frac{1}{2}\chi_0 \omega_L t_{ss} \left[\frac{(\omega_L - \omega)\, t_{ss}}{1 + (\omega_L - \omega)^2 t_{ss}^2}\right] \tag{21.70}$$

$$\chi''_M(\omega) = \frac{1}{2}\chi_0 \omega_L t_{ss} \left[\frac{1}{1 + (\omega_L - \omega)^2 t_{ss}^2}\right] \tag{21.71}$$

The susceptibilities $\chi'_M(\omega)$ and $\chi''_M(\omega)$ are plotted in Fig. 21.7 as a function of $(\omega_L - \omega)t_{ss}$. Fig. 21.7A shows that $\chi''_M(\omega)$ exhibits resonance absorption at $\omega = \omega_L$ and the peak has Lorentzian shape. Eq. (21.71) can be written as

$$\chi''_M(\omega) = \frac{(1/2)\chi_0 \omega_L / t_{ss}}{(\omega_L - \omega)^2 + 1/t_{ss}^2} \tag{21.72}$$

From the above equation we see that the halfwidth $\Delta\omega$ of resonance at half the height is given by

$$\Delta\omega = \frac{1}{t_{ss}} \tag{21.73}$$

The real part of the dynamical magnetic susceptibility $\chi'_M(\omega)$, shown in Fig. 21.7B, exhibits dispersion, which accompanies the absorption.

21.6.1 Free Precession in Static Magnetic Field

One can solve the Bloch equations for the free precession of the spin system in a static (dc) magnetic field $\mathbf{H} = \hat{\mathbf{i}}_3 H$. In a static magnetic field the equilibrium magnetization is $M_z = M_0$. In this case the Bloch equations (21.54) – (21.56) reduce to

$$\frac{dM_x}{dt} = \gamma_I H M_y - \frac{M_x}{t_{ss}} \tag{21.74}$$

$$\frac{dM_y}{dt} = -\gamma_I H M_x - \frac{M_y}{t_{ss}} \tag{21.75}$$

$$\frac{dM_z}{dt} = 0 \tag{21.76}$$

For Eqs. (21.74)–(21.76), we seek damped-oscillator-like solutions of the form

$$M_x = M e^{-t/t'} \cos \omega t \tag{21.77}$$

$$M_y = -M e^{-t/t'} \sin \omega t \tag{21.78}$$

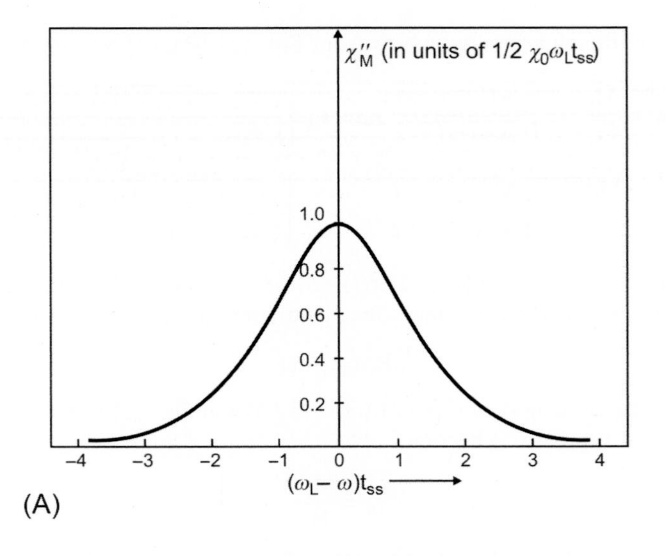

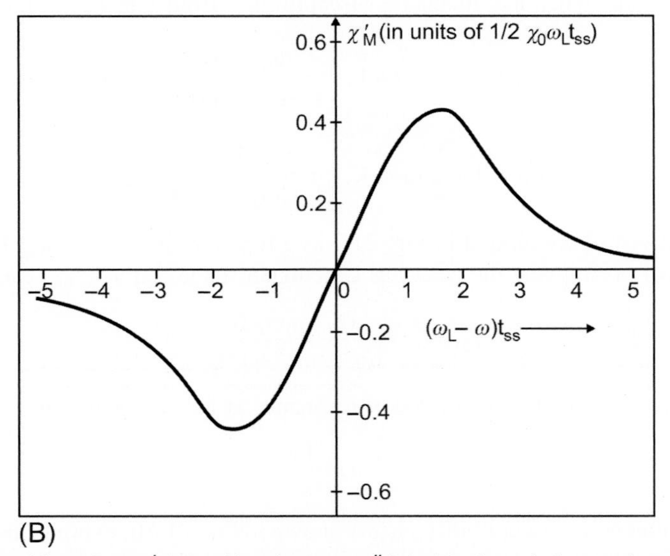

FIG. 21.7 Schematic representation of the real part $\chi'_M(\omega)$ and imaginary part $\chi''_M(\omega)$ of the dynamical magnetic susceptibility $\chi_M(\omega)$ as a function of the frequency ω in a solid.

where t' is the damping time. Substituting Eqs. (21.77), (21.78) into Eqs. (21.74)–(21.76), one can write

$$\left(\frac{1}{t'}-\frac{1}{t_{ss}}\right)\cos\omega t=(\omega_L-\omega)\sin\omega t \tag{21.79}$$

$$\left(\frac{1}{t'}-\frac{1}{t_{ss}}\right)\sin\omega t=-(\omega_L-\omega)\cos\omega t \tag{21.80}$$

Eqs. (21.79), (21.80) will be valid only if

$$t'=t_{ss} \tag{21.81}$$

and

$$\omega=\omega_L \tag{21.82}$$

Eq. (21.81) shows that the damping time is equal to the spin-spin relaxation time. Further, using the analogy of the precession of spins in a static magnetic field with the motion of a damped harmonic oscillator suggests that the spin system exhibits resonance absorption of energy near the Larmor frequency. The frequency width of the response of the system to the driving field will be $\Delta\omega = 1/t_{ss}$.

21.7 MAGNETIC BROADENING OF RESONANCE LINES

It has been observed that resonance absorption of an rf magnetic field is not very sharp, but rather exhibits a peak with finite width called the linewidth. There are a number of physical phenomena that contribute to the linewidth of resonance absorption.

1. In a solid, an inhomogeneous magnetic field gives rise to a broadening of the resonance line. Inhomogeneity in a magnetic field may arise either because of the presence of impurities or due to different values of $\mathbf{H}_{loc}$ at different nuclei (crystal structure effect).
2. If a nucleus possesses a finite magnetic quadrupole moment, it gives rise to what are known as quadrupole effects. These effects destroy the degeneracy of resonance frequencies between nuclear levels with different m_I values (Zeeman effect). Quadrupole effects may give rise to resolved or unresolved splitting of a resonance line. If the quadrupole effects are very small (which may be the case in a number of solids), one gets unresolved splitting of the resonance line, which amounts to broadening.
3. Spin-lattice interactions also give rise to a broadening of the resonance lines. This is because the spin-lattice interactions produce an equilibrium population by balancing the rates of transitions. This puts a limit on the lifetime of the Zeeman states and hence results in a broadening of the resonance lines.

   The coupling between the nuclear magnetic moments in a solid makes a major contribution to the broadening of resonance lines and is considered here. In addition to the applied magnetic field $\mathbf{H}$, a local magnetic field $\mathbf{H}_{loc}(\mathbf{r}_i)$ is produced at the position $\mathbf{r}_i$ of a nuclear magnetic moment (see Eq. 21.19). In a crystalline solid $\mathbf{H}_{loc}(\mathbf{r}_i)$ is different at different lattice positions, yielding an inhomogeneous total magnetic field $\mathbf{H}_{tot}$ given by

$$\mathbf{H}_{tot} = \mathbf{H} + \mathbf{H}_{loc} \tag{21.83}$$

Considering the interactions between the 1NNs, it is evident from Eq. (21.19) that $\mathbf{H}_{loc} \approx \vec{\mu}_I / r_1^3$, where r_1 is the 1NN distance. For a nuclear magnetic moment $\mu_I = 10^{-23}$ erg/gauss and $r_1 = 2$ Å, one obtains $H_{loc} \cong 1$ gauss. Therefore, if $\mathbf{H}_{tot}$ is different at different nuclei, then the various nuclei will exhibit resonance at different frequencies. If $\Delta\mathbf{H}_{loc}$ is the spread in the value of $\mathbf{H}_{loc}$ seen by the nuclei, then it will also be the spread in $\mathbf{H}_{tot}$. Thus, the spread in the resonance frequency $\Delta\omega$, that is, $\omega_L \to \omega_L + \Delta\omega$, is given by

$$\Delta\omega = \gamma_I \Delta H_{loc} \tag{21.84}$$

Knowing the values of $\Delta\omega$ and ΔH_{loc}, one can also find the spin-spin relaxation time t_{ss}. For example, in CaF_2, $\Delta H_{loc} = 2$ gauss and $\Delta\omega = 5 \times 10^4\ s^{-1}$, therefore,

$$t_{ss} = \frac{1}{\Delta\omega} = \frac{1}{5 \times 10^4} = 2 \times 10^{-5}\ s \tag{21.85}$$

21.8 EFFECT OF MOLECULAR MOTION ON RESONANCE

The broadening of resonance lines depends on the nature of the material. The linewidth in the case of liquids is very small compared with solids: the linewidth in a liquid can be as narrow as 0.5 cps, but in solids it is on the order of 5000 cps or more at room temperature. It has further been observed experimentally that the resonance peaks are narrower for a solid in which the atoms (nuclei) are in rapid motion. Further, with an increase in the temperature T of a solid, the motion of the atoms becomes faster and the resonance peaks become narrower. The effect of the motion of the atoms on the width of resonance peaks is usually called *motional narrowing*. This effect is much more pronounced in liquids as the atoms can move much more quickly in random directions. The motional narrowing of resonance peaks can be explained in terms of the local magnetic field $\mathbf{H}_{loc}$ as follows.

It has already been explained that the width of a resonance peak arises due to the inhomogeneous nature of $\mathbf{H}_{loc}$, the crystal structure effect. All of the measurements in a laboratory are made at finite temperatures. At finite temperatures, an atom jumps from one lattice position to another, a process similar to the diffusion of atoms. Let us suppose that an atom remains at a single lattice position for an average time of τ_m seconds. The atoms in a solid move randomly in different directions, as a result of which $\mathbf{H}_{loc}$ at a particular lattice point changes very quickly. The time-averaged value of the local magnetic field $\langle \mathbf{H}_{loc} \rangle$ becomes nearly constant and with much smaller spread. In other words, one can say that the random motion of atoms in different directions makes the solid, and hence $\langle \mathbf{H}_{loc} \rangle$, nearly homogeneous, yielding only a very small spread in $\mathbf{H}_{loc}$. The decrease in the spread of $\mathbf{H}_{loc}$ yields a smaller width of the resonance absorption line. With an increase in temperature, the motion of the atoms increases, yielding a smaller value of ΔH_{loc}, which in turn makes the resonance lines narrower. In other words, one can say that an increase in the motion (velocity) of atoms decreases the value of τ_m, which causes a narrowing of the resonance peaks.

Measurements of the spin-lattice relaxation time t_{sl} are based on the competition between resonance absorption (which tends to equalize the population of different levels) and the spin-lattice interaction (which tends to maintain the Boltzmann distribution). The values of t_{sl} obtained experimentally vary between $10^{-5}-10^4$ s: 10^4 s is the value obtained for ice. The value of t_{sl} is larger for solids than for liquids or gases. For solids, t_{sl} is rarely smaller than 10^{-2} s. Rather, it may be very large at low temperatures. On the other hand, for liquids, t_{sl} may be as short as $10^{-2}-10^{-3}$ s and rarely exceeds a few seconds. This shows that as the atoms in a solid become freer the relaxation time t_{sl} decreases.

It is interesting to note that t_{sl} may reduce strongly when paramagnetic ions are present; these ions have an effective magnetic moment that is 10^3 times as large as the nuclear magnetic moments and they form a very efficient medium for establishing heat contact between the nuclear spins and their surroundings. On the other hand, at low temperatures, there is no variation in the spin-spin relaxation time t_{ss}. With an increase in temperature, t_{ss} increases and, at a certain temperature, t_{sl} and t_{ss} may become approximately equal.

21.9 ELECTRON SPIN RESONANCE

In an atom, both the nucleus and the electrons possess intrinsic spin and hence magnetic moment. Let the magnetic moment of a nucleus with spin I be μ_I and that of an electron μ_B. When a magnetic field $\mathbf{H}$ is applied to a solid, it splits the energy levels of both the nuclei and the electrons. The splitting of the energy levels of the nuclei gives rise to NMR and that of the electrons gives rise to electron spin resonance (ESR). One should note that there are some solids that exhibit ESR.

Consider a solid in which each atom contains only one electron. Let a magnetic field $\mathbf{H}$ be applied to the solid in the z-direction, that is, $\mathbf{H}=\hat{\mathbf{i}}_3 H$. Further, an rf magnetic field $\mathbf{H}_0$ is applied perpendicular to $\mathbf{H}$ (say in the x-direction). The magnetic part of the Hamiltonian due to the atomic electron is given by

$$\widehat{H} = -\vec{\mu}_s \cdot \mathbf{H} = g_s \mu_B \mathbf{s} \cdot \mathbf{H} = g_s \mu_B H s_z \quad (21.86)$$

Here we have used Eq. (18.24) for the spin magnetic moment of an electron. The energy of the electron states is obtained by taking the expectation value of the Hamiltonian as

$$E_{m_s} = \langle m_s | \widehat{H} | m_s \rangle = g_s \mu_B H m_s \quad (21.87)$$

According to the above equation the electron states get split up into two substates (see Fig. 21.8). The energy difference between the two states becomes

$$\Delta E = 2\mu_B H \quad (21.88)$$

Eq. (21.88) gives the condition for a resonance. If an rf field is applied in the xy-plane, then the resonance frequency ω_0 is given by

$$\hbar\omega_0 = 2\mu_B H \quad (21.89)$$

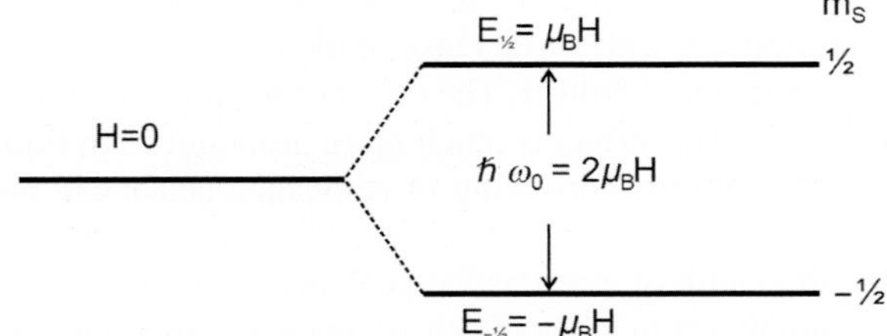

FIG. 21.8 Splitting of an electron energy state with $s_z=1/2$ into two substates in the presence of a dc magnetic field $\mathbf{H}$ applied in the z-direction.

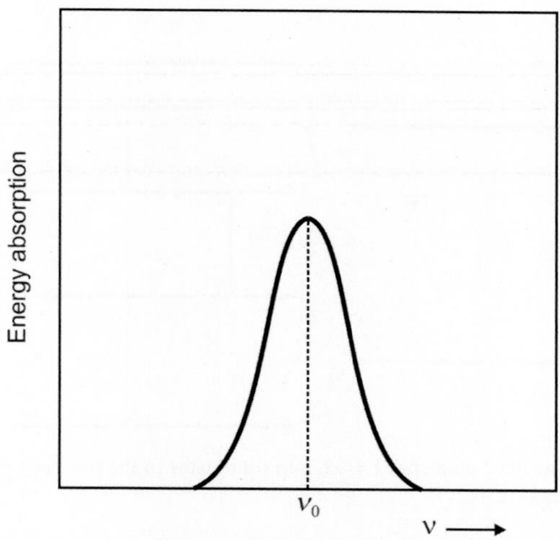

FIG. 21.9 The energy absorption at the resonance frequency ν_0, which broadens into a resonance peak.

So, resonance absorption of the rf field takes place at frequency ω_0. At resonance, the electrons in energy state $E_{-1/2}$ absorb energy and make a transition to the higher energy state $E_{1/2}$. Note that the frequency ω_0 is twice the Larmor frequency ω_L. If we apply $H=1\,\text{weber/m}^2$, then $\omega_0 \approx 30000$ Mc/s.

In the discussion above, only a single electron per atom is assumed, but in general there are a number of electrons in an atom of a solid. Therefore, the resonance frequency is affected by the following interactions:

1. Electron-electron interactions modify the resonance process.
2. The interaction of nuclear magnetic moments with the electronic magnetic moments affects the electron energy levels. One can say that the nuclear spin can take $2I+1$ orientations in the magnetic field and each of these alters the electron energy levels by different amounts (hyperfine interaction).

These interactions, therefore, broaden a resonance line into a resonance peak, as shown in Fig. 21.9. If the electron-electron and electron-nucleus interactions are small compared with the splitting of the energy levels of a single electron, then they can be treated as a perturbation to find the resultant electron states. Thus, one can find the width of the resonance peak. Other effects, such as the inhomogeneous local electric fields arising from impurities and other crystal defects, also change the position of the resonance lines. ESR is observed only in those solids in which each atom has unpaired electrons so as to yield finite electron spin (e.g., paramagnetic substances) because the paired electrons in the valence band produce no effect. Therefore, ESR is often called electron paramagnetic resonance (EPR).

21.10 HYPERFINE INTERACTIONS

So far, we have studied NMR in which the nuclear energy levels are calculated in the presence of mutually perpendicular dc and rf magnetic fields. But in a solid there are electrons revolving around every nucleus that possess finite magnetic moment. Therefore, additional interactions exist between the magnetic moments of electrons and nuclei, usually called *hyperfine interactions,* which cause further splitting of the nuclear energy levels.

The motion of electrons around the nucleus produces a finite magnetic field with the following contributions:

1. The electrons revolving around the nucleus are equivalent to current loops, which produce a finite magnetic field at the nuclear site.
2. The spinning motion of electrons yields spin current about the nucleus. The spin current produces what is called a contact hyperfine interaction. Even if quenching of the orbital angular momentum, and hence of the orbital magnetic moment, occurs, there will be a finite contact hyperfine interaction. Therefore, the contact hyperfine interaction is of special importance in such solids.

The net magnetic field arising from these two contributions interacts with the nucleus to produce the hyperfine interaction. If the hyperfine interaction is small compared with the splitting of the nuclear energy levels, it may be treated as a perturbation in the nuclear resonance phenomenon. The interaction Hamiltonian $\widehat{H}_{sI}$ between the nuclear magnetic moment and the magnetic moment of the electron can be represented as

$$\widehat{H}_{sI} = a_{sI}\,\mathbf{s}\cdot\mathbf{I} \tag{21.90}$$

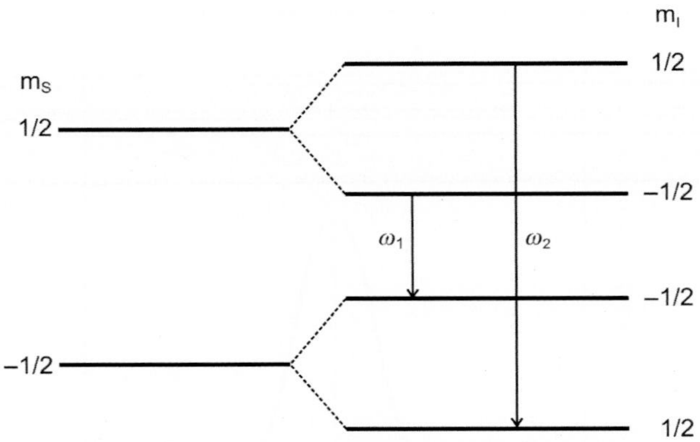

FIG. 21.10 Splitting of an electron state, in a nucleus with spin I = ½, into four states in the presence of hyperfine interactions. Two transitions with frequencies ω_1 and ω_2 are allowed by the selection rules.

where **I** and **s** are the nuclear and electron spins and a_{sI} is the hyperfine constant. The energy eigenvalues of Eq. (21.90) are given by

$$E_{sI} = a_{sI} m_I m_s \tag{21.91}$$

The electron spin quantum number m_s has two values, $\pm 1/2$. For a nucleus with $I = 1/2$ the quantum number m_I also has two values $\pm 1/2$. According to Eq. (21.91), each electron state gets split up into two states with $m_I = \pm 1/2$, giving rise to four states as shown in Fig. 21.10. The transitions between the four states take place by satisfying the following selection rules

$$\Delta m_I = 0, \quad \Delta m_s = \pm 1 \tag{21.92}$$

The allowed transitions yield two resonance lines, as shown in Fig. 21.10.

21.11 KNIGHT SHIFT

Consider a diamagnetic material with nuclear spin **I** to which the magnetic field **H** is applied in the z-direction, that is, $\mathbf{H} = \hat{\mathbf{i}}_3 H$. The magnetic interaction Hamiltonian is given by

$$\widehat{H}_M = -\gamma_I \hbar H I_z \tag{21.93}$$

When an rf magnetic field $\mathbf{H}_0$ is applied in the direction perpendicular to **H**, nuclear resonance absorption is observed at the Larmor frequency $\omega_L = \gamma_I H$. Now consider a metallic solid with the same nuclear spin **I** to which the same magnetic field $\mathbf{H}_0$ is applied in the xy-plane. It is found that nuclear resonance absorption in the metallic solid is observed at a slightly different value of the dc magnetic field. This effect is known as the *Knight shift* and it yields knowledge about the distribution of conduction electrons in a metal. In a metallic solid there exists, in addition, the hyperfine interaction given by

$$\widehat{H}_{sI} = a_{sI} I_z \langle s_z \rangle \tag{21.94}$$

where $\langle s_z \rangle$ is the average value of the conduction electron spin in a metal. The total interaction Hamiltonian in a metallic solid is the sum of the two Hamiltonians given by Eqs. (21.93), (21.94), that is,

$$\widehat{H} = \widehat{H}_M + \widehat{H}_{sI} = [-\gamma_I \hbar H + a_{sI} \langle s_z \rangle] I_z \tag{21.95}$$

$\langle s_z \rangle$ is related to the Pauli spin susceptibility χ_P of the conduction electrons as

$$M_z = \Delta n_s g_s \mu_B \langle s_z \rangle = \chi_P H \tag{21.96}$$

where Δn_s is the difference of the up and down spin densities per unit volume. From the above equation $\langle s_z \rangle$ is given by

$$\langle s_z \rangle = \frac{\chi_P}{\Delta n_s g_s \mu_B} H \tag{21.97}$$

Substituting Eq. (21.97) into Eq. (21.95), we obtain

$$\widehat{H} = -\gamma_I \hbar H [1 - K] I_z \tag{21.98}$$

where

$$K = -\frac{\Delta H}{H} = \frac{a_{sI}\,\chi_P}{\gamma_I \hbar \Delta n_s g_s \mu_B} \tag{21.99}$$

The energy corresponding to the Hamiltonian given by Eq. (21.98) becomes

$$E = -\gamma_I \hbar H[1 - K]\, m_I \tag{21.100}$$

Here the constant K is called the Knight shift and gives the fractional change in magnetic field due to the presence of conduction electrons in a metal. It is also a measure of the shift in energy and hence the resonance frequency. The constant a_{sI} depends on the electron density at the nuclear site, that is, $|\psi(0)|^2$ where $|\psi(\mathbf{r})\rangle$ is the electron wave function and the nucleus is assumed to be situated at the origin. Therefore, by making a reasonable estimate of the hyperfine coupling constant a_{sI}, one can get information about the conduction electron density at the nuclear site. The Knight shift K can be measured experimentally and from it one can find χ_P (see Eq. 21.99).

21.12 QUADRUPOLE INTERACTIONS IN MAGNETIC RESONANCE

If a nucleus is not spherical in shape, which is the case in general, it possesses a multipole character, that is, it has a dipole moment, quadrupole moment, etc. A nucleus having spin **I** greater than or equal to one possesses a quadrupole moment whose $\alpha\beta$-component can be defined as

$$Q_{\alpha\beta} = \int \left(3\, r_\alpha r_\beta - r^2 \delta_{\alpha\beta}\right) \rho_N(\mathbf{r})\, d^3r \tag{21.101}$$

where $\rho_N(\mathbf{r})$ is the nuclear charge density. The diagonal components of the quadrupole moment tensor are defined as

$$Q_{xx} = \int (3x^2 - r^2)\,\rho_N(\mathbf{r})\,d^3r, Q_{yy} = \int (3y^2 - r^2)\,\rho_N(\mathbf{r})\,d^3r \tag{21.102}$$

$$Q = Q_{zz} = \int (3\,z^2 - r^2)\,\rho_N(\mathbf{r})\,d^3r \tag{21.103}$$

The quadrupole moment of a nucleus Q is usually defined by the z-component of the quadrupole tensor.

Problem 21.3

Prove that the quadrupole moment of a nucleus with spherical shape is zero.

Consider a nucleus, situated at the origin, in a solid that experiences crystal potential $V(\mathbf{r})$. If the energy of the nucleus is expanded in terms of multipoles, we get

$$E = qV(0) - \sum_\alpha p_\alpha E_\alpha(0) + \frac{1}{6}\sum_{\alpha,\beta} Q_{\alpha\beta} V_{\alpha\beta}(0) + \cdots \tag{21.104}$$

This equation shows that the nuclear charge interacts with the crystal potential, the nuclear dipole moment interacts with the derivative of the crystal potential (electric field $\mathbf{E} = -\nabla \mathbf{V}$), and the nuclear quadrupole moment interacts with the double derivative of the crystal potential $V_{\alpha\beta}(0)$. The quantity $V_{\alpha\beta}(0)$ is usually called the electric field gradient. From the equation above the quadrupole interaction energy is given by

$$E_{QI} = \frac{1}{6}\sum_{\alpha,\beta} Q_{\alpha\beta} V_{\alpha\beta}(0) \tag{21.105}$$

The quantum mechanical expression for the Hamiltonian of the quadrupole interactions $\widehat{H}_{QI}$ in terms of the principal components of the electric field gradient , that is, $V_{\alpha\alpha}(0)$, is given by (Galsin, 2002)

$$\widehat{H}_{QI} = \frac{eQ}{4I(2I-1)}\left[V_{zz}\,(3I_z^2 - I^2) + \left(V_{xx} - V_{yy}\right)\left(I_x^2 - I_y^2\right)\right] \tag{21.106}$$

If, for simplicity, the electric field gradient is assumed to be cylindrically symmetric, then the field gradient eq is defined as

$$eq = V_{zz}, \quad V_{xx} = V_{yy} \tag{21.107}$$

Therefore, for a cylindrically symmetric field gradient $\widehat{H}_{QI}$ becomes

$$\widehat{H}_{QI} = \frac{e^2qQ}{4I(2I-1)}\left[\left(3I_z^2 - I^2\right)\right] \tag{21.108}$$

The total Hamiltonian of a nucleus in the presence of both a magnetic dipole and a quadrupole moment becomes

$$\begin{aligned}\widehat{H} &= \widehat{H}_M + \widehat{H}_{QI} \\ &= -g_I \mu_{Bp} \mathbf{I}\cdot\mathbf{H} + \frac{e^2qQ}{4I(2I-1)}\left(3I_z^2 - I^2\right)\end{aligned} \tag{21.109}$$

In a number of metals and their alloys, the Zeeman splitting is large, at least 10 times larger than the quadrupole splitting. In such crystalline solids, the quadrupole interactions can be treated as a perturbation. If the magnetic field is applied along the z-direction, that is, along the direction of the principal component of the electric field gradient, then Eq. (21.109) can be written as

$$\widehat{H} = -g_I \mu_{Bp} H I_z + \frac{3e^2qQ}{4I(2I-1)}\left(I_z^2 - \frac{I^2}{3}\right) \tag{21.110}$$

The energy eigenvalue of the nucleus corresponding to the Hamiltonian given by Eq. (21.110) is

$$\begin{aligned}E_{m_I} &= E_{m_I}^0 + E_{m_I}^1 \\ &= -g_I \mu_{Bp} H m_I + \frac{3e^2qQ}{4I(2I-1)}\left[m_I^2 - \frac{1}{3}I(I+1)\right]\end{aligned} \tag{21.111}$$

The first term in Eq. (21.111) gives the equally spaced $2I+1$ nuclear sublevels due to the Zeeman splitting. The second term gives the first-order contribution to the energy due to the quadrupole interactions and further shifts the energy levels. For a nucleus with $I = 1/2$, there are two Zeeman levels with $m_I = \pm 1/2$. The quadrupole interaction term is zero for $m_I = \pm 1/2$. Hence, for a nucleus with $I = 1/2$, there is no splitting due to the quadrupole interactions and the transition frequency remains unchanged, that is, ν_L. For a nucleus with $I = 3/2$, Eq. (21.111) gives

$$E_{m_I} = -g_I \mu_{Bp} H m_I + \frac{e^2qQ}{4}\left[m_I^2 - \frac{5}{4}\right] \tag{21.112}$$

The term $(m_I^2 - 5/4)$ is -1 for $m_I = \pm 1/2$ and $+1$ for $m_I = \pm 3/2$; therefore, Eq. (21.112) can be written as

$$E_{m_I} = -g_I \mu_{Bp} H m_I + (-1)^{|m_I| + 1/2}\frac{e^2qQ}{4} \tag{21.113}$$

The magnetic sublevels for $I = 1/2, 3/2$ in the presence of quadrupole interactions are shown in Fig. 21.11. From the figure it is evident that for $I = 3/2$ the spectrum becomes asymmetrical about the centroid due to the presence of quadrupole interactions.

21.12.1 Nuclear Quadrupole Resonance

If the applied magnetic field is very small, then the Zeeman splitting is quite small compared with the quadrupole splitting. Therefore, the Hamiltonian corresponding to the Zeeman energy can be treated as a perturbation with respect to the Hamiltonian for the quadrupole interactions. In the limit of zero magnetic field, the total Hamiltonian reduces to the quadrupole interaction Hamiltonian H_{QI} and it is called pure nuclear quadrupole resonance (NQR). So, for NQR the Hamiltonian, from Eq. (21.109) becomes

$$\widehat{H} = \widehat{H}_{QI} = \frac{e^2qQ}{4I(I+1)}\left[3I_z^2 - I^2\right] \tag{21.114}$$

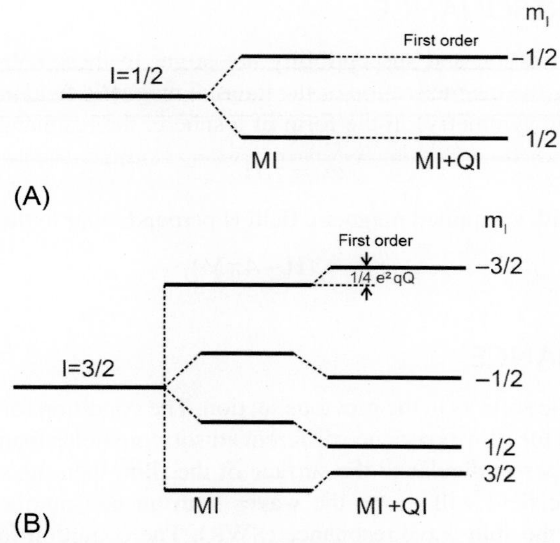

FIG. 21.11 Quadrupole interaction (QI) in the presence of a magnetic interaction (MI) in a nuclear system with symmetry axis parallel to the applied magnetic field **H** for (A) I = 1/2 and (B) I = 3/2.

The energy eigenvalues for the Hamiltonian in Eq. (21.114) are given by

$$E_{m_I} = h\nu_{m_I} = \frac{e^2qQ}{4I(2I-1)}\left[3m_I^2 - I(I+1)\right] \tag{21.115}$$

Eq. (21.115) exhibits twofold degeneracy of energy levels due to the quadrupole interactions: energy levels with $I_z = \pm m_I$ have the same energy. The quadrupole splitting of the energy levels for nuclei with I = 1/2, 3/2, and 5/2 is shown in Fig. 21.12. It is evident from the figure that there is no quadrupole splitting in a nucleus with I = 1/2. It becomes finite if I is equal to or greater than one. Further, if the magnetic field is finite but very weak, the doubly degenerate levels split up but lie very close to each other. Transitions between the split nuclear levels occur according to the selection rule $\Delta m_I = \pm 1$. These transitions can be induced by applying an rf magnetic field of the correct frequency.

There are a number of other resonance processes also, such as ferromagnetic resonance, spin wave resonance, and antiferromagnetic resonance.

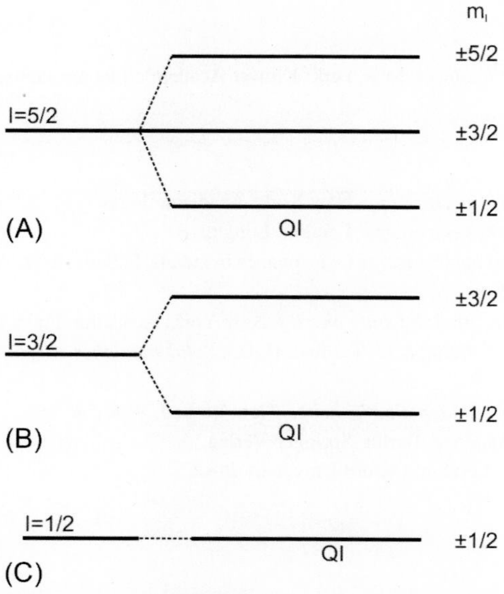

FIG. 21.12 Energy level splitting in the pure quadrupole interaction (QI) for (A) I = 5/2, (B) I = 3/2, and (C) I = 1/2.

21.13 FERROMAGNETIC RESONANCE

In a ferromagnetic solid the magnetization and susceptibility are large. In these solids the shape and structure of the specimen is important in determining the magnetization or the internal magnetic field and hence the resonance frequency. For example, in a solid (having cubic symmetry) in the form of a sphere, the resonance frequency ω_0 is given by

$$\omega_0 = \gamma_I H \tag{21.116}$$

If the solid is in the form of a plate with an applied magnetic field H perpendicular to the surface of the plate, then one gets

$$\omega_0 = \gamma_I (H - 4\pi M) \tag{21.117}$$

21.14 SPIN WAVE RESONANCE

Consider a thin film of a ferromagnetic solid as in the previous section. The condition for a spin wave in the thin film can be achieved if the spins on the surface of the film experience different anisotropic fields than the spins within the film. Suppose a uniform magnetic field is applied perpendicular to the surface of the film, then the spins on the surface are pinned by surface anisotropic interactions. The field will excite the waves with an odd number of half wavelengths within the thickness of the film. This is called the spin wave resonance (SWR). The condition for the SWR is obtained by adding the exchange contribution, represented as DK^2, to Eq. (21.117) to write

$$\omega_0 = \gamma_I (H - 4\pi M) + DK^2 \tag{21.118}$$

Here $K = n\pi/L$ for the mode of a wave with n half wavelengths.

21.15 ANTIFERROMAGNETIC RESONANCE

In an antiferromagnetic solid, one can use the two-sublattice model to study the antiferromagnetic resonance. If the magnetic field is applied in the z-direction (H_A is the magnetic field for sublattice A and $-H_A$ for sublattice B), then the resonance condition becomes

$$\omega_0^2 = \gamma_I^2 H_A (H_A + 2H_E) \tag{21.119}$$

where H_E is the exchange magnetic field given by

$$H_E = \lambda_{AA} M_A \tag{21.120}$$

M_A is the magnetization in the sublattice A.

REFERENCE

Galsin, J. S. (2002). *Impurity scattering in metallic alloys.* New York: Kluwer Academic/Plenum Publishers.

SUGGESTED READING

Bovey, F. A. (1969). *Nuclear magnetic resonance spectroscopy*. New York: Academic Press.

Harris, R. K. (1986). *Nuclear magnetic resonance spectroscopy*. London: Longmans.

Knight, W. D. (1956). Electron paramagnetism and nuclear magnetic resonance in metals. F. Seitz, & D. Turnbull (Eds.), *Solid state physics* (pp. 93–136). Vol. 2(pp. 93–136). New York: Academic Press.

Low, W. (1960). Paramagnetic resonance in solids. In *Solid state physics*. New York: Academic Press. (Suppl. 2).

Pake, G. E. (1956). Nuclear magnetic resonance. F. Seitz, & D. Turnbull (Eds.), *Solid state physics* (pp. 1–91). Vol. 2(pp. 1–91). New York: Academic Press.

Poole, C. P., & Farach, H. A. (1972). *The theory of magnetic resonance*. New York: J. Wiley & Sons.

Slichter, C. P. (1980). *Principles of magnetic resonance*. Berlin: Springer-Verlag.

Winter, J. (1971). *Magnetic resonance in metals*. London: Oxford University Press.

Chapter 22

Superconductivity

Chapter Outline

In normal metals the resistivity depends on two interactions: electron-ion (electron-phonon) and electron-electron interactions. The major contribution to resistivity in metals comes from electron-phonon (e-p) interactions. The e-p interaction decreases linearly with a decrease in temperature due to the linear decrease in the number of phonons with temperature. Therefore, the resistivity in metals decreases linearly with a decrease in temperature and approaches zero at absolute zero. In other words, one can say that the conductivity in metals increases linearly with a decrease in temperature and goes to very high values at absolute zero. Onnes (1911) studied the properties of solids at low temperatures and found that the resistivity of Hg decreased suddenly to very small values at $T \approx 4.2\,K$. In other words, the conductivity of Hg increased suddenly by a very large amount at a particular temperature T_c called the critical temperature. His original measurements on Hg are shown in Fig. 22.1. Due to the large increase in conductivity, the state of the metal below T_c is called the *superconducting state* and the metal in this state is called a *superconductor*. The phenomenon of a sudden and large increase in conductivity at T_c is called superconductivity. Since the discovery of superconductivity, scientists have made efforts to increase the value of T_c. The highest value of T_c that has been reached in pure elements is ~9.26 K in the case of Nb. In an effort to further increase T_c, a number of alloys and compounds that exhibit superconductivity have been synthesized. The highest T_c that had been reached up until 1973 was 23.2 K in the case of Nb_3Ge. Fig. 22.2 shows the increase in T_c over time. As seen in the figure, until 1986 the increase in T_c per year was about 0.25 K. In 1986, a new class of superconductors, usually called high- T_c superconductors, was discovered, in which the value of T_c suddenly increased. These materials then became a subject of great interest for scientists.

22.1 EXPERIMENTAL SURVEY

22.1.1 Electrical Properties

In a superconductor the electrical resistivity decreases greatly at or below T_c and it is not certain whether the electrical resistivity is exactly zero or very close to zero. For this reason, persistent electrical currents have been observed to flow for several years in superconductors. The decay of supercurrents has been studied in a solenoid of $Nb_{0.75}Zn_{0.25}$ using a

Solid State Physics. https://doi.org/10.1016/B978-0-12-817103-5.00022-0

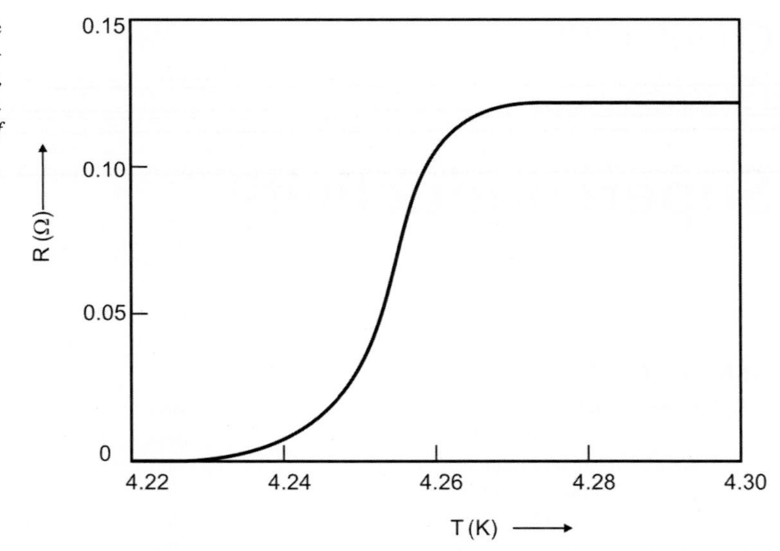

FIG. 22.1 Temperature variation of the resistance R of the metal Hg. The transition from the normal to the superconducting state takes place at $T_c \approx 4.2\,K$. *(Modified from Onnes, H. K. (1911). The superconductivity of mercury (pp. 122–124). Leiden: Communications Physics Laboratory, University of Leiden.)*

precision NMR method in which the magnetic field associated with the supercurrent was measured. It was found that the decay time of the supercurrent is not less than 10^5 years. In some superconducting materials, particularly those used for making superconducting magnets, finite decay times are observed because of an irreversible redistribution of magnetic flux in these materials. The superconducting state is known to be an ordered state of the conduction electrons of the metal and the order is in the formation of loosely bound pairs of electrons below T_c (discussed in detail later).

22.1.2 Magnetic Properties

The magnetic properties shown by superconductors are just as dramatic as their electrical properties. In 1933, Meissner and Ochsenfeld discovered that a superconductor invariably expels all the magnetic flux penetrating it. When a specimen is first placed in a magnetic field and cooled through the T_c for superconductivity, the magnetic flux originally present in the specimen is ejected out of the specimen (see Fig. 22.3). This is called the *Meissner effect*. The more important aspect of the Meissner effect is the discovery of the fact that the magnetic field penetrates a small distance into the specimen near the surface, generally several hundred to several thousand angstroms. Hence, one cannot characterize superconductivity as a state of perfect diamagnetism.

22.1.3 Thermal Properties

22.1.3.1 Entropy

Entropy is a measure of disorder in a system. In a normal metal the entropy decreases linearly with decreasing temperature T. But in a superconductor the entropy decreases markedly on cooling below T_c. The entropy for a metal in the normal state S_n and superconducting state S_s is shown in Fig. 22.4. The lower value of entropy below T_c indicates that the superconducting state is more ordered than the normal state. So, some or all of the electrons thermally excited in a normal metal are ordered in a superconductor.

22.1.3.2 Specific Heat

The specific heat at constant volume of a normal metal C_n is usually the sum of two contributions, one from the lattice C_ℓ and the other from the conduction electrons C_e. At low temperatures it may be expressed as (see Eq. 8.133)

$$C_n = C_e + C_\ell = \gamma_e T + \beta_\ell T^3 \tag{22.1}$$

where γ_e and β_ℓ are constants. The constant γ_e is a measure of the density of electron states at the Fermi energy (see Eq. 9.107). The properties of the lattice are assumed to be the same in both the normal and the superconducting states. Although this assumption has never been proved rigorously, it appears to be valid on the basis of a determination of the Debye-Waller factor in normal and superconducting states of Sn by means of the Mossbauer effect. Therefore, only

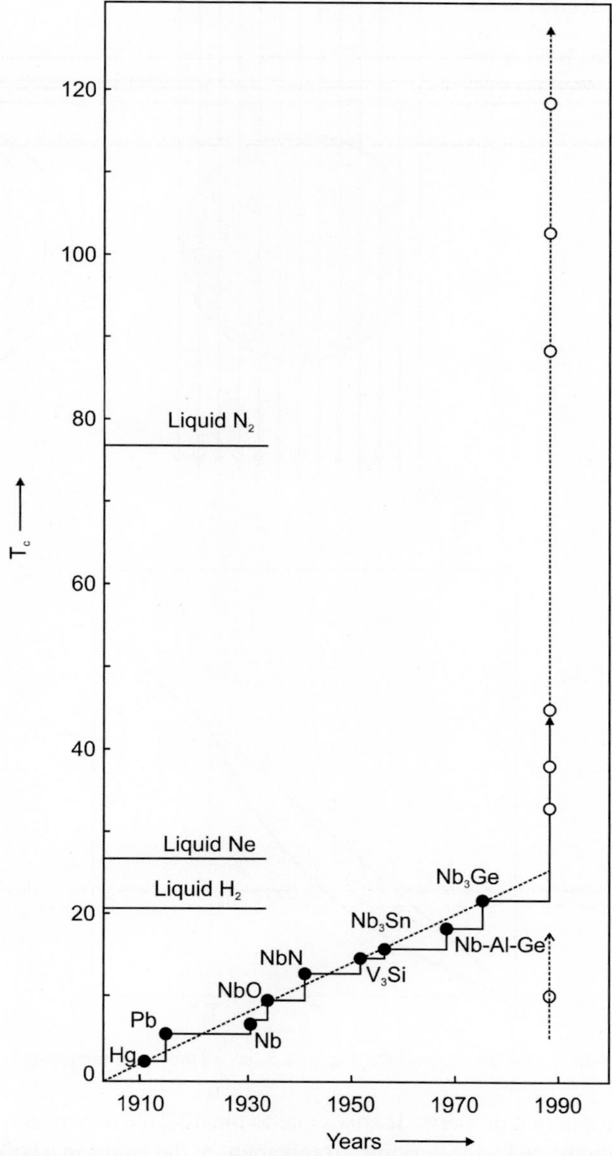

FIG. 22.2 The increase in T_c of a superconductor with time (in years) since the discovery of superconductivity. *(Modified from Bednorz, J. G., & Muller, K. A. (1988). Perovskite type oxides–The new approach to high T_c superconductivity. Reviews of Modern Physics, 60, 585–600.)*

the electronic contribution to the specific heat in the superconducting and the normal states, denoted by C_{es} and C_{en}, respectively, is considered here.

The electronic specific heat shows two striking features when the transition from the normal to the superconducting state occurs in zero magnetic field. First, there is a discontinuous jump in the specific heat at T_c with $C_{es}(T_c) \approx 3\,C_{en}(T_c)$ and then for $T < T_c$, C_{es} decreases exponentially to zero at $T = 0$ (see Fig. 22.5A). The specific heat of a superconductor can be fit into an expression of the form

$$C_{es} = C\,e^{-\Delta/k_B T} \tag{22.2}$$

where Δ is a parameter that is related to the energy gap produced in a superconductor. Fig. 22.5B shows a plot of $\ln(C_{es}/\gamma_e T_c)$ as a function of T_c/T, which is linear with negative slope. This behavior was first clearly demonstrated by Keeson and Kok in Sn and was later observed in other superconductors. It is a transition of the second order because it is characterized by a jump in the specific heat and no latent heat is involved in the transition. If a magnetic field is applied to the material, the transition to the superconducting state occurs at $T < T_c$ and latent heat is associated with the transition, corresponding to the absorption of heat when the sample goes to normal. It is a first-order phase transition.

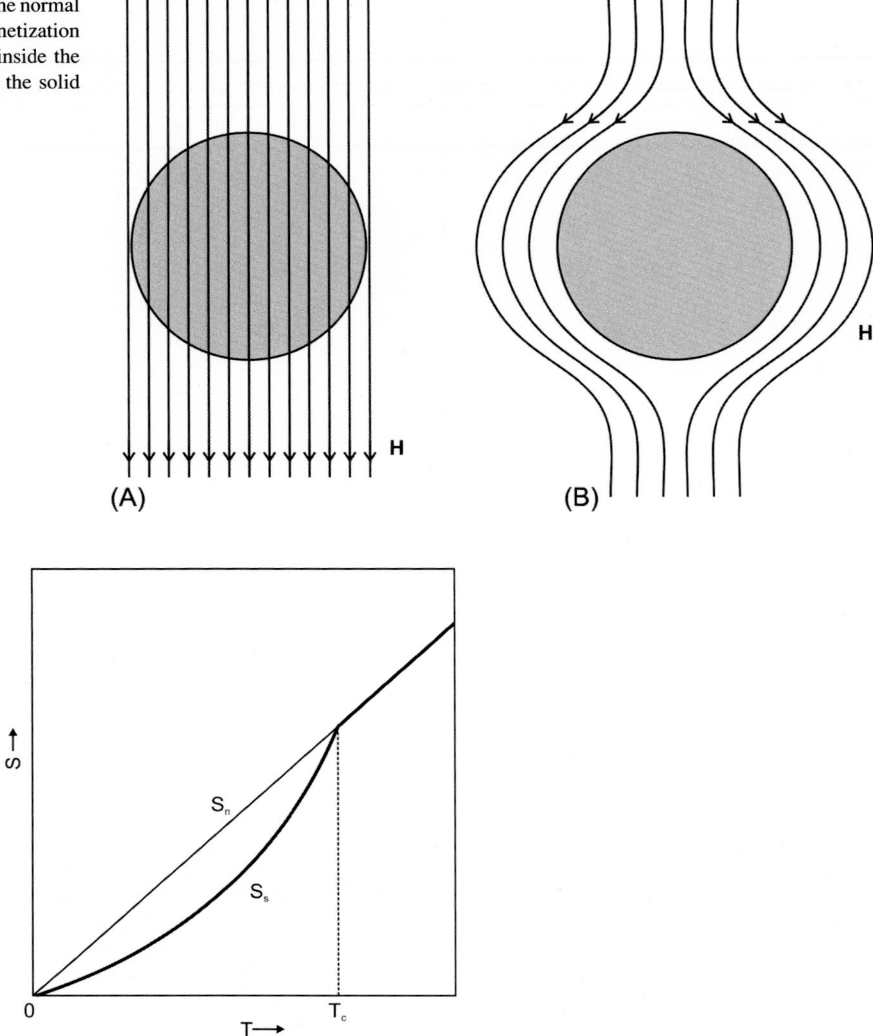

FIG. 22.3 (A) The applied magnetic field **H** in the normal state ($T > T_c$) penetrates a solid, producing magnetization in it and thus generating magnetic induction **B** inside the solid. (B) The magnetic field is expelled out of the solid in the superconducting state with $T < T_c$.

FIG. 22.4 Entropy, both in the normal state S_n and the superconducting state S_s, as a function of temperature T.

In insulators, the energy gap is a lattice property. It arises due to the difference between the binding energies of the two atomic states, which are strongly modified by the periodic arrangement of the atoms in a lattice. In normal metals, the lattice effects completely overwhelm the difference in binding energies of the atomic states, yielding no energy band gap. On the other hand, in superconductors, the energy band gap is an electronic property and is tied to the Fermi electron gas. Just as in the case of insulators, it may be expected that the superconducting energy gap is due to some kind of binding energy between the electrons. The existence of an attractive interaction between two electrons with equal and opposite momenta and opposite spins forms an electron pair. The binding energy required to bind two electrons of opposite spins in a pair is taken from the system itself, thereby decreasing the highest filled energy level of the system. This gives rise to a finite energy gap. One can also argue that the electrons forming bound pairs behave as bosons and, therefore, can be accommodated in fewer states, giving rise to a finite energy gap in the system.

The argument Δ in the exponential of the specific heat (Eq. 22.2) is found to be one-half of the energy gap. The variation of the specific heat of Ga is given by Eq. (22.2) with $\Delta \approx 1.4\, k_B T_c$. Thus, the gap is $E_g = 2\Delta = 2.8\, k_B T_c$, which comes out to be 1.6×10^{-4} eV. The parameter $\Delta(T)$ depends on temperature and its temperature dependence is given as follows:

$$\Delta(T) = \Delta(0)\left[1 - \frac{T}{T_c}\right]^{1/2} \tag{22.3}$$

Fig. 22.6 shows the variation of $\Delta(T)/\Delta(0)$ with T/T_c.

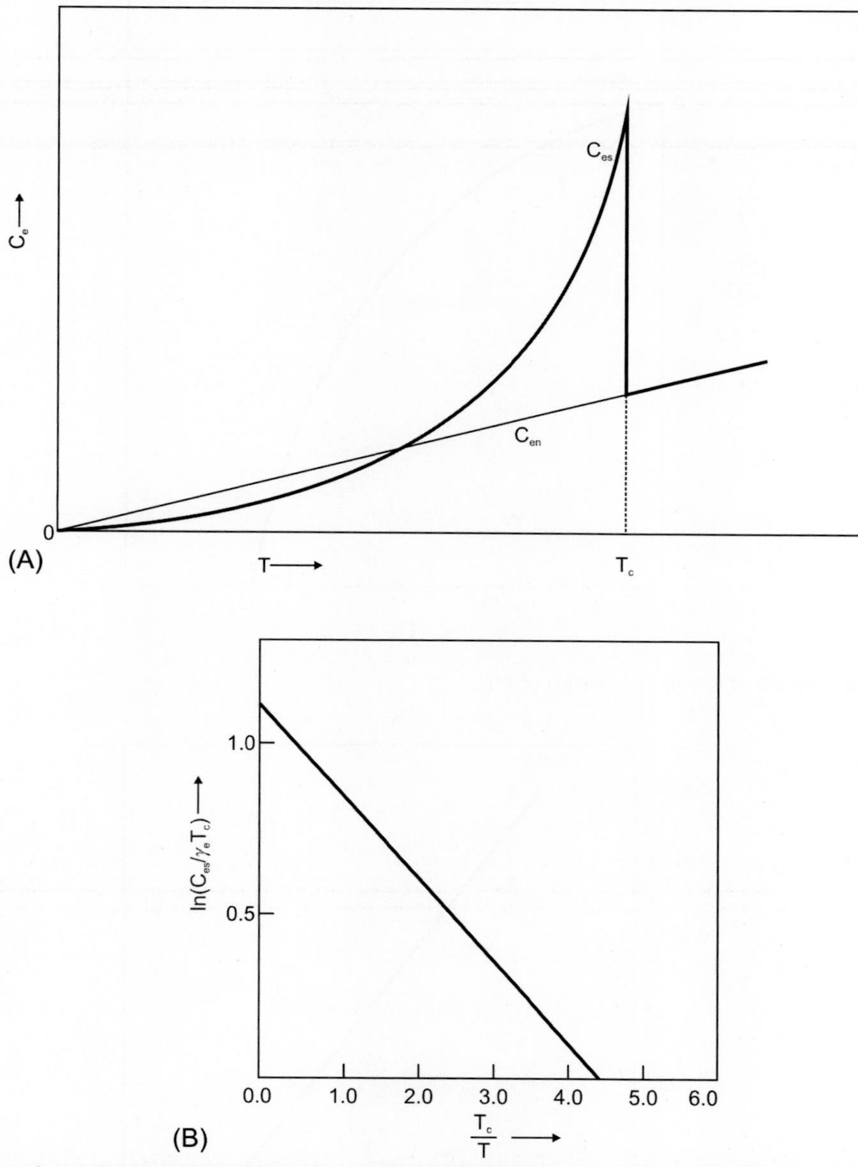

FIG. 22.5 (A) The electronic contribution to the specific heat C_e as a function of temperature both in the normal state (C_{en}) and the superconducting state (C_{es}). (B) $\ln(C_{es}/\gamma_e T_c)$ as a function of T_c/T in the superconducting state.

22.1.4 Isotopic Effect

It has been observed that the T_c of a superconductor varies with isotopic mass M, that is, T_c is different for different isotopes of a superconducting material. In Hg, T_c varies from 4.185 to 4.146K with a variation of atomic mass M from 199.5 to 203.4 amu. Further, T_c changes smoothly when different isotopes of the same element are mixed together. Fig. 22.7 shows that $\log_{10} T_c$ varies linearly with $\log_{10} M$ with a slope of $-\alpha$ for Hg. In other words,

$$\log_{10} T_c = -\alpha \log_{10} M \tag{22.4}$$

Taking the antilogarithm, the above equation gives

$$T_c = M^{-\alpha} \tag{22.5}$$

The dependence of T_c on the isotopic mass M allows us to draw the very important conclusion that lattice vibrations and hence electron-ion interactions may play an important role in superconductivity.

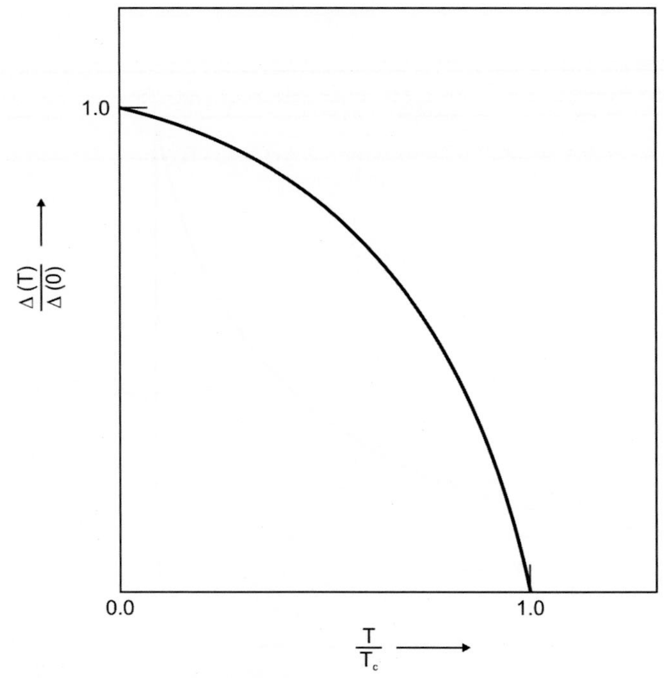

FIG. 22.6 The energy gap parameter $\Delta(T)/\Delta(0)$ as a function of T/T_c.

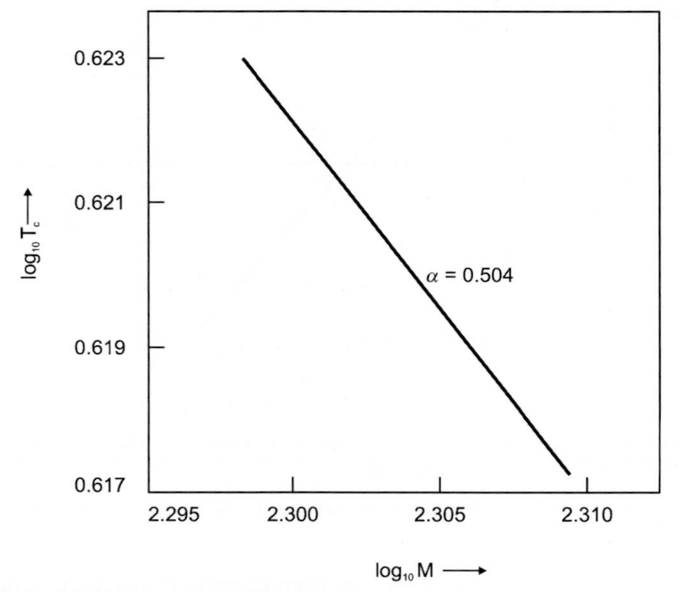

FIG. 22.7 $\log_{10}T_c$ as a function of $\log_{10}M$ of isotopic mass in Hg metal.

22.2 OCCURRENCE OF SUPERCONDUCTIVITY

Superconductivity occurs in many metallic elements, their alloys, and in intermetallic compounds. At present, the range of transition temperature extends from about 23 K for the alloy Nb_3Ge to 0.01 K for the semiconductor $SrTiO_3$, excluding the high-T_c copper oxide superconductors. In many metals, superconductivity has not been found down to a temperature of 1 K. For example, in Li, Na, and K metals, superconductivity has not been found even at 0.08, 0.09, and 0.08 K, respectively. Similarly, Cu, Ag, and Au are normal metals down to 0.05, 0.35, and 0.05 K, respectively. The semiconductors Ge and Si are also found to be in the normal state even at 0.05 and 0.07 K, respectively. Theoretically, it has been predicted that if Na and K are superconductors at all, their T_c will be much less than 10^{-5} K at a pressure of 110 kilobars, after several phase transformations.

It is not known whether every nonmagnetic element will become a superconductor at sufficiently low temperature. Experimentally, it is found that even trace quantities of foreign paramagnetic elements can lower T_c severely. Pure Mo becomes a superconductor at $T_c = 0.92$ K. Experimentally, it has been observed that a few parts per million (ppm) of Fe destroys the superconductivity of Mo completely. Further, one atomic percent of Gd lowers the T_c of La from 5.6 to 0.6 K. Nonmagnetic impurities do not have much effect on T_c, although they may affect the behavior of the superconductor in strong magnetic fields.

None of the monovalent metals, except Cs under pressure, is known to be a superconductor. Further, none of the ferromagnetic metals and none of the rare-earth metals except La (which has an entirely empty 4f electron shell) are superconductors.

22.3 THEORETICAL ASPECTS OF SUPERCONDUCTIVITY

22.3.1 Failure of Ohm's Law in Superconductors

According to Ohm's law

$$\mathbf{E} = \rho \mathbf{J} \tag{22.6}$$

In a superconductor the resistivity ρ goes to zero, while the current density $\mathbf{J}$ is held finite. Therefore, from Eq. (22.6), the electric field $\mathbf{E}$ should go to zero at and below T_c. One of the Maxwell equations relating $\mathbf{E}$ and $\mathbf{B}$ is written as

$$\nabla \times \mathbf{E} = -\frac{1}{c}\frac{\partial \mathbf{B}}{\partial t} \tag{22.7}$$

Substituting $\mathbf{E} = 0$ into Eq. (22.7), one gets

$$\frac{\partial \mathbf{B}}{\partial t} = 0 \tag{22.8}$$

The above equation says that $\mathbf{B}$ is constant in time. This, in turn, implies that the electrodynamic state of a superconductor is a function of its past history. This can be understood through the following examples. Consider two experiments A and B as shown in Fig. 22.8. In experiment A, the superconductor is first cooled below T_c and then placed in a magnetic field. As the magnetic induction within the superconductor must remain the same with time, it follows that no magnetic flux will penetrate into the superconductor. Therefore, the magnetic lines of force must bend away from the superconductor.

In the second experiment B, the opposite situation is presented. Here the specimen in the normal state is first placed in the magnetic field and then cooled below T_c. As a result, magnetic flux will now be frozen inside the superconductor. Thus,

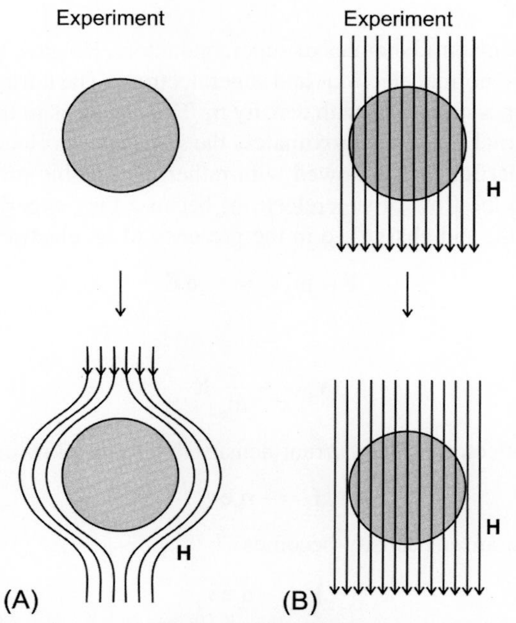

FIG. 22.8 According to Ohm's law, the behavior of a magnetic field in a solid exhibiting superconductivity when (A) the temperature of the solid is first reduced below T_c and then the magnetic field $\mathbf{H}$ is applied. (B) the solid is first placed in the magnetic field $\mathbf{H}$ and then the temperature is reduced below T_c.

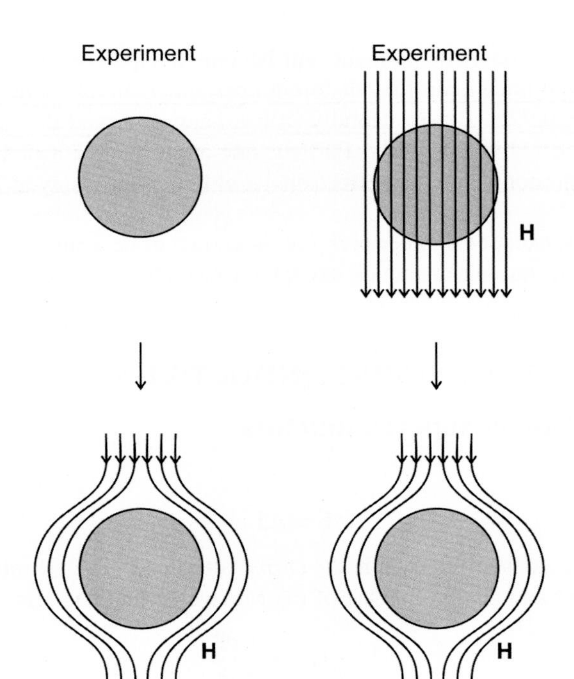

FIG. 22.9 The behavior of the magnetic field in the two cases presented in Fig. 22.8 according to the Meissner effect.

according to Ohm's law, the specimen can exist in two different states with the same external conditions, which is not allowed physically as the system must have a unique equilibrium state. In other words, this means that Ohm's law is not valid in a superconductor. It is for this reason that many theorists felt, prior to the discovery of the Meissner effect, that a successful theory of superconductivity would not be forthcoming.

Meissner and Ochsenfeld proved that the magnetic flux expulsion is reversible, that is, experiments A and B result in the same final state, as demonstrated in Fig. 22.9. Therefore, the Meissner effect showed that superconductivity is a state of true thermodynamic equilibrium, that is, the superconducting state is a single-valued function of **H** and T.

22.3.2 London Theory

London, for the first time, studied the electrodynamics of superconductors. He gave a two-fluid theory in which the electrons are classified into two categories: normal electrons and superelectrons. The normal electrons form a normal fluid with density n_n and the superelectrons form a superfluid with density n_s. The electrons in the normal fluid possess normal properties, that is, they respond to an external field in approximately the same way as electrons in a normal metal. On the other hand, the electrons locked in the superfluid are endowed with rather remarkable properties. For example, the persistent current in a superconductor is due to the flow of superelectrons because they experience no resistance. The equation of motion of superelectrons with mass m_e and charge $-e$ in the presence of an electric field **E** is

$$\mathbf{F} = m_e \dot{\mathbf{v}}_s = -e\mathbf{E} \tag{22.9}$$

which gives

$$\dot{\mathbf{v}}_s = -\frac{e}{m_e}\mathbf{E} \tag{22.10}$$

where $\mathbf{v}_s$ is the velocity of the superelectrons. The current density due to the flow of superelectrons is given by

$$\mathbf{J}_s = -n_s e \mathbf{v}_s \tag{22.11}$$

Therefore, the derivative of the supercurrent density becomes

$$\dot{\mathbf{J}}_s = -n_s e \dot{\mathbf{v}}_s \tag{22.12}$$

Substituting $\dot{\mathbf{v}}_s$ from Eq. (22.10) into Eq. (22.12), one gets

$$\dot{\mathbf{J}}_s = \frac{n_s e^2}{m_e}\mathbf{E} \tag{22.13}$$

The magnetic field **B**, inside the superconductor, can be expressed in terms of the vector potential **A** as

$$\mathbf{B} = \nabla \times \mathbf{A} \tag{22.14}$$

Substituting Eq. (22.14) into Eq. (22.7), we get

$$\mathbf{E} = -\frac{1}{c}\frac{\partial \mathbf{A}}{\partial t} \tag{22.15}$$

Substituting the value of **E** from Eq. (22.15) into Eq. (22.13), we get

$$\dot{\mathbf{J}}_s = -\frac{n_s e^2}{m_e c}\frac{\partial \mathbf{A}}{\partial t} \tag{22.16}$$

In a stationary frame of reference, the partial derivative of time is the same as the total derivative. Therefore, Eq. (22.16) can be written as

$$\frac{d}{dt}\left(\mathbf{J}_s + \frac{n_s e^2}{m_e c}\mathbf{A}\right) = 0 \tag{22.17}$$

The above equation gives

$$\mathbf{J}_s + \frac{n_s e^2}{m_e c}\mathbf{A} = C \tag{22.18}$$

where C is a constant of integration independent of time. The main assumption of the London theory is that C is taken to be zero for a superconductor, which gives

$$\mathbf{J}_s = -\frac{n_s e^2}{m_e c}\mathbf{A} \tag{22.19}$$

The variation of **B** with distance can be studied from the Maxwell equation given by

$$\nabla \times \mathbf{B} = \frac{4\pi}{c}\mathbf{J}_s \tag{22.20}$$

This is nothing but Ampere's law. Taking the curl of the above equation, we can write

$$\nabla \times \nabla \times \mathbf{B} = \frac{4\pi}{c}\nabla \times \mathbf{J}_s \tag{22.21}$$

In the above equation one can use the standard identity defined by

$$\nabla \times \nabla \times \mathbf{a} = \nabla(\nabla \cdot \mathbf{a}) - \nabla^2 \mathbf{a} \tag{22.22}$$

where **a** is a vector. So, Eq. (22.21) becomes

$$\nabla^2 \mathbf{B} = -\frac{4\pi}{c}\nabla \times \mathbf{J}_s \tag{22.23}$$

Here we have used the Maxwell equation defined by

$$\nabla \cdot \mathbf{B} = 0 \tag{22.24}$$

Substituting Eqs. (22.19), (22.14) into Eq. (22.23), we get

$$\nabla^2 \mathbf{B} = \frac{1}{d_L^2}\mathbf{B} \tag{22.25}$$

where

$$d_L = \left(\frac{m_e c^2}{4\pi n_s e^2}\right)^{1/2} \tag{22.26}$$

Consider a superconductor with one of its faces parallel to the y-axis (see Fig. 22.10). If we apply a magnetic field parallel to the face of the superconductor, then its variation along the x-direction, from Eq. (22.25), is given by

$$\frac{d^2\mathbf{B}}{dx^2} = \frac{1}{d_L^2}\mathbf{B} \tag{22.27}$$

The solution of Eq. (22.27) is given by

$$\mathbf{B}(x) = \mathbf{B}(0)\, e^{-x/d_L} \tag{22.28}$$

for $x \geq 0$. At $x = 0$, which corresponds to the face of the superconductor, the magnetic field is equal to the applied field $\mathbf{B}(0) = \mathbf{H}$, but at finite x inside the superconductor, the magnetic field decreases exponentially. At $x = d_L$,

$$\frac{\mathbf{B}(d_L)}{\mathbf{B}(0)} = e^{-1} \tag{22.29}$$

The parameter d_L measures the depth of penetration of the magnetic field and is known as the London penetration depth. Therefore, the London theory explains the Meissner effect. It has been found experimentally that the value of d_L for $Sn_{0.97}I_{0.03}$ alloy increases to nearly twice its value for pure Sn. But according to Eq. (22.26), d_L should change only slightly because the addition of a very small impurity should only change the values of m_e and n_s slightly. Therefore, one of the shortcomings of the London theory is that the expression for d_L does not give any indication at all that it changes significantly in dilute alloys.

Physical insight into the London equation can be obtained from the following simple consideration. Let n and $\mathbf{v}$ be the density and velocity, respectively, of carriers with charge q. Then the current density is given by

$$\mathbf{J} = nq\mathbf{v} \tag{22.30}$$

In the presence of magnetic field $\mathbf{B}$, described by vector potential $\mathbf{A}$, the momentum is given by

$$\mathbf{p} = m\mathbf{v} + \frac{q}{c}\mathbf{A} \tag{22.31}$$

Substituting the value of $\mathbf{v}$ from Eq. (22.31) into Eq. (22.30), one gets

$$\mathbf{J} = \frac{nq}{m}\mathbf{p} - \frac{nq^2}{mc}\mathbf{A} \tag{22.32}$$

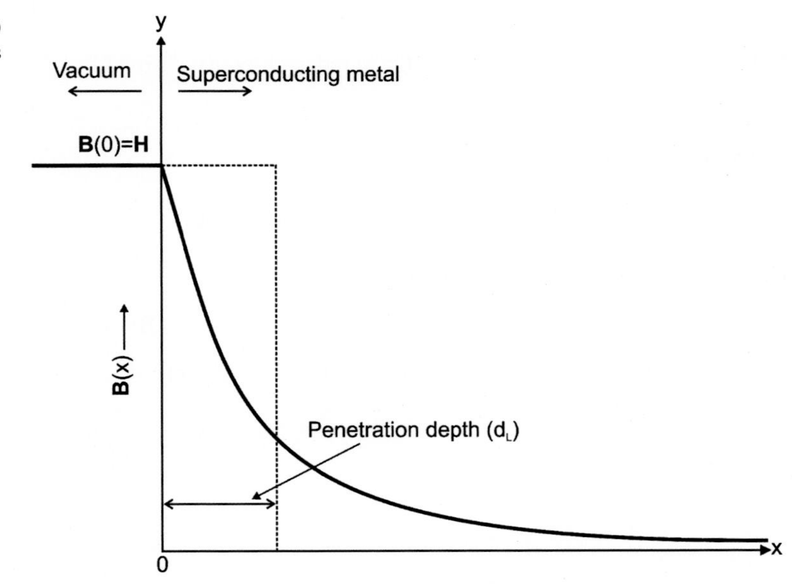

FIG. 22.10 The exponential decay of magnetic field B(x) on entering a superconducting solid. The *dashed line* gives the London penetration depth d_L in the solid.

Eq. (22.32) can be reduced to the London equation by substituting $\mathbf{p}=0$, that is

$$\mathbf{J}=-\frac{nq^2}{mc}\mathbf{A}=-\frac{c}{4\pi}\frac{\mathbf{A}}{d^2} \tag{22.33}$$

where

$$d=\left(\frac{mc^2}{4\pi nq^2}\right)^{1/2} \tag{22.34}$$

If the charge carriers are pairs of electrons with equal and opposite momenta, then these carriers possess charge $q=-2e$ and mass $m=2m_e$, and their density is half that of the conduction electrons, that is, $n=n_s/2$. Substituting these values into Eq. (22.33), one gets the same expression as given by Eq. (22.19) and the parameter d reduces to d_L (Eq. 22.26). Eq. (22.19) can be written as

$$\mathbf{J}_s=-\frac{c}{4\pi}\frac{\mathbf{A}}{d_L^2} \tag{22.35}$$

Hence the London theory predicts the existence of electron pairs with equal and opposite momenta as the charge carriers in a superconductor.

Problem 22.1

A monovalent metal has an electron density of 6.0×10^{22} electrons/cm^3. Find the London penetration depth.

Problem 22.2

Consider a superconducting plate having thickness d such that $d \ll d_L$, the London penetration depth. If a magnetic field H is applied parallel to the surface of the plate, find the magnetization produced inside the plate using the London equation.

22.3.3 Penetration Depth

It has been observed that a superconductor is a diamagnetic material; therefore, it prevents electric current from flowing through the body of a material. But a superconductor cannot be a perfect diamagnetic material because, in this case, the current would be confined to the surface only. If this were so, the current density would become infinite, which is physically impossible. According to the Meissner effect the magnetic field penetrates the superconductor by a small distance near the surface. As a result, the current flows in a very thin surface layer whose thickness is on the order of 10^{-5} cm, although the exact value varies in different materials. Although this thickness is very small, it plays a very important role in determining the properties of superconductors. The depth within which the current flows is called the penetration depth and it is this depth to which the magnetic flux (magnetic field) appears to penetrate in a superconductor. The decrease in the flux density inside a superconductor is shown in Fig. 22.10. The penetration depth in a superconductor is somewhat like the "skin depth" to which high-frequency alternating fields penetrate in a normal metal. The penetration depth can be defined in a number of ways, but the usual definition is given below. If, at a distance x into the metal, the flux density falls to a value B(x), the penetration depth d can be defined as

$$\int_0^\infty B(x)\,dx=d\,B(0) \tag{22.36}$$

where B(0) is the flux density at the surface of the metal. In other words, there would be the same amount of flux inside the superconductor if the flux density of the external field remained constant to a distance d into the metal. Because the penetration depth is very small, we do not notice the flux penetration in magnetic measurements on ordinary-sized specimens.

The penetration depth d(T) does not have a fixed value but varies with temperature T as shown in Fig. 22.11. At low temperatures, it is nearly independent of T with a value d_0 that is characteristic of the particular metal. Above $0.8\,T_c$, d increases rapidly and approaches infinity as T approaches T_c. The variation of d(T) with T is described reasonably well by the relation

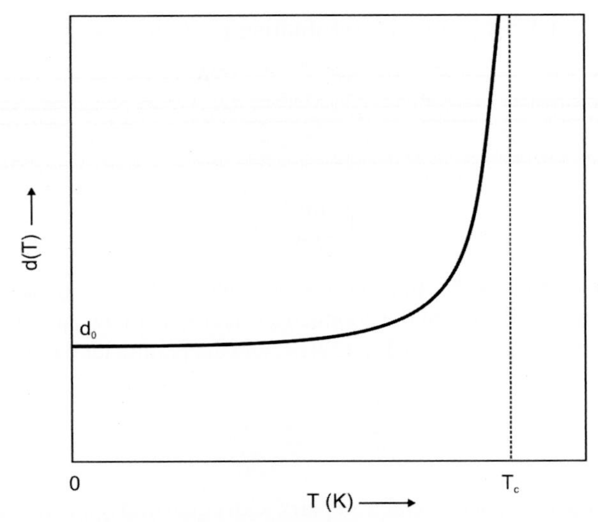

FIG. 22.11 The variation of penetration depth d(T) with T in a superconducting solid. d_0 gives the penetration depth at absolute zero.

$$d(T) = \frac{d_0}{\left[1 - \left(\frac{T}{T_c}\right)^4\right]^{1/2}} \tag{22.37}$$

Perfect diamagnetism, therefore, does not occur in specimens that are very close to their T_c values. The decrease in d(T) is, however, very rapid as the temperature falls below T_c. The small departure from perfect diamagnetism would be extremely difficult to detect in bulk specimens because of the difficulty of holding the temperature sufficiently constant during a measurement.

22.3.4 Coherence Length

It has been seen that when a superconductor is cooled below T_c, some extra order sets in among the conduction electrons. So, the idea that a superconductor can be regarded as consisting of two interpenetrating fluids, a normal fluid and a superfluid, has been introduced. The superelectrons in some ways possess greater order than the normal electrons, and the degree of order can be related to their density n_s. Considering several aspects of the behavior of superconductors, Pippard was led to the idea that n_s cannot change rapidly with position. An appreciable change in n_s can occur within a distance on the order of 10^{-4} cm for pure superconductors. This distance is generally called the Pippard coherence length ξ_P.

An important property of ξ_P is that it depends on the purity of the metal; the value 10^{-4} cm is representative of a pure superconductor. If impurities are present, ξ_P is reduced and, in the case of very impure metals, which are characterized by a very short electron mean free path ℓ_e, ξ_P becomes approximately equal to ℓ_e. The value of ξ_P in a perfectly pure superconductor, which is an intrinsic property, is usually denoted by ξ_0 and is a function of ℓ_e in superconductors containing impurities.

To study the influence of impurity atoms (acting as scattering centers for the electrons) on the coherence length, Pippard measured the penetration depth d_P, generally called the Pippard penetration depth, as a function of ℓ_e and obtained the results shown in Fig. 22.12. A rapid variation of d_P begins when its value is comparable with the mean free path ℓ_e. Pippard modified Eq. (22.35) due to London to write

$$\mathbf{J}_s = -\frac{c}{4\pi}\frac{\mathbf{A}}{d_P^2} \tag{22.38}$$

where d_P is given by

$$\begin{aligned} d_P &= d_L\left(\frac{\xi_0}{\xi_P}\right)^{1/2} \quad \text{if } \xi_P^3 \ll \xi_0 d_L^2 \\ &= d_\infty = d_L\frac{3^{1/6}}{(2\pi)^{2/3}}\left(\frac{\xi_0}{d_L}\right)^{1/3} \quad \text{if } \xi_P^3 \gg \xi_0 d_L^2 \end{aligned} \tag{22.39}$$

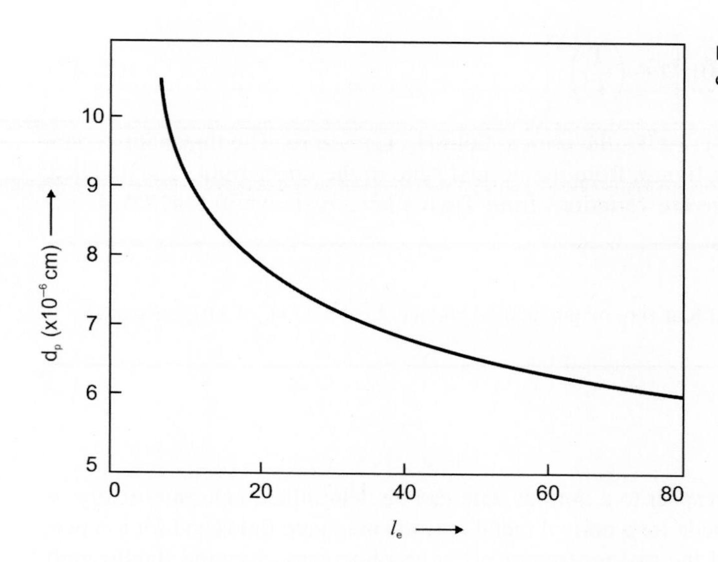

FIG. 22.12 The Pippard's penetration depth d_P as a function of the electron mean free path ℓ_e.

The Pippard's coherence length $\xi_P(\ell_e)$ is given by

$$\frac{1}{\xi_P(\ell_e)} = \frac{1}{\xi_0} + \frac{1}{b_P \ell_e} \tag{22.40}$$

$\xi_P(\ell_e)$ is fitted with the experimental results assuming $\xi_0 = 1.2 \times 10^{-4}$ and $b_P = 0.8$. Thus, Pippard's model solved the long-standing puzzle with d_L, namely, why the measured penetration depths were always a few times larger than those computed from d_L for pure superconductors.

A basic point of the London theory was the absolute rigidity of the superconducting wave function in the presence of a field. Pippard abandoned this point and suggested instead that the perturbing force acting at one point in the superconductor would be felt over a distance $\xi_P(\ell_e)$. Conversely, the response at a point due to a spatially extended perturbation would be obtained by integrating over a finite region surrounding that point. Thus, the coherence length is a measure of the distance within which the gap parameter does not change drastically in a spatially varying magnetic field. Therefore, the coherence length is a measure of the range over which we should average **A** to obtain $\mathbf{J}_s$ (see Eq. 22.38). One can compare the London and Pippard theories by writing down their kernels as follows:

$$K_L(q) = \frac{1}{d_L^2} \tag{22.41}$$

$$\begin{aligned} K_P(q) &= \frac{1}{d_L^2}\frac{\xi_P}{\xi_0}\left[1 - \frac{q^2\xi_P^2}{5} + \cdots\right] \quad \text{for } q\xi_P \ll 1 \\ &= \frac{1}{d_L^2}\frac{\pi}{4q\xi_0}\left[1 - \frac{4}{\pi q \xi_P} + \cdots\right] \quad \text{for } q\xi_P \gg 1 \end{aligned} \tag{22.42}$$

where $q\ (\approx 1/\lambda)$ is the wave vector. $K_L(q)$ and $K_P(q)$ are kernels due to the London and Pippard theories. A comparison of Eqs. (22.41), (22.42) shows that $K_L(q)$ is independent of q, while $K_P(q)$ varies with q. $K_P(q)$ goes over to $K_L(q)$ with a modified penetration depth for $q\xi_P \ll 1$ ($\xi_P \ll \lambda$). This condition may be satisfied for $q \to 0$, that is, at large distances. The other limiting case of $q\xi_P \gg 1$ ($\xi_P \gg \lambda$) is obviously favored by $\ell_e \to \infty$, leading to $\xi_P = \xi_0$. We are thus led to the conclusion that there are two types of superconductors. The first is the Pippard type or type I superconductor having long mean free paths and the second is the London type or type II superconductor with $\xi_P < \lambda$ ($q\xi_P < 1$), exemplified by certain transition metals that have high T_c and, therefore, small ξ_0, or superconducting alloys for which $\xi_P \cong \ell_e < \lambda$.

22.3.5 Destruction of Superconductivity by Magnetic Field

Onnes found that a sufficiently strong magnetic field destroys superconductivity. The threshold or critical value of the applied magnetic field for the destruction of superconductivity is denoted by $H_c(T)$. The temperature variation of $H_c(T)$ follows *Tuyn's law*, defined as

$$H_c(T) = H_c(0)\left[1 - \left(\frac{T}{T_c}\right)^2\right] \tag{22.43}$$

Fig. 22.13 shows the variation of $H_c(T)$ with temperature T. At T_c, the critical field $H_c(T_c)$ is zero. The threshold curves separate the superconducting state, in the lower left of the figure, from the normal state in the upper right. The detailed Bardeen-Cooper-Schrieffer (BCS) theory showed that there are variations from Tuyn's law, as shown in Fig. 22.14.

Problem 22.3

A superconducting metal, Sn has a critical temperature $T_c = 3.7$ K at zero magnetic field and a critical field H_c of 310 gauss at 0 K. Find the critical magnetic field at 2 K.

22.3.6 Stabilization Energy

The stabilization energy of a superconducting state with respect to a normal state can be determined calorimetrically or magnetically. The direct measurement of specific heat is made for a normal metal in finite magnetic fields and for a superconductor in zero magnetic field. The energy difference of the two measurements at absolute zero gives the stabilization

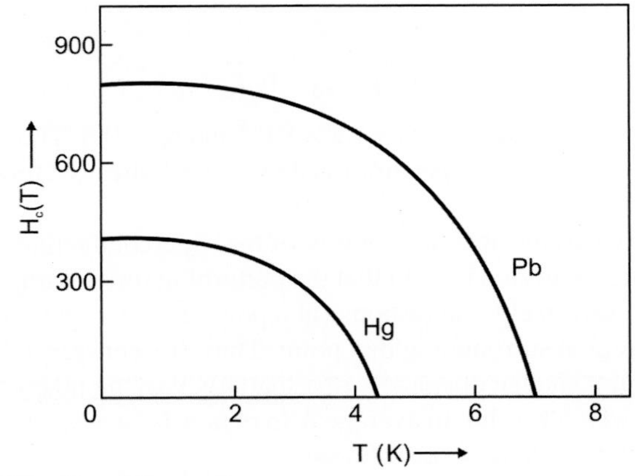

FIG. 22.13 The variation of $\mathbf{H_c}$(T) with T for Pb and Hg metals.

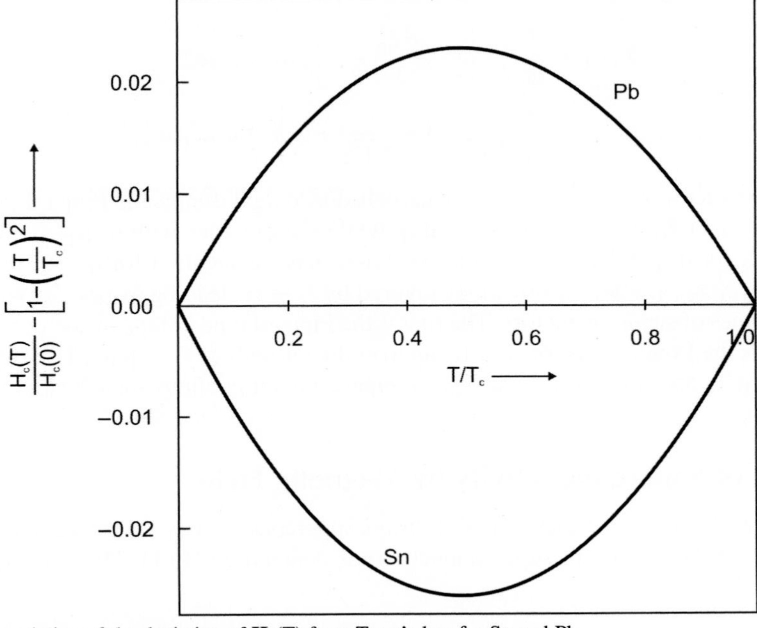

FIG. 22.14 The temperature variation of the deviation of $\mathbf{H_c}$(T) from Tuyn's law for Sn and Pb.

energy of the superconducting state. It is also possible to obtain the stabilization and free energies simply from the critical value of the applied magnetic field H_c that destroys the superconducting state and brings the specimen to the normal state.

The first law of thermodynamics gives the conservation of total energy of the system. If dE is the change in the internal energy of a system and dW is the work done by it, then the heat supplied dQ is given by

$$dQ = dE + dW \tag{22.44}$$

The second law of thermodynamics gives

$$dQ = TdS \tag{22.45}$$

So, from Eqs. (22.44), (22.45), the first law of thermodynamics takes the form

$$dE = TdS - dW \tag{22.46}$$

The work done per unit volume by an applied magnetic field $d\mathbf{H}$ is

$$dW = \mathbf{M} \cdot d\mathbf{H} \tag{22.47}$$

where **M** is the magnetization. The applied field may be due to a permanent magnet, which is brought from infinity to the position **r**. Substituting for dW in Eq. (22.46), we have

$$dE = TdS - \mathbf{M} \cdot d\mathbf{H} \tag{22.48}$$

The magnetic induction **B** in the material is given by

$$\mathbf{B} = \mathbf{H} + 4\pi\mathbf{M} \tag{22.49}$$

In a superconductor **B** is zero, which gives

$$\mathbf{M} = -\frac{\mathbf{H}}{4\pi} \tag{22.50}$$

From Eqs. (22.48), (22.50) one writes

$$dE = TdS + \frac{\mathbf{H} \cdot d\mathbf{H}}{4\pi} \tag{22.51}$$

At absolute zero the increase in energy density of a superconductor in going from infinity to a position **r,** from the above equation, becomes.

$$E_s(\mathbf{H}) - E_s(0) = \frac{\mathbf{H}^2}{8\pi} \tag{22.52}$$

Now consider a normal nonmagnetic metal. If we neglect the small magnetic susceptibility of a metal in the normal state, then $\mathbf{M} = 0$ and the internal energy is independent of the field **H**. Therefore, the value of the internal energy in a normal metal at $\mathbf{H} = 0$ and $\mathbf{H} = \mathbf{H}_c$ are equal, that is,

$$E_n(\mathbf{H}_c) = E_n(0) \tag{22.53}$$

Further, at $\mathbf{H}_c$, the internal energies are equal in both the normal and superconducting states and one can write

$$E_n(\mathbf{H}_c) = E_s(\mathbf{H}_c) = E_s(0) + \frac{H_c^2}{8\pi} \tag{22.54}$$

The specimen is stable in either state when the applied field is equal to $\mathbf{H}_c$. From Eqs. (22.53), (22.54) we have

$$\Delta E = E_n(0) - E_s(0) = \frac{H_c^2}{8\pi} \tag{22.55}$$

ΔE is called the stabilization energy density of the superconducting state of the specimen at absolute zero. As an example, for Al metal $H_c = 105$ gauss at $T = 0$, so the stabilization energy of Al metal is given by

$$\Delta E = \frac{(105)^2}{8 \times 3.142} = 440 \text{ ergs/cm}^3 \tag{22.56}$$

which is in excellent agreement with the experimental result of 430 ergs/cm^3.

22.3.7 Classification of Superconductors

Normal metals (excluding the ferromagnetic metals, such as iron) are virtually nonmagnetic so the magnetic induction **B** inside them is linearly proportional to the applied field, that is,

$$\mathbf{B}=\mu\mathbf{H} \tag{22.57}$$

Fig. 22.15 shows **B** as a function of **H**. The magnetic behavior of a superconductor can also be described in terms of the magnetization **M**. A bulk superconductor behaves, in the presence of an externally applied magnetic field, as if the magnetic induction **B** is zero. In this case, the magnetization **M** is given by Eq. (22.50) and, therefore, the magnetic susceptibility χ_M becomes

$$\chi_M=\frac{\mathbf{M}}{\mathbf{H}}=-\frac{1}{4\pi} \tag{22.58}$$

The magnetic behavior of a superconductor can be described in another way. Substituting Eq. (22.49) into Eq. (22.57), we find

$$\mu=1+4\pi\chi_M \tag{22.59}$$

Substituting χ_M from Eq. (22.58) into Eq. (22.59), the permeability reduces to zero, that is,

$$\mu=0 \tag{22.60}$$

Therefore, the superconducting material can be considered to have zero permeability and hence the flux density inside the material is zero. One can show that these two ways of describing perfect diamagnetism are entirely equivalent. Fig. 22.16 shows the variation of $-4\pi\mathbf{M}$ with **H**. Below the critical magnetic field $\mathbf{H}_c$, $-4\pi\mathbf{M}$ is finite and varies linearly with the magnetic field H as given by Eq. (22.58). At magnetic fields greater than $\mathbf{H}_c$, the material is in the normal state with virtually no magnetization. It is noteworthy that the magnetization curve is reversible. A superconductor that exhibits such behavior is called a type I superconductor or a soft superconductor or an ideal superconductor or a pure superconductor. Type I superconductors possess a sharply defined value of $\mathbf{H}_c$. Type I superconductivity is shown by a specimen in the form of a long solid cylinder placed in a longitudinal magnetic field. In other geometries, the field may not be homogeneous around the specimen and it may penetrate below $\mathbf{H}_c$. For example, in a sphere, the field penetrates at (2/3) $\mathbf{H}_c$ as a consequence of the nonzero demagnetization factor of the sphere. Pure specimens of many materials exhibit this property.

The other superconducting materials exhibit magnetization curves of the form shown in Fig. 22.17 and are known as type II superconductors or hard superconductors. In these superconductors the critical magnetic field is not sharply defined. The magnetic flux starts penetrating at a field $\mathbf{H}_{c_1}$, which is lower than the thermodynamic critical field $\mathbf{H}_c$. For higher values of magnetic field, the material is in a vortex or mixed state between $\mathbf{H}_{c_1}$ and $\mathbf{H}_{c_2}$, which are the lower and upper critical fields, respectively, and the magnetic flux $\mathbf{B}\neq 0$. One can say that the Meissner effect is incomplete between $\mathbf{H}_{c_1}$ and $\mathbf{H}_{c_2}$, but the material has superconducting electrical properties up to $\mathbf{H}_{c_2}$. Above $\mathbf{H}_{c_2}$, the material is a normal

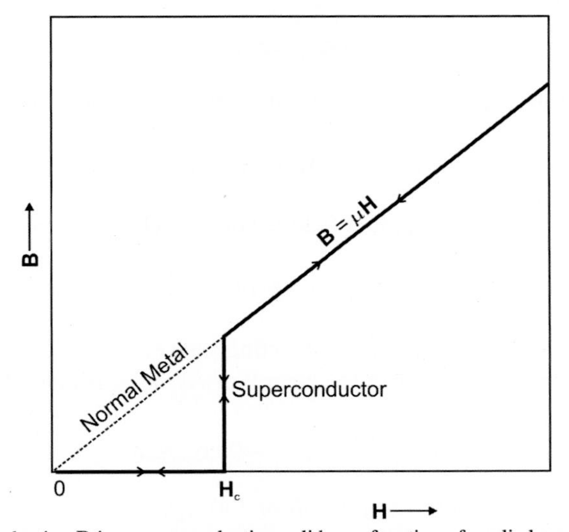

FIG. 22.15 The variation of magnetic induction **B** in a superconducting solid as a function of applied magnetic field **H**.

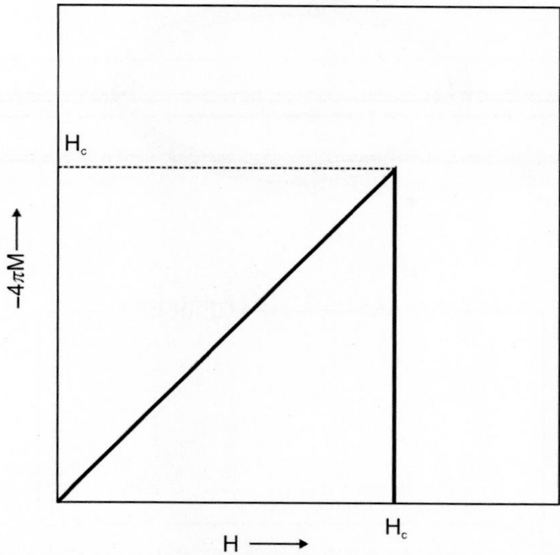

FIG. 22.16 A plot of $-4\pi\mathbf{M}$ as a function of applied magnetic field $\mathbf{H}$ in a type I superconductor. $\mathbf{H}_c$ is the critical magnetic field.

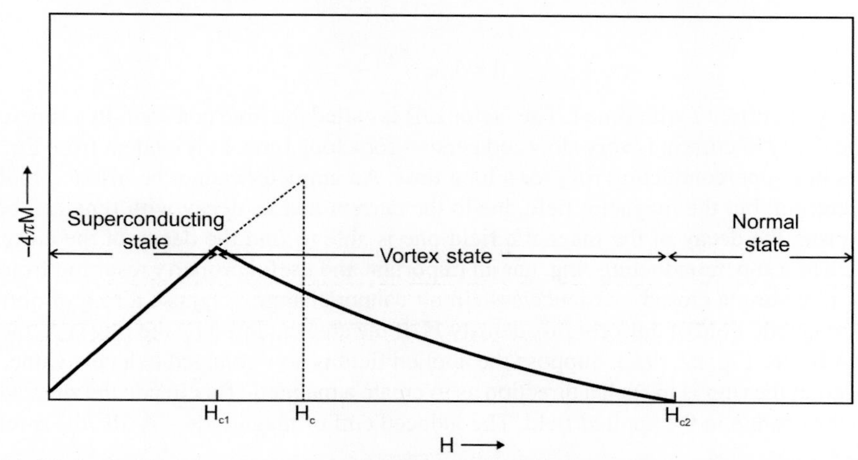

FIG. 22.17 A plot of $-4\pi\mathbf{M}$ as a function of applied magnetic field $\mathbf{H}$ in a type II superconductor. $\mathbf{H}_{c_1}$ and $\mathbf{H}_{c_2}$ are the critical magnetic fields.

conductor in every respect, except for possible surface effects. The value of $\mathbf{H}_{c_2}$ may be 100 times or more than the value of $\mathbf{H}_c$. Furthermore, the magnetization is not reversible and gives rise to what is called hysteresis. When the applied magnetic field is switched off, there remains some finite magnetization in the material, giving rise to a residual flux density $\mathbf{B}_R$ and magnetization $\mathbf{M}_R$. Therefore, the material has trapped some magnetic flux and the superconductor behaves like a permanent magnet. The transition metals or alloys with high values of electrical resistivity in the normal state usually exhibit type II superconductivity. A field $\mathbf{H}_{c_2}$ of 410 kG has been attained in an alloy of Nb, Al, and Ge at the boiling point of He.

22.3.8 Persistent Currents

It has been observed experimentally that the electrical current in a superconducting material may persist for hundreds or thousands of years. Several arguments can be given for the stability of the persistent current, but here we explain this fact through analogy with electricity. Let us take a material in the form of a circular ring, as shown in Fig. 22.18A. It has got both resistance R and inductance L and is equivalent to an electrical circuit, as shown in Fig 22.18B. Suppose current I_0 is allowed to flow through the ring at time $t=0$ and then the current source is switched off. The equation for the flow of current I at time t in the ring is given by

$$L\frac{dI}{dt}+RI=0 \tag{22.61}$$

Integration of the above equation gives

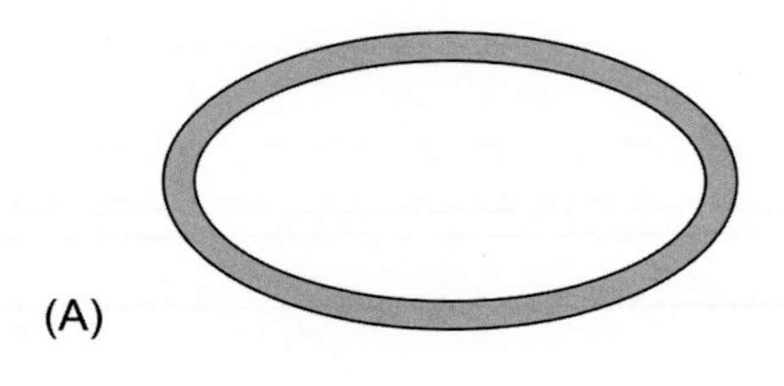

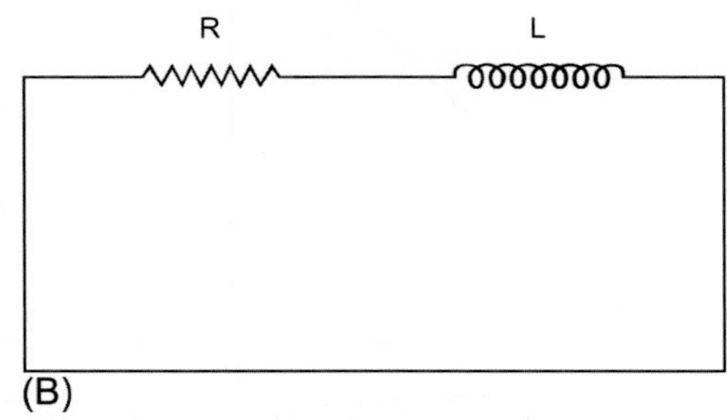

FIG. 22.18 (A) Conducting wire in the form of a ring. (B) Equivalent circuit of the conducting ring with R and L as its resistance and inductance, respectively.

$$I = I_0 e^{-\frac{R}{L}t} \tag{22.62}$$

Eq. (2.62) gives the decay of current I with time t. The factor L/R is called the time constant. In a superconductor, R is very small and, therefore, the decay of current is very slow and persists for a long time, as is evident from Eq. (22.62). Hence, the persistent current exists in a superconducting ring for a long time. An ammeter cannot be inserted in the superconducting circuit to measure the current, but the magnetic field due to the current and its decay with time can be measured without disturbing the circuit. From the decay of the magnetic field one is able to find the decay of the current in the circuit.

A closed circuit, such as a superconducting ring, has an important and useful property resulting from its zero resistance: the total magnetic flux threading a closed resistanceless circuit cannot change. Suppose a ring of normal metal is cooled below T_c in an applied magnetic field of uniform flux density **H**. If the area enclosed by the ring is A, then the magnetic flux linked with the ring is A**H** (see Fig. 22.19A). Suppose the applied field is now changed to a new value. By Lenz's law, the induced current produced in the ring is in such a direction as to create a magnetic flux inside the ring, which tends to cancel the flux change due to the change in the applied field. The induced emf of magnitude $-A\,(dH/dt)$ is related to the induced current I as

$$-A\frac{dH}{dt} = RI + L\frac{dI}{dt} \tag{22.63}$$

In a superconducting circuit, however, $R = 0$ and so

$$A\frac{dH}{dt} + L\frac{dI}{dt} = 0 \tag{22.64}$$

Integrating the above equation, we find

$$LI + AH = 0 \tag{22.65}$$

Here the constant of integration is assumed to be zero. Here LI is the flux linked with the ring due to the induced current I and AH is the flux due to the applied magnetic field. So, according to Eq. (22.65), the total magnetic flux linked with the superconducting ring is constant with time. If the applied magnetic field strength is changed, an induced current is set up of such a magnitude that it creates a flux that exactly compensates the change in the flux from the applied magnetic field. Because the superconducting ring is resistanceless, the induced current flows forever and the original amount of flux is maintained indefinitely. Note that although the total amount of flux enclosed in a resistanceless circuit remains constant, there can be a change in the flux density **H** at any point due to a redistribution of the flux within the circuit. In Fig. 22.19B the flux density has become stronger near the wire and weaker in the center of the enclosed space compared with the uniform distribution in Fig. 22.19A. In both cases, however, the total flux ($\int \mathbf{H}\cdot d\mathbf{A}$) is the same. This property can be used to produce constant magnetic fields with the help of solenoids made from superconducting wires. The property of persistent currents can be used to make superconducting switches.

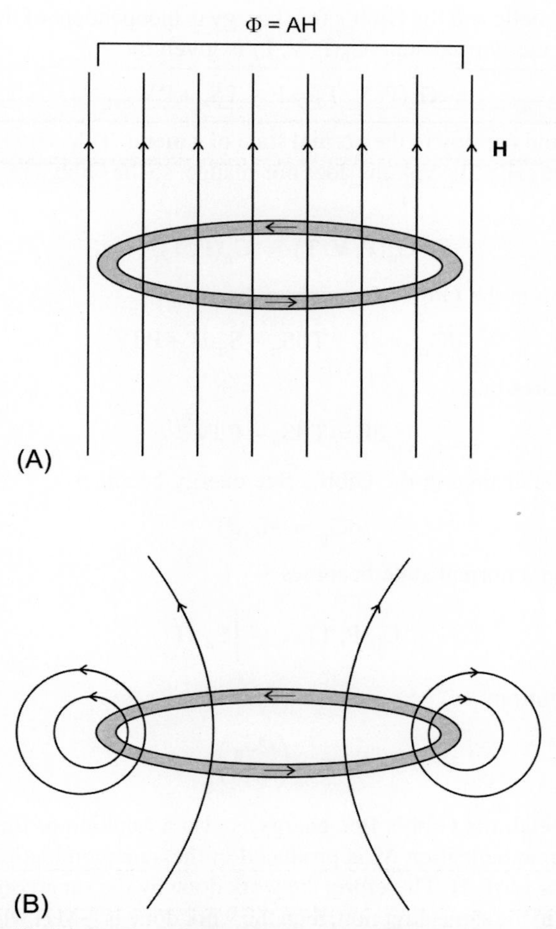

FIG. 22.19 (A) Magnetic flux Φ linked with the normal conducting ring in the presence of a uniform applied magnetic field **H**. (B) Magnetic flux redistribution in the ring in the superconducting state.

One can also consider a second case as follows. Suppose no magnetic field is applied initially and then the ring is cooled down below T_c to become a superconductor. So, the magnetic flux linked with the superconducting ring is zero to start with. Now, if an external magnetic field is applied, subsequently, the net internal flux remains zero in spite of the presence of the external field. This property enables us to use hollow superconducting cylinders to shield enclosures from external magnetic fields. The shielding is perfect only in the case of a long hollow cylinder, in which case the induced currents generate a uniform compensating flux density throughout the interior. For other cases, such as a short ring, it is only the total flux that is maintained at zero and the local magnetic flux density generated by the induced current will not be uniform within the ring. Hence, the flux density due to the persistent current will be stronger than that due to the applied field in some places and weaker in other places, and there will not be an exact cancellation everywhere. In mathematical language one can write

$$\int_A \mathbf{H} \cdot d\mathbf{A} = 0 \tag{22.66}$$

where **H** itself may not necessarily be zero everywhere.

22.3.9 Thermodynamics of Superconductors

It has already been pointed out that the thermal properties, such as the entropy and specific heat, show anomalous features as a normal metal makes the transition to the superconducting state. It has been observed that the state of magnetization of a superconductor depends on the values of the applied magnetic field and temperature and not the way that they are applied. This fact implies that whether or not there is an applied magnetic field, the transition from the superconducting to the normal state is reversible in the thermodynamic sense. Therefore, one can apply thermodynamic arguments to a superconductor taking temperature and magnetic field strength as the thermodynamic variables.

In a normal sate, metal is nonmagnetic and the Gibb's free energy is independent of the strength of the applied magnetic field. So, the Gibb's free energy in the normal state $G_n(P, V, T)$ is given by

$$G_n(P, V, T) = E - TS_n + PV \tag{22.67}$$

where E and S_n are internal energy and entropy in the normal state of a metal. T, V, and P are the temperature, volume, and pressure of the metal. In a solid, practically, the volume does not change, so the Gibb's free energy is independent of volume and one can write

$$G_n(P, V, T) = G_n(P, T) \tag{22.68}$$

At constant pressure, a small change in the Gibb's free energy becomes

$$dG_n = dE - TdS_n - S_n dT + PdV \tag{22.69}$$

The first law of thermodynamics states that

$$dE = TdS_n - PdV \tag{22.70}$$

Using Eq. (22.70) in Eq. (22.69), the change in the Gibb's free energy becomes

$$dG_n = -S_n dT \tag{22.71}$$

Therefore, the Gibb's free energy in a normal state becomes

$$G_n(P, T) = -\int S_n dT \tag{22.72}$$

Eq. (22.71) allows us to write the entropy as

$$S_n = -\left(\frac{\partial G_n}{\partial T}\right)_P \tag{22.73}$$

In the superconducting state of a metal, the Gibb's free energy is also a function of the applied magnetic field. With the application of a magnetic field **H**, magnetization **M** is produced in the superconducting material. The work done by the magnetic field on the superconductor is $\mathbf{M} \cdot \mathbf{H}$. Therefore, the work done by the superconducting material is the negative of this, that is, $-\mathbf{M} \cdot \mathbf{H}$. If **M** and **H** are in the same direction, then the work done is $-MH$ where M and H are the magnitudes of **M** and **H**. Including the magnetic energy, one gets the Gibb's free energy in the superconducting state as (see Eq. 22.67)

$$G_s(P, T, H) = E - TS_s + PV - MH \tag{22.74}$$

Here S_s is the entropy in the superconducting state. At constant pressure, the change in G_s is given by

$$dG_s = dE - TdS_s - S_s dT + PdV - MdH - HdM \tag{22.75}$$

According to the first law of thermodynamics, the internal energy is given by

$$dE = TdS_s - PdV + HdM \tag{22.76}$$

In Eq. (22.76), the work done due to the increase in the magnetic field in vacuum has been neglected. Using Eq. (22.76) in Eq. (22.75), we get

$$dG_s = -S_s dT - MdH \tag{22.77}$$

From the above equation the entropy in the superconducting state is given by

$$S_s = -\left(\frac{\partial G_s}{\partial T}\right)_P - M\frac{\partial H}{\partial T} \tag{22.78}$$

In a superconductor, the value of magnetization is given by Eq. (22.50). Substituting the value of M from Eq. (22.50) into Eq. (22.77), we obtain

$$dG_s = -S_s dT + \frac{1}{4\pi} HdH \tag{22.79}$$

The Gibb's free energy in a superconducting material is obtained by integrating the above equation, that is,

$$G_s(P,T,H) = -\int S_s dT + \frac{H^2}{8\pi} \tag{22.80}$$

The first term gives the Gibb's free energy of a superconductor in the absence of a magnetic field, written as $G_s(P,T,0)$, while the second term is the contribution due to an applied magnetic field. Therefore, Eq. (22.80) can be written as

$$G_s(P,T,H) = G_s(P,T,0) + \frac{H^2}{8\pi} \tag{22.81}$$

where

$$G_s(P,T,0) = -\int S_s dT \tag{22.82}$$

The magnetic field dependence of G_s is shown in Fig. 22.20. At absolute zero a transition occurs from the superconducting to the normal state at $H = H_c$. Therefore, at $T = 0\,K$, $G_s(P,0,H_c) = G_n(P,0)$ and $S_s = S_n$, that is, the entropy in the superconducting state becomes the same as in the normal state. It is noteworthy that H_c is a function of temperature. So, the difference in the Gibb's energy at $H = 0$ and at $H = H_c$ at $T = 0\,K$ is given by

$$\begin{aligned} \Delta G &= G_s(P,0,H_c) - G_n(P,0) \\ &= \left\{-\int S_n dT + \frac{H_c^2}{8\pi}\right\} - \left\{-\int S_n dT\right\} \end{aligned} \tag{22.83}$$

which gives

$$\Delta G = \frac{H_c^2}{8\pi} \approx 10^{-7}\ \text{eV/electron} \tag{22.84}$$

($k_B T_c \approx 10^{-4}$ eV). It is, therefore, difficult to calculate ΔG because the difference is very small. Note that Eq. (22.84) is equal in magnitude to the stabilization energy at $\mathbf{H}_c$ with $T = 0\,K$ (Section 22.3.6).

22.3.9.1 Entropy

With the knowledge of the Gibb's free energy, one can calculate entropy for both the normal and superconducting states. In the normal state, the entropy S_n is given by Eq. (22.73), while in the superconducting state it is given by Eq. (22.78). But here our interest lies in the discontinuity (abrupt change) in entropy at the normal-superconducting phase boundary. At the phase boundary, $G_n = G_s$ and $H = H_c$, therefore,

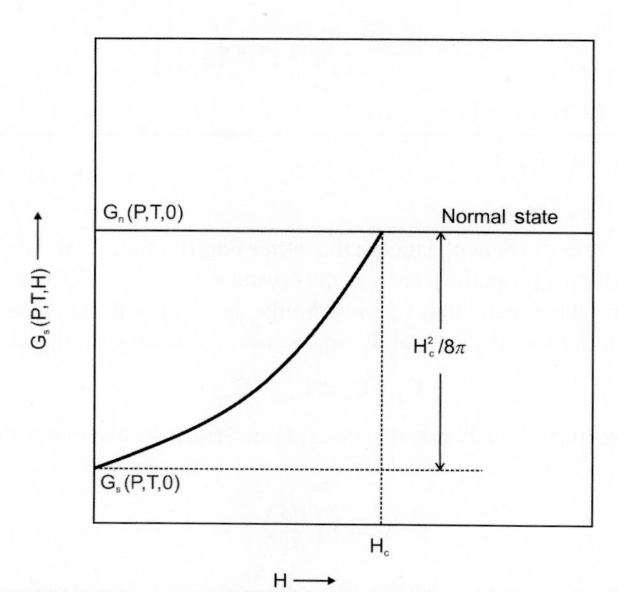

FIG. 22.20 Gibb's free energy of a superconductor $G_s(P,T,H)$ as a function of applied magnetic field $\mathbf{H}$. At absolute zero the solid makes a transition from the superconducting to the normal state at $\mathbf{H} = \mathbf{H}_c$.

$$[dG_n]_{H=H_c} = [dG_s]_{H=H_c} \tag{22.85}$$

Substituting Eqs. (22.71), (22.79) into the above equation, we get

$$S_s - S_n = \frac{1}{8\pi}\frac{d}{dT}H_c^2 \tag{22.86}$$

Because the slope dH_c/dT (see Fig. 22.13) is negative, $S_s - S_n$ is negative. Therefore, the entropy in the normal state is always greater than the entropy in the superconducting state. This shows that the superconducting phase is more ordered than the normal phase.

Alternate Proof

The expression for entropy can also be derived in a slightly different way from Eq. (22.81) for the Gibb's free energy of a superconductor. At $H = H_c$, the Gibb's free energy of a superconductor becomes

$$G_s(P,T,H_c) = G_s(P,T,0) + \frac{H_c^2}{8\pi} \tag{22.87}$$

which is equal to $G_n(P,T)$. Therefore, the difference between $G_s(P,T,H)$ and $G_n(P,T)$ becomes

$$G_s(P,T,H) - G_n(P,T) = \frac{1}{8\pi}\left[H^2 - H_c^2\right] \tag{22.88}$$

With the help of Eq. (22.73), the difference in entropies between the superconducting and normal states can be written as

$$S_s - S_n = -\left[\frac{\partial}{\partial T}(G_s - G_n)\right]_{P,H} \tag{22.89}$$

Substituting Eq. (22.88) into Eq. (22.89), one can write

$$S_s - S_n = \frac{1}{8\pi}\frac{d}{dT}H_c^2 \tag{22.90}$$

which is the same as Eq. (22.86). Here H is assumed to be independent of T.

Problem 22.4

With the help of Eq. (22.86), prove that the temperature dependence of the change in entropy of a superconductor is described by the expression

$$S_s - S_n = -\frac{[H_c(0)]^2}{2\pi T_c^2}T\left[1 - \left(\frac{T}{T_c}\right)^2\right]$$

Further, calculate the value of $S_s - S_n$ at $T = T_c/2$.

22.3.9.2 *Specific Heat*

There are two contributions to the specific heat of a metal: the lattice contribution C_ℓ and the electronic contribution C_e (see Eq. 22.1). One can prove that the lattice properties, such as the crystal structure and Debye temperature, do not change as a metal becomes a superconductor. This means that C_ℓ must be the same in both the superconducting and normal states. Hence, the discontinuous jump in the specific heat at T_c arises from the change in the electronic specific heat, that is,

$$C_s - C_n = C_{es} - C_{en} \tag{22.91}$$

The discontinuous jump in the specific heat at T_c can also be explained from the behavior of the entropy. The specific heat is defined by

$$C = T\left(\frac{\partial S}{\partial T}\right)_V \tag{22.92}$$

From Fig. 22.4 it is evident that there is a sudden change in the entropy at T_c and that the entropy below T_c decreases faster with respect to T in the superconducting state. In other words, the slope dS/dT increases as the transition from the normal to the superconducting state takes place. From this fact it is evident that some extra electron order must appear in the superconducting state.

Here our interest lies in the electronic specific heat at the normal-superconducting phase boundary. The change in electronic specific heat is given by

$$C_s - C_n = T\frac{\partial}{\partial T}(S_s - S_n) \tag{22.93}$$

Using Eq. (22.86) in the above expression, one gets

$$C_s - C_n = \frac{T}{4\pi}\left[\left(\frac{dH_c}{dT}\right)^2 + H_c\frac{d^2H_c}{dT^2}\right] \tag{22.94}$$

Eq. (22.94) gives the general expression for the change in specific heat in the presence of a magnetic field. Note that here the temperature T is less than T_c. From Eq. (22.43), it is straightforward to write

$$\frac{dH_c}{dT} = -2H_c(0)\frac{T}{T_c^2}, \quad \frac{d^2H_c}{dT^2} = -\frac{2H_c(0)}{T_c^2} \tag{22.95}$$

Substituting the above derivatives along with the value of $H_c(T)$ from Eq. (22.43) into Eq. (22.94), we get

$$C_s - C_n = \frac{[H_c(0)]^2}{2\pi T_c}\left[3\frac{T^3}{T_c^3} - \frac{T}{T_c}\right] \tag{22.96}$$

Eq. (22.96) gives the temperature dependence of the change in electronic specific heat below T_c. At $T = T_c$, the above expression reduces to

$$C_s - C_n = \frac{[H_c(0)]^2}{\pi T_c} \tag{22.97}$$

To estimate the jump in specific heat at T_c consider Al metal for which $\gamma_e = 1.35 \times 10^4\ \mathrm{ergs/mol\,K^2}$ in the normal state and $T_c = 1.18\,\mathrm{K}$. At T_c, in the presence of magnetic field $H_c(0) = 99\,\mathrm{gauss}$, one gets

$$\frac{\Delta C}{C_n} = \frac{C_s - C_n}{C_n} = \frac{[H_c(0)]^2/\pi T_c}{\gamma_e T_c} = 0.17 \tag{22.98}$$

22.3.9.3 *First-Order and Second-Order Phase Transitions*

It has been found that S_n is equal to S_s at T_c. Using Eq. (22.73) for entropy, one can write for the superconducting-normal phase transition at T_c,

$$\frac{\partial G_n}{\partial T} = \frac{\partial G_s}{\partial T} \tag{22.99}$$

A phase transition is second order if both G and $\partial G/\partial T$ are continuous at the transition temperature. A second-order transition exhibits two important characteristics: first there is no latent heat involved and second there is a jump in the specific heat at the transition. According to the second law of thermodynamics, heat is given by $dQ = T\,dS$ in a reversible process. Therefore, a small change in latent heat dL for the superconducting-normal transition is given as

$$\begin{aligned} dL &= (dQ)_n - (dQ)_s \\ &= T(dS_n - dS_s) \end{aligned} \tag{22.100}$$

Therefore, the latent heat L is given by

$$L = T(S_n - S_s) \tag{22.101}$$

The first characteristic follows from the fact that $S_s = S_n$ at T_c, yielding $L = 0$ at the transition. The second characteristic is already proved by the fact that there is a discontinuous jump in specific heat at T_c. Therefore, the superconducting-normal state transition is of second order.

Though there is no latent heat when a metal undergoes the superconducting-normal transition in the absence of a magnetic field, the latent heat is finite in the presence of a magnetic field. Substituting the value of $S_n - S_s$ from Eq. (22.86) into Eq. (22.101), we find

$$L = -\frac{1}{4\pi} T \left(H_c \frac{dH_c}{dT} \right) \tag{22.102}$$

In the absence of a magnetic field, the transition occurs at T_c and $H_c(T_c) = 0$; hence, no latent heat is involved. In the presence of an applied magnetic field the transition occurs at some lower temperature T at which $H_c(T) > 0$. Thus, latent heat arises because at temperatures between 0 K and T_c, the entropy of the normal state is greater than that of the superconducting state. So, latent heat must be supplied if the transition is to take place from the superconducting-normal state at constant temperature T. In the presence of an applied magnetic field, therefore, the superconducting-normal transition is first order, that is, although G is continuous, $\partial G/\partial T$ is not.

22.3.10 Bardeen-Cooper-Schrieffer (BCS) Theory

A rigorous mathematical treatment of the Bardeen-Cooper-Schrieffer (BCS) theory (Bardeen, Cooper, & Schrieffer, 1957) is not in the scope of this book as it requires knowledge of quantum field theory. Therefore, the BCS theory will be described qualitatively. Let us first restate the problem of superconductivity. A microscopic theory of superconductivity must be able to explain the following facts:

1. In superconducting materials there is a drastic change in the behavior of the conduction electrons, which is marked by the appearance of long-range order (evident from the existence of positive surface energy) and a gap in their energy spectrum[1] ($\approx 10^{-4}$ eV).
2. The crystal lattice does not show any change in its properties, but nevertheless plays an important role in establishing superconductivity because T_c depends on the atomic mass (isotopic effect).
3. The transition from the superconducting to normal state is a phase transition of second order.

The long-range order noted in point 1 clearly means that the electrons must behave as if there were no interaction between them and that they are randomly oriented. But the conduction electrons in a metal are known to interact strongly through the Coulomb repulsive forces. It is surprising to note that the ordinary free-electron theory, in which the Coulomb repulsion is neglected, works well in metals and semiconductors. Moreover, there is no known mechanism by which the repulsive interaction can lead to an energy gap. On the other hand, an attractive interaction between the electrons can lead to an energy gap. Further, to yield an energy gap of the correct order ($\approx 10^{-4}$ eV), the attractive interaction between the electrons should be weak. The apparent lack of any mechanism for a weak attractive interaction between electrons is the main stumbling block in the way of any microscopic theory of superconductivity.

22.3.10.1 Electron-Phonon Interactions

Electrons in a perfect lattice may be represented by waves, which, at absolute zero, propagate freely through the lattice without any attenuation in the same way as an electric wave can pass along a lossless periodic filter without attenuation. However, if the periodicity of the lattice were destroyed by thermal vibrations, then the electrons would interact with the lattice, yielding a finite electron-ion (electron-lattice) interaction. It is this electron-ion interaction that determines the resistivity of pure metals and semiconductors at room temperature. Because both the energy and momentum must be conserved when an electron is scattered, one of the vibrational modes of the lattice must be excited in the scattering process. This vibrational mode is quantized and an emission (or absorption) of a phonon takes place. Therefore, the electron-ion interaction is usually called the electron-phonon (e-p) interaction.

In 1950, Frohlich postulated the concept of an e-p interaction that is able to couple two electrons together in such a way that they behave as if there were a direct interaction between them. In this interaction one electron emits a phonon, which is then immediately absorbed by another electron. Frohlich was able to show that such an electron-phonon interaction could give rise to a weak attraction between the electrons, which may lead to an energy gap of the right order of magnitude. The Frohlich interaction is schematically represented in Fig. 22.21 in which the straight lines represent electron paths and the

1. It should be noted, however, that there is a very special category of the superconductor indium containing about 1% iron, which does not possess any energy gap as revealed by tunneling experiments. It is believed that in these superconductors, the density of states at the Fermi level is minimum, but there is no actual gap.

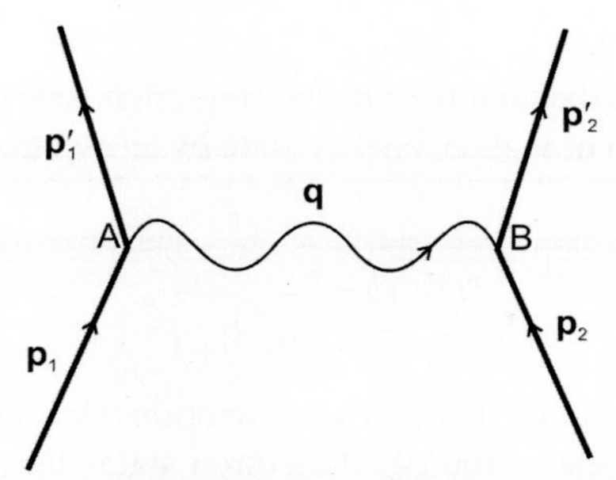

FIG. 22.21 The Feynman diagram for the e-p interaction in a superconductor. $\mathbf{p}_1$ and $\mathbf{p}_2$ represent the momenta for two electrons before interaction, while $\mathbf{p}_1'$ and $\mathbf{p}_2'$ are the corresponding momenta after interaction. Here $\mathbf{q}$ is the momentum (in units of ħ) of a phonon emitted at A and then instantaneously absorbed at B.

wavy line represents a phonon. During the process of emission of a phonon, momentum is conserved, and this can be written as

$$\mathbf{p}_1 = \mathbf{p}_1' + \mathbf{q} \tag{22.103}$$

where $\mathbf{p}_1$ and $\mathbf{p}_1'$ are the momenta of an electron at A before and after scattering and $\mathbf{q}$ is the phonon momentum (in units of $\hbar$), which is given in magnitude by $q = h\nu_{\mathbf{q}}/v_s$ where $\nu_{\mathbf{q}}$ is the frequency of the phonon with momentum $\mathbf{q}$ and v_s the velocity of sound. In the same way, when the emitted phonon is absorbed by the second electron at B, the momentum of the electron changes from $\mathbf{p}_2$ to $\mathbf{p}_2'$ such that

$$\mathbf{p}_2 + \mathbf{q} = \mathbf{p}_2' \tag{22.104}$$

Adding Eqs. (22.103), (22.104), we get

$$\mathbf{p}_1 + \mathbf{p}_2 = \mathbf{p}_1' + \mathbf{p}_2' \tag{22.105}$$

Eq. (22.105) shows that momentum is conserved between the initial and final states, as expected. But the energy does not have to be conserved between the initial and intermediate states, that is, the state in which the first electron has emitted a phonon but the second electron has not yet absorbed it, or between the intermediate and final states. This is because there is an uncertainty relationship between energy and time defined as

$$\Delta E \Delta t \approx \hbar \tag{22.106}$$

The lifetime of the intermediate state Δt is very short, which gives a very large uncertainty in energy ΔE. Therefore, the energy does not have to be conserved in the emission and absorption processes. Such processes, in which energy is not conserved, are known as *virtual processes*. The virtual emission of a phonon is possible only if there is a second electron ready to absorb it almost instantaneously.

Let E_1 and E_1' be the energies of the first electron before and after the virtual emission of a phonon. The quantum mechanical treatment of the process shows that if

$$E_1 - E_2 = h\nu_{\mathbf{q}} \tag{22.107}$$

then the process of emission of a phonon and its subsequent absorption gives rise to an attractive interaction between the two electrons. One can also view the whole process as follows: An electron interacts with an ion of the lattice through an attractive electron-ion interaction. In this process an electron loses energy by emitting a phonon, which is immediately absorbed by the lattice ion producing a distortion in the lattice. This lattice distortion is seen by another electron via the electron-ion interaction and the lattice ion under consideration loses energy by emitting a phonon, which is immediately absorbed by the second electron. This indirect electron-electron interaction via the ion is attractive in nature. Of course, the Coulomb repulsive interaction between the two electrons also exists. But if the indirect electron-electron attractive interaction exceeds the repulsive Coulomb interaction, it yields a weak attractive interaction.

Using the e-p interaction, Frohlich predicted the isotopic effect before it had been discovered experimentally. The fact that the e-p interaction is responsible for superconductivity also explains why superconductors are not good conductors in the normal state, namely, because the e-p interaction is a measure of resistivity. For example, Pb with a reasonably high T_c must have a fairly strong e-p interaction and, therefore, be a poor conductor at room temperature. On the other hand, noble metals (Cu, Ag, and Au) are very good conductors at room temperature and, therefore, must be characterized by a weak e-p interaction. It has been found that the noble metals do not become superconductors, even at the lowest temperature yet attained.

22.3.10.2 Cooper Pairs

Let us first consider a normal metal in which the conduction electrons are nearly free and individually possess energy E and momentum **p**. The probability of occupation of a given energy state by an electron is given by the Fermi-Dirac distribution function.

$$f(E, T) = \frac{1}{e^{\frac{E-E_F}{k_B T}} + 1} \tag{22.108}$$

where E_F is the Fermi energy. At absolute zero, the Fermi-Dirac function takes the form of a step function, as shown by the *dashed line* in Fig. 9.6. In the momentum space, the filled electron states lie in the Fermi sphere with radius p_F where $p_F = (2m_e E_F)^{1/2}$. Suppose two extra electrons are added to a metal at absolute zero. They are forced by the Pauli exclusion principle to occupy states with $p > p_F$, as shown in Fig. 22.22. Cooper has shown that if an attractive interaction between the two electrons exists, however weak, they are able to form a bound state with a total energy less than $2E_F$. To prove this fact some elementary ideas from quantum mechanics will be used here without performing a rigorous mathematical derivation.

Consider two noninteracting electrons with positions $\mathbf{r}_1$ and $\mathbf{r}_2$ and momenta $\mathbf{p}_1$ and $\mathbf{p}_2$. Let $|\psi(\mathbf{r}_1, \mathbf{p}_1; \mathbf{r}_2, \mathbf{p}_2)\rangle$ be the two electron wave function that gives the probability that one electron is at $\mathbf{r}_1$, $\mathbf{p}_1$, while the other at $\mathbf{r}_2$, $\mathbf{p}_2$. $|\psi(\mathbf{r}_1, \mathbf{p}_1; \mathbf{r}_2, \mathbf{p}_2)\rangle$ may be taken as the product of the single-electron wave functions $|\psi(\mathbf{r}_1, \mathbf{p}_1)\rangle$ and $|\psi(\mathbf{r}_2, \mathbf{p}_2)\rangle$ and can be written mathematically as

$$|\psi(\mathbf{r}_1, \mathbf{p}_1; \mathbf{r}_2, \mathbf{p}_2)\rangle = |\psi(\mathbf{r}_1, \mathbf{p}_1)\rangle\, |\psi(\mathbf{r}_2, \mathbf{p}_2)\rangle \tag{22.109}$$

As the electrons are free, the wave functions will simply be plane waves or, more precisely, Bloch waves. If there is an interaction between the pair of electrons, it causes repeated scattering of the electrons accompanied by changes in their momenta. Therefore, the two-electron wave function becomes a mixture of the wave functions obtained after each scattering, which comprise a wide range of momenta. The two-electron wave function then has the form

$$\begin{aligned} |\psi(\mathbf{r}_1, \mathbf{r}_2)\rangle &= \sum_{i,j} a_{ij} \left|\psi\left(\mathbf{r}_1, \mathbf{p}_i; \mathbf{r}_2, \mathbf{p}_j\right)\right\rangle \\ &= \sum_{i,j} a_{ij} |\psi(\mathbf{r}_1, \mathbf{p}_i)\rangle \left|\psi\left(\mathbf{r}_2, \mathbf{p}_j\right)\right\rangle \end{aligned} \tag{22.110}$$

The wave function $|\psi(\mathbf{r}_1, \mathbf{r}_2)\rangle$ gives the probability of finding an electron at $\mathbf{r}_1$ when there is another electron at $\mathbf{r}_2$ regardless of their momenta. Further, $|\psi(\mathbf{r}_1, \mathbf{r}_2)\rangle$ contains wave functions $|\psi(\mathbf{r}_1, \mathbf{p}_i)\rangle$ and $|\psi(\mathbf{r}_2, \mathbf{p}_j)\rangle$ such that in each electron scattering the individual momenta $\mathbf{p}_i$ and $\mathbf{p}_j$ are constantly changing but the total momentum is conserved, that is,

$$\mathbf{p}_i + \mathbf{p}_j = \mathbf{P} \tag{22.111}$$

where **P** is constant (Fig. 22.22). The element $|a_{ij}|^2$ gives the probability of finding the electrons at any instant with individual momenta $\mathbf{p}_i$ and $\mathbf{p}_j$. In the actual scattering process, the electrons experience a mutual interaction potential. If the

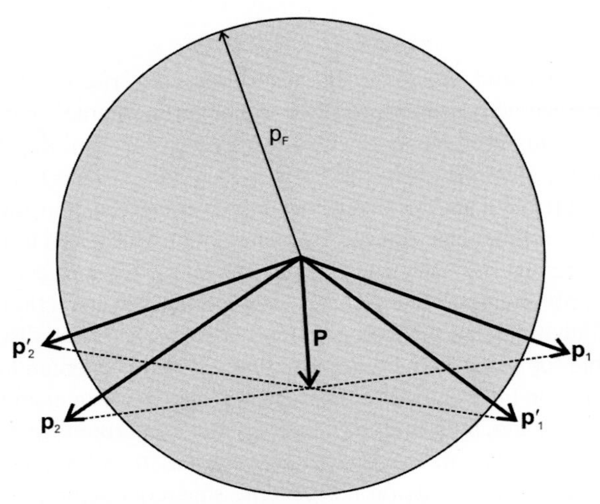

FIG. 22.22 The Fermi sphere with radius p_F in which all the states are filled. Two electrons are added to the free electron gas whose momenta $\mathbf{p}_1$ and $\mathbf{p}_2$ lie outside the Fermi sphere. These two electrons suffer interactions in which their momenta get changed to $\mathbf{p}_1'$ and $\mathbf{p}_2'$ in such a way that the total momentum of the two electrons **P** remains unchanged.

mutual interaction is attractive, then the potential energy resulting from it is negative. Hence, over a period of time during which there are many scattering events, the energy of the two electrons decreases by the time average of this negative potential energy and the amount of decrease is proportional to the number of scattering events that take place, that is, proportional to the number of ways in which two terms can be chosen from the wave function $|\psi(\mathbf{r}_1, \mathbf{r}_2)\rangle$. It is a good approximation to assume that each scattering event contributes an equal amount of $-V$ to the potential energy. In quantum mechanical language, $-V$ is the matrix element of the attractive interaction potential connecting the two electron states, which have same total momentum **P** that is assumed to be independent of the individual momenta of the electrons.

So far nothing has been said about the nature of the interaction, apart from requiring it to be attractive. If it is the electron-ion interaction arising from the actual emission and absorption of a phonon, it turns out from the detailed theory that the probability of scattering is appreciable only if the energy deficit between the initial and intermediate states $(E_1' + h\nu_{\mathbf{q}} - E_1)$ is small, that is,

$$E_1 - E_1' \approx h\nu_{\mathbf{q}} \tag{22.112}$$

If two electrons are added to a metal at absolute zero, then they will individually possess energy greater than E_F as all the states up to E_F are already filled: E_1 and E_1' are both greater than E_F and, at the same time, $E_1 - E_1' \approx h\nu_{\mathbf{q}}$. So, the lower energy state between E_1 and E_1' lies within energy $h\nu_{\mathbf{q}}$ of E_F where $\nu_{\mathbf{q}}$ is an "average" phonon frequency and is about half the Debye frequency (see Fig. 22.23). The difference in energy is given by

$$E_1 - E_1' = \frac{p_1^2}{2m_e} - \frac{p_1'^2}{2m_e} = h\nu_{\mathbf{q}} \tag{22.113}$$

The above equation can be simplified to give

$$\Delta p = \frac{m_e h\nu_{\mathbf{q}}}{p_F} \tag{22.114}$$

where

$$\Delta p = p_1 - p_1' \tag{22.115}$$

$$p_1 = p_1' \approx p_F \tag{22.116}$$

Therefore, the momenta p_1 and p_1' must lie within Δp of the Fermi momentum p_F. Because the allowed values of $\mathbf{p}_i$ and $\mathbf{p}_j$ satisfy Eq. (22.111), the allowed values of $\mathbf{p}_i$ and $\mathbf{p}_j$ will be in the shaded region of Fig. 22.24, that is, these momenta begin or end in a ring whose cross section is the shaded region. The number of such pairs is proportional to the volume of this ring. The volume (and hence the number of such pairs) becomes maximum when $\mathbf{P} = 0$, in which case the ring becomes a spherical shell of thickness Δp and Eq. (22.111) gives

$$\mathbf{p}_i = -\mathbf{p}_j \tag{22.117}$$

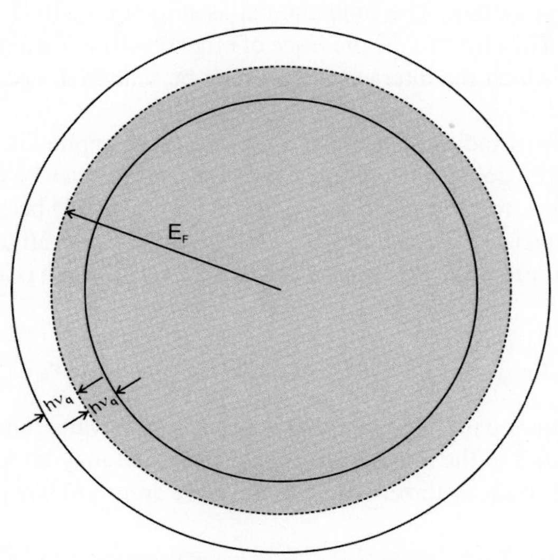

FIG. 22.23 The Fermi sphere with radius E_F is shown by the *dashed line*. The difference between the energies before interaction E_1 and after interaction E_1' lie within energy $h\nu_{\mathbf{q}}$ of the Fermi energy E_F where $\nu_{\mathbf{q}}$ is the phonon frequency.

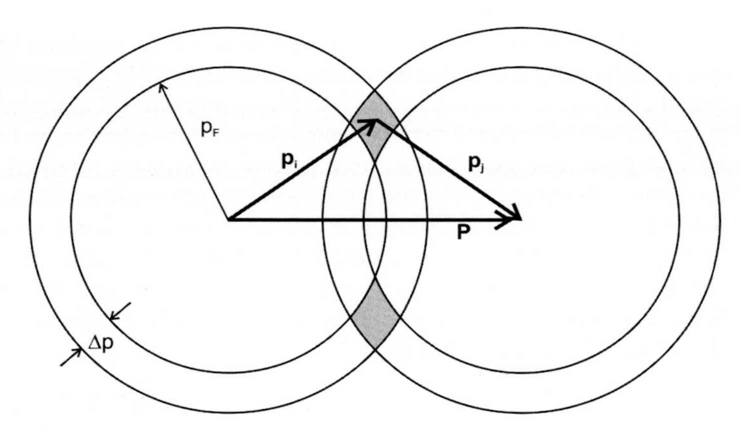

FIG. 22.24 $\mathbf{p}_i$ and $\mathbf{p}_j$ are the momenta of two interacting electrons, when both lie in states within a shell of thickness Δp above p_F. The total momentum of the two electrons is given by $\mathbf{P}=\mathbf{p}_i+\mathbf{p}_j$ (Eq. 22.111).

Thus, the largest number of allowed scattering processes, yielding the maximum decrease of the energy, is obtained by pairing electrons with equal and opposite momenta. Further, a quantum mechanical treatment shows that if the two electrons have opposite spins, then the matrix elements V are largest and, therefore, the decrease in the energy is maximum. With the condition (22.117) the wave function $|\psi(\mathbf{r}_1,\mathbf{r}_2)\rangle$ with the lowest possible energy is given by (from Eq. 22.110)

$$\begin{aligned} |\psi(\mathbf{r}_1,\mathbf{r}_2)\rangle &= \sum_i a_i |\psi(\mathbf{r}_1,\mathbf{p}_i\uparrow;\mathbf{r}_2,-\mathbf{p}_i\downarrow)\rangle \\ &= \sum_i a_i |\psi(\mathbf{r}_1,\mathbf{p}_i\uparrow)\rangle |\psi(\mathbf{r}_2,-\mathbf{p}_i\downarrow)\rangle \end{aligned} \tag{22.118}$$

Here we have written a_i instead of a_{ii}. The wave function $|\psi(\mathbf{r}_1,\mathbf{r}_2)\rangle$ describes what is known as a *Cooper pair*.

The total energy of an electron pair is the sum of the kinetic and potential energies. As the two electrons added to a metal have $p > p_F$, the kinetic energy of the electron pair is $> 2E_F$ (each electron has kinetic energy greater than E_F). If the interaction between the electron pair is attractive, the resulting potential energy will be negative, thereby lowering the total energy in comparison with the kinetic energy. In the case of a Cooper pair, the decrease in energy due to the attractive interaction exceeds the amount by which the kinetic energy is in excess of $2E_F$. Thus, the total energy E_{cp} of a Cooper pair is less than $2E_F$, which results in a small energy gap of magnitude equal to $(1/2)(2E_F - E_{cp})$.

22.3.10.3 Generalization of Cooper Pair Formation

The problem treated by Cooper is a somewhat unrealistic one in the sense that it involves only two interacting electrons, whereas in a metal, there are about 10^{23} conduction electrons per cm^3. Cooper's theory is clearly an improvement over those that completely ignore interactions among the electrons. In the generalization of the Cooper theory, one should also include interactions between three or more electrons. Bardeen, Cooper, and Schrieffer showed that Cooper's simple idea could be applied to a many-electron interacting system. The fundamental assumption of BCS theory is that only the interactions between electron pairs are important. The effect of the presence of other electrons on any one pair is simply to limit, through the Pauli principle, those states into which the interacting pair may be scattered, because some of the states are already occupied.

In a metal, electrons lie in a sphere of radius p_F. Cooper's result may be applied to electrons in a metal with momenta infinitesimally below p_F as these may be transformed into a Cooper pair. If this can be done for one electron pair, the same can also be done for many electron pairs, resulting in lower energy. This is possible because more than one pair of electrons can be represented by the same function $|\psi\rangle$, as given by Eq. (22.118). In this case, all the superconducting electrons can be represented together by the many-electron wave function $|\psi_{n_p}(\mathbf{r}_1,\mathbf{r}_2,\ldots,\mathbf{r}_{n_p})\rangle$, which is a product of the pair wave functions, that is,

$$\left|\psi_{n_p}\left(\mathbf{r}_1,\mathbf{r}_2,\ldots,\mathbf{r}_{n_p}\right)\right\rangle = |\psi(\mathbf{r}_1,\mathbf{r}_2)\rangle|\psi(\mathbf{r}_3,\mathbf{r}_4)\rangle\ldots\left|\psi\left(\mathbf{r}_{n_p-1},\mathbf{r}_{n_p}\right)\right\rangle \tag{22.119}$$

where $n_p/2$ is the total number of Cooper pairs. The fact that many-electron wave functions can be written in the form of Eq. (22.119) shows that there is no limit to the number of Cooper pairs that may be represented by a wave function. The Cooper pair behaves as a composite particle with zero momentum, zero spin, and having a mass twice that of the electron.

Thus, a Cooper pair behaves as a Boson particle obeying the Bose-Einstein statistics. The property that all of the Cooper pairs are in the same quantum state with the same energy will prove to be of great importance.

It is well established now that the electrons that have momenta within the range Δp, given by Eq. (22.114), about p_F, are coupled together to form Cooper pairs. So, the electrons with momenta $p < p_F$ may be raised to states with $p > p_F$ to form Cooper pairs, resulting in a decrease in the total energy. The electrons may also go from $p > p_F$ to states with $p < p_F$. There is a limit to the number of electrons that can be raised from $p < p_F$ to states with $p > p_F$ to form Cooper pairs and it can be understood as follows.

A pair of electrons may be scattered from $(\mathbf{r}_1, \mathbf{p}_i \uparrow; \mathbf{r}_2, -\mathbf{p}_i \downarrow)$ to $(\mathbf{r}_1, \mathbf{p}_j \uparrow; \mathbf{r}_2, -\mathbf{p}_j \downarrow)$ if the states $(\mathbf{r}_1, \mathbf{p}_i \uparrow; \mathbf{r}_2, -\mathbf{p}_i \downarrow)$ are occupied and the states $(\mathbf{r}_1, \mathbf{p}_j \uparrow; \mathbf{r}_2, -\mathbf{p}_j \downarrow)$ are empty. As more and more electrons with $p > p_F$ form Cooper pairs, the chance of finding the states $(\mathbf{r}_1, \mathbf{p}_j \uparrow; \mathbf{r}_2, -\mathbf{p}_j \downarrow)$ empty becomes progressively smaller and smaller. So, the number of scattering processes that may take place is reduced with a constant decrease in the magnitude of the negative potential energy. Eventually a condition is reached in which the decrease in potential energy is insufficient to outweigh the increase in kinetic energy, and it is no longer possible to further lower the total energy of the electrons by forming Cooper pairs. There will be an optimum arrangement, which gives the lowest overall energy, and this arrangement can be described by specifying the probability h_i of the pair state $(\mathbf{r}_1, \mathbf{p}_i \uparrow; \mathbf{r}_2, -\mathbf{p}_i \downarrow)$, which is related to the coefficient a_i occurring in Eq. (22.118). The Pauli principle as applied to Cooper pairs requires that $h_i \leq 1$ and, according to the BCS theory, it is given by

$$h_i = \frac{1}{2}\left[1 - \frac{E_i - E_F}{\left\{(E_i - E_F)^2 + \Delta^2\right\}^{1/2}}\right] \tag{22.120}$$

where $E_i = p_i^2/2m_e$ and the positive square root is taken. The quantity Δ, which has the dimensions of energy, turns out to be of fundamental importance and is given by

$$\Delta = 2h\nu_{\mathbf{q}} e^{-\left\{\frac{1}{N_e(E_F)V}\right\}} \tag{22.121}$$

where $N_e(E_F)$ is the density of electron states at E_F and $-V$ are the matrix elements of the scattering interaction.

The probability $h_i(E)$, as given by the BCS theory for the state $|\psi_{n_p}(\mathbf{r}_1, \mathbf{r}_2, \ldots, \mathbf{r}_{n_p})\rangle$, is shown in Fig. 22.25 and gives the lowest energy (the ground state). The *dashed line* shows the probability of a single-electron state at 0K. The important feature of the figure is that even at absolute zero, the momentum distribution of the electrons in a superconductor does not show an abrupt discontinuity as it does in the case of normal metals. The ground state in a superconductor is the state of lowest energy, in which all the electrons with momenta within the range of Δp about p_F are coupled to form Cooper pairs. This state is often referred to as a condensed state because the electrons are bound together to form a state of lower energy, as happens to the atoms of a gas when they condense to form a liquid.

22.3.10.4 The Energy Gap

A superconductor may be excited to a higher state either by raising the temperature or by illuminating it with light of an appropriate wavelength. In a superconductor, there are Cooper pairs in the momentum range of Δp around p_F (E_F) and each pair has zero total momentum and zero spin. The energy of a pair cannot be increased simply by increasing the momenta of the electrons and at the same time maintaining the condition that their momenta are equal and opposite. However, a pair may break up with the constituent electrons no longer having equal and opposite momenta, in which case the attractive potential energy resulting from their interaction becomes almost negligible. The electrons behave almost like free electrons and for this reason are referred to as "quasiparticles." Furthermore, it is not meaningful to talk of the momenta of the individual electrons before the pair is broken up because the momenta of the individual electrons cannot be specified. The second question is how much energy is necessary to break up a pair so as to produce electrons, or more precisely quasiparticles with unequal momenta $\mathbf{p}_i$ and $\mathbf{p}_j$. According to the BCS theory, the amount of energy required is

$$E = E_i + E_j = \left[(E_i - E_F)^2 + \Delta^2\right]^{1/2} + \left[\left(E_j - E_F\right)^2 + \Delta^2\right]^{1/2} \tag{22.122}$$

Hence, the minimum energy required is 2Δ when $p_i = p_j = p_F$ or $E_i = E_j = E_F$. There is, thus, an energy gap of magnitude 2Δ in the excitation spectrum of a superconductor, and radiation of frequency ν is absorbed if $h\nu > 2\Delta$. The BCS theory is also able to explain macroscopic properties, such as critical temperature, critical magnetic field, and latent heat, among others.

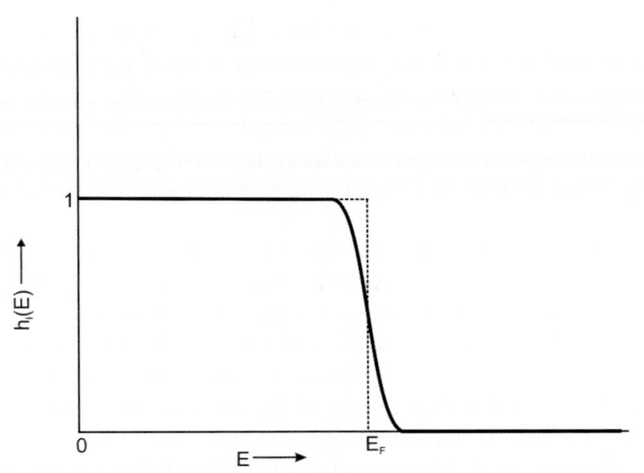

FIG. 22.25 The probability $h_i(E)$ as a function of energy E for the ground state of a superconductor as given by the BCS theory.

22.3.11 Criterion for the Existence of Superconductivity

An important question is whether all metals would exhibit superconductivity if cooled to a low enough temperature. According to the BCS theory this may not necessarily be so. Metals will show superconducting behavior only if the net interaction resulting from the competition between the repulsive Coulomb and e-p interactions is attractive. This is the reason that good conductors, such as Cu, Ag, and Au, which have a weak e-p interaction, do not exhibit superconductivity.

22.3.12 Why Do Magnetic Impurities Lower T_c?

The sharp decrease in T_c due to the presence of magnetic impurities can be explained in terms of the exchange interactions between the conduction electrons and the impurity atoms. The exchange interactions produce spin-flip scattering, which destroys the time reversal correlation of the Cooper pairs. The scattering time for a spin-flip process is obtained by assuming that the impurity spin $\mathbf{s}_I$ is coupled to the conduction electron spin $\mathbf{s}$ by an exchange interaction integral $J(\mathbf{s}_I \cdot \mathbf{s})$. In contrast, with nonmagnetic impurities, the spin-exchange scattering is not time-reversal invariant. This results in a finite lifetime for the Cooper pairs and, therefore, severely reduces T_c. Furthermore, a remarkable conclusion is arrived that there exists a range of magnetic impurity concentrations above which the energy gap in the excitation spectrum becomes zero, yet the specimen remains a superconductor in the sense of having pair correlation and nonzero T_c. The possibility of a *gapless superconductor* was surprising, at first, because the BCS theory contains the energy gap parameter Δ.

22.4 SUPERCONDUCTING QUANTUM TUNNELING

Consider two metals M_1 and M_2 separated by a thin oxide (insulating) layer between them (see Fig. 22.26A). Such a junction is usually called a metal-oxide-metal (MOM) junction. Quantum mechanical theory gives a finite probability for an electron of metal M_1 to be found on the other side of the oxide layer if the thickness of the oxide layer is on the order of or less than the mean free path of the electron. In other words, under such conditions, an electron may tunnel through the oxide layer at constant energy. Tunneling can, however, occur if there are empty states available for an electron to occupy on the other side of the junction. Electron tunneling can also take place when a thin oxide layer separates a superconductor and a normal metal or two superconductors. Tunneling experiments play an important role in determining the following properties of a superconductor:

1. The energy gap 2Δ in a superconductor.
2. The empirical basis for the e-p interaction spectrum as a function of energy.

22.4.1 Single-Electron Superconducting Tunneling

Superconducting single particle tunneling was discovered by Giaever (1960, 1974). Here the junction consists essentially of two superconductors (namely S_1 and S_2) separated by a thin oxide layer (30 Å) as shown in Fig. 22.26B. The length and width of the junction are denoted by L and W, respectively. To understand the tunneling process the energy band diagrams

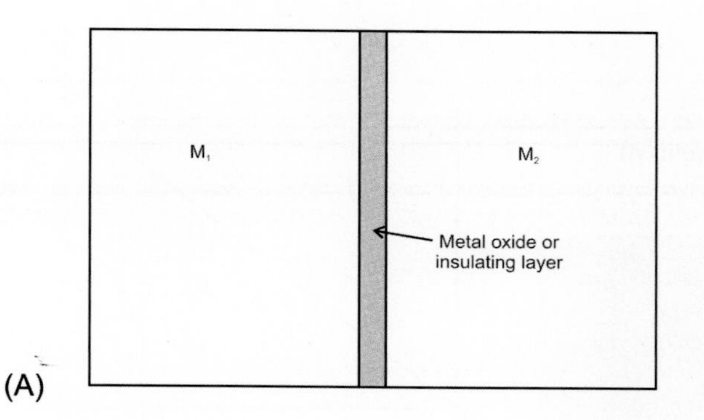

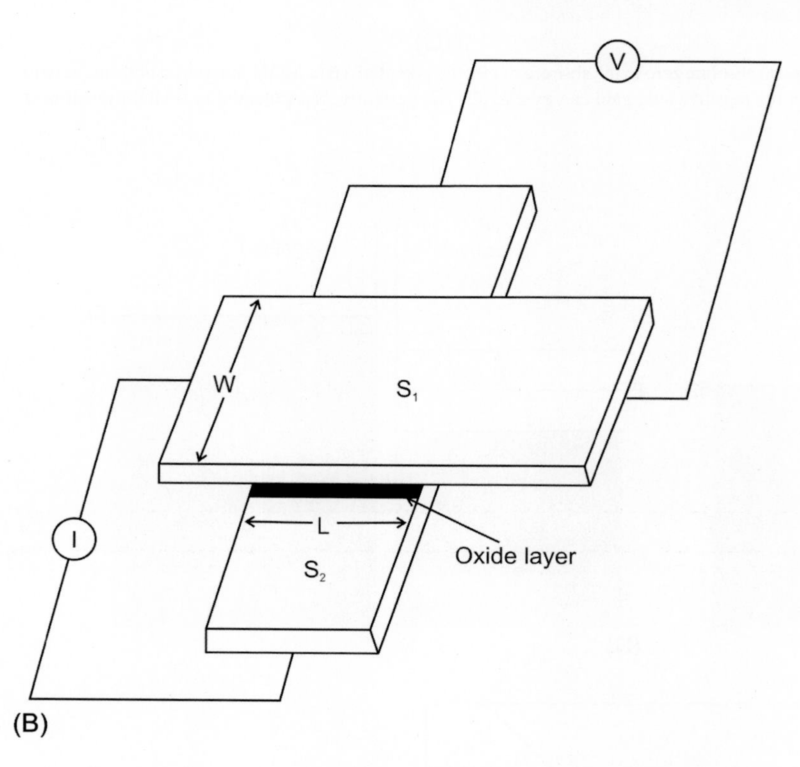

FIG. 22.26 (A) Metal-oxide-metal (MOM) junction where M_1 and M_2 represent two metallic elements. (B) The geometry of a superconductor-oxide-superconductor (SOS) junction. S_1 and S_2 represent two superconductors.

for the different systems will be considered. Fig. 22.27 represents the energy band diagram for a MOM junction. In the equilibrium state the Fermi energy of both the metals is equal (Fig. 22.27A). Were it not so, the electrons would flow from the higher Fermi energy to the lower one to equalize them. When a voltage V is applied to the junction, the electrons on the negative side gain energy eV and, therefore, they tunnel in the direction of the arrow in Fig. 22.27B. The current versus voltage (I-V) characteristic in this case will be linear (see Fig. 22.27C).

The second type of tunneling junction is the metal-oxide-superconductor (MOS) junction for which the energy band diagram is shown in Fig. 22.28A. The Fermi energies for both the superconductor and metal are assumed to be the same in the equilibrium state. The Fermi energy E_F of a superconductor is in the middle of the energy gap. At $T=0\,K$, the energy states at the same energy are either completely empty or filled on the two sides. However, when a voltage V is applied across the oxide junction, the potential energy of the electrons on the negatively charged side is raised; thus, the energy eV separates the Fermi levels on the two sides (Fig. 22.28B). Therefore, at 0 K, a tunneling current flows for $eV \geq \Delta$, which then increases rapidly (see Fig. 22.28C) because a large number of states are available for occupation just above Δ. But when $T > 0\,K$, a few electrons are already in states above E_F on both sides (Fig. 22.29A and B), so some tunnel current flows even for $eV < \Delta$ (see Fig. 22.29C). The variation of current with voltage above the value Δ/e depends on the density of electron states and the tunneling probability.

A superconductor-oxide-superconductor (SOS) junction in the equilibrium state is shown in Fig. 22.30A in which the Fermi energy on both sides has the same value. The application of a voltage raises the potential energy of electrons on the

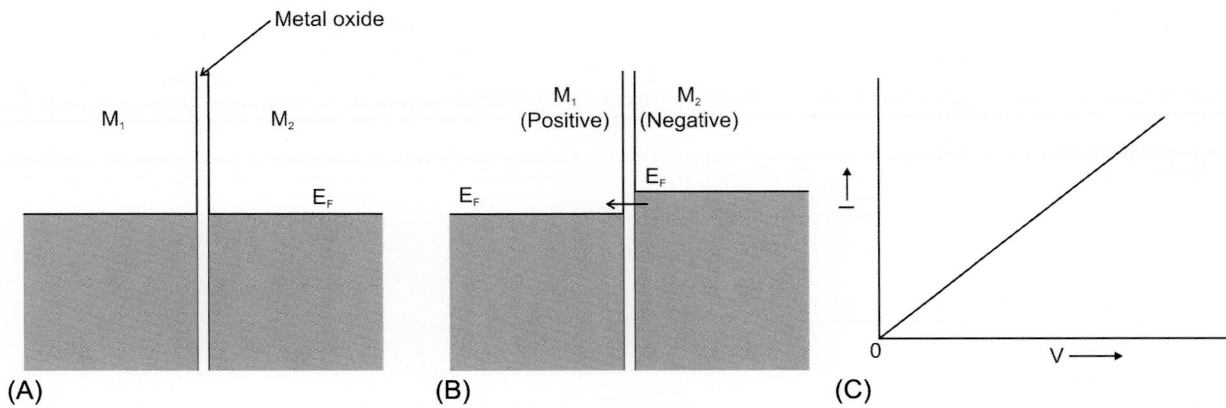

FIG. 22.27 The energy band diagram for (A) a MOM junction at absolute zero in the absence of electric potential. (B) a MOM junction at absolute zero in the presence of electric potential V such that the electrons on the negative side gain energy eV. (C) The current versus potential in a MOM junction at absolute zero, which exhibits linear behavior.

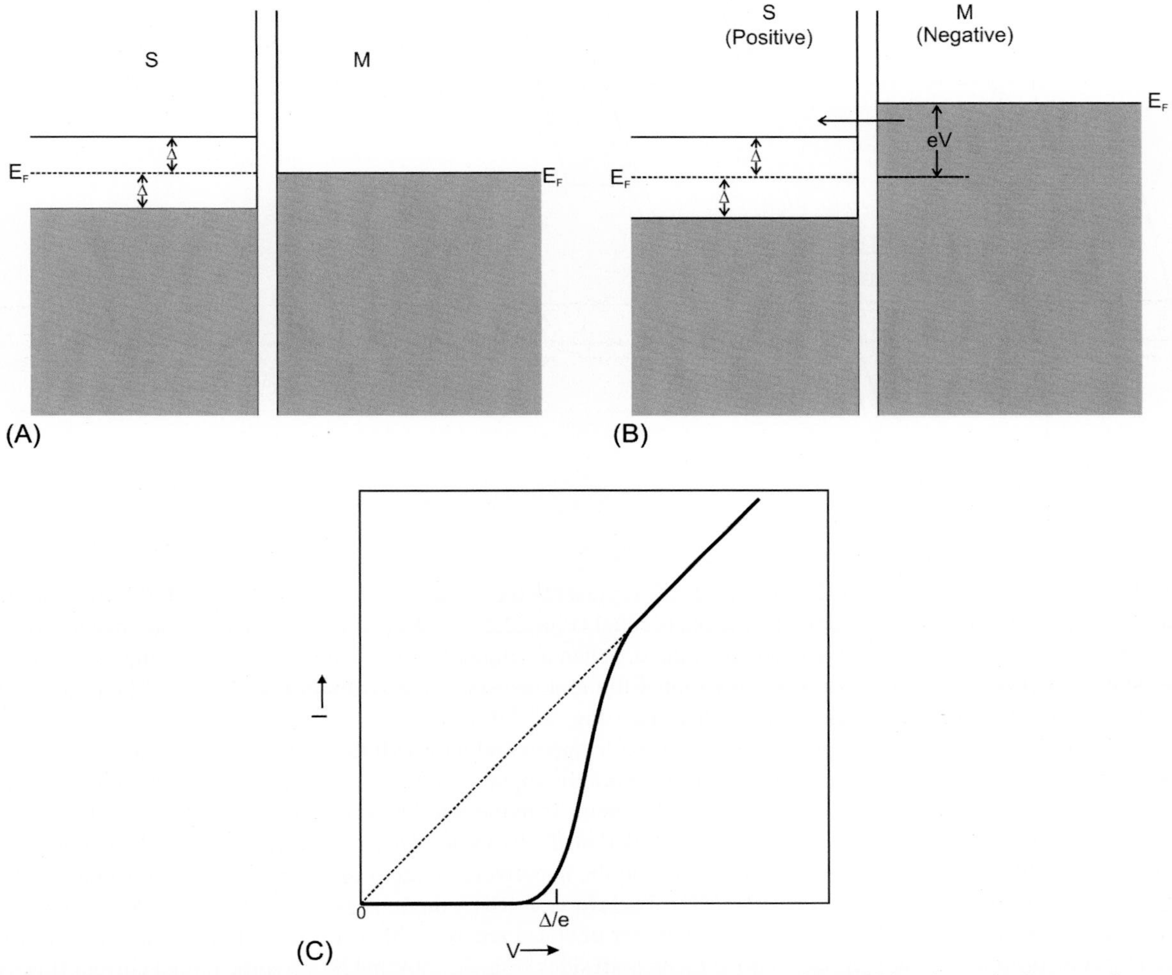

FIG. 22.28 (A) Energy band diagram for a MOS junction at absolute zero. (B) Energy band diagram for a MOS junction at absolute zero when an electric potential V is applied, which raises E_F by eV on the negative side. (C) The current-voltage characteristics of a MOS junction at absolute zero.

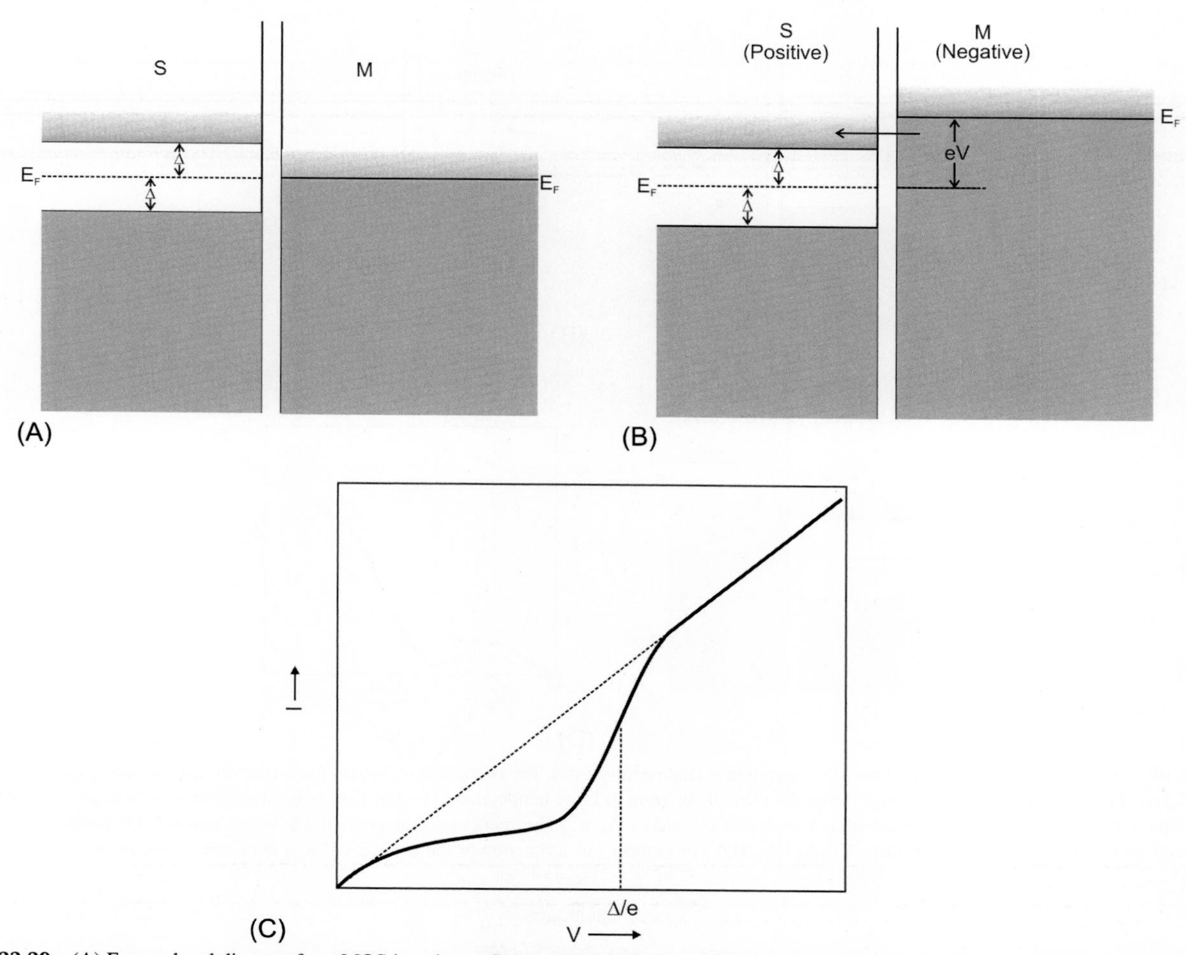

FIG. 22.29 (A) Energy band diagram for a MOS junction at finite temperature. The shade above E_F on the metal side and above the band gap on the superconductor side show excited electrons at finite temperature. (B) MOS junction at finite temperature with applied electric potential V in which the Fermi energy E_F is raised by eV on the negative side. The shaded region above E_F represents the excited electrons at finite temperature. (C) The current-voltage characteristics of a MOS junction.

negatively charged side. At $T > 0\,K$, there are some conduction electrons above E_F in both the superconductors. Therefore, the tunneling current is weak for small voltages and it reaches a maximum for

$$eV_1 = \Delta_1 - \Delta_2 \tag{22.123}$$

(see Fig. 22.30B). In this situation the upper edges of the energy gaps of both the superconductors coincide (when the maxima of the density of states [DOS] of two distributions coincide). When we increase the voltage further, the tunneling current increases much faster for

$$eV_2 = \Delta_1 + \Delta_2 \tag{22.124}$$

and for larger voltages. When the condition given by Eq. (22.124) is satisfied, the bottom of the energy gap of one superconductor coincides with the upper edge of the energy gap of the second superconductor (see Fig. 22.30C). At absolute zero the tunneling current will begin only when Eq. (22.124) is satisfied. Fig. 22.30D shows the I-V characteristics of the SOS junction. From Eqs. (22.123), (22.124) one gets

$$\begin{aligned} \Delta_1 &= \frac{1}{2} e(V_1 + V_2) \\ \Delta_2 &= \frac{1}{2} e(V_2 - V_1) \end{aligned} \tag{22.125}$$

Thus, the tunneling experiment provides the simplest and most direct method for determining the energy gap in superconductors and has been widely used. With the help of tunneling experiments, one can also study the temperature variation of

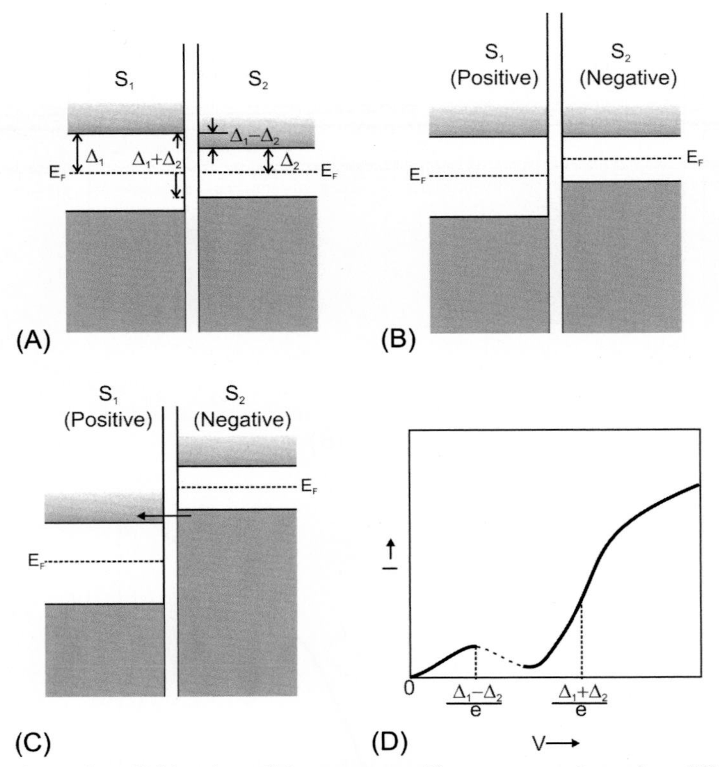

FIG. 22.30 (A) The energy band diagram for a SOS junction at finite temperature. The two superconductors have different energy band gaps with values $2\Delta_1$ and $2\Delta_2$. The shade above the band gaps shows the excited electrons at finite temperature. (B) The energy band diagram of a SOS junction at finite temperature in the presence of an applied potential V such that $eV = \Delta_1 - \Delta_2$. (C) The energy band diagram of a SOS junction at finite temperature in the presence of an applied potential V such that $eV > \Delta_1 + \Delta_2$. (D) The current I as a function of the potential V in a SOS junction at finite temperature.

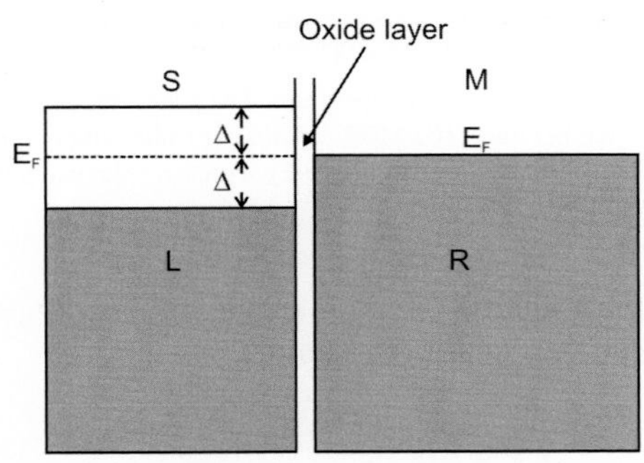

FIG. 22.31 MOS junction in which the material on the right (R) side is a metal and on the left (L) side is a superconductor.

the energy gap in a superconductor. The energy gap in a superconductor can also be measured by other methods, such as ultrasonic attenuation and optical absorption.

Let us evaluate the general expression for the tunneling current in the junction shown in Fig. 22.31 in which L denotes the left-hand side and R the right-hand side of the junction separated by an oxide layer (the two sides may be metals, one may be metal and the other a superconductor, or both may be superconductors). In the superconducting state, the BCS theory gives the energy of a quasiparticle as

$$E = \left[E_{\mathbf{k}}^2 + \{\Delta(E)\}^2\right]^{1/2} \tag{22.126}$$

where $E_{\mathbf{k}} = \hbar^2 k^2/2m_e$ is the energy of each electron in a pair. When the material goes to the superconducting state, the electron states remain the same, that is, there is a one-to-one correspondence between E and $E_{\mathbf{k}}$. Therefore,

$$N_s(E)\,dE = N(E_{\mathbf{k}})\,dE_{\mathbf{k}} \tag{22.127}$$

where $N_s(E)$ and $N(E_{\mathbf{k}})$ are the density of states (DOS) of the superconducting and normal states. To simplify the notation, here and in this remaining chapter we use $N(E_{\mathbf{k}})$ for the electron DOS in place of $N_e(E_{\mathbf{k}})$. Eq. (22.127) gives

$$N_s(E) = N(E_{\mathbf{k}})\frac{dE_{\mathbf{k}}}{dE} \tag{22.128}$$

From Eq. (22.126) one can write

$$E_{\mathbf{k}} = \left[E^2 - \{\Delta(E)\}^2\right]^{1/2} \tag{22.129}$$

Differentiating the above equation, we find

$$\frac{dE_{\mathbf{k}}}{dE} = \frac{E - \Delta(E)\dfrac{d\Delta(E)}{dE}}{\left[E^2 - \{\Delta(E)\}^2\right]^{1/2}} \tag{22.130}$$

Substituting Eq. (22.130) into Eq. (22.128), we get

$$N_s(E) = N(0)\frac{E - \Delta(E)\dfrac{d\Delta(E)}{dE}}{\left[E^2 - \{\Delta(E)\}^2\right]^{1/2}} \tag{22.131}$$

Here N(0) is the value of N(E) at $E=0$ and is assumed to be constant. If Δ is independent of energy, then

$$N_s(E) = N(0)\frac{E}{\left(E^2 - \Delta^2\right)^{1/2}}. \tag{22.132}$$

The electronic DOS for the MOS junction at finite temperature with and without the application of voltage is shown in Fig. 22.32.

The tunneling current can be calculated in terms of the tunneling probability. The probability of an electron tunneling from side L to side R is represented by $p_{L\to R}$ and depends on the following parameters:

1. $p_{L\to R}$ depends on the probability of occupation of state L, given by $f_L(E)$, where $f_L(E)$ is the Fermi-Dirac distribution function for the L state.
2. $p_{L\to R}$ depends on $1 - f_R(E)$, which gives the probability that state R is vacant. Here $f_R(E)$ is the Fermi-Dirac distribution function for the R state.
3. $p_{L\to R}$ depends on the square of the tunneling matrix element T.

The Fermi distribution functions $f_L(E)$ and $f_R(E)$ are given as

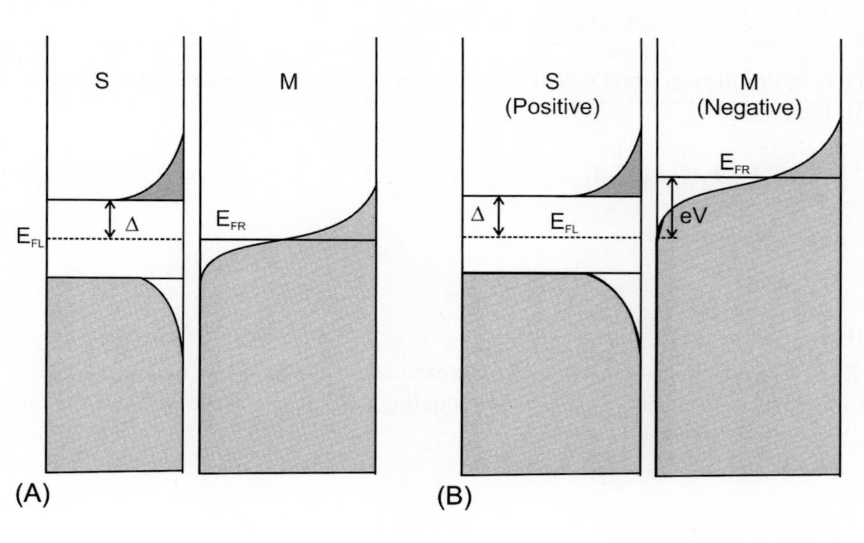

FIG. 22.32 (A) The electronic density of states (DOS) at $T>0\,K$ in a MOS junction in the absence of applied potential. (B) The electronic DOS at $T>0\,K$ when voltage V is applied across the MOS junction.

$$f_L(E) = \left[e^{\left(\frac{E - E_{FL}}{k_B T}\right)} + 1 \right]^{-1} \tag{22.133}$$

$$f_R(E) = \left[e^{\frac{E - E_{FR}}{k_B T}} + 1 \right]^{-1} \tag{22.134}$$

Hence, the probability of an electron tunneling from left to right is given as.

$$p_{L \to R} = \frac{2\pi}{\hbar} |T|^2 f_L(E) N_R(E) [1 - f_R(E)] \tag{22.135}$$

$2\pi/\hbar$ is the constant of proportionality and $N_R(E)$ is the electron DOS on the right side. An electron from the left can occupy any of the vacant states on the right side and the density of vacant states on the right side is given by $N_R(E)\,[1 - f_R(E)]$. The total probability of tunneling from L to R, denoted by $P_{L \to R}$, is obtained by multiplying Eq. (22.135) by $N_L(E)$ and integrating over all energies. The factor $N_L(E)\, f_L(E)$ allows an electron from all possible occupied states on the left to tunnel toward the right. Therefore, one can write

$$P_{L \to R} = \frac{2\pi}{\hbar} \int_{-\infty}^{\infty} |T|^2 N_L(E) N_R(E) f_L(E) [1 - f_R(E)]\, dE \tag{22.136}$$

For simplicity, the matrix element T can be assumed to be a constant, which gives

$$P_{L \to R} = \frac{2\pi}{\hbar} |T|^2 \int_{-\infty}^{\infty} N_L(E) N_R(E) f_L(E) [1 - f_R(E)]\, dE \tag{22.137}$$

In exactly the same way, one can calculate the probability of tunneling from right to left and this is given by

$$P_{R \to L} = \frac{2\pi}{\hbar} |T|^2 \int_{-\infty}^{\infty} N_L(E) N_R(E) f_R(E) [1 - f_L(E)]\, dE \tag{22.138}$$

The tunneling current is obtained by multiplying the probability by the electron charge to write

$$I_{L \to R} = \frac{2\pi e}{\hbar} |T|^2 \int_{-\infty}^{\infty} N_L(E) N_R(E) f_L(E) [1 - f_R(E)]\, dE \tag{22.139}$$

$$I_{R \to L} = \frac{2\pi e}{\hbar} |T|^2 \int_{-\infty}^{\infty} N_L(E) N_R(E) f_R(E) [1 - f_L(E)]\, dE \tag{22.140}$$

Hence, the net tunneling current from right to left is obtained by subtracting Eq. (22.139) from Eq. (22.140), that is,

$$\begin{aligned} I &= I_{R \to L} - I_{L \to R} \\ &= \frac{2\pi e}{\hbar} |T|^2 \int_{-\infty}^{\infty} N_L(E) N_R(E) [f_R(E) - f_L(E)]\, dE \end{aligned} \tag{22.141}$$

The important feature of this formula is that it contains the electron DOS of both sides of the tunneling junction. If voltage V is applied across the junction (Fig. 22.32B), then

$$I = \frac{2\pi e}{\hbar} |T|^2 \int_{-\infty}^{\infty} N_L(E) N_R(E + eV) [f_R(E + eV) - f_L(E)]\, dE \tag{22.142}$$

Let us calculate the tunneling current I for different cases.

22.4.1.1 MOM Tunneling Junction

In the case of a MOM junction, $N_L(E)$ and $N_R(E)$ are constants at the energies of interest and equal to $N(E_F)$, that is, $N_L(E) = N_R(E + eV) = N(E_F)$ and $f_L = f_R = f$ (Fig. 22.27). Hence, from Eq. (22.142), the tunneling current across the MOM junction becomes

$$I_{NN} = \frac{2\pi e}{\hbar}|T|^2[N(E_F)]^2 \int_{-\infty}^{\infty}[f(E+eV) - f(E)]dE$$
$$= \frac{2\pi e}{\hbar}|T|^2[N(E_F)]^2 eV \int_{-\infty}^{\infty}\frac{\partial f}{\partial E}dE \tag{22.143}$$

The above equation can also be written as

$$I_{NN} = \sigma_N V \tag{22.144}$$

where

$$\sigma_N = \frac{2\pi e^2}{\hbar}|T|^2[N(E_F)]^2 \int_{-\infty}^{\infty}\frac{\partial f}{\partial E}dE \tag{22.145}$$

The constant σ_N can be regarded as the normal conductance. Eq. (22.144) gives a linear relation between I and V (Fig. 22.27C), that is, Ohm's law.

22.4.1.2 MOS Tunneling Junction

In a MOS junction, a metal is on the right side (see Fig. 22.28), which has a constant electron DOS, that is, $N_R(E+eV)=C$, a constant. So, Eq. (22.142) gives

$$I_{SN} = \frac{2\pi e}{\hbar}|T|^2 C \int_{-\infty}^{\infty} N_{SL}(E)[f_R(E+eV) - f_L(E)]dE \tag{22.146}$$

Here $N_{SL}(E)$ is the electron DOS of a superconductor on the left side. We are interested in $N_{SL}(E)$ in a small region of energy, on the order of eV. Therefore,

$$I_{SN} = \frac{2\pi e}{\hbar}|T|^2 C \int_{0}^{eV} N_{SL}(E)[f_R(E+eV) - f_L(E)]\,dE \tag{22.147}$$

The density of states $N_{SL}(E)$ is independent of energy E in the range of interest (a few meV). Hence, one can approximately write

$$I_{SN} \propto \int_{0}^{eV} N_{SL}(E)dE \tag{22.148}$$

From Eq. (22.148) one can write

$$\frac{dI_{SN}}{dV} \propto N_{SL}(eV) \tag{22.149}$$

But

$$\frac{dI_{NN}}{dV} \propto N_{NN} \tag{22.150}$$

N_{NN} is the electron DOS in a normal metal. From Eqs. (22.149), (22.150) one can write

$$\frac{(dI_{SN}/dV)}{(dI_{NN}/dV)} = \frac{(dI/dV)_{SN}}{(dI/dV)_{NN}} = \frac{N_{SL}}{N_{NN}} \tag{22.151}$$

Eqs. (22.149)–(22.151) show that the conductance in a MOS junction is proportional to the electronic DOS in a superconductor, while in a MOM junction it is proportional to the electronic DOS in the normal metal. Eq. (22.151) is the central result of single electron tunneling.

22.4.1.3 SOS Tunneling Junction

When the two materials in a junction are superconductors the situation is greatly altered. Eq. (22.142) gives

$$
\begin{aligned}
I_{SS} &= \frac{2\pi e}{\hbar}|T|^2 \int_{-\infty}^{\infty} N_{SL}(E) N_{SR}(E+eV)[f_R(E+eV) - f_L(E)]dE \\
&= C_1 \int \frac{|E|}{\left[E^2 - \Delta_L^2\right]^{1/2}} \frac{|E+eV|}{\left[(E+eV)^2 - \Delta_R^2\right]^{1/2}} [f_R(E+eV) - f_L(E)]\, dE
\end{aligned}
\tag{22.152}
$$

where C_1 is a constant. The integral in Eq. (22.152) can be solved numerically with the help of a computer. It is found that for $T \neq 0\,K$, I_{SS} exhibits a logarithmic singularity at $V_1 = \pm|\Delta_L - \Delta_R|/e$ and a finite discontinuity at $V_2 = \pm|\Delta_L + \Delta_R|/e$ (see Fig. 22.33).

22.4.2 Josephson Tunneling

Consider two superconductors S_L and S_R separated by a thin metal oxide (insulating) layer (see Fig. 22.34). When the thickness of the oxide layer is on the order of the electron mean free path (about 30 Å or more), the Cooper pairs break into quasiparticles (electrons) that can tunnel from one superconductor to the other through the oxide layer. If the thickness of the oxide layer is reduced down to, say, 10 Å, then the Cooper pairs as such can tunnel from one superconductor to the other. The tunneling of the Cooper pairs through the thin potential barrier is called *Josephson tunneling* (Josephson 1962). In this case, the long-range order is transmitted across the boundary. Therefore, the whole system consisting of two superconductors separated by a thin (10 Å) oxide layer behaves, to some extent, as a single superconductor. Unlike ordinary superconductivity, this phenomenon is often called "weak superconductivity" because of the much lower values of the critical parameters involved.

Consider an SOS junction with ψ_L and ψ_R as the pair wave functions for the superconductors on the left and right sides (see Fig. 22.34). The wave functions ψ_L and ψ_R are associated with a macroscopic number of electrons, which are assumed to condense in the same quantum state. In this sense, the superconducting state ψ_L (ψ_R) can be regarded as a macroscopic quantum state, so that $|\psi_L|^2$ ($|\psi_R|^2$) represents the actual Cooper pair density ρ_L (ρ_R). The wave functions ψ_L and ψ_R can be represented as

$$
\begin{aligned}
\psi_L &= \rho_L^{1/2} e^{\imath\phi_L} \\
\psi_R &= \rho_R^{1/2} e^{\imath\phi_R}
\end{aligned}
\tag{22.153}
$$

where φ_L (φ_R) is the phase common to all of the particles on the L (R) side. ρ_L (ρ_R) represents the actual density of the Cooper pairs in the basis state $|L\rangle$ ($|R\rangle$), that is,

$$
\begin{aligned}
\rho_L &= \langle L|\psi_L^* \psi_L|L\rangle \\
\rho_R &= \langle R|\psi_R^* \psi_R|R\rangle
\end{aligned}
\tag{22.154}
$$

$|L\rangle$ and $|R\rangle$ are two orthonormal basis states. Suppose there exists a weak coupling between the two superconductors, then the two macroscopic wave functions ψ_L and ψ_R overlap with each other to a small extent (see Fig. 22.34). Therefore,

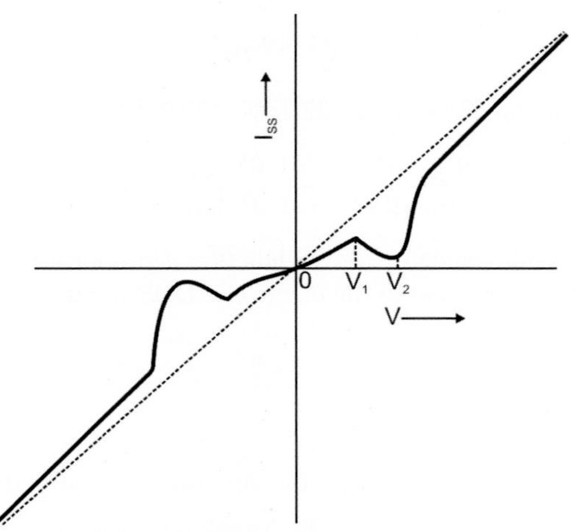

FIG. 22.33 The current-voltage (I-V) characteristics of a SOS tunneling junction.

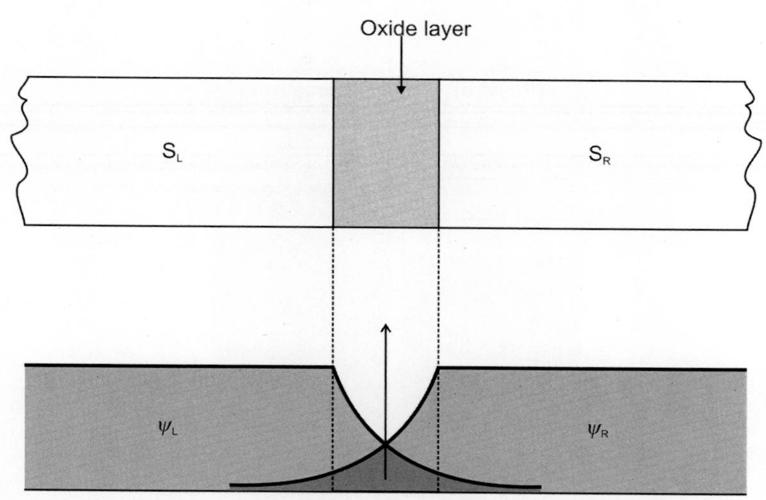

FIG. 22.34 SOS junction with very thin oxide layer. S_L and S_R are the superconductors on the left and right sides and ψ_L and ψ_R are the corresponding electron-pair wave functions.

transitions between the states $|L\rangle$ and $|R\rangle$ can occur. The general state vector with $|L\rangle$ and $|R\rangle$ as the basis states can be written as

$$|\psi\rangle = \psi_L |L\rangle + \psi_R |R\rangle \tag{22.155}$$

Eq. (22.155) says that the particle can either be in the "left" or "right" state with amplitudes ψ_L and ψ_R, respectively. It can be easily proved that

$$\langle\psi|\psi\rangle = \rho_L + \rho_R \tag{22.156}$$

The Schrodinger wave equation for the system is defined by

$$\imath\hbar \frac{\partial}{\partial t}|\psi\rangle = \widehat{H}|\psi\rangle \tag{22.157}$$

with the Hamiltonian given by

$$\widehat{H} = \widehat{H}_L + \widehat{H}_R + \widehat{H}_T \tag{22.158}$$

where

$$\widehat{H}_L = E_L |L\rangle\langle L| \tag{22.159}$$

$$\widehat{H}_R = E_R |R\rangle\langle R| \tag{22.160}$$

$$\widehat{H}_T = K[|L\rangle\langle R| + |R\rangle\langle L|] \tag{22.161}$$

$\widehat{H}_L$ and $\widehat{H}_R$ are the Hamiltonians for the left and right sides and $\widehat{H}_T$ gives the interaction term, which mixes the $|L\rangle$ and $|R\rangle$ states. $\widehat{H}_T$ is also called the tunneling Hamiltonian. E_L and E_R are the ground state energies of superconductors S_L and S_R and K is the coupling amplitude of the two-state system. K is a measure of the interaction energy between S_L and S_R and depends on the specific junction structure (electrode geometry and tunneling barrier, etc.). Substituting Eqs. (22.155), (22.158) into Eq. (22.157) and using the orthonormality condition between $|L\rangle$ and $|R\rangle$, we get

$$\imath\hbar \frac{\partial\psi_L}{\partial t}|L\rangle + \imath\hbar \frac{\partial\psi_R}{\partial t}|R\rangle = E_L \psi_L |L\rangle + E_R \psi_R |R\rangle + \psi_R K|L\rangle + \psi_L K|R\rangle \tag{22.162}$$

Comparing the terms with basis states $|L\rangle$ and $|R\rangle$ on both sides of Eq. (22.162) separately, we find

$$\imath\hbar \frac{\partial\psi_L}{\partial t} = E_L \psi_L + K\psi_R \tag{22.163}$$

$$\imath\hbar \frac{\partial\psi_R}{\partial t} = E_R \psi_R + K\psi_L \tag{22.164}$$

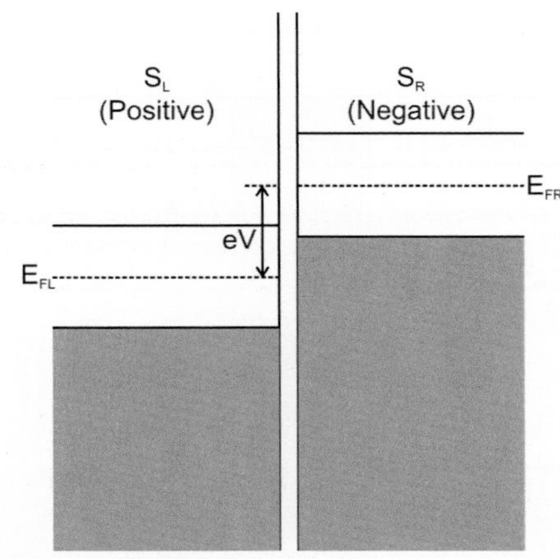

FIG. 22.35 SOS junction with electric voltage V applied across the junction.

The binding energy in a Cooper pair is $2E_F$ as the two electrons in it have energies corresponding to $-\mathbf{k}_F$ and $\mathbf{k}_F$. If the two superconductors are isolated, the energy terms are given by $E_R = 2E_{FR}$ and $E_L = 2E_{FL}$ where E_{FR} and E_{FL} are the Fermi energies of superconductors S_R and S_L. When voltage V is applied to the SOS junction, the Fermi energies shift by eV, that is, $E_{FR} - E_{FL} = eV$ (see Fig. 22.35). Therefore,

$$E_R - E_L = 2eV \tag{22.165}$$

If the zero of energy (reference level of energy) is chosen to be halfway between the two values E_L and E_R, then

$$E_L = -eV, \ \ E_R = eV \tag{22.166}$$

Substituting the values of E_L and E_R from Eq. (22.166) into Eqs. (22.163), (22.164), one gets

$$\imath\hbar\frac{\partial\psi_L}{\partial t} = -eV\psi_L + K\psi_R \tag{22.167}$$

$$\imath\hbar\frac{\partial\psi_R}{\partial t} = eV\psi_R + K\psi_L \tag{22.168}$$

Substituting ψ_L and ψ_R from Eq. (22.153) into Eq. (11.167) and comparing the real and imaginary parts separately, we get

$$-\frac{1}{2}\frac{\partial\rho_L}{\partial t}\sin\phi_L - \rho_L\frac{\partial\phi_L}{\partial t}\cos\phi_L = -\frac{eV}{\hbar}\rho_L\cos\phi_L + \frac{K}{\hbar}(\rho_L\rho_R)^{1/2}\cos\phi_R \tag{22.169}$$

$$\frac{1}{2}\frac{\partial\rho_L}{\partial t}\cos\phi_L - \rho_L\frac{\partial\phi_L}{\partial t}\sin\phi_L = -\frac{eV}{\hbar}\rho_L\sin\phi_L + \frac{K}{\hbar}(\rho_L\rho_R)^{1/2}\sin\phi_R \tag{22.170}$$

Eqs. (22.169), (22.170) are the real and imaginary parts, respectively. Multiplying Eq. (22.169) by $\sin\varphi_L$ and Eq. (22.170) by $\cos\varphi_L$ and subtracting, we obtain

$$\frac{\partial\rho_L}{\partial t} = -\frac{2K}{\hbar}(\rho_L\rho_R)^{1/2}\sin\phi \tag{22.171}$$

where

$$\phi = \phi_L - \phi_R \tag{22.172}$$

Similarly, multiplying Eq. (22.169) by $\cos\varphi_L$ and Eq. (22.170) by $\sin\varphi_L$ and adding, we get

$$\frac{\partial\phi_L}{\partial t} = \frac{eV}{\hbar} - \frac{K}{\hbar}\left(\frac{\rho_R}{\rho_L}\right)^{1/2}\cos\phi \tag{22.173}$$

In exactly the same manner ψ_L and ψ_R can be substituted from Eq. (22.153) into Eq. (22.168) and the real and imaginary parts can be separated. Further, they can be solved for $\partial\rho_R/\partial t$ and $\partial\varphi_R/\partial t$, which are given by

$$\frac{\partial\rho_R}{\partial t} = \frac{2K}{\hbar}(\rho_L\rho_R)^{1/2}\sin\phi \tag{22.174}$$

$$\frac{\partial\phi_R}{\partial t}=-\frac{eV}{\hbar}-\frac{K}{\hbar}\left(\frac{\rho_L}{\rho_R}\right)^{1/2}\cos\phi \tag{22.175}$$

From Eqs. (22.173), (22.175) the relative phase φ is given by

$$\frac{\partial\phi}{\partial t}=\frac{\partial}{\partial t}(\phi_L-\phi_R)=\frac{2eV}{\hbar}+\frac{K}{\hbar}\left[\left(\frac{\rho_L}{\rho_R}\right)^{1/2}-\left(\frac{\rho_R}{\rho_L}\right)^{1/2}\right]\cos\phi \tag{22.176}$$

The current density J is given by

$$\begin{aligned}J&=-\frac{\partial\rho_L}{\partial t}=\frac{\partial\rho_R}{\partial t}\\&=\frac{2K}{\hbar}(\rho_L\rho_R)^{1/2}\sin\phi\end{aligned} \tag{22.177}$$

In a SOS junction, if the tunnel geometry of both the superconductors is the same, then $\rho_L=\rho_R=\rho_0$. Therefore, Eqs. (22.176), (22.177) give

$$J=J_0\sin\phi \tag{22.178}$$

where

$$J_0=\frac{2K\rho_0}{\hbar} \tag{22.179}$$

$$\frac{\partial\phi}{\partial t}=\frac{2eV}{\hbar} \tag{22.180}$$

Eqs. (22.178), (22.180) are the basic relations of Josephson tunneling. For $V=0$ the phase difference φ becomes constant and not necessarily zero. Therefore, the Cooper pairs can tunnel even when no voltage is applied to the SOS junction, yielding a finite current density J. This is the central result of the dc Josephson effect and is different from single-electron tunneling in which the current flows only when the voltage is applied. The I-V characteristic curve in a Sn-SnO-Sn Josephson junction is shown in Fig. 22.36, which clearly shows a finite current at zero voltage. When the current flowing through the junction at zero voltage becomes maximum, a finite voltage suddenly appears across the junction. Indeed, a switching from the zero voltage state to the quasiparticle branch of the I-V curve occurs.

If a constant voltage is applied, then from Eq. (22.180) the relative phase can be written as

$$\phi=\phi_0+\frac{2eV}{\hbar}t \tag{22.181}$$

where φ_0 is the initial phase difference. Substituting Eq. (22.181) into Eq. (22.178), we get

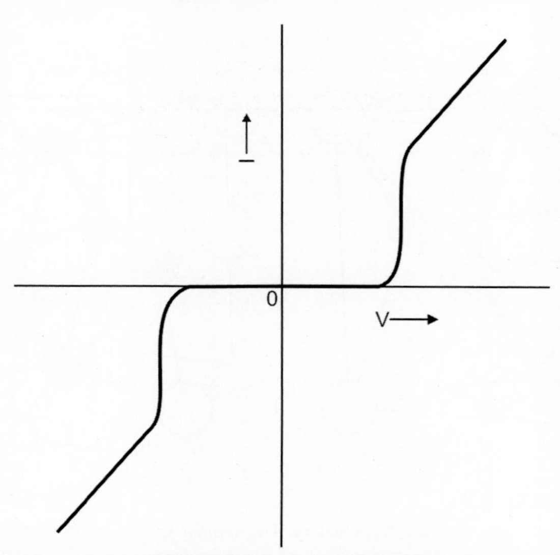

FIG. 22.36 The current-voltage (I-V) characteristics for the Josephson effect in a Sn-SnO-Sn junction.

$$J = J_0 \sin\left(\phi_0 + \frac{2eV}{\hbar} t\right) \tag{22.182}$$

with a frequency $\omega = 2\pi\nu = 2eV/\hbar$. The current density J given by Eq. (22.182) shows a sinusoidal variation and it is called the *ac Josephson effect*. From Eq. (22.182)

$$\nu = \frac{2eV}{h} = 483.6 \text{ THz/V} \tag{22.183}$$

The relationship between the voltage and frequency of the phonon emitted when an electron pair crosses the barrier allows the determination of e/h.

22.5 HIGH-T_C SUPERCONDUCTIVITY

Since the discovery of superconductivity efforts have been made to increase the value of T_c. A family of ternary sulfides, usually called Chevrel phases was discovered, which exhibit a reasonably high value of T_c. Later, high-T_c superconductivity was found in materials with perovskite structure. Bednorz and Muller (1986, 1988) discovered oxide superconductors with $T_c \sim 35$ K. The high-T_c superconducting cuprates (copper oxides) evolved from materials related to the perovskite family. The interesting aspect of the oxide superconductors is that all these materials contain copper oxide layers. In this section, the various high-T_c superconducting materials are described in brief.

22.5.1 Chevrel Phases and Superconductivity

Chevrel phases are described by the general chemical formulae $M_yMo_6X_8$ and $M_yMo_6X_{12}$ in which y is usually one. Here M is one of the metallic elements and X is usually a chalcogenide (S, Se, or Te) or occasionally a heavy and strongly polarizable halide (Br or I). The most common Chevrel phases are $PbMo_6S_8$ and $SnMo_6S_8$. The novel feature of these materials is the presence of one or two fundamental cubic structural units, such as Mo_6X_8 or Mo_6X_{12}, as shown in Fig. 22.37. The internal bonding of Mo_6X_8 or Mo_6X_{12} is very strong, which is due to the fact that the energy E of the valence electrons is far below E_F. These building blocks contribute most of the valence electrons per formula unit (e.g., 95% in the case of $PbMo_6S_8$ with $T_c = 15.2$K). The clusters Mo_6X_8 or Mo_6X_{12} are then combined with M atoms to form the CsCl structure in which M occupies the positions of Cs and Mo_6X_8 or Mo_6X_{12} occupies the positions of Cl. But the structure of MMo_6X_8 or MMo_6X_{12} is slightly uniaxially distorted along its [111] axis compared with the CsCl structure. The structure of $PbMo_6S_8$ is shown in Fig. 22.38. Here Pb provides a soft mechanical link between the clusters of Mo_6S_8. $PbMo_6S_{12}$ has a similar structure.

Because Mo_6X_8 contributes most of the valence electrons, one expects that most of the physical properties, such as the electron DOS at the Fermi energy $N(E_F)$, the electrical resistivity $\rho(T)$, the transition temperature T_c, and the critical field $\mathbf{H}_{c_2}$, depend mainly on Mo_6X_8 with M playing a passive role. But if one carefully examines the Chevrel phases, it is found

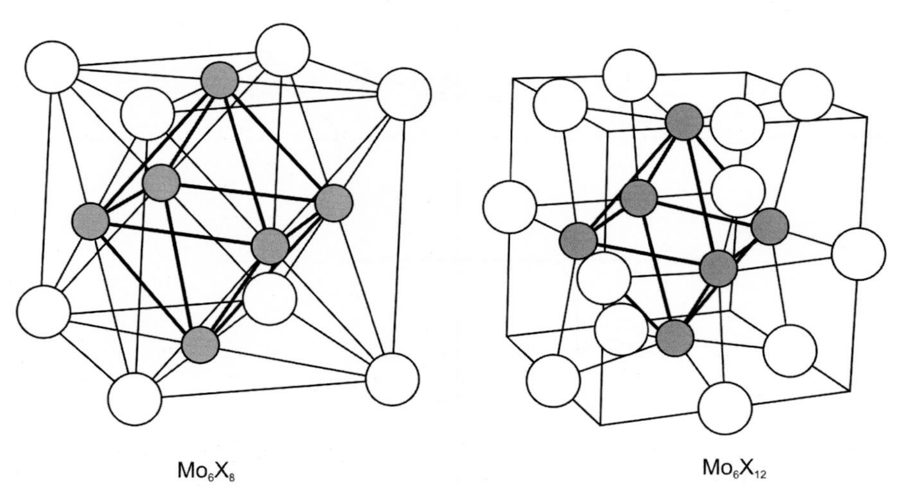

FIG. 22.37 The nearest neighbors of Mo in Mo_6X_8 and Mo_6X_{12} clusters. The *shaded spheres* represent Mo atoms while the *white spheres* represent X atoms. *(Modified from Phillips, J. C. (1989). Physics of high temperature superconductors (p. 61). San Diego: Academic Press.)*

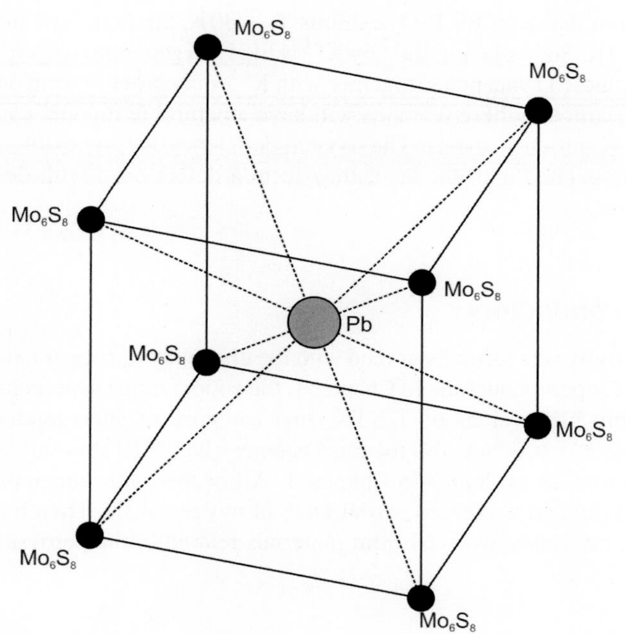

FIG. 22.38 The structure of $PbMo_6S_8$.

that the valence structure of M is more important in determining all the physical properties. In these phases, regardless of the valence electrons in M, M^{+2} is required in forming the structure and the remaining valence electrons with energy near E_F play an important role in producing an electrical bridge between the clusters of Mo_6X_8. For example, Sn and Pb have four valence electrons each. Two electrons are detached to give Sn^{+2} or Pb^{+2} ions and the remaining two electrons have energy near E_F. It is found that for Mo_6S_8, $T_c = 1.8\,K$ so that direct cluster-cluster electron transfer and intracluster e-p interactions alone produce a low T_c. With the introduction of Pb metal in $PbMo_6S_8$, the intercluster Mo - Mo spacing in Mo_6S_8 increases by ~6% and T_c increases to 15.2 K. The bridge element serves two functions: first it produces an electrical bridge and second it provides a mechanical soft link with a strong local e-p interaction.

An interesting aspect of the Chevrel phases is that as X goes from S to Te through Se, band narrowing takes place, which increases $N(E_F)$, all other things being equal, and hence increases T_c. Such behavior is attributed to random internal strains. Further, the Chevrel phases form the first examples of open structure in which M atoms are so weakly bound that they can diffuse readily in and out of the samples, much as ions do in solid electrolytes. This characteristic is used to prepare the high-T_c compounds MMo_6S_8 or MMo_6Se_8 with M = Hg, In, and Tl by low-temperature diffusion ($T < 500^\circ C$). Here Hg^{+2}, In^{+2}, and Tl^{+2} retain only one valence electron (not two as in the case of Sn^{+2} and Pb^{+2}).

22.5.2 Perovskite Superconductivity

The perovskite structure with formula ABX_3 was discussed in Chapter 1 (see Fig. 1.33). Most of these materials exhibit ferroelectric behavior and, therefore, are important from a technological point of view. The most common examples are $BaTiO_3$, $BaBiO_3$, and $CaTiO_3$ and these materials generally have low carrier density. In ABO_3, when the metallic ion A or B is partially replaced by another metallic ion C, the material exhibits high-T_c superconductivity: for example, $BaPb_{1-y}Bi_yO_3$ is a superconductor with $T_c \sim 13\,K$. The conductivity of these elements depends upon the concentration y. In the normal state, with a decrease in T, the sample becomes less metallic as $d\rho/dT \to 0$. For $y > 0.35$ these alloys are semiconducting in nature and for $y < 0.35$ they are metallic in nature: the data reported for $BaPb_{1-y}Bi_yO_3$ alloy refer to powder samples. In the preparation of these alloys it is very difficult to control the parameter y. The resultant samples exhibit a large variation in T_c with y. But the variation in T_c is the smallest near $y = 0.25$, which suggests that the internal stress is minimum for this composition.

There are many differences between the old metallic superconductors and the new high-T_c superconductors, but one factor that remains the same is the presence of lattice instabilities. In the vibrational spectrum of a superconductor, the optical branches in the $[0\,\xi\,\xi]$ and $[\xi\,\xi\,\xi]$ directions exhibit a great deal of softening of the phonons near the M and R points at the BZ faces. This softening is assigned to the rotation of MO_6 octahedra with M as Pb or Bi.

In Chapter 1, it was pointed out that $K_{1-y}Ba_yBiO_3$ exhibits $T_c \sim 30\,K$, but here 30% or less of the sample exhibits the Meissner effect beginning near 30 K. So, replacing Ba^{+2} by K^{+1} is likely to generate oxygen vacancies O. In a small fraction of the sample volume (say 30%), these O vacancies, together with K^{+1}, may order to form domains with typical dimensions on the order of 100 Å. The superlattice in these domains will have a natural tetragonal configuration with an xy-plane of partial O vacancies bounded by K-enriched sheets. The oxygen deficiency is responsible for the enhancement of the e-p interactions. The defect states associated with the layer may form a defect band with defect-enhanced e-p interactions, which may yield high-T_c behavior.

22.5.3 Cu-Oxide Superconductors

The phrase high-T_c superconductivity was formally coined with the discovery of copper oxide superconductors by Bednorz and Muller (1986): a new class of superconductors. In Chapter 1, the copper oxide superconductors were grouped into three categories that exhibit reasonably high values of T_c. The first category of superconductors has K_2NiO_4-type crystal structure, the second category has 123 structure and the third category has 2212 structure (see Fig. 1.34). All of these categories are high-T_c superconductors, as explained in Chapter 1. All of the oxide superconductors with a perovskite-like structure show metallic conductivity and a nonzero partial DOS of oxygen at E_F. Therefore, historically one can say that the high-T_c superconducting cuprates have evolved from materials related to the perovskite family ABX_3.

22.5.4 A_2BX_4 Superconductors

There are several families of high-T_c superconductors that possess A_2BX_4 (AX. ABX_3) structure and among these the most famous are the ferrites having a magnetic cubic spinel structure. Here A and B are cations that are small in size compared with the X anions and they form a close-packed lattice. A and B cations occupy 1/8th of the tetrahedral and 1/2 of the octahedral interstitial positions. Here the B atoms lie in planes and are octahedrally coordinated with the X atoms, while the A atoms are approximately nine-fold coordinated. In such an interstitial ionic compound, the lattice constant depends primarily on the anion size and the anion-anion contacts. The A_2BX_4 superconductor has the tetragonal structure of K_2NiO_4 or K_2NiF_4. Examples of the A_2BX_4 superconductors are La_2CuO_4 and $La_{2-y}Ba_yCuO_4$. Fig. 22.39 shows a comparison of the ABX_3 and A_2BX_4 structures.

In the tetragonal ionic superconductors, the packing along the $\mathbf{a}_3$-axis is usually very tight as shown by the abnormally short A - X distance in all of them. The shape of the octahedra around the B atoms is usually nearly regular, with a very small distortion. The distortion of the octahedra, as expected, is greatest for $Cu^{+2}(d^9)$ because an odd number of d-electrons gives the largest Jahn-Teller effect. However, the distortion can have either sign when X is replaced by F (ionic bonding) or by O (covalent bonding). The distortion also occurs when $Cu^{+2}(d^9)$ is replaced by $Ni^{+2}(d^8)$.

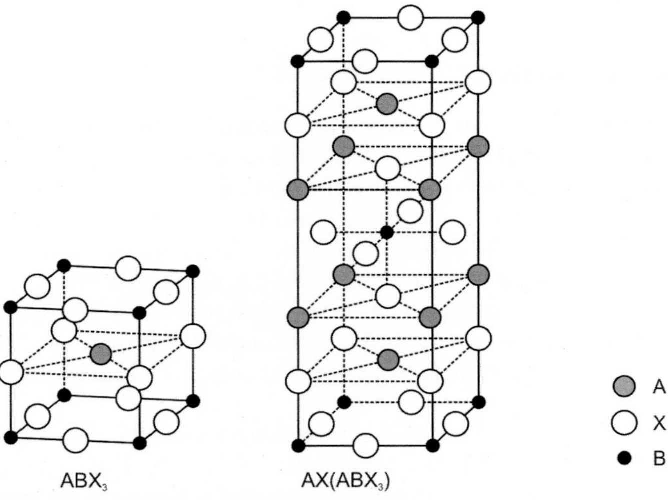

FIG. 22.39 A comparison of ABX_3 and $AX(ABX_3)$ structures in high-T_c superconductors.

22.5.5 Quaternary Copper Oxides

The discovery of $La_{2-y}Ba_yCuO_4$ with $T_c > 30$ K started a world-wide search for related oxides with higher T_c. Chu et al. (1987) and Wu et al. (1987) made the very important observation that the pressure derivative of the transition temperature, that is, dT_c/dP, is positive and very large in $(La, Ba)_2CuO_4$. This suggested the possibility of immensely increasing the value of T_c through the internal pressure generated by the large A - B size difference in ABCuO compounds, with possibly a new crystal structure. The above mentioned possibility became a reality with the discovery of $YBa_2Cu_3O_7$ (123 superconductor) having $T_c = 93$ K and $La(La_{2-y}Ba_y)Cu_3O_7$ having $T_c = 80$ K and possessing an entirely different structure. This is apparently the first true quaternary metallic structure, all previous metallic compounds being binaries, pseudobinaries, ternaries, and pseudoternaries. All of these compounds belong to the 123 family of high-T_c superconductors. Another compound of interest is $YBa_2Cu_3O_{7-y}$, which can have both orthorhombic and tetragonal structures.

A very important difference between $(La, Sr)_2CuO_4$ and $YBa_2Cu_3O_7$ is the role played by the orthorhombic and tetragonal phases. The orthorhombic phase of $(La, Sr)_2CuO_4$ is a semiconductor and becomes a superconductor by doping with Sr, Ba, and Ca, which eventually produces a tetragonal structure. Just the reverse is true in$YBa_2Cu_3O_{7-y}$: the tetragonal phase for $y > 0.7$ is a semiconductor and T_c is maximized by reducing y to <0.1 and obtaining a fully orthorhombic phase. In these superconductors the following points are noteworthy:

1. Fig. 1.34A and B shows the ideal crystallographic structures of $(La, Sr)_2CuO_4$ and $YBa_2Cu_3O_7$, but the actual structure in either superconductor does not necessarily exhibit complete order in the atomic structure. There are regions of the samples that are less ordered (or distorted) and they form a substantial volume fraction of the sample. These regions contribute only a small background to the measured neutron spectra. With three metallic components, there is no reason to expect such a perfect ordering.
2. An additional effect in $YBa_2Cu_3O_{7-y}$ is that T_c is very sensitive to the concentration of oxygen vacancies: T_c changes drastically by adding or removing oxygen to these superconductors.
3. The most remarkable aspect of $YBa_2Cu_3O_{7-y}$ is the easy diffusion of oxygen into the sample and also its evaporation. This facile anion diffusion is correlated with high-T_c superconductivity, especially if the diffused O ions form part of the metallic and superconductive regions embedded in the complex structure of these materials.

22.5.6 Bismates and Thallates

The third type of oxide superconductor includes bismates and thallates. One of the members of the bismates is $Bi_2Sr_2CaCu_2O_8$ with 2212 structure (Fig. 1.34C). The symmetry of the 2212 structure is more evident from Fig. 22.40. Here the CuO_2 layers are separated by Ca layers alternating with a group of SrO - BiO - BiO - SrO layers. The composition of the cation and the O ion can change from sample to sample. For example, one of the superconductors has the composition $Bi_{2.2}Sr_2Ca_{0.8}Cu_2O_{8+y}$. In this structure every cation plane contains nearly coplanar O ions, except for the $Ca_{0.8}Bi_{0.2}$ plane. Actually, the cation planes are strongly bent or bulged by the modulation displacements. Normal to the planes, these displacements are 0.2 Å (Bi,Sr) and 0.3 Å (Cu).

In a bismate compound, Bi is trivalent, that is, Bi^{+3}. This naturally suggests that replacing Bi by Tl results in similar compounds with small but interesting differences in structure. Buckling of the CuO_2 sheets decreases in going from $YBa_2Cu_3O_7$ to $Bi_2Sr_2CaCu_2O_8$ and then to $Tl_2Ba_2CaCu_2O_8$. However, at the same time, the O ions in the Tl plane appear to have moved off the center (along the x-axis) and the Tl atom apparently has a large in-plane vibrational amplitude. It is hard to say whether this is evidence for anharmonicity in Tl vacancies. In that case, it might be evidence for increasing electronic localization and marginal two-dimensionality in the CuO_2 planes.

In bismates and thallates the stacking of superlattices leads to the grouping of CuO_2 planes separated by Ca^{+2} layers bounded by BaO layers. These superlattices form an interesting homologous family written as $(Bi, Tl)_2(Sr, Ca, Ba)_{n+1}Cu_nO_{2(n+2)\pm y}$. This general formula gives rise to the following families of superconductors:

(a) For $n = 1$ one gets the family of superconductors with chemical formula $(Bi, Tl)_2(Sr, Ca, Ba)_2CuO_{6\pm y}$. The familiar members of the series are $Bi_2(Bi, Sr)_2CuO_6$, which is a semiconductor, and $Tl_2Ba_2CuO_6$, which is a superconductor with $T_c = 80$ K.

(b) For $n = 2$, the series has the general formula $(Bi, Tl)_2(Sr, Ca, Ba)_3Cu_2O_{8\pm y}$. The high-$T_c$ superconducting family members of the series are $Bi_2Sr_2CaCu_2O_8$ with $T_c = 85$ K and $Tl_2Ba_2CaCu_2O_8$ with $T_c = 100$ K.

(c) For $n = 3$, the general formula for high-T_c superconductors becomes $(Bi, Tl)_2(Sr, Ca, Ba)_4Cu_3O_{10\pm y}$. The important family members of this series are $(Bi, Tl)_2(Sr, Ca)_4Cu_3O_{10}$ with $T_c = 110$ K and $Tl_2Ba_2Ca_2Cu_3O_{10}$ with $T_c = 125$ K.

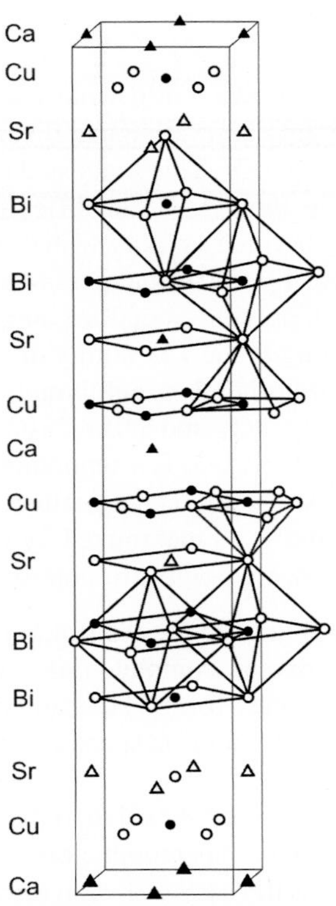

FIG. 22.40 The structure of the high-T_c superconductor $Bi_2Sr_2CaCu_2O_8$. *(Modified from Phillips, J. C. (1989). Physics of high temperature superconductors. New York: Academic Press.)*

It is evident that a greater number of CuO_2 layers (with no CuO planes) results in a higher value of T_c. For example, a greater number of CuO_2 layers in Tl compared with Bi increases the value of T_c. All of the oxide superconductors show metallic conductivity and nonzero partial DOS of oxygen at E_F. The common structure of all oxide superconductors is a CuO_2 layer, which is formed by antibonding bonds between 3d states of Cu having $x^2 - y^2$ symmetry and 2p states of O in the x- and y-directions. The structure of CuO_2 planes in high-T_c materials is more or less the same and is shown in Fig. 22.41. The other layers are also important because:

(a) They provide a polarizable medium that determines the lattice constant and separates the CuO_2 layers.
(b) They can be doped with other atoms to increase the value of T_c.

T_c and other physical properties depend very strongly on the oxygen deficiency denoted by the parameter y, which gives the deviation from the stoichiometry in the structure formula. The layer structure is responsible for strong anisotropies in electrical conductivity and in the parameters of superconductivity.

The exact mechanism of high-T_c superconductivity is still unclear. It has been shown experimentally that electron pairs are responsible for high-T_c superconductivity. In comparison with the known metallic systems, the finite DOS for O sites at E_F is an important new feature. It implies that oscillations of the light O atoms contribute to the e-p interactions. It has been observed that the isotopic effect in high-T_c superconductors is much weaker than in conventional superconductors. Therefore, it has been proposed that nonphononic mechanisms, such as the anisotropy and anharmonicity of lattice vibrations, may also contribute to high-T_c superconductivity. It has been found that the T_c of the La_2CuO_4 structure may be understood within these assumptions. However, the experimental results from 123 superconductors cannot be explained with the help of these assumptions. At present, no single theory exists that is able to explain high-T_c superconductivity in the various systems discussed above. The high-T_c superconductors are of great practical importance because they can be operated with liquid nitrogen as the cooling medium (cooling temperature ≈ 77 K), which reduces the cost considerably. Usually it is assumed that for technical applications T_c should exceed the cooling temperature by a factor of 1.5 at least ($T_c > 115$ K). To use the oxide superconductors in technical applications, the following problems must be solved.

1. Processing of flexible wires or tapes from the brittle oxide superconductors.
2. The density of critical current in the high-T_c superconductors produced by ceramic technology should be increased to values above 10^5 A/cm^2.

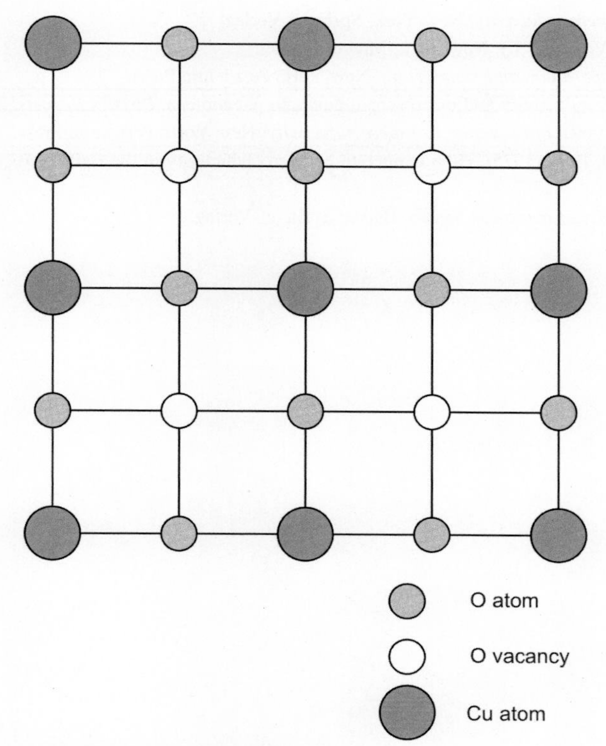

FIG. 22.41 The structure of the CuO_2 plane in a high-T_c superconductor.

In the end we want to point out that an enormous amount of development is still required in the field of high-T_c superconductivity.

Problem 22.5

The London penetration depth is found to depend on temperature as follows:

$$d_L(T) = d_L(0)\left[1 - \left(\frac{T}{T_c}\right)^4\right]^{-1/2} \tag{22.184}$$

If the penetration depth for Pb is 396 Å at 3 K and 1730 Å at 7.1 K, calculate the critical temperature for Pb.

REFERENCES

Bardeen, J., Cooper, L. N., & Schrieffer, J. R. (1957). Theory of superconductivity. *Physical Review*, *108*, 1175–1204.

Bednorz, J. G., & Muller, K. A. (1986). Possible high T_c superconductivity in the Ba-Ca-Cu-O system. *Zeitschrift fur Physik*, *B64*, 189–193.

Bednorz, J. G., & Muller, K. A. (1988). Perovskite type oxides—the new approach to high T_c superconductivity. *Reviews of Modern Physics*, *60*, 585–600.

Chu, C. W., Hor, P. H., Meng, R. L., Gao, L., Huang, Z. J., & Wang, Y. Q. (1987). Evidence for superconductivity above 40 °K in the La-Ba-Cu-O compound system. *Physical Review Letters*, *58*, 405.

Giaever, I. (1960). Energy gap in superconductors measured by electron tunneling. *Physical Review Letters*, *5*, 147.

Giaever, I. (1974). Electron tunneling and superconductivity. *Reviews of Modern Physics*, *46*, 245–250.

Onnes, H. K. (1911). The resistance of platinum at helium temperature. *Communications from the Physical Laboratory at the University of Leiden*, *119b*, 19–26.

Wu, M. K., Ashburn, J. R., Torng, C. J., Hor, P. H., Meng, R. L., Gao, L., et al. (1987). Superconductivity at 93 °K in a new mixed phase Y-Ba-Cu-O compound system at ambient pressure. *Physical Review Letters*, *58*, 908.

SUGGESTED READING

de Gennes, P. G. (1992). *Superconductivity of metals and alloys*. Reading: Addison-Wesley Publishing Co.

Halley, J. W. (1988). *Theories of high temperature superconductors*. New York: Addison-Wesley Publishing Co.

Lynn, J. W. (1990). *High temperature superconductivity*. New York: Springer-Verlag.
Parks, R. D. (1969). *Superconductivity*. (Vols. I & II). New York: Marcel Dekker.
Phillips, J. C. (1989). *Physics of high temperature superconductors*. New York: Academic Press.
Rogovin, D., & Scully, M. (1976). Superconductivity and macroscopic quantum phenomena. *Physics Reports*, *25C*, 175–291.
Rose-Innes, A. C., & Rhoderick, E. H. (1969). *Introduction to superconductivity*. New York: Pergamon Press.
Sleight, A. W., Gillson, J. L., & Bierstedt, P. E. (1975). High temperature superconductivity in the $BaPb_{1-x}Bi_xO_3$ system. *Solid state communications* (p. 27)Vol. 17, (p. 27). .
Vonsovsky, S. V. (1982). *Superconductivity of transition metals*. Berlin: Springer-Verlag.

Chapter 23

Defects in Crystalline Solids

Chapter Outline

The most important feature of crystalline solids is their regular atomic arrangement, but no crystalline solid actually exhibits perfect order: real crystalline solids contain defects to some extent. The defects can be classified into two categories: *electronic defects* and *atomic defects*. Electronic defects are related to the excess electrons in the conduction band and the holes in the valence band. In materials with low conduction electron density, such as semiconductors and insulators, a small change in electron density has a pronounced influence on the electronic properties. A familiar example is an extrinsic semiconductor in which a small amount of either pentavalent or trivalent impurities are added to increase the conductivity. The effect of the electronic defects in semiconductors has already been discussed in Chapter 14. But in metallic solids, the conduction electron density is nearly equal to the density of atoms. Therefore, the addition of a small number of extra electrons is not expected to bring about any appreciable change in the electronic properties of metallic solids. On the other hand, atomic defects play an important role in the electronic properties of metallic solids. In this chapter, we shall discuss atomic defects, usually called *lattice defects*, in crystalline solids. A lattice defect is a state in which the atomic arrangement has departed from regularity in a small region. Here a small region means small in comparison with the dimensions of the crystalline solid: a lattice defect may extend to a few lattice constants. The lattice defects can be classified into three categories.

1. ***Point defects***: A lattice defect that spreads out very little in all three dimensions is called a point defect. Examples include an atomic vacancy, an interstitial atom, and a substitutional atom.
2. ***Line defects***: If a lattice defect is confined to a small region in only one dimension, it is called a line defect. Dislocations are a good example of a line defect.
3. ***Planar defects***: If a lattice defect is confined to a small region in two dimensions, it is called a planar defect. A twin boundary and an extended dislocation are examples of planar defects.

Defects influence the properties of crystalline solids in many ways. For example, they scatter the conduction electrons in a metal, thereby increasing its electrical resistance. This increase is several percent at most in many pure metals, but it may be as much as several tens of percent in the case of alloys.

23.1 POINT DEFECTS IN SOLIDS

A point defect is the simplest imperfection that involves only a single lattice point in a crystalline solid. *Solid solutions* are good examples of point defects in a crystalline solid. Therefore, before discussing the various types of point defects, let us first introduce the concept of solid solutions.

Solid State Physics. https://doi.org/10.1016/B978-0-12-817103-5.00023-2

23.1.1 Solid Solutions

A solid solution is formed by mixing a foreign element B (called an impurity or solute) with a perfect crystalline element A (called the host or solvent) such that the atoms of B share the various crystal sites of element A. Such a solid solution is written as $\underline{A}B$ where the first underlined symbol denotes the host element and the second symbol the impurity. For example, if we add atoms of the element Zn to a pure Al crystal, a solid solution of $\underline{Al}Zn$ is formed. A solid solution is prepared as follows: Elements A and B are mixed in definite proportions and then heated to melt. The crystal is allowed to grow by slowly cooling the molten mixture. The solid solution so prepared acquires some crystal structure depending on the concentration of the impurity. In preparing the solid solution some heat may be required, called the heat of mixing, and there may also be a change in the molar volume. Metallic alloys can be considered as special solid solutions in which the electrons are easily excited.

23.1.1.1 Types of Solid Solutions

Solid solutions can be categorized on the basis of some of the physical parameters, such as structure, concentration, solubility and order.

Structure: A solid solution $\underline{A}B$ may acquire the same crystal structure as that of the host element. Such a solid solution is called a *primary solid solution*. But if the solid solution acquires a crystal structure quite different from either of the elements A or B, then one obtains what is called an *intermediate solid solution*.

Concentration: If the concentration of the impurity atoms B is very low, we obtain the so-called *dilute solid solution*. If the sizes of atoms A and B are nearly equal, then the heat of mixing is zero in the limit of very small concentration and the molar volume remains unchanged. Such a solid solution is called an *ideal solid solution*.

If the concentration of the impurity atoms is large enough, one obtains a *concentrated solid solution*. With an increase in concentration of the impurity, structural changes may appear: a phase change may occur in a solid solution. Consider for example a solid solution $\underline{Al}Fe$: Al has fcc structure, while Fe has bcc structure. For a very small concentration of Fe impurity, the lattice of the solid solution has fcc structure and this is designated as the α-phase or fcc phase. As the concentration of Fe is increased, the lattice becomes distorted but it is still recognizable as the fcc phase and consists primarily of Al atoms with some Fe atoms. At a particular composition, however, the energy of the solid solution becomes so large that a further addition of Fe atoms changes the phase from fcc to bcc. This is designated the β-phase and consists primarily of Fe atoms with some Al atoms in the solid solution.

Solubility: If the components of a solid solution $\underline{A}B$ are soluble at all concentrations, it is called a *completely soluble solid solution*. For example, Ni dissolves in Cu at all concentrations without a change in structure, giving rise to complete solubility and a continuum of solid solutions. It usually occurs if the sizes of the atoms of the two components are nearly equal (differing in size by less than 15%). If two or more elements dissolve only for restricted concentrations, we get what is called a *restricted solid solution*.

Ordering: Solid solutions are sometimes categorized according to the order among the distribution of the atoms of the different elements on the lattice sites. Consider a solid solution $\underline{A}B$ with an equal number of A and B atoms. First possibility is that the atoms A and B are distributed randomly on the lattice. Such a solid solution is called a *random solid solution*. The second possibility is that the distribution of A and B atoms on the lattice may be partially or completely ordered. This is called an *ordered solid solution*. In an ordered solid solution A and B atoms have regular periodic structure with respect to each other, forming a superlattice in which the two types of atoms have a preference for being the 1NNs of one another. In a partially ordered solid solution one can define the degree of order. One should note that a solid solution is completely ordered at absolute zero. With an increase in temperature it becomes less ordered until a transition temperature is reached, above which the structure is fully disordered. If a solid solution in the molten state is cooled rapidly to a temperature below the transition temperature, a metastable state may be produced in which a nonequilibrium disorder is frozen into the lattice. A solid solution may also be named after the nature of the impurity, which will be discussed in the coming section.

23.1.2 Types of Point Defects

As regards point defects, one can ask the following two questions:

1. What is the nature and position of a point defect in a crystalline solid?
2. What is the electronic structure of a crystalline solid for a given nature and position of the point defect?

The different types of atomic defects and their geometry are described below.

23.1.2.1 Substitutional Point Defects

Consider a crystalline solid A in which impurity atoms B knock out the atoms A and take their positions, thus forming a solid solution $\underline{A}B$. The positions of atoms B, which were earlier occupied by atoms A, are called substitutional positions and the impurity is called a substitutional impurity (substitutional point defect). It is found that the impurity atoms that have a size comparable to that of the host atoms occupy the substitutional positions. The reason is that much larger impurities do not have sufficient space to occupy the substitutional positions, while much smaller impurities are not able to knock out the host atoms. Substitutional impurities in the fcc and bcc structures are shown in Fig. 23.1. The positions and distances of the 1NNs and 2NNs of a substitutional impurity are the same as given in Table 1.1. For the formation of substitutional impurities certain conditions called the Hume-Rothery rules should be satisfied. These conditions are stated below:

1. Elements corresponding to both the host and impurity should have the same structure.
2. The radii of the two types of atoms should be approximately the same (within about 15%).
3. Besides the geometrical factors, other factors, such as the valence, should be the same.
4. The two components of the solid solution $\underline{A}B$ should have nearly the same electronegativity.

If the Hume-Rothery rules are not satisfied, an intermediate phase having a different crystal structure than either of the two elements is formed.

23.1.2.2 Vacancies

In a perfect crystal, if one of the atoms is removed from the lattice and taken out of the crystal or to its surface, then the defect formed is called a vacancy. The geometry of the vacancy in fcc and bcc structures is shown in Fig. 23.2. As a substitutional point defect, the positions and distances of the 1NNs and 2NNs of a vacancy are the same as for a substitutional impurity. In an ionic crystal, there are an equal number of negatively and positively charged ions. Therefore, the number of vacancies created by the positively and negatively charged ions must be the same. This is so because, if the positive ions migrate out of the crystal to the surface, the surface becomes positively charged and opposes the migration of additional positive ions out to the crystal surface. On the other hand, the excess negative charge created inside the crystal is favorable for the formation of negative vacancies. In the absence of external forces, therefore, the number of oppositely charged vacancies inside an ionic crystal tends to be equal, forming pairs of vacancies.

23.1.2.3 Interstitial Point Defects

A crystalline solid consists of a periodic array of atoms with finite empty spacings between them called voids (Chapter 1). In most of the solids the atoms are closely packed. Out of the large number of possible close packings, only a few are found to occur in nature, such as cubic and hexagonal close packings, among others. In all types of close packings, the voids can be

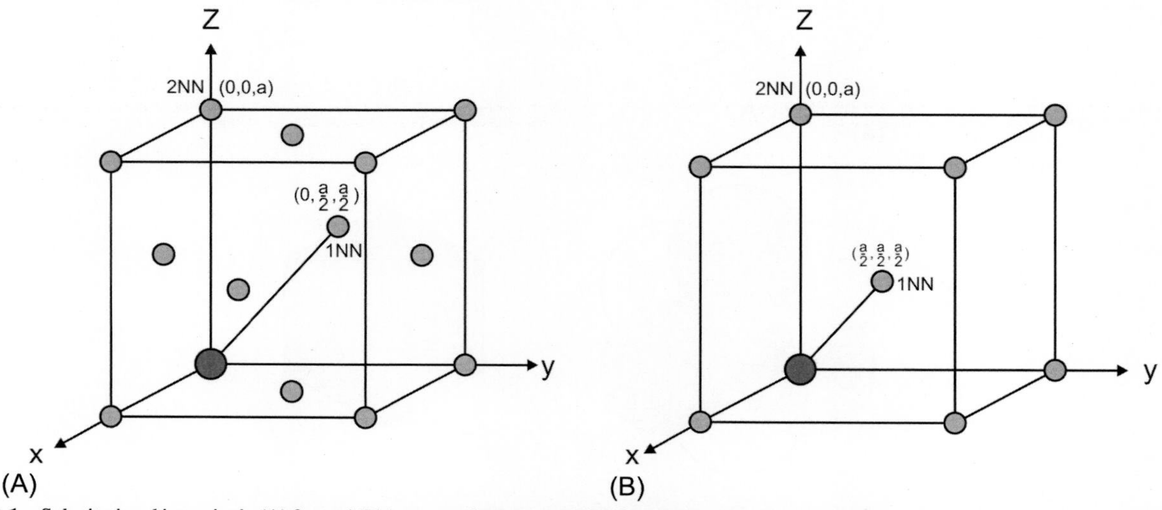

FIG. 23.1 Substitutional impurity in (A) fcc and (B) bcc crystal structures. The bigger atom with the *dark shade* represents an impurity at the origin, *while smaller shaded* atoms represent the host atoms. *(Modified from Galsin, J. S. (2002). Impurity scattering in metallic alloys (p. 28). New York: Kluwer Academic/Plenum Publishers.)*

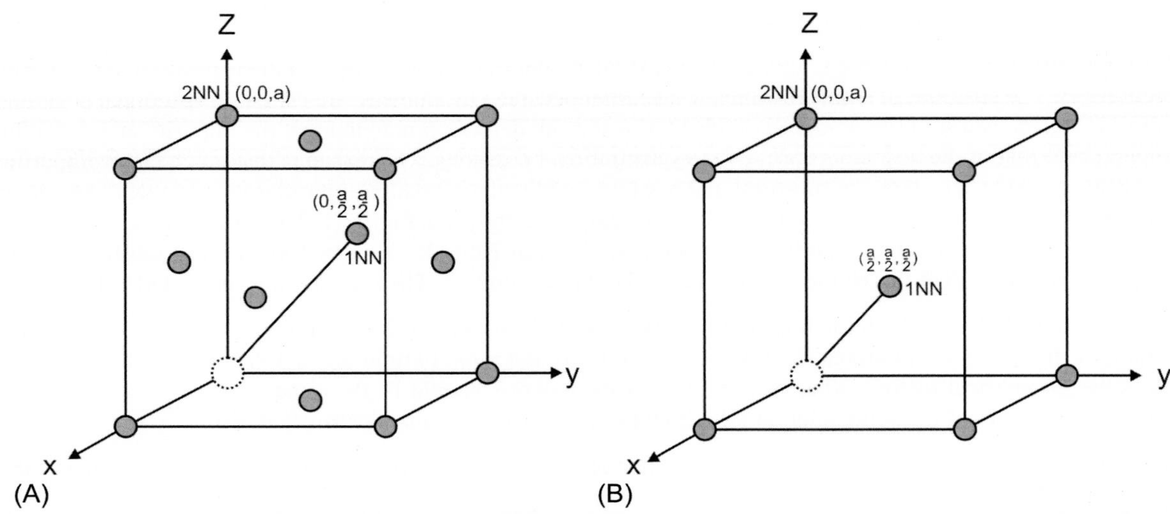

FIG. 23.2 A vacancy in (A) fcc and (B) bcc crystal structures. The *hollow dashed sphere* at the origin represents the vacancy and the *shaded spheres* represent the host atoms. *(Modified from Galsin, J. S. (2002). Impurity scattering in metallic alloys (p. 29). New York: Kluwer Academic/Plenum Publishers.)*

classified into only two categories: tetrahedral and octahedral voids. If the atoms are considered to be hard spheres, then triangular empty spaces exist in each close-packed layer of atoms (Fig. 23.3). If there is a sphere directly over a triangular space, then there results a void with four atoms around it. The four atoms can be arranged at the corners of a regular tetrahedron (Fig. 23.3a), resulting in a void called a tetrahedral void. The center of the tetrahedral void is called the *tetrahedral interstitial position* or *tetrahedral interstice*. Each tetrahedral interstitial position is surrounded by four atoms and each atom is surrounded by eight tetrahedral interstitial positions. Therefore, on average there are two tetrahedral interstitial positions per atom. If a triangular void pointing up in one close-packed layer is covered by a triangular void pointing down in the adjacent layer, then the void that is formed is called an octahedral void which is surrounded by six atoms

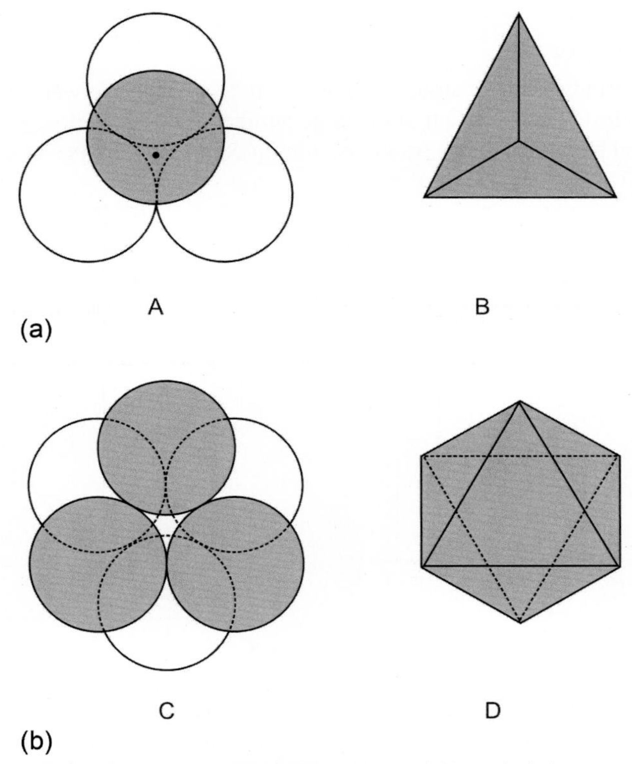

FIG. 23.3 Two types of voids in a crystal structure with (a) tetrahedral symmetry and (b) octahedral symmetry.

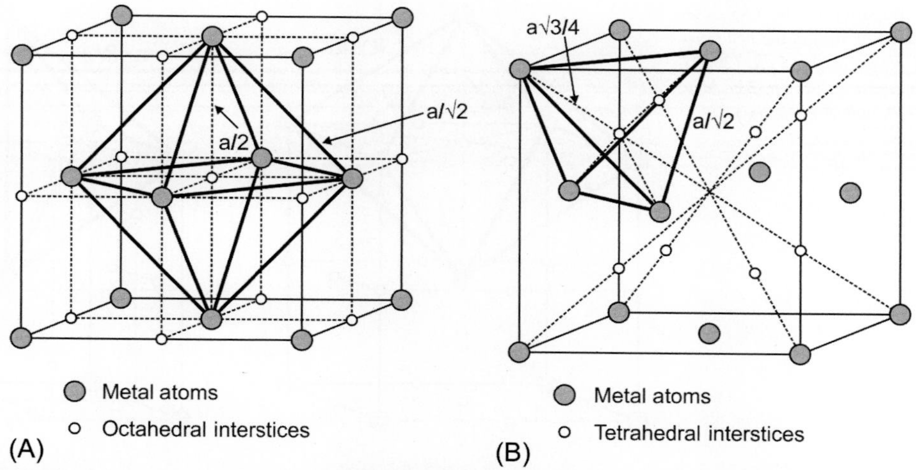

FIG. 23.4 Interstitial impurity in the (A) octahedral void and (B) tetrahedral void in the fcc crystal structure. *(Modified from Galsin, J. S. (2002). Impurity scattering in metallic alloys (p. 31). New York: Kluwer Academic/Plenum Publishers.)*

(see Fig. 23.3b). The center of the octahedral void is called the *octahedral interstitial position* or *octahedral interstice*. Each octahedral interstitial position is surrounded by six atoms and each atom is surrounded by six octahedral interstitial positions. Therefore, the average number of octahedral interstitial positions per atom is one.

It is found that the light elements, such as H, O, N, B, and C, when mixed with metals, take either the tetrahedral or octahedral interstitial positions. This is because the light atoms are not able to knock out the heavy atoms of the metal but can easily accommodate themselves in the voids due to their small size. Fig. 23.4 shows two types of interstitial voids in the case of an fcc structure. The octahedral interstitial positions are the midpoints of the edges of the unit cell, that is, at (0, 0, 1/2), (0, 1/2, 0), (1/2, 0,0), and at (1/2, 1/2, 1/2), which is the center of the unit cell. The tetrahedral interstitial positions are at (1/4, 1/4, 1/4), (1/4, 1/4, 3/4), (3/4, 3/4, 1/4), and (3/4, 3/4, 3/4). The 1NNs and 2NNs of the octahedral and tetrahedral positions in a crystal with fcc structure are given in Table 23.1. The two types of voids in a crystal with bcc structure

TABLE 23.1 Positions, Distances, and Number of 1NNs and 2NNs of Octahedral and Tetrahedral Interstitial Positions in a Host With fcc Crystal Structure

nNN	Position[a]	Number	Distance
Octahedral site			
1NN	(±1/2, 0, 0)a	6	a/2
	(0, ±1/2, 0)a		
	(0, 0, ±1/2)a		
2NN	(±1/2, ±1/2, ±1/2)a	8	$\sqrt{3}a/2$
Tetrahedral site			
1NN	(1/4, 1/4, 1/4)a	4	$\sqrt{3}a/4$
	(−1/4, −1/4, −1/4)a		
	(1/4, −1/4, −1/4)a		
	(−1/4, 1/4, −1/4)a		
2NN	(±1/4, ±3/4, ±1/4)a	12	$\sqrt{11}a/4$
	(±1/4, ±1/4, ±3/4)a		
	(±3/4, ±1/4, ±1/4)a		

[a]*We choose ± such that the parity is odd.*

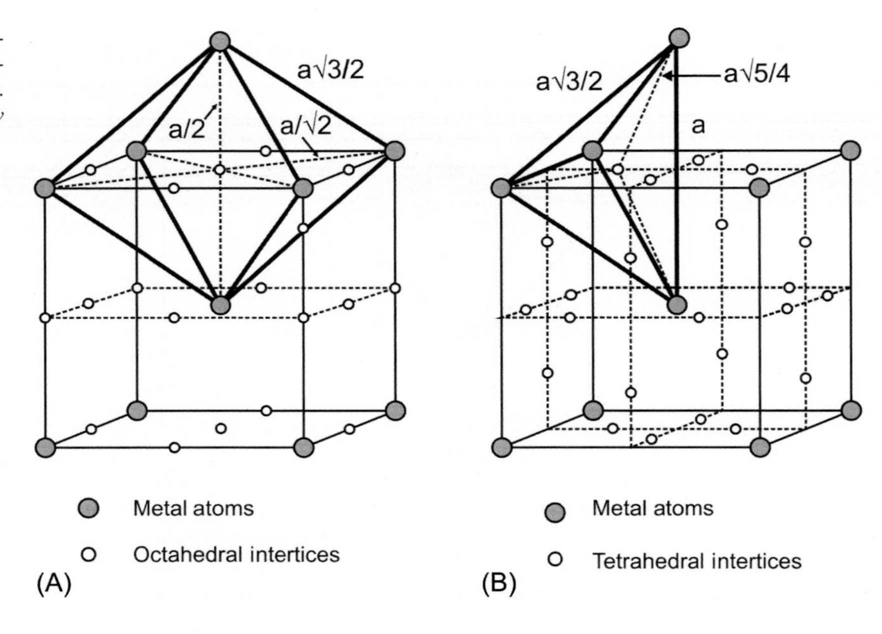

FIG. 23.5 Interstitial impurity in the (A) octahedral void and (B) tetrahedral void in the bcc crystal structure. *(Modified from Galsin, J. S. (2002). Impurity scattering in metallic alloys (p. 33). New York: Kluwer Academic/Plenum Publishers.)*

TABLE 23.2 Positions, Distances, and Number of 1NNs and 2NNs of Octahedral and Tetrahedral Interstitial Positions in a Host With bcc Crystal Structure

nNN	Position	Number	Distance
Octahedral site			
1NN	$(0, 0, \pm 1/2)a$	2	a/2
2NN	$(\pm 1/2, \pm 1/2, 0)a$	4	$a/\sqrt{2}$
Tetrahedral site			
1NN	$(\pm 1/2, -1/4, 0)a$	4	$\sqrt{5}a/4$
	$(0, 1/4, \pm 1/2)a$		
2NN	$(\pm 1/2, 3/4, 0)a$	4	$\sqrt{13}a/4$
	$(0, -3/4, \pm 1/2)a$		

are shown in Fig. 23.5. The 1NNs and 2NNs of the tetrahedral and octahedral voids in a bcc crystal are given in Table 23.2. It is noteworthy that in an fcc crystal, an octahedral void is larger than a tetrahedral void, while the reverse is true in a crystal with bcc structure.

One should note that every symmetry operation of a cube is not a symmetry operation of a regular tetrahedron. For example, a rotation through $\pi/2$ about an axis passing through the center of the cube and parallel to one of its edges takes the cube into itself, but not for a tetrahedron. On the other hand, one can show that all the symmetry operations of a cube are also symmetry operations of a regular octahedron and vice versa. One can classify solid solutions with respect to the nature of the impurity. For example, *substitutional solid solutions* are formed by the presence of substitutional impurities, while *interstitial solid solutions* are formed by the presence of interstitial impurities in a host element.

23.1.2.4 The Frenkel Defects

A Frenkel defect in a crystalline solid arises due to the migration of an atom at the lattice point to a nearby interstitial position (Fig. 23.6). When the interstitial does not fall back into the vacancy so produced, either the vacancy or the interstitial or both may migrate farther away from the point of creation. Ultimately the two components of the Frenkel defect are free from each other's influence. In a pure crystalline solid a Frenkel defect can be regarded as a self-interstitial atomic defect.

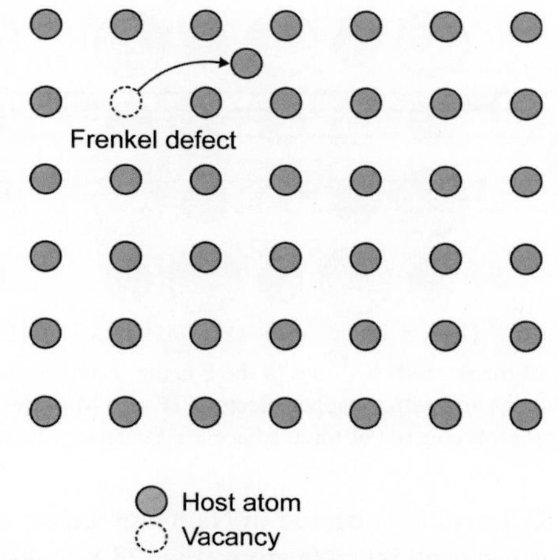

FIG. 23.6 A Frenkel defect in a square lattice. *Shaded spheres* represent host atoms, while the *hollow sphere* represents the vacancy.

23.1.2.5 Color Centers

Ionic crystals, such as NaCl, KCl, and LiF, have forbidden energy gaps on the order of 6 eV and, therefore, these crystals are expected to be transparent to visible light. It is found that the ionic crystals become colored due to the presence of point defects. For example, NaCl crystal becomes yellow in color due to the presence of point defects. A *color center* is a lattice defect that absorbs visible light. There are a number of ways in which an ionic crystals become colored.

1. Crystals become colored with the addition of suitable impurities, such as transition metal ions.
2. The ionic crystals can acquire color by heating in the presence of an alkali metal. The point defects so produced introduce nonstoichiometry in the ionic solid and are called F-centers.
3. The crystals become colored or their color becomes darker by exposing them to high-energy radiation, such as X-rays or γ-rays, or by bombarding them with high-energy electrons or neutrons.

The simplest and most studied color center is the F-center. F-centers are generally produced by heating a crystal in an excess of alkali vapors or by irradiating the crystal by X-rays. For example, when NaCl crystal is heated in the presence of vapors of Na metal, some Na atoms are deposited on the NaCl crystal. These Na atoms lose their outermost electron, forming Na^{+1} ions and occupying the lattice positions. Corresponding to each Na^{+1} ion, there exists an empty position for the negative ion Cl^{-1} and the electron released by the Na atom is bound to the vacancy of the negative ion (see Fig. 23.7), thus maintaining local charge neutrality. This electron is shared by six positive Na^{+1} ions adjacent to the vacant negative site. The excess electron captured at the negative ion site in an alkali halide is called an F-center. The excess electron forms an absorption band in NaCl at about 4650 Å, which is called the F-band. The F-band is in the frequency range of the color blue, which is absorbed and, therefore, is responsible for the yellow color of NaCl crystal. It is noteworthy that the F-band is characteristic of the crystal and not of the alkali vapor in which it is heated.

There also exist more complex color centers. One of the simplest complex color centers is called the F_A center. In general, the F_A center is produced if one of the six 1NN atoms of the F-center in an alkali halide crystal is replaced

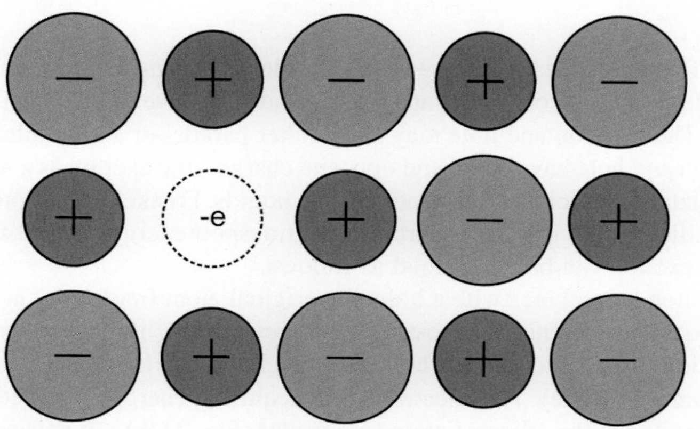

FIG. 23.7 An F-center is a negative ion vacancy with one excess electron bound at the vacancy. The distribution of the excess electron charge is largely on the positive metal ions adjacent to the vacant lattice site.

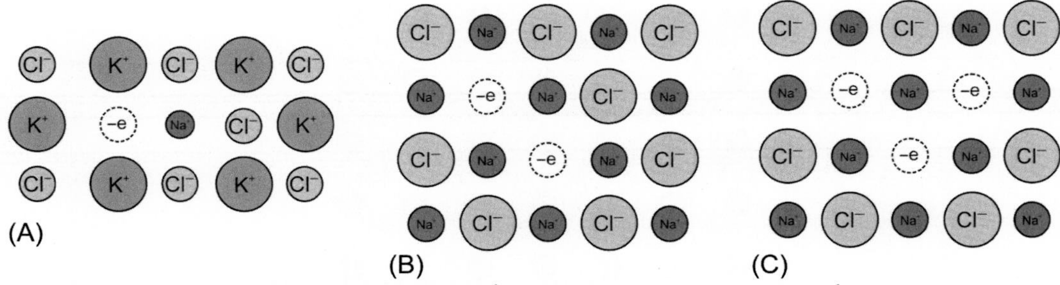

FIG. 23.8 (A) An F_A-center in a KCl crystal. One of the six 1NN K^{+1} ions of the F-center is replaced by Na^{+1} or in general by another alkali ion. The *dashed circle* is used here to denote a vacant negative ion site with a trapped electron. (B) An M-center in a NaCl crystal. It consists of two adjacent F-centers in the crystal. (C) An R-center in a NaCl crystal. It consists of three adjacent F-centers in the crystal.

by a different alkali ion. For example, if KCl crystal is heated in Na metal vapor, then it may happen that one of the first nearest neighbor K atoms of the F-center is replaced by a Na atom (Fig. 23.8A). Further, there is a finite possibility of the formation of a cluster of F-centers. If two adjacent F-centers are formed in an alkali halide crystal, then the color center so formed is called an M-center. But if three adjacent F-centers are formed, then the color center is called an R-center. The different color centers are distinguished by their different optical absorption frequencies. Fig. 23.8B and C show the M- and R-centers in KCl crystal.

Color centers can also be created by trapping holes at a positive ion vacancy. But the hole centers are not as simple as the electron centers. The formation of a hole is described below. The chlorine ion Cl^{-1} has the following electron configuration:

$$Cl^{-1} : 1s^2 2s^2 2p^6 3s^2 3p^6$$

The above electronic configuration is usually called the $3p^6$ configuration and gives a spherically symmetric ion. The creation of a hole in the Cl^{-1} ion gives us the following configuration:

$$Cl : 1s^2 2s^2 2p^6 3s^2 3p^5$$

The $3p^5$ configuration gives rise to an asymmetric ion and has the same configuration as that of a Cl atom. If one more electron is added to Cl^{-1}, the electronic configuration of Cl^{-2} becomes

$$Cl^{-2} : 1s^2 2s^2 2p^6 3s^2 3p^6 4s^1$$

The $3p^6 4s^1$ configuration yields a spherically symmetric ion. Therefore, the states represented by $3p^5$ and $3p^6 4s^1$ are different: one is spherically asymmetric, while the other is spherically symmetric. The asymmetric state will immediately distort the surroundings of the crystal. One can form the molecule of Cl^{-1} as follows:

$$Cl^{-1} + Cl^{-1} = Cl_2^{-1} + e^{-1}$$

Therefore, a hole is trapped in the molecule Cl_2^{-1} as is evident from the configuration of Cl^{-1}.

23.1.3 Excitons

The concept of an exciton was first proposed by Frenkel in 1931 while describing the excitation of atoms in the lattice of an insulator. An exciton in a crystalline solid consists of an electron and a positive hole, which attract each other by electrostatic forces (see Fig. 23.9A). The electron and hole may have either parallel or antiparallel spins, which are coupled by exchange forces. As the electron and hole have equal and opposite charges, the exciton as a whole is an electrically neutral quasiparticle that exists in insulators, semiconductors, and in some liquids. Frenkel further proposed that the exciton is free to move throughout a nonmetallic crystal, just like a particle, and transport energy without transporting electrical charge. The transport of energy by an exciton can be understood as follows.

When an electron in an exciton recombines with a hole, the original atom (molecule) is restored and the exciton vanishes. In the recombination process some energy is lost, which gives a finite lifetime to the excitons. The energy of the exciton during the recombination of the electron and hole may be emitted as light energy, which may get transferred to the electron in the adjacent atom (molecule). This electron, after acquiring energy, is forced to move away from its atom (molecule), producing a new exciton at the adjacent atom (molecule) (Fig. 23.9A). It appears as if the exciton has moved

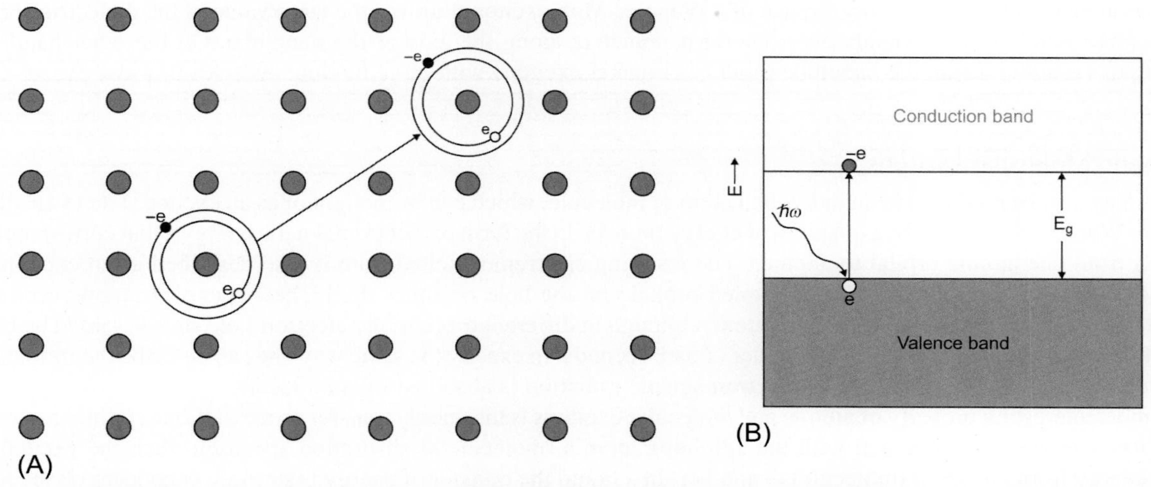

FIG. 23.9 (A) A representation of an exciton as a combination of an electron and hole. The electron and hole have different energies and so revolve about the ion in circles with different radii. The exciton can move from one atom to another in the lattice. (B) Schematic energy band representation of a bound electron-hole pair or an exciton in a semiconductor.

from one atom (molecule) to another along with its energy. In this way the exciton may propagate through the atomic (molecular) solid. There have been several mechanisms proposed for the transfer of energy but two of them are the most important.

1. First is the dissipation of exciton energy due to interactions with the lattice of the crystal.
2. The second method is that the exciton energy is carried away by radiation.

The combination of these two processes has also been used. The excitation of an electron from the valence band to the conduction band in a semiconductor through absorption of a photon provides a simple example of an exciton (Fig. 23.9B). An electron in the conduction band is attracted by a hole in the valence band by Coulomb forces, thus forming an exciton. As some of the energy is spent in binding the electron and hole, the exciton has slightly less energy than an unbound electron and hole in a solid. The wave function of an exciton is said to be hydrogenic, as the excited atomic state is like that of a hydrogen atom. However, the binding energy of an exciton is much smaller and the particle size is much larger than in a hydrogen atom.

23.1.3.1 Types of Excitons

Excitons may be categorized with respect to the attractive interaction between the electron and hole, which in turn depends on the nature of the solid. Excitons are of two types.

Frenkel Excitons

If the Coulomb attraction between the electron and hole is strong enough, one obtains what is called a Frenkel exciton. This is the case in solids with small values of dielectric constant, such as insulators. Frenkel excitons are localized excitations, that is, localized on the same atom or molecule, with a binding energy on the order of 0.1–1.0 eV. Frenkel excitons are usually found in alkali halides and organic molecular crystals composed of aromatic molecules.

Wannier-Mott Excitons

If the attractive Coulomb interaction between the electron and hole is weak, one gets a Wannier-Mott exciton. For example, in semiconductors, the dielectric constant is large, which immensely reduces the Coulomb interaction between the electron and hole. Therefore, the distance between the electron and hole is large, yielding a large radius for the Wannier-Mott exciton: larger than the lattice spacing. The binding energy of a Wannier-Mott exciton is much larger than that of a hydrogen atom: it is typically on the order of 0.01 eV. Wannier-Mott excitons are found in semiconductors with small energy band gaps and high values of dielectric constant and in liquids, such as xenon.

Both types of excitons can occur in a single-walled carbon nanotube. This is because of the peculiar nature of the Coulomb interactions between the electron and hole in one dimension. The dielectric constant inside the carbon nanotube is large enough, thus favoring the occurrence of a Wannier-Mott exciton. Further, the large value of the dielectric constant yields a wave function that extends over several nanometers along the axis of the nanotube. On the other hand, poor screening in vacuum outside the nanotube produces Frenkel excitons with large binding energies (0.4–1.0 eV).

Atomic and Molecular Excitons

Excitons may also be produced in an individual atom or molecule, which can be thought of as an excited state of an atom or molecule. When an atom absorbs a quantum of energy (may be in the form of electromagnetic energy), that corresponds to a transition from one atomic orbital to another. The resulting electronic excited state is also described as an exciton. The excited electron occupies the lowest unoccupied orbital and the hole occupies the highest occupied atomic orbital. As the electron and hole both occur in the same atom, although in different orbitals, the electron-hole state is said to be bound. The lifetime of an atomic exciton is on the order of nanoseconds. In exactly the same way one can describe the formation of a molecular exciton when a quantum of electromagnetic radiation is absorbed by a molecule.

The most interesting property of atomic and molecular excitons is the energy transfer process. If one atomic (molecular) exciton has a proper energy match with the adjoining atom's (molecule's) absorption spectrum, then the exciton may transfer energy from one atom (molecule) to another. In a liquid the transfer of energy is strongly dependent on the interatomic (intermolecular) distance.

23.1.4 Statistical Distribution of Point Defects

The statistical distribution of point defects can be studied using thermodynamics of solids. The thermodynamic study involves the process of minimizing the energy of a crystal with respect to the formation of point defects. The energy of a crystalline solid is given by Gibb' s free energy, defined as

$$G(E, P, V, T, S) = E + PV - TS \tag{23.1}$$

Here P, V, T, S are the pressure, volume, temperature, and entropy. E is the energy of formation of the point defect and depends on its nature. In a crystalline solid P and V are nearly constant; therefore, the Gibb's free energy can be replaced by the Helmholtz free energy F, defined as

$$F = E - TS \tag{23.2}$$

Suppose N_P is the number of point defects introduced in a crystalline solid with N_L lattice sites. The change in Helmholtz free energy ΔF of the crystal due to N_P point defects at a particular temperature becomes

$$\Delta F = N_P E - T\Delta S \tag{23.3}$$

where ΔS is the change in entropy. The entropy has two contributions: configurational (or mixing) entropy ΔS_c and vibrational entropy ΔS_v. Therefore,

$$\Delta S = \Delta S_c + N_P \Delta S_v \tag{23.4}$$

ΔS_v arises from the vibrational motion of each point defect about its equilibrium position at finite temperature, but we are neglecting it in the present discussion. ΔS_c arises due to the distribution of the point defects on the lattice, which produces disorder in the atomic distribution. From Eqs. (23.3), (23.4) one can write

$$\Delta F = N_P E - T\Delta S_c \tag{23.5}$$

Let the point defects be distributed on the lattice in a number W of ways. From statistical considerations, ΔS_c can be related to W as

$$\Delta S_c = k_B \ln(W) \tag{23.6}$$

The equilibrium state of the crystal demands that ΔF be minimum with respect to the number of point defects, that is,

$$\frac{\partial}{\partial N_P}(\Delta F) = E - T\frac{\partial}{\partial N_P}(\Delta S_c) = 0 \tag{23.7}$$

From Eq. (23.7) one can calculate the distribution of different point defects on the crystal lattice.

23.1.4.1 Substitutional Point Defects

Let there be N_s substitutional point defects (which include vacancies) in a crystalline solid with N_L atoms or lattice points. If E_s is the energy required to create a substitutional point defect, then Eq. (23.7) becomes

$$E_s - T\frac{\partial}{\partial N_s}(\Delta S_c) = 0 \tag{23.8}$$

The number of ways in which N_s substitutional point defects can be distributed on N_L lattice points is given by

$$W = \frac{N_L!}{(N_L - N_s)!N_s!} \tag{23.9}$$

From Eqs. (23.6), (23.8), (23.9) one can write

$$E_s - k_B T\frac{\partial}{\partial N_s} \ln\left(\frac{N_L!}{(N_L - N_s)!N_s!}\right) = 0 \tag{23.10}$$

If N_s and N_L are large, which is the case in a crystalline solid, then one can apply the Stirling formula defined as

$$\ln(N!) = N \ln(N) - N \tag{23.11}$$

Using the Stirling formula in Eq. (23.10) and simplifying, one gets

$$E_s + k_B T \ln\left(\frac{N_s}{N_L - N_s}\right) = 0 \tag{23.12}$$

The above equation can be solved immediately for N_s to give

$$N_s = N_L \exp\left(-\frac{E_s}{k_B T}\right) \tag{23.13}$$

A vacancy is also a substitutional point defect so the number of vacancies N_v can be obtained in exactly the same way as was done above, that is,

$$N_v = N_L \exp\left(-\frac{E_v}{k_B T}\right) \tag{23.14}$$

where E_v is the energy of formation of a vacancy.

It has already been discussed that the vacancies occur in pairs in ionic crystals: an equal number of positive and negative vacancies. Consider N_{vp} pairs of vacancies in an ionic crystal. The number of ways in which each kind of vacancy (N_{vp} in number) can be arranged is the same as given by Eq. (23.9). Hence the total number of ways in which N_{vp} pairs of vacancies can be arranged on the lattice is given by

$$W = \left[\frac{N_L!}{\left(N_L - N_{vp}\right)!N_{vp}!}\right]^2 \tag{23.15}$$

Substituting Eq. (23.15) into Eq.(23.6), one can calculate ΔS_c. The ΔS_c so obtained is then substituted in the equation

$$E_{vp} - T\frac{\partial}{\partial N_{vp}}(\Delta S_c) = 0 \tag{23.16}$$

to evaluate N_{vp}, which is given by

$$N_{vp} = N_L \exp\left(-\frac{E_{vp}}{2k_B T}\right) \tag{23.17}$$

Here E_{vp} is the energy of formation of a pair of vacancies.

23.1.4.2 Interstitial Point Defects

In a crystalline solid there are N_L lattice points and N_L' interstitial positions. Let E_I be the energy of formation of an interstitial point defect. The number of ways in which N_I interstitial point defects can be distributed on the N_L' interstitial positions is given by

$$W = \frac{N_L'!}{(N_L' - N_I)!N_I!} \tag{23.18}$$

Substituting Eq. (23.18) into Eq. (23.6), we obtain

$$\Delta S_c = k_B \ln\left(\frac{N_L'!}{(N_L' - N_I)!N_I!}\right) \tag{23.19}$$

The equation of motion for an interstitial point defect, from Eq. (23.7), can be written as

$$E_I - T\frac{\partial}{\partial N_I}(\Delta S_c) = 0 \tag{23.20}$$

Using the Stirling formula in Eq. (23.19) to obtain ΔS_c and then substituting into Eq. (23.20), we get

$$E_I + k_B T \ln\left(\frac{N_I}{N_L' - N_I}\right) = 0 \tag{23.21}$$

Solving Eq. (23.21) for N_I, we obtain

$$N_I = N_L' \exp\left(-\frac{E_I}{k_B T}\right) \tag{23.22}$$

23.1.4.3 The Frenkel Defects

Consider a crystalline solid with N_f Frenkel defects. In other words, there are N_f vacancies and N_f interstitial point defects. The number of ways in which N_f vacancies can be distributed on the N_L lattice points is given by

$$W' = \frac{N_L!}{(N_L - N_f)!N_f!} \tag{23.23}$$

Further, the number of ways in which N_f interstitial point defects can be distributed on N_L' interstitial positions is given by

$$W'' = \frac{N_L'!}{(N_L' - N_f)!N_f!} \tag{23.24}$$

The total number of ways W in which N_f Frenkel defects can be distributed in the crystalline solid is given by the multiplication of W' and W'', giving

$$W = W'W'' = \frac{N_L!}{(N_L - N_f)!N_f!}\frac{N_L'!}{(N_L' - N_f)!N_f!} \tag{23.25}$$

Substituting Eq. (23.25) into Eq. (23.6), the mixing entropy due to the Frenkel defects becomes

$$\Delta S_c = k_B \ln\left(\frac{N_L!}{(N_L - N_f)!N_f!}\frac{N_L'!}{(N_L' - N_f)!N_f!}\right) \tag{23.26}$$

The distribution of the Frenkel defects, from Eq. (23.7) is defined as

$$E_f - T\frac{\partial}{\partial N_f}(\Delta S_c) = 0 \tag{23.27}$$

where E_f is the energy of formation of a Frenkel defect. If N_f is large, then ΔS_c can be simplified using the Stirling formula. The ΔS_c so obtained is then substituted into Eq. (23.27) to get

$$-\frac{E_f}{k_B T} = \ln\left[\frac{N_f^2}{(N_L - N_f)\,(N_L' - N_f)}\right] \tag{23.28}$$

If the number of Frenkel defects is much lower than the number of lattice (interstitial) sites, that is, $N_f \ll N_L$ and $N_f \ll N_L'$, then Eq. (23.28) reduces to

$$-\frac{E_f}{k_B T} = \ln\left[\frac{N_f^2}{N_L N_L'}\right] \tag{23.29}$$

Eq. (23.29) can be solved for N_f to give

$$N_f = (N_L N_L')^{1/2} \exp\left(-\frac{E_f}{2k_B T}\right) \tag{23.30}$$

One should note that any of the above defects, at finite temperature, do not remain at one location. After acquiring thermal energy, the point defects can move from one site to another in the crystalline solid.

Problem 23.1

Energy of 1.1 eV is required to move an atom from within a crystal to its surface. What fraction of vacancies is present in the crystal at 1000 and 500 K?

Problem 23.2

Consider a crystal with N_L lattice points (atoms) and N_P point defects distributed in the lattice. Using the Gibb's free energy, derive the equation for the equilibrium state of the crystal. Further, derive an expression for the distribution of substitutional point defects in the crystal.

23.2 DISLOCATIONS

23.2.1 Plastic Deformation of Crystals

With the application of weak stress, a crystal is deformed elastically and returns to its original state upon removal of the stress. However, if the applied stress is sufficiently large (larger than some critical value), a certain amount of deformation remains even after the removal of the stress. Under such conditions the crystal is said to be plastically deformed. It will be seen that an atomic interpretation of the plastic flow of crystals requires the presence of dislocations.

23.2.2 Definition of Dislocation

Let us consider a crystalline solid, in rectangular form, divided into two halves. The upper half of the crystal is slipped over the lower half by a distance **b** (see Fig. 23.10). Suppose the part of the plane in the upper half of the crystal AEFD has slipped by a distance AA$'$ (= DD$'$) above the plane of the lower part in the direction shown by the arrow and acquires the position A$'$EFD$'$, while the remaining part EBCF of this plane is fixed. The plane ABCD is called the *slip plane* and the vector **b** gives the direction and magnitude of the slip, usually called the *Burgers vector*. The line EF is the boundary

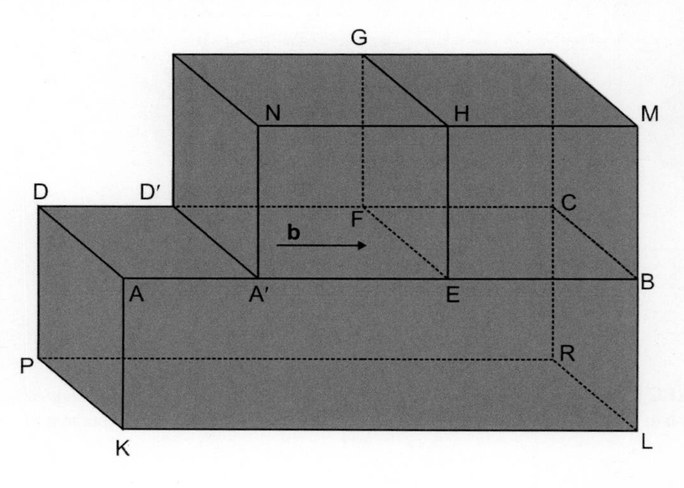

FIG. 23.10 A crystalline solid in which the upper half of the solid has slipped by length AA$'$ with ABCD as the slip plane. The rectangular part AEFD of the slip plane has already slipped, but the part EBCFis unslipped. The line EF is an edge dislocation.

between the slipped and unslipped parts of the plane ABCD. If the distance AA′ (= DD′) is equal to the lattice constant "a" of the crystal, that is, the distance between consecutive atoms, then the slipped part has recovered the regular atomic arrangement of the crystal. So, the wrinkles caused by the slip converge entirely on the line EF. Because the part of the crystal along the line EF has shifted from the regular atomic arrangement, this is a line defect. *The line defect, which constitutes the boundary between the slipped and unslipped parts of the slip plane, is called a dislocation.* When the dislocation line moves in the direction of **b**, the area of the slipped part of the slip plane increases. Ultimately, when the dislocation line reaches the end of the crystal, the whole of the plane in the upper part of the crystal has slipped over the slip plane and it is said that a slip of magnitude **b** has occurred. When the displacement of one lattice plane over the other equals one lattice constant in the slip plane, the slip is said to be a *unit slip* and the dislocation is said to have unit strength.

The slip of a crystal when observed by an optical microscope appears as a *slip band* on the surface of a single crystal. A slip band exhibits a fine structure, which consists of a group of unit slips. A realistic slip, of course, comprises n unit slips taking place successively with length $\mathbf{b}=n\mathbf{a}$. Therefore, one can always analyze a slip of length **b** in terms of unit slips.

Dislocations may not necessarily be limited to a straight line or to a slip plane. For example, let PQRS be a slip plane dividing the crystal into two halves, as shown in Fig. 23.11A. In the slip plane, consider an arbitrary closed curve ABC that encloses the shaded region. This plane, as seen from above, is shown in Fig. 23.11B. Suppose that, by applying a shear stress, the material located over the shaded area in the upper half of the crystal is displaced by an amount **b** relative to the lower half of the crystal. At the same time, the material in the upper half of the crystal lying outside the area ABC is left in place. Therefore, only a fraction of the upper half of the crystal has slipped relative to the lower half. Let us define a parameter f_s as the ratio of the area slipped ABC to the total area of the slip plane PQRS, that is,

$$f_s = \frac{\text{Area enclosed by curve ABC}}{\text{Area of slip plane PQRS}} \tag{23.31}$$

The ratio f_s is referred to as the fraction of slip that has occurred in this plane. If the area ABC is made to grow, f_s increases and, ultimately, approaches unity. In the case of $f_s = 1$, the whole of the upper half of the crystal would be displaced by an amount **b** relative to the lower half, which, in other words, means that a slip of magnitude **b** has occurred. For $f_s < 1$, the average displacement of the upper half relative to the lower half is $f_s |\mathbf{b}|$.

The line ABC marks the boundary, in the slip plane, between the slipped and unslipped material and by definition it constitutes the dislocation line. The vector **b** defines the magnitude and direction of the slip and is the Burgers vector. Because the atoms always acquire the minimum energy position, the vector **b** must connect two atomic equilibrium positions and its value is determined by the crystal structure. The direction of the dislocation line at any given point is described by a unit vector $\hat{\mathbf{t}}$, which is tangent to the dislocation line. The direction of $\hat{\mathbf{t}}$ changes continuously as one moves along the dislocation loop (Fig. 23.11C): $\hat{\mathbf{t}}$ is continuous around the dislocation loop. The directions of the vector $\hat{\mathbf{t}}$ will be opposite on opposite sides of the loop. The direction of the unit vector $\hat{\mathbf{t}}$ (clockwise or counterclockwise) is immaterial as it will only result in a change of sign (plus or minus).

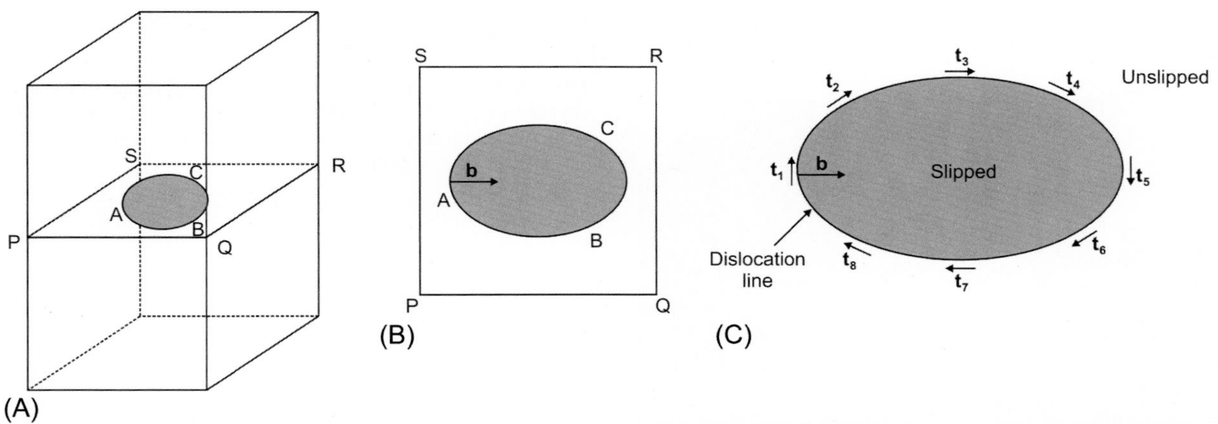

FIG. 23.11 (A) Schematic representation of a closed curved dislocation ABC in the slip plane PQRS in a solid. The slip has occurred only in the *shaded area* in the curved dislocation. (B) The slip plane PQRS with Burgers vector **b** in the slipped portion ABC. (C) The directions of the unit vector $\hat{\mathbf{t}}$ tangent to the dislocation line at different points.

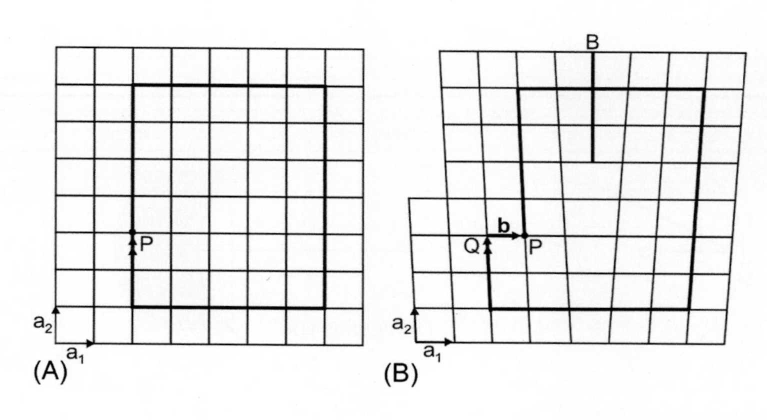

FIG. 23.12 Burgers circuit (a) in a pure crystal and (b) in a real crystal containing an edge dislocation. The starting point is shown by a *black dot* and the end point by a *double arrow*. In a real crystal containing an edge dislocation, the starting and end points are separated by the Burgers vector **b** as shown. Here $\mathbf{a}_1$ and $\mathbf{a}_2$ represent the primitive translation vectors.

Any model of dislocations must answer a number of questions about the plastic flow:

1. It must explain how the dislocations tend to grow (how the size of the slipped region increases) leading to the motion of a slip. Moreover, the calculated critical shear stress should agree quantitatively with its observed value.
2. An observed slip may correspond to displacements of the order of 1000 Å. But the dislocation line ABC, after sweeping through the whole of the slip plane, may produce a slip with $|\mathbf{b}| \approx 2$ Å and then disappear. Therefore, the model must account for the large number of dislocations taking part in the slip process and also the sources that supply such dislocations.
3. Besides the plastic flow, the model should explain other physical properties reasonably well in the presence of dislocations.

To have more insight into the characteristics of dislocations, it is convenient to define dislocations purely geometrically as follows. In a crystal, the parts having regular atomic arrangement are called good regions, while those having irregular atomic arrangement are called bad regions. First, we define a *Burgers circuit*, which is a loop obtained by moving one atomic distance at a time along the primitive lattice vectors of the crystal structure (Fig. 23.12a). An important property of the Burgers circuit is that it encloses a bad region, although it itself is always in a good region of the crystal. To understand the Burgers circuit, consider a pure crystal having no bad regions as a reference crystal (Fig. 23.12a). Start from an arbitrary reference atom at P and move atom by atom to complete a circuit in the crystal. This is usually called a Burgers circuit and is a closed circuit in a pure crystal. Now introduce a line defect in the same crystal, as a result of which the crystal is distorted and, therefore, contains both good and bad regions. If one proceeds from the same atom at P in the same way, connecting the corresponding atoms as in the reference crystal, one ends up at the position Q (see Fig. 23.12b). Therefore, the Burgers circuit in a crystal with a lattice defect is not closed. The vector **b** drawn from the end point Q to the starting point P of the Burgers circuit gives the Burgers vector. The line defect enclosed by the Burgers circuit is called a dislocation. The above definition of dislocation indicates that the Burgers vector is the most important physical concept. The magnitude of the Burgers vector gives the strength of the dislocation.

23.2.3 Force Acting on Dislocations

Suppose a uniform shear stress $\vec{\sigma}_s$ is applied to a crystal, which may not be in the direction of the Burgers vector **b**. This leads to a force on the dislocation line such that the slipped area tends to grow. Consider element dl of a dislocation line (see Fig. 23.13). Suppose this element is displaced in the outward direction by an amount $\mathrm{dl}_\perp$ along the direction perpendicular to dl. The area swept out by the line element is $\mathrm{dA} = \mathrm{dl}\,\mathrm{dl}_\perp$. Now the average shear displacement of the upper part of the crystal relative to the lower part is

$$f_s \mathbf{b} = \frac{\mathrm{dl}\,\mathrm{dl}_\perp}{A}\mathbf{b} = \frac{\mathrm{dA}}{A}\mathbf{b} \tag{23.32}$$

where A is the area of the slip plane. The work done by the shear stress is equal to the total shear force $\mathbf{F} = \vec{\sigma}_s A$ times the average shear displacement, that is,

$$\mathrm{dW} = \mathbf{F} \cdot (f_s \mathbf{b}) = \left(\vec{\sigma}_s \cdot \mathbf{b}\right) \mathrm{dl}\,\mathrm{dl}_\perp \tag{23.33}$$

Hence, the magnitude of the force acting on the element dl in the normal direction is given by

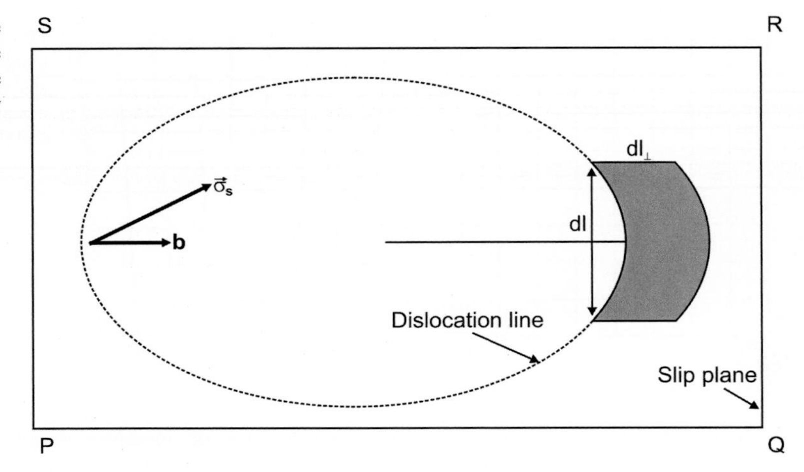

FIG. 23.13 The *dashed line* is the dislocation line in the slip plane PQRS. The small element dl of the dislocation line moves, in the outward direction, through a distance $dl_\perp$ due to the application of shear stress $\vec{\sigma}_s$. The shear stress $\vec{\sigma}_s$ may not be in the direction of the Burgers vector **b** as shown.

$$F_s = \frac{dW}{dl_\perp} = \left(\vec{\sigma}_s \cdot \mathbf{b}\right) dl \tag{23.34}$$

Now the magnitude of the force per unit length acting on the dislocation line is

$$F_d = \frac{dW}{dl\,dl_\perp} = \left(\vec{\sigma}_s \cdot \mathbf{b}\right) \tag{23.35}$$

Thus, the applied shear stress produces a force per unit length $\vec{\sigma}_s \cdot \mathbf{b}$, which is perpendicular to the dislocation line everywhere. If the force is large enough to make the dislocation line move in the direction of F_d, then the slipped area in Fig. 23.13 will grow and slip will occur under the shear stress.

23.2.4 Critical Shear Stress

The dislocation model described above yields a very small value of the critical shear stress σ_{sc} for a slip. Let us consider the regions near the dislocation line somewhat more closely. The boundary between the slipped and unslipped regions is not sharp, but is rather vague due to the interatomic forces extending over several atomic distances. Atoms near the dislocation line, on its inner side, have nearly completed the slip process, but those near the dislocation line, but on the outer side, are just beginning to slip. As a result of the periodic nature of the potential, the atoms outside but near the dislocation line tend to push the dislocation line inward, because this would allow them to occupy their initial equilibrium positions (see Fig. 23.14). On the other hand, the atoms inside but near to the dislocation line tend to push the line outward because this would make it possible for them to occupy their new equilibrium positions. Far away from the dislocation line, on either side of it, the atoms occupy their normal lattice positions and are not affected by the dislocation. Thus, to a first-order

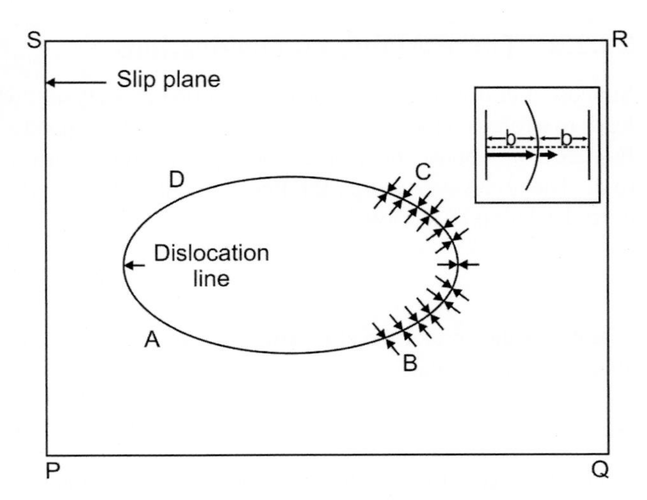

FIG. 23.14 The *dislocation line* ABCD in the slip plane PQRS. The inward and outward forces acting on the part BC of the dislocation line are shown. The inset diagram shows a magnified view of the near completion (startup) of the slip on the inner (outer) side of a small elemental portion of the dislocation line.

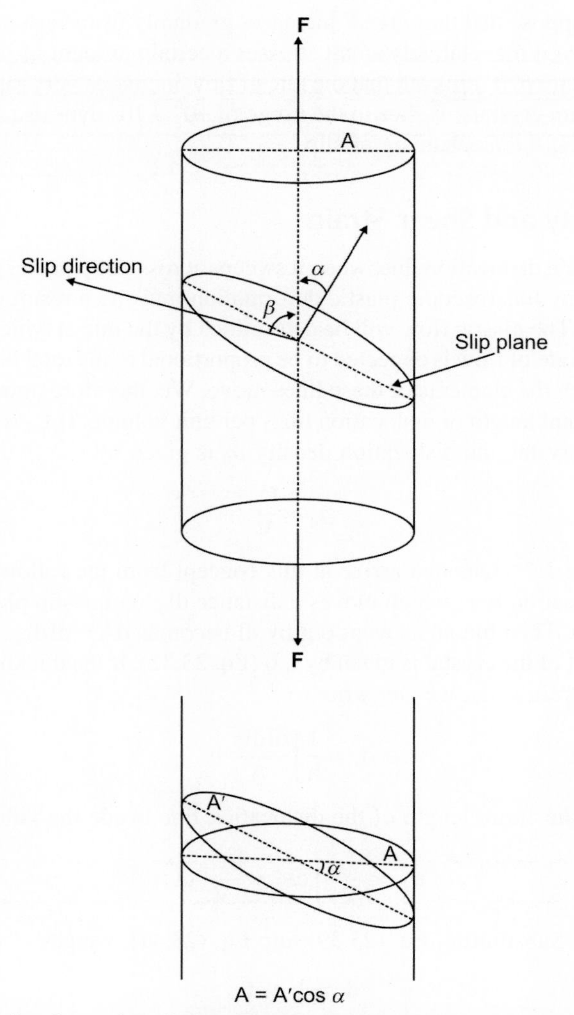

FIG. 23.15 Geometry of slip plane, slip direction, and tensile force **F** acting on a solid in cylindrical form.

approximation, the forces on both sides of the dislocation line balance each other and it should start to move under very small shear forces. Therefore, a first-order approximation will give very low values of $\vec{\sigma}_{sc}$. But in a second-order approximation one finds that the value of $\vec{\sigma}_{sc}$ calculated from this model is of the same order of magnitude as the observed values.

Fig. 23.15 shows a cylindrical crystal with cross sectional area A to which is applied the tensile force **F** with the slip plane shown in the figure. Let the normal to the slip plane make an angle α with **F** and the angle between the slip direction and **F** be β. The force acting on the slip plane per unit area $\mathbf{F}_s$ is given by

$$\mathbf{F}_s = \frac{\mathbf{F}}{A/\cos\alpha} = \frac{\mathbf{F}}{A}\cos\alpha \tag{23.36}$$

The direction of $\mathbf{F}_s$ is the same as that of **F** and can be resolved into two components: one parallel and the other perpendicular to the slip plane. The shear stress is the force acting per unit area (of the slip plane) in the slip direction and is given by

$$\vec{\sigma}_s = \frac{\mathbf{F}}{A/\cos\alpha}\cos\beta = \frac{\mathbf{F}}{A}\cos\alpha\cos\beta \tag{23.37}$$

The tensile stress is the force per unit area normal to the slip plane and is given by

$$\vec{\sigma}_n = \frac{\mathbf{F}}{A/\cos\alpha}\cos\alpha = \frac{\mathbf{F}}{A}\cos^2\alpha \tag{23.38}$$

For particular values of α and β, suppose that the force **F** increases gradually from zero, thereby increasing the magnitudes of the shear and tensile stresses. Even for relatively small stresses a certain amount of plastic flow occurs, but the rate of flow is so small that one speaks of creep. It turns out that the rate of flow increases very rapidly whenever the shear stress $\vec{\sigma}_s$ reaches a critical value $\vec{\sigma}_{sc}$. For pure crystals, $\vec{\sigma}_s$ lies in the range of $10^6 - 10^7$ dynes/sq. cm. At the same time, the results indicate that $\vec{\sigma}_n$ has no influence on the mechanism of slip.

23.2.5 Dislocation Density and Shear Strain

It has already been stated that a single dislocation line, when it sweeps across a slip plane, gives rise to a displacement on the order of a few angstroms. Thus, any macroscopic plastic deformation must be a result of a large number of dislocations sweeping across many slip planes. The plastic flow will be determined by the rate at which dislocation lines sweep through the slip planes. In other words, the rate of flow is expected to be proportional to the total length of all active dislocation lines and the average velocity with which the elements of these lines move. We, therefore, introduce the concept of "dislocation density," which is defined as the total length of dislocation lines per unit volume. If L_d is the total length of the dislocation lines and V is the volume of the crystal, the dislocation density ρ_d is given by

$$\rho_d = \frac{L_d}{V} \tag{23.39}$$

Note that ρ_d has the dimensions of L^{-2}. One can arrive at this concept from the following reasoning.

Let dl be an element of a dislocation line, which moves a distance $dl_\perp$ on the slip plane perpendicular to dl, but in the outward direction (see Fig. 23.13). Then the area swept out by dl becomes $dA = dl\,dl_\perp$. The average displacement of the upper part relative to the lower part of the crystal is given by $f_s\mathbf{b}$ (Eq. 23.32). If the thickness of the crystal perpendicular to the slip plane is h and the shear strain is $d\gamma$, we can write

$$d\gamma = \frac{1}{h}\int \frac{dl\,dl_\perp}{A} b \tag{23.40}$$

The integral over dl gives us L_d, the entire length of the dislocation line inside the volume $V = Ah$ of the crystal. So,

$$d\gamma = \frac{bL_d}{V}\int dl_\perp = \frac{bL_d}{V} d\bar{l}_\perp \tag{23.41}$$

if $d\bar{l}_\perp$ is the average value of $dl_\perp$. Substituting Eq. (23.39) into Eq. (23.41), we get

$$d\gamma = b\rho_d d\bar{l}_\perp \tag{23.42}$$

This is the relation between the average distance moved by the dislocation and the macroscopic strain. The rate of change of shear strain can be written as

$$\frac{d\gamma}{dt} = b\rho_d \frac{d\bar{l}_\perp}{dt} = b\rho_d \bar{v} \tag{23.43}$$

where $\bar{v}$ is the average velocity of the element dl in a direction normal to itself. The dislocation density ρ_d is an extremely important quantity and is related to various phenomena.

23.2.6 Types of Dislocations

Basically, the dislocations are characterized with respect to the directions of the Burgers vector and the dislocation line. The dislocations can be divided into three categories.

1. Edge dislocations
2. Screw dislocations
3. Mixed dislocations

23.2.6.1 *Edge Dislocations*

An edge dislocation is defined as a dislocation for which the Burgers vector **b** is everywhere perpendicular to the dislocation line. The simplest edge dislocation is that in which the dislocation line is a straight line. The formation of a straight-line edge dislocation may be visualized in terms of the slip process. Fig. 23.10 shows a solid in which the upper half is pushed

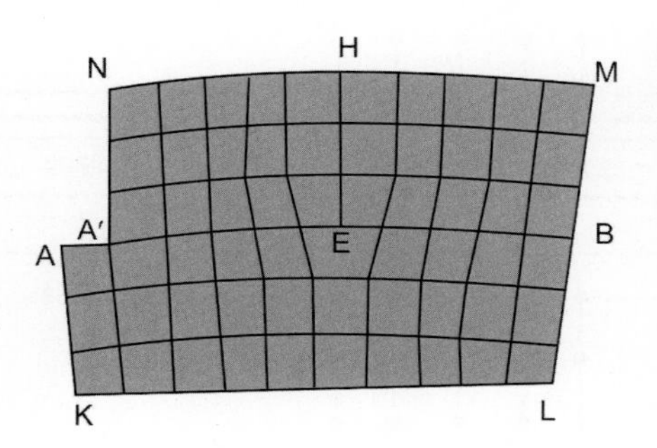

FIG. 23.16 The strain pattern on the front face of the crystal KLMN with straight line edge dislocation HE shown in Fig. 23.10. Here the upper part of the crystal has slipped by AA′. *(Modified from Cottrell, A. H. (1953). Dislocations and plastic flow in crystals (p. 22). Oxford, New York.)*

sideways such that the line A′D′, which initially coincided with AD, is shifted by an amount **b** as indicated. If in this position the two halves are glued together, an edge dislocation is produced. The upper half of the block will clearly be under compression, while the lower half will be under tension. Before the operation, a square network of lines of atoms exists in the solid. But after the operation, the square network gets distorted and the front face AKLMNA′ appears as shown in Fig. 23.16. The strained pattern immediately suggests an alternate method by which an edge dislocation may be produced.

Suppose that the intersection points of the lines of the network represent the positions of atoms in the lattice. Therefore, each point in Fig. 23.16 represents a row of atoms perpendicular to the plane of the paper. The edge dislocation may then be obtained by cutting the block along the plane EFGH (Fig. 23.10) and putting the half plane of atoms initially above AD inside the cut. This gives rise to the "extra" half plane of atoms corresponding to HE in Fig. 23.16, which is typical of an edge dislocation. If the extra half plane is displaced to the right, the slip will progress, and when HE finally reaches the right-hand side of the block, the upper half of the block completes the slip by an amount **b**. The slip process resulting from the motion of edge dislocation is illustrated in Fig. 23.17A.

The edge dislocations for which the extra half plane lies above the slip plane are called positive edge (⊥) dislocations. If the extra half plane lies below the slip plane, it is a negative edge dislocation (T). Fig. 23.17A and B show that the slip resulting from a positive edge dislocation moving to the right is equivalent to the slip resulting from a negative edge dislocation of the same strength moving to the left. The extra half plane may sometimes have an irregular boundary, which gives rise to what is known as an irregular edge dislocation. The definition of an edge dislocation does not necessarily imply

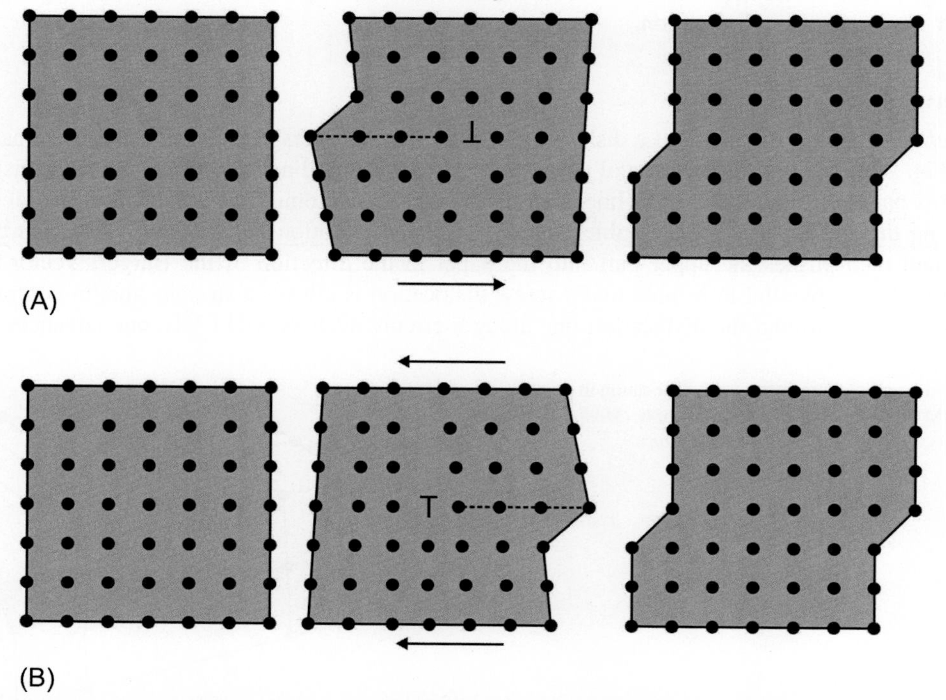

FIG. 23.17 Motion of (A) a positive edge dislocation to the right and (B) negative edge dislocation to the left, both leading to a slip.

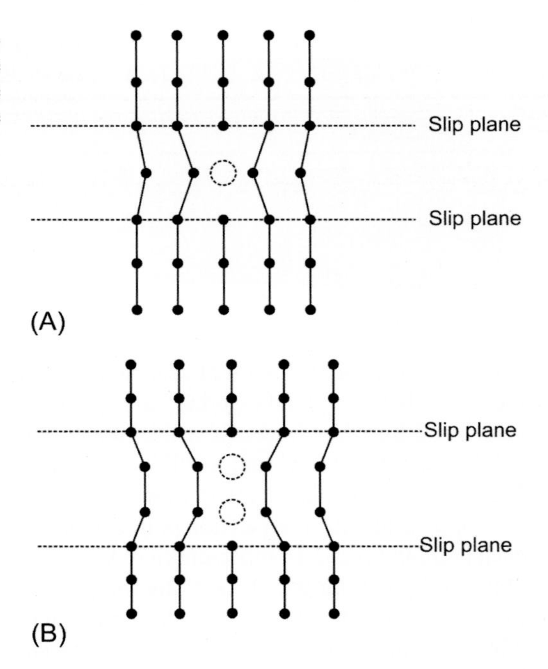

FIG. 23.18 (A) The formation of a row of vacancies, represented by a *dashed circle*, perpendicular to the plane of the paper, upon the recombination of a positive and a negative edge dislocation; the dislocation lines are also perpendicular to the plane of the paper. (B) The formation of two rows of vacancies perpendicular to the plane of the paper. The figure caption is the same as that of part (A) of the figure.

that the dislocation is a straight line. In fact, any curved line can represent an edge dislocation so long as it is perpendicular to the Burgers vector. For example, in Fig. 23.11B and C, the vertical portions of the dislocation line denote edge-type dislocations.

Interstitials or vacancies can be produced as a byproduct of the recombination of a positive and a negative edge dislocation. Consider, for example, the case in which the slip plane of a positive edge dislocation is parallel to that of a negative edge dislocation and suppose that the slip plane of the former lies two interatomic distances above that of the latter. When these dislocations meet, a row of vacancies is left after recombination, as shown in Fig. 23.18A. Fig. 23.18B shows the formation of two rows of vacancies in the crystal when the two slip planes are separated by three interatomic distances. Similarly, if the half planes overlap each other, one or more rows of interstitials become available. Edge dislocation is also sometimes called Taylor-Orowan dislocation.

23.2.6.2 Screw Dislocations

A screw or Burgers dislocation is defined as a dislocation in which the Burgers vector **b** is everywhere parallel to the dislocation line. In Fig. 23.11B and C the horizontal portions of the dislocation line represent screw-type dislocations, as the Burgers vector **b** is parallel to the dislocation line. Fig. 23.19 shows the atomic configuration in the vicinity of a screw dislocation piercing the surface of a simple cubic lattice. A screw dislocation is produced if one cuts the block across the area BFHM and then pushes the upper part into the paper in the direction of the Burgers vector **b**, as indicated. The dislocation line BM is parallel to **b**; note that a screw dislocation is always a straight line, in contrast with an edge dislocation. As one moves around the dislocation line along a circuit, such as AKLCDE, one advances in the direction

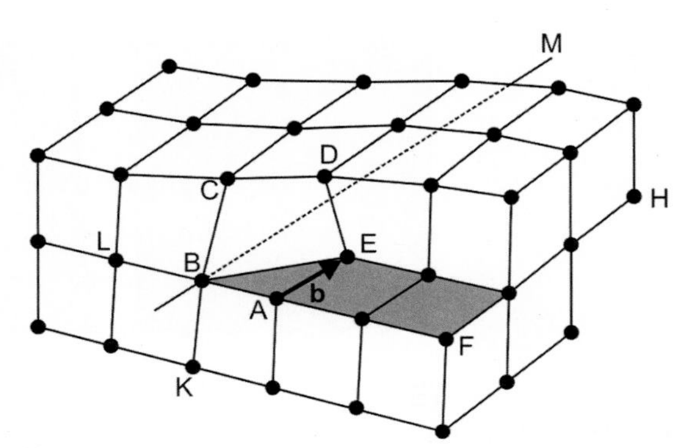

FIG. 23.19 Schematic representation of a screw dislocation in a simple cubic lattice: the dislocation line BM is parallel to the Burgers vector **b**. *(Modified from Dekker, A. J. (1971). Solid state physics (p. 91). London: MacMillan).*

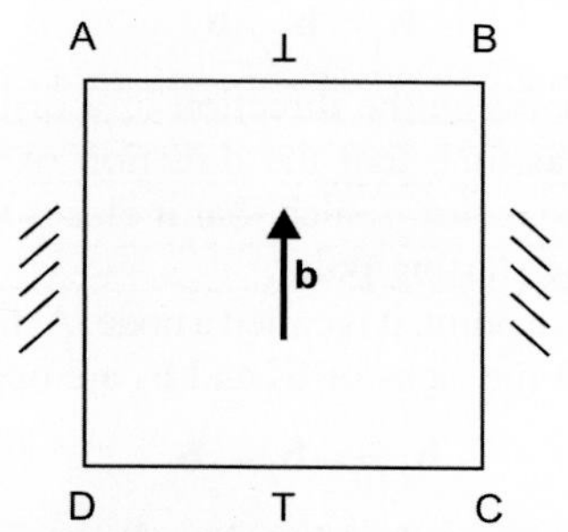

FIG. 23.20 ABCD is a rectangular loop of a dislocation line in the slip plane. It shows both the edge and screw dislocations in a simple way.

of BM by an amount **b** for every turn, hence the term "screw dislocation." Note that no extra half plane appears in the screw dislocation. Thus, the motion of a screw dislocation is more free than that of an edge dislocation. If a screw dislocation moves along any cylindrical surface, its Burgers vector is along its axis. If, in Fig. 23.19 the dislocation line moves to the left, then slip proceeds. Thus, screw dislocations, like edge dislocations, can produce plastic flow. Depending on the motion, screw dislocations can be classified into two categories:

1. If the motion is counterclockwise, it is called a *left-handed screw dislocation* and is denoted by ///.
2. If the motion is clockwise, it is called a *right-handed screw dislocation* and is denoted by \\\\\\.

Let us consider a simple example of edge and screw dislocations. Fig. 23.20 shows a rectangular dislocation loop ABCD produced when the upper part of a crystal slips above the rectangular slip plane. The portion of the crystal above the slip plane but inside the rectangle ABCD is slipped along the Burgers vector **b**. In Fig. 23.20, AB and CD are edge dislocations: AB is a positive edge dislocation and CD is a negative edge dislocation. On the other hand, BC and AD are screw dislocations: BC is a clockwise screw dislocation (\\\\\\) and AD is a counterclockwise screw dislocation (///).

23.2.6.3 Mixed Dislocations

A *mixed dislocation* is defined as a dislocation in which the angle between the Burgers vector and the dislocation line lies between 0° and 90° and possesses, therefore, components of both edge and screw dislocations. Such dislocations are also called compound dislocations. In Fig. 23.11C the curved portions of the dislocation line, which are neither horizontal nor vertical, form mixed dislocations.

23.2.7 Conservation of the Burgers Vector

Let a Burgers circuit be moved gradually only in the good region of a crystal and then deformed gradually only inside the good region. If this Burgers circuit happens to overlap with another Burgers circuit, then these Burgers circuits are said to be equivalent and they possess the same Burgers vector. If a Burgers circuit is moved along a dislocation line (a straight line or a loop), the new Burgers circuits formed are always equivalent to each other as they overlap. So, the Burgers vector remains the same everywhere on the dislocation line. This is called the conservation of the Burgers vector. An important thing that can be inferred from the conservation of the Burgers vector is that the dislocation cannot have an end point inside the crystal. The dislocation line either forms a closed loop inside the crystal or comes out to the surface of the crystal. The closure failure of a Burgers circuit surrounding more than one dislocation line is equal to the sum of the Burgers vectors of the various dislocations.

Consider a dislocation line with the Burgers vector $\mathbf{b}_1$, which is split into two dislocation lines (or branches) with the Burgers vectors $\mathbf{b}_2$ and $\mathbf{b}_3$, as shown in Fig. 23.21A. Then the conservation of the Burgers vector demands

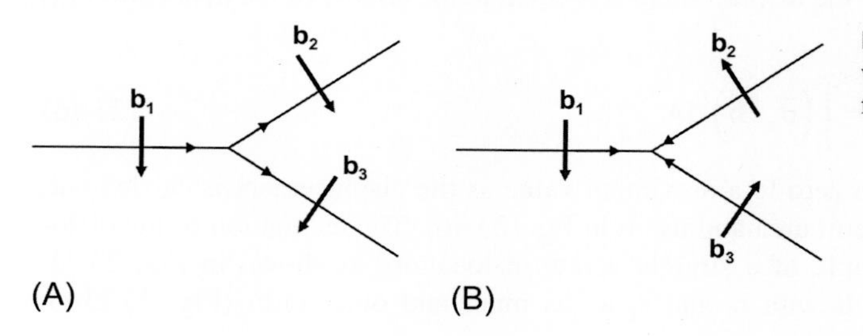

FIG. 23.21 (A) The splitting of the Burgers vector $\mathbf{b}_1$ into two Burgers vectors $\mathbf{b}_2$ and $\mathbf{b}_3$. (B) Three Burgers vectors $\mathbf{b}_1$, $\mathbf{b}_2$, and $\mathbf{b}_3$ meet at a point forming a node of dislocation lines.

$$\mathbf{b}_1 = \mathbf{b}_2 + \mathbf{b}_3 \tag{23.44}$$

It is important to determine the relationship between the direction of a dislocation and the Burgers vector. Construct a Burgers circuit in a clockwise direction and assume that the direction of a dislocation is that of an advancing screw (see Fig. 23.21B). The direction of the Burgers vector is such that it closes the Burgers circuit (see Fig. 23.12b), starting from the end point of the Burgers circuit to the starting point.

When more than two dislocation lines meet at a point, it is called a node. At the node the dislocation lines flow either toward the node or away from the node. In Fig. 23.21B the signs of $\mathbf{b}_2$ and $\mathbf{b}_3$ are opposite to that of $\mathbf{b}_1$. One can, therefore, write

$$\mathbf{b}_1 = -\mathbf{b}_2 - \mathbf{b}_3$$

$$\text{or} \quad \mathbf{b}_1 + \mathbf{b}_2 + \mathbf{b}_3 = 0 \tag{23.45}$$

Kirchhoff s law of dislocations states that the vector sum of the Burgers vectors of all the dislocations flowing into or out of the node vanishes. It is clear from the above discussion that a dislocation either forms a loop in the crystal or comes out at the surface of the crystal. The dislocation lines in a node may not necessarily be in the same plane. For example, in the (111) slip plane of an fcc lattice, the solid can slip along the OX_1, OX_2, and OX_3 directions. So, in an fcc lattice there are three dislocation lines whose Burgers vectors correspond to $OX_1 = \mathbf{b}_1$, $OX_2 = \mathbf{b}_2$, and $OX_3 = \mathbf{b}_3$ (Fig. 23.22A). If the sum of $\mathbf{b}_1$ and $\mathbf{b}_3$ is equal to $\mathbf{b}_2$, then these Burgers vectors can intersect to form a node, as shown in Fig. 23.22B. These dislocation lines may not necessarily be in the same plane. In real crystals, a number of nodes may exist that form a dislocation network. The most stable form of dislocation distribution is found to be a hexagonal network (see Fig. 23.22C) in view of the dislocation energy. Dislocation networks are easily observed by means of transmission electron microscopy in the case of metallic foils. A hexagonal network has been observed in microscopic photographs of AgBr crystal. It is noteworthy that the dislocations are not thermodynamically stable, but they can be mechanically stable.

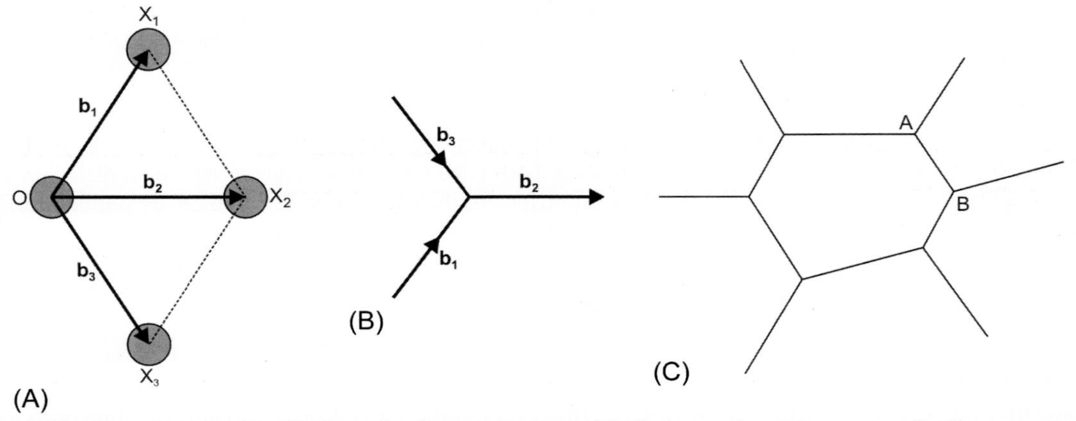

FIG. 23.22 (A) Three Burgers vectors in a slip plane of an fcc crystal network. (B) The intersection of the corresponding Burgers vectors of (A). (C) Hexagonal dislocation network.

23.2.8 Dislocation Energy

To produce a dislocation, work is required to be done, which is stored as the energy of the dislocation. The dislocation energy can be estimated by assuming the crystal to be an elastic solid during the process of creation of the dislocation. Suppose that a uniform shear stress $\vec{\sigma}_s$ is applied to the crystal, which produces the Burgers vector $\mathbf{b}$ not necessarily in the direction of $\vec{\sigma}_s$ (see Fig. 23.13). The shear stress will exert force on the dislocation line causing it to move and thereby increasing the slipped area. The work done by $\vec{\sigma}_s$ in causing the displacement $\mathbf{b}$ is equal to the energy of the dislocation E_D given from Eq. (23.33) as

$$E_D = \int dW = \int \left(\vec{\sigma}_s \cdot \mathbf{b}\right) dA \tag{23.46}$$

In real systems, the force at a point builds up linearly from zero to a maximum value as the displacement is carried out. Therefore, one must calculate the average shear force per unit area and use it in Eq. (23.46). The evaluation of the dislocation energy can be illustrated by taking a simple example of a straight screw dislocation, as shown in Fig. 23.23. Consider a cylindrical shell with length l and thickness dr with r_0 and r_1 as its inner and outer radii (Fig. 23.23A).

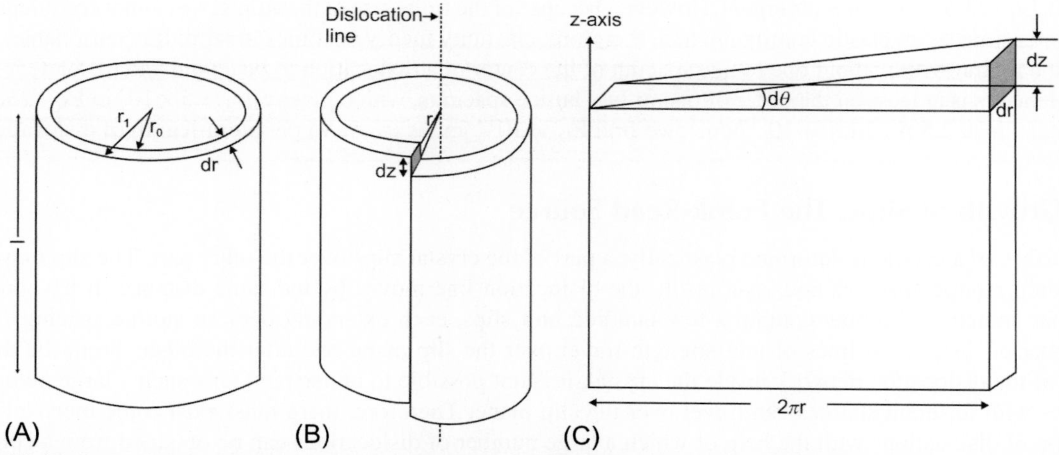

FIG. 23.23 (A) A hollow cylinder with length l and inner and outer radii r_0 and r_1. (B) By applying the shear stress $\vec{\sigma}_s$ the cylinder is cut along the length, thereby producing a dislocation line along the axis of the cylinder. (C) The cylindrical shell is opened out into a flat plate.

Let the cylindrical shell be cut lengthwise by applying a force along the cut plane, thereby producing the screw dislocation (Fig. 23.23B). The magnitude of the shear stress $\vec{\sigma}_s$ can be obtained by the following simplification. Let the crystal be considered to be a series of concentric cylindrical shells with a dislocation line along their axis. Each shell is cut along its cylindrical length where the cut surface intersects the shell. The shell on one side is displaced with respect to the other side by a distance **b**. If the thickness of the shell is small, the geometrical configuration of the shell is not important for calculating the resistive force during the displacement. In particular, the force will be the same if the shell is opened out into a flat plate. Then the problem is reduced to the calculation of the shearing displacement of the plate, as shown in Fig. 23.23C. The shear strain θ is given by

$$\theta = \int d\theta = \frac{1}{2\pi r}\int dz = \frac{b}{2\pi r} \tag{23.47}$$

The integration over dz gives the total displacement b. For a small displacement, according to Hooke's law, the modulus of rigidity η is given by

$$\eta = \frac{\sigma_s}{\theta} = \frac{2\pi r}{b}\sigma_s \tag{23.48}$$

So, the shear stress from the above equation is given by

$$\sigma_s = \frac{\eta b}{2\pi r} \tag{23.49}$$

The shear stress σ_s builds up from zero to its maximum value given by Eq. (23.49), so the average value of stress is

$$\sigma_{av} = \frac{1}{2}\left[0 + \frac{\eta b}{2\pi r}\right] = \frac{\eta b}{4\pi r} \tag{23.50}$$

From Eqs. (23.46), (23.50), the energy of dislocation becomes

$$E_D = \frac{\eta b^2}{4\pi}\int dz \int \frac{dr}{r} \tag{23.51}$$

The integral over dz gives the length of the cylinder l (see Fig. 23.23). Therefore, finally, the dislocation energy becomes

$$E_D = \frac{\eta b^2}{4\pi} l \ln\left(\frac{r_1}{r_0}\right) \tag{23.52}$$

The energy calculated using Eq. (23.52) depends upon the values taken for the integration limits on r. For an infinite crystal ($r_1 \to \infty$), the energy of dislocation is infinite. However, an ordinary-sized finite crystal (say 1 cm on edge) contains many dislocations, which are randomly distributed. Experimental observations show that the mean distance between any two dislocations in a crystal is about 10^4 atomic spacings and hence r_1 is on this order.

As $r_0 \rightarrow 0$, Eq. (23.52) becomes divergent. However, because of the finite size of the atoms, we cannot consider any region of atomic dimensions as an elastic continuum and, therefore, elasticity theory becomes invalid. It is reasonable, therefore, to consider the region within about one lattice spacing of the center of a dislocation to be a void and to delete it from consideration. Hence r_0 is at least on the order of one or two lattice spacings, which gives $r_1/r_0 \approx 5 \times 10^3$ in Eq. (23.52). If the Burgers vector $|\mathbf{b}| \cong 2.5$ Å and $\eta = 10^{11}\,\mathrm{N/m^2}$, we find $E_D = 10^{19}$ joules ($\approx 6\,\mathrm{eV}$) per atom-length of dislocation line.

23.2.9 Growth of Slips: The Frank-Read Source

When the surface of a crystal is deformed plastically, a part of the crystal slips over the other part. The slip may extend to several hundred atomic spacings and, as a result, the dislocation line moves by the same distance. It has already been explained that an actual slip may contain a few hundred unit slips, each extending over an atomic spacing. Therefore, a large number of dislocation lines of unit strength travel over the slip plane one after the other. From the distribution and density of the dislocation network inside the crystal, it is not possible to understand how such a large number of dislocation lines with different sources can travel over the slip plane. Therefore, there must exist some mechanism for the multiplication of dislocations with the help of which a large number of dislocations can be obtained from a single source and that travel one after the other to constitute the observed slip. A number of mechanisms for the dislocation multiplication have been proposed, but one of the most common is the Frank-Read source mechanism. To understand the Frank-Read mechanism, consider part AB of the dislocation network, which always lies in the slip plane P, while the other parts, such as AC, BE, AD, and BF, lie outside the plane P (see Fig. 23.24). If the slip plane is in the page of the book, then AB appears as shown in Fig. 23.25A. If a shear stress $\vec{\sigma}_s$ is applied along the slip plane, the part AB tends to move along with the plane, while the other part outside the plane tends to remain stationary. In other words, AB is pinned at A and B but the rest of the dislocation line expands and becomes distorted, as shown in Fig. 23.25B. If the shear stress is applied externally, then the force acting on the dislocation line per unit length is $\vec{\sigma}_s b$ and this force acts perpendicularly to the dislocation line. Thus, the dislocation line continues to expand, as shown in Fig. 23.25B–D.

The expansion of the dislocation line depends on the sign of each part of it. If the dislocation line in Fig 23.25A is assumed to be a positive edge dislocation, then the sign of each part of the dislocation line when the state labelled d is

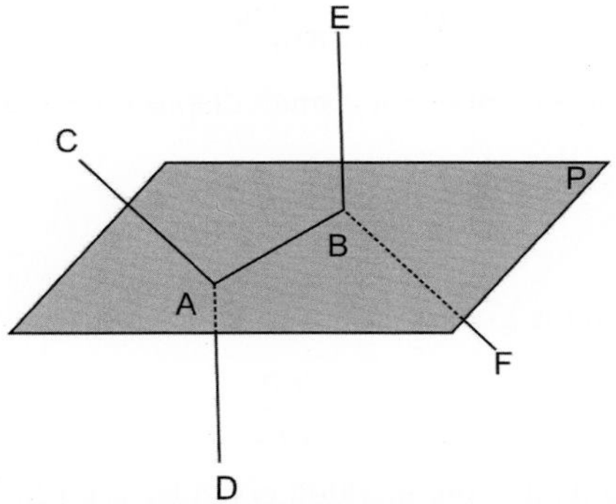

FIG. 23.24 The dislocation network consisting of dislocations CAD and EBF connected with dislocation AB. The dislocation line AB lies in the slip plane P and is called a sweep dislocation. The other parts CAD and EBF lie outside the slip plane and are called pole dislocations.

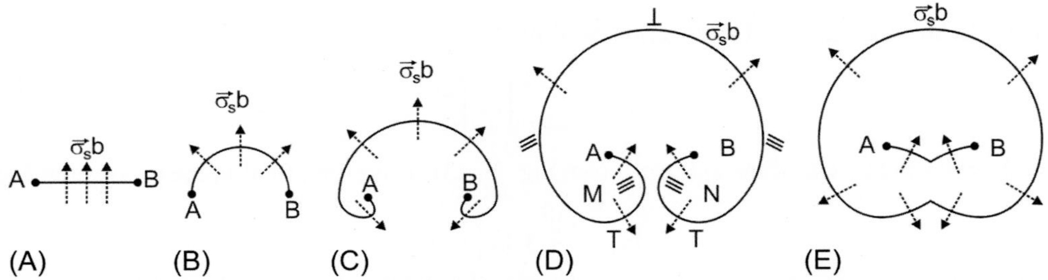

FIG. 23.25 The figure shows the growth of a slip due to the Frank-Read mechanism. (A) The surface of the page is the slip plane and only the part AB of the dislocation line lies in the slip plane. The shear force $\vec{\sigma}_s b$ per unit length always acts perpendicularly to the dislocation line AB. (B) The dislocation line starts expanding in the slip plane but perpendicular to the shear force. (C) and (D) The continued expansion of the dislocation line. (E) Externally expanding dislocation loop is formed along with AB dislocation line which becomes straight. *(Modified from Kubo, R., & Nagamiya, T. (1969). Solid state physics (p. 783). New York: McGraw-Hill Book Co.)*

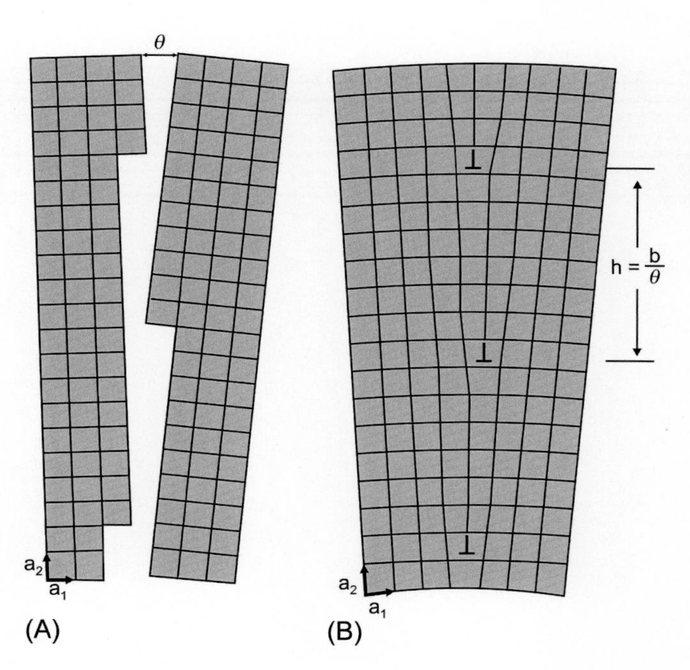

FIG. 23.26 (A) Two crystallites with simple cubic structure are placed close to each other and are inclined to each other by an angle θ. (B) Two crystallites are joined, which forms a set of three positive dislocation lines. The spacing between the two edge dislocations is given by $h = b/\theta$ where b is the Burgers vector of the dislocations. *(Modified from Kubo, R., & Nagamiya, T. (1969).* Solid state physics *(p. 771). New York: McGraw-Hill Book Co.)*

reached is shown in Fig. 23.25D. The parts M and N of the dislocation line face each other and are oppositely wound screw dislocations. On further expansion, the parts M and N come in contact with each other and vanish, leaving the shape indicated in Fig. 23.25E. Thus, when the parts M and N vanish, a heart-shaped loop along with the mountain-shaped part AB is left. The part AB becomes straight, acquiring the same shape as in Fig. 23.25A, while the loop is an externally expanding one. The part AB of the dislocation is called a sweeping dislocation, while the parts CAD and EBF (see Fig. 23.24) are called pole dislocations. If stress is applied continuously from outside, then, again, the shapes of Fig. 23.25B–E are repeated and loops are continuously formed. This is the basic concept of the Frank-Read source.

23.2.10 Grain Boundary

A commercially available crystal comprises very small crystals, usually called crystallites or grains. The crystallites join together to form a bigger crystal and the boundary between the crystallites is called a grain boundary. The formation of a grain boundary can be understood in a simple way by considering two crystallites in which the mutual inclination angle is small. Fig. 23.26A shows two crystallites placed side by side having θ as the small inclination angle. These are simple cubic crystals in which the axes perpendicular to the page are parallel. To join the crystallites, they are rotated about the perpendicular axes in the clockwise and counterclockwise directions by an angle $\theta/2$. The boundary plane formed is the simplest and contains a common crystal axis for both the crystallites. Further, the crystal orientation is symmetric on both sides of the boundary plane. Such a boundary has a vertical arrangement of edge dislocations of like sign, as is clear from Fig. 23.26B. There are three stable vertical edge dislocations of the positive sign at the boundary.

SUGGESTED READING

Farge, Y., & Fontana, M. P. (1979). *Electronic and vibrational properties of point defects in ionic crystals*. Amsterdam: North-Holland Publ. Co.

Flynn, C. P. (1972). *Point defects and diffusion*. London: Clarendon Press.

Fowler, W. B. (1968). *Physics of color centers*. New York: Academic Press.

Friedel, J. (1964). *Dislocations*. New York: Pergamon Press.

Henderson, B. (1972). *Defects in crystalline solids*. New York: Crane, Russak & Co.

Hirth, J. P., & Lothe, J. (1982). *Theory of dislocations* (2nd ed.). New York: J. Wiley & Sons.

Liebfried, G., & Breuer, N. (1978). *Point defects in metals*. (Vols. I & II). Berlin: Springer-Verlag.

Markham, J. J. (1966). *Solid state physics (suppl. 8): F-centers in alkali halides*. New York: Academic Press.

Pines, D. (1963). *Elementary excitations in solids*. New York: W.A. Benjamin.

Seeger, A., Schumacher, D., Schilling, W., & Dielh, J. (1970). *Vacancies and interstitials in metals*. Amsterdam: North-Holland & Publ. Co.

Stoneham, H. M. (1975). *Theory of defects in solids*. London: Oxford University Press.

Chapter 24

Amorphous Solids and Liquid Crystals

Chapter Outline

Solids can be classified broadly into two categories:

1. *Crystalline solids*: In crystalline solids the atoms are periodically arranged in the form of an array and exhibit long-range order. A number of structural and electronic properties of these solids have been discussed in the previous chapters.
2. *Amorphous solids*: Amorphous solids do not exhibit long-range order. In these solids the atoms are not purely randomly distributed, but exhibit a short-range order in the form of a regular coordination number. Long-chain molecules, such as polymers, also fall into the category of amorphous solids. Sometimes the amorphous solids are referred to as glasses. Some common examples of amorphous solids are oxide glasses, polymers of high molecular weight, and a few other inorganic compounds.

To distinguish between crystalline and amorphous solids, one can study the cooling curve of the vapor of a material. Initially, let the material be in the gaseous state and enclosed in a box with finite volume V. As the vapor is cooled, the volume decreases with decreasing temperature T and there is a sharp break in the V-versus- T curve, which marks a change in phase from the gas to the liquid state at the boiling temperature T_B (see Fig. 24.1). As the cooling continues, the volume decreases in a continuous fashion. One can define the coefficient of thermal expansion Γ_{TH} as

$$\Gamma_{TH} = \frac{1}{V}\left(\frac{dV}{dT}\right)_P \tag{24.1}$$

With a decrease in temperature, the liquid-to-solid phase transition takes place (with the exception of He, which remains liquid as $T \rightarrow 0$ at zero pressure). The liquid-to-solid phase transition can take place in two ways:

1. If the cooling is carried out slowly, the liquid usually goes to the crystalline solid phase through path 1. There is a discontinuity in the slope of the V-versus-T curve at the freezing temperature T_F and the whole liquid goes to the crystalline solid phase at a constant temperature T_F.
2. If the cooling is carried out at a very fast rate, there is no discontinuity in volume. Instead the V-versus- T curve acquires a smaller slope, as shown in path 2. There is a narrow temperature range about T_G (shown by shaded region) in which the liquid goes to the amorphous solid phase. T_G is usually called the *glass transition temperature* because amorphous solids were called glasses in the early days. One should note that $T_G < T_F$, therefore, in path 2 the material remains in the liquid state even at temperatures lower than T_F. The narrow range of temperature around T_G in which the liquid goes to the amorphous state can be explained in terms of the bond energies. All the subunits in an amorphous solid do not possess identical surroundings, as a result of which they do not have the same bond energies: the spread in bond

Solid State Physics. https://doi.org/10.1016/B978-0-12-817103-5.00024-4

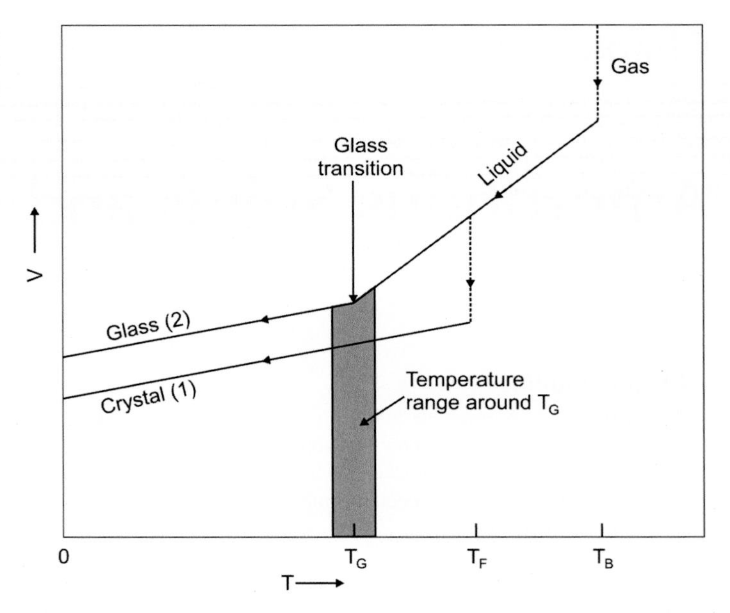

FIG. 24.1 Variation in volume V as a function of temperature T as the material is cooled from the gaseous state to the solid state via the liquid state. Curve 1 is for slow cooling, while curve 2 is for rapid cooling.

energies may be very small. Therefore, the solidification process is associated with a range of energies for the bonds between the subunits. As the liquid is cooled, the lowest energy bonds are formed first and the subunits begin to stick together locally. As the temperature is lowered further, more bonds, in order of increasing energy, are gradually formed until the material is completely hard. Therefore, the process of solidification is completed over a narrow range of temperature around T_G.

It has been found that T_G depends on the cooling rate in the transition from the liquid to the amorphous solid state: T_G shifts to lower temperatures when the cooling rate is decreased. For example, in the organic glass polyvinyl acetate ($CH_2CHOOCCH_3$), if the cooling rate is increased by a factor of 5000, the shift in T_G is only 8 K. The weak dependence of T_G on the cooling rate is due to the temperature dependence of the typical molecular relaxation time.

It can be explained physically why the liquid goes to the amorphous solid state at high cooling rates. In the liquid state, the molecules are randomly distributed and are in constant motion. As the liquid is cooled slowly, the atoms or molecules have sufficient time to occupy their lowest energy states (equilibrium states), yielding a crystalline state in the solid. During the cooling process some nucleation centers are produced, around which the crystal grows in a few hours. On the other hand, if the liquid is cooled rapidly, the atoms or molecules do not have sufficient time to occupy their equilibrium positions. Rather, the random distribution of molecules is frozen into the solid state, yielding an amorphous solid. Therefore, the amorphous state of a solid is not the minimum energy state, but it is still highly stable. The essential aspect of an amorphous solid is the absence of long-range order.

There is a new class of materials, which possess mechanical and symmetry properties in between those of a liquid and a crystalline solid. Such materials are called liquid crystals. In going from the liquid to the solid state, certain materials show a cascade of transitions involving new phases. Therefore, a more appropriate name for a liquid crystal is a mesomorphic phase (mesomorphic means intermediate form). Liquid crystals exhibit partial periodicity (say in one or two Cartesian directions), so it is appropriate to discuss these materials in this chapter.

24.1 STRUCTURE OF AMORPHOUS SOLIDS

It is observed that in an amorphous solid there is a continuous random distribution of atoms or molecules with a short-range order. Three continuous random models are used to describe the structure of amorphous solids.

1. *Continuous random network*: appropriate to the structure of covalent solids.
2. *Random close packing*: appropriate to the structure of simple metallic solids.
3. *Random coil model*: appropriate to the structure of polymers.

Although these models are, to some extent, ideal ones, they represent the best available pictures of the structures of amorphous solids on the atomic scale.

24.1.1 Continuous Random Network Model

Fig. 24.2A shows a honeycomb lattice in two dimensions corresponding to covalently bonded carbon atoms in graphite. In this structure each atom has three 1NNs. Fig. 24.2B shows a graph representing a binary compound in which each black atom is covalently bonded to three white atoms and each white atom is covalently bonded to two black atoms. This is called a "decorated honeycomb lattice." This structure can be derived from the honeycomb lattice by replacing each bond by a pair of bonds with a bridging atom at the center of the original bond. The decorated honeycomb lattice of Fig. 24.2B corresponds to the graph of the layers that make up crystalline As_2S_3 and As_2Se_3. The honeycomb lattices of Fig. 24.2A and B are periodic structures with the following features:

1. The coordination number z for each atom is 3.
2. The 1NN distances (bond lengths) are constant.
3. Both structures are ideal, assuming no dangling bonds.

In analogy with a honeycomb structure, the continuous random network (CRN) model was proposed to explain the atomic arrangement in glasses. Fig. 24.3 shows a graph of the atomic arrangement in the CRN model. Each of the noncrystalline structures has the same short-range order as its crystalline counterpart (Fig. 24.2). There are two fundamental ways in which the crystalline solids and CRNs distinctly differ from each other:

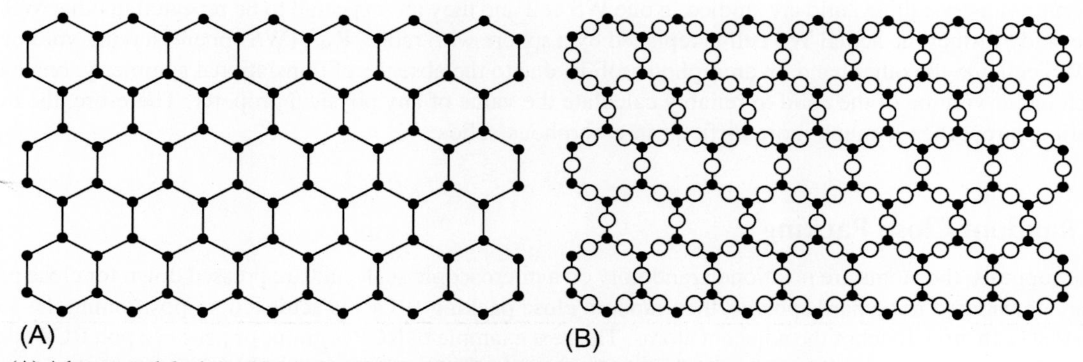

FIG. 24.2 (A) A honeycomb lattice, which is similar to a graphite layer. (B) A decorated honeycomb lattice.

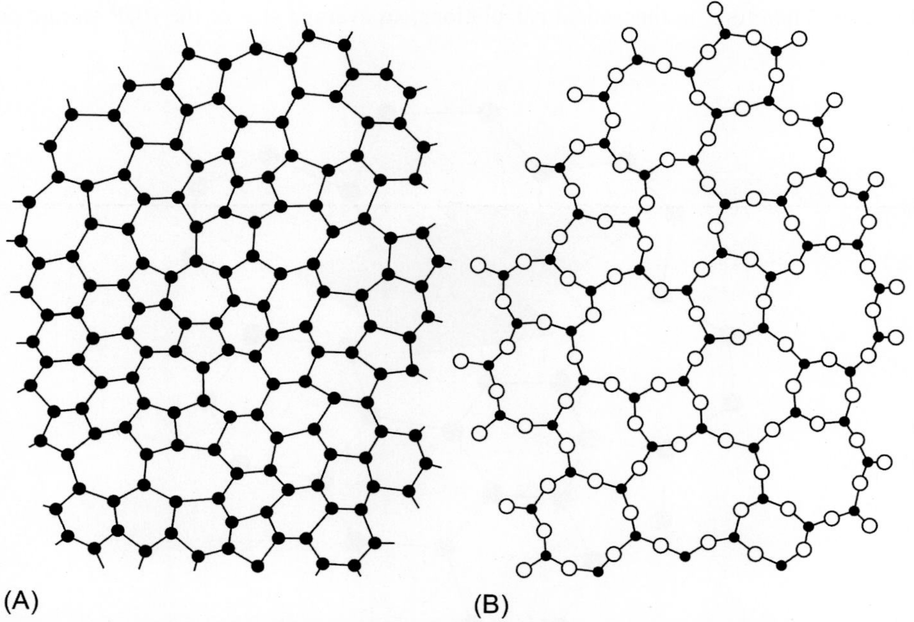

FIG. 24.3 Schematic diagram of a two-dimensional continuous random network (CRN). (A) Showing the CRN of a threefold coordinated honeycomb lattice. (B) Zachariasen diagram for the CRN of a decorated honeycomb lattice.

1. In the CRN model there is a significant spread in bond lengths and bond angles that are not permitted in a crystalline solid.
2. In the CRN model, the long-range order is absent.

In the decorated honeycomb lattice there is an additional spread in the bond angle that occurs at the bridging atom. Further, the maximum spread in the bond angle is at the twofold coordinated atom, which is expected to be much softer than that at the threefold coordinated atom. The spread in the bond lengths is greater in structures with soft bond angles than in structures with stiff bond angles. For example, in Fig. 24.3A the spread in bond lengths is very small compared with that in Fig. 24.3B.

In a crystalline solid the WS cell has a definite shape and volume. In an amorphous solid with CRNs one can also construct WS cells. Fig. 24.4 shows an irregular array of lattice points, which forms the CRN of a solid. The dark dots show the positions of atoms (which are the same as the lattice points) and each atom has three 1NNs. The WS cells of three atoms in the CRN model are shown in Fig. 24.4. We see that WS cells containing different lattice points (or atoms) have different shapes and volumes. A WS cell in an amorphous solid is also called a *Voronoi polyhedron*. Therefore, in going from a crystalline to an amorphous solid, the single atomic polyhedron characteristic is replaced by a statistical distribution of distinct polyhedra. The network formed by the lattice points and the bonds (dots and lines) is called a *simplical graph* of the array and each cell of the CRN is called a *simplex*. The division of space into irregular simplexes is called the *Delaunay division*.

All the WS cells in a crystalline solid are exactly the same (translational symmetry). Therefore, the electronic and vibrational properties of a crystalline solid are studied in one WS cell and they are expected to be repeated in other WS cells. To simplify the study further the actual WS cell is replaced by a sphere with radius R_{WS} (WS sphere) having volume equal to that of the WS cell. On the other hand, in amorphous solids, due to the absence of translational symmetry, one has to consider a much larger volume of the solid to reliably calculate the value of any physical property. Therefore, the theoretical study of various properties is much more difficult in amorphous solids.

24.1.2 Random Close Packing

As the name suggests, the atoms are positioned randomly on a microscopic scale and are pressed down for close packing. If the atoms are considered to be hard spheres, then random close packing (RCP) is achieved by positioning the atoms randomly such that each atom touches the adjacent atoms. The best example of RCP is grains or peas in a pot. RCP gives one of the most satisfactory models for the structure of amorphous metals on a microscopic scale.

The fcc structure is a close-packed structure with a packing fraction of 0.74. In fcc structure, the WS cell is a single polyhedron with definite shape and size. On the other hand, RCP is also a close-packed structure but with polyhedra of different sizes and shapes. Therefore, in theoretical calculations, an average size of the RCP atomic polyhedron is used

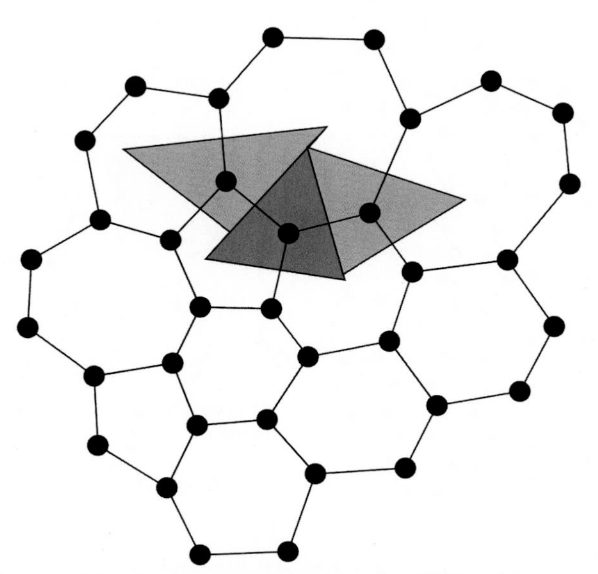

FIG. 24.4 WS cells having different shapes and sizes for three consecutive atoms in a CRN model.

whose size is a little bigger than the size of the fcc polyhedron. The packing fraction for RCP is found to be 0.637. This shows that for atoms of the same size, RCP is about 86% as dense as fcc close packing. Therefore, close packing in crystalline solids corresponds to an absolute potential energy minimum in comparison with RCP because it provides the maximum packing density in amorphous solids.

24.1.3 Long-Chain Molecular Compounds

Compounds in which the smallest units are long-chain molecules instead of atoms are called *polymers* (or occasionally resins or plastics). A long-chain molecule consists of atoms distributed along the length of a chain. In addition, there are side groups (bulky or light) of atoms along the length of the chain. The bonds with which atoms along the length of the main chain are bound are called primary bonds, while those with which atoms of the side groups are bound are called secondary bonds. Long-chain molecules are flexible and get tangled on their own, which results in the development of amorphous structures relatively easily (Fig. 24.5). Polymers are rather loosely packed structures in the solid state because the primary bonds are completely satisfied within the long-chain molecules but the side groups of atoms interfere with the close packing. Any atomic arrangement that contributes to the loose packing of molecular chains favors the formation of amorphous structures. The most important features that favor the formation of amorphous solids are:

1. Long and branched molecular chains.
2. Random arrangement of large side groups along the chains.
3. Copolymer chains: molecular chains that are actually a combination of two or more polymers.
4. Plasticizers: low molecular weight additives that separate the chains from one another.

Small chain paraffins form perfectly crystalline solids, but long-chain paraffins, with molecular weight from a few thousand to a few million, form partially crystalline solids. They are called linear paraffins or polyethylene. On the other hand, branched-polyethylene has side molecular chains that are attached to the main molecular chain at positions normally occupied by a hydrogen atom. As the branches interfere with the crystallization process, the crystallization is partial in branched polyethylene. Further, a greater number of branches results in a higher degree of noncrystallinity in the polyethylene.

The effect of the side group arrangements can be seen by considering the structure of vinyl polymers. The vinyl polymers have a repeating unit of the type:

$$\begin{array}{cc} \mathrm{H} & \mathrm{H} \\ | & | \\ -\mathrm{C}- & \mathrm{C}- \\ | & | \\ \mathrm{H} & \mathrm{X} \end{array} \tag{24.2}$$

where X is some monovalent side group. There are three possible arrangements of side groups in vinyl polymers.

1. Polymers with random distribution of side groups are called *atactic*.
2. Polymers with side groups on the same side of the chain are called *isotactic*.
3. Polymers with side groups arranged alternately from one side to the other are called *syndiotactic*.

If the side groups are small (e.g., X=OH in polyvinyl alcohol), the atomic chains are linear and one obtains a crystalline polymer. But if the side groups are large (e.g., X=Cl in polyvinyl chloride) and randomly distributed along the atomic chain, a polymer with amorphous structure results. In contrast the isotactic and syndiotactic polymers usually crystallize, even when the side groups are large.

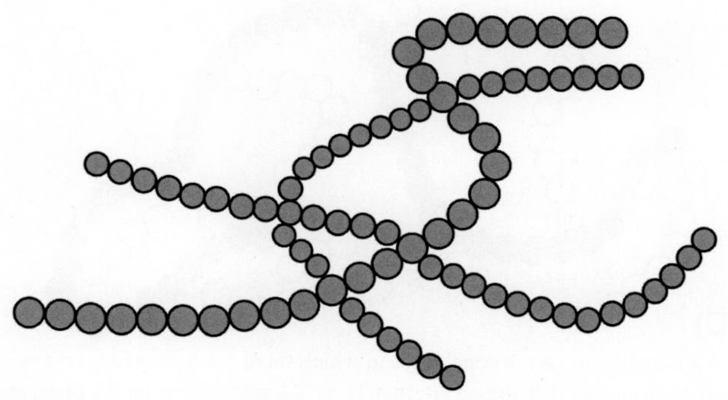

FIG. 24.5 Long-chain molecules in a polymer.

24.1.4 Copolymers

A copolymer consists of two or more polymers arranged on a molecular chain. Copolymers can be made in a number of ways and the simplest ones are illustrated in Fig. 24.6. The process of synthesizing a copolymer is called copolymerization. Copolymerization always decreases the degree of crystallinity of polymer chains and, therefore, promotes the formation of amorphous solids. Quite often a copolymer is developed because a certain amount of noncrystallinity results in better properties. The randomness in the distribution of side groups (regardless of their size) in a molecular chain or in the distribution of atoms in a copolymer is related to the degree of noncrystallinity in a polymer. A greater irregularity in the distribution of atoms in a molecular chain results in a greater degree of noncrystallinity in the formation of an amorphous solid.

24.1.5 Plasticizers

Plasticizers are substances that prevent crystallization of polymers by keeping the chains separated from one another. It is the oldest method to produce amorphous polymers from crystalline polymers. Celluloid is made of nitrocellulose and it is one of the first synthetic crystalline polymers. Celluloid is plasticized with camphor. Cellophane is another common plastic composed of cellulose chains with glycerol as the plasticizer. The disadvantage of this process is that plasticizers are usually of quite low molecular weight and, therefore, they diffuse through the solid and eventually evaporate. Therefore, the introduction of a plasticizer decreases the pliability and increases the tendency to crack with time.

24.1.6 Elastomers

Elastomers are polymers that exhibit a large and reversible extensibility at room temperature. They can be stretched by at least a hundred or a thousand times by applying an external force and they regain their original dimensions when the force

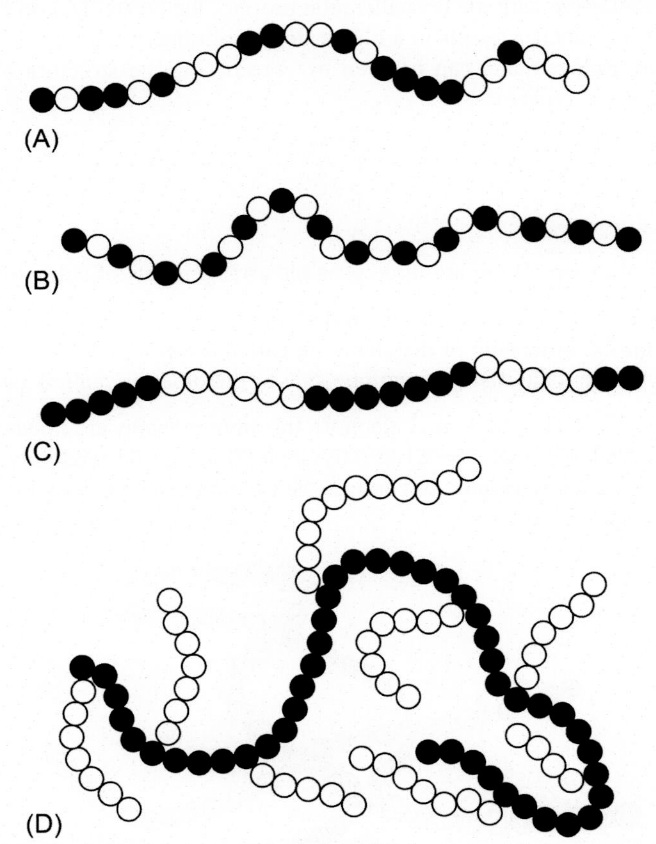

FIG. 24.6 The different arrangements in a copolymer. (A) A copolymer in which there are units of two types of atoms distributed randomly along the chain. (B) A copolymer in which two types of atoms are distributed alternately. (C) A polymer in which there are blocks of two types of atoms repeated alternately. (D) A graft polymer in which the long chain contains the same type of atoms, while the side groups are again polymers but of the other type of atoms.

is switched off. Elastomers are all amorphous polymers at room temperature. Actually, they are intermediate between long-chain molecules and three-dimensional networks. Elastomers satisfy the following criteria in addition to being amorphous:

1. The molecular chains must be very long with many bends in them.
2. At room temperature the chain segments are in a state of constant motion due to the thermal energy.
3. The chains must be connected to one another after every few hundred atoms by crosslinks, which may consist of atoms or a group of atoms that form primary bonds between the chains (Fig. 24.7).

It is important to note that the crosslinks are necessary to explain the reversibility in the extension of an elastomer. If there are no crosslinks, the elastomer will not return to its original shape and size when the external force is switched off. The crosslinks act as pinning points; without them the polymer would deform permanently. The best known elastomer is natural rubber. The molecular chains in rubber are not only long, but tangled and bent (coiled) rather than straight, and are in a state of constant motion at room temperature. The glass temperature T_G in elastomers is quite low, lower than room temperature. If the temperature is lowered below T_G, the elastomer becomes brittle.

Some elastomers, such as S and Se, both of which belong to column VI of the periodic table, also exhibit long-chain structures. In S and Se an amorphous structure can be formed by quenching a viscous melt to room temperature. The bonding in both of these elements is primarily covalent due to the overlapping of p-orbitals, and this leads to long chains of atoms. In S and Se, the atomic chains become so tangled in the liquid state that an amorphous structure develops when the material is quickly cooled. In so-called fibrous S, the long chain S molecules are actually mixed with S_8 ring molecules and the one type of molecule prevents the other type from crystallizing.

24.2 CHARACTERISTICS OF AMORPHOUS SOLIDS

1. Amorphous materials do not possess a sharp solidification temperature as crystalline materials do. They gradually become more viscous over a narrow range of temperature about T_G.
2. Many amorphous materials, such as ordinary window glass, are transparent both in the liquid and solid states. Their transparency arises because there are no inclusions, holes, or internal surfaces with the right properties to scatter light. Further, they have no free electrons or ions that can absorb or emit light by changing their energy states.
3. Fig. 24.8A and B shows C_P versus T for the amorphous solid As_2S_3 and the metallic solid glass $Au_{0.8}Si_{0.1}Ge_{0.1}$, respectively. In As_2S_3, a sharp increase in C_P is seen at T_G and it can be followed continuously from a low temperature up through T_G and well into the liquid phase to T_F and beyond. In the metallic glass $Au_{0.8}Si_{0.1}Ge_{0.1}$, C_P increases sharply at T_G. The dashed line shows the extrapolated value of C_P from just above T_G to just below T_F. Fig. 24.8B gives the first historic evidence of the glass transition. The variation of the coefficient of thermal expansion Γ_{TH} with T is the same as that of C_P (it arises due to the kink in the V versus T curve in Fig. 24.1 at T_G). The behavior of C_P versus T arises due to the bend in the entropy S versus T curve near T_G.

In a second-order thermodynamic transition there is an abrupt rise in the curves for $C_P(T)$ and $\Gamma_{TH}(T)$ as a function of T at the transition temperature. Fig. 24.8A and B shows that a glass transition closely resembles a second-order transition. The curves for $C_P(T)$ and $\Gamma_{TH}(T)$ definitely change their values appreciably in passing through T_G. However, these changes are

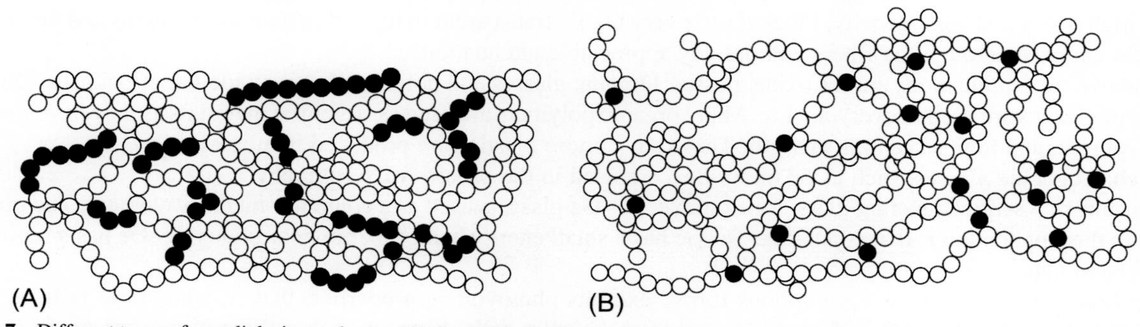

FIG. 24.7 Different types of crosslinks in a polymer. (A) Crosslinked polymer chains in which small chains act as crosslinks. (B) Crosslinked polymer chains in which foreign atoms or molecules form the crosslinks.

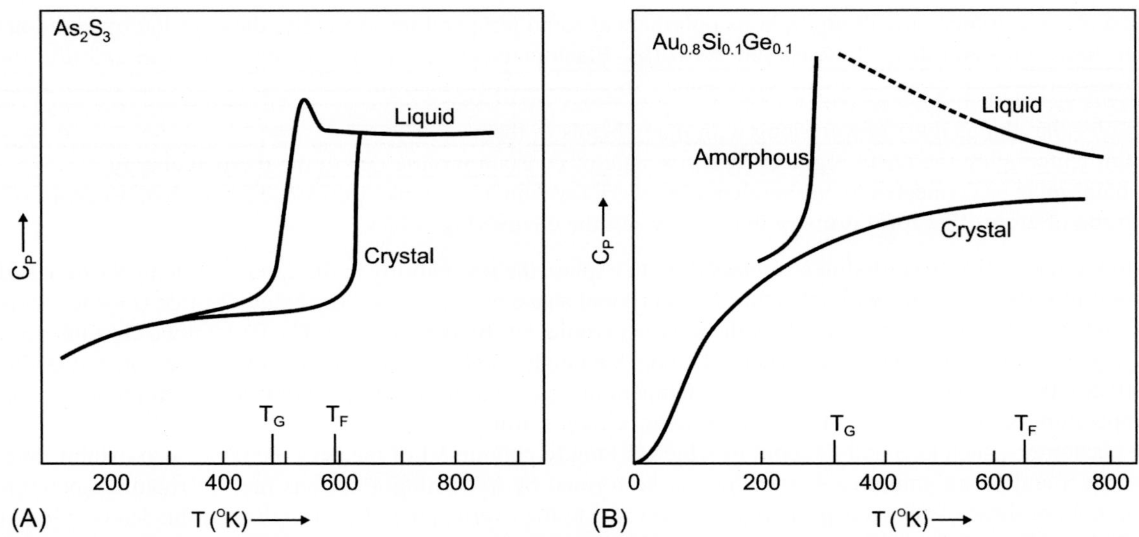

FIG. 24.8 C_P is plotted as a function of temperature T for (A) an amorphous solid As_2S_3, and (B) a metallic glass solid $Au_{0.8}Si_{0.1}Ge_{0.1}$. *(Modified from* Blachnik, R., & Hoppe, A. (1979). *Journal of Non-Crystalline Solids, 34,* 191; Chen, H. S., & Turnbull, D. (1968). *The Journal of Chemical Physics, 48, 2560.)*

not as sharp as they should be in a second-order transition, but are instead spread over a small temperature range. Therefore, one can characterize the liquid ⇔ glass transition as an apparent diffused second-order transition. We know that

$$C_P = T\left(\frac{dS}{dT}\right)_P \tag{24.3}$$

From Eqs. (24.1), (24.3) it is evident that the kinks or bends in the curves for S and V as a function of T are reflected as an abrupt increase in $C_P(T)$ and $\Gamma_{TH}(T)$.

24.3 APPLICATIONS OF AMORPHOUS SOLIDS

Amorphous solids are of immense use in science and technology. Some of the uses of amorphous solids are listed below:

1. Glasses are used as structural materials, such as ordinary window glasses. A large number of ordinary glasses are based on fused silica (SiO_2). As regards the structure, there are two types of glasses:
   (a) *Amorphous glasses*: Amorphous glasses are isotropic as far as the scattering and passage of light is concerned. These glasses are used as ordinary window glasses. It is very easy and cheap to fabricate amorphous glasses.
   (b) *Crystalline glasses*: The crystalline glasses, such as crystalline quartz glasses, are anisotropic as far as the scattering and passage of light is concerned. The crystalline glasses have special symmetry directions that allow more light to pass through compared with other directions. The anisotropy in crystalline glasses is an undesirable property for their use as window glasses. Further, it is not easy to synthesize large crystals as problems associated with the polycrystallinity arise. Therefore, it is very expensive to fabricate large crystalline glasses out of SiO_2.
2. Now a days, fiber glasses have been developed for communication purposes. Fiber is a glass in the form of a fine fiber of very high purity and homogeneity. Fiber glass is very highly transparent to light of certain wavelengths and these wavelengths can pass through the fibers without any appreciable attenuation.
3. The above two applications are for oxide glasses. Organic glasses and polymers have a wide variety of uses. Different kinds of plastics are used in everyday life. Many organic polymers are used as structural materials as they have low cost, light weight, and high structural strength. These days, more plastics are produced than steel.
4. The chalcogenide glasses, such as Se or As_2Se_3, are used in the process of xerography.
5. Tellurium-rich semiconducting glass, for example, Te-Ge glass, is used as a computer memory element. The property used in this application is that crystalline Te-Ge has a small energy band gap, but amorphous Te-Ge has a reasonably large band gap.
6. Si, in both the crystalline and amorphous forms, exhibits photovoltaic properties, that is, when light falls on Si, an electric voltage develops. Therefore, Si is used to make solar cells. Large area thin films of amorphous Si ($\approx 1\ \mu m$

thickness), which absorb most of the solar light falling on them, can be synthesized at very low cost. Crystalline Si is also used in making solar cells: it is usually used in space probes. But much thicker films of crystalline Si (>50 μm) are required for making solar cells. Therefore, the cost of production of a crystalline layer is far more than the cost of amorphous Si. In connection with solar cell technology, the material of interest is amorphous silicon hydride (a - SiH): here a means amorphous. The role of H is to eliminate electronic defects, which are intrinsic to elemental a - Si.

7. Crystalline ferromagnets are both mechanically and magnetically soft (low coercivity and easily magnetized by small magnetic fields). On the other hand, ferromagnetic glasses, such as $Fe_{0.8}B_{0.2}$, $Fe_{0.7}P_{0.2}Co_{0.1}$, and $Co_{0.8}Fe_{0.1}B_{0.1}$, have high saturation magnetization that is isotropic in nature. These ferromagnetic glasses are mechanically quite hard but magnetically soft. The high electrical resistivity of amorphous metals is also helpful in this regard. For these reasons, ferromagnetic glasses are very useful in producing the magnetic cores of power transformers, in which their low-loss properties are very important. Other important applications of amorphous magnets are in magnetic-disc memories and read/write recorder heads.

24.4 LIQUID CRYSTALS

In a crystal the atoms or molecules are arranged on a three-dimensional periodic array of lattice points. In a liquid the molecules are randomly distributed and are in constant motion. These two states of matter differ most obviously in their mechanical properties; for example, a liquid flows easily. The most fundamental difference between a crystal and a liquid is given by its X-ray diffraction pattern: A crystalline solid exhibits sharp Bragg reflection peaks characteristic of the periodicity of the lattice, while a liquid shows a diffused X-ray diffraction pattern. Liquid crystal is a state that is intermediate between the liquid and crystal states. In other words, partial order exists in the lattice of a liquid crystal. Therefore, liquid crystals are also called mesomorphic phases. Liquid crystals can be obtained in two different ways:

1. Liquid crystals can be obtained by imposing positional order in one or two dimensions. If the positional order is imposed only in the z-direction, the solid can be viewed as a set of two-dimensional liquid layers stacked on each other with a well-defined spacing along the z-direction (Fig. 24.9). Each layer has disorder in the arrangement of atoms/molecules in the xy-plane. Such mesophases are usually called *smectic*.
2. Liquid crystals can also be obtained by introducing degrees of freedom that are different from those of the lattice points. For example, in nonspherical molecules, the orientation of the molecules may change in a crystal or in a liquid phase or even in a smectic phase. The orientational change may be of two types.
   - **(a)** Many crystals show a transition from a strongly ordered state to a phase in which each molecule can acquire several equivalent orientations. The high-temperature phase is positionally ordered but orientationally disordered. Such a phase is sometimes called *plastic crystal*. Examples of orientational transitions exist in solid hydrogen, ammonium halides, and also in certain types of organic molecules.
   - **(b)** At low temperatures, certain organic liquids exhibit phases in which the molecules are positionally disordered but orientationally ordered. These are anisotropic liquids called *nematics*. At higher temperatures they undergo a transition to the conventional liquid phase, which is isotropic in nature.

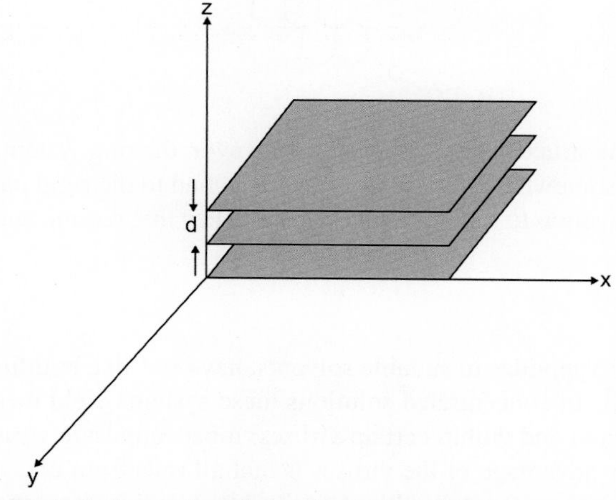

FIG. 24.9 Liquid crystal with positional disorder in the xy-plane. Order exists in the z-direction as the adjacent xy-planes are separated from each other by the distance d.

The term liquid crystal commonly applies to smectic and nematic phases. These types of liquid crystals are found in materials in which the constituent molecules or groups of molecules (usually called *building blocks*) are strongly elongated. From the above discussion it is evident that liquid crystals are quite different from conventional liquids in the sense that they are anisotropic in nature and are more ordered phases.

24.4.1 The Building Blocks

As explained above, in order to synthesize a liquid crystal, one should have elongated objects and these can have the following forms:

1. Small organic molecules.
2. Long helical rods, either naturally occurring or artificially made.
3. More complex units, such as associated structures of molecules and ions.

We present below some familiar examples of the building blocks of liquid crystals.

24.4.1.1 *Small Organic Molecules*

The classical example of a building block is *p*-azoxyanisole (PAA) with formula

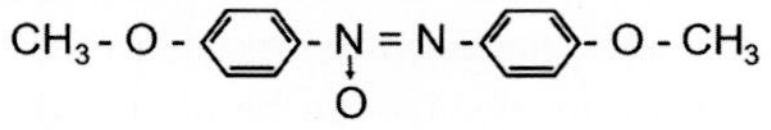

$$CH_3 - O - C_6H_4 - \underset{\downarrow \atop O}{N} = N - C_6H_4 - O - CH_3$$

In this molecule the two benzene rings are nearly coplanar. Therefore, it can be viewed as a rigid rod of length ~20 Å and width ~5 Å. Another example of practical interest is N-(p-methoxybenzylidene)-p-butylaniline (MBBA) with formula

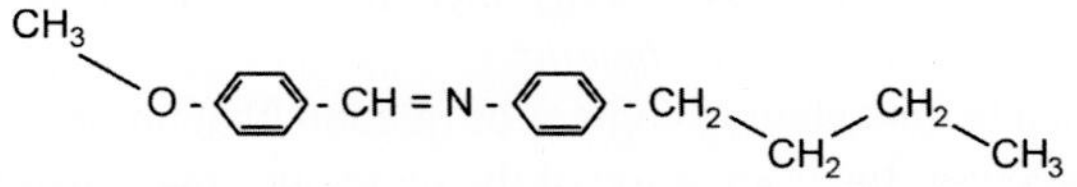

$$CH_3 - O - C_6H_4 - CH = N - C_6H_4 - CH_2 - CH_2 - CH_2 - CH_3$$

Both PAA and MBBA are nematogens, which means they give rise to the nematic type of mesophase.
A broad class of organic molecules with the following general chemical formula also gives mesophases

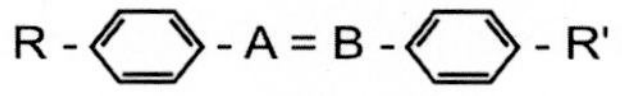

$$R - C_6H_4 - A = B - C_6H_4 - R'$$

The two benzene rings are rigidly bound by a double or triple A - B bond. R and R′ are short chains, which are partly flexible. Another favorable class comprises cholesterol esters, of the general formula

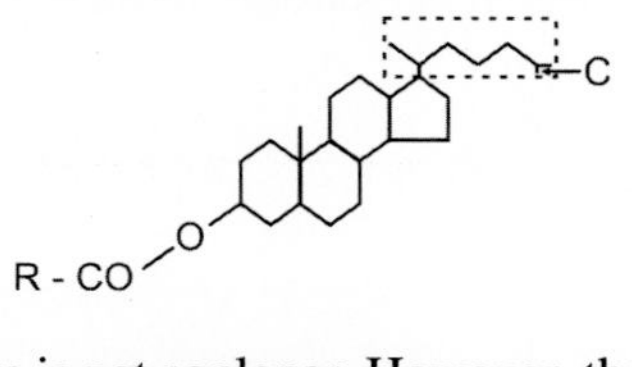

R - CO - O - [steroid ring system] - C

Here the rings are not aromatic and the structure is not coplanar. However, the ring system is rigid, while the saturated chain C and the radical R behave like two somewhat more flexible tails attached to the rigid part. In the pure systems above, the temperature is varied to induce a transition to the liquid crystal state. For this reason, such solids are called *thermotropic*.

24.4.1.2 *Long Helical Rods*

Some materials, such as synthetic polypeptides in suitable solvents, have rod-like building blocks with typical rod lengths on the order of 300 Å and width 20 Å. In concentrated solutions these systems yield mesophases. Similar phases are also found in DNA (deoxyribonucleic acids) and within certain viruses: tobacco mosaic virus with length ~3000 Å and width ~200 Å is the familiar example. One advantage of the viruses is that all rods from one virus species are exactly the same size. In all of these systems the transitions can be induced easily by changing the concentration of rods rather than the temperature. For this reason, they are commonly known as *lyotropic*.

24.4.1.3 Associated Structures

Typical examples of such structures are found in soap-water solutions. Here we have an aliphatic anion $CH_3-(CH_2)_{n-2}-CO_2^{-1}$ (with n in the range of 12–20) and a positive ion (Na^{+1}, K^{+1}, NH_4^{+1}, etc.). The group $-CO_2^{-1}$ is the polar head of the acid, which tends to be in close contact with water molecules, while the nonpolar aliphatic chain avoids water. These two requirements are typical of *amphiphilic materials*. A single chain in a solution can't satisfy these two opposite requirements, but a cluster of chains can do, as shown in Fig. 24.10. The resulting objects (rods or leaflets) become building blocks of mesomorphic solids.

24.4.2 Nematics and Cholesterics

24.4.2.1 Proper Nematics

A schematic representation of the order in a nematic phase is shown in Fig. 24.11A, which shows the following features:

1. The centers of gravity of the long molecules exhibit no positional order in a nematic. In other words, the positions of molecules are randomly distributed in nematics and they flow like liquids. Because of this, there is no sharp Bragg's peak in the X-ray diffraction pattern. For a typical nematic, such as PAA, the viscosities are of order 0.1 Poise.
2. There is some order, however, in the orientation of the molecules. They tend to align parallel to some common axis labeled by a unit vector $\hat{\mathbf{n}}$. Therefore, a nematic is a uniaxial medium with optical axis along $\hat{\mathbf{n}}$. The difference in the refractive indices measured with polarization parallel and normal to $\hat{\mathbf{n}}$ is quite large, typically on the order of 0.2 for PAA. In all known cases, there appears to be a complete rotational symmetry around the $\hat{\mathbf{n}}$ axis.
3. The direction of $\hat{\mathbf{n}}$ is arbitrary in space and is governed by minor forces. For example the direction of $\hat{\mathbf{n}}$ is governed by the guiding effect of the walls of the container.
4. The states along $\hat{\mathbf{n}}$ and $-\hat{\mathbf{n}}$ are indistinguishable. For example, if the individual molecules carry a permanent electric dipole moment, then the number of molecules with dipole moment up or down are equal and hence the material does not exhibit any electric polarization (see Fig. 24.11B).
5. Nematic phases occur only in materials that do not distinguish between right and left, that is, each molecule must be identical to its mirror image. If it is not so, then the system must be *racemic*, that is, a 1:1 mixture of right- and left-handed species.

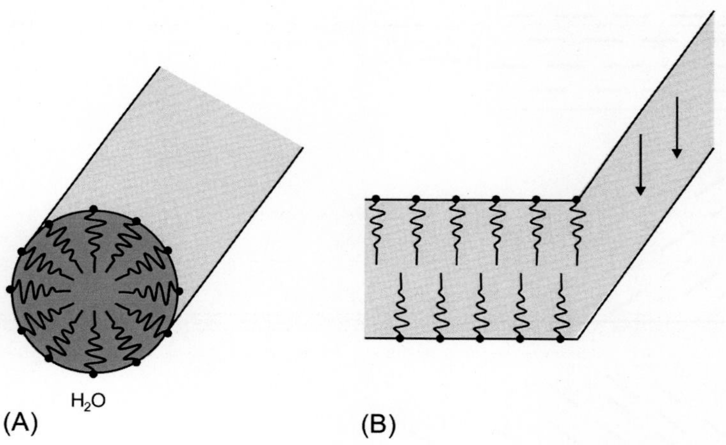

FIG. 24.10 Typical building blocks for amphiphilic materials (A) represents rods, and (B) represents sheets for the fatty acids-water system.

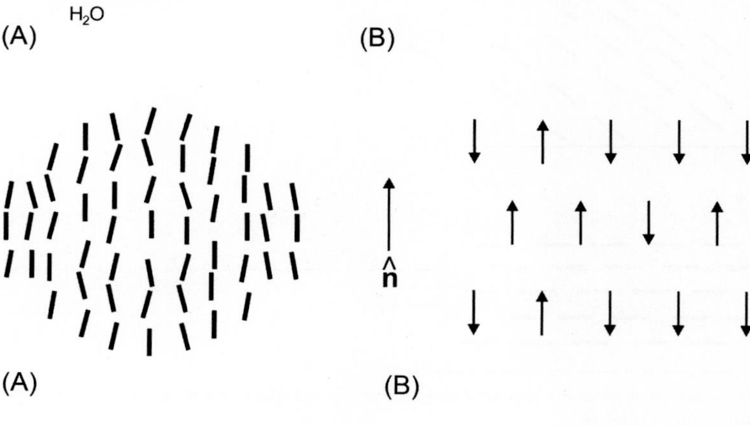

FIG. 24.11 (A) Schematic representation of a nematic phase. The molecules are just like rigid rods whose centers of gravity do not show any positional order. But the orientation of the molecules exhibits some order as they point approximately along the unit vector $\hat{\mathbf{n}}$. (B) Schematic representation of the intrinsic electric dipole moments of molecules in a nematic phase.

24.4.2.2 Cholesterics

If in a nematic liquid, a molecule that is *chiral* (different from its mirror image) is dissolved, it is found that the structure undergoes a helical distortion. The same type of distortion is found in pure cholesterol esters, which are also chiral. For this reason, the helical phase is called *cholesteric* and is a distorted form of the nematic phase. Locally, a cholesteric is very similar to a nematic material. The positions of molecules exhibit no long-range order, but the molecular orientation shows a preferred direction labeled by a unit vector $\hat{\mathbf{n}}$. However, $\hat{\mathbf{n}}$ is not constant in space but varies like a helical about the z-axis (see Fig. 24.12). Here the z-axis is called the helical axis. The Cartesian components of $\hat{\mathbf{n}}$ can be represented as

$$\begin{aligned} n_x &= \cos(q_0 z+\varphi) \\ n_y &= \sin(q_0 z+\varphi) \\ n_z &= 0 \end{aligned} \tag{24.4}$$

Both the helical axis and the value of φ are arbitrary. In Fig. 24.12, the structure is periodic along the z-axis as the states along $\hat{\mathbf{n}}$ and $-\hat{\mathbf{n}}$ are equivalent and the spatial period L is equal to one half of the pitch, that is,

$$L=\frac{\pi}{|q_0|} \tag{24.5}$$

The value of L is ~ 3000 Å and is much larger than the molecular dimensions. Because L is comparable to an optical wavelength, the periodicity results in Bragg scattering of light beams. Both the magnitude and sign of q_0 are meaningful. The sign distinguishes between right- and left-handed helices; a given sample at a given temperature T always produces helices of the same sign. If we change T, q_0 changes and in some particular cases even the sign of $q_0(T)$ may change at a particular temperature T^*. Such materials exhibit interesting features as stated below:

1. At $T=T^*$ the material is found to behave like a conventional nematic.
2. When the temperature crosses T^*, it is found that the physical properties, such as the specific heat, remain quite smooth.

Both properties above show that the local molecular arrangement is indeed similar in the nematic and cholesteric states.

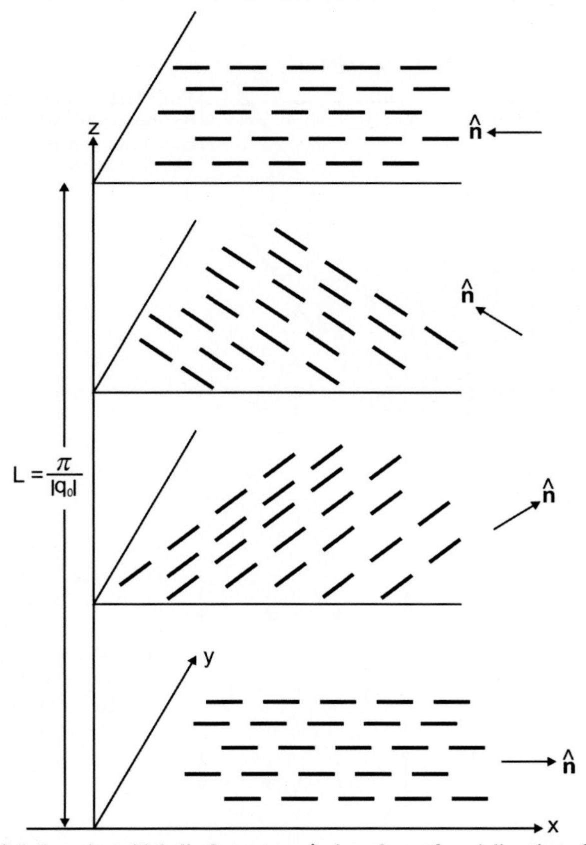

FIG. 24.12 Schematic representation of cholesterics with helical structure. $\hat{\mathbf{n}}$ gives the preferred direction of the rod-like molecules in a particular xy-plane and this direction changes gradually as one moves from one plane to another.

24.4.3 Smectics

Smectic is a Greek word that means soap. Therefore, smectic applies to those mesophases that have mechanical properties similar to those of soaps. From a structural point of view, all smectics are layered structures with a well-defined interlayer spacing that can be measured by X-ray diffraction. Smectics are thus more ordered than nematics. For a given material, the smectic phases always occur at temperatures below the nematic domain. Smectics can be classified mainly into three categories labeled by the letters A, B, and C and these categories can be distinguished by optical techniques.

24.4.3.1 Smectic A

The molecular arrangement in smectic A is shown in Fig. 24.13. It has the following major characteristics:

1. It has a layered structure with layer thickness equal to the length of the constituent molecules.
2. Inside each layer there is no positional long-range order. But there is remarkable one-dimensional orientational order.
3. The system is optically uniaxial, the optical axis being normal to the plane of the layers. Further, there is a complete rotational symmetry around the optical axis.
4. The z-axis is normal to the plane of the layer and the z and –z directions are equivalent.

24.4.3.2 Smectic C

The structure of a smectic C is defined as follows:

1. Each layer is a two-dimensional liquid.
2. The material is optically biaxial.

The axis of the long molecules is tilted with respect to the z-axis (see Fig. 24.14A). This information has been verified by a number of X-ray experiments. The layer thickness d is given as

$$d = \ell \cos\theta \tag{24.6}$$

where ℓ is the length of the molecule and θ is the tilt angle.

3. A simple smectic C is obtained from properties 1 and 2 when the constituent molecules are optically inactive (or with a racemic mixture). If an optically active molecule is added to smectic C, its structure gets distorted. The direction of tilt precesses around the z-axis and a helical configuration (smectic C*) is obtained (see Fig. 24.14B).

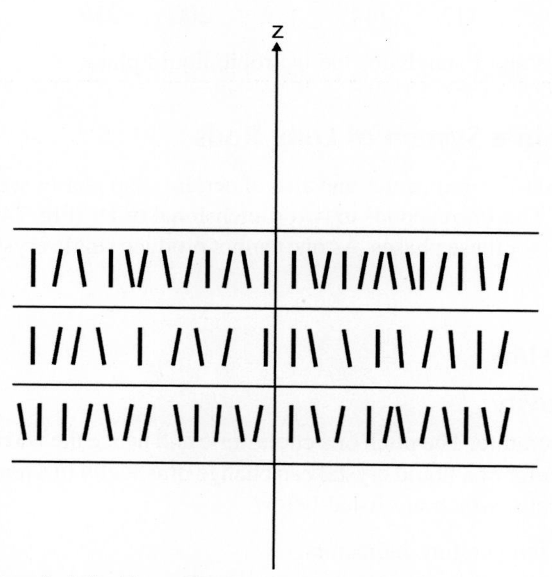

FIG. 24.13 Schematic representation of smectic A liquid crystal.

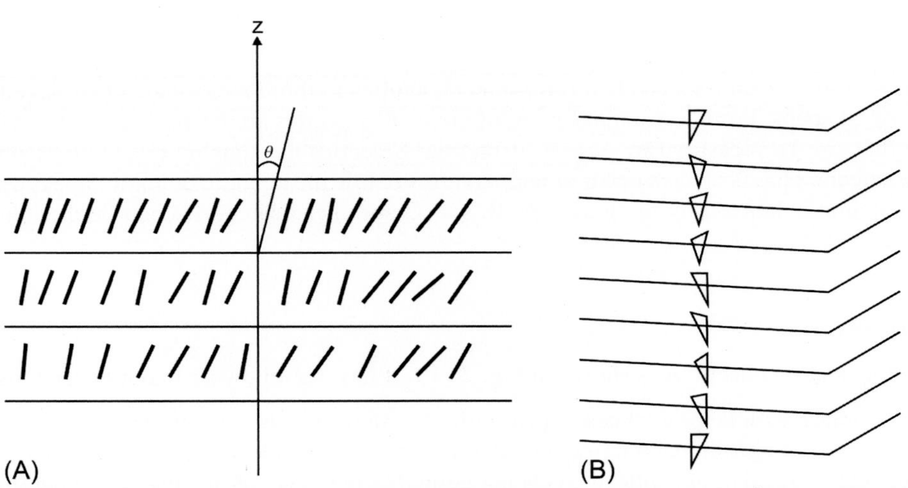

FIG. 24.14 (A) Schematic representation of smectic C liquid crystal. (B) Schematic representation of smectic C* liquid crystal with helical configuration.

24.4.3.3 Smectic B

In smectics A and C each layer behaves as a two-dimensional liquid. In smectic B, however, the layers appear to have the periodicity and rigidity of a two-dimensional solid. The order inside each layer is confirmed by X-ray reflections. The structure of smectic B liquid crystals can be studied by optical methods using an optical microscope. The texture of the smectic B phase (the so-called mosaic texture) shows domains inside each layer that are quite flat. This is in contradiction with A and C smectics in which the textures that are observed most often involve a strong curvature of the layers. Thus, phase B appears as the most ordered one among the A, B, and C phases. If a material is able to display all of the three phases, then the sequence of phase change with increasing temperature is always found to be

$$\mathrm{S} \Leftrightarrow \mathrm{B} \Leftrightarrow \mathrm{C} \Leftrightarrow \mathrm{A}$$

Here S stands for the solid phase. In the above sequence the degree of order decreases with an increase in temperature. A typical material showing all of the phases is terephthal-bis-p-butylaniline) (TBBA) with the formula

$$C_4H_9-\langle\bigcirc\rangle-N=CH-\langle\bigcirc\rangle-CH=N-\langle\bigcirc\rangle-C_4H_9$$

It gives the following set of transition temperatures (°C)

$$\begin{array}{ccccccccccc} \mathrm{S} & \Leftrightarrow & \mathrm{B} & \Leftrightarrow & \mathrm{C} & \Leftrightarrow & \mathrm{A} & \Leftrightarrow & \mathrm{N} & \Leftrightarrow & \mathrm{I} \\ & 113 & & 144 & & 172 & & 200 & & 236 & \end{array}$$

Here N stands for the nematic phase and I stands for the isotropic liquid phase.

24.4.4 Long-Range Order in a System of Long Rods

In the study of polymers with long rod-like molecules and also of certain soap phases we find a set of X-ray reflections that indicate hexagonal packing of rods. This corresponds to two-dimensional order (Fig. 24.15). Frank has proposed the name *canonic* (a Greek word meaning rod) for these phases. As one cannot produce single crystals of such materials, further study of such systems is limited.

24.4.5 Uses of Liquid Crystals

24.4.5.1 Temperature Sensitivity

Liquid crystals are sensitive to temperature. The pitch of a cholesteric and hence the wavelength of the Bragg reflected light depends on temperature. Thus, the color of a liquid crystal can change drastically in a temperature interval of a few degrees. This leads to a number of applications, which are listed below:

(a) Liquid crystals can be used as temperature indicators.
(b) They can be used for the detection of hot points in microcircuits.

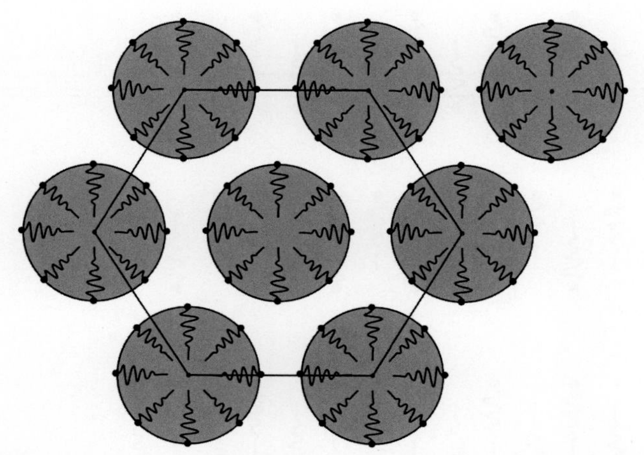

FIG. 24.15 Lipid in water with hexagonal symmetry of rods. The figure shows the circular cross sections of the lipid rods.

(c) They are used to locate fractures and tumors in the human body.
(d) They are used for the conversion of infrared images.

24.4.5.2 Optical Properties

The liquid crystals have unusual optical properties as they are uniaxial. Nematics and cholesterics are extremely sensitive to weak external fields. A change in applied field causes a change in reflectivity of the liquid crystals, making them extremely suitable for use as passive display devices. Nowadays liquid crystal films are used as display devices. The advantage of these display systems is that they work under low voltage and low power and are inexpensive. These devices are usually named liquid crystal display (LCD) devices. Light emitting diodes (LED) are also used for display systems, but they consume more power as they draw more current. The only disadvantage of LCDs is the speed of response. An LCD is a relatively slow device, taking at least tens and sometimes hundreds of milliseconds to respond. The seven-segment display (Fig. 24.16), obtained from a liquid crystal, is used as a numerical display in digital watches and pocket calculators.

On the other hand, smectics have higher viscosity and, therefore, have attracted less attention. However, the amount of work done on smectics is also increasing steadily in various directions.

24.4.5.3 Membrane Biophysics

Biological membranes are thin sheets (80 Å) of lipids and proteins. A lipid is an organic material that is sticky and insoluble in water. Lipids are soluble in either alcohol or ether. They play a crucial role in many living processes, but very little is known about their structure. Most physical techniques cannot be used for a single membrane as the amount of matter available is too small.

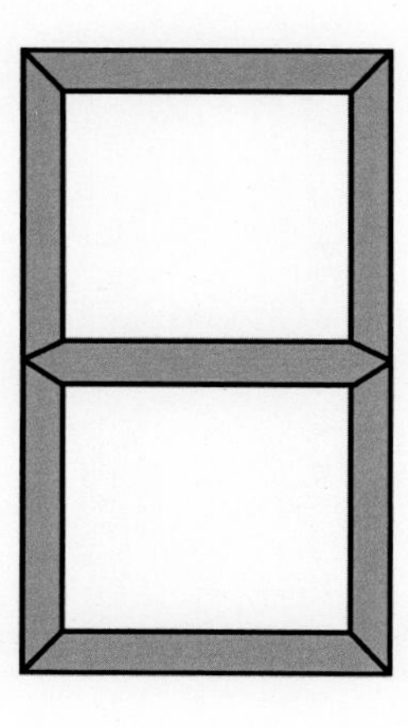

FIG. 24.16 Seven-segment liquid crystal display.

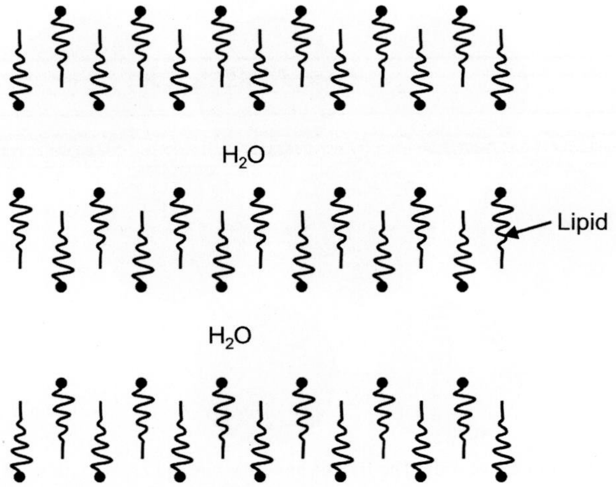

H_2O

Lipid

H_2O

FIG. 24.17 The structure of soap with alternate sheets of water and lipid.

24.4.5.4 Physics of Detergents

Soaps and nonionic detergents show a number of mesophases. The so-called "neat soaps" for instance, correspond to a lamellar phase with successive sheets of water and lipid (Fig. 24.17).

SUGGESTED READING

de Gennes, P. G. (1974). *The physics of liquid crystals*. London: Clarendon Press.

de Gennes, P. G. (1979). *Scaling concepts in polymer physics*. New York: Cornell University Press.

Elliott, S. R. (1990). *Physics of amorphous materials* (2nd ed.). Chichester: J. Wiley & Sons.

Waseda, Y. (1980). *The structure of non-crystalline materials: Liquids and amorphous solids*. New York: McGraw-Hill Book Co.

Zallen, R. (1973). *Physics of amorphous solids*. New York: J. Wiley & Sons.

Chapter 25

Physics of Nanomaterials

Chapter Outline

There are 103 stable elements in the periodic table. The different elements possess different properties, such as electrical and thermal conductivities, magnetism, and superconductivity. All of these properties have been discussed in the previous chapters and they are understood to a reasonable extent. Over the past few decades, scientists have made serious efforts in the miniaturization of machines, particularly electronic devices, or in discovering new materials. One of the ways to do this is to make alloys of different elements, but ultimately scientists discovered nanomaterials. Nano is a Greek word meaning dwarf or extremely small and one nanometer (nm) is equal to 10^{-9} m. The diameter of human hair is about 10^5 times larger than 1 nm. Nanomaterials are solids that are very small in size, in the range of 1–100 nm, and the technology involved in producing these materials and making different nanomachines from them is called nanotechnology. Nanomaterials have actually been produced and used by humans for hundreds of years. For example:

1. The beautiful red color of glass in some cathedrals is due to gold particles in the glass matrix. The color in stained glass windows is due to the presence of metal-oxide clusters in the glass.
2. The decorative glaze known as luster, found in some medieval pottery, contains spherical metallic nanoparticles that give special optical colors.
3. Small colloidal particles of silver are used in image formation in photography.

The techniques used to produce these materials were considered to be trade secrets at the time and are not fully understood even now. The concept of nanomaterials was raised by Richard Feynman. He delivered a talk to the American Physical Society in 1959 entitled, **"There is plenty of room at the bottom."** In this talk, he said that there are no fundamental physical reasons why materials cannot be fabricated by maneuvering individual atoms. In this chapter the reader will be made familiar with different types of nanomaterials, such as nanoparticles, quantum dots, and carbon nanotubes.

25.1 REDUCTION IN DIMENSIONALITY

A solid in which all three dimensions are very large (ideally infinite) is usually called a bulk material (Fig. 25.1A). In conventional solid state physics, an experimental study on a bulk material yields some sort of average value of the property under investigation, which may be quite different from the local behavior inside the solid. The concept of nanomaterials can

Solid State Physics. https://doi.org/10.1016/B978-0-12-817103-5.00025-6

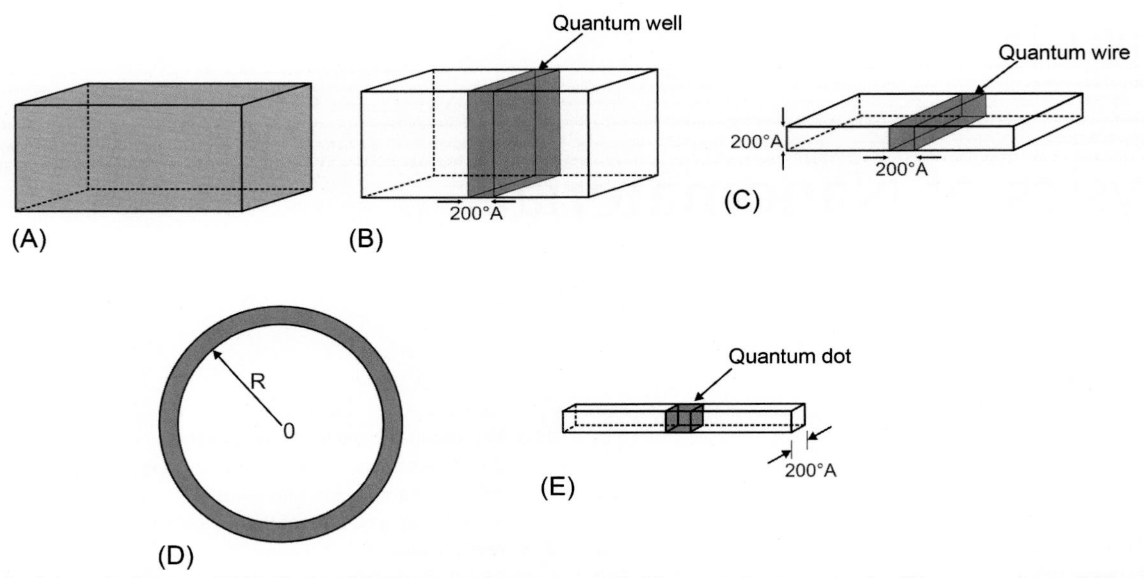

FIG. 25.1 Schematic diagram of (A) bulk material, (B) rectangular quantum well, (C) rectangular quantum wire, (D) quantum ring, and (E) rectangular quantum dot.

be well understood if the dimensions of a bulk material are reduced one by one to the nm range. If one of the dimensions of a bulk material is reduced to nm size, one obtains what is called a *quantum well* (Fig. 25.1B). Further, if two dimensions of the bulk material are reduced to nm size, a *quantum wire* (*nanowire or nanotube*) is obtained (Fig. 25.1C). Nanotubes can be made of any material but the most common are carbon nanotubes, which find immense applications in the electronic industry: carbon nanotubes are commonly used as connecting wires in nanoelectronic circuits. If a quantum wire is bent into the form of a ring, one gets a *quantum ring* (Fig. 25.1D). If all three dimensions of a bulk material are reduced to nm size, one ultimately gets what is called a *quantum dot* (Fig. 25.1E). Quantum dots are usually made of semiconductors and contain tiny droplets of free electrons. Precise control over the shape and size of a quantum dot allows us to have control over the number of electrons: it can contain from a single electron to several thousand electrons. The motion of an electron in a quantum dot is confined in all three directions, thus reducing its kinetic energy to a negligible value, which results in sharp energy levels like those found in atoms. It is for this reason that quantum dots are sometimes called *artificial atoms*, as they are much larger than actual atoms. Fig. 25.2 shows the various types of nanosolids with curvilinear geometry.

25.1.1 Quantum Well

Fig. 25.3A shows a solid with its z-direction reduced to nm size. In this case an electron is free to move in the infinite-dimensional xy-plane, just like the bulk material, but its motion is restricted along the z-direction. Let us define the electron wave vector **k** as

FIG. 25.2 Schematic representation of (A) spherical bulk material, (B) disk-shaped quantum well, (C) cylindrical quantum wire, (D) spherical quantum dot.

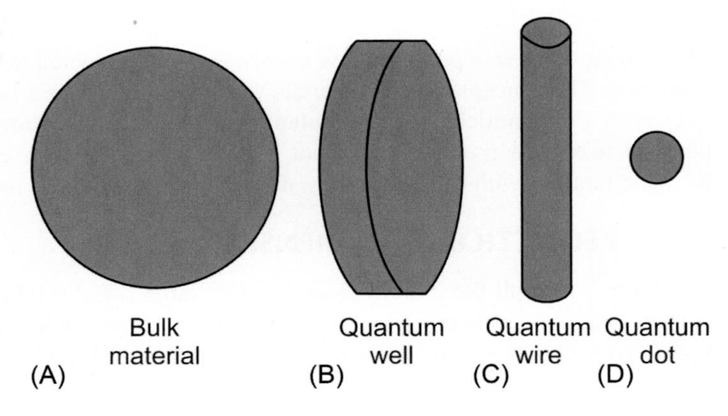

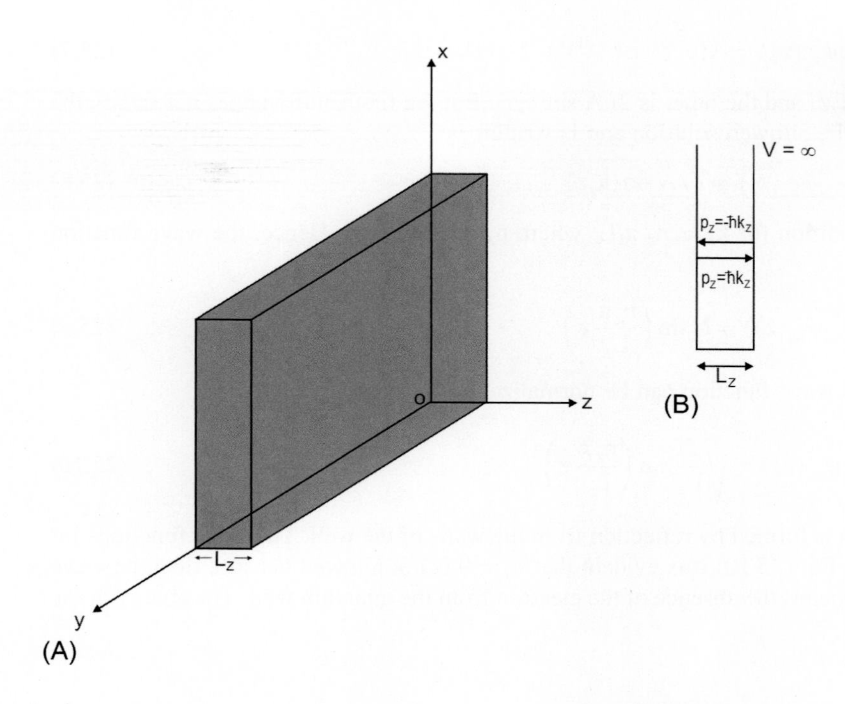

FIG. 25.3 (A) Rectangular solid with width L_z in the nanorange. (B) Rectangular quantum well with infinite potential well depth and having width L_z in the nanorange.

$$\mathbf{k} = \hat{\mathbf{i}}_3 k_z + \mathbf{k}_\perp \tag{25.1}$$

Here $\mathbf{k}_\perp$ is the wave vector in the xy-plane and is given by

$$\mathbf{k}_\perp = \hat{\mathbf{i}}_1 k_x + \hat{\mathbf{i}}_2 k_y = \hat{\mathbf{i}}_1 k_1 + \hat{\mathbf{i}}_2 k_2 \tag{25.2}$$

The motion of the electron in the z-direction is independent of its motion in the xy- plane. Therefore, the electron wave function $|\psi_{\mathbf{k}}(\mathbf{r})\rangle$ can be written as the product of the wave functions in the z-direction $|\psi_{k_z}(z)\rangle$ and in the xy-plane $|\psi_{\mathbf{k}_\perp}(\mathbf{r}_\perp)\rangle$, that is,

$$|\psi_{\mathbf{k}}(\mathbf{r})\rangle = |\psi_{k_z}(z)\rangle\, |\psi_{\mathbf{k}_\perp}(\mathbf{r}_\perp)\rangle \tag{25.3}$$

where $\mathbf{r}_\perp$ is the position vector of the electron in the xy-plane, which is defined as

$$\mathbf{r}_\perp = \hat{\mathbf{i}}_1 x + \hat{\mathbf{i}}_2 y = \hat{\mathbf{i}}_1 r_1 + \hat{\mathbf{i}}_2 r_2 \tag{25.4}$$

Let us first consider the motion of an electron along the z-direction. The electron can be assumed to move in a quantum well of width L_z with walls having very large or infinite potential (see Fig. 25.3B). An electron moving in the positive z-direction suffers reflection from the wall of the well and reverses its direction. It is for this reason that a solid with one of the dimensions reduced to nm size is usually called a quantum well. The electron wave function outside the quantum well is always zero due to the confinement of the electron in the well. Therefore, the electron wave function must go to zero at the walls, that is, at $z=0, L_z$ (boundary conditions). Otherwise there would be a discontinuity in the wave function at the boundary. The wave function of an electron moving in the positive z-direction is given by.

$$\left|\psi^+_{k_z}(z)\right\rangle = A\, e^{\imath k_z z} \tag{25.5}$$

Eq. (25.5) gives the wave function of an electron moving with momentum $\hbar k_z$. One cannot take the wave function given by Eq. (25.5) because it never goes to zero (in $e^{\imath k_z z}$ when $\cos k_z z$ goes to zero, $\sin k_z z$ is maximum and vice versa). But the wave function of an electron moving in the negative z-direction is given by

$$\left|\psi^-_{k_z}(-z)\right\rangle = A\, e^{-\imath k_z z} \tag{25.6}$$

Eq. (25.6) gives momentum $-\hbar k_z$ for the electron. Therefore, the most general wave function is obtained by taking the linear combination of wave functions given by Eqs. (25.5), (25.6), that is,

$$|\psi_{k_z}(z)\rangle = A\left(e^{\imath k_z z} \pm e^{-\imath k_z z}\right) \tag{25.7}$$

Eq. (25.7) gives two solutions: one is $2A\cos(k_z z)$ and the other is $2\imath A\sin(k_z z)$. But the first solution does not satisfy the boundary condition. Therefore, the second is the allowed solution and is written as

$$|\psi_{k_z}(z)\rangle = 2\imath A\sin(k_z z) \tag{25.8}$$

The solution (25.8) satisfies the boundary condition for $k_z = n_z\,\pi/L_z$ where n_z is an integer. Hence, the wave function becomes.

$$|\psi_{k_z}(z)\rangle = N\sin\left(\frac{n_z\pi}{L_z}z\right) \tag{25.9}$$

with N as the normalization factor. The above wave function can be normalized to unity to give

$$|\psi_{k_z}(z)\rangle = \sqrt{\frac{2}{L_z}}\sin\left(\frac{n_z\pi}{L_z}z\right) \tag{25.10}$$

The standing wave associated with the electron is formed by reflection from the walls of the well. The wave functions for $n_z = 1$, 2, and 3 are shown in Fig. 25.4. From Eq. (25.10) it is evident that $n_z = 0$ is not allowed because then the wave function would be zero for all z values, which means the absence of the electron from the quantum well. The energy of the electron in the z-direction is given by.

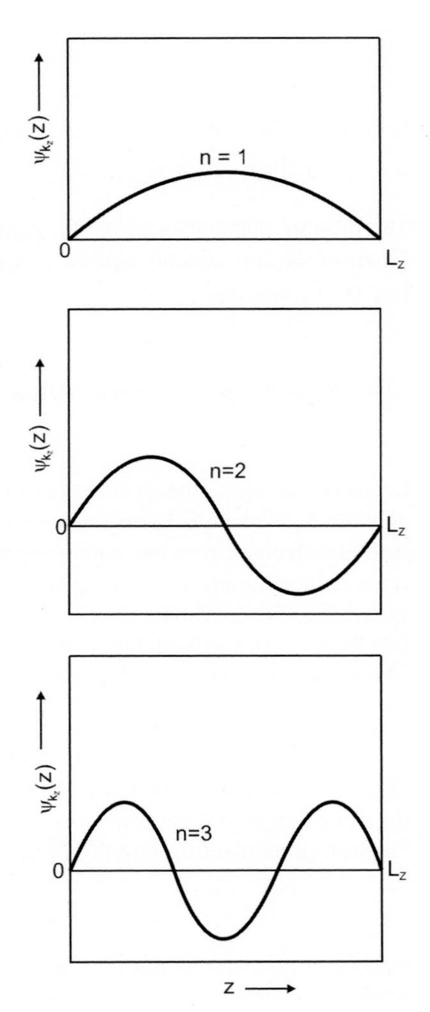

FIG. 25.4 The first three wave functions in increasing order of energy with n = 1, 2, and 3 for a rectangular quantum well.

$$E_{k_z} = \frac{\hbar^2 k_z^2}{2m_e} \tag{25.11}$$

Here m_e is the mass of the electron. Substituting the value of k_z, we get

$$E_{n_z} = \frac{\hbar^2}{2m_e}\left(\frac{n_z \pi}{L_z}\right)^2 = \frac{n_z^2 h^2}{8m_e L_z^2} \tag{25.12}$$

Eq. (25.12) shows that the energies are discrete and the minimum energy is given by.

$$E_{min} = \frac{h^2}{8m_e L_z^2} \tag{25.13}$$

In the x-y plane, an electron is free to move and its wave function is written as.

$$\left|\psi_{\mathbf{k}_\perp}(\mathbf{r}_\perp)\right\rangle = \frac{1}{\sqrt{A}} e^{\imath \mathbf{k}_\perp \cdot \mathbf{r}_\perp} \tag{25.14}$$

where $A = L_x L_y$ is the area of the crystal in the xy-plane. The values of k_x and k_y are given by the periodic boundary conditions as.

$$k_x = \frac{2\pi n_x}{L_x} \text{ and } k_y = \frac{2\pi n_y}{L_y} \tag{25.15}$$

where n_x, $n_y = 0, \pm 1, \pm 2, \ldots$ The energy of a free electron moving in the xy-plane is given by.

$$E_{\mathbf{k}} = \frac{\hbar^2}{2m_e}\left(k_x^2 + k_y^2\right) = \frac{\hbar^2 k_\perp^2}{2m_e} \tag{25.16}$$

From Eqs. (25.12), (25.16), the total energy of the quantum well becomes

$$E_{n_z}\left(k_x, k_y\right) = \frac{h^2 n_z^2}{8m_e L_z^2} + \frac{\hbar^2}{2m_e}\left(k_x^2 + k_y^2\right) \tag{25.17}$$

If $E_{n_z}^0$ is the minimum energy in the z-direction, then Eq. (25.17) can be written as.

$$E_{n_z}\left(k_x, k_y\right) = E_{n_z}^0 + \frac{\hbar^2}{2m_e}\left[\left(\frac{n_z \pi}{L_z}\right)^2 + k_x^2 + k_y^2\right] \tag{25.18}$$

Thus, in a quantum well, corresponding to each value of E_{n_z} there is a subband due to xy-motion (see Fig. 25.5). The Fermi energy E_F varies with the thickness L_z of the slab: the smaller the value of L_z, the higher the value of E_F. But in a slab, E_F will always be higher than it is in the bulk material. It is noteworthy that for an electron wave to be confined within the well, the width of the potential well must have dimensions comparable to the de Broglie wavelength of the electron, which is given by

$$\lambda_{db} = \frac{2\pi\hbar}{\sqrt{2m_e^* k_B T}} \tag{25.19}$$

where m_e^* is the effective electron mass and $k_B T$ gives thermal energy. Taking $m_e^* = 0.4m_e$ and $T = 4\,K$, λ_{db} comes out to be 1000 Å or 100 nm. Therefore, if a potential well is created in a material with a width on the order of nanometers, the electron waves are confined in the well with discrete energy levels.

It is also of interest to evaluate the density of electron states in a quantum well. Suppose, the number of electron states per unit energy (around energy E) is denoted as $N_e(E)$, then the number of allowed electron states dN_e between energy E and $E+dE$ is given by $N_e(E)\,dE$, that is,

$$dN_e = N_e(E)\,dE \tag{25.20}$$

The number dN_e depends directly on the number of allowed wave vectors within some range $\mathbf{k}$ and $\mathbf{k}+d\mathbf{k}$. If $N_e(\mathbf{k})$ is the number of allowed electron states per unit wave vector (around value $\mathbf{k}$), then $N_e(\mathbf{k})\,d\mathbf{k}$ gives the number of

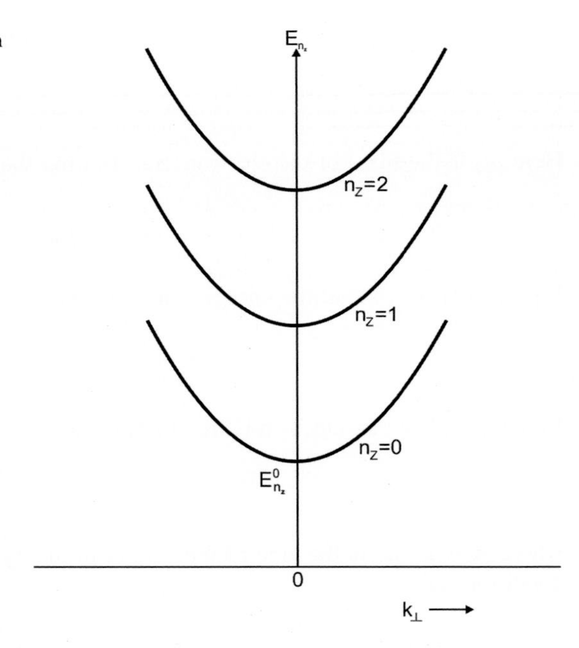

FIG. 25.5 The parabolic bands described by $E_{n_z}(k_x, k_y)$ in the xy-plane in a quantum well for $n_z = 0$, 1, and 2.

allowed states between **k** and **k**+d**k** values. So, $N_e(E)$ dE must be equal to $N_e(\mathbf{k})\,d\mathbf{k}$, assuming one electron to be in one state, that is,

$$N_e(E)\,dE = N_e(\mathbf{k})\,d\mathbf{k} \tag{25.21}$$

In the xy-plane of the quantum well, the allowed values of $\mathbf{k}_\perp$ between two circles with radii $k_\perp$ and $k_\perp + dk_\perp$ is given by.

$$N_e(k_\perp)\,dk_\perp = 2\frac{L_x L_y}{(2\pi)^2}\,2\pi\,k_\perp\,dk_\perp \tag{25.22}$$

using spherical coordinates. Here the factor of two accounts for the spin degeneracy. From the above equation.

$$N_e(k_\perp) = \frac{L_x L_y}{\pi}k_\perp \tag{25.23}$$

From Eq. (25.21) one can write

$$N_e(E) = \frac{N(k_\perp)}{dE/dk_\perp} \tag{25.24}$$

Using Eqs. (25.16), (25.23) in the above expression, we can write.

$$N_e(E) = \frac{1}{2\pi}L_x L_y\left(\frac{2m_e}{\hbar^2}\right) \tag{25.25}$$

Hence the number of electron states per unit energy per unit area in the xy-plane of the quantum well becomes.

$$g_e(E) = \frac{1}{2\pi}\left(\frac{2m_e}{\hbar^2}\right) \tag{25.26}$$

which is independent of energy as in a two-dimensional crystal. The density of electron states for a thin slab becomes constant and can be represented by a step function as shown in Fig. 25.6. The height of each step is given by E_{k_z}, which depends inversely on L_z. As the value of L_z becomes smaller, the height of each step becomes larger, that is, the thickness of each band becomes larger. In a bulk material, $g_e(E)$ as a function of E is a parabola. Therefore, if the bulk material is transformed into a slab (keeping the number of electrons constant), the parabolic band changes into a step function. From Fig. 25.6 it is clear that all of the electrons of the bulk material cannot be accommodated up to E_F if it is transformed into a slab as the electrons in the empty region are left out. To accommodate these electrons, we need to fill the higher energy states, yielding

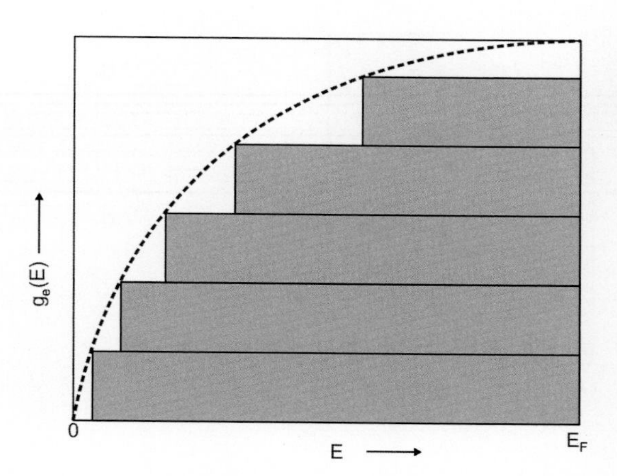

FIG. 25.6 The density of electron states per unit energy per unit volume $g_e(E)$ for a thin slab with width in the nanometer range. The dashed line shows $g_e(E)$ for a bulk material.

a higher value of E_F. As the value of L_z increases, the thickness of the subbands become smaller and the step function approaches closer to the parabolic band. In the limit of very large L_z, the step function coincides exactly with the parabolic band, as expected in a bulk material.

Quantum wells are usually made of semiconducting materials and have many potential applications in modern electronic devices. They can be realized by producing heterojunctions, which are sandwich structures made from semiconducting materials. As an example, consider semiconducting materials that have the same structure and exactly (or nearly) the same lattice constant, but different energy gaps, for example, GaAs and $Al_xGa_{1-x}As$ with $x = 0.3$. Here the band gap of GaAs is less than that of $Al_xGa_{1-x}As$. One of the materials, say GaAs, is taken as the substrate and a thin layer of $Al_xGa_{1-x}As$ is grown over it by using molecular beam epitaxy. The junction created between the two materials is called a heterojunction and the method is called the heteroepitaxy method. Fig. 25.7A shows one heterojunction between GaAs and $Al_xGa_{1-x}As$ and Fig 25.7B shows the conduction and valence bands along with the band gaps of the two materials. Fig. 25.7C shows the change in potential at the heterojunction for both the conduction and valence bands of the two materials. One can further deposit a GaAs layer on the $Al_xGa_{1-x}As$ layer to create a second heterojunction. Fig. 25.8 shows a double heterojunction in which a thin layer of GaAs is created between two layers of $Al_xGa_{1-x}As$. Here GaAs acts as a quantum well for the conduction electrons. Nearly free electrons exist in these semiconductors, which have a much higher potential energy in $Al_xGa_{1-x}As$ than in pure GaAs. Thus, the conduction electrons moving in the z-direction in the GaAs layer become trapped between the two potential walls, just like particles in a one-dimensional square well. But the electrons can move freely in the x- and y-directions. Here the potential well has finite walls instead of an idealized well with infinite walls. This fact affects the wave functions and energies to a small extent, but the equations derived for infinite walls still give reasonably accurate values for the energies. We want to point out here that the holes in the valence band experience a square barrier potential (see Fig. 25.8B). If a series of heterojunctions are constructed at regular intervals, then the potential

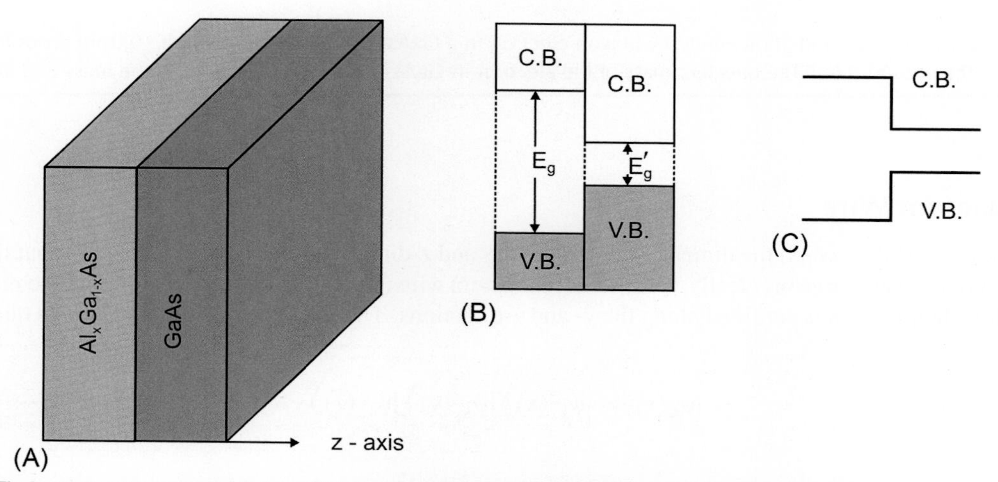

FIG. 25.7 (A) The heterojunction between GaAs and $Al_xGa_{1-x}As$. (B) The valence band (V.B.) and conduction band (C.B.) for GaAs and $Al_xGa_{1-x}As$ in the heterojunction (A). (C) The change in potential at the heterojunction (A) for both the V.B. and C.B. of the two materials.

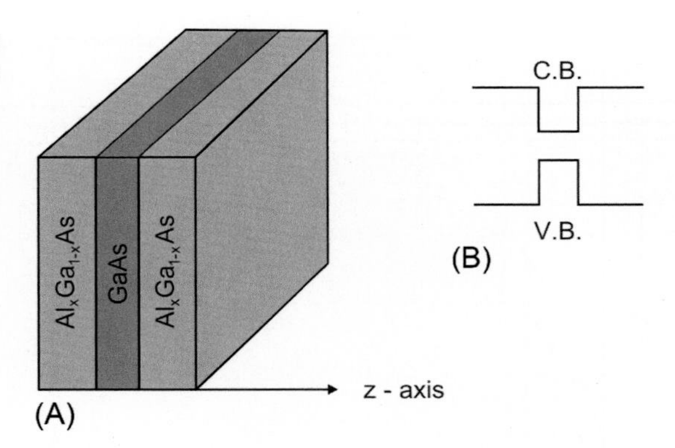

FIG. 25.8 (A) The double heterojunction between GaAs and $Al_xGa_{1-x}As$. (B) The quantum well for the electrons in the C.B. and the potential barrier for the holes in the V.B. in a double heterojunction (A).

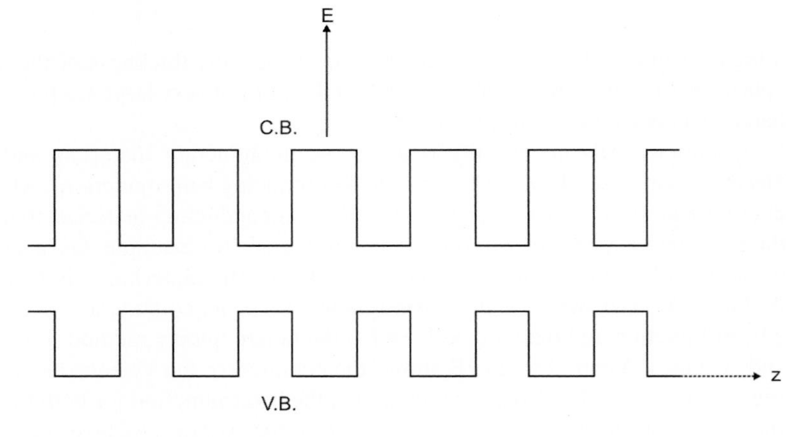

FIG. 25.9 The potential experienced by the electrons in the C.B. and holes in the V.B. in the case of a series of heterojunctions created at regular intervals.

variation in the conduction and valence bands will be periodic, as shown in Fig. 25.9. The conduction electrons experience a periodic square-well potential, while the holes experience a periodic square-barrier potential.

The existence of discrete energy levels in a quantum well is confirmed by the observation of the selective absorption of laser light at certain frequencies, which corresponds to the transition of an electron from one energy level to another. The combination of GaAs and $Al_xGa_{1-x}As$ forms the basis of the semiconductor laser used in compact disc players. The value of the effective mass m_e^* of an electron is found to be much smaller than the free electron mass m_e: in the GaAs quantum well, $m_e^* = 0.067 m_e$ approximately.

Problem 25.1

What is the wavelength of the radiation emitted when an electron in a GaAs quantum well of width 10.0 nm drops from the first excited state to the ground state? The effective mass of the electron in GaAs is $0.067\, m_e$ where m_e is the mass of a free electron.

25.1.2 Quantum Wire

Fig. 25.10 shows a crystal in which the dimensions along the y- and z-directions are reduced to nm size, but the dimension along the x-direction is very large or, ideally, infinite. In a quantum wire (nanowire) the electrons are free to move along the x-direction, while their motion is confined along the y- and z-directions. Therefore, the wave function in a quantum wire is given by.

$$|\psi_{\mathbf{k}}(\mathbf{r})\rangle = \left|\psi_{k_x}(x)\right\rangle \left|\psi_{k_y}(y)\right\rangle \left|\psi_{k_z}(z)\right\rangle \tag{25.27}$$

where

$$\left|\psi_{k_x}(x)\right\rangle = \frac{1}{\sqrt{L_x}} e^{\imath k_x x} \tag{25.28}$$

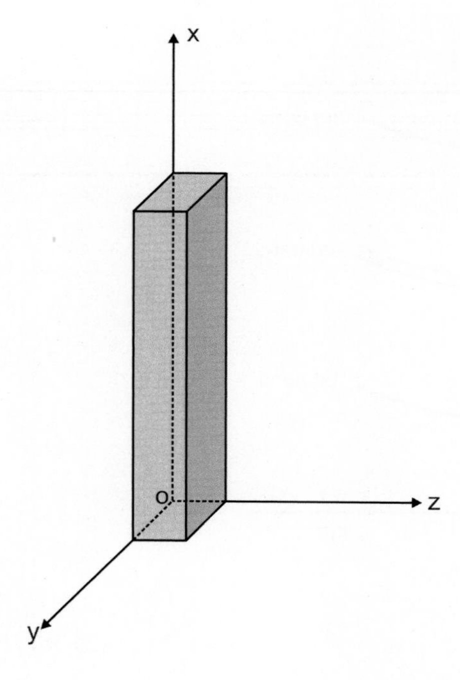

FIG. 25.10 Quantum wire (nanowire) with dimensions L_y and L_z along the y- and z-directions on the order of nm size. The length of the quantum wire along the x-direction is infinite.

$$|\psi_{k_y}(y)\rangle = \sqrt{\frac{2}{L_y}} \sin\left(\frac{n_y \pi}{L_y} y\right) \tag{25.29}$$

$$|\psi_{k_z}(z)\rangle = \sqrt{\frac{2}{L_z}} \sin\left(\frac{n_z \pi}{L_z} z\right) \tag{25.30}$$

The boundary conditions along the x-direction yield the values of k_x as.

$$k_x = \frac{2\pi n_x}{L_x} \tag{25.31}$$

where $n_x = 0, \pm 1, \pm 2, \ldots$. The energies in the y- and z-directions are quantized due to the confinement of the electron motion, while in the x-direction the energy is the same as that of a free electron. Proceeding exactly in the same way as in the quantum well, one gets the x-, y-, and z-components of energy as.

$$E_{k_x} = \frac{\hbar^2 k_x^2}{2m_e} \tag{25.32}$$

$$E_y = \frac{n_y^2 h^2}{8 m_e L_y^2} \tag{25.33}$$

$$E_z = \frac{n_z^2 h^2}{8 m_e L_z^2} \tag{25.34}$$

Hence the total energy of a quantum wire becomes.

$$E_{\mathbf{k}} = \frac{\hbar^2 k_x^2}{2m_e} + \frac{\hbar^2}{2m_e}\left[\left(\frac{n_y \pi}{L_y}\right)^2 + \left(\frac{n_z \pi}{L_z}\right)^2\right] \tag{25.35}$$

Fig. 25.11 shows the energy $E_{\mathbf{k}}$ as a function of k_x. The lowest energy band has $n_y = n_z = 1$ and is parabolic in nature along the k_x direction. The next band is doubly degenerate (with $n_y = 1$, $n_z = 2$ and $n_y = 2$, $n_z = 1$) for $L_y = L_z$. The two bands differ in energy slightly if $L_y \neq L_z$ (see Fig. 25.11). Further, the third band is a single band with $n_y = n_z = 2$ and so on. In one dimension (along the x-direction) Eq. (25.21) becomes

FIG. 25.11 The energy band structure of a quantum wire.

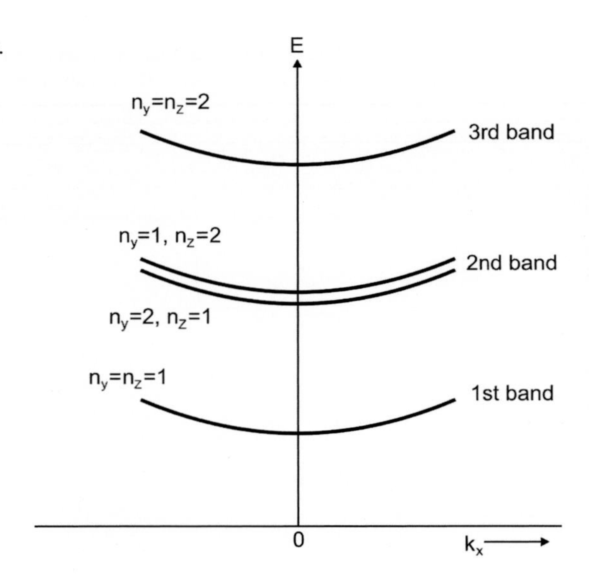

$$N_e\left(E_{k_x}\right) dE_{k_x} = N_e(k_x)\, dk_x \tag{25.36}$$

According to Eq. (25.31) there is one electron state in a wave vector of length $2\pi/L_x$, therefore, the number of allowed states per unit wave vector is given by.

$$N_e(k_x) = 2\frac{L_x}{2\pi} \tag{25.37}$$

From Eq. (25.36) one can write

$$N_e\left(E_{k_x}\right) = \frac{N_e(k_x)}{dE_{k_x}/dk_x} \tag{25.38}$$

Substituting Eqs. (25.32), (25.37) into the above expression, one gets the density of electron states per unit energy as

$$N_e\left(E_{k_x}\right) = \frac{L_x}{2\pi}\left(\frac{2m_e}{\hbar^2}\right)^{1/2} E_{k_x}^{-1/2} \tag{25.39}$$

The density of electron states per unit energy per unit length becomes.

$$g_e\left(E_{k_x}\right) = \frac{1}{2\pi}\left(\frac{2m_e}{\hbar^2}\right)^{1/2} E_{k_x}^{-1/2} \tag{25.40}$$

Fig. 25.12 shows that $g_e(E_{k_x})$ decreases with an increase in energy. Therefore, in a quantum wire most of the electrons lie in the lower energy states.

One important property of a quantum wire is that its conductance is quantized, which can be understood in terms of the above energy bands. Imagine a quantum wire with large metal crystals (three-dimensional electron gas) at each end. Then an electron reaching one end of the quantum wire can escape into the metal crystal. If, for simplicity, we assume 100% transmission, then an electron in the metal crystal at one end flows through the quantum wire and reaches the other end. The lowest energy state of this composite system will have all of the energy levels filled up to the Fermi energy E_F. So, the electrons in the wire at any energy below E_F will be flowing into the metal crystal, but they are certainly flowing into the wire at the same rate because there is no current flow in the ground state.

In a wire with a unit area of cross section, the current density due to an electron moving with velocity v_x and crossing a particular cross section per unit time is.

$$J_x = -e\, v_x \tag{25.41}$$

Using Eq. (25.32), one can write.

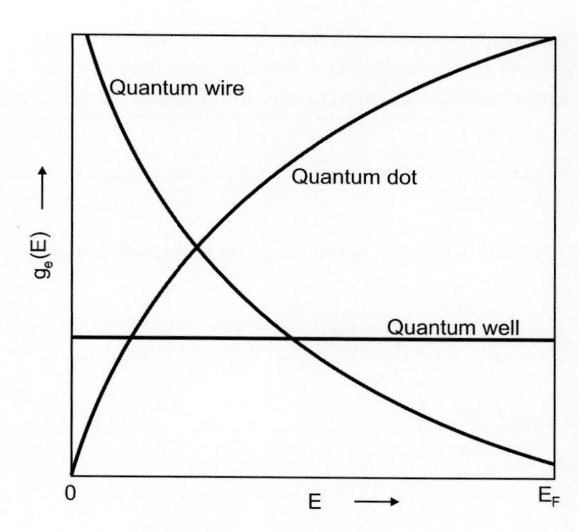

FIG. 25.12 The density of electron states for a quantum wire, quantum well and quantum dot.

$$v_x = \frac{1}{\hbar}\frac{dE_{k_x}}{dk_x} \tag{25.42}$$

Therefore, the general expression for the current density per unit length is

$$J_x = 2\sum_{k_x}(-e)\frac{1}{\hbar}\frac{dE_{k_x}}{dk_x}\frac{1}{L_x} \tag{25.43}$$

The factor of 2 includes the spin degeneracy of the bands. Here we sum over all positive k_x to include all electrons that contribute to the current. Changing the summation into integration, one gets.

$$J_x = \frac{L_x}{2\pi}\frac{2(-e)}{\hbar}\int\frac{dE_{k_x}}{dk_x}\frac{1}{L_x}dk_x$$

which can be written as

$$J_x = \frac{-e}{\pi\hbar}\int dE_{k_x} \tag{25.44}$$

If a voltage V is applied across the quantum wire, it produces the required energy difference for the flow of the current, that is,

$$\int dE_{k_x} = E_{k_x} = -eV \tag{25.45}$$

From Eqs. (25.44), (25.45) we get

$$J_x = \frac{2e^2}{h}V \tag{25.46}$$

which is to be added to every subband at E_F. So, the conductance contribution to every band at E_F is

$$\sigma = \frac{2e^2}{h} \tag{25.47}$$

Eq. (25.47) shows that the conductance in a quantum wire is quantized. The above analysis predicts a minimum conductivity of $2e^2/h$, which corresponds to a maximum resistance of.

$$R_{max} = \frac{h}{2e^2} = \frac{6.63\times10^{-34}}{2\times(1.6\times10^{-19})^2} = 12.95\,\text{kilo}-\text{ohms}. \tag{25.48}$$

Above this value, the resistance becomes infinite and the material becomes an insulator. In chapter 10 it has already been proved that the conductance in a two-dimensional solid (quantum Hall effect) is also quantized, but the quantum of conductance is e^2/h. This shows that the quantized conductance is an important concept in microscopic devices.

25.1.3 Quantum Dot

In a quantum dot, all three dimensions are reduced to nm size (see Fig. 25.13) and hence the wave function of a quantum dot becomes.

$$|\psi_{\mathbf{k}}(\mathbf{r})\rangle = \left|\psi_{k_x}(x)\right\rangle \left|\psi_{k_y}(y)\right\rangle \left|\psi_{k_z}(z)\right\rangle$$
$$= \sqrt{\frac{8}{L_x L_y L_z}} \sin\left(\frac{n_x \pi}{L_x} x\right) \sin\left(\frac{n_y \pi}{L_y} y\right) \sin\left(\frac{n_z \pi}{L_z} z\right) \tag{25.49}$$

Here the values of k_x, k_y and k_z are given by.

$$k_x = \frac{n_x \pi}{L_x}, k_y = \frac{n_y \pi}{L_y} \text{ and } k_z = \frac{n_z \pi}{L_z} \tag{25.50}$$

The energy eigenvalues in a quantum dot become.

$$E_{\mathbf{k}} = E_{k_x} + E_{k_y} + E_{k_z} \tag{25.51}$$

$$= \frac{\hbar^2}{2m_e}\left[\left(\frac{n_x \pi}{L_x}\right)^2 + \left(\frac{n_y \pi}{L_y}\right)^2 + \left(\frac{n_z \pi}{L_z}\right)^2\right] = \frac{\hbar^2 k^2}{2m_e} \tag{25.52}$$

If the quantum dot has all three dimensions the same, that is, $L_x = L_y = L_z = L$ (cubic quantum dot), then.

$$E_n(=E_{\mathbf{k}}) = \frac{\hbar^2}{2m_e}\frac{n^2\pi^2}{L^2} \tag{25.53}$$

where

$$n^2 = n_x^2 + n_y^2 + n_z^2 \tag{25.54}$$

Here E_n can be written as $E(n_x, n_y, n_z)$ for labeling the states. In a cubic quantum dot, the energy of the ground state of a system $E(1,1,1)$ has two-fold spin degeneracy. The first excited state, described by $E(2,1,1)$), $E(1,2,1)$, and $E(1,1,2)$, is six-fold degenerate. Interestingly, a cubic dot exhibits the same multiplicity of states as are found in an atom: a two-fold degenerate s-state and six-fold degenerate p-state, etc.. For this reason, quantum dots are often referred to as *artificial atoms*. Because the electronic structure of quantum dots can be tuned by changing their size or shape, they are particularly attractive building blocks for the development of new nanomaterials and nanotechnologies.

From Eq. (25.50) it is evident that in a volume of $\pi^3/L_xL_yL_z = \pi^3/V$, there is only one allowed **k**-state where V is the volume of the quantum dot. Therefore, the number of states in a spherical shell of radius k and thickness dk is given by.

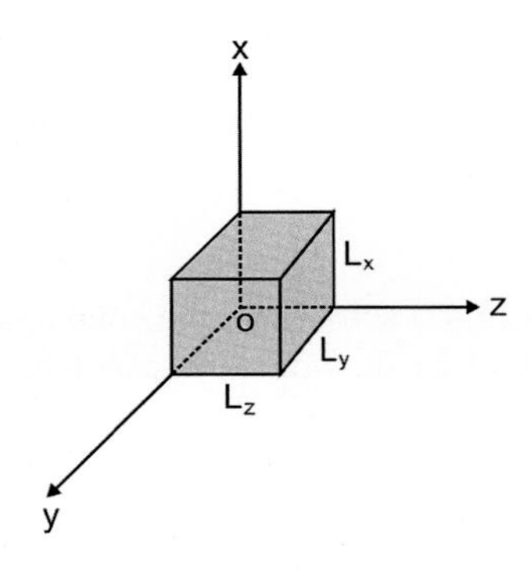

FIG. 25.13 Quantum dot with dimensions L_x, L_y and L_z along the Cartesian directions in the nm range.

$$N_e(\mathbf{k})\,d\mathbf{k} = 2\frac{V}{\pi^3}4\pi k^2 dk \tag{25.55}$$

From Eq. (25.21) one can write

$$N_e(E) = \frac{N_e(\mathbf{k})}{dE/dk} \tag{25.56}$$

Substituting Eqs. (25.52) and (25.55) into the above equation, we get

$$N_e(E) = \frac{4V}{\pi^2}\left(\frac{2m_e}{\hbar^2}\right)^{3/2} E^{1/2} \tag{25.57}$$

Hence the density of electron states per unit energy per unit volume is given by.

$$g_e(E) = \frac{4}{\pi^2}\left(\frac{2m_e}{\hbar^2}\right)^{3/2} E^{1/2} \tag{25.58}$$

Eq. (25.58) is similar to the expression for the density of electron states in a three-dimensional free-electron gas except for the constant factor. Fig. 25.12 shows that in a quantum dot the maximum number of electrons lie near E_F. Comparison of Fig. 9.4 and Fig. 25.12 shows that the density of electron states in nanomaterials is similar to that in a free-electron gas.

25.1.4 Quantum Ring

Consider a quantum wire of length L bent into a circular loop of radius R (see Fig. 25.1D). In this case the system repeats itself after a distance of $L = 2\pi R$ (periodic boundary condition). So, a quantum ring is a one-dimensional solid with a periodic boundary condition. If an electron with wave function $|\psi_k(r)\rangle$ is moving in a circular ring, then its wave function satisfies the periodicity condition, that is,

$$|\psi_k(r)\rangle = |\psi_k(r + 2\pi R)\rangle \tag{25.59}$$

The Schrodinger wave equation for a free electron is given by.

$$-\frac{\hbar^2}{2m_e}\nabla^2|\psi_k(r)\rangle = E_k\,|\psi_k(r)\rangle \tag{25.60}$$

and the wave function by.

$$|\psi_k(r)\rangle = \frac{2}{\sqrt{L}}e^{\imath kr} \tag{25.61}$$

Substituting Eq. (25.61) into Eq. (25.59), we get

$$e^{\imath k(2\pi R)} = 1 = e^{2\pi\imath n} \tag{25.62}$$

The above equation gives the values of the wave vector as.

$$kR = n$$

with $n = 0, \pm 1, \pm 2, \ldots$. One can also write.

$$k_n = \frac{n}{R} \tag{25.63}$$

Here we have labeled the wave vector by the subscript n for convenience. The energy of a free particle moving in the circular ring becomes

$$E_n = \frac{\hbar^2 k_n^2}{2m_e} = \frac{\hbar^2}{2m_e R^2}n^2 \tag{25.64}$$

From the above expression it is clear that the energy of an electron moving in a circular ring is different from that of an electron moving along a particular Cartesian direction and this change is brought about by the periodic boundary condition.

Problem 25.2

Show that the density of electron states in a quantum ring is given by.

$$N_e(E_n) = \frac{m_e R^2}{n\hbar^2}$$

25.2 QUANTUM TUNNELING

It is evident from Fig. 25.9 that a rectangular barrier separates two rectangular quantum wells. If the barrier is quite thin there is a significant probability that electrons will tunnel from one square well into the other. This fact influences the allowed energy levels in a quantum well. Let us study the probability of tunneling from a one-dimensional quantum well into another one (see Fig. 25.14). Let an electron in region 1 with wave vector k be incident from the left side. On reaching the barrier at $z=0$, one part of the wave function is reflected back and the other is transmitted into region 2. Further, on reaching the other end of the barrier at $z=b$, a part of the transmitted wave function may enter region 3, which is called the tunneling component of the wave function. At the two boundaries ($z=0$, b) the wave function and its derivatives should be continuous.

Region 1

If $|\psi_1(z)\rangle$ is the wave function of a free electron in region 1, then the Schrodinger wave equation can be written as.

$$-\frac{\hbar^2}{2m_e}\frac{d^2}{dz^2}|\psi_1(z)\rangle = E|\psi_1(z)\rangle \tag{25.65}$$

or

$$\left(\frac{d^2}{dz^2}+k_0^2\right)|\psi_1(z)\rangle = 0 \tag{25.66}$$

where

$$k_0^2 = \frac{2m_e}{\hbar^2}E \tag{25.67}$$

The solution of Eq. (25.66) comprises incident and reflected waves and, therefore, the most general solution is the linear combination of the two, that is,

$$|\psi_1(z)\rangle = Ae^{\imath k_0 z} + Be^{-\imath k_0 z} \tag{25.68}$$

If the intensity of the incident wave is taken as unity ($A = 1$), then

$$|\psi_1(z)\rangle = e^{\imath k_0 z} + Be^{-\imath k_0 z} \tag{25.69}$$

Region 2

Let $|\psi_2(z)\rangle$ be the wave function of the electron in region 2. In this region, the electron experiences a constant potential V_0, therefore, its Schrodinger equation becomes.

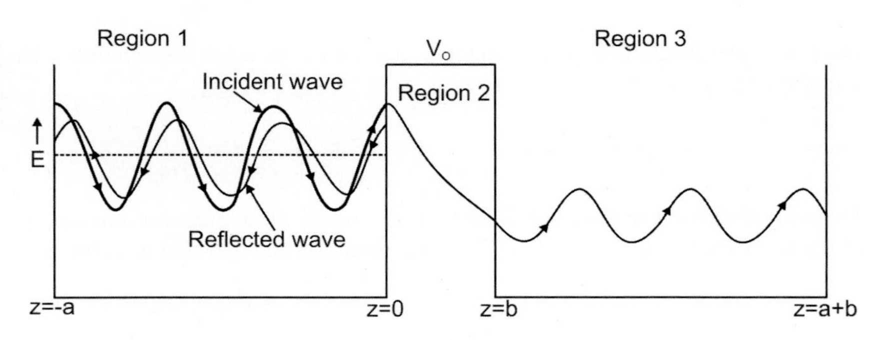

FIG. 25.14 A potential barrier with width b, height V_0, and a as the distance between two consecutive barriers. The tunneling of the incident wave through the potential barrier is depicted in the diagram.

$$\left[\frac{-\hbar^2}{2m_e}\frac{d^2}{dz^2}+V_0\right]|\psi_2(z)\rangle = E|\psi_2(z)\rangle \tag{25.70}$$

which can be written as

$$\left(\frac{d^2}{dz^2}-k^2\right)|\psi_2(z)\rangle = 0 \tag{25.71}$$

where

$$k^2 = \frac{2m_e}{\hbar^2}(V_0 - E) \tag{25.72}$$

The solution of Eq. (25.71) is given by

$$|\psi_2(z)\rangle = Ce^{-kz} + De^{kz} \tag{25.73}$$

The wave function $|\psi_2(z)\rangle$ must go to zero in the asymptotic limit, which is so if $D = 0$. Therefore, $|\psi_2(z)\rangle$ becomes

$$|\psi_2(z)\rangle = Ce^{-kz} \tag{25.74}$$

The equality of the wave functions $|\psi_1(z)\rangle$, $|\psi_2(z)\rangle$ and their derivatives at $z=0$ gives the continuity condition, which from Eqs. (25.69), (25.74) yield.

$$1 + B = C \tag{25.75}$$

$$\imath k_0(1 - B) = -kC \tag{25.76}$$

The values of B and C obtained from Eqs. (25.75), (25.76) are

$$B = \frac{k_0 - \imath k}{k_0 + \imath k}, \quad C = \frac{2k_0}{k_0 + \imath k} \tag{25.77}$$

The above equation gives

$$|B|^2 = 1$$

which means that there is total reflection at $z=0$ as in the classical case. However, the relative probability of finding the electron in region 2 is also finite and is given by.

$$\begin{aligned} P_2(z) &= \left|Ce^{-kz}\right|^2 \\ &= \frac{4k_0^2}{k_0^2 + k^2}e^{-2kz} \end{aligned} \tag{25.78}$$

The probability $P_2(z)$ is appreciable near the barrier edge at $z=0$ and then decreases exponentially (see Fig. 25.14). The probability becomes negligible over a distance that is large compared with $1/k$. If the thickness of the barrier is small (smaller than $1/k$), then we obtain finite and significant probability for finding the electron at its other end (at $z=b$). It will then propagate to the right as a free electron. From Eq. (25.78) the probability of finding the electron at $z=b$ becomes.

$$P_3(b) = \frac{4k_0^2}{k_0^2 + k^2}e^{-2kb} \tag{25.79}$$

The relative probability of finding the electron at $z=b$ is

$$T = \frac{P_3(b)}{P_2(0)} = e^{-2kb} = e^{-2b\sqrt{\frac{2m_e}{\hbar^2}(V_0 - E)}} \tag{25.80}$$

T gives the coefficient of transmission and is an approximate result for large b. For a narrow barrier there is a considerable mixture of rising and falling exponentials.

Region 3

In this region the potential is zero, that is, $V(z) = 0$, so the wave equation is the same as in region 1 but with different amplitude. The wave equation is written as.

$$-\frac{\hbar^2}{2m_e}\frac{d^2}{dz^2}|\psi_3(z)\rangle = E|\psi_3(z)\rangle \quad (25.81)$$

This equation has a solution similar to that of Eq. (25.65) but the difference is that the wave moves toward the right after tunneling through the barrier, that is, in the positive direction. The wave function can therefore, be written as

$$|\psi_3(z)\rangle = De^{\imath k_0 z} \quad (25.82)$$

Here D is the amplitude of the wave in region 3, which is less than the amplitude of the incident wave in region 1, but the frequency and wave vector are the same. The total wave function is obtained by joining together the wave functions from the three regions. Fig. 25.14 shows the real parts of the wave functions in the three regions.

Problem 25.3

Consider a barrier having width $b = 10^{-6}$ cm and barrier height $V_o = 0.3$ eV. Calculate the probability of tunneling through the barrier by an electron having energy (i) 0.1 eV, (ii) 0.2 eV, and (iii) 0.28 eV. Comment on the probability of tunneling with an increase in the energy of the electron.

25.3 NANOPARTICLES

The smallest particles in a solid that can exist independently are atoms. Therefore, there is a fundamental limit to making anything arbitrarily small, which is obviously the size of an atom (0.1 nm). In reducing the size of a solid, one is able to produce material particles with size in the nm range. *A nanoparticle is considered to be a cluster of atoms (molecules) bonded together within a diameter of 100 nm.* A nanoparticle is denoted as A_n where A is the symbol of the element and n is the number of atoms (molecules) in it. For example, Al_{12} denotes a nanoparticle of Al metal with 12 atoms in it. Note that a cluster is an aggregate of atoms or molecules, with a size somewhere between microscopic and macroscopic particles. Fig. 25.15 presents the classification of material into molecules, nanoparticles, and bulk material. It is evident that nanoparticles lie between molecules and bulk materials. This classification is somewhat arbitrary because there are many biological organic molecules containing a large number of atoms that do not fit into this classification. Therefore, one has to evolve some working definition of nanoparticles. It has already been explained in the previous chapters that some sort of critical length characterizes different properties of materials. For example, electrical conductivity is characterized by the mean free path and superconductivity by the coherence length. *So, a nanoparticle can be defined as a cluster of atoms (molecules) with dimensions ranging from 1 to 100 nm forming a very small part of the bulk material with dimensions less than the characteristic length of the phenomenon to be studied.* A quantum dot can be considered as a nanoparticle with a small number of atoms (molecules).

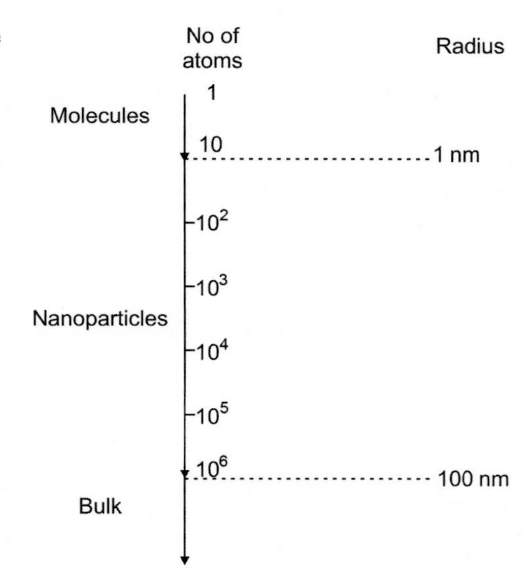

FIG. 25.15 Classification of molecules, nanoparticles, and bulk material as regards the radius and number of atoms.

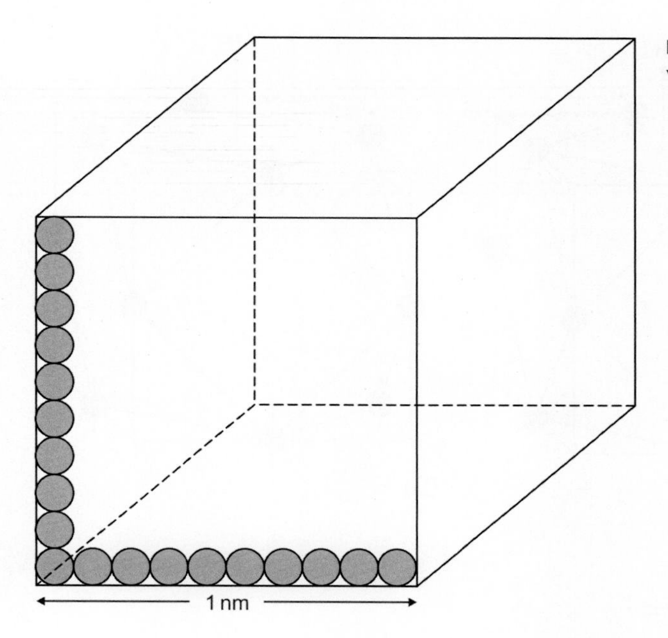

FIG. 25.16 A cube with 1 nm edge and having sc structure being filled with atoms.

To have an idea about the number of atoms in a nanoparticle, one can consider the simple example of a solid in the form of a cube with an edge of 1 nm (see Fig. 25.16) and possessing sc structure. The diameter of an atom is usually taken as 10^{-8} cm (0.1 nm). So, in the case of close packing, the number of atoms in a 1-nm cube comes out to be 1000. The number of atoms in a 1 nm cube with fcc and bcc structures should be still higher due to the higher packing fraction in these structures. But in an actual crystal there are 3–5 atoms lined up in 1 nm of distance. So, a real cubic nanoparticle with a 1-nm edge contains 25–125 atoms. On the other hand, an actual spherical nanoparticle with 1-nm diameter contains 12–65 atoms. This fact shows that the atoms in a nanoparticle are bonded together with some finite bond length.

25.3.1 Magnetic Nanoparticles

There are a large number of paramagnetic and ferromagnetic elements in which each atom possesses a finite intrinsic magnetic dipole moment. Paramagnetic elements exhibit magnetization only in the presence of a magnetic field, while ferromagnetic elements exhibit finite magnetization even in the absence of a magnetic field below the critical temperature T_c. In a bulk paramagnetic element, the intrinsic atomic magnetic moments are randomly oriented, yielding zero magnetization in the absence of an applied magnetic field. As the size of the material is reduced to nm size, the distribution of the atomic magnetic moments may not remain completely random and may yield a finite, though small, magnetic moment on a nanoparticle. As the size of the nanoparticle is reduced further, the degree of randomness in the distribution of a smaller number of atomic magnetic moments may decrease further, thus resulting in a possible increase in the total magnetic moment of the nanoparticle. If an external magnetic field is applied, the magnetic moment of each atom in the nanoparticle experiences two competing forces: thermal and magnetic. The result of these competing forces decides the direction of the magnetic moment of each atom. The vector sum of the magnetic moments of all the atoms gives rise to the net magnetic moment of a nanoparticle (see Fig. 25.17A). Therefore, in the presence of an external magnetic field, the magnetic moment of a nanoparticle may increase further. Fig. 25.17B shows a perfect alignment of all the atomic magnetic moments in a nanoparticle along the direction of the magnetic field, yielding the maximum possible magnetic moment. An example worth mentioning here is that of Rhenium (Re), which is a paramagnetic element in bulk form. However, its nanoparticle Re_n shows interesting behavior with respect to a variation in n. It has been found that Re_n exhibits a significant increase in net magnetic moment for $n < 20$. This example shows that there is a possibility to produce magnetic materials from nanoparticles made of paramagnetic materials. Investigation of the electronic band structure of crystalline solids shows that ferromagnetic behavior depends on two factors:

1. Density of electron states at the Fermi energy E_F denoted by $N(E_F)$.
2. The spin-spin exchange interactions denoted by J.

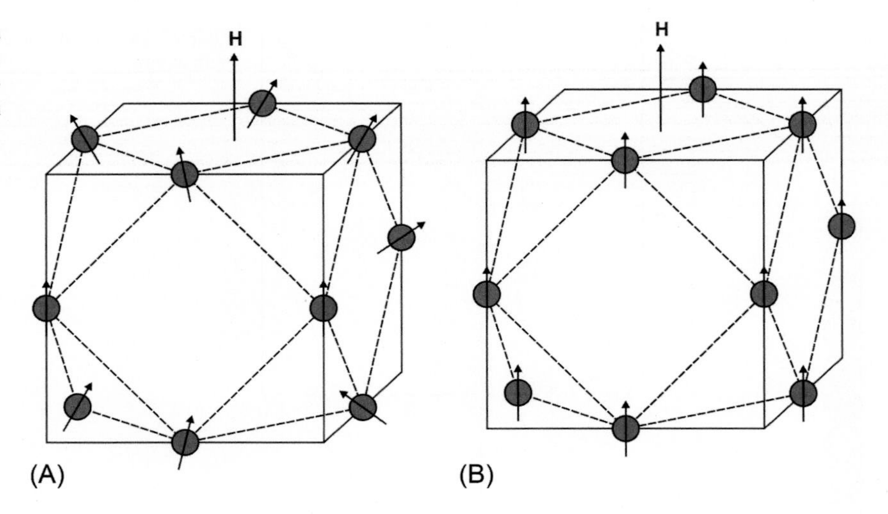

FIG. 25.17 (A) Magnetic moments of atoms in arbitrary directions in a nanoparticle. (B) Magnetic moments of all the atoms in a nanoparticle are aligned in one direction, the direction of the applied magnetic field.

The condition for the existence of ferromagnetism is

$$JN(E_F) \geq 1$$

The formation of energy bands has already been discussed in chapter 12. As the size of a solid is decreased, the electronic energy bands are expected to become narrower, which may cause a possible increase in the density of electronic states in the-band. It is necessary to explore how the spin-spin exchange interactions change with a decrease in the size of a solid. In a bulk material with a large number of atoms or molecules, the probability of spin flipping is large. But with a decrease in the size of the material (or with a decrease in the number of atoms), the probability of a spin flip may decrease significantly and this may be due to an increase in the spin-spin exchange interactions. If the above explanation is accepted, then those nanoparticles made of elements that are nearly ferromagnetic, such as Pd, may exhibit ferromagnetism.

The magnetic moment of a nanoparticle can be measured by the famous Stern-Gerlach experiment (see Fig. 25.18). In this experiment a narrow beam of nanoparticles passes through a nonuniform magnetic field. The magnetic moment of each nanoparticle interacts with the applied magnetic field, which splits the beam into two parts, with the nanoparticles of one magnetic moment orientation moving in the opposite direction to those of the other orientation. An impression of the beam can be taken either on a photographic plate or on a fluorescent screen. From knowledge of the beam separation and the strength of the magnetic field, one can estimate the magnetic moment of a nanoparticle.

25.3.2 Structure of Nanoparticles

The ratio of the surface S to the volume V of a solid is defined by the parameter ζ_{sv} as

$$\zeta_{sv} = S/V \tag{25.83}$$

If the solid is in the form of a sphere of radius r, then $S = 4\pi r^2$ and $V = (4/3)\,\pi r^3$. Hence the parameter ζ_{sv} becomes.

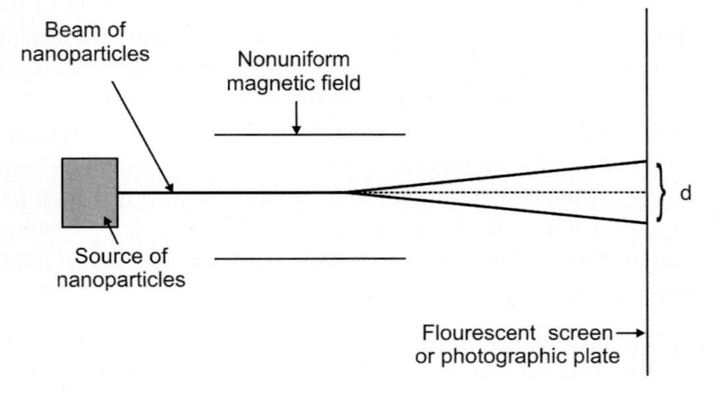

FIG. 25.18 The Stern-Gerlach experiment. A collimated beam of nanoparticles passing through a nonuniform magnetic field is split into two beams. The two beams striking the photographic plate are separated by a distance d.

$$\zeta_{sv} = \frac{4\pi r^2}{(4/3)\pi r^3} = \frac{3}{r} \tag{25.84}$$

Similarly, for a solid in the form of a cube of edge r, the parameter ζ_{sv} becomes.

$$\zeta_{sv} = \frac{6}{r} \tag{25.85}$$

From Eqs. (25.84), (25.85) it is evident that ζ_{sv} is inversely proportional to r: ζ_{sv} increases with a decrease in the size of the material. This amounts to an increase in the surface forces with a decrease in the size of the material. The effect of an increase in surface with a decrease in the size of the solid can also be realized physically: for a smaller number of atoms in a solid, a greater fraction of the atoms is expected to lie on the surface. In a very small nanoparticle, ζ_{sv} approaches 1 as most of the atoms lie on its surface. Therefore, in nanomaterials, the surface forces influence the electronic structure significantly and must be incorporated in the development of a theoretical model.

In the case of large nanoparticles, the structure is the same as that of the bulk material but with some different lattice parameter "a". For example, nanoparticles of Al metal with a size of 80 nm possess fcc structure (X-ray diffraction). It has been observed that the electrostatic interactions between ions and electrons in crystals give rise to a size-dependent intracrystalline pressure (ICP): ICP may increase or decrease with a decrease in the size of a nanoparticle. For example, in Cu, Ag, and Au, the ICP increases with a decrease in the size of the nanoparticle, resulting in a decrease in the lattice parameter. But in Ni, ICP decreases with a decrease in size and hence the lattice parameter increases. The experimental determination of the structure of small nanoparticles is difficult. In small nanoparticles, the surface energy is very large and they acquire the structure having minimum surface energy, which naturally is a sphere in the case of central Coulomb interactions between the atoms. In Al_{13} there are three possible arrangements of atoms (see Fig. 25.19). It is noteworthy that all three possible arrangements of atoms in Al_{13} produce a nearly spherical shape with most of the atoms at the surface. The spherical shape significantly affects the properties of nanoparticles, such as their vibrational structure, stability, and reactivity. Molecular orbital calculations for Al_{13} in conjunction with density functional theory show that the icosahedral structure yields the lowest energy compared with the other two structures. The structure of nanoparticles can also be studied with the help of computer simulation techniques. The above example shows that the structure of a nanoparticle may undergo a significant change with a decrease in its size. One can also argue the reverse. Nanoparticles containing a few atoms possess a geometrical shape that has most of the atoms on the surface. With an increase in the size of a nanoparticle, crystalline order starts to appear and, ultimately, the atoms exhibit the crystal structure of the bulk material.

Some nanoparticles are highly reactive in air, depending on the nature of the material. For example, if an isolated nanoparticle of Al is exposed to air, its surface atoms react immediately with air to form an oxide of Al, that is, Al_2O_3. Thus, a coating of Al_2O_3 is expected to cover a nanoparticle of Al metal, which will protect it from reacting further with oxygen. Such metallic nanoparticles are said to be passivated (here Al nanoparticles are oxygen passivated). Nanoparticles can also be formed in a solution without exposure to air. For example, the decomposition of AlH_3 in a heated solution containing oleic acid produces nanoparticles of Al. In this case the molecules of the oleic acid are bonded to the surface of the nanoparticles. Oleic acid is called a *surfactant*, which will coat the nanoparticles and prevent them from aggregating and oxidizing.

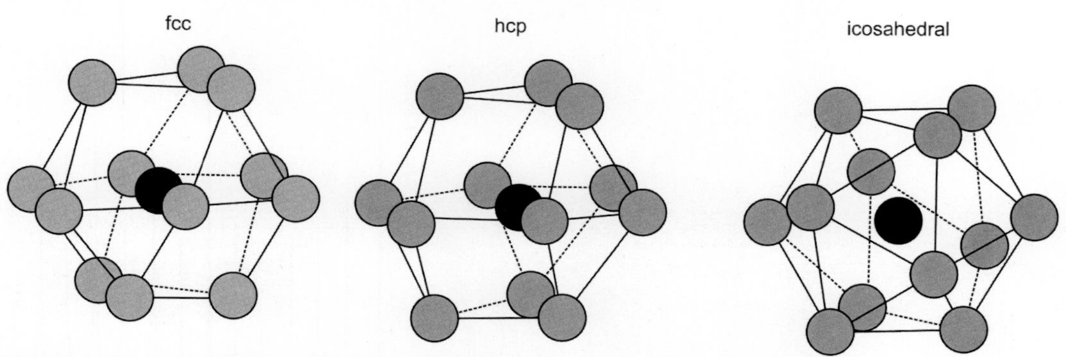

FIG. 25.19 Three possible structures of Al_{13} nanoparticles. *(Modified from Poole, C. P., & Owens, F. J. (2006). Introduction to nanotechnology (p. 78). Asia: John Wiley & Sons.)*

25.3.3 Methods of Synthesis of Nanoparticles

There are a number of methods to synthesize nanoparticles with different sizes. An important aspect of all the methods is to provide some arrangement for the stability of the nanoparticles as regards the shape and size, which can be achieved by the passivation and surfactance of the particles. Here we shall concentrate on the synthesis of metallic nanoparticles. Some commonly used methods to produce metallic nanoparticles are described below.

25.3.3.1 Laser Beam Methods

The basic principle involved in the synthesis of nanoparticles is to evaporate the metal by any means to get constituent atoms and then to allow them to form clusters under favorable conditions. The various methods to evaporate the metal give rise to different fabrication methods. Fig. 25.20 shows the experimental set up for producing nanoparticles of metals using a strong laser beam. It consists of a chamber in which He gas is allowed to enter under pressure from one side (left side). A circular disc of the desired metal is placed in the chamber. A strong laser beam is allowed to fall on the metal disc, which causes evaporation of atoms from its surface. The atoms of the metal are carried away, along with He gas, through a narrow tube into an evacuated container. The sudden expansion of the gas causes cooling, which leads to condensation and hence the formation of nanoparticles (clusters of atoms) with nm size. The nanoparticles are first ionized by ultraviolet radiation and then passed through a skimmer into a mass spectrometer for measurement of their charge/mass ratio, which provides information about the size of the nanoparticles. It is found that nanoparticles A_n with different values of n are formed. Fig. 25.21 shows the number of nanoparticles of lead metal Pb_n as a function of n. From the figure one concludes that Pb_7 and Pb_{10} are more likely to form than other nanoparticles. The numbers 7 and 10 are called *structural magic numbers*.

Metallic nanoparticles can also be produced from a solution of their salts by using a pulsed laser beam. For example, nanoparticles Ag_n can be produced from an aqueous solution of $AgNO_3$ by applying periodic laser pulses. The pulsed laser method employs a container fitted with a rotating sample holder in the middle (see Fig. 25.22) on which is placed a solid disc. A solution of $AgNO_3$ and a reducing agent are put in the container such that the solid disc dips into the solution.

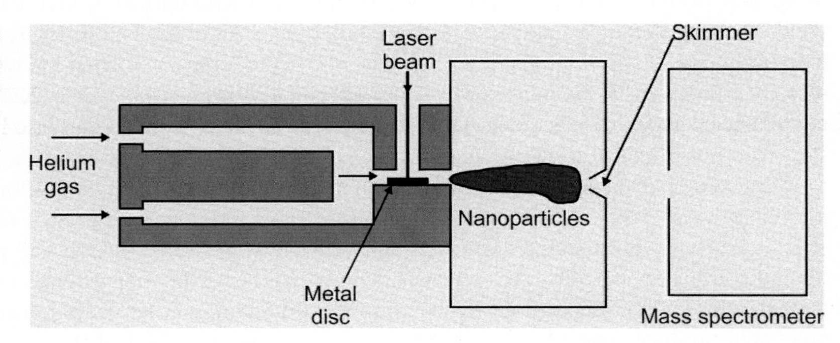

FIG. 25.20 Experimental set up to produce metallic nanoparticles by laser-induced evaporation of atoms from the surface of a metal. *(Modified from Owens, F. J., & Poole, C. P. (1999).* New Superconductors. *New York: Plenum Press.)*

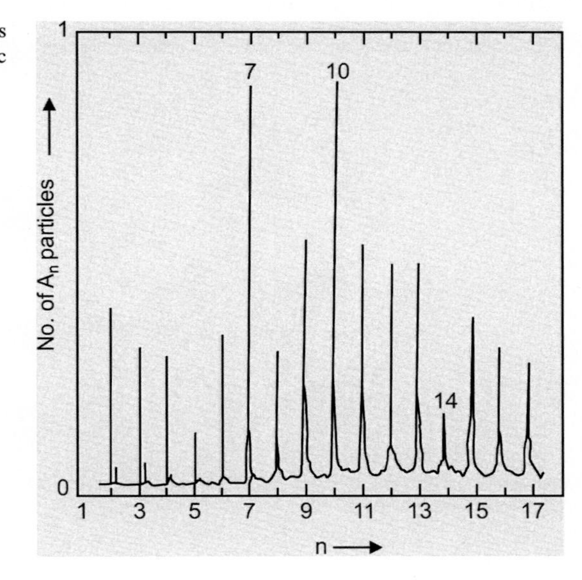

FIG. 25.21 Number of A_n particles of Pb metal as a function of n obtained from mass spectrometry. *(Modified from Duncan, M. A., & Rouvray, D. H. (1989).* Scientific American, *110.)*

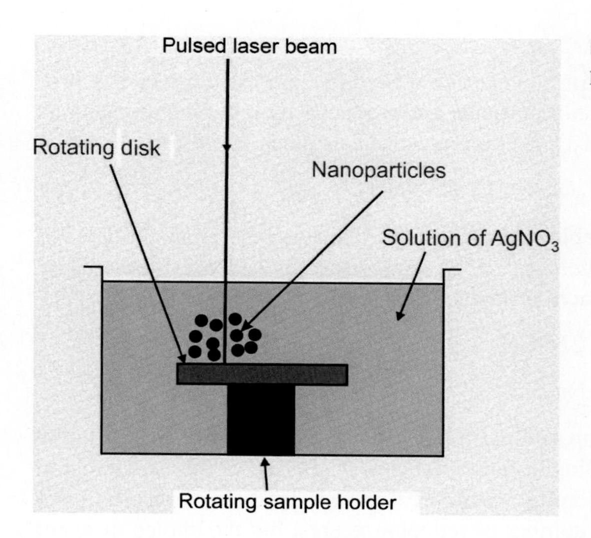

FIG. 25.22 Experimental set up for producing metallic nanoparticles using the pulsed-laser-beam method. *(Modified from Singh, J. (2001).* Materials Today, 2, *10.)*

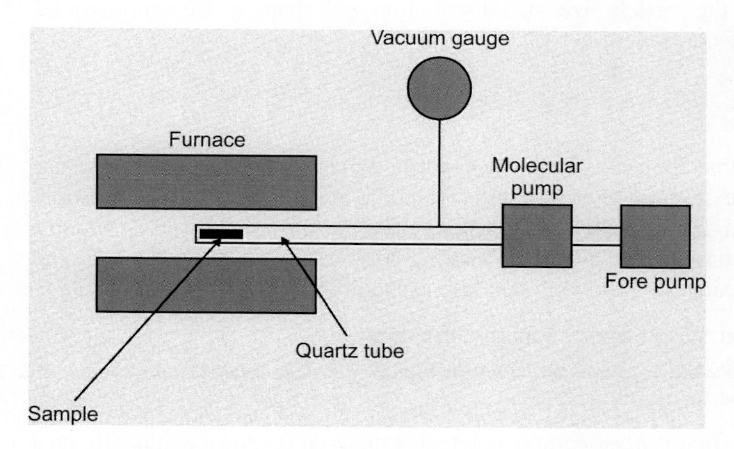

FIG. 25.23 Experimental set up for producing metallic nanoparticles using the method of thermal evaporation. *(Modified from Poole, C. P., & Owens, F. J. (2007).* Introduction to Nanotechnology. *New York: J. Wiley & Sons.)*

The solid disc is rotated with a uniform speed and periodic pulses of a strong laser beam are allowed to fall on it, which create hot spots on the surface of the disc. At the hot spots, $AgNO_3$ and the reducing agent react to release Ag atoms, which further combine to form Ag_n, which are then separated from the solution. One advantage of this method is that by varying the energy of the laser beam and the speed of rotation of the solid disc, the size of the nanoparticles can be varied.

25.3.3.2 Thermal Decomposition

At high temperatures a solid is evaporated into molecules and further heating causes thermal decomposition of molecules into metal cations and molecular anions. The metallic cations then start aggregating, thereby producing metallic nanoparticles. On the other hand, the anions combine to form a gas that escapes into the air. This process is similar to electrolysis in which the passage of an electric current through an aqueous solution of a salt splits the molecules into cations and anions. Therefore, the process of thermal decomposition is sometimes called thermolysis in analogy with electrolysis. The experimental arrangement used for thermal decomposition is shown in Fig. 25.23. It consists of an evacuated quartz tube in which the sample LiN_3 is placed near one end. The other end of the quartz tube is connected to a molecular pump and a fore pump. A vacuum gauge is connected to the tube to measure the pressure. LiN_3 is heated by means of a furnace to 400°C. At this temperature, LiN_3 decomposes into Li cations and N anions and the latter form N_2 gas, which can be detected by the vacuum gauge. Then the N_2 gas is removed from the quartz tube. On the other hand, Li cations aggregate to form nanoparticles (⟨5nm in dimension), which are passivated by a suitable gas. The formation of nanoparticles can be detected by the EPR spectrum of the conduction electrons of the metallic nanoparticles.

In addition to the above method, radiofrequency plasma is also used for producing metallic nanoparticles. In this method metal is heated above the evaporation point using high-voltage rf coils and then the atoms are allowed to condense to form nanoparticles.

25.3.3.3 Chemical Methods

Chemical methods are most useful and convenient for the production of nanoparticles. Metallic nanoparticles can be produced from their salts in the presence of suitable reducing agents. Gold particles are produced by reducing an aqueous solution of chloroauric acid ($HAuCl_4$) with trisodiumcitrate ($Na_3C_6H_5O_7$). The reaction takes place as follows:

$$HAuCl_4 + Na_3C_6H_5O_7 \rightarrow Au^+ + C_6H_5O_7^- + HCl + 3NaCl$$

The gold particles are then stabilized by using thiol or some other suitable surfactant. In the same manner, nanoparticles of other metals, such as Cu, Ag, and Pd, can be synthesized. The nanoparticles of ZnS, a semiconductor, can be synthesized by adding Na_2S to an aqueous solution of $ZnCl_2$ (or $ZnNO_3$) and the reaction proceeds as:

$$ZnCl_2 + Na_2S \rightarrow ZnS + 2NaCl$$

$$ZnNO_3 + Na_2S \rightarrow ZnS + Na_2NO_3$$

ZnS molecules then aggregate to form nanoparticles of ZnS. The nanoparticles of ZnS tend to aggregate further due to the attractive forces between them. So, they are surface passivated by adding liquid thiophenol (C_6H_5SH) or mercaptoethanol (C_2H_5OSH) to the aqueous solution and heating it to the desired temperature. Similarly, one can produce nanoparticles of ZnO by adding NaOH to an aqueous solution of $ZnCl_2$. There are a number of reducing agents, but the choice of agent depends on the metallic salt used. It is noteworthy that the particle size, size distribution, and shape of the nanoparticles depend on the reaction parameters, which can be easily controlled.

25.3.3.4 Self-Assembly Techniques

One of the self-assembly techniques is *epitaxy*, a word that comprises two Greek words. *Epi* means upon and *taxis* means ordered. Therefore, epitaxy is a term applied to the processes used to grow a thin crystalline layer on a crystalline substrate and the growth of such crystals is called epitaxial growth. In this process the substrate serves as the seed crystal on which a new material grows. In epitaxy, crystals can be grown considerably below the melting point of the substrate. For epitaxial growth of a thin crystalline layer on a crystalline substrate, the following conditions should be satisfied:

1. The crystal structures of the deposited materials and the substrate should be the same.
2. The lattice parameters of the two materials should be either exactly or nearly equal.
3. The energy gaps of the two materials should also be nearly the same.

There are two types of epitaxial growth techniques. The first is *homoepitaxy* in which a material is grown epitaxially on a substrate of the same material. For example, Si is deposited epitaxially on a Si wafer. In homoepitaxy the conditions outlined above are automatically satisfied. If the material to be grown is different than the material of the substrate, it is called *heteroepitaxy*. For example, $Al_xGa_{1-x}As$ is grown epitaxially on a substrate of GaAs (see Figs. 25.7 and 25.8). In this system the lattice parameters of the two materials nearly match each other and they possess the same structure. Further, the band gap of the material to be deposited ($Al_xGa_{1-x}As$) is larger than that of the substrate. In heteroepitaxy, the layer grows with a lattice parameter in compliance with that of the substrate. If the lattice parameter of the crystalline thin layer, to be grown epitaxially, is slightly different from that of the substrate, then one obtains a strained layer, which may be in compression or tension along the surface.

The epitaxial growth of alternating layers of crystals with slightly mismatched lattice parameters yields a strained-layer superlattice in which alternate layers either are in tension or compression. The lattice parameter of a strained-layer superlattice is an average of the lattice parameters of the two bulk materials. A number of methods are used to provide the appropriate atoms for growing a crystalline layer. They include chemical vapor deposition (CVD) and molecular beam epitaxy (MBE). Here we shall describe only the latter method. In the MBE method, atoms of an element are evaporated in vacuum and deposited in the form of a layer on a substrate. The substrate may be a crystalline material or an amorphous one. Both homoepitaxial and heteroepitaxial growths can be obtained using the MBE method. First, in situ cleaning of the surface of the substrate is performed in one of the following two ways. In the first method, high-temperature baking, between 1000°C and 1250°C, is carried out for 30 min. This decomposes the native oxide and removes other adsorbed atoms (notably carbon) by evaporation or diffusion into the layer. A better approach is to use a low-energy beam of an inert gas to sputter clean the surface. A short anneal at 800°C to 900°C is sufficient to reorder the surface. The substrate should have an ultraclean and perfectly smooth surface. A schematic diagram for the MBE method is shown in Fig. 25.24. Bringing atoms of a material in contact with the surface of a substrate may do three things:

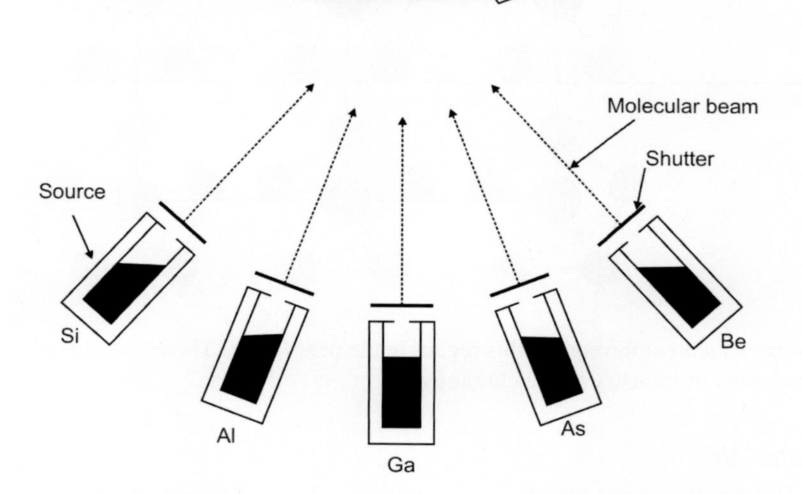

FIG. 25.24 Schematic diagram of the molecular beam epitaxy method.

1. The atoms are either desorbed and thereby leave the surface or they are adsorbed and diffuse on the surface until they join with other adatoms to form an island. Small islands can continue to grow, migrate to other positions, or evaporate. There is a critical size at which the islands become stable and do not experience much evaporation. Thus, there is an initial nucleation stage in which the number of islands increases with the coverage.
2. Second is an aggregation state, in which the number of islands level off and the existing ones grow in size.
3. Finally, there is a coalescence stage, which involves the merger of existing islands with each other to form larger islands. These islands can be converted into quantum dots by covering them with an epitaxial layer. A three-dimensional array of quantum dots can be produced by repeating the deposition sequence just described.

In practice in situ growth techniques, such as MBE and the metal-organic CVD technique, are used to obtain the requisite ultraclean conditions and exquisite control over the deposition. Heteroepitaxy is one of the epitaxial growth techniques that have been widely used for research, as well as in the fabrication of many semiconductor devices. The other methods to fabricate nanoparticles of semiconductors are lithography, colloidal chemistry, and the radiofrequency plasma method. However, self-assembly quantum dots have smaller sizes and stronger confinement potentials than lithographically prepared quantum dots.

25.3.4 Nanostructured Materials

Nanostructured materials are solids in bulk form, with nanoparticles as the basic unit of structure, instead of the atoms or molecules in conventional elements of the periodic table. Nanostructured materials can be classified into two categories.

25.3.4.1 Crystalline Nanostructured Materials

In crystalline nanostructured materials the nanoparticles are the building blocks and they form a three-dimensional periodic array with some symmetry: the nanoparticles have the same size and orientation. Fig. 25.25 shows a two-dimensional ordered lattice of hypothetical nanoparticles. There are two fundamental problems in the synthesis of crystalline nanostructured materials.

1. No method or technique exists with which one can synthesize nanoparticles of a particular size of an element.
2. Secondly, there is no technique available to arrange nanoparticles of a particular size on a periodic lattice to form a crystalline material.

Once techniques are developed to perform the two steps outlined above, one will be able to synthesize many nanostructured crystalline solids from a single element of the periodic table. Hence, we shall be able to produce a large number of nanostructured materials from the existing elements. Computer calculations predict the possibility of crystalline nanostructured materials and some of them may possess extraordinary properties, allowing their use in such applications as super strong

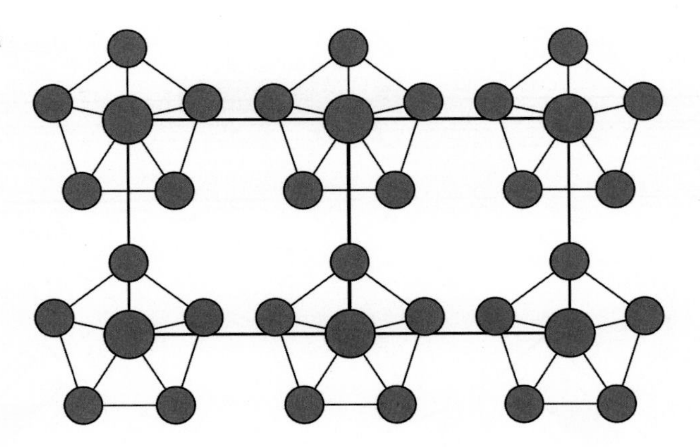

FIG. 25.25 Hypothetical nanoparticles are arranged on a square lattice with perfect order to give a crystalline nanostructured material.

magnets and high-T_c superconductors. Researchers expect a breakthrough in this regard in the near future. Therefore, there is immense scope for crystalline nanostructured materials in industry and technology.

25.3.4.2 Amorphous Nanostructured Materials

In an amorphous nanostructured material, the arrangement of nanoparticles does not exhibit translational periodicity: nanoparticles may have the same size but different orientation, or different size but the same orientation, or both the size and orientation may be different. Fig. 25.26 shows a two-dimensional disordered nanostructure of hypothetical nanoparticles. There are a few methods for the production of amorphous nanostructured materials.

The most commonly used method to make amorphous alloys is the *compaction and consolidation* method. In this method the different component elements of an alloy are ball-milled to nm size. The mixture is then compacted in a tungsten carbide die under high pressure (≈GPa) for many hours. To bring about consolidation the mixture is further subjected to compaction under high pressure and at high temperature. For example, in $Fe_{1-x}Cu_x$ alloy one gets an amorphous nanomaterial with nanoparticles ranging in size from 20 to 70nm. Another commonly used method is the *gas atomization* method. In this method molten metal is made to come out from a fine nozzle under high pressure and this is impacted with a high-velocity beam of inert gas. The molten material produces a fine dispersion of metallic particles with nanometer size in the form of a powder. The powder is then subjected to compaction and consolidation to form bulk samples.

25.3.5 Computer Simulation Technique

In computer simulations a nanoparticle of N atoms with some initial structure is considered. The interaction potential $V(\mathbf{r})$ between the atoms is guessed intuitively. The 3N equations of motion of the N atoms are set. Then, all of the atoms are allowed to move by a small distance at the same time and the 3N equations are solved simultaneously. The energy of the nanoparticle corresponding to the new positions of the atoms is found. The atoms are moved in such a way that the energy of the nanoparticle decreases. The same process is repeated again to further decrease the energy. The atoms are moved bit by bit until a minimum energy for the nanoparticle is obtained. The final positions of the atoms will yield the actual structure of the nanoparticle. This method is quite laborious, as is evident from the following example. For a nanoparticle with 100 atoms one has to solve 300 equations simultaneously, which is a tedious task for an ordinary computer. The main

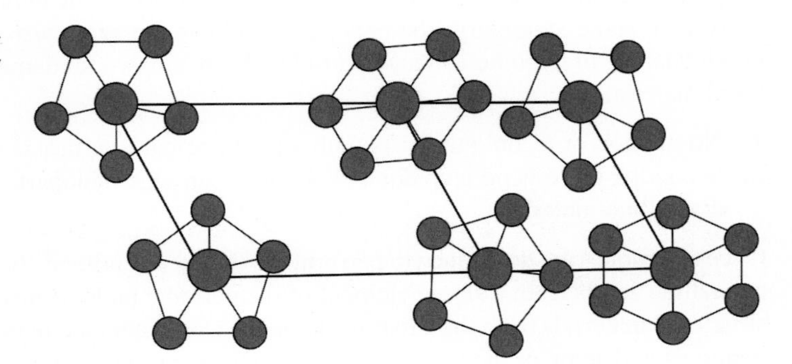

FIG. 25.26 Hypothetical nanoparticles with different sizes and orientations are arranged on a lattice with no translational periodicity yielding an amorphous nanostructured material.

shortcoming of the computer simulation technique is the choice of the interaction potential. If an incorrect interaction potential V(**r**) is chosen, one will end up with incorrect results. Therefore, the intelligence of the research worker lies in the correct choice of the interaction potential.

25.4 NANOMATERIALS OF CARBON

Carbon (C) is one of the most abundant elements on earth and it exists in a number of different solid forms.

1. Crystalline form: brilliant diamond, gray graphite, and its well-known allotropes.
2. Amorphous form: coal and soot.

Kroto, Heath, O'Brien, Curl, and Smalley (1985, 1986) studied the structure of carbon-rich giant stars and found special spectral lines corresponding to some type of long-chain molecules of carbon or nitrogen. Later they produced clusters of carbon atoms in the laboratory by the vaporization of solid carbon. The analysis by mass spectrometer exhibited a number of peaks, but the strongest peak was observed at 60 carbon atoms and the next strongest at 70 carbon atoms. This shows that carbon clusters (nanoparticles) with 60 and 70 atoms, represented by C_{60} and C_{70}, are most stable and they are called C_{60} and C_{70} molecules. The C_{60} molecules are also called buckyballs or fullerenes after the American architect and futurist Buckminster Fuller who designed a geodesic spherical dome using sticks at the Montreal World Exhibition, which looked quite similar to the structure of a C_{60} molecule. Another very interesting class of carbon nanostructure is a carbon nanotube, which is an example of a quantum wire of carbon.

25.4.1 Nanoparticles of Carbon

25.4.1.1 Structure of C_{60} Molecule

NMR experiments have established that C_{60} molecules possess a truncated icosahedral structure, which is like a spherical cage. A regular truncated icosahedron has 90 edges of equal length, 60 vertices, 20 hexagonal faces, and 12 pentagonal faces to form a closed shell. In this structure C atoms have no dangling bonds and the pentagons on the C_{60} molecule are isolated from one another, thereby creating greater chemical and electronic stability. A typical structure of C_{60} is shown in Fig. 25.27. For a chemist it corresponds to a three-dimensional aromatic system in which single and double bonds are alternated. The atomic positions in a C_{60} molecule can be determined by conventional methods, such as neutron or X-ray diffraction experiments. The average 1NN distance between two carbon atoms is 1.44 Å, which is almost identical to that in graphite (1.42 Å). Further, in graphite, each C atom is trigonally bonded to three other C atoms in a sp^2-derived bonding configuration. In a C_{60} molecule, there are single bonds along the hexagon-pentagon edges with bond length around 1.46 Å, whereas there are double bonds along the hexagon-hexagon edges having length around 1.40 Å. Therefore, there is a slight deviation from a regular truncated icosahedron.

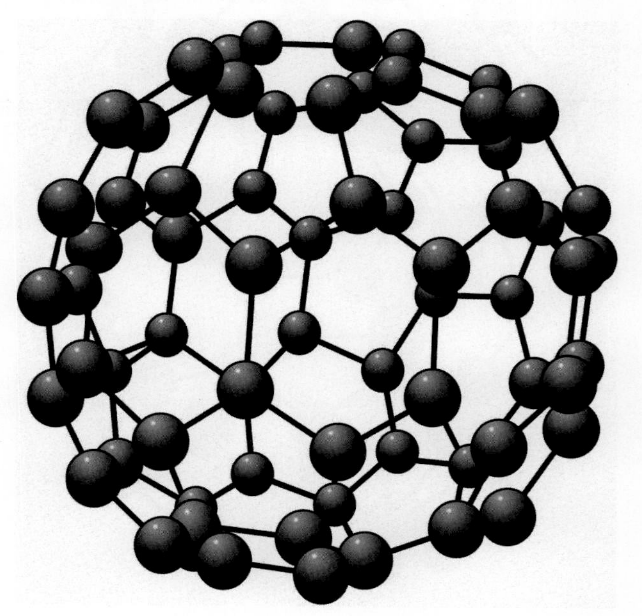

FIG. 25.27 Structure of a C_{60} molecule with spheres representing the C atoms.

It is found that C_{60} is a slowly reacting molecule. In general, it has been found that all the carbon nanoparticles with an even number of carbon atoms ranging from 40 to 80 react equally slowly. Analogously to C_{60}, all these carbon nanoparticles correspond to entirely closed structures that resemble cage-like structures. The bonding requirements of all the valence electrons in C_{60} are satisfied, therefore, it is expected that C_{60} has filled molecular levels. Because all the C atoms in C_{60} are sp^2-hybridized, C_{60} could be like Benzene C_6H_6, which is an aromatic molecule, or like ethane, which is a typical alkene. Note that C_{60} molecules are very hard. If C_{60} molecules are shot onto a steel plate with a velocity of 15,000 mph, they will bounce back unharmed. If a C_{60} molecule is compressed, it will become twice as hard as diamond.

The truncated icosahedron structure of the C_{60} molecule possesses the I_h symmetry group, which consists of the following symmetry operations:

1. Identity operation.
2. Six 5-fold axes through the centers of the 12 pentagonal faces.
3. Ten 3-fold axes through the centers of the 20 hexagonal faces.
4. Fifteen 2-fold axes through the centers of 30 edges joining two hexagons.
5. Each of the 60 rotational symmetry operations can be compounded with the inversion operation to yield 120 symmetry operations in the icosahedral point group I_h.

From the known lengths of single and double bonds, the diameter of a C_{60} molecule comes out to be 7.09 Å. Experimental measurements using the NMR technique yield 7.0 Å as the diameter of a C_{60} molecule, which agrees closely with the theoretical value. There is also an electron cloud of π-electrons, with an estimated thickness of about 3.35 Å, surrounding the C_{60} molecule. Hence, accounting for the electron cloud, the outer diameter of a C_{60} molecule becomes $7.09+3.35=10.44$ Å.

25.4.1.2 Structure of C_{70} Molecule

A C_{70} molecule can be obtained from a C_{60} molecule by adding a ring of 10 carbon atoms, or equivalently, adding a belt of five hexagons around the equatorial plane of the C_{60} molecule, which is normal to one of the five-fold axes (see Fig. 25.28). The two hemispheres of C_{60} molecules are rotated so that they fit continuously on to the belt of hexagons (Dresselhaus, Dresselhaus, & Eklund, 1996). In a C_{70} molecule, there are five inequivalent carbon sites and eight distinct bond lengths. The bond lengths are measured by neutron inelastic scattering experiments and are found to range from 1.356 Å to 1.475 Å (Nilolaev, Dennis, Prassides, & Soper, 1994).

A C_{70} molecule is not spherical in shape but is more like a rugby ball. Knowing the atomic positions in C_{60} one can find the atomic positions in a C_{70} molecule. From the coordinates so obtained the distance of the nearest and farthest C atoms

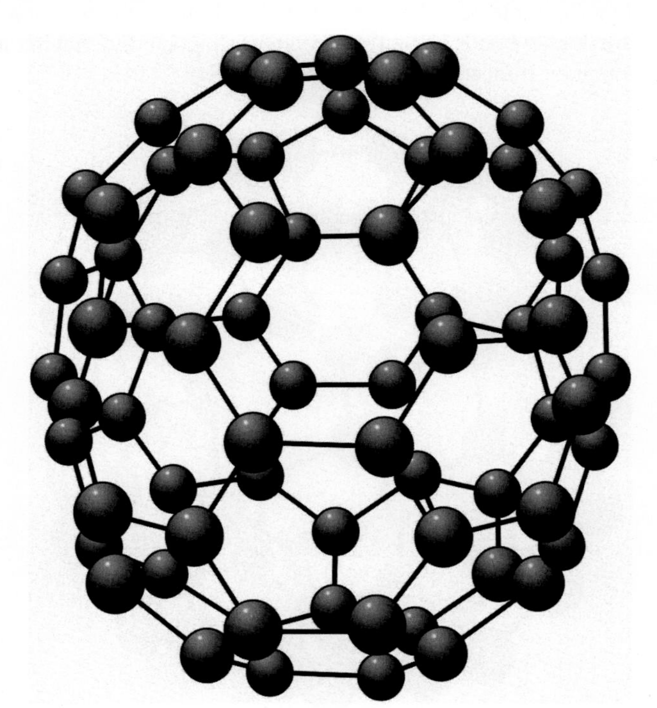

FIG. 25.28 Structure of a C_{70} molecule with spheres representing the C atoms.

from the center of a C_{70} molecule are a = 3.498 Å and b = 4.112 Å and these are taken as the semiminor and semimajor axes of the C_{70} molecule. The reader is referred to Ramirez (1994) who compared some physical properties of C_{60} and C_{70} molecules.

25.4.1.3 *Crystalline C_{60} Solid*

C_{60} and C_{70} molecules are produced in a simple experiment consisting of two graphite rods in a container filled with an inert gas He at a pressure of 100 Torr. An arc is passed through the graphite rods with a current of about 200 amps and a black carbon smoke is produced, which contains fullerenes. Note that fullerenes are best produced at a pressure of 100 Torr in an inert gas. The current is passed intermittently, for a period lasting for about 15 s during each step to avoid excessive heating. Benzene or toluene solution dissolves the fullerene component of the soot to form a colored solution. The color of the solution depends on the concentration and the type of the fullerenes produced.

A toluene solution of concentrated C_{60} molecules will have a dark red color, while that of concentrated C_{70} molecules has a magenta color. So, the color of a toluene solution can be anywhere from pink to dark red to yellow depending on the type of fullerenes present. The major problem in the synthesis of a crystalline solid of C_{60} molecules is that the toluene solution contains a 9:1 mixture of C_{60} and C_{70} molecules and a small number of fullerenes with a higher number of carbon atoms. Thus, there is a need to separate the C_{60} molecules from the rest of the molecules, which can be done with the help of chromatography. Then a single crystal of C_{60} molecules can be grown by the slow evaporation of the pure C_{60} toluene solution. The above procedure indicates that the synthesis and purification of C_{60} fullerenes is difficult and expensive.

At room temperature C_{60} molecules are found to crystallize in the fcc structure with lattice parameter $a \approx 14.2$ Å and it is an insulator. In chapter 1 it was pointed out that spherical atoms are found to crystallize in the fcc structure. The C_{60} molecules have a cage-like structure, which is nearly spherical in nature. Further, the free spinning of C_{60} molecules at high temperatures makes them resemble spheres. X-ray diffraction studies show that as the temperature is lowered crystalline C_{60} solid exhibits a structural transition to sc structure at $T_c = 260$ K, indicating a hindrance to the free rotation of C_{60} molecules.

25.4.1.4 *Alkali-Doped Crystalline C_{60} Solid*

In a crystalline C_{60} solid, 26% of the volume is empty space in the form of voids as the packing fraction of fcc structure is 0.74. The voids are so large that the alkali atoms can easily settle in them. To synthesize alkali-doped crystalline C_{60} solid, C_{60} crystal, along with K metal, is placed in an evacuated tube and then heated to 400°C. At this temperature, K metal vaporizes and its atoms diffuse into the voids of C_{60} crystal to form a compound K_3C_{60}. Fig. 25.29 shows the tetragonal and octahedral positions occupied by the K atoms. The doping of the C_{60} solid with K atoms changes its nature from an insulating to a conducting state. This is because, in K_3C_{60}, each K atom is ionized by releasing an electron (K^{+1}) and the electrons of all the three K atoms become associated with a C_{60} molecule to give C_{60}^{-3}, that is,

$$K_3C_{60} = 3K^{+1} + C_{60}^{-3} \tag{25.86}$$

The ion C_{60}^{-3} has three loosely bound electrons, which get detached and move through the lattice, making C_{60} solid a conductor.

K_3C_{60} also exhibits superconductivity with $T_c = 18$ K, which is quite high compared with the pure elements. Therefore, it constitutes a new class of superconductors that have cubic structure and contain only two components, in contrast with the conventional high-T_c oxide superconductors. Crystalline C_{60} solid can also be doped with other alkali metals and it has been found that T_c increases with an increase in the size of the alkali atoms. Fig. 25.30 shows T_c as a function of the lattice parameter ‘a’ in angstroms. The value of T_c in Cs_2RbC_{60} is 33 K and in $Tl_{2.2}Rb_{2.7}C_{60} \approx 45$ K.

25.4.2 Carbon Nanotubes

Another very interesting class of nanomaterial is the carbon nanotube (CNT), which constitutes a very good example of a quantum wire. One can imagine a CNT as a sheet of graphite rolled into a tube. Fig. 25.31 shows the structures of the CNT formed by rolling a graphite sheet with the axis of the tube along different directions. There are two types of CNT: the single-walled carbon nanotube (SWCNT) and the multiwalled carbon nanotube (MWCNT). Within a MWCNT there can be a number of coaxial CNTs.

CNTs can be synthesized by the evaporation of graphite, which can be accomplished by a number of methods. The advantage of using a laser beam for the evaporation of graphite is that the nanotubes formed are invariably SWCNTs with

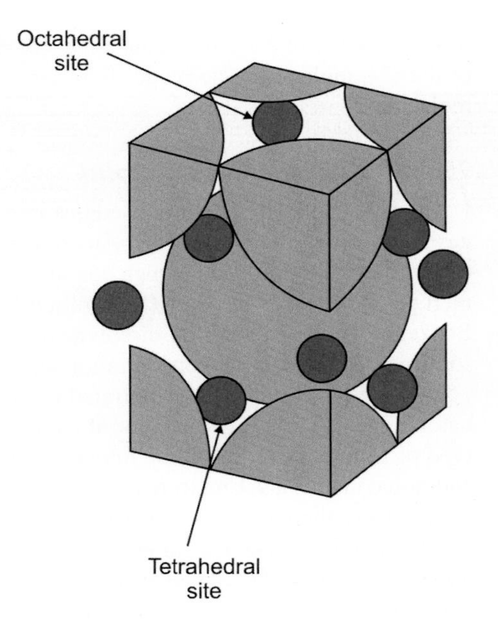

FIG. 25.29 The structure of alkali-doped crystalline C_{60} solid. The dark spheres represent alkali atoms at octahedral and tetrahedral sites of the unit cell of crystalline C_{60} solid. *(Modified from Owens, F. J., & Poole, C. P. (1999).* New Superconductors. *New York: Plenum Press.)*

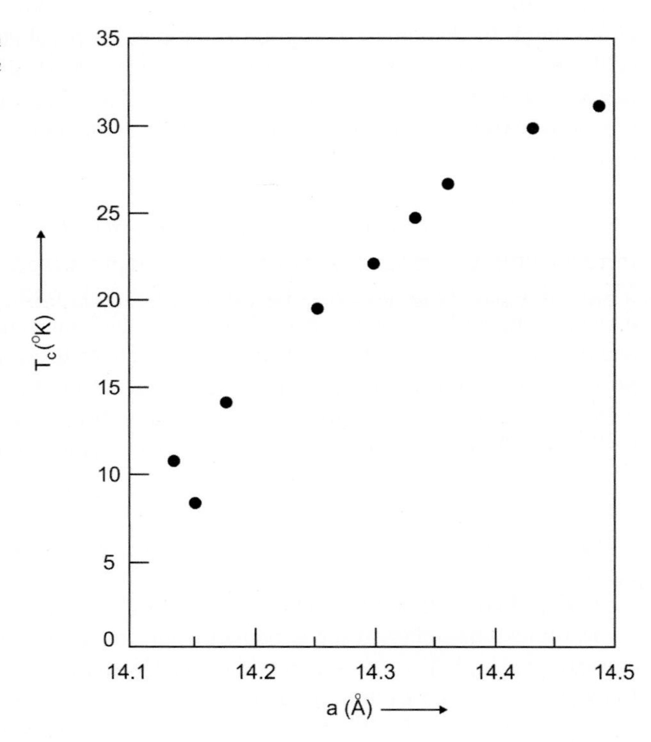

FIG. 25.30 Variation of the superconducting transition temperature T_c with the lattice parameter a in alkali-doped crystalline C_{60} solid. *(Modified from Hebard, A. F. (1992).* Physics Today*, 29.)*

a diameter ranging from 10 to 20 nm and a length of about 100 μm. Fig. 25.32 shows an experimental set up for the synthesis of CNTs. It consists of a quartz tube containing argon (Ar) gas at a pressure of 500 Torr and a graphite sample at its center. The quartz tube is surrounded by a furnace to heat it to very high temperatures (1200°C). At one end of the quartz tube is placed a strong Nd:YAG laser, while at the other end there is a water-cooled Cu collector. The graphite sample contains small numbers of Co, Ni, and Fe atoms, which help in the formation of CNTs. A strong laser beam is allowed to fall on the graphite sample, which evaporates carbon atoms from the surface. These carbon atoms are swept by Ar gas from the high-temperature zone to the cold Cu collector on which the carbon atoms condense to form CNTs. CNTs can also be produced by passing an electric arc between two carbon electrodes. In this method MWCNTs are mostly formed with a diameter in the range 0.7–1.5 nm and with a length of a few μm.

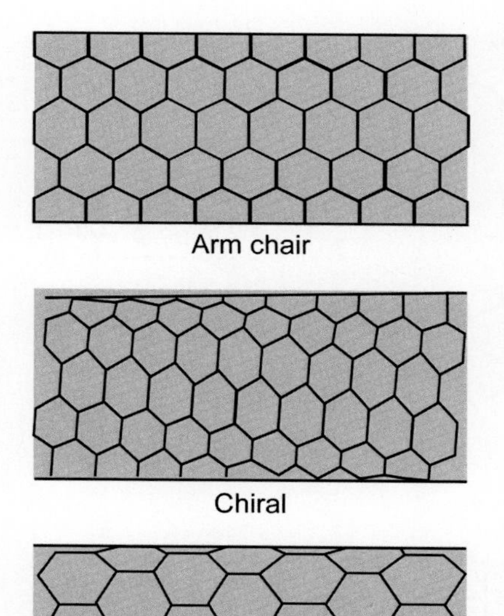

FIG. 25.31 The structure of armchair, chiral, and zigzag CNTs.

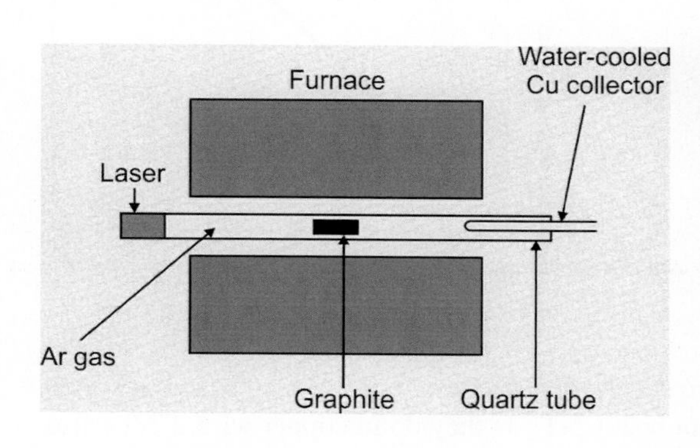

FIG. 25.32 Experimental set up for producing carbon nanotubes by the laser evaporation method. *(Modified from Poole, C. P., & Owens, F. J. (2007). Introduction to nanotechnology. New Delhi: Wiley India.)*

The growth mechanism of CNTs is not fully understood. At such high synthesis temperatures one cannot imagine that long sheets of graphite are really released and folded. It is probably an atom-by-atom or molecule-by-molecule addition under favorable conditions that form the CNTs. In the method above it was proposed that atoms of Co, Ni, and Fe may act as a catalyst in the formation of CNTs.

There are a variety of structures of CNTs, which exhibit different properties. Broadly speaking CNTs can be divided into three categories:

1. Armchair structure,
2. Chiral structure, and.
3. Zigzag structure.

Fig. 25.31 shows all three structures, which are generally closed at both ends where they are like half of a large fullerene. These nanotube structures are obtained by rolling the graphite sheet in different ways. Fig. 25.33 shows a graphite sheet in which $\mathbf{A}_{\text{axis}}$ is a vector about which the graphite sheet is rolled and $\mathbf{B}_{\text{cf}}$ is a circumferential vector perpendicular to $\mathbf{A}_{\text{axis}}$. When $\mathbf{A}_{\text{axis}}$ is parallel to the C—C bonds of carbon hexagons, the structure obtained is called an armchair structure (see Fig. 25.31). But if $\mathbf{A}_{\text{axis}}$ is not parallel to the C—C bonds, then it yields zigzag and chiral structures (Fig. 25.31). As the vector $\mathbf{A}_{\text{axis}}$ can have a number of orientations with respect to the C—C bonds, therefore, a number of chiral and zigzag structures can be obtained.

The most interesting property of CNTs is that they are either metallic or semiconducting in nature depending on the diameter and chirality of the tube. CNTs with armchair structure are metallic in nature with a very high conductivity: they

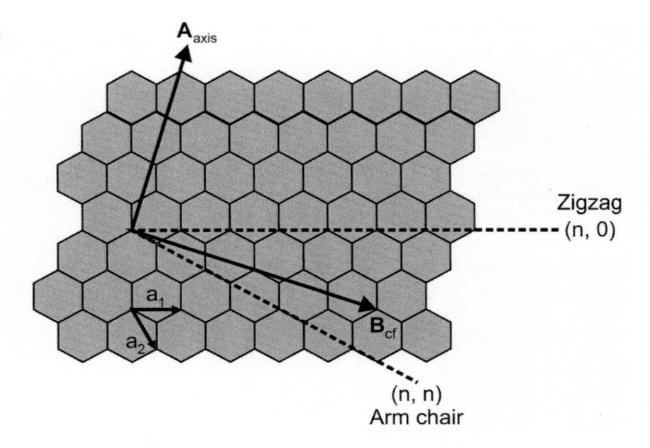

FIG. 25.33 The graphite sheet used in making carbon nanotubes. $\mathbf{A}_{axis}$ represents the axis about which the sheet is rolled and $\mathbf{B}_{cf}$ is the circumferential vector.

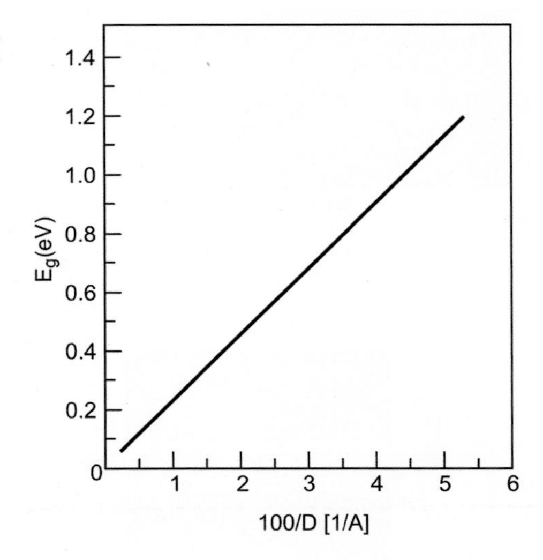

FIG. 25.34 The variation of the energy band gap E_g of a semiconducting chiral carbon nanotube with the diameter D of the tube. *(Modified from Dresselhaus et al. (1994)* Molecular Materials, *4, 27.)*

can carry a current of 10^9 amp/sq.cm. On the other hand, a copper wire can carry a maximum of 10^6 amp/sq.cm at which point it melts because of resistive heating. Further, high currents do not heat CNTs as they do in copper wire. CNTs also have high thermal conductivity with a value that is about two or more times that of a diamond. They are good conductors of heat. CNTs with finite chirality are semiconducting in nature. Fig. 25.34 shows the energy band gap E_g as a function of diameter D of a semiconducting CNT. It increases linearly with a decrease in the diameter. It is an important property that allows semiconducting nanotubes with the desired band gap to be synthesized. This fact allows us to have full control over the energy band gap by varying the diameter of the tube, whereas in crystalline semiconductors Ge and Si, the band gap is fixed (0.70 eV in Ge and 1.02 eV in Si).

From the above discussion it is evident that a single carbon element can exhibit a variety of properties at the nanoscale. One can produce insulating fullerenes, superconducting doped fullerenes, metallic and semiconducting CNTs, and others.

25.5 MICROSCOPES USED FOR NANOMATERIALS

To view nanomaterials special microscopes are needed, which have atomic-scale resolution. A transmission electron microscope is one such instrument, which is mostly used to study the structure of materials. Recently, a number of probe-type microscopes have been invented with which one can study the structure of nanomaterials. In all of these microscopes a probe is placed very close to the surface of the material to measure a local property, such as height, optical absorption, or magnetization. Both the scanning tunneling microscope (STM) and atomic force microscope (AFM) are probe-type microscopes that yield images at the atomic scale. STM and AFM are used as the foremost tools for the manipulation of matter at the nanoscale. The special feature of AFM is that it can examine any rigid surface, either in air or immersed in liquid and can resolve even single atoms that were previously unseen.

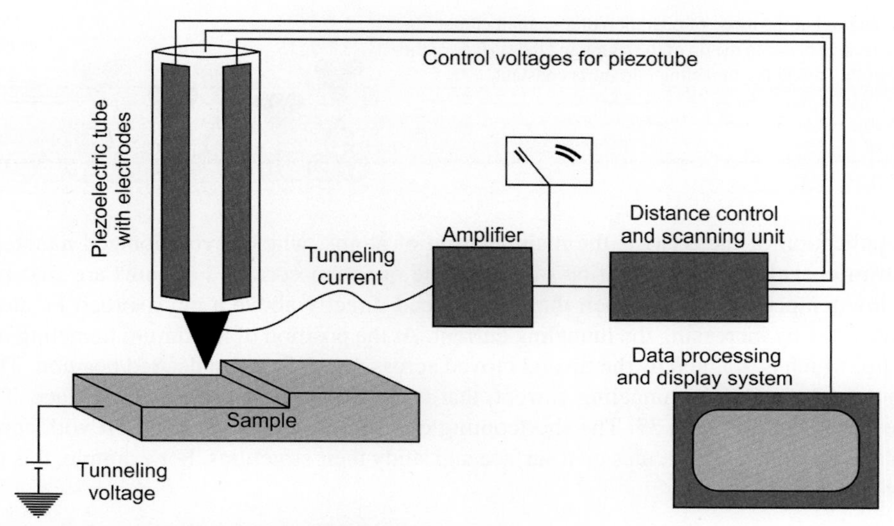

FIG. 25.35 Schematic diagram of a scanning tunneling microscope.

25.5.1 Scanning Tunneling Microscope

An STM works on the principle of quantum tunneling: there is always a finite probability for an electron to tunnel through a potential barrier with potential greater than the energy of the electron (see Fig. 25.14). Fig. 25.35 shows a schematic diagram of an STM. It consists of a stylus with a very fine metallic tip. The tip consists of a cluster of atoms with a single atom at the end. It is this atom that is responsible for most of the tunneling current. The tip is attached to a piezoelectric crystal, which either expands or contracts depending on the direction of the current or direction of the electric field. The tip is attached to a current amplifier, which is connected to a distance control and scanning unit. The distance control and scanning unit provide a control voltage to the piezoelectric crystal, and is also connected to the display unit. A small voltage ($\approx 1.5\,\mathrm{V}$), generally called the tunneling voltage, is imparted to the sample.

An STM is used to study conducting or semiconducting surfaces and has important applications in semiconductor physics and microelectronics. The stylus is lowered toward the sample surface and moved over it. A special robotic arm does the scanning of the surface. When the separation between the tip and the surface is very small (of the order of an atomic diameter), a finite tunneling current ($\approx$ nA) flows either from the tip to the surface or from the surface to the tip depending on the nature of sample surface. From Eq. (25.80) the probability of tunneling of an electron from the tip to the sample or vice versa, is given by.

$$P = e^{-2L\left[\frac{2m_e}{\hbar^2}(V-E)\right]^{1/2}} \tag{25.87}$$

where L is the distance between the tip and the surface, V is the potential barrier, and E the energy of an electron. It is evident from Eq. (25.87) that the probability of tunneling, and hence the tunneling current, depends on the separation L. It also depends on the work function φ_W of the sample surface, which lies in the range of 4–5 eV. The computer software is then used to translate the scanned tunneling current into an image, which is displayed on a monitor. When the tip is negatively biased the electrons tunnel from the occupied states of the tip to the unoccupied states of the surface and an image of the surface is obtained. But if the tip is positively biased, then the electrons tunnel from the occupied states of the surface to the unoccupied states of the tip and an image of the tip is obtained. It should be noted that the STM does not probe the nuclear position but probes only the electron density around the nucleus. So, the image does not always show the position of the atoms.

In a normal mode of operation, the height of the tip is adjusted so as to keep the tunneling current constant. In other words, the tip is moved over the surface keeping the distance between the tip and the surface constant (see Fig. 25.36). In this mode one obtains a topographical map of the surface. In the other mode of operation, the tip is moved horizontally, thereby producing varying tunneling current. The tunneling current is then translated into a topographical map of the surface. But if the surface is rough, then the tip of the STM may collide with the surface atoms in the horizontal-motion mode, thereby destroying either the tip or the surface under investigation. The STM gives a very accurate profile of the surface with a resolution on the order of the size of an atom.

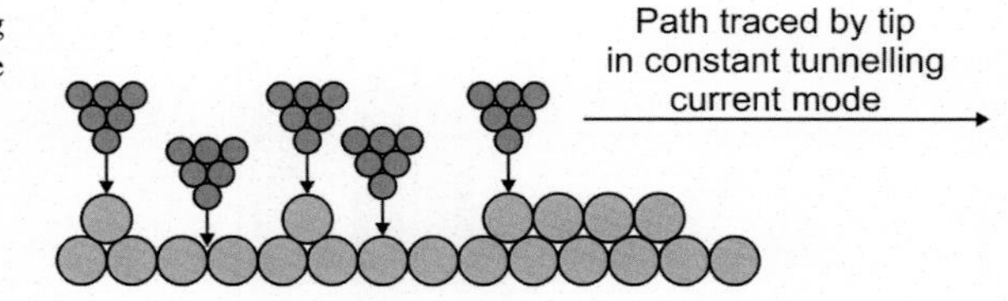

FIG. 25.36 Schematic diagram of the mode of operation of a scanning tunneling microscope in which the distance between the tip of the stylus and the surface of the sample is kept constant. In this mode the tunneling current is constant.

An innovative application of the STM is the manipulation of atoms, which revolutionized nanotechnology: the formation of a "**quantum corral**." In the formation of one of the quantum corrals, Fe atoms are first physisorbed on the Cu surface at very low temperature (4 K). Then the tip is placed directly above a physisorbed Fe atom and lowered to increase the attractive force by increasing the tunneling current. At the position of maximum tunneling current, an Fe atom gets attached to the tip, which is dragged by the tip and moved across the surface to a desired position. Then the atom is left at the desired position by lowering the tunneling current, that is, by lifting the tip above the surface. The quantum corral formed by Fe atoms is shown in Fig. 25.37. The shortcoming of an STM is that it works best with conducting materials. But it is also possible to fix organic molecules on a surface and study their structures. For example, this technique has been used in the study of DNA molecules.

25.5.2 Atomic Force Microscope

An AFM is a very powerful microscope for imaging the topology of a surface as it gives a resolution of 10 picometers. A schematic diagram of an AFM is shown in Fig. 25.38. It consists of a cantilever with a sharp tip at its end. The tip is brought very close to the surface of the sample under study and scans the whole surface. The electrostatic force between the tip and the surface leads to deflection of the cantilever according to Hooke's law. The deflection is measured using a laser beam reflected from the top of the cantilever. The angular deflection of the cantilever causes a two-fold larger angular deflection of the laser beam, which is then reflected from a plane mirror. The reflected laser beam strikes a position-sensitive photo detector consisting of two photodiodes placed side by side. The difference between the two photodiode signals indicates the position of the cantilever tip.

Primarily, there are two modes of operation of an AFM.

1. Contact mode.
2. Dynamic mode.

25.5.2.1 Contact Mode

If the tip of the cantilever is fixed at a constant height, then there is a risk that the tip may collide with the surface, causing damage to it. Hence, usually a feedback mechanism is provided to adjust the tip-to-surface distance so that the force between the two is constant. This can be achieved by mounting the sample on a piezoelectric crystal. The tip is then made to scan the sample surface and the vertical distance $h(x, y)$ needed to keep the force constant is recorded. The resulting map

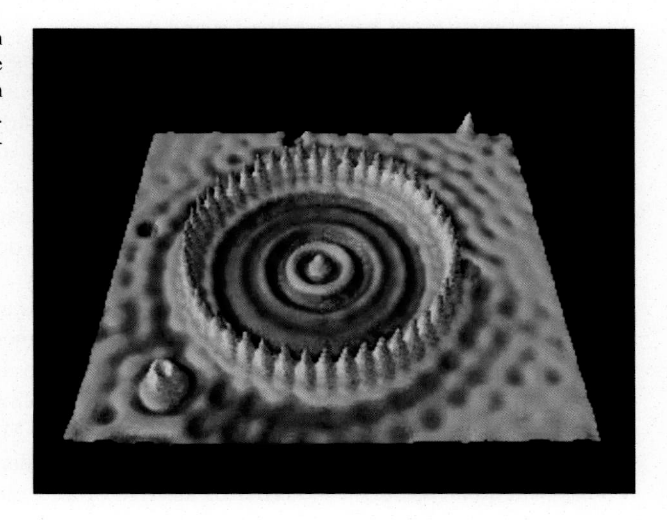

FIG. 25.37 A circular array of Fe atoms placed on a copper surface using an STM tip forming a quantum corral. The ripples inside the corral represent the surface distribution of the electron density that arises from the quantum mechanical energy levels of a circular two-dimensional potential well. *(Created by Don Eigler, IBM Almaden Research Center and taken from Wikipedia with link http://nisenet.org/.../scientific image-quantum Corral.)*

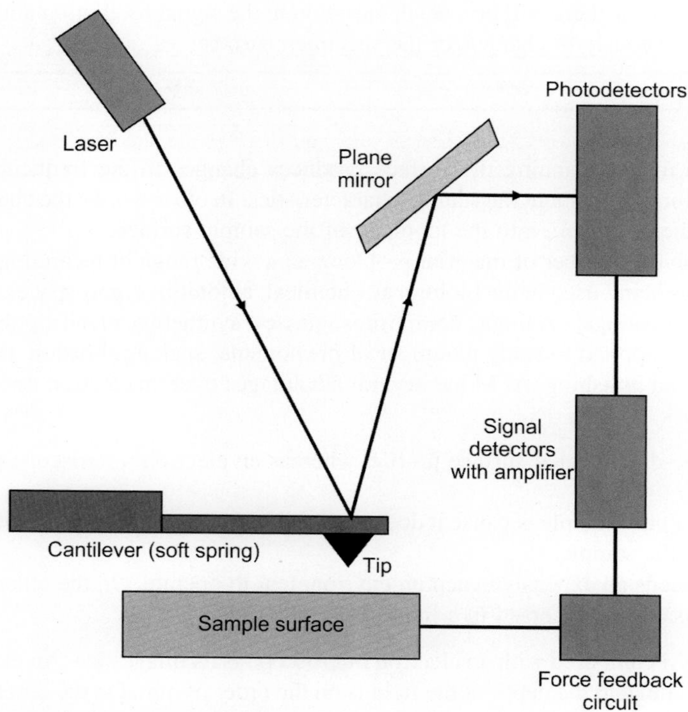

FIG. 25.38 Schematic diagram of an atomic force microscope.

of h(x, y) represents the topology of the surface. The force of attraction between the tip and the surface also depends on the shape of the tip. A normal tip is a 3μm-tall pyramid with an end radius on the order of 30 nm. An electron-beam-deposited (EBD) tip or super tip is also used. Here carbonaceous material is deposited on a normal tip to make it highly pointed (it is long and thin tip).

Contact mode is the most common method of operation of an AFM. As the name suggests, the tip and sample surface remain in close contact as the scanning proceeds. In contact mode, the distance between the tip and the surface is in the repulsive regime of the interatomic force curve (see Fig. 4.3). But sometimes sample surfaces are covered with a layer of absorbed gas, which is 10–30 monolayers thick. As a result, meniscus forces, capillary forces, and van der Wall forces increase the force of friction between the tip and the surface substantially, which affects the sample. In other words, it can sometimes be a destructive mode of imaging.

25.5.2.2 Dynamic Mode

In this mode, the cantilever is made to oscillate close to its resonance frequency (≈50–500 KHz) with the help of an external force. When the tip is brought closer to the sample surface, it interacts with the surface and the oscillations get modified. These changes in oscillations with respect to the external reference oscillation provide information about the characteristics of the surface. Further, there are two possible dynamic modes: amplitude modulation and frequency modulation.

Amplitude Modulation

In the amplitude modulation mode, due to the oscillation of the cantilever, the tip touches the surface for a very short time and is then raised. This is also known as intermittent contact mode or tapping mode. In tapping mode the oscillation amplitude changes, which yields topographical information about the surface of the sample. The amplitude of oscillations increases in valleys and decreases at humps. In this mode there is contact with the sample for a very short time. As the tip scans the surface the short contact reduces the lateral forces drastically. When imaging soft samples (e.g., biological samples) tapping mode may be a far better choice than contact mode. One of the most important factors influencing the resolution is the sharpness of the scanning tip.

The other interesting methods of obtaining image contrast are also possible with tapping mode. In constant force mode, the feedback loop is adjusted so that the amplitude of the cantilever oscillation remains nearly constant. An image can be formed from this signal amplitude as there will be a small variation in the signal oscillation amplitude due to the electronic control not responding instantaneously to changes or the specimen surface.

Frequency Modulation

In the frequency modulation mode, scanning the surface produces changes in the frequency of an external reference frequency, which provides information about the sample characteristics. In other words, the changes in frequency are translated, with the help of computer software, into the topology of the sample surface.

AFM is widely used to solve a number of material problems in a wide range of technologies, such as electronics and telecommunications, as well as being used in the biological, chemical, automotive, aerospace, and energy industries. It can investigate thin and thick film coatings, ceramics, composites, glasses, synthetics, metals, polymers, semiconductors, and biological membranes. AFM is applied to study a number of phenomena, such as abrasion, adhesion, corrosion, etching, friction, lubrication, plating, and polishing. AFM has several advantages over an electron microscope and some are mentioned below:

1. AFM provides a true three-dimensional surface profile, whereas an electron microscope provides a two-dimensional image of the sample.
2. With AFM one can study a pure sample because it does not require any special sample treatment, which either destroys or prevents further use of the sample.
3. An electron microscope needs an expensive vacuum environment to operate. On the other hand, AFM can examine a sample in air or with the sample immersed in a liquid.

The main disadvantage of AFM compared with an electron microscope is its image size. An electron microscope can show an area on the order of mm $\times$ mm and the depth of the field is on the order of mm. On the other hand, an AFM can image a maximum area of around 100 μm $\times$ 100 μm and a depth on the order of μm.

25.5.3 Magnetic Force Microscope

Magnetic materials are very important for making information storage devices. Nanosized magnetic materials also exhibit magnetic domains, just as the bulk materials do. Therefore, a study of the magnetic structure of nanosized materials is important. A magnetic force (MF) microscope is used to study the micromagnetic structure of nanomaterials with lateral resolution down to 30 nm, just as an AFM is used to study chemical structure. This technique is usually called MF microscopy. One of the main advantages of MF microscopy is that thinning or polishing the samples is not necessary. Moreover, MF microscopy yields information on both the chemical and magnetic structures of the surface of the sample under investigation. Therefore, the topology and the magnetic domain structure of a sample may be correlated on the nanometer scale.

The construction and operation of an MF microscope are similar to those of an AFM. A schematic representation of the tip and the sample in an MF microscope is shown in Fig. 25.39. A microfabricated Si tip (about 15 μm in length), which is coated with a ferromagnetic material, is used in an MF microscope. Before each experiment, the tip is magnetically saturated along the axis of the tip (taken as the z-axis) with the help of an external magnetic field. The tip is then mounted back on the microscope in its remnant magnetic state. Here the actual tilt of the tip by an angle of about 10 degrees against the z-axis is neglected.

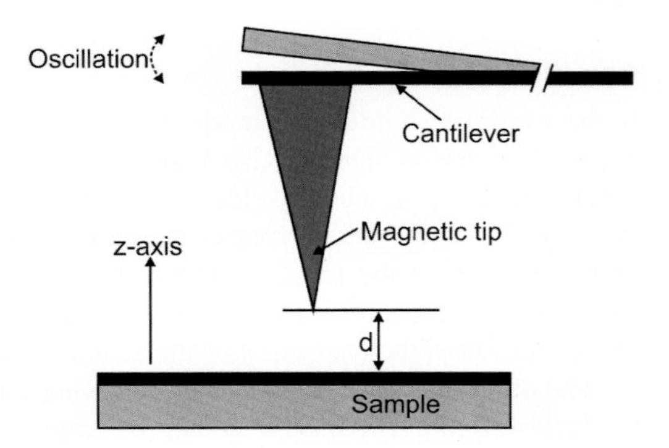

FIG. 25.39 Schematic representation of the principle of operation of a magnetic force microscope, showing the oscillating tip at a distance d above the sample surface.

An MF microscope is used in both the tapping and lift modes. The oscillating tip is raster-scanned across the surface of a given sample, that is, each single line is scanned twice, but at two different distances between the tip and the surface of the sample. In the first scan, a small distance d_1 (≈ 10 nm) is chosen. Because the tip oscillates with an amplitude comparable to d_1, it periodically touches the sample (tapping mode). In this way, the topography of the surface is obtained (the AFM image). In the second scan (called lift mode), a larger distance d_2 (> 40 nm) is chosen between the tip and the sample surface and, therefore, the tip oscillates without touching the sample. In the lift mode only the long-range magnetic interactions between the tip and the field of the sample are important. Depending on the direction of the sample field relative to the direction of the magnetization of the tip (z-direction), the interaction between the tip and field is either attractive or repulsive. Thus, one obtains information about the distribution of the field at d_2 above the surface of the sample, which is called an MF microscope image.

With the help of an MF microscope one can estimate the magnetic domain structure within the surface of the sample. This has been proved with a number of experimental investigations on both magnetic layer systems and nanostructures.

25.6 APPLICATIONS

Nanotechnology is an interdisciplinary subject, which essentially combines physics, chemistry, bioinformatics, and biotechnology. Therefore, it holds the promise of exciting applications in basic sciences, biosciences, medicine, environment, electronics, and a host of other related fields. Some of the common applications of nanomaterials are briefly described below:

25.6.1 Basic Sciences

Quantum dots behave as artificial atoms and electrodes can be connected to them, so one can use them to study atom-like properties. The effect of a magnetic field on a quantum dot can be studied to investigate the magnetic field and its quantization in an atom. The study of a magnetic field in a quantum dot requires a field of 1 Tesla, while an atom requires a field of 10^6 Tesla. Quantum dots can act as a unique laboratory in which fundamental laws of quantum mechanics can be tested.

25.6.2 Nanoelectronics

Nanotubes behave like waveguides for electrons, permitting only a few modes just like fiber optics. Nanotubes that are good conductors can be used as the conducting wires in electronic circuits. There is a possibility of making nanotransistors from carbon nanotubes and their size is expected to be 1/500th of the size of a conventional transistor. The main hurdle in inventing a nanotransistor is in separating semiconducting nanotubes from metallic ones. This invention will revolutionize the electronics industry and will be the ultimate size reduction in electronic components. Further, field-effect transistors made of carbon nanotubes can act as sensitive sensors for some gases. The additional advantage of carbon nanotubes is that they have higher strength and are thermally stable up to about 2800 K.

Nanolasers can be produced from quantum dots because of their discrete energy levels. The color of the light emitted varies with the size of the quantum dot. Nanocomputers and nanoscale memory chips can be produced from nanomaterials. DNA, which is about 2.5 nm wide, has been used to produce a biological computer. Organic light emitting diodes (OLED) have been constructed by Kodak, which are made of nanostructured polymer films, and they are used in car stereos and in cell phones.

25.6.3 Smart Materials

Nanotechnology may create smart or intelligent nanomaterials that can sense and respond to temperature, pressure, light, and electricity. Biosensors can be employed to check body temperature, pulse rate, heart rhythm, blood pressure, and sugar level. Nanosensors can also be used for security and surveillance systems.

25.6.4 Nanocomposite Materials

Nanocomposites of polymers are important from a technological point of view. Dispersing a small number of nanoparticles in a polymer matrix forms a polymer composite. The following properties of materials are improved:

1. Mechanical properties: For example, 2% (by volume) of silicate nanoparticles in polyimide resin increases its strength by 100%.
2. Thermal stability: The thermal stability of the above polymer also increases. The working temperature of the polymer is raised by 100°C.
3. The degree of flammability decreases.

Aircraft, space ships, and other vehicles require lightweight materials with high strength and stiffness (such as polymer nanocomposites) among other properties. The polymer-based nanocomposites have been used for anticorrosion coatings. Nanocomposites have been used in making car bumpers that make them 60% lighter and twice as resistant to denting and scratching, as well as increasing longevity. Glass coated with nanoparticles makes self-cleaning windows. They have been used in catalytic converters in automobiles that help to remove pollution. Certain sunscreens and cosmetics that transparently block harmful radiation have been produced using nanopowders of zinc oxide or titanium oxide.

25.6.5 Nanopharmaceuticals

Nanomaterials are a thousand times smaller than the cells of the human body. Nanodevices are being developed that can slip inside a cell without being recognized by the immune system. Pharmacological agents can be put into buckyballs to deliver medicine more effectively inside the cell. Nanomachines might be possible, which could be directed to correct defects in the cells, to kill cancerous cells without using radiotherapy and chemotherapy, or to repair cell damage.

Radioactive elements can be put into a buckyball, which can then travel through the blood stream and emit radiation. Because buckyballs are excreted intact, the radiation source is removed from the body, reducing the complication of radioactive toxicity. Fluorescent dyes are used to tag cells in genetic research. A light source of the same color is required to illuminate the molecules of the dye, which light up for just for a few seconds. Moreover, dyes have side effects. Quantum dots can be used in the place of a dye with the color of the quantum dot depending on its size. Quantum dots can be lit for a longer time, that is, from a few hours to a few days. Quantum dots allow us to tag different biological components, such as cells and proteins.

Medical monitoring systems, which sound an alert when a diseased organism strikes or appears, can be embedded into a human body. Artificial bone paste, which shows considerable promise for bone repair, is made with ceramic nanoparticles. Some general applications of nanomaterials are listed below. Carbon nanotubes can act as electrodes to contain Li in fuel cells. Nanomembranes can filter toxins from air and water. Nanocoatings on swimsuits repel water, thereby reducing friction with the water and allowing a swimmer to go faster. Nanomaterials can be used to produce low-density insulation, such as light-weight bullet-proof jackets.

In general, there is a need to investigate the toxicity of the various nanomaterials. If the nanoparticles of a particular material happen to be really toxic, then they can create a severe health hazard. Nanoparticles are so small that they can directly penetrate our skin. The release of toxic nanoparticles in the air can be a serious threat to the life of all living animals and human beings. In that case, the production of nanomaterials must be performed under strict environmental control.

25.7 FUTURE THRUST

1. Theoretical understanding of the different types of nanomaterials is lacking. Therefore, there is a need to develop simplified models and theories for these quantum systems.
2. The main emphasis in nanotechnology will be in the fabrication of crystalline nanostructured materials. Once we are able to fabricate these materials, it will open new horizons for both theoretical and experimental studies due to the availability of a few hundred new nanostructured materials.
3. Nanoscience is still in its infancy and no nanoscale machines exist in practice, not even microscale machines. Indeed, only mm-scale machines exist at present. But nanotechnology has great potential to produce nanomachines. The future imperative, therefore, will be to produce nanostructured crystalline bulk materials on one side and nanomachines on the other. It would not be wrong to say that the 21st century will be the century of nanotechnology.

REFERENCES

Dresselhaus, M. S., Dresselhaus, G., & Eklund, P. C. (1996). *Science of fullerenes and carbon nanotubes*. New York: Academic Press.

Kroto, H. W., Heath, J. R., O'Brien, S. C., Curl, R. F., & Smalley, R. E. (1985). *Nature (London)*, *318*, 162.

Kroto, H. W., Heath, J. R., O'Brien, S. C., Curl, R. F., & Smalley, R. E. (1986). *Science*, *242*, 1139.

Nilolaev, A. V., Dennis, T. J. S., Prassides, K., & Soper, A. K. (1994). *Chemical Physics Letters*, *223*, 143.

Ramirez, A. P. (1994). *Condensed Matter News*, *3*, 9.

SUGGESTED READING

Ali Mansoori, G. (2005). *Principles of nanotchnology*. Singapore: World Scientific.
Bruus, H. (2004). *Introduction to nanotechnology*. Denmark: Lyngby spring.
Nalwa, H. S. (2000). *Handbook of nanotechnology*. New York: Academic Press.
Nalwa, H. S. (2002). *Magnetic nanostructures*. California: American Scientific Publishers.
Poole, C. P., & Owens, F. J. (2003). *Introduction to nanotechnology*. New York: J. Wiley & Sons.
Reich, S., Tomsen, C., & Maultzsch, J. (2004). *Carbon nanotubes: Basic concepts and physical properties*. New York: J. Wiley & Sons.
Saito, R., Dresselhaus, G., & Dresselhaus, M. S. (1998). *Physical properties of carbon nanotubes*. Singapore: Imperial College Press.
Schmid, G. (2004). *Nanoparticles: From theory to applications*. New York: J. Wiley & Sons.
Wolf, E. L. (2004). *Nanophysics and nanotechnology*. Weinheim: Wiley-VCH Verlag GmbH & Co.

Appendix A

A.1 VAN DER WAALS-LONDON INTERACTION

In an inert gas crystal, the individual atoms are neutral and the attractive interaction is due to the Van der Waals-London interaction. To understand its origin, consider two inert gas atoms labeled as 1 and 2 at a distance R (Fig. A.1). In the inert gas atom 1 the centers of negative and positive charges coincide, yielding an electrically neutral atom. But the negatively charged electrons revolving around the nucleus of atom 1 possess a finite instantaneous electric dipole moment $\mathbf{p}_1$ whose time-average is zero. The dipole moment $\mathbf{p}_1$ produces an instantaneous electric field $\mathbf{E}_1$ at the position of atom 2 given by

$$\mathbf{E}_1 = \frac{2\mathbf{p}_1}{R^3} \tag{A.1}$$

The field $\mathbf{E}_1$ will induce an instantaneous electric dipole moment $\mathbf{p}_2$ on atom 2, which is linearly proportional to $\mathbf{E}_1$ and is given by

$$\mathbf{p}_2 = \alpha^a \mathbf{E}_1 = \frac{2\alpha^a \mathbf{p}_1}{R^3} \tag{A.2}$$

where α^a is the atomic polarizability (see Chapter 15). The interaction energy $U_a(R)$ between the two electric dipole moments is given by

$$U_a(R) = -\frac{2\mathbf{p}_1 \cdot \mathbf{p}_2}{R^3} = -\frac{4\alpha^a p_1^2}{R^6} \tag{A.3}$$

Eq. (A.3) gives the attractive dipole-dipole interaction energy, which varies as $1/R^6$. Hence, in general, the Van der Waals interaction is written as

$$U_a(R) = -\frac{A}{R^6} \tag{A.4}$$

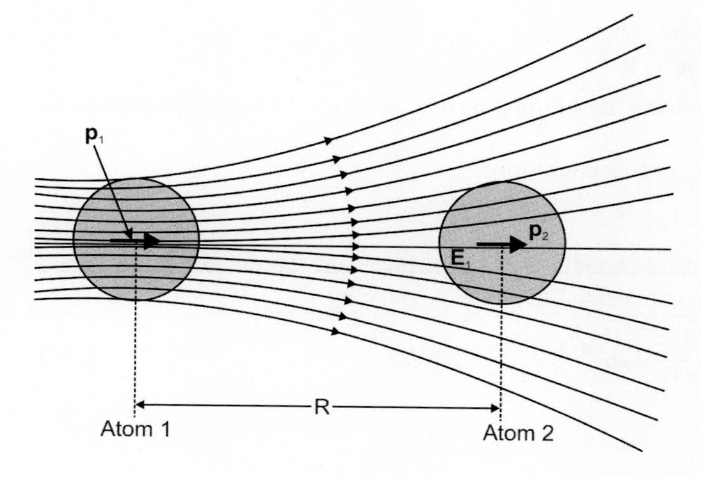

FIG. A.1 Two inert gas atoms labeled as 1 and 2 are separated by distance R. $\mathbf{E}_1$ is the electric field at the position of atom 2 produced by the instantaneous dipole moment $\mathbf{p}_1$ of atom 1.

where A is a constant given by

$$A = 4\alpha^{a} p_1^2 \tag{A.5}$$

From Eq. (A.2) it is evident that α^a has the dimensions of $(\text{length})^3$, that is, r_0^3, where $r_0(\approx 10^{-8}$ cm) is the atomic radius and the dipole moment of an atom is given by $p \approx e\, r_0$. Substituting these values into Eq. (A.5), one can obtain

$$\begin{aligned} A &\approx 4\left(r_0^3\right)(e r_0)^2 = 4e^2 r_0^5 \\ &= 4\left(5\times 10^{-10}\right)^2\left(10^{-8}\right)^5 \approx 10^{-58}\text{erg} - \text{cm}^6 \end{aligned} \tag{A.6}$$

Because of the R^{-6} dependence, the Van der Waals interaction is very small at large distances but increases rapidly as the distance decreases.

Estimation of $\alpha^a p_1^2$

From Eqs. (A.4) to (A.6) one can write

$$U_a(R) = -\frac{10^{-58}}{R^6} \tag{A.7}$$

In the inert gas element Kr, the separation between the atoms is on the order of 4 Å. Therefore, for Kr, the interaction energy becomes

$$U_a(R) = \frac{10^{-58}}{\left(4\times 10^{-8}\right)^6} = 2\times 10^{-14}\text{erg}$$

The temperature T corresponding to $U_a(R)$ given by the above equation becomes

$$T = \frac{U_a(R)}{k_B} = 100^{\circ}\text{K} \tag{A.8}$$

where k_B is the Boltzmann constant.

A.2 REPULSIVE INTERACTION

The derivation of the expression for the repulsive interaction is not in the scope of the present study, but the experimental data obtained from inert gas crystals can be well fit using the following empirical formula

$$U_r(R) = \frac{B}{R^{12}} \tag{A.9}$$

where B is a positive constant. Because of the R^{-12} dependence, it is a very short-ranged interaction: its range is much shorter than the Van der Waals interaction. Therefore, the total interaction potential U(R) is the sum of the Van der Waals and repulsive interactions, that is,

$$\begin{aligned} U(R) &= U_a(R) + U_r(R) \\ &= -\frac{A}{R^6} + \frac{B}{R^{12}} \end{aligned} \tag{A.10}$$

Therefore, the interaction potential in an inert gas crystal is obtained by substituting $m = 6$ and $n = 12$ in the general form of the interaction potential given by Eq. (4.1).

The repulsive interaction can also be fit to the following exponential form:

$$U_r(R) = \lambda_R e^{-R/\rho} \tag{A.11}$$

Here λ_R and ρ are constants: λ_R gives the strength of the repulsive potential while ρ is a measure of its range. So, the second form of the interaction potential in inert gas crystals has an exponential term and is given by

$$U(R) = -\frac{A}{R^6} + \lambda_R e^{-R/\rho} \tag{A.12}$$

Appendix B

Consider an ionic solid with Ze and $-Ze$ as the charges on the positively and negatively charged ions. The repulsive overlap interaction is very short-ranged and, therefore, it is appreciable only with the 1NN ions. On the other hand, the electrostatic interaction is a long-range one and decreases with an increase in distance, but the dominant interaction is again with the 1NN ions. Therefore, the net electrostatic interaction of an ion with all other ions can be represented in terms of the 1NN interaction as $-A_c Z^2 e^2/R$ where A_c is a constant that takes care of the effect of all the other ions. The total interaction potential of an ion can, therefore, be written as

$$U_i(R) = -A_c \frac{Z^2 e^2}{R} + n_0 \frac{B}{R^n} \tag{B.1}$$

Here R is the 1NN distance and n_0 is the number of 1NNs. If there are N ions in the solid, then the total interaction potential energy of the solid becomes

$$U(R) = NU_i(R) = N\left[n_0 \frac{B}{R^n} - A_c \frac{Z^2 e^2}{R}\right] \tag{B.2}$$

In the equilibrium state, Eq. (4.2) of Chapter 4 must be satisfied, which gives the value of the constant B as

$$B = A_c \frac{Z^2 e^2}{n n_0} R_0^{n-1} \tag{B.3}$$

Substituting the value of constant B from Eq. (B.3) into Eq. (B.2), the interaction potential energy of the solid, in the equilibrium state, becomes

$$U(R_0) = -NA_c \frac{Z^2 e^2}{R_0}\left[1 - \frac{1}{n}\right] \tag{B.4}$$

The constant A_c is to be determined. Comparing Eq. (B.4) with Eq. (4.70), one immediately gets $A_c = \alpha_M$, that is, A_c is the Madelung constant, the value of which can be evaluated from the structure of the solid.

Appendix C

An alternate method for evaluating the eigenfunctions of an elastic wave propagating in the [111] direction is as follows. The equations of motion for the displacement of elastic waves in the [111] direction, from Eqs. (5.116), (5.118), (5.119), can be written in matrix form as

$$\begin{pmatrix} C_{11}+C_{44}-3\Lambda & C_{12}+\frac{1}{2}C_{44} & C_{12}+\frac{1}{2}C_{44} \\ C_{12}+\frac{1}{2}C_{44} & C_{11}+C_{44}-3\Lambda & C_{12}+\frac{1}{2}C_{44} \\ C_{12}+\frac{1}{2}C_{44} & C_{12}+\frac{1}{2}C_{44} & C_{11}+C_{44}-3\Lambda \end{pmatrix}\begin{pmatrix} u_1 \\ u_2 \\ u_3 \end{pmatrix}=0 \tag{C.1}$$

For the eigenvalue Λ_2, given by Eq. (5.128), Eq. (C.1) becomes

$$\begin{pmatrix} C_{11}+C_{44}-3\Lambda_2 & C_{12}+\frac{1}{2}C_{44} & C_{12}+\frac{1}{2}C_{44} \\ C_{12}+\frac{1}{2}C_{44} & C_{11}+C_{44}-3\Lambda_2 & C_{12}+\frac{1}{2}C_{44} \\ C_{12}+\frac{1}{2}C_{44} & C_{12}+\frac{1}{2}C_{44} & C_{11}+C_{44}-3\Lambda_2 \end{pmatrix}\begin{pmatrix} u_1 \\ u_2 \\ u_3 \end{pmatrix}=0 \tag{C.2}$$

Substituting the value of Λ_2 from Eq. (5.128) into Eq. (C.2), we obtain

$$\begin{pmatrix} -2\left(C_{12}+\frac{1}{2}C_{44}\right) & C_{12}+\frac{1}{2}C_{44} & C_{12}+\frac{1}{2}C_{44} \\ C_{12}+\frac{1}{2}C_{44} & -2\left(C_{12}+\frac{1}{2}C_{44}\right) & C_{12}+\frac{1}{2}C_{44} \\ C_{12}+\frac{1}{2}C_{44} & C_{12}+\frac{1}{2}C_{44} & -2\left(C_{12}+\frac{1}{2}C_{44}\right) \end{pmatrix}\begin{pmatrix} u_1 \\ u_2 \\ u_3 \end{pmatrix}=0 \tag{C.3}$$

The above matrix equation gives the following three equations

$$\begin{aligned} -2u_1+u_2+u_3&=0 \\ u_1-2u_2+u_3&=0 \\ u_1+u_2-2u_3&=0 \end{aligned} \tag{C.4}$$

Solving Eq. (C.4), one immediately gets

$$u_1=u_2=u_3 \tag{C.5}$$

According to Eq. (C.5) the eigenfunction corresponding to Λ_2 is in the [111] direction. Hence the displacement is in the direction of propagation (**K**) of the wave, which means that it corresponds to a longitudinal wave.

Now for the first solution given by Λ_1, Eq. (C.1) becomes

$$\begin{pmatrix} C_{11}+C_{44}-3\Lambda_1 & C_{12}+\frac{1}{2}C_{44} & C_{12}+\frac{1}{2}C_{44} \\ C_{12}+\frac{1}{2}C_{44} & C_{11}+C_{44}-3\Lambda_1 & C_{12}+\frac{1}{2}C_{44} \\ C_{12}+\frac{1}{2}C_{44} & C_{12}+\frac{1}{2}C_{44} & C_{11}+C_{44}-3\Lambda_1 \end{pmatrix} \begin{pmatrix} u_1 \\ u_2 \\ u_3 \end{pmatrix} = 0 \tag{C.6}$$

Substituting the value of Λ_1 from Eq. (5.134) into Eq. (C.6), one gets

$$\begin{pmatrix} C_{12}+\frac{1}{2}C_{44} & C_{12}+\frac{1}{2}C_{44} & C_{12}+\frac{1}{2}C_{44} \\ C_{12}+\frac{1}{2}C_{44} & C_{12}+\frac{1}{2}C_{44} & C_{12}+\frac{1}{2}C_{44} \\ C_{12}+\frac{1}{2}C_{44} & C_{12}+\frac{1}{2}C_{44} & C_{12}+\frac{1}{2}C_{44} \end{pmatrix} \begin{pmatrix} u_1 \\ u_2 \\ u_3 \end{pmatrix} = 0 \tag{C.7}$$

The three equations given by Eq. (C.7) yield

$$u_1+u_2+u_3=0 \tag{C.8}$$

Let one of the eigenfunctions corresponding to Λ_1 be given by $[1\bar{1}0]$, which satisfies Eq. (C.8). The other eigenfunction must be perpendicular to both [111] and $[1\bar{1}0]$ and can be shown to be $[11\bar{2}]$.

Appendix D: Bose-Einstein Statistics

Bose-Einstein statistics is applicable to bosons: particles with zero or integral spin. Atoms in a solid can be considered to be quantum harmonic oscillators with no spin. Harmonic oscillators are assumed to obey the Maxwell-Boltzmann distribution, according to which the probability of occupation of the nth state P_n with energy E_n is given by

$$P_n \propto e^{-\frac{E_n}{k_B T}} \tag{D.1}$$

Here k_B is the Boltzmann constant and T is the temperature in absolute degrees. The energy of a quantum oscillator is given by

$$E_n = \left(n + \frac{1}{2}\right)\hbar\omega \tag{D.2}$$

where n is an integer or zero and $(1/2)\hbar\omega$ is the zero point energy. The average number of quantum oscillators in the nth state, denoted by $\langle n \rangle$, is given by

$$\langle n \rangle = \frac{\sum_{n=0}^{\infty} n P_n}{\sum_{n=0}^{\infty} P_n} \tag{D.3}$$

Using Eqs. (D.1), (D.2) in Eq. (D.3), one gets

$$\langle n \rangle = \frac{\sum_{n=0}^{\infty} n\, e^{-n\beta_0\hbar\omega}}{\sum_{n=0}^{\infty} e^{-n\beta_0\hbar\omega}} \tag{D.4}$$

where

$$\beta_0 = \frac{1}{k_B T} \tag{D.5}$$

Expanding the series in Eq. (D.4), one can write

$$\langle n \rangle = \frac{d}{dx} \ln\left(1 + e^x + e^{2x} + e^{3x} + \cdots\right) \tag{D.6}$$

where

$$x = -\beta_0\hbar\omega \tag{D.7}$$

Using the standard identity

$$1 + x + x^2 + x^3 + \cdots = \frac{1}{1 - x} \tag{D.8}$$

one can write Eq. (D.6) as

$$\langle n \rangle = \frac{d}{dx} \ln\left(\frac{1}{1 - e^{-\beta_0\hbar\omega}}\right) \tag{D.9}$$

After simplification, Eq. (D.9) becomes

$$\langle n \rangle = \frac{1}{e^{\frac{\hbar\omega}{k_B T}} - 1} \tag{D.10}$$

Eq. (D.10) gives the average occupation number in Bose-Einstein statistics.

Appendix E: Density of Phonon States

E.1 THREE-DIMENSIONAL SOLID

Consider a three-dimensional lattice in the form of a cube of side L and volume V. At finite temperature, atoms of the lattice vibrate in different normal modes and each normal mode of vibration is a function of the wave vector **K**. The dispersion relation $\omega(\mathbf{K})$ for a crystal exhibits different branches as has been seen in Chapters 6 and 7. Imposing the cyclic boundary condition on a crystalline solid shows that there is one **K**-state in a volume of $(2\pi)^3/V$. So, the number of phonon states in an elemental volume d^3K is given by

$$g_p(\mathbf{K})\, d^3K = \frac{V}{(2\pi)^3} d^3K \tag{E.1}$$

$g_p(\mathbf{K})\, d^3K$ is the number of phonon states lying between **K** and **K**+d**K**. Let the volume element d^3K lie between two constant-frequency surfaces with frequencies ω and $\omega + d\omega$, as shown in Fig. E.1. The elemental volume d^3K can then be written as

$$d^3K = d\mathbf{S}_\omega \cdot d\mathbf{K} = dS_\omega \hat{\mathbf{n}} \cdot d\mathbf{K} \tag{E.2}$$

where $\hat{\mathbf{n}}$ is a unit vector perpendicular to the constant-frequency surface with frequency ω. If $dK_\perp$ is the component of the wave vector **K** in the direction of $\hat{\mathbf{n}}$, then

$$d^3K = dS_\omega dK_\perp \tag{E.3}$$

In this approximation, $g_p(\mathbf{K})\, d^3K$ can also be interpreted as the number of phonon states lying between frequencies ω and $\omega + d\omega$, which is represented as $g_p(\omega)\, d\omega$. So, Eq. (E.1) can be written as

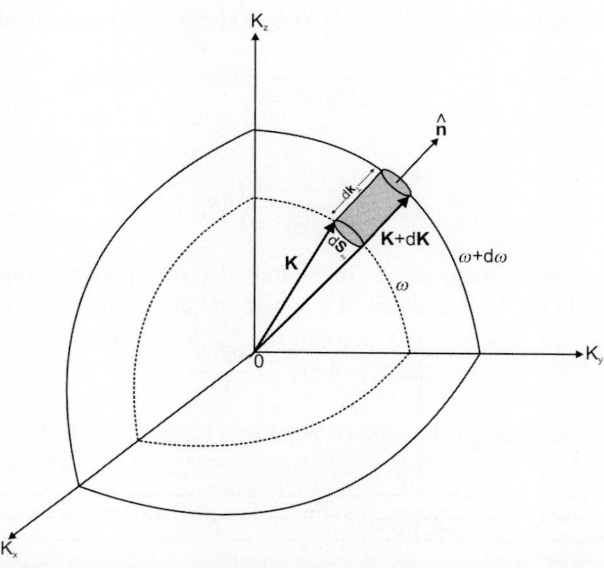

FIG. E.1 The two constant-frequency surfaces with frequencies ω and $\omega + d\omega$ in the **K**-space of a three-dimensional solid. d^3K is an elemental volume lying between the two constant-frequency surfaces having wave vectors **K** and **K**+d**K**. $d\mathbf{S}_\omega$ is an elemental surface on the constant-frequency surface with frequency ω and wave vector **K**.

$$g_p(\omega)\,d\omega = \frac{V}{(2\pi)^3}\int d^3K = \frac{V}{(2\pi)^3}\int dS_\omega\, dK_\perp \tag{E.4}$$

The integral dS_ω is over the surface of constant frequency ω. The dispersion relation makes the frequency a function of the wave vector **K**, that is, $\omega(\mathbf{K})$; therefore,

$$\begin{aligned} d\omega &= \frac{\partial\omega}{\partial K_x}dK_x + \frac{\partial\omega}{\partial K_y}dK_y + \frac{\partial\omega}{\partial K_z}dK_z \\ &= \nabla_{\mathbf{K}}\omega\cdot d\mathbf{K} \end{aligned} \tag{E.5}$$

The different values of **K** may give the same frequency ω, so $\nabla_{\mathbf{K}}\omega$ defines a surface in the **K**-space for which ω is constant. So,

$$d\omega = |\nabla_{\mathbf{K}}\omega|\, dK_\perp \tag{E.6}$$

which gives

$$dK_\perp = \frac{d\omega}{|\nabla_{\mathbf{K}}\omega|} \tag{E.7}$$

Substituting Eq. (E.7) into Eq. (E.4), one can write

$$g_p(\omega)\,d\omega = \frac{V}{(2\pi)^3}\int_{S_\omega}\frac{dS_\omega}{|\nabla_{\mathbf{K}}\omega|}\,d\omega \tag{E.8}$$

From the above expression one can immediately write

$$g_p(\omega) = \frac{V}{(2\pi)^3}\int_{S_\omega}\frac{dS_\omega}{|\nabla_{\mathbf{K}}\omega|} \tag{E.9}$$

Hence the density of phonon states per unit frequency per unit volume becomes

$$g_p(\omega) = \frac{1}{(2\pi)^3}\int_{S_\omega}\frac{dS_\omega}{|\nabla_{\mathbf{K}}\omega|} \tag{E.10}$$

Eqs. (E.9), (E.10) are the general expressions for any type of constant-frequency surface. Because the energy $E_{\mathbf{K}} = \hbar\omega(\mathbf{K})$, therefore, a constant-frequency surface represents a constant-energy surface. For a three-dimensional solid the density of phonon states $g_p(\omega)$ can be calculated from Eq. (E.9) in the Debye approximation defined by Eq. (8.76). From Eq. (8.76) one gets

$$|\nabla_{\mathbf{K}}\omega| = v \tag{E.11}$$

Substituting Eq. (E.11) into Eq. (E.9), one gets

$$g_p(\omega) = \frac{V}{(2\pi)^3}\frac{1}{v}\oint dS_\omega \tag{E.12}$$

The integral is over the surface of constant frequency ω. In the free-electron approximation, the surface of constant energy or frequency is spherical in shape. Let K be the radius of a sphere having frequency ω on its surface, then

$$\oint dS_\omega = 4\pi K^2 = \frac{4\pi\omega^2}{v^2} \tag{E.13}$$

Here we have used the dispersion relation defined by Eq. (8.76). From Eqs. (E.12), (E.13) one can write the phonon density of states with frequency ω as

$$g_p(\omega) = \frac{3V}{2\pi^2 v^3}\omega^2 = A\omega^2 \tag{E.14}$$

where

$$A = \frac{3V}{2\pi^2 v^3} \tag{E.15}$$

The factor of 3 comes from the fact that each atom has associated with it three modes of vibration, one longitudinal, and two transverse modes. For all modes, the velocity of propagation is assumed to be the same. But in reality, the velocities for the longitudinal and transverse modes are different and, therefore, give different contributions to $g_p(\omega)$. It is worthwhile to point out here that real crystals are anisotropic in nature, giving rise to constant energy surfaces that are not spherical in shape. Therefore, an evaluation of $g_p(\omega)$ involves the ab initio calculation of constant-energy surfaces in a crystalline solid.

E.2 TWO-DIMENSIONAL SOLID

Consider a two-dimensional crystalline solid having a surface with area A_0. To find the density of phonon states, one can proceed in the same way as in a three-dimensional solid. Fig. E.2 shows an elemental area d^2K between vectors **K** and **K**+d**K** and bounded by contours of constant frequency ω and $\omega+d\omega$. The number of states in a unit area in the **K**-space is $A_0/(2\pi)^2$. Therefore, the number of phonon states lying between the frequencies ω and $\omega+d\omega$ is given by

$$g_p(\omega)\,d\omega = \frac{A_0}{(2\pi)^2}\int_{l_\omega} d^2K = \frac{A_0}{(2\pi)^2}\int_{l_\omega} dl_\omega\, dK_\perp \tag{E.16}$$

Here the integration is along the contour l_ω of constant frequency ω. As the frequency is a function of the wave vector, that is, $\omega(\mathbf{K})$, therefore,

$$\begin{aligned} d\omega &= \frac{\partial\omega}{\partial K_x}dK_x + \frac{\partial\omega}{\partial K_y}dK_y = \nabla_{\mathbf{K}}\omega\cdot d\mathbf{K} \\ &= |\nabla_{\mathbf{K}}\omega|\,\hat{\mathbf{n}}\cdot d\mathbf{K} = |\nabla_{\mathbf{K}}\omega|\,dK_\perp \end{aligned} \tag{E.17}$$

From the above expression we get

$$dK_\perp = \frac{d\omega}{|\nabla_{\mathbf{K}}\omega|} \tag{E.18}$$

Substituting the value of $dK_\perp$ from the above equation into Eq. (E.16), we get

$$g_p(\omega) = \frac{A_0}{(2\pi)^2}\int_{l_\omega}\frac{dl_\omega}{|\nabla_{\mathbf{K}}\omega|} \tag{E.19}$$

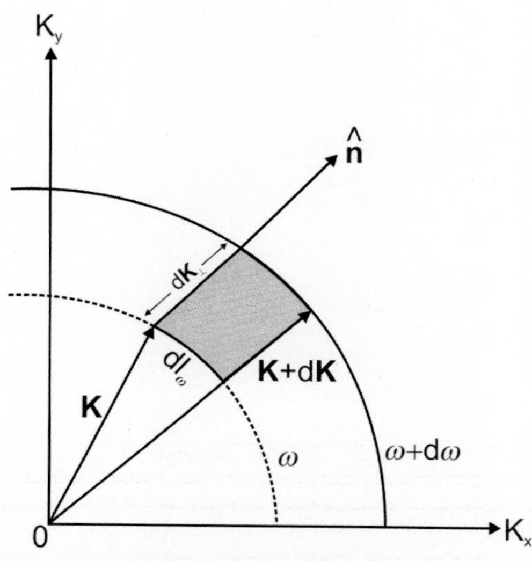

FIG. E.2 The two constant-frequency contours with frequencies ω and $\omega+d\omega$ in the **K**-space of a two-dimensional solid. d^2K is an elemental surface lying between the two constant frequency contours having wave vectors **K** and **K**+d**K**. dl_ω is an elemental line on the constant frequency contour with frequency ω and wave vector **K**.

The density of phonon states in two dimensions can be calculated in the Debye approximation. In a two-dimensional free-electron gas, the integral of dl_ω over a circle of constant frequency gives a length of $2\pi K$. Therefore, Eq. (E.19) gives

$$g_p(\omega) = \frac{A_0}{(2\pi)^2}\frac{2\pi K}{v} = \frac{A_0}{2\pi v^2}\omega \tag{E.20}$$

The density of phonon states in a one-dimensional solid has already been evaluated in Chapter 8 on the specific heat of solids.

Appendix F: Density of Electron States

F.1 THREE-DIMENSIONAL SOLID

Consider a three-dimensional lattice in the form of a cube of side L and volume V. Imposing the cyclic boundary condition on a crystalline solid shows that there is one **k**-state in a volume of $(2\pi)^3/V$. So, the number of electron states in an elemental volume d^3k is given by

$$N_e(\mathbf{k})\,d^3k = 2\frac{V}{(2\pi)^3}\,d^3k \tag{F.1}$$

$N_e(\mathbf{k})\,d^3k$ is the number of electron states lying between **k** and **k**+d**k**. Here the factor of 2 takes account of the spin degeneracy of the electron states. Let the volume element d^3k lie between two constant-energy surfaces with energies $E_\mathbf{k}$ and $E_\mathbf{k}+dE_\mathbf{k}$, as shown in Fig. F.1. The elemental volume d^3k can then be written as

$$d^3k = d\mathbf{S}_{E_k}\cdot d\mathbf{k} = dS_{E_k}\,\hat{\mathbf{n}}\cdot d\mathbf{k} \tag{F.2}$$

where $\hat{\mathbf{n}}$ is a unit vector perpendicular to $d\mathbf{S}_{E_k}$, a small element of surface with constant energy $E_\mathbf{k}$. If $dk_\perp$ is the component of the wave vector d**k** in the direction of $\hat{\mathbf{n}}$, then

$$d^3k = dS_{E_k}\,dk_\perp \tag{F.3}$$

In this approximation, $N_e(\mathbf{k})\,d^3k$ can also be interpreted as the number of electron states lying between energies $E_\mathbf{k}$ and $E_\mathbf{k}+dE_\mathbf{k}$, which is represented as $N_e(E_\mathbf{k})\,dE_\mathbf{k}$. So, Eq. (F.1) can be written as

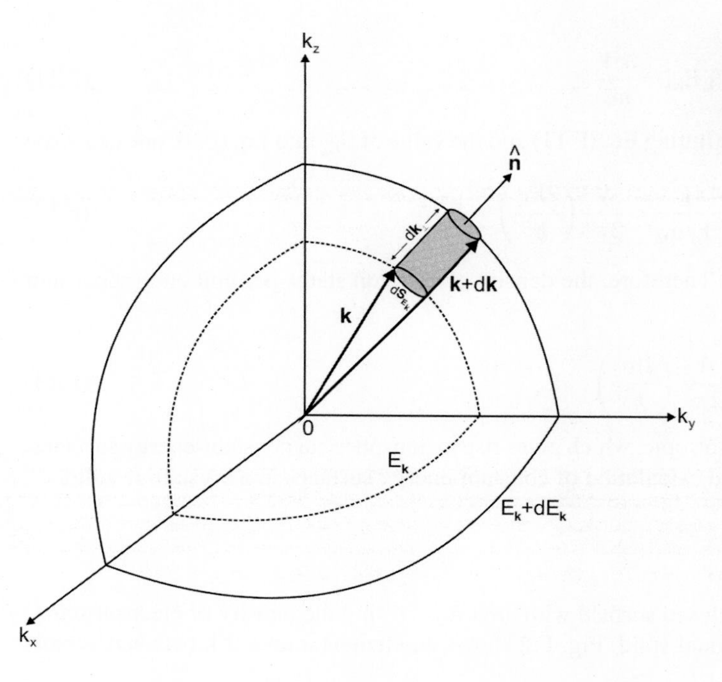

FIG. F.1 The two constant energy surfaces with energies $E_\mathbf{k}$ and $E_\mathbf{k}+dE_\mathbf{k}$ in the **k**-space of a three-dimensional solid. d^3k is an elemental volume lying between the two constant-energy surfaces having wave vectors **k** and **k**+d**k**. $d\mathbf{S}_{E_k}$ is an elemental surface on the constant-energy surface with energy $E_\mathbf{k}$ and wave vector **k**.

$$N_e(E_\mathbf{k})\,dE_\mathbf{k} = 2\frac{V}{(2\pi)^3}\int\limits_{S_{E_k}} d^3k = 2\frac{V}{(2\pi)^3}\int\limits_{S_{E_k}} dS_{E_k}\,dk_\perp \tag{F.4}$$

The integral dS_{E_k} is over the surface of constant energy $E_\mathbf{k}$. The energy derivative is given by

$$\begin{aligned} dE_\mathbf{k} &= \frac{\partial E_\mathbf{k}}{\partial k_x}dk_x + \frac{\partial E_\mathbf{k}}{\partial k_y}dk_y + \frac{\partial E_\mathbf{k}}{\partial k_z}dk_z \\ &= (\nabla_\mathbf{k} E_\mathbf{k})\cdot d\mathbf{k} = |\nabla_\mathbf{k} E_\mathbf{k}|\,dk_\perp \end{aligned} \tag{F.5}$$

From the above equation one can write

$$dk_\perp = \frac{dE_\mathbf{k}}{|\nabla_\mathbf{k} E_\mathbf{k}|} \tag{F.6}$$

Substituting Eq. (F.6) into Eq. (F.4), one can write

$$N_e(E_\mathbf{k})\,dE_\mathbf{k} = 2\frac{V}{(2\pi)^3}\int\limits_{S_{E_k}} \frac{dS_{E_k}}{|\nabla_\mathbf{k} E_\mathbf{k}|}dE_\mathbf{k} \tag{F.7}$$

Therefore, the density of electron states per unit energy $N_e(E_\mathbf{k})$ is given by

$$N_e(E_\mathbf{k}) = 2\frac{V}{(2\pi)^3}\int\limits_{S_{E_k}} \frac{dS_{E_k}}{|\nabla_\mathbf{k} E_\mathbf{k}|} \tag{F.8}$$

Hence the density of electron states per unit energy per unit volume becomes

$$g_e(E_\mathbf{k}) = \frac{2}{(2\pi)^3}\int\limits_{S_{E_k}} \frac{dS_{E_k}}{|\nabla_\mathbf{k} E_\mathbf{k}|} \tag{F.9}$$

Eqs. (F.8), (F.9) are the general expressions for any type of constant energy surface. In the free-electron approximation, the energy of an electron is given by

$$E_\mathbf{k} = \frac{\hbar^2 k^2}{2m_e} \tag{F.10}$$

Therefore,

$$|\nabla_\mathbf{k} E_\mathbf{k}| = \frac{\hbar^2 k}{m_e} \tag{F.11}$$

The area of the constant-energy surface S_{E_k} is $4\pi k^2$. Substituting Eq. (F.11) and the value of S_{E_k} into Eq. (F.8), one can write

$$N_e(E_\mathbf{k}) = 2\frac{V}{(2\pi)^3}\frac{4\pi k^2}{\hbar^2 k/m_e} = \frac{V}{2\pi^2}\left(\frac{2m_e}{\hbar^2}\right)^{3/2} E_\mathbf{k}^{1/2} \tag{F.12}$$

which is the same expression as obtained in Eq. (9.22). Therefore, the density of electron states per unit energy per unit volume becomes

$$g_e(E_\mathbf{k}) = \frac{1}{2\pi^2}\left(\frac{2m_e}{\hbar^2}\right)^{3/2} E_\mathbf{k}^{1/2} \tag{F.13}$$

It is worthwhile to point out here that real crystals are anisotropic, which gives rise to nonspherical constant-energy surfaces. Therefore, the evaluation of $N_e(E_\mathbf{k})$ involves the ab initio calculation of constant-energy surfaces in a crystalline solid.

F.2 TWO-DIMENSIONAL SOLID

Consider a two-dimensional crystalline solid having a closed surface with area A_0. To find the density of electron states, one can proceed in the same way as in a three-dimensional solid. Fig. F.2 shows an elemental area d^2k between vectors

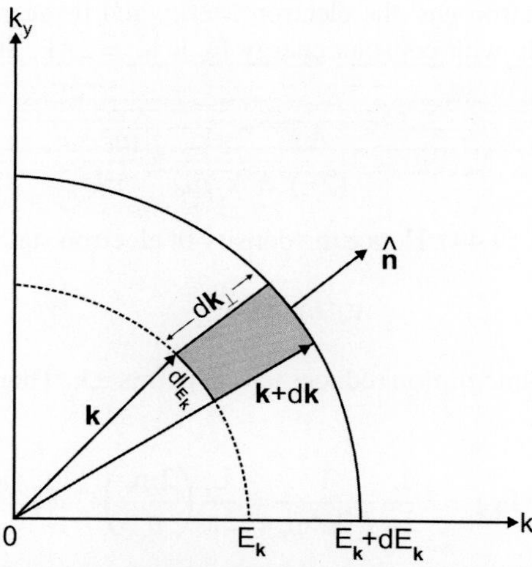

FIG. F.2 The two constant-energy contours with energies $E_{\mathbf{k}}$ and $E_{\mathbf{k}}+dE_{\mathbf{k}}$ in the **k**-space of a two-dimensional solid. d^2k is an elemental surface lying between the two constant-energy contours having wave vectors **k** and **k**+**dk**. $dl_{E_{\mathbf{k}}}$ is an elemental portion of the constant-energy contour with energy $E_{\mathbf{k}}$ and wave vector **k**.

k and **k**+**dk** and bounded by contours of constant energy $E_{\mathbf{k}}$ and $E_{\mathbf{k}}+dE_{\mathbf{k}}$. The number of electron states in a unit area in the **k**-space is $A_0/(2\pi)^2$. Therefore, the number of electron states in d^2k is given by

$$\begin{aligned} N_e(E_{\mathbf{k}})dE_{\mathbf{k}} &= 2\frac{A_0}{(2\pi)^2}\int_{l_{E_{\mathbf{k}}}} d^2k \\ &= 2\frac{A_0}{(2\pi)^2}\int_{l_{E_{\mathbf{k}}}} dl_{E_{\mathbf{k}}}\, dk_\perp \end{aligned} \tag{F.14}$$

Here the integration is along the contour of constant energy $E_{\mathbf{k}}$. As the energy is a function of wave vector, therefore,

$$\begin{aligned} dE_{\mathbf{k}} &= \frac{\partial E_{\mathbf{k}}}{\partial k_x}dk_x + \frac{\partial E_{\mathbf{k}}}{\partial k_y}dk_y \\ &= (\nabla_{\mathbf{k}}E_{\mathbf{k}})\cdot \mathbf{dk} = |\nabla_{\mathbf{k}}E_{\mathbf{k}}|dk_\perp \end{aligned}$$

From the above equation one can write

$$dk_\perp = \frac{dE_{\mathbf{k}}}{|\nabla_{\mathbf{k}}E_{\mathbf{k}}|} \tag{F.15}$$

Substituting the value of $dk_\perp$ from the above equation into Eq. (F.14), we get

$$N_e(E_{\mathbf{k}})dE_{\mathbf{k}} = 2\frac{A_0}{(2\pi)^2}\int_{l_{E_{\mathbf{k}}}} \frac{dl_{E_{\mathbf{k}}}}{|\nabla_{\mathbf{k}}E_{\mathbf{k}}|}dE_{\mathbf{k}} \tag{F.16}$$

Therefore, the density of electron states per unit energy is given by

$$N_e(E_{\mathbf{k}}) = 2\frac{A_0}{(2\pi)^2}\int_{l_{E_{\mathbf{k}}}} \frac{dl_{E_{\mathbf{k}}}}{|\nabla_{\mathbf{k}}E_{\mathbf{k}}|} \tag{F.17}$$

The density of electron states per unit energy per unit area is given by

$$g_e(E_{\mathbf{k}}) = 2\frac{1}{(2\pi)^2}\int_{l_{E_{\mathbf{k}}}} \frac{dl_{E_{\mathbf{k}}}}{|\nabla_{\mathbf{k}}E_{\mathbf{k}}|} \tag{F.18}$$

In the case of a two-dimensional free-electron gas, the electron energy and its derivative are given by Eqs. (F.10), (F.11), respectively. The circumference of a circle with constant energy $E_{\mathbf{k}}$ is $l_{E_{\mathbf{k}}} = 2\pi k$. Substituting these values into Eq. (F.17), we get

$$N_e(E_{\mathbf{k}}) = 2\frac{A_0}{(2\pi)^2}\frac{2\pi k}{\hbar^2 k/m_e} = \frac{m_e A_0}{\pi\hbar^2} \tag{F.19}$$

which is the same result as obtained in Eq. (9.44). Hence the density of electron states per unit energy per unit area becomes

$$g_e(E_{\mathbf{k}}) = \frac{m_e}{\pi\hbar^2} \tag{F.20}$$

In a one-dimensional solid the contour of integration reduces to two points $\pm k$. Therefore, the free electron density of states become

$$N_e(E_{\mathbf{k}}) = 2\frac{L}{2\pi}\frac{1}{\hbar^2 k/m_e} = \frac{L}{2\pi}\left(\frac{2m_e}{\hbar^2}\right)^{1/2} E_{\mathbf{k}}^{-1/2} \tag{F.21}$$

which is the same expression as Eq. (9.51). Hence the density of electron states per unit energy per unit length becomes

$$g_e(E_{\mathbf{k}}) = \frac{1}{2\pi}\left(\frac{2m_e}{\hbar^2}\right)^{1/2} E_{\mathbf{k}}^{-1/2} \tag{F.22}$$

Appendix G: Mean Displacement

The average value of x, that is, $\overline{x}$, can be calculated by using the Maxwell-Boltzmann distribution function as

$$\overline{x} = \frac{\int_{-\infty}^{\infty} x e^{-\frac{1}{k_B T} V(x)} dx}{\int_{-\infty}^{\infty} e^{-\frac{1}{k_B T} V(x)} dx} \tag{G.1}$$

The anharmonic potential in one dimension is given by

$$V(x) = \frac{1}{2}\alpha_F x^2 - \frac{1}{3}\gamma_F x^3 - \frac{1}{4}\delta_F x^4 + \cdots \tag{G.2}$$

(see Eq. 8.145). Substituting Eq. (G.2) into Eq. (G.1), one can write the integral

$$\int_{-\infty}^{\infty} x e^{-\frac{1}{k_B T} V(x)} dx = \int_{-\infty}^{\infty} x e^{-\frac{1}{k_B T}\left[\frac{1}{2}\alpha_F x^2 - \frac{1}{3}\gamma_F x^3 - \frac{1}{4}\delta_F x^4\right]} dx \tag{G.3}$$

Assuming the anharmonic terms to be small and so using the binomial expansion, the above integral can be written as

$$\int_{-\infty}^{\infty} x e^{-\frac{1}{k_B T} V(x)} dx = \int_{-\infty}^{\infty} \left[x + \frac{1}{3k_B T}\gamma_F x^4 + \frac{1}{4k_B T}\delta_F x^5\right] e^{-\frac{1}{2k_B T}\alpha_F x^2} dx$$

The first and the third terms in the above integral are zero, as the integrands are odd in x, but the second term is finite. So,

$$\int_{-\infty}^{\infty} x e^{-\frac{1}{k_B T} V(x)} dx = \frac{2\gamma_F}{3k_B T}\int_{0}^{\infty} x^4 e^{-\frac{1}{2k_B T}\alpha_F x^2} dx = (2\pi)^{1/2}\gamma_F \frac{(k_B T)^{3/2}}{\alpha_F^{5/2}} \tag{G.4}$$

Similarly, one can prove that

$$\int_{-\infty}^{\infty} e^{-\frac{1}{k_B T} V(x)} dx = 2\int_{0}^{\infty} e^{-\frac{1}{2k_B T}\alpha_F x^2} dx = \left(2\pi \frac{k_B T}{\alpha_F}\right)^{1/2} \tag{G.5}$$

Because the integral in Eq. (G.5) is in the denominator of Eq. (G.1), only the harmonic term is retained as the anharmonic terms are small. The integrals (G.4) and (G.5) are evaluated using the following standard integrals.

$$\int_0^\infty x^{2n} e^{-px^2} dx = \frac{(2n-1)!!}{2(2p)^n} \sqrt{\frac{\pi}{p}} \quad \text{for } p > 0 \tag{G.6}$$

$$\int_0^\infty e^{-p^2 x^2} dx = \frac{\sqrt{\pi}}{2p} \quad \text{for } p > 0 \tag{G.7}$$

Substituting the integrals given by Eqs. (G.4), (G.5) into Eq. (G.1), one can immediately write

$$\overline{x} = \gamma_F \frac{k_B T}{\alpha_F^2} = \gamma_F \frac{\overline{E}}{\alpha_F^2} \tag{G.8}$$

where $\overline{E} = k_B T$ is the average energy. Note that Eq. (G.8) is the same as Eq. (8.152).

Appendix H

H.1 BOUND STATES FOR ONE-DIMENSIONAL FREE-ELECTRON GAS

Consider a one-dimensional free-electron gas of N_e electrons confined along a line (say along the x-direction) of length L with infinite potential barriers at the two ends (see Fig. H.1). The electrons occupy different states in the system. Let $|\psi_n(x)\rangle$ be the wave function for the nth state of the system with energy E_n. The Schrodinger equation for the nth state in a one-dimensional free-electron gas is given by

$$-\frac{\hbar^2}{2m_e}\frac{d^2}{dx^2}|\psi_n(x)\rangle = E_n|\psi_n(x)\rangle \tag{H.1}$$

The boundary condition of the system demands that the amplitude of the wave at the two ends of the system be zero, that is,

$$|\psi_n(0)\rangle = |\psi_n(L)\rangle = 0 \tag{H.2}$$

As the system has finite length L with impenetrable walls, its solution will give stationary states (bound states). The Schrodinger equation given by Eq. (H.1) for the one-dimensional free-electron gas can be written as a differential equation of the form

$$\frac{d^2}{dx^2}|\psi_n(x)\rangle + k_n^2|\psi_n(x)\rangle = 0 \tag{H.3}$$

where

$$k_n^2 = \frac{2m_e E_n}{\hbar^2} \tag{H.4}$$

The energy E_n, therefore, can be written as

$$E_n = \frac{\hbar^2 k_n^2}{2m_e} \tag{H.5}$$

The solution of Eq. (H.3) is given by

$$|\psi_n(x)\rangle = A_n \sin k_n x + B_n \cos k_n x \tag{H.6}$$

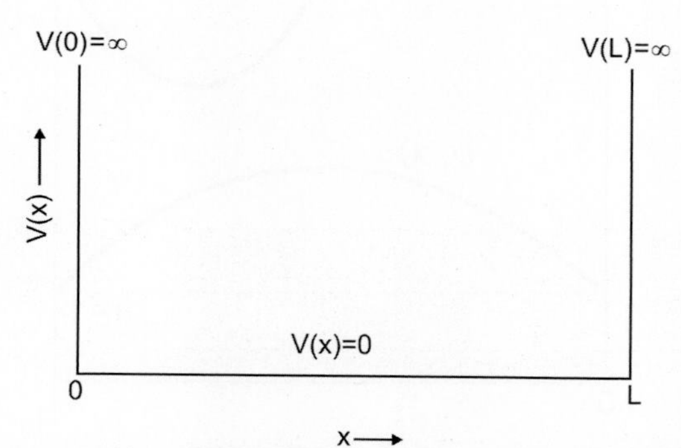

FIG. H.1 The potential function V(x) in a one-dimensional free-electron gas, which is confined to length L along the x-direction.

where A_n and B_n are constants. The boundary condition at $x = 0$ demands $B_n = 0$ yielding

$$|\psi_n(x)\rangle = A_n \sin k_n x \tag{H.7}$$

Applying the boundary condition at $x = L$ to Eq. (H.7), we get

$$\sin k_n L = 0$$

which is satisfied if the electron wave vector k_n has the following values

$$k_n = \frac{n\pi}{L} \tag{H.8}$$

where $n = 0, \pm 1, \pm 2, \ldots$. The constant A_n in Eq. (H.7) can be obtained by normalizing the wave function $|\psi_n(x)\rangle$ to unity, that is,

$$\langle\psi_n(x)|\psi_n(x)\rangle = 1 \tag{H.9}$$

Substituting $|\psi_n(x)\rangle$ from Eq. (H.7) into Eq. (H.9), we get

$$A_n = \left(\frac{2}{L}\right)^{1/2} \tag{H.10}$$

Therefore, the normalized wave function is given by

$$|\psi_n(x)\rangle = \left(\frac{2}{L}\right)^{1/2} \sin\left(\frac{n\pi}{L}x\right) \tag{H.11}$$

The wave functions for the lowest three states with $n = 1, 2,$ and 3 give the stationary states (Fig. H.2). The energy of the nth state is obtained by substituting k_n from Eq. (H.8) into Eq. (H.5) to write

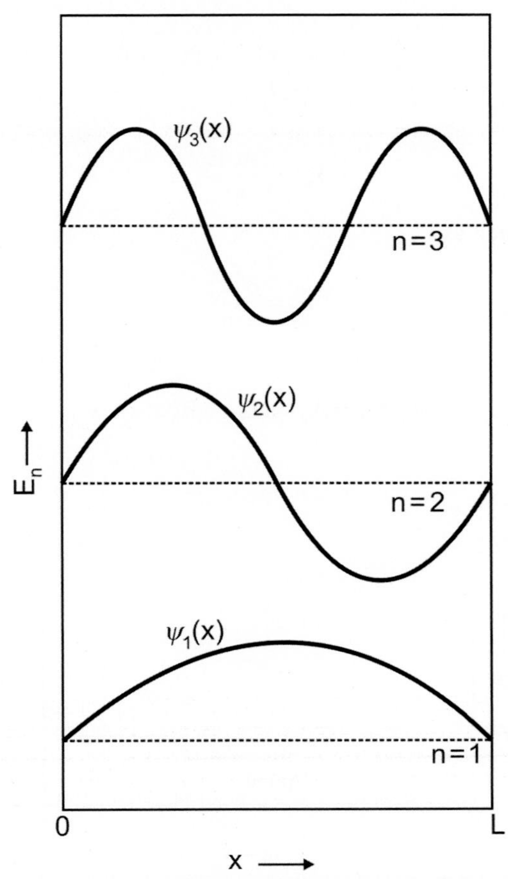

FIG. H.2 The wave functions of the first three bound states with $n = 1, 2,$ and 3, in the one-dimensional free-electron gas, are shown by the *dark lines*. The *dashed lines* show the energies of these states.

$$E_n = \frac{\hbar^2}{2m_e}\left(\frac{n\pi}{L}\right)^2 \tag{H.12}$$

The above expression for energy can also be obtained by substituting Eq. (H.11) into Eq. (H.3). It shows that the energy in a one-dimensional gas is a quadratic function of n. The electrons can be arranged in the various states numbered by n until all the electrons are exhausted. One can define the Fermi level, Fermi energy, Fermi velocity, and the Fermi surface in exactly the same way as in Section 9.2. Let n_F denote the number corresponding to the Fermi state, then

$$2n_F = N_e \tag{H.13}$$

Here the number of electrons N_e in the one-dimensional free-electron gas is assumed to have an even value. The Fermi energy becomes

$$E_F = \frac{\hbar^2}{2m_e}\left(\frac{n_F\pi}{L}\right)^2 = \frac{h^2}{2m_e}\left(\frac{n_F}{2L}\right)^2 = \frac{h^2}{2m_e}\left(\frac{N_e}{4L}\right)^2 \tag{H.14}$$

The total energy E_T in the ground state is given by

$$E_T = 2\sum_{n=1}^{n_F} E_n = 2\sum_{n=1}^{N_e/2} \frac{\hbar^2}{2m_e}\left(\frac{n\pi}{L}\right)^2 \tag{H.15}$$

Here the factor of 2 takes care of the spin degeneracy of the **k**-states. The above summation can be evaluated by using the standard series

$$\sum_{n=1}^{p} n^2 = \frac{1}{6}p\left(2p^2 + 3p + 1\right) \cong \frac{1}{3}p^3 \tag{H.16}$$

for $p \gg 1$. The ground state energy is now straightway given by

$$E_T = \frac{1}{3}N_e E_F \tag{H.17}$$

The number of electron states with wave vector less than or equal to k_n is given by

$$n(k_n) = \frac{L}{\pi}k_n \tag{H.18}$$

Therefore, the density of electron states per unit energy becomes

$$N_e(E_n) = 2\frac{dn(k_n)}{dE_n} = 2\frac{dn(k_n)}{dk_n}\frac{dk_n}{dE_n} \tag{H.19}$$

Substituting Eqs. (H.18), (H.12) into Eq. (H.19), we get

$$N_e(E_n) = \frac{L}{\pi}\left(\frac{2m_e}{\hbar^2}\right)^{1/2} E_n^{-1/2} \tag{H.20}$$

Therefore, the density of electron states per unit length per unit energy is given by

$$g_e(E_n) = \frac{1}{\pi}\left(\frac{2m_e}{\hbar^2}\right)^{1/2} E_n^{-1/2} \tag{H.21}$$

H.2 BOUND STATES FOR TWO- AND THREE-DIMENSIONAL FREE-ELECTRON GAS

For a two-dimensional free-electron gas the Schrodinger equation is given by

$$-\frac{\hbar^2}{2m_e}\left(\frac{\partial^2}{\partial x^2} + \frac{\partial^2}{\partial y^2}\right)|\psi_n(x, y)\rangle = E_n|\psi_n(x, y)\rangle \tag{H.22}$$

If the free-electron gas is confined to a square of side L, then the boundary conditions are

$$\begin{aligned} |\psi_n(x, y)\rangle &= |\psi_n(x+L, y)\rangle \\ |\psi_n(x, y)\rangle &= |\psi_n(x, y+L)\rangle \end{aligned} \tag{H.23}$$

Eq. (H.22) yields two differential equations: one for the x-coordinate and the other for the y-coordinate. These differential equations can be solved in exactly the same manner as for a one-dimensional solid. The total normalized wave function can be written as

$$|\psi_n(x, y)\rangle = \frac{2}{L} \sin\left(\frac{n_x \pi}{L} x\right) \sin\left(\frac{n_y \pi}{L} y\right) \tag{H.24}$$

where n_x and n_y are integers; negative, positive or zero. The energy of the nth state can be obtained by substituting Eq. (H.24) into Eq. (H.22) to write

$$E_n = \frac{\hbar^2}{2m_e}\left[\left(\frac{n_x \pi}{L}\right)^2 + \left(\frac{n_y \pi}{L}\right)^2\right] \tag{H.25}$$

Consider a three-dimensional free-electron gas confined to a cubical box of side L. The Schrodinger wave equation can be written as

$$-\frac{\hbar^2}{2m_e}\left(\frac{\partial^2}{\partial x^2} + \frac{\partial^2}{\partial y^2} + \frac{\partial^2}{\partial z^2}\right)|\psi_n(x, y, z)\rangle = E_n|\psi_n(x, y, z)\rangle \tag{H.26}$$

The boundary conditions on the wavefunction are

$$\begin{aligned} |\psi_n(x, y, z)\rangle &= |\psi_n(x+L, y, z)\rangle \\ |\psi_n(x, y, z)\rangle &= |\psi_n(x, y+L, z)\rangle \\ |\psi_n(x, y, z)\rangle &= |\psi_n(x, y, z+L)\rangle \end{aligned} \tag{H.27}$$

Eq. (H.26) can be solved exactly in the same way as in the two-dimensional free-electron gas and the normalized wave function becomes

$$|\psi_n(x, y, z)\rangle = \left(\frac{8}{L^3}\right)^{1/2} \sin\left(\frac{n_x \pi}{L} x\right) \sin\left(\frac{n_y \pi}{L} y\right) \sin\left(\frac{n_y \pi}{L} z\right) \tag{H.28}$$

The energy of the nth state can be found in the same way as for the two-dimensional free-electron gas and is given by

$$E_n = \frac{\hbar^2}{2m_e}\left[\left(\frac{n_x \pi}{L}\right)^2 + \left(\frac{n_y \pi}{L}\right)^2 + \left(\frac{n_z \pi}{L}\right)^2\right] \tag{H.29}$$

Eqs. (H.12), (H.25), (H.29) show that the energies of the bound states are sharp and discrete in a free-electron gas.

Appendix I: The Fermi Distribution Function Integral

The general Fermi-Dirac distribution function is given by

$$f(E,\mu,T)=\frac{1}{e^{\frac{E-\mu}{k_B T}}+1} \tag{I.1}$$

which can be written as

$$f(x-y)=\frac{1}{e^{x-y}+1} \tag{I.2}$$

where

$$x=\frac{E}{k_B T} \quad \text{and} \quad y=\frac{\mu}{k_B T} \tag{I.3}$$

Here μ is the chemical potential. In studying the temperature dependent properties of solids, one usually comes across an integral of the form

$$I(y)=\int_0^{\infty} h(x)\,f(x-y)\,dx \tag{I.4}$$

where h(x) is assumed to be a slowly varying function of x such that its integral over x gives a finite function H(x), that is,

$$H(x)=\int_0^{x} h(x)\,dx \tag{I.5}$$

Integrating Eq. (I.4) by parts, we can write

$$I(y)=H(x)\,f(x-y)\,|_0^{\infty}-\int_0^{\infty} H(x)\frac{\partial}{\partial x}f(x-y)\,dx \tag{I.6}$$

The first term in Eq. (I.6) is zero because

$$\text{Lim}_{x=\infty} f(x-y)=0 \quad \text{and} \quad \text{Lim}_{x=0} H(x)=0 \tag{I.7}$$

Therefore, Eq. (I.6) reduces to

$$I(y)=-\int_0^{\infty} H(x)\frac{\partial}{\partial x}f(x-y)\,dx \tag{I.8}$$

With suitable substitution the above expression can be written as

$$I(y) = -\int_{-y}^{\infty} H(x+y)\frac{\partial}{\partial x}f(x)\,dx \tag{I.9}$$

$H(x+y)$ can be expanded around y using the Taylor series as

$$\begin{aligned} H(x+y) &= H(y) + x\frac{\partial H}{\partial y} + \frac{1}{2!}x^2\frac{\partial^2 H}{\partial y^2} + \cdots \\ &= e^{x\delta}H(y) \end{aligned} \tag{I.10}$$

where

$$\delta = \frac{\partial}{\partial y} \tag{I.11}$$

Using Eqs. (I.10), (I.11), the integral (I.9) becomes

$$I(y) = -\int_{-y}^{\infty} dx\, e^{x\delta}\frac{\partial}{\partial x}f(x)H(y) \tag{I.12}$$

At low temperatures ($T \ll T_F$), the region with negative values of y contributes negligibly to the above integral. This is due to the fact that $-\partial f/\partial x$ or $-\partial f/\partial E$ is very small for negative values of E. In this approximation Eq. (I.12) can be written as

$$I(y) = -\int_{-\infty}^{\infty} dx\, e^{x\delta}\frac{\partial}{\partial x}f(x)H(y) \tag{I.13}$$

The derivative of the Fermi distribution function f(x) is given by

$$\frac{\partial}{\partial x}f(x) = -\frac{e^x}{(e^x+1)^2}$$

Using the above expression in Eq. (I.13), we write

$$I(y) = \int_{-\infty}^{\infty} dx\, e^{x\delta}\frac{e^x}{(e^x+1)^2}H(y) \tag{I.14}$$

Putting $z = e^x$ in the above integral, we get

$$I(y) = \int_{0}^{\infty} dz\frac{z^\delta}{(z+1)^2}H(y) \tag{I.15}$$

This is an integral of a β-function and its value is given by

$$I(y) = \frac{\pi\delta}{\sin(\pi\delta)}H(y) \tag{I.16}$$

Expanding $\sin(\pi\delta)$ in terms of $\pi\delta$ in the above expression, one finally gets

$$I(y) = \left[1 + \frac{\pi^2}{6}\frac{\partial^2}{\partial y^2} + \frac{7\pi^4}{3(5!)}\frac{\partial^4}{\partial y^4} + \cdots\right]H(y) \tag{I.17}$$

The form of δ from Eq. (I.11) has been substituted in the above expression.

Appendix J: Electron Motion in Magnetic Field

Consider an electron, with momentum **k**, moving in a magnetic field **H**. The force acting on the electron is

$$\begin{aligned} \mathbf{F} &= \frac{d}{dt}(\hbar \mathbf{k}) = -\frac{e}{c}\mathbf{v} \times \mathbf{H} \\ \frac{d\mathbf{k}}{dt} &= -\frac{e}{c\hbar}\mathbf{v} \times \mathbf{H} \end{aligned} \tag{J.1}$$

This means that the wave vector **k** changes in a direction perpendicular to both **v** and **H**. Hence, **k** must be confined to the orbit defined by the intersection of the Fermi surface with a plane normal to **H** (see Fig. J.1). The magnetic field makes a representative **k** point move in an orbit (a circular path) without changing its energy.

If the electron is not scattered, the time period of the electron in the orbit, from Eq. (J.1), is

$$T = \int dt = \oint \frac{c\hbar}{e} \frac{1}{v_\perp H} dk \tag{J.2}$$

Here $\mathbf{v}_\perp$ is the velocity perpendicular to **H**. We know that $T = 2\pi/\omega_c$ where ω_c is the angular frequency of the electron in the orbit, usually called the cyclotron frequency. So,

$$\frac{2\pi}{\omega_c} = \frac{c\hbar}{eH}\oint \frac{dk}{v_\perp} \tag{J.3}$$

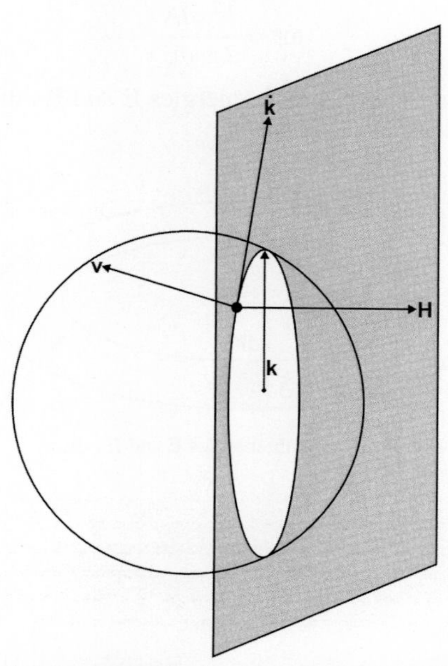

FIG. J.1 The spherical Fermi surface is cut by a shaded plane perpendicular to the magnetic field **H**. The circular orbit of the electron in the **k**-space is in the shaded plane.

Now $\hbar k_\perp = m_e v_\perp$, therefore, from Eq. (J.3) we write

$$\frac{2\pi}{\omega_c} = \frac{m_e c}{eH} \oint \frac{dk}{k_\perp} \tag{J.4}$$

The integral in Eq. (J.4) gives 2π from the elementary geometry of free electrons. So, Eq.(J.4) gives

$$\omega_c = \frac{eH}{m_e c} \tag{J.5}$$

It is customary to define an effective mass m_c^* in the magnetic field, which is usually called the effective cyclotron mass. Therefore,

$$\omega_c = \frac{eH}{m_c^* c} \tag{J.6}$$

Note that m_c^* is not the dynamical mass of the electron, rather it is the property of an orbit, not of a particular electronic state. Substituting Eq. (J.6) into Eq. (J.3), we can write

$$m_c^* = \frac{\hbar}{2\pi} \oint \frac{dk}{v_\perp} \tag{J.7}$$

We know that the velocity v is given in terms of the free electron energy $E = \hbar^2 k^2 / 2 m_e$ as

$$v = \frac{1}{\hbar} \frac{dE}{dk}$$

Therefore, $v_\perp$ can be written as

$$v_\perp = \frac{1}{\hbar} \frac{dE}{dk_\perp} \tag{J.8}$$

Substituting the value of $v_\perp$ from Eq. (J.8) into Eq. (J.7), we get

$$m_c^* = \frac{\hbar^2}{2\pi} \oint \frac{dk_\perp}{dE} dk \tag{J.9}$$

From Fig. J.2 we see that $dk\, dk_\perp$ gives the elemental area between two orbits with energy E and E + dE. Hence the integral in the above equation gives us the change in area per unit energy. Therefore, m_c^* can be written as

$$m_c^* = \frac{\hbar^2}{2\pi} \frac{\partial A}{\partial E} \tag{J.10}$$

∂A is the area in the annular strip between the orbits with energies E and E + dE.

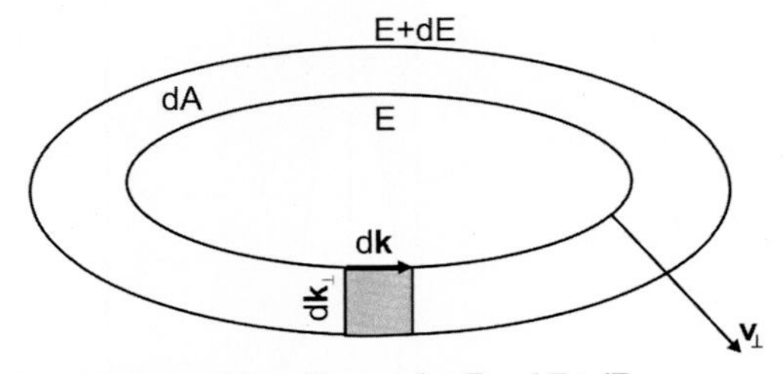

FIG. J.2 Elemental area dA between two constant-energy orbits with energies E and E + dE.

Appendix K

K.1 ONE-DIMENSIONAL SOLID

The Boltzmann transport equation can be derived considering the motion of particles in detail. The phase space for a one-dimensional velocity distribution is shown in Fig. K.1 in which the elemental volume is given by

$$d\tau_2 = dx\,dv_x \tag{K.1}$$

(same as Eq. 8.9). $f(x, v_x, t)$ is the distribution function in one dimension. Further, it is assumed that the time of interaction is very small compared with the relaxation time. Let F_x be the force acting on an electron of mass m_e. In time dt the distance moved by an electron is $v_x\,dt$. Hence, in time dt, all the electrons in area ADEF $= v_x dt dv_x$ enter from the side AD into the elemental area $dx\,dv_x$ of the phase space, and the number of electrons is given by

$$dN_e^{in}(x) = f(x, v_x, t) v_x dv_x dt \tag{K.2}$$

The number of electrons leaving the elemental area $dx\,dv_x$ (ABCD) in time dt through the side BC can be written from Eq. (K.2) as

$$dN_e^{out} = \left[f(x, v_x, t) v_x + \frac{\partial}{\partial x} \{ f(x, v_x, t) v_x \} dx \right] dv_x\,dt \tag{K.3}$$

Hence, the number of electrons entering into the elemental area $dx\,dv_x$ due to the drift parallel to the x-coordinate is obtained by subtracting Eq. (K.3) from Eq. (K.2) to get

$$dN_e^{dx} = -v_x \frac{\partial}{\partial x} [f(x, v_x, t)]\,dx\,dv_x\,dt \tag{K.4}$$

In exactly the same manner, one can calculate the number of electrons entering into the elemental area ABCD due to drift along the velocity axis. The number of electrons entering into the elemental volume $dx\,dv_x$ in time dt from the side AB are contained in area ABGH and are given by

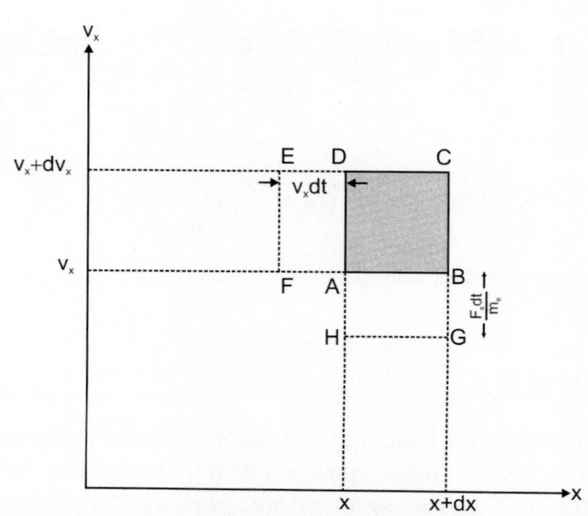

FIG. K.1 The phase space with a one-dimensional velocity distribution function. *The shaded region* ABCD shows the volume element $d\tau_2$.

$$dN_e^{in}(v_x) = f(x, v_x, t)\frac{F_x}{m_e}dt\,dx \tag{K.5}$$

The number of electrons leaving the elemental area $dx\,dv_x$ in time dt through the side CD is given by

$$dN_e^{out}(v_x + dv_x) = \left[f(x, v_x, t)\frac{F_x}{m} + \frac{\partial}{\partial v_x}\left\{f(x, v_x, t)\frac{F_x}{m_e}\right\}dv_x\right]dx\,dt \tag{K.6}$$

Hence, the number of electrons entering into the elemental area $dx\,dv_x$ due to the drift parallel to the velocity axis is obtained by subtracting Eq. (K.6) from Eq. (K.5) and is given by

$$dN_e^{dv_x} = -\frac{F_x}{m_e}\frac{\partial}{\partial v_x}[f(x, v_x, t)]dv_x\,dx\,dt \tag{K.7}$$

where F_x is assumed to be independent of v_x. Now $\frac{\partial}{\partial t}f(x, v_x, t)$ is the rate of increase of the density of electrons in the phase space. Therefore, the increase in the number of electrons in the elemental area $dx\,dv_x$ in time dt becomes $\frac{\partial}{\partial t}f(x, v_x, t)dx\,dv_x dt$. Hence, from Eqs. (K.4), (K.7), one can write

$$\left.\frac{\partial}{\partial t}f(x, v_x, t)\right|_{drift} dx\,dv_x dt = dN_e^{dx} + dN_e^{dv_x} = -\left[v_x\frac{\partial}{\partial x}f(x, v_x, t) + \frac{F_x}{m_e}\frac{\partial}{\partial v_x}f(x, v_x, t)\right]dv_x\,dx\,dt$$

Hence, the rate of change in the distribution function due to the drift produced by the external field is

$$\left.\frac{\partial}{\partial t}f(x, v_x, t)\right|_{drift} = -v_x\frac{\partial}{\partial x}f(x, v_x, t) - \frac{F_x}{m_e}\frac{\partial}{\partial v_x}f(x, v_x, t) \tag{K.8}$$

In time dt, the particles in the range x and x + dx collide with one another only once and their velocities are changed. Some of the electrons in the range x and x + dx with velocities in the range v_x and $v_x + dv_x$ collide with one another in such a way that they go out of the velocity range v_x and $v_x + dv_x$. On the other hand, there are some electrons in the position range x and x + dx whose initial velocities change during the collision process to enter into the velocity range v_x and $v_x + dv_x$. Therefore, as a result of the collision interactions, there is a net increase in the number of electrons in the elemental area ABCD of the phase space. This increase in the number of electrons is proportional to the time interval dt and the elemental area $dx\,dv_x$ of the phase space. The increase in the number of electrons in the elemental area ABCD due to the collision process is given by

$$dN_e^{cell} = \left.\frac{\partial f}{\partial t}\right|_{coll} dx\,dv_x dt \tag{K.9}$$

Hence the total change in the distribution function $f(x, v_x, t)$ is

$$\frac{\partial f(x, v_x, t)}{\partial t}dx\,dv_x dt = \left.\frac{\partial f(x, v_x, t)}{\partial t}\right|_{drift} dx\,dv_x dt + \left.\frac{\partial f(x, v_x, t)}{\partial t}\right|_{coll} dx\,dv_x dt$$

Using Eq. (K.8) in the above equation, we get

$$\frac{\partial f(x, v_x, t)}{\partial t} = -v_x\frac{\partial}{\partial x}f(x, v_x, t) - \frac{F_x}{m_e}\frac{\partial}{\partial v_x}f(x, v_x, t) + \left.\frac{\partial f(x, v_x, t)}{\partial t}\right|_{coll} \tag{K.10}$$

In the equilibrium condition $\partial f/\partial t = 0$, so

$$\left.\frac{\partial f(x, v_x, t)}{\partial t}\right|_{coll} = v_x\frac{\partial}{\partial x}f(x, v_x, t) + \frac{F_x}{m_e}\frac{\partial}{\partial v_x}f(x, v_x, t) \tag{K.11}$$

which gives the Boltzmann transport equation in one dimension (see Eq. 11.17).

K.2 THREE-DIMENSIONAL SOLID

A similar procedure can be adopted to deduce the Boltzmann transport equation in three dimensions. In a three-dimensional solid the volume element in the six-dimensional phase space is given by Eq. (8.12). First consider the motion of electrons moving with velocity v_x in the x-direction. The cross-sectional area of elemental volume $d^3r\,d^3v$ perpendicular to the x-axis is dy $dzdv_xdv_ydv_z$. Hence, the number of electrons entering the elemental volume d^3rd^3v through the side perpendicular to the x-axis at x in time dt is given by

$$dN_e^{in}(x) = [f(\mathbf{r}, \mathbf{v}, t)\, v_x]dy\, dzdv_xdv_ydv_zdt \tag{K.12}$$

The number of electrons leaving the elemental volume d^3rd^3v in time dt through the cross-sectional area perpendicular to the x-axis at x + dx is given by

$$dN_e^{out}(x+dx) = \left[f(\mathbf{r}, \mathbf{v}, t)\, v_x + \frac{\partial}{\partial x}\{f(\mathbf{r}, \mathbf{v}, t)\, v_x\}dx\right] dydzdv_xdv_ydv_z \tag{K.13}$$

Hence, the net number of electrons entering into the elemental volume d^3rd^3v in time dt with velocity distribution along the x-direction is obtained by subtracting Eq. (K.13) from Eq. (K.12) to give

$$dN_e(dx) = -v_x\frac{\partial}{\partial x}\{f(\mathbf{r}, \mathbf{v}, t)\}d^3rd^3vdt \tag{K.14}$$

where the velocity v_x is assumed to be independent of x. Similar expressions can be obtained for the number of electrons entering the elemental volume d^3rd^3v in time dt with velocity distributions only along the y- and z-directions. Hence, the total number of electrons entering into the elemental volume in time dt and possessing a three-dimensional velocity distribution is obtained by combining these three expressions and is written as

$$dN_e(d\mathbf{r}) = -(\mathbf{v}\cdot\nabla_\mathbf{r})f(\mathbf{r}, \mathbf{v}, t)\, d^3rd^3vdt \tag{K.15}$$

Let us now calculate the number of electrons entering into the volume element d^3rd^3v due to the drift in velocity through the side perpendicular to the velocity axis. The cross-sectional area of d^3rd^3v perpendicular to the v_x axis is dx dy $dzdv_ydv_z$. Therefore, the number of electrons entering into the volume element d^3rd^3v in time dt through the side perpendicular to the v_x-direction at v_x is given by

$$dN_e^{in}(v_x) = \left[f(\mathbf{r}, \mathbf{v}, t)\frac{F_x}{m_e}\right] dx\, dy\, dzdv_ydv_zdt \tag{K.16}$$

The number of electrons leaving the elemental volume in time dt through the face perpendicular to the v_x-direction at v_x+dv_x is given by

$$dN_e^{out}(v_x+dv_x) = \left[f(\mathbf{r}, \mathbf{v}, t)\frac{F_x}{m_e} + \frac{\partial}{\partial v_x}\left\{f(\mathbf{r}, \mathbf{v}, t)\frac{F_x}{m_e}\right\}dv_x\right] dxdydzdv_ydv_z\, dt \tag{K.17}$$

Hence, the net number of electrons entering the elemental volume in time dt through the face of the elemental volume perpendicular to the v_x-direction is obtained by subtracting Eq. (K.17) from Eq. (K.16) and is given by

$$dN_e(dv_x) = -\frac{F_x}{m_e}\frac{\partial}{\partial v_x}\{f(\mathbf{r}, \mathbf{v}, t)\}d^3rd^3vdt \tag{K.18}$$

Here it is assumed that the force is independent of the electron velocity. Similar expressions can be obtained for the number of electrons entering into the elemental volume d^3rd^3v in time dt through the faces perpendicular to the v_y- and v_z-directions. Hence the total number of electrons entering into the elemental volume in time dt through the faces perpendicular to the v_x-, v_y- and v_z-directions is obtained by combining these expressions and, therefore, is given by

$$dN_e(dv) = -\left(\frac{\mathbf{F}}{m_e}\cdot\nabla_v\right) f(\mathbf{r}, \mathbf{v}, t)\, d^3rd^3vdt \tag{K.19}$$

The total number of electrons entering into the volume element d^3rd^3v in phase space in time dt due to the drift produced by the external field is obtained by adding Eqs. (K.15), (K.19) to write

$$dN_e^{drift} = \left.\frac{\partial f}{\partial t}\right|_{drift} d^3rd^3vdt = -\left[(\mathbf{v}\cdot\nabla_\mathbf{r})f(\mathbf{r}, \mathbf{v}, t) + \frac{\mathbf{F}}{m_e}\cdot\nabla_v f(\mathbf{r}, \mathbf{v}, t)\right] d^3rd^3vdt \tag{K.20}$$

Due to the collision interactions between the electrons, there is also a net gain in the number of electrons in the elemental volume in time dt, which is given by

$$dN_e^{coll} = \left.\frac{\partial f}{\partial t}\right|_{coll} d^3rd^3vdt \tag{K.21}$$

The total increase in the number of electrons dN_e due to both the external field and the collision interactions is given by

$$dN_e = dN_e^{drift} + dN_e^{coll} = \frac{\partial f}{\partial t} d^3r d^3v dt \tag{K.22}$$

Substituting Eqs. (K.20), (K.21) into Eq. (K.22), we can write

$$\frac{\partial f}{\partial t} = -(\mathbf{v} \cdot \nabla_{\mathbf{r}}) f(\mathbf{r}, \mathbf{v}, t) - \frac{\mathbf{F}}{m_e} \cdot \nabla_{\mathbf{v}} f(\mathbf{r}, \mathbf{v}, t) + \left.\frac{\partial f}{\partial t}\right|_{coll} \tag{K.23}$$

which is the three-dimensional Boltzmann transport equation. In the equilibrium state, $\partial f/\partial t = 0$, that is, the number of electrons per unit volume remains constant in time. Therefore, in the equilibrium state, the Boltzmann transport equation (K.23) assumes the form

$$\left.\frac{\partial f}{\partial t}\right|_{coll} = -\left.\frac{\partial f}{\partial t}\right|_{drift} = (\mathbf{v} \cdot \nabla_{\mathbf{r}}) f(\mathbf{r}, \mathbf{v}, t) + \frac{\mathbf{F}}{m_e} \cdot \nabla_{\mathbf{v}} f(\mathbf{r}, \mathbf{v}, t) \tag{K.24}$$

Eq. (K.24) is the same as Eq. (11.18).

Appendix L: Atomic Magnetic Dipole Moment

Consider an electron of mass m_e moving in a circular orbit of radius r around a nucleus (see Fig. L.1). If ω_0 is the angular velocity and T is the time period of revolution of the electron, then the orbital current produced by the electron is given by

$$I_L = -\frac{e}{T} = -\frac{e\omega_0}{2\pi} \tag{L.1}$$

The area of the circular loop is given by

$$A = \pi r^2 \tag{L.2}$$

From elementary electricity, the magnetic moment produced by the current loop is given by

$$\mu_L = \frac{I_L A}{c} \tag{L.3}$$

where c is the velocity of light. Substituting the values of I_L and A from Eqs. (L.1), (L.2) into Eq. (L.3), one gets

$$\mu_L = -\frac{e}{2m_e c} p_\phi \tag{L.4}$$

with

$$p_\phi = m_e \omega_0 r^2 \tag{L.5}$$

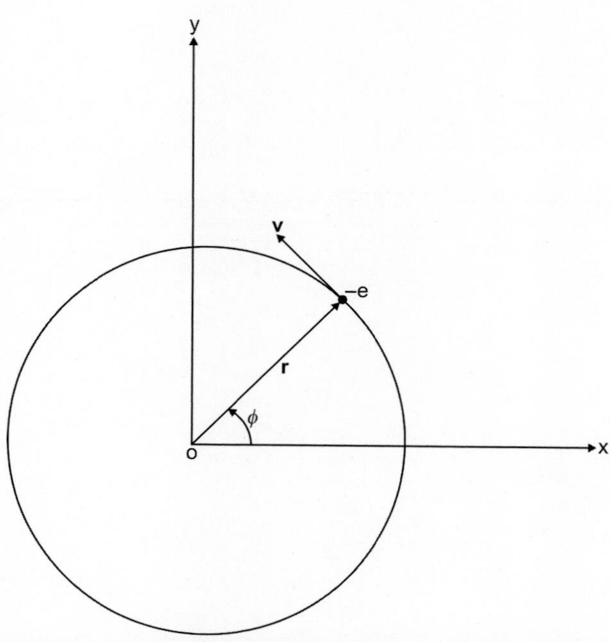

FIG. L.1 An electron of mass m_e is moving with velocity **v** in a circular orbit of radius r around the nucleus in the xy-plane.

Here p_ϕ is the angular momentum of the electron (multiplication of moment of inertia $m_e r^2$ and the angular velocity ω_0). From Bohr's quantization rule for orbits, one can write the orbital angular momentum as

$$\mathbf{p}_\phi = \hbar \mathbf{L} \tag{L.6}$$

where **L** is the orbital angular momentum (in units of ħ) and has integral values as 1, 2, 3,…. From Eqs. (L.4) , (L.6) one can write

$$\vec{\mu}_L = -\mu_B \mathbf{L} \tag{L.7}$$

Here μ_B is called the Bohr magnetron and is defined as

$$\mu_B = \frac{e\hbar}{2 m_e c} \tag{L.8}$$

The negative sign in Eq. (L.7) indicates that the orbital magnetic moment is in a direction opposite to the orbital angular momentum. The above expression is valid only for the orbital motion.

Appendix M: Larmor Precession

To understand Larmor precession, consider the Bohr atom in which an electron with mass m_e is moving with velocity $\mathbf{v}_0$ in a circular orbit of radius r_0 (in the xy-plane) around a nucleus containing only one proton (see Fig. M.1). The equation of motion of the electron is

$$\frac{m_e v_0^2}{r_0} = \frac{e^2}{r_0^2} \tag{M.1}$$

The angular velocity of the electron is given by

$$\omega_0 = \frac{v_0}{r_0} \tag{M.2}$$

From Eqs. (M.1), (M.2) one can write

$$\omega_0^2 = \frac{e^2}{m_e r_0^3} \tag{M.3}$$

Apply magnetic field $\mathbf{H}$ in a direction perpendicular to the plane of the electron orbit. The electron will experience a magnetic force given by

$$\mathbf{F} = -\frac{e}{c}\mathbf{v} \times \mathbf{H} = -\frac{e}{c} vH \tag{M.4}$$

Here $\mathbf{v}$ is different from $\mathbf{v}_0$ because, in the presence of a magnetic field, the velocity of the electron changes and is given by

$$\omega = \frac{v}{r_0} \tag{M.5}$$

Therefore, the equation of motion in the presence of a magnetic field becomes

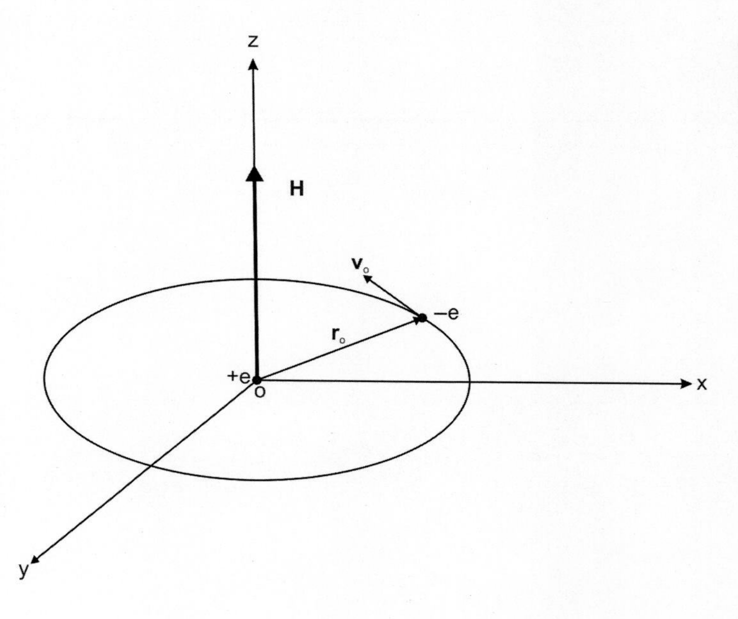

FIG. M.1 The Bohr atom in which an electron with mass m_e and charge $-e$ is revolving in the xy-plane around the nucleus, which contains only one proton having charge e.

$$\frac{m_e v^2}{r_0} = \frac{e^2}{r_0^2} - \frac{e}{c} vH \tag{M.6}$$

Using Eq. (M.5) in Eq. (M.6), one can write

$$\omega^2 + \frac{eH}{m_e c}\omega - \frac{e^2}{m_e r_0^3} = 0 \tag{M.7}$$

The above equation is quadratic in ω and its solution gives

$$\omega = -\omega_L \pm \sqrt{\omega_L^2 + \omega_0^2} \tag{M.8}$$

where

$$\omega_L = \frac{eH}{2m_e c} \tag{M.9}$$

According to Eq. (M.8) the motion of an electron around the nucleus in the presence of a magnetic field is the same as that in the absence of a magnetic field except for the superposition of a precessional angular frequency ω_L as shown in Fig. 18.4. We know that

$$\frac{e}{2m_e c} = 1.4 \times 10^6 \, \text{gauss}^{-1}\,\text{s}^{-1} \tag{M.10}$$

Therefore, even for an applied magnetic field of 10^5 gauss, the Larmor frequency $\omega_L = 10^{11}\,\text{s}^{-1}$ is very small compared with the frequency of an electron in its orbit ($\approx 10^{14} - 10^{15}\,\text{s}^{-1}$). Therefore, to a first-order approximation, Eq. (M.8) can be written as

$$\omega = -\omega_L \pm \omega_0 \tag{M.11}$$

According to Eq. (M.11) the angular frequency decreases in some electrons, which suffer retardation, while it increases in other electrons due to their acceleration in the presence of the magnetic field. Even if the average electron current around the nucleus is zero in the absence of the magnetic field, it becomes finite after the application of the field, yielding a finite induced magnetic dipole moment.

Further Reading

Abragam, A. (1961). *Principles of nuclear magnetism*. Oxford, London: Oxford University Press.
Anderson, P. W. (1963). *Concepts in solids*. New York: Addison-Wesley Publishing Co.
Anderson, P. W. (1984). *Basic notions of condensed matter physics*. New York: Addison-Wesley Publishing Co.
Animalu, A. O. E. (1977). *Intermediate quantum theory of crystalline solids*. New York: Prentice-Hall.
Ashcroft, N. W., & Mermin, N. D. (1976). *Solid state physics*. New York: Holt-Saunders International Edition or Holt, Rinehart and Winston.
Azaroff, L. V., & Brophy, J. J. (1963). *Electronic processes in materials*. New York: McGraw-Hill Book Co.
Barber, D. J., & Loudon, R. (1989). *An introduction to the properties of condensed matter*. Cambridge, London: Cambridge University Press.
Basssani, F., Parravicini, G. P., & Ballinger, R. A. (1975). *Electronic states and optical transitions in solids*. New York: Pergamon Press.
Blakemore, J. S. (1974). *Solid state physics*. London: WB Saunders Co.
Bragg, L. (1949). *The crystalline state: A general survey*. Vol. 1. London: G Bell & Sons.
Brandt, N. B., & Chudinov, S. M. (1975). *Electronic structure of metals*. Moscow: Mir Publishers.
Brick, R. M., & Phillips, A. (1949). *Structure and properties of alloys*. New York: McGraw-Hill Book Co.
Brown, F. C. (1967). *The physics of solids*. New York: WA Benjamin.
Bube, R. H. (1973). *Electronic properties of crystalline solids*. New York: Academic Press.
Buerger, M. J. (1956). *Elementary crystallography*. New York: J. Wiley & Sons.
Busch, G., & Schade, H. (1976). *Lectures on solid state physics*. Oxford, London: Pergamon Press.
Chaikin, P., & Lubensky, T. C. (1995). *Principles of condensed matter physics*. Cambridge, London: Cambridge University Press.
Clark, H. (1957). *Solid state physics*. London: MacMillan Press.
Cochran, J. F., & Hearing, R. R. (1968). *Solid state physics*. (Vols. 1 & 2). New York: Gordon and Breach.
Coles, B. R., & Caplin, A. D. (1976). *The electronic structure of solids*. London: Edward Arnold Publishers.
Cottrell, A. H. (1988). *Introduction to the modern theory of metals*. London: The Institute of Metals.
Davies, J. H. (1998). *The physics of low-dimensional semiconductors: An introduction*. Cambridge, London: Cambridge University Press.
de Gennes, P. G. (1992). Simple views on condensed matter. In Vol. 4. *Series in modern condensed matter physics*. Singapore: World Scientific.
Dekker, A. J. (1971). *Solid state physics*. London: MacMillan Press.
Ehrenreich, H., & Schwartz, L. M. (1976). *Solid state physics*. Vol. 31. New York: Academic Press.
Elliot, R. J., & Gibson, A. F. (1978). *An introduction to solid state physics and its applications*. London: English Language Book Society & MacMillan Press.
Enns, R. H., & Haering, R. R. (1969). *Modern solid state physics*. (Vols. 1 & 2). New York: Gordon and Breach Science Publishers.
Epifanov, G. I. (1979). *Solid state physics*. Moscow: Mir Publishers.
Fan, H. Y. (1987). *Elements of solid state physics*. New York: J. Wiley & Sons.
Galsin, J. S. (2002). *Impurity scattering in metallic alloys*. New York: Kluwer Academic/Plenum Publishers.
Ginzberg, D. M. (1989). *Physical properties of high temperature superconductors*. Singapore: World Scientific.
Goldsmid, H. J. (1968). *Problems in solid state physics*. New York: Academic. Press.
Gorkov, L. P., & Gruner, G. (1989). *Charge density waves in solids*. Amsterdam: North-Holland Publishing Co.
Grimvall, G. (1981). *The electron-phonon interaction in metals*. Amsterdam: North-Holland Publishing Co.
Hall, H. E. (1978). *Solid state physics*. New York: J. Wiley & Sons.
Harrison, W. A. (1966). *Pseudopotentials in the theory of metals*. Massachusetts: WB Benjamin.
Harrison, W. A. (1970). *Solid state theory*. New York: McGraw-Hill Publishing Co.
Harrison, W. A. (1980). *Electronic structure and properties of solids: The physics of chemical bond*. San Francisco: W H Freeman & Co.
Harrison, W. A. (1999). *Elementary electronic structure*. Singapore: World Scientific.
Harrison, W. A. (2000). *Applied quantum mechanics*. Singapore: World Scientific.
Hook, J. R., & Hall, H. E. (1991). *Solid state. physics* (2nd ed.). Chichester: J. Wiley & Sons.
Hume-Rothery, W. (1944). *Structure of metals and alloys*. London: Institute of Metals.
Hume-Rothery, W. (1963). *Electrons, atoms, metals and alloys*. New York: Dover Publishers.
Ibach, H., & Luth, H. (1992). *Solid state physics: An introduction to theory and experiment*. Berlin: Springer-Verlag.
Jiles, D. (1991). *Introduction to magnetism and magnetic materials*. London: Chapman and Hall.
Jones, H. (1960). *The theory of Brillouin zones and electron states in crystals*. Amsterdam: North-Holland Publishing Co.

Jones, W., & March, N. H. (1973). *Theoretical solid state physics*. (Vols. I & II). New York: J. Wiley-Interscience.
Kittel, C. (1963). *Quantum theory of solids*. New York: J. Wiley & Sons.
Kittel, C. (1971). *Introduction to solid state physics* (2nd–8th ed.). New York: J. Wiley & Sons.
Kubo, R., & Nagamiya, T. (1969). *Solid state physics*. New York: McGraw-Hill Book Co.
Landsberg, P. T. (1969). *Solid state theory: Methods and applications*. New York: Wiley-Interscience.
Levy, R. A. (1970). *Principles of solid state physics*. New York: Academic Press.
Loucks, T. L. (1967). *Augmented plane wave method*. New York: WB Benjamin.
Madelung, O. (1978). *Introduction to solid state theory*. Berlin: Springer-Verlag.
Malvino, A., & Bates, D. J. (2007). *Electronic principles*. New Delhi: Tata McGraw-Hill Book Co.
Mandl, F. (1971). *Statistical physics*. New York: J Wiley & Sons.
Maradudin, A. A., Montroll, E. W., Weiss, G. H., & Ipatova, I. P. (1971). Solid state physics. In *Theory of lattice dynamics in harmonic approximation*. New York: Academic Press [Suppl. 3].
Marder, M. P. (2010). *Condensed matter physics*. New York: J Wiley & Sons.
Marion, J. B., & Heald, M. A. (1980). *Classical electromagnetic radiation*. New York: Academic Press.
Marshall, W. (Ed.), (1967). *Theory of magnetism in transition metals*. New York: Academic Press.
Martin, R. M. (2004). *Electronic structure: Basic theory and practical methods*. Cambridge, London: Cambridge University Press.
McKelvey, J. P. (1966). *Solid state and semiconductor physics*. New York: Harper and Row.
Morrison, M. A., Estle, T. L., & Lane, N. F. (1976). *Quantum states of atoms, molecules and solids*. Englewood Cliffs, NJ: Prentice-Hall.
Morse, P. M., & Feshbach, H. (1953). *Method of theoretical physics (parts I and II)*. New York: McGraw-Hill Book Co.
Mott, N. F., & Gurney, R. W. (1957). *Electronic processes in ionic crystals* (2nd ed.). Oxford, London: Oxford University Press.
Myers, H. P. (1997). *Introductory solid state physics* (2nd ed.). London: Taylor & Francis.
Nakajima, S., Toyozawa, Y., & Abe, R. (1980). *The physics of elementary excitations*. Berlin: Springer-Verlag.
Omar, M. A. (1999). *Elementary solid state physics*. Delhi: Pearson Education.
Parr, R. G., & Young, W. (1989). *Density functional theory of atoms and molecules*. Oxford, London: Oxford University Press.
Patterson, J. D. (1971). *Introduction to the theory of solid state physics*. Boston, MA: Addison-Wesley Publ. Co.
Pearson, W. B. (1958). *A handbook of lattice spacings and structure of metals and alloys*. (Vols. 1 & 2). New York: Pergamon Press.
Peierls, R. E. (1955). *Quantum theory of solids*. Oxford, London: Oxford University Press.
Phillips, F. C. (1951). *An introduction to crystallography*. London: Longmans, Green & Co.
Pines, D. (1963). *Elementary excitations in solids*. New York: WB Benjamin.
Raimes, S. (1970). *The wave mechanics of electrons in metals*. Amsterdam: North-Holland Publ. Co.
Reilly, E. O. (2002). *Quantum theory of solids*. London: Taylor and Francis.
Sham, L. J., & Schluter, M. (1991). *Principles and applications of density functional theory*. Singapore: World Scientific.
Sinha, S. K. (1980). Phonons in transition metals. In G. K. Horton, & A. A. Maradudin (Eds.), Vol. 3. *Dynamical properties of solids* (pp. 1–93). Amsterdam: North-Holland Publ. Co.
Slater, J. C. (1974). *Self-consistent field for molecules and solids*. New York: McGraw-Hill Book Co.
Smith, R. A. (1969). *Wave mechanics of crystalline solids* (2nd ed.). London: Chapman and Hall.
Snoke, D. W. (2009). *Solid state physics: Essential concepts*. New Delhi: Pearson Education.
Stringer, J. (1967). *An introduction to the electron theory of solids*. New York: Pergamon Press.
Tanner, B. K. (1995). *Introduction to the physics of electrons in solids*. Cambridge, London: Cambridge University Press.
Taylor, P. L. (1970). *A quantum approach to the solid state*. New Jersey: Prentice-Hall.
Turnbull, R. M. (1979). *The structure of matter*. London: Blackie.
Vainshtein, B. K. (1994). *Modern crystallography*. Vol. 1. Berlin: Springer-Verlag.
Wallis, R. F. (1968). *Localized excitations in solids*. New York: Plenum Publishers.
Wallis, R. F. (1977). *Interaction of radiation with condensed matter*. (Vol. 1). Vienna: International Atomic Energy Agency.
Wanniers, G. H. (1959). *Elements of solid state theory*. Cambridge, London: Cambridge University Press.
Weiss, R. J. (1963). *Solid state physics for metallurgists*. New York; Boston, MA: Pergamon Press; Addison-Wesley Publ. Co.
Wert, C. A., & Thomson, R. M. (1970). *Physics of solids*. New York: McGraw-Hill Book Co.
Zangwill, A. (1987). *Physics at surfaces*. Cambridge, London: Cambridge University Press.
Ziman, J. M. (1969). *The physics of metals: 1 electrons*. Cambridge, London: Cambridge University Press.
Ziman, J. M. (1972a). *Electrons and phonons*. Oxford, London: Oxford University Press.
Ziman, J. M. (1972b). *Principles of the theory of solids*. Cambridge, London: Cambridge University Press.
Ziman, J. M. (1979). *Models of disorder*. Cambridge, London: Cambridge University Press.

Advanced Topics

Abrikosov, A. A. (1972). *Solid state physics (suppl. 12): Introduction to the theory of normal metals*. New York: Academic Press.
Amelinckx, S. (1964). *Solid state physics (suppl. 6): The direct observation of dislocations*. New York: Academic Press.
Anderson, P. W. (1997). *The theory of superconductivity in the high-T_c cuprates*. New York: Princeton University Press.
Bastard, G. (1988). *Wave mechanics applied to semiconductor heterojunctions*. Courtaboeuf: Les Editions de Physique.
Beer, A. C. (1963). *Solid state physics (suppl. 4): Galvanomagnetic effects in semiconductors*. New York: Academic Press.

Beyers, R., & Shaw, T. M. (1989). The structure of $Y_1Ba_2Cu_3O_{7-y}$ and its derivatives. In H. Ehrenreich, & D. Turnbull (Eds.), Vol. 42. *Solid state physics* (pp. 135–212). New York: Academic Press.

Bradley, C. J., & Cracknell, A. P. (1972). *The mathematical theory of symmetry of solids: Representation theory for point defects and space groups*. Oxford, London: Clarendon Press.

Callaway, J. (1974). *Quantum theory of the solid state (parts A & B)*. New York: Academic Press.

Cardona, M. (1969). *Solid state physics (suppl. 11); modulation spectroscopy*. New York: Academic Press.

Cohen, M. L., & Chelikowsky, J. R. (1989). *Electronic structure and optical properties of semiconductors* (2nd ed.). Berlin: Springer-Verlag.

Cohen, M. L., & Heine, V. (1970). The fitting of pseudopotentials to experimental data and their subsequent applications. In H. Ehrenreich, F. Seitz, & D. Turnbull (Eds.), Vol. 24. *Solid state physics* (pp. 37–248). New York: Academic Press.

Cohen, M. H., & Reif, F. (1957). Quadrupole effects in nuclear magnetic resonance studies of solids. In F. Seitz, & D. Turnbull (Eds.), Vol. 5. *Solid state physics* (pp. 321–438). New York: Academic Press.

Compton, W. D., & Rabin, H. (1964). F-aggregate centers in alkali halide crystals. In F. Seitz, & D. Turnbull (Eds.), Vol. 16. *Solid state physics* (pp. 121–226). New York: Academic Press.

Cooper, B. R. (1968). Magnetic properties of rare-earth metals. In F. Seitz, D. Turnbull, & H. Ehrenreich (Eds.), Vol. 21. *Solid state physics* (pp. 393–490). New York: Academic Press.

Das, T. P., & Hahn, E. L. (1958). *Solid state physics (suppl. 1): Nuclear quadrupole resonance spectroscopy*. New York: Academic Press.

Delerue, C., & Lanno, M. (2004). *Nanostructures: Theory and modelling*. New York: Springer-Verlag.

Dimmock, J. O. (1972). The calculation of electronic energy bands by the augmented plane wave method. In H. Ehrenreich, F. Seitz, & D. Turnbull (Eds.), Vol. 26. *Solid state physics* (pp. 103–274). New York: Academic Press.

Douglass, D. H. (1976). *Superconductivity in d- and f-band metals*. New York: Plenum Publishers.

Dresselhaus, M. S., Dresselhaus, G., & Avouris, P. (2001). Carbon nanotubes. *Topics in applied physics*. Vol. 80. Berlin: Springer-Verlag.

Fan, H. Y. (1955). Valence semiconductors Ge and Si. In F. Seitz, & D. Turnbull (Eds.), Vol. 1. *Solid state physics* (pp. 283–365). New York: Academic Press.

Gyorffy, B. L., Kollar, J., Pindor, A. J., Stocks, G. M., Staunton, J., & Winter, H. (1984). *On the theory of ferromagnetism of transition metals at finite temperatures*. London: University of Bristol.

Heine, V. (1970). The pseudopotential concept. In H. Ehrenreich, F. Seitz, & D. Turnbull (Eds.), Vol. 24. *Solid state physics* (pp. 1–36). New York: Academic Press.

Heine, V., & Weaire, D. (1970). Pseudopotential theory of cohesion and structure. In H. Ehrenreich, F. Seitz, & D. Turnbull (Eds.), Vol. 24. *Solid state physics* (pp. 249–463). New York: Academic Press.

Horton, G. K., & Maradudin, A. A. (1974). *Dynamical properties of solids*. (Vols. 1–5). Amsterdam: North-Holland Publishing Co.

Huebener, R. P. (1972). Thermoelectricity in metals and alloys. In H. Ehrenreich, F. Seitz, & D. Turnbull (Eds.), Vol. 27. *Solid state physics* (pp. 63–134). New York: Academic Press.

Isihara, A. (1989). Electron correlations in two dimensions. In H. Ehrenreich & D. Turnbull (Eds.), Vol. 42. *Solid state physics* (pp. 271–402). New York: Academic Press.

Jensen, J., & Mackintosh, A. R. (1991). *Rare earth magnetic structures and excitations*. Oxford, London: Oxford University Press.

Kahn, A. H., & Frederikse, H. P. R. (1959). Oscillatory behavior of magnetic susceptibility and electronic conductivity. In F. Seitz, & D. Turnbull (Eds.), Vol. 9. *Solid state physics* (pp. 257–291). New York: Academic Press.

Keldysh, L. V., Kirzhnitz, D. A., & Maradudin, A. A. (1989). *The dielectric function of condensed systems*. Amsterdam: North-Holland Publishing Co.

Kelly, M. J. (1995). *Low-dimensional semiconductors: Materials physics, technology, devices*. Oxford, London: Clarendon Press.

Knox, R. S. (1963). *Solid state physics (suppl. 5): Theory of excitations*. New York: Academic Press.

Kopfermann, H., & Schneider, E. E. (1958). *Nuclear moments*. New York: Academic Press.

Koster, G. F. (1957). Space groups and their representation. In F. Seitz, & D. Turnbull (Eds.), Vol. 5. *Solid state physics* (pp. 174–256). New York: Academic Press.

Landau, D. P. (1990). *Computer simulation studies in condensed matter physics*. Vol. 1. Berlin: Springer-Verlag.

Landau, D. P. (1991). *Computer simulation studies in condensed matter physics*. Vol. 2. Berlin: Springer-Verlag.

Leibfried, G., & Ludwig, W. (1961). Theory of anharmonic effects in crystals. In F. Seitz, & D. Turnbull (Eds.), Vol. 12. *Solid state physics* (pp. 275–444). New York: Academic Press.

Lowdin, P. O. (1966). *Quantum theory of atoms, molecules and the solid state*. New York: Academic Press.

Moriya, T. (1981). *Electron correlation and magnetism in narrow-band systems*. Berlin: Springer-Verlag.

Mott, N. F. (1974). *Metal-insulator transitions*. London: Taylor & Francis.

Mott, N. F., & Davis, E. A. (1979). *Electronic processes in non-crystalline materials* (2nd ed.). Oxford, London: Clarendon Press.

Nussbaum, A. (1966). Crystal symmetry, group theory and band structure calculations. In F. Seitz, & D. Turnbull (Eds.), Vol. 18. *Solid state physics* (pp. 165–272). New York: Academic Press.

Parr, R. G., & Yang, W. (1989). *Density functional theory of atoms and molecules*. Oxford, London: Clarendon Press.

Phillips, P. (2008). *Advanced solid state physics*. New Delhi: Overseas Press.

Pines, D., & Nozieres, P. (1966). *The theory of quantum liquids*. Vol. 1. New York: WA Benjamin.

Schafroth, M. R. (1960). Theoretical aspects of superconductivity. In F. Seitz, & D. Turnbull (Eds.), Vol. 10. *Solid state physics* (pp. 293–488). New York: Academic Press.

Sham, L. J., & Ziman, J. M. (1963). Electron-phonon interactions. In F. Seitz, & D. Turnbull (Eds.), Vol. 15. *Solid state physics* (pp. 223–298). New York: Academic Press.

Slater, J. C. (1963). *Quantum theory of molecules and solids: Electronic structure of molecules*. Vol. 1. New York: McGraw-Hill Book Co.

Slater, J. C. (1974). *Quantum theory of molecules and solids: The self-consistent field for molecules and solids*. Vol. 4. New York: McGraw-Hill Book Co.

Slater, J. C. (1979). *The calculation of molecular orbits*. New York: J Wiley & Sons.

Sutton, A. P. (1993). *Electronic structure of materials*. Oxford, London: Clarendon Press.

Taylor, K. N., & Darby, M. I. (1972). *Physics of rare-earth solids*. London: Chapman and Hall.

Tinkham, M., & Lobb, C. J. (1989). Physical properties of new superconductors. In H. Ehrenreich, & D. Turnbull (Eds.), Vol. 42. *Solid state physics* (pp. 91–134). New York: Academic Press.

Tyablikov, S. V., Tybulewicz, A., & Mattis, D. C. (1967). *Methods in the quantum theory of magnetism*. New York: Plenum Publishers.

Ziman, J. M. (1972). The calculation of Bloch functions. In H. Ehrenreich, F. Seitz, & D. Turnbull (Eds.), Vol. 26. *Solid state physics* (pp. 1–101). New York: Academic Press.

Index

Note: Page numbers followed by *f* indicate figures, *t* indicate tables, and *b* indicate boxes.

E

F

G

M

Q

R

S

T

U

V

W

X

Z

Printed in the United States
By Bookmasters